Mechanisms in Cellular
Signaling

Transduction Mechanisms in Cellular Signaling

Editors-in-Chief

Edward A. Dennis
Department of Chemistry and Biochemistry,
Department of Pharmacology, School of Medicine,
University of California, San Diego,
La Jolla, California

Ralph A. Bradshaw
Department of Pharmaceutical Chemistry,
University of California, San Francisco,
San Francisco, California

AMSTERDAM • BOSTON • HEIDELBERG • LONDON • NEW YORK • OXFORD
PARIS • SAN DIEGO • SAN FRANCISCO • SINGAPORE • SYDNEY • TOKYO

Academic Press is an imprint of Elsevier

Academic Press is an imprint of Elsevier
525 B Street, Suite 1900, San Diego, CA 92101-4495, USA
30 Corporate Drive, Suite 400, Burlington, MA 01803, USA
32 Jamestown Road, London NW1 7BY, UK
360 Park Avenue South, New York, NY 10010-1710, USA

First edition 2011

Library of Congress Cataloging in Publication Data
Transduction mechanisms in cellular signaling / editors-in-chief, Ralph A. Bradshaw, Edward A. Dennis. – 1st ed.
        p. ; cm.
Summary: "Cell signaling, which is also often referred to as signal transduction or, in more specialized cases, transmembrane signaling, is the process
by which cells communicate with their environment and respond temporally to external cues that they sense there. All cells have the capacity to achieve
this to some degree, albeit with a wide variation in purpose, mechanism, and response. At the same time, there is a remarkable degree of similarity over
quite a range of species, particularly in the eukaryotic kingdom, and comparative physiology has been a useful tool in the development of this field.
The central importance of this general phenomenon (sensing of external stimuli by cells) has been appreciated for a long time, but it has truly become a
dominant part of cell and molecular biology research in the past three decades, in part because a description of the dynamic responses of cells to external
stimuli is, in essence, a description of the life process itself. This approach lies at the core of the developing fields of proteomics and metabolomics, and
its importance to human and animal health is already plainly evident"–Provided by publisher.
    Includes bibliographical references and index.
    ISBN 978-0-12-383862-9 (alk. paper)
    1. Cellular signal transduction. I. Bradshaw, Ralph A., 1941- II. Dennis, Edward A.
    [DNLM: 1. Signal Transduction–physiology. QU 375]
    QP517.C45T72 2011
    571.7'4–dc22
                            2011001757

British Library Cataloging in Publication Data
A catalog record for this book is available from the British Library

ISBN : 978-0-12-383862-9

For information on all Academic Press publications
visit our website at www.elsevierdirect.com

Printed and bound by CPI Group (UK) Ltd, Croydon, CR0 4YY

Working together to grow
libraries in developing countries

www.elsevier.com | www.bookaid.org | www.sabre.org

ELSEVIER   BOOK AID
            International   Sabre Foundation

Contents

Since cell signaling is a major area of biomedical/biological research and continues to advance at a very rapid pace, scientists at all levels, including researchers, teachers, and advanced students, need to stay current with the latest findings, yet maintain a solid foundation and knowledge of the important developments that underpin the field. Carefully selected articles from the 2nd edition of the *Handbook of Cell Signaling* offer the reader numerous, up-to-date views of intracellular signal processing, including membrane receptors, signal transduction mechanisms, the modulation of gene expression/translation, and cellular/organotypic signal responses in both normal and disease states. In addition to material focusing on recent advances, hallmark papers from historical to cutting-edge publications are cited. These references, included in each article, allow the reader a quick navigation route to the major papers in virtually all areas of cell signaling to further enhance his/her expertise.

The Cell Signaling Collection consists of four independent volumes that focus on *Functioning of Transmembrane Receptors in Cell Signaling, Transduction Mechanisms in Cellular Signaling, Regulation of Organelle and Cell Compartment Signaling,* and *Intercellular Signaling in Development and Disease.* They can be used alone, in various combinations or as a set. In each case, an overview article, adapted from our introductory chapter for the Handbook, has been included. These articles, as they appear in each volume, are deliberately overlapping and provide both historical perspectives and brief summaries of the material in the volume in which they are found. These summary sections are not exhaustively referenced since the material to which they refer is.

The individual volumes should appeal to a wide array of researchers interested in the structural biology, biochemistry, molecular biology, pharmacology, and pathophysiology of cellular effectors. This is the ideal go-to books for individuals at every level looking for a quick reference on key aspects of cell signaling or a means for initiating a more in-depth search. Written by authoritative experts in the field, these papers were chosen by the editors as the most important articles for making the Cell Signaling Collection an easy-to-use reference and teaching tool. It should be noted that these volumes focus mainly on higher organisms, a compromise engendered by space limitations.

We wish to thank our Editorial Advisory Committee consisting of the editors of the Handbook of Cell Signaling, 2nd edition, including Marilyn Farquhar, Tony Hunter, Michael Karin, Murray Korc, Suresh Subramani, Brad Thompson, and Jim Wells, for their advice and consultation on the composition of these volumes. Most importantly, we gratefully acknowledge all of the individual authors of the articles taken from the Handbook of Cell Signaling, who are the 'experts' upon which the credibility of this more focused book rests.

Ralph A. Bradshaw, San Francisco, California
Edward A. Dennis, La Jolla, California
January, 2011

Anjon Audhya (23), Weill Institute for Cell and Molecular Biology, Department of Molecular Biology and Genetics, Cornell University, Ithaca, New York

Jesús Balsinde (33), Institute of Molecular Biology and Genetics, Spanish National Research Council (CSIC) and University of Valladolid School of Medicine, Valladolid, Spain

David Barford (12), Section of Structural Biology, Institute of Cancer Research, Chester Beatty Laboratories, London, England, UK

Joseph A. Beavo (38), Department of Pharmacology, University of Washington School of Medicine, Seattle, Washington

Vincent A. Bielinski (18), Department of Pharmacology, University of Texas Southwestern Medical Center, Dallas, Texas

Lutz Birnbaumer (46), Laboratory of Neurobiology, National Institute of Environmental Health Sciences, Research Triangle Park, North Carolina

Ralph A. Bradshaw (1), Department of Pharmaceutical Chemistry, University of California, San Francisco, San Francisco, CA

Ross I. Brinkworth (41), School of Chemistry and Molecular Biosciences and Institute for Molecular Bioscience, University of Queensland, Brisbane, Queensland, Australia

Felricia Brown (40), Department of Physiology, College of Medicine, University of South Alabama, Mobile, Alabama

Janice E. Buss (53), Department of Pharmaceutical Sciences, University of Kentucky, Lexington, Kentucky

Graham Carpenter (31), Department of Biochemistry, Vanderbilt University School of Medicine, Nashville, Tennessee

Patrick J. Casey (49), Departments of Pharmacology and Cancer Biology, Biochemistry, Duke University Medical Center, Durham, North Carolina

Richard A. Cerione (56), Department of Molecular Medicine, Department of Chemistry and Chemical Biology, Cornell University, Ithaca, New York

Gordon Chan (13), Ontario Cancer Institute, Toronto, Ontario, Canada

Chung-Sik Choi (40), Department of Physiology, College of Medicine, University of South Alabama, Mobile, Alabama

Patricia T.W. Cohen (11), Medical Research Council Protein Phosphorylation Unit, College of Life Sciences, University of Dundee, Dundee, UK

Daniela Corda (50), Department of Cell Biology and Oncology, Consorzio Mario Negri Sud, Santa Maria Imbaro, Chieti, Italy

Adrienne D. Cox (54), Departments of Pharmacology, Radiation Oncology, Lineberger Comprehensive Cancer Center, Curriculum in Genetics and Molecular Biology, University of North Carolina at Chapel Hill, Chapel Hill, North Carolina

Mary Dasso (55), Laboratory of Gene Regulation and Development/National Institute of Child Health and Human Development, Bethesda, Maryland

Roger J. Davis (7), Howard Hughes Medical Institute and Program in Molecular Medicine, University of Massachusetts Medical School, Worcester, Massachusetts

Anthony J. Davis (18), Department of Pharmacology, University of Texas Southwestern Medical Center, Dallas, Texas

Edward A. Dennis (1, 33), Department of Chemistry and Biochemistry and Department of Pharmacology, School of Medicine, University of California, San Diego, La Jolla, CA

Channing J. Der (54), Departments of Pharmacology, Lineberger Comprehensive Cancer Center, Curriculum in Genetics and Molecular Biology, University of North Carolina at Chapel Hill, Chapel Hill, North Carolina

Nupur Dey (40), Department of Physiology, College of Medicine, University of South Alabama, Mobile, Alabama

Salim Dhanji (13), Ontario Cancer Institute, Toronto, Ontario, Canada

Zhao Ding (27), Department of Pharmacology, University of Cambridge, Cambridge, England, UK

Jack E. Dixon (29), Pharmacology/Cellular and Molecular Medicine/Chemistry and Biochemistry, University of California, San Diego, La Jolla, California

Wolfgang R. Dostmann (43), Department of Pharmacology, University of Vermont, College of Medicine, Burlington, Vermont

Christian D. Ellson (22), Koch Institute for Integrated Cancer Research, Massachusetts Institute of Technology, Cambridge, Massachusetts

Scott D. Emr (23), Weill Institute for Cell and Molecular Biology, Department of Molecular Biology and Genetics, Cornell University, Ithaca, New York

Edward D. Esplin (18), Department of Pharmacology, University of Texas Southwestern Medical Center, Dallas, Texas

Qiyu Feng (56), Department of Molecular Medicine, Department of Chemistry and Chemical Biology, Cornell University, Ithaca, New York

Garret A. FitzGerald (34), Institute for Translational Medicine and Therapeutics, University of Pennsylvania, Philadelphia, Pennsylvania

Michael A. Frohman (32), Center for Developmental Genetics, Graduate Program in Molecular and Cellular Pharmacology, and the Department of Pharmacology, Stony Brook University, Stony Brook, New York, New York

David A. Fruman (25), Department of Molecular Biology & Biochemistry, and Center for Immunology, University of California, Irvine, California

Maria Di Girolamo (50), Department of Cell Biology and Oncology, Consorzio Mario Negri Sud, Santa Maria Imbaro, Chieti, Italy

Matthew G. Gold (9), Howard Hughes Medical Institute, University of Washington, School of Medicine, Department of Pharmacology, Seattle, Washington

Christopher J. Greenhalgh (8), Division of Molecular Medicine, Walter and Eliza Hall Institute of Medical Research, Parkville, Victoria, Australia

Jesper Z. Haeggström (35), Department of Medical Biochemistry and Biophysics, Division of Chemistry II, Karolinska Institutet, Stockholm, Sweden

Ariella B. Hanker (54), Lineberger Comprehensive Cancer Center, Curriculum in Genetics and Molecular Biology, University of North Carolina at Chapel Hill, Chapel Hill, North Carolina

Carl A. Hansen (51), Department of Biological and Allied Health Sciences, Bloomsburg University, Bloomsburg, Pennsylvania

Douglas J. Hilton (8), Division of Molecular Medicine, Walter and Eliza Hall Institute of Medical Research, Parkville, Victoria, Australia

K.A. Hinchliffe (24), Department of Pharmacology University of Cambridge, Cambridge, England, UK

Rainbo C. Hultman (49), Biochemistry, Duke University Medical Center, Durham, North Carolina

Tony Hunter (2), Molecular and Cell Biology Laboratory, The Salk Institute, La Jolla, California

R.F. Irvine (24), Department of Pharmacology University of Cambridge, Cambridge, England, UK

Benjamin C. Jennings (48), Department of Cell Biology and Physiology, Washington University School of Medicine, St Louis, Missouri

Bruce E. Kemp (6, 41), St Vincent's Institute, Fitzroy, Victoria, Australia, St Vincent's Institute & Department of Medicine, University of Melbourne, St Vincent's Hospital, Fitzroy, Victoria, Australia

Norman J. Kennedy (7), Howard Hughes Medical Institute and Program in Molecular Medicine, University of Massachusetts Medical School, Worcester, Massachusetts

Stephen M. Keyse (17), Cancer Research UK Stress Response Laboratory, Biomedical Research Institute, Ninewells Hospital and Medical School, Dundee, Scotland, UK

Soo-A Kim (29), The Life Science Institute and Department of Biological Chemistry, University of Michigan Medical School, Ann Arbor, Michigan

Michelle E. Kimple (49), Departments of Pharmacology and Cancer Biology, Duke University Medical Center, Durham, North Carolina.

Claude B. Klee (14), Laboratory of Biochemistry, National Cancer Institute, National Institutes of Health, Bethesda, Maryland

Bostjan Kobe (6, 41), School of Chemistry Molecular Biosciences, University of Queensland, Brisbane, Queensland, Australia, School of Chemistry and Molecular Biosciences and Institute for Molecular Bioscience, University of Queensland, Brisbane, Queensland, Australia

Adam J. Kuszak (37), Department of Pharmacology, University of Michigan Medical School, Ann Arbor, Michigan

Lorene K. Langeberg (44), Howard Hughes Medical Institute, University of Washington, Seattle, Washington

Mark A. Lemmon (21), Department of Biochemistry and Biophysics, University of Pennsylvania Medical Center, Philadelphia, Pennsylvania

Alexander Levitzki (5), Unit of Cellular Signaling, Department of Biological Chemistry, The Alexander Silberman Institute of Life Sciences, The Hebrew University of Jerusalem, Israel

Hong-Jun Liao (31), Department of Biochemistry, Vanderbilt University School of Medicine, Nashville, Tennessee

Thomas M. Lincoln (40), Department of Physiology, College of Medicine, University of South Alabama, Mobile, Alabama

Maurine E. Linder (19, 48), Department of Cell Biology and Physiology, Washington University School of Medicine, St Louis, Missouri

Michael Maceyka (36), Department of Biochemistry and Molecular Biology, Virginia Commonwealth University School of Medicine, Richmond, Virginia

George Magklaras (42), Centre for Molecular Medicine Norway, Nordic EMBL Partnership, University of Oslo, Oslo, Norway

Nadir A. Mahmood (18), Department of Pharmacology, University of Texas Southwestern Medical Center, Dallas, Texas

Gerard Manning (2), Razavi Newman Center for Bioinformatics, The Salk Institute, La Jolla, California

Knut Martin Torgersen (29), The Life Science Institute and Department of Biological Chemistry, University of Michigan Medical School, Ann Arbor, Michigan

Jodi McKay (53), Department of Biochemistry, Biophysics and Molecular Biology, Iowa State University, Ames, Iowa

Natalia Mitin (54), Departments of Pharmacology, Lineberger Comprehensive Cancer Center, University of North Carolina, Chapel Hill, North Carolina

Janine Mok (10), Department of Genetics, Yale University School of Medicine, New Haven, Connecticut

Helen R. Mott (57), Department of Biochemistry, University of Cambridge, Cambridge, England, UK

Marc C. Mumby (18), Department of Pharmacology, University of Texas Southwestern Medical Center, Dallas, Texas

James M. Murphy (8), Division of Molecular Medicine, Walter and Eliza Hall Institute of Medical Research, Parkville, Victoria, Australia

Piers Nash (3), Ben May Department for Cancer Research, University of Chicago, Chicago, Illinois

Benjamin G. Neel (13), Ontario Cancer Institute, Toronto, Ontario, Canada

Christian K. Nickl (43), Department of Pharmacology, University of Vermont, College of Medicine, Burlington, Vermont

Nikolaus G. Oberprieler (42), Biotechnology Centre of Oslo, Centre for Molecular Medicine Norway, Nordic EMBL Partnership, University of Oslo, Oslo, Norway

Darerca Owen (57), Department of Biochemistry, University of Cambridge, Cambridge, England, UK

Todd R. Palmby (47), Oral and Pharyngeal Cancer Branch, National Institute of Dental and Craniofacial Research, National Institutes of Health, Bethesda, Maryland, *Current address: Food and Drug Administration, Silver Spring, Maryland

Tony Pawson (3), Samuel Lunenfeld Research Institute, Mt Sinai Hospital, Toronto, Ontario, Canada

Steve M. Prescott (30), Oklahoma Medical Research Foundation, Oklahoma City, Oklahoma

Elzbieta Radzio-Andzelm (39), Department of Pharmacology, Howard Hughes Medical Institute, University of California at San Diego, La Jolla, California

Gretchen A. Repasky (54), Birmingham-Southern College, Birmingham, Alabama

Janet D. Robishaw (51), Weis Center for Research, Geisinger Clinic, Danville, Pennsylvania

Hans Rosenfeldt (47), Oral and Pharyngeal Cancer Branch, National Institute of Dental and Craniofacial Research, National Institutes of Health, Bethesda, Maryland, *Current address: Food and Drug Administration, Silver Spring, Maryland

Anja Ruppelt (42), Biotechnology Centre of Oslo, Centre for Molecular Medicine Norway, Nordic EMBL Partnership, University of Oslo, Oslo, Norway

Neil F. W. Saunders (41), School of Chemistry and Molecular Biosciences and Institute for Molecular Bioscience, University of Queensland, Brisbane, Queensland, Australia

William F. Schwindinger (51), Weis Center for Research, Geisinger Clinic, Danville, Pennsylvania

John D. Scott (9, 44), Howard Hughes Medical Institute, University of Washington, School of Medicine, Department of Pharmacology, Seattle, Washington, Howard Hughes Medical Institute, University of Washington, Seattle, Washington

Hassan Sellak (40), Department of Physiology, College of Medicine, University of South Alabama, Mobile, Alabama

John B. Shabb (45), Department of Biochemistry and Molecular Biology, University of North Dakota School of Medicine and Health Sciences, Grand Forks, North Dakota

Stephen B. Shears (26), Laboratory of Signal Transduction, National Institute of Environmental Health Sciences, Research Triangle Park, North Carolina

Kazuhiro Shiozaki (15), Department of Microbiology, College of Biological Sciences, University of California, Davis, California

Adam M. Silverstein (18), Department of Pharmacology, University of Texas Southwestern Medical Center, Dallas, Texas

J. Silvio Gutkind (47), Oral and Pharyngeal Cancer Branch, National Institute of Dental and Craniofacial Research, National Institutes of Health, Bethesda, Maryland, *Current address: Food and Drug Administration, Silver Spring, Maryland

Stephen J. Smerdon (4), Division of Molecular Structure, National Institute for Medical Research, The Ridgeway, Mill Hill, London, England, UK

Emer M. Smyth (34), Institute for Translational Medicine and Therapeutics, University of Pennsylvania, Philadelphia, Pennsylvania

Michael Snyder (10), Department of Molecular, Cellular, and Developmental Biology, Yale University, New Haven, Connecticut

Sarah Spiegel (36), Department of Biochemistry and Molecular Biology, Virginia Commonwealth University School of Medicine, Richmond, Virginia

Christopher Stefan (23), Weill Institute for Cell and Molecular Biology, Department of Molecular Biology and Genetics, Cornell University, Ithaca, New York

Wenjuan Su (32), Center for Developmental Genetics, Graduate Program in Molecular and Cellular Pharmacology, and the Department of Pharmacology, Stony Brook University, Stony Brook, New York, New York

Roger K. Sunahara (37), Department of Pharmacology, University of Michigan Medical School, Ann Arbor, Michigan

James Surapisitchat (38), Department of Pharmacology, University of Washington School of Medicine, Seattle, Washington

Gillian M. Tannahill (8), Division of Molecular Medicine, Walter and Eliza Hall Institute of Medical Research, Parkville, Victoria, Australia

Kjetil Taskén (42), Biotechnology Centre of Oslo, Centre for Molecular Medicine Norway, Nordic EMBL Partnership, University of Oslo, Oslo, Norway

Hisashi Tatebe (15), Department of Microbiology, College of Biological Sciences, University of California, Davis, California

Colin W. Taylor (27), Department of Pharmacology, University of Cambridge, Cambridge, England, UK

Susan S. Taylor (39), Department of Chemistry and Biochemistry, Department of Pharmacology, Howard Hughes Medical Institute, University of California at San Diego, La Jolla, California

Matthew K. Topham (30), The Huntsman Cancer Institute and Department of Internal Medicine, University of Utah, Salt Lake City, Utah

Lloyd C. Trotman (28), Cancer Center, Cold Spring Harbor Laboratory, Cold Spring Harbor, New York

Philip B. Wedegaertner (52), Department of Biochemistry and Molecular Biology, Kimmel Cancer Center, Thomas Jefferson University, Philadelphia, Pennsylvania

Anders Wetterholm (35), Department of Medical Biochemistry and Biophysics, Division of Chemistry II, Karolinska Institutet, Stockholm, Sweden

Roger L. Williams (20), Medical Research Council, Laboratory of Molecular Biology, Cambridge, England, UK

Michael B. Yaffe (4, 22), Center for Cancer Research, Massachusetts Institute of Technology, Cambridge, Massachusetts, Koch Institute for Integrated Cancer Research, Massachusetts Institute of Technology, Cambridge, Massachusetts

Seun-Ah Yang (14), Division of Hematology-Oncology, University of Pennsylvania School of Medicine, Philadelphia, Pennsylvania

Zhong-Yin Zhang (16), Department of Biochemistry and Molecular Biology, Indiana University School of Medicine, Indianapolis, Indiana

Overview

Chapter 1

Intracellular Signaling*

Edward A. Dennis[1] and Ralph A. Bradshaw[2]

[1]*Department of Chemistry and Biochemistry and Department of Pharmacology, School of Medicine, University of California, San Diego, La Jolla, CA*

[2]*Department of Pharmaceutical Chemistry, University of California, San Francisco, San Francisco, CA*

Cell signaling, which is also often referred to as signal transduction or, in more specialized cases, transmembrane signaling, is the process by which cells communicate with their environment and respond temporally to external cues that they sense there. All cells have the capacity to achieve this to some degree, albeit with a wide variation in purpose, mechanism, and response. At the same time, there is a remarkable degree of similarity over quite a range of species, particularly in the eukaryotic kingdom, and comparative physiology has been a useful tool in the development of this field. The central importance of this general phenomenon (sensing of external stimuli by cells) has been appreciated for a long time, but it has truly become a dominant part of cell and molecular biology research in the past three decades, in part because a description of the dynamic responses of cells to external stimuli is, in essence, a description of the life process itself. This approach lies at the core of the developing fields of proteomics and metabolomics, and its importance to human and animal health is already plainly evident.

ORIGINS OF CELL SIGNALING RESEARCH

Although cells from polycellular organisms derive substantial information from interactions with other cells and extracellular structural components, it was humoral components that first were appreciated to be intercellular messengers. This idea was certainly inherent in the 'internal secretions' initially described by Claude Bernard in 1855 and thereafter, as it became understood that ductless glands, such as the spleen, thyroid, and adrenals, secreted material into the bloodstream. However, Bernard did not directly identify hormones as such. This was left to Bayliss and Starling and their description of secretin in 1902 [1].

Recognizing that it was likely representative of a larger group of chemical messengers, the term *hormone* was introduced by Starling in a Croonian Lecture presented in 1905. The word, derived from the Greek word meaning 'to excite or arouse,' was apparently proposed by a colleague, W. B. Hardy, and was adopted, even though it did not particularly connote the messenger role but rather emphasized the positive effects exerted on target organs via cell signaling (see Wright [2] for a general description of these events). The realization that these substances could also produce inhibitory effects, gave rise to a second designation, 'chalones,' introduced by Schaefer in 1913 [3], for the inhibitory elements of these glandular secretions. The word autocoid was similarly coined for the group as a whole (hormones and chalones). Although the designation chalone has occasionally been applied to some growth factors with respect to certain of their activities (e.g., transforming growth factor), autocoid has essentially disappeared. Thus, if the description of secretin and the introduction of the term hormone are taken to mark the beginnings of molecular endocrinology and the eventual development of cell signaling, then we have passed the hundredth anniversary of this field.

The origins of endocrinology, as the study of the glands that elaborate hormones and the effect of these entities on target cells, naturally gave rise to a definition of hormones as substances produced in one tissue type that traveled systemically to another tissue type to exert a characteristic response. Of course, initially these responses were couched in organ and whole animal responses, although they increasingly were defined in terms of metabolic and

---

Portions of this article were adapted from Bradshaw RA, Dennis EA. Cell signaling: yesterday, today, and tomorrow. In Bradshaw RA, Dennis EA, editors. Handbook of cell signaling. 2nd ed. San Diego, CA: Academic Press; 2008; pp 1–4.

other chemical changes at the cellular level. The early days of endocrinology were marked by many important discoveries, such as the discovery of insulin [4], to name one, that solidified the definition, and a well-established list of hormones, composed primarily of three chemical classes (polypeptides, steroids, and amino acid derivatives), was eventually developed. Of course, it was appreciated even early on that the responses in the different targets were not the same, particularly with respect to time. For example, adrenalin was known to act very rapidly, while growth hormone required a much longer time frame to exert its full range of effects. However, in the absence of any molecular details of mechanism, the emphasis remained on the distinct nature of the cells of origin versus those responding and on the systemic nature of transport, and this remained the case well into the 1970s. An important shift in endocrinological thinking had its seeds well before that, however, even though it took about 25 years for these 'new' ideas that greatly expanded endocrinology to be enunciated clearly.

Although the discovery of polypeptide growth factors as a new group of biological regulators is generally associated with nerve growth factor (NGF), it can certainly be argued that other members of this broad category were known before NGF. However, NGF was the source of the designation *growth factor* and has been, in many important respects, a Rosetta stone for establishing principles that are now known to underpin much of signal transduction. Thus, its role as the progenitor of the field and the entity that keyed the expansion of endocrinology, and with it the field of cell signaling, is quite appropriate. The discovery of NGF is well documented [5] and how this led directly to identification of epidermal growth factor (EGF) [6], another regulator that has been equally important in providing novel insights into cellular endocrinology, signal transduction and, more recently, molecular oncology. However, it was not till the sequences of NGF and EGF were determined [7, 8] that the molecular phase of growth factor research began in earnest. Of particular importance was the postulate that NGF and insulin were evolutionarily related entities [9], which suggested a similar molecular action (which, indeed, turned out to be remarkably clairvoyant), and was the first indication that the identified growth factors, which at that time were quite limited in number, were like hormones. This hypothesis led quickly to the identification of receptors for NGF on target neurons, using the tracer binding technology of the time (see Raffioni *et al.* [10] for a summary of these contributions), which further confirmed their hormonal status. Over the next several years, similar observations were recorded for a number of other growth factors, which in turn, led to the redefinition of endocrine mechanisms to include paracrine, autocrine, and juxtacrine interactions [11]. These studies were followed by first isolation and molecular characterization using various biophysical methods and then cloning of

their cDNAs, initially for the insulin and EGFR receptors [12–14] and then many others. Ultimately, the powerful techniques of molecular biology were applied to all aspects of cell signaling and are largely responsible for the detailed depictions we have today. They have allowed the broad understanding of the myriad of mechanisms and responses employed by cells to assess changes in their environment and to coordinate their functions to be compatible with the other parts of the organism of which they are a part.

INTRACELLULAR SIGNALING MECHANISMS

At the same time that the growth factor field was undergoing rapid development, major advances were also occurring in studies on hormonal mechanisms. In particular, Sutherland and colleagues [15] were redefining hormones as messengers and their ability to produce second messengers. This was, of course, based primarily on the identification of cyclic AMP (cAMP) and its production by a number of classical hormones. However, it also became clear that not all hormones produced this second messenger nor was it stimulated by any of the growth factors known at that time. This enigma remained unresolved for quite a long time until tyrosine kinases were identified [16, 17] and it was shown, first with the EGF receptor [18], that these modifications were responsible for initiating signal transduction for many of those hormones and growth factors that did not stimulate the production of cAMP.

Aided by the tools of molecular biology, it was a fairly rapid transition to the cloning of most of the receptors for hormones and growth factors and the subsequent development of the main classes of signaling mechanisms. These data allowed the six major classes of cell surface receptors for hormones and growth factors to be defined, which included, in addition to the receptor tyrosine kinases (RTKs) described previously, the G-protein coupled receptors (GPCRs) (including the receptors that produce cAMP) that constitute the largest class of cell surface receptors; the cytokine receptors, which recruit the soluble JAK tyrosine kinases and directly activate the STAT family of transcription factors; serine/threonine kinase receptors of the TGFβ superfamily; the tumor necrosis factor (TNF) receptors that activate nuclear factor kappa B (NFκB) via TRAF molecules, among other pathways; and the guanylyl cyclase receptors. Structural biology has not maintained the same pace, and there are still both ligands and receptors for which we do not have three-dimensional information.

In parallel with the development of our understanding of ligand/receptor organization at the plasma membrane, a variety of experimental approaches have also revealed the general mechanisms of transmembrane signal transduction in terms of the major intracellular events that are induced by these various receptor classes. There are three

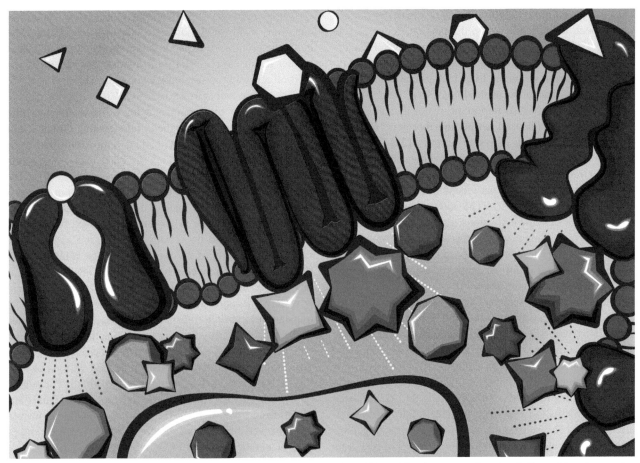

FIGURE 1.1 Intracellular events following receptor activation include the activation of kinase/phosphatase cascades, phospholipases liberating a variety of lipid mediators, cyclic nucleotide production and their downstream events, and numerous G-protein triggered pathways. The immediate signaling pathways often amplify their signals by a series of posttranslational events that in turn release various messengers, often lipids or ions, which over time ultimately result in the modulation of transcriptional events in the nucleus.

principal means by which intracellular signals are propagated: protein posttranslational modifications (PTMs), lipid messengers, and ion fluxes (see Figure 1.1). There are also additional moieties that play significant roles, such as cyclic nucleotides, but their effects are generally manifested in downstream PTMs. There is considerable interplay between the three, particularly in the more complex pathways.

By far the most significant of the PTMs is phosphorylation of serine, threonine, and tyrosine residues (although phosphorylation of several other residues is also known albeit that these modifications are usually found in lower organisms). As already noted, the RTK and cytokine receptors initiate their responses with tyrosine phosphorylation, and there are more than 30 additional nonreceptor tyrosine kinases that also have significant roles in many signaling responses that can be activated by these and other types of receptors. However, the vast bulk of protein phosphorylation occurs downstream from the receptors and is mainly on serine and threonine residues in a ratio of about 20:1. These are produced by a myriad of protein kinases that are

activated themselves through PTMs or through the production of lipid messengers (see below). Indeed, there are over 500 protein kinases in the human genome with more than 100 phosphatases, which emphasizes the investment that has been made in this modification by higher eukaryotes [19]. Therefore, it is perhaps not surprising that through the agency of substantive technological advances in proteomic analyses, mainly in the area of mass spectrometry and its quantitative applications, it has become clear that the level of this PTM, both in terms of type and amount, is significantly greater than originally envisioned [20]. It could not have been readily anticipated from the pioneering studies of Krebs and Fischer in the 1960s [21] when they observed the regulation of muscle phosphorylase activity by protein phosphorylation that this modification would occur essentially universally in cells and that hundreds, if not thousands, of enzyme activities and protein–protein interactions would be regulated by it. Nonetheless, thousands of phosphorylation events have indeed been detected in cellular paradigms that have been appropriately stimulated by one or another growth factor (see, e.g., Olsen *et al.* [22]) and in no case has the complete set of modifications been identified. Clearly, the new challenges are to determine which of these modifications are physiologically meaningful and which kinases (or another type of modifying enzyme) are responsible for which alterations. Further findings using proteomic methodology have demonstrated that other PTMs are also important, if not as widespread. O-GlyNAcylation (also on serine and threonine residues) [23] and N^ε-acetylation of lysine residues [24] are examples of modifications that are receiving increasing attention.

As intracellular signaling was being unraveled, it became increasingly clear that receptor activation and subsequent activations through PTM additions were inducing more than just enzyme activations. Rather, many modifications were providing new, specific sites for forming protein complexes. These were appropriately designated as 'docking sites,' and it introduced the concept of both adaptors and scaffolds, with activated enzymes being called 'effectors.' Adaptors, such as Grb or Shc proteins, and the larger, multisite scaffolds, such as the insulin receptor substrate (IRS), recognize newly formed sites through specific motifs and as the process is repeated, successively build up multicomponent signaling structures [25]. There has now emerged a significant number of binding motifs, recognizing, in addition to PTMs, phospholipids and proline-rich peptide segments to name a few, that are quite widely scattered through the substantial repertoire of signaling molecules and that are activated by different types of receptors in a variety of cell types.

The elucidation of cell signaling mechanisms and the variety of molecules that are employed in these myriad of processes is particularly well exemplified by the lipid messengers. With the exception of steroid hormones, lipids have long been thought to function mainly in energy metabolism and membrane structure. Experimental work for the last two decades has revealed a broad recognition that membrane phospholipids provide many of the important cell signaling molecules via phospholipases and lipid kinases. Key is the role of phospholipase C of which there are four subtypes that are activated by various receptor systems to hydrolyze phosphatidylinositol bisphosphate (PIP_2) to release diglyceride that activates protein kinase C (PKC) and inositol triphosphate (IP_3), which mobilizes intracellular Ca^{2+} central to so many regulatory processes. The phosphorylation of PIP_2 at the 3-position to produce PIP_3 promotes vesicular trafficking and other cellular processes. Phospholipase D releases phosphatidic acid, and phospholipase A2 provides arachidonic acid, which is converted into prostaglandins, leukotrienes, lipoxins, and various P450 products; these ligands in turn bind to unique families of receptors as does platelet activating factor (PAF). The more recent recognition, in the last decade, of the importance of sphingolipids and ceramide in signaling and the discoveries of the unique lysophosphatidic acid and sphingosine phosphate families of receptors have sparked the search for

other new lipid messengers and their receptors. The newly emerging field of lipidomics (see www.lipidmaps.org) holds the promise of expanding our ability to interrogate in greater detail the specificity of agonists and receptors and their effects on lipid signaling events [26–28].

The extent and complexity of GPCRs, in terms of both the ligands that bind them and the effectors they in turn activate, is unparalleled in the other signaling systems. The receptors of this family, with their seven transmembrane segments, function by linking to heteromeric G protein complexes composed of three subunits: α, β, and γ. The α-subunit binds GTP and the receptor-G-protein complex functions as a guanine nucleotide exchange factor (GEF). The ligand induces the G-protein to split into two components – α-GTP and βγ – both of which are active in the further propagation of the signal. When the GTP is hydrolyzed to GDP, it recycles back to the GTP form so it is ready to be reutilized. This type of biochemical 'switch' is widely encountered in biological systems ranging from translation to vesicle transport and is also utilized (as Ras) in the major pathway leading to ERK activation by RTKs. GPCRs are utilized as sensors of peptide/protein hormones, neurotransmitters, amino acids, lipids, and various physiological processes such as light, taste, and smell. The adenylyl cyclases are a major effector for the GPCR signals and are affected by both the α-GTP and βγ subunits. However, they also activate some of the nonreceptor tyrosine kinases, PI-3-K and the β-type of PLC among others.

cAMP, the product of adenylyl cyclase activation, was of course the discovery of Sutherland and colleagues [15], and it exerts much of its effects by the activation of PKA. This is one of the most important mediators in signal transduction pathways. It is composed of two regulatory and two catalytic subunits and is activated when cAMP binds to the regulatory subunits, causing dissociation of the heterotetramer and the concomitant activation of the catalytic subunits. In addition to its multiple cellular roles, it has been an important model for understanding the structure–function relationships of the protein kinase superfamily.

FOCUS AND SCOPE OF THIS VOLUME

The chapters of this volume have been selected from a larger collection [29] and have been organized to emphasize receptor organization and transduction mechanisms functioning in cell signaling. They have been contributed by recognized experts and they are authoritative to the extent that size limitations allow. It is our intention that this survey will be useful in teaching, particularly in introductory courses, and to more seasoned investigators new to this area.

It is not possible to develop any of the areas covered in this volume in great detail, and expansion of any topic is left to the reader. The references in each chapter provide an excellent starting point, and greater coverage can also be found in the parent work [29]. It is important to realize that this volume does not cover other aspects of cell signaling such as the structure and role of cell surface receptors in signaling activities, transcriptional activation and responses in other organelles, and organ-level manifestations, including disease correlates. These can be found in other volumes in this series [30–32].

REFERENCES

1. Bayliss WM, Starling EH. The mechanism of pancreatic secretion. *J Physiol* 1902;**28**:325–53.
2. Wright RD. The origin of the term "hormone". *Trends Biochem. Sci* 1978;**3**:275.
3. Schaefer EA. *The Endocrine Organs*. London: Longman & Green; 1916; p. 6.
4. Banting FG, Best CH. The internal secretion of the pancreas. *J Lab Clin Med* 1922;**7**:251–66.
5. Levi-Montalcini R. The nerve growth factor 35 years later. *Science* 1987;**237**:1154–62.
6. Cohen S. Origins of Growth Factors: NGF and EGF. *J Biol Chem* 2008;**283**:33793–7.
7. Angeletti RH, Bradshaw RA. Nerve growth factor from mouse submaxillary gland: amino acid sequence. *Proc Natl Acad Sci USA* 1971;**68**:2417–20.
8. Savage CR, Inagami T, Cohen S. The primary structure of epidermal growth factor. *J Biol Chem* 1972;**247**:7612–21.
9. Frazier WA, Angeletti RH, Bradshaw RA. Nerve growth factor and insulin. *Science* 1972;**176**:482–8.
10. Raffioni S, Buxser SE, Bradshaw RA. The receptors for nerve growth factor and other neurotrophins. *Annu Rev Biochem* 1993;**62**:823–50.
11. Bradshaw RA, Sporn MB. Polypeptide growth factors and the regulation of cell growth and differentiation: introduction. *Fed Proc* 1983;**42**:2590–1.
12. Ullrich A, Bell JR, Chen EY, Herrera R, Petruzzelli LM, Dull TJ, et al. Human insulin receptor and its relationship to the tyrosine kinase family of oncogenes. *Nature* 1985;**313**:756–61.
13. Ullrich A, Coussens L, Hayflick JS, Dull TJ, Gray A, Tam AW, et al. Human epidermal growth factor receptor cDNA sequence and aberrant expression of the amplified gene in A431 epidermoid carcinoma cells. *Nature* 1985;**309**:418–25.
14. Ebina Y, Ellis L, Jarnagin K, Edery M, Graf L, Clauser E, et al. The human insulin receptor cDNA: the structural basis for hormone transmembrane signalling. *Cell* 1985;**40**:747–58.
15. Robison GA, Butcher RW, Sutherland EW. *Cyclic AMP*. San Diego: Academic Press; 1971.
16. Eckert W, Hutchinson MA, Hunter T. An activity phosphorylating tyrosine in polyoma T antigen immunoprecipitates. *Cell* 1979;**18**:925–33.
17. Hunter T, Sefton BM. Transforming gene product of Rous sarcoma virus phosphorylates tyrosine. *Proc Natl Acad Sci USA* 1980;**77**:1311–5.
18. Ushiro H, Cohen S. Identification of phosphotyrosine as a product of epidermal growth factor-activated protein kinase in A-431 cell membranes. *J Biol Chem* 1980;**255**:8363–5.
19. Manning G, Whyte DB, Martinez R, Hunter T, Sudarsanam S. The protein kinase complement of the human genome. *Science* 2002;**298**:1912–34.
20. Choudhary C, Mann M. Decoding signalling networks by mass spectrometry-based proteomics. *Nature Rev Mol Cell Biol* 2010;**11**:427–39.
21. Krebs EG, Fischer EH. Phosphorylase and related enzymes of glycogen metabolism. *Vitam Horm* 1964;**22**:399–410.
22. Olsen JV, Blagoev B, Gnad F, Macek B, Kumar C, Mortensen P, et al. Global, in vivo, and site-specific phosphorylation dynamics in signaling networks. *Cell* 2006;**127**:635–48.
23. Zeidan Q, Hart GW. The intersections between O-GlcNAcylation and phosphorylation: Implications for multiple signaling pathways. *J Cell Sci* 2010;**123**:13–22.
24. Kouzarides T. Acetylation: a regulatory modification to rival phosphorylation? *EMBO J* 2000;**19**:1176–9.
25. Pawson T. Regulation and targets of receptor tyrosine kinases. *Eur J Cancer* 2002;**38**(Suppl 5):S3–10.
26. Buczynski MW, Stephens DL, Bowers-Gentry RC, Grkovich A, Deems RA, Dennis EA. TLR-4 and sustained calcium agonists synergistically produce eicosanoids independent of protein synthesis in RAW264.7 cells. *J Biol Chem* 2007;**282**:22834–47.
27. Dennis EA. Lipidomics joins the omics evolution. *Proc Natl Acad Sci USA* 2009;**106**:2089–90.
28. Dennis EA, Harkewicz R, Quehenberger O, Brown HA, Milne SB, Myers DS, et al. A mouse macrophage lipidome. *J Biol Chem* 2010;**285**: [PMID20923771].
29. Bradshaw RA, Dennis EA, editors. *Handbook of cell signaling*. 2nd ed. San Diego, CA: Academic Press; 2008.
30. Bradshaw RA, Dennis EA, editors. *Functioning of transmembrane receptors in cell signaling mechanisms*. San Diego, CA: Academic Press; 2011.
31. Bradshaw RA, Dennis EA, editors. *Regulation of organelle and cell compartment signaling*. San Diego, CA: Academic Press; 2011.
32. Dennis EA, Bradshaw RA, editors. *Intercellular signaling in development and disease*. San Diego, CA: Academic Press; 2011.

Phosphorylation/ Dephosphorylation

Part 1

Kinases

Chapter 2

Eukaryotic Kinomes: Genomics and Evolution of Protein Kinases

Gerard Manning[1] and Tony Hunter[2]

[1]*Razavi Newman Center for Bioinformatics, The Salk Institute, La Jolla, California*
[2]*Molecular and Cell Biology Laboratory, The Salk Institute, La Jolla, California*

INTRODUCTION

Ever since the discovery 50 years ago that reversible phosphorylation regulates the activity of glycogen phosphorylase [1], there has been intense interest in the role of protein phosphorylation in regulating protein function. With the advent of DNA cloning and sequencing in the mid-1970s it rapidly became apparent that a large family of eukaryotic protein kinases exists, and the burgeoning numbers of protein kinases led to the speculation that a vertebrate genome might encode as many as 1001 protein kinases [2]. Since then, the importance of protein phosphorylation as a regulatory mechanism has continued to grow, and recent phosphoproteomic analyses suggest that the majority of intracellular proteins can be phosphorylated at one or more sites under an appropriate condition. Protein phosphorylation not only regulates enzymatic activity through inducing conformational changes or through direct steric effects, but also modulates the function of structural proteins through conformational and charge effects. In addition, a major function of protein-linked phosphates is to provide docking sites for other proteins, thus promoting inducible protein–protein association [3].

The catalytic domains of eukaryotic serine/threonine- and tyrosine-specific protein kinases are related in sequence, and belong to the eukaryotic protein kinase (ePK) superfamily, which in turn is a subset of PKL (protein-kinase like) kinases that share a common fold and catalytic mechanism [4]. A few structurally unrelated proteins also have reported protein kinase activity, and a wide variety of protein families can phosphorylate non-protein substrates [5]. Non-ePK protein kinases are termed aPKs, or atypical protein kinases. In addition to Ser, Thr, and Tyr, several other amino acids in proteins can be phosphorylated, including Lys, Arg, and His. The provenance of the responsible

protein kinases remains unclear, although NDPK-B has recently been reported to be a *bona fide* mammalian histidine kinase [6]. The prokaryotic two-component protein kinases, commonly known as "histidine" kinases, form yet another distinct family. These autophosphorylate on histidine, and then transfer the phosphate to an aspartate on a substrate protein. These kinases are also found in plants and protists, but are absent from animals, apart from the unusual mitochondrial PDHK family members, which phosphorylate Ser/Thr.

The ~270 amino acid ePK catalytic domain is characterized by a series of conserved sequence motifs, which define 11 subdomains, and serve as key catalytic elements of the kinase domain [7, 8]. These motifs in combination with the overall catalytic domain sequence can be used to identify other protein kinases through pairwise and HMM profile sequence searches. aPKs can be found using similar approaches. Using this strategy, we have surveyed a series of sequenced eukaryotic genomes to define the protein kinase complement (kinome) of each organism [9–16] and used this as a basis to explore the evolution and global functions of all protein kinases.

THE HUMAN KINOME

In our original survey completed in 2002 [11], we predicted that the human genome has 518 protein kinase genes (2.3 percent of all ~22,500 genes) (Figure 2.1). Of these, 478 encode ePKs, with the others divided between 9 small aPK families, which include the PIKK (PI3 kinase-like kinase), the PDHK (pyruvate dehydrogenase kinase) and alpha kinase (E2F kinase) families. There are 90 tyrosine kinase genes (16 percent of all protein kinases). The complexity of the kinome is further increased by alternative splicing of

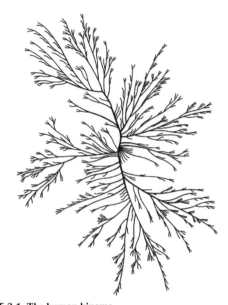

FIGURE 2.1 The human kinome.
A stylized phylogenetic tree represents the 492 ePK domains in the human genome, shaded by group classification. For further details, see http://kinase.com/human/kinome/. Reproduced by permission of *Science* magazine and *Cell Signaling Technologies*.

over half of all kinases [17], which in many cases is known to modulate function, as well as by the existence of regulatory subunits and differential targeting within the cell through association with scaffolding proteins.

Since the 2002 catalog, we have added 7 Ser/Thr kinases (6 four-jointed kinases and a second copy of PITSLRE/CDC2L2), lost 2 Tyr kinases (twinfilin/A6 family), and added NDPK-B (and 8 homologs) as a *bona fide* histidine kinase, to give a total of 532. The list may grow further, as humans have members of other classes of PKLs that are usually thought of as small-molecule kinases but have members known to phosphorylate proteins (e.g., ACAD10/11 from the CAK family [4]) Other ATPases or structurally distinct proteins may also emerge to have kinase activity (as seen in the MinD family of bacterial tyrosine kinases [18]). A few kinases currently in the catalog, such as SgK424 and PRKY are dubious, and may be relegated to pseudogenes.

The major control functions of protein kinases are reflected in their involvement in disease. Thirty-five percent of the kinome (175 genes) has been directly implicated in human disease, through mutation, mis-expression, or copy number changes [19]; 121 protein kinases are implicated in cancer, including 51 of the 90 tyrosine kinases. Many more protein kinases are weakly implicated and are emerging from genome-scale studies, including recent efforts to re-sequence the entire kinome in a wide variety of human cancers to pinpoint driver mutations involved in carcinogenesis. For instance, 164 kinases were mapped to common amplicons [11], and many more such data are now emerging.

Protein kinase catalytic function is often dependent on additional domains in the protein, which regulate activity,

localize, and recruit regulatory proteins/second messengers and substrates. About half the protein kinases are predicted to have additional domains, many of which are implicated in signaling. Of the tyrosine kinases, 25 have P.Tyr binding SH2 domains that play a cardinal role in establishing tyrosine-phosphorylation based signaling networks. In contrast, perhaps surprisingly, only one serine kinase contains a P.Ser/Thr binding domain (an FHA domain in CHK2). In addition, 46 protein kinases have domains that interact with other proteins (e.g., SH3); 55 tyrosine and serine kinases have lipid interaction domains (e.g., PH); 38 have domains linked to small GTPase signaling; and 28 serine kinases have domains linked to calcium signaling. Generally, most members of a protein kinase family have the same set of ancillary domains, but there are some exceptions, and alternative splicing is often used to generate distinct domain combinations from a single gene. A complete listing of additional domains found in human protein kinases is given at http://kinase.com/.

The kinase catalytic domain itself often has ancillary functions. In fact, close to 10 percent of all kinase domains are predicted to have lost enzymatic function, but are retained for non-catalytic reasons. Of the 492 human ePK catalytic domains, 48 are predicted to be inactive, based on loss of key catalytic residues (Lys72/Asp166/Asp184 in PKA) and review of experimental data [20]. These "pseudokinase domains" may serve as docking platforms or scaffolds (e.g., ErbB3 and ILK), structural elements (receptor guanylyl cyclase kinase homology domains), and/or regulatory domains, which might bind and sense ATP levels [21]. Alternatively, they can act as regulators of protein kinases, mimicking mechanisms used by active protein kinases. Most human pseudokinase domains are conserved in all vertebrates, and several are even more ancient: CCK4 is inactive in all metazoans, and the inactive second kinase domain of GCN2 is found in almost all eukaryotes, suggesting that these domains play vital biological roles [11].

There were predicted to be 106 kinase pseudogenes in the human genome. These have sequence similarity to protein kinases, but have stop codons or frameshifts within their sequence, and in many cases (75) lack introns, indicating that they are retrotransposed copies of expressed kinases. For reasons that are unclear, some protein kinase families have a very high ratio of pseudogenes to functional genes (e.g., MARK 28:4). The mouse genome has a similar count of 97 kinase pseudogenes. None of these are orthologous to human, and the families that have high pseudogene counts are distinct from human, implying that no cryptic function remains for most kinase pseudogenes. On the other hand, retrotransposition appears to be the origin of several recently-derived functional kinases, such as the primate-specific TAF1L, CK1α2, and PKACγ genes [14]. Pseudogenes are rare in most invertebrate kinomes, although *C. elegans* has 24 kinase pseudogenes, mostly in recently expanded families [22].

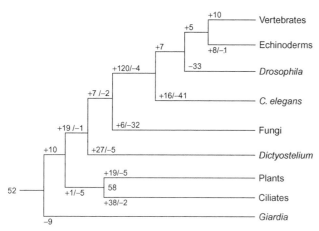

FIGURE 2.2 Gain and loss of kinase classes in eukaryotic evolution. A parsimonious comparison of kinomes suggests an early eukaryotic kinome of 52 distinct subfamilies, with gains and loss (+ and −) occurring at each branch of the tree. The positions of nematodes and insects are not well resolved. Major losses of kinases (−32) were seen in fungi, and a major gain (+120) in the emergence of metazoans. Unclassifiable kinases were omitted from these counts.

COMPARATIVE KINOMICS

Kinome catalogs provide a context for every kinase, allowing the definitive identification of related kinases in the same organism, and equivalent (orthologous) kinases in others. This allows integration of experimental data, and inference from the patterns of evolutionary gain and loss of kinases. To enable these comparisons, we created a hierarchical classification of all protein kinases into Groups, Families, and Subfamilies. In general, groups share overall substrate site preference (e.g., the "proline directed" CMGC group), families share similar biochemical functions (e.g., the MAPK family), and subfamilies are evolutionarily conserved variations within the families, including, for instance, the Erk, Jnk, p38, Erk5, and Erk7 MAPK subfamilies. The full classification scheme and kinomes are available at the KinBase database (http://kinase.com), and extend the original Hanks and Hunter classification [8]. Family and subfamily sizes can vary from a single, deeply conserved member to expansions of up to 10 or more members. Although most protein kinases fall into major groups, many are still outliers ("Other" group), either singly or as small families.

Comparison of several kinomes at the subfamily level shows a rich primordial diversity of protein kinases, and their dynamic gain and loss throughout evolution (Figure 2.2). Remarkably, the common ancestor of all sequenced eukaryotes appears to have had 52 distinct protein kinase subfamilies, of which 28 are still present in all non-parasitic genomes surveyed. By contrast, only three kinases are shared with archaeae, and while bacteria have ePK-like kinases, none are evolutionary widespread, indicating that an elaborate kinase apparatus emerged early in eukaryotic evolution. These pan-eukaryotic protein kinases include well-known kinases involved in the cell

cycle, DNA repair and replication, lipid signaling, the MAPK cascade, and splicing, as well as several other kinases whose functions are still quite obscure, despite being essential for eukaryotic life (http://kinase.com/evolution/).

The size of the kinome grows with genomic and organismal complexity, and is usually ∼2 percent of the full gene count. Exceptions include the large expansion of immunity-related TKL kinases in higher plants, accounting for over 60 percent of the ∼1100 Arabidopsis protein kinases (4 percent of the proteome), the recent expansions in *C. elegans* (below), and the record 2921 kinases of *Paramecium* (7 percent of all genes), largely due to recent whole genome duplications. Curiously, the non-ePK PKL families are all pan-eukaryotic, but are rarely expanded. Thus, it is the ePK superfamily that is most dynamic by far, perhaps due to its specific adaptations for regulated activity and diversity of substrates [4].

The largest increase in kinase diversity comes with the emergence of multicellular animals (metazoans), where 80 new subfamilies emerge, including the canonical tyrosine kinases (TK group), the TGFβ-like receptor serine/threonine kinases, and a wide variety of cytoplasmic kinases. In particular, the TKs are largely cell surface receptors or receptor-associated, and so correlate with the need for intercellular communication in the leap to multicellularity. Metazoans have a concomitant expansion of the P.Tyr binding domains (SH2 and PTB) that transduce TK signals, and the tyrosine-specific PTP phosphatases. Curiously, the compact genome (∼9000 genes) of a unicellular close relative of the metazoans, the choanoflagellate *Monosiga brevicollis*, recently revealed the presence of 128 tyrosine kinase genes, and a similar huge expansion in PTP and SH2 domains [23]. Apart from a mature set of Src-subgroup kinases, none had distinct orthology to metazoans, suggesting that a unicellular common ancestor had a limited TK signaling network, which was independently elaborated in both metazoan and choanoflagellate lineages.

While the TK group is restricted to choanoflagellates and animals, tyrosine phosphorylation is found in many other species, including fungi, where dual-specificity kinases such as GSK3 and DYRK can tyrosine autophosphorylate, and MEKs tyrosine-phosphorylate MAPKs. The Swe1 gene also acts through tyrosine phosphorylation of Cdc2. In plants and *Dictyostelium*, TKL kinases are likely to also phosphorylate on tyrosine, and several such kinases are receptors, or associated with SH2 domains indicating an analogous role in intercellular communication.

The three main invertebrate model systems have characteristic kinomes, which we will explore in more depth.

Saccharomyces cerevisiae

This was first eukaryotic genome to be sequenced, with ∼6200 genes, of which we believe 130 encode protein kinases (2.1 percent of all genes) [5, 7]. These include members of ~25 percent of all subfamilies found in human. Analysis of the *Schizosaccharomyces pombe* genome and drafts of other fungal kinomes show that most fungi have a similar array of kinases. These include seven families that are unique to fungi, and whose functions are particular to the fungal lifestyle, such as cell wall biosynthesis, osmosis, and other stresses. Just as notable is the apparent loss of 26 more ancient kinase classes in *S. cerevisiae*, many of which are also lost in other fungi, possibly during the transition to a sedentary lifestyle (Figure 2.2). Some of these, such as the DYRK2 and PRP4 splicing kinases, are found in other fungi, correlating with their greater use of splicing and gradual loss of kinase functions. *S. cerevisiae* lacks any receptor kinases: it has lost the ancient TKL group, and predates the invention of TKs.

Caenorhabditis elegans

The nematode *C. elegans* is a complex multicellular organism which undergoes a deterministic developmental program. Of its ~20,000 genes, 440 (2.2 percent) encode protein kinases, of which 422 are ePKs [10, 12, 22]. These include almost 75 percent of the kinase subfamilies found in human. There is a striking elaboration of the TK group, with 86 genes (20 percent of the kinome). Compared to *Drosophila*, the relatively large number of TK genes is in part accounted for by the expansion of two families; the Fer non-receptor tyrosine kinase (38 genes), and the Kin16 receptor tyrosine kinases (RTKs) (16 genes). Large expansions are also seen in selected other families (e.g., there are 85 CK1 genes in *C. elegans* compared with 12 in humans and 10 in *Drosophila*). Some of these expansions appear to be very recent, and are not seen in other *Caenorhabditis sp.* genomes; several have degenerated into pseudogenes. There are also a number of nematode-specific kinase subfamilies, largely of unknown function.

Drosophila melanogaster

The dipteran insect *D. melanogaster* has ~13,600 genes – significantly fewer than *C. elegans*, even though by most criteria the fly is a more complex organism. *Drosophila* has 240 protein kinase genes (1.8 percent of all genes), of which 223 are ePKs, with 32 TK genes (14 percent of all protein kinases) [10]. In fact, the percentage of protein kinase genes in *Drosophila* is very similar to that of *C. elegans*, if one trims away the highly expanded protein kinase families in *C. elegans*. Eighteen new protein kinase subfamilies emerge in *Drosophila*, which have functions in immunity (e.g., JAK), morphogenesis (e.g., LIMK), and the nervous system (e.g., MuSK); in contrast, eight subfamilies, including the Met and Trk RTKs are found in nematodes but absent from flies. In fact, preliminary analysis of early metazoan genomes, such as the sea anemone, shows several kinase classes found in vertebrates but not in nematodes or *Drosophila*, suggesting that substantial gene loss occurred in both these lineages, making the timing of invention of many protein kinases uncertain. The various shared and unique protein kinase families in the yeasts, worms, flies, and humans can be explored online at http://kinase.com/.

EMERGENCE AND DIVERSITY OF VERTEBRATE KINOMES

Most vertebrate kinomes are highly similar to that of human, allowing both experimental and computational studies of other vertebrate kinases to be applied to their unique human orthologs. The mouse kinome has single orthologs for 510 of the original 518 human kinases, and most of the differences are due to individual cases of gene loss and duplication by retrotransposition. The divergence increases in more distant vertebrates, but our current estimate is that all vertebrates share a core of ~470 protein kinases (E. D. Scheeff, pers. comm.). The analysis of the sea urchin kinome [13] and availability of invertebrate chordate genomes from amphioxus and *Ciona sp.* show that the major advance in vertebrate kinomes was the proposed four-fold genome duplication, leading to increases in the number of kinases in most subfamilies. By contrast, the only vertebrate-specific subfamilies are the somewhat diffuse atypical kinases FASTK and H11.

CODA

The elucidation of the kinomes of both unicellular and multicellular eukaryotes provides us with rich insights into the evolutionary history of protein kinases, and also defines the precise number of protein kinases that can participate in phosphorylation reactions in a given eukaryotic cell type. On the other side of the coin, we have to remember that phosphorylation cannot regulate protein function unless there are protein phosphatases to remove the regulatory phosphate moieties. Genomics has also had a major impact on the world of protein phosphatases, and the number of protein phosphatases in the three major superfamilies (the "phosphatome") is also very large, with ~150 protein phosphatase catalytic subunit genes being present in the human genome. In combination, the protein kinase and phosphatase genes account for nearly 2.5 percent of all genes in most eukaryotic species. The increasing rate of kinome-wide experimental approaches, including RNAi screens, global expression profiles and phosphoproteomics, promise a greater systems level understanding of the kinome and how it modulates cell behavior. In parallel, protein kinases have become a major therapeutic target, both in addressing kinase-driven diseases and in epistatically correcting other defects, and over 10 kinase inhibitors have now been approved for clinical use.

REFERENCES

1. Walsh DA, Perkins JP, Krebs EG. An adenosine 3′,5′-monophosphate-dependent protein kinase from rabbit skeletal muscle. *J Biol Chem* 1968;**243**:3763–5.

2. Hunter T. A thousand and one protein kinases. *Cell* 1987;**50**:823–9.

3. Hunter T. Signaling –2000 and beyond. *Cell* 2000;**100**:113–27.

4. Kannan N, Taylor SS, Zhai Y, Venter JC, Manning G. Structural and functional diversity of the microbial kinome. *PLoS Biol* 2007;**5**:e17.

5. Cheek S, Ginalski K, Zhang H, Grishin NV. A comprehensive update of the sequence and structure classification of kinases. *BMC Struct Biol* 2005;**5**:6.

6. Srivastava S, Li Z, Ko K, Choudhury P, Albaqumi M, Johnson AK, Yan Y, Backer JM, Unutmaz D, Coetzee WA, Skolnik EY. Histidine phosphorylation of the potassium channel KCa3.1 by nucleoside diphosphate kinase B is required for activation of KCa3.1 and CD4 T cells. *Mol Cell* 2006;**24**:665–75.

7. Hanks SK, Quinn AM, Hunter T. The protein kinase family: conserved features and deduced phylogeny of the catalytic domains. *Science* 1988;**241**:42–52.

8. Hanks SK, Hunter T. Protein kinases 6. The eukaryotic protein kinase superfamily: kinase (catalytic) domain structure and classification. *FASEB J* 1995;**9**:576–96.

9. Hunter T, Plowman GD. The protein kinases of budding yeast: six score and more. *Trends Biochem Sci* 1997;**22**:18–22.

10. Manning G, Plowman GD, Hunter T, Sudarsanam S. Evolution of protein kinase signaling from yeast to man. *Trends Biochem Sci* 2002;**27**:514–20.

11. Manning G, Whyte DB, Martinez R, Hunter T, Sudarsanam S. The protein kinase complement of the human genome. *Science* 2002;**298**:1912–34.

12. Plowman GD, Sudarsanam S, Bingham J, Whyte D, Hunter T. The protein kinases of *Caenorhabditis elegans*: a model for signal transduction in multicellular organisms. *Proc Natl Acad Sci USA* 1999;**96**:13,603–10.

13. Bradham CA, Foltz KR, Beane W, S. Arnone MI, Rizzo F, Coffman JA, Mushegian A, Goel M, Morales J, Geneviere AM, Lapraz F, Robertson AJ, Kelkar H, Loza-Coll M, Townley IK, Raisch M, Rouz MM, Legpage T, Gache C, McClay DR, Manning G. The sea urchin kinome: a first look. *Dev Biol* 2006;**300**:180–93.

14. Caenepeel S, Charydczak G, Sudarsanam S, Hunter T, Manning G. The mouse kinome: discovery and comparative genomics of all mouse protein kinases. *Proc Natl Acad Sci USA* 2004;**101**:11,707–12.

15. Eisen JA, Coyne RS, Wu M, Wu D, Thiagarajan M, Wortman JR, Badger JH, Ren Q, Amedeo P, Jones KM, Tallon LJ, Delcher AL, Salzberg SL, Silva JC, Haas BJ, Majoros WH, Farzad M, Carlton JM, Smith RK, Garg J, Pearlman RE, Karrer KM, Sun L, Manning G, Elde NC, Turkewitz AP, Asai DJ, Wilkes DE, Wang Y, Cai H, Collins K, Stewart BA, Lee SR, Wilamowska K, Weinberg Z, Ruzzo WL, Wloga D, Gaertig J, Frankel J, Tsao CC, Gorovsky MA, Keeling PJ, Waller RF, Patron NJ, Cherry JM, Stover NA, Krieger CJ, Del Toro C, Ryder HF, Williamson SC, Barbeau RA, Hamilton EP, Orias E. Macronuclear genome sequence of the ciliate T*etrahymena thermophila*, a model eukaryote. *PLoS Biol* 2006;**4**:e286.

16. Goldberg JM, Manning G, Liu A, Fey P, Pilcher KE, Xu Y, Smith JL. The *Dictyostelium* kinome – analysis of the protein kinases from a simple model organism. *PLoS Genet* 2006;**2**:e38.

17. Forrest AR, Taylor DF, Crowe ML, Chalk AM, Waddell NJ, Kolle G, Faulkner GJ, Kodzius R, Katayama S, Wells C, Kai C, Kawai J, Carninci P, Hayashizaki Y, Grimmond SM. Genome wide review of transcriptional complexity in mouse protein kinases and phosphatases. *Genome Biol* 2006;**7**:R5.

18. Grangeasse C, Cozzone AJ, Deutscher J, Mijakovic I. Tyrosine phosphorylation: an emerging regulatory device of bacterial physiology. *Trends Biochem Sci* 2007;**32**:86–94.

19. Manning, G. (2005). Protein kinases in human disease. In: *2005–06 Catalog and Technical Reference*, Beverly, MA: Cell Signaling Technologies, pp. 402–9.

20. Scheeff E, Eswaran J, Bunkoczi G, Knapp S, Manning G. Structure of the pseudokinase VRK3 reveals a degraded catalytic site, a highly conserved kinase fold, and a putative regulatory binding site. *Structure* 2009;**17**:128–38.

21. Boudeau J, Miranda-Saavedra D, Barton GJ, Alessi DR. Emerging roles of pseudokinases. *Trends Cell Biol* 2006;**16**:443–52.

22. Manning, G. (2005). Genomic overview of protein kinases. In: *WormBook, C. elegans Research Community* (available at http://wormbook.org).

23. Manning G, Young SL, Miller WT, Zhai Y. The protist, *Monosiga brevicollis*, has a tyrosine kinase signaling network more elaborate and diverse than found in any known metazoan. *Proc Natl Acad Sci USA* 2008;**105**:9674–9.

24. The Arabidopsis Genome Initiative. Analysis of the genome sequence of the flowering plant Arabidopsis thaliana. *Nature* 2000;**408**:796–815.

25. Simillion C, Vandepoele K, Van Montagu MC, Zabeau M, Van de Peer Y. The hidden duplication past of Arabidopsis thaliana. *Proc Natl Acad Sci USA* 2002;**99**:13,627–32.

Chapter 3

Modular Protein Interaction Domains in Cellular Communication

Tony Pawson[1] and Piers Nash[2]

[1]*Samuel Lunenfeld Research Institute, Mt Sinai Hospital, Toronto, Ontario, Canada*
[2]*Ben May Department for Cancer Research, University of Chicago, Chicago, Illinois*

INTRODUCTION

Signal transduction is accomplished by networks of interacting proteins, which function in concert to produce a coherent response to extracellular and internal cues. Intracellular signaling proteins generally contain modular domains that either have a catalytic function (such as kinase activity), or mediate the interactions of proteins with one another or with other biomolecules, including phospholipids, nucleic acids, or small molecules. These latter interaction domains typically fold in such a way that their N and C termini are juxtaposed in space, while their ligand binding site is located on the opposing surface; therefore, they are ideally configured for incorporation into a host polypeptide while retaining their intrinsic binding properties. Interaction domains play a critical role in the selective activation of signaling pathways through their ability to recruit target proteins to activated receptors, and to regulate the ensuing assembly of signaling complexes. Such interaction domains can control not only the specificity of signal transduction but also the kinetics with which cells react to external and intrinsic stimuli, and they can thereby generate complex cellular behaviors. This chapter outlines these general themes; more detail is provided in a number of recent papers [1–7].

PHOSPHOTYROSINE-DEPENDENT PROTEIN–PROTEIN INTERACTIONS

To understand how protein kinases regulate intracellular functions, it is critical to appreciate the mechanisms by which kinases are controlled, and the means by which phosphorylation alters the biochemical properties of their target proteins. Interaction domains present in the protein kinase itself can control catalytic activity through contacts with the kinase domain, and also modulate substrate recognition, either by localizing the kinase to a specific subcellular site or through direct binding to potential targets. The biological activities of protein kinases are, by definition, exerted through their ability to modify substrate proteins by phosphorylation, most commonly in eukaryotic cells on the hydroxyamino acids serine, threonine, and tyrosine. An important and general consequence of phosphorylation is the creation of binding sites for the interaction domains of downstream proteins that recognize specific phosphorylated peptide motifs. Such phosphorylation-dependent protein–protein interactions induce the formation of heteromeric complexes that localize signaling proteins to their sites of action within the cell, and juxtapose polypeptides that act within the same pathways. Interaction domains that recognize phosphorylated sites can also induce conformational changes that regulate the enzymatic and binding properties of their host proteins, for example by mediating intramolecular interactions that result in allosteric regulation.

A principal mechanism by which protein-tyrosine kinases engage downstream targets is through the ability of Src homology 2 (SH2) domains of cytoplasmic proteins to recognize specific phosphotyrosine-containing peptide motifs of four to eight residues, such as those found on activated receptor tyrosine kinases [8–12]. SH2 domains contain approximately 100 amino acids, which normally form an N-terminal α-helix, a central β-sheet, followed by a smaller β-sheet, a second α-helix, and a C-terminal β-strand. This results in a bipartite structure in which the central β-sheet of the SH2 domain separates a conserved phosphotyrosine binding pocket from a more variable surface that often engages residues C-terminal to the phosphotyrosine [13–15]. Commonly, the phosphopeptide ligand binds as an extended strand that crosses the central β-sheet to engage both the phosphotyrosine binding and specificity pockets (Figure 3.1). Phosphopeptide binding requires a conserved arginine residue in the SH2 domain that anchors the phosphorylated tyrosine residue through a buried ionic bond. This phosphotyrosine recognition provides about half of the binding energy in the interaction of an SH2 domain with a phosphopeptide, and for this reason SH2 domains generally bind phosphorylated sites with about 1000-fold higher affinity than their non-phosphorylated counterparts [16, 17]. Because the dissociation constants of SH2 domains associated with optimal phosphorylated sites are commonly in the 0.5–1 μM range, this means that phosphorylation effectively serves as a switch for the recruitment of SH2-containing proteins to phosphotyrosine-containing motifs [18]. The ability of SH2 domains to recognize residues flanking the phosphotyrosine can differ from one SH2 domain to another, and this discrimination provides an element of specificity in tyrosine kinase signaling [10, 19, 20]. For example, the SH2 domain of the Grb2 adaptor protein binds preferentially to phosphotyrosine sites followed two amino acids to the C-terminus (the +2 position) by an asparagine, whereas the SH2 domains of phosphatidyinositol 3'-kinase (PI3K) bind preferentially to phosphorylated sites with a methionine at the +3 position. As an example of such phosphotyrosine-dependent interactions in signal transduction, activated receptor tyrosine kinases become autophosphorylated on multiple sites, for example in their C-terminal tail or juxtamembrane region, each of which binds preferentially to specific SH2-containing proteins. The sequence contexts of a receptor's autophosphorylation sites therefore influence which SH2 proteins it recruits, and thus which cytoplasmic signaling pathways it stimulates in the cell, and with what kinetics [21].

In addition to the canonical binding mode described above, SH2 domains can show considerable flexibility in their binding properties. For example, in several cases SH2 domains recognize residues both N- and C-terminal to the phosphotyrosine, and thus bind extended phosphopeptide motifs with increased specificity [12, 22–23]. SH2 domains can also bind selectively to peptides containing two phosphotyrosine sites, and recognize motifs that adopt conformations that differ from the typical extended strand [24, 25] (Figure 3.1). Furthermore, the SH2 domain of SH2D1A (also called SAP) binds with relatively high affinity to non-phosphorylated sites on SLAM family transmembrane receptors in lymphoid cells, although binding strength increases upon phosphorylation [26, 27]. Moreover, several SH2 domains have secondary recognition sites that are quite distinct from the conventional phosphopeptide binding surface. These can allow for SH2 domain homodimerization, as in the case of the APS protein [26], or the recruitment of atypical ligands. In this latter class, the SH2D1A SH2 domain associates with the

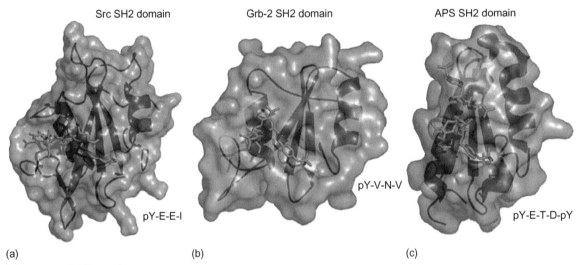

FIGURE 3.1 Modes of SH2 binding.
(a) Structure of the Src SH2 domain bound to a pTyr–Glu–Glu–Ile peptide (PDB: 1SPS) [13]. The surface of the SH2 domain is shown as a translucent skin, and the secondary structural elements of the SH2 domain in ribbons format as an underlying skeleton. The αA helix is to the right and αB helix to the left. The Arg βB5 residue critical for phosphotyrosine binding is in blue. The N-terminal phosphotyrosine of the peptide (stick model) occupies the phosphotyrosine binding pocket. The peptide runs over the central β sheet of the SH2 domain, the +1 and +2 glutamates contact the surface of the domain, and the side-chain of the +3 Ile (to the left) fits in a hydrophobic pocket. (b) The Grb2 SH2 domain in complex with a pYVNV peptide (PBD: 1BMB). A tryptophan in the SH2 domain (light gray) stabilizes the beta-turn conformation essential for high-affinity binding. (c) Two phosphotyrosine binding pockets are present on a single SH2 domain of APS. A single APS SH2 molecule is bound to a pYETDpY peptide (stick model) as part of the activation loop of the activated insulin-receptor. The pTyr-1158 of the insulin-receptor interacts with the conserved SH2 residue Arg-1161 (light gray), while the Lys-455 and Lys-457 of the SH2 domain create a second binding pocket for pTyr-1162 of the insulin-receptor (PDB: 1RQQ). Note that in each case the trajectory of the phosphopeptide with respect to the SH2 domain is different. Ribbon and surface diagrams were generated using Pymol. Adapted from Liu *et al.* [12], with permission.

SH3 domain of the Src family tyrosine kinase Fyn through a basic surface centered on Arg-78 of the SH2 domain [28, 29]. The SH2D1A SH2 domain therefore has an intrinsic adaptor function, since it can simultaneously associate with a (phospho)tyrosine-based motif on the SLAM receptor and the SH3 domain of the Fyn tyrosine kinase, which consequently phosphorylates novel sites in SLAM that recruit additional SH2-containing effectors. These observations make the general point that SH2 domains, in common with many interaction and catalytic domains, have likely been selected in the course of evolution due to their plasticity.

Distinct interaction domains in protein kinases often function in concert to regulate catalytic function. For example, both the affinity and specificity of SH2 domain interactions can be increased by the presence of two tandem SH2 domains in a single protein that together recognize a doubly phosphorylated ligand, as in the case of the ZAP-70 cytoplasmic tyrosine kinase binding to the signaling subunits of the T-cell antigen receptor [30]. SH2 domains are also often linked to distinct interaction modules, such as SH3 domains that typically bind proline-rich motifs that adopt a polyproline type II helix (i.e., Pro–X–X–Pro) [31, 32], as in the case of the c-Src cytoplasmic tyrosine kinase. In the inactive state, when c-Src is phosphorylated on a C-terminal tyrosine residue, its SH2 and SH3 domains both make intramolecular interactions that repress kinase activity. In this autoinhibited conformation, the SH2 domain binds a phosphorylated tyrosine (Y527) in the C-terminal tail, while the SH3 domain associates with SH2-kinase linker [33, 34]. However, upon dephosphorylation of the tail, both the SH2 and SH3 domains are liberated to bind other proteins, such as substrates for phosphorylation, through the coordinate recognition of both phosphotyrosine- and proline-based motifs [35, 36].

Despite their involvement in the autoinhibition of cytoplasmic tyrosine kinases such as Src and Abl, the combination of SH2 and kinase domains likely evolved to promote kinase activity and substrate recognition. For example, in the human Fps/Fes cytoplasmic tyrosine kinase, the SH2 domain interacts with the small lobe of the kinase domain to stabilize the αC helix of the kinase in an active configuration, and is therefore essential for efficient catalytic activity [37]. The Fps/Fes SH2 domain can also bind pre-phosphorylated (primed) substrates, which are thereby presented for further phosphorylation at the active site. In addition, binding of a ligand to the phosphotyrosine binding pocket of the SH2 domain apparently stabilizes the conformation of the SH2 domain, and thus its stimulatory interaction with the kinase domain, raising the possibility that the kinase is only fully active in the presence of an appropriately primed substrate. Mutations in the Fps/Fes SH2 domain can entirely inactivate the adjacent kinase domain, or alter substrate selection, consistent with the close integration of the SH2-kinase unit observed at the structural level [37–39]. In the case of the Abl cytoplasmic tyrosine kinase, the SH2

and SH3 domains interact with the backside of the kinase domain to promote the autoinhibited state [40], but upon activation the SH2 domain forms a new interaction with the small lobe of the kinase that stimulates catalytic activity [37, 41]. These data reinforce the flexibility of the SH2 domain, and its intimate connection with the catalytic domain of cytoplasmic tyrosine kinases in phosphotyrosine signaling.

Although many receptor tyrosine kinase targets have SH2 domains, these proteins otherwise have a variety of different biochemical and biological functions. These include the regulation of small Ras-like GTPases, control of phosphoinositide metabolism, tyrosine phosphorylation and dephosphorylation, transcriptional regulation (STAT proteins), organization of the cytoskeleton, and protein ubiquitination (SOCS box proteins and RING domain proteins such as c-Cbl) (Figure 3.2). SH2-containing proteins, therefore, couple tyrosine kinases to a broad range of regulatory biochemical pathways within the cell. In addition, SH2 domains are often found in adaptor proteins that are composed exclusively of SH2 domains and other interaction modules, such as SH3 domains; these adaptors act as a bridge to physically link phosphotyrosine signaling to multiple targets [42]. These observations suggest that during the course of evolution, the acquisition of SH2 domains may have allowed an otherwise diverse set of signaling proteins to associate with, and be activated by, receptor tyrosine kinases, as we discuss in more detail below.

In addition to SH2 domains, a quite distinct interaction module, the phosphotyrosine binding (PTB) domain, can also recognize proteins in a phosphotyrosine-dependent fashion [43]. The PTB domains of proteins such as Shc, IRS-1, and FRS2 generally recognize Asn–Pro–X–pTyr (NPXpY) motifs in activated receptor tyrosine kinases, such as the epidermal growth factor receptor, the insulin receptor, or the Trk nerve growth factor receptor [44–48]. Additional selectivity in PTB domain recognition is provided by residues flanking the NPXpY motif; for example, the Shc PTB domain binds preferentially to NPXpY motifs with a large aliphatic residue at the -5 position relative to the phosphotyrosine. Upon binding activated receptors, PTB domain proteins themselves become phosphorylated on multiple tyrosine residues corresponding to SH2 domain binding sites, and consequently recruit specific sets of SH2-containing effectors. Proteins with PTB domains therefore function as scaffolds to amplify and expand the range of receptor signaling. For example, once recruited to the activated insulin receptor by its PTB domain, the IRS-1 docking protein is phosphorylated at multiple Tyr–X–X–Met motifs, which consequently bind the SH2 domains of the p85 adaptor subunit of PI3K [49]. IRS proteins are therefore essential for the ability of the insulin receptor to stimulate PI3K signaling, which in turn is crucial for the cellular responses such as increased glucose transport and growth. A subset of C2 domains, such as that found in PKCδ, also bind specific phosphotyrosine sites [50]. The fact that SH2,

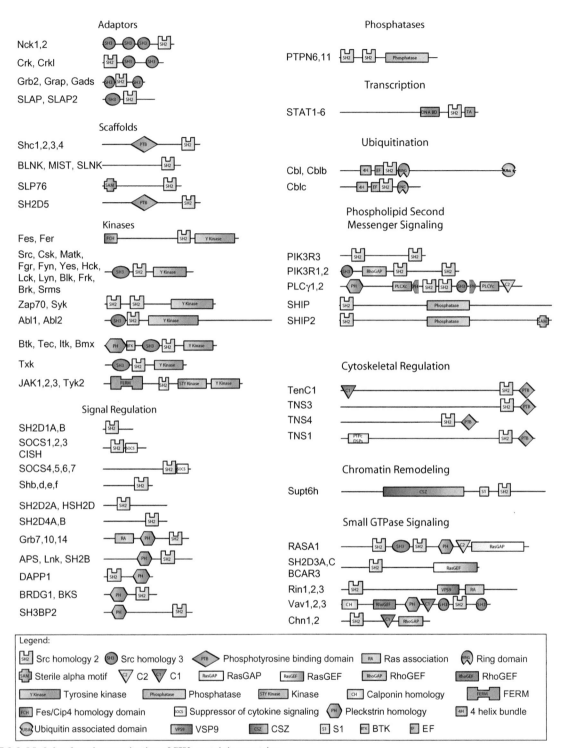

FIGURE 3.2 Modular domain organization of SH2-containing proteins.
Classification and domain composition of the 110 non-redundant human SH2 domain-containing proteins identified in human and mouse by Pfam [108].
More information on SH2-containing proteins can be found at http://sh2.uchicago.edu/ and at SMART (http://smart.embl-heidelberg.de/). Adapted from
Liu et al. [12], with permission.

PTB, and C2 domains have separately evolved the ability to selectively bind phosphotyrosine-containing motifs suggests that such phosphotyrosine-dependent protein–protein interactions are widely exploited by signaling proteins.

These observations indicate that tyrosine kinase signaling requires a successive series of modular, phosphorylation-dependent protein–protein interactions. Interestingly, conventional tyrosine kinases and SH2 domains are absent from yeast, but make a concomitant appearance in the immediate predecessors of multicellular animals. The emergence of SH2 domains to mediate the effect of tyrosine phosphorylation, and the evolution of dedicated tyrosine kinases, represents an evolutionary step that likely facilitated intercellular signaling required for the formation of metazoan species. The human genome encodes some 120 distinct SH2 domains, in 110 proteins [12]; 56 PTB domains are found in 52 human proteins [51], although only a subset of PTB domains have phosphotyrosine binding activity. Knowing the complete set of tyrosine kinases and proteins with phosphotyrosine recognition domains helps to establish the wiring circuitry of phosphotyrosine signaling.

INTERACTION DOMAINS: A COMMON THEME IN SIGNALING

Interaction domains are essential in signaling from many different types of cell surface receptors, as well as in cellular events such as the cell cycle, protein and vesicle trafficking, targeted protein degradation, DNA repair, oxygen sensing, and control of the cytoskeleton. Thus, the SH2 domain serves as a prototype for a large family of protein-interaction modules (Table 3.1), some of which specifically recognize peptide motifs with different forms of posttranslational modification, in a fashion akin to the selective binding of SH2 and PTB domains to phosphotyrosine-containing sequences [7, 52]. A number of proteins or their constituent domains can bind phosphothreonine/phosphoserine-containing motifs (i.e., 14-3-3, FHA, MH2, polo box, BRCT, WD40, WW) and thereby mediate the effects of protein- serine/threonine kinases [53]. For example, in a series of interactions analogous to receptor tyrosine kinase signaling, the activated type I TGFβ receptor serine/threonine kinase becomes autophosphorylated within its juxtamembrane region, thereby creating binding sites for the MH2 domain of a regulatory (R-) SMAD protein, which recognizes pSer–X–pSer motifs [54, 55]. Subsequent phosphorylation of the R-SMAD itself leads to binding to the MH2 domain of SMAD4 and translocation of the R-SMAD/SMAD-4 complex to the nucleus, where it acts to regulate gene expression [56]. In contrast, bromo- and chromodomains (amongst others) bind specific lysine-based motifs (notably in histones) in a fashion dependent on acetylation or methylation, respectively, of the lysine residue, and thereby play an important role in chromatin organization and transcriptional control [57, 58]. In a similar fashion, the VHLβ domain of the VHL protein, a component of an E3 protein ubiquitin ligase complex, selectively binds motifs containing a hydroxyl-proline residue in the transcription factor HIF1α [59, 60] As a consequence, at normal oxygen conditions, when Pro402 and Pro546 of HIF1α are hydroxylated, HIF1α is recognized by VHL and is therefore ubiquitinated and degraded. As the oxygen tension falls, HIF1α

hydroxylation, and thus binding to VHL, is lost, and HIF1α is stabilized to induce the expression of genes such as vascular endothelial growth factor.

A number of interaction domains recognize monoubiquitin or polyubiquitin chains attached through an isopeptide linkage to lysine residues of target proteins [61, 62]. Ubiquitin binding domains differ from domains that recognize other forms of posttranslational modification, such as phosphorylation, in the sense that they only interact with the ubiquitin tag and not with residues in the modified protein itself. However, there are some data to suggest that different ubiquitin binding domains preferentially recognize distinct forms of polyubiquitin, depending on the identity of the lysine residue in ubiquitin that is involved in the extended polyubiquitin chain. As one example of a ubiquitin binding domain involved in signal transduction, ubiquitin interaction motifs (UIM) are a common feature of endocytic proteins, which bind mono- or polyubiquitinated sites and thereby regulate protein trafficking at endosomes [63]. In the context of receptor tyrosine kinase signaling, they act in series with the SH2 domain of Cbl, an E3 protein ubiquitin ligase that binds to specific phosphotyrosine-containing sites on activated receptors, which it consequently ubiquitinates at sites that are then recognized by the UIM domains of proteins involved in trafficking and localization to endosomes [64]. Indeed, the coupling of ubiquitination to protein phosphorylation is a common theme in signal transduction, such that phosphorylation of a protein on tyrosine or serine/threonine marks it for recognition by the phosphorylation-dependent interaction domain of an E3 protein ubiquitin ligase; the phosphorylated protein is consequently ubiquitinated and recognized by ubiquitin binding domains.

In addition to protein–protein interaction domains that selectively recognize sites of posttranslational modification, there are numerous domains that bind unmodified peptide motifs, such as proline-rich sequences (SH3, WW, EVH1, and GYF domains) [32] or the extreme C-terminal residues of target proteins (PDZ domains) [65, 66]. In principle such interactions are constitutive, but they can in fact be regulated, for example by intramolecular interactions that preclude binding to exogenous proteins (as in the case of the SH3 domain of the autoinhibited Src tyrosine kinase), or by phosphorylation of the peptide ligand, which can antagonize binding to modules such as SH3 domains [67]. In the cases described thus far, the peptide ligands for interaction domains are typically in unstructured regions of larger polypeptides. In some cases, however, two interaction domains can oligomerize in a fashion that depends on the folded structures of both domains. For example, the cytoplasmic tail of the Fas receptor contains a death domain, which upon receptor activation heterodimerizes with the death domain of an adaptor protein, FADD. FADD also possesses a death effector domain that associates with the death effector domains of a specialized

TABLE 3.1 Functions of selected protein interaction modules

Interaction module or domain	Example	Binding functions
Phospho-recognition		
SH2	Grb2 & Gads adaptors, Src family kinases, phospholipase C-γ	Phosphotyrosine
PTB	Shc, IRS-1, FRS2	Phosphotyrosine
FHA	Rad53	Phosphothreonine/ phosphoserine
14-3-3	14-3-3 proteins	Phosphothreonine/phosphoserine
WD40	Cdc4, β-TRCP (F-box proteins)	Phosphothreonine/phosphoserine
WW	Pin1	Phosphothreonine/phosphoserine
Proline recognition		
SH3	Src, Crk	Pro–X–X–Pro
WW	YAP, Nedd4	Pro–Pro–X–Tyr
EVH1	Mena, homer	Pro-rich
Gyf	CD2	PPPPGHR motif
Phospholipid recognition		
PH	PKB, mSos, phospholipase C-(C)	$PI(3,4)P_2$, $PI(4,5)P_2$, $PI(3,4,5)P_3$
FYVE	SARA, Hrs, EEA1	$PI(3)P$
PX	p40phox, p47phox	$PI(3)P$
FERM	ERM, Radixin	$PI(4,5)P_2$
Tubby	TULP1, tubby	$PI(4,5)P_2$
Methylated or acetylated residue recognition		
Bromo	P/CAF	Acetylated lysine
Chromo	HP1	Methylated lysine
Other motif recognition		
SH3	Gads & Grb2	Arg–X–X–Lys
	C-terminal SH3	
PDZ	PSD-95	C-terminal motifs
PDZ	Neural nitric oxide synthase (nNos)	PDZ
SAM	—	SAM

protease, pro-caspase 8 [68]. The resulting oligomerization of the caspase leads to its activation and autocleavage, and consequent activation of a pathway leading to apoptotic cell death . In a similar fashion, SAM domains, which are found in a wide range of signaling proteins from receptor tyrosine kinases to cytoplasmic serine/threonine protein kinases and transcriptional regulators, form dimers or more extended oligomers that control a variety of cellular responses [69].

Although intensively studied in the context of proten recognition, interaction domains can bind a variety of biomolecules, including phospholipids, nucleic acids, second messengers, and metabolites. A separate class of interaction domains (i.e., PH, FYVE, PX, ENTH, FERM, GRAM, Tubby) recognizes specific phospholipids, particularly phosphoinositides, and therefore directs proteins to regions in membranes enriched for the appropriate phospholipid [70]. These phospholipid binding domains mediate the effects of lipid kinases and phosphatases, and thus function to spatially restrict the intacellular activities of signaling proteins; they also act in synchrony with protein–protein interaction domains to coordinate intracellular signaling pathways. For example, autophosphorylated receptor tyrosine kinases bind the p85 SH2-containing subunit of PI3K, thereby stimulating PI3K to produce PI(3,4,5)P$_3$. This phospholipid engages the PH domains of intracellular targets such as the serine/threonine protein kinases PKB/Akt and phosphoinositide-dependent protein kinase (PDK1), which consequently are recruited to the plasma membrane, resulting in PKB activation [71]. PKB, in turn, phosphorylates targets that include the pro-apoptotic protein BAD and the transcription factor FKHRL at Ser residues, which subsequently bind 14-3-3 proteins [6]. The 14-3-3 binding represses the ability of the phosphorylated proteins to induce apoptosis by sequestering them in the cytoplasm away from their sites of action. PKB also phosphorylates and inhibits TSC2, a component of a GTPase activating protein that inactivates the Rheb GTPase, and thereby allows Rheb to stimulate the mTOR protein kinase, which in turn increases protein synthesis and cell growth [72]. A dominant mutation in the AKT1/PKB PH domain found in some human cancers increases the affinity of the PH domain for phosphoinositides, resulting in enhanced PKB association with the plasma membrane and aberrant activation of downstream PKB signaling [73]. Thus, a signaling pathway can be constructed from a series of protein and lipid kinases and a succession of phosphorylation-dependent protein–protein and protein–phospholipid interactions. Furthermore, mutations that modulate the affinity of these interactions can lead to ectopic, oncogenic signaling.

BAR and F-BAR domains also associate with phospholipid membranes, but with the additional feature that they typically bind selectively to curved membranes and can themselves induce membrane tubulation [74]. This depends on the ability of BAR domains to form curved helical-bundle dimers that recognize phospholipid membranes through a basic concave surface [75, 76]. Such domains have been implicated in the formation of endocytic vesicles, and are found in proteins with a variety of other interaction domains (e.g., SH3, SH2) and catalytic modules (e.g., tyrosine kinase, RhoGAP). For example, the Fps/Fes cytoplasmic tyrosine kinase has an N-terminal F-BAR domain, followed by the SH2-kinase unit discussed above. The F-BAR domain is involved in localization of the kinase to membranes, although it is not necessary for kinase activity *per se* [77]; a speculative model suggests that binding of the F-BAR domain to membrane sites with an appropriate curvature concentrates the kinase, leading to autophosphorylation of the kinase activation segment, and also positions the SH2-kinase cassette for interaction with phosphorylated substrates or scaffolds.

ADAPTORS, PATHWAYS, AND NETWORKS

As noted above, SH3 domains can be linked to SH2 domains to create adaptor proteins that connect a tyrosine kinase upstream signal to pathways that are engaged by the adaptor's SH3 domains. The Grb2 family of adaptor proteins, for example, links tyrosine kinase signals to Ras and mitogen-activated protein (MAP) kinase pathways. Grb2 contains an SH2 domain flanked on either side by an SH3 domain. As discussed previously, the SH2 domain of Grb2 binds preferentially to pTyr–X–Asn motifs on activated receptors or cytoplasmic docking proteins [19], while the N-terminal SH3 domain associates with a Pro–X–X–Pro motif on Sos, a Ras GDP-GTP exchange factor (GEF) (Figure 3.3) [78, 79]. Genetic data from invertebrates and mammals have shown that this pathway is important for receptor tyrosine kinases to activate the Ras-MAP kinase pathway *in vivo*, leading to cell growth and differentiation [80–82]. Indeed, Grb2 is a key component of a signaling pathway that controls a very early cell fate decision in the developing mouse embryo, by promoting the induction of genes required for the differentiation of primitive endoderm, while inhibiting the expression of genes required for the epiblast lineage [83]. While the the C-terminal SH3 domain of Grb2 can also associate with Sos, it has the additional property of binding to the Gab docking proteins through an atypical Arg–X–X–Lys motif [84]. Thus, Grb2 recruits Gab1 to activated receptor tyrosine kinases, such as the epidermal growth factor (EGF) receptor. The ensuing phosphorylation of Gab1 creates binding sites for additional SH2 proteins, particularly the p85 subunit of PI3K, resulting in the localized production of PI(3,4,5)P$_3$ and activation of survival pathways through the PH-containing serine/threonine kinases PKB and PDK1. Grb2 therefore coordinates the activation of at least two distinct signaling pathways. Of interest, there is direct cross-talk between the Ras and PI3K pathways, since Ras-GTP can bind a Ras association (RA) domain in the catalytic p110 subunit of PI3K, and stimulate PI3K activity both in cells and in animals [85]. Taken together, these data suggest that intracellular signaling likely operates as a network of interlinked pathways, established in part from the reiterated use of rather simple interaction modules and protein/lipid covalent modifications [86].

The interaction domains of adaptor proteins do not necessarily function as beads on a string, but in at least some cases can be are potentially controlled through intramolecular allosteric interactions similar to those described for the Src kinase, as exemplified by the Crk adaptor. The CrkII protein has an SH2–SH3–SH3 domain organization, and binds specific phosphotyrosine motifs through its SH2 domain and proteins that activate the Rac and Rap GTPases through its more N-terminal (n) SH3 domain. CrkII therefore links tyrosine phosphorylated receptors and scaffolds to dynamic regulation of the cytoskeleton and cell adhesion [87]. Structural analysis of human CrkII shows that in the absence of external

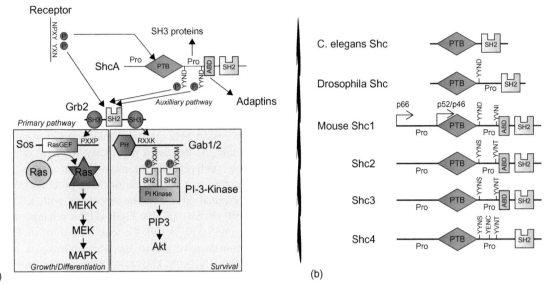

FIGURE 3.3 (a) The Grb2 SH3–SH2–SH3 adaptor couples a pTyr–X–Asn docking site to multiple downstream targets through a series of protein–protein and protein–phospholipid interactions. One pathway to cell growth is assembled through the N-terminal SH3 domain of Grb2 interacting with Pro–X–X–Pro motifs on the Ras-guanine nucleotide exchange factor Sos, leading to activation of Sos and the Erk MAP kinase pathway. A second pathway to cell survival is linked through the C-terminal SH3 domain of Grb2 binding to an Arg–X–X–Lys motif within the scaffolding protein Gab1. Ancillary control over this pathway is generated by the ShcA/Shc1 docking protein. ShcA acts, in part, to extend or amplify the binding and signaling properties of a receptor complex. (b) The Shc docking protein serves as an example of evolved complexity in signal transduction. An ancient form of Shc present in *C. elegans* has SH2 and PTB domains. Shc has gained in complexity concomitantly with evolution from simple to complex multicellular organisms, by acquiring proline-rich sequences, an adaptin binding domain and increasing numbers of tyrosine residues that are substrates for phosphorylation, and act as binding sites for SH2 domains of other proteins such as Grb2. In mammals, Shc has expanded to a four-gene family with additional isoforms created by alternate RNA splicing and alternate use of translation initiation codons. Mammalian Shc2-4 are predominantly localized in the brain.

ligands, the SH2 and SH3 domains participate in a series of intramolecular interactions that block the PXXP binding site of the nSH3 domain [88]. Binding of the CrkII SH2 domain to phosphotyrosine sites in docking proteins such as p130^{cas} may allow the nSH3 domain to recruit downstream targets and initiate signaling. However, once part of this active signaling complex, CrkII, becomes phosphorylated by cytoplasmic tyrosine kinases such as Abl at Tyr221, in the linker region between the two SH3 domains. This is turn induces an intramolecular interaction between the SH2 domain and the phosphorylated Tyr221 site that displaces CrkII from p130^{cas} and thereby terminates signaling. CrkII can therefore adopt two autoinhibited states – a basal conformation in which the binding activity of the nSH3 domain is occluded, and a conformation that terminates active signaling in which the phosphopeptide binding pocket of the SH2 domain is occupied by an internal ligand. Taken together, these data indicate that SH2/SH3 adaptors can adopt a variety of conformations that regulate their signaling properties, and this is likely a rather widespread property of modular signaling proteins.

INTERACTION DOMAINS IN THE EVOLUTION OF SIGNALING PATHWAYS

The evolution of multicellular animals required new mechanisms through which cells could communicate with one another, and transmit intracellular signals. As we have already introduced, the modular organization of interaction domains, and their cognate binding motifs, may have provided a relatively simple basis for the development of new signaling pathways, by the joining of domains and motifs in new combinations. According to this scheme, proteins with quite different biochemical properties can potentially acquire a common mode of regulation by the incorporation of a related domain through genetic rearrangement. For example, the proximal targets of receptor tyrosine kinases have numerous distinct functions, but, because they typically possess an SH2 domain, they have a shared ability to interact with autophosphorylated receptors. The emergence of protein kinases that specifically phosphorylate tyrosine, and of interaction domains that recognize phosphorylated sites, may therefore have been critical for the evolution of metazoan species. The acquisition of an SH2 domain by a pre-existing protein may have been sufficient to physically couple this polypeptide to phosphotyrosine signaling, and the novel chimeric protein would then have undergone positive Darwinian selection if it had useful attributes.

Consistent with this view, unicellular eukaryotes such as yeast lack conventional tyrosine kinases and phosphotyrosine-binding SH2 domains. However, the unicellular protozoan *Monsiga brevicollis*, which is considered to be an immediate predecessor of multicellular animals, has numerous predicted tyrosine kinases and SH2 domains

[89, 90]. Similarly, *Dictyostelium discoideum*, which has a unicellular lifestyle but aggregates into a multicellular form upon exposure to cAMP, has proteins with phosphotyrosine-binding SH2 domains such as STAT and Cbl proteins, as well as protein kinases in which an SH2 domain is linked to a tyrosine kinase-like (TKL) domain with dual specificity [91–93]. *Dictyostelium* may therefore represent an early stage in the evolution of phosphotyrosine signaling, which was employed by protozoa and social amoebae to facilitate their interactions with their environment, and was then co-opted to mediate cell–cell interactions required for multicellular life.

This hypothesis can also be tested by the creation of artificial chimeric proteins, with non-natural combinations of interaction domains and motifs, and investigating whether these have new biological functions. Indeed, a chimeric adaptor protein with an SH2 domain linked to a death effector domain aberrantly recruits pro-caspase 8 to an activated ErbB2 receptor tyrosine kinase, and consequently redirects the mitogenic ErbB2 to induce apoptosis [94].

As animals have become more complex, signaling proteins appear to have typically evolved through the acquisition of new interaction or catalytic domains, or their cognate binding sites or substrate motifs. For example, Shc is a docking protein in which an N-terminal PTB domain and a C-terminal SH2 domain flank a central region containing a proline-rich section and an adaptin binding motif (Figure 3.3). The distinct binding properties of SH2 and PTB domains allow Shc to interact with multiple phosphotyrosine-containing motifs on cell surface receptors. The Shc proteins from *Caenorhabditis elegans* contain this basic organization, but are simpler than their *Drosophila* counterpart [95] in the sense that they lack any tyrosine phosphorylation sites in their central region, whereas *Drosophila* Shc has a central Tyr–X–Asn site that binds the SH2 domain of the Grb2 adaptor upon phosphorylation [96]. In mammals, the situation has become significantly more complex. Mice and humans contain four distinct Shc genes (Shc1–4, also called ShcA–D), of which Shc 2–4 are most prominently expressed in the nervous sytem. Furthermore, mammalian Shc proteins have yet additional tyrosine phosphorylation sites, and all have at least two Tyr–X–Asn motifs that, upon phosphorylation, bind Grb2 and potentially other SH2-containing proteins [97]. As an added complexity, mammalian ShcA is expressed as three protein isoforms, the largest of which (ShcA^{p66}) contains an additional proline-rich N-terminal extension implicated as a factor in cellular response to oxidative stress and in longevity [98]. These data suggest that Shc proteins have multiple functions in signaling that have evolved through the acquisition of new phosphorylation sites and interaction motifs. The tyrosine phosphorylation sites of mammalian Shc potentially allow members of this protein family to function as increasingly complex docking subunits of activated receptor tyrosine kinases, and thereby to enhance the range of receptor signaling (Figure 3.3).

Analysis of mice containing a series of mutant Shc1/ShcA mutant alleles suggests that the more ancient PTB domain is essential for the role of ShcA in mediating embryonic heart development, likely by acting downstream of ErbB receptor tyrosine kinases. In contrast, the more recently acquired Tyr–X–Asn phosphorylation sites are dispensable for embryonic development, but together with the PTB and SH2 domains are required for the development and function of muscle spindles – key components of the monosynaptic stretch reflex circuit that are necessary for limb coordination [99]. These data are consistent with the premise that the acquisition of new interaction domains and motifs yields proteins capable of making new molecular connections; this increased signaling complexity may then have set the stage for the evolution of new cellular activities and the formation of more sophisticated tissues.

EMERGENT PROPERTIES OF MODULAR PROTEIN INTERACTION DOMAIN-DRIVEN SIGNALING NETWORKS

In addition to laying a foundation for rapid evolutionary adaptation and the development of novel connections within signaling networks, the reiterated use of modular protein interaction domains can potentially generate non-linear responses to upstream signals. For instance, binary interactions assembled through modular interaction domains may, under certain circumstances, be used to integrate signals or produce an all-or-none response. A typical enzymatic event, such as a kinase phosphorylating its substrate, or a simple binding event, such as an SH2 domain binding to a phosphorylated tyrosine site, conforms to Michaelis-Menton kinetics, and therefore produces a graded response. In other words, the response to a given stimulus is initially linear and then tapers off in a hyperbolic manner. This contrasts with digital switches in which a certain amount of stimulus converts the system from zero to a complete response. Some signaling pathways have steps at which noise is filtered out, signals are integrated, and all-or none decisions are made [35]. This is particularly important in key decisions such as progression through the cell cycle, when the cell must exercise precise control to avoid catastrophic events such as premature intiation of DNA replication. One mechanism by which a signaling cascade can create a switch-like response (referred to as an ultrasensitive biological switch) is through the requirement for multiple, independent phosphorylation events in order to sanction a requisite protein–protein interaction. Under these conditions, the response varies as a higher order of the kinase concentration, such that three independent phosphorylation events create a stimulus-response that responds to the third order of kinase concentration (modeled with a Hill coefficient of three). This has been observed biologically in a number of situations. In the maturation response of *Xenopus* oocytes, two independent phosphorylation events within the MAP kinase pathway set up conditions for an all-or-none activation of the ERK MAP kinase [100]. In a related example, degradation of the yeast cyclin-dependent kinase (CDK) inhibitor Sic1, a key event required for the G1 to S transition (or START in the cell cycle), requires six of nine serine/threonine phosphorylation sites on Sic1 to be phosphorylated by the CDK activity present in the G1 phase of the cell cycle in order for Sic1 to be targeted for ubiquitination and degradation. Such multiply phosphorylated Sic1 is bound by the WD40 repeat domain of an F-box protein, Cdc4, that serves as the substrate docking subunit of an E3 protein–ubiquitin ligase complex. Thus, Sic1 acts to monitor G1 CDK activity, setting a threshold for kinase activity that must be met in order for START to occur [101]. Multisite phosphorylation, coupled with a simple binary interaction, therefore creates an ultrasensitive response that ensures an orderly transition into the S phase of the cell cycle [102, 103]. The phosphorylated motifs on Sic1 are located in an intrinsically disordered region of the protein, and each phosphosite only has a weak affinity for the Cdc4 WD40 domain; however, experimental analysis indicates that the multiple Sic1 phospho-sites exist in a dynamic equilibrium with the Cdc4 WD40 domain [104]. Despite the fact that it is disordered, the phosphorylated region of Sic1 is compact and shows evidence of transient structure [104], which may create a high local concentration of interaction motifs that suffer minimal steric constraints [105]. Combining short-range specific contact interactions with less specific, spatially long-range polyelectrostatic interactions between the binding partners may then provide a basis for the observed ultrasensitive binding behavior [106]. A switch-like response is by no means a universal outcome of multisite protein phosphorylation, as this can also result in graded regulation of an interaction, producing an effect analogous to a rheostat. For instance, multiple phosphorylations within an unstructured region of the ETS-1 transcription activator results in a graded output, as measured by DNA binding affinity [107].

ACKNOWLEDGEMENTS

Tony Pawson is a Distinguished Scientist of the Canadian Institutes for Health Research.

REFERENCES

1. Pawson T, Scott JD. Signaling through scaffold, anchoring, and adaptor proteins. *Science* 1997;**278**:2075–80.
2. Pawson T, Nash P. Protein–protein interactions define specificity in signal transduction. *Genes Dev* 2000;**14**:1027–47.
3. Kuriyan J, Cowburn D. Modular peptide recognition domains in eukaryotic signaling. *Annu Rev Biophys Biomol Struct* 1997;**26**:259–88.
4. Schlessinger J. Cell signaling by receptor tyrosine kinases. *Cell* 2000;**103**:211–25.
5. Yaffe MB. Phosphotyrosine-binding domains in signal transduction. *Nature Rev Mol Cell Biol* 2002;**3**:177–86.
6. Yaffe MB, Elia AE. Phosphoserine/threonine-binding domains. *Curr Opin Cell Biol* 2001;**13**:131–8.
7. Seet BT, Dikic I, Zhou MM, Pawson T. Reading protein modifications with interaction domains. *Nat Rev Mol Cell Biol* 2006;**7**:473–83.
8. Moran MF, Koch CA, Anderson D, Ellis C, England L, Martin GS, Pawson T. Src homology region 2 domains direct protein–protein interactions in signal transduction. *Proc Natl Acad Sci USA* 1990;**87**:8622–6.
9. Anderson D, Koch CA, Grey L, Ellis C, Moran MF, Pawson T. Binding of SH2 domains of phospholipase C gamma 1, GAP, and Src to activated growth factor receptors. *Science* 1990;**250**:979–82.
10. Songyang Z, Shoelson SE, Chaudhuri M, Gish G, Pawson T, Haser WG, King F, Roberts T, Ratnofsky S, Lechleider RJ, et al. SH2 domains recognize specific phosphopeptide sequences. *Cell* 1993;**72**:767–78.
11. Bradshaw JM, Waksman G. Molecular recognition by SH2 domains. *Adv Protein Chem* 2002;**61**:161–210.
12. Liu BA, Jablonowski K, Raina M, Arce M, Pawson T, Nash PD. The human and mouse complement of SH2 domain proteins–establishing the boundaries of phosphotyrosine signaling. *Mol Cell* 2006;**22**:851–68.
13. Waksman G, Shoelson SE, Pant N, Cowburn D, Kuriyan J. Binding of a high affinity phosphotyrosyl peptide to the Src SH2 domain: crystal structures of the complexed and peptide-free forms. *Cell* 1993;**72**:779–90.
14. Eck MJ, Shoelson SE, Harrison SC. Recognition of a high-affinity phosphotyrosyl peptide by the Src homology-2 domain of p56lck. *Nature* 1993;**362**:87–91.
15. Pascal SM, Singer AU, Gish G, Yamazaki T, Shoelson SE, Pawson T, Kay LE, Forman-Kay JD. Nuclear magnetic resonance structure of an SH2 domain of phospholipase C-gamma 1 complexed with a high affinity binding peptide. *Cell* 1994;**77**:461–72.
16. Piccione E, Case RD, Domchek SM, Hu P, Chaudhuri M, Backer JM, Schlessinger J, Shoelson SE. Phosphatidylinositol 3-kinase p85 SH2 domain specificity defined by direct phosphopeptide/SH2 domain binding. *Biochemistry* 1993;**32**:3197–202.
17. Bradshaw JM, Mitaxov V, Waksman G. Investigation of phosphotyrosine recognition by the SH2 domain of the Src kinase. *J Mol Biol* 1999;**293**:971–85.
18. Ladbury JE, Lemmon MA, Zhou M, Green J, Botfield MC, Schlessinger J. Measurement of the binding of tyrosyl phosphopeptides to SH2 domains: a reappraisal. *Proc Natl Acad Sci USA* 1995;**92**:3199–203.
19. Marengere LE, Songyang Z, Gish GD, Schaller MD, Parsons JT, Stern MJ, Cantley LC, Pawson T. SH2 domain specificity and activity modified by a single residue. *Nature* 1994;**369**:502–5.
20. Maina F, Pante G, Helmbacher F, Andres R, Porthin A, Davies AM, Ponzetto C, Klein R. Coupling Met to specific pathways results in distinct developmental outcomes. *Mol Cell* 2001;**7**:1293–306.
21. Heldin CH, Ostman A, Ronnstrand L. Signal transduction via platelet-derived growth factor receptors. *Biochim Biophys Acta* 1998;**1378**:F79–F113.
22. Frese S, Schubert WD, Findeis AC, Marquardt T, Roske YS, Stradal TE, Heinz DW. The phosphotyrosine peptide binding specificity of Nck1 and Nck2 Src homology 2 domains. *J Biol Chem* 2006;**281**:18,236–18,245.
23. Babon JJ, McManus EJ, Yao S, DeSouza DP, Mielke LA, Sprigg NS, Willson TA, Hilton DJ, Nicola NA, Baca M, Nicholson SE, Norton RS. The structure of SOCS3 reveals the basis of the extended SH2

domain function and identifies an unstructured insertion that regulates stability. *Mol Cell* 2006;**22**:205–16.

24. Groesch TD, Zhou F, Mattila S, Geahlen RL, Post CB. Structural basis for the requirement of two phosphotyrosine residues in signaling mediated by Syk tyrosine kinase. *J Mol Biol* 2006;**356**:1222–36.

25. Hu J, Liu J, Ghirlando R, Saltiel AR, Hubbard SR. Structural basis for recruitment of the adaptor protein APS to the activated insulin receptor. *Mol Cell* 2003;**12**:1379–89.

26. Poy F, Yaffe MB, Sayos J, Saxena K, Morra M, Sumegi J, Cantley LC, Terhorst C, Eck MJ. Crystal structures of the XLP protein SAP reveal a class of SH2 domains with extended, phosphotyrosine-independent sequence recognition. *Mol Cell* 1999;**4**:555–61.

27. Li SC, Gish G, Yang D, Coffey AJ, Forman-Kay JD, Ernberg I, Kay LE, Pawson T. Novel mode of ligand binding by the SH2 domain of the human XLP disease gene product SAP/SH2D1A. *Curr Biol* 1999;**9**:1355–62.

28. Latour S, Gish G, Helgason CD, Humphries RK, Pawson T, Veillette A. Regulation of SLAM-mediated signal transduction by SAP, the X-linked lymphoproliferative gene product. *Nat Immunol* 2001;**2**:681–90.

29. Latour S, Roncagalli R, Chen R, Bakinowski M, Shi X, Schwartzberg PL, Davidson D, Veillette A. Binding of SAP SH2 domain to FynT SH3 domain reveals a novel mechanism of receptor signalling in immune regulation. *Nat Cell Biol* 2003;**5**:149–54.

30. Ottinger EA, Botfield MC, Shoelson SE. Tandem SH2 domains confer high specificity in tyrosine kinase signaling. *J Biol Chem* 1998;**273**:729–35.

31. Mayer BJ. SH3 domains: complexity in moderation. *J Cell Sci* 2001;**114**:1253–63.

32. Zarrinpar A, Bhattacharyya RP, Lim WA. The structure and function of proline recognition domains. *Sci STKE 2003* 2003;**RE8**.

33. Sicheri F, Moarefi I, Kuriyan J. Crystal structure of the Src family tyrosine kinase Hck. *Nature* 1997;**385**:602–9.

34. Xu W, Harrison SC, Eck MJ. Three-dimensional structure of the tyrosine kinase c-Src. *Nature* 1997;**385**:595–602.

35. Pellicena P, Stowell KR, Miller WT. Enhanced phosphorylation of Src family kinase substrates containing SH2 domain binding sites. *J Biol Chem 273* 1998:15,325–15,328.

36. Pellicena P, Miller WT. Processive phosphorylation of p130Cas by Src depends on SH3-polyproline interactions. *J Biol Chem* 2001;**276**:28,190–28,196.

37. Filippakopoulos P, Kofler M, Hantschel O, Gish GD, Grebien F, Salah E, Neudecker P, Kay LE, Turk BE, Superti-Furga G, Pawson T, Knapp S. Structural coupling of SH2-kinase domains links Fes and Abl substrate recognition and kinase activation. *Cell* 2008;**134**:793–803.

38. Sadowski I, Stone JC, Pawson T. A noncatalytic domain conserved among cytoplasmic protein-tyrosine kinases modifies the kinase function and transforming activity of Fujinami sarcoma virus P130gag-fps. *Mol Cell Biol* 1986;**6**:4396–408.

39. Koch CA, Moran M, Sadowski I, Pawson T. The common src homology region 2 domain of cytoplasmic signaling proteins is a positive effector of v-fps tyrosine kinase function. *Mol Cell Biol* 1989;**9**:4131–40.

40. Nagar B, Hantschel O, Young MA, Scheffzek K, Veach D, Bornmann W, Clarkson B, Superti-Furga G, Kuriyan J. Structural basis for the autoinhibition of c-Abl tyrosine kinase. *Cell* 2003;**112**:859–71.

41. Nagar B, Hantschel O, Seeliger M, Davies JM, Weis WI, Superti-Furga G, Kuriyan J. Organization of the SH3–SH2 unit in active and inactive forms of the c-Abl tyrosine kinase. *Mol Cell* 2006;**21**:787–98.

42. Pawson T, Gish GD, Nash P. SH2 domains, interaction modules and cellular wiring. *Trends Cell Biol* 2001;**11**:504–11.

43. Forman-Kay JD, Pawson T. Diversity in protein recognition by PTB domains. *Curr Opin Struct Biol* 1999;**9**:690–5.

44. Kavanaugh WM, Turck CW, Williams LT. PTB domain binding to signaling proteins through a sequence motif containing phosphotyrosine. *Science* 1995;**268**:1177–9.

45. Batzer AG, Blaikie P, Nelson K, Schlessinger J, Margolis B. The phosphotyrosine interaction domain of Shc binds an LXNPXY motif on the epidermal growth factor receptor. *Mol Cell Biol* 1995;**15**:4403–9.

46. van der Geer P, Wiley S, Lai VK, Olivier JP, Gish GD, Stephens R, Kaplan D, Shoelson S, Pawson T. A conserved amino-terminal Shc domain binds to phosphotyrosine motifs in activated receptors and phosphopeptides. *Curr Bioll* 1995;**5**:404–12.

47. Zhou MM, Ravichandran KS, Olejniczak EF, Petros AM, Meadows RP, Sattler M, Harlan JE, Wade WS, Burakoff SJ, Fesik SW. Structure and ligand recognition of the phosphotyrosine binding domain of Shc. *Nature* 1995;**378**:584–92.

48. Uhlik MT, Temple B, Bencharit S, Kimple AJ, Siderovski DP, Johnson GL. Structural and evolutionary division of phosphotyrosine binding (PTB) domains. *J Mol Biol* 2005;**345**:1–20.

49. White MF. The insulin signalling system and the IRS proteins. *Diabetologia* 1997;**40**(2):S2–17.

50. Benes CH, Wu N, Elia AE, Dharia T, Cantley LC, Soltoff SP. The C2 domain of PKCdelta is a phosphotyrosine binding domain. *Cell* 2005;**121**:271–80.

51. Smith MJ, Hardy WR, Murphy JM, Jones N, Pawson T. Screening for PTB domain binding partners and ligand specificity using proteome-derived NPXY peptide arrays. *Mol Cell Biol* 2006;**26**:8461–74.

52. Taverna SD, Li H, Ruthenburg AJ, Allis CD, Patel DJ. How chromatin-binding modules interpret histone modifications: lessons from professional pocket pickrs. *Nat Struct Mol Biol* 2007;**14**:1025–40.

53. Yaffe MB, Smerdon SJ. PhosphoSerine/threonine binding domains: you can't pSERious?. *Structure* 2001;**9**:R33–8.

54. Wu JW, Hu M, Chai J, Seoane J, Huse M, Li C, Rigotti DJ, Kyin S, Muir TW, Fairman R, Massague J, Shi Y. Crystal structure of a phosphorylated Smad2. Recognition of phosphoserine by the MH2 domain and insights on Smad function in TGF-beta signaling. *Mol Cell* 2001;**8**:1277–89.

55. Huse M, Muir TW, Xu L, Chen YG, Kuriyan J, Massague J. The TGF beta receptor activation process: an inhibitor- to substrate-binding switch. *Mol Cell* 2001;**8**:671–82.

56. Attisano L, Wrana JL. Signal transduction by the TGF-beta super-family. *Science* 2002;**296**:1646–7.

57. Owen DJ, Ornaghi P, Yang JC, Lowe N, Evans PR, Ballario P, Neuhaus D, Filetici P, Travers AA. The structural basis for the recognition of acetylated histone H4 by the bromodomain of histone acetyl-transferase gcn5p. *EMBO J* 2000;**19**:6141–9.

58. Nielsen PR, Nietlispach D, Mott HR, Callaghan J, Bannister A, Kouzarides T, Murzin AG, Murzina NV, Laue ED. Structure of the HP1 chromodomain bound to histone H3 methylated at lysine 9. *Nature* 2002;**416**:103–7.

59. Hon WC, Wilson MI, Harlos K, Claridge TD, Schofield CJ, Pugh CW, Maxwell PH, Ratcliffe PJ, Stuart DI, Jones EY. Structural basis for the recognition of hydroxyproline in HIF-1 alpha by pVHL. *Nature* 2002;**417**:975–8.

60. Min Jr. JH, Yang H, Ivan M, Gertler F, Kaelin WG, Pavletich NP Structure of an HIF-1alpha -pVHL complex: hydroxyproline recognition in signaling. *Science* 2002;**296**:1886–9.

61. Hicke L, Schubert HL, Hill CP. Ubiquitin-binding domains. *Nat Rev Mol Cell Biol* 2005;**6**:610–21.

62. Pickart CM, Eddins MJ. Ubiquitin: structures, functions, mechanisms. *Biochim Biophys Acta* 2004;**1695**:55–72.

63. Polo S, Sigismund S, Faretta M, Guidi M, Capua MR, Bossi G, Chen H, De Camilli P, Di Fiore PP. A single motif responsible for ubiquitin recognition and monoubiquitination in endocytic proteins. *Nature* 2002;**416**:451–5.

64. Schmidt MH, Dikic I. The Cbl interactome and its functions. *Nat Rev Mol Cell Biol* 2005;**6**:907–18.

65. Tonikian R, Zhang Y, Sazinsky SL, Currell B, Yeh JH, Reva B, Held HA, Appleton BA, Evangelista M, Wu Y, Xin X, Chan AC, Seshagiri S, Lasky LA, Sander C, Boone C, Bader GD, Sidhu SS. A specificity map for the PDZ domain family. *PLoS Biol* 2008;**6**. e239

66. Stiffler MA, Chen JR, Grantcharova VP, Lei Y, Fuchs D, Allen JE, Zaslavskaia LA, MacBeath G. PDZ domain binding selectivity is optimized across the mouse proteome. *Science* 2007;**317**:364–9.

67. Groemping Y, Lapouge K, Smerdon SJ, Rittinger K. Molecular basis of phosphorylation-induced activation of the NADPH oxidase. *Cell* 2003;**113**:343–55.

68. Carrington PE, Sandu C, Wei Y, Hill JM, Morisawa G, Huang T, Gavathiotis E, Wei Y, Werner MH. The structure of FADD and its mode of interaction with procaspase-8. *Mol Cell* 2006;**22**:599–610.

69. Qiao, F., and Bowie, J.U. (2005). The many faces of SAM. *Sci. STKE 2005*, re7.

70. Cullen PJ, Cozier GE, Banting G, Mellor H. Modular phosphoinositide-binding domains – their role in signalling and membrane trafficking. *Curr Biol* 2001;**11**:R882–93.

71. Milburn CC, Deak M, Kelly SM, Price NC, Alessi DR, Van Aalten DM. Binding of phosphatidylinositol 3,4,5-trisphosphate to the pleckstrin homology domain of protein kinase B induces a conformational change. *Biochem J* 2003;**375**:531–8.

72. Li Y, Inoki K, Yeung R, Guan KL. Regulation of TSC2 by 14-3-3 binding. *J Biol Chem* 2002;**277**:44,593–44,596.

73. Carpten JD, Faber AL, Horn C, Donoho GP, Briggs SL, Robbins CM, Hostetter G, Boguslawski S, Moses TY, Savage S, Uhlik M, Lin A, Du J, Qian YW, Zeckner DJ, Tucker-Kellogg G, Touchman J, Patel K, Mousses S, Bittner M, Schevitz R, Lai MH, Blanchard KL, Thomas JE. A transforming mutation in the pleckstrin homology domain of AKT1 in cancer. *Nature* 2007;**448**:439–44.

74. Dawson JC, Legg JA, Machesky LM. Bar domain proteins: a role in tubulation, scission and actin assembly in clathrin-mediated endocytosis. *Trends Cell Biol* 2006;**16**:493–8.

75. Peter BJ, Kent HM, Mills IG, Vallis Y, Butler PJ, Evans PR, McMahon HT. BAR domains as sensors of membrane curvature: the amphiphysin BAR structure. *Science* 2004;**303**:495–9.

76. Frost A, Perera R, Roux A, Spasov K, Destaing O, Egelman EH, De Camilli P, Unger VM. Structural basis of membrane invagination by F-BAR domains. *Cell* 2008;**132**:807–17.

77. McPherson, V.A., Everingham, S., Karisch, R., Smith, J.A., Udell, C.M., Zheng, J., Jia, Z., and Craig, A.W. (2008). Contributions of F-BAR and SH2 domains of Fes protein-tyrosine kinase for coupling to the Fcε RI pathway in mast cells. *Mol. Cell. Biol.* 2008 Nov 10 (epub ahead of print).

78. Rozakis-Adcock M, Fernley R, Wade J, Pawson T, Bowtell D. The SH2 and SH3 domains of mammalian Grb2 couple the EGF receptor to the Ras activator mSos1. *Nature* 1993;**363**:83–5.

79. Li N, Batzer A, Daly R, Yajnik V, Skolnik E, Chardin P, Bar-Sagi D, Margolis B, Schlessinger J. Guanine-nucleotide-releasing factor hSos1 binds to Grb2 and links receptor tyrosine kinases to Ras signalling. *Nature* 1993;**363**:85–8.

80. Olivier JP, Raabe T, Henkemeyer M, Dickson B, Mbamalu G, Margolis B, Schlessinger J, Hafen E, Pawson T. A Drosophila SH2-SH3 adaptor protein implicated in coupling the sevenless tyrosine kinase to an activator of Ras guanine nucleotide exchange. *Sos Cell* 1993;**73**:179–91.

81. Simon MA, Dodson GS, Rubin GM. An SH3-SH2-SH3 protein is required for p21Ras1 activation and binds to sevenless and Sos proteins *in vitro*. *Cell* 1993;**73**:169–77.

82. Cheng AM, Saxton TM, Sakai R, Kulkarni S, Mbamalu G, Vogel W, Tortorice CG, Cardiff RD, Cross JC, Muller WJ, Pawson T. Mammalian Grb2 regulates multiple steps in embryonic development and malignant transformation. *Cell* 1998;**95**:793–803.

83. Chazaud C, Yamanaka Y, Pawson T, Rossant J. Early lineage segregation between epiblast and primitive endoderm in mouse blastocysts through the Grb2-MAPK pathway. *Dev Cell* 2006;**10**:615–24.

84. Lock LS, Royal I, Naujokas MA, Park M. Identification of an atypical Grb2 carboxyl-terminal SH3 domain binding site in Gab docking proteins reveals Grb2-dependent and -independent recruitment of Gab1 to receptor tyrosine kinases. *J Biol Chem* 2000;**275**:31,536–31,545.

85. Gupta S, Ramjaun AR, Haiko P, Wang Y, Warne PH, Nicke B, Nye E, Stamp G, Alitalo K, Downward J. Binding of ras to phosphoinositide 3-kinase p110alpha is required for ras-driven tumorigenesis in mice. *Cell* 2007;**129**:957–68.

86. Pawson T. Protein modules and signalling networks. *Nature* 1995;**373**:573–80.

87. Feller SM. Crk family adaptors-signalling complex formation and biological roles. *Oncogene* 2001;**20**:6348–71.

88. Kobashigawa Y, Sakai M, Naito M, Yokochi M, Kumeta H, Makino Y, Ogura K, Tanaka S, Inagaki F. Structural basis for the transforming activity of human cancer-related signaling adaptor protein CRK. *Nat Struct Mol Biol* 2007;**14**:503–10.

89. Pincus D, Letunic I, Bork P, Lim WA. Evolution of the phospho-tyrosine signaling machinery in premetazoan lineages. *Proc Natl Acad Sci USA* 2008;**105**:9680–4.

90. Manning G, Young SL, Miller WT, Zhai Y. The protest, Monosiga brevicollis, has a tyrosine kinase signaling network more elaborate and diverse than found in any known metazoan. *Proc Natl Acad Sci USA* 2008;**105**:9674–9.

91. Kawata T, Shevchenko A, Fukuzawa M, Jermyn KA, Totty NF, Zhukovskaya NV, Sterling AE, Mann M, Williams JG. SH2 signaling in a lower eukaryote: a STAT protein that regulates stalk cell differentiation in dictyostelium. *Cell* 1997;**89**:909–16.

92. Moniakis J, Funamoto S, Fukuzawa M, Meisenhelder J, Araki T, Abe T, Meili R, Hunter T, Williams J, Firtel RA. An SH2-domain-containing kinase negatively regulates the phosphatidylinositol-3 kinase pathway. *Genes Dev* 2001;**15**:687–98.

93. Langenick J, Araki T, Yamada Y, Williams JG. A Dictyostelium homologue of the metazoan Cbl proteins regulates STAT signalling. *J Cell Sci* 2008;**121**:3524–30.

94. Howard PL, Chia MC, Del Rizzo S, Liu FF, Pawson T. Redirecting tyrosine kinase signaling to an apoptotic caspase pathway through chimeric adaptor proteins. *Proc Natl Acad Sci USA* 2003;**100**:11,267–11,272.

95. Luzi L, Confalonieri S, Di Fiore PP, Pelicci PG. Evolution of Shc functions from nematode to human. *Curr Opin Genet Dev* 2000;**10**:668–74.

96. Lai KM, Olivier JP, Gish GD, Henkemeyer M, McGlade J, Pawson T. A Drosophila shc gene product is implicated in signaling by the DER receptor tyrosine kinase. *Mol Cell Biol* 1995;**15**:4810–18.

97. van der Geer P, Wiley S, Gish GD, Pawson T. The Shc adaptor protein is highly phosphorylated at conserved, twin tyrosine residues (Y239/240) that mediate protein–protein interactions. *Curr Biol* 1996;**6**:1435–44.

98. Migliaccio E, Giorgio M, Mele S, Pelicci G, Reboldi P, Pandolfi PP, Lanfrancone L, Pelicci PG. The p66shc adaptor protein controls oxidative stress response and life span in mammals. *Nature* 1999;**402**:309–13.

99. Hardy WR, Li L, Wang Z, Sedy J, Fawcett J, Frank E, Kucera J, Pawson T. Combinatorial ShcA docking interactions support diversity in tissue morphogenesis. *Science* 2007;**317**:251–6.

100. Guadagno Jr. TM, Ferrell JE Requirement for MAPK activation for normal mitotic progression in Xenopus egg extracts. *Science* 1998;**282**:1312–15.

101. Nash P, Tang X, Orlicky S, Chen Q, Gertler FB, Mendenhall MD, Sicheri F, Pawson T, Tyers M. Multisite phosphorylation of a CDK inhibitor sets a threshold for the onset of DNA replication. *Nature* 2001;**414**:514–21.

102. Deshaies Jr RJ, Ferrell JE. Multisite phosphorylation and the countdown to S phase. *Cell* 2001;**107**:819–22.

103. Harper JW. A phosphorylation-driven ubiquitination switch for cell-cycle control. *Trends Cell Biol* 2002;**12**:104–7.

104. Mittag T, Orlicky S, Choy WY, Tang X, Lin H, Sicheri F, Kay LE, Tyers M, Forman-Kay JD. Dynamic equilibrium engagement of a polyvalent ligand with a single-site receptor. *Proc Natl Acad Sci USA* 2008;**105**:17,772–17,777.

105. Klein P, Pawson T, Tyers M. Mathematical modeling suggests cooperative interactions between a disordered polyvalent ligand and a single receptor site. *Curr Biol* 2003;**13**:1669–78.

106. Borg M, Mittag T, Pawson T, Tyers M, Forman-Kay JD, Chan HS. Polyelectrostatic interactions of disordered ligands suggest a physical basis for ultrasensitivity. *Proc Natl Acad Sci USA* 2007;**104**:9650–5.

107. Pufall MA, Lee GM, Nelson ML, Kang HS, Velyvis A, Kay LE, McIntosh LP, Graves BJ. Variable control of Ets-1 DNA binding by multiple phosphates in an unstructured region. *Science* 2005;**309**:142–5.

108. Bateman A, Coin L, Durbin R, Finn RD, Hollich V, Griffiths-Jones S, Khanna A, Marshall M, Moxon S, Sonnhammer EL, Studholme DJ, Yeats C, Eddy SR. The Pfam protein families database. *Nucleic Acids Res* 2004;**32**:D138–41.

Chapter 4

Recognition of Phospho-Serine/Threonine Phosphorylated Proteins by Phospho-Serine/Threonine-Binding Domains

Stephen J. Smerdon[1] and Michael B. Yaffe[2]

[1]*Division of Molecular Structure, National Institute for Medical Research, The Ridgeway, Mill Hill, London, England, UK*

[2]*Center for Cancer Research, Massachusetts Institute of Technology, Cambridge, Massachusetts*

INTRODUCTION

The finding that Src homology domain 2 (SH2) and phosphotyrosine binding (PTB) domains could bind to Tyr-phosphorylated motifs on proteins, but not to Ser- or Thr-phosphorylated sequences, suggested that modular domain-mediated phospho-regulation of protein-protein complexes might be a feature unique to tyrosine kinase signaling cascades. However, the subsequent discovery of a diverse group of molecules and domains that specifically recognize phosphorylated serine- and threonine-based motifs has dispelled this idea, and is leading to a more general appreciation of the role of protein phosphorylation and phosphopeptide binding domains in regulating the reversible assembly of multiprotein complexes [1, 2].

14-3-3 PROTEINS

The first phosphoserine/threonine-binding molecules that were identified were members of a family of dimeric proteins called 14-3-3 that were first identified as abundant polypeptides of unknown function in brain [3]; they were later identified as activators of tryptophan and tyrosine hydroxylase [4, 5], and as inhibitors or activators of PKCs [6]. Mammalian cells contain seven distinct 14-3-3 gene products (denoted β, γ, ε, η, σ, τ, and ζ), while plants and fungi contain between 2 and 15. The initial observation that 14-3-3-binding might be regulated by ligand phosphorylation emerged from studies of tryptophan hydroxylase [7] and Raf, the upstream activator of the classical mitogen-activated protein (MAP) kinase pathway [8]. Detailed investigation of the 14-3-3 binding sites on Raf

[9], together with oriented peptide library screening on all mammalian 14-3-3s [10], led to the identification of two optimal pSer/threonine-containing motifs, RSX[pS/pT]XP and RXXX[pS/pT]XP, that are recognized by all 14-3-3 isotypes (pS denotes both pSer and pThr, and X denotes any amino acid, although there are preferences for particular amino acids in different X positions). A third motif, found at the extreme C-terminus of a small number of 14-3-3-binding membrane channel proteins, was discovered in a genetic screen, and corresponds to the sequence SWpTX [11]. To date, several hundred 14-3-3 binding proteins have been identified through a combination of traditional biochemistry, yeast two-hybrid screening, and 14-3-3-directed proteomic screens in yeast [12], higher plants [13, 14] and mammalian cells [15-21]. Many, though not all, of the 14-3-3-binding proteins that have been studied in depth use phosphorylated sequences that closely match the optimal 14-3-3 consensus motifs.

The 14-3-3 proteins participate in, and appear to regulate, a wide range of cellular processes including metabolism, stress and nutritional-deprivation responses, development and cell differentiation, migration, cell division, DNA damage responses, receptor recycling and trafficking, protein degradation, and programmed cell death. In many cases, the mechanistic roles that 14-3-3 binding plays in these processes is not clear, though for a smaller subset of ligands detailed studies are beginning to uncover general mechanisms through which 14-3-3 may regulate their function. For some ligands, 14-3-3 proteins can directly regulate their catalytic activity—as observed, for example, with Raf, exoenzyme S, and serotonin N-acetyltransferase [22–24, 41]. For others, 14-3-3 appears to regulate interactions between its bound protein and other molecules within the cell. For example, growth-factor-mediated phosphorylation of the pro-apoptotic Bcl-2 family protein BAD at Ser-136 and/or Ser-112 facilitates its interaction with 14-3-3 proteins and blocks its association with anti-apoptotic Bcl-2 family members at the mitochondrial membrane [25–27]. In addition, 14-3-3 binding alters the conformation of BAD to stimulating an additional inhibitory phosphorylation at Ser-155 that further blocks programmed cell death [28]. Finally, in a number of cases, 14-3-3 proteins appear to play an important role in controlling the nuclear versus cytoplasmic localization of bound ligands such as Cdc25 and Forkhead family transcription factors [18–21, 29–32], or the endoplasmic reticulum versus plasma membrane distribution of receptors and ion channels [11, 33] through masking of nuclear import sequences, exposure of nuclear export sequences, or interference with endoplasmic reticulum localizing signals.

The X-ray structures of 14-3-3τ and ζ revealed that the molecule is a cup-shaped dimer [34, 35] in which each monomeric subunit consists of nine α-helices (Figure 4.1). The dimer interface is formed from helices αA, αC, and αD, creating a 35-Å$\times$35-Å$\times$20-Å central channel where binding to peptide and protein ligands occurs. Through the structural genomics program, the structures of all mammalian 14-3-3 proteins have been solved [36], as well as that of a plant 14-3-3 [37], revealing a highly conserved global architecture and key features involved in phospho-ligand recognition [38]. Ligand-bound 14-3-3 structures have shown that peptides bind within an amphipathic groove along each edge of the central channel [10, 39, 40] with the entire phosphopeptide main chain in a highly extended conformation until two residues after the pSer, when there is a sudden sharp change in peptide chain direction required

to exit the 14-3-3 binding cleft. The only structure of 14-3-3 proteins in complex with a near-full length *bona fide* protein ligand is 14-3-3ζ bound to serotonin N-acetyltransferase [41]. In this structure, the 14-3-3 binding portion of the enzyme displays a conformation very similar to that seen in isolated phosphopeptide:14-3-3 complexes, including the extended conformation and sudden alteration in chain direction. Furthermore, 14-3-3-binding at least partially restructures the substrate binding site on serotonin N-acetyltransferase, rationalizing some of the 14-3-3 effects on enzyme activity. In addition, a combined X-ray crystal structure of the ternary complex of 14-3-3, plant plasma membrane H$^+$-ATPase, and the fungal toxin fusicoccin revealed that the toxin occupies part of the peptide binding groove, stabilizing the 14-3-3:ATPase interaction, and leading to the formation of active ATPase hexamers [42]. Together, these findings are consistent with the idea that 14-3-3 proteins act as "molecular anvils" to induce conformational changes in their bound ligands [43], though additional X-ray structures of 14-3-3-bound complexes are required before global mechanistic conclusions about 14-3-3 function can be unequivocally stated. At least three sites on several of the mammalian 14-3-3 isotypes are subject to phosphorylation. Phosphorylation of one site, Ser-58, appears to disrupt dimer formation [44, 45], while phosphorylation at either Thr-233 or Ser-184 has been shown to abrogate ligand binding [46–49]. Other potential posttranslational modifications of 14-3-3, as well as their physiological effects, have not yet been well explored.

FHA DOMAINS

Forkhead-associated (FHA) domains are a pThr-binding module found in several prokaryotic and eukaryotic proteins, including kinases, phosphatases, transcription factors, kinesin-like motors, and regulators of small G proteins. FHA domains are ~140 amino acids in length, and extend significantly beyond the core homology region first identified by sequence profiling [50–55].

Recognition that FHA domains were pThr-binding modules came from findings that the FHA domain of KAPP, a protein phosphatase in *Arabidopsis*, was critical for its interaction with phosphorylated receptor-like kinases [53, 56] and that the FHA domains within the *Saccharomyces cerevisiae* cell-cycle checkpoint kinase Rad53p were essential for interaction with the phosphorylated DNA damage control protein Rad9p [57, 58]. Data regarding the specificity of different FHA domains for pThr-based sequence motifs come from peptide library experiments that show sequence-specific binding involving amino acids from the pT$-$3 to the pT$+$3 position [50, 59]. Curiously, substitution of pSer (pS) for pThr (pT) almost completely eliminates phosphopeptide binding, presumably due to a structurally conserved Van der Waals interaction

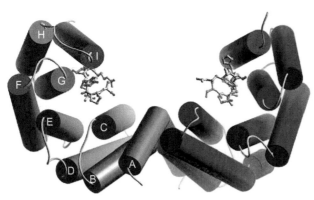

FIGURE 4.1 14-3-3/phosphopeptide interactions.
Dimerization of two 14-3-3 monomers, each of which is composed of nine α-helices, forms a cleft within which phosphoserine-containing ligands bind (shown in ball-and-stick representation). A single, multiply phosphorylated protein ligand may bind simultaneously to both available sites. Alternatively, two singly phosphorylated proteins can bind, one to each monomer, allowing 14-3-3 to act as a molecular scaffold for the assembly of diverse signaling complexes. Reprinted from Yaffe and Smerdon [160], ©Elsevier 2001, with permission.

with the threonine γ-methyl group or to entropic constraints that are unique to phosphothreonine. Tsai and co-workers found that the C-terminal FHA domain of Rad53 binds to pTyr- containing peptides *in vitro* [54, 55], although the *in vivo* relevance of pTyr-dependent signaling mechanisms in budding yeast remains unclear.

The *in vivo* binding partners for most FHA domain proteins are unknown, though a number of studies and clinical observations involving naturally occurring or engineered mutations or deletions within the FHA domains of key signaling molecules have verified their functional importance. Mutations that impair the ability of the N-terminal FHA domain of the yeast checkpoint kinase Rad53p to bind to phosphopeptides also result in increased sensitivity to DNA damage, whereas mutations within the FHA domain of Chk2, the human homolog of Rad53p, have been implicated in a variant form of the human cancer-prone Li-Fraumeni syndrome [60]. In addition, other mutations in FHA-domain-containing proteins including p95/Nbs1 and Chfr also appear to contribute to human tumor formation [61–63]. Chfr may function as a subunit of an E3 ubiquitin ligase, targeting Polo kinase for degradation to establish a cell-cycle checkpoint [64], although the role of the FHA domain in this process is not yet known. The structural

basis of the effect of disease-associated mutations in Nbs1 is also unclear. The FHA domain of Nbs1 precedes one or, more possibly, two BRCT repeat motifs that may also have a phospho-independent binding function. Cellular binding partners of these BRCT domain remain unknown. However, it is clear that the Nbs1 FHA domain mediates an interaction with a series of casein-kinase 2-like phosphorylation sites, possibly in combination with the BRCTs, to maintain Nbs1 (and the other components of the MRN complex, Rad50 and Mre11) at sites of DNA damage [65–68].

Structures of both FHA domains from Rad53p [50, 54, 55], together with FHA domains from Chk2 [69] and Chfr [70], have been determined by nuclear magnetic resonance (NMR) spectroscopy or X-ray crystallography (Figure 4.2). The FHA domain consists of an 11-stranded β-sandwich, with a topology essentially identical to that of the MH2 domain from Smad tumor suppressor molecules [50, 54]. This structural relationship, together with the remarkable similarities in phospho-dependent binding interactions of FHA and MH2 domains, suggests the existence of a superfamily of FHA-like phospho-binding domains [69, 71, 72]. Interestingly, the Chfr FHA domain structure shows a segment-swapped dimer with the C-terminal half of β7 and β8-10 exchanged between monomers. Whether or not these

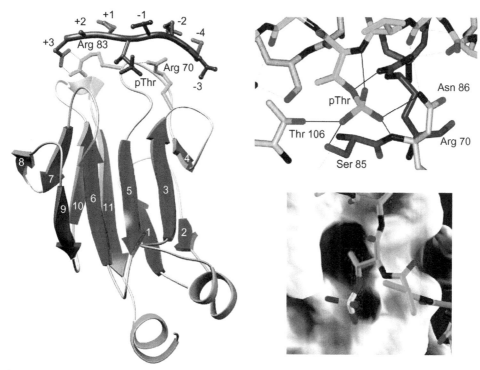

FIGURE 4.2 FHA/phosphopeptide interactions.
The FHA domain architecture as exemplified by the X-ray structure of Rad53p FHA1 in complex with a Rad9p phosphopeptide (left panel). The peptide binds at one end of the β-sandwich domain in an extended conformation. The highly conserved arginine (Arg70) makes a crucial contact with the phosphate group, while a non-conserved arginine (Arg83) acts to select an Asp at the pT+3 position (right panel, top). The phosphate group interacts with a constellation of conserved and semi-conserved FHA domain side-chains (top panel), while the γ-methyl group of the phosphothreonine binds in a pocket on the domain surface (right panel, bottom), likely explaining the observed preference for pThr over pSer in peptide-selection experiments. Reprinted from Yaffe and Smerdon [160], ©Elsevier 2001, with permission.

dimers exist *in vivo* or contribute to Chfr function is not known. Nevertheless, in all pThr peptide complexes, binding occurs at one end of the domain, through interactions between selected residues in the phosphopeptide and loops connecting the β3/4, β4/5, and β6/7 β-strands. Of the seven most highly conserved residues in the FHA family, three make direct interactions with the peptide (two bind directly to the pThr residue), while the remainder form the structural core or stabilize loop regions of the β-sandwich structure. Although this canonical mode of interaction is likely to prove the most common, exceptions and variations are beginning to emerge, including examples of interactions involving highly extended peptide motifs [73], and interactions that are modulated by multi-site phosphorylation [74]. These observations uncover considerably more versatility in FHA function than previously suspected, and we are sure the future holds more surprises!

WW DOMAINS

WW domains contain ~40 amino acids with two invariant tryptophan residues (labeled W in single-letter amino acid code, hence the name *WW domain*) that bind to short proline-rich sequences containing PPXY, PPLP, or PPR motifs [75]. A small subclass of WW domains within the proline isomerases Pin1/Ess1 and their homologs, the splicing factor Prp40 and the ubiquitin ligase Rsp5, show specific binding to phosphoserine-proline motifs within mitotic phosphoproteins [76] and the phosphorylated C-terminal domain (CTD) of RNA polymerase II [77–79].

Phospho-specific WW domain function is best understood for the proline isomerase Pin1, a protein that slows progress through mitosis [80] but is also required for mitotic exit and for the DNA replication checkpoint [81]. In addition to its WW domain, Pin1 contains a proline isomerase (rotamase) domain at its C terminus that catalyzes the specific *cis-trans* isomerization of pSer/Thr-Pro bonds [82, 83]. Both the WW-domain-mediated pS-P binding and the rotamase-catalyzed pS-P isomerization are necessary for Pin1 biological activity [81, 83]. Pin1 also enhances the dephosphorylation of substrates by protein phosphatases, all of which requires the pSer-Pro bond to be in *trans*. Because the WW domain can only bind to the *trans* geometric isomer, its major role may be to stabilize the *trans*-isomerase product for dephosphorylation [84, 85]. WW-domain-facilitated substrate dephosphorylation is likely to be a general mechanism for WW domain function in both cell-cycle progression and regulation of transcriptional elongation.

WW domains fold into three anti-parallel β-strands, forming a single groove that recognizes proline-rich ligands in the context of a type II polyproline helix (Figure 4.3). Specificity for different proline-rich motifs is determined largely by residues within the loop regions that connect the β1/β2 and β2/β3 strands, somewhat akin to the mechanism of ligand binding utilized by FHA domains. The structure of the Pin1 WW domain in complex with a Y*p*SPT*p*SPS peptide from the CTD of RNA polymerase II [86] revealed that all of the phosphate contacts were made between the second pSer and two residues of the peptide in the β1/β2 loop (Ser-16 and Arg-17), along with one in the β2 strand (Tyr-23). These findings explain why only a few WW domains are competent to bind phosphorylated sequence motifs, as the majority of WW domains lack an Arg residue within loop 1.

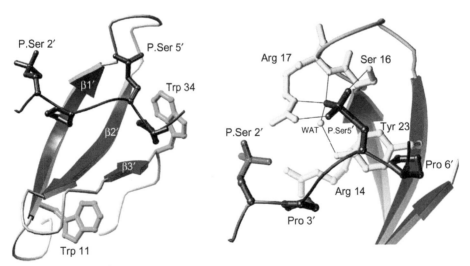

FIGURE 4.3 WW/phosphopeptide interactions.
WW domains are named after two conserved Trp residues that form the structural core of the three-stranded domain (left panel). The pSer-Pro containing phosphopeptide (foreground) derived from RNA polymerase II binds in an extended conformation across the β-sheet (right panel). The oxygen atoms from only one of the two phosphates binds to residues located on the β1-β2 loop.
Reprinted from Yaffe and Smerdon [160], ©Elsevier 2001, with permission.

POLO-BOX DOMAINS

Polo-like kinases (Plk) are a family of serine/threonine protein kinases that play important roles throughout the cell cycle. Named from the phenotype observed in *Drosophila* mutants, the Polo gene in flies was found to encode a protein kinase whose mutation resulted in abnormal spindle poles [87, 88]. Flies, budding yeast, and fission yeast contain a single Polo family member (Polo, CDC5, and Plo1 respectively), while humans, mice, and frogs have four Plk family members, denoted Plk1/Plx1 [89–91], Plk2/Plx2/Snk [92, 93], Plk3/Plx3/Fnk/Prk [95–95] and Plk4/Plx4/Sak, respectively [93–95]. Plk1, which has been most extensively studied, performs a variety of critical functions in essentially all sub-phases of mitosis, and is required for normal centrosome maturation, mitotic spindle formation, chromosome congression followed by efficient separation, cleavage furrow formation, and completion of cytokinesis, as well as for re-entry of G2-arrested cells into mitosis following recovery from DNA damage [96]. Plk2 and Plk3 stimulate S-phase progression [97, 98], while Plk4 is required for proper centriole duplication [99–101]. Plk3 also appears to be involved in several stress-response pathways, including those activated by DNA damage and spindle disruption [95, 102-104]. Exactly how Polo-like kinases select their substrates to control so many distinct aspects of cell division was clarified, in part, by the discovery that a portion of the Plk1 and Plk2 molecule encodes a pSer/pThr-binding domain [105].

All Polo-like kinases contain an N-terminal kinase domain and a conserved C-terminal region containing two Polo-boxes in Plk1, 2, and 3, and a single Polo-box in Plk4 [106, 107]. This C-terminal region, which we generically refer to as the *Polo-box domain* (PBD), is critical for both Plk localization and function, since mutations in this region of human Plk1, for example, result in loss of Plk1 localization in mammalian cells, and an inability of the mutant mammalian protein to bind to spindle poles and the bud neck in yeast complementation experiments [108], causing failure of mitotic progression. Conversely, overexpression of the isolated PBD causes pre-anaphase arrest in mammalian cells [106], and a cytokinesis block in budding yeast [109]. How the PBD functions in subcellular targeting and substrate selection was unknown until the PBD emerged as a serendipitous hit in an unbiased proteomic screen for phosphoserine/threonine-binding domains [105]. That study, and two subsequent detailed X-ray crystallographic studies that followed this discovery [110, 111], showed that the entire PBD of Plk1, including both Polo-boxes, the region between them, and a portion of the linker between the end of the kinase domain and the first Polo-box, function as a single pSer/pThr-recognition module. Intriguingly, the optimal sequence motif recognized by the Plk1 PBD is Ser-[pSer/pThr]-[Pro/X], suggesting that Cdks, MAP kinases, and/or other mitotic kinases, including Plk1 itself, might generate "priming" phosphorylations on substrates or docking proteins to both (1) localize Plks to their substrates [110], and (2) enhance kinase activity by relieving an inhibitory interaction between the kinase domain and Polo-box domain [112]. It appears that in early mitosis Cdks and MAPKs are the likely priming kinases, while in mid-to-late mitosis Plk1 self-priming seems to be the dominant process [113]. The Polo-box domain of Plk2 appears to function in a similar manner [114]. In contrast, the Polo-box domain of Plk3 appears to have only a weak affinity for phosphorylated peptides, and binds to non-phosphorylated peptides equally or better, while the single Polo-box in Plk4, which has been shown to dimerize in solution, also fails to bind to phosphopeptides – a finding consistent with the fact that many of the key residues involved in phospho-specific binding are facing away from the putative binding pocket at the Plk4 Polo-box domain dimer interface [115].

A number of Plk1 substrates at centrosomes, kinetochores, centromeres, the mitotic spindle, and the midbody have been shown to engage Plk1 through direct phospho-specific interactions with the PBD, including hCenexin 1 [116], BubR1 [117–119], PRC1 [120], PICH [121], Mklp-1 [122], Kizuna [123], and NudC [124]. Mutation of the PBD binding sites on these proteins, or mutation of the PBD itself, in many cases, disrupts these interactions and results in aberrant mitosis. A proteomic screen fro PBD-binding ligands revealed over 500 putative interaction partners, including all three Rho-activated protein kinases involved in cytokinesis, although many of the identified interactions could be indirect [125]. Likewise, the PBD in Plk2 binds to the Cdk5-phosphorylated form of SPAR, a RapGAP important for synaptic plasticity in the nervous system [126].

The crystal structures of the Plk1 PBD, both unliganded and in complex with an optimal pSer-containing peptide, revealed that each Polo-box is composed of a six-stranded anti-parallel β-sheet with an α-helix on one side [110, 111]. The phosphopeptide sits in a shallow cleft where the two β-sheets come together (Figure 4.4a) with protein-peptide interactions arising from a zipper-like contribution of residues alternatively arising from each Polo-box. The phosphoserine phosphate itself is directly chelated by a conserved His and Lys residue, along with a network of ordered water molecules. The hydroxyl group of the peptide pSer/pThr-1 Ser, a residue which plays a particularly important role in phosphopeptide selection, participates in three direct hydrogen bonds, two of which involve a critical Trp residue (Trp-414) whose mutation eliminates Polo-box domain binding to ligands [108]. To date there have been no structures of the Polo-box domain in complex with an intact full-length phosphorylated protein ligand, and neither has the structure of full-length Plk1, including both the PBD and the kinase domain, been reported. Both of these types of structures, when available, should shed additional light on the mechanistic basis for Plk1 regulation and potential conformational changes in substrates following their recruitment.

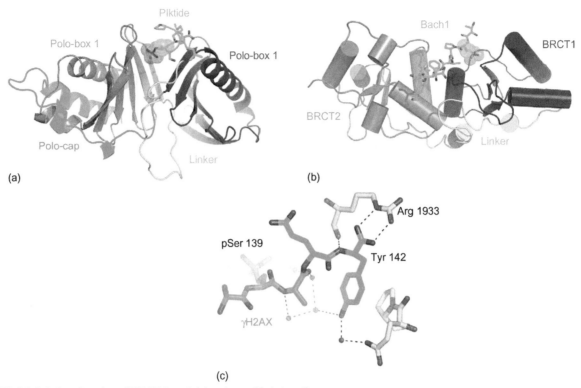

FIGURE 4.4 Polo-box domain and BRCT domain/phosphopeptide interactions.
(a) Each Polo-box in Plk1 (labeled) forms a β6α structure with binding of the pSer-containing peptide at the interface of the two Polo-boxes. The linker connecting the Polo-boxes is indicated. The region N-terminal to the first Polo-box, the Polo-cap, stabilizes the overall structure. (b) Structure of the tandem BRCT domains (labeled BRCT1 and 2) from BRCA1 bound to a phosphopeptide from Bach1. In a generally similar manner to ligand binding by the Polo-box domain shown in (a), the phosphopeptide binds to a groove formed at the interface of the two BRCT domains. The interdomain linker is indicated. In both (a) and (b), the pSer side-chain is shown in CPK representation. (c) A subtle variation in the mechanism of BRCT-phosphopeptide binding is revealed by the structure of the BRCT domains in MDC1 bound to a phosphopeptide from the C-terminus of H2AX corresponding to the sequence pSer-Gln-Glu-Tyr. (Only the β carbon of the peptide Gln is shown). The unique C-terminal recognition arises, in part, from a bivalent interaction of the conserved Arg residue in the tandem BRCT domains with the carboxylate anion of the peptide.

TANDEM BRCT-REPEAT DOMAINS

As their name implies, BRCT (Brca1 C-terminal) motifs were first discovered as tandem, repeated sequences of around 80–100 residues in the C-terminal region of Brca1 [127], and have since been noted in a variety of other proteins, many of which appear to function as adaptor or scaffolding molecules in DNA-damage responses. A role as protein-protein interaction modules was supported by structures of p53 in complex with the BRCT repeat domain of 53BP1 [128, 129]. However, it is now clear that many such domains function through binding to serine- and threonine-phosphorylated motifs that are generally characterized by the presence of a Phe or Tyr in +3 position [130–132]. Structurally, each BRCT motif forms a mixed α−β fold in which a central β-sheet is flanked on both faces by α-helices. BRCT motifs can function individually in some cases, but they most often are active in the tandemly repeated configuration first revealed in the X-ray structure of the Brca1 C-terminal region [133]. In 2004, the molecular basis of phospho-dependent binding of the Brca1 BRCT-repeat domain to the pSer$_{990}$-X-X-Phe motif

of the Bach1 helicase and an *in vitro* selected high-affinity site was demonstrated by X-ray crystallography [134–136]. These structures showed how the peptide binds in a largely extended conformation within a shallow cleft at and across the BRCT-BRCT interface (Figure 4.4b). In this sense, the binding somewhat resembles that observed for interactions with the tandem repeats of the Polo motif in the Plk1 PBD complex described above. In Brca1, the phosphopeptide forms bipartite interactions whereby the pSer is held by hydrogen bonds with Lys 1702 and Ser 1655 from BRCT1, while the Phe +3 is bound by main-chain interactions with a conserved arginine (Arg 1699) and non-polar side-chain interactions with Met 1775. The +3 binding surface is not obviously conserved in other known phospho-interacting BRCT domains. However, Met 1775 in Brca1 is often mutated to arginine or lysine in patients with breast and ovarian cancer, and comparison of the ligand-bound structures with that of a M1775R mutant suggested that a steric clash between the substituted Arg 1775 side-chain and the +3 phenylalanine is likely a major contributor to Brca1 dysfunction in these cases. In contrast to the +3 site, the apparent conservation of the Lys/Thr phosphate clamp in many candidate phosphobinding BRCT repeat domains indicates that the presence of these residues may be predictive of a similar activity in other systems. Indeed, this appears to be the case, and similar interactions have now been observed in the complex between the BRCT domain of human Mdc1 (Mediator of the DNA-damage checkpoint 1) with ATM-phosphorylated H2AX, a variant histone H2A that serves as a primary marker of double-stranded DNA breaks in damaged chromatin [137]. This structure also showed how the remarkable selectivity of Mdc1 for the C-terminal pSer-Gln-Glu-Tyr-COOH motif is mediated through specific salt-bridging interactions with the terminal carboxyl group (Figure 4.4c). Phospho-H2AX (γH2AX) has been proposed to bind to BRCT repeat domains from a number of other DNA-damage response proteins, including Brca1, 53BP1, Nbs1, and microcephalin (MCPH1). However, it now appears that these proteins do not bind γH2AX directly, and instead are recruited to damaged regions through Mdc1 itself, which seems to act as a *bona fide* scaffold [137, 138]. The exception is MCPH1, which can bind directly to γH2AX *in vitro* and whose localization is not apparently reliant on Mdc1 [139]. It will be interesting to see whether the MCPH1 interaction exhibits a specificity for the H2AX terminal carboxyl group as observed for Mdc1, or whether it adopts a more Brca1-like binding mode, as has recently been seen in structures of fission yeast Crb2/γH2A complexes [140].

LEUCINE-RICH REPEATS AND WD40 DOMAINS

In addition to 14-3-3 proteins, FHA domains, WW domains, Polo-box domains, and BRCT-repeats, several other modular signaling domains have been shown to bind to their substrates following serine/threonine phosphorylation, most notably leucine-rich regions, and WD40 repeats. These phosphospecific-binding modules are key participants in phosphorylation-dependent ubiquitin conjugation reactions catalyzed by Skp1-Cdc53-F-box protein (SCF) protein complexes that target the ubiquitinated substrates for proteosome-mediated degradation [141–145]. The SCF complexes are E3 ubiquitin ligases that consist of Skp1, Cul1/Cdc53, Roc1, and an F-box-containing protein that confers substrate specificity in a phosphospecific manner [141–144]. The leucine-rich repeats of WD40 domains within the F-box-containing protein appear to directly mediate phosphospecific binding. The yeast F-box protein Cdc4, for example binds to the phosphorylated form of the yeast Cdk inhibitors Sic1 [144, 146, 147] and Far1 [147, 148], while the mammalian F-box protein Skp2 binds to the phosphorylated form of the Cdk inhibitors p27, p21, and p57 [149–151]. In both yeast and mammalian cells, these interactions promote the ubiquitin-mediated destruction of the inhibitor, leading to activation of G_1 cyclin–Cdk

complexes. Later in mitosis, these proteins, along with an alternative set of F-box proteins is involved in the opposing process of inactivating the cyclin-Cdk complexes through ubiquitin-mediated proteolysis of the cyclin subunit [141–145, 152]. Thus, temporally regulated substrate phosphorylation coupled with combinatorial interchange of different F-box proteins with the core SCF complex can result in waves of proteolysis that drive the cell cycle [151, 152]. Similarly, phosphorylation-dependent recognition of other cell cycle regulators and transcription factors including, Wee1, Cdc25A and B, Bora, PDCD4, claspin, IκBα, and β-catenin, by the WD40 repeat of the F-box protein β-TrCP targets these substrates for ubiquitin-mediated proteolysis to control cell cycle transitions, DNA damage responses, and patterns of gene expression [151, 153, 154].

The WD40 repeat of Cdc4 is sufficient to bind tightly to a singly threonine-phosphorylated peptide from cyclin E1 containing the consensus motif L/I-L/I/P-pT-P-X, where X denotes all amino acids except R and K. However, Sic1, the physiological substrate of Cdc4, does not contain a perfect match to the consensus motif, and relies instead on the phosphorylation of multiple suboptimal motifs to provide cooperative binding [147]. Thus, by using a series of weak distributed phospho-dependent motifs to bind Sic1, Cdc4 ensures that a threshold of phosphorylation must be overcome prior to the initiation of DNA replication [147, 155].

Leucine-rich repeats adopt a C-shaped structure built of single α-helix/β-strand repeats [156], while WD40 repeats form β-propeller structures [157]. The structures of the WD40 repeats of β-TrCP in complex with a di-phosphorylated β-catenin peptide [158], and the WD40 repeats from Cdc4 in complex with a monophosphorylated human cyclin E peptide [159] have been solved. In the β-TrCP structure, the WD40 repeats form a traditional seven-bladed β-propeller, while in the Cdc4 structure, the WD40 repeats form a variant eight-bladed structure. In both cases, there is extensive recognition of multiple amino acids in the phosphopeptide-binding motif by residues in several blades of the propeller, as well as extensive chelation of the phosphoserine phosphates by 1, 2, and 3 Arg residues, respectively. Currently, there is no structure of an LRR domain bound to a phosphopeptide.

CONCLUDING REMARKS

The identification of several families of pSer/Thr-binding modules has provided fascinating insights into how cellular signaling events are regulated by protein–protein interactions mediated by Ser/Thr phosphorylation. The structural diversity of these domains suggests that, in general, their phosphobinding functions have evolved through convergent evolution. Importantly, the phosphorylated motifs recognized by different phosphoserine/threonine binding modules differ somewhat from the optimal phosphorylated motifs generated by various protein kinase families, which provides specificity in assembly of signaling complexes by requiring a unique overlap motif between the activating kinase and the binding domain and also allows abrupt on-and-off activation states to be created through cooperative binding between multiple weak motifs.

Acknowledgements

We are grateful to Lewis C. Cantley and Tony Pawson for helpful discussions. The SJS laboratory is funded by the Medical Research Council, UK, and the Association for International Cancer Research. The MBY laboratory is funded by grants GM60594, CA112967, and ES015339 from the NIH, and a Career Development Award from the Burroughs-Wellcome Fund. We apologize to the many investigators whose work was not cited due to space limitations.

Figures 4.1-4.3 are from Yaffe, M. B. and Smerdon, S. J. (2001), PhosphoSerine/threonine binding domains: you can't be pSERious. *Structure*, 9, R33-38. ©Elsevier 2001, reprinted with permission from Elsevier Science.

REFERENCES

1. Yaffe MB, Elia AE. Phosphoserine/threonine-binding domains. *Curr Opin Cell Biol* 2001;**13**:131–8.
2. Yaffe MB, Smerdon SJ. The use of *in vitro* peptide-library screens in the analysis of phosphoserine/threonine-binding domain structure and function. *Annu Rev Biophys Biomol Struct* 2004;**33**:225–44.
3. Moore B, Perez VJ. Specific acidic proteins of the nervous system. In: Carlson FD, editor. *Physiological and Biochemical Aspects of Nervous Integration*. Prentice-Hall: Englewood Cliffs; 1967. p. 343–59.
4. Ichimura T, Isobe T, Okuyama T, Takahashi N, Araki K, Kuwano R, Takahashi Y. Molecular cloning of cDNA coding for brain-specific 14-3-3 protein, a protein kinase-dependent activator of tyrosine and tryptophan hydroxylases. *Proc Natl Acad Sci USA* 1988;**85**:7084–8.
5. Ichimura T, Isobe T, Okuyama T, Yamauchi T, Fujisawa H. Brain 14-3-3 protein is an activator protein that activates tryptophan 5-monooxygenase and tyrosine 3-monooxygenase in the presence of Ca²⁺ calmodulin-dependent protein kinase II. *FEBS Letts* 1987;**219**:79–82.
6. Toker A, Ellis CA, Sellers LA, Aitken A. Protein kinase C inhibitor proteins. Purification from sheep brain and sequence similarity to lipocortins and 14-3-3 protein. *Eur J Biochem* 1990;**191**:421–9.
7. Furukawa Y, Ikuta N, Omata S, Yamauchi T, Isobe T, Ichimura T. Demonstration of the phosphorylation-dependent interaction of tryptophan hydroxylase with the 14-3-3 protein. *Biochem Biophys Res Commun* 1993;**194**:144–9.
8. Michaud NR, Fabian JR, Mathes KD, Morrison DK. 14-3-3 is not essential for Raf-1 function: identification of Raf-1 proteins that are biologically activated in a 14-3-3- and Ras- independent manner. *Mol Cell Biol* 1995;**15**:3390–7.
9. Muslin AJ, Tanner JW, Allen PM, Shaw AS. Interaction of 14-3-3 with signaling proteins is mediated by the recognition of phosphoserine. *Cell* 1996;**84**:889–97.
10. Yaffe MB, Rittinger K, Volinia S, Caron PR, Aitken A, Leffers H, Gamblin SJ, Smerdon SJ, Cantley LC. The structural basis for 14-3-3: phosphopeptide binding specificity. *Cell* 1997;**91**:961–71.
11. Coblitz B, Shikano S, Wu M, Gabelli SB, Cockrell LM, Spieker M, Hanyu Y, Fu H, Amzel LM, Li M. C-terminal recognition by 14-3-3 proteins for surface expression of membrane receptors. *J Biol Chem* 2005;**280**:36,263–36,272.
12. Kakiuchi K, Yamauchi Y, Taoka M, Iwago M, Fujita T, Ito T, Song SY, Sakai A, Isobe T, Ichimura T. Proteomic analysis of *in vivo* 14-3-3 interactions in the yeast *Saccharomyces cerevisiae*. *Biochemistry* 2007;**46**:7781–92.
13. Alexander RD, Morris PC. A proteomic analysis of 14-3-3 binding proteins from developing barley grains. *Proteomics* 2006;**6**:1886–96.
14. Moorhead G, Douglas P, Cotelle V, Harthill J, Morrice N, Meek S, Deiting U, Stitt M, Scarabel M, Aitken A, MacKintosh C. Phosphorylation-dependent interactions between enzymes of plant metabolism and 14-3-3 proteins. *Plant J* 1999;**18**:1–12.
15. Benzinger A, Muster N, Koch HB, Yates JR III, Hermeking H. Targeted proteomic analysis of 14-3-3 sigma, a p53 effector commonly silenced in cancer. *Mol Cell Proteomics* 2005;**4**:785–95.
16. Wilker EW, van Vugt MA, Artim SA, Huang PH, Petersen CP, Reinhardt HC, Feng Y, Sharp PA, Sonenberg N, White FM, Yaffe MB. 14-3-3σ controls mitotic translation to facilitate cytokinesis. *Nature* 2007;**446**:329–32.
17. Meek SE, Lane WS, Piwnica-Worms H. Comprehensive proteomic analysis of interphase and mitotic 14-3-3-binding proteins. *J Biol Chem* 2004;**279**:32,046–32,054.
18. Pozuelo Rubio M, Geraghty KM, Wong BH, Wood NT, Campbell DG, Morrice N, Mackintosh C. 14-3-3-affinity purification of over 200 human phosphoproteins reveals new links to regulation of cellular metabolism, proliferation and trafficking. *Biochem J* 2004;**379**:395–408.
19. Jin J, Smith FD, Stark C, Wells CD, Fawcett JP, Kulkarni S, Metalnikov P, O'Donnell P, Taylor P, Taylor L, Zougman A, Woodgett JR, Langeberg LK, Scott JD, Pawson T. Proteomic, functional, and domain-based analysis of in vivo 14-3-3 binding proteins involved in cytoskeletal regulation and cellular organization. *Curr Biol* 2004;**14**:1436–50.
20. Fu H, Subramanian RR, Masters SC. 14-3-3 proteins: structure, function, and regulation. *Annu Rev Pharmacol Toxicol* 2000;**40**:617–47.
21. van Hemert MJ, Steensma HY, van Heusden GP. 14-3-3 proteins: key regulators of cell division, signalling and apoptosis. *Bioessays* 2001;**23**:936–46.
22. Roy S, McPherson RA, Apolloni A, Yan J, Lane A, Clyde-Smith J, Hancock JF. 14-3-3 facilitates Ras-dependent Raf-1 activation *in vitro* and *in vivo*. *Mol Cell Biol* 1998;**18**:3947–55.
23. Tzivion G, Luo Z, Avruch J. A dimeric 14-3-3 protein is an essential cofactor for Raf kinase activity. *Nature* 1998;**394**:88–92.
24. Henriksson ML, Troller U, Hallberg B. 14-3-3 proteins are required for the inhibition of Ras by exoenzyme S. *Biochem J* 2000;**349**:697–701.
25. Hsu SY, Kaipia A, Zhu L, Hsueh AJ. Interference of BAD (Bcl-xL/Bcl-2-associated death promoter)-induced apoptosis in mammalian cells by 14-3-3 isoforms and P11. *Mol Endocrinol* 1997;**11**:1858–67.
26. Zha J, Harada H, Yang E, Jockel J, Korsmeyer SJ. Serine phosphorylation of death agonist BAD in response to survival factor results in binding to 14-3-3 not BCL-X(L). *Cell* 1996;**87**:619–28.
27. Hirai I, Wang HG. Survival-factor-induced phosphorylation of Bad results in its dissociation from Bcl-x(L) but not Bcl-2. *Biochem J* 2001;**359**:345–52.

28. Datta SR, Katsov A, Hu L, Petros A, Fesik SW, Yaffe MB, Greenberg ME. 14-3-3 proteins and survival kinases cooperate to inactivate BAD by BH3 domain phosphorylation. *Mol Cell* 2000;**6**:41–51.

29. Yang J, Winkler K, Yoshida M, Kornbluth S. Maintenance of G2 arrest in the Xenopus oocyte: a role for 14-3-3-mediated inhibition of Cdc25 nuclear import. *EMBO J* 1999;**18**:2174–83.

30. Zeng Y, Forbes KC, Wu Z, Moreno S, Piwnica-Worms H, Enoch T. Replication checkpoint requires phosphorylation of the phosphatase Cdc25 by Cds1 or Chk1. *Nature* 1998;**395**:507–10.

31. Brunet A, Kanai F, Stehn J, Xu J, Sarbassova D, Frangioni JV, Dalal SN, DeCaprio JA, Greenberg ME, Yaffe MB. 14-3-3 transits to the nucleus and participates in dynamic nucleocytoplasmic transport. *J Cell Biol* 2002;**156**:817–28.

32. Kumagai A, Dunphy WG. Binding of 14-3-3 proteins and nuclear export control the intracellular localization of the mitotic inducer Cdc25. *Genes Dev* 1999;**13**:1067–72.

33. Rajan S, Preisig-Muller R, Wischmeyer E, Nehring R, Hanley PJ, Renigunta V, Musset B, Schlichthorl G, Derst C, Karschin A, Daut J. Interaction with 14-3-3 proteins promotes functional expression of the potassium channels TASK-1 and TASK-3. *J Physiol* 2002;**545**:13–26.

34. Liu D, Bienkowska J, Petosa C, Collier RJ, Fu H, Liddington R. Crystal structure of the zeta isoform of the 14-3-3 protein. *Nature* 1995;**376**:191–-4.

35. Xiao B, Smerdon SJ, Jones DH, Dodson GG, Soneji Y, Aitken A, Gamblin SJ. Structure of a 14-3-3 protein and implications for coordination of multiple signalling pathways. *Nature* 1995;**376**:188–91.

36. Yang X, Lee WH, Sobott F, Papagrigoriou E, Robinson CV, Grossmann JG, Sundstrom M, Doyle DA, Elkins JM. Structural basis for protein-protein interactions in the 14-3-3 protein family. *Proc Natl Acad Sci USA* 2006;**103**:17,237–17,242.

37. Wurtele M, Jelich-Ottmann C, Wittinghofer A, Oecking C. Structural view of a fungal toxin acting on a 14-3-3 regulatory complex. *EMBO J* 2003;**22**:987–94.

38. Gardino AK, Smerdon SJ, Yaffe MB. Structural determinants of 14-3-3 binding specificities and regulation of subcellular localization of 14-3-3-ligand complexes: a comparison of the X-ray crystal structures of all human 14-3-3 isoforms. *Semin Cancer Biol* 2006;**16**:173–82.

39. Petosa C, Masters SC, Bankston LA, Pohl J, Wang B, Fu H, Liddington RC. 14-3-3ζ binds a phosphorylated Raf peptide and an unphosphorylated peptide via its conserved amphipathic groove. *J Biol Chem* 1998;**273**:16,305–16,310.

40. Rittinger K, Budman J, Xu J, Volinia S, Cantley LC, Smerdon SJ, Gamblin SJ, Yaffe MB. Structural analysis of 14-3-3 phosphopeptide complexes identifies a dual role for the nuclear export signal of 14-3-3 in ligand binding. *Mol Cell* 1999;**4**:153–66.

41. Obsil T, Ghirlando R, Klein DC, Ganguly S, Dyda F. Crystal structure of the 14-3-3ζ:serotonin N-acetyltransferase complex. a role for scaffolding in enzyme regulation. *Cell* 2001;**105**:257–67.

42. Ottmann C, Marco S, Jaspert N, Marcon C, Schauer N, Weyand M, Vandermeeren C, Duby G, Boutry M, Wittinghofer A, Rigaud JL, Oecking C. Structure of a 14-3-3 coordinated hexamer of the plant plasma membrane H+ -ATPase by combining X-ray crystallography and electron cryomicroscopy. *Mol Cell* 2007;**25**:427–40.

43. Yaffe MB. How do 14-3-3 proteins work? Gatekeeper phosphorylation and the molecular anvil hypothesis. *FEBS Letts* 2002;**513**:53–7.

44. Powell DW, Rane MJ, Joughin BA, Kalmukova R, Hong JH, Tidor B, Dean WL, Pierce WM, Klein JB, Yaffe MB, McLeish KR. Proteomic identification of 14-3-3ζ as a mitogen-activated protein kinase-activated protein kinase 2 substrate: role in dimer formation and ligand binding. *Mol Cell Biol* 2003;**23**:5376–87.

45. Woodcock JM, Murphy J, Stomski FC, Berndt MC, Lopez AF. The dimeric versus monomeric status of 14-3-3ζ is controlled by phosphorylation of Ser58 at the dimer interface. *J Biol Chem* 2003;**278**:36,323–36,327.

46. Dubois T, Rommel C, Howell S, Steinhussen U, Soneji Y, Morrice N, Moelling K, Aitken A. 14-3-3 is phosphorylated by casein kinase I on residue 233. Phosphorylation at this site *in vivo* regulates Raf/14-3-3 interaction. *J Biol Chem* 1997;**272**:28,882–28,888.

47. Tsuruta F, Sunayama J, Mori Y, Hattori S, Shimizu S, Tsujimoto Y, Yoshioka K, Masuyama N, Gotoh Y. JNK promotes Bax translocation to mitochondria through phosphorylation of 14-3-3 proteins. *EMBO J* 2004;**23**:1889–99.

48. Sunayama J, Tsuruta F, Masuyama N, Gotoh Y. JNK antagonizes Akt-mediated survival signals by phosphorylating 14-3-3. *J Cell Biol* 2005;**170**:295–304.

49. Yoshida K, Yamaguchi T, Natsume T, Kufe D, Miki Y. JNK phosphorylation of 14-3-3 proteins regulates nuclear targeting of c-Abl in the apoptotic response to DNA damage. *Nat Cell Biol* 2005;**7**:278–85.

50. Durocher D, Taylor IA, Sabassova D, Haire LF, Westcott SL, Jackson SP, Smerdon SJ, Yaffe MB. The molecular basis of FHA domain: phosphopeptide binding specificity and implications for phosphodependent signaling mechanisms. *Mol Cell* 2000;**6**:1169–82.

51. Hammet A, Pike BL, Mitchelhill KI, Teh T, Kobe B, House CM, Kemp BE, Heierhorst J. FHA domain boundaries of the dun1p and rad53p cell cycle checkpoint kinases. *FEBS Letts* 2000;**471**:141–6.

52. Hofmann K, Bucher P. The FHA domain: a putative nuclear signalling domain found in protein kinases and transcription factors [letter]. *Trends Biochem Sci* 1995;**20**:347–9.

53. Li J, Smith GP, Walker JC. Kinase interaction domain of kinase-associated protein phosphatase, a phosphoprotein-binding domain. *Proc Natl Acad Sci USA* 1999;**96**:7821–6.

54. Liao H, Byeon IJ, Tsai MD. Structure and function of a new phosphopeptide-binding domain containing the FHA2 of Rad53. *J Mol Biol* 1999;**294**:1041–9.

55. Wang P, Byeon IJ, Liao H, Beebe KD, Yongkiettrakul S, Pei D, Tsai MD. II. Structure and specificity of the interaction between the FHA2 domain of rad53 and phosphotyrosyl peptidesdagger [In Process Citation].. *J Mol Biol* 2000;**302**:927–40.

56. Stone JM, Collinge MA, Smith RD, Horn MA, Walker JC. Interaction of a protein phosphatase with an Arabidopsis serine-threonine receptor kinase. *Science* 1994;**266**:793–5.

57. Sun Z, Hsiao J, Fay DS, Stern DF. Rad53 FHA domain associated with phosphorylated Rad9 in the DNA damage checkpoint [see comments]. *Science* 1998;**281**:272–4.

58. Durocher D, Henckel J, Fersht AR, Jackson SP. The FHA domain is a modular phosphopeptide recognition motif. *Mol Cell* 1999;**4**:387–94.

59. Liao H, Yuan C, Su MI, Yongkiettrakul S, Qin D, Li H, Byeon IJ, Pei D, Tsai MD. Structure of the FHA1 domain of yeast Rad53 and identification of binding sites for both FHA1 and its target protein Rad9. *J Mol Biol* 2000;**304**:941–51.

60. Bell DW, Varley JM, Szydlo TE, Kang DH, Wahrer DC, Shannon KE, Lubratovich M, Verselis SJ, Isselbacher KJ, Fraumeni JF, Birch JM, Li FP, Garber JE, Haber DA. Heterozygous germ line hCHK2 mutations in Li-Fraumeni syndrome. *Science* 1999;**286**:2528–31.

61. Lee SB, Kim SH, Bell DW, Wahrer DC, Schiripo TA, Jorczak MM, Sgroi DC, Garber JE, Li FP, Nichols KE, Varley JM, Godwin AK, Shannon KM, Harlow E, Haber DA. Destabilization of CHK2 by a missense mutation associated with Li-Fraumeni Syndrome. *Cancer Res* 2001;**61**:8062–7.

62. Scolnick DM, Halazonetis TD. Chfr defines a mitotic stress checkpoint that delays entry into metaphase [see comments]. *Nature* 2000;**406**:430–5.

63. Tauchi H, Kobayashi J, Morishima K, Matsuura S, Nakamura A, Shiraishi T, Ito E, Masnada D, Delia D, Komatsu K. The forkhead-associated domain of NBS1 is essential for nuclear foci formation after irradiation but not essential for hRAD50•hMRE11•NBS1 complex DNA repair activity. *J Biol Chem* 2001;**276**:12–15.

64. Kang D, Chen J, Wong J, Fang G. The checkpoint protein Chfr is a ligase that ubiquitinates Plk1 and inhibits Cdc2 at the G2 to M transition. *J Cell Biol* 2002;**156**:249–59.

65. Chapman JR, Jackson SP. Phospho-dependent interactions between NBS1 and MDC1 mediate chromatin retention of the MRN complex at sites of DNA damage. *EMBO Rep* 2008. in press.

66. Melander F, Bekker-Jensen S, Falck J, Bartek J, Mailand N, Lukas J. Phosphorylation of SDT repeats in the MDC1 N terminus triggers retention of NBS1 at the DNA damage-modified chromatin. *J Cell Biol* 2008;**181**:213–26.

67. Spycher C, Miller ES, Townsend K, Pavic L, Morrice NA, Janscak P, Stewart GS, Stucki M. Constitutive phosphorylation of MDC1 physically links the MRE11-RAD50-NBS1 complex to damaged chromatin. *J Cell Biol* 2008;**181**:227–40.

68. Wu L, Luo K, Lou Z, Chen J. MDC1 regulates intra-S-phase checkpoint by targeting NBS1 to DNA double-strand breaks. *Proc Natl Acad Sci USA* 2008;**105**:11,200–11,205.

69. Li J, Williams BL, Haire LF, Goldberg M, Wilker E, Durocher D, Yaffe MB, Jackson SP, Smerdon SJ. Structural and functional versatility of the FHA domain in DNA-damage signaling by the tumor suppressor kinase Chk2. *Mol Cell* 2002;**9**:1045–54.

70. Stavridi ES, Huyen Y, Loreto IR, Scolnick DM, Halazonetis TD, Pavletich NP, Jeffrey PD. Crystal structure of the FHA domain of the Chfr mitotic checkpoint protein and its complex with tungstate. *Structure (Camb)* 2002;**10**:891–9.

71. Qin BY, Chacko BM, Lam SS, de Caestecker MP, Correia JJ, Lin K. Structural basis of Smad1 activation by receptor kinase phosphorylation. *Mol Cell* 2001;**8**:1303–12.

72. Wu JW, Hu M, Chai J, Seoane J, Huse M, Li C, Rigotti DJ, Kyin S, Muir TW, Fairman R, Massague J, Shi Y. Crystal structure of a phosphorylated Smad2. Recognition of phosphoserine by the MH2 domain and insights on Smad function in TGF-β signaling. *Mol Cell* 2001;**8**:1277–89.

73. Li H, Byeon IJ, Ju Y, Tsai MD. Structure of human Ki67 FHA domain and its binding to a phosphoprotein fragment from hNIFK reveal unique recognition sites and new views to the structural basis of FHA domain functions.. *J Mol Biol* 2004;**335**:371–81.

74. Lee H, Yuan C, Hammet A, Mahajan A, Chen ES, Wu MR, Su MI, Heierhorst J, Tsai MD. Diphosphothreonine-specific interaction between an SQ/TQ cluster and an FHA domain in the Rad53-Dun1 kinase cascade. *Mol Cell* 2008;**30**:767–78.

75. Sudol M, Hunter T. NeW wrinkles for an old domain. *Cell* 2000;**103**:1001–4.

76. Shen M, Stukenberg PT, Kirschner MW, Lu KP. The essential mitotic peptidyl-prolyl isomerase Pin1 binds and regulates mitosis-specific phosphoproteins. *Genes Dev* 1998;**12**:706–20.

77. Chang A, Cheang S, Espanel X, Sudol M. Rsp5 WW domains interact directly with the carboxyl-terminal domain of RNA polymerase II. *J Biol Chem* 2000;**275**:20,562–20,571.

78. Morris DP, Greenleaf AL. The splicing factor, Prp40, binds the phosphorylated carboxyl-terminal domain of RNA polymerase II. *J Biol Chem* 2000;**275**:39,935–39,943.

79. Morris DP, Phatnani HP, Greenleaf AL. Phospho-carboxyl-terminal domain binding and the role of a prolyl isomerase in pre-mRNA 3′-End formation. *J Biol Chem* 1999;**274**:31,583–31,587.

80. Lu KP, Hanes SD, Hunter T. A human peptidyl-prolyl isomerase essential for regulation of mitosis. *Nature* 1996;**380**:544–7.

81. Winkler KE, Swenson KI, Kornbluth S, Means AR. Requirement of the prolyl isomerase Pin1 for the replication checkpoint. *Science* 2000;**287**:1644–7.

82. Ranganathan R, Lu KP, Hunter T, Noel JP. Structural and functional analysis of the mitotic rotamase Pin1 suggests substrate recognition is phosphorylation dependent. *Cell* 1997;**89**:875–86.

83. Yaffe MB, Schutkowski M, Shen M, Zhou XZ, Stukenberg PT, Rahfeld JU, Xu J, Kuang J, Kirschner MW, Fischer G, Cantley LC, Lu KP. Sequence-specific and phosphorylation-dependent proline isomerization: a potential mitotic regulatory mechanism. *Science* 1997;**278**:1957–60.

84. Zhou XZ, Kops O, Werner A, Lu PJ, Shen M, Stoller G, Kullertz G, Stark M, Fischer G, Lu KP. Pin1-dependent prolyl isomerization regulates dephosphorylation of Cdc25C and tau proteins. *Mol Cell* 2000;**6**:873–83.

85. Kops O, Zhou XZ, Lu KP. Pin1 modulates the dephosphorylation of the RNA polymerase II C-terminal domain by yeast Fcp1. *FEBS Lett* 2002;**513**:305–11.

86. Verdecia MA, Bowman ME, Lu KP, Hunter T, Noel JP. Structural basis for phosphoserine-proline recognition by group IV WW domains. *Nat Struct Biol* 2000;**7**:639–43.

87. Llamazares S, Moreira A, Tavares A, Girdham C, Spruce BA, Gonzalez C, Karess RE, Glover DM, Sunkel CE. polo encodes a protein kinase homolog required for mitosis in Drosophila. *Genes Dev* 1991;**5**:2153–65.

88. Sunkel CE, Glover DM. Polo, a mitotic mutant of *Drosophila* displaying abnormal spindle poles. *J Cell Sci* 1988;**89**:25–38.

89. Clay FJ, Ernst MR, Trueman JW, Flegg R, Dunn AR. The mouse Plk gene: structural characterization, chromosomal localization and identification of a processed Plk pseudogene. *Gene* 1997;**198**:329–39.

90. Golsteyn RM, Schultz SJ, Bartek J, Ziemiecki A, Ried T, Nigg EA. Cell cycle analysis and chromosomal localization of human Plk1, a putative homologue of the mitotic kinases *Drosophila* polo and *Saccharomyces cerevisiae* Cdc5. *J Cell Sci* 1994;**107**:1509–17.

91. Kumagai A, Dunphy WG. Purification and molecular cloning of Plx1, a Cdc25-regulatory kinase from *Xenopus* egg extracts. *Science* 1996;**273**:1377–80.

92. Simmons DL, Neel BG, Stevens R, Evett G, Erikson RL. Identification of an early-growth-response gene encoding a novel putative protein kinase. *Mol Cell Biol* 1992;**12**:4164–9.

93. Duncan PI, Pollet N, Niehrs C, Nigg EA. Cloning and characterization of Plx2 and Plx3, two additional Polo-like kinases from *Xenopus laevis*. *Exp Cell Res* 2001;**270**:78–87.

94. Li B, Ouyang B, Pan H, Reissmann PT, Slamon DJ, Arceci R, Lu L, Dai W. Prk, a cytokine-inducible human protein serine/threonine kinase whose expression appears to be down-regulated in lung carcinomas. *J Biol Chem* 1996;**271**:19,402–19,408.

95. Donohue PJ, Alberts GF, Guo Y, Winkles JA. Identification by targeted differential display of an immediate early gene encoding a putative serine/threonine kinase. *J Biol Chem* 1995;**270**:10,351–10,357.

96. van Vugt MA, Bras A, Medema RH. Polo-like kinase-1 controls recovery from a G2 DNA damage-induced arrest in mammalian cells. *Mol Cell* 2004;**15**:799–811.

97. Ma S, Charron J, Erikson RL. Role of Plk2 (Snk) in mouse development and cell proliferation. *Mol Cell Biol* 2003;**23**:6936–43.

98. Zimmerman WC, Erikson RL. Polo-like kinase 3 is required for entry into S phase. *Proc Natl Acad Sci USA* 2007;**104**:1847–52.

99. Bettencourt-Dias M, Rodrigues-Martins A, Carpenter L, Riparbelli M, Lehmann L, Gatt MK, Carmo N, Balloux F, Callaini G, Glover DM. SAK/PLK4 is required for centriole duplication and flagella development. *Curr Biol* 2005;**15**:2199–207.

100. Habedanck R, Stierhof YD, Wilkinson CJ, Nigg EA. The Polo kinase Plk4 functions in centriole duplication. *Nat Cell Biol* 2005;**7**:1140–6.

101. Kleylein-Sohn J, Westendorf J, Le Clech M, Habedanck R, Stierhof YD, Nigg EA. Plk4-induced centriole biogenesis in human cells. *Dev Cell* 2007;**13**:190–202.

102. Bahassi el M, Conn CW, Myer DL, Hennigan RF, McGowan CH, Sanchez Y, Stambrook PJ. Mammalian Polo-like kinase 3 (Plk3) is a multifunctional protein involved in stress response pathways. *Oncogene* 2002;**21**:6633–40.

103. Xie S, Wu H, Wang Q, Cogswell JP, Husain I, Conn C, Stambrook P, Jhanwar-Uniyal M, Dai W. Plk3 functionally links DNA damage to cell cycle arrest and apoptosis at least in part via the p53 pathway. *J Biol Chem* 2001;**276**:43,305–43,312.

104. Xie S, Wu H, Wang Q, Kunicki J, Thomas RO, Hollingsworth RE, Cogswell J, Dai W. Genotoxic stress-induced activation of Plk3 is partly mediated by Chk2. *Cell Cycle* 2002;**1**:424–9.

105. Elia AE, Cantley LC, Yaffe MB. Proteomic screen finds pSer/pThr-binding domain localizing Plk1 to mitotic substrates. *Science* 2003;**299**:1228–31.

106. Seong YS, Kamijo K, Lee JS, Fernandez E, Kuriyama R, Miki T, Lee KS. A spindle checkpoint arrest and a cytokinesis failure by the dominant-negative polo-box domain of Plk1 in U-2 OS cells. *J Biol Chem* 2002;**277**:32,282–32,293.

107. Sonnhammer EL, Eddy SR, Birney E, Bateman A, Durbin R. Pfam: multiple sequence alignments and HMM-profiles of protein domains. *Nucleic Acids Res* 1998;**26**:320–2.

108. Lee KS, Grenfell TZ, Yarm FR, Erikson RL. Mutation of the polo-box disrupts localization and mitotic functions of the mammalian polo kinase Plk. *Proc Natl Acad Sci USA* 1998;**95**:9301–6.

109. Song S, Lee KS. A novel function of *Saccharomyces cerevisiae* CDC5 in cytokinesis. *J Cell Biol* 2001;**152**:451–69.

110. Elia EAH, Rellos P, Haire LF, Chao JW, Ivins FJ, Hoepker K, Mohammad D, Cantley LC, Smerdon SJ, Yaffe MB. The molecular basis for phosphodependent substrate targeting and regulation of Plks by the Polo-box Domain. *Cell* 2003;**115**:83–95.

111. Cheng KY, Lowe ED, Sinclair J, Nigg EA, Johnson LN. The crystal structure of the human polo-like kinase-1 polo box domain and its phospho-peptide complex. *EMBO J* 2003;**22**:5757–68.

112. Jang YJ, Lin CY, Ma S, Erikson RL. Functional studies on the role of the C-terminal domain of mammalian polo-like kinase. *Proc Natl Acad Sci USA* 2002;**99**:1984–9.

113. Neef R, Gruneberg U, Kopajtich R, Li X, Nigg EA, Sillje H, Barr FA. Choice of Plk1 docking partners during mitosis and cytokinesis is controlled by the activation state of Cdk1. *Nat Cell Biol* 2007;**9**:436–44.

114. van de Weerdt BC, Littler DR, Klompmaker R, Huseinovic A, Fish A, Perrakis A, Medema RH. Polo-box domains confer target specificity to the Polo-like kinase family. *Biochim Biophys Acta* 2008;**1783**:1015–22.

115. Leung GC, Hudson JW, Kozarova A, Davidson A, Dennis JW, Sicheri F. The Sak polo-box comprises a structural domain sufficient for mitotic subcellular localization. *Nat Struct Biol* 2002;**9**:719–24.

116. Soung NK, Kang YH, Kim K, Kamijo K, Yoon H, Seong YS, Kuo YL, Miki T, Kim SR, Kuriyama R, Giam CZ, Ahn CH, Lee KS. Requirement of hCenexin for proper mitotic functions of polo-like kinase 1 at the centrosomes. *Mol Cell Biol* 2006;**26**:8316–35.

117. Elowe S, Hummer S, Uldschmid A, Li X, Nigg EA. Tension-sensitive Plk1 phosphorylation on BubR1 regulates the stability of kinetochore microtubule interactions. *Genes Dev* 2007;**21**:2205–19.

118. Matsumura S, Toyoshima F, Nishida E. Polo-like kinase 1 facilitates chromosome alignment during prometaphase through BubR1. *J Biol Chem* 2007;**282**:15,217–15,227.

119. Wong OK, Fang G. Loading of the 3F3/2 antigen onto kinetochores is dependent on the ordered assembly of the spindle checkpoint proteins. *Mol Biol Cell* 2006;**17**:4390–9.

120. Kang YH, Park JE, Yu LR, Soung NK, Yun SM, Bang JK, Seong YS, Yu H, Garfield S, Veenstra TD, Lee KS. Self-regulated Plk1 recruitment to kinetochores by the Plk1-PBIP1 interaction is critical for proper chromosome segregation. *Mol Cell* 2006;**24**:409–22.

121. Baumann C, Korner R, Hofmann K, Nigg EA. PICH, a centromere-associated SNF2 family ATPase, is regulated by Plk1 and required for the spindle checkpoint. *Cell* 2007;**128**:101–14.

122. Liu X, Zhou T, Kuriyama R, Erikson RL. Molecular interactions of Polo-like-kinase 1 with the mitotic kinesin-like protein CHO1/MKLP-1. *J Cell Sci* 2004;**117**:3233–46.

123. Oshimori N, Ohsugi M, Yamamoto T. The Plk1 target Kizuna stabilizes mitotic centrosomes to ensure spindle bipolarity. *Nat Cell Biol* 2006;**8**:1095–101.

124. Zhou T, Aumais JP, Liu X, Yu-Lee LY, Erikson RL. A role for Plk1 phosphorylation of NudC in cytokinesis. *Dev Cell* 2003;**5**:127–38.

125. Lowery DM, Clauser KR, Hjerrild M, Lim D, Alexander J, Kishi K, Ong SE, Gammeltoft S, Carr SA, Yaffe MB. Proteomic screen defines the Polo-box domain interactome and identifies Rock2 as a Plk1 substrate. *EMBO J* 2007;**26**:2262–73.

126. Seeburg DP, Feliu-Mojer M, Gaiottino J, Pak DT, Sheng M. Critical role of CDK5 and Polo-like kinase 2 in homeostatic synaptic plasticity during elevated activity. *Neuron* 2008;**58**:571–83.

127. Koonin EV, Altschul SF, Bork P. BRCA1 protein products. Functional motifs. *Nat Genet* 1996;**13**:266–8.

128. Derbyshire DJ, Basu BP, Serpell LC, Joo WS, Date T, Iwabuchi K, Doherty AJ. Crystal structure of human 53BP1 BRCT domains bound to p53 tumour suppressor. *EMBO J* 2002;**21**:3863–72.

129. Joo WS, Jeffrey PD, Cantor SB, Finnin MS, Livingston DM, Pavletich NP. Structure of the 53BP1 BRCT region bound to p53 and its comparison to the Brca1 BRCT structure. *Genes Dev* 2002;**16**:583–93.

130. Manke IA, Lowery DM, Nguyen A, Yaffe MB. BRCT repeats as phosphopeptide-binding modules involved in protein targeting. *Science* 2003;**302**:636–9.

131. Rodriguez M, Yu X, Chen J, Songyang Z. Phosphopeptide binding specificities of BRCA1 COOH-terminal (BRCT) domains. *J Biol Chem* 2003;**278**:52,914–52,918.

132. Yu X, Chini CC, He M, Mer G, Chen J. The BRCT domain is a phospho-protein binding domain. *Science* 2003;**302**:639–42.

133. Williams RS, Green R, Glover JN. Crystal structure of the BRCT repeat region from the breast cancer-associated protein BRCA1. *Nat Struct Biol* 2001;**8**:838–42.

134. Clapperton JA, Manke IA, Lowery DM, Ho T, Haire LF, Yaffe MB, Smerdon SJ. Structure and mechanism of BRCA1 BRCT domain recognition of phosphorylated BACH1 with implications for cancer. *Nat Struct Mol Biol* 2004;**11**:512–18.

135. Shiozaki EN, Gu L, Yan N, Shi Y. Structure of the BRCT repeats of BRCA1 bound to a BACH1 phosphopeptide: implications for signaling. *Mol Cell* 2004;**14**:405–12.

136. Williams RS, Lee MS, Hau DD, Glover JN. Structural basis of phosphopeptide recognition by the BRCT domain of BRCA1. *Nat Struct Mol Biol* 2004;**11**:519–25.

137. Stucki M, Clapperton JA, Mohammad D, Yaffe MB, Smerdon SJ, Jackson SP. MDC1 directly binds phosphorylated histone H2AX to regulate cellular responses to DNA double-strand breaks. *Cell* 2005;**123**:2034–47.

138. Huen MS, Grant R, Manke I, Minn K, Yu X, Yaffe MB, Chen J. RNF8 transduces the DNA-damage signal via histone ubiquitylation and checkpoint protein assembly. *Cell* 2007;**131**:901–14.

139. Wood JL, Singh N, Mer G, Chen J. MCPH1 functions in an H2AX dependent but MDC1 independent pathway in response to DNA damage. *J Biol Chem* 2007;**282**:35,416–35,423.

140. Kilkenny ML, Doré AS, Roe SM, Nestoras K, Ho JC, Watts FZ, Pearl LH. Structural and functional analysis of the Crb2-BRCT2 domain reveals distinct roles in checkpoint signaling and DNA damage repair. *Genes Dev* 2008;**22**:2034–47.

141. Craig KL, Tyers M. The F-box: a new motif for ubiquitin dependent proteolysis in cell cycle regulation and signal transduction. *Prog Biophys Mol Biol* 1999;**72**:299–328.

142. Deshaies RJ. SCF and Cullin/Ring H2-based ubiquitin ligases. *Annu Rev Cell Dev Biol* 1999;**15**:435–67.

143. Patton EE, Willems AR, Tyers M. Combinatorial control in ubiquitin-dependent proteolysis: don't Skp the F-box hypothesis. *Trends Genet* 1998;**14**:236–43.

144. Skowyra D, Craig KL, Tyers M, Elledge SJ, Harper JW. F-box proteins are receptors that recruit phosphorylated substrates to the SCF ubiquitin-ligase complex [see comments]. *Cell* 1997;**91**:209–19.

145. Willems AR, Goh T, Taylor L, Chernushevich I, Shevchenko A, Tyers M. SCF ubiquitin protein ligases and phosphorylation-dependent proteolysis. *Phil Trans R Soc Lond B Biol Sci* 1999;**354**:1533–50.

146. Feldman RM, Correll CC, Kaplan KB, Deshaies RJ. A complex of Cdc4p, Skp1p, and Cdc53p/cullin catalyzes ubiquitination of the phosphorylated CDK inhibitor Sic1p. *Cell* 1997;**91**:221–30.

147. Nash P, Tang X, Orlicky S, Chen Q, Gertler FB, Mendenhall MD, Sicheri F, Pawson T, Tyers M. Multisite phosphorylation of a CDK inhibitor sets a threshold for the onset of DNA replication. *Nature* 2001;**414**:514–21.

148. Blondel M, Galan JM, Chi Y, Lafourcade C, Longaretti C, Deshaies RJ, Peter M. Nuclear-specific degradation of Far1 is controlled by the localization of the F-box protein Cdc4. *EMBO J* 2000;**19**:6085–97.

149. Carrano AC, Eytan E, Hershko A, Pagano M. SKP2 is required for ubiquitin-mediated degradation of the CDK inhibitor p27. *Nat Cell Biol* 1999;**1**:193–9.

150. Tsvetkov LM, Yeh KH, Lee SJ, Sun H, Zhang H. p27(Kip1) ubiquitination and degradation is regulated by the SCF(Skp2) complex through phosphorylated Thr187 in p27. *Curr Biol* 1999;**9**:661–4.

151. Frescas D, Pagano M. Deregulated proteolysis by the F-box proteins SKP2 and beta-TrCP: tipping the scales of cancer. *Nat Rev Cancer* 2008;**8**:438–49.

152. King RW, Deshaies RJ, Peters JM, Kirschner MW. How proteolysis drives the cell cycle. *Science* 1996;**274**:1652–9.

153. Hart M, Concordet JP, Lassot I, Albert I, del los Santos R, Durand H, Perret C, Rubinfeld B, Margottin F, Benarous R, Polakis P. The F-box protein beta-TrCP associates with phosphorylated beta-catenin and regulates its activity in the cell. *Curr Biol* 1999;**9**:207–10.

154. Yaron A, Hatzubai A, Davis M, Lavon I, Amit S, Manning AM, Andersen JS, Mann M, Mercurio F, Ben-Neriah Y. Identification of the receptor component of the IκBα-ubiquitin ligase. *Nature* 1998;**396**:590–4.

155. Klein P, Pawson T, Tyers M. Mathematical modeling suggests cooperative interactions between a disordered polyvalent ligand and a single receptor site. *Curr Biol* 2003;**13**:1669–78.

156. Kobe B, Deisenhofer J. A structural basis of the interactions between leucine-rich repeats and protein ligands. *Nature* 1995;**374**:183–6.

157. ter Haar E, Harrison SC, Kirchhausen T. Peptide-in-groove interactions link target proteins to the beta-propeller of clathrin [see comments]. *Proc Natl Acad Sci USA* 2000;**97**:1096–100.

158. Wu G, Xu G, Schulman BA, Jeffrey PD, Harper JW, Pavletich NP. Structure of a β-TrCP1-Skp1-β-catenin complex: destruction motif binding and lysine specificity of the SCF(β-TrCP1) ubiquitin ligase. *Mol Cell* 2003;**11**:1445–56.

159. Orlicky S, Tang X, Willems A, Tyers M, Sicheri F. Structural basis for phosphodependent substrate selection and orientation by the SCFCdc4 ubiquitin ligase. *Cell* 2003;**112**:243–56.

160. Yaffe MB, Smerdon SJ. PhosphoSerine/threonine binding domains: you can't pSERious? *Structure* 2001;**9**:R33–8.

Chapter 5

Protein Kinase Inhibitors

Alexander Levitzki

Unit of Cellular Signaling, Department of Biological Chemistry, The Alexander Silberman Institute of Life Sciences, The Hebrew University of Jerusalem, Israel

SIGNAL TRANSDUCTION THERAPY AND PROTEIN KINASE INHIBITORS

The realization that aberrations in signal transduction networks involve protein kinases has given impetus to researchers to target these kinases for various therapies, mainly of cancers. Therapy aimed at tampering with the signaling pathways of the cancer cell has been coined "signal transduction therapy" [1]. Protein kinase inhibitors are the front runners in signal transduction therapy, because the catalytic sites of enzymes are easier targets for drug design than protein–protein interaction domains. This picture will change, however, as the structures of other signaling proteins are deciphered, creating new opportunities for drug design.

The kinase map, the kinome, describes all the families of protein kinases [2, 3]. In the human genome, 409 serine/threonine (Ser/Thr) kinases, 59 receptor tyrosine kinases (RPTKs), and 32 non-receptor protein tyrosine kinases (PTKs) can currently be identified. With the addition of splice variants, the actual number of protein kinases is slightly higher. In contrast to the Ser/Thr kinases, which are mostly involved in cellular housekeeping, PTKs are almost exclusively involved in cellular signaling. Since protein phosphorylation affects ~30 percent of cellular proteins, and many kinases are involved in signaling, we stand to see accelerated development of protein kinase inhibitors as therapeutic agents. Needless to say, these agents are also useful for the study of signaling pathways.

The search for protein kinase inhibitors remains mostly in the field of cancer, since the relevance of certain kinases to specific cancers has been shown. The altered signal transduction network in cancer cells allows them to utilize their environment to their advantage without obeying the full network of signals that regulate the normal cell [1, 4]. Cancer cells sprout due to mutations in their growth signaling pathways. These mutations induce stress, which the mutated cell evades by further mutations that enhance

survival signals. These two sets of mutations enhance the proliferation of tumor cells, which now resist apoptotic messages [4]. During its evolution, the cancer cell becomes highly dependent on its abnormal signaling network. It loses a significant portion of its signaling genes and thrives on fewer, more intense signals. Thus, the few enhanced pathways on which the cancer cell depends are actually its potential Achilles' heel. Depriving the cancer cell of a few of these enhanced signaling elements may sensitize it to stress and even induce its demise. The neighboring normal cells, which retain their robustness due to their more elaborate signaling network, suffer little harm. In the case of the chronic phase of chronic myelogenous leukemia (CML), deprivation of a single signaling pathway, namely Bcr-Abl signaling, is sufficient to induce cell death.

Since protein tyrosine phosphorylation is mainly involved in signaling, it is not surprising that tyrosine kinase inhibitors (TKIs) were the first to enter the clinic. It was recognized early on that protein tyrosine kinases comprise a major fraction of the signaling elements whose activities are enhanced in the cancer cell, and on whose activities the survival of the cancer cell is highly dependent [4]. As well as targeting the aberrant signaling elements within the cancer cell, signal transduction therapy also includes anti-angiogenic therapy, which targets the newly dividing endothelial cells lining the fresh blood vessels that are generated in response to VEGF secreted by the tumor [5].

PROTEIN TYROSINE KINASE INHIBITORS

Although protein kinases have been known since the discovery of protein phosphorylation in the 1950s, no one turned to them as drug targets until tyrosine phosphorylation was discovered [6]. The identification of tyrosine kinase activity as the hallmark of the oncogenic activity of $pp60^{c-Src}$, and soon thereafter of dozens of other oncoproteins, drew the attention of researchers and clinicians alike towards these proteins as potential targets for drugs. This realization led, in the late 1980s, to the development of tyrosine kinase inhibitors (TKIs; or "tyrphostins," for "*ty*rosine *phos*phorylation *in*hibitors") as a strategy to combat cancer and other proliferative diseases [7, 8].

The enhanced activities of PTKs are associated with enhanced proliferation and strong survival signals, the two most prominent traits of cancer cells. The development of PTK inhibitors pioneered the approach of signal transduction therapy [9], aimed at eradicating cancer cells as well as managing other proliferative diseases and inflammatory conditions. Research in this area, which began in the late 1980s, has demonstrated that it is possible to generate small molecules with a high degree of selectivity against different PTKs, even closely related ones like EGFR and Her-2/neu. Currently, there are seven TKIs in the clinic (Table 5.1), and a few dozen in pre-clinical development.

A number of serine/threonine kinases have also been found to play central roles in cancer. These include the cyclin dependent kinases (Cdks), Erks, Raf, and PKB/Akt, which play key roles in cell proliferation, cell division, and anti-apoptotic signaling. Mutations in various signaling components regulating these enzymes or within the enzymes themselves lead to their constitutive activation and oncogenic behavior. No Ser/Thr kinase inhibitor has been approved yet for clinical use, although the search for clinical candidates to inhibit PKC was begun in the early 1980s, soon after PKC was discovered.

Similarly B-Raf/C-Raf and PKB/Akt inhibitors are in development. The most likely reason why Ser/Thr inhibitors have not made it so far, is their toxicity. Early on it was noted that tyrphostins are much less toxic than chemotherapeutic agents, probably due to their much narrower specificity. For example, the JAK2 inhibitor, AG 490, was found to possess excellent anti-tumor activity but was not inhibitory to normal B or T cells [10]. The diseased Pre-B ALL cells or multiple myeloma cells depend for their survival on the persistently active JAK2, whereas normal B and T cells, as well as other immune cells, are oblivious to the inhibition of JAK2. This pioneering study validated the hypothesis outlined above, and is considered an important milestone in signal transduction therapy [11]. The relatively mild toxicity of AG 490 allowed its combination with immunotherapy [12]. This is not possible with cytotoxic agents, since they harm the immune system. AG490 however never made it to the clinic due to its pharmacokinetic properties.

Chemistry of tyrosine kinase inhibitors

A number of naturally occurring compounds have been found to be potent inhibitors of PTKs. Despite initial promise, these compounds have all been found to be too toxic for administration as drugs. Probably, these compounds or their metabolites hit crucial off-target proteins, causing toxicity.

The first group of synthetic tyrphostins/TKIs synthesized was the benzene malononitrile tyrphostins [9, 13]. These compounds were found to be substrate-competitive ATP competitive "mixed-competitive" – i.e., their interaction with the EGFR [14] or PDGFR [15] reduces the binding affinity for both ATP and substrate, causing a reduction in the catalytic activity of the enzyme, without binding to either site. This type of behavior suggests that the agent qualifies as an allosteric inhibitor.

Most of the inhibitors generated are based on a scaffold structured around two or more aromatic rings with a number of nitrogen atoms, as in AG 1478 and AG 1295.

Quinaozolines like ZD1839/Iressa and OSI774/Tarceva made it to the clinic; Tarceva is still used, while Iressa is largely unused, mainly because of its inferior pharmacokinetic properties. Quinaxalines such as AG 1296 [16] or AGL 2043 [17, 18] were found to block PDGFR, c-Kit and Flt-3.

As tyrphostins became cyclized (Figure 5.1)[13, 16, 19], incorporating the nitrile nitrogen into a second ring, most of the compounds became ATP competitive [16, 19–21]. In fact, since 1994 the main thrust in the development of PTK inhibitors, especially by pharmaceutical companies, has been towards the generation of ATP mimics (ATP competitive kinase inhibitors) [22]. In spite of the high degree of conservation between the ATP scaffolds, it is possible to generate relatively selective inhibitors. As early as 1993 it was demonstrated that ATP competitive tyrphostins such as AG 825 can discriminate between the kinase domains of Her-2/neu and EGFR with almost two orders of magnitude difference in affinity [23], in spite of the almost 80 percent identity in the kinase domains of the two related PTKs.

The crystal structure of the inactive form of Hck with the Pfizer inhibitor PP1 [24] and of the active form of Lck with PP2 [25] clarified why these ATP mimics bind better

TABLE 5.1 Protein tyrosine kinase inhibitors currently in use in the clinic, and their targets

Agent	Cancer	Target
Gleevec/STI571 (imatinib)	CML, GIST	Bcr-Abl
Iressa/ZD1839 (gefitinib)	NSCLC	EGFR
Nilotib	CML	Bcr-Abl
Tarceva/OSI774 (erlotinib)	NSCLC	EGFR
Sutent/SU11248 (sunitinib)	GIST	Kit, PDGFR, FGFR
Sprycel/BMS (dasatinib)	CML	Bcr-Abl/Src
Tykerb/GW2016 (lapatinib)	Breast cancer	ErbB 2, EGFR
Nexavar (Bay 439006)	VEGFR2/ FGFR1	VEGFR2/B-Raf and many more protein kinases

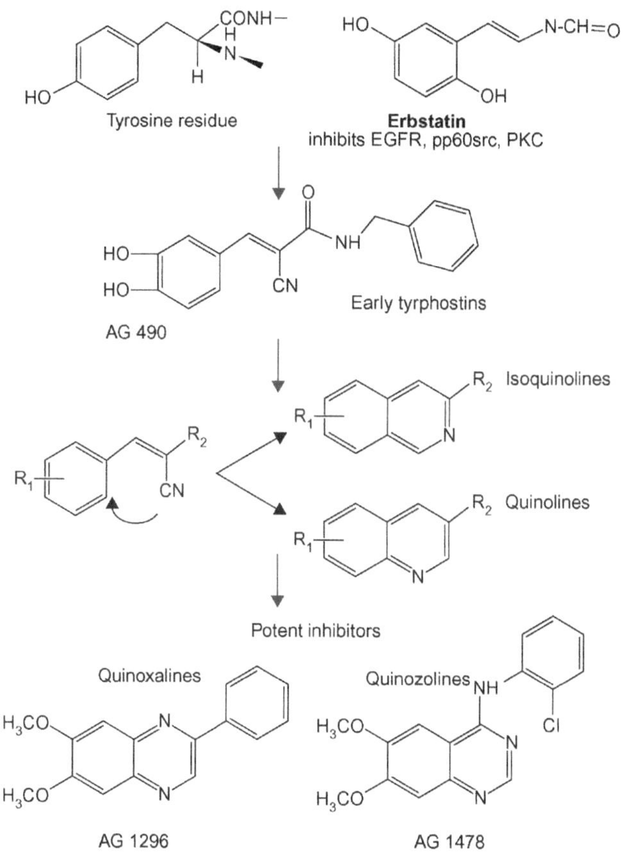

FIGURE 5.1 The evolution of tyrphostins.

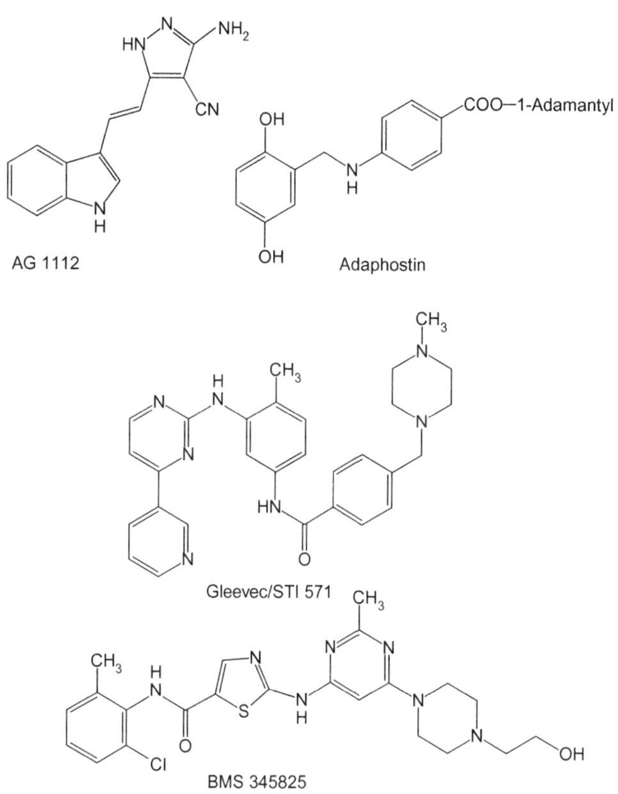

FIGURE 5.2 Gleevec, its predecessors and follow-up.

to the Src family kinase than to EGFR, and much better than to a number of other tyrosine kinases and PKAs [24]. Nonetheless, although PP1 has been presumed to be a selective Src family inhibitor, it actually also inhibits PDGFR [26] and, even more potently RIP2, a serine kinase! Even Csk, p38, Erk and CK 1, and MAPK were inhibited with an affinity less than one order of magnitude below that required to inhibit Src family kinases [27]. Furthermore, PP1 inhibits the active form of Src in an ATP non-competitive manner [28]. These findings strongly suggest that highly selective ATP-mimics will be very difficult to come by. Of the seven TKIs that have got as far as clinical use (Table 5.1), all are ATP competitors. Similarly, all of the close to several dozen compounds in pre-clinical and clinical development, were designed such that the ATP binding site is always the anchor for inhibitor binding, although neighboring residues participate as well. No pure substrate competitive inhibitors or allosteric inhibitors have reached the clinic yet, since the main effort is still focused on ATP competitive inhibitors.

From tyrphostins to Gleevec

In 1993 it was demonstrated that tyrphostin AG 1112 (Figure 5.2), which targets Bcr-Abl, induces the terminal differentiation of K562 cells [29]. Another Bcr-Abl selective tyrphostin, AG 957, is substrate competitive [30, 31]. AG

957 synergizes with anti-Fas receptor antibody to induce the purging of cells carrying the *BCR-ABL* "Philadelphia chromosome" (Ph$^+$), the hallmark of CML [32]. Druker and colleagues followed up on these studies utilizing the PDGFR kinase inhibitor CGP 57148 [33] (renamed STI 571/Gleevec/Glivec), which was found to block Bcr-Abl and induce the death of CML cells. With the advent of Gleevec, the 5-year survival rate in newly diagnosed CML patients improved to > 95 percent. Prior to Gleevec, ~50 percent of patients progressed to advanced disease within 3–5 years.

Interestingly, patients who take Gleevec (400–800 mg daily) suffer only minor side effects and tolerate the drug well. This is rather surprising, given that Gleevec blocks c-Abl, PDGFR, and c-Kit, which play important roles in normal cells. The most likely explanation is that normal cells can get by, even when over 90 percent of these targets are blocked, since they also utilize the alternative pathways which all normal cells possess [4]. CML cells, on the other hand, are highly dependent on the enhanced signaling of Bcr-Abl for their survival ("addicted"), and therefore die when Bcr-Abl is blocked. Thus, the principle of the enhanced sensitivity of the cancer cell to an inhibitor that targets the element whose signaling is enhanced, and on which the cancer cell depends for its survival, has now been validated in the clinic.

The findings with Bcr-Abl have been re-enforced by the remarkable activity of Gleevec on a subpopulation of gastrointestinal stromal tumor (GIST) patients [34, 35]. The common denominator of the patients who respond to the drug is that their tumors express Kit receptors carrying mutations in exon 11, which convert the receptor to a persistently active kinase. As in chronic CML, it seems that the survival of the tumor cells is highly dependent on the signaling of the mutated Kit receptor kinase, whereas normal cells can compensate for c-Kit blockade. Again, GIST patients on Gleevec suffer few side effects, probably because normal cells utilize alternative pathways when c-Kit is blocked. Detailed structural studies on the interaction of the Gleevec with Abl kinase shed light on the selectivity demonstrated by this anti-CML agent ref. Schinder et al. *Science* 2000, 289, 1938.

EGFR family kinase inhibitors

Interest in EGFR-targeted therapy is intense, since EGFR family members are overexpressed and/or activated through an autocrine mechanism in many tumors – up to 90 percent of certain kinds of tumor. Of the dozen tyrphostins currently in the clinic, Iressa [21] and Tarceva [36] target EGFR, and Lapatinib [37] targets both EGFR and Her-2 (Figure 5.3). Of the three antibodies targeting PTK signaling, two – Erbitux and Herceptin – target EGFR and Her-2, respectively. Erbitux and Tarceva, in combination with chemotherapy, extend survival by weeks in non-small-cell-lung carcinoma and colon cancer, with isolated cases of better outcome. Although combinations with chemotherapy gives slightly better results, none of these anti-EGFR agents brings about a cure or long remission. Gleevec is able to bring about long-term remission of CML patients, because Bcr-Abl is the principal survival factor for the CML cell in its early phase. The rather weak performance of anti-EGFR modalities implies that EGFR is not a primary survival factor for the cancers where it has been tested, despite its frequent overexpression. Erbitux has a better therapeutic effect, probably because antibodies induce bystander effects in which the host immune system is triggered to act against the tumor – a feature small kinase inhibitors do not possess.

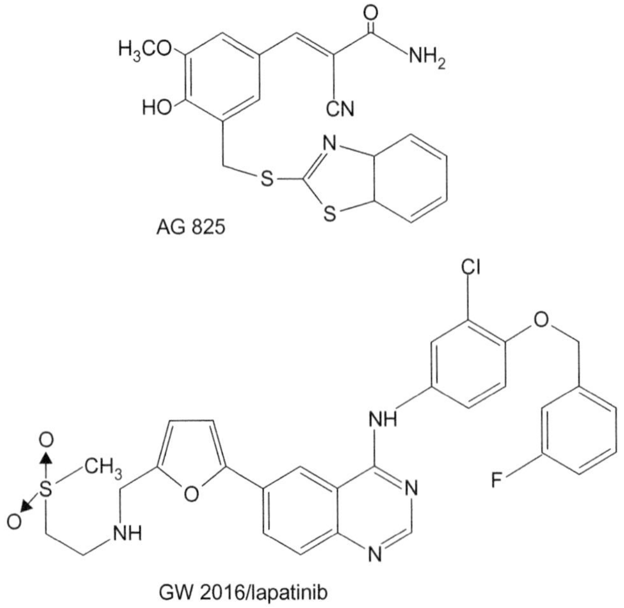

Covalent EGFR kinase affinity labels

The covalent attachment of a selective inhibitor to the EGFR kinase domain abolishes completely the catalytic activity of the receptor, and is therefore believed by some to possess better clinical potential. Indeed, the Parke-Davis compound CI-1033 [38] is highly effective *in vivo* as an EGFR kinase inhibitor. CI-1033 gives a long-lasting effect and seems to possess higher efficacy than its reversible analogs in animal tumor models. The covalent label binds to cysteine-773, close to the ATP binding domain, and most probably targets the receptor for degradation, which leads to cell death. So far, CI-1033 has not reached the clinic. This is most likely for two reasons: first, EGFR is not a sole survival element, so inhibiting it has only a weak effect on cell survival; secondly, the persistence of CI-1033 in body fluids may be limited, because of the chemical reactivity of the side-chain.

Anti-angiogenic kinase inhibitors: from AG 1433 to SUTENT/Sunitinib

Angiogenesis is mediated by the activity of various receptors, mainly VEGFR but also PDGFR and FGFR. Following the initial agent AG1433 [39], Sugen developed SU 11248/Sutent/Sunitinib [40], as a multi-targeted agent, inhibiting PDGFR, VEGFR2, and FGFR. This agent has been approved for the treatment of Gleevec-resistant GIST, since it also inhibits Kit, and for the treatment of renal cell carcinoma (RCC), probably because of its anti-angiogenic activity.

PTK inhibitors synergize with pro-apoptotic agents

Transformed cells possess a heightened state of sensitivity to stress/apoptotic signals [41, 42]. Thus, cancer cells are susceptible to agents like cis-Platin (CDDP) and to pro-apoptotic ligands like FasL. As a cancer progresses, the potentiated sensitivity to stress is shielded by a massive anti-apoptotic signaling network. To resensitize the cancer cells to pro-apoptotic stress, it is necessary to remove the anti-apoptotic shield. This is probably why PTK inhibitors synergize with cytotoxic drugs or pro-apoptotic proteins such as FasL in tissue culture and in animal experiments.

The first demonstration of this principle was reported for Her-2/neu expressing lung cancer cell lines. The degree

FIGURE 5.3 Her-2 kinase inhibitors.

of synergism between a Her-2 kinase inhibitor (AG 825) (Figure 5.3) and the cytotoxic agent CDDP, etoposide, or doxorubicin increased with the level of expression of Her-2/neu in a series of "isogenic" patient-derived NSLC cell lines [43]. Similarly, an EGFR-directed tyrphostin (AG 1478) and cis-platinum synergized in tissue culture and in a tumor model *in vivo* [44].

Synergy between Tarceva or Iressa and cytotoxic agents has not been observed in human trials. This is at least partially due to the fact that EGFR is not actually a survival factor, so it does not provide an anti-apoptotic shield, which is necessary for the synergy to take place. However, even in those patients in which mutant EGFR confers sensitivity to Tarceva and Iressa and in which the mutant EGFR is anti-apoptotic, such synergy has not yet been reported. These findings highlight the limitations of animal models in predicting the behavior of human cancer.

PTK inhibitors for the treatment of indications other than cancer

The enhanced activity of PTKs is also related to diseases other than cancer. The enhanced activity of EGFR is a hallmark of psoriasis and papilloma. In both instances, the EGFR is overexpressed and the diseased cells produce EGFR stimulating ligands through an autocrine mechanism. This is the reason why EGFR kinase inhibitors inhibit the growth of both psoriatic keratinocytes [45] and keratinocytes immortalized with HPV 16 [46]. It has been suggested that EGFR kinase inhibitors be used as topical agents to treat these conditions.

PDGFR is the key player in re-stenosis following balloon angioplasty, and therefore PDGFR kinase inhibitors have been tested as inhibitors of balloon-induced stenosis. Local application of AG 1295 [47] and AGL 2043 [18] during balloon angioplasty has indeed been shown to be highly effective against the development of stenosis after balloon angioplasty. One can envisage the utilization of PTK inhibitors for other indications, like pulmonary fibrosis, in which the enhanced activity of a PTK seems to play an important role in the pathophysiology of the disease.

PTK inhibitors as cancer prevention agents

Interest has increased over the past few years in anti-cancer agents that are natural components of foodstuffs [48]. It has been suggested that Far Eastern males have low rates of prostate cancer because they consume food rich in genistein, a non-selective PTK inhibitor which may prevent the growth of metastatic cancer [49, 50]. Dietary genistein downregulates the expression of the androgen receptor and estrogen receptors α and β in the rat prostate, at concentrations comparable to those found in humans on a soy diet [51]. Thus, downregulated sex steroid receptor expression

may be responsible for the lower incidence of prostate cancer in populations whose diet contains high levels of phytoestrogens. Also, it has been suggested that polyphenols, which are components of wine, can prevent cancer. Although largely attributed to their antioxidant properties, it is possible that inhibition of PTKs mediate some of the anti-neoplastic effects of polyphenols.

PTK inhibitors for diagnostic purposes

It is essential to establish a treatment modality according to the presence of the drug target in the diseased tissue. Thus, for example, it is extremely important to know, ahead of treatment with an anti-EGFR agent, if the tumor indeed overexpresses the receptor. This can be achieved by imaging the tumor by positron emission tomography (PET), utilizing an EGFR kinase inhibitor that is a positron emitter. Recent studies have shown that, for such purposes, an irreversible EGFR kinase inhibitor is superior to a reversible one [52]. PET imaging of the EGFR is extremely important for the determination of which patients with non-small cell lung carcinoma are eligible for treatment with Iressa. Since many tumors express either EGFR or Her-2, or both, a dual PET imager based on Lapatinib/GW 2016 may also be useful.

SER/THR KINASE INHIBITORS

Among the few hundred Ser/Thr kinases, a handful are involved in transmitting the signals of upstream PTKs. Figure 5.4 shows Ser/Thr kinases that have essential roles in cell proliferation and anti-apoptotic signaling and are known to be involved in cancer (the figure does not consider additional kinases that are involved in other proliferative diseases, including diabetes and inflammatory diseases.) The activities of these kinases can be enhanced in cancerous cells by the synergistic action of enhanced upstream PTKs combined with the inactivation of downstream negative regulators.

A few examples are appropriate: The cyclin-dependent kinases (Cdks), which execute the cell cycle, can be hyperactivated not only as a result of enhanced upstream signaling but also as a result of the overexpression of cyclin D1 and the deletion of Cdk inhibitors, such as p15 and p16. Thus Cdk2, Cdk1/Cdc2, and Cdk4 have become targets for new anti-neoplastic agents. Similarly, PI-3' kinase signaling has been found to be enhanced by the deletion of its negative regulator, PTEN. This deletion, characteristic of high-grade tumors, potentiates the already strong positive regulation of the PI-3' kinase/PDK1/PKB/mTor pathway by PTKs. Inhibitors of PDK1, PKB, and mTor are in development as anti-tumor agents. Although mTor is a protein kinase, its catalytic domain more closely resembles members of the PI3K family but also ATM- and DNA-dependent

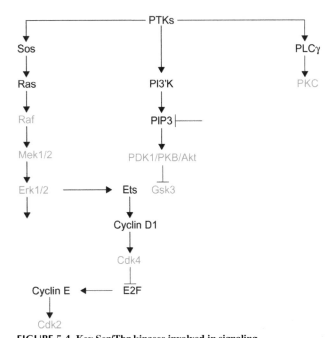

FIGURE 5.4 Key Ser/Thr kinases involved in signaling.
A small fraction of Ser/Thr kinases are valid targets for drug development. These are shown in bold black print and boxed. Crosses appear on tumor suppressors whose activities are compromised in cancer, and which therefore augment the activation of various kinases.

kinase. The Ras/Raf/Mek/Erk pathway is also downstream of PTKs.

Numerous inhibitors against the targets depicted in Figure 5.4 have been prepared. The first small molecular weight Ser/Thr inhibitor was H89, which is relatively specific for PKA. However, it also inhibits other kinases, and possibly other non-kinase off-target proteins.

PKB/Akt inhibitors

Progress has been made towards the generation of PKB inhibitors (sometimes called Aktstatins). NL 71-101 is a derivative of H89, the classical PKA inhibitor [53]. Attempts to widen the gap in selectivity towards PKB/Akt based on NL 71-101 failed when the compound was found to have unwarranted toxicity *in vivo* at doses close to those required for anti-tumor activity. The PKB inhibitor A-443654 [54] was found to be non-specific. AktI-1,2 [55] is a specific, mixed competitive inhibitor which blocks the activation of PKB by binding to the pleckstrin homology (PH) domain of PKB in its inactive conformation. The affinity of the agent, however is rather low (IC50~2.6–4 μM). Another PKB/Akt activation inhibitor is triciribine/API-2 [56]. This compound was identified by screening the NCI library. Triciribine/API-2 does not block a large family of other Ser/Thr kinases, and its *in vivo* anti-tumor activity is very impressive. It is likely, however, that it has other targets, since this compound has been found to

possess effective anti-HIV activity and has been known for ~25 years. Another interesting PKB/Akt inhibitor is the peptide substrate-mimic, PTR6164. This inhibitor has a Ki for PKB of ~90 nM [57], with a 10-fold lower affinity for PKA and PKC, and even lower affinities for other Ser/Thr kinases. PTR 6164 is highly non-toxic, and shows excellent anti-tumor activity.

Inhibitors of the Ras pathway: Raf and Mek inhibitors

Inhibitors of the Ras/Raf/Mek/Erk pathway are of great potential, since mutated Ras is the hallmark of many cancers, occurring in 30 percent of all human tumors, ~ 80 percent of pancreatic cancers, 50 percent of colorectal cancers, ~40 percent of lung cancers, and ~20 percent of hematopoietic malignancies. So far no real success has been achieved with Ras inhibitors, but promising developments have been reported for Raf and Mek inhibitors. Mutations that activate the kinase activity of B-Raf occur in ~66 percent of human melanomas, suggesting that Raf kinase inhibitors such as BAY 43-9006/Nexavar [58] (Figure 5.5) could be utilized to treat metastatic melanoma. However, Raf inhibitors also inhibit feedback regulation, and thus their utility will be limited the feedback arm can be stopped preferentially. Furthermore, Nexavar is highly promiscuous inhibitor and blocks quite a few PTKs as well as Ser/Thr kinases. Due to the potency of Nexavar against VEGFR2 [59], and probably of other PTKs, it has been approved for clinical use against renal cell carcinoma (RCC). The Mek inhibitors PD 184352/CI 1040 [60] and a second generation compound, PD 0325901(Figure 5.5) have satisfactory toxicity profiles. Both PD 184352 and PD0325901 are highly selective MKK1/2 inhibitors, binding allosterically to the inactive forms of the enzyme, which accounts for their effective cellular activities. These promising compounds are still in clinical development.

Cdks

Inhibitors of Cdks are being developed as anti-cancer agents [61–66]. Flavopiridol inhibits Cdk4/cyclin D and Cdk1/CyclinB1 with IC_{50} values of 200 nM as compared to PKA and PKC, which it inhibits with an IC_{50} of 960 nM and 10 μM, respectively [67]. This inhibitor is also still in clinical development.

p38 and Jnk inhibitors

The classical p38 inhibitor SB 203580 has been found to be rather non-selective and hits many other targets – some, such as RIP2, even better than p38. Similarly, the Jnk inhibitors SP 600125 and AS 601245 are non-specific and inhibit many Ser/Thr kinases. For a comprehensive review on the non-selectivity of these agents, see [27].

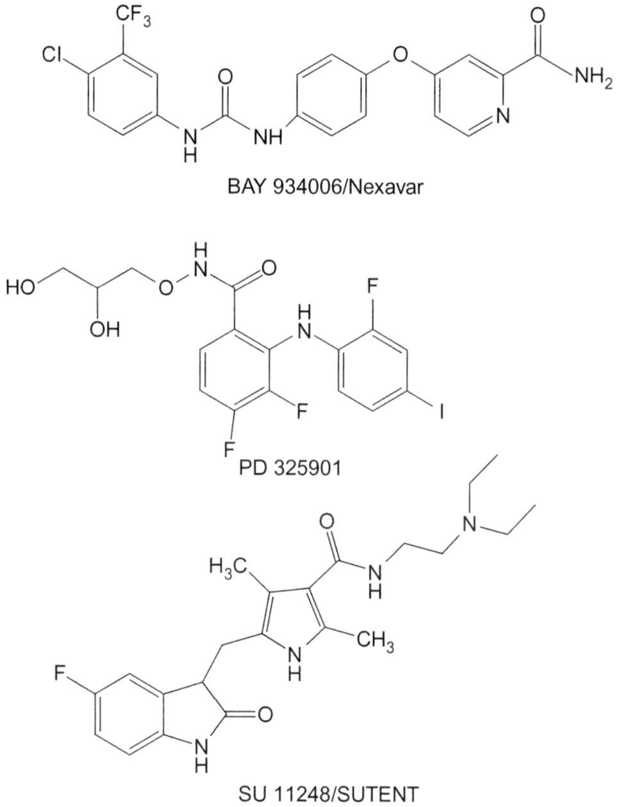

BAY 934006/Nexavar

PD 325901

SU 11248/SUTENT

FIGURE 5.5 Nexavar, Mek inhibitor (PD 0325901) and SUTENT.

PKC inhibitors

Although PKC isozymes have been known for a long time, little progress has been made in the utilization of PKC inhibitors as therapeutic agents. Some agents are in development as anti-cancer drugs [62, 68–70]. PKCβ inhibitors are also potential agents against vascular dysfunction [65] and are being evaluated as agents against vascular retinopathy, a complication of diabetes. No PKC inhibitor has entered the clinic yet, most probably due to toxicity issues.

Rapamycin

Rapamycin, an inhibitor of mTor, is effective as an inhibitor of angiogenesis, and therefore is a potential anti-cancer drug [72]. Rapamycin also is effective in the inhibition of re-stenosis when applied on coated stents [73]. Rapamycin is actually not a Ser/Thr kinase inhibitor, but inhibits Tor complex 1 (TORC1) by binding to FK506-binding protein. The activity of TORC1 is positively regulated by PKB/Akt, which is also activated by TORC1, placing TORC1 both upstream and downstream of PKB/Akt. Inhibition of mTor abolishes the feedback mechanism elicited by PKB/Akt, leading to increased levels of IRS1 and reinstatement of IGF1R signaling. Therefore, rapamycin has performed rather weakly as an anti-tumor agent in the clinic. One way

to achieve better efficacy may be to combine it with IGF1R and/or PI-kinase inhibitors. This reasoning may also affect how to use PKB/Akt inhibitors.

ATP mimics *vis-à-vis* substrate mimics

It is an interesting dilemma, whether to pursue pure ATP competitors or to opt for substrate competitive inhibitors or allosteric inhibitors that bind to unique areas of the target protein. Almost all existing protein kinase inhibitors indeed bind to the ATP site. Exceptions are: the Mek1/2 inhibitor PD 184352, which binds to an allosteric site of Mek1/2; AktI-1,2, which binds to the inactive form of PKB/Akt; PTR 6164, which is a substrate competitive PKB/Akt inhibitor; and rapamycin, which binds to a TORC1 subunit and not to the kinase domain. Some of the ATP binders, like AG 1296, change their mode of binding from ATP-competitive to "mixed competitive" upon PDGFR activation [15]. It is unclear as yet whether the weak performance of ATP competitive inhibitors (except Gleevec) is due to the fact that they are ATP mimics, hitting too many targets and requiring high dosages due to the high intracellular levels of ATP. Gleevec is obviously the most effective protein kinase inhibitor, but this is mainly due to the exceptional nature of chronic CML. When CML progresses to a more advanced stage, Gleevec loses its stardom. Advanced CML is no longer "addicted" only to Bcr-Abl, but rather resembles other hematological malignancies, which are driven by a plethora of oncogenic pathways, and a single Bcr-Abl inhibitor is no longer able to contain the disease.

Multi-targeted protein kinase inhibitors

The promiscuity of a kinase inhibitor does not necessarily affect its efficacy in the clinic. Sunitinib/Sutent/SU11248 (Figure 5.5) is a multi-targeted agent, inhibiting VEGFR2, FGFR, and PDGFR. Sutent is effective against renal cell carcinoma (RCC), since all these receptors are involved in angiogenesis. Sutent is reported to prolong median progression-free survival as compared to interferon-α therapy by a factor of more than two. Sutent is also effective against c-Kit, and therefore is effective in the treatment of Gleevec-resistant GIST. The median progression-free survival is enhanced more than four-fold, as compared to placebo, in GIST patients. Its favorable toxicity profile and pharmacokinetic properties enable the use of therapeutic doses. Similar results were obtained for another multi-targeted agent, BAY 439006/Nexavar, which blocks VEGFR2, FGFR, and B-Raf. These findings suggest that multi-targeted agents should probably be used to treat other forms of cancer, as long as the toxicity profile is acceptable.

On the other hand, Gleevec has very narrow specificity and has excellent efficacy against early CML and a subclass of GIST patients. In both cases there is one protein tyrosine kinase on which the targeted cancer cell depends for survival, and therefore its blockade is effective. Another interesting case is Lapatinib/Tykerb, which is a potent blocker of EGFR and Her-2. Tykerb prolonged the median time to progression when combined with chemotherapy, compared to chemotherapy alone, in a group of patients who progressed after treatment with other therapeutic regimens, including herceptin. Lapatinib is a more selective inhibitor than Sutent, Nexavar, and even Gleevec, not inhibiting any of the tested kinases other than EGFR, Her-2, and Her-4.

Bain and colleagues investigated the specificity of many protein kinase inhibitors, including most of the clinically approved agents, against 70 protein kinases. These investigators found that the inhibitory profiles did not correlate strongly with the originally intended target [27]. It is likely that if that study had been performed against the whole repertoire of ~500 protein kinases, the correlation would have been even weaker. This study reinforces the conclusion that the focus should be on toxicity issues and pharmacological properties more than on selectivity.

REFERENCES

1. Levitzki A. Signal-transduction therapy A novel approach to disease management. *Eur J Biochem* 1994;**226**(1):1–13.
2. Manning G, Plowman GD, Hunter T, Sudarsanam S. Evolution of protein kinase signaling from yeast to man. *Trends Biochem Sci* 2002;**27**(10):514–20.
3. Manning G, Whyte DB, Martinez R, Hunter T, Sudarsanam S. The protein kinase complement of the human genome. *Science* 2002;**298**(5600):1912–34.
4. Klein S, Levitzki A. Targeted cancer therapy promise and reality. *Adv Cancer Res* 2007;**97**:295.
5. Sorafenib. new drug. Second-line treatment of kidney cancer: better evaluated than sunitinib. *Prescrire Intl* 2007;**16**(90):141–3.
6. Eckhart W, Hutchinson MA, Hunter T. An activity phosphorylating tyrosine in polyoma T antigen immunoprecipitates. *Cell* 1979;**18**(4):925–33.
7. Levitzki A, Gazit A. Tyrosine kinase inhibition: an approach to drug development. *Science* 1995;**267**(5205):1782–8.
8. Levitzki A, Mishani E. Tyrphostins and other tyrosine kinase inhibitors. *Annu Rev Biochem* 2006;**75**:93–109.
9. Yaish P, Gazit A, Gilon C, Levitzki A. Blocking of EGF-dependent cell proliferation by EGF receptor kinase inhibitors. *Science* 1988;**242**(4880):933–5.
10. Meydan N, Grunberger T, Dadi H, Shahar M, Arpaia E, Lapidot Z, Leeder JS, Freedman M, Cohen A, Gazit A, Levitzki A, Roifman CM. Inhibition of acute lymphoblastic leukaemia by a Jak-2 inhibitor. *Nature* 1996;**379**(6566):645–8.
11. Ito T, May WS. Drug development train gathering steam. *Nature Med* 1996;**2**(4):403–4.
12. Burdelya L, Catlett-Falcone R, Levitzki A, Cheng F, Mora LB, Sotomayor E, Coppola D, Sun J, Sebti S, Dalton WS, Jove R, Yu H. Combination therapy with AG-490 and interleukin 12 achieves greater antitumor effects than either agent alone. *Mol Cancer Ther* 2002;**1**(11):893–9.
13. Gazit A, Yaish P, Gilon C, Levitzki A. Tyrphostins I: synthesis and biological activity of protein tyrosine kinase inhibitors. *J Med Chem* 1989;**32**(10):2344–52.
14. Posner I, Levitzki A. Kinetics of phosphorylation of the SH2-containing domain of phospholipase C gamma 1 by the epidermal growth factor receptor. *FEBS Letts* 1994;**353**(2):155–61.
15. Kovalenko M, Ronnstrand L, Heldin CH, Loubtchenkov M, Gazit A, Levitzki A, Bohmer FD. Phosphorylation site-specific inhibition of platelet-derived growth factor beta-receptor autophosphorylation by the receptor blocking tyrphostin AG1296. *Biochemistry* 1997;**36**(21):6260–9.
16. Kovalenko M, Gazit A, Bohmer A, Rorsman C, Ronnstrand L, Heldin CH, Waltenberger J, Bohmer FD, Levitzki A. Selective platelet-derived growth factor receptor kinase blockers reverse sis-transformation. *Cancer Res* 1994;**54**(23):6106–14.
17. Gazit A, Yee K, Uecker A, Bohmer FD, Sjoblom T, Ostman A, Waltenberger J, Golomb G, Banai S, Heinrich MC, Levitzki A. Tricyclic quinoxalines as potent kinase inhibitors of PDGFR kinase, Flt3 and Kit. *Bioorg Med Chem* 2003;**11**(9):2007–18.
18. Banai S, Gertz SD, Gavish L, Chorny M, Perez LS, Lazarovichi G, Ianculuvich M, Hoffmann M, Orlowski M, Golomb G, Levitzki A. Tyrphostin AGL-2043 eluting stent reduces neointima formation in porcine coronary arteries. *Cardiovasc Res* 2004;**64**(1):165–71.
19. Osherov N, Levitzki A. Epidermal-growth-factor-dependent activation of the src-family kinases. *Eur J Biochem* 1994;**225**(3):1047–53.
20. Ward WH, Cook PN, Slater AM, Davies DH, Holdgate GA, Green LR. Epidermal growth factor receptor tyrosine kinase. Investigation of catalytic mechanism, structure-based searching and discovery of a potent inhibitor. *Biochem Pharmacol* 1994;**48**(4):659–66.
21. Wakeling AE, Barker AJ, Davies DH, Brown DS, Green LR, Cartlidge SA, Woodburn JR. Specific inhibition of epidermal growth factor receptor tyrosine kinase by 4-anilinoquinazolines. *Breast Cancer Res Treat* 1996;**38**(1):67–73.
22. Levitzki A. Protein tyrosine kinase inhibitors as novel therapeutic agents. *Pharmacol Ther* 1999;**82**(2–3):231–9.
23. Osherov N, Gazit A, Gilon C, Levitzki A. Selective inhibition of the epidermal growth factor and HER2/neu receptors by tyrphostins. *J Biol Chem* 1993;**268**(15):11,134–11,142,.
24. Schindler T, Sicheri F, Pico A, Gazit A, Levitzki A, Kuriyan J. Crystal structure of Hck in complex with a Src family-selective tyrosine kinase inhibitor. *Mol Cell* 1999;**3**(5):639–48.
25. Zhu X, Kim JL, Newcomb JR, Rose PE, Stover DR, Toledo LM, Zhao H, Morgenstern KA. Protein, structure, structural analysis of the lymphocyte-specific kinase Lck in complex with non-selective and Src family selective kinase inhibitors. *Structure Fold Des* 1999;**7**(6):651–61.
26. Waltenberger J, Uecker A, Kroll J, Frank H, Mayr U, Bjorge JD, Fujita D, Gazit A, Hombach V, Levitzki A, Bohmer FD. A dual inhibitor of platelet-derived growth factor beta-receptor and Src kinase activity potently interferes with motogenic and mitogenic responses to PDGF in vascular smooth muscle cells. A novel candidate for prevention of vascular remodeling. *Circ Res* 1999;**85**(1):12–22.
27. Bain J, Plater L, Elliott M, Shpiro N, Hastie J, McLauchlan H, Klevernic I, Arthur S, Alessi D, Cohen P. The selectivity of protein kinase inhibitors; a further update. *Biochem J* 2007;**408**(3):297–315.
28. Karni R, Mizrachi S, Reiss-Sklan E, Gazit A, Livnah O, Levitzki A. The pp60c-Src inhibitor PP1 is non-competitive against ATP. *FEBS Letts* 2003;**537**(1–3):47–52.
29. Anafi M, Gazit A, Zehavi A, Ben-Neriah Y, Levitzki A. Tyrphostin-induced inhibition of p210bcr-abl tyrosine kinase activity induces K562 to differentiate. *Blood* 1993;**82**(12):3524–9.

30. Kaur G, Gazit A, Levitzki A, Stowe E, Cooney DA, Sausville EA. Tyrphostin induced growth inhibition: correlation with effect on p210bcr-abl autokinase activity in K562 chronic myelogenous leukemia. *Anticancer Drugs* 1994;**5**(2):213–22.

31. Anafi M, Gazit A, Gilon C, Ben-Neriah Y, Levitzki A. Selective interactions of transforming and normal abl proteins with ATP, tyrosine-copolymer substrates, and tyrphostins. *J Biol Chem* 1992;**267**(7):4518–23.

32. Carlo-Stella C, Regazzi E, Sammarelli G, Colla S, Garau D, Gazit A, Savoldo B, Cilloni D, Tabilio A, Levitzki A, Rizzoli V. Effects of the tyrosine kinase inhibitor AG957 and an Anti-Fas receptor antibody on CD34(+) chronic myelogenous leukemia progenitor cells. *Blood* 1999;**93**(11):3973–82.

33. Druker BJ, Tamura S, Buchdunger E, Ohno S, Segal GM, Fanning S, Zimmermann J, Lydon NB. Effects of a selective inhibitor of the Abl tyrosine kinase on the growth of Bcr-Abl positive cells. *Nature Med* 1996;**2**(5):561–6.

34. Tuveson DA, Willis NA, Jacks T, Griffin JD, Singer S, Fletcher CD, Fletcher JA, Demetri GD. STI571 inactivation of the gastrointestinal stromal tumor c-KIT oncoprotein: biological and clinical implications. *Oncogene* 2001;**20**(36):5054–8.

35. Joensuu H, Roberts PJ, Sarlomo-Rikala M, Andersson LC, Tervahartiala P, Tuveson D, Silberman S, Capdeville R, Dimitrijevic S, Druker B, Demetri GD. Effect of the tyrosine kinase inhibitor STI571 in a patient with a metastatic gastrointestinal stromal tumor. *N Engl J Med* 2001;**344**(14):1052–6.

36. Ciardiello F, Tortora G. Anti-epidermal growth factor receptor drugs in cancer therapy. *Expert Opin Invest Drugs* 2002;**11**(6):755–68.

37. Xia W, Mullin RJ, Keith BR, Liu LH, Ma H, Rusnak DW, Owens G, Alligood KJ, Spector NL. Anti-tumor activity of GW572016: a dual tyrosine kinase inhibitor blocks EGF activation of EGFR/erbB2 and downstream Erk1/2 and AKT pathways. *Oncogene* 2002;**21**(41):6255–63.

38. Smaill JB, Rewcastle GW, Loo JA, Greis KD, Chan OH, Reyner EL, Lipka E, Showalter HD, Vincent PW, Elliott WL, Denny WA. Tyrosine kinase inhibitors. 17. Irreversible inhibitors of the epidermal growth factor receptor: 4-(phenylamino)quinazoline- and 4-(phenylamino)pyrido[3,2-d]pyrimidine-6-acrylamides bearing additional solubilizing functions. *J Med Chem* 2000;**43**(7):1380–97.

39. Strawn LM, McMahon G, App H, Schreck R, Kuchler WR, Longhi MP, Hui TH, Tang C, Levitzki A, Gazit A, Chen I, Keri G, Orfi L, Risau W, Flamme I, Ullrich A, Hirth KP, Shawver LK. Flk-1 as a target for tumor growth inhibition. *Cancer Res* 1996;**56**(15):3540–5.

40. Mendel DB, Laird AD, Xin X, Louie SG, Christensen JG, Li G, Schreck RE, Abrams TJ, Ngai TJ, Lee LB, Murray LJ, Carver J, Chan E, Moss KG, Haznedar JO, Sukbuntherng J, Blake RA, Sun L, Tang C, Miller T, Shirazian S, McMahon G, Cherrington JM. *In vivo* antitumor activity of SU11248, a novel tyrosine kinase inhibitor targeting vascular endothelial growth factor and platelet-derived growth factor receptors: determination of a pharmacokinetic/pharmacodynamic relationship. *Clin Cancer Res* 2003;**9**(1):327–37.

41. Benhar M, Engelberg D, Levitzki A. ROS, stress-activated kinases and stress signaling in cancer. *EMBO Rep* 2002;**3**(5):420–5.

42. Benhar M, Dalyot I, Engelberg D, Levitzki A. Enhanced ROS production in oncogenically transformed cells potentiates c-Jun N-terminal kinase and p38 mitogen-activated protein kinase activation and sensitization to genotoxic stress. *Mol Cell Biol* 2001;**21**(20):6913–26.

43. Tsai CM, Levitzki A, Wu LH, Chang KT, Cheng CC, Gazit A, Perng RP. Enhancement of chemosensitivity by tyrphostin AG825 in high-p185(neu) expressing non-small cell lung cancer cells. *Cancer Res* 1996;**56**(5):1068–74.

44. Nagane M, Narita Y, Mishima K, Levitzki A, Burgess AW, Cavenee WK, Huang HJ. Human glioblastoma xenografts overexpressing a tumor-specific mutant epidermal growth factor receptor sensitized to cisplatin by the AG1478 tyrosine kinase inhibitor. *J Neurosurg* 2001;**95**(3):472–9.

45. Powell TJ, Ben-Bassat H, Klein BY, Chen H, Shenoy N, McCollough J, Narog B, Gazit A, Harzstark Z, Chaouat M, Levitzki R, Tang C, McMahon J, Shawver L, Levitzki A. Growth inhibition of psoriatic keratinocytes by quinazoline tyrosine kinase inhibitors. *Br J Dermatol* 1999;**141**(5):802–10.

46. Ben-Bassat H, Rosenbaum-Mitrani S, Hartzstark Z, Shlomai Z, Kleinberger-Doron N, Gazit A, Plowman G, Levitzki R, Tsvieli R, Levitzki A. Inhibitors of epidermal growth factor receptor kinase and of cyclin-dependent kinase 2 activation induce growth arrest, differentiation, and apoptosis of human papilloma virus 16-immortalized human keratinocytes. *Cancer Res* 1997;**57**(17):3741–50.

47. Banai S, Wolf Y, Golom G, Pearle A, Waltenberger J, Fishbein I, Schneider A, Gazit A, Perez L, Huber R, Lazarovichi G, Rabinovich L, Levitzki A, Gertz SD. PDGF-receptor tyrosine kinase blocker AG1295 selectively attenuates smooth muscle cell growth *in vitro* and reduces neointimal formation after balloon angioplasty in swine. *Circulation* 1998;**97**(19):1960–9.

48. Lamartiniere CA, Cotroneo MS, Fritz WA, Wang J, Mentor-Marcel R, Genistein chemoprevention EA. Genistein chemoprevention: timing and mechanisms of action inmurine mammary and prostate. *J Nutrition* 2002;**132**(3):5528–88.

49. Mentor-Marcel R, Lamartiniere CA, Eltoum IE, Greenberg NM, Elgavish A. Genistein in the diet reduces the incidence of poorly differentiated prostatic adenocarcinoma in transgenic mice (TRAMP). *Cancer Res* 2001;**61**(18):6777–82.

50. Bergan RC, Waggle DH, Carter SK, Horak I, Slichenmyer W, Meyers M. Tyrosine kinase inhibitors and signal transduction modulators: rationale and current status as chemopreventive agents for prostate cancer. *Urology* 2001;**57**:77–80.

51. Fritz WA, Wang J, Eltoum IE, Lamartiniere CA. Dietary genistein down-regulates androgen and estrogen receptor expression in the rat prostate. *Mol Cell Endocrinol* 2002;**186**(1):89–99.

52. Mishani E, Abourbeh G, Jacobson O, Dissoki S, Ben Daniel R, Rozen Y, Shaul M, Levitzki A. High-affinity epidermal growth factor receptor (EGFR) irreversible inhibitors with diminished chemical reactivities as positron emission tomography (PET)-imaging agent candidates of EGFR overexpressing tumors. *J Med Chem* 2005;**48**(16):5337–48.

53. Reuveni H, Livnah N, Geiger T, Klen S, Ohne O, Cohen I, Benhar M, Gellerman G, Levitzki A. Towards a PKB inhibitor: Modification of a selective PKA inhibitor by rational design. *Biochemistry* 2002;**41**(32):10,304–10,314,.

54. Luo Y, Shoemaker AR, Liu X, Woods KW, Thomas SA, de Jong R, Han EK, Li T, Stoll VS, Powlas JA, Oleksijew A, Mitten MJ, Shi Y, Guan R, McGonigal TP, Klinghofer V, Johnson EF, Leverson JD, Bouska JJ, Mamo M, Smith RA, Gramling-Evans EE, Zinker BA, Mika AK, Nguyen PT, Oltersdorf T, Rosenberg SH, Li Q, Giranda VL. Potent and selective inhibitors of Akt kinases slow the progress of tumors *in vivo*. *Mol Cancer Ther* 2005;**4**(6):977–86.

55. Barnett SF, Defeo-Jones D, Fu S, Hancock PJ, Haskell KM, Jones RE, Kahana JA, Kral AM, Leander K, Lee LL, Malinowski J, McAvoy EM, Nahas DD, Robinson RG, Huber HE. Identification and characterization of pleckstrin-homology-domain-dependent and isoenzyme-specific Akt inhibitors. *Biochem J* 2005;**385**(Pt 2):399–408.

56. Yang L, Dan HC, Sun M, Liu Q, Sun XM, Feldman RI, Hamilton AD, Polokoff M, Nicosia SV, Herlyn M, Sebti SM, Cheng JQ. Akt/protein

kinase B signaling inhibitor-2, a selective small molecule inhibitor of Akt signaling with antitumor activity in cancer cells overexpressing Akt. *Cancer Res* 2004;**64**(13):4394–9.

57. Litman P, Ohne O, Ben-Yaakov S, Shemesh-Darvish L, Yechezkel T, Salitra Y, Rubnov S, Cohen I, Senderowitz H, Kidron D, Livnah O, Levitzki A, Livnah N. A novel substrate mimetic inhibitor of PKB/Akt inhibits prostate cancer tumor growth in mice by blocking the PKB pathway. *Biochemistry* 2007;**46**(16):4716–24.

58. Lyons JF, Wilhelm S, Hibner B, Bollag G. Discovery of a novel Raf kinase inhibitor. *Endocr Relat Cancer* 2001;**3**:219–25.

59. Adnane L, Trail PA, Taylor I, Wilhelm SM. Sorafenib (BAY 43-9006, Nexavar®), a dual-action inhibitor that targets RAF/MEK/ERK pathway in tumor cells and tyrosine ,inases VEGFR/PDGFR in tumor vasculature. *Methods Enzymol* 2005;**407**:597–612.

60. Sebolt-Leopold JS, Dudley DT, Herrera R, Van Becelaere K, Wiland A, Gowan RC, Tecle H, Barrett SD, Bridges A, Przybranowski S, Leopold WR, Saltiel AR. Blockade of the MAP kinase pathway suppresses growth of colon tumors in vivo. *Nat Med* 1999;**5**(7):810–16.

61. Sausville EA, Johnson J, Alley M, Zaharevitz D, Senderowicz AM. Inhibition of CDKs as a therapeutic modality. *Ann NY Acad Sci* 2000:221–2. discussion.

62. Kaubisch A, Schwartz GK. Cyclin-dependent kinase and protein kinase C inhibitors: a novel class of antineoplastic agents in clinical development. *Cancer J* 2000;**6**(4):192–212.

63. Mani S, Wang C, Wu K, Francis R, Pestell R. Cyclin-dependent kinase inhibitors: novel anticancer agents. *Expert Opin Invest Drugs* 2000;**9**(8):1849–70.

64. Murthi KK, Dubay M, McClure C, Brizuela L, Boisclair MD, Worland PJ, Mansuri MM, Pal K. Structure–activity relationship studies of flavopiridol analogues. *Bioorg Med Chem Letts* 2000;**10**(10):1037–41.

65. Roy KK, Sausville EA. Early development of cyclin dependent kinase modulators. *Curr Pharm Des* 2001;**7**(16):1669–87.

66. Sausville EA. Complexities in the development of cyclin-dependent kinase inhibitor drugs. *Trends Mol Med.* 2002;**8**(4):S32–7.

67. Kelland LR. Flavopiridol, the first cyclin-dependent kinase inhibitor to enter the clinic: current status. *Expert Opin Invest Drugs* 2000;**9**(12):2903–11.

68. da Rocha AB, Mans DR, Regner A, Schwartsmann G. Targeting protein kinase C: new therapeutic opportunities against high-grade malignant gliomas?. *Oncologist* 2002;**7**(1):17–33.

69. Goekjian PG, Jirousek MR. Protein kinase C inhibitors as novel anticancer drugs. *Expert Opin Invest Drugs* 2001;**10**(12):2117–240.

70. Teicher BA, Alvarez E, Menon K, Esterman MA, Considine E, Shih C, Faul MM. Antiangiogenic effects of a protein kinase Cbeta-selective small molecule. *Cancer Chemother Pharmacol* 2002;**49**(1):69–77.

71. Ishii H, Jirousek MR, Koya D, Takagi C, Xia P, Clermont A, Bursell SE, Kern TS, Ballas LM, Heath WF, Stramm LE, Feener EP, King GL. Amelioration of vascular dysfunctions in diabetic rats by an oral PKC beta inhibitor. *Science* 1996;**272**(5262):728–31.

72. Guba M, von Breitenbuch P, Steinbauer M, Koehl G, Flegel S, Hornung M, Bruns CJ, Zuelke C, Farkas S, Anthuber M, Jauch KW, Geissler EK. Rapamycin inhibits primary and metastatic tumor growth by antiangiogenesis: involvement of vascular endothelial growth factor. *Nature Med* 2002;**8**(2):128–35.

73. Morice MC, Serruys PW, Sousa JE, Fajadet J, Ban Hayashi E, Perin M, Colombo A, Schuler G, Barragan P, Guagliumi G, Molnar F, Falotico R. A randomized comparison of a sirolimus-eluting stent with a standard stent for coronary revascularization. *N Engl J Med* 2002;**346**(23):1773–80.

Chapter 6

Principles of Kinase Regulation

Bostjan Kobe[1] and Bruce E. Kemp[2]

[1] School of Chemistry Molecular Biosciences, University of Queensland, Brisbane, Queensland, Australia

[2] St Vincent's Institute, Fitzroy, Victoria, Australia

INTRODUCTION

Protein kinases represent one of the largest protein families, corresponding to ~2 percent of eukaryotic proteins [1]. Their importance is further reflected in the fact that phosphorylation is the most abundant type of cellular regulation, affecting essentially every cellular process, including metabolism, growth, differentiation, motility, membrane transport, learning, and memory [2]. To function as switches controlling all these various processes, kinases must be tightly regulated. Improper regulation leads to cancer and various other diseases. Regulation is an integral part of protein kinase function that controls the timing of the catalytic activity and substrate specificity.

Protein kinase can be regulated in diverse ways, ranging from transcriptional control through subcellular localization and recruitment of substrates (using anchoring, adaptor and scaffold proteins or domains [3]), to structural and chemical modifications of the proteins themselves. The specificity in the cell is determined by a combination of "peptide specificity" of the kinase (the molecular recognition of the sequence surrounding the phosphorylation site), substrate recruitment and phosphatase activity [4–6]. This short review focuses on the principles of regulation of protein kinase activity at the protein level (for further details, see related recent reviews, [7–11]).

PROTEIN KINASE STRUCTURE

Eukaryotic protein kinase domains segregate into two large groups, phosphorylating either serine/threonine or tyrosine residues on target proteins. However, both groups have similar 3D structures, comprising two lobes with the active site in a cleft between the lobes [12–15]. The smaller, N-terminal (N-) lobe contains mainly β structure, and one important helix termed helix C. The larger C-terminal (C-) lobe is mainly α-helical with helices E and F in the core especially conserved. Important structural motifs include the phosphate-binding (P-) loop, which contains a glycine-rich motif and is involved in ATP binding, and the activation (A-) loop, often containing phosphorylation sites, which provides a platform for peptide substrate binding (Figure 6.1). The two lobes are connected through a flexible hinge.

Eukaryotic protein kinases typically catalyze the same reaction, the transfer of the gamma-phosphate from ATP to the hydroxyl group of a Ser, Thr, or Tyr. All kinases also adopt strikingly similar structures when they are in the active form [12–15]. The active structure ensures that the substrates and the kinase catalytic machinery are properly aligned. By contrast, the inactive structures of different kinases show remarkable diversity. The mechanisms of control range from allosteric to intrasteric and everything in between [16], and are used to modulate the conformations of the A-loop and the P-loop, the position of helix C, the access to ATP and substrate binding sites, and the relative orientation of the two lobes (Figures 6.1, 6.2). The control can be exerted by internal regions of the kinase catalytic domains, by segments flanking the catalytic domain, and/or additional subunits or proteins; these regions or proteins can respond to second messengers, their expression may be controlled by the functional state of the cell, they can target the kinase to different substrates or subcellular locations, or inhibit the kinase activity. These possible regulatory sites affect each other, resulting in a rich spectrum of possible regulatory pathways. We first review allosteric and intrasteric behaviors in protein kinases, and then address how individual sites are regulated.

GENERAL PRINCIPLES OF CONTROL

Allosteric Regulation

Allosteric regulation is a widespread mechanism of control of protein function; effectors bind to regulatory sites

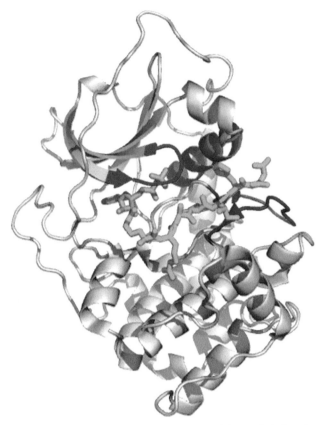

FIGURE 6.1 Ribbon diagram of the structure of the catalytic domain of the cAMP-dependent protein kinase (PKA) [54].
The bound ADP and peptide substrate (only seven residues are shown) are shown in stick representation in dark and light gray, respectively. The various regulatory regions are indicated in dark gray: (A-loop, below the substrate; P-loop, above ADP; helix C, above the substrate.). The online color version of this figure is at http://www.sciencedirect.com/science/book/9780123741455.The figure was generated using PyMol (DeLanoScientific LLC)

distinct from the active site, usually inducing conformational changes that influence the activity [17]. Allosteric effectors generally bear no structural resemblance to their target protein's substrate. For example, end-products of metabolic pathways can act at early steps of the pathway to exert feedback control. In protein kinases, allosteric control can be exerted by flanking sequences or separate subunits/ proteins. Allostery can be mediated through a number of mechanisms, but most commonly by affecting the conformation of the A-loop. Some representative examples of allosteric control in protein kinases are listed below.

A classical example of allosteric control in protein kinases is cyclin binding to cyclin-dependent kinases (CDKs), where cyclin binding induces a reformation of the ATP binding site [18]. The protein kinases Aurora A and B, involved in mitosis, are activated by binding of TPX2 (a microtubule-associated protein) and INCENP (the inner centromere protein), respectively. In both cases, allosteric activator binding induces the A-loop to adopt an active conformation [19, 20]. Epidermal growth factor (EGF) activates the receptor tyrosine kinase (EGFR) through dimerization, which turns out to be an asymmetric homodimerization, involving the C-lobe of one molecule interacting with the N-lobe of the other, and stabilizing the active conformation [21]. The stress-activated kinases PKR and GCN2 are under allosteric control by RNA, in both cases through dimerization, but by distinct mechanisms. In PKR, dimerization triggers autophosphorylation, which in turn remodels the substrate (eukaryotic elongation factor 2α) binding site [22]. In GCN2, on the other hand, dimerization induces an inactive form, which is activated by tRNA binding through opening of the domains, as suggested by structural studies of active mutant proteins [23]. In MAP kinases, the A-loop and domain orientations are affected by binding of proteins containing docking motifs, resulting in activation of kinases ERK2 [24] and Fus3 [25], or inhibition of JNK1 through binding of a scaffolding protein [26]. The SH3 and SH2 domains flanking the kinase domain in tyrosine kinases c-Src and Hck hold the kinase

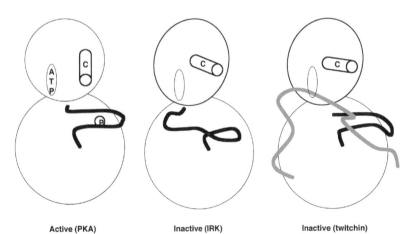

Active (PKA)   Inactive (IRK)   Inactive (twitchin)

FIGURE 6.2 A schematic diagram showing prototypic kinase active structure (PKA) and inactive structures (IRK and twitchin), highlighting the regulatory regions.
The two lobes of the catalytic domain, A-loop (thick line) with the phosphorylation site in the loop indicated (P), helix C (cylinder indicated with "C"), the ATP binding site (ATP), and the autoregulatory sequence (gray thick line) are shown.

lobes in a rigid inactive state, which is relieved by tyrosine phosphorylation in the C-terminal region of the kinase [27, 28]. In stem cell factor receptor c-Kit, the juxta-membrane region N-terminal to the kinase domain holds the A-loop in an inactive state until activated by phosphorylation of two tyrosines in this region [29].

Intrasteric Regulation

The term *intrasteric regulation* was first introduced to describe autoregulation of protein kinases and phosphatases by internal sequences that resembled substrates ("pseudosubstrates"), and acted directly at the active site [30]. It is now clear that this form of control extends to diverse enzyme classes and beyond enzymes to receptors and protein targeting domains, and acts through an intrasteric autoregulatory sequences (IARseq) that resemble the specific protein's substrates or ligands to varying extents [16]. Intrasteric interactions typically result in inhibition of protein function, and diverse mechanisms can be used for activation, including protein activators or ligands, phosphorylation, proteolysis, reduction of disulfide bonds, or combinations of these. A large subfamily of intrasterically regulated kinases is activated by calcium-binding proteins (e.g., calmodulin-dependent kinases (CaMKs), titin, twitchin), or calcium directly in the case of plant calcium-dependent protein kinases (these contain a calcium-binding domain contiguous with the kinase domain). In twitchin, a C-terminal IARseq threads through the active site cleft between the two lobes of the protein kinase domain, making a plethora of contacts with the peptide substrate binding site and other regions essential for catalysis and ATP-binding [31]. The binding site of the activator S100A1 has been mapped to one portion of the autoinhibitory sequence [31, 32]. A very similar mechanism of inhibition occurs in the related giant kinase titin; however, here a combination of phosphorylation and calmodulin (CaM) binding (to a site analogous to S100A1 binding site in twitchin) is required to activate the enzyme [33]. The more distantly related CaMK-1 is also activated by CaM binding to an IARseq, but the structure shows a modified mode of inhibition where the IARseq exits the active site in the opposite direction to twitchin kinase [34]. In p21-activated kinase (PAK), the autoregulatory sequence is located on a distinct "inhibitory switch domain" binding to the C-lobe of the kinase, which both blocks the substrate binding site and causes various distortions to the kinase domain [35]. PAK is activated by the GTP form of the Rho family of G proteins. Other protein kinase families predicted to be intrasterically regulated include glycogen synthase kinase 3β, which is proposed to be autoinhibited by a phosphorylated N-terminus acting as a pseudosubstrate [36, 37], and the protein kinase C family and cGMP-dependent protein kinases, where the IARseqs reside N-terminal to the catalytic domain and the activators are phospholipids and diacylglycerol, and cGMP, respectively [38, 39].

Insulin receptor tyrosine kinase (IRK) revealed a more subtle autoinhibitory mechanism with a tyrosine residue bound to the active site [13]. This tyrosine may be considered a transient pseudosubstrate, and is ultimately autophosphorylated in response to insulin binding to the extracellular part of the receptor. Phosphorylation of this and two other tyrosine residues results in a rearrangement, allowing access to the active site [40]. A similar blocking of the active site by a tyrosine is also observed in the inactive structure of the MAP kinase ERK2 [41]. In this case, the tyrosine is one of the two residues phosphorylated by a distinct upstream kinase to yield the active enzyme [42].

REGULATORY SITES IN PROTEIN KINASE DOMAINS

Activation Loop

The activation (A-) loop, which follows the C-terminal to the so-called "DFG" motif involved in Mg^{2+} and ATP phosphate coordination, represents the most complex element of protein kinase structure, and shows a great variety of behaviors. The loop has evolved a remarkable ability to rearrange in response to phosphorylation. Many kinases that require activation by phosphorylation in the A-loop contain an arginine residue immediately preceding the catalytic aspartate, and have therefore been termed *RD kinases* [43]. The A-loop represents a part of the active site, and also has an influence on the position of helix C; it needs to be in an open and extended conformation to allow substrate binding. The most dramatic examples of the modulation of activity by phosphorylation of the A-loop include the previously mentioned IRK and MAPK ERK2. The phosphorylation even modulates the oligomerization and nuclear localization of ERK2 [44]. The inactive conformation of the A-loop has been exploited medically as a specific binding site for the anticancer drug Gleevec in Bcr-Abl [45].

Helix C

Helix C modulates kinase activity because it is coupled through intramolecular contacts to both the ATP binding site and the A-loop, by moving as a rigid body in response to intra- and intermolecular regulators. The best understood example is CDK2, where cyclin binding directly to the helix and its vicinity induces a rotation that reconstitutes the ATP binding site [18]. In Src family tyrosine kinases, the binding of the adjacent SH3 domain from the protein holds helix C in a conformation similar to the inactive state of CDK2; ligands to the SH3 domain (and SH2 domain also present in the protein) allow helix C to assume the active conformation [46–48]. Both CDKs and Src family kinases simultaneously require phosphorylation in the A-loop for full activity.

P-loop

The conformation of the P-loop differs subtly between the active kinases. The classical P-loop motif is GXGXXG, but some kinases have only two of the glycines. It is likely flexible so it can both accommodate ATP binding despite subtle changes in the orientation of the two lobes, and respond to regulators. Inhibitory phosphorylation of the P-loop can also occur, as in the case WEE1 phosphorylating the cell cycle kinase Cdc2 on tyrosine in its P-loop and blocking activity.

ATP Binding Site

Blocking the conserved ATP binding site is a common mechanism to regulate protein kinase activity, and the specificity stems from additional protein–protein interfaces (e.g., p16–CDK6 interaction [49, 50]) or complex intramolecular interactions (e.g., IRK, twitchin kinase).

Substrate Binding Site

The substrate peptide-binding site is in the groove between the N- and C-lobes, with the A-loop constituting a part of it. It is most often blocked via intrasteric mechanisms, or modulated via the conformation of the A-loop.

Flanking Segments

Polypeptide segments flanking the protein kinase domain either N- or C-terminally are responsible for autoinhibition in most kinases exhibiting intrasteric regulation (e.g., twitchin, CaMKI, PAK). In some cases the flanking regions are not directed to the active site, but inhibit the catalytic activity allosterically through conformational change alone. EphB2 kinase is activated by phosphorylation in both the A-loop and the N-terminal flanking sequence. In an unphosphorylated state, this N-terminal sequence binds to the N-lobe and stabilizes helix C and the A-loop in a catalytically unproductive conformation [50]. Phosphorylation of the N-terminal segment also activates the type I TGFβ receptor (TGFβ-IR). This N-terminal "GS" region inhibits the kinase activity when bound to the inhibitory protein FKBP12, distorting the N-lobe and particularly helix C [51]. The activators of EphB2 and TGFβ-IR are various SH2 domains and Smad proteins, respectively.

CONCLUSIONS

The structures of different protein kinases in active and inactive states have revealed some fundamental principles of kinase regulation, but also many variations and combinations of the major themes. It is expected that many further variations exist, and new structural information continues to illustrate new variations. Further insights into the mechanisms of regulation will come from combining structural analysis with other approaches, such as computational [52] and biophysical [53] methods.

REFERENCES

1. Manning G, Whyte DB, Martinez R, Hunter T, Sudarsanam S. The protein kinase complement of the human genome. *Science* 2002;**298**:1912–34.
2. Hunter T. Signaling–2000 and beyond. *Cell* 2000;100:113–27.
3. Pawson T, Scott JD. Signalling through scaffold, anchoring, and adaptor proteins. *Science* 1997;**278**:2075–80.
4. Zhu G, Liu Y, Shaw S. Protein kinase specificity. A strategic collaboration between kinase peptide specificity and substrate recruitment. *Cell Cycle* 2005;**4**:52–6.
5. Kobe B, Kampmann T, Forwood J, Listwan P, Brinkworth RI. Substrate specificity of protein kinases and computational prediction of substrates. *Biochim Biophys Acta* 2005;1754:200–9.
6. Ubersax Jr. JA, Ferrell JE Mechanisms of specificity in protein phosphorylation. *Nat Rev Mol Cell Biol* 2007;**8**:530–41.
7. Cowan-Jacob SW. Structural biology of protein tyrosine kinases. *Cell Mol Life Sci* 2006;**63**:2608–25.
8. Pellicena P, Kuriyan J. Protein–protein interactions in the allosteric regulation of protein kinases. *Curr Opin Struct Biol* 2006;**16**:702–9.
9. Gold MG, Barford D, Komander D. Lining the pockets of kinases and phosphatases. *Curr Opin Struct Biol* 2006;**16**:693–701.
10. Shi Z, Resing KA, Ahn NG. Networks for the allosteric control of protein kinases. *Curr Opin Struct Biol* 2006;**16**:686–92.
11. Remenyi A, Good MC, Lim WA. Docking interactions in protein kinase and phosphatase networks. *Curr Opin Struct Biol* 2006;**16**:676–85.
12. Knighton DR, Zheng J, Ten Eyck LF, Ashford VA, Xuong N-H, Taylor SS, Sowadski JM. Crystal structure of the catalytic subunit of cyclic adenosine monophosphate-dependent protein kinase. *Science* 1991;**253**:407–14.
13. Hubbard SR, Wei L, Ellis L, Hendrickson WA. Crystal structure of the tyrosine kinase domain of the human insulin receptor. *Nature* 1994;**372**:746–54.
14. Brinkworth RI, Breinl RA, Kobe B. Structural basis and prediction of substrate specificity in protein serine/threonine kinases. *Proc Natl Acad Sci USA* 2003;**100**:74–9.
15. Nolen B, Taylor S, Ghosh G. Regulation of protein kinases; controlling activity through activation segment conformation. *Mol Cell* 2004;**15**:661–75.
16. Kobe B, Kemp BE. Active site-directed protein regulation. *Nature* 1999;**402**:373–6.
17. Monod J, Changeux JP, Jacob F. Allosteric proteins and cellular control systems. *J Mol Biol* 1963;**6**:306–29.
18. Jeffrey PD, Russo AA, Polyak K, Gibbs E, Hurwitz G, Massague J, Pavletich NP. Mechanism of CDK activation revealed by the structure of a cyclinA–CDK2 complex. *Nature* 1995;**376**:313–20.
19. Bayliss R, Sardon T, Vernos I, Conti E. Structural basis of Aurora-A activation by TPX2 at the mitotic spindle. *Mol Cell* 2003;**12**:851–62.
20. Sessa F, Mapelli M, Ciferri C, Tarricone C, Areces LB, Schneider TR, Stukenberg PT, Musacchio A. Mechanism of Aurora B activation by INCENP and inhibition by hesperadin. *Mol Cell* 2005;**18**:379–91.

21. Zhang X, Gureasko J, Shen K, Cole PA, Kuriyan J. An allosteric mechanism for activation of the kinase domain of epidermal growth factor receptor. *Cell* 2006;**125**:1137–49.

22. Dar AC, Dever TE, Sicheri F. Higher-order substrate recognition of eIF2α by the RNA-dependent protein kinase PKR. *Cell* 2005;**122**:887–900.

23. Padyana AK, Qiu H, Roll-Mecak A, Hinnebusch AG, Burley SK. Structural basis for autoinhibition and mutational activation of eukaryotic initiation factor 2α protein kinase GCN2. *J Biol Chem* 2005;**280**:29,289–29,299.

24. Zhou T, Sun L, Humphreys J, Goldsmith EJ. Docking interactions induce exposure of activation loop in the MAP kinase ERK2. *Structure* 2006;**14**:1011–19.

25. Bhattacharyya RP, Remenyi A, Good MC, Bashor CJ, Falick AM, Lim WA. The Ste5 scaffold allosterically modulates signaling output of the yeast mating pathway. *Science* 2006;**311**:822–6.

26. Heo YS, Kim SK, Seo CI, Kim YK, Sung BJ, Lee HS, Lee JI, Park SY, Kim JH, Hwang KY, et al. Structural basis for the selective inhibition of JNK1 by the scaffolding protein JIP1 and SP600125. *EMBO J* 2004;**23**:2185–95.

27. Schindler T, Sicheri F, Pico A, Gazit A, Levitzki A, Kuriyan J. Crystal structure of Hck in complex with a Src family-selective tyrosine kinase inhibitor. *Mol Cell* 1999;**3**:639–48.

28. Xu W, Doshi A, Lei M, Eck MJ, Harrison SC. Crystal structures of c-Src reveal features of its autoinhibitory mechanism. *Mol Cell* 1999;**3**:629–38.

29. Mol CD, Dougan DR, Schneider TR, Skene RJ, Kraus ML, Scheibe DN, Snell GP, Zou H, Sang BC, Wilson KP. Structural basis for the autoinhibition and STI-571 inhibition of c-Kit tyrosine kinase. *J Biol Chem* 2004;**279**:31,655–31,663.

30. Kemp BE, Pearson RB. Intrasteric regulation of protein kinases and phosphatases. *Biochim Biophys Acta* 1991;**1094**:67–76.

31. Kobe B, Heierhorst J, Feil SC, Parker MW, Benian GM, Weiss KR, Kemp BE. Giant protein kinases: domain interactions and structural basis of autoregulation. *EMBO J* 1996;**15**:6810–21.

32. Heierhorst J, Kobe B, Feil SC, Parker MW, Benian GM, Weiss KR, Kemp BE. Ca^{2+}/S100 regulation of giant protein kinases. *Nature* 1996;**380**:636–9.

33. Mayans O, van der Ven PFM, Wilm M, Mues A, Young P, Furst D, Wilmanns M, Gautel M. Structural basis for activation of the titin kinase domain during myofibrillogenesis. *Nature* 1998;**395**:863–9.

34. Goldberg J, Nairn AC, Kuriyan J. Structural basis for the auto-inhibition of calcium/calmodulin-dependent protein kinase I. *Cell* 1996;**84**:875–87.

35. Lei M, Lu W, Meng W, Parrini MC, Eck MJ, Mayer BJ, Harrison SC. Structure of PAK1 in an autoinhibited conformation reveals a multistage activation switch. *Cell* 2000;**102**:387–97.

36. Dajani R, Fraser E, Roe SM, Young N, Good V, Dale TC, Pearl LH. Crystal structure of glycogen synthase kinase 3 beta: structural basis for phosphate-primed substrate specificity and autoinhibition. *Cell* 2001;**105**:721–32.

37. ter Haar E, Coll JT, Austen DA, Hsiao HM, Swenson L, Jain J. Structure of GSK3beta reveals a primed phosphorylation mechanism. *Nat Stuct Biol* 2001;**8**:593–6.

38. Kemp BE, Faux MC, Means AR, House C, Tiganis T, Hu S-H, Mitchelhill KI. Structural aspects: pseudosubstrate and substrate interactions. In: Woodgett JR, editor. *Protein Kinases*. 1st edn Oxford: IRL Press; 1994. p. 30–67.

39. Kheifets V, Mochly-Rosen D. Insight into intra- and inter-molecular interactions of PKC: design of specific modulators of kinase function. *Pharmacol Res* 2007;**55**:467–76.

40. Hubbard SR. Crystal structure of the activated insulin receptor tyrosine kinase in complex with peptide substrate and ATP analog. *EMBO J* 1997;**16**:5572–81.

41. Zhang F, Strand A, Robbins D, Cobb MH, Goldsmith EJ. Atomic structure of the MAP kinase ERK2 at 2.3-Å resolution. *Nature* 1994;**367**:704–11.

42. Canagarajah BJ, Khokhlatchev A, Cobb MH, Goldsmith EJ. Activation mechanism of the MAP kinase ERK2 by dual phosphorylation. *Cell* 1997;**90**:859–69.

43. Johnson LN, ŒNoble MEM, Owen DJ. Active and inactive protein kinases: structural basis for regulation. *Cell* 1996;**85**:149–58.

44. Khokhlatchev AV, Canagarajah B, Wilsbacher J, Robinson M, Atkinson M, Goldsmith E, Cobb MH. Phosphorylation of the MAP kinase ERK2 promotes its homodimerization and nuclear translocation. *Cell* 1998;**93**:605–15.

45. Schindler T, Bornmann W, Pellicena P, Miller WT, Clarkson B, Kuriyan J. Structural mechanism for STI-571 inhibition of abelson tyrosine kinase. *Science* 2000;**289**:1938–42.

46. Sicheri F, Moarefi I, Kuriyan J. Crystal structure of the Src family tyrosine kinase Hck. *Nature* 1997;**385**:602–53.

47. Xu W, Harrison SC, Eck MJ. Three-dimensional structure of the tyrosine kinase c-Src. *Nature* 1997;**385**:595–602.

48. Moarefi I, LaFevre-Bernt M, Sicheri F, Huse M, Lee CH, Kuriyan J, Miller WT. Activation of the Src-family tyrosine kinase Hck by SH3 domain displacement. *Nature* 1997;**385**:650–3.

49. Russo AA, Tong L, Lee JO, Jeffrey PD, Pavletich NP. Structural basis for inhibition of the cyclin-dependent kinase Cdk6 by the tumour suppressor p16INK4a. *Nature* 1998;**395**:237–43.

50. Brotherton DH, Dhanaraj V, Wick S, Brizuela L, Domaille PJ, Volyanik E, Xu X, Parisini E, Smith BO, Archer SJ, et al. Crystal structure of the complex of the cyclin D-dependent kinase Cdk6 bound to the cell-cycle inhibitor p19INK4d. *Nature* 1998;**395**:244–50.

51. Huse M, Chen YG, Massague J, Kuriyan J. Crystal structure of the cytoplasmic domain of the type I TGF beta receptor in complex with FKBP12. *Cell* 1999;**96**:425–36.

52. Young MA, Gonfloni S, Superti-Furga G, Roux B, Kuriyan J. Dynamic coupling between the SH2 and SH3 domains of c-Src and Hck underlies their inactivation by C-terminal tyrosine phosphorylation. *Cell* 2001;**105**:115–26.

53. Li F, Gangal M, Juliano C, Gorfain E, Taylor SS, Johnson DA. Evidence for an internal entropy contribution to phosphoryl transfer: a study of domain closure, backbone flexibility, and the catalytic cycle of cAMP-dependent protein kinase. *J Mol Biol* 2002;**315**:459–69.

54. Madhusudan., Trafny EA, Xuong N-H, Adams JA, Ten Eyck LF, Taylor SS, Sowadski JM. cAMP-dependent protein kinase: crystallographic insights into substrate recognition and phosphotransfer. *Protein Sci* 1994;**3**:176–87.

Chapter 7

Mammalian MAP Kinases

Norman J. Kennedy and Roger J. Davis

Howard Hughes Medical Institute and Program in Molecular Medicine, University of Massachusetts Medical School, Worcester, Massachusetts

INTRODUCTION

Mitogen-activated protein (MAP) kinases are critical mediators of signal transduction in mammalian cells [1–5]. The MAP kinases are encoded by a group of genes that are related in sequence. These enzymes are also functionally related. For example, MAP kinases phosphorylate substrate proteins on conserved Ser–Pro and Thr–Pro motifs [6]. However, the substrate specificity of individual MAP kinases is different for individual MAP kinase isoforms. This substrate specificity is mediated, in part, by the selective docking of MAP kinases to substrate proteins [7]. A second functional similarity between MAP kinases is the mechanism of activation. MAP kinases are activated within conserved protein kinase signaling modules composed of a MAP kinase, a MAP kinase kinase, and a MAP kinase kinase kinase. Phosphorylation and activation of MAP kinases by MAP kinase kinases occurs on two residues within a tripeptide motif (Thr–Xaa–Tyr) located within the T-loop that controls the conformation of the MAP kinase active site [8].

There are three major groups of mammalian MAP kinases: the extracellular signal regulated protein kinases (ERK), the p38 MAP kinases, and the c-Jun NH$_2$-terminal kinases (JNK) (Table 7.1). These groups of MAP kinases can be distinguished by the sequence of the dual phosphorylation motif that mediates MAP kinase activation: Thr–Glu–Tyr (ERK), Thr–Gly–Tyr (p38), and Thr–Pro–Tyr (JNK). In addition, there are several protein kinase that are related to the MAP kinases with similar dual phosphorylation motifs: Thr–Glu–Tyr (MAK and MOK), Thr–Asp–Tyr (ICK, KKIAMRE, KKIALRE), and Thr–His–Glu (NLK).

Here we review the properties of the three major groups of MAP kinases: ERK, p38, and JNK.

THE ERK GROUP OF MAP KINASES

ERK1 and ERK2

The ERK1 and ERK2 protein kinases are activated by phosphorylation of the Thr–Glu–Tyr motif located in the T-loop by the MAP kinase kinases MEK1 and MEK2 (Table 7.2). Structural analysis of non-phosphorylated and inactive ERK2 [9] and phosphorylated and activated ERK2 [10] by X-ray crystallography reveals that the mechanism of activation involves conformational changes that remodel the T-loop and the active site. The activated form of ERK2 has been identified as a homodimer [11], and it is likely that this dimeric form of ERK2 is critical for nuclear accumulation [12].

The ERK1 and ERK2 protein kinases are major targets of the Ras signaling pathway. Substrates of these MAP kinases include a wide array of proteins, including the epidermal growth factor receptor [6], cytoplasmic phospholipase A2 [13], and Ets family transcription factors [14]. The ERK1 and ERK2 protein kinases are implicated in proliferation [15], tumorigenesis [16], and differentiation [17]. It appears that the timecourse of ERK1 and ERK2 activation may be critical for specifying the biological outcome of signal transduction: transient activation correlates with growth; and sustained activation correlates with growth arrest and differentiation [17].

Recent studies indicate that the targeted disruption of the *Erk1* gene in mice causes defects in thymocyte development [18] and causes increased synaptic plasticity associated with improved striatal-mediated learning and memory [19]. In contrast, *Erk2*$^{-/-}$ mice died during embryonic development because of defects in placental development [20, 21]. It is likely that ERK1 and ERK2 exhibit partial functional redundancy. However, studies of compound mutant mice lacking both *Erk1* and *Erk2* have not yet been reported (Table 7.3).

ERK3 and ERK4

The ERK3/4 subgroup of MAP kinases is encoded by two genes [22, 23]. These protein kinases are not true members of the ERK group because they contain the sequence Ser–Glu–Gly instead of the dual phosphorylation motif

TABLE 7.1 Nomenclature of mammalian MAP kinases[1]

Name	Alternative name	Gene name	Chromosomal location	UniGene ID
ERK1	p44 MAPK	MAPK3	16p11.2	Hs.861
ERK2	p42 MAPK	MAPK1	22q11.21	Hs.431850
ERK3	p97 MAPK	MAPK6	15q21	Hs.411847
ERK4	p63 MAPK	MAPK4	18q12-q21	Hs.433728
ERK5	Big MAP kinase	MAPK7	17p11.2	Hs.150136
ERK7		ERK7	7q34	Rn.42898
ERK8		MAPK15	8q24.3	Hs.493169
p38α	CSBP, RK SAPK2A	MAPK14	6p21.3-p21.2	Hs.588289
p38β	P38-2, p38β2 SAPK2B	MAPK11	22q13.33	Hs.57732
p38γ	SAPK3, ERK6	MAPK12	22q13.33	Hs.432642
p38δ	SAPK4	MAPK13	6p21.31	Hs.178695
JNK1	SAPKγ, SAPK1C	MAPK8	10q11.22	Hs.138211
JNK2	SAPKα, SAPK1A	MAPK9	5q35	Hs.654461
JNK3	SAPKβ, SAPK1B	MAPK10	4q22.1-q23	Hs.125503

[1]*Human MAP kinases corresponding to the ERK group (ERK1, ERK2, ERK3, ERK4, ERK5, ERK7, and ERK8), the p38 MAP kinase group (p38α, p38β, p38γ, p38δ), and the JNK group (JNK1, JNK2, and JNK3) are presented with their alternative names, the gene name, the chromosomal localization, and the UniGene identification number.*

TABLE 7.2 Nomenclature of mammalian MAP kinase kinases[1]

Name	Alternative name	Gene name	Chromosomal location	UniGene ID
MEK1	MAPKK1, MKK1	MAP2K1	15q22.1 – q22.33	Hs.145442
MEK2	MAPKK2, MKK2	MAP2K2	19p13.3	Hs.465627
MKK3	MEK3, SKK2	MAP2K3	17q11.2	Hs.514012
MKK4	MEK4, SEK1, JNKK1, SKK1	MAP2K4	17p11.2	Hs.514681
MEK5	MKK5	MAP2K5	15q23	Hs.114198
MKK6	MEK6, SKK3	MAP2K6	17q24.3	Hs.463978
MKK7	MEK7, SEK2, JNKK2, SKK4	MAP2K7	19p13.3 – p13.2	Hs.531754

[1]*Human MAP kinase kinases are presented with their alternative names, the gene name, the chromosomal localization, and the UniGene identification number.*

TABLE 7.3 Targeted disruption of genes in mice[1]

	Phenotype	Reference
ERK1	Viable. Defects in thymocyte maturation. Increased synaptic plasticity and improved striatal-mediated learning and memory.	18, 19
ERK2	Lethal. Defects in placental development.	20, 21
ERK5	Lethal. Embryos die during mid-gestation. Defects in cardiovascular development.	35
p38α	Lethal. Defects in formation of the placenta and expression of erythropoietin	55–57
p38β	Viable. No reported defects.	169
p38γ	Viable. No reported defects.	62
p38δ	Viable. No reported defects.	62
p38γ + p38δ	Viable. No reported defects.	62
JNK1	Viable. Defects in CD4 effector T cell differentiation and function. Defects in CD8 T cell activation.	76, 80, 81, 84
JNK2	Viable. Defects in CD4 effector T cell differentiation and function. Increased CD8 T cell activation.	79–81, 84
JNK1 + JNK2	Lethal. Embryos die during mid-gestation with neural tube closure defects and increased fore-brain apoptosis. Primary fibroblasts are resistant to stress-induced apoptosis.	71, 82, 83
JNK3	Viable. Defects in neuronal apoptosis in response to excitotoxic stress and ischemia.	77, 170
MEK1	Lethal. Defects in placental development *in vivo* and cell migration *in vitro*.	171
MEK2	Viable. No reported defects.	172
MKK3	Viable. Defects in cytokine secretion by CD4 T cells and macrophages. Defects in activation-induced cell death of peripheral CD4 T cells.	173–175
MKK4	Lethal. Embryos die during mid-gestation with liver apoptosis.	66, 176–182
MEK5	Lethal. Embryos die during mid-gestation with defects in cardiac development.	36
MKK6	Viable. Defects in thymocyte apoptosis.	173
MKK7	Lethal. Embryos die during mid-gestation with defective liver development. Defects in TNF-stimulated JNK activation and pre-mature senescence in cultured fibroblasts.	66, 84, 183, 184
MKK4 + MKK6	Lethal. Embryos die during mid-gestation with defects in placental formation and vascular development. Defects in TNF-stimulated p38 activation in cultured fibroblasts.	43
MKK4 + MKK7	Lethal. Embryos die during mid-gestation. JNK is not activated in cultured fibroblasts.	66

[1]*The effect of disruption of the murine genes that encode MAP kinases and MAP kinase kinases is summarized.*

Thr–Glu–Tyr. A MAP kinase kinase activity for ERK3 has been described, but the regulation of this protein kinase activity is not understood [5]. ERK3, but not ERK4, is rapidly processed by ubiquitin-mediated degradation and has a short half-life *in vivo* [24, 25]. The physiological function of ERK3 and ERK4 remains unclear. However, recent studies have indicated that one target of this signaling pathway is MAPKAPK5 [26–29].

ERK5

The ERK5 protein kinase is activated by the MAP kinase kinase MEK5 by dual phosphorylation on the conserved T-loop motif Thr–Glu–Tyr [30, 31]. Substrates of ERK5 include members of the MEF2 family of transcription factors [32]. The signaling pathways that lead to the activation of the ERK5/MEK5 module are poorly understood. However, ERK5 is critical for survival signal transduction by retrograde NGF receptors localized in endosomes [33], and for proliferation caused by epidermal growth factor [34]. Recent studies indicate that ERK5 is essential for viability during murine embryonic development because ERK5-deficiency causes defects in cardiovascular development [35]. Similarly, MEK5 deficiency causes embryonic death that is associated with decreased ERK5 activity [36].

ERK7 and ERK8

The ERK7 and ERK8 protein kinases form a subgroup of ERK-related protein kinases that contain the dual phosphorylation motif Thr–Glu–Tyr [37, 38]. The ERK7 isoform has only been detected in the rat. In contrast, only the ERK8 isoform has been detected in humans. These observations suggest that ERK7 and ERK8 may represent orthologs in different species. ERK7 and ERK8 are not activated by MEK1 and MEK2. Recent studies indicate that ERK7 is constitutively activated by autophosphorylation [39], and the amount of ERK7 expression is regulated by ubiquitin-mediated degradation [40]. In contrast, it has been reported that ERK8 has a low basal activity that is increased in cells with activated Src or following serum-stimulation [38]. The function of ERK7 and ERK8 is unclear. However, ERK7 may interact with an intracellular chloride channel [41] and may regulate the ubiquitin-mediated degradation of the estrogen receptor [42]. Further studies of these novel MAP kinase isoforms to define their physiological function are warranted.

THE P38 GROUP OF MAP KINASES

Regulation of the p38 group of MAP kinases differs from the ERK group of MAP kinases [1, 3]. The ERK1 and ERK2 protein kinases are activated by many growth factors by a Ras-dependent mechanism. In contrast, the p38 MAP kinases are activated by inflammatory cytokines and by the exposure of cells to environmental stress. It is thought that members of the Rho family of small GTPases, rather than the Ras family of GTPases, are critical mediators of p38 MAP kinase activation. The p38 group of MAP kinases is activated by three different MAP kinase kinases [4]. Two of these MAP kinase kinases specifically activate the p38 MAP kinases (MKK3 and MKK6). In contrast, the third MAP kinase kinase (MKK4) activates both the JNK and p38 MAP kinases. Studies of $Mkk3^{-/-}$, $Mkk4^{-/-}$, and $Mkk6^{-/-}$ mice and compound mutants lacking these MKK isoforms have confirmed this specificity *in vivo* [43].

p38α and p38β MAP Kinases

The p38α and p38β MAP kinases are structurally related protein kinases that contain the dual phosphorylation motif Thr–Gly–Tyr [1,3]. The structure of p38α MAP kinase has been determined by X-ray crystallography [44, 45]. This structure is similar to ERK2. However, there are significant differences that distinguish p38α MAP kinase from ERK2. Included among these differences is the structure of the active site of p38α MAP kinase. This difference has allowed the discovery of active site-directed inhibitors that are extremely selective for p38α and p38β MAP kinases [46]. These drugs are useful for functional dissection of the p38α and p38β MAP kinase signaling pathway. In addition, it is likely that such drugs may be useful for therapeutic intervention. In particular, a major role for p38α and p38β MAP kinases in inflammatory responses has been identified. The p38α and p38β MAP kinases appear to be critical for the expression of several inflammatory cytokines, including interleukin-1, interleukin-6, and tumor necrosis factor [47]. These p38 MAP kinases regulate several steps in cytokine expression, including transcription, mRNA stability, and translation. The transcriptional effects of the p38α and p38β MAP kinases are mediated by phosphorylation of several transcription factors, including ATF2 and members of the MEF2, and Ets families [48–50]. The mechanisms that account for the effects of p38α and p38β MAP kinases on mRNA stability and translation have not been defined. However, recent studies have implicated proteins that bind to the 3′ ARE elements of regulated mRNAs and MAPKAP2, a protein kinase that is phosphorylated and activated by p38α and p38β MAP kinases [51–54].

Mice with targeted disruption of the p38α MAP kinase gene have been described [55–57]. These mice die during mid-gestation because of defects in the formation of the placenta. p38α MAP kinase-deficient cells derived from these embryos exhibit severe defects in the expression of inflammatory cytokines [56]. Studies using conditional ablation of the *p38α* gene have demonstrated that p38 MAPK is a negative regulator of myocyte development [58], cardiomyocyte proliferation [59], lung stem cell function [60], and tumorigenesis [60].

p38γ and p38δ MAP Kinases

The p38γ and p38δ MAP kinases are related enzymes that represent a separate subgroup of p38 MAP kinases [1, 3]. Like other members of the p38 family, the p38γ and p38δ MAP kinases contain a Thr–Gly–Tyr dual phosphorylation motif. However, the p38γ and p38δ MAP kinases are not inhibited by drugs that selectively inhibit the p38α and p38β MAP kinases [46]. The regulation of the p38γ and p38δ MAP kinases is similar to the p38α/β MAP kinases, and all four p38 MAP kinases are activated by inflammatory cytokines and exposure to environmental stress. Interestingly, p38γ MAP kinase interacts with the PDZ domain of α1-syntrophin, although the functional significance of this interaction is unclear [61]. Possible roles for p38γ MAP kinase in the regulation of PSD-95 and dDlg have also been described [62, 63]. Recent studies have demonstrated an important role for p38δ MAP kinase in the regulation of translation by eEF2 kinase [64]. Knockout mice that lack expression of p38γ or p38δ and compound mutants lacking both p38γ and p38δ have been described, but no obvious phenotype was detected [62].

THE JNK GROUP OF MAP KINASES

The JNK group of MAP kinases is activated by many of the same stimuli that cause activation of the p38 MAP kinases, including the exposure of cells to inflammatory cytokines and environmental stress [65]. This similarity in regulation indicates that the JNK and p38 MAP kinases may be functionally related. Indeed, both the JNK and p38 MAP kinases are collectively named stress-activated protein kinases (SAPK). However, the mechanism of JNK activation differs from the p38 MAP kinases. First, the dual phosphorylation motif located in the T-loop of JNK is Thr–Pro–Tyr. Second, the signaling module that activates JNK includes the MAP kinase kinases MKK4 and MKK7 [65]. The protein kinase MKK7 is a specific activator of the JNK pathway, while MKK4 can activate both p38 MAP kinase and JNK [65]. Biochemical studies demonstrate that MKK4 and MKK7 can cooperate to activate JNK by selectively phosphorylating JNK on Tyr and Thr, respectively, and gene disruption experiments in mice demonstrate functional cooperation of MKK4 and MKK7 in vivo [66–68].

The JNK protein kinases are encoded by three genes. Two of the genes (JNK1 and JNK2) are expressed ubiquitously, and one gene (JNK3) is expressed in a limited number of tissues, including the brain, heart, and testis [65]. Alternative splicing creates ten different JNK protein kinases: four JNK1 isoforms; four JNK2 isoforms; and two JNK3 isoforms [69]. Analysis of JNK3 by X-ray crystallography indicates that this MAP kinase is structurally related to both ERK2 and p38α MAP kinase [70].

The physiological role of JNK appears to be complex. It is established that JNK is required for the normal regulation of AP-1 transcription activity [65]. This is mediated, in part, by the phosphorylation of the transcription factors ATF2, c-Jun, JunB, and JunD on two sites within the NH2-terminal activation domain [65]. JNK can also regulate other transcription factors by phosphorylation. Nevertheless, although it is known that JNK causes increased AP-1-dependent gene expression, the physiological consequence of JNK activation appears to be both cell-type and context dependent. JNK can cause apoptosis by a mechanism that involves the mitochondrial pathway [71, 72]. In addition, JNK is required for the generation of reactive oxygen species during TNF-induced necrosis [73]. Nevertheless, JNK can also signal cell survival [74, 75]. The mechanism that accounts for these markedly divergent cellular responses to JNK activation is unclear. However, it is likely that the cellular response to JNK activation reflects the activation state of other signal transduction pathways within the cell. A reasonable hypothesis is that the program of gene expression that is induced by JNK-stimulated AP-1 activity depends on the cooperation of AP-1 with other transcription factors bound to the promoters of relevant genes. Further studies will be required to test this hypothesis.

Gene disruption studies in mice demonstrate that JNK1, JNK2, and JNK3 are not essential for viability [76–81]. However, these mice exhibit defects in apoptosis and immune responses. Compound mutations in JNK1 and JNK2 cause early embryonic lethality associated with neural tube defects and markedly increased apoptosis in the developing forebrain [82, 83]. Studies of $Jnk1^{-/-}$ $Jnk2^{-/-}$ CD4 T cells isolated from $Rag1^{-/-}$ chimeric mice indicate that JNK is required for effector CD4 T cell function, but is not required for CD4 T cell activation [84]. Primary embryo fibroblasts isolated from $Jnk1^{-/-}$ $Jnk2^{-/-}$ embryos exhibit severe growth defects with premature senescence associated with increased expression of ARF and p53. These JNK-deficient fibroblasts also exhibit marked resistance to stress-induced apoptosis [71]. This resistance to apoptosis was also observed in $Mkk4^{-/-}$ $Mkk7^{-/-}$ fibroblasts that contain JNK, but lack a mechanism to allow JNK activation [66]. Together, these data indicate that JNK contributes to multiple physiological processes, including differentiation, survival, and apoptosis.

MAP KINASE DOCKING INTERACTIONS

Although MAP kinases can phosphorylate Ser-Pro and Thr-Pro sites on substrate proteins, the substrate specificities of individual MAP kinases are distinct. One mechanism that can dictate the substrate specificity of a MAP kinase is the requirement for a docking site on the substrate [7]. The docking site is physically separate from the site of phosphorylation and is required for efficient substrate phosphorylation by MAP kinases. Mutational removal of the docking site prevents substrate phosphorylation by MAP kinases. Examples of docking sites include the δ domain on c-Jun that interacts with JNK [65], the D domain on the Elk-1 transcription factor that binds ERK and JNK [85, 86], the D domain on MEF transcription factors that binds p38α MAP kinase [87], and the FXF motif on the SAP-1 transcription factor that binds ERK and p38α MAP kinase [88, 89]. The δ and D domains are similar and consist of a Leu–Xaa–Leu motif separated from several basic residues [7]; mutational analysis indicates that these motifs bind MAP kinases at a common site [90–92]. In contrast, the FXF domain contains the motif Phe–Xaa–Phe–(Pro) [7], and the site of interaction on MAP kinases has not been fully defined.

Interestingly, these docking domains are conserved in many proteins that interact with MAP kinases, including MKK isoforms, MKP isoforms, scaffold proteins, and substrates [7]. Thus, the NH$_2$-terminal region of many MKK isoforms contains a D domain [93]. Disruption of this MAP kinase docking site on MAP kinase kinases is caused by the anthrax lethal factor protease and prevents MAP kinase activation [94]. Similarly, dysregulation of MAP kinase function can occur by disruption of MAP kinase interactions with other proteins with conserved interaction motifs, including MKP isoforms, scaffold proteins, and substrates [7].

Recent studies have provided structural insight into the mechanism of protein docking to MAP kinases. First, hydrogen exchange mass spectroscopy demonstrates that separate binding pockets on MAP kinases interact with D domain and FXFP ligands [95]. Second, several X-ray crystal structures of MAP kinase docking interactions with D domain ligands have been solved, including: ERK2 with the D domains of MKP3 [96], MEK2 [97], and HePTP [97]; p38α MAP kinase with the D domains of MKK3 and MEF2 [44]; JNK1 with the D domain of JIP1 [98]; and also the yeast MAPK Fus3 with the D domains of the MKK isoform Ste7, the MKP isoform Msg5, and the substrate Far1 [99]. This structural analysis demonstrated that the Leu–Xaa–Leu motif is directly involved in the protein–protein contact and that the basic residues in the D-domain may interact with acidic residues in the common docking site [100]. The site of interaction on the surface of these MAP kinases is not located close to the T loop or the active site. However, the interaction induces conformational changes in the MAP kinases that leads to exposure of the T-loop. Thus, the active site and the T-loop of MAP kinases are available for interaction with docked proteins, including MAP kinase kinases, MAP kinase phosphatases, and substrates.

SCAFFOLD PROTEINS

The protein kinases that form MAP kinase signaling modules can interact via a series of sequential binary interactions to create a protein kinase cascade. One example is represented by MKK4, which can dock to both an upstream kinase (MEKK1) and to a down–stream MAP kinase (JNK) [101]. Alternatively, the protein kinases may simultaneously interact with a common component of the cascade. Thus, MEKK2 can synergistically interact with both MKK4 and JNK [102] and MEKK1 can bind c-Raf-1, MEK1, and ERK2 [103]. MEKK1 can also bind JNK in a phosphorylation-dependent manner [104]. These interactions may lead to the assembly of a functional signaling module [105].

Functional MAP kinase signaling modules can also be created by the interaction of the protein kinases with scaffold molecules that serve to assemble the protein kinases [105]. These putative scaffold proteins include KSR, MP1, the JIP proteins, POSH, MORG1, OSM, β-arrestin-2, SKRP1 and MKPX (Table 7.4). In addition, several other molecules have been proposed to function as MAP kinase scaffolds, including Filamin [106], Crk II [107], IKAP [108], and Paxillin [109].

The molecular mechanism that accounts for the role of these scaffold proteins in the regulation of MAP kinase activation is poorly defined [110]. However, studies in yeast have established that functional scaffold proteins can be created by assembling protein interaction modules [111, 112] and X-ray crystal structure analysis has provided insight into the function of the Ste5 scaffold-mediated activation of the Fus3 MAP kinase [113]. Further studies are required to understand the function of mammalian MAP kinase scaffold proteins.

KSR

Kinase suppressor of Ras (KSR) shares structural similarity with the MAP kinase kinase kinase c-Raf-1 [114, 115]. One major difference between KSR and c-Raf-1 is that while KSR does have a protein kinase-like domain, KSR does not function as a protein kinase [116]. KSR binds to c-Raf-1, MEK1/2 and ERK1/2 and appears to function as a scaffold for the activation of the ERK1/2 signaling module that is activated by growth factors [114, 115]. Following the activation of growth factor receptors, KSR is recruited to the cell surface by a phosphorylation-dependent mechanism that involves the C-TAK1 protein kinase [117]. Gene disruption studies in mice have examined the requirement of KSR for growth factor-stimulated ERK1/2 activation. KSR-deficient cells derived from these mice were found to exhibit partial defects in ERK activation [118]. This partial defect may reflect a specialized role for KSR under certain conditions. Alternatively, it is possible that the functions of KSR are redundant with the product of another gene that encodes the related protein KSR-2 (Table 7.4). Further studies are required to define the role of these KSR proteins. Nevertheless, it is very likely that the mammalian KSR proteins do function as an essential scaffold for ERK1/2 activation because RNAi experiments

TABLE 7.4 Nomenclature of mammalian MAP kinase scaffold proteins[1]

Name	Alternative name	Gene name	Chromosomal location	UniGene ID
KSR-1	KSR	*KSR1*	17q11.1	Hs.133534
KSR-2		*KSR2*	12q24.22-q24.23	Hs.375836
MP-1		*MAP2K1IP1*	4q22.3	Hs.696082
JIP-1	IB1	*MAPK8IP1*	11p12 – p11.2	Hs.234249
JIP-2	IB2	*MAPK8IP2*	22q13.33	Hs.558180
JIP-3	JSAP1, SYD2	*MAPK8IP3*	16p13.3	Hs.207763
JIP-4	JLP, SYD1	*SPAG9*	17q21.33	Hs.463439
POSH	Sh3md2, R75531	*SH3RF1*	4q32.3-q33	Hs.301804
POSH2		*C17orf64*	17q23.2	Hs.129312
MORG1		*MORG1*	19p13.13	Hs.657204
OSM	CCM2, Malcavernin	*CCM2*	7p13	Hs.148272
β-Arrestin-2		*ARRB2*	17p13	Hs.435811
SKRP1	JKAP, DUSP19	*SKRP1*	2q32.1	Hs.132237
MKPX	VHX, JSP-1	*DUSP22*	6p25.3	Hs.29106

[1]*Human MAP kinase scaffold proteins are presented with their alternative names, the gene name, the chromosomal localization, and the UniGene identification number.*

have demonstrated an important role for KSR in the activation of ERK in *Drosophila* [119].

MP-1

The MP-1 scaffold protein binds to MEK1 and ERK1 [120]. Transfection assays demonstrate that MP-1 potentiates the activation of ERK1 caused by MEK1 [120]. Studies demonstrate that MP1 also binds to the late endosomal protein p14 [121]. MP-1 therefore localizes a MEK1/ERK1 signaling module on the cytoplasmic surface of late endosomes. It is possible that MP1 contributes to ERK activation following ligand-induced endocytosis of growth factor receptors. This role of MP-1 on late endosomes serves to distinguish MP-1 from the KSR scaffold proteins which appear to function at the cell surface. Recent studies show that PAK1-dependent ERK activation during cell adhesion and spreading involves MP1 [122]. The possible functional interaction between the MP-1 and KSR scaffold proteins warrants further study. In addition, gene disruption studies are required to establish the physiological function of MP-1 in ERK activation *in vivo*.

MORG1

The MORG1 scaffold protein interacts with the MP-1 scaffold protein and binds members of the ERK cascade, including cRaf-1, MEK1/2, and ERK1/2 [123]. MORG1 also appears to interact with hypoxia-inducible factor prolyl hydroxylase 3 [124]. Transfection assays indicate MORG1 enhances ERK activation by LPA and PMA, but not that caused by EGF [123]. RNAi studies demonstrate that MORG1 gene suppression decreases ERK activation [123]. However, MORG1-deficient mice have not yet been described.

JIP

Four genes encode the JIP group of scaffold proteins [65]. The JIP1 and JIP2 proteins are structurally similar and contain an SH3 domain and a PTB domain in the COOH-terminal region [125–128]. The PTB domain can interact with p190 RhoGEF and with members of the low density lipoprotein receptor family, including ApoER2 and LRP1

[129–131]. In addition, JIP1 and JIP2 have been reported to bind the Rac exchange factor Tiam1 and Ras-GRF while JIP2 interacts with members of the fibroblast growth factor homologous protein family [132–134]. The JIP3 and JIP4 proteins are structurally distinct and consist of an extended coiled-coil domain [135–138]. All of these JIP proteins share several common properties. Each JIP isoform binds to JNK, MKK7, and members of the mixed-lineage group of MAP kinase kinase kinases. An interaction of JIP3 with MEKK1 and MKK4 has also been described [136]. Transfection studies demonstrate that the JIP proteins potentiate the activation of JNK [125, 126, 135, 136] and that both the Notch [139] and AKT [140] signaling pathways can suppress this function of JIP proteins. Some studies have demonstrated that JIP2 and JIP4 can also activate p38 MAP kinases under some circumstances [132, 133, 137, 138].

The JIP1, JIP2, JIP3, and JIP4 proteins bind to kinesin light chain and are transported by the microtubule motor protein kinesin [138, 141–144]. This interaction with motor proteins accounts for the accumulation of the JIP proteins in the growth cones of developing neurons. In mature neurons, the JIP1 and JIP2 proteins accumulate at synapses while JIP3 and JIP4 are mostly localized to perinuclear vesicular structures. The JIP proteins may act as adaptor molecules for the transport of cargo by the kinesin motor protein. In addition, the JIP proteins may act to locally regulate JNK activation in response to specific stimuli.

Three groups have reported the phenotype of mice with targeted disruption of the *Jip1* gene. One group reported that the JIP1-deficiency causes very early embryonic lethality prior to implantation [145]. In contrast, two groups reported that JIP1-deficient mice are viable [142, 146]. It is possible that the difference in viability reflects an effect of the mouse strain background. The viable JIP1-deficient mice were found to exhibit defects in stress-induced JNK activation in hippocampal neurons following exposure to stress *in vivo* and *in vitro* [142, 146]. In addition, JIP1-deficient mice exhibit marked defects in JNK activation, insulin resistance, and diabetes in mice fed a high fat diet [147]. These data indicate that JIP1 is required for a limited spectrum of stress-stimuli that activate JNK.

Mice with targeted ablation of the *Jip2* gene and compound mutant mice with loss of expression of both JIP1 and JIP2 have recently been reported [148]. These JIP-deficient mice exhibit major defects in stress-induced JNK activation in neurons. In addition, neurons from these mice exhibit major defects in NMDA receptor function [148].

Two groups have reported the construction of mice with targeted ablation of the *Jip3* gene [149, 150]. These mice die shortly after birth because of a failure to breathe, most likely because of severe neurological defects. Indeed, these mice exhibit defects in axonal guidance during development, including the failure of morphogenesis of commissures in the central nervous system. Interestingly, this

defect in JIP3-deficient mice can be suppressed by transgenic overexpression of JIP1 [150], suggesting that JIP1 and JIP3 may serve partially redundant functions in mice. Studies of JIP4-deficient mice have not yet been reported.

Structural studies of JIP scaffold proteins have only been reported for the isoform JIP1. These studies have documented partial structures of JIP1, including the interaction of JIP1 with JNK1 [98] and the homodimerization of JIP1 that is mediated by the SH3 domain [151].

POSH

The POSH group of scaffold proteins includes two isoforms, POSH and POSH2 [152, 153]. POSH can interact with mixed-lineage kinases (MLK), the MKK isoforms MKK4 and MKK7, and with JNK [154]. POSH and POSH2 contain a functional RING domain that serves to promote the ubiquitin-mediated degradation of both POSH proteins [155]. The mechanism of POSH function to increase JNK activity is thought to be mediated by the interaction of POSH with JIP scaffold proteins, and it is possible that JIP proteins may mediate the effects of POSH on JNK activation [156]. The role of POSH and POSH2 may be to stabilize JNK pathway components by preventing their degradation, including JIP proteins, MLK isoforms, and JNK [153, 155].

β-Arrestin

Ligand-induced activation of seven transmembrane spanning receptors causes receptor phosphorylation, recruitment of arrestin molecules, and subsequent downregulation of heterotrimeric G protein signaling [157]. The arrestin molecules can also serve as a platform for the recruitment of additional molecules to the receptor [157]. Evidence has been presented that indicates that the ubiquitously expressed isoform β-arrestin-1 may serve as a scaffold for components of the ERK pathway [158]. More detailed studies have been performed on β-arrestin-2. This scaffold protein contains a D domain that selectively binds to JNK3 [159, 160]. Interestingly, both β-arrestin-2 and JNK3 are selectively expressed in the brain and the heart. The β-arrestin-2 scaffold also binds the MAP kinase kinase kinase ASK1 [160]. The signaling module assembled by β-arrestin-2 also contains MKK4, which interacts with both JNK3 and ASK1, but does not directly contact β-arrestin-2 [160]. Biochemical assays demonstrate that β-arrestin-2 is essential for the activation of JNK3 caused by the angiotensin II receptor [160]. Interestingly, the activated receptor bound to the β-arrestin-2 scaffold complex is localized to endosomal structures. The β-arrestin-2 scaffold provides a mechanism for the activation of JNK by seven transmembrane spanning receptors. In addition, the signaling module assembled by β-arrestin-2 provides a mechanism for the selective activation of the JNK3 isoform of JNK. Studies of β-arrestin-2 -deficient mice have been reported [161]. Further studies of these mice to investigate defects in the activation of JNK3 are warranted.

OSM

The OSM scaffold protein, also known as CCM2/malcavernin, is implicated in cerebral cavernous malformation syndrome [162]. Biochemical studies indicate that OSM interacts with components of the p38 pathway, including MEKK3, MKK3, and p38 MAP kinase [163]. RNAi-mediated depletion studies indicate that the OSM scaffold complex is required for p38 MAP kinase activation caused by hyperosmotic shock [163].

SKRP1/MKPX1

SKRP1 and MKPX1 are two related small phosphatases that belong to the MAP kinase phosphatase (MKP) family. Like other MKPs, these phosphatases can inactivate MAP kinases [164, 165]. However, it appears that the normal function of these phosphatases is to activate JNK [166–168]. Interestingly, the phosphatase catalytic activity is required for these MKPs to activate JNK, but the physiologically relevant substrates have not been identified. It appears that these phosphatases interact with MKK7 [166, 167] and may also interact with ASK1 [166]. Gene disruption studies in mice demonstrated that SKRP1 is not an essential gene, but SKRP1 is required for JNK activation caused by tumor necrosis factor-α and transforming growth factor-β, but not for JNK activation caused by ultraviolet radiation [167]. Together, these data suggest that these MKPs may act as scaffold proteins for the JNK signaling pathway.

REFERENCES

1. Kyriakis JM, Avruch J. Mammalian mitogen-activated protein kinase signal transduction pathways activated by stress and inflammation. *Physiol Rev* 2001;**81**:807–69.
2. Lewis TS, Shapiro PS, Ahn NG. Signal transduction through MAP kinase cascades. *Adv Cancer Res* 1998;**74**:49–139.
3. Schaeffer HJ, Weber MJ. Mitogen-activated protein kinases: specific messages from ubiquitous messengers. *Mol Cell Biol* 1999;**19**:2435–44.
4. Ip YT, Davis RJ. Signal transduction by the c-Jun N-terminal kinase (JNK) –from inflammation to development. *Curr Opin Cell Biol* 1998;**10**:205–19.
5. Cobb MH. MAP kinase pathways. *Prog Biophys Mol Biol* 1999;**71**:479–500.
6. Davis RJ. The mitogen-activated protein kinase signal transduction pathway. *J Biol Chem* 1993;**268**:14,553–14,556.
7. Enslen H, Davis RJ. Regulation of MAP kinases by docking domains. *Biol Cell* 2001;**93**:5–14.
8. Cobb MH, Goldsmith EJ. How MAP kinases are regulated. *J Biol Chem* 1995;**270**:14,843–14,846.
9. Wang Z, Harkins PC, Ulevitch RJ, Han J, Cobb MH, Goldsmith EJ. The structure of mitogen-activated protein kinase p38 at 2.1-A resolution. *Proc Natl Acad Sci USA* 1997;**94**:2327–32.
10. Canagarajah BJ, Khokhlatchev A, Cobb MH, Goldsmith EJ. Activation mechanism of the MAP kinase ERK2 by dual phosphorylation. *Cell* 1997;**90**:859–69.
11. Khokhlatchev AV, Canagarajah B, Wilsbacher J, Robinson M, Atkinson M, Goldsmith E, Cobb MH. Phosphorylation of the MAP kinase ERK2 promotes its homodimerization and nuclear translocation. *Cell* 1998;**93**:605–15.
12. Cobb MH, Goldsmith EJ. Dimerization in MAP-kinase signaling. *Trends Biochem Sci* 2000;**25**:7–9.
13. Lin LL, Wartmann M, Lin AY, Knopf JL, Seth A, Davis RJ. cPLA2 is phosphorylated and activated by MAP kinase. *Cell* 1993;**72**:269–78.
14. Marais R, Wynne J, Treisman R. The SRF accessory protein Elk-1 contains a growth factor-regulated transcriptional activation domain. *Cell* 1993;**73**:381–93.
15. Whitmarsh AJ, Davis RJ. A central control for cell growth. *Nature* 2000;**403**:255–6.
16. Mansour SJ, Matten WT, Hermann AS, Candia JM, Rong S, Fukasawa K, Vande Woude GF, Ahn NG. Transformation of mammalian cells by constitutively active MAP kinase kinase. *Science* 1994;**265**:966–70.
17. Cowley S, Paterson H, Kemp P, Marshall CJ. Activation of MAP kinase kinase is necessary and sufficient for PC12 differentiation and for transformation of NIH 3T3 cells. *Cell* 1994;**77**:841–52.
18. Pages G, Guerin S, Grall D, Bonino F, Smith A, Anjuere F, Auberger P, Pouyssegur J. Defective thymocyte maturation in p44 MAP kinase (Erk 1) knockout mice. *Science* 1999;**286**:1374–7.
19. Mazzucchelli C, Vantaggiato C, Ciamei A, Fasano S, Pakhotin P, Krezel W, Welzl H, Wolfer DP, Pages G, Valverde O, Marowsky A, Porrazzo A, Orban PC, Maldonado R, Ehrengruber MU, Cestari V, Lipp HP, Chapman PF, Pouyssegur J, Brambilla R. Knockout of ERK1 MAP kinase enhances synaptic plasticity in the striatum and facilitates striatal-mediated learning and memory. *Neuron* 2002;**34**:807–20.
20. Hatano N, Mori Y, Oh-hora M, Kosugi A, Fujikawa T, Nakai N, Niwa H, Miyazaki J, Hamaoka T, Ogata M. Essential role for ERK2 mitogen-activated protein kinase in placental development. *Genes Cells* 2003;**8**:847–56.
21. Saba-El-Leil MK, Vella FD, Vernay B, Voisin L, Chen L, Labrecque N, Ang SL, Meloche S. An essential function of the mitogen-activated protein kinase Erk2 in mouse trophoblast development. *EMBO Rep* 2003;**4**:964–8.
22. Boulton TG, Nye SH, Robbins DJ, Ip NY, Radziejewska E, Morgenbesser SD, DePinho RA, Panayotatos N, Cobb MH, Yancopoulos GD. ERKs: a family of protein-serine/threonine kinases that are activated and tyrosine phosphorylated in response to insulin and NGF. *Cell* 1991;**65**:663–75.
23. Gonzalez FA, Raden DL, Rigby MR, Davis RJ. Heterogeneous expression of four MAP kinase isoforms in human tissues. *FEBS Letts* 1992;**304**:170–8.
24. Coulombe P, Rodier G, Bonneil E, Thibault P, Meloche S. N-Terminal ubiquitination of extracellular signal-regulated kinase 3 and p21 directs their degradation by the proteasome. *Mol Cell Biol* 2004;**24**:6140–50.
25. Coulombe P, Rodier G, Pelletier S, Pellerin J, Meloche S. Rapid turnover of extracellular signal-regulated kinase 3 by the ubiquitin-proteasome pathway defines a novel paradigm of mitogen-activated

protein kinase regulation during cellular differentiation. *Mol Cell Biol* 2003;**23**:4542–58.

26. Aberg E, Perander M, Johansen B, Julien C, Meloche S, Keyse SM, Seternes OM. Regulation of MAPK-activated protein kinase 5 activity and subcellular localization by the atypical MAPK ERK4/MAPK4. *J Biol Chem* 2006;**281**:35,499–35,510.

27. Seternes OM, Mikalsen T, Johansen B, Michaelsen E, Armstrong CG, Morrice NA, Turgeon B, Meloche S, Moens U, Keyse SM. Activation of MK5/PRAK by the atypical MAP kinase ERK3 defines a novel signal transduction pathway. *EMBO J* 2004;**23**:4780–91.

28. Kant S, Schumacher S, Singh MK, Kispert A, Kotlyarov A, Gaestel M. Characterization of the atypical MAPK ERK4 and its activation of the MAPK-activated protein kinase MK5. *J Biol Chem* 2006;**281**:35,511–35,519.

29. Schumacher S, Laass K, Kant S, Shi Y, Visel A, Gruber AD, Kotlyarov A, Gaestel M. Scaffolding by ERK3 regulates MK5 in development. *EMBO J* 2004;**23**:4770–9.

30. English JM, Vanderbilt CA, Xu S, Marcus S, Cobb MH. Isolation of MEK5 and differential expression of alternatively spliced forms. *J Biol Chem* 1995;**270**:28,897–28,902.

31. Zhou G, Bao ZQ, Dixon JE. Components of a new human protein kinase signal transduction pathway. *J Biol Chem* 1995;**270**:12,665–12,669.

32. Kato Y, Kravche KK, Tapping RI, Han J, Ulevitch RJ, Lee JD. BMK1/ERK5 regulates serum-induced early gene expression through transcription factor MEF2C. *EMBO J* 1997;**16**:7054–66.

33. Watson FL, Heerssen HM, Bhattacharyya A, Klesse L, Lin MZ, Segal RA. Neurotrophins use the Erk5 pathway to mediate a retrograde survival response. *Nat Neurosci* 2001;**4**:981–8.

34. Kato Y, Tapping RI, Huang S, Watson MH, Ulevitch RJ, Lee JD. Bmk1/Erk5 is required for cell proliferation induced by epidermal growth factor. *Nature* 1998;**395**:713–16.

35. Regan CP, Li W, Boucher DM, Spatz S, Su MS, Kuida K. Erk5 null mice display multiple extraembryonic vascular and embryonic cardiovascular defects. *Proc Natl Acad Sci USA* 2002;**99**:9248–53.

36. Wang X, Merritt AJ, Seyfried J, Guo C, Papadakis ES, Finegan KG, Kayahara M, Dixon J, Boot-Handford RP, Cartwright EJ, Mayer U, Tournier C. Targeted deletion of mek5 causes early embryonic death and defects in the extracellular signal-regulated kinase 5/myocyte enhancer factor 2 cell survival pathway. *Mol Cell Biol* 2005;**25**:336–45.

37. Abe MK, Kuo WL, Hershenson MB, Rosner MR. Extracellular signal-regulated kinase 7 (ERK7), a novel ERK with a C- terminal domain that regulates its activity, its cellular localization, and cell growth. *Mol Cell Biol* 1999;**19**:1301–12.

38. Abe MK, Saelzler MP, Espinosa R, Kahle KT, Hershenson MB, Le Beau MM, Rosner MR. ERK8, a new member of the mitogen-activated protein kinase family. *J Biol Chem* 2002;**277**:16,733–16,743.

39. Abe MK, Kahle KT, Saelzler MP, Orth K, Dixon JE, Rosner MR. ERK7 is an autoactivated member of the MAPK family. *J Biol Chem* 2001;**276**:21,272–21,279.

40. Kuo WL, Duke CJ, Abe MK, Kaplan EL, Gomes S, Rosner MR. ERK7 expression and kinase activity is regulated by the ubiquitin-proteosome pathway. *J Biol Chem* 2004;**279**:23,073–23,081.

41. Qian Z, Okuhara D, Abe MK, Rosner MR. Molecular cloning and characterization of a mitogen-activated protein kinase-associated intracellular chloride channel. *J Biol Chem* 1999;**274**:1621–7.

42. Henrich LM, Smith JA, Kitt D, Errington TM, Nguyen B, Traish AM, Lannigan DA. Extracellular signal-regulated kinase 7, a regulator of hormone-dependent estrogen receptor destruction. *Mol Cell Biol* 2003;**23**:5979–88.

43. Brancho D, Tanaka N, Jaeschke A, Ventura JJ, Kelkar N, Tanaka Y, Kyuuma M, Takeshita T, Flavell RA, Davis RJ. Mechanism of p38 MAP kinase activation in vivo. *Genes Dev* 2003;**17**:1969–78.

44. Chang CI, Xu BE, Akella R, Cobb MH, Goldsmith EJ. Crystal structures of MAP kinase p38 complexed to the docking sites on its nuclear substrate MEF2A and activator MKK3b. *Mol Cell* 2002;**9**:1241–9.

45. Wilson KP, Fitzgibbon MJ, Caron PR, Griffith JP, Chen W, McCaffrey PG, Chambers SP, Su MS. Crystal structure of p38 mitogen-activated protein kinase. *J Biol Chem* 1996;**271**:27,696–27,700.

46. English JM, Cobb MH. Pharmacological inhibitors of MAPK pathways. *Trends Pharmacol Sci* 2002;**23**:40–5.

47. Lee JC, Laydon JT, McDonnell PC, Gallagher TF, Kumar S, Green D, McNulty D, Blumenthal MJ, Heys JR, Landvatter SW, et al. A protein kinase involved in the regulation of inflammatory cytokine biosynthesis. *Nature* 1994;**372**:739–46.

48. Han J, Jiang Y, Li Z, Kravchenko VV, Ulevitch RJ. Activation of the transcription factor MEF2C by the MAP kinase p38 in inflammation. *Nature* 1997;**386**:296–9.

49. Whitmarsh AJ, Yang SH, Su MS, Sharrocks AD, Davis RJ. Role of p38 and JNK mitogen-activated protein kinases in the activation of ternary complex factors. *Mol Cell Biol* 1997;**17**:2360–71.

50. Raingeaud J, Whitmarsh AJ, Barrett T, Derijard B, Davis RJ. MKK3- and MKK6-regulated gene expression is mediated by the p38 mitogen-activated protein kinase signal transduction pathway. *Mol Cell Biol* 1996;**16**:1247–55.

51. Kotlyarov A, Neininger A, Schubert C, Eckert R, Birchmeier C, Volk HD, Gaestel M. MAPKAP kinase 2 is essential for LPS-induced TNF-alpha biosynthesis. *Nat Cell Biol* 1999;**1**:94–7.

52. Neininger A, Kontoyiannis D, Kotlyarov A, Winzen R, Eckert R, Volk HD, Holtmann H, Kollias G, Gaestel M. MK2 targets AU-rich elements and regulates biosynthesis of tumor necrosis factor and interleukin-6 independently at different post- transcriptional levels. *J Biol Chem* 2002;**277**:3065–8.

53. Winzen R, Kracht M, Ritter B, Wilhelm A, Chen CY, Shyu AB, Muller M, Gaestel M, Resch K, Holtmann H. The p38 MAP kinase pathway signals for cytokine-induced mRNA stabilization via MAP kinase-activated protein kinase 2 and an AU-rich region-targeted mechanism. *EMBO J* 1999;**18**:4969–80.

54. Mahtani KR, Brook M, Dean JL, Sully G, Saklatvala J, Clark AR. Mitogen-activated protein kinase p38 controls the expression and posttranslational modification of tristetraprolin, a regulator of tumor necrosis factor alpha mRNA stability. *Mol Cell Biol* 2001;**21**:6461–9.

55. Adams RH, Porras A, Alonso G, Jones M, Vintersten K, Panelli S, Valladares A, Perez L, Klein R, Nebreda AR. Essential role of p38alpha MAP kinase in placental but not embryonic cardiovascular development. *Mol Cell* 2000;**6**:109–16.

56. Allen M, Svensson L, Roach M, Hambor J, McNeish J, Gabel CA. Deficiency of the stress kinase p38alpha results in embryonic lethality: characterization of the kinase dependence of stress responses of enzyme-deficient embryonic stem cells. *J Exp Med* 2000;**191**:859–70.

57. Tamura K, Sudo T, Senftleben U, Dadak AM, Johnson R, Karin M. Requirement for p38alpha in erythropoietin expression: a role for stress kinases in erythropoiesis. *Cell* 2000;**102**:221–31.

58. Perdiguero E, Ruiz-Bonilla V, Serrano AL, Munoz-Canoves P. Genetic deficiency of p38alpha reveals its critical role in myoblast cell cycle exit: the p38alpha-JNK connection. *Cell Cycle* 2007;**6**:1298–303.

59. Engel FB, Schebesta M, Duong MT, Lu G, Ren S, Madwed JB, Jiang H, Wang Y, Keating MT. p 38 MAP kinase inhibition enables proliferation of adult mammalian cardiomyocytes. *Genes Dev* 2005;**19**:1175–87.

60. Ventura JJ, Tenbaum S, Perdiguero E, Huth M, Guerra C, Barbacid M, Pasparakis M, Nebreda AR. p38alpha MAP kinase is essential in lung stem and progenitor cell proliferation and differentiation. *Nat Genet* 2007;**39**:750–8.

61. Hasegawa M, Cuenda A, Spillantini MG, Thomas GM, Buee-Scherrer V, Cohen P, Goedert M. Stress-activated protein kinase-3 interacts with the PDZ domain of alpha1-syntrophin. A mechanism for specific substrate recognition. *J Biol Chem* 1999;**274**:12,626–12,631.

62. Sabio G, Arthur JS, Kuma Y, Peggie M, Carr J, Murray-Tait V, Centeno F, Goedert M, Morrice NA, Cuenda A. p38gamma regulates the localisation of SAP97 in the cytoskeleton by modulating its interaction with GKAP. *EMBO J* 2005;**24**:1134–45.

63. Sabio G, Reuver S, Feijoo C, Hasegawa M, Thomas GM, Centeno F, Kuhlendahl S, Leal-Ortiz S, Goedert M, Garner C, Cuenda A. Stress- and mitogen-induced phosphorylation of the synapse-associated protein SAP90/PSD-95 by activation of SAPK3/p38gamma and ERK1/ERK2. *Biochem J* 2004;**380**:19–30.

64. Knebel A, Morrice N, Cohen P. A novel method to identify protein kinase substrates: eEF2 kinase is phosphorylated and inhibited by SAPK4/p38delta. *EMBO J* 2001;**20**:4360–9.

65. Davis RJ. Signal transduction by the JNK group of MAP kinases. *Cell* 2000;**103**:239–52.

66. Tournier C, Dong C, Turner TK, Jones SN, Flavell RA, Davis RJ. MKK7 is an essential component of the JNK signal transduction pathway activated by proinflammatory cytokines. *Genes Dev* 2001;**15**:1419–26.

67. Lawler S, Fleming Y, Goedert M, Cohen P. Synergistic activation of SAPK1/JNK1 by two MAP kinase kinases in vitro. *Curr Biol* 1998;**8**:1387–90.

68. Fleming Y, Armstrong CG, Morrice N, Paterson A, Goedert M, Cohen P. Synergistic activation of stress-activated protein kinase 1/c-Jun N-terminal kinase (SAPK1/JNK) isoforms by mitogen-activated protein kinase kinase 4 (MKK4) and MKK7. *Biochem J* 2000;**352**(1):145.

69. Gupta S, Barrett T, Whitmarsh AJ, Cavanagh J, Sluss HK, Derijard B, Davis RJ. Selective interaction of JNK protein kinase isoforms with transcription factors. *EMBO J* 1996;**15**:2760–70.

70. Xie X, Gu Y, Fox T, Coll JT, Fleming MA, Markland W, Caron PR, Wilson KP, Su MS. Crystal structure of JNK3: a kinase implicated in neuronal apoptosis. *Structure* 1998;**6**:983–91.

71. Tournier C, Hess P, Yang DD, Xu J, Turner TK, Nimnual A, Bar-Sagi D, Jones SN, Flavell RA, Davis RJ. Requirement of JNK for stress-induced activation of the cytochrome c- mediated death pathway. *Science* 2000;**288**:870–4.

72. Lei K, Nimnual A, Zong WX, Kennedy NJ, Flavell RA, Thompson CB, Bar-Sagi D, Davis RJ. The Bax subfamily of Bcl2-related proteins is essential for apoptotic signal transduction by c-Jun NH(2)-terminal kinase. *Mol Cell Biol* 2002;**22**:4929–42.

73. Ventura Jr JJ, Cogswell P, Flavell RA, Baldwin AS, Davis RJ. JNK potentiates TNF-stimulated necrosis by increasing the production of cytotoxic reactive oxygen species. *Genes Dev* 2004;**18**:2905–15.

74. Hess P, Pihan G, Sawyers CL, Flavell RA, Davis RJ. Survival signaling mediated by c-Jun NH(2)-terminal kinase in transformed B lymphoblasts. *Nat Genet* 2002;**32**:201–5.

75. Potapova O, Gorospe M, Dougherty RH, Dean NM, Gaarde WA, Holbrook NJ. Inhibition of c-Jun N-terminal kinase 2 expression suppresses growth and induces apoptosis of human tumor cells in a p53-dependent manner. *Mol Cell Biol* 2000;**20**:1713–22.

76. Dong C, Yang DD, Wysk M, Whitmarsh AJ, Davis RJ, Flavell RA. Defective T cell differentiation in the absence of Jnk1. *Science* 1998;**282**:2092–5.

77. Yang DD, Kuan CY, Whitmarsh AJ, Rincon M, Zheng TS, Davis RJ, Rakic P, Flavell RA. Absence of excitotoxicity-induced apoptosis in the hippocampus of mice lacking the Jnk3 gene. *Nature* 1997;**389**:865–70.

78. Yang DD, Conze D, Whitmarsh AJ, Barrett T, Davis RJ, Rincon M, Flavell RA. Differentiation of CD4+T cells to Th1 cells requires MAP kinase JNK2. *Immunity* 1998;**9**:575–85.

79. Sabapathy K, Hu Y, Kallunki T, Schreiber M, David JP, Jochum W, Wagner EF, Karin M. JNK2 is required for efficient T-cell activation and apoptosis but not for normal lymphocyte development. *Curr Biol* 1999;**9**:116–25.

80. Sabapathy K, Kallunki T, David JP, Graef I, Karin M, Wagner EF. c-Jun NH2-terminal kinase (JNK)1 and JNK2 have similar and stage-dependent roles in regulating T cell apoptosis and proliferation. *J Exp Med* 2001;**193**:317–28.

81. Conze D, Krahl T, Kennedy N, Weiss L, Lumsden J, Hess P, Flavell RA, Le Gros G, Davis RJ, Rincon M. c-Jun NH(2)-terminal kinase (JNK)1 and JNK2 have distinct roles in CD8(+) T cell activation. *J Exp Med* 2002;**195**:811–23.

82. Kuan CY, Yang DD, Samanta Roy DR, Davis RJ, Rakic P, Flavell RA. The Jnk1 and Jnk2 protein kinases are required for regional specific apoptosis during early brain development. *Neuron* 1999;**22**:667–76.

83. Sabapathy K, Jochum W, Hochedlinger K, Chang L, Karin M, Wagner EF. Defective neural tube morphogenesis and altered apoptosis in the absence of both JNK1 and JNK2. *Mech Dev* 1999;**89**:115–24.

84. Dong C, Yang DD, Tournier C, Whitmarsh AJ, Xu J, Davis RJ, Flavell RA. JNK is required for effector T-cell function but not for T-cell activation. *Nature* 2000;**405**:91–4.

85. Yang SH, Whitmarsh AJ, Davis RJ, Sharrocks AD. Differential targeting of MAP kinases to the ETS-domain transcription factor Elk-1. *EMBO J* 1998;**17**:1740–9.

86. Yang SH, Yates PR, Whitmarsh AJ, Davis RJ, Sharrocks AD. The Elk-1 ETS-domain transcription factor contains a mitogen-activated protein kinase targeting motif. *Mol Cell Biol* 1998;**18**:710–20.

87. Yang SH, Galanis A, Sharrocks AD. Targeting of p38 mitogen-activated protein kinases to MEF2 transcription factors. *Mol Cell Biol* 1999;**19**:4028–38.

88. Galanis A, Yang SH, Sharrocks AD. Selective targeting of MAPKs to the ETS domain transcription factor SAP- 1. *J Biol Chem* 2001;**276**:965–73.

89. Jacobs D, Glossip D, Xing H, Muslin AJ, Kornfeld K. Multiple docking sites on substrate proteins form a modular system that mediates recognition by ERK MAP kinase. *Genes Dev* 1999;**13**:163–75.

90. Tanoue T, Maeda R, Adachi M, Nishida E. Identification of a docking groove on ERK and p38 MAP kinases that regulates the specificity of docking interactions. *EMBO J* 2001;**20**:466–79.

91. Tanoue T, Yamamoto T, Nishida E. Modular structure of a docking surface on MAPK phosphatases. *J Biol Chem* 2002;**277**:22,942–22,949.

92. Tanoue T, Adachi M, Moriguchi T, Nishida E. A conserved docking motif in MAP kinases common to substrates, activators and regulators. *Nat Cell Biol* 2000;**2**:110–16.

93. Bardwell L, Thorner J. A conserved motif at the amino termini of MEKs might mediate high affinity interaction with the cognate MAPKs. *Trends Biochem Sci* 1996;**21**:373–4.

94. Duesbery NS, Webb CP, Leppla SH, Gordon VM, Klimpel KR, Copeland TD, Ahn NG, Oskarsson MK, Fukasawa K, Paull KD, Vande Woude GF. Proteolytic inactivation of MAP-kinase-kinase by anthrax lethal factor. *Science* 1998;**280**:734–7.

95. Lee T, Hoofnagle AN, Kabuyama Y, Stroud J, Min X, Goldsmith EJ, Chen L, Resing KA, Ahn NG. Docking motif interactions in MAP kinases revealed by hydrogen exchange mass spectrometry. *Mol Cell* 2004;**14**:43–55.

96. Liu S, Sun JP, Zhou B, Zhang ZY. Structural basis of docking inter-actions between ERK2 and MAP kinase phosphatase 3. *Proc Natl Acad Sci USA* 2006;**103**:5326–31.

97. Zhou T, Sun L, Humphreys J, Goldsmith EJ. Docking interac-tions induce exposure of activation loop in the MAP kinase ERK2. *Structure* 2006;**14**:1011–19.

98. Heo YS, Kim SK, Seo CI, Kim YK, Sung BJ, Lee HS, Lee JI, Park SY, Kim JH, Hwang KY, Hyun YL, Jeon YH, Ro S, Cho JM, Lee TG, Yang CH. Structural basis for the selective inhibition of JNK1 by the scaffold-ing protein JIP1 and SP600125. *EMBO J* 2004;**23**:2185–95.

99. Remenyi A, Good MC, Bhattacharyya RP, Lim WA. The role of docking interactions in mediating signaling input, output, and dis-crimination in the yeast MAPK network. *Mol Cell* 2005;**20**:951–62.

100. Weston CR, Lambright DG, Davis RJ. Signal transduction. MAP kinase signaling specificity. *Science* 2002;**296**:2345–7.

101. Xia Y, Wu Z, Su B, Murray B, Karin M. JNKK1 organizes a MAP kinase module through specific and sequential interactions with upstream and downstream components mediated by its amino-termi-nal extension. *Genes Dev* 1998;**12**:3369–81.

102. Cheng J, Yang J, Xia Y, Karin M, Su B. Synergistic interaction of MEK kinase 2, c-Jun N-terminal kinase (JNK) kinase 2, and JNK1 results in efficient and specific JNK1 activation. *Mol Cell Biol* 2000;**20**:2334–42.

103. Karandikar M, Xu S, Cobb MH. MEKK1 binds Raf-1 and the ERK2 cascade components. *J Biol Chem* 2000;**275**:40,120–40,127.

104. Gallagher ED, Xu S, Moomaw C, Slaughter CA, Cobb MH. Binding of JNK/SAPK to MEKK1 is regulated by phosphorylation. *J Biol Chem* 2002;**12**:12.

105. Whitmarsh AJ, Davis RJ. Structural organization of MAP-kinase signaling modules by scaffold proteins in yeast and mammals. *Trends Biochem Sci* 1998;**23**:481–5.

106. Marti A, Luo Z, Cunningham C, Ohta Y, Hartwig J, Stossel TP, Kyriakis JM, Avruch J. Actin-binding protein-280 binds the stress-activated protein kinase (SAPK) activator SEK-1 and is required for tumor necrosis factor-alpha activation of SAPK in melanoma cells. *J Biol Chem* 1997;**272**:2620–8.

107. Girardin SE, Yaniv M. A direct interaction between JNK1 and CrkII is critical for Rac1- induced JNK activation. *EMBO J* 2001;**20**:3437–46.

108. Holmberg C, Katz S, Lerdrup M, Herdegen T, Jaattela M, Aronheim A, Kallunki T. A novel specific role for I kappa B kinase com-plex-associated protein in cytosolic stress signaling. *J Biol Chem* 2002;**277**:31,918–31,928.

109. Ishibe S, Joly D, Zhu X, Cantley LG. Phosphorylation-dependent paxillin-ERK association mediates hepatocyte growth factor-stimu-lated epithelial morphogenesis. *Mol Cell* 2003;**12**:1275–85.

110. Morrison DK, Davis RJ. Regulation of MAP kinase signaling mod-ules by scaffold proteins in mammals. *Annu Rev Cell Dev Biol* 2003;**19**:91–118.

111. Park SH, Zarrinpar A, Lim WA. Rewiring MAP kinase path-ways using alternative scaffold assembly mechanisms. *Science* 2003;**299**:1061–4.

112. Harris K, Lamson RE, Nelson B, Hughes TR, Marton MJ, Roberts CJ, Boone C, Pryciak PM. Role of scaffolds in MAP kinase pathway specificity revealed by custom design of pathway-dedicated signal-ing proteins. *Curr Biol* 2001;**11**:1815–24.

113. Bhattacharyya RP, Remenyi A, Good MC, Bashor CJ, Falick AM, Lim WA. The Ste5 scaffold allosterically modulates signaling output of the yeast mating pathway. *Science* 2006;**311**:822–6.

114. Morrison DK. KSR: a MAPK scaffold of the Ras pathway?. *J Cell Sci* 2001;**114**:1609–12.

115. Roy F, Therrien M. MAP Kinase Module: The Ksr Connection. *Curr Biol* 2002;**12**:R325–7.

116. Michaud NR, Therrien M, Cacace A, Edsall LC, Spiegel S, Rubin GM, Morrison DK. KSR stimulates Raf-1 activity in a kinase-inde-pendent manner. *Proc Natl Acad Sci USA* 1997;**94**:12,792–12,796,.

117. Muller J, Ory S, Copeland T, Piwnica-Worms H, Morrison DK. C-TAK1 regulates Ras signaling by phosphorylating the MAPK scaffold, KSR1. *Mol Cell* 2001;**8**:983–93.

118. Nguyen A, Burack WR, Stock JL, Kortum R, Chaika OV, Afkarian M, Muller WJ, Murphy KM, Morrison DK, Lewis RE, McNeish J, Shaw AS. Kinase suppressor of Ras (KSR) is a scaffold which facili-tates mitogen- activated protein kinase activation *in vivo*. *Mol Cell Biol* 2002;**22**:3035–45.

119. Roy F, Laberge G, Douziech M, Ferland-McCollough D, Therrien M. KSR is a scaffold required for activation of the ERK/MAPK module. *Genes Dev* 2002;**16**:427–38.

120. Schaeffer HJ, Catling AD, Eblen ST, Collier LS, Krauss A, Weber MJ. MP1: a MEK binding partner that enhances enzymatic activa-tion of the MAP kinase cascade. *Science* 1998;**281**:1668–71.

121. Wunderlich W, Fialka I, Teis D, Alpi A, Pfeifer A, Parton RG, Lottspeich F, Huber LA. A novel 14-kilodalton protein interacts with the mitogen-activated protein kinase scaffold mp1 on a late endo-somal/lysosomal compartment. *J Cell Biol* 2001;**152**:765–76.

122. Pullikuth A, McKinnon E, Schaeffer HJ, Catling AD. The MEK1 scaffolding protein MP1 regulates cell spreading by integrating PAK1 and Rho signals. *Mol Cell Biol* 2005;**25**:5119–33.

123. Vomastek T, Schaeffer HJ, Tarcsafalvi A, Smolkin ME, Bissonette EA, Weber MJ. Modular construction of a signaling scaffold: MORG1 interacts with components of the ERK cascade and links ERK signaling to specific agonists. *Proc Natl Acad Sci USA* 2004;**101**:6981–6.

124. Hopfer U, Hopfer H, Jablonski K, Stahl RA, Wolf G. The novel WD-repeat protein Morg1 acts as a molecular scaffold for hypoxia-inducible factor prolyl hydroxylase 3 (PHD3). *J Biol Chem* 2006;**281**:8645–55.

125. Yasuda J, Whitmarsh AJ, Cavanagh J, Sharma M, Davis RJ. The JIP group of mitogen-activated protein kinase scaffold proteins. *Mol Cell Biol* 1999;**19**:7245–54.

126. Whitmarsh AJ, Cavanagh J, Tournier C, Yasuda J, Davis RJ. A mammalian scaffold complex that selectively mediates MAP kinase activation. *Science* 1998;**281**:1671–4.

127. Bonny C, Nicod P, Waeber G. IB1, a JIP-1-related nuclear protein present in insulin-secreting cells. *J Biol Chem* 1998;**273**:1843–6.

128. Negri S, Oberson A, Steinmann M, Sauser C, Nicod P, Waeber G, Schorderet DF, Bonny C. cDNA cloning and mapping of a novel islet-brain/JNK-interacting protein. *Genomics* 2000;**64**:3.

129. Meyer D, Liu A, Margolis B. Interaction of c-Jun amino-terminal kinase interacting protein-1 with p190 rhoGEF and its localization in differentiated neurons. *J Biol Chem* 1999;**274**:35,113–35,118.

130. Gotthardt M, Trommsdorff M, Nevitt MF, Shelton J, Richardson JA, Stockinger W, Nimpf J, Herz J. Interactions of the low density lipo-protein receptor gene family with cytosolic adaptor and scaffold pro-teins suggest diverse biological functions in cellular communication and signal transduction. *J Biol Chem* 2000;**275**:25,616–25,624.

131. Stockinger W, Brandes C, Fasching D, Hermann M, Gotthardt M, Herz J, Schneider WJ, Nimpf J. The reelin receptor ApoER2 recruits JNK-interacting proteins-1 and -2. *J Biol Chem* 2000;**275**:25,625–25,632.

132. Schoorlemmer J, Goldfarb M. FGF homologous factors and the islet brain-2 scaffold protein regulate activation of a stress-activated protein kinase. *J Biol Chem* 2002;**18**:18.

133. Schoorlemmer J, Goldfarb M. Fibroblast growth factor homologous factors are intracellular signaling proteins. *Curr Biol* 2001;**11**:793–7.

134. Buchsbaum RJ, Connolly BA, Feig LA. Interaction of Rac exchange factors Tiam1 and Ras-GRF1 with a scaffold for the p38 mitogen-activated protein kinase cascade. *Mol Cell Biol* 2002;**22**:4073–85.

135. Kelkar N, Gupta S, Dickens M, Davis RJ. Interaction of a mitogen-activated protein kinase signaling module with the neuronal protein JIP3. *Mol Cell Biol* 2000;**20**:1030–43.

136. Ito M, Yoshioka K, Akechi M, Yamashita S, Takamatsu N, Sugiyama K, Hibi M, Nakabeppu Y, Shiba T, Yamamoto KI. JSAP1, a novel jun N-terminal protein kinase (JNK)-binding protein that functions as a Scaffold factor in the JNK signaling pathway. *Mol Cell Biol* 1999;**19**:7539–48.

137. Lee CM, Onesime D, Reddy CD, Dhanasekaran N, Reddy EP. JLP: A scaffolding protein that tethers JNK/p38MAPK signaling modules and transcription factors. *Proc Natl Acad Sci USA* 2002;**99**:14,189–14,194.

138. Kelkar N, Standen CL, Davis RJ. Role of the JIP4 scaffold protein in the regulation of mitogen-activated protein kinase signaling pathways. *Mol Cell Biol* 2005;**25**:2733–43.

139. Kim JW, Kim MJ, Kim KJ, Yun HJ, Chae JS, Hwang SG, Chang TS, Park HS, Lee KW, Han PL, Cho SG, Kim TW, Choi EJ. Notch interferes with the scaffold function of JNK–interacting protein 1 to inhibit the JNK signaling pathway. *Proc Natl Acad Sci USA* 2005;**102**:14,308–14,313.

140. Kim AH, Yano H, Cho H, Meyer D, Monks B, Margolis B, Birnbaum MJ, Chao MV. Akt1 regulates a JNK scaffold during excitotoxic apoptosis. *Neuron* 2002;**35**:697–709.

141. Verhey KJ, Meyer D, Deehan R, Blenis J, Schnapp BJ, Rapoport TA, Margolis B. Cargo of kinesin identified as JIP scaffolding proteins and associated signaling molecules. *J Cell Biol* 2001;**152**:959–70.

142. Whitmarsh AJ, Kuan CY, Kennedy NJ, Kelkar N, Haydar TF, Mordes JP, Appel M, Rossini AA, Jones SN, Flavell RA, Rakic P, Davis RJ. Requirement of the JIP1 scaffold protein for stress-induced JNK activation. *Genes Dev* 2001;**15**:2421–32.

143. Bowman AB, Kamal A, Ritchings BW, Philp AV, McGrail M, Gindhart JG, Goldstein LS. Kinesin-dependent axonal transport is mediated by the sunday driver (SYD) protein. *Cell* 2000;**103**:583–94.

144. Nguyen Q, Lee CM, Le A, Reddy EP. JLP associates with kinesin light chain 1 through a novel leucine zipper-like domain. *J Biol Chem* 2005;**280**:30,185–30,191.

145. Thompson NA, Haefliger JA, Senn A, Tawadros T, Magara F, Ledermann B, Nicod P, Waeber G. Islet-brain1/JNK-interacting protein-1 is required for early embryogenesis in mice. *J Biol Chem* 2001;**276**:27,745–27,748.

146. Im JY, Lee KW, Kim MH, Lee SH, Ha HY, Cho IH, Kim D, Yu MS, Kim JB, Lee JK, Kim YJ, Youn BW, Yang SD, Shin HS, Han PL. Repression of phospho-JNK and infarct volume in ischemic brain of JIP1-deficient mice. *J Neurosci Res* 2003;**74**:326–32.

147. Jaeschke A, Czech MP, Davis RJ. An essential role of the JIP1 scaffold protein for JNK activation in adipose tissue. *Genes Dev* 2004;**18**:1976–80.

148. Kennedy NJ, Martin G, Ehrhardt AG, Cavanagh-Kyros J, Kuan C-Y, Rakic P, Flavell RA, Treistman SN, Davis RJ. *Requirement of JIP scaffold proteins for NMDA-mediated signal transduction.* 2007;**21**:2336–46.

149. Kelkar N, Delmotte MH, Weston CR, Barrett T, Sheppard BJ, Flavell RA, Davis RJ. Morphogenesis of the telencephalic commissure requires scaffold protein JNK-interacting protein 3 (JIP3). *Proc Natl Acad Sci USA* 2003;**100**:9843–8.

150. Ha HY, Cho IH, Lee KW, Lee KW, Song JY, Kim KS, Yu YM, Lee JK, Song JS, Yang SD, Shin HS, Han PL. The axon guidance defect of the telencephalic commissures of the JSAP1-deficient brain was partially rescued by the transgenic expression of JIP1. *Dev Biol* 2005;**277**:184–99.

151. Kristensen O, Guenat S, Dar I, Allaman-Pillet N, Abderrahmani A, Ferdaoussi M, Roduit R, Maurer F, Beckmann JS, Kastrup JS, Gajhede M, Bonny C. A unique set of SH3-SH3 interactions controls IB1 homodimerization. *EMBO J* 2006;**25**:785–97.

152. Tapon N, Nagata K, Lamarche N, Hall A. A new rac target POSH is an SH3-containing scaffold protein involved in the JNK and NF-kappaB signalling pathways. *EMBO J* 1998;**17**:1395–404.

153. Wilhelm M, Kukekov NV, Xu Z, Greene LA. Identification of POSH2, a novel homologue of the c-Jun N-terminal kinase scaffold protein POSH. *Dev Neurosci* 2007;**29**:355–62.

154. Xu Z, Kukekov NV, Greene LA. POSH acts as a scaffold for a multiprotein complex that mediates JNK activation in apoptosis. *EMBO J* 2003;**22**:252–61.

155. Xu Z, Kukekov NV, Greene LA. Regulation of apoptotic c-Jun N-terminal kinase signaling by a stabilization-based feed-forward loop. *Mol Cell Biol* 2005;**25**:9949–59.

156. Kukekov NV, Xu Z, Greene LA. Direct interaction of the molecular scaffolds POSH and JIP is required for apoptotic activation of JNKs. *J Biol Chem* 2006;**281**:15,517–15,524.

157. Miller WE, Lefkowitz RJ. Expanding roles for beta-arrestins as scaffolds and adapters in GPCR signaling and trafficking. *Curr Opin Cell Biol* 2001;**13**:139–45.

158. Luttrell LM, Roudabush FL, Choy EW, Miller WE, Field ME, Pierce KL, Lefkowitz RJ. Activation and targeting of extracellular signal-regulated kinases by beta-arrestin scaffolds. *Proc Natl Acad Sci USA* 2001;**98**:2449–54.

159. Miller WE, McDonald PH, Cai SF, Field ME, Davis RJ, Lefkowitz RJ. Identification of a motif in the carboxyl terminus of beta -arrestin2 responsible for activation of JNK3. *J Biol Chem* 2001;**276**:27,770–27,777.

160. McDonald PH, Chow CW, Miller WE, Laporte SA, Field ME, Lin FT, Davis RJ, Lefkowitz RJ. Beta-arrestin 2: a receptor-regulated MAPK scaffold for the activation of JNK3. *Science* 2000;**290**:1574–7.

161. Bohn LM, Lefkowitz RJ, Gainetdinov RR, Peppel K, Caron MG, Lin FT. Enhanced morphine analgesia in mice lacking beta-arrestin 2. *Science* 1999;**286**:2495–8.

162. Zawistowski JS, Stalheim L, Uhlik MT, Abell AN, Ancrile BB, Johnson GL, Marchuk DA. CCM1 and CCM2 protein interactions in cell signaling: implications for cerebral cavernous malformations pathogenesis. *Hum Mol Genet* 2005;**14**:2521–31.

163. Uhlik MT, Abell AN, Johnson NL, Sun W, Cuevas BD, Lobel-Rice KE, Horne EA, Dell'Acqua ML, Johnson GL. Rac-MEKK3-MKK3 scaffolding for p38 MAPK activation during hyperosmotic shock. *Nat Cell Biol* 2003;**5**:1104–10.

164. Alonso A, Merlo JJ, Na S, Kholod N, Jaroszewski L, Kharitonenkov A, Williams S, Godzik A, Posada JD, Mustelin T. Inhibition of T cell antigen receptor signaling by VHR-related MKPX (VHX), a new dual specificity phosphatase related to VH1 related (VHR). *J Biol Chem* 2002;**277**:5524–8.

165. Zama T, Aoki R, Kamimoto T, Inoue K, Ikeda Y, Hagiwara M. A novel dual specificity phosphatase SKRP1 interacts with the MAPK kinase MKK7 and inactivates the JNK MAPK pathway. Implication for the precise regulation of the particular MAPK pathway. *J Biol Chem*, 2002;**277**:23,909–23,918.

166. Zama T, Aoki R, Kamimoto T, Inoue K, Ikeda Y, Hagiwara M. Scaffold role of a mitogen-activated protein kinase phosphatase, SKRP1, for the JNK signaling pathway. *J Biol Chem* 2002;**277**:23,919–23,926.

167. Chen AJ, Zhou G, Juan T, Colicos SM, Cannon JP, Cabriera-Hansen M, Meyer CF, Jurecic R, Copeland NG, Gilbert DJ, Jenkins NA, Fletcher F, Tan TH, Belmont JW. The dual specificity JKAP specifically activates the c-Jun N-terminal kinase pathway. *J Biol Chem* 2002;**277**:36,592–36,601.

168. Shen Y, Luche R, Wei B, Gordon ML, Diltz CD, Tonks NK. Activation of the Jnk signaling pathway by a dual-specificity phosphatase, JSP-1. *Proc Natl Acad Sci USA* 2001;**98**:13,613–13,618.

169. Beardmore VA, Hinton HJ, Eftychi C, Apostolaki M, Armaka M, Darragh J, McIlrath J, Carr JM, Armit LJ, Clacher C, Malone L, Kollias G, Arthur JS. Generation and characterization of p38beta (MAPK11) gene-targeted mice. *Mol Cell Biol* 2005;**25**:10,454–10,464.

170. Kuan CY, Whitmarsh AJ, Yang DD, Liao G, Schloemer AJ, Dong C, Bao J, Banasiak KJ, Haddad GG, Flavell RA, Davis RJ, Rakic P. A critical role of neural-specific JNK3 for ischemic apoptosis. *Proc Natl Acad Sci USA* 2003;**100**:15,184–15,189.

171. Giroux S, Tremblay M, Bernard D, Cardin-Girard JF, Aubry S, Larouche L, Rousseau S, Huot J, Landry J, Jeannotte L, Charron J. Embryonic death of Mek1-deficient mice reveals a role for this kinase in angiogenesis in the labyrinthine region of the placenta. *Curr Biol* 1999;**9**:369–72.

172. Belanger LF, Roy S, Tremblay M, Brott B, Steff AM, Mourad W, Hugo P, Erikson R, Charron J. Mek2 is dispensable for mouse growth and development. *Mol Cell Biol* 2003;**23**:4778–87.

173. Tanaka N, Kamanaka M, Enslen H, Dong C, Wysk M, Davis RJ, Flavell RA. Differential involvement of p38 mitogen-activated protein kinase kinases MKK3 and MKK6 in T-cell apoptosis. *EMBO Rep* 2002;**3**:785–91.

174. Lu HT, Yang DD, Wysk M, Gatti E, Mellman I, Davis RJ, Flavell RA. Defective IL-12 production in mitogen-activated protein (MAP) kinase kinase 3 (Mkk3)-deficient mice. *EMBO J* 1999;**18**:1845–57.

175. Wysk M, Yang DD, Lu HT, Flavell RA, Davis RA. Requirement of mitogen-activated protein kinase kinase 3 (MKK3) for tumor necrosis factor-induced cytokine expression. *Proc Natl Acad Sci USA* 1999;**96**:3763–8.

176. Yang D, Tournier C, Wysk M, Lu HT, Xu J, Davis RJ, Flavell RA. Targeted disruption of the MKK4 gene causes embryonic death, inhibition of c-Jun NH2-terminal kinase activation, and defects in AP-1 transcriptional activity. *Proc Natl Acad Sci USA* 1997;**94**:3004–9.

177. Nishina H, Vaz C, Billia P, Nghiem M, Sasaki T, De la Pompa JL, Furlonger K, Paige C, Hui C, Fischer KD, Kishimoto H, Iwatsubo T, Katada T, Woodgett JR, Penninger JM. Defective liver formation and liver cell apoptosis in mice lacking the stress signaling kinase SEK1/MKK4. *Development* 1999;**126**:505–16.

178. Nishina H, Fischer KD, Radvanyi L, Shahinian A, Hakem R, Rubie EA, Bernstein A, Mak TW, Woodgett JR, Penninger JM. Stress-signalling kinase Sek1 protects thymocytes from apoptosis mediated by CD95 and CD3. *Nature* 1997;**385**:350–3.

179. Nishina H, Bachmann M, Oliveira-dos-Santos AJ, Kozieradzki I, Odermatt B, Wakeham A, Shahinian A, Takimoto H, Bernstein A, Mak TW, Woodgett JR, Ohashi PS, Penninger JM. Impaired CD28-mediated interleukin 2 production and proliferation in stress kinase SAPK/ERK1 kinase (SEK1)/mitogen-activated protein kinase kinase 4 (MKK4)-deficient T lymphocytes. *J Exp Med* 1997;**186**:941–53.

180. Nishina H, Radvanyi L, Raju K, Sasaki T, Kozieradzki I, Penninger JM. Impaired TCR-mediated apoptosis and Bcl-XL expression in T cells lacking the stress kinase activator SEK1/MKK4. *J Immunol* 1998;**161**:3416–20.

181. Ganiatsas S, Kwee L, Fujiwara Y, Perkins A, Ikeda T, Labow MA, Zon LI. SEK1 deficiency reveals mitogen-activated protein kinase cascade crossregulation and leads to abnormal hepatogenesis. *Proc Natl Acad Sci USA* 1998;**95**:6881–6.

182. Swat W, Fujikawa K, Ganiatsas S, Yang D, Xavier RJ, Harris L, Davidson NL, Ferrini L, Davis RJ, Labow MA, Flavell RA, Zon LI, Alt FW. SEK1/MKK4 is required for maintenance of a normal peripheral lymphoid compartment but not for lymphocyte development. *Immunity* 1998;**8**:625–34.

183. Sasaki T, Wada T, Kishimoto H, Irie-Sasaki J, Matsumoto G, Goto T, Yao Z, Wakeham A, Mak TW, Suzuki A, Cho SK, Zuniga-Pflucker JC, Oliveira-dos-Santos AJ, Katada T, Nishina H, Penninger JM. The stress kinase mitogen-activated protein kinase kinase (MKK)7 is a negative regulator of antigen receptor and growth factor receptor- induced proliferation in hematopoietic cells. *J Exp Med* 2001;**194**:757–68.

184. Wada T, Joza N, Cheng HY, Sasaki T, Kozieradzki I, Bachmaier K, Katada T, Schreiber EM, Wagner F, Nishina H, Penninger JM. MKK7 couples stress signalling to G2/M cell-cycle progression and cellular senescence. *Nat Cell Biol* 2004;**6**:215–26.

Chapter 8

The Negative Regulation of JAK/STAT Signaling

James M. Murphy, Gillian M. Tannahill, Douglas J. Hilton and Christopher J. Greenhalgh

Division of Molecular Medicine, Walter and Eliza Hall Institute of Medical Research, Parkville, Victoria, Australia

INTRODUCTION

Cytokines are a large family of protein messengers, which are classified into four families based on their structures: hematopoietin, interferon, chemokine, and tumor necrosis factor (TNF). These proteins elicit diverse cellular responses upon binding to cognate transmembrane receptors that are expressed on the surface of target cells. The hematopoietin cytokines, characterized structurally by a four-helical bundle arranged in an up–up–down–down topology [1], are of particular interest to our laboratory, and thus their signaling will be the focus of this chapter. The recognition of hematopoietin cytokines by their cognate receptors induces oligomerization of receptor subunits (or reorientation of receptor subunits within a constitutive oligomer) to transmit a signal from outside to within the cell. Typically, the intracellular portions of hematopoietin receptors lack any intrinsic kinase activity and are therefore dependent on receptor-associated effectors, such as Janus kinases (JAKs), to initiate signaling pathways. Receptor reorientation or oligomerization leads to transphosphorylation and thus activation of JAK molecules that are constitutively associated with the receptor subunit intracellular domains. Activated JAKs phosphorylate tyrosine residues within the intracellular portion of the receptor to generate docking sites for numerous phosphotyrosine recognition domain-containing signaling effectors. In particular, the signal transducers and activators of transcription (STAT) proteins are well-characterized as downstream effectors of JAK signaling. STATs are recruited to the phosphorylated receptor *via* their Src Homology-2 (SH2) phosphotyrosine recognition domain, localizing the STAT in proximity of the activated JAK. JAK-mediated STAT phosphorylation facilitates STAT dimerization [2] or reorientation of constitutive STAT dimers [3] to enable transport to the cell nucleus where they serve as transcription factors that regulate transcription of a suite of genes.

Aberrant cytokine receptor signaling is the basis of many mammalian diseases, as exemplified by the implication of the constitutively-activated V617F mutant JAK2 in the human leukemia, polycythemia vera [4, 5]. Normal cytokine signaling critically depends on a finely-tuned balance between cytokine-dependent signals and mechanisms to antagonize these signals to allow a cell to be resensitized to further extracellular cues. Studies over the past decade have elucidated several mechanisms by which cytokine signaling through the JAK/STAT pathway is negatively regulated. Herein, we review the current knowledge of how cytokine signaling is attenuated by negative regulation of JAK signaling by phosphorylation/dephosphorylation of JAKs, inhibition of STATs by Protein Inhibitors of Activated STAT (PIAS) proteins, and the inhibitory role of Suppressor of Cytokine Signaling (SOCS) family proteins.

Attenuation of JAK/STAT Signaling by Phosphatases

Tyrosine phosphorylation is the primary process by which the signals arising from cytokine receptor activation are conveyed within the cell. The engagement of cytokine orientates receptor subunits to facilitate transphosphorylation of the JAKs, affecting 13 tyrosine residues over JAK Homology (JH) domains JH1, JH2, JH6, and JH7 in the case of JAK2 [6]. Critically, a key pair of tyrosines within the activation loops of JAKs, which are conserved throughout the JAK family (Y1022/Y1023 in JAK1; Y1007/Y1008 in JAK2; Y980/Y981 in JAK3; Y1054/Y1055 in Tyk2), are subject to transphosphorylation. Phosphorylation of the former of these two tyrosines upregulates the tyrosine kinase activity of the JAKs [7–10], presumably owing to (1) the introduced phosphate moiety precluding tyrosine occupation of the JAK active site, thereby relieving autoinhibition, and (2) the phosphotyrosine mediating ionic interactions with helix αC residues in the N-terminal lobe of the kinase domain to maintain an active conformation [11, 12]. The N-terminal FERM (band 4.1, ezrin, radixin and moesin) domain (JH4–7) of JAKs has been implicated in receptor association, since JAK3 FERM domain mutations identified in severe combined immunodeficiency (SCID) patients prevent JAK3:receptor interaction and subsequent JAK3 activation [13, 14]. Recent studies have deduced that auto- or trans-phosphorylation of Tyr119 within the JAK2 FERM domain, following activation of the erythropoietin (EPO) receptor, imposed another level of JAK2 negative regulation by inducing dissociation from the receptor and subsequent JAK2 degradation [15].

Due to the importance of JAK tyrosine phosphorylation for signal transduction – especially within the activation loop – it follows that dephosphorylation of tyrosine residues is a mechanism employed within the cell to downregulate JAK kinase activity. To this end, several protein phosphatases have been identified as modulators of JAK/STAT signaling over the past decade: the SH2

domain-containing tyrosine phosphatases (SHPs); protein tyrosine phosphatase 1B (PTP1B) and T-cell protein tyrosine phosphatase (TC-PTP); protein tyrosine phosphatase basophil-like (PTP-BL); and the transmembrane protein, CD45 (Figure 8.1, left panel). Each of these phosphatases exhibits constitutive activity with the capacity to dephosphorylate the JAK activation segment and, in the case of the SHPs and PTP1B, the ability to dephosphorylate STATs.

The first phosphatase modulators of JAK/STAT signaling to be identified were the SHPs: SHP-1, also known as SHP, hematopoietic cell phosphatase (HCP), or PTP1C; and SHP-2, also known as Syp, PTP2C, or PTP1D. SHP-1 is primarily expressed in hematopoietic cells and some epithelial tissues [16], whilst SHP-2 is ubiquitously expressed [17].

These phosphatases are named for the two SH2 domains located N-terminal to the tyrosine phosphatase domain. The SH2 domains play a central role in SHP recruitment to targets, and recent studies using degenerate peptide libraries have provided insights into the respective binding specificities of the SHP-1 and SHP-2 N- and C-terminal SH2 domains [18–21]. Predominantly, the SH2 domains mediate SHP recruitment to tyrosine-phosphorylated intracellular portions of many receptors (reviewed in [21]), in addition to phospho-STAT5 [22], consistent

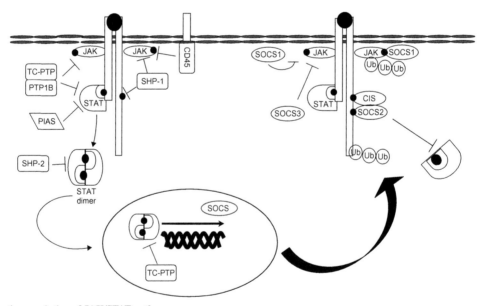

FIGURE 8.1 Negative regulation of JAK/STAT pathway.
JAKS (JAK1–3 and TYK2) are constitutively associated with a proline-rich membrane-proximal domain of these receptors, often referred to as Box1/Box2 region. Upon ligand stimulation, receptors undergo conformational changes and oligomerize. This in turn activates JAKs, which subsequently transphosphorylate each other and the cytoplasmic domain of the receptor. The JAKs mediate phosphorylation of specific receptor tyrosine residues to create docking sites for the Src-homology 2 (SH2) domains of the signal transducer and activator of transcription (STATs). STAT recruitment is followed by tyrosine (and serine) phosphorylation on key residues, by JAKs (and other closely associated kinases). The activated STATs (STAT1–5, and STAT6A and STAT6B) translocate to the nucleus, where they function as transcription factors to regulate gene expression. There are many biochemical processes, which regulate JAK/STAT signaling. *Left panel*: The phosphatases, SHP-1, SHP-2, CD45, PTP-1B, and TC-PTP, control the phosphorylation status and consequent activity of receptors, JAKs and STATs. The PIAS family proteins modulate STAT activity via many mechanisms, including sumoylation/sequestration, and occluding DNA binding. *Right panel*: The Suppressor of Cytokine Signaling (SOCS) genes were identified in 1997 (102–105). These genes are upregulated by STATs and encode the SOCS proteins, which act in a negative feedback loop to suppress further cytokine signaling. SOCS proteins are known to inhibit signalling by (1) inhibiting JAK activity, (2) competing for binding of the receptor with STATs, and (3) causing proteasomal degradation of signaling intermediates.

with the notion that SHPs antagonize JAK/STAT signaling both by occluding interactions with other effectors and by catalyzing the elimination of the phosphotyrosine motifs that serve as docking sites for effectors. Interestingly, the SHP-2 SH2 domains exhibit overlapping binding specificities with the SH2 domain of another JAK/STAT signaling antagonist, SOCS3, suggesting both negative regulators compete for binding sites on phosphorylated receptors, including those for the growth factors, interleukin-6, leptin, and EPO [19, 23–26].

The interaction of the SHP SH2 domains with their binding partners is a key step in the activation of the intrinsic phosphatase activity, since the SH2 domains occupy the phosphatase active site in an intramolecular fashion and thereby negatively regulate the enzyme [27]. SHP-1 is a key negative regulator of receptor signaling in hematopoietic and epidermal cells, and has been implicated in dephosphorylating the receptors for EPO [28], interleukin-3 (IL-3) [29, 30], granulocyte colony-stimulating factor (G-CSF) [31], colony-stimulating factor-1 (CSF-1) [32], Kit ligand/stem cell factor (SCF) [33], and epidermal growth factor (EGF) [34], and the JAK family members, JAK1 [35] and JAK2 [28, 36]. In contrast, SHP-2 exhibits different substrate specificities, which have been attributed to intrinsic differences in its catalytic domain [34, 37]. The role of SHP-2 in growth factor signaling appears to be much more complicated, and is therefore less well understood. As for SHP-1, SHP-2 has been shown to associate with several growth factor receptors [38–40] and JAKs [41, 42], although, in contrast to SHP-1, few of these interactors appear to be SHP-2 substrates. One of few examples to date is the IL-3 β-receptor, which was identified as a target capable of recruiting SHP-2 via its SH2 domains and was subject to Y612 dephosphorylation by the SHP-2 phosphatase domain [38, 43]. SHP-2 appears to play an additional role as an adaptor and, in addition to the dual SH2 domains, features several phosphotyrosine motifs capable of recruiting downstream effectors. In fact, JAK1 and JAK2 were reported to phosphorylate Y304 and Y327 of SHP-2 [41], although it appears more likely that JAK/STAT signaling activates a downstream kinase that is responsible for SHP-2 phosphorylation [42]. Interestingly, in contrast to SHP-1, SHP-2 does not dephosphorylate JAKs but serves as a STAT5 phosphatase [44, 45], and thus negative regulates JAK/STAT signaling in a complementary manner.

The critical biological roles of SHP-1 and SHP-2 have been elucidated by analyses of mouse mutant and knockout models. Spontaneous mutations in the *Shp-1* gene give rise to the moth-eaten phenotype, in which mice have patchy hair and exhibit immunodeficiency, autoimmunity, elevated erythropoiesis, and tissue infiltration of granulocytes, macrophages, and lymphocytes [46, 47]. A targeted mouse mutant of *SHP-2*, lacking exon 3, resulted in embryonic lethality at midgestation in homozygotes [48], whilst embryonic stem cells – homozygous for the mutant *SHP-2*

lacking exon 3 – exhibited suppressed differentiation into erythroid/myeloid progenitors *in vitro* [49]. Intriguingly, the $Shp\text{-}1^{-/-}/Shp\text{-}2^{-/-}$ double knockout was embryonic lethal, although analysis of yolk sacs indicated an elevation in erythroid and erythroid-myeloid lineage progenitors compared to the $Shp\text{-}2^{-/-}$ single knockout [50]. Overall, SHP-1 has been described predominantly as a negative regulator of hematopoietic cell signaling. However, further studies will be necessary to definitively define the role of SHP-2 in hematopoiesis.

The phosphatases, PTP1B and TC-PTP, share 74 percent identity between their catalytic domains, yet, intriguingly, appear to possess complementary specificities for dephosphorylation of substrate JAKs and STATs. The crystal structure of PTP1B in complex with an insulin receptor peptide identified the importance of a motif composed of two consecutive phosphotyrosines for active site binding, where the first phosphotyrosine occupies the catalytic site and the second occupies an adjacent phosphotyrosine recognition pocket [51]. Analogous motifs occur within the JAK activation loops, and subtle differences between the JAK1/JAK3 or JAK2/Tyk2 sequences dictate whether these proteins are substrates of TC-PTP or PTP1B, respectively [52–54]. Tyk2 and JAK2 contain the (E/D)-pY-pY-(R/K) motif in their activation loops, which was shown to be the preferred substrate of PTP1B [52]. By comparison, JAK1 and JAK3 activation loops contain a Thr or Val residue C-terminal to the tandem phosphotyrosines, rather than R/K, and this difference appears to be critical for recognition by TC-PTP in preference to PTP1B [55]. PTP1B and TC-PTP have also been implicated in STAT dephosphorylation [56–60], although the STATs do not contain consensus substrate motifs and consequently the sites that are dephosphorylated remain unidentified. Recently, a 45-kD, nuclear variant of TC-PTP has been demonstrated to dephosphorylate nuclear STATs [57–60], and thus suggests a mechanism for terminating STAT transcription factor activities to enable their relocalization to the cytoplasm to participate further in JAK/STAT signaling.

The distinct biological roles of the PTP1B and TC-PTP phosphatases *in vivo* have been defined in mouse knockout studies. $PTP1B^{-/-}$ mice exhibit hypersensitivity to the growth factors, growth hormone, insulin, and leptin [54, 61, 62], and are resistant to diet-induced diabetes and obesity [62, 63]. In contrast, $TC\text{-}PTP^{-/-}$ mice die neonatally from a systemic inflammatory disease characterized by mononuclear cell infiltration [64, 65]. These mice also exhibit defects in immune cell responses and hematopoiesis, especially erythropoiesis [64].

Recently, another phosphatase, PTP basophil-like or PTP-BL, was identified as a physiologically-important negative regulator of JAK-STAT signaling owing to its capacity to dephosphorylate STAT1, STAT3, STAT4, STAT5, and STAT6 *in vitro* and *in vivo* [66]. Obtaining a more complete understanding of the physiological function (s) of PTP-BL awaits the availability of a mouse knockout model.

The transmembrane protein phosphatase, CD45, is an abundant, well-characterized regulator of Src family tyrosine kinases, such as Lck [67, 68] and Fyn [69], in B and T cells. Unexpectedly, CD45 was found to have a broader negative regulatory role as a hematopoietic JAK phosphatase, where CD45 antagonized stimulation by IL-3, IL-4, and EPO *in vitro* by decreasing STAT3 tyrosine phosphorylation, but not serine phosphorylation [70]. CD45 was shown to directly dephosphorylate the tandem phosphotyrosines of the JAK1, JAK2, JAK3, and Tyk2 activation loops [70]. However, to date it has not been established whether sites extraneous to the JAK activation loops are CD45 substrates, and whether CD45 dephosphorylates STATs. The importance of CD45 in attenuating JAK/STAT signaling has been validated using $CD45^{-/-}$ mice, where bone-marrow cells cultured from these mice exhibit extended phosphorylation of JAK1 and STAT5 following stimulation with IL-7 [71].

MODULATION OF JAK KINASE ACTIVITY BY EXTRINSIC KINASES

Phosphorylation of sites extraneous to the JAK activation loops by extrinsic kinases has recently been described as another level of regulation of JAK activity. Intrinsic JAK kinase activity is attenuated by intramolecular interactions between the catalytic domain (JH1) and the adjacent, catalytically-inactive pseudokinase domain (JH2) [72]. Consequently, modulation of this interaction will influence the catalytic activity of the JAKs. Recently, phosphorylation of Tyr570 in the pseudokinase domain of JAK2 was revealed to inhibit kinase activity, whereas the Y570F mutant JAK2 exhibited constitutive activity [73]. Whilst the mechanism of JAK2 inactivation and the kinase responsible for Tyr570 phosphorylation are currently unknown, Feener *et al.* [73] speculated that Tyr570 phosphorylation may sterically hinder activation or substrate accessibility to the JAK2 kinase domain. Subsequent studies have identified another pseudokinase domain residue, Ser523 of JAK2, as a site for serine phosphorylation [74, 75] in response to cytokine receptor activation, leading to down-regulation of catalytic activity. JAK2 was mutated to eliminate Ser523 phosphorylation, leading to enhanced JAK2 catalytic activity. As Ser523 resides in the pseudokinase domain of JAK2, like Tyr570, it is possible that negative regulation of JAK catalytic activity may occur via a mechanism analogous to that postulated for Tyr570 phosphorylation (above). Mazurkiewicz-Munoz *et al.* [74] reported that ERK1 or ERK2 are capable of phosphorylating Ser523 of JAK2, although this does not preclude Ser523 serving as a substrate for other serine/threonine kinases. Additionally, it is likely that as yet unidentified phosphatases which target phosphorylated Ser523 or Tyr570 may provide another level of regulation of the JAK kinase activity.

PROTEIN INHIBITORS OF ACTIVATED STATS (PIAS)

The PIAS family of proteins comprises four proteins, PIAS1, PIASx (or PIAS2), PIAS3, and PIASy (or PIAS4), originally identified as candidate STAT interactors using yeast two-hybrid assays [76]. PIAS3 was initially characterized based on its interaction STAT3 [76]. Subsequently, PIAS1 and PIASy were shown to inhibit STAT1 [77, 78], and PIASx to inhibit STAT4 [79]. Despite their name, PIAS proteins neither act only as inhibitors nor show specificity for STATs. Rather, they positively or negatively regulate numerous (>60) proteins in addition to STATs, many of which are transcription factors (reviewed in [80]).

The domain architecture of PIAS proteins comprises an N-terminal SAP (scaffold-attachment factor A and B), apoptotic chromatin-condensation inducer in the nucleus (ACINUS) and PIAS) domain [81]; the "PINIT" motif [82]; a C3HC4-motif-type RING finger-like zinc binding domain [83]; an acidic domain [84]; and a variable serine/threonine-rich domain (Figure 8.2). The SAP domain, PINIT motif and RING domain all appear to contribute to the nuclear localization of PIAS proteins [82]. A number of mechanisms by which PIAS proteins could potentially modulate JAK/STAT signaling have been postulated, consistent with the attributes of the component domains. In particular, the recent discovery that the RING finger-like domain functions as a SUMO (small ubiquitin-like modifier)-E3 ligase has greatly enhanced our understanding of PIAS mechanisms of action [85]. PIAS proteins have subsequently been shown to sumoylate (that is covalently, posttranslationally attach SUMO to lysine residues in a manner analogous to ubiquitylation) many targets, such as STAT1 [86, 87], c-Jun and p53 [88].

PIAS-mediated sumoylation is one mechanism by which PIAS may attenuate transcription factor activity. In the case of STAT1, a conserved residue, Lys703, is subject to PIAS1-mediated sumoylation [86, 87]. Sumoylation was inferred to serve a negative regulatory role since mutation of STAT1 Lys703 to Arg led to increased interferon (IFN)-γ-mediated transactivation [86, 87]. In a recent study, MAPK-induced phosphorylation of Ser727 in STAT1 was shown to enhance PIAS1 binding and SUMO conjugation to STAT1, suggesting a mechanism by which stress or growth stimuli transmitted *via* MAPK may stimulate PIAS1 activity to repress STAT1-induced transcription [89]. Phosphorylation of PIAS1 at Ser90 by IKKα was recently identified as another level of PIAS1 fine-tuning [90]. Ser90 phosphorylation was found to be critical for PIAS1 to repress STAT1- or NFκB-induced gene transcription, even though Ser90 mutations did not disrupt the PIAS1 : NFκB interaction [90]. The molecular basis for the role of PIAS1 Ser90 phosphorylation in repressing gene transcription has not yet been established.

Insights into the role PIAS-mediated sumoylation may play in regulating transcriptional activities can be gleaned

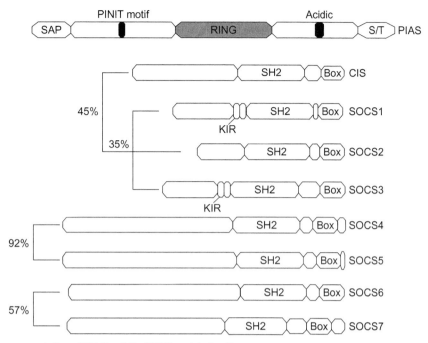

FIGURE 8.2 Schematic representation of PIAS and the SOCS protein family.
PIAS proteins are composed of an N-terminal SAP (scaffold-attachment factor A and B), apoptotic chromatin-condensation inducer in the nucleus (ACINUS), and PIAS) domain; the "PINIT" motif; a C3HC4-motif-type RING finger-like zinc binding domain; an acidic domain; and a variable serine/threonine-rich domain.
There are eight members of the SOCs family, CIS and SOCS1–SOCS7. The eight members all contain an SH2 domain, for phosphotyrosine binding, and a SOCS box which has been implicated in proteasomal degradation. SOCS1 and SOCS3 contain a Kinase Inhibitory Region (KIR) thought to be necessary for high-affinity binding to JAKs. They also contain a highly variable N-terminal domain.

from studies of other transcription factors. PIASy-mediated sumoylation was found to repress the transcriptional activity of LEF1 by altering the distribution of LEF1 from whole nuclear to punctate subnuclear structures [85]. Redistribution of the protein complex has been attributed to the N-terminal SAP domain, which, in *in vitro* studies, has been shown to bind the nuclear matrix attachment region DNA [85] or A/T-rich DNA surrogates [91, 92]. However, even though sumoylation was required for PIASy-mediated LEF1 redistribution [85], the role of SUMO remains unclear. It has been speculated that SUMO may hone the complex to nuclear matrix SUMO receptors or play a role in organizing the complex [83], since LEF1 SUMOylation did not alter LEF1 DNA binding or binding to the transcriptional coactivator, β-catenin [85].

PIAS proteins are known to modulate transcription factor activities by mechanisms that are independent of SUMO ligase activity, as exemplified by the wild-type inhibition of androgen receptor transcriptional activity by the SUMO ligase-dead, W363A mutant PIASy [93]. One such mechanism is the blocking DNA-binding of a transcription factor. The initial characterization of PIAS3 revealed that its interaction with STAT3 inhibited the DNA binding activity of STAT3 [76]. Similarly, PIAS1 was shown to inhibit the DNA binding activities of STAT1 [77], with this function attributed to a C-terminal region [94]. In contrast, PIAS3-mediated STAT3 DNA binding inhibition was attributed to

an N-terminal region of PIAS3 [95]. The molecular basis for the inhibition of DNA binding by PIAS proteins, and thus why DNA binding is inhibited in only some of the PIAS:transcription factor complexes, remains unclear. Nevertheless, it is plausible that DNA binding is inhibited by PIAS occluding the DNA binding site of the transcription factor within the PIAS:transcription factor complex or PIAS engagement of a transcription factor causing a conformational change or dissociation of a transcription complex (STAT monomerization, for example).

Another distinct mechanism by which PIAS proteins may modulate transcription is by serving as scaffolds for the recruitment of co-regulators to a protein complex. This function is implied by the observation that PIAS proteins interact with histone deacetylases (HDACs), enzymes that catalyze the removal of acetyl moieties from lysine residues in histones and other proteins. PIASy was shown to constitutively interact with HDAC1 and HDAC2, and treating cells the HDAC inhibitor, trichostatin A, was found to alleviate transcriptional repression in cells transfected with reporter gene constructs [93, 96]. Similarly, PIASx-β was shown to interact with HDAC3 [97], and trichostatin A treatment likewise abrogated IL-12-stimulated STAT4-dependent gene activation [79].

Studies of mice individually lacking one of the four PIAS genes have failed to shed light on the key physiological functions of each PIAS protein. These knockout mice exhibit only mild phenotypes, thus suggesting that functional redundancy exists between family members. For instance: *PIASy*$^{-/-}$ mice appear phenotypically normal and show only slight perturbations in Wnt and IFNγ signaling [98, 99]; *PIAS1*$^{-/-}$ mice are runts and show increased IFNγ- and IFNβ-mediated innate immune responses [100]. More recently, studies using PIASy/PIAS1 double-knockout mice have found that PIAS1 and PIASy synergistically regulate the induction of STAT1- and NFκB-dependent genes [101]. Overall, these studies emphasize the complexities of studying PIAS function *in vivo*, and indicate that truly understanding the molecular mechanisms of PIAS-mediated negative regulation of JAK/STAT signaling will rely on multidisciplinary experimental approaches in the future.

SUPPRESSORS OF CYTOKINE SIGNALING

The final family of inhibitors that we will discuss is the Suppressor of Cytokine Signaling (SOCS) proteins, which, unlike the above-described phosphatases and PIAS proteins, are not constitutively expressed.

SOCS genes encode a family of eight proteins (cytokine-inducible SH2 protein (CIS) and SOCS1–7) [102–105], which contain two highly conserved domains: a central SH2 domain that determines binding specificity, and a C-terminal domain termed a SOCS box implicated in proteasomal degradation (Figure 8.2). A 24 amino acid kinase inhibitory region (KIR) is also present in the N-terminal domain of SOCS1 and SOCS3 [106]. CIS, SOCS1, SOCS2, and SOCS3 are the best characterized members of the SOCS family, and consequently this chapter will predominantly focus on these family members as a paradigm for SOCS-mediated signal regulation. In contrast, the biological roles of SOCS4–SOCS7 remain poorly understood, probably due to their limited induction in response to cytokine stimuli. CIS, SOCS1, SOCS2, and SOCS3 have been readily characterized, as their mRNA and protein levels are low in unstimulated cells, but are rapidly induced following cytokine stimulation to act in a negative feedback loop to inhibit signaling [107].

Numerous *in vitro* overexpression structure/function studies of SOCS proteins have found they inhibit the action of a wide range of cytokines that utilize JAK/STAT signaling via a number of different mechanisms.

SOCS1 and SOCS3 are Kinase Inhibitors with Important Mutually Exclusive Roles

Initial phosphopeptide binding studies suggested SOCS1 inhibits JAK2 by binding to a key tyrosine (Tyr-1007) and inserts its KIR into the activation loop of its kinase domain, which acts as a pseudosubstrate, preventing substrates from accessing the catalytic pocket [108–110]. However, subsequent work demonstrated that SOCS1 abrogated tyrosine phosphorylation of STAT1 via direct binding to phosphorylated type 1 (IFNα/β) and 2 (IFNγ) interferon receptors [111, 112]. Consequently, a model of SH2 domain : receptor interaction and KIR-derived inhibition has been proposed which is supported by studies showing that a SOCS1-KIR mimetic peptide that binds directly to the autophosphorylation site of JAK2 inhibits STAT1 activation [113].

The physiological importance of SOCS1 is evident from studies of *SOCS1*$^{-/-}$ mice, which die neonatally from a range of autoinflammatory disease symptoms arising from dysregulation of IFNγ signaling [114, 115]. However, the story is considerably more complicated, as animals lacking IFNγ die at later time points from a different subset of inflammatory conditions, indicating that other SOCS1 targets are present. This hypothesis is supported by studies showing that the lethal phenotype can be ameliorated by blocking other signaling pathways or intermediates, such as the interleukin-4 receptor (IL-4R) [116], IFNα/β [112], IL-12R [117], IL-15 [118], and common γ-chain (γc) receptor signaling [119], demonstrating clear and discrete roles for SOCS1 in controlling cytokine signaling for homeostasis and in response to an immune challenge. These defined regulatory roles in distinct pathways have contributed to our understanding of how SOCS1 plays a role in controlling LPS responses [120], restrains arthritic development [121], and protects β cells from cytokine-mediated destruction [122]. The absence of SOCS1 expression due to gene methylation is suspected to cause the development of a range of primary cancers, including hepatocellular carcinoma (HCC), breast and ovarian cancer lines, acute myeloid leukemia, and multiple myeloma [123–127].

SOCS3 is structurally most closely related to SOCS1, and the mechanism of action of these proteins involves first binding to the receptor, followed by inhibition of JAK activity via a KIR-mediated JAK inhibition [128]. SOCS3 inhibits the signaling propagated by numerous stimuli that induce its expression, including signals from IL-2, IL-6, IL-9, IFNs, EPO, GM-CSF, GH, PRL, insulin, and leptin (reviewed by [129]). This inhibition is predominantly achieved by SOCS3 engagement of the activated receptors of these cytokines. Accordingly, SOCS3 has been shown specifically to bind to phosphorylated tyrosine 759 (Tyr759) of the gp130 receptor [24], Tyr800 of IL-12 receptor β2 [130], Tyr985 of the leptin receptor [25, 131], Tyr401 and the tandem-phosphorylated motif Tyr429/Tyr431 of the EPO receptor [26, 132], and Tyr729 of the G-CSF receptor [133], via an its extended SH2 domain. Interestingly, a number of these receptor-docking sites are also binding sites for another signaling regulator, SHP2 (described above), although the relative contributions of SOCS and SHP-2 to signal modulation and pathway cross-talk remains a matter of ongoing debate. Surprisingly, SOCS3 is not phosphorylated by JAKs, but relies on other kinases such as Src kinases and receptor-tyrosine kinases (e.g., EGF receptor or PDGF receptor) for activation, suggesting a role for SOCS3 as a mediator of cross-talk between signaling pathways [134].

Due to the embryonic lethality of SOCS3 knockout mice, in vivo investigations have been performed using conditional $SOCS3^{-/-}$ mice. These studies demonstrated that SOCS3 performs physiological functions that are largely distinct from SOCS1, including regulation of leptin [135], IL-6 [136–138], leukemia inhibitory factor (LIF) (139), and G-CSF signaling (140, 141). The critical role of SOCS3 in attenuating gp130 signaling is highlighted when deletion of SOCS3 results in prolonged STAT activation, thereby affecting the functional outcome of IL-6 signaling [136–138]. In $SOCS3^{-/-}$ macrophages, IL-6 behaves more like the immunosuppressive cytokine IL-10 [138]. Both IL-6 and IL-10 act through STAT3 and, despite the rapid induction of SOCS1 and SOCS3 by IFNγ and IL-6, there appears to be little functional redundancy. IL-10-induced STAT3 activation was not enhanced in $SOCS3^{-/-}$ macrophages, suggesting that, at least in the myeloid lineage, SOCS3 specifically targets gp130 family cytokines. These studies also provide insights into how signaling systems that use the same signaling intermediates can elicit distinct biological outcomes.

Given the importance of SOCS3 in regulating a number of inflammatory cytokines, it is not surprising that the absence of SOCS3 has been implicated susceptibility to chronic inflammatory diseases such as inflammatory bowel disease (IBD), rheumatoid arthritis (RA), and Crohn's disease (CD) [142, 143]. For example, a mutation in gp130 (Y759F) renders SOCS3 (and SHP-2) unable to bind and resulted in mice with autoimmune arthritis (143), while deletion of SOCS3 in the hematopoietic and endothelial cell compartment was associated with particularly severe IL-1 dependent inflammatory arthritis [144].

There is also a growing body of evidence suggesting that a lack of SOCS3 correlates with the development and progression of malignancies. Mice in which $SOCS3$ is deleted from hepatocytes develop HCC at an accelerated rate, presumably because of dysregulation of cytokine signaling through the IL-6 and EGF receptors, accelerating cell proliferation after hepatectomy [145]. The notion that a lack of SOCS3 expression results in enhanced cell growth and migration in response to cytokines and growth factors is supported by detection of methylation-associated silencing of the SOCS3 gene in a large number of cancers and cancerous cell lines, including lung cancer [146], melanoma [147], head and neck squamous cell cancer [148], adenocarcinoma [149], and hepatocellular cancers [150].

CIS and SOCS are Binding Competitors of STATs that Affect a Number of Developmental Pathways

The inhibitory mechanisms of CIS and SOCS2 differ from those described for SOCS1 and SOCS3 (above), as neither CIS nor SOCS2 act upon JAKs. CIS associates with the STAT5 binding sites on the EPO receptor (EPOR) and growth hormone receptor (GHR), preventing STAT5 recruitment and activation [151–154]. The biological role of CIS has remained elusive, since knockout studies have reportedly failed to detect any significant phenotype [155]. In comparison, the CIS transgenic mouse showed suppressed STAT5A/B activity, and exhibited stunted growth, defective lactation, and T cell defects, presumably due to defects in growth hormone, prolactin, and IL-2 signaling, respectively [156]. Overall, these studies suggest a functional redundancy for CIS, which will hopefully be resolved in future studies by the generation of compound knockout mice simultaneously lacking another member of the SOCS family.

The physiological role of SOCS2 was revealed when $SOCS2^{-/-}$ mice were generated, as these mice exhibit a gigantism phenotype, presumed to arise from the dysregulation of the growth hormone (GH) axis [157]. This phenotype was recapitulated in mice lacking both SOCS2 and GH upon administration of exogenous GH [158], thereby implicating SOCS2 as a modulator of GH sensitivity in vivo. Subsequent studies indicate that SOCS2 also regulates the activity of IGF-I signaling, at least in the gastrointestinal tract [159], plays a significant role in prolactin-driven mammary gland development [160], and promotes neurogenic differentiation by blocking the actions of GH [161]. The only reported departure from SOCS2 acting in developmental processes came from strong evidence that SOCS2 plays a crucial regulatory role of limiting the anti-inflammatory actions of aspirin-induced lipoxins in dendritic cells [162].

Several biochemical studies have provided insights into the molecular basis for SOCS2 action. SOCS2 is thought to partially block STAT5 recruitment by binding phosphorylated Tyr487 and Tyr595 on the GHR, since mutations of these sites result in prolongation of GH-induced STAT5 activation [158]. Previous in vitro studies have shown that overexpression of SOCS2 had biphasic effects on signaling: low levels partially inhibited STAT5 action, while high levels enhanced STAT5 activity [163]. However, the molecular mechanisms underpinning this phenomenon are currently unclear.

A number of studies have linked SOCS2 to cancer. Mice lacking one allele of $SOCS2$ spontaneously developed multiple hyperplastic and lymphoid polyps when crossed to GH transgenic mice [164], and, interestingly, it was not possible to generate GH transgenic mice that lacked both $SOCS2$

alleles. Some breast cancer tumors show methylation of the *SOCS2* gene [124], and it has been recently shown that SOCS2 has significant prognostic value, since breast cancer patients that have high SOCS2 and IGF-I expression in their tumors live significantly longer [165].

SOCS are Ubiquitin Ligases that Promote the Degradation of their Partners

X-ray crystallography and biochemical studies have defined that the C-terminal 40 amino acid domain of the SOCS proteins, termed the SOCS box, interacts with Elongin B and Elongin C, Cullin-5 and RING-box-2 (RBX2) to form an E3 ligase complex termed ECS (Elongin B/C–Cullin–SOCS box protein) [166, 167] (Figure 8.3). The ECS binds to a ubiquitin-activating enzyme (E1) and a ubiquitin-conjugating enzyme (E2), allowing the ECS complex to recognize and bind target proteins for polyubiquitination and degradation via the 26S proteasome [168]. Proteasomal degradation has emerged as a key mechanism of signal attenuation, and the nuclear protein, SLIM, was recently characterized as a ubiquitin E3 ligase which targets STATs for degradation via the proteasome [169].

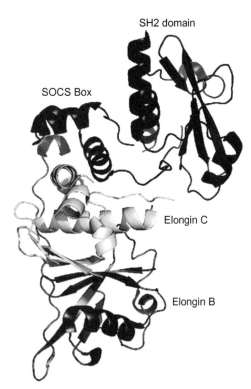

SH2 domain

SOCS Box

Elongin C

Elongin B

FIGURE 8.3 X-ray crystal structure of the SOCS2–Elongin C–Elongin B complex.
This crystal structure reveals that the SOCS box of SOCS2 is a tri-helical fold that binds to Elongin C, which is an adaptor for binding to Elongin B (166). The SH2 domain and SOCS box of SOCS2 are colored black; Elongin C is light gray; Elongin B is dark gray.
This figure was drawn using PyMol from the Protein Data Bank file, 2C9W.

The interaction of the SOCS box with the ECS complex provides a clear link between SOCS proteins and proteasomal degradation, which is central to attenuating the levels of SOCS proteins within the cell [170–175]. However, it remains a point of contention whether SOCS proteins can serve as adaptors to bridge their binding partners to the proteasome and instigate their degradation. Evidence supporting this hypothesis has accumulated from a mass of overexpression studies over the past few years: (1) SOCS1 was shown to target JAK2 and TEL-JAK2 fusion protein for ubiquitination and proteasomal degradation [176–178]; (2) CIS was required for EPOR and GHR degradation [154, 179]; (3) SOCS3 was required for ubiquitination of the G-CSFR on a membrane-proximal lysine, which led to lysosomal-routing following receptor activation [180]; and (4) SOCS2 was found to enhance IL-2- and IL-3-dependent signaling by stimulating the degradation of SOCS3 in a SOCS box-dependent manner [181], providing a model for cross-talk between different members of the SOCS family.

Proteasomal degradation plays a more intricate role in regulating the stability of SOCS3 via SOCS box-independent mechanisms. The N-terminus of SOCS3 contains a residue subject to ubiquitination, lysine 6, and the deletion of this residue in a truncated isoform lacking the first 12 N-terminal amino acids stabilizes SOCS3 [182]. Additionally, SOCS3 contains a PEST (proline-, glutamic acid-, serine- and threonine-rich) motif within the SH2 domain, which is not required for STAT inhibition, but regulates the half-life of SOCS3 [183].

Clear physiological evidence for the importance of the SOCS box has been gleaned from *in vivo* studies, as mice lacking the SOCS box of SOCS1 present a hypomorphic phenotype and develop a fatal inflammatory disease [184], and mice lacking the SOCS box of SOCS3 exhibit hyper-responsiveness to G-CSF [185]. In addition, mutations within the SOCS box of SOCS1, which interfere with the degradation of phospho-JAK2, have been detected in mediastinal lymphoma cell lines [186].

FUTURE OUTLOOK

In this chapter, we have described our current understanding of the mechanisms underlying JAK-STAT signaling. The vast array of posttranslational modifications, such as phosphorylation, dephosphorylation, sumoylation, and ubiquitination, that are known to control JAK-STAT signaling underscores the intrinsic complexity of this pathway and the difficulties associated with studying individual regulators in isolation. In the future, a more comprehensive understanding of the redundancy and cross-talk between the regulators of JAK-STAT signaling may be accessible using a systems biology approach. This approach will rely on comprehensive cataloging of modulators of JAK-STAT signaling, which, based on the recent discoveries of novel attenuators such as SLIM [169] and PTP-BL [66], is far from complete.

ACKNOWLEDGEMENTS

The authors gratefully acknowledge financial support from the National Health and Medical Research Council of Australia (NHMRC; Program Grant 461219) and the National Institutes of Health (Grant ROI-CA-22556). James M. Murphy is the recipient of a CJ Martin (Biomedical) fellowship from the NHMRC; Christopher J. Greenhalgh is the recipient of an NHMRC Industry Fellowship; and Douglas J. Hilton is the recipient of an NHMRC Australia Fellowship.

REFERENCES

1. Bazan JF. Hematopoietic receptors and helical cytokines. *Immunol Today* 1990;**11**:350–4.

2. Shuai K, Horvath CM, Huang LHT, Qureshi SA, Cowburn D, Darnell JE. Interferon activation of the transcription factor Stat91 involves dimerization through SH2-phosphotyrosyl peptide interactions. *Cell* 1994;**76**:821–8.

3. Mao X, Ren Z, Parker GN, et al. Structural bases of unphosphorylated STAT1 association and receptor binding. *Mol Cell* 2005;**17**:761–71.

4. Kralovics R, Passamonti F, Buser AS, et al. A gain-of-function mutation of JAK2 in myeloproliferative disorders. *N Engl J Med* 2005;**352**:1779–90.

5. Jones AV, Kreil S, Zoi K, et al. Widespread occurrence of the JAK2 V617F mutation in chronic myeloproliferative disorders. *Blood* 2005;**106**:2162–8.

6. Matsuda T, Feng J, Witthuhn BA, Sekine Y, Ihle JN. Determination of the transphosphorylation sites of Jak2 kinase. *Biochem Biophys Res Comm* 2004;**325**:586–94.

7. Gauzzi MC, Velazquez L, McKendry R, Mogensen KE, Fellous M, Pellegrini S. Interferon-α-dependent activation of Tyk2 requires phosphorylation of positive regulatory tyrosines by another kinase. *J Biol Chem* 1996;**271**:20,494–20,500.

8. Zhou YJ, Hanson EP, Chen YQ, et al. Distinct tyrosine phosphorylation sites in JAK3 kinase domain positively and negatively regulate its enzymatic activity. *Proc Natl Acad Sci USA* 1997;**94**:13,850–13,855.

9. Liu KD, Gaffen SL, Goldsmith MA, Greene WC. Janus kinases in interleukin-2-mediated signaling: JAK1 and JAK3 are differentially regulated by tyrosine phosphorylation. *Curr. Biol.* 1997;**7**:817–26.

10. Feng J, Witthuhn BA, Matsuda T, Kohlhuber F, Kerr IM, Ihle JN. Activation of Jak2 catalytic activity requires phosphorylation of Y1007 in the kinase activation loop. *Mol Cell Biol* 1997;**17**:2497–501.

11. Boggon TJ, Li YQ, Manley PW, Eck MJ. Crystal structure of the Jak3 kinase domain in complex with a staurosporine analog. *Blood* 2005;**106**:996–1002.

12. Lucet IS, Fantino E, Styles M, et al. The structural basis of Janus kinase 2 inhibition by a potent and specific pan-Janus kinase inhibitor. *Blood* 2006;**107**:176–83.

13. Cacalano NA, Migone TS, Bazan F, et al. Autosomal SCID caused by a point mutation in the N-terminus of Jak3: mapping of the Jak3-receptor interaction domain. *EMBO J* 1999;**18**:1549–58.

14. Zhou YJ, Chen M, Cusack NA, et al. Unexpected effects of FERM domain mutations on catalytic activity of Jak3: Structural implication for Janus kinases. *Mol Cell* 2001;**8**:959–69.

15. Funakoshi-Tago M, Pelletier S, Matsuda T, Parganas E, Ihle JN. Receptor specific downregulation of cytokine signaling by autophosphorylation in the FERM domain of Jak2. *EMBO J* 2006;**25**:4763–72.

16. Plutzky J, Neel BG, Rosenberg RD. Isolation of a src homology 2-containing tyrosine phosphatase. *Proc Natl Acad Sci* 1992;**89**:1123–7.

17. Feng GS, Hui CC, Pawson T. SH2-containing phosphotyrosine phosphatase as a target of protein-tyrosine kinases. *Science* 1993;**259**:1607–11.

18. Beebe KD, Wang P, Arabaci G, Pei DH. Determination of the binding specificity of the SH2 domains of protein tyrosine phosphatase SHP-1 through the screening of a combinatorial phosphotyrosyl peptide library. *Biochemistry* 2000;**39**:13,251–13,260.

19. De Souza D, Fabri LJ, Nash A, Hilton DJ, Nicola NA, Baca M. SH2 domains from suppressor of cytokine signaling-3 and protein tyrosine phosphatase SHP-2 have similar binding specificities. *Biochemistry* 2002;**41**:9229–36.

20. Sweeney MC, Wavreille AS, Park J, Butchar JP, Tridandapani S, Pei D. Decoding protein–protein interactions through combinatorial chemistry: sequence specificity of SHP-1, SHP-2, and SHIP SH2 domains. *Biochemistry* 2005;**44**:14,932–14,947.

21. Imhof D, Wavreille AS, May A, Zacharias M, Tridandapani S, Pei DH. Sequence specificity of SHP-1 and SHP-2 Src homology 2 domains – critical roles of residues beyond the pY + 3 position. *J Biol Chem* 2006;**281**:20,271–20,282.

22. Chughtai N, Schimchowitsch S, Lebrun JJ, Ali S. Prolactin induces SHP-2 association with Stat5, nuclear translocation, and binding to the β-casein gene promoter in mammary cells. *J Biol Chem* 2002;**277**:31,107–31,114.

23. Nicholson SE, De Souza D, Fabri LJ, et al. Suppressor of cytokine signaling-3 preferentially binds to the SHP-2-binding site on the shared cytokine receptor subunit gp130. *Proc Natl Acad Sci USA* 2000;**97**:6493–8.

24. Schmitz J, Weissenbach M, Haan S, Heinrich PC, Schaper F. SOCS3 exerts its inhibitory function on interleukin-6 signal transduction through the SHP2 recruitment site of gp130. *J Biol Chem* 2000;**275**:12,848–12,856.

25. Bjorbaek C, Lavery HJ, Bates SH, et al. SOCS3 mediates feedback inhibition of the leptin receptor via Tyr(985). *J Biol Chem* 2000;**275**:40,649–40,657.

26. Sasaki A, Yasukawa H, Shouda T, Kitamura T, Dikic I, Yoshimura A. CIS3/SOCS-3 suppresses erythropoietin (EPO) signaling by binding the EPO receptor and JAK2. *J Biol Chem* 2000;**275**:29,338–29,347.

27. Hof P, Pluskey S, Dhe-Paganon S, Eck MJ, Shoelson SE. Crystal structure of the tyrosine phosphatase SHP-2. *Cell* 1998;**92**:441–50.

28. Klingmuller U, Lorenz U, Cantley LC, Neel BG, Lodish HF. Specific recruitment of SH-PTP1 to the erythropoietin receptor causes inactivation of JAK2 and termination of proliferative signals. *Cell* 1995;**80**:729–38.

29. Yi TL, Mui ALF, Krystal G, Ihle JN. Hematopoietic-cell phosphatase associates with the interleukin-3 (IL-3) receptor-β chain and downregulates IL-3-induced tyrosine phosphorylation and mitogenesis. *Mol Cell Biol* 1993;**13**:7577–86.

30. Pei DH, Lorenz U, Klingmuller U, Neel BG, Walsh CT. Intramolecular regulation of protein-tyrosine-phosphatase SH-PTP1 – a new function for src homology-2 domains. *Biochemistry* 1994;**33**:15,483–15,493.

31. Ward AC, Oomen S, Smith L, et al. The SH2 domain-containing protein tyrosine phosphatase SHP-1 is induced by granulocyte colony-stimulating factor (G-CSF) and modulates signaling from the G-CSF receptor. *Leukemia* 2000;**14**:1284–91.

32. Chen HE, Chang S, Trub T, Neel BG. Regulation of colony-stimulating factor 1 receptor signaling by the SH2 domain-containing tyrosine phosphatase SHPTP1. *Mol Cell Biol* 1996;**16**:3685–97.

33. Yi TL, Ihle JN. Association of hematopoietic-cell phosphatase with c-Kit after stimulation with c-Kit ligand. *Mol Cell Biol* 1993;**13**:3350–8.

34. Tenev T, Keilhack H, Tomic S, et al. Both SH2 domains are involved in interaction of SHP-1 with the epidermal growth factor receptor but cannot confer receptor-directed activity to SHP-1/SHP-2 chimera. *J Biol Chem* 1997;**272**:5966–73.

35. David M, Chen HYE, Goelz S, Larner AC, Neel BG. Differential regulation of the alpha/beta interferon-stimulated JAK/STAT pathway by the SH2 domain-containing tyrosine phosphatase SHPTP1. *Mol Cell Biol* 1995;**15**:7050–8.

36. Jiao H, Berrada K, Yang W, Tabrizi M, Platanias LC, Yi T. Direct association with and dephosphorylation of Jak2 kinase by the SH2-domain-containing protein tyrosine phosphatase SHP-1. *Mol Cell Biol* 1996;**16**:6985–92.

37. O'Reilly AM, Neel BG. Structural determinants of SHP-2 function and specificity in Xenopus mesoderm induction. *Mol Cell Biol* 1998;**18**:161–77.

38. Bone H, Dechert U, Jirik F, Schrader JW, Welham MJ. SHP1 and SHP2 protein-tyrosine phosphatases associate with βc after interleukin-3-induced receptor tyrosine phosphorylation-Identification of potential binding sites and substrates. *J Biol Chem* 1997;**272**:14,470–14,476.

39. Kim HK, Hawley TS, Hawley RG, Baumann H. Protein tyrosine phosphatase 2 (SHP-2) moderates signaling by gp130 but is not required for the induction of acute-phase plasma protein genes in hepatic cells. *Mol Cell Biol* 1998;**18**:1525–33.

40. Tauchi T, Feng GS, Shen R, et al. Involvement of SH2-containing phosphotyrosine phosphatase SYP in erythropoietin receptor signal-transduction pathways. *J Biol Chem* 1995;**270**:5631–5.

41. Yin TG, Shen R, Feng GS, Yang YC. Molecular characterization of specific interactions between SHP-2 phosphatase and JAK tyrosine kinases. *J Biol Chem* 1997;**272**:1032–7.

42. Schaper F, Gendo C, Eck M, et al. Activation of the protein tyrosine phosphatase SHP2 via the interleukin-6 signal transducing receptor protein gp130 requires tyrosine kinase Jak1 and limits acute-phase protein expression. *Biochem J* 1998;**335**:557–65.

43. Wheadon H, Paling NRD, Welham MJ. Molecular interactions of SHP1 and SHP2 in IL-3-signalling. *Cell Signal* 2002;**14**:219–29.

44. Yu CL, Jin YJ, Burakoff SJ. Cytosolic tyrosine dephosphorylation of STAT5 – Potential role of SHP-2 in STAT5 regulation. *J Biol Chem* 2000;**275**:599–604.

45. Chen J, Yu WM, Qu CK. SHP-2 tyrosine phosphatase negatively regulates hematopoietic cell survival by dephosphorylation of Stat5. *Blood* 2003;**102**:836A.

46. Tsui HW, Siminovitch KA, de Souza L, Tsui FWL. Moth-eaten and viable moth-eaten mice have mutations in the hematopoietic-cell phosphatase gene. *Nature Genetics* 1993;**4**:124–9.

47. Shultz LD, Schweitzer PA, Rajan TV, et al. mutations at the murine moth-eaten locus are within the hematopoietic-cell protein-tyrosine-phosphatase (HCPH) gene.. *Cell* 1993;**73**:1445–54.

48. Saxton TM, Henkemeyer M, Gasca S, et al. Abnormal mesoderm patterning in mouse embryos mutant for the SH2 tyrosine phosphatase Shp-2. *EMBO J* 1997;**16**:2352–64.

49. Qu CK, Shi ZQ, Shen R, Tsai FY, Orkin SH, Feng GS. A deletion mutation in the SH2-N domain of Shp-2 severely suppresses hematopoietic cell development. *Mol Cell Biol* 1997;**17**:5499–507.

50. Qu CK, Nguyen S, Chen JZ, Feng GS. Requirement of Shp-2 tyrosine phosphatase in lymphoid and hematopoietic cell development. *Blood* 2001;**97**:911–14.

51. Salmeen A, Andersen JN, Myers MP, Tonks NK, Barford D. Molecular basis for the dephosphorylation of the activation segment of the insulin receptor by protein tyrosine phosphatase 1B. *Mol Cell* 2000;**6**:1401–12.

52. Myers MP, Andersen JN, Cheng A, et al. TYK2 and JAK2 are substrates of protein-tyrosine phosphatase 1B. *J Biol Chem* 2001;**276**:47,771–47,774.

53. Simoncic PD, Lee-Loy A, Barber DL, Tremblay ML, McGlade CJ. The T cell protein tyrosine phosphatase is a negative regulator of janus family kinases 1 and 3. *Curr. Biol.* 2002;**12**:446–53.

54. Gu F, Dube N, Kim JW, et al. Protein tyrosine phosphatase 1B attenuates growth hormone-mediated JAK2-STAT signaling. *Mol Cell Biol* 2003;**23**:3753–62.

55. Espanel X, Huguenin-Reggiani M, van Huijsduijnen RH. The SPOT technique as a tool for studying protein tyrosine phosphatase substrate specificities. *Protein Sci* 2002;**11**:2326–34.

56. Aoki N, Matsuda T. A cytosolic protein-tyrosine phosphatase PTP1B specifically dephosphorylates and deactivates prolactin-activated STAT5a and STAT5b. *J Biol Chem* 2000;**275**:39,718–39,726.

57. ten Hoeve J, Ibarra-Sanchez MD, Fu YB, et al. Identification of a nuclear Stat1 protein tyrosine phosphatase. *Mol Cell Biol* 2002;**22**:5662–8.

58. Yamamoto T, Sekine Y, Kashima K, et al. The nuclear isoform of protein-tyrosine phosphatase TC-PTP regulates interleukin-6-mediated signaling pathway through STAT3 dephosphorylation. *Biochem Biophys Res Comm* 2002;**297**:811–17.

59. Aoki N, Matsuda T. A nuclear protein tyrosine phosphatase TC-PTP is a potential negative regulator of the PRL-mediated signaling pathway: dephosphorylation and deactivation of signal transducer and activator of transcription 5a and 5b by TC-PTP in nucleus. *Mol Endocrinol* 2002;**16**:58–69.

60. Lu XQ, Chen J, Sasmono RT, et al. T-cell protein tyrosine phosphatase, distinctively expressed in activated-B-cell-like diffuse large B-cell lymphomas, is the nuclear phosphatase of STAT6. *Mol Cell Biol* 2007;**27**:2166–79.

61. Elchebly M, Payette P, Michaliszyn E, et al. Increased insulin sensitivity and obesity resistance in mice lacking the protein tyrosine phosphatase-1B gene. *Science* 1999;**283**:1544–8.

62. Klaman LD, Boss O, Peroni OD, et al. Increased energy expenditure, decreased adiposity, and tissue-specific insulin sensitivity in protein-tyrosine phosphatase 1B-deficient mice. *Mol Cell Biol* 2000;**20**:5479–89.

63. Cheng A, Uetani N, Simoncic PD, et al. Attenuation of leptin action and regulation of obesity by protein tyrosine phosphatase 1B. *Dev Cell* 2002;**2**:497–503.

64. YouTen KE, Muise ES, Itie A, et al. Impaired bone marrow microenvironment and immune function in T cell protein tyrosine phosphatase-deficient mice. *J Exp Med* 1997;**186**:683–93.

65. Heinonen KM, Nestel FP, Newell EW, et al. T-cell protein tyrosine phosphatase deletion results in progressive systemic inflammatory disease. *Blood* 2004;**103**:3457–64.

66. Nakahira M, Tanaka T, Robson BE, Mizgerd JP, Grusby MJ. Regulation of signal transducer and activator of transcription signaling by the tyrosine phosphatase PTP-BL. *Immunity* 2007;**26**:163–76.

67. Mustelin T, Coggeshall KM, Altman A. Rapid activation of the T-cell tyrosine protein-kinase PP56Lck by the CD45 phosphotyrosine phosphatase. *Proc Natl Acad Sci USA* 1989;**86**:6302–6.

68. Ostergaard HL, Shackelford DA, Hurley TR, et al. Expression of CD45 alters phosphorylation of the Lck-encoded tyrosine protein-kinase in murine lymphoma T-cell lines. *Proc Natl Acad Sci USA* 1989;**86**:8959–63.

69. Shiroo M, Goff L, Biffen M, Shivnan E, Alexander D. CD45-tyrosine phosphatase-activated P59Fyn couples the T-cell antigen receptor to pathways of diacylglycerol production, protein-kinase-C activation and calcium influx. *EMBO J* 1992;**11**:4887–97.

70. Irie-Sasaki J, Sasaki T, Matsumoto W, et al. CD45 is a JAK phosphatase and negatively regulates cytokine receptor signalling. *Nature* 2001;**409**:349–54.

71. Fleming HE, Milne CD, Paige CJ. CD45-deficient mice accumulate pro-B cells both in vivo and in vitro. *J Immunol* 2004;**173**:2542–51.

72. Saharinen P, Silvennoinen O. The pseudokinase domain is required for suppression of basal activity of Jak2 and Jak3 tyrosine kinases and for cytokine-inducible activation of signal transduction. *J Biol Chem* 2002;**277**:47,954–47,963.

73. Feener EP, Rosario F, Dunn SL, Stancheva Z, Myers MG. Tyrosine phosphorylation of Jak2 in the JH2 domain inhibits cytokine signaling. *Mol Cell Biol* 2004;**24**:4968–78.

74. Mazurkiewicz-Munoz A, Argetsinger LS, Kouadio JK, Stensballe A, Jensen ON, Cline JM, Carter-Su C. Phosphorylation of JAK2 at serine 523: a negative regulator of JAK2 that is stimulated by growth hormone and epidermal growth factor. *Mol Cell Biol* 2006;**26**:4052–62.

75. Ishida-Takahashi R, Rosario F, Gong Y, Kopp K, Stancheva Z, Chen X, Feener EP, Myers MG. Phosphorylation of Jak2 on Ser523 inhibits Jak2-dependent leptin receptor signaling. *Mol Cell Biol* 2006;**26**:4063–73.

76. Chung CD, Liao JY, Liu B, et al. Specific inhibition of Stat3 signal transduction by PIAS3. *Science* 1997;**278**:1803–5.

77. Liu B, Liao JY, Rao XP, et al. Inhibition of Stat1-mediated gene activation by PIAS1. *Proc Natl Acad Sci USA* 1998;**95**:10,626–10,631.

78. Liu B, Gross M, ten Hoeve J, Shuai K. A transcriptional corepressor of Stat1 with an essential LXXLL signature motif. *Proc Natl Acad Sci USA* 2001;**98**:3203–7.

79. Arora T, Liu B, He HC, et al. PIASx is a transcriptional co-repressor of signal transducer and activator of transcription 4. *J Biol Chem* 2003;**278**:21,327–21,330,.

80. Shuai K, Liu B. Regulation of gene-activation pathways by PIAS proteins in the immune system. *Nature Rev Immunol* 2005;**5**:593–605.

81. Aravind L, Koonin EV. SAP – a putative DNA-binding motif involved in chromosomal organization. *Trends Biochem Sci* 2000;**25**:112–14.

82. Duval D, Duval G, Kedinger C, Poch O, Boeuf H. The 'PINIT' motif, of a newly identified conserved domain of the PIAS protein family, is essential for nuclear retention of PIAS3L. *FEBS Letts* 2003;**554**:111–18.

83. Jackson PK. A new RING for SUMO: wrestling transcriptional responses into nuclear bodies with PIAS family E3 SUMO ligases. *Genes Dev* 2001;**15**:3053–8.

84. Jimenez-Lara AM, Heine MJS, Gronemeyer H. PIAS3 (protein inhibitor of activated STAT-3) modulates the transcriptional activation mediated by the nuclear receptor coactivator TIF2. *FEBS Letts* 2002;**526**:142–6.

85. Sachdev S, Bruhn L, Sieber H, Pichler A, Melchior F, Grosschedl R. PIASy, a nuclear matrix-associated SUMO E3 ligase, represses LEF1 activity by sequestration into nuclear bodies. *Genes Dev* 2001;**15**:3088–103.

86. Ungureanu D, Vanhatupa S, Kotaja N, et al. PIAS proteins promote SUMO-1 conjugation to STAT1. *Blood* 2003;**102**:3311–13.

87. Rogers RS, Horvath CM, Matunis MJ. SUMO modification of STAT1 and its role in PIAS-mediated inhibition of gene activation. *J Biol Chem* 2003;**278**:30,091–30,097.

88. Schmidt D, Muller S. Members of the PIAS family act as SUMO ligases for c-Jun and p53 and repress p53 activity. *Proc Natl Acad Sci USA* 2002;**99**:2872–7.

89. Vanhatupa S, Ungureanu D, Paakkunainen M, Silvennoinen O. MAPK-induced Ser(727) phosphorylation promotes SUMOylation of STAT1. *Biochem J* 2008;**409**:179–85.

90. Liu B, Yang YH, Chernishof V, et al. Proinflammatory stimuli induce IKK α-mediated phosphorylation of PIAS1 to restrict inflammation and immunity. *Cell* 2007;**129**:903–14.

91. Tan JA, Hall SH, Hamil KG, Grossman G, Petrusz P, French FS. Protein inhibitors of activated STAT resemble scaffold attachment factors and function as interacting nuclear receptor coregulators. *J Biol Chem* 2002;**277**:16,993–17,001.

92. Okubo S, Hara F, Tsuchida Y, et al. NMR structure of the N-terminal domain of SUMO ligase PIAS1 and its interaction with tumor suppressor p53 and A/T-rich DNA oligomers. *J Biol Chem* 2004;**279**:31,455–31,461.

93. Gross M, Yang R, Top I, Gasper C, Shuai K. PIASy-mediated repression of the androgen receptor is independent of sumoylation. *Oncogene* 2004;**23**:3059–66.

94. Liao JY, Fu YB, Shuai K. Distinct roles of the NH2- and COOH-terminal domains of the protein inhibitor of activated signal transducer and activator of transcription (STAT) 1 (PIAS1) in cytokine-induced PIAS1–Stat1 interaction. *Proc Natl Acad Sci USA* 2000;**97**:5267–72.

95. Sonnenblick A, Levy C, Razin E. Interplay between MITF, PIAS3, and STAT3 in mast cells and melanocytes. *Mol Cell Biol* 2004;**24**:10,584–10,592.

96. Long JY, Matsuura I, He DM, Wang GN, Shuai K, Liu F. Repression of Smad transcriptional activity by PIASy, an inhibitor of activated STAT. *Proc Natl Acad Sci USA* 2003;**100**:9791–6.

97. Tussie-Luna MI, Bayarsaihan D, Seto E, Ruddle RH, Roy AL. Physical and functional interactions of histone deacetylase 3 with TFII-I family proteins and PIAS alpha beta. *Proc Natl Acad Sci USA* 2002;**99**:12,807–12,812.

98. Wong KA, Kim R, Christofk H, Gao J, Lawson G, Wu H. Protein inhibitor of activated STAT y (PIASy) and a splice variant lacking exon 6 enhance sumoylation but are not essential for embryogenesis and adult life. *Mol Cell Biol* 2004;**24**:5577–86.

99. Roth W, Sustmann C, Kieslinger M, et al. PIASy-deficient mice display modest defects in IFN and Wnt signaling. *J Immunol* 2004;**173**:6189–99.

100. Liu B, Mink S, Wong KA, et al. PIAS1 selectively inhibits interferon-inducible genes and is important in innate immunity. *Nature Immunol* 2004;**5**:891–8.

101. Tahk S, Liu B, Chernishof V, Wong KA, Wu H, Shuai K. Control of specificity and magnitude of NF-kappa B and STAT1-mediated gene activation through PIASy and PIAS1 cooperation. *Proc Natl Acad Sci USA* 2007;**104**:11,643–11,648.

102. Starr R, Willson TA, Viney EM, et al. A family of cytokine-inducible inhibitors of signalling. *Nature* 1997;**387**:917–21.

103. Naka T, Narazaki M, Hirata M, et al. Structure and function of a new STAT-induced STAT inhibitor. *Nature* 1997;**387**:924–9.

104. Minamoto S, Ikegame K, Ueno K, et al. Cloning and functional analysis of new members of STAT induced STAT inhibitor (SSI) family: SSI-2 and SSI-3. *Biochem Biophys Res Comm* 1997;**237**:79–83.

105. Endo TA, Masuhara M, Yokouchi M, et al. A new protein containing an SH2 domain that inhibits JAK kinases. *Nature* 1997;**387**:921–4.

106. Nicholson SE, Willson TA, Farley A, et al. Mutational analyses of the SOCS proteins suggest a dual domain requirement but distinct mechanisms for inhibition of LIF and IL-6 signal transduction. *EMBO J* 1999;**18**:375–85.

107. Hilton DJ, Richardson RT, Alexander WS, et al. Twenty proteins containing a C-terminal SOCS box form five structural classes. *Proc Natl Acad Sci USA* 1998;**95**:114–19.

108. Narazaki M, Fujimoto M, Matsumoto T, et al. Three distinct domains of SSI-1/SOCS-1/JAB protein are required for its suppression of interleukin 6 signaling. *Proc Natl Acad Sci USA* 1998;**95**:13,130–13,134.

109. Nicholson SE, Hilton DJ. The SOCS proteins: a new family of negative regulators of signal transduction. *J Leukocyte Biol* 1998;**63**:665–8.

110. Yasukawa H, Misawa H, Sakamoto H, et al. The JAK-binding protein JAB inhibits Janus tyrosine kinase activity through binding in the activation loop. *EMBO J* 1999;**18**:1309–20.

111. Qing Y, Costa-Pereira AP, Watling D, Stark G. Role of tyrosine 441 of interferon-gamma receptor subunit 1 in SOCS-1-mediated attenuation of STAT1 activation. *J Biol Chem* 2005;**280**:1849–53.

112. Fenner JE, Starr R, Cornish AL, et al. Suppressor of cytokine signaling 1 regulates the immune response to infection by a unique inhibition of type I interferon activity. *Nature Immunol* 2006;**7**:33–9.

113. Waiboci LW, Ahmed CM, Mujtaba MG, et al. Both the suppressor of cytokine signaling 1 (SOCS-1) kinase inhibitory region and SOCS-1 mimetic bind to JAK2 autophosphorylation site: implications for the development of a SOCS-1 antagonist. *J Immunol* 2007;**178**:5058–68.

114. Alexander WS, Starr R, Metcalf D, et al. Suppressors of cytokine signaling (SOCS): negative regulators of signal transduction. *J Leukocyte Biol* 1999;**66**:588–92.

115. Starr R, Metcalf D, Elefanty AG, et al. Liver degeneration and lymphoid deficiencies in mice lacking suppressor of cytokine signaling-1. *Proc Natl Acad Sci USA* 1998;**95**:14,395–14,399.

116. Dickensheets H, Vazquez N, Sheikh F, et al. Suppressor of cytokine signaling-1 is an IL-4-inducible gene in macrophages and feedback inhibits IL-4 signaling. *Genes Immunity* 2007;**8**:21–7.

117. Eyles JL, Metcalf D, Grusby MJ, Hilton DJ, Starr R. Negative regulation of interleukin-12 signaling by suppressor of cytokine signaling-1. *J Biol Chem* 2002;**277**:43,735–43,740.

118. Davey GM, Starr R, Cornish AL, et al. SOCS-1 regulates IL-15-driven homeostatic proliferation of antigen-naive CD8 T cells, limiting their autoimmune potential. *J Exp Med* 2005;**202**:1099–108.

119. Cornish AL, Chong MM, Davey GM, et al. Suppressor of cytokine signaling-1 regulates signaling in response to interleukin-2 and other γc-dependent cytokines in peripheral T cells. *J Biol Chem* 2003;**278**:22,755–22,761.

120. Gingras S, Parganas E, de Pauw A, Ihle JN, Murray PJ. Re-examination of the role of suppressor of cytokine signaling 1 (SOCS1) in the regulation of toll-like receptor signaling. *J Biol Chem* 2004;**279**:54,702–54,707.

121. Egan PJ, Lawlor KE, Alexander WS, Wicks IP. Suppressor of cytokine signaling-1 regulates acute inflammatory arthritis and T cell activation. *J Clin Invest* 2003;**111**:915–24.

122. Chong MMW, Chen Y, Darwiche R, et al. Suppressor of cytokine signaling-1 overexpression protects pancreatic β cells from CD8(+) T cell-mediated autoimmune destruction. *J Immunol* 2004;**172**:5714–21.

123. Yoshikawa H, Matsubara K, Qian GS, et al. SOCS-1, a negative regulator of the JAK/STAT pathway, is silenced by methylation in human hepatocellular carcinoma and shows growth-suppression activity. *Nature Genet* 2001;**28**:29–35.

124. Sutherland KD, Lindeman GJ, Choong DYH, et al. Differential hypermethylation of SOCS genes in ovarian and breast carcinomas. *Oncogene* 2004;**23**:7726–33.

125. Chen CY, Tsay W, Tang JL, et al. SOCS1 methylation in patients with newly diagnosed acute myeloid leukemia. *Genes Chromosomes Cancer* 2003;**37**:300–5.

126. Watanabe D, Ezoe S, Fujimoto M, et al. Suppressor of cytokine signalling-1 gene silencing in acute myeloid leukemia and human haematopoietic cell lines. *Br J Haematol* 2004;**126**:726–35.

127. Galm O, Yoshikawa H, Esteller M, Osieka R, Herman JG. SOCS-1, a negative regulator of cytokine signaling, is frequently silenced by methylation in multiple myeloma. *Blood* 2003;**101**:2784–8.

128. Sasaki A, Yasukawa H, Suzuki A, et al. Cytokine-inducible SH2 protein-3 (CIS3/SOCS3) inhibits Janus tyrosine kinase by binding through the N-terminal kinase inhibitory region as well as SH2 domain. *Genes Cells* 1999;**4**:339–51.

129. Fujimoto M, Naka T. Regulation of cytokine signaling by SOCS family molecules. *Trends Immunol* 2003;**24**:659–66.

130. Yamamoto K, Yamaguchi M, Miyasaka N, Miura O. SOCS-3 inhibits IL-12-induced STAT4 activation by binding through its SH2 domain to the STAT4 docking site in the IL-12 receptor β2 subunit. *Biochem Biophys Res Comm* 2003;**310**:1188–93.

131. Eyckerman S, Broekaert D, Verhee A, Vandekerckhove J, Tavernier J. Identification of the Y985 and Y1077 motifs as SOCS3 recruitment sites in the murine leptin receptor. *FEBS Letts* 2000;**486**:33–7.

132. Hortner M, Nielsch U, Mayr LM, Heinrich PC, Haan S. A new high affinity binding site for suppressor of cytokine signaling-3 on the erythropoietin receptor. *Eur J Biochem* 2002;**269**:2516–26.

133. Hortner M, Nielsch U, Mayr LM, Johnston JA, Heinrich PC, Haan S. Suppressor of cytokine signaling-3 is recruited to the activated granulocyte-colony stimulating factor receptor and modulates its signal transduction. *J Immunol* 2002;**169**:1219–27.

134. Sommer U, Schmid C, Sobota RM, et al. Mechanisms of SOCS3 phosphorylation upon interleukin-6 stimulation – contributions of Src- and receptor-tyrosine kinases. *J Biol Chem* 2005;**280**:31,478–31,488.

135. Mori H, Hanada R, Hanada T, et al. Socs3 deficiency in the brain elevates leptin sensitivity and confers resistance to diet-induced obesity. *Nature Med* 2004;**10**:739–43.

136. Lang R, Pauleau AL, Parganas E, et al. SOCS3 regulates the plasticity of gp130 signaling. *Nature Immunol* 2003;**4**:546–50.

137. Croker BA, Krebs DL, Zhang JG, et al. SOCS3 negatively regulates IL-6 signaling in vivo. *Nature Immunol* 2003;**4**:540–5.

138. Yasukawa H, Ohishi M, Mori H, et al. IL-6 induces an anti-inflammatory response in the absence of SOCS3 in macrophages. *Nature Immunol* 2003;**4**:551–6.

139. Takahashi Y, Carpino N, Cross JC, Torres M, Parganas E, Ihle JN. SOCS3: an essential regulator of LIF receptor signaling in trophoblast giant cell differentiation. *EMBO J* 2003;**22**:372–84.

140. Croker BA, Metcalf D, Robb L, et al. SOCS3 is a critical physiological negative regulator of G-CSF signaling and emergency granulopoiesis. *Immunity* 2004;**20**:153–65.

141. Kimura A, Kinjyo I, Matsumura Y, et al. SOCS3 is a physiological negative regulator for granulopoiesis and granulocyte colony-stimulating factor receptor signaling. *J Biol Chem* 2004;**279**:6905–10.

142. Suzuki A, Hanada T, Mitsuyama K, et al. CIS3/SOCS3/SSI3 plays a negative regulatory role in STAT3 activation and intestinal inflammation. *J Exp Med* 2001;**193**:471–81.

143. Atsumi T, Ishihara K, Kamimura D, et al. A point mutation of Tyr-759 in interleukin 6 family cytokine receptor subunit gp130 causes autoimmune arthritis. *J Exp Med* 2002;**196**:979–90.

144. Wong PKK, Egan PJ, Croker BA, et al. SOCS-3 negatively regulates innate and adaptive immune mechanisms in acute IL-1-dependent inflammatory arthritis. *J Clin Invest* 2006;**116**:1571–81.

145. Riehle KJ, Campbell JS, McMahan RS, et al. Regulation of liver regeneration and hepatocarcinogenesis by suppressor of cytokine signaling 3. *J Exp Med* 2008;**205**:91–103.

146. He B, You L, Uematsu K, et al. SOCS-3 is frequently silenced by hypermethylation and suppresses cell growth in human lung cancer. *Proc Natl Acad Sci USA* 2003;**100**:14,133–14,138.

147. Tokita T, Maesawa C, Kimura T, et al. Methylation status of the SOCS3 gene in human malignant melanomas. *Intl J Oncol* 2007;**30**:689–94.

148. Weber A, Hengge UR, Bardenheuer W, et al. SOCS-3 is frequently methylated in head and neck squamous cell carcinoma and its precursor lesions and causes growth inhibition. *Oncogene* 2005;**24**:6699–708.

149. Tischoff I, Hengge UR, Vieth M, et al. Methylation of SOCS-3 and SOCS-1 in the carcinogenesis of Barrett's adenocarcinoma. *Gut* 2007;**56**:1047–53.

150. Niwa Y, Kanda H, Shikauchi Y, et al. Methylation silencing of SOCS-3 promotes cell growth and migration by enhancing JAK/STAT and FAK signalings in human hepatocellular carcinoma. *Oncogene* 2005;**24**:6406–17.

151. Matsumoto A, Masuhara M, Mitsui K, et al. CIS, a cytokine inducible SH2 protein, is a target of the JAK-STAT5 pathway and modulates STAT5 activation. *Blood* 1997;**89**:3148–54.

152. Yoshimura A, Ohkubo T, Kiguchi T, et al. A novel cytokine-inducible gene CIS encodes an SH2-containing protein that binds to tyrosine-phosphorylated interleukin 3 and erythropoietin receptors. *EMBO J* 1995;**14**:2816–26.

153. Ram PA, Waxman DJ. SOCS/CIS protein inhibition of growth hormone-stimulated STAT5 signaling by multiple mechanisms. *J Biol Chem* 1999;**274**:35,553–35,561.

154. Verdier F, Rabionet R, Gouilleux F, et al. A sequence of the CIS gene promoter interacts preferentially with two associated STAT5A dimers: a distinct biochemical difference between STAT5A and STAT5B. *Mol Cell Biol* 1998;**18**:5852–60.

155. Marine J-C, McKay C, Wang D, et al. SOCS3 is essential in the regulation of fetal liver erythropoiesis. *Cell* 1999;**98**:617–27.

156. Matsumoto A, Seki Y, Kubo M, et al. Suppression of STAT5 functions in liver, mammary glands, and T cells in cytokine-inducible SH2-containing protein 1 transgenic mice. *Mol Cell Biol* 1999;**19**:6396–407.

157. Metcalf D, Greenhalgh CJ, Viney E, et al. Gigantism in mice lacking suppressor of cytokine signalling-2. *Nature* 2000;**405**:1069–73.

158. Greenhalgh CJ, Rico-Bautista E, Lorentzon M, et al. SOCS2 negatively regulates growth hormone action *in vitro* and *in vivo*. *J Clin Invest* 2005;**115**:397–406.

159. Michaylira CZ, Simmons JG, Ramocki NM, et al. Suppressor of cytokine signaling-2 limits intestinal growth and enterotrophic actions of IGF-I *in vivo*. *Am J Physiol Gastro Liver Physiol* 2006;**291**:G472–81.

160. Harris J, Stanford PM, Sutherland K, et al. Socs2 and Elf5 mediate prolactin-induced mammary gland development. *Mol Endocrinol* 2006;**20**:1177–87.

161. Turnley AM, Faux CH, Rietze RL, Coonan JR, Bartlett PF. Suppressor of cytokine signaling 2 regulates neuronal differentiation by inhibiting growth hormone signaling. *Nature Neurosci* 2002;**5**:1155–62.

162. Machado FS, Johndrow JE, Esper L, et al. Anti-inflammatory actions of lipoxin A(4) and aspirin-triggered lipoxin are SOCS-2 dependent. *Nature Med* 2006;**12**:330–4.

163. Favre H, Benhamou A, Finidori J, Kelly PA, Edery M. Dual effects of suppressor of cytokine signaling (SOCS-2) on growth hormone signal transduction. *FEBS Letts* 1999;**453**:63–6.

164. Michaylira CZ, Ramocki NM, Simmons JG, et al. Haplotype insufficiency for suppressor of cytokine signaling-2 enhances intestinal growth and promotes polyp formation in growth hormone-transgenic mice. *Endocrinology* 2006;**147**:1632–41.

165. Haffner MC, Petridou B, Peyrat JP, et al. Favorable prognostic value of SOCS2 and IGF-I in breast cancer. *BMC Cancer* 2007;**7**:9.

166. Bullock AN, Debreczeni JE, Edwards AM, Sundstrom M, Knapp S. Crystal structure of the SOCS2-elongin C-elongin B complex defines a prototypical SOCS box ubiquitin ligase. *Proc Natl Acad Sci USA* 2006;**103**:7637–42.

167. Bullock AN, Rodriguez MC, Debreczeni JE, Songyang Z, Knapp S. Structure of the SOCS4–ElonginB/C complex reveals a distinct SOCS box interface and the molecular basis for SOCS-dependent EGFR degradation. *Structure* 2007;**15**:1493–504.

168. Kamura T, Maenaka K, Kotoshiba S, et al. VHL-box and SOCS-box domains determine binding specificity for Cul2-Rbx1 and Cul5-Rbx2 modules of ubiquitin ligases. *Genes Dev* 2004;**18**:3055–65.

169. Tanaka T, Soriano MA, Grusby MJ. SLIM is a nuclear ubiquitin E3 ligase that negatively regulates STAT signaling. *Immunity* 2005;**22**:729–36.

170. Narazaki M, Fujimoto M, Matsumoto T, et al. Three distinct domains of SSI-1/SOCS-1/JAB protein are required for its suppression of interleukin 6 signaling. *Proc Natl Acad Sci USA* 1998;**95**:13,130–13,134.

171. Kamura T, Sato S, Haque D, et al. The Elongin BC complex interacts with the conserved SOCS-box motif present in members of the SOCS, ras, WD-40 repeat, and ankyrin repeat families. *Genes Dev* 1998;**12**:3872–81.

172. Hanada T, Yoshida T, Kinjyo I, et al. A mutant form of JAB/SOCS1 augments the cytokine-induced JAK/STAT pathway by accelerating degradation of wild-type JAB/CIS family proteins through the SOCS-box. *J Biol Chem* 2001;**276**:40,746–40,754.

173. Haan S, Ferguson P, Sommer U, et al. Tyrosine phosphorylation disrupts elongin interaction and accelerates SOCS3 degradation. *J Biol Chem* 2003;**278**:31,972–31,979.

174. Zhang JG, Farley A, Nicholson SE, et al. The conserved SOCS box motif in suppressors of cytokine signaling binds to elongins B and C and may couple bound proteins to proteasomal degradation. *Proc Natl Acad Sci USA* 1999;**96**:2071–6.

175. Chen XP, Losman JA, Rothman P. SOCS proteins, regulators of intracellular signaling. *Immunity* 2000;**13**:287–90.

176. Kamizono S, Hanada T, Yasukawa H, et al. The SOCS box of SOCS-1 accelerates ubiquitin-dependent proteolysis of TEL-JAK2. *J Biol Chem* 2001;**276**:12,530–12,538.

177. Frantsve J, Schwaller J, Sternberg DW, Kutok J, Gilliland DG. Socs-1 inhibits TEL-JAK2-mediated transformation of hematopoietic cells through inhibition of JAK2 kinase activity and induction of proteasome-mediated degradation. *Mol Cell Biol* 2001;**21**:3547–57.

178. Ungureanu D, Saharinen P, Junttila I, Hilton DJ, Silvennoinen O. Regulation of Jak2 through the ubiquitin-proteasome pathway involves phosphorylation of Jak2 on Y1007 and interaction with SOCS-1. *Mol Cell Biol* 2002;**22**:3316–26.

179. Landsman T, Waxman DJ. Role of the cytokine-induced SH2 domain-containing protein CIS in growth hormone receptor internalization. *J Biol Chem* 2005;**280**:37,471–37,480.

180. Irandoust MI, Aarts LH, Roovers O, Gits J, Erkeland SJ, Touw IP. Suppressor of cytokine signaling 3 controls lysosomal routing of G-CSF receptor. *EMBO J* 2007;**26**:1782–93.

181. Tannahill GM, Elliott J, Barry AC, Hibbert L, Cacalano NA, Johnston JA. SOCS2 can enhance interleukin-2 (IL-2) and IL-3 signaling by accelerating SOCS3 degradation. *Mol Cell Biol* 2005;**25**:9115–26.

182. Sasaki A, Inagaki-Ohara K, Yoshida T, et al. The N-terminal truncated isoform of SOCS3 translated from an alternative initiation AUG codon under stress conditions is stable due to the lack of a major ubiquitination site Lys-6. *J Biol Chem* 2003;**278**:2432–6.

183. Babon JJ, McManus EJ, Yao S, et al. The structure of SOCS3 reveals the basis of the extended SH2 domain function and identifies an unstructured insertion that regulates stability. *Mol Cell* 2006;**22**:205–16.

184. Zhang JG, Metcalf D, Rakar S, et al. The SOCS box of suppressor of cytokine signaling-1 is important for inhibition of cytokine action in vivo. *Proc Natl Acad Sci USA* 2001;**98**:13,261–13,265.

185. Boyle K, Egan P, Rakar S, et al. The SOCS box of suppressor of cytokine signaling-3 contributes to the control of G-CSF responsiveness *in vivo*. *Blood* 2007;**110**:1466–74.

186. Melzner I, Bucur AJ, Bruderlein S, et al. Biallelic mutation of SOCS-1 impairs JAK2 degradation and sustains phospho-JAK2 action in the MedB-1 mediastinal lymphoma line. *Blood* 2005;**105**:2535–42.

Chapter 9

Protein Proximity Interactions

Matthew G. Gold and John D. Scott

Howard Hughes Medical Institute, University of Washington, School of Medicine, Department of Pharmacology, Seattle, Washington

The cell interior is an architectural triumph in which specific protein–protein interactions provide the mortar. Such "protein proximity" interactions increase rates of signal transduction, insulate different signaling pathways, and enable sophisticated dynamic cellular processes to proceed. This section of the handbook introduces the mechanisms that contribute to the compartmentalization of signaling enzymes and the assembly of multiprotein signaling networks, in three areas:

1. Advances in the analysis of protein–protein interactions
2. Subcellular structures and multiprotein complexes that contribute to cell signaling
3. Kinase and phosphatase targeting.

ADVANCES IN THE ANALYSIS OF PROTEIN–PROTEIN INTERACTIONS

Mass spectrometry (MS) is now a mainstay in identifying protein–protein interactions. In the opening chapter (Chapter 154 of Handbook of Cell Signaling, Second Edition), Timothy Haystead discusses an advanced MS-based approach for the analysis of *in vivo* phosphorylation sites and the kinases and phosphatases that regulate them. Peter Verveer and Philippe Bastiaens (Chapter 155 of Handbook of Cell Signaling, Second Edition) introduce Fluorescence Resonance Energy Transfer (FRET) and optical techniques that provide sensitive means to quantify and detect protein interactions in living cells. New to this edition of the *Handbook*, Ian Donaldson reviews the data resources for investigating protein interactions (Chapter 170 of Handbook of Cell Signaling, Second Edition).

SUBCELLULAR STRUCTURES AND MULTIPROTEIN COMPLEXES THAT CONTRIBUTE TO CELL SIGNALING

In the next section, we consider protein proximity interactions in the context of multiprotein complexes and subcellular structures. Christopher Turner and co-authors describe the network of intracellular interactions that comprise the focal adhesion (Chapter 156 of Handbook of Cell Signaling, Second Edition), before Frank Mason and Scott Soderling discuss WASP and WAVE protein complexes (Chapter 157 of Handbook of Cell Signaling, Second Edition). Mary Kennedy describes the synaptic NMDA-receptor complex (Chapter 158 of Handbook of Cell Signaling, Second Edition). Yann Hyvert and Jean-Luc Imler describe the network of intracellular interactions downstream of Toll family receptors (Chapter 159 of Handbook of Cell Signaling, Second Edition), and Andrey Shaw and Emanuele Giurisato introduce signaling at the immunological synapse (Chapter 160 of Handbook of Cell Signaling, Second Edition). Mark Hochstrasser discusses the ubiquitin proteasome system (Chapter 161 of Handbook of Cell Signaling, Second Edition), and in the final chapter Guy Salvesen reviews cell signaling by caspases (Chapter 162 of Handbook of Cell Signaling, Second Edition).

KINASE AND PHOSPHATASE TARGETING

Seven chapters introduce some of the anchoring, adapter and scaffolding proteins that organize broad specificity protein kinases and phosphoprotein phosphatases. Elaine Elion and co-authors provide a historical perspective by describing the identification and analyses of scaffolding proteins that maintain Mitogen Activated Protein (MAP) kinase cascades in yeast (Chapter 163 of Handbook of Cell Signaling, Second Edition). A complementary chapter by Norman Kennedy and Roger Davis defines MAP kinase and Jun kinase scaffolds in mammalian cells (Chapter 164 of Handbook of Cell Signaling, Second Edition), while Matthew Pink and Mark Dell'Acqua introduce A-kinase Anchoring Proteins (AKAPs) that localize the cAMP-dependent protein kinase at specific sites inside cells in Chapter 165 of Handbook of Cell Signaling, Second Edition. The context-dependent assembly of AKAP signal transduction units is then explored by Lorene Langeberg and John Scott in Chapter 166 of Handbook of Cell Signaling, Second Edition. Anthony Baucum and Roger Colbran then review the extensive literature on protein phosphatase localization with emphasis on PP1 and calcineurin targeting in dendrites (Chapter 167 of Handbook of Cell Signaling, Second Edition). In Chapter 168 of Handbook of Cell Signaling, Second Edition, Marc Mumby and co-authors catalogue the numerous families of targeting subunits that compartmentalize and modulate the activity of protein phosphatase 2 A. Finally, Hubert Hondermarck introduces the key role of 14-3-3 proteins in protein proximity interactions (Chapter 169 of Handbook of Cell Signaling, Second Edition).

Chapter 10

Global Analysis of Phosphoregulatory Networks

Janine Mok[1] and Michael Snyder[2]

[1] *Department of Genetics, Yale University School of Medicine, New Haven, Connecticut*

[2] *Department of Molecular, Cellular, and Developmental Biology, Yale University, New Haven, Connecticut*

INTRODUCTION

The activity of most, if not all, cellular proteins is regulated on some level by posttranslational modifications. Hundreds of posttranslational modifications have been described to date, and all function to amplify the complexity of the eukaryotic proteome [1]. The most extensively studied, however, is that of reversible protein phosphorylation. Reversible protein phosphorylation, through the action of protein kinases and phosphatases, has been implicated to play a role in nearly every basic cellular process [2]. These processes range from DNA replication, gene transcription, and protein translation to cell differentiation, cell growth, and intercellular communication. In fact, abnormal phosphorylation, such as that resulting from mutations in kinases or phosphatases, has been recognized as a cause or a consequence of many human diseases such as cancer [3, 4]. Notably, many different mechanisms of regulation have been described for phosphorylation. Phosphorylation can not only directly activate or deactivate a protein target, but also affect the rate at which a protein is degraded, its ability to translocate from one subcellular compartment to another, and its capacity to bind to other proteins.

Phosphorylation remains one of the most widespread posttranslational modifications present in all eukaryotes. Nearly one-third of the eukaryotic proteome is predicted to be phosphorylated [5]. Furthermore, kinases are encoded by roughly 2 percent of the eukaryotic genome, and thus represent one of the largest classes of proteins [6]. There are predicted to be 518 protein kinases in humans [7], 540 in mice [8], and 122 in yeast [9].

Insights into phosphorylation signaling pathways have traditionally been revealed through single gene studies. However, recent years have witnessed a revolution in the way phosphorylation is studied. More and more global studies have emerged that look at phosphorylation on a proteomic scale. These studies have provided tremendous insight into which proteins are phosphorylated, which kinases are responsible for these phosphorylation events, and how the various signaling pathways are connected.

GLOBAL MAPPING OF PHOSPHORYLATION

Previous methods to identify phosphorylated residues *in vivo* involve the use of ^{32}P to metabolically label the protein under investigation. Typically, protease digestion of the protein, followed by high performance liquid chromatography (HPLC) or two-dimensional gel electrophoresis separation of the peptide mixture, and collection of the radioactive fractions or bands is coupled with Edman sequencing to determine the site of phosphorylation [10–12]. However, Edman sequencing is limited by its requirement to purify the phosphorylated protein, making this method difficult to adapt to high-throughput studies. Consequently, mass spectrometry (MS), which dispenses the need for a purification step, has emerged as the preferred tool for the global mapping of phosphorylation sites. The use of MS to characterize phosphorylation on a large scale is nevertheless not without its own problems. Signals of phosphopeptides in the MS spectra of complex samples are, in most cases, suppressed by the relatively high concentration of non-phosphopeptides present in the proteolytic digests of samples. Furthermore, substoichiometric phosphorylation often occurs, and phosphopeptides may show inefficient ionization or may be lost preferentially during handling by adsorption to plastics. Recent studies have addressed these issues by developing techniques for phosphopeptide enrichment and optimizing ionization. These advances have resulted in the identification of thousands of *in vivo* phosphorylation sites, and have set the stage for emerging technologies of quantitative mapping of phosphorylation sites in complex samples.

Many techniques have been developed to enrich for phosphopeptides. One of these techniques is affinity chromatography using immobilized antibodies specific for phosphorylated residues. This technique has most successfully been applied using anti-phosphotyrosine antibodies to analyze tyrosine phosphorylation. Rush *et al.* used this technique in conjunction with liquid chromatography-tandem mass spectrometry (LC-MS/MS) to identify 194 phosphotyrosine sites in pervanadate-treated Jurkat T cells, 185 phosphotyrosine sites in NIH3T3 cells expressing a constitutively active c-Src allele, and 180 phosphotyrosine sites in SU-DHL-1 cells – a cancer line derived from anaplastic large cell lymphomas [13]. Rikova *et al.* similarly used this technique to identify 4552 phosphotyrosine sites on over 2700 proteins in 41 non-small cell lung cancer (NSCLC) cell lines and over 150 NSCLC tumors [14], and Ballif *et al.* used the technique to analyze murine brains and identified 414 phosphotyrosine sites on 303 proteins [15]. Anti-phosphoserine and anti-phosphothreonine antibodies are, however, seldom used for affinity chromatography, as non-specific binding with these antibodies tends to be high. Moreover, because phosphoserine- and phosphothreonine-containing peptides are often lost prior to elution from the immobilized antibodies, flow-through and washing steps need to be conscientiously monitored for the presence of phosphoproteins.

A technique that can be used to enrich for all phosphopeptides is immobilized metal affinity chromatography (IMAC) [16, 17]. IMAC relies on the affinity of phosphate groups for certain metal ions (namely Fe^{3+} and Ga^{3+}) that are bound to tethered chelating reagents present on solid-phase supports. When combined with the esterification of carboxylic groups to prevent non-specific binding, this technique is highly selective for phosphorylated peptides [18]. The effectiveness of IMAC to identify phosphorylation sites within the context of a complex mixture has been demonstrated by a number of studies. Ficarro *et al.* identified 383 phosphorylation sites on 216 proteins in *Saccharomyces cerevisiae* [19], Kim *et al.* identified 213 phosphorylation sites and 25 more for which the precise site of phosphorylation was ambiguous on 116 proteins in HT-29 human colon adenocarcinoma cells [20], and Moser *et al.* identified 339 phosphorylation sites on over 200 proteins in rat liver [21]. Analogous to IMAC is the use of titanium dioxide and zirconium dioxide columns for phosphopeptide enrichment [22,23]. These compounds adsorb phosphopeptides through bidentate interactions. Still relatively new, these methods need to be further developed, but show promise in being more efficient than IMAC as significantly less column preparation time is required.

Another technique used for phosphopeptide enrichment is strong cation exchange (SCX) [24]. When performed at low pH, SCX separates phosphorylated from non-phosphorylated peptides on the basis of the charge difference associated with the negatively charged phosphate group. Singly phosphorylated peptides display a net charge state of +1 at low pH, which is different from non-phosphopeptides, which have a net charge of +2. Ballif *et al.* used SCX LC-MS/MS to profile the phosphoproteome of developing murine brain, and identified 460 phosphorylation sites and 86 more for which the precise site of phosphorylation was ambiguous on 351 proteins [25]. This technique was also used by Beausoleil *et al.* to analyze nuclear phosphoproteins in HeLa cells [26]. Remarkably, this study identified 2002 phosphorylation sites on 967 proteins. However, a concern with this approach is that only singly phosphorylated peptides are enriched by SCX; all singly phosphorylated peptides with basic residues or multiply phosphorylated peptides have a net charge other than +1 and therefore will be missed in the analysis. Furthermore, as with IMAC, the multiple fractionation steps used during the enrichment process present the need for large amounts of starting material and extensive analysis time.

Other techniques make use of chemical tagging of phosphate groups through covalent modification. Phosphoserine and phosphothreonine residues have reactive chemistries in which base-catalyzed β-elimination reactions result in a Cα=Cβ bond that is susceptible to Michael addition by various nucleophilic tags, including those containing reactive groups that can be further adapted for solid-phase capture [27]. Oda *et al.* developed a method in which phosphate groups are replaced by biotinylated moieties, enabling phosphopeptides to be selectively purified using immobilized avidin [28]. In a similar approach, McLachlin *et al.* used dithiothreitol (DTT) for nucleophilic addition to the Cα=Cβ bond [29]. Here, DTT peptides are covalently attached via a disulfide exchange to an immobilized thiolate resin for purification, and released for MS analysis by eluting with an excess of unconjugated DTT. Jalili *et al.* have analogously derivatized phosphoserine- and phosphothreonine-containing proteins using a histidine tag that has a thiol-containing cysteine residue [30]. The resulting histidine-tagged peptides are purified by Ni^{2+}-IMAC and are either eluted with imidazole or cleaved with Factor Xa. Lastly, Knight *et al.* have used the reactive chemistries of phosphoserine and phosphothreonine residues to transform them into lysine analogs (aminoethylcysteine and β-methylaminoethylcysteine) and in turn generate trypsin recognition sites at each previously phosphorylated residue [31].

There are a few caveats to the chemical tagging approaches described above. The first is that Michael addition may occur at both Cα and Cβ, and in turn result in heterogeneous products [32]. Secondly, β-elimination of other posttranslational modifications, such as O-glycosylation, can occur. This event will likely yield products that can be misinterpreted as a signature of phosphorylation. Labeling of some protein samples is also substoichiometric [33].

Most notable, however, is the fact that these approaches rely on reactive chemistries specific to phosphoserine and phosphothreonine residues.

Consequently, an alternative approach has been developed that can be used to label phosphoserine, phosphothreonine, and phosphotyrosine residues. This approach involves reacting phosphorylated residues with N-2-dimethylaminpropyl-N′-ethylcarbodiimide (EDC) to produce direct linkages to the phosphate group [34]. The resulting phosphoramidate adduct can then be coupled with dendrimers for solid phase capture, or captured onto functionalized glass beads. Tao *et al.* conjugated protease-digested peptides, initially purified using anti-phosphotyrosine antibodies, to dendrimers for enrichment by solid phase capture and used them to map phosphorylated tyrosine residues in human T cells [35]. In addition to all the known tyrosine phosphorylation sites within the immunoreceptor tyrosine-based activation motifs and the T cell receptor CD3 chains, they identified several novel tyrosine phosphorylation sites on a total of 97 proteins. Using similar chemistry, Bodenmiller *et al.* derivatized phosphorylated peptides with cystamine for capture onto maleimide glass beads to profile the phosphoproteome of *Drosophila melanogaster* Kc167 cells [36]. They identified 571 phosphorylation sites, many of which mapped to proteins involved in growth and development, such as *unkempt*, and the related *Saccharomyces cerevisiae* septin protein *peanut.*

These phosphopeptide enrichment techniques have all been instrumental in reducing sample complexity and enabling the successful mapping of phosphorylation sites. However, a recent study demonstrates that each phosphopeptide enrichment technique differs in its specificity in isolation, suggesting that the comprehensive mapping of phosphorylation sites by MS is likely only achievable by using multiple phosphopeptide enrichment techniques in parallel [37]. In this study, Bodenmiller *et al.* directly compared phosphoramidate chemistry, IMAC, and titanium dioxide phosphopeptide enrichment methods using *Drosophila melanogaster* Kc167 cells. They showed that while each exhibited similar reproducibility, there was at best a 34 percent overlap between any two of the three methods. Consistent with this observation is the work by Villen *et al.* [38]. Using SCX, in combination with IMAC, this group was able to identify 5250 phosphorylation sites in murine liver. However, by preparing additional samples using affinity chromatography with anti-phosphotyrosine antibodies, they were able to identify an additional 363 phosphorylation sites that were not previously identified by SCX-IMAC. Table 10.1 provides a summary of the references large-scale MS studies.

The successful mapping of phosphorylation sites by MS is also largely dependent upon efficient peptide ionization. Typically, MS/MS is performed using low-energy collisionally activated dissociation (CAD) in positive ion mode. In CAD, ions acquire positive charge by the addition of protons, and fragmentation primarily occurs by nucleophilic reactions [39]. Sites of cleavage are thus strongly influenced by peptide sequences and the distribution of protons across backbone and side-chain atoms. CAD ionization of phosphopeptides often promotes the β-elimination of phosphoric acid from phosphorylated residues without breaking the amide bonds along the peptide backbone – a phenomenon called neutral loss. Answering the need to improve the ionization of phosphopeptides, other ionization techniques have emerged for MS mapping of posttranslational modifications. These techniques include electron capture dissociation (ECD) and electron transfer dissociation (ETD) [27]. In ECD and ETD, peptides are fragmented through interactions with low-energy electrons (ECD) or radical anions (ETD), and in turn form peptide radicals that readily undergo backbone cleavage [40,41]. ECD and ETD have advantages over CAD in detecting phosphorylation, because peptide fragmentation is less influenced by peptide sequence, and neutral loss reactions are reduced. Moreover, MS of negatively charged ions tends to be more sensitive than positive-mode MS for detecting phosphopeptides, since negative-mode MS of phosphorylated serine, threonine, and tyrosine residues yield fragments of $-79\,\mathrm{Da}$ (PO_{3-}) or $-63\,\mathrm{Da}$ (PO_{2-}). A recent study conducted by Chi *et al.* demonstrated the importance of efficient ionization of phosphopeptides in mapping phosphorylation sites [42]. Using ETD, these authors identified 1252 phosphorylation sites on 629 proteins in *Saccharomyces cerevisiae* tryptic digests enriched by IMAC, whereas the same laboratory, previously using CAD, only identified 383 phosphorylation sites on 216 proteins from similarly prepared samples.

The development of techniques to profile phosphorylation qualitatively on a proteomic scale has paved the way for more quantitative approaches to take the stage. Quantitative information is essential for the accurate modeling of signal transduction pathways. In fact, most changes resulting from a targeted perturbation are only detectable if some quantitative information is already known. Moreover, quantitative information ensures that the failure to identify or detect a phosphopeptide by MS analysis is the result of the absence of that particular peptide, and not the result of its presence below the threshold of detection. Quantitative information can be expressed both in absolute and in relative terms. Absolute quantitation of phosphorylation is the actual abundance of phosphopeptides in a sample, whereas relative quantitation of phosphorylation is the difference in phosphopeptide abundance between two or more samples. Several approaches that quantify phosphorylation have been described to date. The most successful of them involve coding phosphopeptides with stable isotopes, and are described below.

The simplest approach for quantifying phosphorylation involves the chemical synthesis of isotope-coded peptides that mimic native peptides formed by proteolysis, complete with any posttranslational modifications. These synthetic

TABLE 10.1 Summary of referenced large-scale MS studies

Author	Number identified	Cell/tissue type	Phosphopeptide enrichment strategy
Rush et al. [13]	194 phosphosites	Jurkat T cells	Anti-phosphotyrosine IP
	185 phosphosites	NIH3T3-Src cells	
	180 phosphosites	SU-DHL-1 cells	
Rikova et al. [14]	4551 phosphosites on 2700 proteins	Non-small cell lung cancer cells lines and tumors	Anti-phosphotyrosine IP
Ballif et al. [15]	414 phosphosites on 303 proteins	Adult murine brain	Anti-phosphotyrosine IP
Ficarro et al. [19]	383 phosphosites on 216 proteins	Saccharomyces cerevisiae	IMAC
Kim et al. [20]	213 phosphosites + 25 ambiguous sites on 116 proteins	HT-29 human colon adenocarcinoma cells	IMAC
Moser and White [21]	339 phosphosites on >200 proteins	Rat liver	IMAC
Ballif et al. [25]	460 phosphosites + 86 ambiguous sites on 351 proteins	Developing murine brain	SCX
Beausoleil et al. [26]	2002 phosphosites on 967 proteins	HeLa cells	SCX
Tao et al. [35]	97 tyrosine phosphoproteins	Jurkat T cells	Anti-phosphotyrosine IP, Phosphoramidate chemistry
Bodenmiller et al. [36]	571 phosphosites	Drosophila melanogaster Kc167 cells	Phosphoramidate chemistry
Villen et al. [38]	5635 phosphosites on 2328 proteins	Murine liver	SCX, Anti-phosphotyrosine IP, IMAC
Chi et al. [42]	1252 phosphosites on 629 proteins	Saccharomyces cerevisiae	IMAC
Qian et al. [44]	86 phosphopeptides	Human breast cancer MCF-7 cells	β-elimination chemistry
Zhang et al. [47]	104 tyrosine phosphosites on 76 proteins	Human mammary epithelial cells	iTRAQ, Anti-phosphotyrosine IP, IMAC
Gruhler et al. [52]	729 phosphosites on 503 proteins	Saccharomyces cerevisiae	SCX, IMAC
Olsen et al. [53]	6600 phosphosites on 2244 proteins	HeLa cells	SCX, TiO_2

peptides are spiked into the proteolyzed sample in known quantities before MS analysis, and used as internal standards to determine the absolute abundance of their native counterparts. Stemmann et al. used this approach, termed AQUA (Absolute QUAntitation), to quantify the cell cycle dependent phosphorylation of separase in Xenopus oocytes [43]. They showed that inactive separase is present in a phosphorylated state during metaphase, and rapidly becomes dephosphorylated during the cell's transition to anaphase when separase is active. The main limitation of this approach is that it requires the ability to identify and

synthesize suitable internal peptide standards. Thus, AQUA is best suited for hypothesis-driven studies.

Other approaches quantifying phosphorylation are more amenable to analysis of phosphorylation on a proteomic scale. These approaches, however, measure relative phosphorylation, and involve pooling two or more differentially labeled samples for MS analysis (Figure 10.1). One such approach involves labeling peptides using isotopically coded chemical tags. Qian et al. made use of β-elimination and Michael addition chemistry to derivatize phosphoserine and phosphothreonine residues with an isotope-coded solid

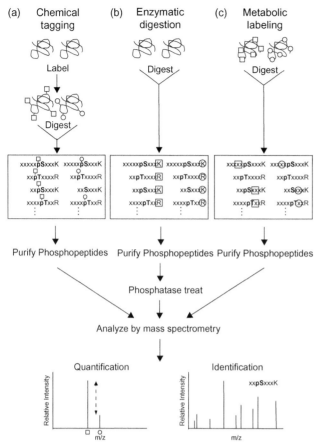

FIGURE 10.1 Isotopic labeling methods for relative quantitation of phosphorylation.
(a) Proteins are labeled through covalent modifications with isotopically coded chemical tags prior to proteolysis [35, 44, 47]. (b) Phosphopeptides are labeled at their carboxy-terminal ends during proteolysis with isotopically coded H₂O [48, 49]. (c) Proteins are metabolically labeled by growing cells in media containing isotopically coded amino acids [51–53].

phase reagent that contained either light ([^{12}C$_6$ ^{14}N]leucine) or heavy ([^{13}C$_6$ ^{15}N]leucine) stable isotopes and a photocleavable linker [44]. UV photocleavage of the captured labeled phosphopeptides followed by LC-MS/MS analysis identified 86 phosphopeptides in human breast cancer MCF-7 cells, nearly 70 percent of which could be equally be detected as light and heavy peptides. Esterification of carboxylic groups (present both as side-chains of aspartic and glutamic acids and as carboxy-termini) using deuterated methanol is another strategy used to label peptides with isotopically coded chemical tags. The study described above by Tao *et al.* used this approach to label pervanadate treated and untreated Jurkat T cells before phosphopeptide purification to quantify the dynamic changes in phosphorylation of the 97 tyrosine phosphorylated proteins in response to activation of the T cell receptor [35]. Primary amine groups have also been targeted for isotopically coded chemical tagging. Trypsin will not cleave at derivatized lysines, so peptides will generally have one label for the amino terminus

as well as one for each internal lysine [45]. iTRAQ (isobaric Tag for Relative and Absolute Quantitation) makes use of this chemistry to derivatize peptides with a tag that generates a specific reported ion in fragmentation spectra [46]. There are four isoforms of the iTRAQ tag that produce fragment ions of mass 114, 115, 116, or 117 Da, and, through a carbonyl balance group, are all isobaric. Zhang *et al.* used iTRAQ to follow the dynamics of tyrosine phosphorylation in response to epidermal growth factor (EGF) in epithelial cells [47]. Proteolytic digests of cells stimulated with EGF for 0, 5, 10, and 30 minutes were differentially labeled with the four isoforms of the iTRAQ tag, pooled, and subjected to anti-phosphotyrosine immunoprecipitation and IMAC. They identified and quantified 104 tyrosine phosphorylation sites on 76 proteins.

Enzymatic digestion has also been used to isotopically code phosphopeptides. In this approach, phosphopeptides are labeled during proteolysis that is performed in the presence of either normal (H₂^{16}O) or heavy water (H₂^{18}O). This encoding strategy labels every carboxy-terminus that is produced during proteolysis. Typically, equal portions of the differentially labeled peptides are pooled, subjected to some form of phosphopeptide purification, and dephosphorylated before being analyzed by MS to measure the ratios of phosphopeptides from each pool, which are distinguishable by a four Dalton shift. Bonenfant *et al.* demonstrated that this labeling technique could be used to detect the phosphorylation of a GST fusion of the yeast protein kinase Npr1 (nitrogen permease reactivator protein 1) in wild-type cells relative to cells treated with rapamycin and to cells lacking Sit4, the phosphatase known to be responsible for the dephosphorylation of Npr1 [48]. More recently, Smith *et al.* showed ^{18}O labeling could be extended to quantify phosphorylation in complex mixtures [49]. Yeast cell lysates were spiked with a synthetic peptide, present in both a phosphorylated and an unphosphorylated form. After labeling, the sample labeled with ^{18}O was dephosphorylated using a cocktail of phosphatases, and the sample labeled with ^{16}O was left untreated. The samples were then pooled and subjected to LC-MS/MS. Phosphorylated peptides could then be identified using an algorithm that deconvolved the MS spectra to identify pairs of peaks that exhibited different intensities. While this approach benefits from the fact that only the unphosphorylated peptide versions of the peptides are measured, careful attention must be paid to ensure that both carboxyl oxygens are labeled. If only one ^{18}O is incorporated, the mass offset of two Daltons is not sufficient to separate the isotopic envelopes and thus, as noted by Smith *et al.*, complicates the quantification.

Another approach for quantifying relative phosphorylation is SILAC, or stable isotope labeling by amino acids in cell cultures. In SILAC, proteomic samples are differentially labeled metabolically before lysis [50]. Cells are grown in media lacking a standard essential amino acid but supplemented with different non-radioactive isotopically labeled forms of that amino acid. As cells double, the isotopically coded amino acid is incorporated in a sequence-specific manner into each protein of the cell's proteome. Zhu *et al.* demonstrated the utility of SILAC labeling by encoding human skin fibroblast cells with isotopic serine and showing that the phosphorylation of the histone protein H2A.X in response to low-dose radiation could be detected [51]. Gruhler *et al.* used SILAC labeling with arginine and lysine in combination with SCX and IMAC to conduct a global study of phosphorylation in yeast [52]. They identified 729 phosphorylation sites on 503 proteins, and observed 139 sites that were differentially regulated at least two-fold in response to mating pheromone. Olsen *et al.* also used SILAC labeling with arginine and lysine, in combination instead with SCX and titanium dioxide chromatography, to monitor phosphorylation in HeLa cells [53]. They identified 6600 phosphorylation sites on 2244 proteins, with 14 percent of these sites modulated at least two-fold upon stimulation with EGF. The obvious disadvantage to SILAC is that it cannot be used to isotopically label non-cultured cell samples, such as primary tissue or blood. However, recent work by Ishihama *et al.* suggests that SILAC may still be applicable to these types of studies. They developed a technique, termed the culture-derived isotope tags (CDIT) approach, which uses SILAC labeled cells as the bridging internal standard between two tissue samples [54]. Using this technique, they quantified more than 900 proteins identified from murine brain with a SILAC labeled Neuro2A cell line as internal standards.

The developments in MS methods described above have sparked the widespread use of MS to map phosphorylation sites, and in doing so have provided a wealth of information detailing the extent to which phosphorylation is used as a regulatory mechanism in eukaryotes. However, while phosphopeptide enrichment strategies such as IMAC, SCX, and chemical tagging of phosphorylated residues have significantly reduced sample complexity, and optimized ionization strategies have improved sensitivity in detecting phosphopeptides, it is important to note that, in general, only the most abundant sites are being identified. Furthermore, quantitative methods are still in their infancy,

and have only recently been applied to whole proteomes. Thus, there continues to be a need for the development of more sensitive and efficient methods that can be applied to study specific protein and whole proteomes alike.

GLOBAL IDENTIFICATION OF KINASE–SUBSTRATE PAIRS

In addition to identifying and quantifying phosphorylation, determining the kinases responsible for these events is equally important for our understanding of phosphorylation networks. Traditionally, the ability of a kinase to phosphorylate a substrate has been assayed by focused experiments where the kinase and particular candidate substrate are purified separately and then mixed together in an *in vitro* solution kinase assay. Specific kinase-substrate pairs have also been assayed *in vivo* by immunoblotting to detect gel shifts among kinase active and inactive cells. Recently, however, a number of technologies have emerged that allow high-throughput, system-wide identification of substrates for a given kinase.

One of these technologies is that of protein microarrays (Figure 10.2) [55–58]. Using *Saccharomyces cerevisiae* as a model system, our lab has developed protein microarrays that consist of ~4400 of the ~6000 yeast proteins spotted in duplicate at high spatial density onto modified glass slides [57]. To assay a kinase for substrates, one of these glass slides is incubated in kinase buffer containing radiolabeled ATP and purified active kinase. Following thorough washing, the slide is then exposed to autoradiography film. *In vitro* substrates of a kinase are identified by determining the amount of radiolabeled phosphate incorporated at each pair of spots relative to that of the corresponding pair on an autophosphorylation (no kinase) control slide. Using this method, Ptacek *et al.* assayed 87 different yeast protein kinases for their *in vitro* substrates, and observed ~4200 phosphorylation events on more than 1300 proteins. The obvious advantage of protein microarrays is the ease with which one can simultaneously assay the *in vitro* phosphorylation of thousands of proteins by a kinase. Protein microarrays are currently commercially available for yeast

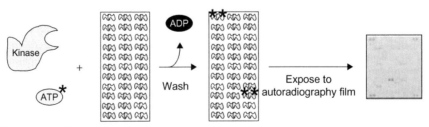

FIGURE 10.2 *In vitro* **kinase assays using protein microarrays.**
A glass slide containing ~4400 yeast GST-fusion proteins spotted in duplicate is incubated with kinase and radiolabeled ATP, washed thoroughly to remove unincorporated radiolabel, and exposed to autoradiography film. *In vitro* substrates are identified by computational analysis of the quantitated autoradiogram [57].

and human, containing ~4100 and ~8000 proteins, respectively, and have already been tested for *Caenorhabditis elegans* [59] and *Arabidopsis* [60, 61]. Thus, this approach will likely become even more robust as cloning efforts enable the entire proteomes of these organisms to be included on the protein microarray.

Kinase assays using analog-sensitive alleles have also contributed to the identification of kinase-substrate pairs (Figure 10.3). Several kinases, including the cyclin-dependent kinases Cdk1 and Pho85, have been mutated at a conserved bulky residue in their ATP binding pockets so that they preferentially bind an ATP analog (N^6-benzyl-ATP)

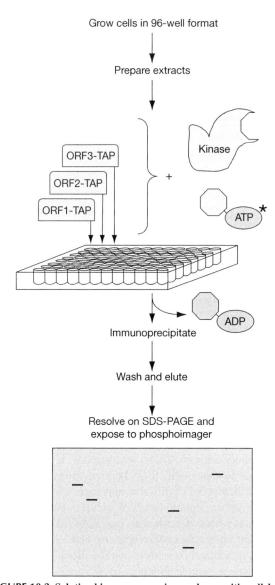

Grow cells in 96-well format

Prepare extracts

ORF3-TAP

ORF2-TAP

ORF1-TAP

Kinase

+

ATP *

Immunoprecipitate

ADP

Wash and elute

Resolve on SDS-PAGE and
expose to phosphoimager

FIGURE 10.3 Solution kinase assays using analog sensitive alleles. Extracts prepared from TAP-tagged strains are incubated in the presence of a mutant analog sensitive kinase and a radiolabeled bulky ATP analog. Candidate substrates are subsequently purified using the TAP tag, resolved on a SDS-PAGE gel, and exposed using a phosphoimager. In this assay, all radiolabeled proteins are necessarily phosphorylated by the mutant kinase as no other kinase can efficiently use the bulky ATP analog [62–64].

that cannot be accommodated by wild-type kinases [62–64]. Radiolabeling of this ATP analog thus allows detection of substrates that are specifically phosphorylated by the analog-sensitive (*as*) kinase. In this method, lysates are prepared individually from strains containing candidate substrates epitope-tagged at their endogenous locus and mixed with purified *as* kinase in the presence of the radiolabeled ATP analog. The candidate substrates are then immunoprecipitated and resolved by gel electrophoresis before being exposed to a phosphoimager. Positive substrates are identified as those having incorporated the radiolabeled phosphate. This method has been used by Ubersax *et al.* to identify 181 candidate substrates of Cdk1 among 385 proteins that contained multiple Cdk1 consensus phosphorylation sites [62], and by Loog *et al.* to reveal the specificity of ~40 of these 181 substrates for Clb5-Cdk1 over Clb2-Cdk1 [63]. This method has furthermore been adapted by Dephoure *et al.* to identify substrates of kinases with no known target sequences by pooling multiple epitope-tagged strains and then deconvoluting those pools that show positive phosphorylation to identify specific substrates [64]. Dephoure *et al.* screened 4250 yeast proteins and identified 24 strong candidate substrates of Pho85-Pcl1. Included among these 24 candidate substrates was the known Pho85 substrate, Rvs167. The major strength of this approach is that the kinase reactions are carried out with near physiological levels of exogenous kinase in native whole cell extracts. However, this approach is still *in vitro* in nature, and can only be used for kinases in which an analog sensitive mutation can be introduced that does not affect functionality.

Recently, reverse in-gel kinase assay (RIKA) has emerged as yet another method to identify substrates for a given kinase (Figure 10.4). This method, developed by Li *et al.*, involves the immobilization of the kinase of interest by polymerization in a denaturing polyacrylamide gel that is used to resolve a tissue or cell protein extract [65]. Following two-dimensional electrophoresis, the gel is subjected to a series of washes to remove the SDS detergent and to refold the kinase and candidate substrates into their native form. An *in situ* kinase reaction is then performed by incubating the gel in kinase buffer containing radiolabeled ATP. Substrates phosphorylated by the kinase of interest are identified on an autoradiogram as those proteins that incorporated radiolabeled phosphate, and the identities of these novel substrates are determined by mass spectrometry analysis of the corresponding spots excised from a silver-stained gel run in parallel without kinase. Using this method, Li *et al.* were able to identify 10 novel *in vitro* substrates for casein kinase 2 (CK2) and two novel *in vitro* substrates for protein kinase A (PKA) [65]. The main advantage of RIKA is its ability to query all physiologic isoforms present in the proteome. Many proteins are alternatively spliced or covalently modified posttranslationally, either in specific cell types or under different conditions, and these substrates are likely missed by other methods, such as protein microarrays,

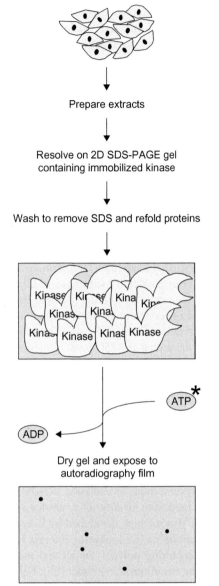

FIGURE 10.4 Reverse in-gel kinase assays.
In this assay, purified kinase is immobilized in a SDS-PAGE gel by polymerization that is in turn used to resolve cell extracts in two-dimensions. After a series of washes to remove the SDS detergent and refold the proteins, the gel is incubated in kinase buffer in the presence of radiolabeled ATP, dried, and exposed to autoradiography film. Substrates are identified by MS analysis of the corresponding bands cut from a silver-stained gel run in parallel without kinase [65].

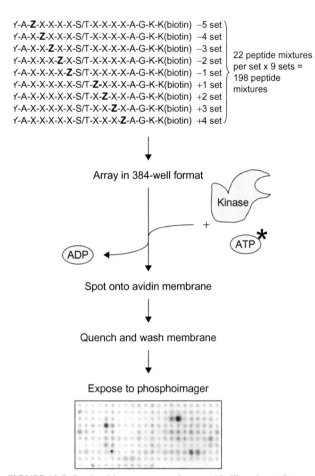

FIGURE 10.5 *In vitro* kinase assays using peptide libraries to determine consensus phosphorylation motifs.
A positional scanning peptide library consisting of 198 distinct peptide mixtures is designed such that each of the 20 amino acids, phospho-threonine, and phosphotyrosine are fixed at each of the nine positions surrounding a central serine/threonine residue. Each peptide is also biotinylated at its carboxy-terminus, thus allowing immobilization on an avidin-impregnated membrane following incubation with kinase and radiolabeled ATP. Membranes are then washed thoroughly, dried, and exposed using a phosphoimager. Motifs read from the phosphoimage are analyzed by computer algorithms to identify candidate phosphorylation sites [75].

that only assay the major isoform of each cellular protein. RIKA is, however, dependent upon proper renaturation of both the kinase and the lysate proteins, as improperly folded proteins can result in the phosphorylation of non-native substrates. RIKA also requires the kinase of interest to be active in gel. It will be interesting to see whether co-polymerization of a kinase with cofactors and/or adaptors is feasible.

The utility of peptide libraries in identifying substrates of a kinase has also been demonstrated (Figure 10.5).

Peptide libraries have been used to determine the consensus phosphorylation motif that is targeted by a given kinase and, based on this motif, to make predictions of possible substrates. A number of different approaches which involve screening either immobilized or solution-phase peptide libraries have been described to date [66–75]. One of the most recent approaches, developed by Hutti *et al.*, makes use of a positional scanning peptide library consisting of 198 distinct peptide mixtures as substrates in solution kinase assays performed in 384-well format [75]. In this method, each peptide is biotinylated at its carboxy-terminus, and the library is designed such that each mixture fixes one of the 20 naturally occurring amino acids, phosphothreonine, or phosphotyrosine at one of the nine positions surrounding a central serine/threonine residue, leaving the remaining positions degenerate. Following incubation with radiolabeled ATP and active kinase, the peptides are arrayed onto an avidin-coated membrane. The membrane is then washed to remove the unincorporated radiolabel, and exposed to a phosphoimager. Using this technique, Hutti *et al.* demonstrated the ability of this approach to rapidly profile the sequence preferences for phosphorylation of seven different serine/threonine kinases. They not only recapitulated known motifs for a selection of basic, proline-directed, and acidophilic kinases, but also determined the specificity of Pim2, a mammalian kinase encoded by a proto-oncogene involved in mediating cell survival for which its consensus phosphorylation motif was not previously known [76, 77]. While motif-based predictions can still be difficult for those kinases that exhibit little selectivity in their consensus phosphorylation motifs, this approach has additional utility in filtering the hitlists generated by other phosphoproteomic methods. Furthermore, this approach provides important information for mapping actual phosphorylation sites in known substrates, which tend to fall within a sequence context specific to the phosphorylating kinase.

A comparative proteomics approach has also shown to be useful in identifying kinase-substrate pairs. Budovskaya *et al.* predicted candidate substrates of yeast PKA by systematically analyzing the evolutionary conservation of PKA consensus sites (RRx[S/T]ϕ motifs where ϕ is any hydrophobic residue) in the *Saccharomyces cerevisiae* proteome across a group of related budding yeasts (five *Saccharomyces* species, and *Candida albicans*) [78]. Guided by the premise that a higher degree of conservation is likely to identify functional sites *in vivo*, they were able to identify 44 candidate PKA substrates, 5 of which were already known to be *bona fide* substrates. As a proof of principle, they validated Atg1, an autophagy-related protein kinase, as a novel *in vivo* target of PKA by metabolically labeling wild-type and *pka* mutant strains with ^{32}P-orthophosphate. What is particularly interesting about this approach is that the candidate substrates identified for *Saccharomyces cerevisiae* PKA are equally likely to be substrates of *Candida albicans* PKA. Thus, this method focuses narrowly on the dissection of phosphorylation signaling pathways that have withstood natural selection. It should be pointed out, however, that this method relies on knowing the consensus phosphorylation motif targeted by the kinase, and thus underscores the need for methods that identify kinase consensus phosphorylation motifs, such as the peptide array approach described above.

These technologies have changed the way novel kinase-substrate pairs are identified such that whole proteomes can now be assayed for candidate substrates of a kinase in a single experiment. Nonetheless, the kinases responsible for a large portion of the many thousands of phosphorylation events identified by recent MS efforts are still not known. Moreover, these technologies, though high-throughput, remain *in vitro* in nature and, consequently,

follow-up experiments are needed to validate the identified candidates as *bona fide in vivo* targets. A recent study by Matsuoka *et al.*, however, suggests that candidate substrates can be identified globally *in vivo* by applying information obtained from these *in vitro* technologies to current MS approaches [79]. Matsuoka *et al.* took advantage of the fact that the consensus phosphorylation motif ([S/T]-Q) targeted by ATM and ATR, two related protein kinases involved in the DNA damage response, is known. They used antibodies that recognized phosphorylated forms of these motifs and combined this purification strategy with SILAC labeling with arginine and lysine to identify proteins by LC-MS/MS that are likely to be phosphorylated *in vivo* by ATM and ATR in response to DNA damage. They observed more than 900 phosphorylation sites containing the consensus ATM/ATR phosphorylation motif in 700 proteins that were inducibly phosphorylated in 293T human embryonic kidney cells in response to infrared radiation. Interestingly, these 700 proteins are implicated in a number of other pathways, including those involved in DNA replication, DNA repair, and cell control, and thus uncover an extensive network of pathways previously not known to be responsive to DNA damage.

CONCLUSION

Our knowledge of phosphorylation-based signaling networks has advanced enormously over the past 5 years. With existing large-scale phosphoproteomic studies providing a solid foundation, there is much promise for gaining a more complete understanding of phosphorylation and reaching the point where detailed mathematical models of cell signaling pathways can be made. Obtaining a perfect model will, however, be difficult. No screen identifies all substrates, and it is likely that the readily identifiable ones have already been found. There will also always be problem substrates, which are present in low abundance, and problem kinases, which require special conditions or adaptor proteins for activation. Nevertheless, the array of technologies described in this chapter remains a powerful toolkit with which to continue our investigation of the different aspects of phosphorylation-based signal transduction. Different technologies have different limitations; thus, integration of studies using bioinformatics will likely be the key to accelerating our understanding of eukaryotic phosphorylation networks.

REFERENCES

1. Agris PF. Decoding the genome: a modified view. *Nucleic Acids Res* 2004;**32**:223–38.
2. Hunter T. Signaling – 2000 and beyond. *Cell* 2000;**100**:113–27.
3. Blume-Jensen P, Hunter T. Oncogenic kinase signalling. *Nature* 2001;**411**:355–65.

4. Yoo LI, Chung DC, Yuan J. LKB1–a master tumour suppressor of the small intestine and beyond. *Nat Rev Cancer* 2002;**2**:529–35.

5. Cohen P. The origins of protein phosphorylation. *Nat Cell Biol* 2002;**4**:E127–30.

6. Manning G, Plowman GD, Hunter T, Sudarsanam S. Evolution of protein kinase signaling from yeast to man. *Trends Biochem Sci* 2002;**27**:514–20.

7. Manning G, Whyte DB, Martinez R, Hunter T, Sudarsanam S. The protein kinase complement of the human genome. *Science* 2002;**298**:1912–34.

8. Caenepeel S, Charydczak G, Sudarsanam S, Hunter T, Manning G. The mouse kinome: discovery and comparative genomics of all mouse protein kinases. *Proc Natl Acad Sci USA* 2004;**101**:11,707–712.

9. Hunter T, Plowman GD. The protein kinases of budding yeast: six score and more. *Trends Biochem Sci* 1997;**22**:18–22.

10. Wettenhall RE, Aebersold RH, Hood LE. Solid-phase sequencing of 32P-labeled phosphopeptides at picomole and subpicomole levels. *Methods Enzymol* 1991;**201**:186–99.

11. MacDonald JA, Mackey AJ, Pearson WR, Haystead TA. A strategy for the rapid identification of phosphorylation sites in the phosphoproteome. *Mol Cell Proteomics* 2002;**1**:314–22.

12. Coull JM, Pappin DJ, Mark J, Aebersold R, Koster H. Functionalized membrane supports for covalent protein microsequence analysis. *Anal Biochem* 1991;**194**:110–20.

13. Rush J, Moritz A, Lee KA, et al. Immunoaffinity profiling of tyrosine phosphorylation in cancer cells. *Nat Biotechnol* 2005;**23**:94–101.

14. Rikova K, Guo A, Zeng Q, et al. Global survey of phosphotyrosine signaling identifies oncogenic kinases in lung cancer. *Cell* 2007;**131**:1190–203.

15. Ballif BA, Carey GR, Sunyaev SR, Gygi SP. Large-scale identification and evolution indexing of tyrosine phosphorylation sites from murine brain. *J Proteome Res* 2008;**7**:311–18.

16. Andersson L, Porath J. Isolation of phosphoproteins by immobilized metal (Fe^{3+}) affinity chromatography. *Anal Biochem* 1986;**154**:250–4.

17. Posewitz MC, Tempst P. Immobilized gallium(III) affinity chromatography of phosphopeptides. *Anal Chem* 1999;**71**:2883–92.

18. Brill LM, Salomon AR, Ficarro SB, Mukherji M, Stettler-Gill M, Peters EC. Robust phosphoproteomic profiling of tyrosine phosphorylation sites from human T cells using immobilized metal affinity chromatography and tandem mass spectrometry. *Anal Chem* 2004;**76**:2763–72.

19. Ficarro SB, McCleland ML, Stukenberg PT, et al. Phosphoproteome analysis by mass spectrometry and its application to Saccharomyces cerevisiae. *Nat Biotechnol* 2002;**20**:301–5.

20. Kim JE, Tannenbaum SR, White FM. Global phosphoproteome of HT-29 human colon adenocarcinoma cells. *J Proteome Res* 2005;**4**:1339–46.

21. Moser K, White FM. Phosphoproteomic analysis of rat liver by high capacity IMAC and LC-MS/MS. *J Proteome Res* 2006;**5**:98–104.

22. Larsen MR, Thingholm TE, Jensen ON, Roepstorff P, Jorgensen TJ. Highly selective enrichment of phosphorylated peptides from peptide mixtures using titanium dioxide microcolumns. *Mol Cell Proteomics* 2005;**4**:873–86.

23. Kweon HK, Hakansson K. Selective zirconium dioxide-based enrichment of phosphorylated peptides for mass spectrometric analysis. *Anal Chem* 2006;**78**:1743–9.

24. Schmelzle K, White FM. Phosphoproteomic approaches to elucidate cellular signaling networks. *Curr Opin Biotechnol* 2006;**17**:406–14.

25. Ballif BA, Villen J, Beausoleil SA, Schwartz D, Gygi SP. Phosphoproteomic analysis of the developing mouse brain. *Mol Cell Proteomics* 2004;**3**:1093–101.

26. Beausoleil SA, Jedrychowski M, Schwartz D, et al. Large-scale characterization of HeLa cell nuclear phosphoproteins. *Proc Natl Acad Sci USA* 2004;**101**:12,130–135.

27. Witze ES, Old WM, Resing KA, Ahn NG. Mapping protein post-translational modifications with mass spectrometry. *Nat Methods* 2007;**4**:798–806.

28. Oda Y, Nagasu T, Chait BT. Enrichment analysis of phosphorylated proteins as a tool for probing the phosphoproteome. *Nat Biotechnol* 2001;**19**:379–82.

29. McLachlin DT, Chait BT. Improved beta-elimination-based affinity purification strategy for enrichment of phosphopeptides. *Anal Chem* 2003;**75**:6826–36.

30. Jalili PR, Sharma D, Ball HL. Enhancement of ionization efficiency and selective enrichment of phosphorylated peptides from complex protein mixtures using a reversible poly-histidine tag. *J Am Soc Mass Spectrom* 2007;**18**:1007–17.

31. Knight ZA, Schilling B, Row RH, Kenski DM, Gibson BW, Shokat KM. Phosphospecific proteolysis for mapping sites of protein phosphorylation. *Nat Biotechnol* 2003;**21**:1047–54.

32. Byford MF. Rapid and selective modification of phosphoserine residues catalysed by Ba2+ ions for their detection during peptide microsequencing. *Biochem J* 1991;**280**:261–5.

33. Goshe MB. Characterizing phosphoproteins and phosphoproteomes using mass spectrometry. *Brief Funct Genomic Proteomic* 2006;**4**:363–76.

34. Zhou H, Watts JD, Aebersold R. A systematic approach to the analysis of protein phosphorylation. *Nat Biotechnol* 2001;**19**:375–8.

35. Tao WA, Wollscheid B, O'Brien R, et al. Quantitative phosphoproteome analysis using a dendrimer conjugation chemistry and tandem mass spectrometry. *Nat Methods* 2005;**2**:591–8.

36. Bodenmiller B, Mueller LN, Pedrioli PG, et al. An integrated chemical, mass spectrometric and computational strategy for (quantitative) phosphoproteomics: application to Drosophila melanogaster Kc167 cells. *Mol Biosyst* 2007;**3**:275–86.

37. Bodenmiller B, Mueller LN, Mueller M, Domon B, Aebersold R. Reproducible isolation of distinct, overlapping segments of the phosphoproteome. *Nat Methods* 2007;**4**:231–7.

38. Villen J, Beausoleil SA, Gerber SA, Gygi SP. Large-scale phosphorylation analysis of mouse liver. *Proc Natl Acad Sci USA* 2007;**104**:1488–93.

39. Wysocki VH, Resing KA, Zhang Q, Cheng G. Mass spectrometry of peptides and proteins. *Methods* 2005;**35**:211–22.

40. Stensballe A, Jensen ON, Olsen JV, Haselmann KF, Zubarev RA. Electron capture dissociation of singly and multiply phosphorylated peptides. *Rapid Commun Mass Spectrom* 2000;**14**:1793–800.

41. Syka JE, Coon JJ, Schroeder MJ, Shabanowitz J, Hunt DF. Peptide and protein sequence analysis by electron transfer dissociation mass spectrometry. *Proc Natl Acad Sci USA* 2004;**101**:9528–33.

42. Chi A, Huttenhower C, Geer LY, et al. Analysis of phosphorylation sites on proteins from Saccharomyces cerevisiae by electron transfer dissociation (ETD) mass spectrometry. *Proc Natl Acad Sci USA* 2007;**104**:2193–8.

43. Stemmann O, Zou H, Gerber SA, Gygi SP, Kirschner MW. Dual inhibition of sister chromatid separation at metaphase. *Cell* 2001;**107**:715–26.

44. Qian WJ, Goshe MB, Camp DG 2nd, Yu LR, Tang K, Smith RD. Phosphoprotein isotope-coded solid-phase tag approach for enrichment and quantitative analysis of phosphopeptides from complex mixtures. *Anal Chem* 2003;**75**:5441–50.

45. Ong SE, Mann M. Mass spectrometry-based proteomics turns quantitative. *Nat Chem Biol* 2005;**1**:252–62.

46. Ross PL, Huang YN, Marchese JN, et al. Multiplexed protein quantitation in Saccharomyces cerevisiae using amine-reactive isobaric tagging reagents. *Mol Cell Proteomics* 2004;**3**:1154–69.

47. Zhang Y, Wolf-Yadlin A, Ross PL, et al. Time-resolved mass spectrometry of tyrosine phosphorylation sites in the epidermal growth factor receptor signaling network reveals dynamic modules. *Mol Cell Proteomics* 2005;**4**:1240–50.

48. Bonenfant D, Schmelzle T, Jacinto E, et al. Quantitation of changes in protein phosphorylation: a simple method based on stable isotope labeling and mass spectrometry. *Proc Natl Acad Sci USA* 2003;**100**:880–5.

49. Smith JR, Olivier M, Greene AS. Relative quantification of peptide phosphorylation in a complex mixture using 18O labeling. *Physiol Genomics* 2007;**31**:357–63.

50. Blagoev B, Mann M. Quantitative proteomics to study mitogen-activated protein kinases. *Methods* 2006;**40**:243–50.

51. Zhu H, Hunter TC, Pan S, Yau PM, Bradbury EM, Chen X. Residue-specific mass signatures for the efficient detection of protein modifications by mass spectrometry. *Anal Chem* 2002;**74**:1687–94.

52. Gruhler A, Olsen JV, Mohammed S, et al. Quantitative phosphoproteomics applied to the yeast pheromone signaling pathway. *Mol Cell Proteomics* 2005;**4**:310–27.

53. Olsen JV, Blagoev B, Gnad F, et al. Global, in vivo, and site-specific phosphorylation dynamics in signaling networks. *Cell* 2006;**127**:635–48.

54. Ishihama Y, Sato T, Tabata T, et al. Quantitative mouse brain proteomics using culture-derived isotope tags as internal standards. *Nat Biotechnol* 2005;**23**:617–21.

55. MacBeath G, Schreiber SL. Printing proteins as microarrays for high-throughput function determination. *Science* 2000;**289**:1760–3.

56. Zhu H, Bilgin M, Bangham R, et al. Global analysis of protein activities using proteome chips. *Science* 2001;**293**:2101–5.

57. Ptacek J, Devgan G, Michaud G, et al. Global analysis of protein phosphorylation in yeast. *Nature* 2005;**438**:679–84.

58. Gelperin DM, White MA, Wilkinson ML, et al. Biochemical and genetic analysis of the yeast proteome with a movable ORF collection. *Genes Dev* 2005;**19**:2816–26.

59. Reboul J, Vaglio P, Rual JF, et al. C. elegans ORFeome version 1.1: experimental verification of the genome annotation and resource for proteome-scale protein expression. *Nat Genet* 2003;**34**:35–41.

60. Feilner T, Kersten B. Phosphorylation studies using plant protein microarrays. *Methods Mol Biol* 2007;**355**:379–90.

61. Popescu SC, Popescu GV, Bachan S, et al. Differential binding of calmodulin-related proteins to their targets revealed through high-density Arabidopsis protein microarrays. *Proc Natl Acad Sci USA* 2007;**104**:4730–5.

62. Ubersax JA, Woodbury EL, Quang PN, et al. Targets of the cyclin-dependent kinase Cdk1. *Nature* 2003;**425**:859–64.

63. Loog M, Morgan DO. Cyclin specificity in the phosphorylation of cyclin-dependent kinase substrates. *Nature* 2005;**434**:104–8.

64. Dephoure N, Howson RW, Blethrow JD, Shokat KM, O'Shea EK. Combining chemical genetics and proteomics to identify protein kinase substrates. *Proc Natl Acad Sci USA* 2005;**102**:17,940–945.

65. Li X, Guan B, Srivastava MK, Padmanabhan A, Hampton BS, Bieberich CJ. The reverse in-gel kinase assay to profile physiological kinase substrates. *Nat Methods* 2007;**4**:957–62.

66. Tegge WJ, Frank R. Analysis of protein kinase substrate specificity by the use of peptide libraries on cellulose paper (SPOT-method). *Methods Mol Biol* 1998;**87**:99–106.

67. Wu JJ, Phan H, Lam KS. Comparison of the intrinsic kinase activity and substrate specificity of c-Abl and Bcr-Abl. *Bioorg Med Chem Letts* 1998;**8**:2279–84.

68. Rodriguez M, Li SS, Harper JW, Songyang Z. An oriented peptide array library (OPAL) strategy to study protein–protein interactions. *J Biol Chem* 2004;**279**:8802–7.

69. Rychlewski L, Kschischo M, Dong L, Schutkowski M, Reimer U. Target specificity analysis of the Abl kinase using peptide microarray data. *J Mol Biol* 2004;**336**:307–11.

70. Songyang Z, Blechner S, Hoagland N, Hoekstra MF, Piwnica-Worms H, Cantley LC. Use of an oriented peptide library to determine the optimal substrates of protein kinases. *Curr Biol* 1994;**4**:973–82.

71. Songyang Z, Lu KP, Kwon YT, et al. A structural basis for substrate specificities of protein Ser/Thr kinases: primary sequence preference of casein kinases I and II. NIMA, phosphorylase kinase, calmodulin-dependent kinase II, CDK5, and Erk1. *Mol Cell Biol* 1996;**16**:6486–93.

72. Nishikawa K, Toker A, Johannes FJ, Songyang Z, Cantley LC. Determination of the specific substrate sequence motifs of protein kinase C isozymes. *J Biol Chem* 1997;**272**:952–60.

73. Obata T, Yaffe MB, Leparc GG, et al. Peptide and protein library screening defines optimal substrate motifs for AKT/PKB. *J Biol Chem* 2000;**275**:36,108–115.

74. Fujii K, Zhu G, Liu Y, et al. Kinase peptide specificity: improved determination and relevance to protein phosphorylation. *Proc Natl Acad Sci USA* 2004;**101**:13744–9.

75. Hutti JE, Jarrell ET, Chang JD, et al. A rapid method for determining protein kinase phosphorylation specificity. *Nat Methods* 2004;**1**:27–9.

76. Yan B, Zemskova M, Holder S, et al. The PIM-2 kinase phosphorylates BAD on serine 112 and reverses BAD-induced cell death. *J Biol Chem* 2003;**278**:45,358–367.

77. Fox CJ, Hammerman PS, Cinalli RM, Master SR, Chodosh LA, Thompson CB. The serine/threonine kinase Pim-2 is a transcriptionally regulated apoptotic inhibitor. *Genes Dev* 2003;**17**:1841–54.

78. Budovskaya YV, Stephan JS, Deminoff SJ, Herman PK. An evolutionary proteomics approach identifies substrates of the cAMP-dependent protein kinase. *Proc Natl Acad Sci USA* 2005;**102**:13,933–938.

79. Matsuoka S, Ballif BA, Smogorzewska A, et al. ATM and ATR substrate analysis reveals extensive protein networks responsive to DNA damage. *Science* 2007;**316**:1160–6.

Part 2

Phosphatases

Phosphatase Families Dephosphorylating Serine and Threonine Residues in Proteins

Patricia T.W. Cohen

Medical Research Council Protein Phosphorylation Unit, College of Life Sciences, University of Dundee, Dundee, UK.

CURRENT CLASSIFICATION OF PROTEIN SERINE/THREONINE PHOSPHATASES

Protein serine/threonine phosphatases are enzymes that reverse the actions of protein kinases by cleaving phosphate from serine and threonine residues in proteins. They are structurally distinct from the superfamily of protein phosphatases, which encompasses the tyrosine and dual specificity (tyrosine and serine/threonine) phosphatases. They are also functionally distinct from the acid and alkaline phosphatases that cleave phosphate from other molecules, although in some cases they may be distantly structurally related through a common ancestor.

The current classification of protein serine/threonine phosphatases is based upon the amino acid sequences of the catalytic subunits, which fall into three families (Table 11.1). The two major families have been termed the PPP (phosphoprotein phosphatase) family, of which the prototypic member is Ppp1c/PP1, and the PPM (protein phosphatase, Mg^{2+}/Mn^{2+}-dependent) family, of which the prototypic member is Ppm1c/PP2C. These two families show no similarities in amino acid sequences, and have different consensus amino acid sequences and topology surrounding the active sites, but the three-dimensional structures of the catalytic subunits and mechanism of catalysis show some similarities, suggesting that the two families may have undergone convergent evolution [1]. The PPP family is thought to have evolved from a distant ancestral metallophosphoesterase that also gave rise to the purple acid phosphatases [2]. The FCP family of protein serine/threonine phosphatases was recognized more recently from its founding member, TFIIF-interacting CTD phosphatase 1 (FCP1), which is a transcription factor IIF-interacting protein that dephosphoylates the carboxy-terminal domain (CTD) of RNA polymerase II [3]. The FCP family

possesses a distinct amino acid consensus sequence and catalytic mechanism from those of the PPP and PPM families, and forms an arm of the haloacid dehalogenase (HAD) superfamily, which includes phosphoesterases acting on a diverse set of non-protein substrates [4].

BACKGROUND

The discovery that protein phosphatases regulate cellular functions originates from studies on the dephosphorylation of a single serine residue in glycogen phosphorylase [5]. Many different high molecular weight protein phosphatase activities in a variety of protein substrates were identified subsequently. A classification based on sensitivities to inhibitor proteins and activation by metal ions proposed that most known eukaryotic cytosolic phosphatase activities could be accounted for by four distinct types of protein serine/threonine phosphatase catalytic subunits: PP1, PP2A, PP2B, and PP2C [6]. PP1 in type 1 protein phosphatases dephosphorylated the β subunit of phosphorylase kinase (phosphorylated by PKA) and was potently inhibited by the proteins inhibitor-1 and inhibitor-2. PP2A in type 2 protein phosphatases preferentially dephosphorylated the α subunit of phosphorylase kinase (phosphorylated by PKA) and was insensitive to the inhibitor proteins. PP1 and PP2A were unaffected by metal ions, PP2B (found to be identical with a major calcium binding protein, calcineurin, isolated from brain [7]) was dependent on calcium ions and calmodulin, while PP2C was dependent on magnesium ions for activity. The primary structures for mammalian PP1 and PP2A, deduced from the complementary DNA, demonstrated that these catalytic subunits belonged to the same family [8]. In contrast, analysis of cDNA encoding PP2C revealed an unrelated amino acid sequence [9]. These studies therefore

TABLE 11.1 Families of protein serine/threonine phosphatases

Family	PPP	PPM	FCP
Active site "signature" (mammals[1])	$-GDxHG(x)_{-23}GDxVDRG(x)_{-25}GNHE-$	$-(E/Q)D(x)_nDGH(A/G)(x)_nD(N/D)-$	$-DxDx(T/V/I)L-$
Active site	Bimetal (Fe^{3+} + Zn^{2+} in native PPP3)	Bimetal (both Mn^{2+} in expressed Ppm1)	Aspartic acid
Catalytic mechanism	Metal-activated water molecule initiates single step dephosphorylation	Metal-activated water molecule initiates single step dephosphorylation	Aspartic acid based hydrolysis; phosphoaspartate intermediate formed
Other characteristics	Some active in absence of metal ions, others activated by Ca^{2+}. Some members inhibited by naturally occurring toxins	Dependent on Mg^{2+} or Mn^{2+} for activity where tested. Some members are also activated by Ca^{2+}.	Dependent on Mg^{2+}, Mn^{2+}, or Ca^{2+} for activity.
Human genes	≥13	≥17	≥8
Common names of some members	Ppp1c/protein phosphatase1/PP1 Ppp2c/protein phosphatase 2A/PP2A Ppp3c/protein phosphatase 2B/PP2B calcineurin/Ca^{2+}-calmodulin regulated protein phosphatase	Ppm1/protein phosphatase 2C PDP/pyruvate dehydrogenase phosphatase S. cerevisiae PTC	RNA polymerase II CTD phosphatase Dullard
Evolution	Present before divergence of eukaryotes and prokaryotes	Present in eukaryotes; in some prokaryotes probably by horizontal gene transfer	Present in eukaryotes and some prokaryotes
Proteins with a similar domain that are not known to have protein Ser/Thr phosphatase activity	Diadenosine tetraphosphatase (E. coli) and related phosphatases Purple acid phosphatases have weak sequence similarities to PPPs	TAK-1 binding protein N-terminal domain of adenyl cyclase (S. cerevisiae)	Other proteins (including phosphoesterases, sugar phosphomutases, ATPases, dehalogenases) in the haloacid dehalogenase (HAD) superfamily.

The nomenclature for the human genes and encoded proteins of the PPP and PPM families proposed by a committee at the FASEB Conference on Protein Phosphatases in 1992 is used. Other terms and synonyms in current usage for the proteins are also included.
[1]*Variations in the active site sequences occur in lower organisms, but all the aspartic acid (D) residues are invariant all three families. In the PPP "signature" asparagine (N) and histidine (H) residues are also invariant.*

identified two distinct gene families encoding the protein serine/threonine phosphatases, termed the PPP family, which includes PP1, PP2A, and PP2B, and the PPM family, of which PP2C is the founding member.

EVOLUTION AND CONSERVED FEATURES OF THE PPP FAMILY

Members of the PPP family of protein phosphatases are widely distributed in all eukaryotic phyla. A truncated PPP domain with protein serine phosphatase activity was detected in bacteriophages, cyanobacteria, and other (but not all) eubacteria, and an entire PPP domain was discovered in archeabacteria [10], indicating that PPPs were present before the divergence of prokaryotes and eukaryotes.

The PPP catalytic domain, which spans ~270 amino acids in mammals, possesses the active site "signature" –GDxHG(x)~23GDxVDRG(x)~25GNHE–, which is virtually invariant among eukaryotes and shows little variation even in prokaryotes. The motif is located in the amino-terminal half of the eukaryotic catalytic domain (Figure 11.1). Amino acids in the conserved sections I, II, and III play cru-

cial roles in binding divalent metal ions (probably Fe^{2+} and Zn^{2+} in the native enzymes) located at the catalytic center and the phosphate group of the substrate (Figure 11.2). Activation of a water molecule by the metals facilitates hydroxide-ion initiated cleavage of phosphatate from phosphoserine and phosphothreonine in the protein substrate in a single step, without formation of a phosphoenzyme intermediate. The carboxy-terminal half of the eukaryotic catalytic domain contains motifs IV to VII (–HGG–, –WxD–, –RG–, and –RxH–), which are highly conserved among eukaryotes and archeabacteria [10, 11] and are mainly involved in further interactions with one of the metal ions (Figure 11.2). The –SAxNY– motif VIII, invariant in eukaryotes but absent from prokaryotes including archaebacteria, is in a flexible loop region that has been implicated in the binding of toxins and inhibitor-2 to the eukaryotic PPPs (see Chapter 86 of Handbook of Cell Signaling, Second Edition).

In eukaryotes, the PPP family is found to comprise seven different groups (Ppp1 to Ppp7) by comparison of the amino acid sequences of the catalytic domains (Figure 11.3). The Ppp2, Ppp4, and Ppp6 groups are more closely related to each other (~60 percent identity) than they are to other groups such as PP1, to which they show about 40 percent identity. Ppp5 and Ppp7 groups are also more closely

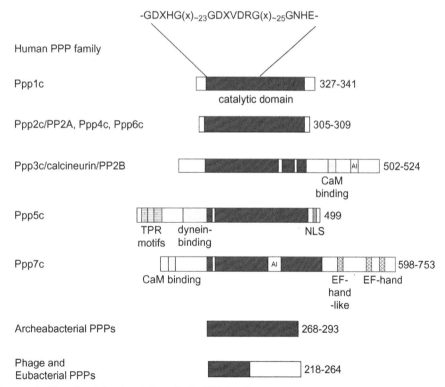

FIGURE 11.1 Domain organization of protein phosphatases in the PPP family.
Bacteriophage and eubacteria Ppps show homology to the eukaryotic and archeabacterial protein phosphatases in the amino-terminal half of the catalytic domain, but the carboxy-terminal halves show little similarity. Autoinhibitory regions (AI), Ca^{2+}-binding EF hands, tetratricopeptide repeats (TPR), putative nuclear localisation signal (NLS), and calmodulin (CaM) binding sites are indicated. The numbers of amino acids in different phosphatases with each type of structure is given on the right. The invariant amino acid motifs found in all PPP family members are shown at the top of the figure.

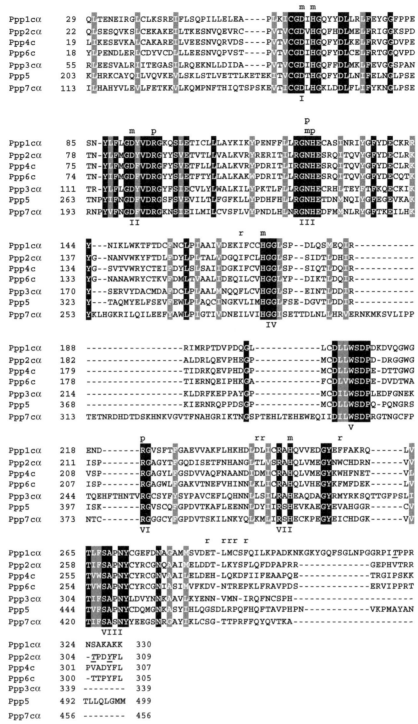

FIGURE 11.2 Comparison of human PPP sequences, showing highly conserved regions I–VIII.
The active site "signature" comprises regions I–III. Bacteriophages and many prokaryotes do not have the conserved regions IV to VIII. m indicates metal ion binding; p indicates phosphate binding; r indicates interactions with the RVxF motif of regulatory subunits with Ppp1c; amino acids that are phosphorylated are in italics and underlined. Ppp1cγ1 (not shown) is phosphorylated at non-homologous sites (T307 and T318) to Ppp1cα T320. The C-terminal leucine of Ppp2cα and β, Ppp4c and probably Ppp6c are carboxymethylated. Conserved region VIII and the cysteine immediately following it in several PPPs are crucial for the binding of several toxins. The alignment was made at http://www.ebi.ac.uk/Tools/clustalw2/ and shaded at http://www.ch.embnet.org/software/Box-form.html.

related to each other than to other PPPs; they show the most similarities to the highly divergent prokaryotic Ppps (not shown). Members of the PPP subgroups Ppp1–Ppp7 are present in mammals and *Drosophila melanogaster*, but Ppp7 orthologs are absent from yeasts (Figure 11.3), and Ppp3/calcineurin/PP2B orthologs have not been identified

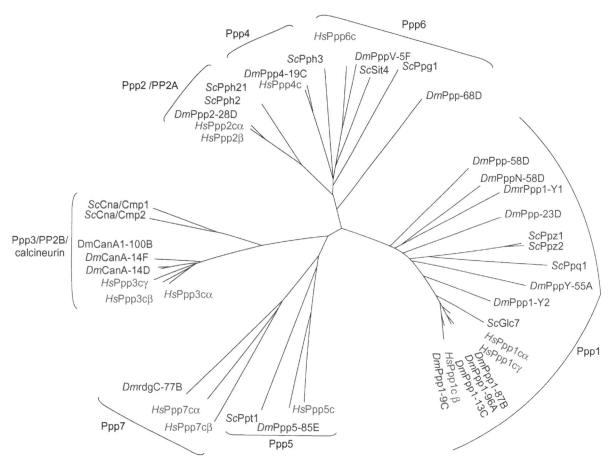

FIGURE 11.3 The PPP phosphome, a phylogenetic tree, depicting the relationships between *Homo sapiens* (Hs, red), *Drosophila melanogaster* (Dm, blue) and budding yeast *Saccharomyces cerevisiae* (Sc, brown) protein serine/threonine phosphatases in the PPP family.
The chromosomal location is included for the *Drosophila* PPPs. The unrooted tree is derived from multiple alignment (http://www.clustalw.genome. ad.jp/) of the phosphatase catalytic domain [starting ~30 amino acids prior to the invariant GDXHG motif and terminating ~30 amino acids following the conserved SAXNY motif [11] and corresponding to residues 29–301 of Ppp1cα_1]. Ppp1c and Ppp2c show approximately 40 percent sequence identity. Isoforms of Ppp1c show >85 percent sequence identity, while novel *Drosophila* phosphatases in Ppp1 subfamily show <65 percent sequence identity. Similar values occur in the PPP2A subfamily. Genbank/Swiss-Prot accession numbers for human PPP sequences are in Table 11.2.; *Drosophila* PPP sequences, Ppp1-13C (Q05547), Ppp1-87B/*Su-var* (P12982), Ppp1-96A (P48461), Ppp1-9C/*flw* (P48462), Ppp1Y2 (AF427494), PppY-55A (P11612), Ppp-23D/PpD6 (Q9VQL9), Ppp-58B/PpD5 (Q9W2A5), Ppp1Y1 (AF427493), PppN-58A (O77294), Ppp2- 28D/*mts* (P23696), Ppp4-19C/*cmm* (O76932), PppV-5F (Q27884), Ppp-68D (AY058490), CanA1- 100B(P48456), CanA-14F (Q9VXF1) CanA-14D (Q27889), Ppp5-85E/PpD3 (Q9VH81), rdgC-77B(P40421); *S. cerevisiae* PPP sequences Glc7 (P32598), Ppz1 (P26570), Ppz2 (P33329), Ppq1 (P32945), Pph21 (P23594), Pph22 (P23595), Pph3 (P32345), Sit4 (P20604), Ppg1 (P32838), Ppt1(P53043) Cna/Cmp1 (P23287), Cna/Cmp2 (P14747).

in plants [12, 13]. Twelve PPP family members are known in *S. cerevisiae* and 13 in the human species [14, 15], indicating that there has not been a significant increase in the number of PPP catalytic subunits from yeast to man. Interestingly, *Drosophila* possesses 19 PPP family members. Two are located on the Y chromosome (Ppp1-Y1 and Ppp1-Y2), and at least two others are related to male-specific functions (PppY-55A and PppN-58A), all of which appear to lack homologs in *Homo sapiens*. Some Ppps in *S. cerevisiae* (Ppq1, Ppz1, Ppz2, Ppg1) also have no homologs in humans. In plants, at least 8 Ppp1c isoforms have been produced by gene duplication. Thus, it is possible that gene duplication within different eukaryotes may be a mechanism used to produce regulation by Ser/Thr dephosphorylation of processes specific to particular organisms.

HOLOENZYME STRUCTURES OF PPP FAMILY MEMBERS

The main diversity of the eukaryotic protein serine/threonine phosphatases in the PPP family resides in the fact that most are high molecular mass complexes containing one or more regulatory subunits (Table 11.2). The regulatory subunits or interactors of Ppp1c include regulatory proteins, phosphatase inhibitor proteins, and proteins originally identified by their own intrinsic properties or functions, such as the retinomablastoma protein (Rb). Most Ppp1c interactors have no readily discernible sequence similarities, but bind in a mutually exclusive manner to the Ppp1c catalytic subunit [16, 17]. The interactions allow a single PPP catalytic subunit to form numerous complexes, which participate

TABLE 11.2 Structures and properties of the human PPP catalytic subunits

Human gene name	Gene locus	Protein names	Protein size (kDa)[2]	Protein Accession number[3]	Tissue distribution	Holoenzyme structure and/or interactions	Activators/inhibitors	Reviews/references
PPP1CA	11q13.3	Ppp1cα/PP1α	α 37	NP_002699 P62136	Widely expressed	Mainly heterodimeric; >70 different regulatory subunits or interactors; some heterotrimeric and oligomeric complexes	*inhibitor-1 inhibitor-2 okadaic acid microcystin tautomycetin*	[16, 17]
PPP1CB	2p23.2	Ppp1cβ/PP1β/PP1δ	β 37	NP_002700 P62140	Widely expressed			
PPP1CC	12q24	Ppp1cγ/PP1γ/PP1cγ	γ1 37 γ2 38	NP_002701 P36873-1 P36873-2	γ1 widely expressed; γ2 mainly in testis			
PPP2CA	5q31.1	Ppp2cα/PP2Acα/PP2Acα	α 36	NP_002706 P67775	Widely expressed	Mainly heterotrimeric, core regulatory A (PR65) subunit +≥14 different B regulatory subunits; dimeric complex with α4 or TIP; some oligomeric	*okadaic acid microcystin fostriecin*	[20, 23] Chapter 182 of Handbook of Cell Signaling, Second Edition
PPP2CB	8p12	Ppp2cβ/PP2Aβ/PP2Acβ	β 36	NP_004147 P62714	Widely expressed			
PPP4C	16p11.2	Ppp4c/PP4/PPX	35	NP_002711 P60510	Widely expressed	Heterotrimeric, core regulatory R2 subunit and third subunit R3; heterodimeric, complex with R1, α4 or TIP subunits; oligomeric with R2 and Gemins	*okadaic acid microcystin fostriecin*	[21]
PPP6C	9q33.3	Ppp6c/PP6	35	NP_002712 O00743	Widely expressed	Heterodimeric, regulatory subunits, R1, R2, R3 (orthologues of yeast SAPs), α4 or TIP; heterotrimeric, third subunit Ankrd28	*okadaic acid microcystin*	[22, 49]

								Chapter 98 of Handbook of Cell Signaling, Second Edition
PPP3CA	4q24	Ppp3cα/PP2Bα/ CNA/CALNA	α$_1$ 59 α$_2$ 58	NP_000935 Q08209-1 Q08209-2	Widely expressed, abundant in brain	Heterodimer. Fused calmodulin binding and autoinhibitory domains; heterodimer with regulatory 19-kDa Ca^{2+} binding B subunit; interacts with calmodulin; some oligomeric complexes	Ca^{2+}/calmodulin *FKBP12-FK506* *Cyclosporin/-cyclophilin*	
PPP3CB	10q22.2	Ppp3cβ/ PP2Bβ/CNA2/ CALNA2/ CALNB	β$_1$ 59 β$_2$ 58 β$_3$ 58	NP_066955 P16298-1 P16298-2 P16298-3	Widely expressed, abundant in brain			
PPP3CC	8p21.3	Ppp3cγ/PP2Bγ/ CANγ	57	NP_005596 P48454	Mainly in testis			
PPP5C	19q13.3	Ppp5c/PP5/PPT	57	NP_006238 P53041	Widely expressed	monomeric with fused TPR domain, associates with Hsp90 and Hsp70.	*okadaic acid* *microcystin*	[27, 51, 52]
PPP7CA[1]	Xp22	Ppp7cα/PPEF1/ PP7α	α 76	NP_006231 O14829	Mainly in retina and sensory neurons	monomeric, with fused Ca^{2+}-binding EF hand domains, interacts with calmodulin	Ca^{2+}/calmodulin	[15, 30, 57]
PPP7CB[1]	4q21.1	Ppp7cβ/PPEF2/ PP7β	β$_1$ 86 β$_2$ 69	NP_006230 O14830-1 O14830-2	Mainly in retina and sensory neurons			

The nomenclature for the human genes and encoded proteins of the PPP and PPM families proposed by a committee at the FASEB Conference on Protein Phosphatases in 1992 is used. Ppp1c–Ppp6c are present in the nucleus and cytoplasm. Enrichment at some subcellular locations is likely to be due to regulatory subunit binding. Ppp7 is cytosolic.
[1] The names PPP7CA and PPP7CB are used here for genes referred to as PPEF1 and PPEF2 in some databases.
[2] Different molecular masses refer to splice variants. A Ppp1cα variant (NP_001008709) has been reported.
[3] Accession sequences are located at http://www.ncbi.nlm.nih.gov/sites/entrez?db=pubmed and http://uniprot.org/.

in many different cellular functions. Over 70 different Ppp1c interactors are known in humans, and a similar additional number are predicted [18]. In contrast, only about 12 Ppp1c interactors have been identified in *S. cerevisiae*, providing an explanation for the increasing diversity of functions regulated by reversible phosphorylation in higher compared to lower eukaryotes. About 70 percent of Ppp1c regulatory subunits bind through a conserved "–RVxF–" motif (with the consensus (R/K)x_1(V/I)x_2(F/W) where x_1 may be absent or any residue except large hydrophobic residues, and x_2 is any amino acid except proline and phospho-amino acids), providing an explanation for mutually exclusive interaction of most of these interacting subunits with Ppp1c. A hydrophobic groove formed by amino acids near the carboxyterminus of Ppp1c binds to the "–RVxF–" motif in the interacting subunit (Figure 11.2). Many secondary binding sites have been identified that may be different for each regulatory subunit [16, 19]. Trimeric PP1 complexes of Ppp1c/PP1 with an associated variable subunit have been identified where the third subunit is commonly an inhibitor protein.

Ppp2/PP2A comprises a "core" complex of the catalytic subunit bound to a core 65-kDa regulatory subunit (termed A or PR65), which interacts with one of many variable B regulatory subunits [20]. Most forms of Ppp2/PP2A have three subunits. Since there are at least 18 variable B subunits encoded by 14 genes, which fall into 4 unrelated groups (2 A subunits and 2 catalytic subunits), over 70 different trimeric PP2A holoenzymes combinations could be generated. The variable B regulatory subunits of Ppp2/PP2A possess two conserved domains that mediate mutually exclusive interaction with the complex of the of Ppp2c/PP2Ac and the A "core" regulatory subunit [21]. However, dimeric complexes of Ppp2c/PP2Ac and the A "core" regulatory subunit as well as Ppp2c/PP2Ac associated with a single regulatory subunit also exist [20].

Ppp4 possesses similar holoenzyme complexes to those of Ppp2/PP2A. Dimeric complexes such as Ppp4c-R1 have been identified, and trimeric holoenzymes, although only three variable subunits (R3A, R3B, and Gemin) are known to bind to the Ppp4c-R2 core complex [22]. Although most holoenzyme structures of Ppp6c (and its *S. cerevisiae* homolog Sit4p) identified to date are dimeric, trimeric complexes of Ppp6c have been recently identified [23]. Despite the 60–65 percent amino acid sequence identity of Ppp2c/PP2Ac, Ppp4c, and Ppp6c, virtually all regulatory subunits are specific to a particular phosphatase except for two (α4 and TIP), both of which interact independently with all three catalytic subunits to form dimeric complexes [24, 25].

Oligomeric complexes of protein serine/threonine phosphatases may facilitate cellular signaling. Interactions of phosphatases and kinases can be mediated by anchoring or adaptor proteins [26]. Alternatively, the interaction may be direct in kinase/phosphatase complexes, which have the ability to act as signaling modules or molecular switches. Ppp2/PP2A holoenzyme has been isolated in association with several different kinases, and a complex of Ppp1c with Nek2 (NIMA-related kinase) has been studied in which the kinase could be regarded as the regulatory subunit of the phosphatase catalytic subunit (or *vice versa*) [16]. Ppp4c-R2 binds to the oligomeric survival of motor neuron complex, probably via Gemin4.

Although Ppp1c and Ppp2c /PP2Ac possess only very short amino and carboxy-terminal regions outside the catalytic region, other PPP family members may possess fused amino and carboxy-terminal domains that, at least in some cases, impart distinct properties to the catalytic domain (Figure 11.1). Ppp5 has an inhibitory amino terminal domain containing three tetratricopeptide repeats (TPR), which might be expected to mediate protein–protein interactions [27, 28], but there is no evidence that Ppp5 forms complexes with specific regulatory subunits. Ppp3c (PP2B, calcineurin A subunit) has a carboxy-terminal region which tightly binds one regulatory B subunit that confers Ca^{2+} dependency, and the subsequent weaker interaction with calmodulin allows the dimer to respond to intracellular Ca^{2+} changes by the binding of Ca^{2+} to calmodulin [29]. The interaction with calmodulin displaces the autoinhibitory pseudo-substrate domain near carboxy-terminus of the catalytic subunit (Figure 11.1). Ppp7c has a C-terminal domain with EF hand-like sequences that confer Ca^{2+} sensitivity, and an amino-terminal domain that interacts with calmodulin through a calmodulin-binding motif that is distinct from that of Ppp3c [15, 30].

The catalytic subunits of Ppp1 and Ppp2/PP2A have been extremely conserved throughout the evolution of multicellular eukaryotes. The full-length amino acid sequences of human Ppp1c and Ppp2c/PP2Ac show greater than 80 percent and 73 percent identity, respectively, to their *S. cerevisiae* orthologs, and these enzymes are among the most slowly evolving proteins known. This may be because they interact with a wide variety of regulatory proteins, and also because of their many crucial roles in cellular processes such as the cell cycle.

CATALYTIC ACTIVITIES OF THE PPP FAMILY MEMBERS

PPPs dephosphorylate phosphoserine and phosphothreonine residues in proteins *in vivo*. *In vitro*, the catalytic subunits and Ppp1 and Ppp2/PP2A will dephosphorylate a wide variety of phosphorylated substrates that they do not dephosphorylate *in vivo*. Although Ppp1c preferentially dephosphorylates glycogen phosphorylase (phosphorylated by phosphorylase kinase), and Ppp2c/PP2Ac and Ppp4c preferentially dephosphorylate casein (phosphorylated by PKA), both phosphatase catalytic subunits will dephosphorylate both protein substrates as well as peptide substrates *in vitro* [31]. PPP catalytic subunits will dephosphorylate serine and threonine residues that lie in unrelated amino acid sequences, implying that, in contrast to many protein kinases, specificity is not determined primarily by the linear sequence of amino acids on either side of the phosphorylated residue [32].

Mammalian Ppp1 and Ppp2 have also been shown to dephosphorylate histidine residues in proteins *in vitro* [33]. Ppp1 expressed in *E. coli* displays properties that are slightly different from the native enzyme, including dependence or partial dependence on divalent metal ions such as Mn^{2+} or Mg^{2+}, and protein-phosphotyrosine phosphatase activity [34]. These features are unlikely to be a property of the native enzyme, because they disappear when Ppp1 is refolded *in vitro* in the presence of inhibitor-2, a process which may also occur *in vivo* [35]. Ppp2/PP2A exhibits some protein tyrosine phosphatase activity in the presence of the protein PTPA [20]. The carboxyl group of the carboxyl terminal leucine residue of Ppp2c/PP2Ac, Ppp4c, and Ppp6c is methylated, a modification regulated by methylesterase and methyltransferase enzymes. Carboxyl methylation regulates Ppp2/PP2A activities by controlling the association of regulatory B subunits [36], a process which may also be modulated by phosphorylation of amino acids (Thr 304 and Tyr 307 in human Ppp2/PP2A) near the C-terminus [37]. Interestingly, recent studies have indicated that PTPA is required to fold Ppp2c/PP2Ac into its active conformation, and that this is coupled to methylation and holoenzyme assembly [38].

The activities of two the PPP members (Ppp3c/PP2B/CNA and Ppp7c) are dependent on Ca^{2+} ions for activity, and are regulated by changes in intracellular Ca^{2+} levels through Ca^{2+}-dependent association of the Ca^{2+}-binding protein calmodulin. There is no evidence for the existence of the free catalytic subunits of Ppp1, Ppp2/PP2A, Ppp4, Ppp6, and Ppp3/calcineurin/PP2B within the cell, and thus the major determinants of substrate specificity are the regulatory subunits of these enzymes that may modify the activity of the catalytic subunits towards different substrates and target the catalytic subunit to different subcellular locations and substrates.

FUNCTIONS OF PPP FAMILY MEMBERS

The functions of Ppp1 and Ppp2/PP2A holoenzymes are extremely diverse and are largely determined by the variable regulatory subunits, which are discussed in detail in other chapters. Briefly, Ppp1 complexes in mammals regulate functions as disparate as glycogen metabolism, muscle contraction, protein synthesis (transcription, mRNA processing, and translation), signal transduction via receptors and ion channels, cell cycle progression, and apoptosis [16, 17]. Deletion of the single PP1 gene in *S. cerevisiae* or the major Ppp1-87B gene in *D. melanogaster* causes

lethality. However, the different isoforms of Ppp1c in higher eukaryotes exhibit some specificity of function. In mammals, Ppp1cβ is the only isoform specifically binding to the myosin targeting subunit [39] and muscle glycogen targeting subunit [40]. Mice homozygous for a targeted mutation of the Ppp1cγ gene, which disrupts the expression of the Ppp1cγ1 and Ppp1cγ2 isoforms, exhibit a defect in spermiogenesis, but are otherwise healthy [41]. This suggests that function of Ppp1cγ1 may be compensated for (most likely by PP1α), but that Ppp1cγ2, the predominant isoform in testis, serves a distinct function. The mammalian Ppp1c isoforms are present in the nucleus and cytoplasm, but show differential abundance at particular subcellular locations, and regulatory subunits may in some cases show binding preferences between the Ppp1α or Ppp1γ1 isoforms [42]. Most forms of regulation of the activities of PP1 complexes in response to cellular signals and intracellular changes are likely to be transmitted through the regulatory subunits. However, it has been shown that Ppp1cα is phosphorylated on Thr 320 by Cdks and inactivated during G1 phase of the cell cycle, and that this is crucial for the phosphorylation of the tumor suppressor protein Rb and progression into S phase [43]. In this function, Ppp1cα may be regarded as a tumor suppressor.

Ppp2/PP2A regulates a wide variety of cellular processes, including metabolism, protein synthesis (transcription, mRNA processing, translation), cellular signaling, G2/M cell cycle progression, apoptosis, stress responses, and DNA replication. Ppp2c/PP2Ac has recently been demonstrated to be a tumor suppressor that dephosphorylates the oncogenic protein c-Myc after displacement of a novel PP2A inhibitor protein, cancerous inhibitor of PP2A (CIP2A), from cMyc [44, 45]. Dephosphorylation of c-Myc on Ser-62 initiates the proteolytic degradation of c-Myc and prevents transformation [44, 45]. Targeted gene deletion of Ppp2cα/PP2Acα in the mouse is lethal, indicating functional differences between the isoforms, and differential subcellular localization of isoforms has been noted the early embryo [20]. It seems likely that the isoform-specific properties of the catalytic subunits are determined by differential binding of the variable regulatory subunits.

Although the activities of Ppp1, Ppp2/PP2A, and Ppp3/PP2B/calcineurin were detected in cellular lysates [31], Ppp4, Ppp5, Ppp6 were not detected – probably because of their low basal activities. Nevertheless, emerging evidence indicates that they all regulate a distinct variety of cellular functions. Ppp4 has been implicated in centrosome maturation, spliceosomal assembly, cellular signaling, and chromatin remodeling [22]. Disruption of Ppp4c in mice causes embryonic lethality and T-cell-specific disruption of Ppp4c blocks thymocyte development [46]. Ppp4c plays a role in the migration of human cells and in centrosome maturation, and may coordinate the two processes via the regulation of Rac1 and Cdc42 [47]. During recovery from DNA damage, human histone 2AX is dephosphorylated by Ppp4c in conjunction with Ppp2c/PP2Ac [48a, 48b].

Ppp6, like its *S. cerevisiae* ortholog Sit4p, regulates G1/S cell cycle progression in human cells, but the regulation appears to be in the stability of cyclin D1, rather by alteration of transcription of the G1 cyclins, which is found in yeast [49]. Ppp6 has been implicated in the regulation of stress response pathways – for example, interleukin-1 signaling via TAK1 [50]. Although most functions of Ppp2/PP2A, Ppp4, and Ppp6 are distinct, regulatory subunits that bind to all of them, $\alpha4$ (ortholog of *S. cerevisiae* TAP42, which regulates the TOR pathway) and TIP (type 2 A interacting protein) are likely to result in some overlapping functions in the nutrient response and the DNA damage response pathways that they may respectively regulate [24, 25].

Ppp5 has been implicated in the regulation of signal transduction, glucocorticoid- and estrogen-mediated cell growth and proliferation, transcription, apoptosis, DNA damage response, cell cycle arrest, and other stress related pathways [27, 51, 52]. Ppp5 associates with glucocorticoid and estrogen receptors though interaction with heat shock protein (Hsp) 90, and has recently been shown to dephosphorylate the glucocorticoid receptor, modulating gene expression [53]. Although interactions with a variety of other proteins have been documented, the major proteins binding to Ppp5 are the chaperones Hsp90 and Hsp70 [54]. Ppp5 binds to the TPR acceptor site at the carboxy terminus of Hsp90 [27], and thus may act as a co-chaperone for Hsp90 in the refolding of client proteins that bind to the amino-terminal region of Hsp90. Deletion of *S. cerevisiae* Ppt1, the yeast ortholog of mammalian Ppp5, leads to hyperphosphorylation of Hsp90 [55]. Mice with an insertional mutation in the Ppp5 gene causing a deficiency of Ppp5 are viable, and embryonic fibroblasts exhibit a decrease in the ataxia telangiectasis mutated (ATM)-mediated DNA damage response [56].

The functions regulated by Ppp7 in mammals are not known [15]. Mutants in the *D. melanogaster* ortholog rdgC show defects in the termination of the light response in photoreceptors, and severe retinal degeneration. RdgC dephosphorylates rhodopsin, thereby uncoupling this receptor from its regulatory G protein, and it has been postulated that Ppp7 and its orthologs may generally regulate dephosphorylation of G-protein-coupled receptors. Although Ppp7c is primarliy expressed in the retina and sensory neurons, mice lacking both isoforms Ppp7c/PPEF show normal light response, photoreceptor integrity, and rhodopsin phosphorylation [57]. However, compensatory mechanisms, such as an increase in PP2A activity, which is known to be able to dephosphorylate rhodopsin, have not been eliminated in this model.

Ppp3/PP2B is the only PPP in most tissues that is responsive to calcium ions. Changes in Ca^{2+} signals allow Ppp3/PP2B to regulate several functions, including transcription and signal transduction via receptors and ion channels (see Chapter 98 of Handbook of Cell Signaling, Second Edition). In T cells, activation of the T cell receptor increases intracellular Ca^{2+}, enabling Ppp3/PP2B to dephosphorylate the cytosolic transcription factor, nuclear factor for activated T cells (NF-AT). The NF-AT–phosphatase complex translocates to the nucleus, where it activates the transcription of the T cell growth factor interleukein-2. Particular interest in this pathway developed when it was realized that NF-AT dephosphorylation is the target of the immunosuppressant drugs cyclosporin and FK506. Ppp3/PP2B activates the Na^+/K^+ ATPase transporter, and inhibition of this enzyme in the kidney by the immunosuppressive drugs may contribute to their nephrotoxicity. However, Ppp3/PP2B is particularly abundant in brain, where it plays a major role in regulating pre- and post-synaptic events in excitory and inhibitory neurons by dephosphorylating ligand-gated ion channels (NMDA) and G-protein-coupled receptors (AMPA), as well as the PP1 inhibitor dopamine and cAMP regulated phosphoprotein, 32 kDa (DARPP-32/PPP1R15B). Overall, both Ppp3/PP2B and Ppp1 limit synaptic transmission and plasticity, and negatively regulate learning and memory processes [58].

MEDICAL IMPORTANCE OF THE PPP FAMILY

The first immunosuppressant drugs, cyclosporin and FK506, which changed the face of transplantation surgery, were found to be Ppp3c (PP2B/calcinceurin) inhibitors. They are still very effective for short-term use, but their toxicity (particularly to kidney function), compared to the newer mTOR inhibitors, currently limits their use as long-term immunosuppressants [59]. The eukaryotic PPPs, Ppp1c, Ppp2c/PP2Ac, Ppp4c, Ppp5c, and Ppp6c, are potently inhibited by naturally occurring toxins and tumor promoters and suppressors, such as okadaic acid and microcystin, which pose public health problems due to their presence in seafood and fresh water reservoirs, respectively (Chapter 87 of Handbook of Cell Signaling, Second Edition). No inhibitors distinguish between Ppp2c/PP2Ac, Ppp4c, and Ppp6c, although fostriecin, identified empirically as an anticancer agent and a potent inhibitor of both Ppp4c and Ppp2c/PP2A, has not yet been tested on Ppp6c. There are data suggesting that specific inhibitors of Ppp4c and Ppp5c may block cell division and cell growth, and thus be useful anticancer agents [22, 60]. Ppp4c has also been implicated in the development of resistance to the widely used platinum-based anticancer drugs (oxaliplatin and related compounds), suggesting the possibility that specific inhibition of Ppp4c may be useful in prevention of the resistance. Interestingly, DNA tumor viruses (SV40 and polyoma virus) have been shown to compromise the tumor suppressor function of PP2A by producing proteins that compete with regulatory subunit CIP2A [45]. The dsRNA virus, herpes simplex, expresses PP1-binding proteins that recruit Ppp1c from host cell complexes in order to enhance its own replication and evade host cell defense mechanisms [16]. Ppp1 has also been found to regulate the transcription of the retrovirus, human immunodeficiency virus-1 [61].

THE PPM FAMILY

The human PPM family has grown to at least 17 recognized members in recent years, outstripping the size of the human PPP family (Table 11.3). However, most are not known to interact with regulatory subunits, and therefore it seems unlikely that their diversity will reach that known in the human PPP family. PPMs are present in other eukaryotes, including plants (where they number 69) and yeasts; phylogenetic analyses suggest that PPMs may have arisen in early eukaryotes and been exported to prokaryotes by horizontal gene transfer. Similarly to the PPPs, the two metal ions at the catalytic center of Ppm1α/PP2Cα are thought to activate an associated water molecule to provide a nucleophile initiating hydrolysis of the substrate phosphomonoester bond [1]. However, unlike the PPPs, Ppm1α/PP2Cα, and Ppm1β/PP2Cβ are dependent on magnesium or manganese ions for activity. This is probably because one of the two metal ions at the active site forms limited contacts with the enzyme and thus binds with low affinity. The low metal ion affinity enables metal chelators such as EDTA to inhibit Ppm1α/PP2Cα and Ppm1β/PP2Cβ, but there are no known naturally occurring inhibitors of the PPMs. There is no evidence that the PPMs are regulated by intracellular Mg^{2+}/Mn^{2+} concentrations, which are not known to vary substantially.

Ppm1α/PP2Cα and Ppm1β/PP2Cβ consist of a catalytic domain with only short amino- and carboxy-terminal regions, while other members have additional domains (Table 11.3) – for example, PP2Cϵ, encoded by PPM1L, has a N-terminal transmembrane domain, which may enable it to dephosphorylate a membrane ceramide transport protein [62]. Others are found to associate with proteins that may be substrates [63].

A mitochondrial pyruvate dehydrogenase phosphatase catalytic subunit (Pdpc) has an amino-terminal Ca^{2+} binding domain and interacts with one regulatory subunit to form a heterodimer. Proteins containing inactive PPM pseudo-phosphatase domains, for example in TAK1, are not included in Table 11.3. Most active PPM family members are involved in the regulation of cell growth and cellular stress signaling. Several may be tumor suppressors (Ppm1A, Ppm1B, Ilkap, Phlpp1, Phlpp2), and Wip1 may be an oncoprotein. Their functions are reviewed in more detail in Chapter 91 of Handbook of Cell Signaling, Second Edition, and in reference [64].

THE FCP FAMILY

The FCP family is a small, but growing, distinct group of protein serine/threonine phosphatases, which now comprises at least eight mammalian members (Table 11.4) and is recognized as a branch of the HAD superfamily. Members of the FCP family Fcp1, Scp1, and Dullard are dependent on Mg^{2+} for activity, and possess a similar catalytic mechanism to other phosphomonoesterases in this superfamily. The two aspartic acid residues at the catalytic center coordinate a Mg^{2+} ion, which binds the substrate phosphate group. The hydroxide ion of the first aspartic acid initiates catalysis of the phosphoserine or phosphothreonine in the protein substrate with formation of a phosphoaspartyl intermediate, which is subsequently protonated by the second aspartic acid. Like other members of the HAD superfamily, Dullard was found to be inhibited by BeF_3 and not by NaF, which inhibits members of the PPP and PPM families.

The founding mammalian member, Fcp1, plays a role in the regulation of transcription by dephosphorylating serine residues in the many short repeat sequences in the carboxy-terminal domain (CTD) of RNA polymerase II [3]. Fcp1 comprises an N-terminal catalytic domain in which two aspartic acid residues have been shown to be essential for function, and a C-terminal domain that interacts with the transcription factor TFIIF. The small CTD phosphatases (Scp1, Scp2, and Scp3) lack the C-terminal domain, and function in growth factor signaling by dephosphorylating Smads [65, 66]. Dephosphorylation of the linker region of Smad 2 and 3 enhances growth factor TGFβ signaling, while dephosphorylation of the linker region and C-terminal tail of Smad 1 resets bone morphogenic protein (BMP) signaling to the basal state. Scp1 has also been shown to repress the transcription of neuronal genes in non-neuronal tissues [67].

More recently, it has been recognized that other members of the FCP family exist that perform entirely different functions. For example, Dullard has an N-terminal membrane binding domain, localizes to the nuclear membrane, and is likely to play a crucial role in nuclear membrane biogenesis [68]. By comparison with its *S. cerevisiae* ortholog, it is believed to dephosphorylate the protein, lipin, which generates diacylglycerol for membrane biogenesis. The orthologs of Dullard and FCP1 in *Drosophila* and in *S. cerevisiae* are encoded by essential genes. It seems likely that the next few years will provide more functional information on this novel family of protein serine/threonine phosphatases.

CONCLUDING REMARKS

The number of protein serine/threonine phosphatase catalytic subunits encoded in the human genome is now a minimum 38 and, given the diversity of Ppp1 and Ppp2/PP2A complexes, together with several Ppp4 and Ppp6 complexes, the protein serine/threonine phosphatase holoenzymes may total about 200. This goes some way to redress the balance for the 418 human protein serine/threonine

TABLE 11.3 Structures and properties of the human PPM (protein phosphatases magnesium or manganese dependent) catalytic subunits

Human gene name	Gene locus	Protein names	Protein Size (kDa)[1]	Protein Accession no.[2]	Tissue distribution Subcellular localization	Holoenzyme structure and/or interactions
PPM1A	14q23.1	Ppm1α/PP2Cα	α₁ 42 α₂ 36	NP_066283 P35813-1 NP_808820 P35813-2	Widely expressed; *cytoplasm, nucleus*	Monomeric
PPM1B	2p21	Ppm1β/PP2Cβ	β₁ 53 β₂ 43	NP_002697 O75688-1 NP_808907 O75688-2	Widely expressed, splice variants show tissue specific expression; *cytoplasm*	Monomeric
PPM1G	2p23.3	Ppm1γ/PP2Cγ/FIN13	59	NP_817092 O15355	Widely expressed; *nucleus*	Monomeric N-terminal CH and acidic domains
ILKAP	2q37.3	ILKAP	43	NP_110395 Q9H0C8	Widely expressed; *cytoplasm*	Monomeric
PPM1D	17q23.2	Wip1/PP2Cδ	67	NP_003611 O15297	Widely expressed; *nucleus*	Monomeric
PPM1E	17q22	POPX1, CaMKP-N	85	NP_055721 Q8WY54-1	Widely expressed, abundant in brain; *nucleus*	N-terminal PIX binding domain
PPM1F	22q11.22	CaMIIKPase/POPX2/ hFEM2/CaMKP	50	NP_055449 P49539	Widely expressed; *cytoplasm*	N-terminal PIX binding domain
PPM1H	12q14.1 -q14.2	NERPP-2C/ARHCL1	56	NP_065751 Q9ULR3	Neurites, growth cones, also widely expressed	Associates with hCAS/CSE1L

						Associates with UBC9 in presence of SUMO
PPM1J	1p13.2	PP2Cζ	55	Testicular germ cells	NP_005158 Q5JR12	Associates with UBC9 in presence of SUMO
PPM1K	4q22.1	PP2Cκ	41	Widely expressed.	NP_689755 Q8N3J5-1	ND
PPM1L	3q26.1	PP2Cε	41	Widely expressed; endoplasmic reticulum	NP_640338 Q5SGD2	Transmembrane N-terminal domain
PPM1M	3p21.1	PP2Cη	30	Widely expressed; nucleus	NP_653242 Q96M16	ND
	12q24.11	TA-PP2C/PPTC7	33	T-cells and other cell types	NP_644812	ND
PHLPP1	18q21.3	SCOP/PLEKHE1/ PHLPP1	α 134 β 185	Cytoplasm, nucleus, membranes	NP_919431 O60346	N-terminal PH-domain
PHLPPL	16q22.3	PHLPP2	147	Cytoplasm, nucleus, membranes	NP_055835 Q6ZVD8	N-terminal PHdomain
PDP1	8q22.1	PDP1/PDP1c/ PPM2C	61	Widely expressed, abundant in skel. muscle; mitochondria	NP_060914 Q9P0J1	Heterodimeric
PDP2	16q22.1	PDP2/PDP2c	60	Widely expressed, abundant in liver; mitochondria	NP_065837 Q9P2J9	Heterodimeric

The nomenclature for the human genes and encoded proteins of the PPP and PPM families proposed by a committee at the FASEB Conference on Protein Phosphatases in 1992 is used.
[1] Different values for the number of amino acids refers to splice variants; ILKAP: Integrin linked kinase associated phosphatase; CaMIIKPase, Ca²⁺-calmodulin–dependent protein kinase II phosphatase, POPX, partner of PIX; Wip1,wild-type p53 induced phosphatase; NERPP-2C, neurite extension related protein phosphatase, related to PP2C; PDPC, pyruvate dehydrogenase phosphatase catalytic subunit; PIX, Pak [p21(Cdc42/Rac)-activated kinase] interacting guanine nucleotide exchange factor; UBC9, ubiquitin conjugating enzyme 9; SUMO, small ubiquitin-related modifier; hCAS/CSE1L, human cellular apoptosis susceptibility protein; CH, collagen homology; PH, pleckstrin homology.
[2] Accession sequences and references are located at http://www.ncbi.nlm.nih.gov/sites/entrez?db=pubmed and http://uniprot.org/.

TABLE 11.4 Structures and properties of the human FCP protein phosphatase catalytic subunits

Human gene name	Gene locus	Protein names	Protein size (kDa)	Protein Accession number[1]	Tissue distribution / *Subcellular localization*	Holoenzyme structure or interactions	Functions/enzymes regulated
CTDP1	18q23	FCP1	104	NP_004706 NP_430255 Q9Y5B0	Widely expressed; *nucleus*	Heterodimer	Dephosphorylates the C-terminus of RNA polymerase II. Mg^{2+} dependent
CTDSP1	2q35	SCP1	29	NP_872580 NP_067021 Q9GZU7	Non-neuronal tissues, several cell lines, embryos (frog); *nucleus*	Associates with Smads	Transcriptional co-repressor inhibiting neuronal gene transcription in non-neuronal cells. Dephosphorylation of Smad1 C-terminal tail, attenuating BMP signaling. Dephosphorylation of Smad1, 2, & 3 linker region increasing TGFβ-induced transcriptional activity.
CTDSP2	12q13-q15	SCP2	31	NP_005721 O14595	Several cell lines, embryos (frog); *nucleus*	Associates with Smads	Amplified in human sarcomas. Attenuates steroid activated transcription. Dephosphorylation of Smad1 C-terminal tail, attenuating BMP signaling. Dephosphorylation of Smad1, 2, & 3 linker region increasing TGFβ-induced transcriptional activity.
CTDSPL	3p21.3	SCP3	30	NP_005799	Several cell lines, embryos (frog); *nucleus*	Associates with Smads	Tumour suppressor

	Location	Protein	aa	Accession[1]	Expression; localization	Domain	Function
DULLARD	17p13	Dullard	28	NP_001008393, O15194	Several cell lines; nuclear membrane	Transmembrane domain	Dephosphorylation of Smad1 C-terminal tail, attenuating BMP signaling. Dephosphorylation of Smad1, 2, & 3 linker region increasing TGFβ-induced transcriptional activity
CTDSPL2	15q15.3	HSPC129/ CTDSPL2	53	NP_056158, O95476	ND	ND	Nuclear biogenesis
				NP_057480, Q8IY19			ND
TIMM50	19q13	TIM50 isoform-1, Tim50a isoform-2	40, 50	Q3ZCQ8-1, NP_001001563, Q3ZCQ8-2	Widely expressed; 1 – mitochondria, 2 – nucleus	1 – subunit of Tim23 complex; 2 – associates with coilin	1 – mitochondrial protein import; 2 – snRNP biogenesis
UBLCP1	5q33.3	UBLCP	37	NP_659486, Q8WVY7	Widely expressed; nucleus	Ubiquitin-like domain	ND

FCP, TFIIF interacting C-terminal domain (CTD) phosphatase 1; SCP, small CTD phosphatase 1; HHPC129/CTDSPL2, CTD small phosphatase like 2; TIMM50, translocase of inner mitochondrial membrane 50; UBLCP1, ubiquitin-like domain containing CTD phosphatase 1.
[1]Accession sequences and references are located at http://www.ncbi.nlm.nih.gov/sites/entrez?db=pubmed and http://www.uniprot.org/.

kinases recognized. Identifying all their functions is a major challenge, but may uncover promising drug targets among the largely pharmacologically untapped protein serine/threonine phosphatases.

ACKNOWLEDGEMENTS

The author thanks the Medical Research Council, UK for financial support.

REFERENCES

1. Das AK, Helps NR, Cohen PTW, Barford D. Crystal structure of the protein serine/threonine phosphatase 2C at 2.0 A resolution. *EMBO J* 1996;**15**:6798–809.

2. Guddat LW, McAlpine AS, Hume D, Hamilton S, de Jersey J, Martin JL. Crystal structure of mammalian purple acid phosphatase. *Structure* 1999;**7**:757–67.

3. Kobor MS, Archambault J, Lester W, Holstege FCP, Gileadi O, Jansma DB, Jennings EG, Kouyoumdjian F, Davidson AR, Young RA, Greenblatt J. An unusual eukaryotic protein phosphatase required for transcription by RNA polymerase II and CTD dephosphorylation in *S-cerevisiae*. *Mol Cell* 1999;**4**:55–62.

4. Burroughs AM, Allen KN, Dunaway-Mariano D, Aravind L. Evolutionary genomics of the HAD superfamily: understanding the structural adaptations and catalytic diversity in a superfamily of phosphoesterases and allied enzymes. *J Mol Biol* 2006;**361**:1003–34.

5. Wosilait WD, Sutherland EW. The relationship of epinephrine and glucagon to liver phosphorylase: II. Enzymatic inactivation of liver phosphorylase. *J Biol Chem* 1956;**218**:469–81.

6. Ingebritsen TS, Cohen P. Protein phosphatases: properties and role in cellular regulation. *Science* 1983;**221**:331–8.

7. Stewart AA, Ingebritsen TS, Manalan A, Klee CB, Cohen P. Discovery of a Ca²⁺-dependent and calmodulin-dependent protein phosphatase – probable identity with calcineurin (CAM-BP80). *FEBS Letts* 1982;**137**:80–4.

8. Berndt N, Campbell DG, Caudwell FB, Cohen P, da Cruz e Silva EF, da Cruz e Silva OB, Cohen PTW. Isolation and sequence analysis of a cDNA clone encoding a type-1 protein phosphatase catalytic subunit: homology with protein phosphatase 2A. *FEBS Letts* 1987;**223**:340–6.

9. Tamura S, Lynch KR, Larner J, Fox J, Yasui A, Kikuchi K, Suzuki Y, Tsuiki S. Molecular cloning of rat type 2C (1A) protein phosphatase mRNA. *Proc Natl Acad Sci* 1989;**86**:1796–800.

10. Kennelly PJ. Protein kinases and protein phosphatases in prokaryotes: a genomic perspective. *FEMS Microbiol Letts* 2002;**206**:1–8.

11. Barton GJ, Cohen PTW, Barford D. Conservation analysis and structure prediction of the protein serine/threonine phosphatases: sequence similarity with diadenosine tetraphosphatase from *E. coli* suggests homology to the protein phosphatases. *Eur J Biochem* 1994;**220**:225–37.

12. Farkas I, Dombradi V, Miskei M, Szabados L, Koncz C. Arabidopsis PPP family of serine/threonine phosphatases. *Trends Plant Sci* 2007;**12**:169–76.

13. Kerk D, Templeton G, Moorhead GB. Evolutionary radiation pattern of novel protein phosphatases revealed by analysis of protein data from the completely sequenced genomes of humans, green algae, and higher plants. *Plant Physiol* 2008;**146**:351–67.

14. Stark MJR. Yeast protein serine/threonine phosphatases: multiple roles and diverse regulation. *Yeast* 1996;**12**:1647–75.

15. Andreeva AV, Kutuzov MA. RdgC/PP5-related phosphatases: novel components in signal transduction. *Cell Signal* 1999;**11**:555–62.

16. Cohen PTW. Protein phosphatase 1-targeted in many directions. *J Cell Sci* 2002;**115**:241–56.

17. Ceulemans H, Bollen M. Functional diversity of protein phosphatase-1, a cellular economizer and reset button. *Physiol Rev* 2004;**84**:1–39.

18. Meiselbach H, Sticht H, Enz R. Structural analysis of the protein phosphatase 1 docking motif: molecular description of the binding specificities identifies interacting proteins. *Chem Biol* 2006;**13**:49–59.

19. Terrak M, Kerff F, Langsetmo K, Tao T, Dominguez R. Structural basis of protein phosphatase 1 regulation. *Nature* 2004;**429**:780–4.

20. Janssens V, Goris J. Protein phosphatase 2A: a highly regulated family of serine/threonine phosphatases implicated in cell growth and signalling. *Biochem J* 2001;**353**:417–39.

21. Li X, Virshup D. Two conserved domains in the regulatory B subunits mediate binding to the A subunit of protein phosphatase 2A. *Eur J Biochem* 2002;**269**:546–52.

22. Cohen PTW, Philp A, Vázquez-Martin C. Protein phosphatase 4 – from obscurity to vital functions. *FEBS Letts* 2005;**579**:3278–86.

23. Stefansson B, Ohama T, Daugherty AE, Brautigan DL. Protein phosphatase 6 regulatory subunits composed of ankyrin repeat domains. *Biochemistry* 2008;**47**:1442–51.

24. Virshup DM. Protein phosphatase 2A: a panoply of enzymes. *Curr Opin Cell Biol* 2000;**12**:180–5.

25. McConnell JL, Gomez RJ, McCorvey LR, Law BK, Wadzinski BE. Identification of a PP2A-interacting protein that functions as a negative regulator of phosphatase activity in the ATM/ATR signaling pathway. *Oncogene* 2007;**26**:6021–30.

26. Pawson T, Scott JD. Signaling through scaffold, anchoring and adaptor proteins. *Science* 1998;**278**:2075–80.

27. Chinkers M. Protein phosphatase 5 in signal transduction. *Trends Endocrinol Metab* 2001;**12**:28–32.

28. Yang J, Roe SM, Cliff MJ, Williams MA, Ladbury JE, Cohen PTW, Barford D. Molecular basis for TPR domain-mediated regulation of protein phosphatase 5. *EMBO J* 2005;**24**:1–10.

29. Klee CB, Ren H, Wang X. Regulation of the calmodulin-stimulated protein phosphatase, calcineurin. *J Biol Chem* 1998;**273**:13,367–370.

30. Kutuzov MA, Solov'eva OV, Andreeva AV, Bennett N. Protein Ser/Thr phosphatases PPEF interact with calmodulin. *Biochem Biophys Res Commun* 2002;**293**:1047–52.

31. Cohen P. The structure and regulation of protein phosphatases. *Annu Rev Biochem* 1989;**58**:453–508.

32. Pinna LA, Donella-Deana A. Phosphorylated synthetic peptides as tools for studying protein phosphatases. *Biochim Biophys Acta* 1994;**1222**:415–31.

33. Kim Y, Huang J, Cohen P, Matthews HR. Protein phosphatases 1, 2A and 2C are protein histidine phosphatases. *J Biol Chem* 1993;**268**:18,513–518.

34. MacKintosh C, Garton AJ, McDonnell A, Barford D, Cohen PTW, Cohen P, Tonks NK. Further evidence that inhibitor-2 acts like a chaperone to fold PP1 into its native conformation. *FEBS Letts* 1996;**397**:235–8.

35. Alessi DR, Street AJ, Cohen P, Cohen PTW. Inhibitor-2 functions like a chaperone to fold three expressed isoforms of mammalian protein phosphatase-1 into a conformation with the specificity and regulatory properties of the native enzyme. *Eur J Biochem* 1993;**213**:1055–66.

36. Tolstykh T, Lee J, Vafai S, Stock JB. Carboxyl methylation regulates phosphoprotein phosphatase 2A by controlling the association of regulatory B subunits. *EMBO J* 2000;**19**:5682–91.

37. Longin S, Zwaenepoel K, Louis JV, Dilworth S, Goris J, Janssens V. Selection of protein phosphatase 2A regulatory subunits is mediated by the C terminus of the catalytic Subunit. *J Biol Chem* 2007;**282**:26,971–980.

38. Hombauer H, Weismann D, Mudrak I, Stanzel C, Fellner T, Lackner DH, Ogris E. Generation of active protein phosphatase 2A is coupled to holoenzyme assembly. *PLoS Biol* 2007;**5**:e155.

39. Moorhead G, Johnson D, Morrice N, Cohen P. The major myosin phosphatase in skeletal muscle is a complex between the β-isoform of protein phosphatase 1 and the MYPT2 gene product. *FEBS Letts* 1998;**438**:141–4.

40. Toole BJ, Cohen PTW. The skeletal muscle-specific glycogen-targeted protein phosphatase 1 plays a major role in the regulation of glycogen metabolism by adrenaline in vivo. *Cell Signal* 2007;**19**:1044–55.

41. Varmuza S, Jurisicova A, Okano K, Hudson J, Boekelheide K, Shipp EB. Spermiogenesis is impaired in mice bearing a targeted mutation in the protein phosphatase 1cgamma gene. *Dev Biol* 1999;**205**:98–110.

42. Trinkle-Mulcahy L, Andersen J, Lam YW, Moorhead G, Mann M, Lamond AI. Repo-Man recruits PP1 gamma to chromatin and is essential for cell viability. *J Cell Biol* 2006;**172**:679–92.

43. Liu CW, Wang RH, Berndt N. Protein phosphatase 1α activity prevents oncogenic transformation. *Mol Carcinog* 2006;**45**:648–56.

44. Junttila MR, Puustinen P, Niemela M, Ahola R, Arnold H, Bottzauw T, Ala-aho R, Nielsen C, Ivaska J, Taya Y, Lu SL, Lin S, Chan EK, Wang XJ, Grenman R, Kast J, Kallunki T, Sears R, Kahari VM, Westermarck J. CIP2A inhibits PP2A in human malignancies. *Cell* 2007;**130**:51–62.

45. Mumby M. PP2A: unveiling a reluctant tumor suppressor. *Cell* 2007;**130**:21–4.

46. Shui JW, Hu MC, Tan TH. Conditional knockout mice reveal an essential role of protein phosphatase 4 in thymocyte development and pre-T-cell receptor signaling. *Mol Cell Biol* 2007;**27**:79–91.

47. Martin-Granados C, Philp A, Oxenham SK, Prescott AR, Cohen PTW. Depletion of protein phosphatase 4 in human cells reveals essential roles in centrosome maturation and Rho GTPases. *Intl J Biochem Cell Biol* 2008;**40**:2315–32.

48a. Chowdhury D, Xu X, Zhong X, Ahmed F, Zhong J, Liao J, Dykxhoorn DM, Weinstock DM, Pfeifer GP and Lieberman J. A PP4-phosphatase complex dephosphorylates gamma-H2AX generated during DNA replication. *Mol Cell* 2008;**31**:33–46.

48b. Nakada S, Chen GI, Gingras AC and Durocher D. PP4 is a gamma H2AX phosphatase required for recovery from the DNA damage checkpoint. *EMBO Rep* 2008;**91**:1019–1026.

49. Stefansson B, Brautigan DL. Protein phosphatase PP6 N terminal domain restricts G1 to S phase progression in human cancer cells. *Cell Cycle* 2007;**6**:1386–92.

50. Kajino T, Ren H, Iemura S, Natsume T, Stefansson B, Brautigan DL, Matsumoto K, Ninomiya-Tsuji J. Protein phosphatase 6 down-regulates TAK1 kinase activation in the IL-1 signaling pathway. *J Biol Chem* 2006;**281**:39,891–896.

51. Hinds Jr TD, Sanchez ER. Protein phosphatase 5. *Intl J Biochem Cell Biol* 2008;**40**:2358–62.

52. Golden T, Swingle M, Honkanen RE. The role of serine/threonine protein phosphatase type 5 (PP5) in the regulation of stress-induced signaling networks and cancer. *Cancer Metastasis Rev* 2008;**27**:169–78.

53. Wang Z, Chen W, Kono E, Dang T, Garabedian MJ. Modulation of glucocorticoid receptor phosphorylation and transcriptional activity by a C-terminal-associated protein phosphatase. *Mol Endocrinol* 2007;**21**:625–34.

54. Zeke T, Morrice N, Vázquez Martin C, Cohen PTW. Human protein phosphatase 5 dissociates from heat shock proteins and is proteolytically activated in response to arachidonic acid and the microtubule depolymerising drug nocodazole. *Biochem J* 2005;**385**:45–56.

55. Wandinger SK, Suhre MH, Wegele H, Buchner J. The phosphatase Ppt1 is a dedicated regulator of the molecular chaperone Hsp90. *EMBO J* 2006;**25**:367–76.

56. Yong W, Bao S, Chen H, Li D, Sanchez ER, Shou W. Mice lacking protein phosphatase 5 are defective in ataxia telangiectasia mutated (ATM)-mediated cell cycle arrest. *J Biol Chem* 2007;**282**:14,690–694.

57. Ramulu P, Kennedy M, Xiong W-H, Williams J, Cowan M, Blesh D, Yau K-W, Hurley JB, Nathans J. Normal light response, photoreceptor integrity and rhodopsin dephosphorylation in mice lacking both protein phosphatases with EF hands (PPEF-1 and PPEF-2). *Mol Cell Biol* 2001;**21**:8605–14.

58. Mansuy IM, Shenolikar S. Protein serine/threonine phosphatases in neuronal plasticity and disorders of learning and memory. *Trends Neurosci* 2006;**29**:679–86.

59. Banner NR, Lyster H, Yacoub MH. Clinical immunosuppression using the calcineurin-inhibitors cyclosporin and tacrolimus. In: Pinna LA, Cohen PTW, editors. *Handbook of experimental pharmacology: inhibitors of protein Kinases and protein phosphatases*: Springer-Verlag Heidelberg; 2005. p. 321–59.

60. Honkanen RE. Serine/threonine protein phosphatase inhibitors with antitumour activity. In: Pinna LA, Cohen PTW, editors. *Handbook of experimental pharmacology: inhibitors of protein kinases and protein phosphatases*: Springer-Verlag Heidelberg; 2005. p. 297–317.

61. Ammosova T, Washington K, Debebe Z, Brady J, Nekhai S. Dephosphorylation of CDK9 by protein phosphatase 2A and protein phosphatase-1 in Tat-activated HIV-1 transcription. *Retrovirology* 2005;**2**:47.

62. Saito S, Matsui H, Kawano M, Kumagai K, Tomishige N, Hanada K, Echigo S, Tamura S, Kobayashi T. Protein Phosphatase 2C{epsilon} Is an endoplasmic reticulum integral membrane protein that dephosphorylates the ceramide transport protein CERT to enhance its association with organelle membranes. *J Biol Chem* 2008;**283**:6584–93.

63. Sugiura T, Noguchi Y, Sakurai K, Hattori C. Protein phosphatase 1H, overexpressed in colon adenocarcinoma, is associated with CSE1L. *Cancer Biol Ther* 2008;**7**:285–93.

64. Lammers T, Lavi S. Role of type 2C protein phosphatases in growth regulation and in cellular stress signaling. *Crit Rev Biochem Mol Biol* 2007;**42**:437–61.

65. Sapkota G, Knockaert M, Alarcon C, Montalvo E, Brivanlou AH, Massague J. Dephosphorylation of the linker regions of Smad1 and Smad2/3 by small C-terminal domain phosphatases has distinct outcomes for bone morphogenetic protein and transforming growth factor-β pathways. *J Biol Chem* 2006;**281**:40,412–419.

66. Wrighton KH, Willis D, Long J, Liu F, Lin X, Feng XH. Small C-terminal domain phosphatases dephosphorylate the regulatory linker regions of Smad2 and Smad3 to enhance transforming growth factor-β signaling. *J Biol Chem* 2006;**281**:38,365–375.

67. Yeo M, Lee SK, Lee B, Ruiz EC, Pfaff SL, Gill GN. Small CTD phosphatases function in silencing neuronal gene expression. *Science* 2005;**307**:596–600.

68. Kim Jr Y, Gentry MS, Harris TE, Wiley SE, Lawrence JC, Dixon JE. A conserved phosphatase cascade that regulates nuclear membrane biogenesis. *Proc Natl Acad Sci USA* 2007;**104**:6596–601.

Chapter 12

The Structure and Topology of Protein Serine/Threonine Phosphatases

David Barford

Section of Structural Biology, Institute of Cancer Research, Chester Beatty Laboratories, London, England, UK

INTRODUCTION

Structural studies of the two families of protein phosphatases responsible for dephosphorylating serine and threonine residues have revealed that, although these families are unrelated in sequence, the architecture of their catalytic domains is remarkably similar and distinct from the protein tyrosine phosphatases. The diversity of structure within the PPP and PPM families is generated by regulatory subunits and domains that function to modulate protein specificity and to localize the phosphatase to particular subcellular locations [1].

PROTEIN SERINE/THREONINE PHOSPHATASES OF THE PPP FAMILY

The protein Ser/Thr phosphatases PP1, PP2A, and PP2B of the PPP family, together with PP2C of the PPM family, account for the majority of the protein serine/threonine phosphatase activity *in vivo*. While PP1, PP2A, and PP2B share a common catalytic domain of 280 residues, these enzymes are divergent within their non-catalytic N and C termini and are distinguished by their associated regulatory subunits to form a diverse variety of holoenzymes. Major members of the PPP family are encoded by numerous isoforms that share a high degree of sequence similarity, especially within their catalytic domains. Greater sequence diversity occurs within the extreme N and C termini of the proteins. Although these isoforms have similar substrate specificities and interact with the same regulatory subunits *in vitro*, the phenotype of a functional loss is isoform specific, indicating they perform distinct functions *in vivo*.

Overall Structure and Catalytic Mechanism

Structural analyses of members of the PPP family have begun to reveal the molecular basis for catalysis, inhibition by toxins, and aspects of their regulatory mechanisms. Crystal structures are available for (1) various isoforms of PP1 in complex with natural toxins [2, 3], the phosphate mimic tungstate [4], and a peptide of the RVxF targeting motif [5]; (2) PP2B in the auto-inhibited state [6] and as a complex with FK506/FKBP [6, 7]; (3) the PR65/A-subunit of PP2A [8]; and (4) a regulatory TPR domain of PP5 [9]. The catalytic domains of the PP1 catalytic subunit (PP1c) and PP2B share a common architecture consisting of a central β-sandwich of two mixed β-sheets surrounded on one side by seven α-helices and on the other by a subdomain comprised of three α-helices and a three-stranded β-sheet (Figure 12.1a). The interface of the three β-sheets at the top of the β-sandwich creates a shallow catalytic site channel. Conserved amino acid residues present on loops emanating from the β-strands of this central β-sandwich are responsible for coordinating a pair of metal ions to form a binuclear metal centre (Figure 12.2a). Crystallographic data on PP1c and PP2B provided the first compelling insight regarding the role of metal ions in the catalytic reaction of the PPP family. The identity of the two metal ions is slightly controversial. Proton-induced X-ray emission spectroscopy performed on PP1c crystals produced from the protein expressed in *Escherichia coli* indicated that the metal ions were Fe^{2+} (or Fe^{3+}) and Mn^{2+} [4], whereas atomic absorption spectroscopy of bovine brain PP2B indicated a stoichiometric ratio of Zn^{2+} and iron [10]. There have also been conflicting reports concerning the iron oxidation states, with both Fe^{2+} and Fe^{3+} being observed. Native

(a)

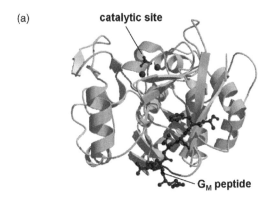

(b)

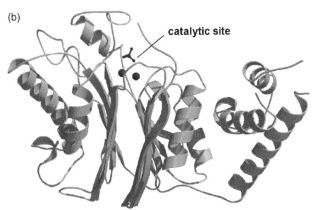

FIGURE 12.1 Ribbon diagrams showing an overview of the structures of protein Ser/Thr phosphatase catalytic domains.
(a) The catalytic subunit of human protein phosphatase 1γ in complex with the RVxF motif PP1 binding peptide. The catalytic site is indicated by the two metal ions and bound sulfate ion. (b) Human PP2Cα the catalytic site is indicated by the two metal ions and bound phosphate ion. Figure 12.1a reproduced from [5] (Egloff et al., 1997), with permission; Figure 12.2 is reproduced from [19] (Das et al., 1996), with permission.

PP2B is most likely to contain Fe^{2+}, explaining the time-dependent inactivation of PP2B that results from the oxidation of the Fe–Zn center [11]. Oxidation of the binuclear metal center and phosphatase inactivation may represent a mechanism for PP2B regulation by redox potential during oxidative stress or as a result of reactive oxygen species generation following receptor tyrosine kinase activation, in a process reminiscent of the inactivation of protein tyrosine phosphatases by oxidation of the catalytic site Cys residue by hydrogen peroxide and perhaps analogous to a possible PP2C regulatory mechanism, discussed later [10].

The structure of PP1c with tungstate and PP2B with phosphate indicated that two oxygen atoms of the oxyanion-substrate coordinate the metal ions (Figure 12.2a) [4, 7]. Two water molecules, one of which is a metal-bridging water molecule, contribute to the octahedral hexa-coordination of the metal ions. The metal coordinating residues (aspartates, histidines, and asparagines) are

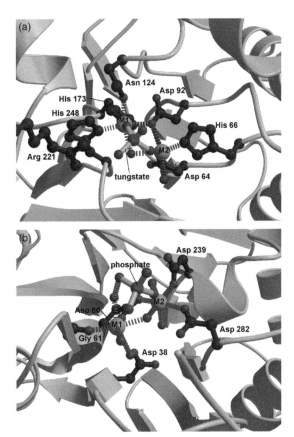

FIGURE 12.2 Comparison of the catalytic sites of the protein Ser/Thr phosphatases, showing the common binuclear metal center, coordinating the phosphate of the phosphorylated protein substrate: (a) human PP1γ (for clarity, Arg 96 is not shown) and (b) human PP2Cα.

invariant among all PPP family members. These residues, together with Arg and His residues that interact with the phosphate group of the phosphorylated residue, occur within five conserved sequence motifs found in other enzymes, including the purple acid phosphatase, whose common function is to catalyze phosphoryl transfer reactions to water [12]. These observations suggest that PPP and purple acid phosphatases evolved by divergent evolution from an ancestral metallophosphoesterase. Consistent with roles in catalysis, mutation of these residues either eliminates or profoundly reduces catalytic activity.

PPPs catalyze dephosphorylation in a single step with a metal-activated water molecule or hydroxide ion. The most convincing evidence for this notion is that the purple acid phosphatase, which is generally related in structure to the PPPs both at the tertiary level and at the catalytic site [13], promote dephosphorylation with inversion of configuration of the oxygen geometry of the phosphate ion [14]. This indicates that a phosphoryl-enzyme intermediate would not occur. The two metal-bound water molecules are within van der Waals distance of the phosphorous atom of the phosphate bound to the catalytic site, and one of them is likely to be metal-activated nucleophile.

Interactions with Regulatory Subunits

PP1 and PP2A are responsible for regulating diverse cellular functions by dephosphorylating multiple and varied protein substrates. This seemingly paradoxical situation was resolved by the discovery that distinct forms of PP1 and PP2A holoenzymes occur *in vivo*, where essentially the same catalytic subunit is complexed to different targeting and regulatory subunits. For PP1 it has been shown that targeting subunits confer substrate specificity by directing particular PP1 holoenzymes to a subcellular location and by enhancing or suppressing activity toward different substrates. The control of PP1 holoenzyme structure and activity by the combinatorial selection of different targeting/regulatory subunits has recently been the subject of numerous reviews [15, 16] and will not be discussed at length here, although the contrasting structural mechanism by which the conserved PP1 and PP2A catalytic subunits are able to form diverse holoenzyme structures will be discussed.

In the case of PP1, it is known that the binding of targeting subunits to its catalytic subunit is mutually exclusive, suggesting that there are one or more common or overlapping binding sites, recognized by all PP1-binding subunits. It is therefore a little surprising that PP1-binding subunits are highly diverse structurally and share little to no overall sequence similarities. The key to understanding this paradox came from the crystal structure of PP1c in complex with a short, 13-residue peptide derived from a region of the PP1-glycogen targeting subunit (G_M) responsible for PP1c interactions (Figure 12.1a). This structure showed that the peptide associated with the phosphatase via two hydrophobic residues (Val and Phe), which engage a hydrophobic groove on the protein surface formed from the interface of the two β-sheets of the central β-sandwich and remote from the catalytic site [5]. Two basic residues of the peptide immediately N-terminal to the Val residue form salt-bridge interactions with Asp and Glu residues at one end of the peptide binding channel. Alanine substitutions of either the Val or Phe residues of the peptide abolish PP1–peptide interactions. Analysis of other PP1-binding subunit sequences revealed the presence of the identical or related sequence motif RRVxF, found to mediate the interactions between the G_M peptide and PP1c [5]. The role of the degenerative RVxF motif in mediating PP1-regulatory subunit interactions is now supported by numerous experimental observations. First, for various regulatory subunits, mutation of either hydrophobic residue of the motif in native proteins weakens or eliminates their association with PP1c. Second, peptides corresponding to the RVxF motif competitively disrupt the interactions of regulatory subunits with PP1c. Third, the use of a common or overlapping PP1c binding site explains why the interactions of regulatory subunits is mutually exclusive. The number of PP1 holoenzyme structures that could be generated in this manner is potentially infinite, and to date over a hundred PP1 regulatory subunits have been characterized.

The residues of PP1 that interact with the RVxF peptide are conserved in all isoforms of PP1 in all eukaryotic species, although not within PP2A and PP2B, explaining why PP1-binding subunits are unique to PP1. The mechanism of combinatorial control of PP2A holoenzyme structure is different from that of PP1 and is mediated by a scaffolding subunit termed the PR65/A subunit, which simultaneously associates with the PP2A catalytic subunit and a variable regulatory B subunit. The ability to recognize a variety of regulatory subunits (perhaps over 50) is conferred by the architecture of the PR65/A subunit, which consists of 15 tandem repeats of a 39 amino-acid sequence termed the HEAT motif and related in structure to ARM repeats. These repeats assemble to create an extended molecule ideally suited for mediating protein–protein interactions [8]. Combinatorial generation of variable PP2A holoenzymes is achieved by the ability of different combinations of HEAT motifs to select different regulatory B subunits [8].

Interactions of Natural Toxins with PP1

PP1 and PP2A are specifically and potently inhibited by a variety of naturally occurring toxins such as okadaic acid, a diarrhetic shellfish poison and powerful tumor promoter, and microcystin, a liver toxin produced by blue-green algae [17]. Whereas PP2B is only poorly inhibited by the toxins that affect PP1 and PP2A, it is known to be the immunosuppressive target of FK506 and cyclosporin in association with their major cellular binding proteins, the cis-trans peptidyl prolyl isomerase FKBP12 and cyclophilin, respectively [18]. The mechanism of inhibition of PP1 by okadaic acid and microcystin LR have been defined by structures of PP1 in complex with these inhibitors. Both inhibitors, although structurally different, bind to a similar region of the phosphatase, occupying the catalytic channel to directly block phosphatase–substrate interactions. Regions of PP1 that contact the toxins include the hydrophobic groove and the β12/β13 loop, the latter undergoing conformational changes to optimize contacts with microcystin [2, 3]. Both toxins disrupt substrate–phosphatase interactions by competing for sites on the protein that coordinate the phosphate group of the substrate. For example, carboxylate and carbonyl groups of microcystin interact with two of the metal-bound water molecules [2], whereas okadaic acid contacts the two phosphate-coordinating arginine residues (Arg-96 and Arg-221) [3]. A similar mechanism of phosphatase inhibition by steric hindrance of substrate binding is observed in the structure of the full-length PP2B holoenzyme, for which the autoinhibitory domain lies over the substrate binding channel of the catalytic domain in such a way that a Glu side chain accepts a hydrogen bond from two of the metal-bound water molecules [6].

PROTEIN SERINE/THREONINE PHOSPHATASES OF THE PPM FAMILY

Protein phosphatases of the PPM family are present in both eukaryotes and prokaryotes for which the defining member is PP2C. Biochemically, the PPM family was distinguished from the PPP family by its requirement for divalent metal ions (Mg^{2+}) for catalytic activity, although it is now known from crystal structures that both PPP and PPM phosphatases catalyze dephosphorylation reactions by means of a binuclear divalent metal center. Within the PPM family, the PP2C domain occurs in numerous structural contexts that reflect structural diversity [19]. For example, the PP2C domain of the Arabidopsis ABI1 gene is fused with EF hand motifs, whereas in KAPP-1, a kinase interaction domain associated with a phosphorylated receptor precedes the phosphatase domain. Other less closely related examples include the Ca^{2+} stimulated mitochondrial pyruvate dehydrogenase phosphatase, which contains a catalytic subunit sharing 22 percent sequence identity with that of mammalian PP2C, and the SpoIIE phosphatase of *Bacillus subtilus*, which has ten membrane-spanning regions preceding the PP2C-like catalytic domain. A surprising homolog is a 300-residue region of yeast adenylyl cyclase present immediately N-terminus to the cyclase catalytic domain that shares sequence similarity with PP2C. This domain may function to mediate Ras-GTP activation of adenylyl cyclase activity and is not known to possess protein phosphatase activity. In eukaryotes, the various isoforms of PP2C have been implicated in diverse functions such as regulation of cell-cycle progression mediated by dephosphorylation of CDKs [20], to regulation of RNA splicing [21], control of p53 activity [22], and regulation of stress response pathways in yeast [23].

The sequences of protein phosphatases of the PPM family share no similarity with those of the PPP family, and the natural toxins that inhibit the PPP family have no affect on PPM family phosphatases. It therefore came as a surprise when the crystal structure of human PP2Cα revealed a striking similarity in tertiary structure and catalytic site architecture to the PPP protein phosphatases (Figure 12.1b) [19]. Mammalian PP2C consists of two domains; an N-terminal catalytic domain common to all members of the PP2C family is fused to a 90-residue C-terminal domain, unique to the mammalian PP2Cs [19]. The catalytic domain is dominated by a central, buried β-sandwich of 11 β-strands formed by the association of two antiparallel β-sheets, both of which are flanked by a pair of antiparallel α-helices inserted between the two central β-strands, with four additional α-helices. The C-terminal domain is formed from three antiparallel α-helices remote from the catalytic site, suggesting a role in defining substrate specificity rather than catalysis.

At the catalytic site of PP2C, two Mn^{2+} ions, separated by 4 Å, form a binuclear metal center and are coordinated by four invariant aspartate residues and a non-conserved Glu residue (Figure 12.2b) [19]. These residues are situated at the top of the central β-sandwich that forms a shallow channel suitable for the dephosphorylation of phosphoserine- and phosphothreonine-containing proteins. Six water molecules coordinate the two metal ions. One of these water molecules bridges the two metal ions and four form hydrogen bonds to a phosphate ion at the catalytic site. Dephosphorylation is probably catalyzed by a metal-activated water molecule that acts as nucleophile in a mechanism similar to that proposed for the PPP family. A recent kinetic analysis of PP2Cα by Denu and colleagues [24] indicated that Mn^{2+} and Fe^{2+} are the most effective divalent metal ions in promoting dephosphorylation reactions, suggesting that at least one of these ions must be present at the catalytic site. In contrast, Zn^{2+} and Ca^{2+} competitively inhibit PP2C Mn^{2+}-dependent activity. An Fe^{2+}-containing catalytic site would be analogous to the PPP protein phosphatases and possibly explains the H_2O_2-mediated inactivation of PP2C [25] consequent on the redox sensitivity of Fe^{2+} and its oxidation to the inactive Fe^{3+} valence state [24]. The finding that an ionizable group with a pKa of 7.0 has to be deprotonated for catalysis was fully consistent with the notion from the crystal structure that a metal activated water molecule acts as a nucleophile [19, 24]. Fluoride has long been used as an inhibitor of both PPP and PPM phosphatases, and the rationale for this inhibition is now clear from the catalytic mechanism of serine/threonine phosphatases revealed by the crystal structures. By substituting for the metal-bound nucleophilic water molecule, fluoride prevents metal-activated nucleophilic attack on the phospho-protein substrate. Substitution of Ala for the Asp residues of the catalytic site in yeast and plant PP2C homologs and in the related phosphatase SpoIIE from B. subtilus abolishes catalytic activity, supporting a role for the metal ions in catalysis [26, 27].

CONCLUSIONS

Although the PPP and PPM phosphatases share a similar tertiary structure, characterized by a central β-sandwich and flanking α-helices, with related catalytic site architectures and mechanisms, two observations suggest that these protein families have evolved from distinct ancestors. First, the secondary structure topology of the PPP and PPM families are unrelated and there is no simple rearrangement of chain connectivities that would allow the PPP topology to be transformed into the PPM topology. Second, the conserved sequence motifs of the PPP family required for metal binding and catalysis, typical of some metallophosphoesterases, are distinct from the conserved sequence motifs of the PPM family. The comparison of the PPP and PPM families of Ser/Thr protein phosphatases provides interesting contrasts with the protein phosphatases of the $C(X)_5R$-motif superfamily composed of three distinct gene families: (1) the conventional protein tyrosine phosphatases (PTPs), including dual-specificity phosphatases and PTEN lipid phosphatases; (2) low-molecular-weight protein tyrosine phosphatases (lmPTPs); and (3) Cdc25 dual-specificity phosphatases [28]. These proteins share a similar overall tertiary structure composed of a central β-sheet surrounded on both sides by α-helices which results in an identical catalytic site architecture in all three families characterized by the nucleophilic Cys-residue located within a phosphate binding cradle. What distinguishes the three protein families are differences in their secondary structure topology; however, the topologies of the PTP and lmPTP families are related by a simple permutation of secondary structure connectivity, suggesting that these families may have evolved from the same ancestor and diverged as a result of exon shuffling events. In recent years, much has been learned of the structural details of Ser/Thr protein phosphatases relating to their overall folds, catalytic mechanisms, and interactions with regulatory subunits and toxins. However, still elusive is the structure of a Ser/Thr protein phosphatase in complex with a phosphoprotein substrate that would explain the basis for the selectivity for Ser/Thr residues and reveal the mechanism of substrate selectivity by regulatory subunits. It is hoped that future studies of Ser/Thr phosphatase will address these questions.

REFERENCES

1. Hubbard MJ, Cohen P. On target with a new mechanism for the regulation of protein phosphorylation. *Trends Biochem Sci* 1993;**18**:172–7.
2. Goldberg J, Huang HB, Kwon YG, Greengard P, Nairn AC, Kuriyan J. Three-dimensional structure of the catalytic subunit of protein serine/threonine phosphatase-1. *Nature* 1995;**376**:745–53.
3. Maynes JT, Bateman KS, Cherney MM, Das AK, Luu HA, Holmes CF, James MN. Crystal structure of the tumor-promoter okadaic acid bound to protein phosphatase-1. *J Biol Chem* 2001;**276**:44,078–44,082.
4. Egloff MP, Cohen PT, Reinemer P, Barford D. Crystal structure of the catalytic subunit of human protein phosphatase 1 and its complex with tungstate. *J Mol Biol* 1995;**254**:942–59.
5. Egloff MP, Johnson DF, Moorhead G, Cohen PT, Cohen P, Barford D. Structural basis for the recognition of regulatory subunits by the catalytic subunit of protein phosphatase 1. *EMBO J* 1997;**16**:1876–87.
6. Kissinger CR, Parge HE, Knighton DR, Lewis CT, Pelletier LA, Tempczyk A, Kalish VJ, Tucker KD, Showalter RE, Moomaw EW, et al. Crystal structures of human calcineurin and the human FKBP12–FK506–calcineurin complex. *Nature* 1995;**378**:641–4.
7. Griffith JP, Kim JL, Kim EE, Sintchak MD, Thomson JA, Fitzgibbon MJ, Fleming MA, Caron PR, Hsiao K, Navia MA. X-ray structure of calcineurin inhibited by the immunophilin-immunosuppressant FKBP12–FK506 complex. *Cell* 1995;**82**:507–22.
8. Groves MR, Hanlon N, Turowski P, Hemmings BA, Barford D. The structure of the protein phosphatase 2A PR65/A subunit reveals the conformation of its 15 tandemly repeated HEAT motifs. *Cell* 1999;**96**:99–110.
9. Das AK, Cohen PW, Barford D. The structure of the tetratricopeptide repeats of protein phosphatase 5: implications for TPR-mediated protein–protein interactions. *EMBO J* 1998;**17**:1192–9.
10. Wang X, Culotta VC, Klee CB. Superoxide dismutase protects calcineurin from inactivation. *Nature* 1996;**383**:434–7.
11. Yu L, Haddy A, Rusnak F. Evidence that calcineurin accommodates an active site binuclear metal center. *J Am Chem Soc* 1995;**117**:10,147–10,148.
12. Lohse DL, Denu JM, Dixon JE. Insights derived from the structures of the Ser/Thr phosphatases calcineurin and protein phosphatase 1. *Structure* 1995;**3**:987–90.
13. Klabunde T, Strater N, Frohlich R, Witzel H, Krebs B. Mechanism of Fe(III)–Zn(II) purple acid phosphatase based on crystal structures. *J Mol Biol* 1996;**259**:737–48.
14. Mueller EG, Crowder MW, Averill BA, Knowles JR. Purple acid phosphatase: a diron enzyme that catalyses a direct phosphogroup transfer to water. *J Am Chem Soc* 1993;**115**:2974–5.
15. Bollen M. Combinatorial control of protein phosphatase-1. *Trends Biochem Sci* 2001;**26**:426–31.
16. Cohen PT. Protein phosphatase 1: targeted in many directions. *J Cell Sci* 2002;**115**:241–56.
17. MacKintosh C, MacKintosh RW. Inhibitors of protein kinases and phosphatases. *Trends Biochem Sci* 1994;**19**:444–8.
18. Liu Jr. J, Farmer JD, Lane WS, Friedman J, Weissman I, Schreiber SL Calcineurin is a common target of cyclophilin-cyclosporin A and FKBP–FK506 complexes. *Cell* 1991;**66**:807–15.
19. Das AK, Helps NR, Cohen PTW, Barford D. Crystal structure of human protein serine/threonine phosphatase 2C at 2.0-Å resolution. *EMBO J* 1996;**15**:6798–809.
20. Cheng A, Ross KE, Kaldis P, Solomon MJ. Dephosphorylation of cyclin-dependent kinases by type 2C protein phosphatases. *Genes Dev* 1999;**13**:2946–57.
21. Murray MV, Kobayashi R, Krainer AR. The type 2C Ser/Thr phosphatase PP2Cγ is a pre-mRNA splicing factor. *Genes Dev* 1999;**13**:87–97.
22. Takekawa M, Adachi M, Nakahata A, Nakayama I, Itoh F, Tsukuda H, Taya Y, Imai K. p53-inducible wip1 phosphatase mediates a negative feedback regulation of p38 MAPK-p53 signaling in response to UV radiation. *EMBO J* 2000;**19**:6517–26.
23. Shiozaki K, Russell P. Counteractive roles of protein phosphatase 2C (PP2C) and a MAP kinase kinase homolog in the osmoregulation of fission yeast. *EMBO J* 1995;**14**:492–502.
24. Fjeld CC, Denu JM. Kinetic analysis of human serine/threonine protein phosphatase 2Cα. *J Biol Chem* 1999;**274**:20,336–20,343.
25. Meinhard M, Grill E. Hydrogen peroxide is a regulator of ABI1, a protein phosphatase 2C from Arabidopsis. *FEBS Letts* 2001;**508**:443–6.
26. Adler E, Donella-Deana A, Arigoni F, Pinna LA, Stragler P. Structural relationship between a bacterial developmental protein and eukaryotic PP2C protein phosphatases. *Mol Microbiol* 1997;**23**:57–62.
27. Sheen J. Mutational analysis of protein phosphatase 2C involved in abscisic acid signal transduction in higher plants. *Proc Natl Acad Sci USA* 1998;**95**:975–80.
28. Barford D, Das AK, Egloff M-P. The structure and mechanism of protein phosphatases. Insights into catalysis and regulation. *Annu Rev Biophys Biomol Struct* 1998;**27**:133–64.

Chapter 13

SH2 Domain-Containing Protein-Tyrosine Phosphatases

Benjamin G. Neel, Gordon Chan and Salim Dhanji

Ontario Cancer Institute, Toronto, Ontario, Canada

HISTORY AND NOMENCLATURE

Shp1 was cloned by four groups, using PCR [1–4]. Later, mammalian Shp2s were cloned using similar approaches [5–9], and Shp orthologs in *Xenopus* [10], chicken [11] and, most recently, zebrafish [12] were reported. A subsequent agreement led to the adoption of single names for each mammalian Shp [13], with "Shp1" replacing PTP1C [1], SH-PTP1 [2], HCP [3], and SHP [4], and SH-PTP2 [5], Syp [6], PTP1D [7], PTP2C [8], and SH-PTP3 [9] now termed "Shp2." Genomic sequencing efforts have re-introduced some confusion. The Human Gene Mapping Nomenclature Committee employs a standardized naming system for PTPs, in which the gene encoding SHP1 is termed *PTPN6* and that for SHP2 is *PTPN11* (the "N" indicates "non-transmembrane; the number specifies the order in which the PTP was reported to the database). Technically, the respective protein products should be termed PTPN6 and PTPN11. For the balance of this chapter, we use the standard nomenclature to refer to each gene, but will refer to the proteins by their more commonly used names, Shp1 and Shp2 (or, for the human versions SHP1 and SHP2), respectively.

Drosophila corkscrew (*csw*) was identified in a screen for modifiers of the Torso receptor tyrosine kinase (RTK) pathway [14]. Although initially believed to be an Shp1 ortholog, sequence analysis [5] and functional studies [15] indicate that Csw actually is the ortholog of Shp2. *C. elegans* has a single Shp, *ptp-2*, whose function also appears most analogous to Shp2 [16]. It remains unclear whether Csw and Ptp-2 also have some functions similar to Shp1, or if, as seems more likely, Shp1 evolved to carry out functions unique to vertebrates. Notably, zebrafish has both Shp1 and Shp2 orthologs [12].

STRUCTURE, EXPRESSION, AND REGULATION

Primary Structure

Shps (Figure 13.1a) have two SH2 domains at their N-termini (hereafter, N-SH2 and C-SH2); a classical protein-tyrosine phosphatase (PTP) catalytic domain and a C-terminal tail ("C-tail"). The Csw PTP domain is split by a cysteine/serine-rich insert (~150 amino acids) with no clear protein motifs [14]. The detailed function of the Csw insert is unknown, although it is conserved in other *Drosophila* species and "insert-less mutants" act as weak *csw* loss of function mutants [15]. Vertebrate Shp2 orthologs lack an insert, so either its function is specific to insect signaling pathway(s), or it is taken over by another protein invertebrate. Furthermore, some *csw* splice variants lack the insert, suggesting that it is important only in some signaling pathways [17].

Shps also differ in their C-tails; indeed, this region differs most between the two Shps (Figure 13.1a, b). Shp1 and Shp2 have two tyrosyl phosphorylation sites in this region, which are phosphorylated differentially by RTKs and non-receptor PTKs [6, 7, 18–25]. The Csw C-tail retains only the more proximal tyrosine (Y542 in Shp2), whereas Ptp-2 lacks both sites. Shp2 and Csw (but again, not Ptp-2) have proline-rich domains that may bind SH3 domain-containing proteins, possibly including Src family PTKs [26]. Shp1 lacks a proline-rich domain, but has a basic sequence that can function as a nuclear localization sequence (NLS) [27, 28].

Expression

Shp2 and its orthologs are expressed ubiquitously, although at variable levels in different tissues [5, 6, 8, 14]. Shp1 is

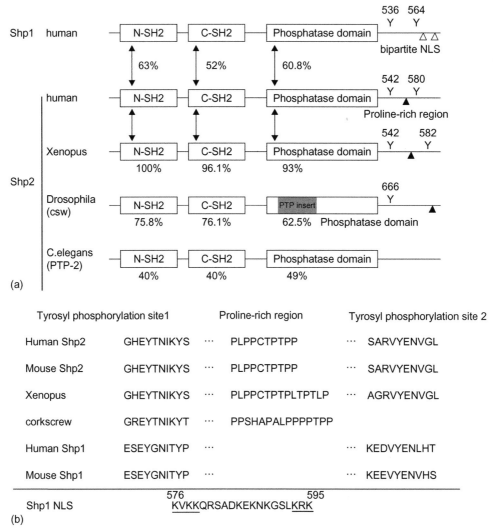

FIGURE 13.1 The Shp family.
(a) Schematic structures of Shp family members. The SH2 and PTP domains and C-tail are indicated, and the relative positions of tyrosyl phosphorylation sites, proline-rich domains, and nuclear localization sequences are indicated. (b) Features of Shp C-tails. Shown are sequences surrounding potential tyrosyl phosphorylation sites and proline-rich domains. Also shown is the potential bi-partite NLS found in Shp1, but not other family members.

more restricted, with high levels in hematopoietic cells, lower amounts in most epithelial and some neuronal cells, and little or none in fibroblasts [2, 4, 29]. The Shp1 gene has two promoters that function in a tissue-specific fashion, generating Shp1 isoforms with slightly different N-termini [30, 31]. In humans, the more 3′ (downstream) of these promoters is expressed only in hematopoietic cells, whereas the upstream promoter is expressed only in epithelia. The murine 3′ promoter may be active in epithelial cells as well, but the upstream promoter retains epithelial-specific expression [31]. A third Shp1 isoform, generated by alternative splicing, has a C-terminal extension [32]. Splice variants within the PTP domain of Shp2 and Csw also have been defined [5, 14, 33]. The Shp2 isoforms reportedly have different PTP activities [33], but their physiological significance has not been determined. Although differential

expression may explain some differences in the roles of the Shps, they clearly are not the whole story. Many cells and tissues, particularly hematopoietic cells, express high levels of both Shps, but the consequences of loss of either molecule differ dramatically (see "Biological functions of Shps," below).

SH2 Domain Function

Not surprisingly, the SH2 domains of Shps target them to phosphotyrosyl-containing (pTyr) proteins. Multiple proteins are known to bind the SH2 domains of mammalian Shp1 and Shp2. Most fall into three distinct categories: receptors (RTKs, cytokine receptors, or integrin cytoplasmic domains), scaffolding adaptors (e.g., IRS, DOS/Gab and FRS proteins), and so-called "immune inhibitory receptors" (commonly termed "inhibitory receptors"). The latter comprise a large number of glycoproteins, first described in immune cells, hence the name [34]. However, several inhibitory receptors are expressed more widely. Some Shp-binding proteins (e.g., Shps1/Sirpα, Pir-B) bind both Shps [35–38], whereas others (e.g., Gab/Dos family proteins) show specificity in Shp binding [39,40].

Unlike for most SH2 domains, residues both N- and C-terminal of pTyr contribute to binding to the Shps (Table 13.1). Early studies showed that Shp SH2 domains bind to "Immuno-receptor Tyrosine-based Inhibitory Motifs"(ITIMs): I/V/L–X–pY–X–X–I/V/L [41–44] (reviewed in [45–47]). However, why some pTyr-peptides that fit the ITIM consensus bind the N- vs C-SH2 of one or both Shps initially was unclear. Subsequently, peptide library-based approaches, including bulk binding assays [48] and one bead–one sequence binding, followed by partial Edman/mass spectrometric sequencing [49–52] have been used to address this issue.

The latter approach, in particular, has led to important new insights into the binding properties of the N- and C-SH2 domains of the Shps (Table 13.1). Unexpectedly, for example, the N-SH2 of Shp2 actually binds to four different classes of pTyr peptides. Of these, class I conforms to the ITIM motif (with some important subtleties), whereas class IV is consistent with previous, controversial reports of Shp2 binding to proteins with a pTyr–X–X–P motif [53] (reviewed in [51]). There are no documented Shp2 interactions with proteins containing class II or III motifs, although these findings suggest obvious possibilities to be tested. Notably, class I peptides are detected far more commonly that those with class IV motifs, but the latter bind the N-SH2 with significantly higher affinity. This probably explains why the class II–IV were missed in earlier, "bulk binding" experiments; they were overwhelmed by the greater number of possible lower affinity class I peptides. In more recent work [52], single bead approaches have been used to uncover further specificity at the +4 to +6 positions of the N- and C-SH2 domains of both Shps (Table 13.1).

In addition to binding Shps, ITIMs often can bind the 5′ inositol phosphatases Ship-1 and Ship-2. Furthermore, Shp2 binding sites can overlap with the binding sites for the SH2 domain of the E3 ubiquitin ligase Socs3 and possibly other Socs proteins [54–58]. The above studies have

TABLE 13.1 pTyr peptide consensus sequences that selectively interact with SH2 domains of Shp1, Shp2, SHIP and SOCS3

	−2	−1	pY	+1	+2	+3	+4	+5	+6
SHP2									
N-SH2									
Class I	I/L/V/m	X	**pY**	T/V/A	X	I/V/L/f	W/y/h	F/H/W	P/G/R/a
Class II	W	M/T/v	**pY**	y/r	I/L	X	N.D		
Class III	I/V	X	**pY**	L/M/T	Y	A/P/T/S/g	N.D		
Class IV	I/L/V	X	**pY**	F/M	X	P	N.D		
C-SH2	T/V/I/y	X	**pY**	A/s/t/v	X	I/v/i	W/R/H	H/R	R/H/y
SHP1									
N-SH2	Similar to SHP2	Similar to SHP2	**pY**			W/R/H/Y/l/k/f	H/R/k/w	R/k/h/g	
C-SH2	T/v/i	X	**pY**	C/A/t	X	L/m/v	R/kl/h/w/y	R/H/y	R/h/y
SHIP	X	X	**pY**	Y/S/T/v	L/y/m/f	L/M/i/v	N.D		
								H/V/I/Y	
SOCS3	N.D.	N.D.	**pY**	S/A/V/Y/F	φ	V/I/L	φ		

Lower-case letters represent less frequently selected residues; X is any amino acid, φ represents hydrophobic residue. N.D., not determined. See text for details.

clarified the specificity determinants for these interactions [48, 51, 59–64], and suggest pTyr peptides that should bind selected SH2 domain-containing proteins from this group. In this way, it has been possible to determine, for example, that Socs3, rather than Shp2, is the major negative regulator of gp130 signaling (see [48] and " Shp signaling and substrates", below).

Still, some reported binding interactions do not exactly fit these consensus sequences. For example, mast cell function-associated antigen (MAFA) contains an ITIM-like motif with Ser at Y-2, and a pTyr-peptide bearing this sequence can bind to both Shps [59]. Also, Shp1 reportedly binds several TNF family (death) receptors via a conserved A–X–pY–X–X–L motif. Even more surprisingly, binding could be competed by a short peptide lacking any residue at Y-2 [65]. These reports stand in marked contrast to earlier studies, which revealed an essential role for positions upstream of pTyr [43, 60, 61], and a specific requirement for hydrophobic residues at the -2 position. Conceivably, some of these non-conforming interactions are indirect; nevertheless, further work is required to resolve these inconsistencies.

Finally, a recent study showed that the N-SH2 domain of Shp2 (but not the other Shp SH2 domains) can bind pTyr-containing peptides via an alternative mechanism that involves only residues N-terminal to pTyr [66]. Although no known Shp2 interacting proteins use this binding mode, this study identified several interesting signaling molecules that fit this consensus and warrant investigation for Shp2 binding *in vivo*.

Regulation of PTP Activity

Shps have very low basal activity, but addition of a pTyr-ligand for the N-SH2 substantially enhances catalysis [67–70] (reviewed in [71]). Bis-phosphorylated (bidentate) pTyr-ligands have even greater effect, resulting in ~50-fold-increased activity [72]. Comparable stimulation results from N-SH2 truncation [69, 70, 73, 74].

The crystal structure of Shp2 lacking its C-tail (i.e., containing the N- and C-SH2 and PTP domains) provides an elegant molecular explanation for these findings [75] (reviewed in [71]) (Figure 13.2a, b). In the structure, the "back-side" of the N-SH2 (the surface opposite the pTyr-peptide binding pocket) is wedged into the PTP domain, physically and chemically inhibiting the catalytic cleft and contorting the N-SH2 pTyr peptide-binding pocket. Thus, in the basal state, the PTP domain is inhibited by the N-SH2, *and* pTyr-peptide binding is incompatible with binding of the N-SH2 back-side to the PTP domain. The C-SH2 has minimal interactions with the PTP domain, and its pTyr-binding pocket is unperturbed in the basal state. Thus, the C-SH2 probably serves a "search" function, surveying the cell for appropriate pTyr targets. If it binds to a bidentate

ligand (one that also has an N-SH2 binding site), the effective increase in local concentration of the N-SH2 ligand can reverse inhibition by the PTP domain, allowing release of the N-SH2 and enzyme activation. Alternatively, high-affinity ligands for the N-SH2 (e.g., Class IV ligands; see "SH2 domain function," below) may be able to cause activation in the absence of C-SH2 binding.

The biological relevance of the Shp2 structure was verified by analyzing mutants of the N-SH2/PTP domain interface. Such "activated" mutants have increased basal activity *in vitro*, retain the ability to bind N-SH2 ligands, and behave as gain of function ("activated") mutants *in vivo* [76]. More dramatic confirmation came with the finding that analogous mutants cause Noonan Syndrome (NS) and various hematologic malignancies in humans (see "Shp signaling and substrates," below). Subsequently, the crystal structure of C-terminally truncated Shp1 has been solved and found to be quite similar to that of Shp2 [77].

Alternative Mechanisms of Regulation; the Role of the C-Tail

The effects of C-tail tyrosyl phosphorylation remain controversial. Insulin receptor-induced tyrosyl phosphorylation of Shp1 (at Y536) stimulates activity [23]. Similar effects were reported for Src-catalyzed phosphorylation [78], although earlier work [21] found no effect of tyrosyl phosphorylation, perhaps due to auto-dephosphorylation. An early report claimed that tyrosyl phosphorylation of Shp2 increases activity [7], but this study could not distinguish the effects of phosphorylation from SH2 domain association.

Protein ligation techniques have been used to replace Y542 or Y580 of Shp2 with a non-hydrolyzable pTyr mimetic [79]. Phosphorylation at either position was found to stimulate catalysis by two- to three-fold (the same-fold stimulation observed with a stimulatory pTyr-peptide in these experiments). Mutagenesis, combined with protease resistance studies, suggested that stimulation by pY-542 involved intramolecular engagement of the N-SH2 domain, whereas pY-580 stimulated activity by binding to the C-SH2. A subsequent study, using an analogous approach, found a marked increase in phophatase activity of Shp1 phosphorylated at Y-536, and a much smaller effect of phosphonate modification of Y-564 [80].

Although these studies are novel and provocative, they are difficult to reconcile completely with previous studies. It is unclear how pY-580, upon engaging the C-SH2, would cause enzyme activation while the N-SH2 remains "wedged" in the PTP domain; notably, Y580-phosphorylated Shp2 is further activated by a pTyr-peptide for the N-SH2, suggesting that the N-SH2/PTP domain interaction mechanism remains intact. The Shp2 crystal structure provides no obvious explanation for how C-SH2 engagement could affect

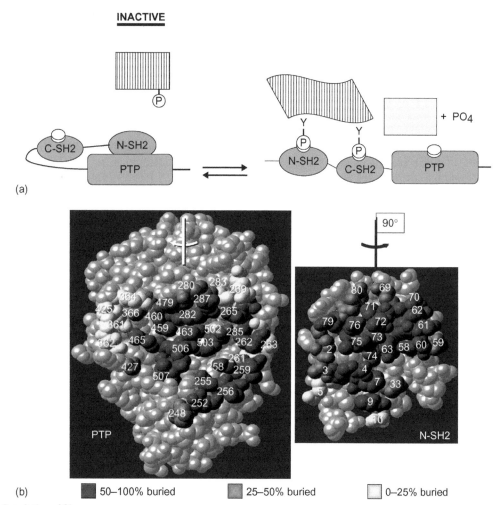

FIGURE 13.2 Regulation of Shps.
(a) Cartoon illustrating regulation of a prototypical Shp by a pTyr peptide. In the basal state the Shp is largely inactive, because the back-side of the N-SH2 is inserted into the catalytic cleft. This results in mutual allosteric inhibition, with the N-SH2 inhibiting the PTP domain and the PTP domain contorting the pTyr-binding pocket of the N-SH2, which is on the opposite surface. The C-SH2 is left essentially unperturbed in the basal state, with its pTyr-peptide binding pocket in a conformation suitable for binding an appropriate ligand. Most likely, the C-SH2 has the primary targeting function to most binding proteins. If an appropriate binding protein has an adjacent pTyr-peptide sequence capable of binding the N-SH2, the increase in local concentration overcomes the mutual allosteric inhibition, resulting in opening of the enzyme and activation. For further details, see text. (b) Location of mutations in Noonan syndrome. Shown is an "open book" representation of the Shp2 crystal structure in which one is looking directly into the surfaces of the N-SH2 back-side loop and the PTP domain that are involved in basal inhibition. The positions of known Noonan syndrome mutations found in these domains are indicated. Note the high correspondence between NS mutations and residues involved in basal inhibition. Initial figure courtesy of S. Shoelson, Joslin Diabetes Center, and modified as above.

catalytic activity, and previous enzymological studies showed no effect of C-SH2 engagement alone [75] (reviewed in [71]). Also, given the "closed" conformation of Shp2 in the absence of pTyr–peptide binding, it is hard to see how binding of pY542 to the N-SH2 leads increased proteolytic susceptibility. Further work is required to resolve these issues.

Serine/threonine phosphorylation of the Shps also has been reported (reviewed in [81]). Shp1 undergoes serine phosphorylation in response to G-protein [82] and PKC agonists [83–85] in various hematopoietic cells. Phosphorylation has been reported to [84, 85], stimulates [82], or no effect [83] on Shp1 catalytic activity. Specific PKC isoforms phosphorylate Shp2 *in vitro* [86, 87]; obviously, further studies are required to resolve these discrepancies and in

transfected cells [87]. Mutagenesis studies suggest that S576 and S591 are the sites of phosphorylation, but mutation of these sites has no apparent effect on catalytic activity or biological function [87]. On the other hand, recent studies suggest that PKC activation in response to Fcγ or Toll receptor (TLR) agonists increases the activity of Shp2 *in vivo* [88]. Erk-dependent phosphorylation of Shp2 was reported to be inhibitory by one group [89], but not another [83]. The site(s) of serine/threonine phosphorylation were not identified, but sequence analysis predicts these are likely to be within the C-tail. Shp2 is required for Erk activation in many signaling pathways, raising the possibility that phosphorylation of Shp2 by Erk is part of a negative feedback loop.

Certain phospholipids, particularly phosphatidic acid, stimulate Shp1 (but not Shp2) [90] by binding to a specific high-affinity site within the C-tail [91]. The physiological significance of phospholipid binding remains to be determined, although one intriguing possibility is that it relates to the reported ability of Shps to be recruited to lipid rafts [92–94]. Recent studies suggest that localization of Shp1 in rafts may be critical for its ability to inhibit TCR signaling [95].

The C-tail of Shp1, but not Shp2, has a functional NLS. Substantial amounts of Shp1 are found in the nuclei in epithelial cells [27] and tissues ([96] (H. Keilhack and B. G. Neel, unpublished observations), whereas under basal (randomly growing) conditions Shp1 is cytoplasmic in hematopoietic cells [27, 28, 83]. However, cytokine stimulation can result in nuclear translocation of Shp1 [28]. Translocation occurs with delayed kinetics (>1 h post-stimulation), which might reflect a requirement for synthesis of a new protein to promote nuclear import or repress nuclear export. Most studies of Shp signaling have focused on immediate events following receptor stimulation; this work argues that attention be paid to later events as well.

Besides these conventional mechanisms of regulation, Shp2 (and other PTP family members) is regulated by regulated by reversible oxidation. Increasing evidence suggests that hydrogen peroxide and other reactive oxygen species (ROS) are generated upon growth factor and cytokine stimulation, and act as second messengers (reviewed in [97, 98]). Shp2 undergoes transient inactivation by ROS in response to PDGF stimulation of Rat 1 fibroblasts, and it was argued that Shp2 inactivation is required for normal PDGFR function in these cells [99]. Importantly, however, other work indicates that Shp2 plays a signal-enhancing role in PDGF signaling ([100–102]; see "Shp2 in RTK signaling," below). More recent work argues for a critical role for ROS-catalyzed inactivation of Shp2 in TCR signaling [103] and ROS-mediated Shp1 inactivation in BCR signaling [104]. Further studies are required to clarify how the phosphorylation, other C-tail functions, and reversible oxidation help orchestrate the precise regulation of Shp activity in space and time.

Shp Inhibitors

As typical members of the PTP superfamily, both Shp1 and Shp2 are inhibited by vanadate and other vanadium-based compounds. Although such agents are non-selective PTP inhibitors, vanadate has been tested in human trials for Type 2 diabetes [105]. Calpeptin is a protease inhibitor with selectivity for calpains. Calpeptin has also been demonstrated to bind to Shp2 (and not PTP1B, Shp1, or PTPα), in Jurkat cell-line extracts and can inhibit its phosphatase activity *in vitro* at doses of $100 \mu g/ml$ and $1 mg/ml$ [106]. However, the lack of specificity between calpeptin and proteases and Shp2 may complicate this type of analysis.

More recently, a screen against the NCI Diversity Set chemical library was performed and the compound, NSC-87877, was found to be a potential Shp2 inhibitor [107]. Computer docking predicts that NSC-87877 forms hydrogen bonds with the R465 residue in the PTP catalytic motif and two other non-conserved residues, K280 and N281, located adjacent to the phosphotyrosine recognition loop. Although this compound shows no selectivity between Shp2 and Shp1, it inhibits both Shps with a lower IC_{50} compared to other related PTPs tested, including PTP1B, HePTP, PTPβ, CD45, and LAR. More recently, Lawrence *et al* reported several compounds that have more selective inhibitory activity on Shp2 compared to Shp1 [108], although these compounds were not assessed in any cell-based assays. Finally, PHPS4 was reported to be a selective inhibitor of Shp2, and to inhibit Shp2-dependent cellular processes such as HGF-evoked ERK activation [109].

BIOLOGICAL FUNCTIONS OF SHPS

Genetic models for murine Shp1 and for Shp2 orthologs in mouse, *Drosophila*, and *C. elegans* have been invaluable for defining the biological functions of the Shps. The phenotypes of Shp-deficient organisms will be described briefly here; more complete descriptions are available in other reviews.

The *Motheaten* Phenotype

Two naturally occurring point mutations exist in the murine Shp1 gene, each of which causes abnormal splicing of Shp1 transcripts [110, 111]. The *motheaten* (*me*) allele generates an early frameshift; consequently, *me/me* mice are protein-null. The *motheaten viable* (*me*^{v}) allele encodes two aberrant Shp1 proteins; one with a small deletion, the other with a small insertion in the PTP domain. Together, these retain only about 20 percent of wild-type Shp1 activity, demonstrating an essential role for PTP activity in Shp1 function.

The phenotypes of *me/me* and *me*^{v}/*me*^{v} mice differ only in severity, with *me/me* mice developing abnormalities earlier (2–3 weeks) than *me*^{v}/*me*^{v} (9–12 weeks) [112–116]. For this reason, we use "**me**" to refer generically to Shp1-deficient mice. The **me** phenotype derives its name from patchy hair loss, which gives the mice a "motheaten" appearance. The hair loss, in turn, results from sterile dermal abscesses comprised of neutrophils. Inflammation is also prominent elsewhere, including the joints, liver, and lungs. The latter leads to the early demise of **me** mice, due to severe interstitial pneumonitis caused by accumulations of alveolar macrophages and neutrophils. The macrophage population in **me** mice is expanded and exhibits abnormal differentiation, with a dramatic increase in CD5+ monocytoid cells and a decrease in cells expressing tissue/marginal zone macrophage markers [117, 118]. Some dendritic cell populations are increased, whereas others are diminished [118], and osteoclast numbers and function are enhanced, leading to osteopenia in **me** mice [119]. Shp1-deficient mice on either *nude* or *Rag* knockout background still develop inflammatory disease [120], and the disease can be reproduced by transplantation of bone marrow cells and prevented by incubation with Mac 1 antibodies [121]. Thus, lymphoid cells are dispensable, and defects in the myeloid lineage are critical, if not sufficient, for development of the **me** inflammatory syndrome.

Although, from the host standpoint, the myeloid defects present the gravest problems, every other hematopoietic lineage is affected by Shp1 deficiency (reviewed in [112–116]). The thymus undergoes premature involution, possibly due to defective homing of a thymic accessory cell [122, 123] Thymocytes and peripheral T cells lacking Shp1 actually exhibit increased mitogenesis in response to T cell antigen receptor (TCR) stimulation [124, 125]. Consistent with enhanced responsiveness, crosses to TCR transgenic mice show that Shp1 deficiency lowers the threshold for thymic selection [126–129]. Normal B (B2 cell) lymphopoiesis is reduced, but there is a marked increase in B1 (CD5+) cells. The remaining B cells appear hyperactivated, and produce autoantibodies [116, 130]. Proliferation [131, 132] and calcium flux [133] in response to B cell antigen receptor (BCR) stimulation are reportedly enhanced in **me** lymphocytes, and the response threshold of a transgenic BCR is lowered [133]. NK cell activity is decreased [134], but the remaining NK cells show enhanced lytic activity [135]. Motheaten mice are anemic, probably due to chronic hemolysis, although their erythroid progenitors are hyper-responsive to erythropoietin (EPO) [136–138]. Increased numbers of certain mast cell populations also have been reported [139, 140].

Very recently, Croker and colleagues, using random mutagenesis, generated a strain in which affected mice displayed spontaneous autoimmunity and inflammatory lesions in several organs [141]. The features of these mice were reminiscent of (although less severe than) the **me** phenotype. The responsible mutation was mapped to the Shp1 (*Ptpn6*) locus, and was found to result in a Shp1 protein with 20–50 percent reduction in specific activity. Furthermore, the inflammatory phenotype was lost when the mutant mice were crossed to IRAK-4$^{-/-}$, MyD88$^{-/-}$, or IL-1 receptor$^{-/-}$ mice, or housed in a completely germ-free environment. These new findings suggest that, even though the inflammatory lesions in **me** mice are sterile, the phenotype is likely initiated by microbial stimuli that result in IL-1 production, which then leads to uncontrolled inflammation. This study demonstrates the importance of Shp1 in controlling inflammation initiated by environmental microbes, either through preventing overactivation of cells

at the site of infection or by dampening the inflammatory feedback loop initiated by microbial recognition. Because the lymphohematopoietic system is highly interactive, identifying which **me** abnormalities are primary (i.e., cell-autonomous) defects, as opposed to secondary consequences of the myeloid defects, poses major (and ongoing) challenges. The recent availability of inducible (floxed) Shp1 knockout mice [142] should facilitate resolution of this key issue. Nevertheless, many of the abnormalities in **me** mice have been ascribed to loss of negative regulation of specific signaling pathways in the absence of Shp1 (see "Shp signaling and substrates," below).

Invertebrate Models of Shp2 Deficiency

Csw is a maternal effect mutation affecting the so-called "terminal class" pathway [14], which is initiated by the RTK Torso and controls embryonic head and tail development [143] (reviewed in [144]). Loss of function mutations in *csw* were found to have a phenotype similar to although less severe than *torso* mutations, which provided the first evidence of a positive (i.e., signal-enhancing) function for an Shp2 ortholog [14]. Csw also is a required positive component of the *sevenless*, *breathless* (FGFR), and *Drosophila* EGFR (DER) pathways [15, 145, 146].

Ptp-2 functions in at least two RTK signaling pathways. In vulval development, which is controlled by the EGFR ortholog Let-23, *ptp-2* mutation alone has no obvious effect. However, *ptp-2* deficiency suppresses the multivulva phenotype induced by mutation of the negative regulator *lin-15*. Interestingly, *lin-15* mutations cause Let-23 activation even in the absence of the EGFR ligand, Lin-3, implying that, Ptp-2 may play an important role in a "lig-and-independent" Let-23 pathway [16]. Ptp-2 also is important for signaling by the FGF receptor (FGFR) ortholog Egl-15 [147], and has an essential role in an as yet unidentified pathway required for oogenesis [16]. The functions of Ptp-2 in both the Let-23 and Egl-15 pathways are likely mediated through the adaptor Soc-1, which is structurally homologous to mammalian Gab proteins [147, 148].

Functions of Vertebrate Shp2

Studies of *Xenopus* embryogenesis provided initial evidence of a role for Shp2 in vertebrate development [10]. Expression of dominant negative Shp2 disrupts gastrulation, causing severe tail truncations reminiscent of, but less severe than, the effects of dominant negative FGFR. Dominant negative Shp2 also blocks FGF-induced mesoderm induction and elongation of ectodermal explants. "Activated" mutants of Shp2 (similar to those found in Noonan syndrome; see "Shp signaling and substrates", below) induce elongation of ectodermal explants in the absence of exogenous FGF. Such mutants do not, by themselves, induce mesodermal gene expression, although they potentiate induction of the Erk pathway by FGF [76].

Targeted mutations of murine Shp2 indicate a key role for Shp2 in mammalian development. Homozyotic deletion of either Exon 2 [149] or Exon 3 [150] results in early embryonic lethality. Exon 3 $(Ex3)^{-/-}$ embryos die between E8.5 and E10.5, with a range of abnormalities consistent with defective gastrulation and mesodermal differentiation [150, 151]. These defects resemble the effects of dominant negative Shp2 (and FGFR) mutants in *Xenopus*, and the effects of vertebrate FGFR mutations [152]. Chimeric analyses using Ex $3^{-/-}$ embryonal stem (ES) cells reveal an essential role for Shp2 in limb development and branchial arch formation, two other pathways controlled by FGFR signaling [153, 154]. Studies of hematopoietic differentiation in $Ex3^{-/-}$ ES cells [155, 156] and in chimeric mice [153] indicate a stringent requirement for Shp2 in the earliest progenitors, consistent with a role for Shp2 in Kit (stem cell factor receptor) signaling [157].

The Ex 3 mutation generates a truncated Shp2 protein that lacks part of its N-SH2 domain and is expressed at ~50 percent of wild-type levels (so, in Ex3$^{-/-}$ cells, it is expressed at ~25 percent normal). Owing to the N-SH2 deletion, however, the Ex3 mutant is activated markedly; consequently, Ex3$^{-/-}$ cells actually have increased Shp2 activity [150, 155], although the mutant protein is defective at localizing correctly in at least some signaling pathways [158]. This raised the possibility that some effects of Ex3 deletion might be neomorphic.

Recent studies of other targeted mutations argue against this possibility. Ex2$^{-/-}$ embryos die earlier (~E6–6.5) than Ex3$^{-/-}$ embryos. Despite earlier reports (which used antibodies against the N-terminus to assess expression), Ex2$^{-/-}$ mice also express an N-terminally truncated protein, so it is unclear why they die earlier than Ex3$^{-/-}$ mice. However, a variant Ex2 mutation (Ex2*), in which a strong splice acceptor sequence was introduced into the targeting construct, is, in fact, protein-null. Ex2*$^{-/-}$ embryos also die pre-implantation [159]. Detailed analysis of such embryos reveals defective trophoblast development. Trophoblast stem (TS) cells isolated from inducible (floxed) Shp2 mutant mice are normal, but excision of the floxed Shp2 allele results in massive cell death, which appears to be mediated, at least in part, through an FGF/FGFR-mediated SFK/Ras/Erk/Bim pathway. However, depletion of Bim with an shRNA only partially impairs TS cell death in the absence of Shp2, arguing for at least one other Shp2-dependent pathway [159]. Although defective FGF4 signaling clearly contributes to peri-implantation lethality in Shp2-null mice, the timing and nature of the lethality of these embryos also are consistent with roles for Shp2 in β1 integrin [160] signaling (see "Signaling by Shp2 and its orthologs," below).

Two groups have generated floxed (fl) alleles of Shp2 and, by crossing them to appropriate Cre recombinase lines, have provided insight into the role of Shp2 in specific cell/tissue types *in vivo*. These findings will be summarized only briefly here; readers are urged to consult the appropriate primary references for details.

As indicated above, deletion of Shp2 in TS cells results in their rapid death [159]. Three different types of neuronal Cre lines, which express Cre at different times and/or different cells, have been used to delete Shp2 in the nervous system, with quite distinct results. Deletion in post-mitotic neurons using CAM kinase-Cre results in early-onset obesity and leptin resistance [161]. This in turn leads to insulin resistance (but normoglycemia), hypertriglyceridemia, and hepatic steatosis. The basis for the increased weight is decreased metabolic rate, as food consumption is normal. These phenotypes correlate with *increased* leptin-evoked JAK2 and STAT3 phosphorylation, which is surprising, given that STAT3 signaling mediates the anti-obesity effects of leptin [162]. On the other hand, leptin-evoked Erk activation is impaired, leading the authors to conclude that deficiency of this pathway is dominant to STAT3 in leptin signaling. Another possibility, though, is that Shp2 deficiency also affects other anorexigenic pathways (e.g., BDNF or CNTF signaling). The hypothalamic/pituitary axis was also dramatically affected in these mice, which could definitely contribute to their obese phenotype.

In contrast, pan-neuronal deletion using the CRE3 line results in both obesity and diabetes, along with a wide array of diabetic complications (vasculitis, glomerular disease, gastric hypomotility, etc) [163]. Also, unlike the CAM kinase-Cre-deleted mice, CRE3-deletion causes hyperphagia. CRE3-deleted mice also show differences in linear growth, neuropeptides, and basal and leptin-evoked signaling responses (STAT and Erk activation).

The reasons for these differences remain to be determined, but could include the more widespread deletion of Shp2 in CRE3-deleted animals (compared to the selective deletion in forebrain neurons in the CAMK-Cre-deleted mice) and/or the earlier onset of deletion in CRE3- (~E11.5) versus CAMK-Cre- (~P5) deleted mice. Given the highly complex nature of neuronal regulation of body weight, deletion of Shp2 in more restricted neuronal populations (e.g., specific hypothalamic neurons) will be required to better understand the role of Shp2 in body-mass regulation.

Deletion of Shp2 in neural precursor cells (which give rise to neurons and glia) using nestin-Cre results in a third set of phenotypes, again distinct from those caused by either CAMK- or CRE-3-evoked deletion [164]. Nestin-Cre-deleted mice are viable at birth, but show severe postnatal growth retardation and perinatal lethality. Cortical differentiation is markedly impaired, with decreased numbers of neurons and oligodendrocytes and a slight increase in astrocytes. Consistent with these findings, there is decreased proliferation and increased apoptosis of neuronal progenitors in Shp2-deleted brains. Defective neurogenesis and increased astrocyte generation are also seen in neuronal stem cell cultures *in vitro* – a result consistent with siRNA knockdown experiments [165] in wild-type neural stem cell cultures. In both Shp2-deleted and -knockdown cells, Erk activation is diminished and STAT3 activation enhanced, presumably accounting for the effects on neurogenesis and gliogenesis, respectively. Together, these findings indicate an instructive role for Shp2 in neuronal fate specification [164, 165]. Interestingly, the converse effects are seen in a mouse model of Noonan syndrome [165], which is, as indicated above, caused by activated mutants of Shp2; these effects likely contribute to the cognitive impairment seen in many NS patients (see "Shp signaling and substrates." below).

Shp2 deletion also has distinct effects in skeletal and cardiac muscle, respectively. Shp2 deficiency in skeletal muscle results in impaired myotube growth, due to defective NFAT-mediated IL4 activation [166]. Lack of Shp2 in cardiomyocytes causes the rapid onset of dilated cardiomyopathy without an intervening hypertrophic phase, and also prevents the normal hypertrophic response to aortic banding [167]. The latter is accompanied by decreased Erk activation in response to multiple growth factors or banding-induced stretch, as well as increased Rho activation.

Mammary-specific deletion of Shp2 results in impaired lobular-alveolar outgrowth during pregnancy, with decreased milk production and accelerated involution [168]. STAT5 activation is decreased during normal pregnancy or following Prl injection, while STAT3 activation is enhanced.

Shp2 is also required for efficient liver regeneration in response to partial hepatectomy [169], as a consequence of defective growth factor-evoked proliferation. The latter, in turn, most likely results from defective Erk activation in the absence of Shp2.

Finally, T cell-specific Shp2 deletion results in defective thymic deletion and impaired mature T cell function [170]. Such mice show decreased thymic cellularity and a partial block in double negative thymocyte development at the DN3 to DN4 stage, consistent with impaired pre-TCR signaling. Surviving mature cells also have impaired TCR-evoked proliferation and decreased activation marker expression. Defective Erk activation probably underlies this phenotype as well.

SHP SIGNALING AND SUBSTRATES

Shp1

Shp1 has been implicated as a regulator of signaling by RTKs, cytokine receptors, multi-chain immune recognition receptors (MIRRs), chemokine receptors, and integrins. Many of these studies utilized cells from *me/me* or *me^v/me^v* mice. Such cells provide the advantage of a genetic model of Shp deficiency, but the reported defects may reflect altered development caused by Shp1 deficiency rather than the effects of Shp1 on a specific signaling pathway *per se*.

Other studies use point mutations of Shp1 binding proteins (e.g., inhibitory receptors, cytokine receptors, RTKs), and infer that phenotypic effects reflect loss of Shp1 binding. This assumption can be dangerous, however, because as discussed above (in "SH2 domain function" and Table 13.1), Shp1 binding sites can overlap with those of other SH2-containing proteins. Several more recent studies have used (or included) Shp1 knockdown (by shRNA or siRNA) approaches, which can circumvent such problems. Nevertheless, despite much progress in defining signaling pathways affected by Shp1, its direct targets in several pathways remain controversial.

Regulation of Myeloid Cell Signaling by Shp1

Bone marrow macrophages (BMMs) from **me** mice were reported to show increased proliferation in response to colony stimulating factor 1 (CSF-1; MCSF) [171]. Subsequent studies found no effect of Shp1 on proliferation *per se* [172, 173], although **me** BMMs required less CSF-1 for survival [172]. The targets of Shp1 in this pathway are also controversial. In one study, the CSF-1R was found to be hyperphosphorylated, albeit only briefly, following stimulation [171]. Others failed to confirm these observations, instead identifying p62Dok (Dok) as the major hyperphosphorylated species [172]. The reason for this discrepancy is not clear. Regardless, because Dok is primarily a negative regulator, acting to recruit RasGap [174–176], Dok hyperphosphorylation presumably cannot explain the lower CSF-1 dependence of **me** BMMs.

Shp1 does not bind directly to either the CSF-1R or Dok. Instead, two inhibitory receptors, Shps1 (Sirpα/BIT/MFR) and PirB (p91A) are its binding partners in BMMs [38,177–179]. Both also are Shp1 substrates [38], but it also does not appear as if dephosphorylation of these proteins plays a major negative regulatory role. Moreover, it not clear if Shps1 and/or PirB regulate RTK signaling in BMMs or have another function, such as in integrin signaling [180].

Three recent reports [181–183] place Shp1 downstream of Toll-like receptors (TLRs) in macrophages and dendritic cells (DCs), and argue that Shp1 suppresses pro-inflammatory cytokine production (Figure 13.3a). One report suggests that Shp1 dephosphorylates Vav1 to effect this function [182]. A second argues that Shp1 binds directly to IRAK1 via an ITIM in its kinase domain, and inhibits IRAK kinase activity [181]. The third indicates that Shp1 associates with TRAF6 [183], a downstream target of IRAK. Surprisingly, inhibition of IRAK1 by Shp1 is reportedly PTP-independent [181]. As IRAK1 is required for pro-inflammatory cytokine production downstream of TLRs [184], the latter model could account for excessive cytokine production by macrophages and DC in **me** mice. Besides inhibiting pro-inflammatory cytokine production, though, Shp1 was reported to promote type I interferon (IFN) production in macrophages and DCs

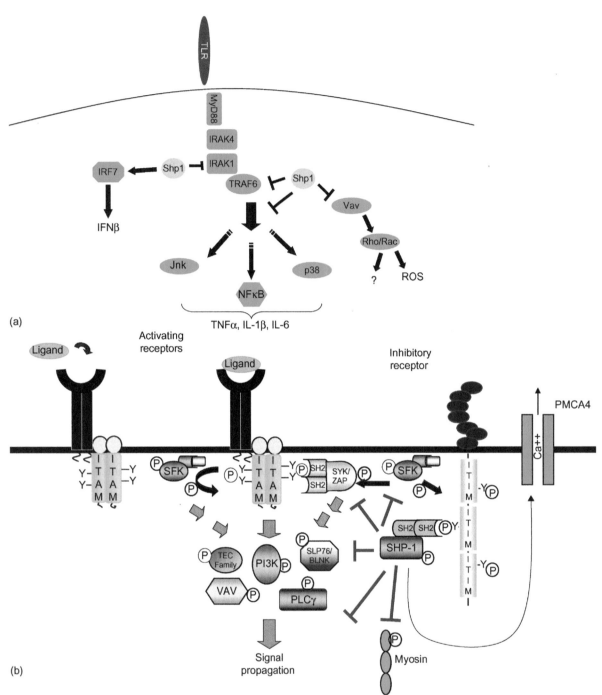

FIGURE 13.3 Signaling by Shp1.

(a) Regulation of TLR signaling. Upon TLR engagement, MyD88, IRAK4, and IRAK1 are recruited to the receptor followed by direct activation of TRAF6. The activation of TRAF6 then leads to the eventual activation of NFκB as well as the MAP kinases p38 and Jnk, resulting in the production of the pro-inflammatory cytokines IL-1β, TNFα, and IL-6. Shp1 regulates this cascade, possibly via the dephosphorylation of TRAF6 or Vav1, resulting in the overproduction of pro-inflammatory cytokines. In addition, Shp1 binds to the kinase domain of IRAK1 and inhibits its activity, in a phosphatase-independent manner, leading to decreased IRF7 phosphorylation and somehow, increased production of type I IFN (see text for details). (b) Regulation of MIRR signaling by inhibitory receptor/Shp complexes. Shown is a typical activating multi-chain immune recognition receptor (e.g., TCR, BCR, FcR, activating NK receptor) and a typical inhibitory receptor. Upon cell activation, signals from the activating MIRR result in tyrosyl phosphorylation of the inhibitory receptor, which in turn recruits Shp1 (and possibly Shp2 and/or Ship). Recruitment both localizes and activates the Shp, which then dephosphorylates one or more targets in the vicinity of the activating receptor complex. Possible direct targets are indicated. However, for most pathways the direct substrates of Shp1 remain unknown or controversial. For many MIRR pathways, crosslinking of inhibitory receptors to activating receptors may be important for inhibitory signaling. (c) Regulation of cytokine signaling. Shown is a model of the EPOR, which couples to JAK2. Upon ligand binding, the receptor-associated PTK becomes activated and phosphorylates the receptor on multiple sites, including the binding site for Shp1. Shp1 may directly dephosphorylate JAK2, leading to its inactivation. However, Socs proteins are important for inactivating cytokine receptors. How Shps and Socs proteins interact to effect negative regulation is unknown. Shown is a highly speculative model in which Shp1 and Socs proteins might collaborate to negatively regulate cytokine receptor signaling. In this model, Shp1 is responsible for dephosphorylating the adjacent pY1008 on JAK2, thereby allowing an appropriate Socs protein to bind at pTyr 1007. For details, see text.

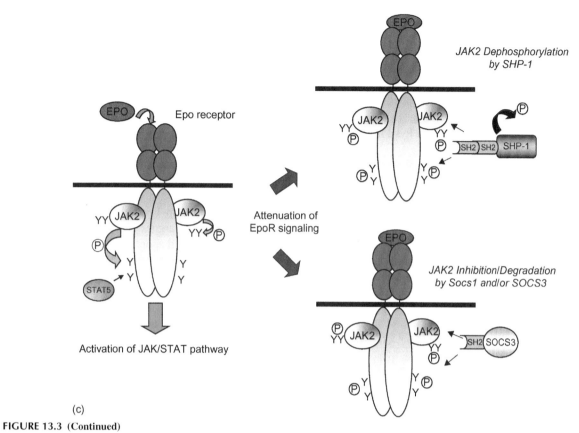

(c)

FIGURE 13.3 (Continued)

through IRAK1 inhibition [181], and consequent prevention of IRF7 phosphorylation. The latter result is surprising, given that IRAK1 is required for the production of IFN downstream of TLRs [185]. Also, if loss of IRAK regulation is a key pathway that leads to fatal inflammation in **me** mice, it remains unclear why the *me^v* allele, which encodes a largely intact Shp1 protein, is hypermorphic. Increased Vav phosphorylation, by promoting Rac activation, could result in ROS production [186], which could directly cause tissue damage and also inhibit other PTPs, leading to a "feed-forward" inflammatory state. Conceivably, both excessive Vav and IRAK activation contribute to the **me** phenotype.

Furthermore, Shp1 negatively regulates cytokine signaling in BMMs. IFNα/β signaling is dramatically enhanced in **me** BM Ms, as shown by increased JAK1 and STAT1 tyrosine phosphorylation [187]. These data are consistent with other studies implicating Shp1 in Janus PTK dephosphorylation (see below). Others have reported enhanced granulocyte-macrophage colony stimulating factor (GM-CSF)-evoked proliferation in **me** BMMs [173]. These workers observed no change in JAK/STAT phosphorylation, but did notice a hyperphosphorylated 126-kDa species, which most likely is Shps1 and/or PirB. As indicated previously, hyperphosphorylation of these proteins alone is unlikely to explain increased GM-CSF responsiveness.

Shp1-deficient BMMs are markedly hyper-adherent to ligands for both β1 and β2 integrins, suggesting a negative regulatory role for Shp1 in integrin signaling [188]. The direct targets of Shp1 in this pathway also remain unclear, although actions on src family PTKs (SFKs) and/or the p85 subunit of phosphatidylinositol 3-kinase (PI3K) have been suggested. If SFKs are, in fact, Shp1 targets, presumably only specific members mediate the increased adhesiveness, because SFK activity also is increased in CD45^−/− BMM SFKs, yet these cells fail to sustain integrin-mediated adhesion [189]. Shps1 (and possibly PirB) probably play an important role in mediating Shp1 action in integrin signaling. Shps1 is rapidly tyrosyl phosphorylated in response to BMM adhesion, most likely by SFKs [180], and recruits Shp1. Shps1 also forms complexes with other proteins that play important roles in integrin signaling. One contains the adaptor proteins Skaphom/R and SLAP130/Fyb (now known as ADAP). ADAP is essential for in inside-out signaling in T cells [190–192], perhaps by virtue of its ability to bind Ena/Vasp family proteins [193, 194]. The other Shps1 complex contains the focal adhesion kinase (FAK)-related PTK Pyk2, and FAK, of course, regulates fibroblast adhesion [195,196]. Skaphom, ADAP, and Pyk2 associate constitutively with Shps1, but all undergo inducible phosphorylation in response to adhesion [180]. It is tempting to speculate that upon recruitment to tyrosine-phosphorylated Shps1, Shp1 dephosphorylates one or more of these associated proteins. Although tyrosyl phosphorylation activates Pyk2, its effect on Skaphom and ADAP remains to be determined, as does whether any of these proteins are direct Shp1 targets. In any case, only some effects of Shp1 on BMMs can be mediated via Shps1, as mice lacking the cytoplasmic domain of Shps1 do not have excessive inflammation [197]. Interestingly, while haplotaxis is defective in **me** BMMs, probably owing to defective de-adhesion, they show markedly increased chemotaxis in response to chemokines [198]. Chemokines signal via G-coupled receptors. It remains unclear what the targets are for a PTP in chemokine signaling, although Pyk2 is a tempting target [199].

Shp1 also regulates phagocytosis. Initial evidence was provided by a hallmark phenotype of me mice: erythrophagocytosis. Moreover, **me** BMM show increased ingestion of IgG-opsonized sheep red blood cells (RBCs), indicating defective negative regulation of FcγR signaling [200]. Complement-mediated phagocytosis, which utilizes β2 integrins, is also enhanced. Both of these negative regulatory pathways involve Shps1 (and most likely, Shps1/Shp1 complexes) and its ligand CD47 [201], a ubiquitously expressed glycoprotein [202]. CD47 on the RBC surface engages macrophage Shps1, leading to its tyrosine phosphorylation, Shp1 recruitment, and inhibition of FcγR signaling [203,204]. A similar mechanism prevents inappropriate phagocytosis of platelets, and absence of this pathway in Shps1$^{-/-}$ mice explains their mild thromobocytopenia [205].

Recently, the CD47/Shps1 pathway was found to be crucial for the superior long-term acceptance of human bone marrow by NOD/SCID mice [206]. Shps1 found in most mouse strains does not recognize human CD47; consequently, macrophages (and/or other myeloid cells from these mice) are not prevented from reacting against human xenografts. However, NOD mice express an Shps1 variant that binds human CD47, resulting in inhibition of these cells and graft acceptance. A similar inability of mouse Shps1 to recognize porcine CD47 is implicated in rejection of these xenografts [207].

Neutrophil signaling is also affected by Shp1 deficiency. The number and size of colonies evoked by G-CSF are increased in bone marrow from **me**, compared with wild-type, mice, and these colonies also differ qualitatively, containing increased numbers of macrophage-like cells [208]. Increased G-CSF responsiveness also is apparent in short-term proliferation assays and is reflected by an increased magnitude of STAT activity in progenitors [208]. Cell lines expressing dominant negative Shp1 [209] and Shp1$^{-/-}$ DT40 B cells expressing the G-CSFR [210] also show increased G-CSF-evoked STAT activation. Under endogenous conditions, no association between the G-CSFR and Shp1 is detected, although an unidentified 92-kDa species coprecipitates [208]. When Shp1 and the G-CSFR are massively overexpressed in 293 T cells, a small amount of co-precipitation is detected. Despite the lack of strong association, the G-CSFR C-terminus is necessary for Shp1 to mediate its effects on G-CSF signaling, although the receptor tyrosines are dispensable [210]. Thus, the mechanism by which Shp1 regulates GCSF signaling, including its direct targets, remains unclear.

A provocative study indicates that Shp1 acts as an effector of death receptors in neutrophils, such as the TNF receptor (TNFR) and Fas [65]. These receptors contain a conserved AXYXXL motif in their cytoplasmic domains, and undergo tyrosine phosphorylation upon activation. Despite its non-canonical nature, pTyr peptides containing this motif were found to bind to Shp1 from neutrophils. Death receptor activation antagonizes cytokine-promoted neutrophil survival and depends on the presence of the pYXXL motif. Moreover, me/+ neutrophils appear relatively resistant to death receptor stimuli. These effects correlate with increased cytokine-evoked Lyn phosphorylation, suggesting that Lyn may be a target for death receptor/Shp-1 complexes.

A recent report also argues that Lyn and Shp1 play opposing roles in neutrophil apoptosis, but via the regulation of caspase-8 [211]. Lyn was found to phosphorylate Tyr-397 and -465 of caspase-8, preventing cleavage and subsequent apoptosis induction. When caspase-8 is phosphorylated on Tyr-310, however, it recruits Shp1, which subsequently dephosphorylates caspase-8, allowing apoptosis to proceed.

Neutrophil function is altered in **me** mice. Oxidant production, surface expression of the integrin subunit CD18, and adhesion to plastic are enhanced in **me** neutrophils, whereas chemotaxis is diminished, perhaps due to increased adhesion [212]. However, it is not certain whether all of these defects reflect direct effects of the absence of Shp1, altered granulocytic differentiation, or the inflammatory milieu in **me** mice.

Finally, Shp1 has been implicated in regulation of the *CYBB* and *NCF2* genes in myeloid cells, in part by antagonizing tyrosine phosphorylation of IRF1 and IRF8 (ICSBP) [213] and HoxA10 [214]. The same group has also proposed a similar function for Shp2 [215, 216] (see "Shp2 in cytokine signaling", below).

Regulation of Lymphocyte Signaling by Shp1

T cells, B cells, and NK cells all show various effects of Shp1-deficiency (see "The motheaten phenotype," above, and [217–220]). In almost all cases, the absence of Shp1 results in enhanced responses to activating receptor engagement; i.e., Shp1, in general, is a negative regulator of lymphocyte signaling. In most cases Shp1 action is mediated by its recruitment to inhibitory receptors, which allows Shp1 to dephosphorylate key signaling molecules proximal to and downstream of the antigen receptors (Figure 13.3b). Inhibitory receptors, by virtue of their ligand-binding ability, also permit cross-regulation of signaling by means of cell–cell interactions. Because the signaling output of activating receptor engagement is crucial for modulating lymphocyte development, Shp1 deficiency also has important effects on T cell, B cell, and NK cell development, as described above.

Genetic and biochemical evidence show that Shp1 plays a key role in regulating signals emanating from the BCR [142, 219, 221, 222]. For example, in **me** mice (see "The motheaten phenotype," above), as well as B cell-specific Shp1 knockout mice [142], there is an increased number of unconventional B-1 B cells, which develop in response to stronger BCR ligation during selection [223]. Studies using **me** mice expressing a transgenic BCR with or without co-expression of the cognate antigen also show that Shp1 deficiency alters BCR signaling threshold [133]. These studies provide strong evidence that the lack of Shp1 specifically in B cells shifts selection from the B-2 to the B-1 lineage [142], but also complicates the interpretation of signaling differences in "B" cells from **me** mice (which consist of different numbers of B1 vs B2 cells).

Shp1 is found in the BCR complex prior to stimulation [104,132], but rapidly dissociates from the complex after BCR ligation [104,132]. Thus, Shp1 may control the threshold for receptor activation by keeping the resting BCR in check (presumably by maintaining Igα/β-ITAMs in a dephosphorylated state). However, inhibitory receptors probably provide the major means by which Shp1 may also attenuate B cell signaling via inhibitory receptors; these include CD22 [224–226], CD72 [227], CD5 [228], PirB/p91A, ILT2, PECAM, and CEACAM [222], and Fc receptor-like 5 [229]. Shp1 was initially suggested to mediate inhibitory signaling by FcγRIIB, the Fc receptor that causes inhibition of B cell activation by immune complexes [131]. Subsequent studies showed that Shp1 is dispensable for FcγRIIB-mediated inhibition, which instead is transmitted via the inositol phosphatase Ship [230–233].

B cells from **me** mice reportedly show enhanced induction of tyrosine phosphorylation and activation of the ERK MAPK pathway, as well as an increased BCR-evoked calcium response [218, 220]. Only the enhanced calcium response is reproduced in B-cell specific Shp1 knockout mice, however [142], indicating that the other effects may be indirect consequences of Shp1 deficiency in other cell types. Most likely, altered calcium responses reflect loss of CD22/Shp1 complexes, as CD22$^{-/-}$ B cells show a similar phenotype [224–226, 234, 235]. CD22 and Shp1 have been linked to calcium mobilization through the membrane ATPase PMCA4 [236]. PMCA4 promotes calcium extrusion following BCR ligation, and tyrosine-phosphorylated CD22 associates with PMCA4 in an Shp1-dependent manner (Figure 13.3b). Shp1 may dephosphorylate a tyrosine residue in PMCA4 known to be phosphorylated in platelets [237–239], resulting in inhibition of PMCA4. Arguing against the sole importance of PMCA4 in CD22/Shp1 inhibition, however, is the observation that chemical inhibition of PMCA4 does not completely reverse the effects of CD22 on B cell calcium responses [236]. Other proposed Shp1 targets that might mediate these effects include BLNK [240], Vav [241–243], SLP-76 [240, 244, 245], the p85 subunit of PI3K [246], Btk [247], Syk [248], 3BP2 [249], and myosin [250] (Figure 13.3b). Other inhibitory receptors also affect B cell function. CD5 on B-1 B cells associates with Shp1 and regulates activation [228]. PirB ligation leads to

inhibition of BCR signal transduction via Shp1, resulting in inactivation of Btk, Syk, and PLCγ [251, 252].

T cells from **me** mice display increased tyrosine phosphorylation and proliferation in response to TCR receptor stimulation [124, 125]. Furthermore, **me** thymocytes or thymocytes expressing a dominant negative form of Shp1 show enhanced positive selection as well as enhanced clonal deletion, likely due to stronger TCR signaling [126, 128, 129]. Shp1 associates with CD5 [125, 253], members of the ILT family (LIR, MIR, CD85) [254], CEACAM [255, 256], BTLA [257], and PD-1 [258, 259]. As for B cells, though, there is disagreement over the precise targets of Shp1 in T cells (Figure 13.3b). There are reports that Shp1 inhibits TCR-proximal events, such as CD3ζ phosphorylation, ZAP-70 activation, and/or Lck and Fyn activation [124, 125, 260, 261]. For example, CD4-associated Lck and Fyn activity are enhanced in **me** thymocytes [124], and Shp1 preferentially dephosphorylates the activating tyrosyl residue on Lck [262, 263]. Regulation of Lck by Shp1 is suggested to be critical for discrimination between strong TCR stimulation by cognate antigen, and weak stimulation by self-antigen or altered peptide ligands [264, 265]. Lck also phosphorylates Shp1 in response to TCR stimulation [21]. This leads to a model in which TCR signaling leads to Lck-mediated phosphorylation of Shp1, but then tyrosyl phosphorylated Shp1 binds to and inactivates Lck. This negative feedback loop can only be broken by stronger stimuli, which result in Erk activation and phosphorylation of Lck on S59, preventing Shp1 binding [264, 265]. Another proposed target of Shp1 in T cells is PI3K [246], although the consequences of PI3K tyrosyl phosphorylation have not been clearly established.

Shp1 deficiency in mature T cells is associated with T cell hyperproliferation due to enhanced TCR signaling. It is unclear, however, whether this is due to stronger or more sustained TCR stimulation owing to the absence of a negative feedback loop (such as that described above), or a lack of inhibitory receptor signaling. In support of the former hypothesis, a recent study suggests that the hyperproliferation of **me** T cells might be due to the ability of the cells to form more stable conjugates with antigen presenting cells, leading to more sustained TCR stimulation [266]. It is still possible, however, that inhibitory receptor signaling plays a role in the hyperproliferation, as BTLA and PD-1, among others, are all expressed on activated T cells, and their ligands are present on antigen presenting cells (reviewed in [267]). Most likely, the discrepancies between the various reports of different T cell targets of Shp1 reflect its binding to different (and possibly multiple) inhibitory receptors, and possibly redundancy with other negative regulatory molecules in different types of T cells. Hopefully, it will be possible to sort out this complexity using conditional Shp1 knockout mice [142].

Shp1 associates with multiple inhibitory receptors in NK cells, including those belonging to the killer immunoglublin-like receptor (KIR), Ly49 receptor (KIR-equivalent in mice), and CD94/NKG2 familes, as well as Siglec-7 and -9 [33, 221, 268, 269]. These receptors bind to major histocompatibility (MHC) class I molecules or, in the case of Siglecs, sialic acid on target cells, and antagonize NK cell activation via Shp1. Engagement of inhibitory receptors leads to phosphorylation by SFKs in NK cells, and the subsequent recruitment of Shp1 to receptor ITIMs, although binding of Shp1 (or Shp2) to KIR2DL1 appears to be facilitated by β-arrestin 2 [270]. Shp1 then dephosporylates the activating receptors themselves, the kinases proximal to those receptors [269], or other substrates such as SLP-76 [244] or Vav1 [243].

Shp1 also regulates cytokine receptor signaling in lymphocytes. In human T cells, Shp1 dephosphorylates IL-2Rβ [271], and thus plays a direct role in the ability of T cells (and NK cells) to respond to IL-2 and IL-15 (which share this receptor chain). The IL-2/IL-15 receptor complex lacks obvious ITIMs, so it is unclear how Shp1 accesses IL2Rβ, although it could involve recruitment of Shp1 to lipid rafts [95, 272–274]. Whether Shp1 regulates mouse IL-2Rβ is unclear, as **me** thymocytes show a normal IL2 dose response [124].

Shp1 also dampens IL-4 signaling, although in this case there is a clear ITIM motif on the IL-4 receptor [275, 276]. Increased IL4 signaling may help account for enhanced TH2 responses in *me/+* mice [277]. IL-12 responsiveness also is altered in *me/+* mice [278, 279]. Thus, Shp1, through its effects on IL-4 and IL-12 signaling, alters both humoral (Th2) and cell-mediated CD4 T cell responses. Recently, the IL-10 receptor has been shown to activate Shp1, resulting in the suppression of co-stimulation in T cells [280]. Another suppressive cytokine, TGFβ, requires Shp1 for at least some of its inhibitory functions, such as the suppression of the transcription factor T-bet [281]. How Shp1 inhibits a signaling pathway initiated by a serine/threonine kinase cascade remains to be determined. The generation of regulatory T (Treg) cells is enhanced in **me** mice [282], but whether this is a consequence of enhanced TCR stimulation, enhanced responsiveness to cytokines that support Treg development or function, or both, remains to be determined.

Finally, there have been reports that Shp1 is required for Fas-induced cell death [283,284], analogous to its role in mediating death receptor signaling in neutrophils (see "Regulation of myeloid cell signaling by Shp1," above). However, two subsequent reports failed to confirm such a requirement [124, 285].

Shp1 Signaling in Erythroid Cells

Shp1 binds directly to the EPO receptor (EpoR) [286, 287], and studies with EpoR mutants suggest that upon recruitment, Shp1 dephosphorylates the receptor-associated kinase, JAK2 [287]. This model (Figure 13.3c) provides a possible molecular explanation for the phenotype of a family that inherits an EpoR truncation and exhibits erythrocytosis [288]. "Knock-in" mice bearing a similar truncation do not have erythocytosis, but do show increased EPO sensitivity [289].

However, the finding that Socs-3 can be recruited to the same region raises the question of whether the effects of these EpoR mutants are due to loss of Shp1 or Socs binding [56]. Conceivably, Shp1 and Socs proteins act in combination to negatively regulate the EpoR and possibly other cytokine receptors. Upon activation, JAK2 (and other Janus PTKs) becomes phosphorylated on two adjacent sites, Y1007 and Y1008. Monophosphorylated Y1007-containing peptides bind Socs3, but it is unclear if bis-phosphorylated (i.e., pY1007/pY1008) peptides can bind. Perhaps Shp1 first dephosphorylates Y1008, thereby allowing a Socs protein to bind via pY1007 and mediate degradation (Figure 13.3c). An analogous model may explain the positive actions of Shp2 in cytokine signaling (see "Signaling by Shp2 and its orthologs," below). Shp1 also binds to [20] and negatively regulates [139,140] Kit. Loss of this regulation may contribute to the enhanced erythogenic potential of progenitors from **me** mice.

Shp1 Signaling in Epithelial Cells

The few studies that have explored Shp1 actions in epithelial cells have yielded surprising results. First, as indicated above ("Alternative mechanisms of regulation; the role of the C-tail"), a significant amount of Shp1 appears to be nuclear in these cell types [27, 96]. Also, in at least some cell types and signaling pathways, Shp1 may play a positive signaling role [290], although how it does so remains to be determined. However, Shp1 clearly plays a negative regulatory role in other epithelial signaling pathways. For example, it binds directly to and dephosphorylates the RTK Ros [291]. Loss of this regulation may help explain the sterility of **me** mice [291]. In addition, Shp1 may negatively regulate β-catenin signaling in intestinal epithelial cells [292], and associate with and mediate the growth inhibitory effects of the sst2 somatostatin receptor [293–295].

Finally, Marette and colleagues [296] have proposed a novel role for Shp1 in regulating insulin action. They found that **me** mice or mice expressing dominant negative *Shp1* or an *Shp1* shRNA in their livers (by adenoviral transduction) show improved glucose tolerance. This was accompanied by increased insulin receptor, IRS-1 and IRS-2 tyrosyl phosphorylation, as well as increased PI3K association with IRS proteins and Akt activation. Shp1 was found to co-immunoprecipitate with IR as well as the inhibitory receptor CEACAM, and to dephosphorylate both of these molecules. In addition to its effects on insulin receptor signaling, Shp1 regulation of CEACAM phosphorylation correlates with, and is proposed to regulate, hepatic clearance of insulin.

Signaling by Shp2 and Its Orthologs

Shp2 also is implicated in a wide variety of signaling pathways, yet, unlike Shp1, in most cases it appears to have a positive role. There are, however, exceptions, as described below.

Shp2 in RTK Signaling

In most, if not all, RTK signaling pathways, Shp2 is required for full activation of the Erk MAP kinase pathway. For some RTKs, and in some cell types (e.g., IGF-1 signaling in fibroblasts), there is virtually no Erk activation in the absence of Shp2; in others, initial Erk activation is normal, but Shp2 is required for sustained signaling [102, 150, 158, 297].

Although Shp2 binds directly to some RTKs (e.g., PDGFR), in many other pathways it binds to one or more scaffolding adaptors. For example, Shp2 binds to Gab1 upon EGFR, HGFR, and FGFR stimulation [298, 299], to IRS family proteins following insulin/IGF stimulation [300], and to FRS2 (also known as SNT) downstream of the FGFR [301]. Studies using fibroblasts derived from Gab1$^{-/-}$ [302, 303] and FRS-2$^{-/-}$ [304] mice, and experiments with chimeric and mutant forms of scaffolding adaptors [305, 306], indicate that these scaffolding adaptor/Shp2 complexes mediate the effects of Shp2 on Erk activation, at least in some RTK pathways. Genetic analyses of the DER, EGL-15, and Btl pathways suggest similar roles for Dos/Csw [146], Dof/Csw [307], and Soc-1/Ptp-2 complexes [147].

How Shp2 mediates Erk activation remains controversial/incompletely understood (Figure 13.4a). There is general agreement that Shp2 acts upstream of Ras. For example, cells expressing dominant negative Shp2 [308], Ex3$^{-/-}$ fibroblasts [102,158], or fibroblasts with complete Shp2 deletion [159,309], show defective Ras activation in response to multiple RTKs. Initial work suggested an "adaptor" model, in which Shp2 becomes tyrosyl phosphorylated in response to RTK stimulation and then binds Grb2/Sos [24,310,311]. Although this mechanism may contribute to Ras activation in some RTK pathways, it is unlikely to be generally required. For one thing, Shp2 is not tyrosyl phosphorylated in all RTK signaling pathways. Moreover, earlier studies of *Xenopus* [312] and *Drosophila* [313] showed no absolute requirement for the C-terminal tyrosyl residues in XFGFR and Sevenless, respectively. These studies involved overexpression of mutant forms of Csw/Shp2, and thus could have been misleading. However, more recent studies, in which Ex3$^{-/-}$ fibroblasts were reconstituted with Shp2 mutants lacking one or both C-terminal phosphorylation sites confirm that tyrosyl phosphorylation is not required at all for EGF or IGF-1-evoked Erk activation. In contrast, tyrosyl phosphorylation is required for maximal restoration of PDGF and FGF responses, although there is no correlation between the ability of individual tyrosyl phosphorylation site mutants to bind Grb2 and to enhance Erk activation [314]. Studies in which the Shp2 C-terminal phosphorylation sites were replaced with non-hydrolyzable phosphotyrosine analogs suggest that these residues can bind differentially to the N- and C-SH2 domains to relieve basal inhibition of Shp2 [79]. However, given that the majority of tyrosyl phosphorylated Shp2 appears to be bound to Grb2, the physiological relevance of this model is unclear. One attractive possibility by which tyrosyl phosphorylation might enhance Shp2 action is by binding to potential substrates that have SH2 domains, helping to "pry" them open for subsequent dephosphorylation [314]; a similar model has been proposed for dephosphorylation of Src by PTPα [315]. Nevertheless, how tyrosyl phosphorylation promotes Erk activation in response to specific growth factors, as well as the physiological consequences of tyrosyl phosphorylation in the whole organism context, remain unknown.

In contrast, multiple studies indicate that the PTP activity of Shp2 is vital for its positive signaling function. For example, it has long been known that mutation or deletion of the PTP domain of Shp2 blocks its ability to enhance Erk activation [40, 316]. More recently, similar effects were reported for small molecule inhibitors of Shp2 catalytic activity [107–109].

The key question, then – and the major area of controversy – is the identity of the critical substrate(s) Shp2 must dephosphorylate to help activate the Ras/Erk pathway. Several candidates, and presumptive mechanisms, have been suggested (Figure 13.4a). Early work indicated that although Shp2 can dephosphorylate most, if not all, PDGFR phosphotyrosyl resides, it preferentially dephosphorylates the Ras-Gap binding site. These findings suggested a model in which Shp2 has a dual function, first preventing precocious recruitment of RasGap and then promoting dephosphorylation of other receptor phosphotyrosines [317]. Analogous findings were reported in studies of Torso (which is structurally related to the PDGFR) signaling in *Drosophila* [318]. More recently, a similar model, by which Shp2 dephosphorylates a presumptive Ras-Gap binding site on the EGFR, was proposed to account for Shp2 action in EGF signaling [319].

Although this model and these observations are attractive, they do not appear to account for all available data. For example, Dos (*daughter of sevenless*), a *Drosophila* Gab ortholog, is required for embryogenesis, presumably in Torso signaling, and Csw binding to Dos appears to be essential for this function [320–322]. Gab1$^{-/-}$ fibroblasts and Shp2 Ex3$^{-/-}$ fibroblasts [102,323] also have defective PDGF-evoked Erk activation [302]. Furthermore, Gab1 and, more specifically, the Shp2 binding sites on Gab1, are required for Ras/Erk activation in EGFR signaling. In the absence of Gab1, EGF-induced Ras-guanine nucleotide exchange (GEF) activity is reduced, whereas Ras-GAP activity appears unaffected [324]. Thus, it would appear

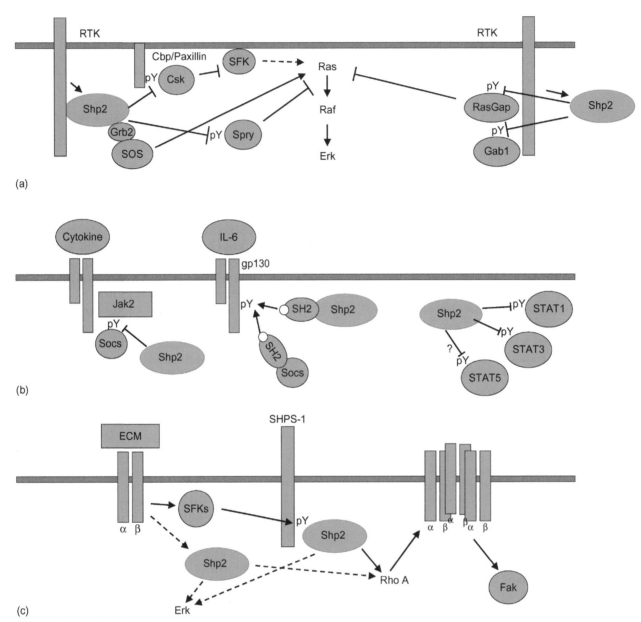

FIGURE 13.4 Signaling by Shp2.
(a) Role in RTK signaling. The precise targets that Shp2 must dephosphorylate to mediate Ras activation are unknown. Shp2 may act as an adapter in the recruitment of Grb2/SOS complex, leading to activation of Ras. It also dephosphorylates Csk binding sites on CBP and Paxillin, thus enhancing the activation of SFKs. Shp2 may mediate the desphosphorylation of Spry, a negative regulator of Ras. Finally, Shp2 catalyzes the dephosphorylation of RasGap, or its binding site on Gab1, and thus prolongs the activation of Ras. For details, see text. (b) Role in cytokine signaling. In cytokine signaling pathways, Shp2 may function analogously to its role in RTK signaling (a). A highly speculative alternative, however, is that Shp2 acts on pY1007 to regulate Socs protein recruitment and consequent degradation. Note that this model is the converse of that in Figure 13.3b. In IL-6 signaling, Shp2, through its SH2 domain, competes for the SOCS-binding site on gp130, and thus prolongs the activation of JAK/STAT pathway. Shp2 may also directly dephosphorylate STAT1, STAT3, and STAT5. (c) Role in integrin signaling. Upon binding of integrin to extracellular matrix (ECM), Shp2 is recruited to lipid rafts. Activation of Shp2 may also be facilitated by binding to SHPS-1. These processes lead to the activation of Erk and RhoA, the latter then triggers cytoskeletal arrangement, integrin clustering, and the activation of Fak. See text for details.

that Shp2 action in the EGFR pathway (or at least, an Shp2 action in the EGFR pathway) requires Gab1 and affects a Ras-GEF, not a Ras-Gap. Although Gab1/Shp2 complexes could have different targets and/or distinct effects downstream of the EGFR and PDGFR, this seems unlikely, because Ras activation is affected in both pathways. In any case, mere recruitment of Shp2/Csw to PDGFR/Torso (and consequent regulation of Ras-Gap binding) does not seem to account for Shp2's role in Ras activation. A recent study suggests that Ras-Gap may also bind to several sites in Gab1 [325], and that Shp2 dephosphorylates these sites. However, endogenous Ras-Gap binding to Gab1 was not demonstrated. Furthermore, studies of Dos mutants indicate that Csw is required even when all other Dos tyrosyl residues (besides the Csw sites) are absent [321]. Moreover, tethering Csw to the membrane can bypass the requirement for Dos [145, 146], as well as rescue activation of the Erk pathway in mammalian cells expressing Shp2-binding defective Gab1 [326]. Finally, as indicated above, Gab1/Shp2 complexes appear to regulate Ras exchange, not Ras inactivation.

Cells expressing a Gab1/Shp2 fusion exhibit enhanced Src activity, and Src inhibitors block the ability of the fusion to activate Erk [327]. Furthermore, several studies indicate that growth factor-evoked SFK activation is defective in Shp2-deficient cells [159, 309, 328–330]. These findings suggest that Shp2 acts upstream of, and even directly on, Src in the Ras/Erk pathway, an attractive idea, since SFKs have negative regulatory C-terminal tyrosyl phosphorylation sites. Shp2 does not appear to dephosphorylate SFK directly, however. Rather, Shp2 dephosphorylates binding proteins for Csk such as Cbp/Pag and Paxillin [309, 331]. However, the generality of this model for other growth factors and cell types remains unclear.

Other suggested substrates for Shp2 include a group of poorly understood inhibitors of the Ras/Erk pathway, the Sprouty and Spred proteins. Transfection studies suggest that Shp2 binds and dephosphorylates Sprouty (Spry) proteins, thereby inactivating their inhibitory potential [332, 333]. Regulation of Spry tyrosyl phosphorylation under endogenous conditions remains to be demonstrated, and the detailed effects of tyrosyl phosphorylation and mechanism of action of this family remains unclear (reviewed in [334]). The related Spred family members (Spred-1, Spred-2, and Spred-3) [335, 336] are also tyrosyl phosphorylated inhibitors implicated in downregulation of Ras/Erk activation by various RTKs, including EGF, FGF, NFG, VEGF, and SCF [335–338]. Although regulation of these proteins by Shp2 has not been reported, they are attractive substrates because homozygous mutation of *SPRED1* produces a variant Neurofibromatosis-like syndrome and, as discussed above ("Shp signaling and substrates"), Shp2 gain-of-function mutants and other Ras/Erk pathway mutations cause related disorders.

Two final candidate substrates are α-catenin [339] and major vault protein [340]. In fibroblasts transformed by a constitutively activated FGFR3 mutant and expressing catalytically impaired Shp2, α-catenin tyrosyl phosphorylation is markedly enhanced, which leads to increased recruitment of α- and β-catenin to the plasma membrane and reversion of transformation. Notably, α-catenin knockout mice have sustained Ras/Erk activation [341], which raised the possibility that altered phosphorylation of α-catenin might also affect this pathway. Indeed, expression of a

phosphomimetic mutant of the α-catenin phosphorylation site in normal fibroblasts significantly impaired Erk activation, consistent with the idea that increased α-catenin tyrosyl phosphorylation might account, at least in part, for defective Erk activation in Shp2-deficient cells. It would be interesting to know whether this mutant acts at the level of Ras, and whether Shp2 depletion in cells expressing the α-catenin phosphomimetic mutant has any further effect on Erk activation. The role of MVP is less clear, although it may act as a scaffold protein for activated Erks and potentiate Ras-induced Elk-1 activation [340].

Although the major role of Shp2 is to regulate Ras/Erk activation, genetic and biochemical evidence indicate other key roles in RTK signaling. For example, in the Torso pathway, Csw mutations have more severe effects than mutation of either Drk (Grb2) or Ras [342]. Epistasis analysis places Csw both upstream and downstream of Raf or in a parallel pathway in Sevenless signaling [145], and upstream and downstream of (or parallel to) Ras in the DER pathway [146]. The latter function may involve regulation of nuclear import, since Csw associates with the importin β ortholog Dim-7, which controls nuclear transport of the Erk ortholog, Rolled [343]. This notion is consistent with the studies of the role of Gab2/Shp2 complexes in cytokine signaling [39].

In mammalian cell lines, Shp2 is required for PI-3K activation [323, 328] downstream of some RTKs (e.g., PDGFR, IGFR, FGFR). However, in the EGFR pathway Shp2 has the converse role, negatively regulating PI3K activation by dephosphorylating the PI3K binding sites on Gab1 [323, 344]. Recent studies indicate that Shp2 also suppresses S6K1 activation in IGF signaling [345]. Shp2 mutant fibroblasts exhibit increased Jnk activation in response to a variety of stress stimuli [102], whereas fibroblasts lacking the Shp2-binding protein Shps1 (see below) show enhanced Jnk activation in response to IGF-1 [197]. Further studies are required to unravel the molecular details of these actions of Shp2.

Shp2 in Cytokine Signaling

Multiple studies indicate that Shp2 also is important for Erk activation in response to a variety of cytokines. It is likely that Shp2 has a similar/analogous target(s) and mechanism of action (i.e., acting upstream of Ras) for regulating Erk activation in cytokine receptor and RTK signaling, although this has not been established directly. Many studies also suggest unique actions of Shp2 in cytokine signaling–e.g., in regulating JAK and/or STAT activation. Both positive and negative effects have been reported (Figure 13.4b), however, and it is unclear whether these discrepancies are pathway or cell-context dependent.

Members of the JAK family of kinases have several autophosphorylation sites [346–351], the best studied of which are two tyrosines in their activation loops (Y1007/8 in JAK2) Mutagenesis studies have shown that Y1007 and Y1008 are essential for kinase activity [351]. Y1007 plays a dual role, however, as it also recruits Socs family proteins (e.g., Socs-1), which both inhibit JAK kinase activity and function as ubiquitin ligases to promote JAK2 degradation [352–355]. Ali and colleagues suggest that in response to prolactin receptor activation, Shp2 dephosphorylates Y1007, thereby preventing Socs recruitment to this site and prolonging activation of JAK2 and STAT5 [356]. This is an intriguing model, but, given that Y1007 phosphorylation is also essential for JAK2 activity, it is unclear how Shp2 (by dephosphorylating pY1007) could prevent Socs recruitment while simultaneously preserving JAK2 activity. Shp2 does, however, play a net positive role in prolactin signaling *in vivo*, as mammary-specific Shp2 deficiency results in impaired Prl-stimulated STAT5 activation [168].

Shp2 may also play a positive signaling role by regulating Socs-recruitment more indirectly. As discussed above, the SH2 domains of Shp2 and Socs3 recognize similar motifs. In IL-6 signaling, Shp2 competes with Socs3 for binding to pY759 of gp130, and thus could positively regulate STAT3 (and possibly JAK1/2) activation by preventing precocious downregulation of the receptor [48, 54, 55, 357–360]. As predicted by this model, Y759F mutation results in sustained STAT3 activation, which was initially interpreted as indicating that Shp2 dephosphorylates STAT3. However, a gp130 mutant with the sequence around Y759 altered so that it retains binding to Shp2 but loses Socs-binding ability still shows sustained STAT3 activation in response to cytokine stimulation [361]. These data suggest that the increased STAT3 activation seen in the Y759F mutant reflects lack of Socs3, not Shp2, recruitment to pY759. This type of Shp2 function is phosphatase-independent, and thus could help explain the action of *PTPN11* mutants that cause LEOPARD syndrome (see "Shp2 and human disease, below"). Conversely, other studies suggest that Shp2 does, in fact, negatively regulate (dephosphorylate) STAT3 (see below).

Studies of other cytokine receptor signaling pathways also are consistent with a positive role for Shp2 in the JAK/STAT pathway, although the detailed mechanism remains unclear. For example, immortalized yolk sac hematopoietic cells from Ex3$^{-/-}$ mice show impaired IL-3 induced JAK2/STAT5 activation [362]. Conversely, Ba/F3 [363] or bone marrow mast cells [364] that overexpress leukemia-associated "activated" mutants of human *PTPN11* (SHP2^{E76K}), show enhanced IL3-evoked JAK2 and STAT5 phosphorylation.

There also are multiple reports of negative regulatory (inhibitory) roles for Shp2 in JAK/STAT signaling, however (Figure 13.4b). Shp2 is constitutively associated with the IFNγ receptor, and IFNγ-evoked JAK1 and STAT1 activation are enhanced in immortalized fibroblasts from Ex3$^{-/-}$ mice. Accordingly, these cells are hypersensitive

to the cytotoxic effects of IFNγ [365]. Increased STAT1 phosphorylation was observed prior to enhanced JAK1 activation in this study, which suggests that Shp2 directly dephosphorylates STAT1, while the effects on JAK1 could be more indirect. Ex3$^{-/-}$ cells also show increased IFNα responsiveness, along with increased STAT1 (but not STAT2) tyrosyl phosphorylation [365], again consistent with a role for Shp2 as a STAT1 phosphatase. Neither Tyk2 nor JAK1 tyrosyl phosphorylation was detected in either wild-type or mutant cells in this study, though, so action at the JAK kinase level cannot be excluded. Biochemical studies also suggest that STAT1 is a direct Shp2 substrate, although this report also claimed that Shp2 dephosphorylates STAT1 serine residues [366], which seems unlikely for structural reasons.

In contrast to the results described above, several studies argue that Shp2 is a STAT5 phosphatase and thereby inhibits STAT5-dependent signaling and biological responses [367–369]. Also, in disagreement with the mutant Y759 gp130 studies [361], there are several reports that Shp2 is a *bona fide* STAT3 phosphatase *in vivo*. For example, while Prl-stimulated STAT5 phosphorylation is decreased in Shp2-deleted mammary cells, STAT3 (which is not normally phosphorylated) shows increased tyrosyl phosphorylation under basal (and stimulated) conditions [168]. Furthermore, LIF-stimulated STAT3 phosphorylation is elevated in Ex3$^{-/-}$ ES cells [156], leptin-stimulated STAT3 phosphorylation is increased in mice with forebrain-specific Shp2 deletion [161], CNTF-evoked STAT3 phosphorylation is enhanced in neuronal progenitors exposed to Shp2-specific shRNAs [165], and STAT3 activation is increased in mice with global deletion of Shp2 in neurons and glial cells [164].

In some settings, Shp2 deficiency seems to have no effect on cytokine-evoked JAK/STAT pathway activation. For example, there is little difference in IFNα-evoked STAT1 or STAT3 activation [88] or in GM-CSF-evoked JAK2/STAT5 activation (W. Yang, G.Chan and B.G. Neel, unpublished observations) in wild-type and Shp2-deficient bone marrow-derived macrophages. Furthermore, IL-6-evoked STAT activation is unaffected in cardiomyocytes from mice with cardiac-specific deletion of Shp2 [167].

How can these apparently discrepant findings be reconciled? It remains possible, of course, that some of these reports and/or their interpretations are erroneous. In particular, whether Shp2 is a physiological JAK phosphatase remains unclear, and warrants further examination. However, considerable biochemical and genetic evidence indicate that Shp2 can act as a physiologically relevant STAT phosphatase-at least for STAT1 and STAT3 (strong evidence that Shp2 dephosphorylates STAT5 *in vivo* is lacking). Whether it does or not – i.e., whether Shp2 deficiency will affect STAT phosphorylation *in vivo* – probably depends on which other PTPs (and possibly other negative regulators) are expressed in the same cells. If sufficient amounts of another STAT1 or STAT3 phosphatase are present, Shp2 deficiency might have no effect on STAT1 or STAT3 tyrosyl phosphorylation because the other PTP can compensate. Such compensation need not even need to occur at the level of the STAT protein itself, but instead could result from enhanced receptor or JAK phosphorylation. Notably, in such situations, the effects of dominant negative or PTP-inactive mutants (such as those seen in LEOPARD syndrome; see "Shp and human diseases," above) on Shp2-deficient pathways might be stronger than Shp2 deficiency: when Shp2 is absent, the other PTP (if present) can access the STAT protein, whereas the PTP-inactive mutant would compete for the same sites and could therefore lead to increased STAT phosphorylation not seen in the null mutants. A further implication of such a model is that whether Shp2 is the "dominant" PTP in a particular signaling pathway – even in the same cell – might be subject to regulation. Indeed, there is evidence that "cross-talk" by other signaling pathways can alter the effects of Shp2 on a cytokine signaling pathway. Activation of macrophages by either Fcγ or Toll-like receptor stimulation markedly impairs their response to IFNα; this, in turn, requires PKC and Shp2. In this context, PKC appears to activate Shp2 so that it now dominantly impairs IFNα-evoked JAK/STAT signaling [88]. How PKC activates Shp2 is unclear, though, given that PKC while phosphorylates Shp2, it does not appear to increase its catalytic activity *in vitro* (see "Alternative mechanisms of regulation; the role of the C-tail," above).

In addition to its effects on cytokine-evoked Erk and JAK/STAT activation, Shp2 appears to inhibit Jnk activation in response to IL3 [370]. This inhibitory role contrasts with a requirement for Shp2 function in Kit-evoked Jnk activation in mast cells [371], but is similar to its potential role in IGF1 and EGF signaling [197] and its effects in stress-response pathways [102]. Shp2 also appears to be required for IL-1 and TNFR-induced NFκB activation [372]. Although the detailed mechanism remains unclear, it is interesting to note that mice lacking α-catenin, a proposed Shp2 substrate, show enhanced NFκB activation [373].

Finally, recent studies have suggested an entirely new ways in which Shp2 might regulate cytokine signaling. Eklund and colleagues have reported that, like Shp1 [213]), Shp2 catalyzes dephosphorylation of the interferon consensus sequence binding protein (ICSBP, also known as IRF8) at Y92 and Y95 in GM-CSF and M-CSF signaling, thereby negatively regulating the transcription of genes such as *Nf1* [215]. Shp1 deficiency causes the **me** phenotype, so if (increased) ICSBP tyrosyl phosphorylation contributes to this phenotype, Shp2, at least at normal endogenous levels of activation, must not be able to dephosphorylate ICSBP. Consistent with this notion, Shp2 deficiency in myeloid cells does not cause any obvious phenotype that could be attributed to increased ICSBP phosphorylation (W. Yang and B.G. Neel, unpublished).

However, when Shp2 activity is dramatically enhanced, as in leukemia-associated *PTPN11* mutants, it could "crossover" and dephosphorylate a substrate like ICSBP that is normally dephosphorylated by Shp1. Consistent with the idea that ICSBP can be a target of activated (although probably not endogenous) Shp2, ICSBP deficiency also causes myeloproliferative disease [374]. However, expressing activated Shp2 in ICSBP$^{-/-}$ mice results in acute myelogenous leukemia [375], indicating that leukemia-associated Shp2 has other key targets besides ICSBP. Notably, the same group suggests that both Shp1 [214] and Shp2 [216] also regulate tyrosyl phosphorylation of HoxA10. Clearly, further investigation and validation of these roles are clearly warranted.

Shp2 in Integrin Signaling

Studies of Ex3$^{-/-}$ fibroblasts and cells expressing dominant negative mutants have established that Shp2 function regulates integrin signaling and biological responses. Yet, as with other Shp2-regulated pathways, there are significant differences between various groups on the molecular and cellular details (Figure 13.4c). There is general consensus that spreading on a variety of integrin ligands (and, therefore, in response to activation of various integrins) requires active Shp2 [94, 376–378], although one group reported increased spreading in cells expressing catalytically inactive Shp2 [379]. Studies of Ex3$^{-/-}$ fibroblasts indicate no effect of Shp2 deficiency on initial adhesion [376–378], yet experiments with dominant negative mutants argue that adhesion is enhanced [379, 380]. All, however, agree that Shp2 is required for integrin-dependent Erk activation, focal contact maturation, cytoskeletal reorganization, migration, and invasion [376] [94,377–380]. The reasons for these discrepancies are unclear, but could include differences in cell type studied (fibroblasts vs epithelial cells), immortalization method (spontaneous, large T antigen), or experimental details (presence or absence of growth factors in the media during plating experiments, times at which adhesion is measured). Notably, Manes and colleagues showed that IGF-1 treatment stimulates adhesion in wild-type MCF-7 cells, but not in those expressing dominant negative Shp2 [380], so the presence of residual serum/growth factors in plating experiments might account for some intergroup differences. Likewise, elegant microscopy studies indicate that initial focal contact formation is normal in Ex 3$^{-/-}$ fibroblasts, but focal adhesion maturation/strengthening is defective [378]. Thus, the apparent effect of Shp2 on adhesion *per se* could be different depending on the precise time of measurement and the particular cell type.

Perhaps not surprisingly, the effects of Shp2 on integrin signaling are also controversial. Several compelling studies argue that Shp2 regulates integrin function by desphosphorylating Fak, either directly upon plating onto integrin ligands, or in response to stimulation by growth factors or the

H. pylori oncogene CagA [376] [378, 380–384]. However, Shp2 appears to promote FAK phosphorylation under some circumstances [94, 309, 377], and still other studies show no effect of Shp2 on FAK phosphorylation [379]. Even those studies that report that Shp2 dephosphorylates Fak differ on the physiological consequences of this regulation, with some arguing that Fak dephosphorylation promotes focal adhesion disassembly [380], while others argue that dephosphorylation of Fak is required for focal adhesion maturation [378] Likewise, some studies report that Shp2 positively regulates integrin-evoked SFK activation [309, 329, 377], whereas others report no effect [94].

The effect of Shp2 on more downstream events in integrin signaling is also a subject of controversy. Although all agree that Shp2 is required for integrin-evoked Erk activation, its effects on activation of the small G protein Rho, which controls stress fiber formation as well as other biological processes [385], are less clear.

Initial studies showed that Rho activation is enhanced in Shp2$^{-/-}$ fibroblasts [106], and more recent studies show that Rho activity is also increased in Shp2-deficient cardiomyocytes [167]. Yet in skeletal myoblasts, Shp2 reportedly promotes Rho activation by dephosphorylating p190RhoGapB [386], and genetic analyses suggest that the same pathway may be required for adipocyte vs skeletal muscle fate determination [387]. Further complexity is introduced by the recent report that Rho kinase-II (RockII), a key downstream affector of Rho (and possibly Rock as well), is inhibited by tyrosyl phosphorylation during adhesion of D2 myeloid leukemia cells to fibronectin, and that Shp2 appears to mediate dephosphorylation of this residue [388]. An important caveat of the above studies is that RockII tyrosyl phosphorylation was detected in pervanadate-treated cells, making it unclear whether this phosphorylation is physiologically relevant.

How Shp2 is recruited and activated to mediate its effects on integrin signaling also remains unclear. Shp2 is reported to co-immunoprecipitate with Fak in response to integrin engagement [376, 380], but the molecular basis for this interaction – whether it is direct – remains unclear. Shp2 may translocate into lipid rafts during integrin-induced adhesion, but again, how this occurs is unknown. Shps1 becomes tyrosine phosphorylated following integrin activation, probably by SFKs [36, 377, 389], and recruits Shp2. Thus, the effects of Shp2 on adhesion could be mediated by Shps1/Shp2 complexes (Figure 13.4c). Yet, studies of fibroblasts from mice lacking the Shps1 cytoplasmic domain reveal a more complicated picture. Whereas Rho activation is enhanced in Shp2 mutant fibroblasts, it is decreased in Shps1 mutants, despite the fact that both types of cells have increased stress fibers and defective migration [197]. Also, whether Shps1/Shp2 complexes are in lipid rafts is unclear. Another transmembrane glycoprotein, PZR, is also tyrosyl phosphorylated in response to adhesion [390], but its role remains poorly understood.

Whether the above disparate observations can cohere into a single model of Shp2 involvement in integrin signaling, or whether instead Shp2 does have distinct integrin- and/or cell-type dependent actions, remains a key challenge for the field. Perhaps, as initially suggested by Lacalle and colleagues [94], the answer is that Shp2 has multiple actions in integrin signaling (Figure 13.4c). These authors suggest that, following initial adhesion to integrin ligands, Shp2 is recruited into lipid rafts, where it promotes Rho activation (conceivably by dephosphorylating and inactivating p190RhoGap [386]). Although the mechanism by which Shp2 is recruited into rafts is unresolved, one attractive possibility is that Shps1 carries out this role, given that SFK activity is required for raft recruitment [94] and, as mentioned above, Shps1 mutants show decreased Rho activation upon integrin engagement [197]. Rho activation, in turn, leads to integrin clustering and Fak activation – effects that could be evoked artificially in non-adherent cells expressing a chimeric Shp2 bearing a lipid raft targeting sequence [380]). Upon plating of these cells, however, Rho is initially inactivated and then reactivated to a higher level in chimera-expressing cells. Remarkably, Shp2 activity is required for Rho deactivation, which might be a consequence of Shp2 dephosphorylation of Fak and/or Rock [388], and chimera-expressing cells also have higher levels of reactivated Rho. If, in fact, Shp2 does act sequentially during several steps in integrin signaling, then the exact times at which specific signaling events were assessed in the various studies of Shp2-deficient or mutant-expressing cells, as well as the nature of the negative feedback pathways in different cell types (or in cells immortalized in different ways), could have major effects on the experimental outcome.

There is one system in which sequential action by Shp2 has been proposed to mediate cross-talk between integrin and RTK signaling pathways. A series of papers by Clemmons' group shows that in vascular smooth muscle cells (VSMC), engagement of integrin αvβ3 is necessary to promote efficient IGF1R signaling and migration [391] (reviewed in [392]). They argue for a "fire brigade" model in which Shp2 is first recruited to tyrsosyl phosphorylated β3 itself, apparently via direct association of Shp2 with p62Dok [393]. IGF-1R stimulation results in Shps-1 tyrosyl phosphorylation, and Shp2 is then "passed" to Shps-1 (Figure 13.4c). Shp2 then dephosphorylates Shps1, resulting in Shp2 release and association with IGF1R itself. Failure to recruit Shp2 to either Dok or Shps1 results in premature association of Shp2 with the IGF1R, precocious IGF1R dephosphorylation, and termination of signaling. This model is intriguing, although it may not apply to all cell types. For example, fibroblasts that express an Shps1 mutant that cannot bind Shp2 show *increased* Erk activation, not the decreased Erk activation predicted by the Clemmons' model. Likewise, Dok deficiency typically results in increased growth factor-evoked Erk activation

[394], whereas the above model would predict decreased Erk activation in response to IGF1. Conceivably, specific integrin/RTK pairing results in quite different interactions with and/or effects of Shp2. Indeed, there is some precedence for this, as plating of endothelial cells on collagen, but not vitronectin, results in Shp2-dependent dephosphorylation of VEGFR2 [330]. Intriguingly, on vitronectin, Shp2 does not associate with VEGFR2, possibly via $\alpha v\beta 3$ binding as in VSMC, whereas collagen promotes VEGFR2/Shp2 interaction. Systematic reinvestigation of these issues in Shp2-null cells (or cells rendered Shp2-deficient by RNAi) would be quite helpful in resolving these issues.

Shp2 in Antigen Receptor Signaling

Upon TCR activation, Shp2 becomes tyrosyl phosphorylated and associates with the scaffolding adaptor Gab2 [39, 395–397]. Shp2 tyrosyl phosphorylation and association with Gab1 have been reported in B cells [398], although Shp2/Gab2 complexes also form in some B cell lines [39]. The functional consequences of these interactions remain somewhat unclear. Stable expression of dominant negative Shp2 in a T cell line inhibits TCR-evoked Erk activation [396]. However, overexpression of Gab2 inhibits TCR-evoked activation of Erk- and NFAT-dependent reporters [399,400]. One group found that only the PI3K binding sites were required for this inhibitory effect [399], whereas another observed a requirement for both Shp2 and PI3K binding [400]. TCR-evoked T cell proliferation is slightly increased in one line of Gab2$^{-/-}$ mice [401], although we have failed to observe consistent effects of Gab2 deficiency on T cells in our Gab2$^{-/-}$ line (H. Gu, J. Pratt, and B. G. Neel., unpublished). Experiments with chimeric mice suggest that Gab1 deficiency has minimal effects on B and T cell development, but, by acting via Shp2, specifically inhibits TI-2 responses in marginal zone B cells [402].

In addition to its proposed role in antigen receptor-evoked Erk activation, Shp2 is implicated in regulation of TCR-evoked adhesion. Experiments using T cell lines indicate that Shp2 dephosphorylates Vav1 and ADAP, which in turn regulate integrin activation by means of "inside-out" signaling [103]. Interestingly, TCR-evoked production of ROS is proposed to transiently inactivate Shp2, thereby facilitating adhesion upon T cell activation. This intriguing idea awaits validation using appropriate genetic models.

Shp2 also may mediate some inhibitory receptor signals. Shp2 was reported to bind the CD28 family member CTLA-4 and dampen TCR activation, most likely by dephosphorylating TCR-associated kinases [53]. The purported binding site for Shp2 on CTLA-4 is quite atypical, raising questions as to whether Shp2 binds directly to CTLA-4. Moreover, subsequent studies indicated that the Shp2 binding site is dispensable for CTLA-4-mediated inhibition [403–405]. Thus, while some role for Shp2 in CTLA-4 action cannot be excluded, Shp2 does not appear to be an essential mediator of CTLA-4 negative regulation. The inhibitory receptor PD-1 contains two potential ITIMs, and a chimera between the extracellular domain of FcγRIIB and the intracellular domain of PD1 inhibits BCR activation upon co-crosslinking. Shp2, but not Shp1 or Ship, bind the chimeric receptor, implicating Shp2 as the mediator of inhibitory signaling [406]. However, subsequent studies showed that PD1 could, in fact, bind Shp1 as well [258]. As Shp2 binding sites can also overlap with those of other inhibitory molecules, such as Socs proteins, confirmation using dominant negative, RNAi, or B-cell specific knockout approaches would be desirable. In this regard, our unpublished studies indicate that B-cell specific deletion of Shp2 has little if any effect on B-lymphoid development or function (W. Yang and B. G. Neel., unpublished). Shp2 is implicated in inhibitory signaling by CD31 (PECAM) in T cells [407, 408], B cells [409], and platelets [410], although its precise role awaits validation by cell-selective knockout approaches. Shp2 binds to many other inhibitory receptors in a wide array of cell types (e.g., NK cells, dendritic cell, macrophages, neutrophils), most of which also bind Shp1[33, 411, 412]. In the case of the NK cell inhibitory receptor KIR3DL1, Shp2 does appear to inhibit target cell conjugation and cytotoxicity [413]. Whether Shp2 has a role in mediating other inhibitory signals remains unknown, and, given its positive regulatory function, answering this question is often a daunting proposition.

Finally, two studies have addressed the role of Shp2 in T cell development and function *in vivo*. T-cell specific expression of catalytically impaired (C/S) Shp2 results in normal development, but T cells from these mice showed increased expression of activation markers and also produced more IL4 upon TCR stimulation [411]. These data contrast with the earlier cell line data indicating a requirement for Shp2 in TCR-evoked Erk activation [396], suggesting instead a minor negative regulatory role. Mice expressing dominant negative Shp2 in T cells did show defective thymic-dependent B cell responses, consistent with a decreased ability to provide T cell help. No obvious signaling defects were seen in these T cells, though.

More recently, T-cell specific Shp2 knockout mice have been generated, with quite different consequences [170]. These mice show a partial defect in thymic development, attributable to defective pre-TCR signaling. Thymocytes from these mice exhibit defective anti-CD3-evoked Erk activation, consistent with the earlier cell line results, indicating that Shp2 is required for TCR signaling as well. Because the developmental defect was only partial, peripheral Shp2-deficient T cells could also be studied, and they too showed defective TCR-evoked Erk activation, IL2 production, and expression of activation markers. Surprisingly, PMA-induced T cell activation was also defective in the absence of Shp2.

Taken together, these two studies suggest that Shp2 may have both positive and negative effects on T cell signaling/

function *in vivo*. The knockout studies clearly demonstrate the former. Yet although dominant negative experiments must always be interpreted with caution, the enhanced activation seen in T-cell specific Shp2 C/S mice could reflect negative regulatory functions obscured by the activation defect in the T-cell specific knockout mice.

Shp2 in Other Signaling Pathways

Recent work suggests that Shp2 negatively regulates Toll-like receptor (TLR) signaling in both peritoneal macrophages and fibroblasts [414]. Remarkably, this appears to involve a PTP activity-independent pathway in which Shp2, via amino acids 273–538, binds directly to the kinase domain of the TLR signaling component TBK1, inhibiting its catalytic activity. In the absence of Shp2, LPS-evoked interferon and pro-inflammatory cytokine signaling is enhanced. These effects of Shp2 on TLR signaling contrast with the recently described negative regulatory role of Shp1, which binds to and inhibits IRAK1 (Figure 13.3a). It will be interesting and important to see whether these PTP-independent functions of Shp2 help explain the phenotype of LEOPARD syndrome, which is caused by PTP-defective Shp2 mutants (see "Shp signaling and substrates," above).

Shp2 also appears to be important in signaling caused by fluid shear in endothelial cells [415, 416]. However, the precise mechanism is unclear. One group reports that these effects are mediated via a Pecam/Tie2 complex [416], whereas the other reports a Gab1/Shp2/PKA complex [415]. Further work will be required to clarify this signaling mechanism and test its physiological relevance *in vivo*.

Shp2, acting via Gab1, is critical for Jnk activation and cell death in response to oxidative stress [417]. Interestingly, Gab1 also sends an independent anti-apoptotic signal via the PI3K/Akt pathway in response to oxidative stress, with the net outcome (survival vs death) dependent on the relative strength of the Gab/Shp2/Jnk and Gab/PI3K/Akt responses. The precise mechanism by which Shp2 regulates Jnk activation in this pathway is not clear, although it is reminiscent of the Rac-Jnk pathway activated downstream of Kit [371].

Finally, Shp2 is implicated in regulation of cell cycle arrest in response to DNA damage. SV40 large T antigen-immortalized $exon3^{-/-}$ fibroblasts show reduced G_2 arrest in response to DNA damaging reagents, and reduced DNA damage-induced apoptosis [418–420]. In contrast, these mutant cells show enhanced epigallocatechin-3-gallate-induced apoptosis [421]. It remains to be determined whether these effects are present in non-transformed cells.

DETERMINANTS OF SHP SPECIFICITY

Shp1/Shp2 chimeras have been used to probe the determinants of Shp specificity in *Xenopus* [312] and mammalian cell [422] systems. These studies show that both the SH2 and the PTP domains contribute, and, surprisingly, the PTP domain contains the more critical specificity determinants. Even more surprisingly, the C-tails, which differ the most amongst Shps, do not confer specificity, at least in these systems. This suggests a combinatorial model of Shp specificity, whereby the SH2 domains direct Shps to appropriate cellular locations and the PTP domain then selects appropriate substrates.

Unfortunately, no structure exists for an Shp2–substrate complex. However, the PTP domain of Shp1 has been co-crystallized with relatively "good" and "poor" substrate peptides [423, 424]. These structures suggest that the $\alpha0$ helix and the $\alpha1/\beta1$, $\alpha5$-loop-$\alpha6$, and $\beta5$-loop-$\beta6$ motifs make contacts with bound peptides. The residues in the $\alpha1/\beta1$ motif involved in binding are identical in Shp1 and Shp2, whereas the two Shps have only relatively minor differences in $\beta5$-loop-$\beta6$ and $\alpha5$-loop-$\alpha6$. Thus, the more divergent $\alpha0$ helices may be of greatest importance in determining differences in substrate specificity.

There are caveats to this interpretation, however. Although the enzyme–peptide contacts extend at least four residues to either side of the pTyr, additional interactions important for specificity might extend even further, and thus would have been missed. Finally, it remains possible that "specificity" is not due to *bona fide* differences in substrate preferences between Shp1 and Shp2, but reflects the higher intrinsic catalytic rate of the Shp1 PTP domain [425–428]. Additional structural, biological, and biochemical studies will be required to resolve these issues.

SHPS AND HUMAN DISEASE

The most exciting recent discoveries about Shps have focused on their involvement in several different human diseases. Most of this work has centered on the role of Shp2 mutations in developmental disorders and neoplasia, and has been the subject of several recent reviews [429–433]. Therefore, these studies are reviewed briefly below, and readers interested in the details are urged to consult these other sources. Shp1 has also been implicated in certain types of malignancy, although by epigenetic, rather than mutational, alteration.

The breakthrough finding was supplied by the laboratory of Bruce Gelb, who used positional cloning to identify the gene for ~50 percent of the cases of Noonan syndrome (NS) as *PTPN11,* which encodes SHP2 [434]. NS is a fairly common (incidence ~1:2000 births) autosomal dominant developmental disorder characterized by abnormal faces, webbed neck resembling that in Turner syndrome, proportionate short stature, and cardiac abnormalities – most frequently pulmonic stenosis and/or other valvuloseptal defects. Chest and spine deformities, mental retardation, delayed puberty, crytorchidism, and bleeding diathesis occur with variable penetrance [435–438]. Also, case reports had noted a possible association between NS and increased incidence of malignancy, particularly juvenile myelomonocytic leukemia [439–442]. More recent studies identified gain-of-function mutation in *KRAS* (~1–2 percent overall) [443, 444], *SOS1* (~10 percent) [445, 446], or *RAF1* (~5 percent) [447, 448] in NS patients without *PTPN11* mutations. These studies strongly suggest that NS is a disease caused by excessive RAS/ERK pathway activation. Although the gene(s) responsible for ~30–40 percent of NS cases remain to be identified, genetic lesions in other components this pathway will likely be uncovered by ongoing sequencing efforts.

Rare NS patients with specific germline *PTPN11* mutations develop juvenile myelomonocytic leukemia (JMML). Moreover, somatic *PTPN11* mutations are found in ~35 percent of sporadic JMML patients [449,450] (see below), and at lower incidence in other myeloid neoplasms, including acute myelogenous leukemia (AML), chronic myelomyelogenous leukemia (CMML), and myelodysplastic syndrome (MDS) [449–452], as well as in B-ALL [453, 454]. *PTPN11* mutations are rare in solid tumors, with the possible exception of neuroblastoma [452, 455], and maybe "passenger mutations" [456] in these neoplasms.

Most NS and leukemia mutations in *PTPN11* affect the N-SH2 or PTP domain residues in SHP2 involved in basal inhibition [434, 449–453, 457, 458] (Figure 13.2b). Consequently, disease-associated alleles are activated, gain-of-function mutants of Shp2 analogous to those discussed above (see "Regulation of PTP activity"). There are some interesting exceptions, though. T42A, a NS mutant, resides in the binding pocket in the N-SH2 domain and enhances its affinity for pTyr peptides. As a result, this mutant demonstrates fairly normal basal PTP activity but enhanced PTyr-peptide-induced activation [459]. Similarly, D106A and E139D show only increased pTyr-peptide induced PTP activity, most likely due to "kinking" of the inter-SH2 domain linker and the C-SH2 domain, respectively [459]. The net result of all of these mutations is that lower levels of pTyr peptide ligands are needed to activate these mutant SHP2 proteins. In contrast, the mutants N308S and Q506P reside in the PTP domain, and show enhanced PTP activity only against some substrates [459]. Interestingly, Hatakayema's group recently identified a T407K mutant in a single case of hepatocellular carcinoma [460]. Although, as discussed above, this could (or is likely to) be a passenger mutation in this tumor, it nonetheless shows unusual properties, including the ability to transform NIH3T3 cells to factor independence (which other disease-associated *PTPN11* mutants cannot do), which the authors attribute to altered substrate specificity. Conceivably, detailed analysis of the substrate specificity of such "specificity switch" mutants may yield important insights into the key targets for SHP2 in RAS/ERK pathway activation.

The precise mechanisms and pathways by which *PTPN11* mutations cause the various NS abnormalities remain to be elucidated, but previous studies yield several clues. The cardiac abnormalities, particularly the valve defects, probably stem from abnormal regulation of the EGFR (and other EGFR family) signaling pathways [461], whereas the facial/skeletal abnormalities are reminiscent of activating mutations in FGFR signaling pathways [462]. Proportionately shortened stature may be due to defective GH and/or IGF-1 signaling. The origins of the cognitive and coagulation abnormalities are less clear, and may reflect the role of Shp2 in more than one signaling pathways.

The effects of NS and leukemia-associated mutations in *PTPN11* have been modeled in the mouse [463] and *Drosophila* [464]. Knock-in mice bearing the NS allele D61G show most of the hallmark features of this disorder, including facial dysmorphia, proportionate short stature, and a variety of valvuloseptal defects, although the latter occur with variable penetrance. They also develop an initially mild, well-tolerated myeloproliferative disease (MPD). Developing D61G/+ embryos show increased Erk activation in areas (cardiac cushions, facial placodes, limb buds) that subsequently show phenotypic abnormalities, although, interestingly, other tissues, as well as fibroblasts isolated from these embryos, show no evidence of increased Erk activation. NS and leukemia mutants have also been introduced into bone marrow cells using retroviral gene transfer. Leukemia-associated mutants can potently transform primary bone marrow cells to cytokine-independence [364, 465, 466], whereas NS alleles are far weaker in this assay [364, 466]. When transplanted into lethally irradiated mice, BM transduced with leukemia-associated alleles causes a fatal MPD [364], although this disease is incompletely (~60 percent) penetrant, occurs only after a significant time lag, and may be strain background-dependent [466]. Hematopoietic cells (bone marrow macrophages or mast cells) expressing leukemogenic mutants show increased cytokine-evoked activation of Erk, Akt, and/or STAT5; similar effects have been observed when these mutants are introduced into some [363, 467] but not all [364] immortalized, factor-dependent cell lines. Flies expressing *csw* engineered to contain NS or leukemia-associated mutations also show genetic (e.g., the ability to be enhanced or suppressed by other mutant alleles in the Ras/Erk pathway) and biochemical evidence of Erk pathway hypermorphism [464]. As in the mouse models, *csw* alleles bearing "leukemia" mutations are also more potent in the *Drosophila* system.

Germline *PTPN11* mutations are also found in at least 90 percent of patients with LEOPARD (Lentigines, Electrocardiogram abnormalities, Ocular hypertelorism, Pulmonic stenosis, Abnormalities of genitalia, Retardation of growth, and Deafness) syndrome (LS) [468,469]. In contrast to the mutations that cause NS and JMML, LS mutants reside in the PTP domain and typically affect residues important for catalysis. Furthermore, LS alleles are PTP-impaired or inactive *in vitro*, and can act as dominant negative mutants that inhibit growth-factor-evoked ERK activation in transient and stable transfection assays [470–472]. The location of LS mutants within the SHP2 structure, their potency as dominant negative mutants, and their increased binding to scaffolding adaptors such as GAB1 under basal and growth factor-stimulated conditions, suggest that they are in the "open" conformation.

These studies argue that LS is associated with impaired RAS/ERK pathway activation. If so, this poses the obvious paradox of how gain- and loss-of-function mutations in the same pathway can have such similar pathological effects. Very recent studies comparing the effects of "NS" and "LS" mutants engineered into zebrafish embryos are consistent with the idea that they have opposite effects on signaling pathways controlling gastrulation. Interestingly, however, their effects on this process were attributed to differential effects on Rho-, not Erk-, dependent pathways [12].

In even more recent work, "LS" mutants in *csw* showed gain-of-functions effects in *Drosophila* wing and eye development [473], which were suppressed by hypomorphic alleles and enhanced by hypermorphic alleles in the Ras/Erk pathway. Thus, it remains controversial how these mutants evoke pathological response in LS patients. Hopefully, clarity will come from ongoing mouse modeling efforts (M. Kontaridis and B.G. Neel, unpublished data).

At least one Shp2 binding protein plays an important role in human disease. In an SHP2-dependent manner, GAB2 overexpression increases the proliferative potential of human MCF10A cells [474, 475]. Furthermore, GAB2 cooperates with HER2/Neu overexpression to evoke an invasive-like phenotype in these cells and to increase HER2-evoked breast carcinogenesis in mice [475]. Both enhanced proliferation and cooperation with HER2/Neu require intact SHP2 binding sites on GAB2 [475]. Gab2 is also a critical determinant of the lineage (myeloid vs lymphoid) and severity (latency) of Bcr-Abl-evoked leukemia [476]. Gab2$^{-/-}$ cells expressing Bcr-Abl exhibit defective activation of the PI3K/Akt and Erk pathways; although the former probably reflects Gab2 binding to PI3K, defective Erk activation reflects inability to recruit Shp2. Consistent with these findings, Shp2-mutant (Ex3) yolk-sac (YS) cells are refractory to Bcr-Abl transformation/leukemogenesis [477], and Gab2 or Shp2 knockdown also impair human CML blasts [478]. In addition to its role in Bcr/Abl-evoked disease, Gab2 is required downstream of the receptor tyrosine kinase Stk for induction of Friend virus-evoked erythroleukemias in mice [479]. Previous work also established a positive role for Gab2 in v-SEA (the chicken-viral ortholog of STK)-evoked transformation of mouse fibroblasts [480]. Whether Shp2/Gab2 signaling is required for human erythroid malignancies remains to be determined.

Inappropriate activation of Shp2 also is critical in human gastric cancer pathogenesis. *H. pylori*, the major cause of gastric ulcers and carcinoma worldwide, encodes a number of virulence determinants, one of which, CagA, becomes tyrosyl phosphorylated by Src family kinases on C-terminal residues that conform to SHP2-binding EPIYA motifs [481,482]. SHP2 knockdown diminishes the "hummingbird phenotype" induced by CagA, suggesting that SHP2 is necessary for CagA to evoke morphological changes in gastric epithelial cells [483]. In very recent studies [484], transgenic mice expressing CagA fairly ubiquitously developed gastric hyperplasia, which progressed to frank adenocarcinoma in ~10 percent of mice after a long (>1 year) latent period. These mice also developed MPD similar to that caused by leukemia-associated alleles of *Shp2*. Neither gastric carcinoma nor MPD developed in transgenic mice expressing CagA lacking its SHP2 binding sites, arguing strongly that, as in Gab2-evoked malignancies, the ability of CagA to activate Shp2 is essential for pathogenesis.

In contrast to the pathogenic effects of increased Shp2 action, deficient Shp1 function plays a role in human disease. EPOR mutations that result in loss of Shp1 binding are associated with familial erythocytosis, although it is unclear whether loss of Shp1 binding (or Shp1 binding alone) is important. Shp1 deficiency was initially implicated in the pathogenesis of polcythemia vera [485], but these findings were susbsequently disputed [486]. Given the finding by multiple groups that JAK2 gain-of-function mutants cause polycythemia vera and related adult MPDs (reviewed in [487]), these initial findings were likely erroneous. There is markedly decreased Shp1 expression in human lymphoma [488], HTLV-1 induced leukemia [271], and Sezary syndrome [489], suggesting a possible role for Shp1 in suppressing these malignancies. Shp1 repression is effected by methylation and, in at least some cases, expression can be restored by methylation inhibitors, suggesting possible therapeutic approaches [490, 491].

Several human pathogens hijack Shp1 to help suppress the host immune response and/or to effect tissue damage. Binding of Shp1 to CD46, the measles virus receptor in macrophages, is associated with the production of IL12 and nitric oxide [492]. The Opa52 protein produced by *N. gonorrheae* binds to CECAM (CD66a) on primary CD4+ T cells. This leads to recruitment of Shp1 and Shp2, and suppression of T cell activation and proliferation [493,494]. Finally, elevated levels of Shp1 and Shp2 are associated with congenital neutropenia [495], although whether the Shps play a causal role in this disorder remains unclear.

SUMMARY AND FUTURE DIRECTIONS

Substantial progress has been made since the discovery of Shps in the early 1990s. We know many of the pathways in which Shps participate, the biological consequences of loss-of-function, and, in the case of Shp2, gain of function mutations in Shps. Shp structures have been solved to atomic resolution, and the basics of regulation of Shp activity are well understood.

Still, many important questions remain. Chief amongst these is the resolution of the proximate targets of each of the Shps. We do not understand Shp specificity in detail, and questions remain about the role of phosphorylation and, indeed, the C-tail (and, for Csw, the PTP insert) domain in general. The precise mechanism by which human disease-associated mutations in Shp2 cause disease remains to be clarified, and the development of potent and specific Shp inhibitors remains a "Holy Grail" of the field. With the available of powerful technologies such as inducible and tissue-specific knockout mice, RNAi, and substrate trapping mutants, answers to these and other questions about the Shp subfamily should emerge soon.

ACKNOWLEDGEMENTS

Work in the authors' laboratory is supported by NIH R37CA49152, R01 CA114945, and HL083273 (to Benjamin G. Neel). Benjamin G. Neel also is a Canada Research Chair, Tier I. Salim Dhanji holds a CIHR fellowship.

REFERENCES

1. Shen S-H, Bastien L, Posner BI, Chrétien P. A protein-tyrosine phosphatase with sequence similarity to the SH2 domain of the protein-tyrosine kinases. *Nature* 1991;**352**:736–9.

2. Plutzky J, Neel B, Rosenberg R, Eddy R, Byers M, Jani-Sait S, et al. Chromosomal localization of an SH2-containing tyrosine phosphatase. *Genomics* 1992;**13**:869–72.

3. Yi TL, Cleveland JL, Ihle JN. Protein tyrosine phosphatase containing SH2 domains: characterization, preferential expression in hematopoietic cells, and localization to human chromosome 12p12–p13. *Mol Cell Biol* 1992;**12**(2):p836–46.

4. Matthews RJ, Bowne DB, Flores E, Thomas ML. Characterization of hematopoietic intracellular protein tyrosine phosphatases: description of a phosphatase containing an SH2 domain and another enriched in proline-, glutamic acid-, serine-, and threonine-rich sequences. *Mol Cell Biol* 1992;**12**:2396–405.

5. Freeman Jr. RM, Plutzky J, Neel BG Identification of a human src homology 2-containing protein-tyrosine-phosphatase: a putative homolog of *Drosophila* corkscrew. *Proc Natl Acad Sci USA* 1992;**89**(23):11,239–43.

6. Feng G-S, Hui C-C, Pawson T. SH2-containing phosphotyrosine phosphatase as a target of protein-tyrosine kinases. *Science* 1993;**259**:1607–11.

7. Vogel W, Lammers R, Huang J, Ullrich A. Activation of a phosphotyrosine phosphatase by tyrosine phosphorylation. *Science* 1993;**259**(5101):1611–14.

8. Ahmad S, Banville D, Zhao Z, Fischer E, Shen S. A widely expressed human protein-tyrosine phosphatase containing src homology 2 domains. *Proc Natl Acad Sci USA* 1993;**90**(6):2197–201.

9. Adachi M, Sekiya m, Miyachi T, Matsuno K, Hinoda y, Imai K, et al. Molecular cloning of novel protein-tyrosine phosphatase SH-PTP3 with sequence similarity to the src-homology region 2. *FEBS Letts* 1992;**314**:335–9.

10. Tang TL, Freeman RM, O'Reilly AM, Neel BG, Sokol SY. The SH2-containing protein tyrosine phosphatase SH-PTP2 is required upstream of MAP kinase for early *Xenopus* development. *Cell* 1995;**80**:473–83.

11. Park C, La MK, Tonks N, Hayman M. Cloning and expression of the chicken protein tyrosine phosphatase SH-PTP2. *Gene* 1996;**177**(1–2):93–7.

12. Jopling C, van Geemen D, den Hertog J. Shp2 knockdown and Noonan/LEOPARD mutant Shp2-induced gastrulation defects. *PLoS Genet* 2007;**3**(12):e225.

13. Adachi M, Fisher EH, Ihle J, Imai K, Jirik F, Neel B, et al. Mammalian SH2-containing protein tyrosine phophatases. *Cell* 1996;**85**:15.

14. Perkins LA, Larsen I, Perrimon N. corkscrew endodes a putative protein tyrosine phosphatase that functions to transduce the terminal signal from the receptor tyrosine kinase torso. *Cell* 1992;**70**:225–36.

15. Perkins LA, Johnson MR, Melnick MB, Perrimon N. The non-receptor protein tyrosine phosphatase Corkscrew functions in multiple receptor tyrosine kinase pathways in *Drosophila*. *Dev Biol* 1996;**180**:63–81.

16. Gutch MJ, Flint AJ, Keller J, Tonks NK, Hengartner MO. The *Caenorhabditis elegans* SH2 domain-containing protein tyrosine phosphatase PTP-2 participates in signal transduction during oogenesis, embryogenesis and vulval development. *Genes Dev* 1997;**12**:571–85.

17. Perkins LA, Melnick M. Corkscrew, an essential protein tyrosine phosphatase required for multiple developmental decisions in *Drosophila*. In: Merlevede W, editor. *Advances of protein phosphatases*. Philadelphia: Coronet Books Inc; 1995. p. 355–85.

18. Yeung Y, Berg KL, Pixley FJ, Angeletti RH, Stanley ER. Protein tyrosine phosphatase-1C is rapidly phosphorylated on tyrosine in macrophages in response to colony stimulating factor-1. *J Biol Chem* 1992;**267**:23,447–50.

19. Lechleider RJ, Freeman RM, Neel BG. Tyrosyl phosphorylation and growth factor receptor association of the human corkscrew homologue, SH-PTP2. *J Biol Chem* 1993;**268**:13,434–8.

20. Yi T, Ihle JN. Association of hematopoietic cell phosphatase with c-Kit after stimulation with c-Kit ligand. *Mol Cell Biol* 1993;**13**(6):3350–8.

21. Lorenz U, Ravichandran KS, Pei D, Walsh CT, Burakoff SJ, Neel BG. Lck-dependent tyrosyl phosphorylation of the phosphotyosine phosphatase SH-PTP1 in murine T cells. *Mol Cell Biol* 1994;**14**:1824–34.

22. Bouchard P, Zhao Z, Banville D, Dumas F, Fischer EH, Shen SH. Phosphorylation and identification of a major tyrosine phosphorylation site in protein tyrosine phosphatase 1C. *J Biol Chem* 1994;**269**(30):19,585–9.

23. Uchida T, Matozaki T, Noguchi T, Yamao T, Horita K, Suzuki T, et al. Insulin stimulates the phosphorylation of Tyr^{538} and the catalytic activity of PTP1C, a protein tyrosine phosphatase with Src homology-2 domains. *J Biol Chem* 1994;**269**:12,220–8.

24. Bennett AM, Tang TL, Sugimoto S, Walsh CT, Neel BG. Protein-tyrosine-phosphatase SHPTP2 couples platelet-derived growth factor receptor beta to Ras. *Proc Natl Acad Sci USA* 1994;**91**(15):7335–9.

25. Vogel W, Ullrich A. Multiple *in vivo* phosphorylated tyrosine phosphatase SHP-2 engages binding to Grb2 via Tyrosine 584. *Cell Growth Diff* 1996;**7**:1589–97.

26. Walter AO, Peng ZY, Cartwright CA. The Shp-2 tyrosine phosphatase activates the Src tyrosine kinase by a non-enzymatic mechanism. *Oncogene* 1999;**18**(11):1911–20.

27. Craggs G, Kellie S. A functional nuclear localization sequence in the C-terminal domain of Shp-1. *J Biol Chem* 2001;**276**:23,719–725.

28. Yang W, Tabrizi M, Yi T. A bipartite NLS at the SHP-1 C-terminus mediates cytokine-induced SHP-1 nuclear localization in cell growth control. *Blood Cells Mol Dis* 2002;**28**:63–74.

29. Yi T, Cleveland JL, Ihle JN. Protein tyrosine phosphatase containing SH2 domains: characterization, preferential expression in hematopoietic

cells, and localization to human chromosome 12p12–13. *Mol Cell Biol* 1992;**12**(2):836–46.

30. Banville D, Stocco R, Shen SH. Human protein tyrosine phosphatase 1C (PTPN6) gene structure: alternative promoter usage and exon skipping generate multiple transcripts. *Genomics* 1995;**27**:165–73.

31. Tsui HW, Hasselblatt K, Martin A, Mok SC, Tsui FWL. Molecular mechanisms underlying *SHP-1* gene expression. *Eur J Biochem* 2002;**269**:3057–64.

32. J.in Y-J, Yu C-L, Burakoff SJ. Human 70-kDa SHP-1L differs from 68-kDa SHP-1 in its C-terminal structure and catalytic activity. *J Biol Chem* 1999;**274**:28,301–7.

33. Ravetch JV, Lanier LL. Immune inhibitory receptors. *Science* 2000;**290**:84–9.

34. Daeron M, Jaeger S, Du Pasquier L, Vivier E. Immunoreceptor tyrosine-based inhibition motifs: a quest in the past and future. *Immunol Rev* 2008;**224**:11–43.

35. Noguchi T, Matozaki T, Fujioka Y, Yamao T, Tsuda M, Takada T, et al. Characterization of a 115-kDa protein that binds to SH-PTP2, a protein-tyrosine phosphatase with Src homology 2 domains, in Chinese hamster ovary cells. *J Biol Chem* 1996;**271**:27,652–8.

36. Fujioka Y, Matozaki T, Noguchi T, Iwamatsu A, Yamao T, Takahashi N, et al. A novel membrane glycoprotein, SHPS-1, that binds the SH2-domain-containing protein tyrosine phosphatase SHP-2 in response to mitogens and cell adhesion. *Mol Cell Biol* 1996;**16**:6887–99.

37. Kharitonenkov A, Chen Z, Sures I, Wang H, Schilling J, Ullrich A. A family of proteins that inhibit signaling though tyrosine kinase receptors. *Nature* 1997;**386**:181–6.

38. Timms JF, Carlberg K, Gu H, Chen H, Kamatkar S, Nadler MJ, et al. Identification of major binding proteins and substrates for the SH2-containing protein tyrosine phosphatase SHP-1 in macrophages. *Mol Cell Biol* 1998;**18**(7):3838–50.

39. Gu H, Pratt JC, Burakoff SJ, Neel BG. Cloning and characterization of the major SHP-2 binding protein in hematopoietic cells (p97) reveals a novel pathway for cytokine-induced gene activation. *Mol Cell* 1998;**2**(6):729–40.

40. Neel BG, Gu H, Pao L. The 'Shp"ing news: SH2 domain-containing tyrosine phosphatases in cell signaling. *Trends Biochem Sci* 2003;**28**:284–93.

41. Burshtyn DN, Scharenberg AM, Wagtmann N, Rajagopalan S, Berrada K, Yi T, et al. Recruitment of tyrosine phosphatase HCP by the killer cell inhibitory receptor. *Immunity* 1996;**4**:77–85.

42. Olcese L, Lang P, Vely F, Cambiaggi A, Marguet D, Blery M, et al. Human and mouse killer-cell inhibitory receptors recruit PTP1C and PTP1D protein tyrosine phosphatases. *J Immunol* 1996;**156**(12):p4531–4.

43. Burshtyn DN, Yang W, Yi T, Long EO. A novel phosphotyrosine motif with a critical amino acid at position-2 for the SH2 domain-mediated activation of the tyrosine phosphatase SHP-1. *J Biol Chem* 1997;**272**:13,066–72.

44. Burshtyn DN, Lam AS, Weston M, Gupta N, Warmerdam PA, Long EO. Conserved residues amino-terminal of cytoplasmic tyrosines contribute to the SHP-1-mediated inhibitory function of killer cell Ig-like receptors. *J Immunol* 1999;**162**(2):897–902.

45. Thomas ML. Of ITAMs and ITIMs: turning on and off the B cell antigen receptor. *J Exp Med* 1995;**181**:1953–6.

46. Cambier JC. Inhibitory receptors abound?. *Proc Natl Acad Sci USA* 1997;**94**:5993–5.

47. Vivier E, Daeron M. Immunoreceptor tyrosine-based inhibition motifs. *Immunol Today* 1997;**18**(6):286–91.

48. De Souza D, Fabri LJ, Nash A, Hilton DJ, Nicola NA, Baca M. SH2 domains from suppressor of cytokine signaling-3 and protein tyrosine phosphatase SHP-2 have similar binding specificities. *Biochemistry* 2002;**41**:9229.

49. Beebe KD, Wang P, Arabaci G, Pei D. Determination of the binding specificity of the SH2 domains of protein tyrosine phosphatase SHP-1 through the screening of a combinatorial phosphotyrosyl peptide library. *Biochemistry* 2000;**39**:13,251–260.

50. Sweeney MC, Wavreille AS, Park J, Butchar JP, Tridandapani S, Pei D. Decoding protein-protein interactions through combinatorial chemistry: sequence specificity of SHP-1, SHP-2, and SHIP SH2 domains. *Biochemistry* 2005;**44**(45):14,932–47.

51. Wavreille AS, Garaud M, Zhang Y, Pei D. Defining SH2 domain and PTP specificity by screening combinatorial peptide libraries. *Methods* 2007;**42**(3):207–19.

52. Imhof D, Wavreille AS, May A, Zacharias M, Tridandapani S, Pei D. Sequence specificity of SHP-1 and SHP-2 Src homology 2 domains. Critical roles of residues beyond the pY+3 position. *J Biol Chem* 2006;**281**(29):20,271–82.

53. Marengere LEM, Waterhouse P, Duncan GS, Mittrucker H-W, Feng G-S, Mak TW. Regulation of T cell receptor signaling by tyrosine phosphatase SYP association with CTLA-4. *Science* 1996;**272**:1170–3.

54. Schmitz J, Weissenbach M, Haan S, Heinrich PC, Schaper F. SOCS3 exerts its inhibitory function on interleukin-6 signal transduction through the SHP2 recruitment site of gp130. *J Biol Chem* 2000;**275**:12,848–56.

55. Nicholson SE, De Souza D, Fabri LJ, Corbin J, Willson TA, Zhang JG, et al. Suppressor of cytokine signaling-3 preferentially binds to the SHP-2-binding site on the shared cytokine receptor subunit gp130. *Proc Natl Acad Sci USA* 2000;**97**(12):6493–8.

56. Sasaki A, Yasukawa H, Shouda T, Kitamura T, Dikic I, Yoshimura A. CIS3/SOCS-3 suppresses erythropoietin (EPO) signaling by binding the EPO receptor and JAK2. *J Biol Chem* 2000;**275**:29,338–47.

57. Bjorbaek C, Lavery HJ, Bates SH, Olson RK, Davis SM, Flier JS, et al. SOCS3 mediates feedback inhibition of the leptin receptor via tyr985. *J Biol Chem* 2000;**275**:40,649–57.

58. Eyckerman S, Broekaert D, Verhee A, Vanderkerckhove J, Tavernier J. Identification of the Y985 and Y1077 motifs as SOCS3 recruitment sites in the murine leptin receptor. *FEBS Letts* 2000;**486**:33–7.

59. Philosof-Oppenheimer R, Hampe CS, Schlessinger K, Fridkin M, Pecht I. An immunoreceptor tyrosine-based inhibitory motif, with serine at site Y-2, binds SH2-domain-containing phosphatases. *Eur J Biochem* 2000;**267**:703–11.

60. Vely F, Olivero S, Olcese L, Moretta A, Damen JE, Liu L, et al. Differential association of phosphatases with hematopoietic co-receptors bearing immunoreceptor tyrosine-based inhibition motifs. *Eur J Immunol* 1997;**27**:1994–2000.

61. Famiglietti SJ, Nakamura K, Cambier JC. Unique features of SHIP, SHP-1 and SHP-2 binding to FCgammaRIIb revealed by surface plasmon resonance analysis. *Immunol Letts* 1999;**68**:35–40.

62. Vely F, Trautmann A, Vivier E. BIAcore analysis to test phosphopeptide-SH2 domain interactions. *Methods Mol Biol* 2000;**121**:313–21.

63. Bruhns P, Marchetti P, Fridman WH, Vivier E, Daeron M. Differential roles of N- and C-terminal immunoreceptor tyrosine-based inhibition motifs during inhibition of cell activation by killer cell inhibitory receptors. *J Immunol* 1999;**162**:3168–75.

64. Baca M. Analysis of SH2 ligands and identification of sites of interaction. *Methods Mol Biol* 2004;**249**:111–20.

65. Daigle I, Yousefi S, Colonna M, Green DR, Simon HU. Death receptors bind SHP-1 and block cytokine-induced anti-apoptotic signaling in neutrophils. *Nat Med* 2002;**8**(1):61–7.

66. Qin C, Wavreille AS, Pei D. Alternative mode of binding to phosphotyrosyl peptides by Src homology-2 domains. *Biochemistry* 2005;**44**(36):12,196–202.

67. Case RD, Piccione E, Wolf G, Benett AM, Lechleider RJ, Neel BG, et al. SH-PTP2/Syp SH2 domain binding specificity is defined by direct interactions with platelet-derived growth factor beta-receptor, epidermal growth factor receptor, and insulin receptor substrate-1-derived phosphopeptides. *J Biol Chem* 1994;**269**(14):10,467–74.

68. Lechleider RJ, Sugimoto S, Bennett AM, Kashishian AS, Cooper JA, Shoelson SE, et al. Activation of the SH2-containing phosphotyrosine phosphatase SH-PTP2 by its binding site 1009, on the human platelet-derived growth factor β. *J Biol Chem* 1993;**268**:21,478–81.

69. Sugimoto S, Wandless TJ, Shoelson SE, Neel BG, Walsh CT. Activation of the SH2-containing protein tyrosine phosphatase, SH-PTP2, by phosphotyrosine containing peptides derived from insulin receptor substrate-1. *J Biol Chem* 1993;**268**:2733–6.

70. Pei D, Lorenz U, Klingmuller U, Neel BG, Walsh CT. Intramolecular regulation of protein tyrosine phosphatase SH-PTP1: a new function for Src homology 2 domains. *Biochemistry* 1994;**33**:15,483–93.

71. Barford D, Neel BG. Revealing mechanisms for SH2 domain mediated regulation of the protein tyrosine phosphatase SHP-2. *Structure* 1998;**6**(3):249–54.

72. Pluskey S, Wandless TJ, Walsh CT, Shoelson SE. Potent stimulation of SH-PTP2 phosphatase activity by simultaneous occupancy of both SH2 domains. *J Biol Chem* 1995;**270**:2900–87.

73. Townley R, Shen S-H, Banville D, Ramachandran C. Inhibition of the activity of protein tyrosine phosphatase 1C by its SH2 domains. *Biochemistry* 1993;**32**:13,414–18.

74. Pregel MJ, Shen S-H, Storer AC. Regulation of protein tyrosine phosphatase 1C: opposing effects of the two src homology 2 domains. *Protein Eng* 1995;**12**:1309–16.

75. Hof P, Pluskey S, Dhe-Paganon S, Eck MJ, Shoelson SE. Crystal structure of the tyrosine phosphatase SHP-2. *Cell* 1998;**92**(4):441–50.

76. O'Reilly AM, Pluskey S, Shoelson SE, Neel BG. Activated mutants of SHP-2 preferentially induce elongation of *Xenopus* animal caps. *Mol Cell Biol* 1999;**20**:299–311.

77. Yang J, Liang X, Niu T, Meng W, Zhao Z, Zhou GW. Crystal structure of the catalytic domain of protein-tyrosine phosphatase SHP-1. *J Biol Chem* 1998;**273**:199–207.

78. Frank C, Burkhardt C, Imhof D, Ringel J, Zschornig O, Wieligmann K, et al. Effective dephosphorylation of Src substrates by SHP-1. *J Biol Chem* 2004;**279**(12):11,375–83.

79. Lu W, Gong D, Bar-Sagi D, Cole PA. Site-specific incorporation of a phosphotyrosine mimetic reveals a role for tyrosine phosphorylation of SHP-2 in cell signaling. *Mol Cell* 2001;**8**:759–69.

80. Zhang Z, Shen K, Lu W, Cole PA. The role of C-terminal tyrosine phosphorylation in the regulation of SHP-1 explored via expressed protein ligation. *J Biol Chem* 2003;**278**(7):4668–74.

81. Poole AW, Jones ML. A SHPing tale: perspectives on the regulation of SHP-1 and SHP-2 tyrosine phosphatases by the C-terminal tail. *Cell Signal* 2005;**17**(11):1323–32.

82. Li RY, Gaits F, Ragab A, Ragab-Thomas JM, Chap H. Tyrosine phosphorylation of an SH2-containing protein tyrosine phosphatase is coupled to platelet thrombin receptor via a pertussis toxin-sensitive heterotrimeric G-protein. *EMBO J* 1995;**14**:2519–26.

83. Zhao Z, Shen S-H, Fischer EH. Phorbol ester-induced expression, phosphorylation, and translocation of protein-tyrosine-phosphatase 1C in HL-60 cells. *Proc Natl Acad Sci USA* 1994;**91**:5007–11.

84. Brumell JH, Chan CK, Butler J, Borregaard N, Siminovitch KA, Grinstein S, et al. Regulation of Src homology 2-containing tyrosine phosphatase 1 during activation of human neutrophils. *J Immunol* 1997;**272**:875–82.

85. Jones ML, Craik JD, Gibbins JM, Poole AW. Regulation of SHP-1 tyrosine phosphatase in human platelets by serine phosphorylation at its C terminus. *J Biol Chem* 2004;**279**(39):40,475–83.

86. Zhao Z, Larocque R, Ho WT, Fischer EH, Shen SH. Purification and characterization of PTP2C, a widely distributed protein tyrosine phosphatase containing two SH2 domains. *J Biol Chem* 1994;**269**:8780–5.

87. Strack V, Krutzfeldt J, Kellerer M, Ullrich A, Lammers R, Haring H-U. The protein-tyrosine phosphatase SHP2 is phosphorylated on serine residues 576 and 591 by protein kinase C isoforms α, β1, β2, and eta. *Biochemistry* 2002;**41**:603–8.

88. Du Z, Shen Y, Yang W, Mecklenbrauker I, Neel BG, Ivashkiv LB. Inhibition of IFN-α signaling by a PKC- and protein tyrosine phosphatase SHP-2-dependent pathway. *Proc Natl Acad Sci USA* 2005;**102**(29):10,267–72.

89. Peraldi P, Zhao Z, Filloux C, Fischer E, Van Obberghen E. Protein-tyrosine-phosphatase 2C is phosphorylated and inhibited by 44-kDa mitogen-activated protein kinase. *Proc Natl Acad Sci USA* 1994;**91**:5002–6.

90. Zhao Z, Shen S-H, Fischer EH. Stimulation by phospholipids of a protein-tyrosine-phosphatase containing two *src* homology 2 domains. *Proc Natl Acad Sci USA* 1993;**90**:4251–5.

91. Frank C, Keilhack H, Opitz F, Zschornig O, Bohmer F-D. Binding of phosphatidic acid to the protein-tyrosine phosphatase SHP-1 as a basis for activity modulation. *Biochemistry* 1999;**38**:11,993–12,002.

92. Kosugi A, Sakakura J, Yasuda K, Ogata M, Hamaoka T. Involvement of SHP-1 tyrosine phosphatase in TCR-mediated signaling pathways in lipid rafts. *Immunity* 2001;**14**:669–80.

93. Su MW-C, Yu C-L, Burakoff SJ, Jin Y-J. Targeting src homology 2 domain-containing tyrosine phosphatase (SHP-1) into lipid rafts inhibits CD3-induced T cell activation. *J Immunol* 2001;**166**:3975.

94. Lacalle RA, Mira E, Gomez-Mouton C, Jimenez-Baranda S, Martinez-A. C, Manes S. Specific SHP-2 partitioning in raft domains triggers integrin activation via Rho activation. *J Cell Biol* 2002;**157**:277–89.

95. Fawcett VC, Lorenz U. Localization of Src homology 2 domain-containing phosphatase 1 (SHP-1) to lipid rafts in T lymphocytes: functional implications and a role for the SHP-1 carboxyl terminus. *J Immunol* 2005;**174**(5):2849–59.

96. Ram PA, Waxman DJ. Interaction of growth hormone-activated STATs with SH2-containing phosphotyrosine phosphatase SHP-1 and nuclear JAK2 tyrosine kinase. *J Biol Chem* 1997;**272**(28):17,694–702.

97. Finkel T, Serrano M, Blasco MA. The common biology of cancer and ageing. *Nature* 2007;**448**(7155):767–74.

98. Tonks NK. Protein tyrosine phosphatases: from genes, to function, to disease. *Nat Rev Mol Cell Biol* 2006;**7**(11):833–46.

99. Meng T-C, Fukada T, Tonks NK. Reversible oxidation and inactivation of protein tyrosine phosphatases *in vivo*. *Mol Cell* 2002;**9**:387–99.

100. Valius M, Kazlauskas A. Phospholipase C-gamma1 and phospho-tidylinositol 3-kinase are the downstream mediators of the PDGF receptor's mitogenic signal. *Cell* 1993;**73**:321–34.

101. Roche S, McGlade J, Jones M, Gish GD, Pawson J, Courtneidge SA. Requirement of phospholipase C†, the tyrosine phosphatase Syp and the adaptor proteins Shc and Nck for PDGF-induced DNA synthesis: evidence for the existence of Ras-dependent and Ras-independent pathways. *EMBO J* 1996;**18**:4940–8.

102. Shi ZQ, Lu W, Feng GS. The Shp-2 tyrosine phosphatase has opposite effects in mediating the activation of extracellular signal-regulated and c-Jun NH2-terminal mitogen-activated protein kinases. *J Biol Chem* 1998;**273**(9):4904–8.

103. Kwon J, Qu CK, Maeng JS, Falahati R, Lee C, Williams MS. Receptor-stimulated oxidation of SHP-2 promotes T-cell adhesion through SLP-76-ADAP. *EMBO J* 2005;**24**(13):2331–41.

104. Singh DK, Kumar D, Siddiqui Z, Basu SK, Kumar V, Rao KV. The strength of receptor signaling is centrally controlled through a cooperative loop between Ca2+ and an oxidant signal. *Cell* 2005;**121**(2):281–93.

105. Goldfine AB, Simonson DC, Folli F, Patti ME, Kahn CR. *In vivo* and *in vitro* studies of vanadate in human and rodent diabetes mellitus. *Mol Cell Biochem* 1995;**153**(1–2):217–31.

106. Schoenwaelder SM, Petch LA, Williamson D, Shen R, Feng GS, Burridge K. The protein tyrosine phosphatase Shp-2 regulates RhoA activity. *Curr Biol* 2000;**10**:1523–6.

107. Chen L, Sung SS, Yip ML, Lawrence HR, Ren Y, Guida WC, et al. Discovery of a novel shp2 protein tyrosine phosphatase inhibitor. *Mol Pharmacol* 2006;**70**(2):562–70.

108. Lawrence HR, Pireddu R, Chen L, Luo Y, Sung SS, Szymanski AM, et al. Inhibitors of Src homology-2 domain containing protein tyrosine phosphatase-2 (Shp2) based on oxindole scaffolds. *J Med Chem* 2008;**51**(16):4948–56.

109. Hellmuth K, Grosskopf S, Lum CT, Wurtele M, Roder N, von Kries JP, et al. Specific inhibitors of the protein tyrosine phosphatase Shp2 identified by high-throughput docking. *Proc Natl Acad Sci USA* 2008;**105**(20):7275–80.

110. Tsui HW, Siminovitch KA, deSouza L, Tsui FWL. *Motheaten* and *viable motheaten* mice have mutations in the haematopoietic cell phosphatase gene. *Nat Genet* 1993;**4**:124–9.

111. Shultz LD, Schweitzer PA, Rajan TV, Yi T, Ihle JN, Matthews RJ, et al. Mutations at the murine motheaten locus are within the hematopoietic cell protein-tyrosine phosphatase (Hcph) gene. *Cell* 1993;**73**:1445–54.

112. Shultz LD. Hematopoiesis and models of immunodeficiency. *Sem Immunol* 1991;**3**:397–408.

113. Neel BG. Structure and function of SH2-domain containing tyrosine phosphatases. *Sem Cell Biol* 1993;**4**:419–32.

114. Tsui FWL, Tsui HW. Molecular basis of the motheaten phenotype. *Immunol Rev* 1994;**136**:185–206.

115. Bignon JS, Siminovitch KA. Identification of PTP1C mutation as a genetic defect in motheaten and viable motheaten mice: a step toward defining the roles of protein tyrosine phosphatases in the regulation of hemopoietic cell differentiation and function. *Clin Immunol Immunopathol* 1994;**73**:168–79.

116. Shultz LD, Rajan TV, Greiner DL. Severe defects in immunity and hematopoiesis caused by SHP-1 protein-tyrosine-phosphatase deficiency. *Trends Biotechnol* 1997;**15**:302–7.

117. Takahashi K, Miyakawa K, Wynn AA, Nakayama K-I, Myint YY, Naito M, et al. Effects of granulocyte/macrophage colony-stimulating factor on the development and differentiation of CD5-positive macrophages and their potential derivation from a CD5-positive B-cell lineage in mice. *Am J Pathol* 1998;**152**:445–56.

118. Nakayama K, Takahashi K, Shultz L, Miyakawa K, Tomita K. Abnormal development and differentiation of macrophages and dendritic cells in viable motheaten mutant mice deficient in haematopoietic cell phosphatase. *Intl J Exp Pathol* 1997;**78**(4):245–57 [published erratum appears in *Intl J Exp Pathol* 1997;**78**(5):364].

119. Umeda S, Beamer W, Takagi K, Naito M, Hayashi S, Yonemitsu H, et al. Deficiency of SHP-1 protein-tyrosine phosphatase activity results in heightened osteoclast function and decreased bone density. *Am J Pathol* 1999;**155**(1):223–33.

120. Yu CC, Tsui HW, Ngan BY, Shulman MJ, Wu GE, Tsui FW. B and T cells are not required for the viable motheaten phenotype. *J Exp Med* 1996;**183**:371–80.

121. Koo GC, Rosen H, Sirotina A, Ma XD, Shultz LD. Anti-CD11b antibody prevents immunopathologic changes in viable motheaten bone marrow chimeric mice. *J Immunol* 1993;**151**:6733–41.

122. Greiner DL, Goldschneider I, Komschlies KL, Medlock ES, Bollum FJ, Shultz L. Defective lymphopoiesis in bone marrow of motheaten (*me/me*) and viable motheaten (*me^v/me^v*) mutant mice. *J Exp Med* 1986;**164**:1129–44.

123. Komschlies KL, Greiner DL, Shultz L, Goldschneider I. Defective lymphopoieses in the bone marrow of motheaten (*me/me*) and viable motheaten (*me^v/me^v*) mutant mice. *J Exp Med* 1987;**166**:1162–7.

124. Lorenz U, Ravichandran KS, Burakoff SJ, Neel BG. Lack of SHPTP1 results in *src*-family kinase hyperactivation and thymocyte hyperresponsiveness. *Proc Natl Acad Sci USA* 1996;**93**:9624–9.

125. Pani G, Fischer K-D, Rascan IM, Siminovitch KA. Signaling capacity of the T cell antigen receptor is negatively regulated by the PTP1C tyrosine phosphatase. *J Exp Med* 1996;**184**:839–52.

126. Plas DR, Willians CB, Jersh GJ, White LS, White JM, Paust S, et al. The tyrosine phosphatase SHP-1 regulates thymocyte positive selection. *J Immunol* 1999;**162**:5680–4.

127. Johnson KG, LeRoy FG, Borysiewicz LK, Matthews RJ. TCR signaling thresholds regulating T cell development and activation are dependent upon SHP-1. *J Immunol* 1999;**162**:3802–13.

128. Carter JD, Neel BG, Lorenz U. The tyrosine phosphatase SHP-1 influences thymocyte selection by setting TCR signaling thresholds. *Intl Immunol* 1999;**11**:1999–2014.

129. Zhang J, Somani AK, Yuen D, Yang Y, Love PE, Siminovitch KA. Involvement of the SHP-1 tyrosine phosphatase in regulation of T cell selection. *J Immunol* 1999;**163**:3012–21.

130. Shultz LD, Sidman CL. Genetically determined murine models of immunodeficiency. *Annu Rev Immunol* 1987;**5**:367–403.

131. D'Ambrosio D, Hippen KL, Minskoff SA, Mellman I, Pani G, Siminovitch KA, et al. Recruitment and activation of PTP1C in negative regulation of antigen receptor signaling by Fc gamma RIIB1. *Science* 1995;**268**:293–7.

132. Pani G, Kozlowski M, Cambier JC, Mills GB, Siminovitch KA. Identification of the tyrosine phosphatase PTP1C as a B cell antigen receptor-associated protein involved in the regulation of B cell signaling. *J Exp Med* 1995;**181**:2077–84.

133. Cyster JG, Goodnow CC. Protein tyrosine phosphatase 1C negatively regulates antigen receptor signaling in B lymphocytes and determines thresholds for negative selection. *Immunity* 1995;**2**:1–20.

134. Clark EA, Shultz LD, Pollack SB. Mutations in mice that influence natural killer (NK) cell activity. *Immunogenetics* 1981;**12**:601–13.

135. Nakamura MC, Niemi EC, Fisher MJ, Shultz LD, Seaman WE, Ryan JC. Mouse Ly-49A interrupts early signaling events in natural

killer cell cytotoxicity and functionally associates with the SHP-1 tyrosine phosphatase. *J Exp Med* 1997;**185**(4):673–84.

136. Shultz LD, Bailer CL, Coman DR. Hematopoietic stem cell function in motheaten mice. *Exp Hematol* 1983;**11**:667–80.

137. Shultz LD, Coman DR, Bailey CL, Beamer WG, Sidman CL. "Viable motheaten," a new allele at the motheaten locus. *Am J Pathol* 1984;**116**:179–92.

138. van Zant G, Shultz L. Hematologic abnormalities of the immunodeficient mouse mutant, viable motheaten (*me^v*). *Exp Hematol* 1989;**17**:81–7.

139. Paulson RF, Vesely S, Siminovitch KA, Bernstein A. Signalling by the W/Kit receptor tyrosine kinase is negatively regulated in vivo by the protein tyrosine phosphatase Shp1. *Nat Genet* 1996;**13**:309–15.

140. Lorenz U, Bergemann AD, Steinberg HN, Flanagan JG, Li X, Galli SJ, et al. Genetic analysis reveals cell type-specific regulation of receptor tyrosine kinase c-Kit by the protein tyrosine phosphatase SHP1. *J Exp Med* 1996;**184**:1111–26.

141. Croker BA, Lawson BR, Berger M, Eidenschenk C, Blasius AL, Moresco EM, et al. Inflammation and autoimmunity caused by a SHP1 mutation depend on IL-1, MyD88, and a microbial trigger. *Proc Natl Acad Sci USA* 2008;**105**(39):15,028–33.

142. Pao LI, Lam KP, Henderson JM, Kutok JL, Alimzhanov M, Nitschke L, et al. B cell-specific deletion of protein-tyrosine phosphatase Shp1 promotes B-1a cell development and causes systemic autoimmunity. *Immunity* 2007;**27**(1):35–48.

143. Duffy JB, Perrimon N. The torso pathway in *Drosophila*: lessons on receptor tyrosine kinase signaling and pattern formation. *Dev Biol* 1994;**166**:380–95.

144. Furriols M, Casanova J. In and out of Torso RTK signalling. *EMBO J* 2003;**22**(9):1947–52.

145. Allard JD, Chang HC, Herbst R, McNeill H, Simon MA. The SH2-containing tyrosine phosphatase corkscrew is required during signaling by sevenless, Ras1 and Raf. *Development* 1996;**122**:1137–46.

146. Hamlet MRJ, Perkins LA. Analysis of Corkscrew signaling in the Drosophila epidermal growth factor receptor pathway during myogenesis. *Genetics* 2001;**159**:1073–87.

147. Schutzman JL, Borland CZ, Newman JC, Robinson MK, Kokel M, Stern MJ. The *Caenorhabditis elegans* EGL-15 signaling pathway implicates a DOS-like multisubstrate adaptor protein in Fibroblast Growth Factor signal transduction. *Mol Cell Biol* 2001;**21**:8104–16.

148. Hopper NA. The adaptor protein soc-1/Gab1 modifies growth factor receptor output in *Caenorhabditis elegans*. *Genetics* 2006;**173**(1):163–75.

149. Arrandale JM, Gore-Willse A, Rocks S, Ren JM, Zhu J, Davis A, et al. Insulin signaling in mice expressing reduced levels of Syp. *J Biol Chem* 1996;**271**:21,353–21,358.

150. Saxton TM, Henkemeyer M, Gasca S, Shen R, Rossi DJ, Shalaby F, et al. Abnormal mesoderm patterning in mouse embryos mutant for the SH2 tyrosine phosphatase Shp-2. *EMBO J* 1997;**16**(9):2352–64.

151. Saxton T, Pawson T. Morphogenetic movements at gastrulation require the SH2 tyrosine phosphatase Shp2. *Proc Natl Acad Sci USA* 1999;**96**(7):3790–5.

152. Yamaguchi TP, Harpal K, Henkemeyer M, Rossant J. Fgfr-1 is required for embryonic growth and mesodermal patterning during mouse gastrulation. *Genes Dev* 1994;**8**:3032–44.

153. Qu CK, Yu WM, Azzarelli B, Cooper S, Broxmeyer HE, Feng GS. Biased suppression of hematopoiesis and multiple developmental defects in chimeric mice containing Shp-2 mutant cells. *Mol Cell Biol* 1998;**18**(10):6075–82.

154. Saxton T, Ciruna B, Holmyard D, Kulkarni S, Harpal K, Rossant J, et al. The SH2 tyrosine phosphatase Shp2 is required for mammalian limb development. *Nat Genet* 2000;**24**:420–3.

155. Qu CK, Shi ZQ, Shen R, Tsai FY, Orkin SH, Feng GS. A deletion mutation in the SH2-N domain of Shp-2 severely suppresses hematopoietic cell development. *Mol Cell Biol* 1997;**17**(9):5499–507.

156. Chan RJ, Johnson SA, Li Y, Yoder MC, Feng GS. A definitive role of Shp-2 tyrosine phosphatase in mediating embryonic stem cell differentiation and hematopoiesis. *Blood* 2003;**102**(6):2074–80.

157. Tauchi T, Feng G-S, Marshall MS, Shen R, Mantel C, Pawson T, et al. The ubiquitously expressed Syp phosphatase interacts with c-kit and Grb2 in hematopoietic cells. *J Biol Chem* 1994;**269**:25,206–11.

158. Shi ZQ, Yu DH, Park M, Marshall M, Feng GS. Molecular mechanism for the Shp-2 tyrosine phosphatase function in promoting growth factor stimulation of Erk activity. *Mol Cell Biol* 2000;**20**:1526–36.

159. Yang W, Klaman LD, Chen B, Araki T, Harada H, Thomas SM, et al. An Shp2/SFK/Ras/Erk signaling pathway controls trophoblast stem cell survival. *Dev Cell* 2006;**10**(3):317–27.

160. Stephens LE, Sutherland AE, Klimanskaya IV, Andrieux A, Meneses J, Pedersen RA, et al. Deletion of b1 integrins in mice results in inner cell mass failure and peri-implantation lethality. *Genes Dev* 1995;**9**:1883–95.

161. Zhang EE, Chapeau E, Hagihara K, Feng GS. Neuronal Shp2 tyrosine phosphatase controls energy balance and metabolism. *Proc Natl Acad Sci USA* 2004;**101**(45):16,064–9.

162. Bates SH, Stearns WH, Dundon TA, Schubert M, Tso AWK, Wang Y, et al. STAT3 signaling is required for leptin regulation of energy balance but not reproduction. *Nature* 2003;**421**:856–9.

163. Krajewska M, Banares S, Zhang EE, Huang X, Scadeng M, Jhala US, et al. Development of diabesity in mice with neuronal deletion of Shp2 tyrosine phosphatase. *Am J Pathol* 2008;**172**(5):1312–24.

164. Ke Y, Zhang EE, Hagihara K, Wu D, Pang Y, Klein R, et al. Deletion of Shp2 in the brain leads to defective proliferation and differentiation in neural stem cells and early postnatal lethality. *Mol Cell Biol* 2007;**27**(19):6706–17.

165. Gauthier AS, Furstoss O, Araki T, Chan R, Neel BG, Kaplan DR, et al. Control of CNS cell-fate decisions by SHP-2 and its dysregulation in Noonan syndrome. *Neuron* 2007;**54**(2):245–62.

166. Fornaro M, Burch PM, Yang W, Zhang L, Hamilton CE, Kim JH, et al. SHP-2 activates signaling of the nuclear factor of activated T cells to promote skeletal muscle growth. *J Cell Biol* 2006;**175**(1):87–97.

167. Kontaridis MI, Yang W, Bence KK, Cullen D, Wang B, Bodyak N, et al. Deletion of Ptpn11 (Shp2) in cardiomyocytes causes dilated cardiomyopathy via effects on the extracellular signal-regulated kinase/mitogen-activated protein kinase and RhoA signaling pathways. *Circulation* 2008;**117**(11):1423–35.

168. Ke Y, Lesperance J, Zhang EE, Bard-Chapeau EA, Oshima RG, Muller WJ, et al. Conditional deletion of Shp2 in the mammary gland leads to impaired lobulo-alveolar outgrowth and attenuated STAT5 activation. *J Biol Chem* 2006;**281**(45):34,374–80.

169. Bard-Chapeau EA, Yuan J, Droin N, Long S, Zhang EE, Nguyen TV, et al. Concerted functions of Gab1 and Shp2 in liver regeneration and hepatoprotection. *Mol Cell Biol* 2006;**26**(12):4664–74.

170. Nguyen TV, Ke Y, Zhang EE, Feng GS. Conditional deletion of Shp2 tyrosine phosphatase in thymocytes suppresses both pre-TCR and TCR signals. *J Immunol* 2006;**177**(9):5990–6.

171. Chen HE, Chang S, Trub T, Neel BG. Regulation of colony-stimulating factor 1 receptor signaling by the SH2 domain-containing tyrosine phosphatse SHPTP1. *Mol Cell Biol* 1996;**16**:3685–97.

172. Berg KL, Siminovitch KA, Stanley ER. SHP-1 regulation of p62(DOK) tyrosine phosphorylation in macrophages. *J Biol Chem* 1999;**274**:35,855–65.

173. Jiao H, Yang W, Berrada K, Tabrizi M, Shultz L, Yi T. Macrophages from motheaten and viable motheaten mutant mice show increased proliferative responses to GM-CSF: detection of potential HCP substrates in GM-CSF signal transduction. *Exp Hematol* 1997;**25**(7): 592–600.

174. Carpino N, Wisniewski D, Strife A, Marshak D, Kobayashi R, Stillman B, et al. p62(dok): a constitutively tyrosine-phosphorylated, GAP-associated protein in chronic myelogenous leukemia progenitor cells. *Cell* 1997;**88**:197–204.

175. Di Cristofano A, Niki A, Zhao M, Karnell FG, Clarkson B, Pear WS, et al. p62^{dok}, a negative regulator of Ras and mitogen-activated protein kinase (MAPK) activity, opposes leukemogenesis by p210$^{bcr-abl}$. *J Exp Med* 2001;**194**:275–84.

176. Yamanashi Y, Baltimore D. Identification of the Abl- and ras-GAP-associated 62-kDa protein as a docking protein, Dok. *Cell* 1997;**88**(2):205–11.

177. Berg KL, Carlberg K, Rohrschneider LR, Siminovitch KA, Stanley ER. The major SHP-1-binding tyrosine phosphorylated protein in macrophages is a member of the KIR/LIR family and an SHP-1 substrate. *Oncogene* 1998;**17**:2535–41.

178. Saginario C, Sterling H, Beckers C, Kobayashi R, Solimena M, Ullu E, et al. MFR, a putative receptor mediating the fusion of macrophages. *Mol Cell Biol* 1998;**18**(11):6213–23.

179. Veillette A, Thibaudeau E, Latour S. High expression of inhibitory receptor Shps-1 and its association with protein-tyrosine phosphatase Shp-1 in macrophages. *J Biol Chem* 1998;**273**:22,719–28.

180. Timms JF, Swanson KD, Marie-Cardine A, Raab M, Rudd CE, Schraven B, et al. SHPS-1 is a scaffold for assembling distinct adhesion-regulated multi-protein complexes in macrophages. *Curr Biol* 1999;**9**:927–30.

181. An H, Hou J, Zhou J, Zhao W, Xu H, Zheng Y, et al. Phosphatase SHP-1 promotes TLR- and RIG-I-activated production of type I interferon by inhibiting the kinase IRAK1. *Nat Immunol* 2008;**9**(5): 542–50.

182. Hardin AO, Meals EA, Yi T, Knapp KM, English BK. SHP-1 inhibits LPS-mediated TNF and iNOS production in murine macrophages. *Biochem Biophys Res Commun* 2006;**342**(2):547–55.

183. Zhang Z, Jimi E, Bothwell AL. Receptor activator of NF-kappa B ligand stimulates recruitment of SHP-1 to the complex containing TNFR-associated factor 6 that regulates osteoclastogenesis. *J Immunol* 2003;**171**(7):3620–6.

184. Gottipati S, Rao NL, Fung-Leung WP. IRAK1: a critical signaling mediator of innate immunity. *Cell Signal* 2008;**20**(2):269–76.

185. Uematsu S, Sato S, Yamamoto M, Hirotani T, Kato H, Takeshita F, et al. Interleukin-1 receptor-associated kinase-1 plays an essential role for Toll-like receptor (TLR)7- and TLR9-mediated interferon-α induction. *J Exp Med* 2005;**201**(6):915–23.

186. Diebold BA, Bokoch GM. Rho GTPases and the control of the oxidative burst in polymorphonuclear leukocytes. *Curr Top Microbiol Immunol* 2005;**291**:91–111.

187. David III M, Petricoin E, Benjamin C, Pine R, Weber MJ, Larner AC. Requirement for MAP kinase (ERK2) activity in interferon alpha- and interferon beta-stimulated gene expression through STAT proteins. *Science* 1995;**269**(5231):1721–3.

188. Roach TI, Slater SE, White LS, Zhang X, Majerus PW, Brown EJ, et al. The protein tyrosine phosphatase SHP-1 regulates integrin-mediated adhesion of macrophages. *Curr Biol* 1998;**8**(18):1035–8.

189. Roach T, Slater S, Koval M, White L, Mcfarland E, Okumura M, et al. CD45 regulates src family member kinase activity associated with macrophage integrin-mediated adhesion. *Curr Biol* 1997;**7**:408–17.

190. Griffiths EK, Krawczyk C, Kong YY, Raab M, Hyduk SJ, Bouchard D, et al. Positive regulation of T cell activation and integrin adhesion by the adapter Fyb/Slap. *Science* 2001;**293**:2260–3.

191. Hunter AJ, Ottoson N, Boerth N, Koretzky GA, Shimizu Y. A novel function for the SLAP-130/FYB adapter protein in b1 integrin-signalling and T lymphocyte migration. *J Immunol* 2000;**164**:1143–7.

192. Peterson EJ, Woods ML, Dmowski SA, Derimanov G, Jordan MS, Wu JN, et al. Coupling of the TCR to integrin activation by SLAP130/Fyb. *Science* 2001;**293**:2263–5.

193. Krause M, Sechi AS, Konradt M, Monner D, Gertler FB, Wehland J. Fyn-binding protein (Fub)/SLP-76-associated protein (SLAP), Ena/Vasodilator-stimulated phosphoprotein (VASP) proteins and the Arp2/3 complex link T cell receptor (TCR) signaling to the actin cytoskeleton. *J Cell Biol* 2000;**149**:181–94.

194. Simeoni L, Kliche S, Lindquist J, Schraven B. Adaptors and linkers in T and B cells. *Curr Opin Immunol* 2004;**16**(3):304–13.

195. Ilic D, Furuta Y, Kanazawa S, Takeda N, Sobue K, Nakatsuji N, et al. Reduced cell motility and enhanced focal adhesion contact formation in cells from FAK-deficient mice. *Nature* 1995;**377**(6549):539–44.

196. Mitra SK, Schlaepfer DD. Integrin-regulated FAK-Src signaling in normal and cancer cells. *Curr Opin Cell Biol* 2006;**18**(5):516–23.

197. Inagaki K, Yamao T, Noguchi T, Matozaki T, Fukunaga K, Takada T, et al. SHPS-1 regulates integrin-mediated cytoskeletal reorganization and cell motility. *EMBO J* 2000;**19**(24):6721–31.

198. Kim CH, Qu C-K, Hangoc G, Cooper S, Anzai N, Feng G-S, et al. Abnormal chemokine-induced responses of immature and mature hematopoietic cells from motheaten mice implicate the protein-tyrosine phosphatase SHP-1 in chemokine responses. *J Exp Med* 1999;**190**:681–90.

199. Dikic I, Tokiwa G, Lev S, Courtneidge SA, Schlessinger J. A role for Pyk2 and Src in linking G-coupled receptors with MAP kinase activation. *Nature* 1996;**383**:547–50.

200. Gresham HD, Dale BM, Potter JW, Chang PW, Vines CM, Lowell CA, et al. Negative regulation of phagocytosis in murine macrophages by the Src kinase family member, Fgr. *J Exp Med* 2000;**191**(3):515–28.

201. Jiang P, Lagenaur CF, Narayanan V. Integrin-associated protein is a ligand for the P84 neural adhesion molecule. *J Biol Chem* 1999;**274**(2):559–62.

202. Lindberg FP, Gresham HD, Schwarz E, Brown EJ. Molecular cloning of integrin–associated protein: an immunoglobulin family member with multiple membrane-spanning domains implicated in avb3-dependent ligand binding. *J Cell Biol* 1993;**123**:485–96.

203. Oldenborg PA, Zheleznyak A, Fang YF, Lagenaur CF, Gresham HD, Lindgerg FP. Role of CD47 as a marker of self on red blood cells. *Science* 2000;**288**:2051–4.

204. Oldenborg P-A, Gresham HD, Lindberg FP. CD47-Signal Regulatory Protein a (SIRPa) regulates Fcg and complement receptor-mediated phagocytosis. *J Exp Med* 2001;**193**:855–61.

205. Yamao T, Noguchi T, Takeuchi O, Nishiyama U, Morita H, Hagiwara T, et al. Negative regulation of platelet clearance and of the macrophage phagocytic response by the transmembrane glycoprotein SHPS-1. *J Biol Chem* 2002;**277**(42):39,833–9.

206. Takenaka K, Prasolava TK, Wang JC, Mortin-Toth SM, Khalouei S, Gan OI, et al. Polymorphism in Sirpa modulates engraftment of human hematopoietic stem cells. *Nat Immunol* 2007;**8**(12):1313–23.

207. Ide K, Wang H, Tahara H, Liu J, Wang X, Asahara T, et al. Role for CD47-SIRPalpha signaling in xenograft rejection by macrophages. *Proc Natl Acad Sci USA* 2007;**104**(12):5062–6.

208. Tapley P, Shevde NK, Schweitzer PA, Gallina M, Christianson SW, Lin IL, et al. Increased G-CSF responsiveness of bone marrow cells from hematopoietic cell phosphatase deficient viable motheaten mice. *Exp Hematol* 1997;**25**(2):p122–31.

209. Ward AC, Oomen SPMA, Smith L, Gits J, van Leeuwen D, Soede-Bobok AA, et al. The SH2 domain-containing protein tyrosine phosphatase SHP-1 is induced by granulocyte colony-stimulating factor (G-CSF) and modulates signaling from the G-CSF receptor. *Leukemia* 2000;**14**:1284–91.

210. Dong F, Qiu Y, Yi T, Touw IP, Larner AC. The carboxyl terminus of the granulocyte colony-stimulating factor receptor, truncated in patients with severe congenital neutropenia/acute myeloid leukemia, is required for SH2-containing phosphatase-1 suppression of STAT activation. *J Immunol* 2001;**167**:6447–52.

211. Jia SH, Parodo J, Kapus A, Rotstein OD, Marshall JC. Dynamic regulation of neutrophil survival through tyrosine phosphorylation or dephosphorylation of caspase-8. *J Biol Chem* 2008;**283**(9):5402–13.

212. Kruger J, Butler JR, Cherapanov V, Dong Q, Ginzberg H, Govindarajan A, et al. Deficiency of Src homology 2-containing phosphatase 1 results in abnormalities in murine neutrophil function: studies in motheaten mice. *J Immunol* 2000;**165**:5847–9.

213. Kautz B, Kakar R, David E, Eklund EA. SHP1 protein-tyrosine phosphatase inhibits gp91PHOX and p67PHOX expression by inhibiting interaction of PU.1, IRF1, interferon consensus sequence-binding protein, and CREB-binding protein with homologous Cis elements in the CYBB and NCF2 genes. *J Biol Chem* 2001;**276**(41):37, 868–78.

214. Eklund EA, Goldenberg I, Lu Y, Andrejic J, Kakar R. SHP1 protein-tyrosine phosphatase regulates HoxA10 DNA binding and transcriptional repression activity in undifferentiated myeloid cells. *J Biol Chem* 2002;**277**(39):36,878–88.

215. Huang W, Saberwal G, Horvath E, Zhu C, Lindsey S, Eklund EA. Leukemia-associated, constitutively active mutants of SHP2 protein tyrosine phosphatase inhibit NF1 transcriptional activation by the interferon consensus sequence binding protein. *Mol Cell Biol* 2006;**26**(17):6311–32.

216. Lindsey S, Huang W, Wang H, Horvath E, Zhu C, Eklund EA. Activation of SHP2 protein-tyrosine phosphatase increases HoxA10-induced repression of the genes encoding gp91(PHOX) and p67(PHOX). *J Biol Chem* 2007;**282**(4):2237–49.

217. Long EO. Regulation of immune responses through inhibitory receptors. *Annu Rev Immunol* 1999;**17**:875–905.

218. Pao LI, Badour K, Siminovitch KA, Neel BG. Nonreceptor protein-tyrosine phosphatases in immune cell signaling. *Annu Rev Immunol* 2007;**25**:473–523.

219. Siminovitch KA, Neel BG. Regulation of B cell signal transduction by SH2-containing protein-tyrosine phosphatases. *Semin Immunol* 1998;**10**(4):329–47.

220. Zhang J, Somani A-K, Siminovitch KA. Role of the Shp-1 tyrosine phosphatase in the negative regulation of cell signaling. *Sem Immunol* 2000;**12**:361–78.

221. Billadeau DD, Leibson PJ. ITAMs versus ITIMs: striking a balance during cell regulation. *J Clin Invest* 2002;**109**:161–8.

222. Tamir I, Dal Porto JM, Cambier JC. Cytoplasmic protein tyrosine phosphatases SHP-1 and SHP-2: regulators of B cell signal transduction. *Curr Opin Immunol* 2000;**12**:307–15.

223. Allman D, Pillai S. Peripheral B cell subsets. *Curr Opin Immunol* 2008;**20**(2):149–57.

224. Nitschke L, Carsetti R, Ocker B, Kohler G, Lamers MC. CD22 is a negative regulator of B cell receptor signaling. *Curr Biol* 1997;**7**:133–43.

225. O'Keefe TL, Williams GT, Davies SL, Neuberger MS. Hyperresponsive B cells in CD22-deficient mice. *Science* 1996;**274**(5288):798–801.

226. Otipoby KL, Andersson KB, Draves KE, Klaus SJ, Farr AG, Kerner JD, et al. CD22 regulates thymus-independent responses and the lifespan of B cells. *Nature* 1996;**384**(6610):634–7.

227. Pan C, Baumgarth N, Parnes JR. CD72-deficient mice reveal non-redundant roles of CD72 in B cell development and activation. *Immunity* 1999;**11**:495–506.

228. Bikah G, Carey J, Ciallella JR, Tarakhovsky A, Bondada S. CD5-mediated negative regulation of antigen receptor-induced growth signals in B-1 B cells. *Science* 1996;**274**:1906–9.

229. Haga CL, Ehrhardt GR, Boohaker RJ, Davis RS, Cooper MD. Fc receptor-like 5 inhibits B cell activation via SHP-1 tyrosine phosphatase recruitment. *Proc Natl Acad Sci USA* 2007;**104**(23):9770–5.

230. Ono M, Bolland S, Tempst P, Ravetch JV. Role of the inositol phosphatase SHIP in negative regulation of the immune system by the receptor FcgRIIB. *Nature* 1996;**383**:263–6.

231. Ono M, Okada H, Bolland S, Yanagi S, Kurosaki T, Ravetch JV. Deletion of SHIP or SHP-1 reveals two distinct pathways for inhibitory signaling. *Cell* 1997;**90**:293–301.

232. Nadler MJ, Chen B, Anderson S, Wortis H, Neel BG. Protein-tyrosine phosphatase SHP-1 is dispensible for FcgRIIB-mediated inhibition of B cell antigen receptor activation. *J Biol Chem* 1997;**272**:20,038–43.

233. Gupta N, Scharenberg AM, Burshtyn DN, Wagtmann N, Lioubin MN, Rohrschneider LA, et al. Negative signaling pathways of the killer cell inhibitory receptor and Fc gamma RIIb1 require distinct phosphatases. *J Exp Med* 1997;**186**:473–8.

234. Sato S, Miller AS, Inaoki M, Bock CB, Jansen PJ, Tang LK, et al. CD22 is both a positive and negative regulator of lymphocyte antigen receptor signal transduction: altered signaling in CD22-deficient mice. *Immunity* 1996;**5**:551–62.

235. Nadler MJS, McLean PA, Neel BG, Wortis HH. B cell antigen receptor-evoked calcium influx is enhanced in CD22-deficient B cell lines. *J Immunol* 1997;**159**:4233–43.

236. Chen J, McLean PA, Neel BG, Okunade G, Shull GE, Wortis HH. CD22 attenuates calcium signaling by potentiating plasma membrane calcium-ATPase activity. *Natur Immunol* 2004;**5**:651–7.

237. Dean WL, Chen D, Brandt PC, Vanaman TC. Regulation of platelet plasma membrane Ca^{2+}-ATPase by cAMP-dependent and tyrosine phosphorylation. *J Biol Chem* 1997;**272**(24):15,113–19.

238. Inoue O, Suzuki-Inoue K, Dean WL, Frampton J, Watson SP. Integrin alpha2beta1 mediates outside-in regulation of platelet spreading on collagen through activation of Src kinases and PLCgamma2. *J Cell Biol* 2003;**160**(5):769–80.

239. Wan TC, Zabe M, Dean WL. Plasma membrane Ca^{2+}-ATPase isoform 4b is phosphorylated on tyrosine 1176 in activated human platelets. *Thromb Haemost* 2003;**89**(1):122–31.

240. Mizuno K, Tagawa Y, Mitomo K, Arimura Y, Hatano N, Katagiri T, et al. Src homology region 2 (SH2) domain-containing phosphatase-1 dephosphorylates B cell linker protein/SH2 domain leukocyte protein of 65 kDa and selectively regulates c-Jun NH2-terminal kinase activation in B cells. *J Immunol* 2000;**165**:1344–51.

241. Sato S, Jansen PJ, Tedder TF. CD19 and CD22 expression reciprocally regulates tyrosine phosphorylation of Vav protein during B lymphocyte signaling. *Proc Natl Acad Sci USA* 1997;**94**:13,158–62.

242. Kon-Kozlowski M, Pani G, Pawson T, Siminovitch KA. The tyrosine phosphatase PTP1C associates with Vav, Grb 2, and mSos1 in hematopoietic cells. *J Biol Chem* 1996;**271**:3856–62.

243. Stebbins CC, Watzl C, Billadeau BD, Leibson PJ, Burshtyn DN, Long EO. Vav1 dephosphorylation by the tyrosine phosphatase SHP-1 as a mechanism for inhibition of cellular cytotoxicity. *Mol Cell Biol* 2003;**23**:6291–9.

244. Binstadt B, Billadeau D, Jevremovic D, Williams B, Fang N, Yi T, et al. SLP-76 is a direct substrate of SHP-1 recruited to killer cell inhibitory receptors. *J Biol Chem* 1998;**273**(42):27,518–23.

245. Mizuno K, Tagawa Y, Watanabe N, Ogimoto M, Yakura H. SLP-76 is recruited to CD22 and dephosphorylated by SHP-1, thereby regulating B cell receptor-induced c-Jun N-terminal kinase activation. *Eur J Immunol* 2005;**35**(2):644–54.

246. Cuevas B, Lu Y, Watt S, Kumar R, Zhang J, Siminovitch KA, et al. SHP-1 regulates Lck-induced phosphatidylinositol 3-kinase phosphorylation and activity. *J Biol Chem* 1999;**274**(39):27,583–9.

247. Maeda A, Scharenberg AM, Tsukada S, Bolen JB, Kinet JP, Kurosaki T. Paired immunoglobulin-like receptor B (PIR-B) inhibits BCR-induced activation of Syk and Btk by SHP-1. *Oncogene* 1999;**18**(14):2291–7.

248. Dustin LB, Plas DR, Wong J, Hu YT, Soto C, Chan AC, et al. Expression of dominant negative src-homology domain 2-containing protein-tyrosine phosphatase-1 results in increased Syk tyrosine kinase activity and B cell activation. *J Immunol* 1999;**162**:2717–24.

249. Yu Z, Maoui M, Zhao ZJ, Li Y, Shen SH. SHP-1 dephosphorylates 3BP2 and potentially downregulates 3BP2-mediated T cell antigen receptor signaling. *FEBS J* 2006;**273**(10):2195–205.

250. Baba T, Fusaki N, Shinya N, Iwamatsu A, Hozumi N. Myosin is an in vivo substrate of the protein tyrosine phosphatase (SHP-1) after mIgM cross-linking. *Biochem Biophys Res Commun* 2003;**304**(1):67–72.

251. Blery M, Kubagawa H, Chen C-C, Vely F, Cooper MD, Vivier E. PIR-B is an inhibitory receptor that recruits SHP-1. *Proc Natl Acad Sci* 1998;**95**(5):2446–51.

252. Hayami K, Fukuta D, Nishikawa Y, Yamashita Y, Inui M, Ohyama Y, et al. Molecular cloning of a novel murine cell-surface glycoprotein homologous to killer cell inhibitory receptors. *J Biol Chem* 1997;**272**:7320–7.

253. Perez-Villar JJ, Whitney GS, Bowen MA, Hewgill DH, Aruffo AA, Kanner SB. CD5 negatively regulates the T-cell antigen receptor signal transduction pathway: involvement of SH2-containing phosphotyrosine phosphatase SHP-1. *Mol Cell Biol* 1999;**19**(4):2903–12.

254. Bellon T, Kitzig F, Sayos J, Lopez-Botet M. Mutational analysis of immunoreceptor tyrosine-based inhibition motifs of the Ig-like transcript 2 (CD85j) leukocyte receptor. *J Immunol* 2002;**168**(7):3351–9.

255. Chen Z, Chen L, Qiao S-W, Nagaishi T, Blumberg RS. Carcinoembryonic antigen-related cell adhesion molecule 1 inhibits proximal TCR signaling by targeting ZAP-70. *J Immunol* 2008;**180**(9):6085–93.

256. Nagaishi T, Pao L, Lin SH, Iijima H, Kaser A, Qiao SW, et al. SHP1 phosphatase-dependent T cell inhibition by CEACAM1 adhesion molecule isoforms. *Immunity* 2006;**25**(5):769–81.

257. Watanabe N, Gavrieli M, Sedy JR, Yang J, Fallarino F, Loftin SK, et al. BTLA is a lymphocyte inhibitory receptor with similarities to CTLA-4 and PD-1. *Nat Immunol* 2003;**4**(7):670–9.

258. Chemnitz JM, Parry RV, Nichols KE, June CH, Riley JL. SHP-1 and SHP-2 associate with immunoreceptor tyrosine-based switch motif of programmed death 1 upon primary human T cell stimulation, but only receptor ligation prevents T cell activation. *J Immunol* 2004;**173**(2):945–54.

259. Parry RV, Chemnitz JM, Frauwirth KA, Lanfranco AR, Braunstein I, Kobayashi SV, et al. CTLA-4 and PD-1 receptors inhibit T-cell activation by distinct mechanisms. *Mol Cell Biol* 2005;**25**(21):9543–53.

260. Plas DR, Johnson R, Pingel JT, Matthews RJ, Dalton M, Roy G, et al. Direct regulation of ZAP-70 by SHP-1 in T cell antigen receptor signaling. *Science* 1996;**272**:1173–6.

261. Sozio MS, Mathis MA, Young JA, Walchli S, Pitcher LA, Wrage PC, et al. PTPH1 is a predominant protein-tyrosine phosphatase capable of interacting with and dephosphorylating the T cell receptor zeta subunit. *J Biol Chem* 2004;**279**(9):7760–9.

262. Chiang GG, Sefton BM. Specific dephosphorylation of the Lck tyrosine kinase at Tyr 394 by the SHP-1 protein tyrosine phosphatase. *J Biol Chem* 2001;**276**:23,173–8.

263. Raab M, Rudd CE. Hematopoietic cell phosphatase (HCP) regulates p56LCK phosphorylation and ZAP-70 binding to T cell receptor zeta chain. *Biochem Biophys Res Commun* 1996;**222**(1):50–7.

264. Stefanova I, Hemmer B, Vergelli M, Martin R, Biddison WE, Germain RN. TCR ligand discrimination is enforced by competing ERK positive and SHP-1 negative feedback pathways. *Nat Immunol* 2003;**4**(3):248–54.

265. Feinerman O, Veiga J, Dorfman JR, Germain RN, Altan-Bonnet G. Variability and robustness in T cell activation from regulated heterogeneity in protein levels. *Science* 2008;**321**(5892):1081–4.

266. Sathish JG, Dolton G, Leroy FG, Matthews RJ. Loss of Src homology region 2 domain-containing protein tyrosine phosphatase-1 increases CD8+ T cell-APC conjugate formation and is associated with enhanced in vivo CTL function. *J Immunol* 2007;**178**(1):330–7.

267. Keir ME, Sharpe AH. The B7/CD28 costimulatory family in autoimmunity. *Immunol Rev* 2005;**204**:128–43.

268. Avril T, Floyd H, Lopez F, Vivier E, Crocker PR. The membrane-proximal immunoreceptor tyrosine-based inhibitory motif is critical for the inhibitory signaling mediated by Siglecs-7 and -9, CD33-related Siglecs expressed on human monocytes and NK cells. *J Immunol* 2004;**173**(11):6841–9.

269. Binstadt BA, Brumbaugh KM, Dick CJ, Scharenberg AM, Williams BL, Colonna M, et al. Sequential involvement of Lck and SHP-1 with MHC-recognizing receptors on NK cells inhibits FcR-initiated tyrosine kinase activation. *Immunity* 1996;**5**:629–38.

270. Yu MC, Su LL, Zou L, Liu Y, Wu N, Kong L, et al. An essential function for beta-arrestin 2 in the inhibitory signaling of natural killer cells. *Nat Immunol* 2008;**9**(8):898–907.

271. Migone TS, Cacalano NA, Taylor N, Yi T, Waldmann TA, Johnston JA. Recruitment of SH2-containing protein tyrosine phosphatase SHP-1 to the interleukin 2 receptor; loss of SHP-1 expression in human T-lymphotropic virus type I-transformed T cells. *Proc Natl Acad Sci USA* 1998;**95**(7):3845–50.

272. Liu Y, Kruhlak MJ, Hao JJ, Shaw S. Rapid T cell receptor-mediated SHP-1 S591 phosphorylation regulates SHP-1 cellular localization and phosphatase activity. *J Leukoc Biol* 2007;**82**(3):742–51.

273. Sankarshanan M, Ma Z, Iype T, Lorenz U. Identification of a novel lipid raft-targeting motif in Src homology 2-containing phosphatase 1. *J Immunol* 2007;**179**(1):483–90.

274. Matko J, Bodnar A, Vereb G, Bene L, Vamosi G, Szentesi G, et al. GPI-microdomains (membrane rafts) and signaling of the multichain interleukin-2 receptor in human lymphoma/leukemia T cell lines. *Eur J Biochem* 2002;**269**(4):1199–208.

275. Huang Z, Coleman JM, Su Y, Mann M, Ryan J, Shultz LD, et al. SHP-1 regulates STAT6 phosphorylation and IL-4-mediated function in a cell type-specific manner. *Cytokine* 2005;**29**(3):118–24.

276. Kashiwada M, Giallourakis CC, Pan P-Y, Rothman PB. Immunoreceptor tyrosine-based inhibitory motif of the IL-4 receptor associates with SH2-containing phosphatases and regulates IL-4-induced proliferation. *J Immunol* 2001;**167**:6382–7.

277. Kamata T, Yamashita M, Kimura M, Murata K, Inami M, Shimizu C, et al. src homology 2 domain-containing tyrosine phosphatase SHP-1 controls the development of allergic airway inflammation. *J Clin Invest* 2003;**111**(1):109–19.

278. Deng C, Minguela A, Hussain RZ, Lovett-Racke AE, Radu C, Ward ES, et al. Expression of the tyrosine phosphatase SRC homology 2 domain-containing protein tyrosine phosphatase 1 determines T cell activation threshold and severity of experimental autoimmune encephalomyelitis. *J Immunol* 2002;**168**(9):4511–18.

279. Yu WM, Wang S, Keegan AD, Williams MS, Qu CK. Abnormal Th1 cell differentiation and IFN-gamma production in T lymphocytes from motheaten viable mice mutant for Src homology 2 domain-containing protein tyrosine phosphatase-1. *J Immunol* 2005;**174**(2): 1013–19.

280. Taylor A, Akdis M, Joss A, Akkoç T, Wenig R, Colonna M, et al. IL-10 inhibits CD28 and ICOS costimulations of T cells via src homology 2 domain-containing protein tyrosine phosphatase 1. *J Allergy Clin Immunol* 2007;**120**(1):76–83.

281. Park I-K, Shultz LD, Letterio JJ, Gorham JD. TGF-β1 inhibits T-bet induction by IFN-γ in murine CD4+ T Cells through the protein tyrosine phosphatase Src homology region 2 domain-containing phosphatase-1. *J Immunol* 2005;**175**(9):5666–74.

282. Carter JD, Calabrese GM, Naganuma M, Lorenz U. Deficiency of the Src homology region 2 domain-containing phosphatase 1 (SHP-1) causes enrichment of CD4+CD25+ regulatory T cells. *J Immunol* 2005;**174**(11):6627–38.

283. Su X, Zhou T, Wang Z, Yang P, Jope RS, Mountz JD. Defective expression of hematopoietic cell protein tyrosine phosphatase (HCP) in lymphoid cells blocks Fas-mediated apoptosis. *Immunity* 1995;**2**(4):p353–62.

284. Su X, Zhou T, Yang PA, Wang Z, Mountz JD. Hematopoietic cell protein-tyrosine phosphatase-deficient motheaten mice exhibit T cell apoptosis defect. *J Immunol* 1996;**156**:4198–208.

285. Takayama H, Lee MH, Shirota-Someya Y. Lack of requirement for SHP-1 in both Fas-mediated and perforin-mediated cell death induced by CTL. *J Immunol* 1996;**157**:3943–8.

286. Yi T, Zhang J, Miura O, Ihle JN. Hematopoietic cell phosphatase associates with erythropoietin (Epo) receptor after Epo-induced receptor tyrosine phosphorylation: identification of potential binding sites. *Blood* 1995;**85**(1):87–95.

287. Klingmuller U, Lorenz U, Cantley LC, Neel BG, Lodish HF. Specific recruitment of SH-PTP1 to the erythropoietin receptor causes inactivation of JAK2 and termination of proliferative signals. *Cell* 1995;**80**:729–38.

288. De La Chappelle A, Traskelin A-L, Juvonen E. Truncated erythropoietin receptor causes dominantly inherited benign human erythrocytosis. *Proc Natl Acad Sci USA* 1993;**90**:4495–9.

289. Zang H, Sato K, Nakajima H, McKay C, Ney PA, Ihle JN. The distal region and receptor tyrosines of the Epo receptor are non-essential for *in vivo* erythropoiesis. *EMBO J* 2001;**20**:3156–66.

290. Su L, Zhao Z, Bouchard P, Banville D, Fischer EH, Krebs EG, et al. Positive effect of overexpressed protein-tyrosine phosphatase PTP1C on mitogen-activated signaling in 293 cells. *J Biol Chem* 1996;**271**(17):10,385–90.

291. Keilhack H, Muller M, Bohmer SA, Frank C, Weidner KM, Birchmeier W, et al. Negative regulation of Ros receptor tyrosine kinase signaling. An epithelial function of the SH2 domain protein tyrosine phosphatase SHP-1. *J Cell Biol* 2001;**152**:325–34.

292. Duchesne C, Charland S, Asselin C, Nahmias C, Rivard N. Negative regulation of beta-catenin signaling by tyrosine phosphatase

SHP-1 in intestinal epithelial cells. *J Biol Chem* 2003;**278**(16): 14,274–83.

293. Lopez F, Esteve JP, Buscail L, Delesque N, Saint-Laurent N, Theveniau M, et al. The tyrosine phosphatase SHP-1 associates with the sst2 somatostatin receptor and is an essential component of sst2-mediated inhibitory growth signaling. *J Biol Chem* 1997;**272**(39):24,448–54.

294. Bousquet C, Delesque N, Lopez F, Saint-Laurent N, Esteve JP, Bedecs K, et al. sst2 somatostatin receptor mediates negative regulation of insulin receptor signaling through the tyrosine phosphatase SHP-1. *J Biol Chem* 1998;**273**(12):7099–106.

295. Lopez F, Ferjoux G, Cordelier P, Saint-Laurent N, Esteve JP, Vaysse N, et al. Neuronal nitric oxide synthase: a substrate for SHP-1 involved in sst2 somatostatin receptor growth inhibitory signaling. *FASEB J* 2001;**15**(12):2300–2.

296. Dubois MJ, Bergeron S, Kim HJ, Dombrowski L, Perreault M, Fournes B, et al. The SHP-1 protein tyrosine phosphatase negatively modulates glucose homeostasis. *Nat Med* 2006;**12**(5):549–56.

297. Zhang SQ, Tsiaras WG, Araki T, Wen G, Minichiello L, Klein R, et al. Receptor-specific regulation of phosphatidylinositol 3'-kinase activation by the protein tyrosine phosphatase Shp2. *Mol Cell Biol* 2002;**22**(12):4062–72.

298. Holgado-Madruga M, Emlet DR, Moscatello DK, Godwin AK, Wong AJ. A Grb2-associated docking protein in EGF- and insulin-receptor signalling. *Nature* 1996;**379**:560–4.

299. Weidner KM, Di Cesare S, Sachs M, Brinkmann V, Behrens J, Birechmeier W. Interaction between Gab1 and the c-Met receptor tyrosine kinase is responsible for epithelial morphogenesis. *Nature* 1996;**384**:173–6.

300. White MF, Yenush L. The IRS-signaling system: a network of docking proteins that mediate insulin and cytokine action. *Curr Top Microbiol Immunol* 1998;**228**:179–208.

301. Kouhara H, Hadari Y, Spivak-Kroizman T, Schilling J, Bar-Sagi D, Lax I, et al. A lipid-anchored Grb2-binding protein that links FGF-Receptor activation to the Ras/MAPK Signaling pathway. *Cell* 1997;**89**:693–702.

302. Itoh M, Yoshida Y, Nishida K, Narimatsu M, Hibi M, Hirano T. Role of Gab1 in heart, placenta and skin development and growth factor- and cytokine-induced extracellular signal-related kinase mitogen-activated protein kinase activation. *Mol Cell Biol* 2000;**20**:2695–704.

303. Sachs M, Brohmann H, Zechner D, Muller T, Hulsken J, Walther I, et al. Essential role of Gab1 for signaling by the c-Met receptor *in vivo*. *J Cell Biol* 2000;**150**:1375–84.

304. Hadari YR, Gotoh N, Kouhara H, Lax I, Schlessinger J. Critical role for the docking-protein FRS2 alpha in FGF receptor-mediated signal transduction pathways. *Proc Natl Acad Sci USA* 2001;**98**:8578–83.

305. Maroun CR, Naujokas MA, Holgado-Madruga M, Wong AJ, Park M. The tyrosine phosphatase SHP-2 is required for sustained activation of extracellular signal-regulated kinase and epithelial morphogenesis downstream from the Met receptor tyrosine kinase. *Mol Cell Biol* 2000;**20**:8513–25.

306. Schaeper U, Gehring NH, Fuchs KP, Sachs M, Kempkes B, Birchmeier W. Coupling of Gab1 to c-Met, Grb2, and Shp2 mediates biological responses. *J Cell Biol* 2000;**149**:1419–32.

307. Petit V, Nussbaumer U, Dossenbach C, Affolter M. Downstream-of-FGFR is a fibroblast growth factor-specific scaffolding protein and recruits Corkscrew upon receptor activation. *Mol Cell Biol* 2004;**24**(9):3769–81.

308. Noguchi T, Matozaki T, Horita K, Fujioka Y, Kasuga M. Role of SH-PTP2, a protein-tyrosine phosphatase with src homology 2 domains, in insulin-stimulated ras activation. *Mol Cell Biol* 1994;**14**: 6674–82.

309. Zhang SQ, Yang W, Kontaridis MI, Bivona TG, Wen G, Araki T, et al. Shp2 regulates SRC family kinase activity and Ras/Erk activation by controlling Csk recruitment. *Mol Cell* 2004;**13**(3):341–55.

310. Li W, Nishimura R, Kashishian A, Batzer AG, Kim WJ, Cooper JA, et al. A new function for a phosphotyrosine phosphatase: linking GRB2-Sos to a receptor tyrosine kinase. *Mol Cell Biol* 1994;**14**(1):509–17.

311. Welham MJ, Dechert U, Leslie KB, Jirik F, Schrader JW. Interleukin (IL)-3 and granulocyte/macrophage colony-stimulating factor, but not IL-4, induce tyrosine phosphorylation, activation, and association of SHPTP2 with Grb2 and phosphatidylinositol 3′-kinase. *J Biol Chem* 1994;**269**:23,764–8.

312. O'Reilly AM, Neel BG. Structural determinants of SHP-2 function and specificity in *Xenopus* mesoderm induction. *Mol Cell Biol* 1998;**18**:161–77.

313. Allard JD, Herbst R, Carroll PM, Simon MA. Mutational analysis of the SRC homology 2 domain protein-tyrosine phosphatase Corkscrew. *J Biol Chem* 1998;**273**(21):13,129–35.

314. Araki T, Nawa H, Neel BG. Tyrosyl phosphorylation of Shp2 is required for normal Erk activation in response to some but not all growth factors. *J Biol Chem* 2003;**278**:41,677–84.

315. Zheng XM, Resnick RJ, Shalloway D. A phosphotyrosine displacement mechanism for activation of Src by PTPalpha. *EMBO J* 2000;**19**(5):964–78.

316. Feng GS. Shp-2 tyrosine phosphatase: signaling one cell or many. *Exp Cell Res* 1999;**253**:47–54.

317. Klinghoffer RA, Kazlauskas A. Identification of a putative Syp substrate, the PDGF beta receptor. *J Biol Chem* 1995;**270**(38):22,208–17.

318. Cleghon V, Feldmann P, Ghiglione C, Copeland TD, Perrimon N, Hughes DA, et al. Opposing actions of CSW and RasGAP modulate the strength of Torso RTK signaling in the *Drosophila* terminal pathway. *Mol Cell* 1998;**2**:719–27.

319. Agazie YM, Hayman MJ. Molecular mechanism for a role of SHP2 in epidermal growth factor receptor signaling. *Mol Cell Biol* 2003;**23**:7875–86.

320. Herbst R, Carroll PM, Allard JD, Schilling J, Raabe T, Simon MA. Daughter of sevenless is a substrate of the phosphotyrosine phosphatase Corkscrew and functions during sevenless signaling. *Cell* 1996;**85**(6):899–909.

321. Herbst R, Zhang X, Qin J, Simon MA. Recruitment of the protein tyrosine phosphatase CSW by DOS is an essential step during signaling by the sevenless receptor tyrosine kinase. *EMBO J* 1999;**18**:6950–1.

322. Bausenwein BS, Schmidt M, Mielke B, Raabe T. In vivo functional analysis of the daughter of sevenless protein in receptor tyrosine kinase signaling. *Mech Dev* 2000;**90**:205–15.

323. Zhang SQ, Tsiaris WG, Araki T, Wen G, Minichiello L, Klein R, et al. Receptor-specific regulation of phosphatidyl 3′-kinase activation by the protein tyrosine phosphatase Shp2. *Mol Cell Biol* 2002;**22**:4062–72.

324. Yamasaki S, Nishida K, Yoshida Y, Itoh M, Hibi M, Hirano T. Gab1 is required for EGF receptor signaling and the transformation by activated ErbB2. *Oncogene* 2003;**22**:1546–56.

325. Montagner A, Yart A, Dance M, Perret B, Salles JP, Raynal P. A novel role for Gab1 and SHP2 in epidermal growth factor-induced Ras activation. *J Biol Chem* 2005;**280**(7):5350–60.

326. Cunnick JM, Mei L, Doupnik cA, Wu J. Phosphotyrosines 627 and 659 of Gab1 constitute a bisphosphoryl tyrosine-based activation motif (BTAM) conferring binding and activation of SHP2. *J Biol Chem* 2001;**276**:24,380–7.

327. Cunnick JM, Meng S, Ren Y, Desponts C, Wang HG, Djeu JY, et al. Regulation of the mitogen-activated protein kinase signaling pathway by SHP2. *J Biol Chem* 2002;**277**(11):9498–504.

328. Wu CJ, O'Rourke DM, Feng GS, Johnson GR, Wang Q, Greene MI. The tyrosine phosphatase SHP-2 is required for mediating phosphatidylinositol 3-kinase/Akt activation by growth factors. *Oncogene* 2001;**20**(42):6018–25.

329. Bertotti A, Comoglio PM, Trusolino L. Beta4 integrin activates a Shp2-Src signaling pathway that sustains HGF-induced anchorage-independent growth. *J Cell Biol* 2006;**175**(6):993–1003.

330. Mitola S, Brenchio B, Piccinini M, Tertoolen L, Zammataro L, Breier G, et al. Type I collagen limits VEGFR–2 signaling by a SHP2 protein-tyrosine phosphatase-dependent mechanism 1. *Circ Res* 2006;**98**(1):45–54.

331. Ren Y, Meng S, Mei L, Zhao ZJ, Jove R, Wu J. Roles of Gab1 and SHP2 in paxillin tyrosine dephosphorylation and Src activation in response to epidermal growth factor. *J Biol Chem* 2004;**279**:8497–505.

332. Hanafusa H, Torii S, Yasunaga T, Matsumoto K, Nishida E. Shp2, an SH2-containing protein-tyrosine phosphatase, positively regulates receptor tyrosine kinase signaling by dephosphorylating and inactivating the inhibitor Sprouty. *J Biol Chem* 2004;**279**(22):22,992–5.

333. Jarvis LA, Toering SJ, Simon MA, Krasnow MA, Smith-Bolton RK. Sprouty proteins are *in vivo* targets of Corkscrew/SHP-2 tyrosine phosphatases. *Development* 2006;**133**(6):1133–42.

334. Mason JM, Morrison DJ, Basson MA, Licht JD. Sprouty proteins: multifaceted negative-feedback regulators of receptor tyrosine kinase signaling. *Trends Cell Biol* 2006;**16**:46–54.

335. Wakioka T, Sasaki A, Kato R, Shouda T, Matsumoto A, Miyoshi K, et al. Spred is a Sprouty-related suppressor of Ras signalling. *Nature* 2001;**412**(6847):647–51.

336. Kato R, Nonami A, Taketomi T, Wakioka T, Kuroiwa A, Matsuda Y, et al. Molecular cloning of mammalian Spred-3 which suppresses tyrosine kinase-mediated Erk activation. *Biochem Biophys Res Commun* 2003;**302**(4):767–72.

337. Nonami A, Kato R, Taniguchi K, Yoshiga D, Taketomi T, Fukuyama S, et al. Spred-1 negatively regulates interleukin-3-mediated ERK/mitogen-activated protein (MAP) kinase activation in hematopoietic cells. *J Biol Chem* 2004;**279**(50):52,543–51.

338. Taniguchi K, Kohno R, Ayada T, Kato R, Ichiyama K, Morisada T, et al. Spreds are essential for embryonic lymphangiogenesis by regulating vascular endothelial growth factor receptor 3 signaling. *Mol Cell Biol* 2007;**27**(12):4541–50.

339. Burks J, Agazie YM. Modulation of alpha-catenin Tyr phosphorylation by SHP2 positively effects cell transformation induced by the constitutively active FGFR3. *Oncogene* 2006;**25**(54):7166–79.

340. Kolli S, Zito CI, Mossink MH, Wiemer EA, Bennett AM. The major vault protein is a novel substrate for the tyrosine phosphatase SHP-2 and scaffold protein in epidermal growth factor signaling. *J Biol Chem* 2004;**279**(28):29,374–85.

341. Vasioukhin V, Bauer C, Degenstein L, Wise B, Fuchs E. Hyperproliferation and defects in epithelial polarity upon conditional ablation of alpha-catenin in skin. *Cell* 2001;**104**(4):605–17.

342. Hou XS, Chou T-B, Melnick MB, Perrimon N. The torso receptor tyrosine kinase can activate Raf in a Ras-independent pathway. *Cell* 1995;**81**:63–71.

343. Lorenzen JA, Baker SE, Denhez F, Melnick MB, Brower DL, Perkins LA. Nuclear import of activated D-ERK by DIM-7, an importin family member encoded by the gene *moleskin*. *Development* 2001;**128**:1403–14.

344. Mattoon DR, Lamothe B, Lax I, Schlessinger J. The docking protein Gab1 is the primary mediator of EGF-stimulated activation of the PI-3K/Akt cell survival pathway. *BMC Biol* 2004;**2**:24.

345. Zito CI, Qin H, Blenis J, Bennett AM. SHP-2 regulates cell growth by controlling the mTOR/S6 kinase 1 pathway. *J Biol Chem* 2007;**282**(10):6946–53.

346. Argetsinger LS, Kouadio JL, Steen H, Stensballe A, Jensen ON, Carter-Su C. Autophosphorylation of JAK2 on tyrosines 221 and 570 regulates its activity. *Mol Cell Biol* 2004;**24**(11):4955–67.

347. Feener Jr. EP, Rosario F, Dunn SL, Stancheva Z, Myers MG Tyrosine phosphorylation of JAK2 in the JH2 domain inhibits cytokine signaling. *Mol Cell Biol* 2004;**24**(11):4968–78.

348. Funakoshi-Tago M, Pelletier S, Matsuda T, Parganas E, Ihle JN. Receptor specific downregulation of cytokine signaling by autophosphorylation in the FERM domain of JAK2. *EMBO J* 2006;**25**(20):4763–72.

349. Godeny MD, Sayyah J, Von Der Linden D, Johns M, Ostrov DA, Caldwell-Busby J, et al. The N-terminal SH2 domain of the tyrosine phosphatase, SHP-2, is essential for JAK2-dependent signaling via the angiotensin II type AT1 receptor. *Cell Signal* 2007;**19**(3):600–9.

350. Matsuda T, Feng J, Witthuhn BA, Sekine Y, Ihle JN. Determination of the transphosphorylation sites of JAK2 kinase. *Biochem Biophys Res Commun* 2004;**325**(2):586–94.

351. Feng J, Witthuhn BA, Matsuda T, Kohlhuber F, Kerr IM, Ihle JN. Activation of JAK2 catalytic activity requires phosphorylation of Y1007 in the kinase activation loop. *Mol Cell Biol* 1997;**17**(5):2497–501.

352. Yasukawa H, Misawa H, Sakamoto H, Masuhara M, Sasaki A, Wakioka T, et al. The JAK-binding protein JAB inhibits Janus tyrosine kinase activity through binding in the activation loop. *Embo J* 1999;**18**(5):1309–20.

353. Frantsve J, Schwaller J, Sternberg DW, Kutok J, Gilliland DG. Socs-1 inhibits TEL-JAK2-mediated transformation of hematopoietic cells through inhibition of JAK2 kinase activity and induction of proteasome-mediated degradation. *Mol Cell Biol* 2001;**21**(10):3547–57.

354. Kamizono S, Hanada t, Yasukawa H, Minoguchi S, Kato R, Minoguchi M, et al. The SOCS box of SOCS-1 accelerates ubiquitin-dependent proteolysis of TEL-JAK2. *J BiolChem* 2001;**276**:12,530–8.

355. Ungureanu D, Saharinen P, Junttila I, Hilton DJ, Silvennoinen O. Regulation of JAK2 through the ubiquitin-proteasome pathway involves phosphorylation of JAK2 on Y1007 and interaction with SOCS-1. *Mol Cell Biol* 2002;**22**(10):3316–26.

356. Ali S, Nouhi Z, Chughtai N, Ali S. SHP-2 regulates SOCS-1-mediated Janus kinase-2 ubiquitination/degradation downstream of the prolactin receptor. *J Biol Chem* 2003;**278**:52,021–31.

357. Lehmann U, Schmitz J, Weissenbach M, Sobota RM, Hortner M, Friederichs K, et al. SHP2 and SOCS3 contribute to Tyr-759-dependent attenuation of interleukin-6 signaling through gp130. *J Biol Chem* 2003;**278**(1):661–71.

358. Tebbutt NC, Giraud AS, Inglese M, Jenkins B, Waring P, Clay FJ, et al. Reciprocal regulation of gastrointestinal homeostasis by SHP2 and STAT-mediated trefoil gene activation in gp130 mutant mice. *Nat Med* 2002;**8**(10):1089–97.

359. Jenkins BJ, Grail D, Nheu T, Najdovska M, Wang B, Waring P, et al. Hyperactivation of STAT3 in gp130 mutant mice promotes gastric hyperproliferation and desensitizes TGF-beta signaling. *Nat Med* 2005;**11**(8):845–52.

360. Jenkins BJ, Roberts AW, Najdovska M, Grail D, Ernst M. The threshold of gp130-dependent STAT3 signaling is critical for normal regulation of hematopoiesis. *Blood* 2005;**105**(9):3512–20.

361. Fairlie WD, De Souza D, Nicola NA, Baca M. Negative regulation of gp130 signalling mediated through tyrosine-757 is not dependent on the recruitment of SHP2. *Biochem J* 2003;**372**(Pt 2):495–502.

362. Yu WM, Hawley TS, Hawley RG, Qu CK. Catalytic-dependent and -independent roles of SHP-2 tyrosine phosphatase in interleukin-3 signaling. *Oncogene* 2003;**22**(38):5995–6004.

363. Yu WM, Daino H, Chen J, Bunting KD, Qu CK. Effects of a leukemia-associated gain-of-function mutation of SHP-2 phosphatase on inter-leukin-3 signaling. *J Biol Chem* 2006;**281**(9):5426–34.

364. Mohi MG, Williams IR, Dearolf CR, Chan G, Kutok JL, Cohen S, et al. Prognostic, therapeutic, and mechanistic implications of a mouse model of leukemia evoked by Shp2 (PTPN11) mutations. *Cancer Cell* 2005;**7**(2):179–91.

365. You M, Yu DH, Feng GS. Shp-2 tyrosine phosphatase functions as a negative regulator of the interferon-stimulated JAK/STAT pathway. *Mol Cell Biol* 1999;**19**(3):2416–24.

366. Wu TR, Hong YK, Wang XD, Ling MY, Dragoi AM, Chung AS, et al. SHP-2 is a dual-specificity phosphatase involved in STAT1 dephosphorylation at both tyrosine and serine residues in nuclei. *J Biol Chem* 2002;**277**(49):47572–80.

367. Yu CL, Jin YJ, Burakoff SJ. Cytosolic tyrosine dephosphorylation of STAT5. Potential role of SHP-2 in STAT5 regulation. *J Biol Chem* 2000;**275**(1):599–604.

368. Chen Y, Wen R, Yang S, Schuman J, Zhang EE, Yi T, et al. Identification of Shp-2 as a STAT5A phosphatase. *J Biol Chem* 2003;**278**(19):16,520–7.

369. Chen J, Yu WM, Bunting KD, Qu CK. A negative role of SHP-2 tyrosine phosphatase in growth factor-dependent hematopoietic cell survival. *Oncogene* 2004;**23**(20):3659–69.

370. Wheadon H, Edmead C, Welham MJ. Regulation of interleukin-3-induced substrate phosphorylation and cell survival by SHP-2 (Src-homology protein tyrosine phosphatase 2). *Biochem J* 2003;**376**(Pt 1):147–57.

371. Yu M, Luo J, Yang W, Wang Y, Mizuki M, Kanakura Y, et al. The scaffolding adapter Gab2, via Shp-2, regulates kit-evoked mast cell proliferation by activating the Rac/JNK pathway. *J Biol Chem* 2006;**281**(39):28,615–26.

372. You M, Flick LM, Yu D, Feng GS. Modulation of the nuclear factor kappa B pathway by Shp-2 tyrosine phosphatase in mediating the induction of interleukin (IL)-6 by IL-1 or tumor necrosis factor. *J Exp Med* 2001;**193**:101–10.

373. Kobielak A, Fuchs E. Links between alpha-catenin, NF-kappaB, and squamous cell carcinoma in skin. *Proc Natl Acad Sci USA* 2006;**103**(7):2322–7.

374. Holtschke T, Lohler J, Kanno Y, Fehr T, Giese N, Rosenbauer F, et al. Immunodeficiency and chronic myelogenous leukemia-like syndrome in mice with a targeted mutation of the ICSBP gene. *Cell* 1996;**87**(2):307–17.

375. Konieczna I, Horvath E, Wang H, Lindsey S, Saberwal G, Bei L, et al. Constitutive activation of SHP2 in mice cooperates with ICSBP deficiency to accelerate progression to acute myeloid leukemia. *J Clin Invest* 2008;**118**(3):853–67.

376. Yu DH, Qu CK, Henegariu O, Lu X, Feng GS. Protein-tyrosine phosphatase Shp-2 regulates cell spreading, migration, and focal adhesion. *J Biol Chem* 1998;**273**(33):21,125–31.

377. Oh E-S, Gu H, Saxton T, Timms J, Hausdorff S, Frevert E, et al. Regulation of early events in integrin signaling by the protein-tyrosine phosphatase SHP-2. *Mol Cell Biol* 1999;**19**:3205–15.

378. von Wichert G, Haimovich B, Feng GS, Sheetz MP. Force-dependent integrin-cytoskeleton linkage formation requires downregulation of focal complex dynamics by Shp2. *EMBO J* 2003;**22**(19):5023–35.

379. Inagaki K, Noguchi T, Matozaki T, Horikawa T, Fukunaga K, Tsuda M, et al. Roles for the protein tyrosine phosphatase SHP-2 in cytoskeletal organization, cell adhesion and cell migration revealed by overexpression of a dominant negative mutant. *Oncogene* 2000;**19**(1):75–84.

380. Manes S, Mira E, Gomez-Mouton C, Zhao ZJ, Lacalle RA, Martinez AC. Concerted activity of tyrosine phosphatase SHP-2 and focal adhesion kinase in regulation of cell motility. *Mol Cell Biol* 1999;**19**(4):3125–35.

381. Miao H, Burnett E, Kinch M, Simon E, Wang B. Activation of EphA2 kinase suppresses integrin function and causes focal-adhesion-kinase dephosphorylation. *Nat Cell Biol* 2000;**2**:62–9.

382. Vadlamudi RK, Adam L, Nguyen D, Santos M, Kumar R. Differential regulation of components of the focal adhesion complex by heregulin: role of phosphatase SHP-2. *J Cell Physiol* 2002;**190**:189–99.

383. Tsutsumi R, Takahashi A, Azuma T, Higashi H, Hatakeyama M. Focal adhesion kinase is a substrate and downstream effector of SHP-2 complexed with *Helicobacter pylori* CagA. *Mol Cell Biol* 2006;**26**(1):261–76.

384. Marin TM, Clemente CF, Santos AM, Picardi PK, Pascoal VD, Lopes-Cendes I, et al. Shp2 negatively regulates growth in cardiomyocytes by controlling focal adhesion kinase/Src and mTOR pathways. *Circ Res* 2008;**103**(8):813–24.

385. Heasman SJ, Ridley AJ. Mammalian Rho GTPases: new insights into their functions from *in vivo* studies. *Nat Rev Mol Cell Biol* 2008;**9**(9):690–701.

386. Kontaridis MI, Eminaga S, Fornaro M, Zito CI, Sordella R, Settleman J, et al. SHP-2 positively regulates myogenesis by coupling to the Rho GTPase signaling pathway. *Mol Cell Biol* 2004;**24**:5340–2.

387. Sordella R, Jiang W, Chen GC, Curto M, Settleman J. Modulation of Rho GTPase signaling regulates a switch between adipogenesis and myogenesis. *Cell* 2003;**113**(2):147–58.

388. Lee HH, Chang ZF. Regulation of RhoA-dependent ROCKII activation by Shp2. *J Cell Biol* 2008;**181**(6):999–1012.

389. Tsuda M, Matozaki T, Fukunaga K, Fujioka Y, Imamoto A, Noguchi T, et al. Integrin-mediated tyrosine phosphorylation of SHPS-1 and its association with SHP-2. Roles of Fak and Src family kinases. *J Biol Chem* 1998;**273**(21):13,223–9.

390. Eminaga S, Bennett AM. Noonan syndrome-associated SHP-2/Ptpn11 mutants enhance SIRPalpha and PZR tyrosyl phosphorylation and promote adhesion-mediated ERK activation. *J Biol Chem* 2008;**283**(22):15,328–38.

391. Ling Y, Maile LA, Clemmons DR. Tyrosine phosphorylation of the beta3-subunit of the alphaVbeta3 integrin is required for embrane association of the tyrosine phosphatase SHP-2 and its further recruitment to the insulin-like growth factor I receptor. *Mol Endocrinol* 2003;**17**(9):1824–33.

392. Clemmons DR, Maile LA. Minireview: integral membrane proteins that function coordinately with the insulin-like growth factor I receptor to regulate intracellular signaling. *Endocrinology* 2003;**144**(5):1664–70.

393. Ling Y, Maile LA, Badley-Clarke J, Clemmons DR. DOK1 mediates SHP-2 binding to the alphaVbeta3 integrin and thereby regulates insulin-like growth factor I signaling in cultured vascular smooth muscle cells. *J Biol Chem* 2005;**280**(5):3151–8.

394. Zhao M, Schmitz AA, Qin Y, Di Cristofano A, Pandolfi PP, Van Aelst L. Phosphoinositide 3-kinase-dependent membrane recruitment of p62(dok) is essential for its negative effect on mitogen-activated protein (MAP) kinase activation. *J Exp Med* 2001;**194**(3):265–74.

395. Frearson JA, Yi T, Alexander DR. A tyrosine-phosphorylated 110-120-kDa protein associates with the C-terminal SH2 domain of phosphotyrosine phosphatase-1D in T cell receptor-stimulated T cells. *Eur J Immunol* 1996;**26**:1539–43.

396. Frearson JA, Alexander DR. The phosphotyrosine phosphatase SHP-2 participates in a multimeric signaling complex and regulates T cell receptor (TCR) coupling to the Ras/mitogen-activated protein kinase (MAPK) pathway in Jurkat T cells. *J Exp Med* 1998;**187**:1417–26.

397. Nishida K, Yoshida Y, Itoh M, Fukada T, Ohtani T, Shirogane T, et al. Gab-family adapter proteins act downstream of cytokine and growth factor receptors and T- and B-cell antigen receptors. *Blood* 1999;**93**(6):1809–16.

398. Ingham RJ, Holgado-Madruga M, Siu C, Wong AJ, Gold MR. The Gab1 protein is a docking site for multiple proteins involved in signaling by the B cell antigen receptor. *J Biol Chem* 1998;**273**(46):30,630–7.

399. Pratt JC, Igras VE, Maeda H, Baksh S, Gelfand EW, Burakoff SJ, et al. Cutting edge: Gab2 mediates an inhibitory phosphatidylinositol 3'-kinase pathway in T cell antigen receptor signaling. *J Immunol* 2000;**165**:4158–63.

400. Yamasaki S, Nishida K, Hibi M, Sakuma M, Shiina R, Takeuchi A, et al. Docking protein Gab2 is phosphorylated by ZAP-70 and negatively regulates T cell receptor signaling by recruitment of inhibitory molecules. *J Biol Chem* 2001;**276**:45,175–83.

401. Yamasaki S, Nishida K, Sakuma M, Berry D, McGlade CJ, Hirano T, et al. Gads/Grb2-mediated association with LAT is critical for the inhibitory function of Gab2 in T cells. *Mol Cell Biol* 2003;**23**:2515–29.

402. Itoh S, Itoh M, Nishida K, Yamasaki S, Yoshida Y, Narimatsu M, et al. Adapter molecule Grb2-associated binder 1 is specifically expressed in marginal zone B cells and negatively regulates thymus-independent antigen-2 responses. *J Immunol* 2002;**168**(10):5110–16.

403. Nakaseko C, Miyatake S, Iida T, Hara S, Abe R, Ohno H, et al. Cytotoxic lymphocyte antigen 4 (CTLA-4) engagement delivers an inhibitory signal through the membrane proximal region in the absence of the tyrosine motif in the cytoplasmic tail. *J Exp Med* 1999;**190**:765–74.

404. Cinek T, Sadra A, Imboden JB. Tyrosine-independent transmission of inhibitory signals by CTLA-4. *J Immunol* 2000;**164**:5–8.

405. Baroja ML, Luxenberg D, Chau T, Ling V, Strathdee CA, Carreno BM, et al. The inhibitory function of CTLA-4 does not require its phosphorylation. *J Immunol* 2000;**164**:49–55.

406. Okazaki T, Maeda A, Nishimura H, Kurosaki T, Honjo T. PD-1 immunoreceptor inhibits B cell receptor-mediated signalling by recruiting src homology 2-domain-containing tyrosine phosphatase 2 to phosphotyrosine. *Proc Natl Acad Sci USA* 2001;**98**:13,866–71.

407. Newton-Nash DK, Newman PJ. A new role for platelet-endothelial cell adhesion molecule-1 (CD31): inhibition of TCR-mediated signal transduction. *J Immunol* 1999;**163**(2):682–8.

408. Prager E, Staffler G, Majdic O, Saemann M, Godar S, Zlabinger G, et al. Induction of hyporesponsiveness and impaired T lymphocyte activation by the CD31 receptor:ligand pathway in T cells. *J Immunol* 2001;**166**(4):2364–71.

409. Newman DK, Hamilton C, Newman PJ. Inhibition of antigen-receptor signaling by platelet endothelial cell adhesion molecule-1 (CD31) requires functional ITIMs, SHP-2, and p56^lck. *Blood* 2001;**97**:2351–7.

410. Cicmil M, Thomas JM, Leduc M, Bon C, Gibbins JM. Platelet endothelial cell adhesion molecule-1 signaling inhibits the activation of human platelets. *Blood* 2002;**99**:137–44.

411. Salmond RJ, Alexander DR. SHP2 forecast for the immune system: fog gradually clearing. *Trends Immunol* 2006;**27**(3):154–60.

412. Simeoni L, Lindquist JA, Smida M, Witte V, Arndt B, Schraven B. Control of lymphocyte development and activation by negative regulatory transmembrane adapter proteins. *Immunol Rev* 2008;**224**: 215–28.

413. Yusa S, Campbell KS. Src homology region 2-containing protein tyrosine phosphatase-2 (SHP-2) can play a direct role in the inhibitory function of killer cell Ig-like receptors in human NK cells. *J Immunol* 2003;**170**(9):4539–47.

414. An H, Zhao W, Hou J, Zhang Y, Xie Y, Zheng Y, et al. SHP-2 phosphatase negatively regulates the TRIF adaptor protein-dependent type I interferon and proinflammatory cytokine production. *Immunity* 2006;**25**(6):919–28.

415. Dixit M, Loot AE, Mohamed A, Fisslthaler B, Boulanger CM, Ceacareanu B, et al. Gab1, SHP2, and protein kinase A are crucial for the activation of the endothelial NO synthase by fluid shear stress. *Circ Res* 2005;**97**(12):1236–44.

416. Tai LK, Zheng Q, Pan S, Jin ZG, Berk BC. Flow activates ERK1/2 and endothelial nitric oxide synthase via a pathway involving PECAM1, SHP2, and Tie2. *J Biol Chem* 2005;**280**(33):29,620–4.

417. Holgado-Madruga M, Wong AJ. Gab1 is an integrator of cell death versus cell survival signals in oxidative stress. *Mol Cell Biol* 2003;**23**(13):4471–84.

418. Yuan L, Yu WM, Qu CK. DNA damage-induced G2/M checkpoint in SV40 large T antigen-immortalized embryonic fibroblast cells requires SHP-2 tyrosine phosphatase. *J Biol Chem* 2003;**278**(44):42,812–20.

419. Yuan L, Yu WM, Xu M, Qu CK. SHP-2 phosphatase regulates DNA damage-induced apoptosis and G2/M arrest in catalytically dependent and independent manners, respectively. *J Biol Chem* 2005;**280**(52):42,701–6.

420. Yuan L, Yu WM, Yuan Z, Haudenschild CC, Qu CK. Role of SHP-2 tyrosine phosphatase in the DNA damage-induced cell death response. *J Biol Chem* 2003;**278**(17):15,208–16.

421. Amin AR, Thakur VS, Paul RK, Feng GS, Qu CK, Mukhtar H, et al. SHP-2 tyrosine phosphatase inhibits p73-dependent apoptosis and expression of a subset of p53 target genes induced by EGCG. *Proc Natl Acad Sci USA* 2007;**104**(13):5419–24.

422. Tenev T, Keilhack H, Tomic S, Stoyanov B, Stein-Gerlach M, Lammers R, et al. Both SH2 domains are involved in interaction of SHP-1 with the epidermal growth factor receptor but cannot confer receptor-directed activity to SHP-1/SHP-2 chimera. *J Biol Chem* 1997;**272**:5966–73.

423. Yang J, Cheng Z, Niu T, Liang X, Zhao ZJ, Zhou GW. Structural basis for substrate specificity of protein-tyrosine phosphatase SHP-1. *J Biol Chem* 2000;**275**:4066–71.

424. Yang J, Cheng Z, Niu T, Liang X, Zhao ZJ, Zhou GW. Protein tyrosine phosphatase SHP-1 specifically recognizes C-terminal residues of its substrates via Helix aO. *J Cell Biochem* 2001;**83**:14–20.

425. Pei D, Neel BG, Walsh CT. Overexpression, purification, and characterization of *Src* homology 2-containing protein tyrosine phosphatase. *Proc Natl Acad Sci, USA* 1993;**90**:1092–6.

426. Sugimoto S, Lechleider RJ, Shoelson SE, Neel BG, Walsh CT. Expression, purification, and characterization of SH2-containing protein tyrosine phosphatase, SH-PTP2. *J Biol Chem* 1993;**268**:22,771–6.

427. Zhao ZY, Shen SH, Fischer EH. Structure, regulation and function of SH2 domain-containing protein tyrosine phosphatases. *Adv Prot Phosphatases* 1995;**9**:301–21.

428. Niu T, Liang X, Yang J, Zhao Z, Zhou GW. Kinetic comparison of the catalytic domains of SHP-1 and SHP-2. *J Cell Biochem* 1999;**72**:145–50.

429. Chan RJ, Feng GS. PTPN11 is the first identified proto-oncogene that encodes a tyrosine phosphatase. *Blood* 2007;**109**(3):862–7.

430. Chan G, Kalaitzidis D, Neel BG. The tyrosine phosphatase Shp2 (PTPN11) in cancer. *Cancer Metast Rev* 2008;**27**(2):179–92.

431. Mohi MG, Neel BG. The role of Shp2 (PTPN11) in cancer. *Curr Opin Genet Dev* 2007;**17**(1):23–30.

432. Gelb BD, Tartaglia M. Noonan syndrome and related disorders: dysregulated RAS-mitogen activated protein kinase signal transduction. *Hum Mol Genet* 2006;**15**:R220–6.

433. Bentires-Alj M, Kontaridis MI, Neel BG. Stops along the RAS pathway in human genetic disease. *Nat Med* 2006;**12**(3):283–5.

434. Tartaglia M, Mehler EL, Goldberg R, Zampino G, Brunner HG, Kremer H, et al. Mutations in PTPN11, encoding protein tyrosine phosphatase SHP-2, cause Noonan syndrome. *Nat Genet* 2001;**29**:465–8.

435. Noonan JA. Hypertelorism with Turner phenotype. A new syndrome with associated congenital heart disease. *Am J Dis Child* 1968;**116**(4):373–80.

436. Noonan JA. Noonan syndrome: an update and review for the primary pediatrician. *Clin Pediatr* 1994;**33**:548–55.

437. Noonan JA. Noonan syndrome revisited. *J Pediatr* 1999;**135**:667–8.

438. Daoud MS, Dahl PR, Su WPD. Noonan syndrome. *Sem Dermatol* 1995;**14**:140–4.

439. Attard-Montalto SP, Kingston JE, Eden T. Noonan's syndrome and acute lymphoblastic leukemia. *Med Pediatr Oncol* 1994;**23**:391–2.

440. Bader-Meunier B, Tchernia G, Mielot F, Fontaine JL, Thomas C, Lyonnet S, et al. Occurrence of myeloproliferative disorder in patients with Noonan syndrome. *J Pediatr* 1997;**130**:885–9.

441. Choong K, Freedman MH, Chitayat D, Kelly EN, Taylor G, Zipursky A. Juvenile myelomonocytic leukemia and Noonan Syndrome. *J Pediatr Hematol Oncol* 1999;**21**:523–7.

442. Klopfenstein KJ, Sommer A, Ruymann FB. Neurofibromatosis-Noonan syndrome and acute lymphoblastic leukemia: a report of two cases. *J Pediatr Hematol Oncol* 1999;**21**:158–60.

443. Schubbert S, Zenker M, Rowe SL, Boll S, Klein C, Bollag G, et al. Germline KRAS mutations cause Noonan syndrome. *Nat Genet* 2006;**38**(3):331–6.

444. Carta C, Pantaleoni F, Bocchinfuso G, Stella L, Vasta I, Sarkozy A, et al. Germline missense mutations affecting KRAS Isoform B are associated with a severe Noonan syndrome phenotype. *Am J Hum Genet* 2006;**79**(1):129–35.

445. Roberts AE, Araki T, Swanson KD, Montgomery KT, Schiripo TA, Joshi VA, et al. Germline gain-of-function mutations in SOS1 cause Noonan syndrome. *Nat Genet* 2007;**39**(1):70–4.

446. Tartaglia M, Pennacchio LA, Zhao C, Yadav KK, Fodale V, Sarkozy A, et al. Gain-of-function SOS1 mutations cause a distinctive form of Noonan syndrome. *Nat Genet* 2007;**39**(1):75–9.

447. Razzaque MA, Nishizawa T, Komoike Y, Yagi H, Furutani M, Amo R, et al. Germline gain-of-function mutations in RAF1 cause Noonan syndrome. *Nat Genet* 2007;**39**(8):1013–17.

448. Pandit B, Sarkozy A, Pennacchio LA, Carta C, Oishi K, Martinelli S, et al. Gain-of-function RAF1 mutations cause Noonan and LEOPARD syndromes with hypertrophic cardiomyopathy. *Nat Genet* 2007;**39**(8):1007–12.

449. Tartaglia M, Niemeyer CM, Fragale A, Song X, Buechner J, Jung A, et al. Somatic mutations in PTPN11 in juvenile myelomonocytic leukemia, myelodysplastic syndromes and acute myeloid leukemia. *Nat Genet* 2003;**34**:148–50.

450. Loh ML, Vattikuti S, Schubbert S, Reynolds MG, Carlson E, Lieuw KH, et al. Mutations in PTPN11 implicate the SHP-2 phosphatase in leukemogenesis. *Blood* 2004;**103**(6):2325–31.

451. Loh ML, Reynolds MG, Vattikuti S, Gerbing RB, Alonzo TA, Carlson E, et al. PTPN11 mutations in pediatric patients with acute myeloid leukemia: results from the Children's Cancer Group. *Leukemia* 2004;**18**(11):1831–4.

452. Bentires-Alj M, Paez JG, David FS, Keilhack H, Halmos B, Naoki K, et al. Activating mutations of the Noonan syndrome-associated SHP2/PTPN11 gene in human solid tumors and adult acute myelogenous leukemia. *Cancer Res* 2004;**64**(24):8816–20.

453. Tartaglia M, Martinelli S, Cazzaniga G, Cordeddu V, Iavarone I, Spinelli M, et al. Genetic evidence for lineage-related and differentiation stage-related contribution of somatic PTPN11 mutations to leukemogenesis in childhood acute leukemia. *Blood* 2004;**104**:307–13.

454. Yamamoto T, Isomura M, Xu Y, Liang J, Yagasaki H, Kamachi Y, et al. PTPN11, RAS and FLT3 mutations in childhood acute lymphoblastic leukemia. *Leuk Res* 2006;**30**(9):1085–9.

455. Martinelli S, Carta C, Flex E, Binni F, Cordisco EL, Moretti S, et al. Activating PTPN11 mutations play a minor role in pediatric and adult solid tumors. *Cancer Genet Cytogenet* 2006;**166**(2):124–9.

456. Sjoblom T, Jones S, Wood LD, Parsons DW, Lin J, Barber TD, et al. The consensus coding sequences of human breast and colorectal cancers. *Science* 2006;**314**(5797):268–74.

457. Tartaglia M, Kalidas K, Shaw A, Song X, Musat DL, van der Burgt I, et al. PTPN11 mutations in Noonan syndrome: molecular spectrum, genotype-phenotype correlation, and phenotypic heterogeneity. *Am J Hum Genet* 2002;**70**:1555–63.

458. Kosaki K, Suzuki T, Muroya K, Hasegawa T, Sato S, Matsuo N, et al. PTPN11 (protein-tyrosine phosphatase, nonreceptor-type 11) mutations in seven Japanese patients with Noonan syndrome. *J Clin Endocrinol Metab* 2002;**87**(8):3529–33.

459. Keilhack H, David FS, McGregor M, Cantley LC, Neel BG. Diverse biochemical properties of Shp2 mutants: implications for disease phenotypes. *J Biol Chem* 2005;**280**:30,984–93.

460. Miyamoto D, Miyamoto M, Takahashi A, Yomogita Y, Higashi H, Kondo S, et al. Isolation of a distinct class of gain-of-function SHP-2 mutants with oncogenic RAS-like transforming activity from solid tumors. *Oncogene* 2008;**27**(25):3508–15.

461. Chen B, Bronson RT, Klaman LD, Hampton TG, Wang JF, Green PJ, et al. Mice mutant for Egfr and Shp2 have defective cardiac semilunar valvulogenesis. *Nat Genet* 2000;**24**:296–9.

462. Webster MK, Donoghue DJ. FGFR activation in skeletal disorders: too much of a good thing. *Trends Genet* 1997;**13**:178–82.

463. Araki T, Mohi MG, Ismat FA, Bronson RT, Williams IR, Kutok JL, et al. Mouse model of Noonan syndrome reveals cell type- and gene dosage-dependent effects of Ptpn11 mutation. *Nat Med* 2004;**10**(8): 849–57.

464. Oishi K, Gaengel K, Krishnamoorthy S, Kamiya K, Kim I-K, Ying H, et al. Transgenic *Drosophila* models of Noonan syndrome causing PTPN11 gain-of-function mutations. *Hum Mol Genet* 2006;**15**:543–53.

465. Chan RJ, Leedy MB, Munugalavadla V, Voorhorst CS, Li Y, Yu M, et al. Human somatic PTPN11 mutations induce hematopoietic-cell hypersensitivity to granulocyte-macrophage colony-stimulating factor. *Blood* 2005;**105**(9):3737–42.

466. Schubbert S, Lieuw K, Rowe SL, Lee CM, Li X, Loh ML, et al. Functional analysis of leukemia-associated PTPN11 mutations in primary hematopoietic cells. *Blood* 2005;**106**(1):311–17.

467. Ren Y, Chen Z, Chen L, Woods NT, Reuther GW, Cheng JQ, et al. Shp2E76K mutant confers cytokine-independent survival of TF-1 myeloid cells by up-regulating Bcl-XL. *J Biol Chem* 2007;**282**(50):36,463–73.

468. Digilio MC, Conti E, Sarkozy A, Mingarelli R, Dottorini T, Marino B, et al. Grouping of multiple-lentigines/LEOPARD and Noonan syndromes on the PTPN11 gene. *Am J Hum Genet* 2002;**71**(2):389–94.

469. Legius E, Schrander-Stumpel C, Schollen E, Pulles-Heintzberger C, Gewillig M, Fryns JP. PTPN11 mutations in LEOPARD syndrome. *J Med Genet* 2002;**39**(8):571–4.

470. Kontaridis MI, Swanson KD, David FS, Barford D, Neel BG. PTPN11 (Shp2) mutations in LEOPARD syndrome have dominant negative, not activating, effects. *J Biol Chem* 2006;**281**(10):6785–92.

471. Tartaglia M, Martinelli S, Stella L, Bocchinfuso G, Flex E, Cordeddu V, et al. Diversity and functional consequences of germline and somatic PTPN11 mutations in human disease. *Am J Hum Genet* 2006;**78**(2):279–90.

472. Hanna N, Montagner A, Lee WH, Miteva M, Vidal M, Vidaud M, et al. Reduced phosphatase activity of SHP-2 in LEOPARD syndrome: consequences for PI3K binding on Gab1. *FEBS Letts* 2006;**580**(10):2477–82.

473. Oishi K, Zhang H, Gault WJ, Wang CJ, Tan CC, Kim IK, et al. Phosphatase-defective LEOPARD syndrome mutations in PTPN11 have gain-of-function effects during *Drosophila* development. *Hum Mol Genet* 2008. in press.

474. Brummer T, Schramek D, Hayes VM, Bennett HL, Caldon CE, Musgrove EA, et al. Increased proliferation and altered growth factor dependence of human mammary epithelial cells overexpressing the Gab2 docking protein. *J Biol Chem* 2006;**281**(1):626–37.

475. Bentires-Alj M, Gil SG, Chan R, Wang ZC, Wang Y, Imanaka N, et al. A role for the scaffolding adapter GAB2 in breast cancer. *Nat Med* 2006;**12**(1):114–21.

476. Sattler M, Mohi MG, Pride YB, Quinnan LR, Malouf NA, Podar K, et al. Critical role for Gab2 in transformation by BCR/ABL. *Cancer Cell* 2002;**1**(5):479–92.

477. Chen J, Yu WM, Daino H, Broxmeyer HE, Druker BJ, Qu CK. SHP-2 phosphatase is required for hematopoietic cell transformation by Bcr-Abl. *Blood* 2007;**109**(2):778–85.

478. Scherr M, Chaturvedi A, Battmer K, Dallmann I, Schultheis B, Ganser A, et al. Enhanced sensitivity to inhibition of SHP2, STAT5, and Gab2 expression in chronic myeloid leukemia (CML). *Blood* 2006;**107**(8):3279–87.

479. Teal HE, Ni S, Xu J, Finkelstein LD, Cheng AM, Paulson RF, et al. GRB2-mediated recruitment of GAB2, but not GAB1, to SF-STK supports the expansion of Friend virus-infected erythroid progenitor cells. *Oncogene* 2006;**25**(17):2433–43.

480. Ischenko I, Petrenko O, Gu H, Hayman MJ. Scaffolding protein Gab2 mediates fibroblast transformation by the SEA tyrosine kinase. *Oncogene* 2003;**22**(41):6311–18.

481. Higashi H, Tsutsumi R, Muto S, Sugiyama T, Azuma T, Asaka M, et al. SHP-2 tyrosine phosphatase as an intracellular target of *Helicobacter pylori* CagA protein. *Science* 2002;**295**:683–6.

482. Tsutsumi R, Higashi H, Higuchi M, Okada M, Hatakeyama M. Attenuation of *Helicobacter pylori* CagA x SHP-2 signaling by interaction between CagA and C-terminal Src kinase. *J Biol Chem* 2003;**278**(6):3664–70.

483. Higuchi M, Tsutsumi R, Higashi H, Hatakeyama M. Conditional gene silencing utilizing the lac repressor reveals a role of SHP-2 in cagA-positive *Helicobacter pylori* pathogenicity. *Cancer Sci* 2004;**95**(5):442–7.

484. Ohnishi N, Yuasa H, Tanaka S, Sawa H, Miura M, Matsui A, et al. Transgenic expression of *Helicobacter pylori* CagA induces gastrointestinal and hematopoietic neoplasms in mouse. *Proc Natl Acad Sci USA* 2008;**105**(3):1003–8.

485. Wickrema A, Chen F, Namin F, Yi T, Ahmad S, Uddin S, et al. Defective expression of the SHP-1 phosphatase in polycythemia vera. *Exp Hematol* 1999;**27**:1124–32.

486. Asimakopoulos FA, Hinshelwood S, Gilbert JG, Delibrias CC, Gottgens B, Fearon DT, et al. The gene encoding hematopoietic cell phosphatase (SHP-1) is structurally and transcriptionally intact in polycythemia vera. *Oncogene* 1997;**14**:1215–22.

487. Morgan KJ, Gilliland DG. A role for JAK2 mutations in myeloproliferative diseases. *Annu Rev Med* 2008;**59**:213–22.

488. Oka T, Yoshino T, Hayashi K, Ohara N, Nakanishi T, Yamaai Y, et al. Reduction of hematopoietic cell-specific tyrosine phosphatase SHP-1 gene expression in nature killer cell lymphoma and various yypes of lymphomas/leukemias: combination analysis with cDNA expression array and tissue microarray. *Am J Pathol* 2001;**159**:1495–505.

489. Leon F, Cespon C, Franco A, Lombardia M, Roldan E, Escribano L, et al. SHP-1 Expression in peripheral T cells from patients with Sezary syndrome and in the T cell line HUT-78: implications in JAK3-mediated Signaling. *Leukemia* 2002;**16**:1470–7.

490. Zhang Q, Raghunath PN, Vonderheid E, Odum N, Wasik MA. Lack of phosphotyrosine phosphatase SHP-1 expression in malignant T-cell lymphoma cells results from methylation of the SHP-1 promoter. *Am J Pathol* 2000;**157**(4):1137–46.

491. Nakase K, Cheng J, Zhu Q, Marasco WA. Mechanisms of SHP-1 P2 promoter regulation in hematopoietic cells and its silencing in HTLV-1-transformed T cells. *J Leukoc Biol* 2008. in press.

492. Kurita-Taniguchi M, Fukui A, Hazeki K, Hirano A, Tsuji S, Matsumoto M, et al. Functional modulation of human macrophages through CD46 (Measles Virus Receptor): production of IL-12 p40 and nitric oxide in association with recruitment of protein-tyrosine thosphatase SHP-1 to CD46. *J Immunol* 2000;**165**:5143–52.

493. Boulton IC, Gray-Owen SD. Neisserial binding to CEACAM1 arrests the activation and proliferation of CD4$^+$ T lymphocytes. *Nat Immunol* 2002;**3**:229–36.

494. Lee HS, Ostrowski MA, Gray-Owen SD. CEACAM1 dynamics during neisseria gonorrhoeae suppression of CD4$^+$ T lymphocyte activation. *J Immunol* 2008;**180**(10):6827–35.

495. Tidow N, Kasper B, Welte K. SH2-containing protein tyrosine phosphatase SHP-1 and SHP-2 are dramatically increased at the protein level in neutrophils from patients with severe congenital neutropenia (Kostmann's Syndrome). *Exp Hematol* 1999;**27**:1038–45.

Chapter 14

Calcineurin

Claude B. Klee[1] and Seun-Ah Yang[2]

[1] *Laboratory of Biochemistry, National Cancer Institute, National Institutes of Health, Bethesda, Maryland*

[2] *Division of Hematology-Oncology, University of Pennsylvania School of Medicine, Philadelphia, Pennsylvania*

INTRODUCTION

Calcineurin (also called PP2B), a protein phosphatase under the control of Ca^{2+} and calmodulin (CaM), is ideally suited to play an important role in modulating cellular responses in response to the second messenger Ca^{2+}. The identification of calcineurin as the target of the immunosuppressive drugs cyclosporin A (CsA) and FK506, complexed with their respective binding proteins cyclophilin A (CypA) and FKBP12 (FK506 binding protein), revealed the key role of calcineurin in the Ca^{2+}-dependent steps of T cell activation [1]. This discovery led to purification of the transcription factor NFAT (nuclear factor of activated T cells) and the first identification of a complete transduction pathway from the plasma membrane to the nucleus [2, 3]. The specific inhibition of calcineurin by FK506 and CsA and the over-expression of the catalytic subunit of a CaM-independent derivative of calcineurin (calcineurin $A\alpha$, residues 1 to 392) have been widely used to identify the roles of calcineurin in the regulation of cellular processes as diverse as gene expression, ion homeostasis, muscle differentiation, embryogenesis, secretion, and neurological functions. It is no wonder that alteration of calcineurin activity has been implicated in the pathogenesis of such diseases as cardiac hypertrophy, congenital heart disease, and immunological and neurological disorders. For further information, the reader is referred to comprehensive reviews and references therein [2–6].

ENZYMATIC PROPERTIES

The serine/threonine phosphatase activity of calcineurin is completely dependent on Ca^{2+} concentrations found in stimulated cells (0.5–1 μM). A 19-residue synthetic peptide containing the phosphorylation site of the RII subunit of cAMP-dependent protein kinase (PKA) is routinely used to measure the phosphatase activity of the purified enzyme.

A small activation is observed upon addition of Ca^{2+} ($K_{act} = 0.5 \mu$M), while addition of an equimolar amount of CaM results in a 50-to 100-fold increase of the V_{max} [4]. The cooperative Ca^{2+} dependence of the CaM stimulation (Hill coefficient of 2.5 to 3) allows calcineurin to respond to narrow Ca^{2+} thresholds following cell stimulation. Because of its high affinity for CaM ($K_{act} \leq 10^{-10}$M), the activation of calcineurin in response to a Ca^{2+} signal can precede the activation of most, if not all, CaM-regulated enzymes.

In crude extracts, calcineurin activity is distinguished from that of PP1, 2A, and 2C by (1) its Ca^{2+} and CaM dependence; (2) its resistance to the endogenous inhibitors (inhibitor-1, DARPP-32, inhibitor-2), okadaic acid, microcystin, and calyculin; and (3) its specific inhibition by FK506 (but not rapamycin) and CsA in the presence of saturating amounts of their respective binding proteins, FKBP12 and CypA [2]. The crude enzyme, with a specific activity 10 to 20 times that of the purified enzyme, is subject to a time- and Ca^{2+}/CaM-dependent inactivation that is prevented by superoxide dismutase and reversed by ascorbate. This observation suggested that *in vivo* calcineurin activity may also be modulated by reactive oxygen species [7]. This reversible inactivation provides a mechanism for the temporal regulation of the protein phosphorylation by CaM-dependent kinases and phosphatases [8]. Determination of calcineurin activity *in vivo* can only be achieved by monitoring the extent of dephosphorylation of endogenous substrates, such as the transcription factor NFAT, if they are present at detectable levels [2].

The substrate specificity of calcineurin depends not only on recognition of the sequence surrounding the phosphorylated residues but to the presence of docking domains. NFAT contains two such domains responsible for its Ca^{2+}-dependent and phosphorylation-independent anchoring to calcineurin [2, 9]. The anchoring of NFAT allows its dephosphorylation, despite its low intracellular concentration, and the nuclear cotranslocation of calcineurin and NFAT.

Handbook of Cell Signaling, Three-Volume Set 2 ed.

STRUCTURE

Calcineurin is also characterized by a unique and highly conserved subunit structure. It is a heterodimer of a 58- to 64-kDa catalytic subunit, calcineurin A (CnA), tightly bound even in the presence of EGTA, ($K_d \leq 10^{-13}$ M), to a regulatory 19-kDa regulatory subunit, calcineurin B (CnB). CnB is an EF-hand Ca^{2+} binding protein of the CaM family. It binds four mol of Ca^{2+}, two with high affinity (K_d^{-7} M) and two with moderate affinity, in the micromolar range [10, 11]. The crystal structure of the recombinant α-isoform of human calcineurin (Figure 14.1a) confirmed the domain organization of CnA predicted by limited proteolysis and site-directed mutagenesis [2, 4]. The N-terminal two-thirds of the molecule contains the catalytic domain, the structure of which is similar to those of PP1 and PP2A. The active site contains an Fe^{3+}-Zn^{2+} dinuclear metal center [12]. Iron and zinc are bound to residues provided by the two faces of a β-sandwich. The last β-sheet extends into a five-turn amphipathic α-helix, the top polar face of which is covered by a 33-Å groove formed by the N- and C-terminal lobes and the C-terminal strand of CnB. With the exception of two short α-helices corresponding to the inhibitory domain blocking the catalytic center, the C-terminal regulatory domain (including the CaM binding domain), not visible in the electron density map, is flexible and sensitive to proteolytic attack in the absence of CaM. The catalytic and CnB binding domain (residues 1 to 392), associated with the fully liganded form of CnB, is resistant to proteolysis. It is fully activated in the absence of CaM but still requires the presence of less than 10^{-7} M Ca^{2+} [2].

The crystal structure of a proteolytic derivative of bovine calcineurin (residues 15 to 392) complexed with FKBP12-FK506 (Fig. 1B) is similar to that of the recombinant protein. Myristic acid, covalently linked to the N-terminal glycine of CnB, lies parallel to the hydrophobic face of the N-terminal helix of CnB. This perfectly conserved post-translational modification of CnB is apparently not involved in membrane association or required for enzymatic activity. It may serve as a stabilizing structural element and is required for interaction with phospholipids [2, 4, 13]. The polar bottom face of the CnB binding helix of CnA and CnB forms the site of interaction with the FKBP-FK506 and CsA-CyP complexes. The key role of CnB in forming the drug binding site provides a molecular basis for the exquisite specificity of FK506 and CsA as calcineurin inhibitors.

REGULATION

Role of CaM

The crystal structure of calcineurin helps to define the different roles and mechanisms of action of the two structurally similar Ca^{2+}-regulated proteins, CaM and CnB, in the regulation of calcineurin. The catalytic center blocked by the inhibitory domain and the flexible calmodulin binding domain, freely accessible for calmodulin binding, is consistent with the widely accepted mechanism of CaM stimulation of CaM-regulated enzymes. According to this mechanism, binding of CaM results in the displacement of the inhibitory domain and exposure of the catalytic center [4]. The requirement for Fe^{2+} (as opposed to Fe^{3+}) for calcineurin activity explains the redox sensitivity of calcineurin activity in crude tissue extracts [14, 15]. Crude and ascorbate-activated purified calcineurin is an Fe^{2+}-Zn^{2+} enzyme with an optimum pH of 6.1 [15]. The Ca^{2+}/CaM-induced exposure of Fe^{2+} facilitates its oxidation, which is responsible for the inactivation of the enzyme. Partial depletion of iron and zinc as well as oxidation of the iron during the purification procedure are responsible

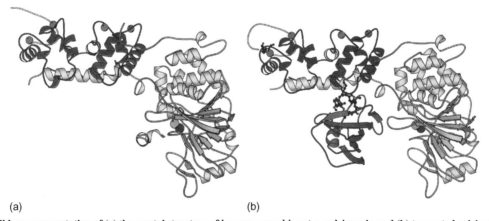

(a)   (b)

FIGURE 14.1 Ribbon representation of (a) the crystal structure of human recombinant α-calcineurin and (b) truncated calcineurin complexed with FKBP12-FK506.
CnA is shown in light gray, CnB in dark gray; iron and zinc are shown as light-gray and black spheres, respectively. The four Ca^{2+} bound to CnB are shown as dark-gray spheres, and FKBP12 is shown in dark gray. Myristic acid, covalently linked to the N-terminal glycine, and FK506 are shown in ball and stick representations (PDB code 1AUI [44] and 1TCO [45]).

for the low activity of the purified enzyme and its stimulation by 0.1-mM Mn^{2+} and 6-mM Mg^{2+} [15].

Role of CnB

CnB serves both a structural and a regulatory role. Ca^{2+}-independent binding of CnB to CnA, mediated by the high-affinity C-terminal sites, ensures the folding of active enzyme [9, 10]. Ca^{2+} binding to the N-terminal sites induces a conformational change of the regulatory domain resulting in the exposure of the drug and CaM binding domains [1, 16].

Endogenous Regulators

The presence of anchoring and inhibitory proteins, for which expression varies from tissue to tissue, adds another level of complexity to calcineurin regulation. The PKA scaffold protein, AKAP79, anchors calcineurin to specific sites of action but also inhibits its activity. Calsarcin-1 and -2, which tether calcineurin to α-actinin at the z-line of the sarcomere in cardiac and skeletal muscle, respectively, have been proposed to couple calcineurin activity to muscle contraction [17]. Calsarcins interact with calcineurin close to its active site and inhibit its activity. What is not clear is how the Ca^{2+}-independent binding of AKAP79 (K_I= 200 nM) and the inhibition of calsarcin are reversed.

Cain/Cabin1, a 240-kDa nuclear protein of yet unknown function, has been identified as a non-competitive inhibitor of calcineurin [18, 19]. It is abundant in brain, kidney, liver, and testis, but is absent in muscle. *In vivo* binding of Cabin1 to calcineurin requires Ca^{2+} and PKC activity and is inhibited by FK506-FKBP, suggesting that it binds at the drug interacting site. A basic domain in the C terminus of Cabin1 has been identified as a calcineurin binding site [18].

A family of 22- to 24-kDa proteins identified in yeast (RCn1p) and mammalian cells (calsuppressins; MCIP1, 2, and 3) are believed to be feedback inhibitors of calcineurin [20, 21]. The mammalian proteins are identical to proteins encoded in the DSCR1 (Down's syndrome critical region 1) gene on chromosome 21 (ZAKI-4, DSCRIL1, and DSCRIL). Their expression in heart and skeletal muscle is upregulated by a calcineurin/NFAT-dependent mechanism. MCPs inhibit both calcineurin activity and expression. They do not compete with CaM or FK506/FKBP but interact with calcineurin in a Ca^{2+}-independent fashion through a highly conserved ISPPxSPP motif, similar to the SP motifs of NFAT; they inhibit calcineurin activity *in vitro* as well as NFAT activation *in vivo*.

DISTRIBUTION AND ISOFORMS

Although found predominantly in neural tissues, calcineurin is present in all eukaryotes and in all tissues examined.

There are three mammalian isoforms (α, β, γ) of calcineurin. The human genes (PPP3CA, PPP3CB, PPP3CC) are located on human chromosomes 4, 10, and 8, respectively. Additional isoforms, products of alternative splicing, have not been characterized at the protein level. An N-terminal polyproline motif is a conserved feature of the β isoform. A C-terminal nuclear localization signal and stretch of basic residues are responsible for the high pI (7.1) of the testis-specific γ isoform, whereas the neural α and the broadly distributed β isoform have pIs of 5.6 and 5.8, respectively [2, 4]. Two mammalian isoforms of calcineurin B - CnB1 (associated with the α and β isoforms of CnA) and CnB2 (expressed only in testis) - are the products of two genes: PPP3R1 (located on chromosome 2) and PPP3R2 [2]. The α and β isoforms have been expressed in SF9 cells and bacteria where coexpression of the two subunits is required to yield a soluble recombinant β isoform [2].

FUNCTIONS

T cell Activation

The T cell Ca^{2+}/calcineurin/NFAT pathway, also shown to be applicable to other cell types [2, 3–5], requires a sustained release of Ca^{2+} from IP_3-sensitive stores [22]. Four NFAT isoforms (NFAT1-4) share a conserved N-terminal regulatory domain composed of a serine-rich motif, a nuclear localization signal, and three SP motifs flanked on both sides by two calcineurin binding motifs [2, 8]. The N-terminal motif, PxIxIT, missing in NFAT5, binds calcineurin (K_d=2.5 μM) at a site tentatively identified as residues 1 to 14 [23], and the C-terminal motif binds calcineurin (K_d=1.3 μM) at a site that may overlap with the drug binding domain [9]. Dephosphorylation of NFAT results in nuclear translocation of the calcineurin-NFAT complex and enhancement of DNA binding and transcriptional activity [2, 3]. Nuclear export depends on NFAT rephosphorylation upon removal of Ca^{2+} or calcineurin inactivation (or dissociation?). Identifying the mechanisms and kinases involved in the process has been elusive. GSK3, casein kinase 1/MEKK1, and Jun N-terminal kinase (JNK) have all been implicated [2, 3]. The specificity of these kinases for different isoforms of NFAT as well as the complexity due to cell background may be responsible for the failure to identify a single kinase responsible for this process [2, 3].

Muscle Differentiation

Calcineurin-mediated dephosphorylation of two transcription factors, NFAT3 and MEF2, plays a critical role in the switch of muscle fiber subtype that follows the onset of innervation and nerve activity [5]. This contractile phenotype transition consists of an increased expression of slow fiber proteins (slow MHC-1, SERCA 2a) and decreased expression of the fast fiber proteins (MHC2a, SERCA1, creatine kinase, citrate synthase). It is achieved by NFAT as well as by MEF2D activation through calcineurin-mediated dephosphorylation and a Ca^{2+}-dependent phosphorylation of MEF2D at specific serines residues [24]. The previously reported role of calcineurin in muscle hypertrophy is controversial. No hypertrophy is observed in transgenic mice overexpressing calcineurin [25]; activation of the mTOR (phosphatidylinositol 3-kinase [PI3K]/AKT) pathway plays a major role in the insulin-like growth factor 1 (IGF-1)-induced hypertrophy of preformed myotubes [26]; and calcineurin inhibitors do not block the increase of fiber size induced by nerve stimulation in regenerating muscle [27].

Cardiac Hypertrophy

Calcineurin-mediated activation of NFAT3, which, in conjunction with the transcription factor GATA4, leads to the induction of fetal cardiac genes along with natriuretic factor, have been shown to play a major role cardiac hypertrophy [5]. Calcineurin can also induce cardiac hypertrophy acting synergistically with a Ca^{2+}-dependent kinase to activate MEF2D. Regardless of its cause (overexpression of CaM-independent calcineurin, pressure overload, induction by hypertrophic agonists), cardiac hypertrophy is prevented by overexpression of MCP1 and the inhibitory domains of Cabin1/Cain or AKAP79 [5, 28, 29]. Less reproducible inhibition by FK506 and CsA may be due to high levels of calcineurin or low levels of binding proteins [5]. Furthermore, the hypertrophic response to calcineurin activation is impaired in transgenic mice expressing a constitutive form of GSK3 [29], and transgenic mice lacking the β-isoform of calcineurin (the predominant isoform in heart) have impaired ability to develop cardiac hypertrophy in response to hypertrophic agonists [30]. The calcineurin-mediated activation of PKC by calcineurin [31, 32] suggests that calcineurin, acting upstream of PKC, can also be implicated in cardiac hypertrophy through the activation of the PKC and JNK in parallel or downstream of MAP kinase pathways.

Cell Death and Differentiation

Emphasizing the general role of calcineurin in the regulation of gene expression is the broad tissue distribution of NFAT and the many genes for which expression is directly induced by NFAT-NFAT2; the cytokines IL-2, -3, -4; TNFα GM-CSF; IFNγ, the chemokines IL-8 and MIP-1a; and the receptors FasL, CD40L, CTLA-4, NF-AT2, Oct2, Egr, NFκB50p - as well as other genes activated by calcineurin, such as Elk-1 and BAD) [2, 3]. Three calcineurin-mediated pathways - NFAT activation of NFκB and FasL, synergistic induction of a member of the steroid/thyroid receptor family, Nur77, by NFAT and MEF2D; and dephosphorylation of BAD -perhaps explain the involvement of calcineurin in Ca^{2+}- and possibly H_2O_2-induced apoptosis [2, 33, 34]. The calcineurin/NFAT pathway is essential for the development of heart valves and the vascular developmental pattern during embryogenesis [35].

Ion Homeostasis

In yeast and fungi, calcineurin plays a major role in the regulation of ion homeostasis by a mechanism similar to the NFAT-mediated transcriptional control in mammalian cells [36–38]. In addition to its regulation of the Na^+/K^+ pump in kidney, recent evidence indicates that the Ca^{2+}/calcineurin/NFAT pathway is also responsible for the regulation of expression of the inositol 1,4,5-triphosphate (IP3) receptors, the plasma membrane Ca^{2+} pumps, and the Ca^{2+} exchanger in mammalian cells [39]. It was also reported that the regulation of Ca^{2+} fluxes from the IP3 and ryanodine channels was mediated by an FKBP-mediated interaction of calcineurin with the receptors acting as endogenous analogs of FKBP, but recent evidence indicates that calcineurin does not interact with either one of these two receptors [40].

Neuronal Functions

Consistent with the high concentration of calcineurin in the brain (1% of total protein), the list of neuronal functions modulated by calcineurin is continuously expanding. A major role of calcineurin in brain is to trigger a protein phosphatase cascade initiated by the dephosphorylation of two endogenous inhibitors of PP-1 (inhibitor-1 and DARPP-32). It is sensitive to PP-1 as well as calcineurin inhibitors. The dephosphorylation of these inhibitors, which do not contain anchoring domains, is not inhibited by specific inhibitors of NFAT [2]. This cascade counteracts the stimulatory effects induced by cAMP- and Ca^{2+}-regulated kinase. It has been shown to explain the antagonistic effects of glutamate binding to the NMDA receptor and dopamine binding to the D1-like dopamine receptors in striatal neurons [41], as well as the complex regulation of synaptic plasticity that includes induction of long-term potentiation (LTP) and long-term memory [42, 43] and the modulation of the activity of the transcription factor CREB [8]. Calcineurin plays an important role in cellular trafficking by dephosphorylating a family of proteins involved in endocytosis and in the release of neurotransmitters [2]. Dephosphorylation of other calcineurin substrates involved in the downregulation of receptor- and voltage-gated channels remains to be identified. Two other potentially important substrates of calcineurin are NO synthase and adenylate cyclase [2].

CONCLUSION

The importance of calcineurin in the regulation of cellular processes and its involvement in the pathogenesis of many diseases is now well established, but the role of other signaling molecules should not be underestimated. To fully assess the contribution of calcineurin in the transduction of so many diverse signals it is evident that we must understand how different pathways interact with each other.

REFERENCES

1. Liu Jr. J, Farmer JD, Lane WS, Friedman J, Weissman I, Schreiber SL. Calcineurin is a common target of cyclophilin-cyclosporin A and FKBP-FK506 complexes. *Cell* 1991;**66**:807–15.
2. Aramburu J, Rao A, Klee CB. Calcineurin: from structure to function. *Curr Top Cell Regul* 2000;**36**:237–95.
3. Clipstone NA, Fiorentino DF, Crabtree GR. Molecular analysis of the interaction of calcineurin with drug-immunophilin complexes. *J Biol Chem* 1994;**269**:26,431–26,437.
4. Klee CB, Draetta GF, Hubbard MJ. *Calcineurin Adv Enzymol* 1988;**61**:149–200.
5. Olson EN, Williams S. Calcineurin signaling and muscle remodeling. *Cell* 2000;**101**:689–92.
6. Rusnak F, Mertz P. Calcineurin: form and function. *Physiol Rev* 2000;**80**:1483–521.
7. Wang X, Culotta VC, Klee CB. Superoxide dismutase protects calcineurin from inactivation. *Nature* 1996;**383**:434–7.
8. Bito H, Deisseroth K, Tsien RW. CREB phosphorylation and dephosphorylation: a Ca^{2+}- and stimulus-duration-dependent switch for hippocampal gene expression. *Cell* 1996;**87**:1203–14.
9. Park S, Uesugi M, Verdine GL. A second calcineurin binding site on the NFAT regulatory domain. *Proc Natl Acad Sci USA* 2000;**97**:7130–5.
10. Feng B, Stemmer PM. Interactions of calcineurin A, calcineurin B, and Ca^{2+}. *Biochemistry* 1999;**38**:12,481–12,489.
11. Gallagher SC, Gao ZH, Li S, Dyer RB, Trewhella J, Klee CB. There is communication between all four Ca^{2+}-bindings sites of calcineurin B. *Biochemistry* 2001;**40**:12,094–12,202.
12. Rusnak F, Mertz P. Calcineurin: form and function. *Physiol Rev* 2000;**80**:1483–521.
13. Perrino BA, Martin BA. Ca^{2+}- and myristoylation-dependent association of calcineurin with phosphatidylserine. *J Biochem (Tokyo)* 2001;**129**:835–41.
14. Namgaladze D, Hofer HW, Ullrich V. Redox control of calcineurin by targeting the binuclear Fe^{2+}-Zn^{2+} center at the enzyme active site. *J Biol Chem* 2002;**277**:5962–9.
15. Ghosh MC, Klee CB. Native calcineurin is a Fe^{2+} enzyme. *FASEB J* 1997;**11**:A1024.
16. Yang S-A, Klee CB. Low affinity $Ca2+$-binding sites of calcineurin B mediate conformational changes of calcineurin A. *Biochemistry* 2000;**39**:16,147–16,154.
17. Frey N, Richardson JA, Olson EN. Calsarcins, a novel family of sarcomeric calcineurin-binding proteins. *Proc Natl Acad Sci USA* 2000;**97**:14,632–14,637.
18. Sun L, Youn HD, Loh C, Stolow M, He W, Liu JO. Cabin 1, a negative regulator for calcineurin signaling in T lymphocytes. *Immunity* 1998;**8**:703–11.
19. Lai MM, Burnett PE, Wolosker H, Blackshaw S, Snyder SH. Cain, a novel physiologic protein inhibitor of calcineurin. *J Biol Chem* 1998;**273**:18,325–18,331.
20. Fuentes JJ, Genesca L, Kingsbury TJ, Cunningham KW, Perez-Riba M, Estivil X, Luna S. DSCR1, overexpressed in Down syndrome, is an inhibitor of calcineurin-mediated signaling pathways. *Hum Mol Genet* 2000;**9**:1681–90.
21. Yang J, Rothermel B, Vega RB, Frey N, McKinsey TA, Olson EN, Bassel-Duby R, Williams RS. Independent signals control expression of the calcineurin inhibitory proteins MCIP1 and MCIP2 in striated muscles. *Circ Res* 2000;**87**:E61–8.
22. Jayaraman T, Marks AR. Calcineurin is downstream of the inositol 1,4,5-trisphosphate receptor in the apoptotic and cell growth pathways. *J Biol Chem* 2000;**275**:6417–20.
23. Tokoyoda K, Takemoto Y, Nakayama T, Arai T, Kubo M. Synergism between the calmodulin-binding and autoinhibitory domains on calcineurin is essential for the induction of their phosphatase activity. *J Biol Chem* 2000;**275**:11,728–11,734.
24. Wu H, Rothermel B, Kanatous S, Rosenberg P, Naya FJ, Shelton JM, Hutcheson KA, DiMaio JM, Olson EN, Bassel-Duby R, Williams RS. Activation of MEF2 by muscle activity is mediated through a calcineurin-dependent pathway. *EMBO J* 2001;**20**:6414–23.
25. Naya FJ, Mercer B, Shelton J, Richardson JA, Williams RS, Olson EN. Stimulation of slow skeletal muscle fiber gene expression by calcineurin in vivo. *J Biol Chem* 2000;**275**:4545–8.
26. Rommel C, Bodine SC, Clarke BA, Rossman R, Nunez L, Stitt TN, Yancopoulos GD, Glass DJ. Mediation of IGF-1-induced skeletal myotube hypertrophy by PI(3)K/Akt/mTOR and PI(3)K/Akt/GSK3 pathways. *Nat Cell Biol* 2001;**3**:1009–13.
27. Serrano AL, Murgia M, Pallafacchina G, Calabria E, Coniglio P, Lomo T, Schiaffino S. Calcineurin controls nerve activity-dependent specification of slow skeletal muscle fibers but not muscle growth. *Proc Natl Acad Sci USA* 2001;**98**:13,108–13,113.
28. Hill JA, Rothermel B, Yoo KD, Cabuay B, Demetroulis E, Weiss RM, Kutschke W, Bassel-Duby R, Williams RS. Targeted inhibition of calcineurin in pressure-overload cardiac hypertrophy. Preservation of systolic function. *J Biol Chem* 2002;**277**:10,251–10,255.
29. Antos CL, McKinsey TA, Frey N, Kutschke W, McAnally J, Shelton JM, Richardson JA, Hill JA, Olson EN. Activated glycogen synthase-3 beta suppresses cardiac hypertrophy *in vivo*. *Proc Natl Acad Sci USA* 2002;**99**:907–12.
30. Bueno OF, Wilkins BJ, Tymitz KM, Glascock BJ, Kimball TF, Lorenz JN, Molkentin JD. Impaired cardiac hypertrophic response in calcineurin Aβ-deficient mice. *Proc Natl Acad Sci USA* 2002;**99**:4586–91.
31. De Windt LJ, Lim HW, Haq S, Force T, Molkentin JD. Calcineurin promotes protein kinase C and c-Jun NH2-terminal kinase activation in the heart. Cross-talk between cardiac hypertrophic signaling pathways. *J Biol Chem* 2000;**275**:13,571–13,579.
32. Zhu W, Zou Y, Shiojima I, Kudoh S, Aikawa R, Hayashi D, Mizukami M, Toko H, Shibasaki F, Yazaki Y, Nagai R, Komuro I. Ca^{2+}/calmodulin-dependent kinase II and calcineurin play critical roles in endothelin-1-induced cardiomyocyte hypertrophy. *J Biol Chem* 2000;**275**:15,239–15,245.
33. Youn D, Chatila TA, Liu JO. Integration of calcineurin and MEF2 signals by the coactivator p300 during T-cell apoptosis. *EMBO J* 2000;**19**:4323–31.
34. Furuke K, Shiraishi M, Mostowski HS, Bloom ET. Fas ligand induction in human NK cells is regulated by redox through a calcineurin-nuclear factors of activated T-cell-dependent pathway. *J Immunol* 1999;**162**:1988–93.

35. Graef IA, Chen F, Chen L, Kuo A, Crabtree G. Signals transduced by Ca^{2+}/calcineurin and NF-ATc3/c4 pattern the developing vasculature. *Cell* 2001;**105**:863–75.

36. Cyert MS. Genetic analysis of calmodulin and its targets in Saccharomyces cerevisiae. *Annu Rev Genet* 2001;**35**:647–72.

37. Hemenway CS, Heitman J. Calcineurin: structure, function, and inhibition. *Cell Biochem Biophys* 1999;**30**:115–51.

38. Kingsbury TJ, Cunningham KW. A conserved family of calcineurin regulators. *Genes Dev* 2000;**14**:1595–604.

39. Li L, Guerini D, Carafoli E. Calcineurin controls the transcription of Na/Ca^{2+} exchanger isoforms in developing cerebellar neurons. *J Biol Chem* 2000;**275**:20,903–20,910.

40. Bultynck G, Rossi D, Callewaert G, Missiaen L, Sorrentino V, Parys JB, De Smedt H. The conserved sites for the FK506-binding proteins in ryanodine receptors and inositol 1,4,5-trisphosphate receptors are structurally and functionally different. *J Biol Chem* 2001;**276**:47,715–47,724.

41. Fienberg AA, Hiroi N, Mermelstein PG, Song W, Snyder GL, Nishi A, Cheramy A, O'Callaghan JP, Miller DB, Cole DG, Corbett R, Haile CN, Cooper DC, Onn SP, Grace AA, Ouimet CC, White FJ, Hyman SE, Surmeier DJ, Girault J, Nestler EJ, Greengard P. DARPP-32: regulator of the efficacy of dopaminergic neurotransmission. *Science* 1998;**281**:838–42.

42. Malleret G, Haditsch U, Genoux D, et al. Inducible and reversible enhancement of learning, memory, and long-term potentiation by genetic inhibition of calcineurin. *Cell* 2001;**104**:675–86.

43. Zeng H, Chattarji S, Barbarosie M, Rondi-Reig L, Philpot BD, Miyakawa T, Bear MF, Tonegawa S. Forebrain-specific calcineurin knockout selectively impairs bidirectional synaptic plasticity and working/episodic-like memory. *Cell* 2001;**107**:617–29.

44. Kissinger CR, Parge HE, Knighton DR, Lewis CT, Pelletier LA, Tempczyk A, Kalish VJ, Tucker KD, Showalter RE, Moomaw EW, et al. Crystal structures of human calcineurin and the human FKBP12-FK506-calcineurin complex. *Nature* 1995;**378**:641–4.

45. Griffith JP, Kim JL, Kim EE, Sintchak MD, Thomson JA, Fitzgibbon MJ, Fleming MA, Caron PR, Hsiao K, Navia MA. X-ray structure of calcineurin inhibited by the immunophilin-immunosuppressant FKBP12-FK506 complex. *Cell* 1995;**82**:507–22.

Chapter 15

Protein Serine/Threonine-Phosphatase 2C (PP2C)

Hisashi Tatebe and Kazuhiro Shiozaki

Department of Microbiology, College of Biological Sciences, University of California, Davis, California

INTRODUCTION

In early biochemical studies of protein phosphatase activities in mammalian tissues, PP2C was defined as a Mg^{2+}- or Mn^{2+}-dependent serine/threonine-specific activity that is insensitive to a phosphatase inhibitor, okadaic acid [1]. Subsequent isolation of PP2C genes from yeast to humans demonstrated that PP2C is an evolutionarily conserved phosphatase family, which shares no apparent similarity in amino acid sequence with the other phosphatase families – PP1, PP2A, and PP2B (Calcineurin). However, the crystal structure of the PP2C catalytic core shows significant resemblance to those of other serine/threonine phosphatases, such as PP1 [2]. PP2C functions as a monomer, and no regulatory subunit has been reported. Because of its cation dependency, PP2C is sometimes referred as PPM (protein phosphatase, magnesium or manganese dependent). Activity of some PP2C isoforms is also affected by oxidation or unsaturated fatty acids *in vitro* [3–6].

Eukaryotic organisms appear to have multiple PP2C genes; even the simplest eukaryotes, the budding yeast *Saccharomyces cerevisiae* and the fission yeast *Schizosaccharomyces pombe*, have seven and five PP2C genes in their genomes, respectively. In humans there are at least 16 PP2C genes, which express at least 22 isoforms by alternative splicing [7]. Multiple genes encoding PP2C-like phosphatases have also been found in the sequenced genomes of the fruit fly *Drosophila melanogaster* (10 genes) and the nematode *Caenorhabditis elegans* (8 genes). *Arabidopsis thaliana*, the most popular plant model system, has an extremely large number (76) of PP2C genes in its genome [8]. Intriguingly, several Mg^{2+}-dependent phosphatases in prokaryotes also possess clear sequence similarities with eukaryotic PP2C enzymes.

PP2Cs and related protein phosphatases are implicated in numerous biological processes, often in stress signaling. In this chapter we will review various signal transduction mechanisms where the molecular functions of PP2C enzymes have been relatively well elucidated. PHLPP, PP2C-like phosphatases implicated in the regulation of protein kinase B (PKB), is discussed in Chapter 103 of Handbook of Cell Signaling, Second Edition.

PP2C FUNCTIONS CONSERVED IN BOTH LOWER AND HIGHER EUKARYOTES

Regulation of the Stress-Activated MAP Kinase Cascades

One of the well-defined roles of PP2C in cell signaling is the downregulation of the stress-activated MAP kinase cascades in eukaryotes. A MAPK cascade is composed of three kinases – MAP kinase (MAPK), MAPK kinase (MAPKK), and MAPKK kinase (MAPKKK) – sequential activation of which transmits extracellular stimuli to the nucleus [9]. Dephosphorylation of any of the kinases in the cascade results in downregulation of the final outcome.

Studies in the budding yeast *S. cerevisiae* showed that the multicopy expression of PP2C genes rescues the mutations that cause hyperactivation of the Hog1 osmosensing MAPK cascade [10]. In the fission yeast *S. pombe*, inactivation of the stress-activated MAPK, Spc1 (also known as Sty1), suppresses the phenotypes of PP2C-deficient cells [11, 12]. These genetic data implied that PP2C negatively regulates the stress MAPK cascades. Subsequent biochemical studies demonstrated that fission yeast PP2C enzymes, Ptc1 and Ptc3, dephosphorylate Thr-171 in the T-loop (activation loop) of Spc1 MAPK [13]. Interestingly, the *ptc1*⁺

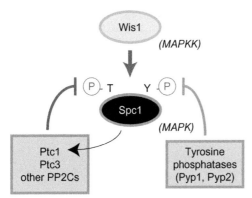

FIGURE 15.1 Downregulation of Spc1 MAP kinase by PP2Cs.
Wis1 MAPKK activated by environmental stress phosphorylates Thr-171 and Tyr-173 in the activation loop of Spc1 MAPK. Subsequently, activated Spc1 phosphorylates downstream transcription factors, which in turn induce stress-resistance genes as well as $ptc1^+$ encoding a PP2C enzyme. Ptc1 dephosphorylates Spc1 Thr-171 to inactivate the Spc1 pathway. Ptc3, another PP2C dephosphorylating Spc1, is expressed constitutively. On the other hand, phosphorylation of Spc1 Tyr-171 is negatively regulated by two tyrosine-specific phosphatases, Pyp1 and Pyp2.

gene is transcriptionally induced by activation of the Spc1 pathway, while the $ptc3^+$ gene is expressed constitutively [14]. Thus, PP2C participates both in suppressing the Spc1 MAPK under normal growth conditions and in a negative feedback regulation of Spc1 activated by stress (Figure 15.1). In budding yeast, a PP2C enzyme, Ptc1, dephosphorylates Thr-174 of Hog1 MAPK, the equivalent of Spc1 Thr-171 in fission yeast, to maintain the low basal activity of Hog1 and inactivate Hog1 during adaptation [15]. Two additional PP2Cs, Ptc2 and Ptc3, appear to limit the maximum activation level of Hog1 during stress [16].

PP2C regulation of stress MAPK cascades is also conserved in higher eukaryotes. Mammalian cells have two stress-activated MAP kinases, p38 and JNK. PP2Cα as well as the PP2Cδ isoform called Wip1 inactivate p38 MAPK by dephosphorylating its Thr-180 in the T-loop [17, 18]. Importantly, p38 MAPK activated by UV stress phosphorylates the tumor suppressor p53, which induces transcription of Wip1. Thus, like Spc1 MAPK and the Ptc1 phosphatase in fission yeast, the mammalian p38 MAPK and Wip1 PP2Cδ form a negative feedback loop. Note that PP2Cα, β, and ε isoforms also inhibit MAPKKs (MKK3, 4, 6, 7) and MAPKKKs (ASK1, TAK1) upstream of the stress-activated MAPKs p38 and JNK [17, 19–21].

In alfalfa plants, a stress-inducible PP2C, MP2C, was identified as a negative regulator of a stress-activated MAPK [22]. Environmental stress stimuli, such as wounding, activate multiple alfalfa MAPKs, among which MP2C physically interacts and inactivates a MAPK called SIMK by directly dephosphorylating Thr-213 in its T-loop [23]. As MP2C expression is induced upon stress stimuli that activate SIMK, SIMK and MP2C may also form a negative feedback loop similar to that in fission yeast and humans.

However, it remains to be determined whether MP2C expression is regulated by SIMK.

Dephosphorylation of the Cyclin-Dependent Kinase (Cdk)

Cdk is the master regulator driving the eukaryotic cell cycle at critical steps. During cell cycle progression, Cdk is regulated by interaction with cyclin subunits and by the inhibitory tyrosine phosphorylation. In addition, like MAPKs, activation of Cdk requires threonine phosphorylation in the T-loop, which is carried out by the Cdk-activating kinase (CAK). In budding yeast, Ptc2 and Ptc3 PP2Cs efficiently dephosphorylate this phosphorylated threonine in the Cdc28 Cdk both *in vivo* and *in vitro* [24]. In humans, the corresponding threonine residue in Cdk2 and Cdk6 is dephosphorylated by PP2Cα and PP2Cβ2 [25]. Binding of cyclin to Cdk inhibits dephosphorylation by PP2C in both organisms, indicating that PP2C dephosphorylates only monomeric Cdk.

Regulation in DNA Checkpoint Through Checkpoint Kinases

Chk1 and Chk2 kinases are key regulators mediating checkpoint signaling in response to genotoxic stress [26]. These kinases are activated through direct phosphorylation by the PI3K-like kinases ATM and ATR to induce cellular responses to damages of chromosomal DNA, including cell cycle arrest, DNA repair, and apoptosis. In budding yeast, a Chk2 homolog, Rad53, is phosphorylated upon DNA damage in a manner dependent on the Mec1 ATR, bringing about cell cycle arrest at G2/M. The following evidence indicates that Ptc2 and Ptc3 PP2Cs dephosphorylate Rad53 to inactivate the DNA-damage checkpoint signaling [27]. First, Ptc2 and Ptc3 physically interact with the Rad53 protein. Second, inactivation of Ptc2 and Ptc3 causes prolonged phosphorylation of Rad53 and checkpoint arrest. Third, ectopic overexpression of Ptc2 and Ptc3 leads to dephosphorylation of Rad53 and rapid adaptation to checkpoint arrest. Also in mammalian cells, a PP2C isoform, Wip1 (PP2Cδ), directly binds, dephosphorylates, and inactivates Chk1 and Chk2 [28, 29] (see below).

PP2C FUNCTIONS SPECIFIC IN HIGHER EUKARYOTES

PP2Cα, a Smad Phosphatase in TGFβ Signaling

In addition to the negative regulation of the JNK and p38 MAPK cascades, PP2Cα also inhibits the Smad proteins in signaling induced by the transforming growth factor β (TGFβ) in metazoan cells [30, 31]. The TGFβ superfamily plays a key role in a variety of biological processes, including early development and pathogenesis of diverse diseases. TGFβ binds to and activates the receptor Ser/Thr kinase at the plasma membrane, which in turn phosphorylates and activates Smad in the cytoplasm. Activated Smad proteins then translocate into the nucleus, where they regulate expression of a set of genes in collaboration with other transcription factors. Nuclear PP2Cα binds and dephosphorylates Smad, facilitating nuclear export and inactivation of Smad [30, 31]. While depletion of PP2Cα enhances the signaling, ectopic expression of PP2Cα abolishes Smad-mediated responses, including transcription.

Wip1 (PP2Cδ), an Oncogenic PP2C

As described above, the human PP2Cδ Wip1 negatively regulates the stress MAPK signaling. Additionally, recent observations clearly point to the oncogenic function of Wip1. Wip1-deficient mice exhibit defects in reproductive organs, immune function, and cell cycle control [32], but are resistant to spontaneous and oncogene-induced tumor formation [33, 34]. Consistently, fibroblasts derived from those mice are resistant to oncogene-induced transformation. Furthermore, the *Wip1* gene is frequently amplified in human breast tumors carrying wild-type p53, suggesting that Wip1 abrogates the p53 tumor-suppressor activity [35, 36].

In addition to inhibition of p53 through inactivation of p38 MAPK [18], Wip1 directly dephosphorylates and regulates p53 as well as its regulators, MDM2 and ATM (Figure 15.2). p53 phosphorylation on Ser-15 by the ATM and ATR kinases is important to induce apoptosis in response to ionizing radiation and UV [37], and Wip1 is responsible for dephosphorylation of this residue [28]. Ser-15 phosphorylation of p53 is implicated in inhibiting the interaction of p53 with MDM2, an E3 ubiquitin ligase priming p53 for degradation [38]. Wip1 also dephosphorylates MDM2 at Ser-395, which is phosphorylated by ATM [39, 40]. The Wip1-dependent dephosphorylation stabilizes MDM2 and increases its affinity to p53, leading to reduction in the p53 protein level and hence activity [40]. Furthermore, Wip1 inhibits the ATM kinase through dephosphorylation of ATM Ser-1981 [41]. Thus, Wip1 negatively regulates the ATM/ATR-p53 pathway at multiple levels.

In addition to p53 and MDM2, Wip1 appears to dephosphorylate a variety of downstream components of the ATM and ATR kinases. Indeed, *in vitro* experiments demonstrated that Wip1 preferentially dephosphorylates phosphoserine/threonine followed by glutamine [42], the consensus motif for phosphorylation by ATM/ATR [43]. For example, Wip1 dephosphorylates and inactivates Chk1 [28] and Chk2 [29] kinases, which are phosphorylated by ATR at $S^{345}Q$ and $T^{68}Q$, respectively.

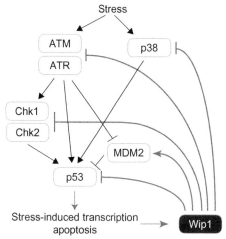

FIGURE 15.2 Feedback regulation of p53 by Wip1 PP2C. Under stressful conditions, p53 is activated by multiple upstream regulators, including ATM/ATR, Chk1/Chk2, p38, and MDM2. Active p53 induces transcription of Wip1, which in turn quenches p53 activity through dephosphorylation of p53 and its regulators. Arrows indicate activation, while T-bars indicate inhibition.

Calmodulin-Dependent Protein Kinase Phosphatase (CaMKP)

Ca^{2+}/calmodulin-dependent kinase II (CaMKII) is activated by autophosphorylation upon Ca^{2+} stimulus, and inactivated when dephosphorylated by protein phosphatases. Multiple protein phosphatases have been biochemically identified as CaMKII phosphatases, one of which is a PP2C enzyme, CaMKP. CaMKP physically interacts with CaMKII and dephosphorylates its autophosphorylation site [44, 45]. CaMKP has also been found to dephosphorylate and inactivate p21-activated protein kinase (PAK) [46].

CaMKP in the nematoda *C. elegans*, Fem-2 [47], is involved in sex determination. *fem-2* was originally identified as one of the genes required for the induction of sexual dimorphism in *C. elegans* [48, 49]. Fem-2 protein exhibits PP2C activity *in vitro*, and the *fem-2* gene complements the budding yeast PP2C mutant [50]. Both the PP2C-like domain and the amino-terminal, non-catalytic extension are essential for Fem-2 function [50].

PP2C in Plant: Abscisic Acid Signaling in *Arabidopsis Thaliana*

The large number of PP2C genes found in the *Arabidopsis thaliana* genome may reflect the importance of PP2C in plant physiology. *Arabidopsis* PP2Cs are classified into ten subgroups based on sequence similarity [8, 51]. Among those is the most extensively studied *Abi1/2* subgroup in the abscisic acid (ABA) signaling transduction [52, 53]. ABA is a plant hormone important for the maintenance of seed dormancy, stomatal closure, and growth inhibition. Loss-of-function mutations in the *abi1/abi2* genes cause hypersensitivity to ABA, indicating that Abi1/2 PP2Cs are likely to be negative regulators of the ABA signaling. Interestingly, the amount of the *ABI1* and *ABI2* mRNAs increases in response to ABA [54], suggesting that Abi PP2Cs are part of a negative feedback loop in the ABA pathway. Several other PP2Cs in the Abi1/2 subgroup have also been found to negatively regulate ABA signaling [55–58].

Small molecule second messengers have emerged in the control of Abi1/2 PP2Cs. In response to ABA, phospholipase D at the plasma membrane produces phosphatidic acid (PA), which binds to Abi1 at a 1 : 1 ratio [59, 60]. Binding of PA recruits Abi1 to the plasma membrane, and also reduces the phosphatase activity of Abi1 [59]. A mutant of phospholipase D as well as an *abi1* mutant defective in interaction with PA fail to promote stomatal closure in response to ABA [60]. Another second messenger implicated in the regulation of Abi PP2Cs is reactive oxygen species (ROS), the amount of which is elevated by ABA. A mutant plant for NADPH oxidase exhibits deficiency in the ABA-mediated signaling, and hydrogen peroxide reduces Abi1/2 phosphatase activity *in vitro* [3, 4, 61].

Multiple protein kinases have been isolated that are possibly regulated by Abi1/2. SnRK2.6, a key regulator of stomatal aperture by ABA, is an SNF1-related kinase which interacts with Abi1 and is activated in response to ABA [62, 63]. There are also several SnRK3-type kinases that are involved in ABA signaling and bind to Abi1/2 PP2Cs [64, 65].

Bacterial PP2Cs: Key Regulators of σ Factors

Bacillus subtilis has five PP2C-like phosphatases, SpoIIE, PrpC, RsbU, RsbX, and RsbP. SpoIIE is a key regulator of the sporulation specific σ^F factor [66], while RsbU, RsbX, and RxbP are involved in the regulation of the general stress-responsive σ^B factor [67]. Until prespore formation takes place, the SpoIIAB anti-σ factor represses activity of the σ^F transcription factor by directly binding to σ^F, as well as by phosphorylating the SpoIIAA anti-anti-σ factor [68–71]. Upon prespore formation, the SpoIIE PP2C dephosphorylates and activates SpoIIAA, which then releases σ^F from SpoIIAB to induce σ^F-dependent transcription in the prespore [72, 73].

The σ^B transcription factor is activated by energy stress (i.e., starvation of carbon, phosphate, or oxygen) and environmental stress (i.e., high salt, heat shock, or ethanol), leading to induction of many stress-response genes [67]. Energy and environmental stresses are transmitted by two distinct signaling cascades that are linked by the RsbV anti-anti-σ factor (Figure 15.3). Energy stress signaling is mediated by the RsbP PP2C, which dephosphorylates RsbV to induce σ^B-dependent gene expression through inhibition of the anti-σ factor, RsbW [74]. On the other

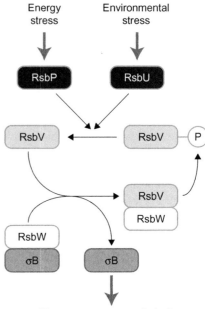

FIGURE 15.3 Stress signaling in *B. subtilis*.
The anti-anti-σ factor RsbV is dephosphorylated by two PP2C-like phosphatases, RsbP and RsbU, which are activated upon energy or environmental stress, respectively. Dephosphorylated RsbV binds to the RsbW anti-σ factor to release σ^B. Free σ^B then induces a set of stress-responsive genes. RsbW also has a protein kinase activity that phosphorylates and inactivates RsbV.

hand, environmental stress signals are conveyed through dephosphorylation of RsbV by the RsbU PP2C, which is activated by upstream regulators including the RsbX PP2C [75]. In the absence of stress, the kinase activity of RsbW represses the pathway by phosphorylating RsbV.

REFERENCES

1. Cohen P. The structure and regulation of protein phosphatases. *Annu Rev Biochem* 1989;**58**:453–508.

2. Barford D, Das AK, Egloff MP. The structure and mechanism of protein phosphatases: insights into catalysis and regulation. *Annu Rev Biophys Biomol Struct* 1998;**27**:133–64.

3. Meinhard M, Grill E. Hydrogen peroxide is a regulator of ABI1, a protein phosphatase 2C from Arabidopsis. *FEBS Letts* 2001;**508**:443–6.

4. Meinhard M, Rodriguez PL, Grill E. The sensitivity of ABI2 to hydrogen peroxide links the abscisic acid-response regulator to redox signalling. *Planta* 2002;**214**:775–82.

5. Baudouin E, Meskiene I, Hirt H. Short communication: unsaturated fatty acids inhibit MP2C, a protein phosphatase 2C involved in the wound-induced MAP kinase pathway regulation. *Plant J* 1999;**20**:343–8.

6. Klumpp S, Selke D, Hermesmeier J. Protein phosphatase type 2C active at physiological Mg^{2+}: stimulation by unsaturated fatty acids. *FEBS Letts* 1998;**437**:229–32.

7. Lammers T, Lavi S. Role of type 2C protein phosphatases in growth regulation and in cellular stress signaling. *Crit Rev Biochem Mol Biol* 2007;**42**:437–61.

8. Schweighofer A, Hirt H, Meskiene I. Plant PP2C phosphatases: emerging functions in stress signaling. *Trends Plant Sci* 2004;**9**:236–43.

9. Widmann C, Gibson S, Jarpe MB, Johnson GL. Mitogen-activated protein kinase: conservation of a three-kinase module from yeast to human. *Physiol Rev* 1999;**79**:143–80.

10. Maeda T, Wurgler-Murphy SM, Saito H. A two-component system that regulates an osmosensing MAP kinase cascade in yeast. *Nature* 1994;**369**:242–5.

11. Shiozaki K, Russell P. Cellular function of protein phosphatase 2C in yeast. *Cell Mol Biol Res* 1994;**40**:241–3.

12. Shiozaki K, Russell P. Counteractive roles of protein phosphatase 2C and a MAP kinase kinase homolog in the osmoregulation of fission yeast. *EMBO J* 1995;**14**:492–502.

13. Nguyen AN, Shiozaki K. Heat shock-induced activation of stress MAP kinase is regulated by threonine- and tyrosine-specific phosphatases. *Genes Dev* 1999;**13**:1653–63.

14. Gaits F, Shiozaki K, Russell P. Protein phosphatase 2C acts independently of stress-activated kinase cascade to regulate the stress response in fission yeast. *J Biol Chem* 1997;**272**:17,873–9.

15. Warmka J, Hanneman J, Lee J, Amin D, Ota I. Ptc1, a type 2C Ser/Thr phosphatase, inactivates the HOG pathway by dephosphorylating the mitogen-activated protein kinase hog1. *Mol Cell Biol* 2001;**21**:51–60.

16. Young C, Mapes J, Hanneman J, Al-Zarban S, Ota I. Role of Ptc2 type 2C Ser/Thr phosphatase in yeast high-osmolarity glycerol pathway inactivation. *Eukaryot Cell* 2002;**1**:1032–40.

17. Takekawa M, Maeda T, Saito H. Protein phosphatase 2Ca inhibits the human stress-responsive p38 and JNK MAPK pathways. *EMBO J* 1998;**17**:4744–52.

18. Takekawa M, Adachi M, Nakahata A, et al. p53-inducible wip1 phosphatase mediates a negative feedback regulation of p38 MAPK-p53 signaling in response to UV radiation. *EMBO J* 2000;**19**:6517–26.

19. Hanada M, Kobayashi T, Ohnishi M, et al. Selective suppression of stress-activated protein kinase pathway by protein phosphatase 2C in mammalian cells. *FEBS Letts* 1998;**437**:172–6.

20. Hanada M, Ninomiya-Tsuji J, Komaki K, et al. Regulation of the TAK1 signaling pathway by protein phosphatase 2C. *J Biol Chem* 2001;**276**:5753–9.

21. Li MG, Katsura K, Nomiyama H, et al. Regulation of the interleukin-1-induced signaling pathways by a novel member of the protein phosphatase 2C family (PP2Cepsilon). *J Biol Chem* 2003;**278**:12,013–21.

22. Meskiene I, Bogre L, Glaser W, et al. MP2C, a plant protein phosphatase 2C, functions as a negative regulator of mitogen-activated protein kinase pathways in yeast and plants. *Proc Natl Acad Sci USA* 1998;**95**:1938–43.

23. Meskiene I, Baudouin E, Schweighofer A, et al. Stress-induced protein phosphatase 2C is a negative regulator of a mitogen-activated protein kinase. *J Biol Chem* 2003;**278**:18,945–52.

24. Cheng A, Ross KE, Kaldis P, Solomon MJ. Dephosphorylation of cyclin-dependent kinases by type 2C protein phosphatases. *Genes Dev* 1999;**13**:2946–57.

25. Cheng A, Kaldis P, Solomon MJ. Dephosphorylation of human cyclin-dependent kinases by protein phosphatase type 2C alpha and beta 2 isoforms. *J Biol Chem* 2000;**275**:34,744–9.

26. Bartek J, Lukas J. Chk1 and Chk2 kinases in checkpoint control and cancer. *Cancer Cell* 2003;**3**:421–9.

27. Leroy C, Lee SE, Vaze MB, et al. PP2C phosphatases Ptc2 and Ptc3 are required for DNA checkpoint inactivation after a double-strand break. *Mol Cell* 2003;**11**:827–35.

28. Lu X, Nannenga B, Donehower LA. PPM1D dephosphorylates Chk1 and p53 and abrogates cell cycle checkpoints. *Genes Dev* 2005;**19**:1162–74.

29. Fujimoto H, Onishi N, Kato N, et al. Regulation of the antioncogenic Chk2 kinase by the oncogenic Wip1 phosphatase. *Cell Death Diff* 2006;**13**:1170–80.

30. Lin X, Duan X, Liang YY, et al. PPM1A functions as a Smad phosphatase to terminate TGFβ signaling. *Cell* 2006;**125**:915–28.

31. Duan X, Liang YY, Feng XH, Lin X. Protein serine/threonine phosphatase PPM1A dephosphorylates Smad1 in the bone morphogenetic protein signaling pathway. *J Biol Chem* 2006;**281**:36,526–32.

32. Choi J, Nannenga B, Demidov ON, et al. Mice deficient for the wild-type p53-induced phosphatase gene (Wip1) exhibit defects in reproductive organs, immune function, and cell cycle control. *Mol Cell Biol* 2002;**22**:1094–105.

33. Bulavin DV, Phillips C, Nannenga B, et al. Inactivation of the Wip1 phosphatase inhibits mammary tumorigenesis through p38 MAPK-mediated activation of the p16(Ink4a)-p19(Arf) pathway. *Nat Genet* 2004;**36**:343–50.

34. Nannenga B, Lu X, Dumble M, et al. Augmented cancer resistance and DNA damage response phenotypes in PPM1D null mice. *Mol Carcinogen* 2006;**45**:594–604.

35. Bulavin DV, Demidov ON, Saito S, et al. Amplification of PPM1D in human tumors abrogates p53 tumor-suppressor activity. *Nat Genet* 2002;**31**:210–15.

36. Rauta J, Alarmo EL, Kauraniemi P, Karhu R, Kuukasjarvi T, Kallioniemi A. The serine-threonine protein phosphatase PPM1D is frequently activated through amplification in aggressive primary breast tumours. *Breast Cancer Res Treat* 2006;**95**:257–63.

37. Sluss HK, Armata H, Gallant J, Jones SN. Phosphorylation of serine 18 regulates distinct p53 functions in mice. *Mol Cell Biol* 2004;**24**:976–84.

38. Shieh SY, Ikeda M, Taya Y, Prives C. DNA damage-induced phosphorylation of p53 alleviates inhibition by MDM2. *Cell* 1997;**91**:325–34.

39. Maya R, Balass M, Kim ST, et al. ATM-dependent phosphorylation of Mdm2 on serine 395: role in p53 activation by DNA damage. *Genes Dev* 2001;**15**:1067–77.

40. Lu X, Ma O, Nguyen TA, Jones SN, Oren M, Donehower LA. The Wip1 phosphatase acts as a gatekeeper in the p53-Mdm2 autoregulatory loop. *Cancer Cell* 2007;**12**:342–54.

41. Shreeram S, Demidov ON, Hee WK, et al. Wip1 phosphatase modulates ATM-dependent signaling pathways. *Mol Cell* 2006;**23**:757–64.

42. Yamaguchi H, Durell SR, Chatterjee DK, Anderson CW, Appella E. The Wip1 phosphatase PPM1D dephosphorylates SQ/TQ motifs in checkpoint substrates phosphorylated by PI3K-like kinases. *Biochemistry* 2007;**46**:12,594–603.

43. Kim ST, Lim DS, Canman CE, Kastan MB. Substrate specificities and identification of putative substrates of ATM kinase family members. *J Biol Chem* 1999;**274**:37,538–43.

44. Kitani T, Ishida A, Okuno S, Takeuchi M, Kameshita I, Fujisawa H. Molecular cloning of Ca^{2+}/calmodulin-dependent protein kinase phosphatase. *J Biochem (Tokyo)* 1999;**125**:1022–8.

45. Harvey BP, Banga SS, Ozer HL. Regulation of the multifunctional Ca^{2+}/calmodulin-dependent protein kinase II by the PP2C phosphatase PPM1F in fibroblasts. *J Biol Chem* 2004;**279**:24,889–98.

46. Koh CG, Tan EJ, Manser E, Lim L. The p21-activated kinase PAK is negatively regulated by POPX1 and POPX2, a pair of serine/threonine phosphatases of the PP2C family. *Curr Biol* 2002;**12**:317–21.

47. Tan KM, Chan SL, Tan KO, Yu VC. The *Caenorhabditis elegans* sex-determining protein FEM-2 and its human homologue, hFEM-2, are Ca^{2+}/calmodulin-dependent protein kinase phosphatases that promote apoptosis. *J Biol Chem* 2001;**276**:44,193–202.

48. Pilgrim D, McGregor A, Jackle P, Johnson T, Hansen D. The *C. elegans* sex-determining gene fem-2 encodes a putative protein phosphatase. *Mol Biol Cell* 1995;**6**:1159–71.

49. Hansen D, Pilgrim D. Molecular evolution of a sex determination protein. FEM-2 (pp2c) in *Caenorhabditis*. *Genetics* 1998;**149**:1353–62.

50. Chin-Sang ID, Spence AM. *Caenorhabditis elegans* sex-determining protein FEM-2 is a protein phosphatase that promotes male development and interacts directly with FEM-3. *Genes Dev* 1996;**10**:2314–25.

51. Kerk D, Bulgrien J, Smith DW, Barsam B, Veretnik S, Gribskov M. The complement of protein phosphatase catalytic subunits encoded in the genome of Arabidopsis. *Plant Physiol* 2002;**129**:908–25.

52. Gosti F, Beaudoin N, Serizet C, Webb AA, Vartanian N, Giraudat J. ABI1 protein phosphatase 2C is a negative regulator of abscisic acid signaling. *Plant Cell* 1999;**11**:1897–910.

53. Merlot S, Gosti F, Guerrier D, Vavasseur A, Giraudat J. The ABI1 and ABI2 protein phosphatases 2C act in a negative feedback regulatory loop of the abscisic acid signalling pathway. *Plant J* 2001;**25**:295–303.

54. Leung J, Merlot S, Giraudat J. The Arabidopsis ABSCISIC ACID-INSENSITIVE2 (ABI2) and ABI1 genes encode homologous protein phosphatases 2C involved in abscisic acid signal transduction. *Plant Cell* 1997;**9**:759–71.

55. Nishimura N, Yoshida T, Kitahata N, Asami T, Shinozaki K, Hirayama T. ABA-Hypersensitive Germination1 encodes a protein phosphatase 2C, an essential component of abscisic acid signaling in Arabidopsis seed. *Plant J* 2007;**50**:935–49.

56. Yoshida T, Nishimura N, Kitahata N, et al. ABA-hypersensitive germination3 encodes a protein phosphatase 2C (AtPP2CA) that strongly regulates abscisic acid signaling during germination among Arabidopsis protein phosphatase 2Cs. *Plant Physiol* 2006;**140**:115–26.

57. Kuhn JM, Boisson-Dernier A, Dizon MB, Maktabi MH, Schroeder JI. The protein phosphatase AtPP2CA negatively regulates abscisic acid signal transduction in Arabidopsis, and effects of abh1 on AtPP2CA mRNA. *Plant Physiol* 2006;**140**:127–39.

58. Saez A, Apostolova N, Gonzalez-Guzman M, et al. Gain-of-function and loss-of-function phenotypes of the protein phosphatase 2C HAB1 reveal its role as a negative regulator of abscisic acid signalling. *Plant J* 2004;**37**:354–69.

59. Zhang W, Qin C, Zhao J, Wang X. Phospholipase D alpha 1-derived phosphatidic acid interacts with ABI1 phosphatase 2C and regulates abscisic acid signaling. *Proc Natl Acad Sci USA* 2004;**101**:9508–13.

60. Mishra G, Zhang W, Deng F, Zhao J, Wang X. A bifurcating pathway directs abscisic acid effects on stomatal closure and opening in Arabidopsis. *Scienc (NY)* 2006;**312**:264–6.

61. Kwak JM, Mori IC, Pei ZM, et al. NADPH oxidase AtrbohD and AtrbohF genes function in ROS-dependent ABA signaling in Arabidopsis. *EMBO J* 2003;**22**:2623–33.

62. Yoshida R, Umezawa T, Mizoguchi T, Takahashi S, Takahashi F, Shinozaki K. The regulatory domain of SRK2E/OST1/SnRK2.6 interacts with ABI1 and integrates abscisic acid (ABA) and osmotic stress signals controlling stomatal closure in Arabidopsis. *J Biol Chem* 2006;**281**:5310–18.

63. Mustilli AC, Merlot S, Vavasseur A, Fenzi F, Giraudat J. Arabidopsis OST1 protein kinase mediates the regulation of stomatal aperture by abscisic acid and acts upstream of reactive oxygen species production. *Plant Cell* 2002;**14**:3089–99.

64. Guo Y, Xiong L, Song CP, Gong D, Halfter U, Zhu JK. A calcium sensor and its interacting protein kinase are global regulators of abscisic acid signaling in Arabidopsis. *Dev Cell* 2002;**3**:233–44.

65. Ohta M, Guo Y, Halfter U, Zhu JK. A novel domain in the protein kinase SOS2 mediates interaction with the protein phosphatase 2C ABI2. *Proc Natl Acad Sci USA* 2003;**100**:11,771–6.

66. Hilbert DW, Piggot PJ. Compartmentalization of gene expression during *Bacillus subtilis* spore formation. *Microbiol Mol Biol Rev* 2004;**68**:234–62.

67. Hecker M, Volker U. General stress response of *Bacillus subtilis* and other bacteria. *Adv Microb Physiol* 2001;**44**:35–91.

68. Najafi SM, Willis AC, Yudkin MD. Site of phosphorylation of SpoIIAA, the anti-anti-σ factor for sporulation-specific sigma F of *Bacillus subtilis*. *J Bacteriol* 1995;**177**:2912–13.

69. Diederich B, Wilkinson JF, Magnin T, Najafi M, Errington J, Yudkin MD. Role of interactions between SpoIIAA and SpoIIAB in regulating cell-specific transcription factor σ F of *Bacillus subtilis*. *Genes Dev* 1994;**8**:2653–63.

70. Min KT, Hilditch CM, Diederich B, Errington J, Yudkin MD. Sigma F, the first compartment-specific transcription factor of *B. subtilis*, is regulated by an anti-σ factor that is also a protein kinase. *Cell* 1993;**74**:735–42.

71. Duncan L, Losick R. SpoIIAB is an anti-sigma factor that binds to and inhibits transcription by regulatory protein sigma F from *Bacillus subtilis*. *Proc Natl Acad Sci USA* 1993;**90**:2325–9.

72. Arigoni F, Duncan L, Alper S, Losick R, Stragier P. SpoIIE governs the phosphorylation state of a protein regulating transcription factor sigma F during sporulation in *Bacillus subtilis*. *Proc Natl Acad Sci USA* 1996:3238–42.

73. Duncan L, Alper S, Arigoni F, Losick R, Stragier P. Activation of cell-specific transcription by a serine phosphatase at the site of asymmetric division. *Science (NY)* 1995;**270**:641–4.

74. Vijay K, Brody MS, Fredlund E, Price CW. A PP2C phosphatase containing a PAS domain is required to convey signals of energy stress to the σB transcription factor of *Bacillus subtilis*. *Mol Microbiol* 2000;**35**:180–8.

75. Yang X, Kang CM, Brody MS, Price CW. Opposing pairs of serine protein kinases and phosphatases transmit signals of environmental stress to activate a bacterial transcription factor. *Genes Dev* 1996;**10**:2265–75.

Chapter 16

Inhibitors of Protein Tyrosine Phosphatases

Zhong-Yin Zhang

Department of Biochemistry and Molecular Biology, Indiana University School of Medicine, Indianapolis, Indiana

INTRODUCTION

Protein tyrosine phosphatases (PTPs) are a large family of enzymes whose structural diversity and complexity rival those of protein tyrosine kinases (PTKs). Unlike PTKs, however, which share sequence identity with protein serine/threonine kinases, the PTPs show no structural similarity with the protein Ser/Thr phosphatases. Not surprisingly, inhibitors of protein Ser/Thr phosphatases such as okadaic acid and microcystin are not effective against PTPs. In addition, these two classes of protein phosphatases have evolved to employ completely different strategies to accomplish the dephosphorylation reaction. Thus, while the protein Ser/Thr phosphatases are metalloenzymes with bimetallic centers containing iron, and effect catalysis by direct attack of an activated water molecule at the phosphorus atom of the substrate [1], the PTPs do not require metals and proceed through a covalent phosphocysteine intermediate during catalytic turnover [2]. The hallmark that defines the PTP superfamily is the active site amino acid sequence $(H/V)C(X)^5R(S/T)$, also called the PTP signature motif, in the catalytic domain. An analysis of the completed human genome sequence suggested that humans have 107 PTPs [3], including both the tyrosine-specific and dual-specific phosphatases, which can utilize protein substrates that contain pTyr, as well as pSer and pThr.

Although all PTPs share a common catalytic mechanism and catalyze the same biochemical reaction – the hydrolysis of phosphoamino acids – they have distinct (and often unique) biological functions *in vivo* [4]. Genetics and biochemical studies indicate that PTPs are involved in a number of disease processes [4, 5]. However, because PTPs can both enhance and antagonize cellular signaling, it is essential to elucidate the physiological context in which PTPs function. One of the major challenges of the PTP field is to establish the exact functional roles for individual PTPs, both in normal cellular physiology and in pathogenic conditions. Potent and specific PTP inhibitors could serve as very powerful tools to delineate the physiological roles of these enzymes. They can also be good starting points for therapeutic developments. This chapter provides a summary and update of currently known PTP inhibitors. Both specific and non-specific small molecule competitive and reversible PTP inhibitors are discussed. In addition, some examples of covalent and time-dependent PTP inactivators are also presented.

COVALENT PTP MODIFIERS

The PTPs employ covalent catalysis, utilizing the thiol group of the active site Cys residue as the attacking nucleophile, to form a thiophosphoryl enzyme intermediate (E–P) during the PTP reaction [2]. Substitutions of the Cys residue completely abrogate PTP activity. The nucleophilic cysteine is housed within the active site architecture specifically designed to bind a negatively charged phosphoryl group. Consequently, the pKa for the sulfhydryl group of the active site Cys is extremely low (~5) [6]. Thus, the PTPs are very sensitive to thiol specific alkylating agents. For example, the PTPs can be irreversibly inactivated by iodoacetate, N-ethylmaleimide, and 5,5′-dithio-2–nitrobenzoic acid [6–8]. In addition, PTPs can also be inactivated by heavy metals including Zn^{2+}, Cu^{2+}, and *p*-(hydroxymercuri)benzoate, possibly through covalent bond formation with the active site thiol group. There were several attempts to design specific, PTP active site directed alkylating agents, taking into consideration the architecture of the PTP catalytic site and the nature of the thiol-mediated phosphate hydrolysis. The 4–fluoromethylphenylphosphate (**1**, Figure 16.1) was designed as a mechanism-based phosphatase inactivator [9, 10], which, upon the cleavage of the phosphate ester bond by the phosphatase, rapidly liberates the fluoride ion and forms a reactive quinone methide intermediate. Subsequent attack by PTP nucleophilic residues would result in formation of covalent adducts. Unfortunately, the lack of selectivity among various phosphatases and the unfavorable kinetics prevent the wide use of this compound in PTP research. The α-halobenzylphosphonate **2** (Figure 16.1) is an irreversible inactivator of the *Yersinia* PTP and PTP1B [11]. Mechanistically, compound **2** would be expected to undergo nucleophilic displacement of the halide without cleavage of the carbon–phosphorus bond. More recently, the α-bromobenzylphosphonate moiety has been exploited to develop both biotin- and rhodamine-tagged PTP probes (**3** and **4** respectively, Figure 16.1) that exhibit extremely high specificity toward PTPs while remaining inert to other proteins, including alkaline phosphatase, acid phosphatases, Ser/Thr protein phosphatases, proteases, the Src kinase, BSA, lysozyme, glyceraldehyde-3–phosphate dehydrogenase, and glutathione S transferase and the whole proteome from *E. coli* [12, 13]. These properties indicate that these α-bromobenzylphosphonate-based PTP probes can be used to interrogate the state of PTP activity in the whole proteome, thereby facilitating the simultaneous activity-based profiling of all PTPs in samples of high complexity. α-Haloacetophenone derivatives (**5**, Figure 16.1) have also been shown to be capable of covalently modifying SHP1 and PTP1B, possibly via nucleophilic displacement of the halide by the active site thiol group [14]. A unique feature of these compounds is that the inhibition can be reversed upon photoactivation. Interestingly, the dipeptide aldehyde calpain inhibitor Calpeptin (**6**, Figure 16.1) was recently shown to preferentially inhibit membrane-associated

PTPs [15]. Although the exact mechanism of inhibition has not been investigated, it is possible that the aldehyde functionality may react with residues in the PTP active site to form covalent adducts. Finally, several vitamin K analogs (e.g., **7**, Figure 16.1) have been shown to be effective PTP inactivators, possibly involving Michael-type nucleophilic addition of the active site Cys to the menadione moiety [16].

Due to the extremely low pKa of the active site thiol group, the PTPs are also prone to metal ion-catalyzed oxidation by O_2 in the air. Thus, it is a common practice to include EDTA and DTT in PTP assay buffers in order to keep the active site Cys in the reduced form. In addition to molecular oxygen, exposure of the PTPs to reactive oxygen species (ROS) can also result in PTP inactivation. For example, it has been shown that treatment of various PTPs with hydrogen peroxide [17, 18], superoxide radical anion [19], and nitric oxide [20] all lead to the oxidation of the active site Cys. For PTP1B and PTPα, oxidation of the catalytic cysteine by hydrogen peroxide leads to the formation of sulfenic acid, which is followed by a rapid condensation reaction to yield a cyclic sulfenamide [21, 22]. Interestingly, oxidation of the active site Cys from PTEN [23], Cdc25C [24], and PRL1 [25] results in the formation of a disulfide bond with a vicinal cysteine within the active site. Both cyclic sulfenamide and disulfide formation protect the active site Cys from further irreversible oxidation by ROS. Since ROS can be generated endogenously in the cell upon stimulation by peptide growth factors, cytokines, GPCR agonists, and stress, and since the oxidation of the active site Cys by ROS is in many cases reversible, it has been suggested that reversible oxidation of PTPs by ROS may provide a general mechanism for temporal negative regulation of PTP activity [26].

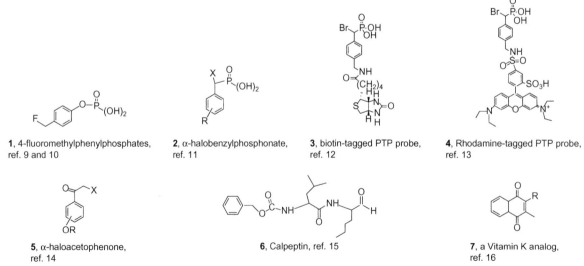

1, 4-fluoromethylphenylphosphates, ref. 9 and 10

2, α-halobenzylphosphonate, ref. 11

3, biotin-tagged PTP probe, ref. 12

4, Rhodamine-tagged PTP probe, ref. 13

5, α-haloacetophenone, ref. 14

6, Calpeptin, ref. 15

7, a Vitamin K analog, ref. 16

FIGURE 16.1 Structures of covalent inactivators of PTP.

OXYANIONS AS PTP INHIBITORS

Inorganic phosphate is a hydrolytic product of the PTP reaction, and serves as a reversible competitive PTP inhibitor (K_i ~5 mM) [27]. Because of the similar physical and chemical properties, many oxyanions, including sulfate, arsenate, molybdate, and tungstate, can competitively inhibit PTPs to various degrees (10 μM to mM). These oxyanions could inhibit PTPs by simply mimicking the tetrahedral geometry of the phosphate ion. The crystal structure of the PTP complexed with tungstate reveals the interactions between the enzyme and the phosphoryl moiety of the substrate [28]. The oxyanions adopt a tetrahedral configuration, and the oxygen atoms of the oxyanion make hydrogen bonds with the guanidinium group of the invariant Arg residue in the active site as well as the NH amides of the peptide backbone making up the PTP signature motif (H/V)$\mathbf{C}$(X)$_5$$\mathbf{R}$(S/T).

Vanadate is by far the most potent oxyanion inhibitor of PTPs, with K_i values of less than 1 μM [29]. Thus, the binding affinity of vanadate for PTPs is several orders of magnitude higher than that of phosphate. Because vanadate is known to be able to readily adopt five-coordinate structures, such observations have led to the hypothesis that vanadate may inhibit the PTPs by forming complexes that resemble the trigonal-bipyramidal geometry of the transition state. The crystal structures of the PTP–vanadate complexes show that indeed the vanadate moiety in the PTP complex is trigonal bipyramidal, with three short equatorial non-bridging V–O bonds, one apical bridging V–O bond, and one long V–S bond to the active site Cys [30–32]. Interestingly, a detailed kinetic and Raman spectroscopic study indicates that the PTP–vanadate complex may not be a true transition state analog for the PTP reaction [33], which occurs via a high dissociative metaphosphate-like transition state [34]. However, even if only a fraction of the transition state energy is captured by vanadate, it may be sufficient to account for its higher potency against PTPs.

It is important to point out that vanadate does not display any selectivity against individual members of the PTP family. Since there are no readily available PTP inhibitors with potencies comparable to that of vanadate, vanadate has been widely used as a pharmacological reagent for global PTP inhibition. Of note is the ability of vanadate to mimic the effects of insulin and to stimulate cellular proliferation. In many of the *in vivo* studies, the effective vanadate concentration used was well into the mM range. This may be caused by the presence of chelating and reducing agents, which decrease the free vanadate concentration. Because vanadate inhibits many classes of phosphoryl transfer enzymes, including phosphatases, ATPases, and nucleases, interpretation of cellular effects by vanadate should always be exercised with care.

Pervanadate, which is the complex of vanadate with hydrogen peroxide, is a more potent general PTP inhibitor than vanadate [35]. Unlike vanadate, which inhibits the PTPs reversibly and competitively, pervanadate inhibits the PTPs by irreversibly oxidizing the catalytic Cys [29]. Because pervanadate retains the vanadyl moiety, which is directed to the phosphate-binding pocket, and at the same time gains a peroxide group, which targets the active site Cys, it serves as a more efficient and specific reagent for global suppression of PTP activity at concentrations significantly lower than those employed for vanadate.

NON-HYDROLYZABLE pTyr SURROGATES AS PTP INHIBITORS

X-ray crystallographic studies have revealed a very similar active site (the pTyr binding site) for all PTPs. In the substrate-bound form of PTP1B, the terminal non-bridge phosphate oxygens of pTyr form an extensive array of hydrogen bonds with the main-chain nitrogens of the PTP signature motif (residues 215–221) and the guanidinium side-chain of Arg221. The phenyl ring of pTyr is engaged in hydrophobic interactions with the active site cavity, formed by the non-polar side-chains of Val49, Ala217, Ile219, and Gln262, and the aryl side-chains of Tyr46 and Phe182, which sandwich the pTyr ring and delineate the boundaries of the pTyr-binding pocket [36, 37]. Consistent with the fact that pTyr is essential for peptide/protein substrate binding, the interactions between pTyr and the PTP active site represent the dominant driving force for pTyr-containing peptide recognition. With this in mind, major efforts have been made to develop non-hydrolyzable pTyr surrogates that contain both a phosphate mimic that substitutes the phosphoryl group, and an aromatic scaffold that can occupy the active site pocket in a manner reminiscent of the benzene ring in pTyr. A variety of non-hydrolyzable pTyr surrogates have been reported (Figure 16.2).

One of the most commonly used non-hydrolyzable pTyr surrogates is phosphonodifluoromethyl phenylalanine (F_2Pmp, **8**) [38], which is over 1000 times more potent than phosphonomethyl phenylalanine (Pmp, **9**) when incorporated into a peptide. This may be attributed to a direct interaction between the fluorine atoms and PTP active site residues [39]. Other non-hydrolyzable pTyr surrogates include sulfotyrosine (**10**) [40], O-malonyltyrosine (**11**) [41], fluoro-O-malonyl tyrosine (**12**) [42], aryloxymethyl phosphonate (**13**) [43], cinnamic acid (**14**) [44], 3–carboxy-4–O–carboxymethyl) tyrosine (**15**) [45], salicylic acid (**16**) [46], aryl difluoromethylene phosphonate (**17**) [47], aryl difluoromethylene sulfonate (**18**) and tetrazole (**19**) [48], 2–oxalylamino)-benzoic acid (**20**) [49], 5–carboxy-2–naphthoic acid (**21**) [50], and α-ketocarboxylic acid (**22**) [51]. Like pTyr, none of these non-hydrolyzable pTyr surrogates alone exhibits high affinity toward PTPs. However, when attached to an appropriate structural scaffold, the pTyr surrogate-containing compounds can be very effective PTP inhibitors.

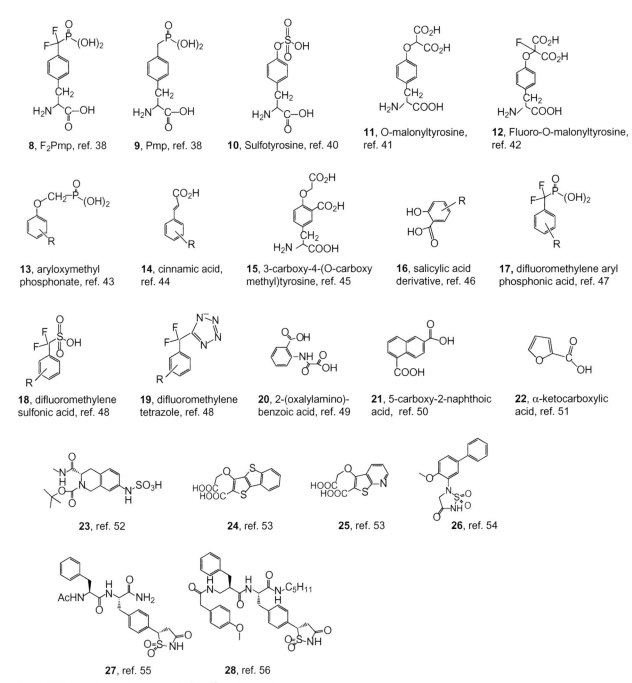

FIGURE 16.2 Structures of nonhydrolyzable pTyr surrogates.

There is continued interest in developing novel pTyr mimetics with more favorable pharmacological properties. Compound **23** (Figure 16.2) was developed based on hits identified from high-throughput screening of a compound collection [52]. With an IC_{50} of 42.5 μM for PTP1B, the single negatively charged compound **23** is a very promising pTyr mimetic worth further optimization. Another recently reported example is compound **24** [53]. Compound **25** ($K_i = 230$ μM) was obtained from an initial screen as an active site binding motif. Optimization efforts guided by crystal structures resulted in compound **24**, with a K_i of

9.2 μM. In addition to high throughput screening methods, rational design also serves as an efficient way to develop a novel active site binding motif. For example, a 1,2,5–thiadiazolidin-3–one-1,1–dioxide group was recently reported to mimic the phosphoryl moiety of pTyr, and the corresponding 1,2,5–thiadiazolidin-3–one-1,1–dioxide containing compound **26** (Figure 16.2) has an IC_{50} of 2.47 μM for PTP1B, indicating that **26** is a very effective pTyr mimetic [54]. Compound **27**, which contains a similar isothiazolidinone group, is also an excellent pTyr mimetic, with a K_i of 0.19 μM for PTP1B [55]. Using the isothiazolidinone group as the pTyr mimetic, a peptide-based inhibitor **28** was synthesized which showed an IC_{50} of 40 nM, demonstrating the utility of the isothiazolidinone to serve as a highly efficacious pTyr mimetic [56].

BIDENTATE PTP INHIBITORS

Although pTyr is essential for peptide/protein substrate recognition, pTyr by itself does not display high affinity for PTPs [57]. This and the fact that the PTP active site (pTyr binding site) is highly conserved among various PTPs present a serious challenge for the development of potent and selective PTP inhibitors targeted primarily to the active site. The discovery of a second aryl phosphate-binding site in PTP1B (defined by residues Arg24, Arg254, Met258, Gly259, and Gln262), which is not conserved among the PTPs and is adjacent to the active site, provides a novel paradigm for the design of tight-binding and specific PTP1B inhibitors that can span both sites [37]. Moreover, kinetic and structural studies have shown that amino acid residues flanking the pTyr are also required for efficient PTP substrate recognition [58]. These results suggest that subpockets adjacent to the PTP active site may also be targeted for inhibitor development. Consequently, an effective strategy for PTP inhibitor design is to attach a non-hydrolyzable pTyr surrogate to a properly functionalized structural element, which interacts with the immediate surroundings beyond the catalytic site. This strategy produces bidentate PTP inhibitors that simultaneously bind both the active site and a unique

adjacent peripheral site, thereby exhibiting both enhanced affinity and specificity.

Initial attempts to exploit this strategy generated several bis-aryldifluorophosphonate inhibitors that display modest selectivity for PTP1B [47, 59, 60]. Medicinal chemistry efforts directed to the optimization of the 3–carboxy-4–(O-carboxymethyl) tyrosine core (**15**, Figure 16.2) and its attached peptide template led to several small molecule peptidomimetics (e.g., **29** and **30** in Figure 16.3) that displayed sub- to micromolar potency against PTP1B, and augmented insulin action in the cell [61, 62]. Using a structure-based approach, the Novo Nordisk group was able to introduce a substituent into the core structure of 2–(oxalylamino)-benzoic acid (**20**) to address the second aryl phosphate-binding pocket in PTP1B [63]. This transformed a general, low-affinity, and non-selective PTP inhibitor into a reasonably potent (K_i=0.6 μM) and selective inhibitor for PTP1B (**31**, Figure 16.3). Using an NMR-based screen, small molecules that could occupy the active site and the second aryl phosphate-binding site were identified. An appropriate linker was then installed to connect the two binding fragments. Compounds **32** (K_i = 22 nM, two-fold selectivity over TC-PTP) [64] and **33** (K_i = 18 nM, four-fold selectivity over TC-PTP) [65] were both obtained using this approach. A completely different approach, namely combinatorial chemistry, was employed to identify bidentate PTP1B inhibitors capable of simultaneously occupying both the active site and a unique peripheral site in PTP1B [66]. This effort resulted in the identification of compound **34** (Figure 16.3), which displays a K_i value of

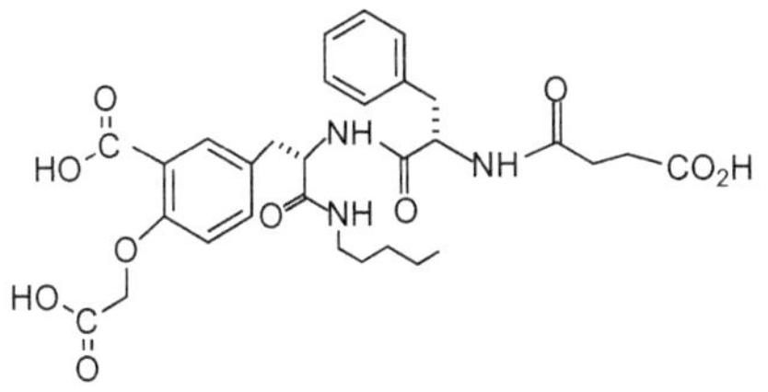

29, ref. 61 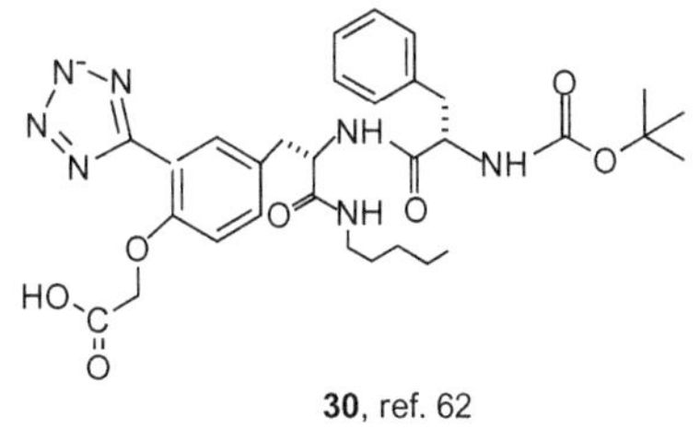   **30**, ref. 62 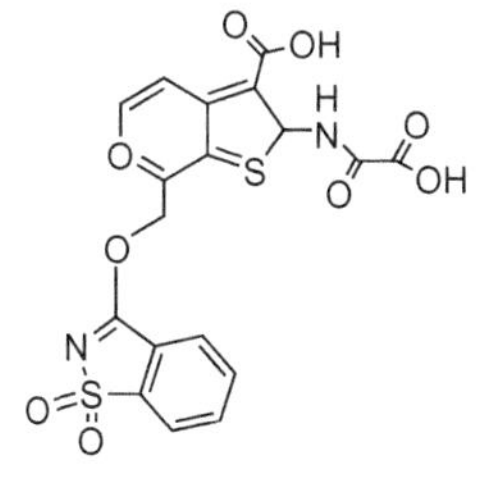   **31**, ref. 63

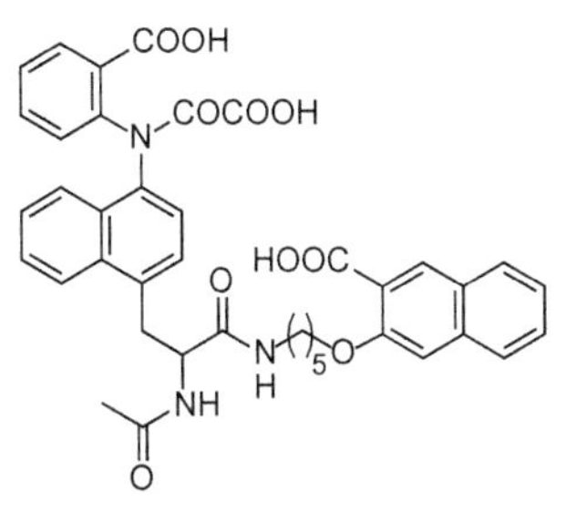

32, ref. 64 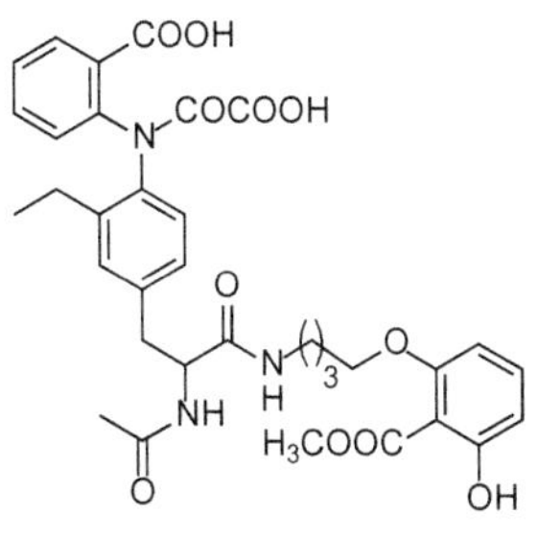   **33**, ref. 65 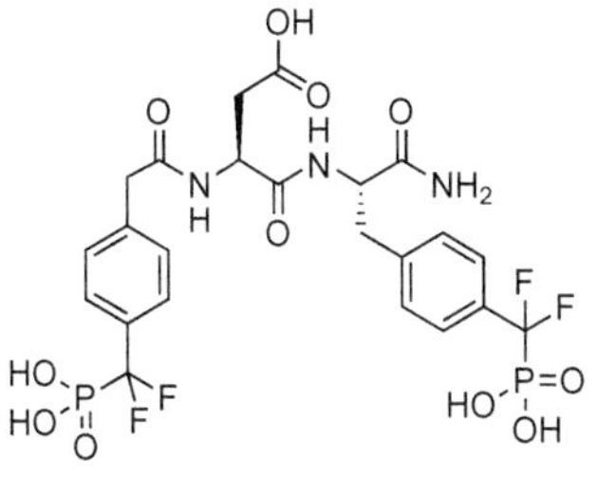   **34**, ref. 66

FIGURE 16.3 Structures of potent and selective bidentate PTP1B inhibitors.

2.4 nM for PTP1B and exhibits several orders of magnitude selectivity in favor of PTP1B against a panel of PTPs. Compound **34** is the most potent and selective PTP1B inhibitor identified to date. Subsequent structural and mutagenesis studies reveal that the distal element in **34** does not interact with the second aryl phosphate-binding pocket, but rather occupies a distinct area involving residues Lys41, Asn44, Tyr46, Arg47, Asp48, Lys116, and Phe182 [67, 68]. The interactions between compound **34** and PTP1B are unique, and provide the molecular basis for its potency and selectivity for PTP1B. Collectively, these results demonstrate that it is feasible to acquire potent, yet highly selective, PTP inhibitory agents.

OTHER PTP INHIBITORS

PTP1B was the founding member of the PTP family, and a large amount of structural and mechanistic information is available for PTP1B. Furthermore, biochemical and genetic data suggest that PTP1B is a negative regulator for both insulin and leptin signaling. Consequently, most of the PTP inhibitors that are reported in the literature are directed to PTP1B. However, other PTPs have also received considerable attention, notably CD45, Cdc25, and YopH. Most of the inhibitors described earlier for CD45 and Cdc25 display only modest potency (~10 μM) with very limited selectivity [58, 69, 70]. Many of them were identified from natural product screens. In most cases, the manner by which these compounds interact with the target PTPs is unclear, rendering structure-based optimization of new analogs difficult. A recent high-throughput evaluation of 10,070 compounds in a publicly available chemical repository of the National Cancer Institute led to the discovery of NSC 95397 (2,3–bis-[2–hydroxyethylsulfanyl]-[1,4]naphthoquinone). (**35**, Figure 16.4), which displayed mixed inhibition kinetics with *in vitro* K_i values for Cdc25A, -B, and -C of 32, 96, and 40 nM, respectively [71]. Compound **35** showed significant growth inhibition against human and murine carcinoma cells, and blocked G2/M phase transition. Medicinal chemistry efforts around the 9,10–phenanthrenedione core resulted in potent CD45 inhibitors (**36**, Figure 16.4), some of which inhibit T cell receptor-mediated proliferation with activities in the low micromolar range [72]. Interestingly, suramin (**37**, Figure 16.4), one of the oldest synthetic therapeutics, which has long been used for the treatment of sleeping sickness and onchocerciasis, has been shown to be a potent reversible and competitive inhibitor of a number of PTPs, including YopH, PTP1B, and Cdc25A [73, 74]. This is consistent with the observation that suramin leads to enhanced levels of tyrosine phosphorylation in several cell lines. More recently, sodium stibogluconate (**38**, Figure 16.4), a pentavalent antimonial used for the treatment of leishmaniasis, has been suggested as a potent inhibitor of PTPs

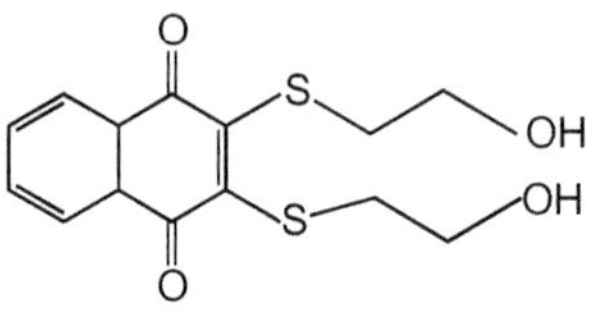

35, ref. 71   **36**, ref. 72

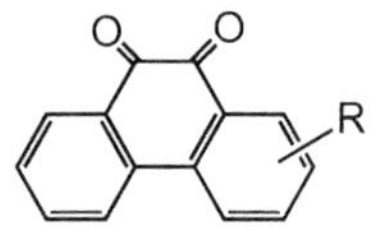

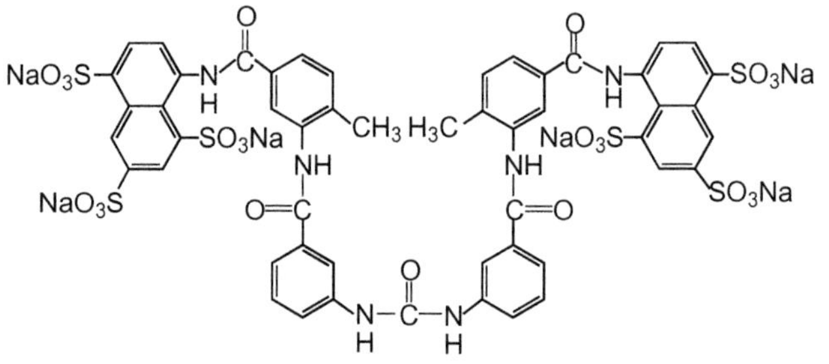

37, ref. 73, 74

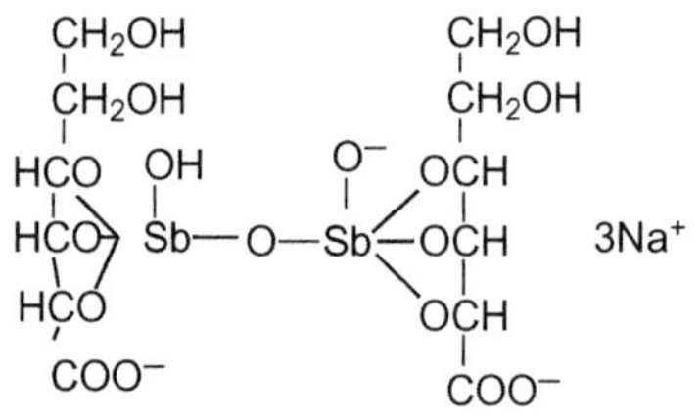

38, ref. 75

FIGURE 16.4 Other PTP inhibitors.

[75]. Although sodium stibogluconate augments cytokine responses in hemopoietic cell lines, its exact mode of action against PTPs is not clear and requires further investigation.

CONCLUDING REMARKS

The importance of PTPs in the regulation of cellular signaling is well established. In spite of the large number of PTPs identified to date, and the emerging roles played by PTPs in human diseases, a detailed understanding of the role played by PTPs in normal physiology and in pathogenic conditions has been hampered by the absence of PTP-specific agents. Such PTP specific inhibitors could potentially serve as useful tools in determining the physiological significance of protein tyrosine phosphorylation in complex cellular signal transduction pathways, and may constitute valuable therapeutics in the treatment of several human diseases. Despite the difficulties in obtaining such compounds, there are now several relatively specific inhibitors for PTP1B. It appears that significant differences exist within the active site and its immediate surroundings of various PTPs, such that selective, tight-binding PTP inhibitors can be developed. In principle, an identical approach (i.e., to create bidentate inhibitors that could span both the active site and a unique adjacent peripheral site) used for PTP1B could also be employed to produce specific small molecule inhibitors for all members of the PTP family that would enable the pharmacological modulation of selected signaling pathways for treatment of various diseases.

Combinatorial solid-phase library synthesis is finding wide applicability throughout the pharmaceutical industry, and, not surprisingly, this technique has begun to yield fruitful results in the area of PTP inhibitors.

ACKNOWLEDGEMENT

Work in the author's laboratory was supported by Grants CA69202, and DK68447 from NIH.

REFERENCES

1. Barford D. Structural studies of reversible protein phosphorylation and protein phosphatases. *Biochem Soc Trans* 1999;**27**:751–66.
2. Zhang Z-Y. Mechanistic studies on protein tyrosine phosphatases. *Prog Nucl Acid Res Mol Biol* 2003;**73**:171–220.
3. Alonso A, Sasin J, Bottini N, Friedberg I, Friedberg I, Osterman A, Godzik A, Hunter T, Dixon J, Mustelin T. Protein tyrosine phosphataes in the human genome. *Cell* 2006;**117**:699–711.
4. Tonks NK. Protein tyrosine phosphatases: from genes, to function, to disease. *Nat Rev Mol Cell Biol* 2006;**7**:833–46.
5. Zhang Z-Y. Protein tyrosine phosphatases: prospects for therapeutics. *Curr Opin Chem Biol* 2001;**5**:416–23.
6. Zhang Z-Y, Dixon JE. Active site labeling of the *Yersinia* protein tyrosine phosphatase: the determination of the pK$_a$ of the active site cysteine and the function of the conserved histidine 402. *Biochemistry* 1993;**32**:9340–5.
7. Tonks NK, Diltz CD, Fischer EH. Characterization of the major protein-tyrosine-phosphatases of human placenta. *J Biol Chem* 1988;**263**:6731–7.
8. Pot DA, Woodford TA, Remboutsika E, Haun RS, Dixon JE. Cloning, bacterial expression, purification, and characterization of the cytoplasmic domain of rat LAR, a receptor-like protein tyrosine phosphatase. *J Biol Chem* 1991;**266**:19,688–96.
9. Myers JK, Widlanski TS. Mechanism-based inactivation of prostatic acid phosphatase. *Science* 1993;**262**:1451–3.
10. Wang Q, Dechert U, Jirik F, Withers SG. Suicide inactivation of human prostatic acid phosphatase and a phosphotyrosine phosphatase. *Biochem Biophys Res Commun* 1994;**200**:577–83.
11. Taylor WP, Zhang Z-Y, Widlanski TS. Quiescent affinity inactivators of protein tyrosine phosphatases. *Bioorg Med Chem* 1996;**4**:1515–20.
12. Kumar S, Zhou B, Liang F, Wang W-Q, Huang Z, Zhang Z-Y. Activity-based probes for protein tyrosine phosphatases. *Proc Natl Acad Sci USA* 2004;**101**:7943–8.
13. Kumar S, Zhou B, Liang F, Yang H, Wang W-Q, Zhang Z-Y. Global analysis of protein tyrosine phosphatase activity with ultra-sensitive fluorescent probes. *J Proteome Res* 2006;**5**:1898–905.
14. Arabaci G, Guo X-C, Beebe KD, Coggeshall KM, Pei D. α-Haloacetophenone derivatives as photoreversible covalent inhibitors of protein tyrosine phosphatases. *J Am Chem Soc* 1999;**121**:5085–6.
15. Schoenwaelder SM, Burridge K. Evidence for a calpeptin-sensitive protein-tyrosine phosphatase upstream of the small GTPase Rho,

A novel role for the calpain inhibitor calpeptin in the inhibition of protein-tyrosine phosphatases. *J Biol Chem* 1999;**274**:14,359–67.
16. Tamura K, Southwick EC, Kerns J, Rosi K, Carr BI, Wilcox C, Lazo JS. Cdc25 inhibition and cell cycle arrest by a synthetic thioalkyl vitamin K analogue. *Cancer Res* 2000;**60**:1317–25.
17. Lee SR, Kwon KS, Kim SR, Rhee SG. Reversible inactivation of protein-tyrosine phosphatase 1B in A431 cells stimulated with epidermal growth factor. *J Biol Chem* 1998;**273**:15,366–72.
18. Denu JM, Tanner KG. Specific and reversible inactivation of protein tyrosine phosphatases by hydrogen peroxide: evidence for a sulfenic acid intermediate and implications for redox regulation. *Biochemistry* 1998;**37**:5633–42.
19. Barrett WC, DeGnore JP, Keng Y-F, Zhang Z-Y, Yim M-B, Chock PB. Roles of superoxide radical anion in signal transduction mediated by reversible regulation of protein tyrosine phosphatase 1B. *J Biol Chem* 1999;**274**:34,543–6.
20. Caselli A, Chiarugi P, Camici G, Manao G, Ramponi G. *In vivo* inactivation of phosphotyrosine protein phosphatases by nitric oxide. *FEBS Letts* 1995;**374**:249–52.
21. Salmeen A, Andersen JN, Myers MP, Meng TC, Hinks JA, Tonks NK, Barford D. Redox regulation of protein tyrosine phosphatase 1B involves a sulphenyl-amide intermediate. *Nature* 2003;**423**:769–73.
22. Yang J, Groen A, Lemeer S, Jans A, Slijper M, Roe SM, den Hertog J, Barford D. Reversible oxidation of the membrane distal domain of receptor PTPα is mediated by a cyclic sulfenamide. *Biochemistry* 2007;**46**:709–19.
23. Leslie NR, Bennett D, Lindsay YE, Stewart H, Gray A, Downes CP. Redox regulation of PI 3-kinase signalling via inactivation of PTEN. *EMBO J* 2003;**22**:5501–10.
24. Savitsky PA, Finkel T. Redox regulation of Cdc25C. *J Biol Chem* 2002;**277**:20,535–40.
25. Sun J-P, Wang W-Q, Yang H, Liu S, Liang F, Fedorov AA, Almo SC, Zhang Z-Y. Structure and biochemical properties of PRL1, a phosphatase implicated in cell growth, differentiation, and tumor invasion. *Biochemistry* 2005;**44**:12,009–21.
26. Tonks NK. Redox redux: revisiting PTPs and the control of cell signaling. *Cell* 2005;**121**:667–770.
27. Zhang Z-Y. Kinetic and mechanistic characterization of a mammalian protein tyrosine phosphatase, PTP1. *J Biol Chem* 1995;**270**:11,199–204.
28. Stuckey JA, Schubert HL, Fauman E, Zhang Z-Y, Dixon JE, Saper MA. Crystal structure of *Yersinia* protein tyrosine phosphatase at 2.5 Å and the complex with tungstate. *Nature* 1994;**370**:571–5.
29. Huyer G, Liu S, Kelly J, Moffat J, Payette P, Kennedy B, Tsaprailis G, Gresser MJ, Ramachandran C. Mechanism of inhibition of protein-tyrosine phosphatases by vanadate and pervanadate. *J Biol Chem* 1997;**272**:843–51.
30. Denu JM, Lohse DL, Vijayalakshmi J, Saper MA, Dixon JE. Visualization of intermediate and transition-state structures in protein-tyrosine phosphatase catalysis. *Proc Natl Acad Sci USA* 1996;**93**:2493–8.
31. Pannifer AD, Flint AJ, Tonks NK, Barford D. Visualization of the cysteinyl-phosphate intermediate of a protein-tyrosine phosphatase by x-ray crystallography. *J Biol Chem* 1998;**273**:10,454–62.
32. Zhang M, Zhou M, Van Etten RL, Stauffacher CV. Crystal structure of bovine low molecular weight phosphotyrosyl phosphatase complexed with the transition state analog vanadate. *Biochemistry* 1997;**36**:15–23.
33. Deng H, Callender R, Huang Z, Zhang Z-Y. Is PTPase-vanadate a true transition state analog? *Biochemistry* 2002;**41**:5865–72.

34. Hengge AC, Sowa G, Wu L, Zhang Z-Y. Nature of the transition state of the protein-tyrosine phosphatase-catalyzed reaction. *Biochemistry* 1995;**34**:13,982–7.

35. Posner BI, Faure R, Burgess JW, Bevan AP, Lachance D, Zhang-Sun G, Fantus IG, Ng JB, Hall DA, Lum BS, Shaver A. Peroxovanadium compounds. A new class of potent phosphotyrosine phosphatase inhibitors which are insulin mimetics. *J Biol Chem* 1994;**269**:4596–604.

36. Jia Z, Barford D, Flint AJ, Tonks NK. Structural basis for phosphotyrosine peptide recognition by protein tyrosine phosphatase 1B. *Science* 1995;**268**:1754–8.

37. Puius YA, Zhao Y, Sullivan M, Lawrence DS, Almo SC, Zhang Z-Y. Identification of a second aryl phosphate-binding site in protein-tyrosine phosphatase 1B: a paradigm for inhibitor design. *Proc Natl Acad Sci USA* 1997;**94**:13,420–5.

38. Burke Jr TR, Smyth M, Nomizu M, Otaka A, Roller PP. Preparation of fluoro- and hydroxy-4-(phosphonomethyl)-D,L-phenylalanine suitable protected for solid-phase synthesis of peptides containing hydrolytically stable analogues of *O*-phosphotyrosine. *J Org Chem* 1993;**58**:1336–40.

39. Chen L, Wu L, Otaka A, Smyth MS, Roller PP, Burke TR, den Hertog J, Zhang Z-Y. Why is phosphono-difluoromethyl phenylalanine a more potent inhibiting moiety than phosphonomethyl phenylalanine toward protein-tyrosine phosphatases? *Biochem Biophys Res Commun* 1995;**216**:976–84.

40. Liotta AS, Kole HK, Fales HM, Roth J, Bernier MA. Synthetic trissulfotyrosyl dodecapeptide analogue of the insulin receptor 1146-kinase domain inhibits tyrosine dephosphorylation of the insulin receptor in situ. *J Biol Chem* 1994;**269**:22,996–23,001.

41. Kole HK, Akamatsu M, Ye B, Yan X, Barford D, Roller PP, Burke Jr TR. Protein-tyrosine phosphatase inhibition by a peptide containing the phosphotyrosyl mimetic, L-O-malonyltyrosine. *Biochem Biophys Res Commun* 1995;**209**:817–22.

42. Roller PP, Wu L, Zhang Z-Y, Burke Jr TR. Potent inhibition of protein-tyrosine phosphatase-1B using the phosphotyrosyl mimetic fluoro-O-malonyl tyrosine (FOMT). *Bioorg Med Chem Letts* 1998;**8**:2149–50.

43. Ibrahimi OA, Wu L, Zhao K, Zhang Z-Y. Synthesis and characterization of a novel class of protein tyrosine phosphatase inhibitors. *Bioorg Med Chem Letts* 2000;**10**:457–60.

44. Moran EJ, Sarshar S, Cargill JF, Shahbaz MM, Lio A, Mjalli AMM, Armstrong RW. Radio frequency tag encoded combinatorial library method for the discovery of tripeptide-substituted cinnamic acid inhibitors of the protein tyrosine phosphatase PTP1B. *J Am Chem Soc* 1995;**117**:10,787–8.

45. Burke Jr TR, Yao ZJ, Zhao H, Milne GWA, Wu L, Zhang Z-Y, Voigt JH. Enantioselective synthesis of nonphosphorous-containing phosphotyrosyl mimetics and their use in the preparation of tyrosine phosphatase inhibitory peptides. *Tetrahedron* 1998;**54**:9981–94.

46. Sarmiento M, Wu L, Keng Y-F, Song L, Luo Z, Huang Z, Wu G-Z, Yuan AK, Zhang Z-Y. Structure-based discovery of small molecule inhibitors targeted to protein tyrosine phosphatase 1B. *J Med Chem* 2000;**43**:146–55.

47. Taing M, Keng Y-F, Shen K, Wu L, Lawrence DS, Zhang Z-Y. Potent and highly selective inhibitors of the protein tyrosine phosphatase 1B. *Biochemistry* 1999;**38**:3793–803.

48. Kotoris CC, Chen M-J, Taylor SD. *Bioorg Med Chem Letts* 1998;**8**:3275–80.

49. Andersen HS, Iversen LF, Jeppesen CB, Branner S, Norris K, Rasmussen HB, Moller KB, Moller NPH. 2-(oxalylamino)-benzoic acid is a general, competitive inhibitor of protein-tyrosine phosphatases. *J Biol Chem* 2000;**275**:7101–8.

50. Gao Y, Voigt J, Zhao H, Pais GCG, Zhang X, Wu L, Zhang Z-Y, Burke Jr TR. Utilization of a peptide lead for the discovery of a novel PTP1B-binding motif. *J Med Chem* 2001;**44**:2869–78.

51. Chen YT, Onaran MB, Doss CJ, Seto CT. α-Ketocarboxylic acid-based inhibitors of protein tyrosine phosphatases. *Bioorg Med Chem Letts* 2001;**11**:1935–8.

52. Klopfenstein SR, Evdokimov AG, Colson AO, Fairweather NT, Neuman JJ, Maier MB, Gray JL, Gerwe GS, Stake GE, Howard BW, Farmer JA, Pokross ME, Downs TR, Kasibhatla B, Peters KG. 1,2,3,4-tetrahydroisoquinolinyl sulfamic acids as phosphatase PTP1B inhibitors. *Bioorg Med Chem Letts* 2006;**16**:1574–8.

53. Moretto AF, Kirincich SJ, Xu WX, Smith MJ, Wan ZK, Wilson DP, Follows BC, Binnun E, Joseph-McCarthy D, Foreman K, Erbe DV, Zhang YL, Tam SK, Tam SY, Lee J. Bicyclic and tricyclic thiophenes as protein tyrosine phosphatase 1B inhibitors. *Bioorg Med Chem* 2006;**14**:2162–77.

54. Black E, Breed J, Breeze AL, Embrey K, Garcia R, Gero TW, Godfrey L, Kenny PW, Morley AD, Minshull CA, Pannifer AD, Read J, Rees A, Russell DJ, Toader D, Tucker J. Structure-based design of protein tyrosine phosphatase-1B inhibitors. *Bioorg Med Chem Letts* 2005;**15**:2503–7.

55. Combs AP, Yue EW, Bower M, Ala PJ, Wayland B, Douty B, Takvorian A, Polam P, Wasserman Z, Zhu W, Crawley ML, Pruitt J, Sparks R, Glass B, Modi D, McLaughlin E, Bostrom L, Li M, Galya L, Blom K, Hillman M, Gonneville L, Reid BG, Wei M, Becker-Pasha M, Klabe R, Huber R, Li Y, Hollis G, Burn TC, Wynn R, Liu P, Metcalf B. Structure-based design and discovery of protein tyrosine phosphatase inhibitors incorporating novel isothiazolidinone heterocyclic phosphotyrosine mimetics. *J Med Chem* 2005;**48**:6544–8.

56. Yue EW, Wayland B, Douty B, Crawley ML, McLaughlin E, Takvorian A, Wasserman Z, Bower MJ, Wei M, Li Y, Ala PJ, Gonneville L, Wynn R, Burn TC, Liu PC, Combs AP. Isothiazolidinone heterocycles as inhibitors of protein tyrosine phosphatases: synthesis and structure-activity relationships of a peptide scaffold. *Bioorg Med Chem* 2006;**14**:5833–49.

57. Zhang Z-Y, Maclean D, McNamara DJ, Sawyer TK, Dixon JE. Protein tyrosine phosphatase substrate specificity: the minimum size of the peptide and the positioning of the phosphotyrosine. *Biochemistry* 1994;**33**:2285–90.

58. Zhang Z-Y. Protein tyrosine phosphatases: structure and function, substrate specificity, and inhibitor development. *Annu Rev Pharmacol Toxicol* 2002;**42**:209–34.

59. Desmarais S, Friesen RW, Zamboni R, Ramachandran C. [Difluoro(phosphono)methyl]phenylalanine-containing peptide inhibitors of protein tyrosine phosphatases. *Biochem J* 1999;**337**:219–23.

60. Jia Z, Ye Q, Dinaut AN, Wang Q, Waddleton D, Payette P, Ramachandran C, Kennedy B, Hum G, Taylor SD. Structure of protein tyrosine phosphatase 1B in complex with inhibitors bearing two phosphotyrosine mimetics. *J Med Chem* 2001;**44**:4584–94.

61. Bleasdale JE, Ogg D, Palazuk BJ, Jacob CS, Swanson ML, Wang XY, Thompson DP, Conradi RA, Mathews WR, Laborde AL, Stuchly CW, Heijbel A, Bergdahl K, Bannow CA, Smith CW, Svensson C, Liljebris C, Schostarez HJ, May PD, Stevens FC, Larsen SD. Small molecule peptidomimetics containing a novel phosphotyrosine bioisostere inhibit protein tyrosine phosphatase 1B and augment insulin action. *Biochemistry* 2001;**40**:5642–54.

62. Liljebris C, Larsen SD, Ogg D, Palazuk BJ, Bleasdale JE. Investigation of potential bioisosteric replacements for the carboxyl groups of peptidomimetic inhibitors of protein tyrosine phosphatase 1B: identification of a tetrazole-containing inhibitor with cellular activity. *J Med Chem* 2002;**45**:1785–98.

63. Iversen LF, Andersen HS, Moller KB, Olsen OH, Peters GH, Branner S, Mortensen SB, Hansen TK, Lau J, Ge Y, Holsworth DD, Newman MJ, Moller NPH. Steric hindrance as a basis for structure-based design of selective inhibitors of protein-tyrosine phosphatases. *Biochemistry* 2001;**40**:14,812–20.

64. Szczepankiewicz BG, Liu G, Hajduk PJ, Abad-Zapatero C, Pei Z, Xin Z, Lubben TH, Trevillyan JM, Stashko MA, Ballaron SJ, Liang H, Huang F, Hutchins CW, Fesik SW, Jirousek MR. Discovery of a potent, selective protein tyrosine phosphatase 1B inhibitor using a linked-fragment strategy. *J Am Chem Soc* 2003;**125**:4087–96.

65. Liu G, Xin Z, Liang H, Abad-Zapatero C, Hajduk PJ, Janowick DA, Szczepankiewicz BG, Pei Z, Hutchins CW, Ballaron SJ, Stashko MA, Lubben TH, Berg CE, Rondinone CM, Trevillyan JM, Jirousek MR. Selective protein tyrosine phosphatase 1B inhibitors: Targeting the second phosphotyrosine binding site with non-carboxylic acid-containing ligands. *J Med Chem* 2003;**46**:3437–40.

66. Shen K, Keng Y-F, Wu L, Guo X-L, Lawrence DS, Zhang Z-Y. Acquisition of a specific and potent PTP1B inhibitor from a novel combinatorial library and screening procedure. *J Biol Chem* 2001;**276**:47,311–19.

67. Guo X-L, Shen K, Wang F, Lawrence DS, Zhang Z-Y. Probing the molecular basis for potent and selective protein tyrosine phosphatase 1B inhibition. *J Biol Chem* 2002;**277**:41,014–22.

68. Sun J-P, Fedorov AA, Lee S-Y, Guo X-L, Shen K, Lawrence DS, Almo SC, Zhang Z-Y. Crystal structure of PTP1B in complex with a potent and selective bidentate inhibitor. *J Biol Chem* 2003;**278**:12,406–14.

69. Burke Jr TR, Zhang Z-Y. Protein tyrosine phosphatases: structure, mechanism and inhibitor discovery. *Biopolymers (Pept Sci)* 1998;**47**:225–41.

70. Ripka WC. Protein tyrosine phosphatase inhibition. *Annu Rep Med Chem* 2000;**35**:231–50.

71. Lazo JS, Nemoto K, Pestell KE, Cooley K, Southwick EC, Mitchell DA, Furey W, Gussio R, Zaharevitz DW, Joo B, Wipf P. Identification of a potent and selective pharmacophore for Cdc25 dual specificity phosphatase inhibitors. *Mol Pharmacol* 2002;**61**:720–8.

72. Urbanek RA, Suchard SJ, Steelman GB, Knappenberger KS, Sygowski LA, Veale CA, Chapdelaine MJ. Potent reversible inhibitors of the protein tyrosine phosphatase CD45. *J Med Chem* 2001;**44**:1777–93.

73. Zhang Y-L, Keng Y-F, Zhao Y, Wu L, Zhang Z-Y. Suramin is an active site-directed, reversible, and tight-binding inhibitor of protein-tyrosine phosphatases. *J Biol Chem* 1998;**273**:12,281–7.

74. McCain DF, Wu L, Nickel P, Kassack MU, Kreimeyer A, Gagliardi A, Collins DC, Zhang Z-Y. Suramin derivatives as inhibitors and activators of protein tyrosine phosphatases. *J Biol Chem* 2004;**279**:14,713–25.

75. Pathak MK, Yi T. Sodium stibogluconate is a potent inhibitor of protein tyrosine phosphatases and augments cytokine responses in hemopoietic cell lines. *J Immunol* 2001;**167**:3391–7.

Chapter 17

MAP Kinase Phosphatases

Stephen M. Keyse

Cancer Research UK Stress Response Laboratory, Biomedical Research Institute, Ninewells Hospital and Medical School, Dundee, Scotland, UK

INTRODUCTION

Mitogen-activated protein kinases (MAPKs) constitute a highly conserved family of enzymes, which regulate a diverse array of physiological processes in mammalian cells and tissues. These include cell proliferation and differentiation, development, immune function, stress responses and cell death [1–4]. MAPK pathways in organisms from yeast to man share a common architecture, consisting of a core three-component kinase cascade or module. MAPKs are activated by phosphorylation of both threonine and tyrosine residues within a signature T–X–Y motif located in the activation loop of the kinase. This is catalyzed by a dual-specificity MAPK kinase (MKK or MEK), which is in turn phosphorylated and activated by a MAPK kinase kinase (MKKK or MEKK) [5]. Genetic and biochemical studies have revealed the existence of multiple distinct MAPK pathways in a variety of organisms, which differ both in their responses to specific agonists, and in their downstream substrates and effectors. In mammalian cells these comprise three main pathways: the p42/p44 MAPKs or extracellular signal regulated kinases (ERKs), the Jun N-terminal kinases (JNKs 1–3), and the p38 MAPKs (α, β, γ, and δ).

One conserved property of MAPK signaling cascades is that the physiological outcome of signaling is critically dependent on both the magnitude and the duration of MAPK activation [6]. This suggests that negative regulatory mechanisms will be an important determinant of pathway output. It is now clear that a major point of control in these pathways occurs at the level of the MAPK itself, and is mediated by the intervention of MAPK-specific protein phosphatases. Because MAPKs require phosphorylation of both threonine and tyrosine residues to become fully active, dephosphorylation of either or both of these residues can lead to MAPK inactivation. Thus MAPKs can be potential substrates for all three major classes of protein phosphatases in cells, namely Ser/Thr phosphatases, tyrosine-specific phosphatases (PTPs), and dual-specificity Thr/Tyr phosphatases. Work in a wide variety of model organisms has now demonstrated that MAPKs are indeed regulated by representative enzymes of all three classes of protein phosphatase. The following sections describe the characterization, roles, and properties of protein phosphatases that participate in the regulation of MAPK activities.

SER/THR PROTEIN PHOSPHATASES AND THE REGULATION OF MAPK ACTIVITY

Protein serine/threonine phosphatases (PSTPs) comprise a group of enzymes in mammalian cells, which includes protein phosphatase 1 (PP1), PP2A, PP2B (calcineurin), PP2C, PP3, PP4, PP5, and others [7]. Studies of the biological functions of PSTPs have been hampered by the fact that a relatively small number of PSTP catalytic subunits function as parts of hetero-oligomeric compexes in which they interact with a large and expanding number of regulatory and targeting subunits. Clearly PSTPs can affect the outcome of MAPK signaling by intervention at a number of points within these signaling pathways, but this chapter will focus only on those PSTPs that have been shown to dephosphorylate the MAPKs themselves.

PP2C Isozymes Regulate MAPK Activity

Much of our understanding of the roles played by different classes of protein phosphatase in the regulation of MAPK activity has come from studies in the budding yeast *S. cerevisiae*. With respect to PSTPs, a synthetic lethal screen identified the *PTC1* gene, which encodes a type 2C PSTP, as essential for the growth of yeast lacking the tyrosine-specific phosphatase encoded by *PTP2*, indicating that these two phosphatases regulate the same signaling pathway [8]. Soon afterwards, both *PTP2* and *PTC1* were identified as important regulators of the osmo-regulatory Hog1 MAPK pathway [9, 10]. Further work identified the related PSTPs encoded by *PTC2* and *PTC3* as also playing a key role in Hog1 regulation. In this context, Ptc1 acts to maintain a low basal level of signaling through the Hog1 pathway, whereas *PTC2* and *PTC3* seem to limit the maximum activity of Hog1 under conditions of osmotic stress [11]. In parallel studies in the fission yeast *S. pombe*, the type 2C PSTPs encoded by *ptc1+* and *ptc3+* were found to be responsible for dephosphorylation of the regulatory threonine phospho-acceptor site within the stress-activated Spc1/Sty1 MAPK – a further indication of the degree of conservation of both MAPK pathway architecture and regulation [12].

While this work implicates PP2C in the regulation of MAPK activity, it does not address the mechanism by which these enzymes are able to specifically recognize MAPK substrates. However, global 2–hybrid protein–protein interaction screens in *S. cerevisiae* identified Nbp2, a Nap-1 binding protein containing an SH3 domain, as an interaction partner of Ptc1. Subsequent genetic and biochemical studies revealed that Nbp2 is involved in the negative regulation of Hog1. The Pbs2 protein acts both as a MAPK kinase towards Hog1, but also as a scaffold to assemble Hog1 and upstream activators including the MKKKs Ste11 and Ssk2/22. Nbp2 binds to Pbs2 via its SH3 domain, thus acting as an adaptor protein to bring Ptc1 to the scaffold complex, where it can exert its phosphatase activity towards Hog1 (Figure 17.1) [13]. Interestingly, both Ptc1 and its binding partner Nbp2 have also been implicated in the regulation of Slt2/Mpk1, a stress-activated MAPK pathway involved in the maintenance of cell wall integrity (Figure 17.1) [14]. This suggests that Ptc1 could control the activities of two distinct stress-activated MAPK pathways, and the recent finding that the ability of yeast cells to survive cell wall stress requires the sequential activation of the Hog1 and Slt2/Mpk1 MAPKs is a further indication that the activities of these two pathways are coordinately regulated [15].

There are 16 different genes encoding PP2C isoforms in the human genome and, taking into account alternative splicing, these encode at least 22 PP2C isozymes [16]. The first indication that type 2C PSTPs might also regulate mammalian MAPKs came with the results of screening human cDNA libraries for genes which downregulated the Hog1 pathway in *S. cerevisiae*. This identified human PP2Cα as a potential MAPK phosphatase, and biochemical studies demonstrated that PP2Cα could selectively dephosphorylate p38 MAPK when expressed in mammalian cells [17]. The effects of PP2Cα on p38 appeared to be direct and also dependent on the phosphorylation of p38, as the two proteins could only be co-immunoprecipitated from extracts when cells were exposed to osmotic stress. Interestingly, in contrast to the situation in yeasts where PP2C regulates only the MAPK, PP2Cα also dephosphorylated and inactivated the MAPK kinases (MKKs) MKK3 and SEK1 [17]. PP2Cβ also regulates both p38 and JNK, but not ERK signaling, in mammalian cells. However, in contrast to PP2Cα, this appears to be entirely due to its ability to modulate upstream MAPKK and MAPKKKs rather than the MAPKs themselves [18, 19].

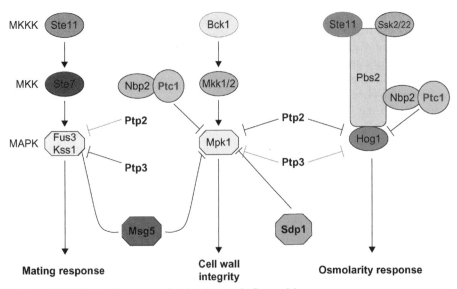

FIGURE 17.1 The regulation of MAPK signaling by protein phosphatases in *S. cerevisiae*.
Positive interactions are indicated by arrows while negative interactions are indicated by bars. Line weights indicate the potency of activities. Note that only those MAPK pathways which have been shown to be regulated by protein phosphatases are shown. For details, see text.

PP2Cδ (PPM1D) was first identified as a human phosphatase which was inducible in a p53-dependent manner by ionizing radiation and named Wip-1 (wild-type p53-induced phosphatase-1) [20]. Wip-1 was later shown to dephosphorylate and inactivate p38 MAPK and thereby attenuate the p38-dependent phosphorylation of p53 on serines 38 and 46. Phosphorylation of these residues is required for p53 transcriptional activity, and Wip-1 was also able to inhibit p53-mediated apoptosis, indicating that it acted as part of a negative feedback loop to control the stress-induced activation of p53 [21, 22]. The characterization of mice lacking PPM1D and cells derived from these animals revealed that as well as suppressing the activation of p53, Wip-1 also regulated the p16 (INK4A)-p19 (ARF) tumor suppressors, and that mouse embryo fibroblasts lacking Wip1 were resistant to the effects of transforming oncogenes. *In vivo* studies then demonstrated that loss of Wip-1 impaired mammary carcinogenesis induced by MMTV-driven expression of either ErbB2 (c-Neu) or H-Ras1, strongly implicating Wip-1 as an oncogenic phosphatase [23]. Although it is clear that Wip-1 has other direct targets within cellular DNA-damage response pathways including the checkpoint kinases Chk1 and Chk2, the ataxia-telangiectasia mutated (ATM) kinase, and uracil DNA glycosylase [24], pharmacological inhibition of p38 MAPK activity was able to reconstitute the sensitivity of Wip-1 null mice to mammary carcinogenesis, indicating that it is the constitutive activation of p38 in the absence of Wip-1 which prevents tumorigenesis [23]. Based on these observations, Wip-1 might also be expected to play some role in human cancers. This was confirmed with the observation that the Wip-1 gene is amplified in breast carcinomas, and is also overexpressed in ovarian cancer and neuroblastoma [22, 25, 26]. Chemical inhibitors of Wip-1 have now been identified; these cause p38-dependent toxicity in cancer cell lines which overexpress Wip-1, and are also able to decrease the proliferation of both breast tumor xenografts and tumors that develop in MMTV-c-Neu transgenic mice [27, 28], These results indicate that Wip-1 may represent a valuable new therapeutic target in the selective treatment of certain solid tumors.

PP2A and the Regulation of ERK Signaling

The various forms of PP2A can account for as much as 30–50 percent of the total PSTP activity in certain mammalian cells and tissues. PP2A is a heterotrimeric protein comprised of a structural A subunit, catalytic C subunit, and highly variable, regulatory B subunit [29]. The structural A subunit is found in two isoforms, α or β, and functions as a scaffold to facilitate the binding of the B and C subunits. The catalytic C subunit also exists in two isoforms, α or β, while the regulatory B subunits comprise four distinct families called B/B55/PPP2R2, B′/PR61/B56/PPP2R5 (referred to as B56 here), B″, and B‴. These B subunits

bind the A subunit in a mutually exclusive manner to form distinct PP2A holoenzymes. Thus, at least 100 different PP2A heterotrimeric complexes can be formed through combinatorial association of these subunits, and studies in several model systems have suggested that particular PP2A holoenzyme complexes mediate specific physiological functions [30, 31].

A key part of the original evidence that the activation of ERK2 MAPK required phosphorylation on both threonine and tyrosine residues was its sensitivity to inactivation by either the tyrosine-specific phosphatase PTP1B or the catalytic subunit of PP2A *in vitro* [32]. Shortly afterwards, it was demonstrated that the inactivation of PP2A, either by exposure of cells to okadaic acid or following transient expression of the small-T antigen of simian virus 40, led to activation of ERK2, and that this was secondary to loss of PP2A activity towards both MAPK kinase (MKK or MEK) and ERK2 itself [33, 34]. Biochemical experiments that exploited the differential sensitivity of protein phosphatases to chemical inhibitors then revealed that PP2A was the major vanadate-insensitive phosphatase activity that acted to dephosphorylate threonine-183 of ERK2 in a variety of cell lines exposed to growth factors [35]. Subsequent work has also implicated PP2A, in complex with the tyrosine-specific phosphatase HePTP, as forming a cholesterol-regulated dual-specificity phosphatase activity capable of inactivating ERK2 in the context of Caveoli / lipid rafts. Loss of this activity on cholesterol depletion has been invoked to explain the mechanism by which ERK activity is coupled to membrane cholesterol levels [36]. Finally, PP2A has been implicated in the regulation of both MAPK kinase and ERK, but it is not clear if these activities are mediated by a specific PP2A holoenzyme. A recent study has used both siRNA-mediated knockdown and overexpression to identify the B56 family of PP2A regulatory subunits as participating in the dephosphorylation of pT-ERK [37].

MAPK REGULATION BY CLASSICAL PROTEIN TYROSINE PHOSPHATASES (PTPs)

Protein tyrosine phosphatases (PTPs) constitute a large and structurally diverse family of cysteine-dependent enzymes, all of which contain the minimal active site consensus sequence Cys–X_5–Arg. PTPs can be broken down into three major classes. These comprise the classical PTPs, which can be further subdivided into transmembrane (receptor-type) PTPs and non-transmembrane (cytosolic) PTPs; the dual-specificity phosphatases (DUSPs); and the low molecular weight PTPs [38, 39]. Within this overall group, two subfamilies of enzymes have been implicated in the regulation of MAPK activity; these are the classical tyrosine-specific PTPs and the dual-specificity protein phosphatases (DUSPs). The following section deals with the former group of proteins and their involvement in the regulation of MAPK pathways in yeasts and higher eukaryotes.

PTPs and the Regulation of MAPK Signaling in Yeasts

The involvement of classical PTPs in the regulation of MAPK signaling was first discovered by genetic and biochemical experiments in yeasts. A genetic screen in *S. cerevisiae* for negative regulators of the Hog1 osmoregulatory MAPK revealed that the tyrosine phosphatases encoded by *PTP2* and *PTP3* were responsible for Hog1 inactivation [40, 41]. Of these two enzymes, Ptp2 is the major regulator of Hog1, while Ptp3 plays a more limited role (Figure 17.1). Interestingly, in addition to regulating the activity of Hog1, these two phosphatases also play a role in determining the subcellular localization of this MAPK. The nuclear retention of Hog1 is impaired in strains lacking Ptp2, while loss of Ptp3 caused Hog1 to accumulate in the nucleus. These results suggest that Ptp2 acts as a nuclear anchor for Hog1, while Ptp3 may function as a cytoplasmic tether for the MAPK [42]. In support of this idea, overexpression of Ptp2 caused nuclear translocation of Hog1 while overexpression of Ptp3 caused cytoplasmic retention of Hog1. In the fission yeast *S. pombe*, the Sty1/Spc1 MAPK, which is the functional homolog of Hog1, is also regulated by two tyrosine-specific phosphatases encoded by *Pyp1+* and *Pyp2+* [43, 44]. Furthermore, *Pyp2+* is transcriptionally upregulated in response to osmotic stress, and this is mediated by the Atf-1 transcription factor which lies downstream of Sty1/Spc1[45–47]. Thus, Pyp2 acts in a negative feedback loop to regulate MAPK activity.

S. cerevisiae possesses a second stress-responsive MAPK in Mpk1/Slt2, which acts in a signaling pathway to maintain cell wall integrity and can be activated in response to heat shock and oxidative stress. Like Hog1, Mpk1/Slt2 is also targeted by Ptp2 and Ptp3, with Ptp2 the major regulator of this MAPK (Figure 17.1) [48]. Finally, in addition to regulating these stress-responsive MAPKs in budding yeast, both Ptp2 and Ptp3 also play a role in the regulation of the pheromone-responsive Fus3 MAPK, which regulates the mating response. However, in this case Ptp3 is the major activity responsible for inactivation of Fus3, with Ptp2 playing the lesser role (Figure 17.1) [49, 50]. This serves to emphasize both the complexity and potential for cross-regulation of MAPK pathways by these phosphatases *in vivo*.

PTPs also Regulate MAPK Signaling in *Drosophila* and Mammals

Given the highly conserved nature of MAPK signaling pathways, it was to be expected that tyrosine-specific phosphatases would also be involved in the regulation of these signaling pathways in higher eukaryotes. The *Drosophila* MAPK encoded by the *Rolled* (*rl*) gene is most homologous to mammalian ERK1 and 2, and lies downstream of *Ras1*, where it can act as a binary switch to trigger cell differentiation in the developing eye. A genetic screen to isolate mutations in genes that act either positively or negatively downstream of Ras1 led to the isolation of tyrosine phosphatase ERK/enhancer of Ras1 (PTP-ER). PTP-ER binds to and inactivates the *rl* MAPK both *in vitro* and *in vivo* [51]. Biochemical studies of the interaction between the yeast Ptp3 phosphatase and its MAPK substrate Fus3 identified a conserved kinase interaction motif (KIM) within the amino-terminal non catalytic domain of Ptp3 [49]. PTP-ER also contains a related motif, which is essential for MAPK binding. At much the same time, a group of highly related PTPs that were able to regulate MAPK signaling in mammalian cells was identified. Striatal enriched phosphatase (STEP) and STEP-like phosphatase (PTP-SL), which are expressed in neuronal cells, were found to bind to the ERK1 and 2 MAPKs [52]. Both STEP and PTP-SL were also able to specifically dephosphorylate the tyrosine residue within the T-E-Y motif of the ERK2 activation loop and, like the *Drosophila* enzyme PTP-ER, both phosphatases interacted with ERK2 via a conserved KIM motif located within the amino-terminal non-catalytic domain. Similar results were reported for the related phosphatase PTPBR7 [53]. Furthermore, in addition to mediating substrate selectivity, the interaction between the KIM motif of PTP-SL and ERK2 also resulted in cytoplasmic retention of this MAPK [54]. In parallel studies, two lymphoid-specific phosphatases, hematopoietic PTP (HePTP) and leukocyte PTP (LC-PTP), were found to be able to block antigen-dependent T-lymphocyte activation [55–57]. Like PTP-SL and STEP, the binding of HePTP to both ERK2 and p38 MAPKs is mediated by a KIM within its amino-terminal non-catalytic domain. Confirmation that ERK1 and 2 are *bona fide* targets of HePTP came with the finding that lymphocytes from HePTP(−/−) mice show enhanced ERK activation after both phorbol myristate acetate (PMA) and anti-CD3-mediated T-cell receptor (TCR) stimulation [58].

An added twist to the regulation of MAPK signaling by these enzymes came with the discovery that PTP-SL, HePTP, and STEP are regulated by the protein kinase A pathway. PKA phosphorylates a serine residue within the KIM of all three enzymes, and phosphorylation greatly reduces the ability of these phosphatases to bind to and dephosphorylate both the ERK1/2 and p38 MAPKs [59–61]. These results suggest a mechanism of protein tyrosine phosphatase-mediated cross-talk between the PKA and MAPK pathways. The physiological significance of this regulatory mechanism was underlined by the observation that the STEP phosphatase is constitutively phosphorylated within its KIM motif in rat neurons. However, in response to glutamate-mediated activation of NMDA receptors, an influx of Ca^{2+} leads to activation of the Ca^{2+}-dependent phosphatase calcineurin and the dephosphorylation and activation of STEP.

This serves to limit the duration of glutamate-induced ERK activation, as well as its translocation to the nucleus and subsequent downstream nuclear signaling [62].

MAPK REGULATION BY DUAL-SPECIFICITY PROTEIN PHOSPHATASES

Unusually, the existence of dual-specificity phosphatases that are dedicated to the regulation of MAPK activity was first demonstrated in mammalian cells, with the discovery of homologous enzymes in yeasts, *Drosophila*, and *C. elegans* following on from this work. *S. cerevisiae* has two enzymes that show sequence homology with the prototypic VH1 dual-specificity phosphatase of vaccinia virus, namely Msg5 and Sdp1. Haploid yeast cells respond to mating pheromone by activating the Fus3 MAPK pathway. Fus3 activation leads to an initial block in the G1 phase of the cell cycle, which, in the absence of a mating partner and continued presence of pheromone, is gradually overcome. Msg5 was isolated in a genetic analysis of this adaptation process as a suppressor of pheromone-induced cell cycle arrest, and found to act by dephosphorylating the regulatory threonine and tyrosine residues of Fus3 [63]. Msg5 is also a prototypic example of negative feedback control by this class of phosphatases, as its pheromone-inducible expression is dependent on the transcription factor Ste12p, which lies downstream of Fus3 itself. Subsequent work has also identified Msg5 as a regulator of the stress-activated Mpk1/Slt2 MAPK pathway involved in maintaining cell wall integrity, where it seems to act to suppress basal Mpk1/Slt2 activity in unstressed cells (Figure 17.1) [64, 65].

Mpk1/Slt2 also appears to be the sole target of the Sdp1 phosphatase which regulates the activation of this pathway in response to heat stress (Figure 17.1) [66, 67]. Interestingly, Sdp1 is the first example of a cysteine-dependent protein phosphatase which is sensitive to the presence of reducing agent and shows increased activity on oxidation [68]. This is mediated by formation of an intramolecular disulfide bond between two cysteine residues, one of which is immediately adjacent to the catalytic cysteine, while the other lies in an essential region of the amino-terminus of Sdp1. This may represent a mechanism by which Sdp1 can overcome the intrinsic limitation of using a thiol-based phosphatase to regulate the activity of a stress-responsive MAPK by allowing continued function under conditions of oxidative stress.

Both *Drosophila* and *C. elegans* Utilize Dual-Specificity Phosphatases to Regulate MAPK Signaling

The first example of a DUSP involved in the physiological regulation of a stress-responsive MAPK pathway came from studies of the *puckered* (*puc*) phosphatase in the process of dorsal closure during *Drosophila* embryogenesis. *Basket* (*bsk*), the fly homolog of JNK, plays a key role in mediating changes in cell migration and morphology of those lateral epithelial cells which eventually cover the dorsal region of the embryo, and mutations in *puc* lead to abnormalities in this process [69]. *Puc* encodes a dual-specificity phosphatase with significant sequence homology to the human MKP encoded by DUSP1/MKP-1 (see below), and inactivates *bsk* by dephosphorylation of the threonine and tyrosine residues within the activation loop of the kinase [70]. *Puc* expression is also dependent on the activity of *bsk*, indicating that, like Msg5 in yeast, *puc* functions as part of a negative feedback control to regulate MAPK activity. Subsequent work has shown that *puc* also promotes cell viability by antagonizing JNK-induced apoptosis, and that heterozygous loss of function mutations in *puc* extend median and maximum lifespan in *Drosophila* [71]. In addition to *puc*, *Drosophila* also express a second MKP which is homologous to the mammalian phosphatase encoded by DUSP6/MKP-3 (see later text). DMKP-3, like its mammalian counterpart, specifically targets the classical ERK MAPK, and plays an essential role in the negative regulation of the Ras/ERK pathway during early *Drosophila* development [72, 73].

Two functional dual-specificity MKPs, *lip-1* and *vhp-1*, have been characterized in *C. elegans*, and play important roles in development and the regulation of stress responses. During vulval development, a signal from the anchor cell stimulates the Ras/MAPK signaling pathway in the closest vulval precursor cell P6.p to induce the primary cell fate. A lateral signal from these P6.p cells then activates the Notch signaling pathway in the neighboring P5.p and P7.p cells to prevent them from adopting the primary fate and to specify the secondary fate. Berset and colleagues [74] discovered that it is the Notch-mediated upregulation of *lip-1* transcription in P5.p and P7.p cells and consequent downregulation of the ERK MAPK pathway which mediates this lateral signaling and hence pattern formation. In addition to this role in vulval development, the regulation of MAPK activity by *lip-1* also controls the extent of germline proliferation in *C. elegans*, where it acts as a pivotal regulator of the decision between cell proliferation and differentiation [75].

The *C. elegans* *vhp-1* phosphatase is most closely related to the mammalian DUSP16/MKP-7 phosphatase (see below). Loss of *vhp-1* causes larval arrest, and the observation that this phenotype is suppressed by expression of the worm JNK homolog *kgb-1* indicates that this MAPK is a physiological target of *vhp-1* [76]. Loss of *vhp-1* also partially suppressed the Cu^{2+}-sensitive phenotype of mutants in both *kgb-1* and its upstream activator *Mek1*, implicating this phosphatase in stress responses triggered by exposure to metal ions. Interestingly, deletion of the *C. elegans* p38 MAPK encoded by *pmk-1* also caused a modest sensitivity to Cu^{2+}, enhanced the metal sensitivity of the *kgb-1* mutant, and, most importantly, suppressed the resistance to Cu^{2+} seen in the *kgb-1*, *vhp-1* double mutant. Taken together these data indicate that the JNK and p38 MAPKs are partially redundant regulators of the heavy metal response, and that both kinases are physiological targets for *vhp-1* [76]. Finally, the *pmk-1* pathway plays a key role in mediating immune function in *C. elegans*, and subsequent work showed that both *mek-1* and *vhp-1* play a role in regulating immunity against bacterial pathogens. Thus, a single MKP is able to integrate signals transduced by distinct MAPK pathways (Figure 17.2) [77].

MAMMALIAN DUAL-SPECIFICITY MAPK PHOSPHATASES

The dual-specificity protein phosphatases (DUSPs), also known as MAPK phosphatases (MKPs), are by far the largest group of protein phosphatases dedicated to the regulation of MAPK activity in mammalian cells (see Table 17.1). These constitute a family of 10 catalytically active enzymes, which share a common structure comprising a catalytic domain that shows significant sequence homology to the prototypic dual-specificity phosphatase VH1 of vaccinia virus, and an amino terminal non-catalytic domain [78–80]. This latter part of the protein contains two regions of sequence similarity with the catalytic domain of the cell cycle regulatory phosphatase Cdc25 [81], and reflects a common evolutionary origin of this domain within the MKPs and Cdc25 with the rhodanese family of sulfotransferases [82]. The N-terminal domain also contains a KIM motif consisting of a cluster of basic amino acids, and this mediates specific protein–protein interactions between the MKPs and their MAPK substrates [83]. Based on sequence similarity, substrate specificity, and subcellular localization, the 10 MKPs can be subdivided into four distinct groups. The first of these contains four inducible nuclear MKPs – DUSP1/MKP-1,

DUSP2/PAC1, DUSP4/MKP-2, and DUSP5. The second group contains three cytoplasmic ERK-specific MKPs – DUSP6/MKP-3, DUSP7/MKP-X, and DUSP9/MKP-4. The final group is made up of three MKPs – DUSP8/hVH5, DUSP10/MKP-5, and DUSP16/MKP-7, all of which show substrate selectivity for the stress-activated MAPKs JNK and p38. These three groups are dealt with in turn below, highlighting recent advances in our understanding of their physiological functions.

The Inducible Nuclear MKPs

The inducible nuclear phosphatases DUSP1/MKP-1 (also known as CL100, Erp, or hVH1) and DUSP2/PAC1 were the first members of this family of enzymes to be identified as specific regulators of MAPK phosphorylation [84–86]. DUSP1/MKP-1 is subject to rapid transcriptional upregulation in response to a wide variety of growth factors and conditions of cellular stress [87–89], while DUSP2/PAC1 was identified as a mitogen-inducible gene in T cells [90]. This family was expanded with the discovery of DUSP4/MKP-2 (which is most highly related to DUSP1/MKP-1) and DUSP5 (also known as hVH3) [91–95]. Both DUSP1/MKP-1 and DUSP2/PAC1 proteins were originally characterized as potent and specific inactivators of the ERK1 and ERK2 MAPKs in mammalian cells. However, it is now clear that DUSP1/MKP-1 is also able to dephosphorylate and inactivate the stress-activated MAPKs JNK and p38, and that all three major classes of MAPK are targets for DUSP1/MKP-1 *in vivo* [96–99]. The ability to interact with and inactivate both mitogen and stress-activated MAPK isoforms is shared by DUSP2/PAC1 and DUSP4/MKP-2 [96]. In contrast, DUSP5/hVH3 appears to act specifically on the ERK1/2 MAPKs [100].

The generation of knockout mice and studies of cultured cells derived from these animals has revealed that the inducible nuclear MKPs play important non-redundant roles in the regulation of MAPK signaling. Mice lacking DUSP1/MKP-1 were originally characterized as having no obvious phenotype or abnormalities in the regulation of the ERK1/2 MAPKs [101]. However, more recent work has revealed that this phosphatase plays a key role in the regulation of stress responses, immunity, and metabolic homeostasis. Mouse embryo fibroblasts (MEFs) derived from mice lacking DUSP1/MKP-1 are sensitive to a wide variety of stresses, including cisplatin, hydrogen peroxide, ionizing radiation, and anisomycin [102–105]. Increased cell death correlates with both elevated levels and duration of JNK and p38 activation, suggesting that DUSP1 plays a key role in cell survival by attenuating the level of signaling though stress-activated MAPK pathways.

The observation that both LPS-induced p38 activation and tumor necrosis factor (TNF) production were greatly increased in mouse macrophages from DUSP1/MKP-1

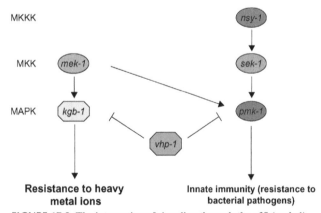

MKKK   nsy-1

MKK   mek-1   sek-1

MAPK   kgb-1   pmk-1

vhp-1

Resistance to heavy metal ions   **Innate immunity (resistance to bacterial pathogens)**

FIGURE 17.2 The integration of signaling through the p38 (pmk-1) and JNK (kgb-1) signaling pathways by the dual-specificity phosphatase vhp-1 in *C. elegans*.
Positive interactions are indicated by arrows; negative interactions are indicated by bars. For details see text.

TABLE 17.1 Nomenclature and key properties of the mammalian dual-specificity MAPK phosphatases (see text for details and literature references)

Gene	MKP	Trivial names	Chromosomal localization	Subcellular localization	Substrate selectivity	Physiological function(s)
DUSP1	MKP-1	CL100, Erp, 3CH134, hVH1	5q34	Nuclear	JNK,p38,ERK	Negative regulation of immune function; regulation of metabolic homeostasis and stress response
DUSP4	MKP-2	Typ1, Sty8, hVH2,	8p11–p12	Nuclear	JNK, p38, ERK	Unclear at present; inducible by many of the same stimuli as DUSP1
DUSP2		PAC1	2q11	Nuclear	ERK, p38	Positive regulator of inflammatory responses; null mice are resistant to autoimmune inflammatory arthritis
DUSP5		hVH3, B23	10q25	Nuclear	ERK	IL-2 inducible gene in T cells; may act as a nuclear anchor for inactive ERK
DUSP6	MKP-3	Pyst1, rVH6	12q22–q23	Cytoplasmic	ERK	Negative feedback regulator of ERK activity downstream of FGFR signalling
DUSP7	MKP-X	Pyst2, B59	3p21	Cytoplasmic	ERK	Function(s) poorly understood at present
DUSP9	MKP-4	Pyst3	Xq28	Cytoplasmic	ERK, p38	Essential for placental development and function (labyrinth formation)
DUSP8		M3/6, hVH5, HB5	11p15.5	Cytoplasmic/ nuclear	JNK, p38	Function(s) poorly understood at present
DUSP10	MKP-5		1q41	Cytoplasmic/ nuclear	JNK, p38	Function(s) poorly understood at present
DUSP16	MKP-7		12p12	Cytoplasmic/ nuclear	JNK, p38	Functions in innate and adaptive immunity

knockout mice suggests a possible role for DUSP1/MKP-1 in the regulation of inflammation during the innate immune response [106]. This was confirmed in a number of studies reporting both the overproduction of cytokines, including TNF, interleukin 6 (IL6), and interleukin 10 (IL10), and acute sensitivity of DUSP1/MKP-1 null mice to lethal endotoxic shock. In addition, these animals also exhibit a significant increase in both the incidence and severity of experimentally induced autoimmune arthritis [107–110]. Thus DUSP1/MKP-1 controls the levels of both pro-inflammatory (TNF) and anti-inflammatory (IL10) cytokines in response to bacterial endotoxins, indicating that a dynamic balance between MAPK activation and the activity of DUSP1/MKP-1 plays a key role in determining the outcome of signaling downstream of toll-like receptor (TLR) in mediating the innate immune response (Figure 17.3). The observation that DUSP1/MKP-1 transcription is increased in response to synthetic glucocorticoids such as dexamethasone raises the possibility that at least part of the anti-inflammatory effects of these drugs could be mediated by DUSP1/MKP-1 itself [111, 112]. This idea is supported by the finding that the suppression

of a subset of pro-inflammatory genes by dexamethasone is reduced in macrophages from DUSP1/MKP-1 knockout mice, and that the anti-inflammatory effects of dexamethasone on zymosan-induced inflammation are also impaired in these animals [113]. However, a more recent study found that pro-inflammatory cytokine (IL-6, TNF) and chemokine (C-C motif ligand 2 or CCL2,) is still downregulated by glucocorticoids in macrophages from DUSP1/MKP-1 null mice. In addition, while these animals show enhanced mast cell degranulation and are highly susceptible to anaphylaxis, these effects were still alleviated by dexamethasone. Finally, glucocorticoids also repressed other inflammatory responses, such as dinitrofluorobenzene-induced contact hypersensitivity and LPS-induced mortality in DUSP1/MKP-1 null mice, indicating that the anti-inflammatory action of these drugs is largely independent of DUSP1 [114].

In addition to its effects in the immune system, DUSP1/MKP-1 has also been implicated in the regulation of MAPK signaling in insulin-responsive tissues, where it appears to play a key role in metabolic homeostasis. DUSP1/MKP-1 null mice gain weight at a slower rate

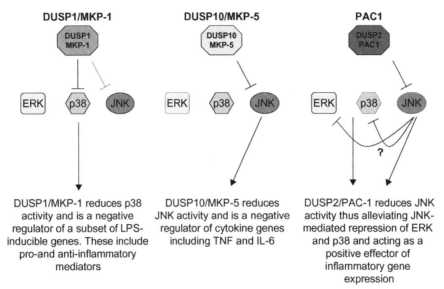

FIGURE 17.3 MKPs are key regulators of MAPK activity in immune effector cells.
For details see text.

than their wild-type littermates when weaned onto a normal chow diet. Furthermore, they are also resistant to the development of obesity when fed on a high fat diet, but remain susceptible to the development of glucose intolerance and hyperinsulinemia. Thus, it appears that DUSP1/MKP-1 is able to regulate body mass independently of the regulation of glucose homeostasis [115]. Examination of MAPK signaling in insulin-responsive tissues reveals complex effects on multiple MAPK pathways on deletion of DUSP1/MKP-1. Abnormally high levels of JNK and p38 activation are seen in white adipose tissue (WAT), skeletal muscle, and liver, while elevated ERK1/2 activation is seen in WAT and muscle, but not in the liver [115]. How, then, do these complex changes in MAPK activities translate into the observed phenotype? Elevated JNK activity has previously been associated with both obesity and insulin resistance, with mice lacking JNK1 exhibiting insulin sensitivity and resistance to diet-induced obesity [116]. Given that loss of DUSP1/MKP-1 causes elevated JNK activation, these observations are difficult to reconcile with the observed phenotype in the DUSP1/MKP-1 null mice. However, the phenotype of animals lacking DUSP1/MKP-1 may be a reflection of two key properties of this phosphatase. First, DUSP1/MKP-1 regulates not only JNK, but also p38 and ERK signaling. p38 MAPK can regulate levels of energy expenditure via activation of peroxisome proliferator-activated receptor-gamma (PPARγ) co-activator 1α (PGC1α), while mice lacking ERK1 display resistance to diet-induced obesity and do not become as insulin resistant as wild-type animals [117, 118]. Thus, the combined effects of DUSP1/MKP-1 loss on all three major MAPK pathways rather than the deregulation of a single MAPK pathway may contribute to the observed phenotype.

Secondly, DUSP1/MKP-1 is a nuclear enzyme and might be expected to preferentially affect nuclear signaling events, leaving cytoplasmic MAPK activities relatively unaffected. In support of this, the phosphorylation of cytoplasmic JNK substrates such as insulin receptor substrate 1 (IRS-1) was not altered in mice lacking DUSP1/MKP-1. In contrast, levels of nuclear phospho-cJun were significantly increased in hepatocytes from the null animals [115]. Therefore, increases in the activity of all three major classes of MAPK in the nucleus may lead to altered transcriptional regulation of key genes which regulate fatty acid metabolism and energy expenditure.

Thus far, DUSP2/PAC1 is the only other nuclear MKP for which the results of gene knockout experiments have been reported. Like DUSP1/MKP-1, deletion of DUSP2/PAC1 produces complex changes in the activities of all three major classes of MAPK in immune effector cells, and reveals an unexpected positive regulatory function for this enzyme in mediating certain inflammatory responses. First, there is a significant decrease in the expression of pro-inflammatory mediators, cytokines, and chemokines in macrophages from DUSP2/PAC1 mice following exposure to LPS. These animals were then assessed using an autoimmune (K/BxN serum-induced) model of inflammatory arthritis. In this model, pathology is driven by effector leukocytes such as mast cells, neutrophils, and macrophages, and DUSP2/PAC1 null animals showed greatly diminished levels of inflammatory cell infiltration, and eventual joint destruction, when compared with wild-type controls [119]. Despite the fact that DUSP2/PAC1 has been reported to dephosphorylate ERK2 and p38 but not JNK in mammalian cells [96], deletion of DUSP2/PAC1 resulted in reduced levels of ERK and p38 activation and increased levels of phospho-JNK in macrophages exposed to LPS [119].

Pharmacological inhibition of JNK in these cells resulted in a significant increase in phospho-ERK levels, indicating that, by elevating JNK activity, loss of DUSP2/PAC1 might suppress ERK activation by an as yet undefined mechanism (Figure 17.3). Whether or not the elevated level of JNK activation seen in DUSP2/PAC1 null cells is a reflection of the direct activity of DUSP2/PAC1 towards this MAPK remains to be determined.

Of the two remaining members of this group of MKPs, rather less is known. DUSP4/MKP-2 is most closely related to DUSP1/MKP-1, and is expressed in many of the same cell lines and tissues where it is inducible in response to both growth factors and cellular stress [91–93]. Like DUSP1/MKP-1, DUSP4/MKP-2 can inactivate both ERK and JNK MAPKs. However, although it binds tightly to the p38 MAPK, it does not appear to be able to cause significant inactivation of this MAPK [96, 120]. DUSP5, the final member of this group, was originally characterized as inducible in response to both growth factors and heat shock [94, 95]. More recently, DUSP5 mRNA was found to be inducible by interleukin 2 (IL-2) in T cells, and DUSP5 was also identified as a direct transcriptional target of the p53 tumor suppressor [121, 122]. Interestingly, unlike DUSP1/MKP-1, DUSP2/PAC1, and DUSP4/MKP-2, DUSP5 is highly specific in its ability to bind to and inactivate ERK1/2, showing no significant activity towards p38 or JNK. Furthermore, overexpression of DUSP5 causes both the inactivation and nuclear translocation of endogenous ERK2, indicating that DUSP5 may play a role as a nuclear anchor for inactive MAPK [100]. Finally in agreement with a possible physiological role for DUSP5 in regulating the outcome of IL-2 signaling in T cells, overexpression of DUSP5 causes a block in thymocyte development at the double positive stage, and DUSP5-expressing mature T cells exhibit decreased IL-2–dependent proliferation and defective IL-2–mediated gene expression [123].

The Cytoplasmic ERK-Specific MKPs

This second group of MKPs comprises three closely related phosphatases: DUSP6/MKP-3, DUSP7/MKP-X, and DUSP9/MKP-4. Of these, the most widely studied and best characterized is DUSP6/MKP-3, a cytosolic MKP, which specifically recognizes and inactivates the ERK1/2 MAPKs *in vitro* and *in vivo* [98, 124]. The ability of DUSP6/MKP-3 to specifically interact with ERK2 and its cytoplasmic localization are determined by a kinase interaction motif (KIM) and nuclear export signal (NES), respectively, both of which are located within the amino terminal non-catalytic domain of the protein [125, 126]. The substrate selectivity of DUSP6/MKP-3 for ERK1/2 is further enhanced by allosteric activation of the catalytic domain on binding to MAPK [127]. A combination of biochemical and structural studies has revealed that in the absence of substrate, key catalytic residues display a distorted geometry.

In particular, a loop containing an essential aspartic acid residue (Asp262) is displaced to such an extent that it cannot participate in catalysis. However, ERK binding causes substrate-induced closure of this general acid loop over the active site, thus repositioning Asp262 for efficient catalysis [128–130].

The first clues as to the physiological function (s) of DUSP6/MKP-3 came with the observation that DUSP6/MKP-3 expression correlated with several sites of fibroblast growth factor (FGF) signaling in both mouse and chicken embryos [131, 132]. Subsequent studies in the developing limbs and nervous tissue of chicken embryos revealed that the expression of DUSP6/MKP-3 was mediated by FGF as part of a negative feedback control to regulate levels of activated ERK2 in these tissues [132, 133]. The use of pharmacological inhibitors of either the Ras/MAPK or PI-3 kinase signaling pathways indicates that the expression of DUSP6/MKP-3 in response to FGF in both limb buds and nervous tissue is absolutely dependent on the former pathway [132, 134]. This is corroborated by recent studies in mouse NIH3T3 fibroblasts in which FGF-dependent DUSP6/MKP-3 transcription was found to be dependent on MAPK signaling and was mediated via the binding of the Ets1 and Ets2 transcription factors to a conserved site within the DUSP6/MKP-3 gene promoter [135]. Confirmation of a role for DUSP6/MKP-3 in the regulation of FGF signaling *in vivo* has now come from studies using mice lacking a functional DUSP6/MKP-3 gene. Embryos lacking DUSP6/MKP-3 show increased levels of ERK phosphorylation in the limb buds, and present with a variably penetrant postnatal lethality, skeletal abnormalities including dwarfism and premature fusion of the cranial sutures (craniosynostoses), and deafness [136]. All of these features are consistent with increased activity of FGF signaling.

Little is known about DUSP7/MKP-X, the second member of this subfamily. Like DUSP6/MKP-3, it shows substrate selectivity for ERK1/2 and its mRNA is inducible in response to serum stimulation [137]. Overexpression of DUSP7/MKP-X can also block the cell transformation by both v-*raf* and H-*ras*, most probably due to its ability to dephosphorylate ERK1 and 2 [138]. DUSP9/MKP-4, the final member of this group to be characterized, is somewhat unusual. Unlike DUSP6/MKP-3, which targets only ERK1/2, DUSP9/MKP-4 also shows significant activity towards the stress-activated MAPKs, and in particular p38 MAPK [139, 140]. Further insights into the physiological function of this MKP came with the finding that mice lacking DUSP9/MKP-4 die *in utero* due to placental insufficiency and, specifically, a failure of labyrinth development [141]. This correlates with the expression pattern of DUSP9/MKP-4 in wild-type animals, and suggests that the regulation of ERK and/or p38 by this phosphatase is critical for placental development and function. This is supported by the observation that deletion of the murine genes encoding either p38alpha or MEK1 (the upstream activator of ERK1/2) gives rise to placental abnormalities that are almost identical to those seen in the DUSP9/MKP-1 knock-out mice [142, 143].

Despite its essential role in the formation of extraembryonic tissues, when tetraploid rescue experiments were performed in order to bypass the effects of this defect, DUSP9/MKP-4 male knockout mice were born. These animals appeared normal and were fertile, indicating that DUSP9/MKP-4 is dispensable for normal embryonic development [141]. The early lethality of the DUSP9/MKP-4 gene deletion and the low efficiency of tetraploid rescue has precluded a more extensive analysis of the function of DUSP9/MKP-4 in adult tissues. However, two recent studies have indicated that the ability of DUSP9/MKP-4 to regulate signaling through more than one MAPK pathway may define its physiological role(s). First, in a clonal model of epidermal carcinogenesis, DUSP9/MKP-4 was downregulated at initiation and lost at malignant conversion. Furthermore, reconstitution of DUSP9/MKP-4, but not DUSP6/MKP-3, expression in malignant tumor cells resulted in cell death and tumor suppression [144]. Secondly, overexpression of DUSP9/MKP-4 in the liver of ob/ob mice decreased ERK and JNK phosphorylation, leading to a reduction in fed and fasted glycemia, improved glucose intolerance, decreased expression of gluconeogenic and lipogenic genes, and reduced hepatic steatosis. These results imply that MKP-4 has a protective effect against the development of insulin resistance through its ability to dephosphorylate and inactivate more than one class of MAPK [145].

MKPs that Specifically Inactivate p38 and JNK

This final group of three MKPs comprises DUSP8 (also known as hVH5 and M3/6), DUSP10/MKP-5, and DUSP16/MKP-7. All three MKPs are able to target p38 and/or JNK, and have very little activity towards ERK1/2 [124, 146–149]. Of these MKPs, most is known about DUSP10/MKP-5 and to date this is the only member of this group which has been the subject of a gene deletion experiment. DUSP10/MKP-5 is inducible in a mouse macrophage cell line in response to LPS and is also expressed in T cells, suggestive of a function in regulating immune system function. Deletion of the murine DUSP10/MKP-5 gene does not cause gross developmental problems, nor is the normal development of the immune system compromised in these animals. However, both T-helper cells and LPS-treated macrophages from DUSP10/MKP-5 knockout mice exhibit higher levels of activated JNK when compared to wild-type controls, suggesting a role for this enzyme in the regulation of innate or adaptive immunity. Interestingly, no apparent abnormalities were noted in the regulation of p38 MAPK in these cells. When peritoneal macrophages from DUSP10/MKP-5 null animals were exposed to LPS they produced increased levels of pro-inflammatory cytokines such as IL-6 and TNF, and the injection of LPS into null mice led to increased serum levels of TNF when compared with wild-type animals [150]. The T cell priming activity of LPS-treated antigen presenting cells (APCs) was also significantly enhanced, indicating that DUSP10/MKP-5 is a negative regulator of innate immunity, and differentiated T helper 1 (Th1) and T helper 2 (Th2) cells produced increased levels of interferon gamma (IFNγ) and interleukin-4 (IL-4), indicating that cytokine expression in effector T cells is also subject to negative regulation by DUSP10/MKP-5.

Finally, DUSP10/MKP-5 null animals were analyzed using immune and autoimmune disease models. Immunization with myelin oligodendrocyte glycoprotein (MOG) induces experimental autoimmune encephalomyelitis (EAE), and animals lacking DUSP10/MKP-5 displayed a much-reduced incidence and severity of disease. This indicates that DUSP10/MKP-5 plays a critical role in either the generation or expansion of the autoreactive T cells which underlie this pathology [150]. To examine T cell-mediated immunity to infection, both DUSP10/MKP-5 null and wild-type animals were infected with lymphocytic choriomeningitis virus (LCMV). The animals lacking DUSP10/MKP-5 cleared virus normally, and exhibited no differences in their primary T cell response to infection. However, a second viral challenge caused markedly elevated serum levels of TNF, leading to the immune-mediated death of the DUSP10/MKP-5 null mice some 2–4 days later [150]. From these results, it is clear that the regulation of JNK activity by DUSP10/MKP-5 plays a key role as a negative regulator of innate and adaptive immunity (Figure 17.3). It remains to be seen if this phosphatase plays any role either in the regulation of p38 MAPK activity or in the cellular stress response. It is hoped that gene targeting experiments for DUSP8 and DUSP16/MKP-7 will provide insights into the physiological functions of the remaining members of this group of JNK/p38-specific MKPs.

SUMMARY

Genetic and biochemical experiments in a wide variety of model organisms have revealed that a complex network of protein phosphatases controls MAPK activities. These studies have revealed important regulatory principles which extend from yeast to man. Studies in *S. cerevisiae* were the first to demonstrate that a single MAPK can be subject to regulation by all three major classes of protein phosphatase, including type 2 serine/threonine phosphatases, classical cysteine-dependent tyrosine phosphatases, and dual-specificity protein phosphatases, and also to suggest that these enzymes could facilitate cross-talk between distinct signaling pathways. Another concept first demonstrated in yeast is that protein phosphatases may regulate the subcellular localization as well as the activity of target

MAPKs, and there is now support for such a role for both tyrosine-specific and dual-specificity phosphatases in mammalian cells. In common with many other components of signal transduction pathways, protein–protein interactions appear to be critical for the specific and differential targeting and inactivation of distinct MAPK isoforms as exemplified by the dual-specificity MKPs in mammalian cells. In addition, certain of these MKPs also undergo substrate-induced catalytic activation, further enhancing their specificity of action. Another concept first demonstrated in yeast is that the genes encoding MKPs can be subject to rapid transcriptional upregulation in response to signals mediated by the MAPK pathway that they target. The ERK-mediated expression of DUSP6/MKP-3 mRNA and protein in response to FGF signaling is a good example of this kind of negative feedback control in the regulation of a mammalian MAPK pathway. Finally, it has been demonstrated that certain MKPs, such as DUSP1/MKP-1 and DUSP2/PAC1, can regulate the activities of multiple MAPK pathways in certain cells and tissues. This coordinated regulation of distinct MAPKs is crucial to the regulation of important physiological endpoints such as immune function and metabolic homeostasis, indicating that MKPs act as key integration points in the regulation of MAPK signaling.

ACKNOWLEDGEMENTS

Work in the author's laboratory is supported by Cancer Research UK.

REFERENCES

1. Chang L, Karin M. Mammalian MAP kinase signalling cascades. *Nature* 2001;**410**:37–40.
2. Pearson G, Robinson F, Beers Gibson T, Xu BE, Karandikar M, Berman K, Cobb MH. Mitogen-activated protein (MAP) kinase pathways: regulation and physiological functions. *Endocr Rev* 2001;**22**:153–83.
3. Dong C, Davis RJ, Flavell RA. MAP kinases in the immune response. *Annu Rev Immunol* 2002;**20**:55–72.
4. Wada T, Penninger JM. Mitogen-activated protein kinases in apoptosis regulation. *Oncogene* 2004;**23**:2838–49.
5. Marshall CJ. MAP kinase kinase kinase, MAP kinase kinase and MAP kinase. *Curr Opin Genet Dev* 1994;**4**:82–9.
6. Marshall CJ. Specificity of receptor tyrosine kinase signaling: transient versus sustained extracellular signal-regulated kinase activation. *Cell* 1995;**80**:179–85.
7. Gallego M, Virshup DM. Protein serine/threonine phosphatases: life, death, and sleeping. *Curr Opin Cell Biol* 2005;**17**:197–202.
8. Maeda T, Tsai AY, Saito H. Mutations in a protein tyrosine phosphatase gene (PTP2) and a protein serine/threonine phosphatase gene (PTC1) cause a synthetic growth defect in *Saccharomyces cerevisiae*. *Mol Cell Biol* 1993;**13**:5408–17.
9. Maeda T, Wurgler-Murphy SM, Saito H. A two-component system that regulates an osmosensing MAP kinase cascade in yeast. *Nature* 1994;**369**:242–5.
10. Warmka J, Hanneman J, Lee J, Amin D, Ota I. Ptc1, a type 2C Ser/Thr phosphatase, inactivates the HOG pathway by dephosphorylating the mitogen-activated protein kinase Hog1. *Mol Cell Biol* 2001;**21**:51–60.
11. Young C, Mapes J, Hanneman J, Al-Zarban S, Ota I. Role of Ptc2 type 2C Ser/Thr phosphatase in yeast high-osmolarity glycerol pathway inactivation. *Eukaryot Cell* 2002;**1**:1032–40.
12. Nguyen AN, Shiozaki K. Heat-shock-induced activation of stress MAP kinase is regulated by threonine- and tyrosine-specific phosphatases. *Genes Dev* 1999;**13**:1653–63.
13. Mapes J, Ota IM. Nbp2 targets the Ptc1-type 2C Ser/Thr phosphatase to the HOG MAPK pathway. *Embo J* 2004;**23**:302–11.
14. Du Y, Walker L, Novick P, Ferro-Novick S. Ptc1p regulates cortical ER inheritance via Slt2p. *EMBO J* 2006;**25**:4413–22.
15. Bermejo C, Rodriguez E, Garcia R, Rodriguez-Pena JM, Rodriguez de la Concepcion ML, Rivas C, Arias P, Nombela C, Posas F, Arroyo J. The sequential activation of the yeast HOG and SLT2 pathways is required for cell survival to cell wall stress. *Mol Biol Cell* 2008. in press.
16. Lammers T, Lavi S. Role of type 2C protein phosphatases in growth regulation and in cellular stress signaling. *Crit Rev Biochem Mol Biol* 2007;**42**:437–61.
17. Takekawa M, Maeda T, Saito H. Protein phosphatase 2Calpha inhibits the human stress-responsive p38 and JNK MAPK pathways. *EMBO J* 1998;**17**:4744–52.
18. Hanada M, Kobayashi T, Ohnishi M, Ikeda S, Wang H, Katsura K, Yanagawa Y, Hiraga A, Kanamaru R, Tamura S. Selective suppression of stress-activated protein kinase pathway by protein phosphatase 2C in mammalian cells. *FEBS Letts* 1998;**437**:172–6.
19. Hanada M, Ninomiya-Tsuji J, Komaki K, Ohnishi M, Katsura K, Kanamaru R, Matsumoto K, Tamura S. Regulation of the TAK1 signaling pathway by protein phosphatase 2C. *J Biol Chem* 2001;**276**:5753–9.
20. Fiscella M, Zhang H, Fan S, Sakaguchi K, Shen S, Mercer WE, Vande Woude GF, O'Connor PM, Appella E. Wip1, a novel human protein phosphatase that is induced in response to ionizing radiation in a p53-dependent manner. *Proc Natl Acad Sci USA* 1997;**94**:6048–53.
21. Takekawa M, Adachi M, Nakahata A, Nakayama I, Itoh F, Tsukuda H, Taya Y, Imai K. p53-inducible wip1 phosphatase mediates a negative feedback regulation of p38 MAPK-p53 signaling in response to UV radiation. *EMBO J* 2000;**19**:6517–26.
22. Bulavin Jr. DV, Demidov ON, Saito S, Kauraniemi P, Phillips C, Amundson SA, Ambrosino C, Sauter G, Nebreda AR, Anderson CW, Kallioniemi A, Fornace AJ, Appella E. Amplification of PPM1D in human tumors abrogates p53 tumor-suppressor activity. *Nat Genet* 2002;**31**:210–15.
23. Bulavin Jr. DV, Phillips C, Nannenga B, Timofeev O, Donehower LA, Anderson CW, Appella E, Fornace AJ. Inactivation of the Wip1 phosphatase inhibits mammary tumorigenesis through p38 MAPK-mediated activation of the p16(Ink4a)-p19(Arf) pathway. *Nat Genet* 2004;**36**:343–50.
24. Lu X, Nguyen T-A, Moon S-H, Darlington Y, Sommer M, Donehower LA. The type 2C phosphatase Wip1: An oncogenic regulator of tumour suppressor and DNA damage pathways. *Cancer Metast Rev* 2008 in press.
25. Hirasawa A, Saito-Ohara F, Inoue J, Aoki D, Susumu N, Yokoyama T, Nozawa S, Inazawa J, Imoto I. Association of 17q21–q24 gain in ovarian clear cell adenocarcinomas with poor prognosis and identification of PPM1D and APPBP2 as likely amplification targets. *Clin Cancer Res* 2003;**9**:1995–2004.

26. Saito-Ohara F, Imoto I, Inoue J, Hosoi H, Nakagawara A, Sugimoto T, Inazawa J. PPM1D is a potential target for 17q gain in neuroblastoma. *Cancer Res* 2003;**63**:1876–83.

27. Rayter S, Elliott R, Travers J, Rowlands MG, Richardson TB, Boxall K, Jones K, Linardopoulos S, Workman P, Aherne W, Lord CJ, Ashworth A. A chemical inhibitor of PPM1D that selectively kills cells overexpressing PPM1D. *Oncogene* 2008;**27**:1036–44.

28. Belova Jr GI, Demidov ON, Fornace AJ, Bulavin DV. Chemical inhibition of Wip1 phosphatase contributes to suppression of tumorigenesis. *Cancer Biol Ther* 2005;**4**:1154–8.

29. Mumby M. The 3D structure of protein phosphatase 2A: new insights into a ubiquitous regulator of cell signaling. *ACS Chem Biol* 2007;**2**:99–103.

30. Arnold HK, Sears RC. A tumor suppressor role for PP2A-B56α through negative regulation of c-Myc and other key oncoproteins. *Cancer Metast Rev* 2008. in press.

31. Sablina AA, Hahn WC. SV40 small T antigen and PP2A phosphatase in cell transformation. *Cancer Metast Rev* 2008;**6**. in press.

32. Anderson NG, Maller JL, Tonks NK, Sturgill TW. Requirement for integration of signals from two distinct phosphorylation pathways for activation of MAP kinase. *Nature* 1990;**343**:651–3.

33. Sontag E, Fedorov S, Kamibayashi C, Robbins D, Cobb M, Mumby M. The interaction of SV40 small tumor antigen with protein phosphatase 2A stimulates the map kinase pathway and induces cell proliferation. *Cell* 1993;**75**:887–97.

34. Haystead TA, Weiel JE, Litchfield DW, Tsukitani Y, Fischer EH, Krebs EG. Okadaic acid mimics the action of insulin in stimulating protein kinase activity in isolated adipocytes. The role of protein phosphatase 2a in attenuation of the signal. *J Biol Chem* 1990;**265**:16,571–80.

35. Alessi DR, Gomez N, Moorhead G, Lewis T, Keyse SM, Cohen P. Inactivation of p42 MAP kinase by protein phosphatase 2A and a protein tyrosine phosphatase, but not CL100, in various cell lines. *Curr Biol* 1995;**5**:283–95.

36. Wang PY, Liu P, Weng J, Sontag E, Anderson RG. A cholesterol-regulated PP2A/HePTP complex with dual specificity ERK1/2 phosphatase activity. *EMBO J* 2003;**22**:2658–67.

37. Letourneux C, Rocher G, Porteu F. B56-containing PP2A dephosphorylate ERK and their activity is controlled by the early gene IEX-1 and ERK. *EMBO J* 2006;**25**:727–38.

38. Andersen JN, Mortensen OH, Peters GH, Drake PG, Iversen LF, Olsen OH, Jansen PG, Andersen HS, Tonks NK, Moller NP. Structural and evolutionary relationships among protein tyrosine phosphatase domains. *Mol Cell Biol* 2001;**21**:7117–36.

39. Alonso A, Sasin J, Bottini N, Friedberg I, Friedberg I, Osterman A, Godzik A, Hunter T, Dixon J, Mustelin T. Protein tyrosine phosphatases in the human genome. *Cell* 2004;**117**:699–711.

40. Wurgler-Murphy SM, Maeda T, Witten EA, Saito H. Regulation of the Saccharomyces cerevisiae HOG1 mitogen-activated protein kinase by the PTP2 and PTP3 protein tyrosine phosphatases. *Mol Cell Biol* 1997;**17**:1289–97.

41. Jacoby T, Flanagan H, Faykin A, Seto AG, Mattison C, Ota I. Two protein-tyrosine phosphatases inactivate the osmotic stress response pathway in yeast by targeting the mitogen-activated protein kinase, Hog1. *J Biol Chem* 1997;**272**:17,749–55.

42. Mattison CP, Ota IM. Two protein tyrosine phosphatases, Ptp2 and Ptp3, modulate the subcellular localization of the Hog1 MAP kinase in yeast. *Gene Dev* 2000;**14**:1229–35.

43. Millar JB, Buck V, Wilkinson MG. Pyp1 and Pyp2 PTPases dephosphorylate an osmosensing MAP kinase controlling cell size at division in fission yeast. *Gene Dev* 1995;**9**:2117–30.

44. Shiozaki K, Russell P. Cell-cycle control linked to extracellular environment by MAP kinase pathway in fission yeast. *Nature* 1995;**378**:739–43.

45. Wilkinson MG, Samuels M, Takeda T, Toone WM, Shieh JC, Toda T, Millar JB, Jones N. The Atf1 transcription factor is a target for the Sty1 stress-activated MAP kinase pathway in fission yeast. *Gene Dev* 1996;**10**:2289–301.

46. Shiozaki K, Russell P. Conjugation, meiosis, and the osmotic stress response are regulated by Spc1 kinase through Atf1 transcription factor in fission yeast. *Gene Dev* 1996;**10**:2276–88.

47. Degols G, Shiozaki K, Russell P. Activation and regulation of the Spc1 stress-activated protein kinase in Schizosaccharomyces pombe. *Mol Cell Biol* 1996;**16**:2870–7.

48. Mattison CP, Spencer SS, Kresge KA, Lee J, Ota IM. Differential regulation of the cell wall integrity mitogen-activated protein kinase pathway in budding yeast by the protein tyrosine phosphatases Ptp2 and Ptp3. *Mol Cell Biol* 1999;**19**:7651–60.

49. Zhan XL, Guan KL. A specific protein–protein interaction accounts for the in vivo substrate selectivity of Ptp3 towards the Fus3 MAP kinase. *Gene Dev* 1999;**13**:2811–27.

50. Zhan XL, Deschenes RJ, Guan KL. Differential regulation of FUS3 MAP kinase by tyrosine-specific phosphatases PTP2/PTP3 and dual-specificity phosphatase MSG5 in Saccharomyces cerevisiae. *Gene Dev* 1997;**11**:1690–702.

51. Karim FD, Rubin GM. PTP–ER, a novel tyrosine phosphatase, functions downstream of Ras1 to downregulate MAP kinase during *Drosophila* eye development. *Mol Cell* 1999;**3**:741–50.

52. Pulido R, Zuniga A, Ullrich A. PTP-SL and STEP protein tyrosine phosphatases regulate the activation of the extracellular signal-regulated kinases ERK1 and ERK2 by association through a kinase interaction motif. *EMBO J* 1998;**17**:7337–50.

53. Ogata M, Oh-hora M, Kosugi A, Hamaoka T. Inactivation of mitogen-activated protein kinases by a mammalian tyrosine-specific phosphatase, PTPBR7. *Biochem Biophys Res Commun* 1999;**256**:52–6.

54. Zuniga A, Torres J, Ubeda J, Pulido R. Interaction of mitogen-activated protein kinases with the kinase interaction motif of the tyrosine phosphatase PTP-SL provides substrate specificity and retains ERK2 in the cytoplasm. *J Biol Chem* 1999;**274**:21,900–7.

55. Saxena M, Williams S, Brockdorff J, Gilman J, Mustelin T. Inhibition of T cell signaling by mitogen-activated protein kinase-targeted hematopoietic tyrosine phosphatase (HePTP). *J Biol Chem* 1999;**274**:11,693–700.

56. Saxena M, Williams S, Gilman J, Mustelin T. Negative regulation of T cell antigen receptor signal transduction by hematopoietic tyrosine phosphatase (HePTP). *J Biol Chem* 1998;**273**:15,340–4.

57. Oh-hora M, Ogata M, Mori Y, Adachi M, Imai K, Kosugi A, Hamaoka T. Direct suppression of TCR-mediated activation of extracellular signal-regulated kinase by leukocyte protein tyrosine phosphatase, a tyrosine-specific phosphatase. *J Immunol* 1999;**163**:1282–8.

58. Gronda M, Arab S, Iafrate B, Suzuki H, Zanke BW. Hematopoietic protein tyrosine phosphatase suppresses extracellular stimulus-regulated kinase activation. *Mol Cell Biol* 2001;**21**:6851–8.

59. Blanco-Aparicio C, Torres J, Pulido R. A novel regulatory mechanism of MAP kinases activation and nuclear translocation mediated by PKA and the PTP-SL tyrosine phosphatase. *J Cell Biol* 1999;**147**:1129–36.

60. Saxena M, Williams S, Tasken K, Mustelin T. Crosstalk between cAMP-dependent kinase and MAP kinase through a protein tyrosine phosphatase. *Nat Cell Biol* 1999;**1**:305–11.

61. Paul S, Snyder GL, Yokakura H, Picciotto MR, Nairn AC, Lombroso PJ. The Dopamine/D1 receptor mediates the phosphorylation and

inactivation of the protein tyrosine phosphatase STEP via a PKA-dependent pathway. *J Neurosci* 2000;**20**:5630–8.

62. Paul S, Nairn AC, Wang P, Lombroso PJ. NMDA-mediated activation of the tyrosine phosphatase STEP regulates the duration of ERK signaling. *Nat Neurosci* 2003;**6**:34–42.

63. Doi K, Gartner A, Ammerer G, Errede B, Shinkawa H, Sugimoto K, Matsumoto K. MSG5, a novel protein phosphatase promotes adaptation to pheromone response in S. cerevisiae. *EMBO J* 1994;**13**:61–70.

64. Flandez M, Cosano IC, Nombela C, Martin H, Molina M. Reciprocal regulation between Slt2 MAPK and isoforms of Msg5 dual-specificity protein phosphatase modulates the yeast cell integrity pathway. *J Biol Chem* 2004;**279**:11,027–34.

65. Martin H, Flandez M, Nombela C, Molina M. Protein phosphatases in MAPK signalling: we keep learning from yeast. *Mol Microbiol* 2005;**58**:6–16.

66. Collister M, Didmon MP, MacIsaac F, Stark MJ, MacDonald NQ, Keyse SM. YIL113w encodes a functional dual-specificity protein phosphatase which specifically interacts with and inactivates the Slt2/Mpk1p MAP kinase in S. cerevisiae. *FEBS Letts* 2002;**527**:186–92.

67. Hahn JS, Thiele DJ. Regulation of the *Saccharomyces cerevisiae* Slt2 kinase pathway by the stress-inducible Sdp1 dual specificity phosphatase. *J Biol Chem* 2002;**277**:21,278–84.

68. Fox GC, Shafiq M, Briggs DC, Knowles PP, Collister M, Didmon MJ, Makrantoni V, Dickinson RJ, Hanrahan S, Totty N, Stark MJ, Keyse SM, McDonald NQ. Redox-mediated substrate recognition by Sdp1 defines a new group of tyrosine phosphatases. *Nature* 2007;**447**:487–92.

69. Ring JM, Martinez Arias A. puckered, a gene involved in position-specific cell differentiation in the dorsal epidermis of the Drosophila larva. *Dev* 1993(Suppl):251–9.

70. Martin-Blanco E, Gampel A, Ring J, Virdee K, Kirov N, Tolkovsky AM, Martinez-Arias A. puckered encodes a phosphatase that mediates a feedback loop regulating JNK activity during dorsal closure in *Drosophila*. *Gene Dev* 1998;**12**:557–70.

71. Wang MC, Bohmann D, Jasper H. JNK signaling confers tolerance to oxidative stress and extends lifespan in *Drosophila*. *Dev Cell* 2003;**5**:811–16.

72. Kim SH, Kwon HB, Kim YS, Ryu JH, Kim KS, Ahn Y, Lee WJ, Choi KY. Isolation and characterization of a *Drosophila* homologue of mitogen-activated protein kinase phosphatase-3 which has a high substrate specificity towards extracellular-signal-regulated kinase. *Biochem J* 2002;**361**:143–51.

73. Kim M, Cha GH, Kim S, Lee JH, Park J, Koh H, Choi KY, Chung J. MKP-3 has essential roles as a negative regulator of the Ras/mitogen-activated protein kinase pathway during *Drosophila* development. *Mol Cell Biol* 2004;**24**:573–83.

74. Berset T, Hoier EF, Battu G, Canevascini S, Hajnal A. Notch inhibition of RAS signaling through MAP kinase phosphatase LIP-1 during *C. elegans* vulval development. *Science* 2001;**291**:1055–8.

75. Lee MH, Hook B, Lamont LB, Wickens M, Kimble J. LIP-1 phosphatase controls the extent of germline proliferation in Caenorhabditis elegans. *EMBO J* 2006;**25**:88–96.

76. Mizuno T, Hisamoto N, Terada T, Kondo T, Adachi M, Nishida E, Kim DH, Ausubel FM, Matsumoto K. The Caenorhabditis elegans MAPK phosphatase VHP-1 mediates a novel JNK-like signaling pathway in stress response. *EMBO J* 2004;**23**:2226–34.

77. Kim DH, Liberati NT, Mizuno T, Inoue H, Hisamoto N, Matsumoto K, Ausubel FM. Integration of *Caenorhabditis elegans* MAPK pathways mediating immunity and stress resistance by MEK-1 MAPK kinase and VHP-1 MAPK phosphatase. *Proc Natl Acad Sci USA* 2004;**101**:10,990–4.

78. Camps M, Nichols A, Arkinstall S. Dual specificity phosphatases: a gene family for control of MAP kinase function. *FASEB J* 2000;**14**:6–16.

79. Keyse SM. Protein phosphatases and the regulation of mitogen-activated protein kinase signalling. *Curr Opin Cell Biol* 2000;**12**:186–92.

80. Theodosiou A, Ashworth A. MAP kinase phosphatases. *Genome Biol* 2002;**3**. REVIEWS3009.

81. Keyse SM, Ginsburg M. Amino acid sequence similarity between CL100, a dual-specificity MAP kinase phosphatase and cdc25. *Trends Biochem Sci* 1993;**18**:377–8.

82. Bordo D, Bork P. The rhodanese/Cdc25 phosphatase superfamily. Sequence-structure-function relations. *EMBO Rep* 2002;**3**:741–6.

83. Tanoue T, Nishida E. Molecular recognitions in the MAP kinase cascades. *Cell Signal* 2003;**15**:455–62.

84. Alessi DR, Smythe C, Keyse SM. The human CL100 gene encodes a Tyr/Thr-protein phosphatase which potently and specifically inactivates MAP kinase and suppresses its activation by oncogenic ras in *Xenopus* oocyte extracts. *Oncogene* 1993;**8**:2015–20.

85. Sun H, Charles CH, Lau LF, Tonks NK. MKP-1 (3CH134), an immediate early gene product, is a dual specificity phosphatase that dephosphorylates MAP kinase *in vivo*. *Cell* 1993;**75**:487–93.

86. Zheng CF, Guan KL. Dephosphorylation and inactivation of the mitogen-activated protein kinase by a mitogen-induced Thr/Tyr protein phosphatase. *J Biol Chem* 1993;**268**:16,116–19.

87. Keyse SM, Emslie EA. Oxidative stress and heat shock induce a human gene encoding a protein-tyrosine phosphatase. *Nature* 1992;**359**:644–7.

88. Charles CH, Abler AS, Lau LF. cDNA sequence of a growth factor-inducible immediate early gene and characterization of its encoded protein. *Oncogene* 1992;**7**:187–90.

89. Noguchi T, Metz R, Chen L, Mattei MG, Carrasco D, Bravo R. Structure, mapping, and expression of erp, a growth factor-inducible gene encoding a nontransmembrane protein tyrosine phosphatase, and effect of ERP on cell growth. *Mol Cell Biol* 1993;**13**:5195–205.

90. Rohan PJ, Davis P, Moskaluk CA, Kearns M, Krutzsch H, Siebenlist U, Kelly K. PAC-1: a mitogen-induced nuclear protein tyrosine phosphatase. *Science* 1993;**259**:1763–6.

91. Misra-Press A, Rim CS, Yao H, Roberson MS, Stork PJ. A novel mitogen-activated protein kinase phosphatase. Structure, expression, and regulation. *J Biol Chem* 1995;**270**:14,587–96.

92. King AG, Ozanne BW, Smythe C, Ashworth A. Isolation and characterisation of a uniquely regulated threonine, tyrosine phosphatase (TYP 1) which inactivates ERK2 and p54jnk. *Oncogene* 1995;**11**:2553–63.

93. Guan KL, Butch E. Isolation and characterization of a novel dual specific phosphatase, HVH2, which selectively dephosphorylates the mitogen-activated protein kinase. *J Biol Chem* 1995;**270**:7197–203.

94. Ishibashi T, Bottaro DP, Michieli P, Kelley CA, Aaronson SA. A novel dual specificity phosphatase induced by serum stimulation and heat shock. *J Biol Chem* 1994;**269**:29,897–902.

95. Kwak SP, Dixon JE. Multiple dual specificity protein tyrosine phosphatases are expressed and regulated differentially in liver cell lines. *J Biol Chem* 1995;**270**:1156–60.

96. Chu Y, Solski PA, Khosravi-Far R, Der CJ, Kelly K. The mitogen-activated protein kinase phosphatases PAC1, MKP-1, and MKP-2 have unique substrate specificities and reduced activity in vivo toward the ERK2 sevenmaker mutation. *J Biol Chem* 1996;**271**:6497–501.

97. Franklin CC, Kraft AS. Conditional expression of the mitogen-activated protein kinase (MAPK) phosphatase MKP-1 preferentially inhibits p38 MAPK and stress-activated protein kinase in U937 cells. *J Biol Chem* 1997;**272**:16,917–23.

98. Groom LA, Sneddon AA, Alessi DR, Dowd S, Keyse SM. Differential regulation of the MAP, SAP and RK/p38 kinases by Pyst1, a novel cytosolic dual-specificity phosphatase. *EMBO J* 1996;**15**:3621–32.

99. Slack DN, Seternes OM, Gabrielsen M, Keyse SM. Distinct binding determinants for ERK2/p38alpha and JNK map kinases mediate catalytic activation and substrate selectivity of map kinase phosphatase-1. *J Biol Chem* 2001;**276**:16,491–500.

100. Mandl M, Slack DN, Keyse SM. Specific inactivation and nuclear anchoring of extracellular signal-regulated kinase 2 by the inducible dual-specificity protein phosphatase DUSP5. *Mol Cell Biol* 2005;**25**:1830–45.

101. Dorfman K, Carrasco D, Gruda M, Ryan C, Lira SA, Bravo R. Disruption of the erp/mkp-1 gene does not affect mouse development: normal MAP kinase activity in ERP/MKP-1-deficient fibroblasts. *Oncogene* 1996;**13**:925–31.

102. Wang Z, Cao N, Nantajit D, Fan M, Liu Y, Li JJ. Mitogen-activated protein kinase phosphatase-1 represses c-Jun NH2-terminal kinase-mediated apoptosis via NF-kappaB regulation. *J Biol Chem* 2008;**283**:21,011–23.

103. Zhou JY, Liu Y, Wu GS. The role of mitogen-activated protein kinase phosphatase-1 in oxidative damage-induced cell death. *Cancer Res* 2006;**66**:4888–94.

104. Wu JJ, Bennett AM. Essential role for mitogen-activated protein (MAP) kinase phosphatase-1 in stress-responsive MAP kinase and cell survival signaling. *J Biol Chem* 2005;**280**:16,461–6.

105. Wang Z, Xu J, Zhou JY, Liu Y, Wu GS. Mitogen-activated protein kinase phosphatase-1 is required for cisplatin resistance. *Cancer Res* 2006;**66**:8870–7.

106. Zhao Q, Shepherd EG, Manson ME, Nelin LD, Sorokin A, Liu Y. The role of mitogen-activated protein kinase phosphatase-1 in the response of alveolar macrophages to lipopolysaccharide: attenuation of proinflammatory cytokine biosynthesis via feedback control of p38. *J Biol Chem* 2005;**280**:8101–8.

107. Chi H, Barry SP, Roth RJ, Wu JJ, Jones EA, Bennett AM, Flavell RA. Dynamic regulation of pro- and anti-inflammatory cytokines by MAPK phosphatase 1 (MKP-1) in innate immune responses. *Proc Natl Acad Sci USA* 2006;**103**:2274–9.

108. Hammer M, Mages J, Dietrich H, Servatius A, Howells N, Cato AC, Lang R. Dual specificity phosphatase 1 (DUSP1) regulates a subset of LPS-induced genes and protects mice from lethal endotoxin shock. *J Exp Med* 2006;**203**:15–20.

109. Zhao Q, Wang X, Nelin LD, Yao Y, Matta R, Manson ME, Baliga RS, Meng X, Smith CV, Bauer JA, Chang CH, Liu Y. MAP kinase phosphatase 1 controls innate immune responses and suppresses endotoxic shock. *J Exp Med* 2006;**203**:131–40.

110. Salojin KV, Owusu IB, Millerchip KA, Potter M, Platt KA, Oravecz T. Essential role of MAPK phosphatase-1 in the negative control of innate immune responses. *J Immunol* 2006;**176**:1899–907.

111. Lasa M, Abraham SM, Boucheron C, Saklatvala J, Clark AR. Dexamethasone causes sustained expression of mitogen-activated protein kinase (MAPK) phosphatase 1 and phosphatase-mediated inhibition of MAPK p38. *Mol Cell Biol* 2002;**22**:7802–11.

112. Kassel O, Sancono A, Kratzschmar J, Kreft B, Stassen M, Cato AC. Glucocorticoids inhibit MAP kinase via increased expression and decreased degradation of MKP-1. *EMBO J* 2001;**20**:7108–16.

113. Abraham SM, Lawrence T, Kleiman A, Warden P, Medghalchi M, Tuckermann J, Saklatvala J, Clark AR. Antiinflammatory effects of dexamethasone are partly dependent on induction of dual specificity phosphatase 1. *J Exp Med* 2006;**203**:1883–9.

114. Maier JV, Brema S, Tuckermann J, Herzer U, Klein M, Stassen M, Moorthy A, Cato AC. Dual specificity phosphatase 1 knockout mice show enhanced susceptibility to anaphylaxis but are sensitive to glucocorticoids. *Mol Endocrinol* 2007;**21**:2663–71.

115. Wu JJ, Roth RJ, Anderson EJ, Hong EG, Lee MK, Choi CS, Neufer PD, Shulman GI, Kim JK, Bennett AM. Mice lacking MAP kinase phosphatase-1 have enhanced MAP kinase activity and resistance to diet-induced obesity. *Cell Metab* 2006;**4**:61–73.

116. Hirosumi J, Tuncman G, Chang L, Gorgun CZ, Uysal KT, Maeda K, Karin M, Hotamisligil GS. A central role for JNK in obesity and insulin resistance. *Nature* 2002;**420**:333–6.

117. Bost F, Aouadi M, Caron L, Even P, Belmonte N, Prot M, Dani C, Hofman P, Pages G, Pouyssegur J, Le Marchand-Brustel Y, Binetruy B. The extracellular signal-regulated kinase isoform ERK1 is specifically required for *in vitro* and *in vivo* adipogenesis. *Diabetes* 2005;**54**:402–11.

118. Fan M, Rhee J, St-Pierre J, Handschin C, Puigserver P, Lin J, Jaeger S, Erdjument-Bromage H, Tempst P, Spiegelman BM. Suppression of mitochondrial respiration through recruitment of p160 myb binding protein to PGC-1alpha: modulation by p38 MAPK. *Gene Dev* 2004;**18**:278–89.

119. Jeffrey KL, Brummer T, Rolph MS, Liu SM, Callejas NA, Grumont RJ, Gillieron C, Mackay F, Grey S, Camps M, Rommel C, Gerondakis SD, Mackay CR. Positive regulation of immune cell function and inflammatory responses by phosphatase PAC-1. *Nat Immunol* 2006;**7**:274–83.

120. Chen P, Hutter D, Yang X, Gorospe M, Davis RJ, Liu Y. Discordance between the binding affinity of mitogen-activated protein kinase subfamily members for MAP kinase phosphatase-2 and their ability to activate the phosphatase catalytically. *J Biol Chem* 2001;**276**:29,440–9.

121. Kovanen PE, Young L, Al-Shami A, Rovella V, Pise-Masison CA, Radonovich MF, Powell J, Fu J, Brady JN, Munson PJ, Leonard WJ. Global analysis of IL-2 target genes: identification of chromosomal clusters of expressed genes. *Intl Immunol* 2005;**17**:1009–21.

122. Ueda K, Arakawa H, Nakamura Y. Dual-specificity phosphatase 5 (DUSP5) as a direct transcriptional target of tumor suppressor p53. *Oncogene* 2003;**22**:5586–91.

123. Kovanen PE, Bernard J, Al-Shami A, Liu C, Bollenbacher-Reilley J, Young L, Pise-Masison C, Spolski R, Leonard WJ. T-cell development and function are modulated by dual specificity phosphatase DUSP5. *J Biol Chem* 2008;**283**:17,362–9.

124. Muda M, Theodosiou A, Rodrigues N, Boschert U, Camps M, Gillieron C, Davies K, Ashworth A, Arkinstall S. The dual specificity phosphatases M3/6 and MKP-3 are highly selective for inactivation of distinct mitogen-activated protein kinases. *J Biol Chem* 1996;**271**:27,205–8.

125. Muda M, Theodosiou A, Gillieron C, Smith A, Chabert C, Camps M, Boschert U, Rodrigues N, Davies K, Ashworth A, Arkinstall S. The mitogen-activated protein kinase phosphatase-3 N-terminal noncatalytic region is responsible for tight substrate binding and enzymatic specificity. *J Biol Chem* 1998;**273**:9323–9.

126. Karlsson M, Mathers J, Dickinson RJ, Mandl M, Keyse SM. Both nuclear-cytoplasmic shuttling of the dual specificity phosphatase MKP-3 and its ability to anchor MAP kinase in the cytoplasm are mediated by a conserved nuclear export signal. *J Biol Chem* 2004;**279**:41,882–91.

127. Camps M, Nichols A, Gillieron C, Antonsson B, Muda M, Chabert C, Boschert U, Arkinstall S. Catalytic activation of the phosphatase MKP-3 by ERK2 mitogen-activated protein kinase. *Science* 1998;**280**:1262–5.

128. Stewart AE, Dowd S, Keyse SM, McDonald NQ. Crystal structure of the MAPK phosphatase Pyst1 catalytic domain and implications for regulated activation. *Nat Struct Biol* 1999;**6**:174–81.

129. Rigas JD, Hoff RH, Rice AE, Hengge AC, Denu JM. Transition state analysis and requirement of Asp-262 general acid/base catalyst for full activation of dual-specificity phosphatase MKP3 by extracellular regulated kinase. *Biochemistry* 2001;**40**:4398–406.

130. Zhou B, Zhang ZY. Mechanism of mitogen-activated protein kinase phosphatase-3 activation by ERK2. *J Biol Chem* 1999;**274**:35,526–34.

131. Dickinson RJ, Eblaghie MC, Keyse SM, Morriss-Kay GM. Expression of the ERK-specific MAP kinase phosphatase PYST1/MKP3 in mouse embryos during morphogenesis and early organogenesis. *Mech Dev* 2002;**113**:193–6.

132. Eblaghie MC, Lunn JS, Dickinson RJ, Munsterberg AE, Sanz-Ezquerro JJ, Farrell ER, Mathers J, Keyse SM, Storey K, Tickle C. Negative feedback regulation of FGF signaling levels by Pyst1/MKP3 in chick embryos. *Curr Biol* 2003;**13**:1009–18.

133. Kawakami Y, Rodriguez-Leon J, Koth CM, Buscher D, Itoh T, Raya A, Ng JK, Esteban CR, Takahashi S, Henrique D, Schwarz MF, Asahara H, Izpisua Belmonte JC. MKP3 mediates the cellular response to FGF8 signalling in the vertebrate limb. *Nat Cell Biol* 2003;**5**:513–19.

134. Smith TG, Karlsson M, Lunn JS, Eblaghie MC, Keenan ID, Farrell ER, Tickle C, Storey KG, Keyse SM. Negative feedback predominates over cross-regulation to control ERK MAPK activity in response to FGF signalling in embryos. *FEBS Letts* 2006; **580**:4242–5.

135. Ekerot M, Stavridis MP, Delavaine L, Mitchell MP, Staples C, Owens DM, Keenan ID, Dickinson RJ, Storey KG, Keyse SM. Negative-feedback regulation of FGF signalling by DUSP6/MKP-3 is driven by ERK1/2 and mediated by Ets factor binding to a conserved site within the DUSP6/MKP-3 gene promoter. *Biochem J* 2008;**412**:287–98.

136. Li C, Scott DA, Hatch E, Tian X, Mansour SL. Dusp6 (Mkp3) is a negative feedback regulator of FGF-stimulated ERK signaling during mouse development. *Development* 2007;**134**:167–76.

137. Dowd S, Sneddon AA, Keyse SM. Isolation of the human genes encoding the pyst1 and Pyst2 phosphatases: characterisation of Pyst2 as a cytosolic dual-specificity MAP kinase phosphatase and its catalytic activation by both MAP and SAP kinases. *J Cell Sci* 1998;**111**:3389–99.

138. Shin DY, Ishibashi T, Choi TS, Chung E, Chung IY, Aaronson SA, Bottaro DP. A novel human ERK phosphatase regulates H-ras and v-raf signal transduction. *Oncogene* 1997;**14**:2633–9.

139. Muda M, Boschert U, Smith A, Antonsson B, Gillieron C, Chabert C, Camps M, Martinou I, Ashworth A, Arkinstall S. Molecular cloning and functional characterization of a novel mitogen-activated protein kinase phosphatase, MKP-4. *J Biol Chem* 1997;**272**:5141–51.

140. Dickinson RJ, Williams DJ, Slack DN, Williamson J, Seternes OM, Keyse SM. Characterization of a murine gene encoding a developmentally regulated cytoplasmic dual-specificity mitogen-activated protein kinase phosphatase. *Biochem J* 2002;**364**:145–55.

141. Christie GR, Williams DJ, Macisaac F, Dickinson RJ, Rosewell I, Keyse SM. The dual-specificity protein phosphatase DUSP9/MKP-4 is essential for placental function but is not required for normal embryonic development. *Mol Cell Biol* 2005;**25**:8323–33.

142. Adams RH, Porras A, Alonso G, Jones M, Vintersten K, Panelli S, Valladares A, Perez L, Klein R, Nebreda AR. Essential role of p38alpha MAP kinase in placental but not embryonic cardiovascular development. *Mol Cell* 2000;**6**:109–16.

143. Giroux S, Tremblay M, Bernard D, Cardin-Girard JF, Aubry S, Larouche L, Rousseau S, Huot J, Landry J, Jeannotte L, Charron J. Embryonic death of Mek1-deficient mice reveals a role for this kinase in angiogenesis in the labyrinthine region of the placenta. *Curr Biol* 1999;**9**:369–72.

144. Liu Y, Lagowski J, Sundholm A, Sundberg A, Kulesz-Martin M. Microtubule disruption and tumor suppression by mitogen-activated protein kinase phosphatase 4. *Cancer Res* 2007;**67**:10,711–19.

145. Emanuelli B, Eberle D, Suzuki R, Kahn CR. Overexpression of the dual-specificity phosphatase MKP-4/DUSP-9 protects against stress-induced insulin resistance. *Proc Natl Acad Sci USA* 2008;**105**:3545–50.

146. Theodosiou A, Smith A, Gillieron C, Arkinstall S, Ashworth A. MKP5, a new member of the MAP kinase phosphatase family, which selectively dephosphorylates stress-activated kinases. *Oncogene* 1999;**18**:6981–8.

147. Tanoue T, Moriguchi T, Nishida E. Molecular cloning and characterization of a novel dual specificity phosphatase, MKP-5. *J Biol Chem* 1999;**274**:19,949–56.

148. Tanoue T, Yamamoto T, Maeda R, Nishida E. A Novel MAPK phosphatase MKP-7 acts preferentially on JNK/SAPK and p38 alpha and beta MAPKs. *J Biol Chem* 2001;**276**:26,629–39.

149. Masuda K, Shima H, Watanabe M, Kikuchi K. MKP-7, a novel mitogen-activated protein kinase phosphatase, functions as a shuttle protein. *J Biol Chem* 2001;**276**:39,002–11.

150. Zhang Y, Blattman JN, Kennedy NJ, Duong J, Nguyen T, Wang Y, Davis RJ, Greenberg PD, Flavell RA, Dong C. Regulation of innate and adaptive immune responses by MAP kinase phosphatase 5. *Nature* 2004;**430**:793–7.

Chapter 18

Protein Phosphatase 2A

Adam M. Silverstein, Anthony J. Davis, Vincent A. Bielinski, Edward D. Esplin, Nadir A. Mahmood and Marc C. Mumby

Department of Pharmacology, University of Texas Southwestern Medical Center, Dallas, Texas

INTRODUCTION

Serine/threonine phosphatases are integral components of many signal transduction pathways. There are eight classes of serine/threonine phosphatases in vertebrates. Protein serine/threonine phosphatases 1, 2 A, 2B/calcineurin, 4, 5, 6, and 7 are members of the PPP gene family that contain a conserved serine/threonine phosphatase domain. Protein phosphatase 2 A (PP2A) is a ubiquitously expressed member of the PPP gene family that accounts for a substantial portion of the total serine/threonine phosphatase activity in many cell types. PP2A is an essential enzyme that functions in fundamental cellular processes, including metabolism and the cell cycle. Like the other signaling molecules discussed in this chapter, proximity interactions play a primary role in regulating PP2A.

Once thought of as a single, broad-specificity phosphatase, PP2A is actually many different enzymes composed of complexes between catalytic subunits, scaffold subunits, regulatory subunits, and interacting proteins [1–3]. The catalytic and scaffold subunits bind tightly to form a core dimer that is the common component of most, but not all, forms of PP2A. The core dimer interacts with an array of regulatory subunits to generate multiple heterotrimeric holoenzymes. Additional interactions between PP2A and a variety of interacting proteins generate additional diversity. The regulatory subunits and interacting proteins target PP2A to specific substrates and intracellular locations. The existence of many different forms of PP2A accounts for the ability of the enzyme to regulate a wide variety of biological processes.

Interaction of the core dimer with regulatory subunits is critical for PP2A function. The regulatory subunits bind to the core dimer through interactions with both the scaffold and the catalytic subunits. The scaffold subunit is composed entirely of 15 copies of a conserved motif termed the HEAT repeat [4]. HEAT repeats 1–10 mediate interactions with regulatory subunits whereas repeats 11–15 mediate interaction with the C subunit [5]. The regulatory subunits must form contacts with both the scaffold and the catalytic subunits to generate stable heterotrimers [5, 6]. The regulatory subunits bind to the core dimer in a mutually exclusive manner. Although some sites of interaction are conserved, there are unique amino acids within the scaffold subunit that are involved in the interaction with individual regulatory subunits [7] (Figure 18.1).

PP2A REGULATORY SUBUNITS MEDIATE PROXIMITY INTERACTIONS

Regulatory subunits play a primary role in specifying the proximity interactions of PP2A. Three families of PP2A regulatory subunits have been identified in vertebrates by biochemical and genetic methods. A list of PP2A subunits is presented in Table 18.1. In order to avoid confusion, we have used a nomenclature for the PP2A subunits derived from their official human gene symbols. In contrast to the scaffold and catalytic subunits, which are ubiquitously expressed, the PP2A regulatory subunits are expressed in a cell- and tissue-specific manner. PP2A regulatory subunits are also differentially expressed during development and have distinct subcellular localizations. Neither the structural basis for interaction of regulatory subunits with the PP2A core dimer nor the biochemical effects of these interactions have been clearly elucidated. The PP2A regulatory subunit families have little overall amino acid sequence similarity. Several regulatory subunits contain WD domains, which have been proposed as a conserved motif responsible for the interaction with the core dimer [3, 8–10]. Recently a

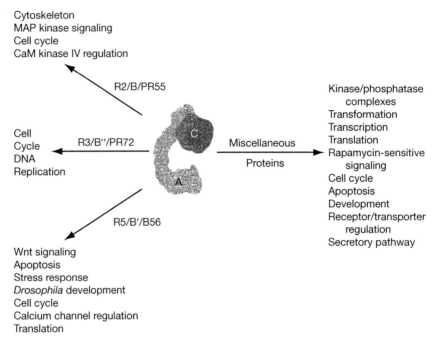

FIGURE 18.1 PP2A is a family of enzymes with multiple cellular functions. The PP2A holoenzyme consists of a common core dimer (AC) that complexes with a wide variety of regulatory molecules to generate a diversity of enzyme forms. These regulatory molecules include three regulatory subunit families (R2, R3, and R5) and a variety of miscellaneous proteins that interact with the core dimer or the free catalytic subunit. The regulatory molecules target PP2A to distinct substrates and intracellular locations, allowing the enzyme to participate in numerous cellular functions. The types of functions targeted by the individual regulatory subunits and miscellaneous proteins are listed.

loosely conserved A-subunit binding domain has been identified in each of the regulatory subunit families [11]. The regulatory subunits have effects on the kinetics of dephosphorylation that are consistent with a role in controlling the binding of substrates to PP2A [12–15]. This model is consistent with the notion that regulatory subunit-mediated proximity interactions play a role in targeting PP2A to phosphoprotein substrates. In contrast to simple enzyme–substrate interactions, the interaction of PP2A with many substrates involves a stable interaction involving regions of the enzyme removed from the active site. These stable interactions serve to maintain a high effective concentration of PP2A in the vicinity of the substrate.

Consistent with roles in defining PP2A specificity, different families of PP2A regulatory subunits have non-overlapping functions. The stress-induced growth arrest caused by mutations in the R5 subunit gene (RTS1) in yeast can be rescued by introduction of wild-type versions of either the yeast R5 gene or the human R5γ gene [16]. In contrast, wild-type R5 cannot rescue the cold-sensitive phenotype resulting from mutations in the yeast R2 subunit gene (CDC55). Knockdown of individual PP2A regulatory subunits in *Drosophila* S2 cells by RNA interference causes distinct defects. Loss of the R2 subunit causes an increase in insulin-dependent MAP kinase signaling, whereas loss of both R5 isoforms induces apoptosis [17]. These data suggest that PP2A holoenzymes containing the R2 subunit play a negative regulatory role in

MAP kinase signaling whereas holoenzymes containing the R5 subunit function in cell survival.

The R2 Family

The R2 family comprises a set of proteins present in a form of PP2A originally designated PP2A$_1$ [18]. This family currently contains four known isoforms (Table 18.1) that are 79–87 percent identical. R2α mRNA is ubiquitously expressed and is the most abundant PP2A regulatory subunit in many cells and tissues. The R2β and R2γ isoforms are only expressed at high levels in brain and testis. Although R2α and R2β are both expressed in the brain, they are present at different levels in different types of neurons [19]. R2α is distributed mainly in neuronal cell bodies and is localized in both the cytosol and nucleus. In contrast, the β isoform is excluded from the nucleus and is localized in axons and dendrites in addition to the cell body. Expression of R2 subunit mRNA is also differentially regulated during development. The differential expression and localization of R2 subunits support the idea that different members of this family play distinct roles in regulating PP2A functions.

Information about the functions of PP2A regulatory subunits has been derived from genetic analysis in yeast, *Drosophila*, and *C. elegans*. The pleiotropic phenotypes of

TABLE 18.1 Nomenclature of mammalian PP2A subunits

Name	Gene symbol	Aliases	Chromosomal location	LocusID/Accession number[a]
Catalytic subunits				
α isoform	PPP2CA	Cα, PP2A$_{Cα}$	5q23–31	5515
β isoform	PPP2CB	Cβ, PP2A$_{Cβ}$	8p21–12	5516
Scaffold subunits				
α isoform	PPP2R1A	Aα, PR65α	19	5518
β isoform	PPP2R1B	Aβ, PR65β	11q23	5519
R2 subunits				
α isoform	PPP2R2A	Bα, PR55α	8	5520
β isoform	PPP2R2B	Bβ, PR55β	5q31–5q33	5521
γ isoform	PPP2R2C	Bγ	4p16	5522
δ isoform		Bδ		AF180350
R3 subunits				
α isoform	PPP2R3	PR72, Bδ	3	5523
β isoform		PR59		AF050165
γ isoform		PR48		28227
R5 subunits				
α isoform	PPP2R5A	B'α, B56α	1q41	5525
β isoform	PPP2R5B	B'β, B56β	11q12	5526
γ isoform	PPP2R5C	B'γ, B56γ	3p21	5527
δ isoform	PPP2R5D	B'δ, B56δ	6p21.1	5528
ε isoform	PPP2R5E	B'ε, B56ε	7p11.1–12	5529

[a] *Entries in this column include the LocusID, when available, for the NCBI LocusLink entry for the corresponding the protein, or the GenBank/EMBL Accession number.*

mutant alleles of the R2 subunit in yeast and its numerous genetic interactions indicate that the R2 (cdc55p) protein plays multiple roles during mitosis, including the bud morphogenetic checkpoint and the mitotic spindle-assembly checkpoint [20–22]. The genetic results suggest that R2/cdc55p is involved in promoting activation of the yeast cell cycle regulatory kinase CDC2 (cyclin B/Cdc28 in S. cerevisiae) via dephosphorylation of the inhibitory tyrosine 19 phosphorylation site. Since PP2A does not directly dephosphorylate tyrosine, a likely target of R2 action is the cdc25 dual-specificity phosphatase, which is responsible for dephosphorylating tyrosine 19 in *S. cerevisiae* cdc28p.

Reduced levels of the R2 subunit in *Drosophila* result in varied phenotypes depending on the severity of the alleles. The aar1 allele (for abnormal anaphase resolution) contains a P-element insertion in the R2 gene [23]. Mutant aar1 flies die as larvae or early adults with overcondensed chromosomes and abnormal anaphase figures in larval brain cells. These defects can be rescued by reintroduction of the wild-type R2 gene. The aar1 phenotype is reminiscent of the mitotic spindle-assembly checkpoint defects seen in the yeast R2 mutants. Another P-element mutant allele of *Drosophila* R2 (twins[P]) causes death at an early pupal stage and shows pattern duplication of wing imaginal discs [24]. Flies harboring a weaker allele, twins[55], survive but have duplicated bristles in sensory neurons [25]. The effects of the twins mutation are consistent with a role for the R2 subunit in *Drosophila* embryonic cell fate determination. Both the aar1and twins[P] mutant larvae have a specific reduction in phosphatase activity toward substrates of cyclin-dependent kinases, suggesting the R2 subunit directs PP2A toward these substrates.

The R2 subunit targets PP2A to pathways that regulate MAP kinase activity. Overexpression of the small-t antigen of SV40 virus disrupts endogenous PP2A complexes containing the R2 subunit. This leads to enhanced activation of MAP kinase in response to growth factors in some but not all cell types [26, 27]. The small-t antigen effects may involve protein kinase C and the PI3 kinase pathway [28]. Depletion of the R2 subunit in *Drosophila* S2 cells via RNA interference also leads to a prolonged activation of MAP kinase in response to insulin [17]. These studies indicate that the R2 subunit plays a negative role in regulating MAP kinase activity, presumably by targeting PP2A to a component that is activated by phosphorylation. The *C. elegans* R2 subunit (sur-6) was isolated as a suppressor of the multivulval phenotype caused by an activated ras mutation [29]. Sur-6 mutations do not cause defects in vulval development by themselves but enhance the effects of weak mutant alleles of the *C. elegans* Raf protein kinase. These genetic interactions indicate that sur-6 mutations reduce signaling through the Ras pathway and may act with the kinase suppressor of raf (KSR) protein in a common pathway to positively regulate signaling through the Ras-Raf-MAP kinase pathway. The PP2A core dimer can associate with the Raf-1 protein kinase (Table 18.2). This interaction appears to mediate the dephosphorylation of inhibitory phosphorylation sites and enhance activation of Raf-1 during mitogenic stimulation [30]. This interaction does not appear to be mediated by the R2 subunit, since neither R2α nor R2β were detected in Raf-1 complexes. These studies suggest that PP2A is targeted to components of signaling pathways that regulate MAP kinase in both positive and negative ways. At least some of this targeting is mediated by R2 subunits. Multiple roles in MAP kinase signaling are consistent with genetic studies showing that mutations in the PP2A catalytic subunit have both positive and negative effects on MAP kinase activation in *Drosophila* [31]. The multiple actions in MAP kinase signaling are likely to be due to different forms of PP2A acting at distinct sites in this regulatory network.

Another R2-mediated proximity interaction regulates the microtubule cytoskeleton. A population of PP2A is associated with microtubules in neuronal and non-neuronal cells [32]. The association of PP2A with microtubules in brain is specific for R2α- and R2β-containing isoforms, and can be enhanced by a heat-labile anchoring factor [15]. PP2A holoenzymes containing R2α or R2β also interact with the neuronal microtubule-associated protein tau (Table 18.2) and act as potent tau phosphatases [33]. The microtubulebinding and organizing activity of tau is regulated by phosphorylation. Hypophosphorylated forms of tau bind to microtubules, leading to increased microtubule stability. In contrast, hyperphosphorylated tau dissociates from microtubules, leading to a decrease in microtubule stability. Tau-dependent stabilization of microtubules is important for formation and maintenance of axons in the central nervous system [34]. Disruption of the PP2A-tau interaction by expression of SV40 small-t antigen (which disrupts interaction of R2 subunits with the core dimer) causes hyperphosphorylation of tau and its dissociation from microtubules [33]. These observations suggest that proximity interactions among R2-containing forms of PP2A, microtubules, and tau play important roles in maintaining tau in a hypophosphorylated state. The targeted dephosphorylation of tau is important for axonal integrity, since inhibition of PP2A leads to tau hyperphosphorylation, loss of organized microtubules, and axonal degeneration in cultured neuronal cells [35]. The R2-mediated interactions of PP2A with microtubules and tau may have implications in neurodegenerative diseases, including Alzheimer's disease, where tau becomes hyperphosphorylated.

Expansion of a novel CAG trinucleotide repeat within the human R2β gene (PPPR2B) is associated with a form of autosomal dominant spinocerebellar ataxia termed SCA12 [36]. SCA12 is caused by neurodegeneration with atrophy of the cortex and cerebellum. The CAG expansion lies near the transcription start site of the R2β gene and could alter expression of this brain-specific isoform. The presence of the CAG expansion in affected individuals and its absence in non-affected family members suggest that altered expression of R2β may cause this disease. Although the mechanism of R2β loss in SCA12 is unknown, these data suggest that R2β may play a role in maintenance of neuronal viability.

The R3 Family

The second family of regulatory subunits identified by molecular cloning was the R3 family (Table 18.1). The R3 subunit was first identified as a 74-kDa protein present in a PP2A holoenzyme termed PCS$_M$ [37]. Current evidence indicates that this family plays a role in targeting PP2A to proteins involved in cell cycle regulation, including Cdc6, p107, and CG-NAP (Table 18.2). The gene encoding the R3 subunit (designated R3α in Table 18.1) produces two alternatively spliced transcripts encoding

TABLE 18.2 PP2A interacting proteins

Protein	Comments	Refs
Signaling proteins/transcription factors		
Adenomatous polyposis coli (APC)	APC binds to R5 subunits in yeast two-hybrid assays. This interaction may target PP2A to the Wnt signaling pathway, but physical complexes between PP2A and APC have not been demonstrated. Overexpression of R5 subunits decrease β-catenin levels and suppress Wnt signaling.	49
Axin	Axin forms complexes with the C and R5 subunits. The interaction targets PP2A to a complex of axin, APC, GSK3, and β-catenin and plays a role in regulating Wnt signaling.	50, 51
Cas (p130 Crk-associated substrate)	Cas is a Src substrate that has increased association with PP2A when Src is activated. PP2A dephosphorylates serine residues on Cas *in vitro*.	60
E-cadherin/β-catenin	The C_α but not the C_β subunit is required for stabilization of E-cadherin/β-catenin complexes at the plasma membrane.	61
Heat shock transcription factor 2 (HSF2)	HSF2 interacts with the A-subunit in two-hybrid and co-immunoprecipitation assays. HSF2 may displace the catalytic subunit from PP2A holoenzymes.	62, 63
HOX11	HOX 11 is homeobox transcription factor that controls development of the spleen. HOX11 binds to the PP2A catalytic subunit and inhibits phosphatase activity. HOX 11 also interacts with protein phosphatase 1.	64
HRX	HRX binds to PP2A through the SET/I_2^{PP2A} inhibitor protein. HRX is commonly mutated in acute leukemias.	65
Sex combs reduced (SCR)	SCR is a *Drosophila* homeobox transcription factor that interacts with the *Drosophila* R5 subunit in two-hybrid assays. SCR is homologous to human HOX5 and HOX6. PP2A may control phosphorylation and DNA binding activity of SCR.	57
RelA	RelA interacts with the scaffold subunit *in vitro*. The association may be transient since cross-linking is required to isolate a PP2A/RelA complex. RelA is dephosphorylated by PP2A *in vitro*.	66
Shc	PP2A associates with the PTB domain of Shc in the basal state and dissociates in response to insulin- and EGF-induced tyrosine phosphorylation. Expression of SV40 small-t antigen also causes dissociation of this complex.	67
Sp1	The Sp1 transcription factor interacts with the catalytic subunit in dividing T lymphocytes.	68, 69
STAT5	STAT5 associates with PP2A in an IL-3-dependent manner in the cytoplasm but not the nucleus.	70
Cell cycle related proteins		
Anaphase-promoting complex/cytosome (APC/C)	APC/C binds to the adenovirus E4orf4-PP2A complex. E4orf4 may target PP2A to APC/C, leading to its inactivation. This interaction may play a role in E4orf4-mediated cell cycle arrest and apoptosis.	71
Cdc6	Cdc6 binds to the R3γ/PR48 subunit and interacts with the AC–R3 heterotrimer. The interaction may regulate Cdc6 phosphorylation and DNA replication. Overexpression of R3γ causes G1 arrest.	43

(Continued)

TABLE 18.2 (Continued)

Protein	Comments	Refs
Cdc25c	Cdc25c co-immunoprecipitates with PP2A following crosslinking of cell lysates. The interaction requires the R2 subunit and results in dephosphorylation of Cdc25c dual-specificity phosphatase. The interaction is enhanced by the HIV-1 Vpr protein, suggesting that dephosphorylation and inactivation of cdc25c is involved in Vpr-mediated G2 arrest.	72
Cyclin G2	The association of cyclin G2 with PP2A catalytic and R5 subunits correlates with its ability to inhibit cell cycle progression.	54
DNA polymerase α-primase	PP2A is recovered with the hypophosphorylated form of DNA polymerase α-primase in G1. PP2A dephosphorylates DNA polymerase α-primase and restores its origin-dependent initiation activity *in vitro*.	73
p107	p107 (a retinoblastoma-related protein) binds the R3β/PR59 subunit-containing holoenzyme. Overexpression of R3β/PR59 causes p107 dephosphorylation and G1 arrest.	41

Membrane receptors/transporters

Protein	Comments	Refs
Beta$_2$-adrenergic receptor	The association of PP2A with this G-protein-coupled receptor is dependent upon agonist stimulation, receptor internalization, and acidification of endosomes. PP2A dephosphorylation is important for receptor resensitization and recycling to plasma membrane.	74
Biogenic amine transporters	Dopamine, norepinephrine, and serotonin transporters associate with PP2A. Transporter phosphorylation results in disruption of the PP2A association. The interaction may be involved in the regulation of the surface expression of transporters.	75
Class C L-type calcium channel (Ca$_v$1.2)	PP2A binds to the pore-forming $_{1C}$ subunit of this channel and reverses PKA-catalyzed serine channel phosphorylation. The interaction is selective for R5γ-containing PP2A complexes.	58
CXCR2 chemokine receptor	The chemokine receptor CXCR2 is a G-protein-coupled receptor involved in chemotaxis. CXCR2 receptor interacts with the AC core dimer. The interaction is dependent on internalization of the receptor following agonist stimulation.	76
NMDA receptor	PP2A forms a stable complex with NR3A subunit of the NMDA receptor. The association increases phosphatase activity and dephosphorylation of the NR1 subunit. Stimulation of the receptor leads to dissociation of PP2A and a reduction in phosphatase activity.	77

Protein kinases

Protein	Comments	Refs
CaM kinase IV (CaMKIV)	CaMKIV binds to the AC-R2 form and is dephosphorylated by PP2A.	78
Casein kinase II (CK2)	CK2 binds to the AC core dimer. CK2 can phosphorylate and stimulate PP2A activity *in vitro*.	79
JAK2	There is a transient association of JAK2 and PP2A upon interleukin-11 stimulation of adipocytes.	80
p21-Activated kinase (PAK1)	PAK1 interacts with and is a substrate of PP2A.	81
p70 S6 kinase	p70 S6 kinase is a PP2A substrate.	81
PKCα	The PP2A catalytic subunit co-immunoprecipitates with PKCα PKCα is dephosphorylated by PP2A. This association may be involved in the regulation of mast cell IL-6 production.	82

(Continued)

TABLE 18.2 (Continued)

Protein	Comments	Refs
PKCδ	PKCδ is a substrate for PP2A.	83
PKR (Double-stranded RNA-dependent protein kinase)	PKR binds to and phosphorylates the R5α regulatory subunit. Phosphorylation of R5α enhances PP2A activity and may alter the activity of the translation initiation factor eIF4.	84
RAF-1	RAF-1 interacts with the AC core dimer. PP2A dephosphorylates inhibitory sites on RAF-1.	30
Src	PP2A binds to the SH2, SH3, and catalytic domains of Src. This interaction decreases Src tyrosine kinase activity.	85
Apoptotic proteins		
Cyclin G1	Cyclin G1 binds to R5 subunits and the association is dependent on the induction of p53. Cyclin G1 plays a role in enhancing apoptosis.	53, 55
Bcl-2	Bcl-2 interacts with the PP2A isoform containing the R5α subunit. PP2A dephosphorylates Bcl-2 and regulates the function of Bcl-2 in apoptosis.	86–88
Cytoskeletal proteins		
CG-NAP (AKAP 350/450/ CG-NAP)	This 450-kDa centrosome and Golgi localized PKN-associated protein coimmunoprecipitates with PP2A in R3α-130 expressing cells. CG-NAP is involved in regulation of centrosome dynamics during the cell cycle.	40
Mid-1	Mid-1 binds to the PP2A interacting protein alpha 4 at a site independent from the C-subunit binding site. This interaction may regulate mid-1 binding to microtubules and formation of the midline during embryonic development.	89
Myosin	PP2A associates with myosin following mast cell activation. This interaction may play a role in regulating cytoskeletal remodeling and mast cell secretion.	90
Neurofilament proteins (NFs)	The AC–R2 complex associates with NF proteins. PP2A dephosphorylates sites in all three NF proteins (NF-L, NF-M, and NF-H). Dephosphorylation by PP2A promotes assembly of NF-L into filaments.	91, 92
Paxillin	Paxillin interacts with C-subunit and R5γ regulatory subunit. R5γ1 co-localizes with paxillin at focal adhesions and may target PP2A to paxillin.	59
Tau	Tau specifically interacts with R2-containing trimers. AC–R2 trimers dephosphorylate tau, promote microtubule binding, and stabilize microtubules.	33, 93
Vimentin	The AC-R2 complex associates with and dephosphorylates vimentin in an interaction mediated by the R2 subunit. Depletion of R2 by antisense RNA causes hyperphosphorylation of vimentin and reorganization of intermediate filaments.	94
Secretory pathway proteins		
Carboxypeptidase D (CPD)	PP2A binds to and dephosphorylates the cytoplasmic tail of this secretory pathway protein. PP2A may play a role in the intracellular trafficking of CPD between the cell surface and the trans-Golgi network.	95
Mannose-6-phosphate receptor (cation-dependent)	PP2A binds to the cytoplasmic tail of this secretory pathway protein.	95

(Continued)

TABLE 18.2 (Continued)

Protein	Comments	Refs
Peptidylglycine-a-amidating mono-oxygenase (PAM)	PP2A binds to the cytoplasmic tail of this secretory pathway protein.	95
TGN38	PP2A binds to the cytoplasmic tail of this secretory pathway protein.	95
Translation		
Eukaryotic termination factor 1 (eRF1)	eRF1 binds to the AC core dimer through C subunit. This interaction may target PP2A to ribosomes.	96
α4/Tap42 (IGBP1)	Alpha 4 interacts directly with the C subunit and decreases phosphatase acitivity toward eIF4E-BP1 that has been phosphorylated by the mTOR kinase.	97–100
Viral proteins		
Adenovirus E4orf4 protein	E4orf4 binds to the AC-R2 and AC-R5 complexes. Formation of a complex with AC–R2 is required for E4orf4-mediated apoptosis.	56, 101, 102
HIV Vpr protein	Vpr binds to AC–R2 complex and mediates Vpr-induced G2 arrest. This interaction regulates the Cdc25 dual-specificity phosphatase and Wee1 kinase.	68, 72, 103
Polyomavirus middle tumor antigen	Middle-T antigen binds to the AC core dimer and targets PP2A to the signaling complex assembled tumor antigen by middle-T antigen. The role of this interaction in middle-T mediated transformation is not clear.	104–106
Polyomavirus small tumor antigen	Similar to SV40 small-t antigen. Binds to the AC core dimer.	104–106
SV40 small tumor antigen	Binds to AC core dimer, displacing the R2 subunits and inhibiting PP2A acitivity toward some substrates. This interaction enhances MAP kinase signaling and viral transformation.	104, 105
Other cellular proteins		
I_1^{PP2A} (PHAP1, mapmodulin)	I_1^{PP2A} can inhibit PP2A activity *in vitro*, but its physiological function is unknown.	107
I_2^{PP2A} (SET)	I_2^{PP2A} can inhibit PP2A *in vitro*, but its function is unknown.	108
Phosphotyrosyl phosphatase activator (PTPA)	PTPA displays a weak interaction with PP2A and can enhance the low activity of the AC core dimer toward phosphotyrosine.	3
Protein phosphatase 5 (PP5)	PP5 interacts with the scaffold subunit of PP2A and may replace the catalytic subunit. The interaction appears to involve the R3α subunit, which co-immunoprecipitates with PP5.	39
Protein phosphatase methylesterase (PME-1)	Associates with catalytically inactive C-subunit point mutants. Demethylates the catalytic subunit *in vitro*.	109
SG2NA	SG2NA binds to the AC core dimer. The protein is localized in nucleus. SG2NA contains WD repeats, such as R2 subunits and striatin, and binds calmodulin. The function of SG2NA is currently unknown.	10
Striatin	Striatin binds to the AC core dimer. The protein contains WD repeats, such as R2 subunits and SG2NA, and binds to calmodulin. The function of striatin is currently unknown.	10

proteins of 72 and 130 kDa [38]. R3α-72 and R3α-130 contain the same C-terminal protein sequence, but PR130 contains a 665-amino-acid N-terminal extension. Both the 72- and 130-kDa variants are selectively but not exclusively expressed in skeletal muscle and heart. *In vitro*, the R3α subunit suppresses the activity of the AC dimer toward exogenous substrates and increases sensitivity of the enzyme to polycations [37]. The functions of R3α-72 or R3α-130 subunits have not been identified. Protein phosphatase 5 (another member of the PPP gene family) can interact with PP2A. Immunoprecipitated PP5 is associated with R3α-72 but not other regulatory subunits [39]. Although the significance of this interaction is not known, the data suggest that PP5 can be present in a PP2A oligomer containing the scaffold and R3α-72 subunits and that PP5 might act as the catalytic subunit in this heterocomplex. R3α-130 interacts with the giant scaffolding protein CG-NAP (centrosome and Golgi localized PKN-associated protein). CG-NAP anchors a signaling complex containing protein kinase-A, protein kinase-N, protein kinase-Cε, PP2A (R3α-130), and protein phosphatase 1 to the centrosome and Golgi apparatus in a cell-cycle-dependent manner [40]. The CG-NAP signaling complex may mediate some of the complex phosphorylation-based regulation of the centrosome that occurs during the cell cycle. One potential substrate for PP2A in this complex is protein kinase-N.

The R3 family contains additional isoforms that function in cell cycle regulation through unique proximity interactions. The R3β (PR59) protein was discovered in a yeast two-hybrid screen via the retinoblastoma-related protein p107 as bait [41]. R3β forms complexes with the PP2A core dimer when expressed in cells. Although R3β shares 56 percent identity with R3α-72, the interaction with p107 is specific. Furthermore, although R3β binds to p107, it fails to interact with the retinoblastoma protein. Forced overexpression of R3β results in dephosphorylation of p107 and cell cycle arrest in the G1 phase. R3β-mediated cell cycle arrest may be the result of hypophosphorylation of p107 (due to increased PP2A targeting) and its association with the E2F transcription factor. Binding of p107 to E2F would repress expression of genes required for entry into S phase. R3β may be targeted to dephosphorylate p107 in response to UV irradiation [42].

The R3γ regulatory subunit (PR48) was discovered in a yeast two-hybrid screen with the Cdc6 protein as bait [43]. Cdc6 is required for formation of pre-replication complexes during DNA replication. Phosphorylation of Cdc6 by S-phase cyclin-dependent kinases is the rate-limiting step for initiation of DNA replication. In mammalian cells, phosphorylation of Cdc6 at the beginning of S phase causes its dissociation from chromatin and triggers replication. In addition, Cdc6 phosphorylation induces its nuclear export and ubiquitin-dependent degradation. R3γ shares 50 and 68 percent sequence identity with R3α and R3β, respectively. R3γ localizes to the nucleus in mammalian cells and, like PR59, forced overexpression of PR48 results in cell cycle arrest at G1.

The R5 Family

The R5 regulatory subunits are a complex family of proteins that are components of a PP2A holoenzyme originally termed PP2A$_0$ [18, 44]. There are at least five isoforms (Table 18.1) that have distinct patterns of expression [45–47]. The α and γ isoforms are expressed predominantly in muscle, the β and δ isoforms in brain, and the ε isoform in brain and testis. In cardiac muscle, nearly all of the PP2A holoenzyme is composed of the R5α subunit [44]. *In vitro*, the R5 subunits suppress phosphatase activity toward multiple substrates [14]. This implies that the R5 subunits target PP2A by disfavoring interactions with some substrates while favoring interactions with others. The R5 family has been subdivided into cytosolic and nuclear types based on localization of transiently expressed proteins [46, 48]. The R5α, R5β, and R5ε isoforms are cytoplasmic whereas R5γ and R5δ are present in both the cytoplasm and nucleus. Ectopically expressed R5 subunits are also phosphorylated in intact cells. Thus, the regulation of PP2A or interaction with other proteins may be modulated by covalent modification of R5 family members.

The R5 subunits mediate interactions between PP2A and components of the Wnt signaling pathway involved in cell growth and transformation. Members of the R5 family were identified in a yeast two-hybrid screen by using the adenomatous polyposis coli (APC) protein as bait [49]. APC forms a signaling complex with axin and glycogen synthase kinase 3β that mediates the phosphorylation and proteasome-dependent degradation of β-catenin. A basal level of β-catenin degradation normally prevents transcription of β-catenin target genes involved in cell growth and transformation. Stimulation of the Wnt pathway causes inhibition of β-catenin phosphorylation and degradation, leading to increased transcription of β-catenin target genes. Ectopic expression of R5 subunits in mammalian cells causes a reduction in β-catenin levels and a decrease in expression of β-catenin target genes. Further supporting a role for PP2A in the Wnt/β-catenin pathway, the catalytic subunit of PP2A interacts with axin in two-hybrid assays and can be co-immunoprecipitated with axin [50]. Subsequent studies have shown that the scaffold subunit, the catalytic subunit, and R5 subunits can be immunoprecipitated with axin from *Xenopus* embryos [51, 52]. Ectopic expression of the PP2A scaffold subunit, the catalytic subunit, or R5 subunits all have ventralizing activity in *Xenopus* embryos, consistent with a negative role in Wnt/β-catenin signaling. The R5 subunits appear to interact directly with axin at a site that is distinct from the sites that interact with APC, GSK-3β, and β-catenin [51]. The data are all consistent with an important role for R5 subunits in targeting PP2A to the axin/GSK-3/APC complex and regulating the Wnt signaling pathway.

The R5 subunits are also linked to cell survival and apoptosis. Cyclin G1, cyclin G2, and cyclin I are members of a unique family of cyclin-related proteins that are expressed in brain and muscle. R5 subunits interact with both cyclin G1 [53] and cyclin G2 [54]. Cyclin G1 and R5 subunits can be co-immunoprecipitated from neurons ,whereas cyclin G2–R5-catalytic subunit complexes can be isolated from cultured cells [54]. Although the function of the cyclin G1 is not known, the p53 tumor suppressor protein regulates its transcription. Ectopic expression of cyclin G1 enhances apoptosis in response to multiple stimuli in cultured cells [55]. Similarly, forced overexpression of cyclin G2 causes formation of aberrant nuclei and cell cycle arrest [54]. These observations raise the possibility that the cyclin G1-PP2A interaction could be involved in cell cycle arrest and apoptosis. The interaction of R5 subunits with the adenovirus E4orf4 protein is essential for E4orf4-mediated apoptosis [56]. Finally, the use of RNA interference in *Drosophila* cells has shown that loss of both of the *Drosophila* R5 subunits results in apoptosis [17].

R5 subunits interact with a variety of other proteins, thus indicating roles for this family in other signaling pathways (Table 18.2). A *Drosophila* homolog of R5 interacts with a homeodomain-containing transcription factor called Sex Combs Reduced. This interaction positively modulates transcriptional activity [57]. The R5γ subunit is associated with L-type calcium channels, where it appears to target PP2A to regulatory sites phosphorylated by protein kinase A [58]. R5α interacts with the double-stranded RNA-dependent protein kinase PKR. PKR phosphorylates R5α, leading to an increase in PP2A phosphatase activity. PKR-enhanced PP2A activity may lead to decreased phosphorylation of eIF4E and altered protein synthesis. R5-containing PP2A may also be targeted to focal adhesions through interaction with paxillin [59].

PP2A-INTERACTING PROTEINS

Proximity interactions are the most important mechanism for regulating the activity of PP2A. Association with interacting proteins mediates many proximity interactions of PP2A, and allows targeting of this phosphatase to a wide variety of signaling pathways. PP2A interacting proteins include phosphoproteins that are PP2A substrates, scaffold proteins, and components of the cytoskeleton. As discussed above, many of these interactions occur with PP2A holoenzymes and are mediated by specific regulatory subunits. However, interacting proteins have been identified that interact directly with the PP2A core dimer and the catalytic subunit. PP2A-interacting proteins include virally encoded proteins and a host of cellular proteins that participate in interesting aspects of sig-

nal transduction. A compilation of the currently identified PP2A-interacting proteins is presented in Table 18.2. Although many of the proteins listed in the table are substrates for PP2A, others act to target PP2A to specific signaling complexes, and some alter signaling by disrupting endogenous PP2A complexes. These proteins have been grouped into categories based on functional similarities. Brief descriptions of individual proteins and their interaction with PP2A are presented in the table.

REFERENCES

1. Millward TA, Zolnierowicz S, Hemmings BA. Regulation of protein kinase cascades by protein phosphatase 2 A. *Trends Biochem Sci* 1999;**24**:186–91.
2. Virshup DM. Protein phosphatase 2 A: a panoply of enzymes. *Curr Opin Cell Biol* 2000;**12**:180–5.
3. Janssens V, Goris J. Protein phosphatase 2 A: a highly regulated family of serine/threonine phosphatases implicated in cell growth and signalling. *Biochem J* 2001;**353**:417–39.
4. Groves MR, Hanlon N, Turowski P, Hemmings BA, Barford D. The structure of the protein phosphatase 2 A PR65/A subunit reveals the conformation of its 15 tandemly repeated HEAT motifs. *Cell* 1999;**96**:99–110.
5. Ruediger R, Hentz M, Fait J, Mumby M, Walter G. Molecular model of the A subunit of protein phosphatase 2 A: interaction with other subunits and tumor antigens. *J Virol* 1994;**68**:123–9.
6. Kamibayashi C, Lickteig RL, Estes R, Walter G, Mumby MC. Expression of the A subunit of protein phosphatase 2 A and characterization of its interactions with the catalytic and regulatory subunits. *J Biol Chem* 1992;**267**:21,864–21,872.
7. Ruediger R, Fields K, Walter G. Binding specificity of protein phosphatase 2 A core enzyme for regulatory B subunits and T antigens. *J Virol* 1999;**73**:839–42.
8. Neer EJ, Schmidt CJ, Nambudripad R, Smith TF. The ancient regulatory-protein family of WD-repeat proteins. *Nature* 1994;**371**:297–300.
9. Griswold-Prenner I, Kamibayashi C, Maruoka EM, Mumby MC, Derynck R. Physical and functional interactions between type I transforming growth factor beta receptors and B-alpha, a WD-40 repeat subunit of phosphatase 2 A. *Mol Cell Biol* 1998;**18**:6595–604.
10. Moreno CS, Park S, Nelson K, Ashby D, Hubalek F, Lane WS, Pallas DC. WD40 repeat proteins striatin and S/G(2) nuclear autoantigen are members of a novel family of calmodulin-binding proteins that associate with protein phosphatase 2 A. *J Biol Chem* 2000;**275**:5257–63.
11. Li X, Virshup DM. Two conserved domains in regulatory B subunits mediate binding to the A subunit of protein phosphatase 2 A. *Eur J Biochem* 2002;**269**:546–52.
12. Chen S-C, Kramer G, Hardesty B. Isolation and partial characterization of an M_r 60,000 subunit of a type 2 A phosphatase from rabbit reticulocytes. *J Biol Chem* 1989;**264**:7267–75.
13. Imaoka T, Imazu M, Usui H, Kinohara N, Takeda M. Resolution and reassociation of three distinct components from pig heart phosphoprotein phosphatase. *J Biol Chem* 1983;**258**:1526–35.
14. Kamibayashi C, Estes R, Lickteig RL, Yang S-I, Craft C, Mumby MC. Comparison of heterotrimeric protein phosphatase 2 A containing different B subunits. *J Biol Chem* 1994;**269**:20,139–20,148.
15. Price NE, Mumby MC. Effects of regulatory subunits on the kinetics of protein phosphatase 2 A. *Biochemistry* 2000;**39**:11,312–11,318.

16. Zhao Y, Boguslawski G, Zitomer RS, DePaoli–Roach AA. Saccharomyces cerevisiae homologs of mammalian B and B′ subunits of protein phosphatase 2 A direct the enzyme to distinct cellular functions. *J Biol Chem* 1997;**272**:8256–62.

17. Silverstein AM, Barrow CA, Davis AJ, Mumby MC. Actions of PP2A on the MAP kinase pathway and apoptosis are mediated by distinct regulatory subunits. *Proc Natl Acad Sci USA* 2002;**99**:4221–6.

18. Tung HYL, Alemany S, Cohen P. The protein phosphatases involved in cellular regulation 2. Purification, subunit structure and properties of protein phosphatases-2A0, 2A1, and 2A2 from rabbit skeletal muscle. *Eur J Biochem* 1985;**148**:253–63.

19. Strack S, Zaucha JA, Ebner FF, Colbran RJ, Wadzinski BE. Brain protein phosphatase 2 A: developmental regulation and distinct cellular and subcellular localization by B subunits. *J Comp Neurol* 1998;**392**:515–27.

20. Healy AM, Zolnierowicz S, Stapelton AE, Goebl M, DePaoli-Roach AA, Pringle JR. CDC55, a *Saccharomyces cerevisiae* gene involved in cellular morphogenesis: identification, characterization, and homology to the B subunit of mammalian type 2 A protein phosphatase. *Mol Cell Biol* 1991;**11**:5767–80.

21. Minshull J, Straight A, Rudner AD, Dernburg AF, Belmont A, Murray AW. Protein phosphatase 2 A regulates MPF activity and sister chromatid cohesion in budding yeast. *Curr Biol* 1996;**6**:1609–20.

22. Wang Y, Burke DJ. Cdc55p, the B-type regulatory subunit of protein phosphatase 2 A, has multiple functions in mitosis and is required for the kinetochore/spindle checkpoint in Saccharomyces cerevisiae. *Mol Cell Biol* 1997;**17**:620–6.

23. Mayer-Jaekel RE, Ohkura H, Gomes R, Sunkel CE, Baumgartner S, Hemmings BA, Glover DM. The 55-kD regulatory subunit of *Drosophila* protein phosphatase 2 A is required for anaphase. *Cell* 1993;**72**:621–33.

24. Uemura T, Shiomi K, Togashi S, Takeichi M. Mutation of twins encoding a regulator of protein phosphatase 2 A leads to pattern duplication in *Drosophila* imaginal disks. *Genes Dev* 1993;**7**:429–40.

25. Shiomi K, Takeichi M, Nishida Y, Nishi Y, Uemura T. Alternative cell fate choice induced by low-level expression of a regulator of protein phosphatase 2 A in the Drosophila peripheral nervous system. *Development* 1994;**120**:1591–9.

26. Sontag E, Fedorov S, Kamibayashi C, Robbins D, Cobb M, Mumby M. The interaction of SV40 small tumor antigen with protein phosphatase 2 A stimulates the MAP kinase pathway and induces cell proliferation. *Cell* 1993;**75**:887–97.

27. Frost JA, Alberts AS, Sontag E, Guan K, Mumby MC, Feramisco JR. SV40 small t antigen cooperates with mitogen activated kinases to stimulate AP-1 activity. *Mol Cell Biol* 1994;**14**:6244–52.

28. Sontag E, Sontag JM, Garcia A. Protein phosphatase 2 A is a critical regulator of protein kinase C zeta signaling targeted by SV40 small t to promote cell growth and NF-kappaB activation. *EMBO J* 1997;**16**:5662–71.

29. Sieburth DS, Sundaram M, Howard RM, Han M. A PP2A regulatory subunit positively regulates Ras-mediated signaling during *Caenorhabditis elegans* vulval induction. *Genes Dev* 1999;**13**:2562–9.

30. Abraham D, Podar K, Pacher M, Kubicek M, Welzel N, Hemmings BA, Dilworth SM, Mischak H, Kolch W, Baccarini M. Raf-1-associated protein phosphatase 2 A as a positive regulator of kinase activation. *J Biol Chem* 2000;**275**:22,300–22,304.

31. Wassarman DA, Solomon NM, Chang HC, Karim FD, Therrien M, Rubin GM. Protein phosphatase 2 A positively and negatively regulates Ras1-mediated photoreceptor development in *Drosophila*. *Genes Dev* 1996;**10**:272–8.

32. Sontag E, Nunbhakdi-Craig V, Bloom GS, Mumby MC. A novel pool of protein phosphatase 2 A is associated with microtubules and is regulated during the cell cycle. *J Cell Biol* 1995;**128**:1131–44.

33. Sontag E, Nunbhakdi-Craig V, Lee G, Bloom GS, Mumby MC. Regulation of the phosphorylation state and microtubule-binding activity of tau by protein phosphatase 2 A. *Neuron* 1996;**17**:1201–7.

34. Billingsley ML, Kincaid RL. Regulated phosphorylation and dephosphorylation of tau protein: effects on microtubule interaction, intracellular trafficking and neurodegeneration. *Biochem J* 1997;**323**:577–91.

35. Merrick SE, Trojanowski JQ, Lee VMY. Selective destruction of stable microtubules and axons by inhibitors of protein serine/threonine phosphatases in cultured human neurons (NT2N cells). *J Neurosci* 1997;**17**:5726–37.

36. Holmes SE, O'Hearn EE, McInnis MG, Gorelick-Feldman DA, Kleiderlein JJ, Callahan C, Kwak NG, Ingersoll-Ashworth RG, Sherr M, Sumner AJ, Sharp AH, Ananth U, Seltzer WK, Boss MA, Vieria-Saecker AM, Epplen JT, Riess O, Ross CA, Margolis RL. Expansion of a novel CAG trinucleotide repeat in the 5′ region of PPP2R2B is associated with SCA12. *Nat Genet* 1999;**23**:391–2.

37. Waelkens E, Goris J, Merlevede W. Purification and properties of polycation-stimulated phosphorylase phosphatases from rabbit skeletal muscle. *J Biol Chem* 1987;**262**:1049–59.

38. Hendrix P, Mayer-Jaekel RE, Cron P, Goris J, Hofsteenge J, Merlevede W, Hemmings BA. Structure and expression of a 72-kDa regulatory subunit of protein phosphatase 2 A.Evidence for different size forms produced by alternative splicing. *J Biol Chem* 1993;**268**:15,267–15,276.

39. Lubert EJ, Hong Y, Sarge KD. Interaction between protein phosphatase 5 and the A subunit of protein phosphatase 2 A: evidence for a heterotrimeric form of protein phosphatase 5. *J Biol Chem* 2001;**276**:38,582–38,587.

40. Takahashi M, Shibata H, Shimakawa M, Miyamoto M, Mukai H, Ono Y. Characterization of a novel giant scaffolding protein, CG-NAP, that anchors multiple signaling enzymes to centrosome and the Golgi apparatus. *J Biol Chem* 1999;**274**:17,267–17,274.

41. Voorhoeve PM, Hijmans EM, Bernards R. Functional interaction between a novel protein phosphatase 2 A regulatory subunit, PR59, and the retinoblastoma-related p107 protein. *Oncogene* 1999;**18**:515–24.

42. Voorhoeve PM, Watson RJ, Farlie PG, Bernards R, Lam EW. Rapid dephosphorylation of p107 following UV irradiation. *Oncogene* 1999;**18**:679–88.

43. Yan Z, Fedorov SA, Mumby MC, Williams RS. PR48, a novel regulatory subunit of protein phosphatase 2 A, interacts with Cdc6 and modulates DNA replication in human cells. *Mol Cell Biol* 2000;**20**:1021–9.

44. Zolnierowicz S, Csortos C, Bondor J, Verin A, Mumby MC, DePaoli-Roach AA. Diversity in the regulatory B-subunits of protein phosphatase 2 A: identification of a novel isoform highly expressed in brain. *Biochemistry* 1994;**33**:11,858–11,867.

45. McCright B, Virshup DM. Identification of a new family of protein phosphatase 2 A regulatory subunits. *J Biol Chem* 1995;**270**:26,123–26,128.

46. Ahmadian-Tehrani M, Mumby MC, Kamibayashi C. Identification of a novel protein phosphatase 2 A regulatory subunit highly expressed in muscle. *J Biol Chem* 1996;**271**:5164–70.

47. Csortos C, Zolnierowicz S, Bako E, Durbin SD, DePaoli-Roach AA. High complexity in the expression of the B′ subunit of protein phosphatase 2A0. Evidence for the existence of at least seven novel isoforms. *J Biol Chem* 1996;**271**:2578–88.

48. McCright B, Rivers AM, Audlin S, Virshup DM. The B56 family of protein phosphatase 2 A (PP2A) regulatory subunits encodes differentiation-induced phosphoproteins that target PP2A to both nucleus and cytoplasm. *J Biol Chem* 1996;**271**:22,081–22,089.

49. Seeling JM, Miller JR, Gil R, Moon RT, White R, Virshup DM. Regulation of beta-catenin signaling by the B56 subunit of protein phosphatase 2 A. *Science* 1999;**283**:2089–91.

50. Hsu W, Zeng L, Costantini F. Identification of a domain of axin that binds to the serine/threonine protein phosphatase 2 A and a self-binding domain. *J Biol Chem* 1999;**274**:3439–45.

51. Yamamoto H, Hinoi T, Michiue T, Fukui A, Usui H, Janssens V, Van Hoof C, Goris J, Asashima M, Kikuchi A. Inhibition of the Wnt signaling pathway by the PR61 subunit of protein phosphatase 2 A. *J Biol Chem* 2001;**276**:26,875–26,882.

52. Li X, Yost HJ, Virshup DM, Seeling JM. Protein phosphatase 2 A and its B56 regulatory subunit inhibit Wnt signaling in *Xenopus*. *EMBO J* 2001;**20**:4122–31.

53. Okamoto K, Kamibayashi C, Serrano M, Prives C, Mumby MC, Beach D. p53-dependent association between cyclin G and the B′ subunit of protein phosphatase 2 A. *Mol Cell Biol* 1996;**16**:6593–602.

54. Bennin DA, Arachchige Don AS, Brake T, McKenzie JL, Rosenbaum H, Ortiz L, DePaoli-Roach AA, Horne MC. Cyclin G2 associates with protein phosphatase 2 A catalytic and regulatory B′ subunits in active complexes and induces nuclear aberrations and a G1/S phase cell cycle arrest. *J Biol Chem* 2002;**277**:27,449–27,467.

55. Okamoto K, Prives C. A role of cyclin G in the process of apoptosis. *Oncogene* 1999;**18**:4606–15.

56. Shtrichman R, Sharf R, Kleinberger T. Adenovirus E4orf4 protein interacts with both B alpha and B′ subunits of protein phosphatase 2 A, but E4orf4-induced apoptosis is mediated only by the interaction with B alpha. *Oncogene* 2000;**19**:3757–65.

57. Berry M, Gehring W. Phosphorylation status of the SCR homeodomain determines its functional activity: essential role for protein phosphatase 2 A,B′. *EMBO J* 2000;**19**:2946–57.

58. Davare MA, Horne MC, Hell JW. Protein phosphatase 2 A is associated with class C L-type calcium channels (Cav1.2) and antagonizes channel phosphorylation by cAMP-dependent protein kinase. *J Biol Chem* 2000;**275**:39,710–39,717.

59. Ito A, Kataoka TR, Watanabe M, Nishiyama K, Mazaki Y, Sabe H, Kitamura Y, Nojima H. A truncated isoform of the PP2A B56 subunit promotes cell motility through paxillin phosphorylation. *EMBO J* 2000;**19**:562–71.

60. Yokoyama N, Miller WT. Protein phosphatase 2 A interacts with the Src kinase substrate p130(CAS). *Oncogene* 2001;**20**:6057–65.

61. Gotz J, Probst A, Mistl C, Nitsch RM, Ehler E. Distinct role of protein phosphatase 2 A subunit C alpha in the regulation of E-cadherin and beta-catenin during development. *Mechan Dev* 2000;**93**:83–93.

62. Hong YL, Sarge KD. Regulation of protein phosphatase 2 A activity by heat shock transcription factor 2. *J Biol Chem* 1999;**274**:12,967–12,970.

63. Hong YL, Lubert EJ, Rodgers DW, Sarge KD. Molecular basis of competition between HSF2 and catalytic subunit for binding to the PR65/A subunit of PP2A. *Biochem Biophys Res Commun* 2000;**272**:84–9.

64. Kawabe T, Muslin AJ, Korsmeyer SJ. Hox11 interacts with protein phosphatases PP2A and PP1 and disrupts a G2/M cell-cycle checkpoint. *Nature* 1997;**385**:454–8.

65. Adler HT, Nallaseth FS, Walter G, Tkachuk DC. HRX leukemic fusion proteins form a heterocomplex with the leukemia-associated protein SET and protein phosphatase 2 A. *J Biol Chem* 1997;**272**:28,407–28,414.

66. Yang J, Fan GH, Wadzinski BE, Sakurai H, Richmond A. Protein phosphatase 2 A interacts with and directly dephosphorylates RelA. *J Biol Chem* 2001;**276**:47,828–47,833.

67. Ugi S, Imamura T, Ricketts W, Olefsky JM. Protein phosphatase 2 A forms a molecular complex with Shc and regulates Shc tyrosine phosphorylation and downstream mitogenic signaling. *Mol Cell Biol* 2002;**22**:2375–87.

68. Elder RT, Yu M, Chen M, Zhu X, Yanagida M, Zhao Y. HIV-1 Vpr induces cell cycle G2 arrest in fission yeast (*Schizosaccharomyces pombe*) through a pathway involving regulatory and catalytic subunits of PP2A and acting on both Wee1 and Cdc25. *Virology* 2001;**287**:359–70.

69. Lacroix I, Lipcey C, Imbert J, Kahn-Perles B. Sp1 transcriptional activity is upregulated by phosphatase 2 A in dividing T lymphocytes. *J Biol Chem* 2002;**277**:9598–605.

70. Yokoyama N, Reich NC, Miller WT. Involvement of protein phosphatase 2a in the interleukin-3-stimulated Jak2–Stat5 signaling pathway. *J Interferon Cytokine Res* 2001;**21**:369–78.

71. Kornitzer D, Sharf R, Kleinberger T. Adenovirus E4orf4 protein induces PP2A-dependent growth arrest in Saccharomyces cerevisiae and interacts with the anaphase-promoting complex/cyclosome. *J Cell Biol* 2001;**154**:331–44.

72. Hrimech M, Yao XJ, Branton PE, Cohen EA. Human immunodeficiency virus type 1 Vpr-mediated G(2) cell cycle arrest: Vpr interferes with cell cycle signaling cascades by interacting with the B subunit of serine/threonine protein phosphatase 2 A. *EMBO J* 2000;**19**:3956–67.

73. Dehde S, Rohaly G, Schub O, Nasheuer HP, Bohn W, Chemnitz J, Deppert W, Dornreiter I. Two immunologically distinct human DNA polymerase alpha-primase subpopulations are involved in cellular DNA replication. *Mol Cell Biol* 2001;**21**:2581–93.

74. Krueger KM, Daaka Y, Pitcher JA, Lefkowitz RJ. The role of sequestration in G protein-coupled receptor resensitization. Regulation of beta2-adrenergic receptor dephosphorylation by vesicular acidification. *J Biol Chem* 1997;**272**:5–8.

75. Bauman AL, Apparsundaram S, Ramamoorthy S, Wadzinski BE, Vaughan RA, Blakely RD. Cocaine and antidepressant-sensitive biogenic amine transporters exist in regulated complexes with protein phosphatase 2 A. *J Neurosci* 2000;**20**:7571–8.

76. Fan GH, Yang W, Sai J, Richmond A. Phosphorylation-independent association of CXCR2 with the protein phosphatase 2 A core enzyme. *J Biol Chem* 2001;**276**:16,960–16,968.

77. Chan SF, Sucher NJ. An NMDA receptor signaling complex with protein phosphatase 2 A. *J Neurosci* 2001;**21**:7985–92.

78. Westphal RS, Anderson KA, Means AR, Wadzinski BE. A signaling complex of Ca^{2+}-calmodulin-dependent protein kinase IV and protein phosphatase 2 A. *Science* 1998;**280**:1258–61.

79. Heriche JK, Lebrin F, Rabilloud T, Leroy D, Chambaz EM, Goldberg Y. Regulation of protein phosphatase 2 A by direct interaction with casein kinase 2α. *Science* 1997;**276**:952–5.

80. Fuhrer DK, Yang YC. Complex formation of JAK2 with PP2A, P13K, and Yes in response to the hematopoietic cytokine interleukin-11. *Biochem Biophys Res Commun* 1996;**224**:289–96.

81. Westphal RS, Coffee RL, Marotta A, Pelech SL, Wadzinski BE. Identification of kinase-phosphatase signaling modules composed of p70 S6 kinase-protein phosphatase 2 A (PP2A) and p21-activated kinase-PP2A. *J Biol Chem* 1999;**274**:687–92.

82. Boudreau RT, Garduno R, Lin TJ. Protein phosphatase 2 A and protein kinase Calpha are physically associated and are involved in *Pseudomonas aeruginosa*-induced interleukin 6 production by mast cells. *J Biol Chem* 2002;**277**:5322–9.

83. Srivastava J, Goris J, Dilworth SM, Parker PJ. Dephosphorylation of PKCdelta by protein phosphatase 2Ac and its inhibition by nucleotides. *FEBS Letts* 2002;**516**:265–9.

84. Xu Z, Williams BR. The B56alpha regulatory subunit of protein phosphatase 2 A is a target for regulation by double-stranded RNA-dependent protein kinase PKR. *Mol Cell Biol* 2000;**20**:5285–99.

85. Yokoyama N, Miller WT. Inhibition of Src by direct interaction with protein phosphatase 2 A. *FEBS Letts* 2001;**505**:460–4.

86. Deng XM, Ito T, Carr B, Mumby M, May WS. Reversible phosphorylation of Bcl2 following interleukin 3 or bryostatin 1 is mediated by direct interaction with protein phosphatase 2 A. *J Biol Chem* 1998;**273**:34,157–34,163.

87. Ruvolo PP, Deng X, Ito T, Carr BK, May WS. Ceramide induces bcl2 dephosphorylation via a mechanism involving mitochondrial PP2A. *J Biol Chem* 1999;**274**:20,296–20,300.

88. Ruvolo PP, Clark W, Mumby M, Gao F, May WS. A functional role for the B56 alpha-subunit of protein phosphatase 2 A in ceramide-mediated regulation of Bcl2 phosphorylation status and function. *J Biol Chem* 2002;**277**:22,847–22,852.

89. Liu J, Prickett TD, Elliott E, Meroni G, Brautigan DL. Phosphorylation and microtubule association of the Opitz syndrome protein mid-1 is regulated by protein phosphatase 2 A via binding to the regulatory subunit alpha 4. *Proc Natl Acad Sci USA* 2001;**98**:6650–5.

90. Holst J, Sim AT, Ludowyke RI. Protein phosphatases 1 and 2 A transiently associate with myosin during the peak rate of secretion from mast cells. *Mol. Biol. Cell* 2002;**13**:1083–98.

91. Saito T, Shima H, Osawa Y, Nagao M, Hemmings BA, Kishimoto T, Hisanaga S. Neurofilament-associated protein phosphatase 2 A: its possible role in preserving neurofilaments in filamentous states. *Biochemistry* 1995;**34**:7376–84.

92. Strack S, Westphal RS, Colbran RJ, Ebner FF, Wadzinski BE. Protein serine/threonine phosphatase 1 and 2 A associate with and dephosphorylate neurofilaments. *Brain Res Mol Brain Res* 1997;**49**:15–28.

93. Sontag III E, Nunbhakdi-Craig V, Lee G, Brandt R, Kamibayashi C, Kuret J, White CL, Mumby MC, Bloom GS. Molecular interactions among protein phosphatase 2 A, tau, and microtubules: implications for the regulation of tau phosphorylation and the development of tauopathies. *J Biol Chem* 1999;**274**:25,490–25,498.

94. Turowski P, Myles T, Hemmings BA, Fernandez A, Lamb NC. Vimentin dephosphorylation by protein phosphatase 2 A is modulated by the targeting subunit B55. *Mol Biol Cell* 1999;**10**:1997–2015.

95. Varlamov O, Kalinina E, Che FY, Fricker LD. Protein phosphatase 2 A binds to the cytoplasmic tail of carboxy– peptidase D and regulates post-trans-Golgi network trafficking. *J Cell Sci* 2001;**114**:311–22.

96. Andjelkovic N, Zolnierowicz S, Van Hoof C, Goris J, Hemmings BA. The catalytic subunit of protein phosphatase 2 A associates with the translation termination factor eRF1. *EMBO J* 1996;**15**:156–67.

97. Chen J, Peterson RT, Schreiber SL. Alpha 4 associates with protein phosphatases 2 A, 4, and 6. *Biochem Biophys Res Commun* 1998;**247**:827–32.

98. Maeda K, Inui S, Tanaka H, Sakaguchi N. A new member of the alpha 4-related molecule (alpha 4-b) that binds to the protein phosphatase 2 A is expressed selectively in the brain and testis. *Eur J Biochem* 1999;**264**:702–6.

99. Nanahoshi M, Nishiuma T, Tsujishita Y, Hara K, Inui S, Sakaguchi N, Yonezawa K. Regulation of protein phosphatase 2 A catalytic activity by alpha4 protein and its yeast homolog tap42. *Biochem Biophys Res Commun* 1998;**251**:520–6.

100. Jiang Y, Broach JR. Tor proteins and protein phosphatase 2 A reciprocally regulate Tap42 in controlling cell growth in yeast. *EMBO J* 1999;**18**:2782–92.

101. Kleinberger T, Shenk T. Adenovirus E4orf4 protein binds to protein phosphatase 2 A, and the complex down regulates E1A-enhanced junB transcription. *J Virol* 1993;**67**:7556–60.

102. Shtrichman R, Sharf R, Barr H, Dobner T, Kleinberger T. Induction of apoptosis by adenovirus E4orf4 protein is specific to transformed cells and requires an interaction with protein phosphatase 2 A. *Proc Natl Acad SciUSA* 1999;**96**:10,080–10,085.

103. Tung HY, De Rocquigny H, Zhao LJ, Cayla X, Roques BP, Ozon R. Direct activation of protein phosphatase-2A0 by HIV-1 encoded protein complex NCp7:vpr. *FEBS Letts* 1997;**401**:197–201.

104. Mumby M. Regulation by tumour antigens defines a role for PP2A in signal transduction. *Sem Cancer Biol* 1995;**6**:229–37.

105. Pallas DC, Shahrik LK, Martin BL, Jaspers S, Miller TB, Brautigan DL, Roberts TM. Polyoma small and middle T antigens and SV40 small t antigen form stable complexes with protein phosphatase 2 A. *Cell* 1990;**60**:167–76.

106. Cayla X, Ballmer-Hofer K, Merlevede W, Goris J. Phosphatase 2 A associated with polyomavirus small-T or middle-T antigen is an okadaic acid-sensitive tyrosyl phosphatase. *Eur J Biochem* 1993;**214**:281–6.

107. Li M, Makkinje A, Damuni Z. Molecular identification of I1PP2A, a novel potent heat-stable inhibitor protein of protein phosphatase 2 A. *Biochemistry* 1996;**35**:6998–7002.

108. Li M, Makkinje A, Damuni Z. The myeloid leukemia-associated protein SET is a potent inhibitor of protein phosphatase 2 A. *J Biol Chem* 1996;**271**:11,059–11,062.

109. Ogris E, Du XX, Nelson KC, Mak EK, Yu XX, Lane WS, Pallas DC. A protein phosphatase methylesterase (PME-1) is one of several novel proteins stably associating with two inactive mutants of protein phosphatase 2 A. *J Biol Chem* 1999;**274**:14,382–14,391.

Lipid Signaling

Chapter 19

Lipid-Mediated Localization of Signaling proteins

Maurine E. Linder

Department of Cell Biology and Physiology, Washington University School of Medicine, St Louis, Missouri

INTRODUCTION

Extracellular signals elicit responses by binding to receptors at the cell surface and activating signal transducers localized at the inner leaflet of the plasma membrane. Signals propagate to intracellular effectors, resulting in changes in enzyme activity, ion channel gating, or protein–protein interactions. The components of a signaling pathway must be oriented spatially and temporally to execute the appropriate response to an extracellular cue. Many effectors are constitutively present at the membrane, but others must be recruited from the cytoplasm in response to signals. Lipids participate in the localization or recruitment of signaling proteins to cellular membranes through several mechanisms. First, a large number of proteins involved in signal transduction and membrane trafficking contain protein modules that bind to specific lipids embedded in cell membranes [1]. Second, lipids play a key role in forming lipid rafts – small, cholesterol-rich subdomains of the plasma membrane that form transiently and facilitate protein interactions [2]. Finally, lipids covalently modify proteins, which impacts their membrane interactions, trafficking, and function. Here we review the structures of the different lipid modifications, describe how lipids are attached to proteins and discuss how these modifications impact the localization of lipidated proteins.

LIPID MODIFICATIONS ON THE CYTOPLASMIC FACE OF MEMBRANES

N-myristoylation

Protein N-myristoylation is the addition of myristic acid to a protein through an amide linkage to an N-terminal glycine residue (Figure 19.1). Numerous signaling proteins are modified with amide-linked myristate, including the Src family of non-receptor tyrosine kinases, members of the G_i family of G-protein α subunits, the catalytic subunit of protein kinase A, and the MARCKS protein, a protein kinase C substrate (Table 19.1). Modification of the protein occurs co-translationally upon removal of the initiator methionine by methionylpeptidase. There is an absolute requirement

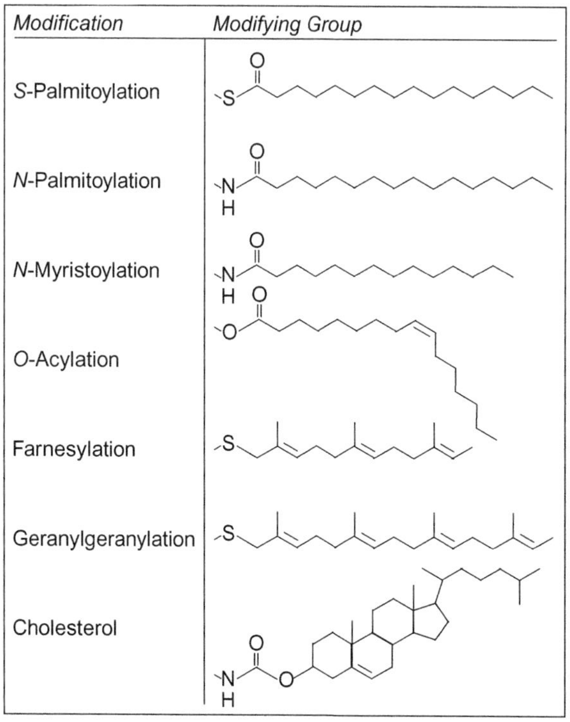

FIGURE 19.1 Structures of covalent lipid modifications.
Nadolski, M.J. and Linder, M.E. (2007), Protein lipidation, *FEBS J.* 274(20):5202–5210; reproduced by permission of Blackwell Publishing.

TABLE 19.1 Lipid modifications of selected signaling proteins[1]

Protein	Modified sequence	Modification
G-protein coupled receptors		
Rhodopsin	-----^{319}TTL**CC**GKN^{326}---	S-palmitoylation
β$_2$-adrenergic receptor	-----^{337}QELL**C**LRR^{344}----	S-palmitoylation
Src family kinases		
Src	^{1}**MG**SNKSKPK---	N-myristoylation
Fyn	^{1}**MGC**VQ**C**KDK---	N-myristoylation, S-palmitoylation
Lck	^{1}**MGC**G**C**SSHP---	N-myristoylation, S-palmitoylation
G-protein α subunits		
αt1	^{1}**MG**AGASAE---	N-myristoylation
αi1, αi2, αi3, αo, αz	^{1}**MGC**---	N-myristoylation, S-palmitoylation
αs	^{1}**MGC**LGNSKT---	N-palmitoylation, S-palmitoylation
αq	^{1}MTLESIMA**CC**LS---	S-palmitoylation
G-protein γ subunits		
γ1	---^{65}KELKGG<u>C</u>VIS-$_{COOH}$	Farnesyl isoprenoid, CM
γ2	---^{63}EKKFF<u>C</u>AIL-$_{COOH}$	Geranylgeranyl isoprenoid, carboxylmethylation
Monomeric GTPases		
K-Ras 4B	---^{175}KKKKKKSKTK<u>C</u>VIM-$_{COOH}$	Farnesyl isoprenoid, CM
H-Ras	---^{181}**C**MS**C**K<u>C</u>VLS-$_{COOH}$	Farnesyl isoprenoid, CM, S-palmitoylation
N-Ras	---^{181}**C**MGLP<u>C</u>VVM-$_{COOH}$	Farnesyl isoprenoid, CM, S-palmitoylation
Rho A	---^{181}ARRGKKKSG<u>C</u>LVL-$_{COOH}$	Geranylgeranyl isoprenoid, CM
Rab3a	---^{216}GD<u>CAC</u>-$_{COOH}$	Geranylgeranyl isoprenoid, CM

[1]Sequences shown are human except for rhodopsin (bovine); N-myristoylated glycines are boldface; S-palmitoylated cysteines are shown in outline; N-palmitoylated cysteines are in bold italics; prenylated cysteines are underlined; CM, carboxyl-methylation.

for a glycine residue as the attachment site. Often, N-myristoylated proteins have Ser or Thr at position five in the amino acid sequence. A computer algorithm that predicts N-myristoylation sequences is available (http://mendel.imp.ac.at/myristate/SUPLmain.htm) [3].

The enzyme responsible for myristoylation is N-myristoyltransferase (NMT) [4, 5]. NMT is encoded by a single gene in *Saccharomyces cerevisiae* that is essential for cell viability. Drosophila lacking NMT have multiple developmental defects. There are two N-myristoyltransferase genes, NMT1 and NMT2, in humans and mice that encode

enzymes with overlapping functions. NMT1 is required for early embryonic development in mice. Human NMT1 appears to be targeted predominately to subcellular fractions enriched in ribosomes, consistent with its role as a co-translational protein modifier [6]. This has been documented by assays of enzyme activity and immunoblots of subcellular fractions derived from mammalian cell lines. Association with ribosomes is mediated by the N-terminal domain of NMT1. A similar domain is present in NMT2.

More recently, posttranslational myristoylation has been characterized for a number of proteins. This so-called "morbid myristoylation" [7] was discovered for the pro-apoptotic protein BID [8]. Caspase 8 cleavage exposes an internal glycine residue in the C-terminal fragment of BID (t-BID). Myristoylation of t-BID facilitates its targeting to the inner mitochondrial membrane, thereby enhancing BID-induced release of cytochrome C and cell death. Subsequent studies have revealed that actin, gelsolin, and caspase-activated p21-activated protein kinase 2 are also posttranslationally myristoylated [9–11]. Posttranslational myristoylation resulting from caspase activation may represent a mechanism to coordinate changes in cytoskeletal dynamics associated with programmed cell death [11].

Prenylation

Prenylated proteins are recognized by their characteristic C-terminal motifs (Table 19.1) [12]. Ras, other monomeric GTPases, and G-protein γ subunits undergo posttranslational processing at a CaaX motif (C = cys, a = generally a small aliphatic amino acid, and X=an uncharged amino acid). The identity of the X residue is important for directing whether a substrate will be modified with a farnesyl or geranylgeranyl group [13, 14] (Figure 19.1). When the protein terminates in serine, methionine, alanine, glutamine, or phenylalanine, the protein is modified with a C15 farnesyl isoprenoid. When the protein terminates in leucine, phenylalanine, and sometimes methionine, the protein is modified with a C20 geranylgeranyl isoprenoid. CaaX-motif proteins are modified by farnesyl transferase or geranylgeranyl transferase type I. Proteins ending with –XXXCC, –XXCXC, –XXCCX, –XCCXX or –CCXXX are typically modified by geranylgeranyltransferase type II (GGTase II). Members of the Rab family of GTPases are among the substrates of GGTase II, and are modified with two geranylgeranyl groups. A computer algorithm to identify prenylation motifs in proteins is available online (http://mendel.imp.ac.at/sat/PrePS/index.html) [15].

Prenylation occurs through a thioether linkage, and is just the first of a series of posttranslational processing steps for CaaX proteins [16]. Following prenylation in the cytoplasm by a prenyltransferase, the protein associates with the endoplasmic reticulum where the –aaX peptide is proteolytically cleaved by Ras converting enzyme, followed by methylation of the carboxyl group on the prenylated cysteine by isoprenylcysteine methyltransferase. Although prenylation is a stable modification, methylation of the prenylated cysteine residue is reversible and may be regulated.

S-palmitoylation

In addition to modification with amide-linked myristate, proteins are also fatty acylated at cysteine residues through reversible thioester linkages [17] (Figure 19.1). This is commonly referred to as *palmitoylation*, although other long-chain fatty acids besides palmitate are incorporated into

proteins through this mechanism. Palmitoylation is a post-translational event that occurs both on intracellular membranes and at the plasma membrane. Palmitate modifies cysteine residues in a variety of sequence motifs (Table 19.1). A computer algorithm to identify palmitoylation motifs in proteins is available online (http://bioinformatics.lcd-ustc.org/css_palm/) [18]. In signal transducers, palmitoylated cysteines are frequently located near N-myristoylated glycine residues or immediately upstream of prenylated C-terminal cysteines. However, some G-protein α subunits are modified exclusively with palmitate. Integral membrane proteins can also be palmitoylated. Many G-protein-coupled receptors are modified at cysteine residues in the C-terminal cytoplasmic domain. Insertion of the fatty acid into the membrane creates a fourth cytoplasmic loop in these serpentine receptors.

The enzymes that mediate palmitoylation eluded identification for many years. Genetic and biochemical analysis studies in yeast led to the discovery of the first protein acyltransferases (PATs) [19]. Erf2/Erf4 is the PAT that palmitoylates yeast Ras. Akr1 is the PAT for yeast casein kinase 2. Erf2 and Akr1 are both integral membrane proteins that share a domain referred to as the DHHC domain, a cysteine-rich domain with a conserved Asp–His–His–Cys signature motif. Evidence suggests that the DHHC domain is directly involved in the palmitoyl transfer reaction. In contrast to the prenyltransferases and N-myristoyltransferases, the DHHC proteins constitute a large family. There are 7 DHHC proteins in *Saccharomyces cerevisiae*, and 23 in humans. In yeast, significant progress has been made in identifying cognate enzyme : substrate pairs, and there is evidence to suggest that DHHC proteins have overlapping substrate specificity [20]. Fewer examples are known at present for mammalian cells. DHHC proteins are expressed in diverse tissues, and reside in different subcellular locations [21]. As discussed below, the localization of DHHC PATs is likely to dictate the localization of protein substrates.

N-palmitoylation

There are a few examples of proteins in which palmitate is attached to the N-terminus of the protein through a stable amide linkage. Several secreted proteins are N-palmitoylated as described below. On the cytoplasmic face of membranes, $G_{\alpha s}$ is unique among the G-protein subunits in being modified with both amide-linked palmitate at its N-terminal glycine and thioester-linked palmitate at the adjacent cysteine [22]. The mechanism of N-palmitoylation of $G_{\alpha s}$ is unknown.

LIPID MODIFICATIONS IN THE LUMEN OF THE SECRETORY PATHWAY

GPI anchors

A number of cell surface proteins are anchored to the plasma membrane through a glycosylphosphatidylinositol (GPI)

[23]. This lipid modification is composed of a phosphatidyli-nositol lipid connected through a carbohydrate linker to the C-terminal carboxyl group of a protein. This process occurs in the lumen of the endoplasmic reticulum. Following addition of the GPI moiety, the protein traffics through the secretory pathway to the cell surface, where the GPI anchor tethers the protein to the extracellular face of the plasma membrane. A computer algorithm is available to identify proteins that are potential candidates for the addition of GPI anchors (http://mendel.imp.ac.at/sat/gpi/gpi_server.html) [24].

Lipid Modifications of Secreted Morphogens, Growth Factors, and Hormones

The secreted morphogens Hedgehog and Wnt have recently gained attention as targets for lipid modification [25]. Like GPI-anchored proteins, these proteins receive their lipid modifications in the lumen of the secretory pathway. Hedgehog is modified both with cholesterol and with palmitate. As the protein undergoes an autoprocessing event that results in cleavage between the glycine and cysteine residues of a Gly–Cys–Phe motif, a cholesterol moiety is added to the now C-terminal glycine residue of the N-terminal cleavage product (Figure 19.1). This same cleavage product is then N-palmitoylated on the cysteine at its extreme N-terminus (Figure 19.1). Lipidation of Hedgehog influences the range and potency of the signals generated.

Dual lipidation is also found on the Wnt protein, but in this case the modifications are thioester- and oxyester-linked fatty acids. The first modification of Wnt discovered was S-palmitoylation. Murine Wnt-3a is S-palmitoylated at the conserved cysteine 77. S-palmitoylation is not required for secretion of the protein, but is important for Wnt's ability to signal [26]. The second modification of Wnt-3a is O-acylation at serine 209. Interestingly, the acyl group identified by mass spec is palmitoleic acid (16:1), a monounsaturated fatty acid (Figure 19.1). As opposed to S-palmitoylation, O-acylation of Wnt is associated with exit from the ER and subsequent secretion [27]. Lipidation of Wnt at both sites is dependent upon the integral membrane protein Porcupine, which is localized in the endoplasmic reticulum [27, 28]. Porcupine shares sequence homology with membrane-bound O-acyltransferases (MBOATs) [29], and is presumed to directly modify Wnt proteins.

The Drosophila epidermal growth factor receptor (EGFR) ligand Spitz is also a substrate for palmitoylation [28]. Similar to Hedgehog, Spitz is N-palmitoylated at a cysteine residue exposed by cleavage of the signal sequence. Palmitoylation of both Spitz and Hedgehog is dependent on Rasp (also known has Skinny Hedgehog, Central Missing, and Sightless) in flies, and Hedgehog acyltransferase (Hhat) in vertebrates. Like Porcupine, Rasp is a member of the MBOAT family of proteins. Hedgehog palmitoylation has been reconstituted in vitro with purified Hhat, providing direct biochemical evidence that Hhat is a bona fide PAT [30].

The appetite-stimulating peptide hormone ghrelin is subject to a unique lipid modification, octanolyation. The eight-carbon fatty acid octanoate is attached to ghrelin at serine 3 through an oxyester linkage (O-acylation), and is necessary for the physiological activity of the peptide [31]. GOAT (ghrelin O-acyltransferase) is the MBOAT protein responsible for ghrelin acylation [32, 33]. Because inhibitors of GOAT could potentially suppress appetite, GOAT is considered a promising therapeutic target to curb obesity.

LOCALIZATION OF LIPID-MODIFIED PROTEINS

Membrane Interactions

Covalent lipid modifications promote membrane association by direct insertion of the fatty acid or isoprenoid into the lipid bilayer [34]. However, N-myristoylated and prenylated proteins are found in the cytoplasm as well as on cellular membranes, whereas S-palmitoylated proteins appear to be exclusively associated with membranes. Measurements of the apparent affinities of lipid-modified peptides for liposomes demonstrate that a farnesyl or myristoyl group is not sufficient to confer stable membrane association. Thus, proteins modified singly by myristate or a farnesyl isoprenoid are predicted to be exquisitely sensitive to other properties of the protein that promote or interfere with membrane association.

Two well-characterized properties that act cooperatively with N-myristoylation or farnesylation to promote membrane binding are electrostatic interactions or additional lipid modifications [34]. Examples of both mechanisms are found in the Src family kinases. All members of this family are N-myristoylated. Positively-charged residues near the N-terminus of Src cooperate with the myristoyl moiety to provide a stable membrane anchor by binding to the negatively-charged phospholipid head groups of the lipid bilayer. Lck and Fyn are palmitoylated at cysteine residues adjacent to or nearby the myristoylated N-terminus. The additional hydrophobicity provided by the thioester-linked fatty acids promotes long-lived interactions with membranes [35]. Similar mechanisms operate in prenylated proteins. Electrostatic interactions promote membrane interactions of the G-protein βγ complex [36] and K-Ras proteins [37]. H- and N-Ras are modified with palmitate subsequent to prenylation and C-terminal processing [16]. In addition, membrane affinity of prenylated proteins is increased approximately 10-fold by carboxyl-methylation [35].

How do lipidation motifs target proteins to specific membranes? Short peptide sequences that direct tandem N-myristoylation and palmitoylation or farnesylation and palmitoylation are sufficient for plasma membrane localization, suggesting that the lipid modifications themselves can provide specificity [38]. The kinetic membrane-trapping model proposes that farnesylation or myristoylation confers rapid exchange of a protein with cellular membranes until it is trapped by palmitoylation. The dually lipidated protein can then exert its effects on the membrane where it was captured, or it can move to other locations by vesicular transport. An important prediction of this model is that the localization and substrate specificity of protein palmitoyltransferases will dictate the localization of palmitoylated proteins in conjunction with other protein–protein interactions. The DHHC family of PATs lends itself to this model, with PATs distributed on endomembranes and the plasma membrane, where they can recognize and "trap" their cognate substrates [21].

Trafficking Through Acylation and Deacylation Cycles

The reversibility of protein palmitoylation distinguishes it from the static lipid modifications prenylation and N-myristoylation. Regulation of protein trafficking by reversible palmitoylation has been best illustrated for palmitoylated forms of Ras using live cell imaging [39, 40]. As described above, all Ras isoforms undergo farnesylation in the cytoplasm and proteolytic removal of the –aaX motif and carboxylmethylation of the farnesylated cysteine at the cytoplasmic face of the ER. N-Ras and H-Ras are further modified with palmitate at one or two cysteines, respectively, upstream of the farnesylated cysteine (Table 19.1). Palmitoylation occurs early in the secretory pathway, either at the ER or Golgi, and the proteins move to the plasma membrane by vesicular transport. Here, H- and N-Ras can transduce signals from the cell surface. At the plasma membrane, H- and N-Ras are depalmitoylated and traffic by a diffusion-limited process through the cytoplasm. At the Golgi, Ras can be repalmitoylated and again be stably anchored to membranes. It appears that the more stable the membrane attachment, the more prominent the plasma membrane localization, as the dually palmitoylated H-Ras accumulates to a greater extent at the plasma membrane than does the mono-palmitoylated N-Ras, which is more prominently distributed to the Golgi. This has important functional consequences, as Ras proteins signal at the Golgi in response to a different repertoire of activators than it sees at the cell surface [41]. Accordingly, the signaling output from the cell will depend on the localization of H- and N-Ras, which in turn are regulated in part by their palmitoylation status.

Protein Lipidation and Lipid Rafts

Lipid rafts have been defined as "small (10–200 nm), heterogeneous, highly dynamic, sterol- and sphingolipid-enriched domains that compartmentalize cellular processes.

Small rafts can sometimes be stabilized to form larger platforms through protein–protein and protein–lipid interactions" [42]. Raft lipids have been proposed to exist in a more ordered state, similar to the liquid-ordered (l_o) phase described in model membranes [2, 43]. Lipid raft formation in the exoplasmic leaflet of the plasma membrane is driven by the tight packing of the long saturated acyl chains of sphingolipids with cholesterol. The structure of raft lipids in the cytoplasmic leaflet is less clear, but these domains are also believed to exist in a state where acyl chains of lipids are tightly packed, highly ordered, and extended. Proteins with a high affinity for an ordered lipid environment are recruited to rafts. These are proteins with GPI anchors (which contain predominantly saturated fatty acids), or proteins such as Src family kinases that are modified with dual fatty acylation motifs. Crosslinking of GPI-anchored proteins induces activation of Src family kinases. The lipid raft domain provides a platform for transduction of a signal from a protein anchored only to the exoplasmic leaflet to proteins associated with the inner leaflet of the plasma membrane.

Trafficking of Integral Membrane Proteins

By far the most frequent lipid modification found on integral membrane proteins is palmitoylation of cysteine residues within the cytoplasmic domain(s). Integral membrane proteins do not require covalent lipid modifications for membrane attachment. Hence, palmitoylation must serve other functions. Evidence is accumulating, for a number of receptors, that palmitoylation is required for cell surface expression [44–47]. The biosynthetic itinerary of an integral membrane protein begins in the ER, and exit from the ER requires that the protein is properly folded. The retention of palmitoylation-defective proteins early in the secretory pathway suggests that palmitoylation is a signal to bypass the quality control machinery and enter the secretory pathway.

Palmitoylation can also influence the trafficking of integral membrane proteins through the endocytic pathway. For example, endocytosis of the anthrax toxin receptor is regulated through two posttranslational modifications, palmitoylation and ubiquitination, which play opposing roles [48]. Palmitoylation and ubiquitination of the anthrax toxin receptors occurs at cysteine residues within the cytoplasmic domain of the receptor. The consequence of mutating the palmitoylated cysteines to alanine appears to be premature targeting of the receptor to lysosomes. A non-palmitoylated mutant receptor had a significantly shorter half-life than the wild-type protein, and could be visualized accumulating in a lysosomal compartment when cells were treated with leupeptin, a lysosomal enzyme inhibitor. The palmitoylation-defective receptor was ubiquitinated, pointing to a protective role for palmitoylation.

SUMMARY

Our concept of covalent lipid modifications as simple membrane anchors has evolved to an appreciation that these moieties contribute to protein localization and trafficking through specific and regulated interactions. Live cell imaging has revolutionized the way protein trafficking is monitored in cells, and led to an appreciation that lipid-modified proteins rapidly move between cellular compartments. Novel chemical biology strategies and proteomics have expanded the list of modified proteins [20, 49–51], and the enzymology of protein lipidation has grown with the discovery of the DHHC and MBOAT family of PATs. Regulation of protein stability by palmitoylation has emerged as a functional consequence of modifying integral membrane proteins. The future is likely to bring a deeper and more sophisticated understanding of how lipids regulate the localization and function of signaling proteins.

ACKNOWLEDGEMENTS

Work in the author's laboratory is supported by the National Institute of General Medical Sciences.

REFERENCES

1. Lemmon MA. Membrane recognition by phospholipid-binding domains. *Nat Rev Mol Cell Biol* 2008;**9**:99–111.

2. Edidin M. The state of lipid rafts: from model membranes to cells. *Ann Rev Bioph Biom* 2003;**32**:257–83.

3. Maurer-Stroh S, Eisenhaber B, Eisenhaber F. N-terminal N-myristoylation of proteins: refinement of the sequence motif and its taxon-specific differences. *J Mol Biol* 2002;**317**:523–40.

4. Farazi T, Waksman G, Gordon J. The biology and enzymology of protein N-myristoylation. *J Biol Chem* 2001;**276**:36,007–501.

5. Resh MD. Trafficking and signaling by fatty-acylated and prenylated proteins. *Nat Chem Bio* 2006;**2**:584–90.

6. Glover C, Hartman K, Felsted R. Human N-myristoyltransferase amino-terminal domain involved in targeting the enzyme to the ribosomal subcellular fraction. *J Biol Chem* 1997;**272**:28,680–9.

7. Mishkind M. Morbid myristoylation. *Trends Cell Biol* 2001;**11**:191.

8. Zha J, Weiler S, Oh K, Wei M, Korsmeyer S. Posttranslational N-myristoylation of BID as a molecular switch for targeting mitochondria and apoptosis. *Science* 2000;**290**:1761–5.

9. Utsumi T, Sakurai N, Nakano K, Ishisaka R. C-terminal 15-kDa fragment of cytoskeletal actin is posttranslationally N-myristoylated upon caspase-mediated cleavage and targeted to mitochondria. *FEBS Letts* 2003;**539**:37–44.

10. Sakurai N, Utsumi T. Posttranslational N-myristoylation is required for the anti-apoptotic activity of human tGelsolin, the C-terminal caspase cleavage product of human gelsolin. *J Biol Chem* 2006; **281**:14,288–95.

11. Vilas GL, Corvi MM, Plummer GJ, Seime AM, Lambkin GR, Berthiaume LG. Posttranslational myristoylation of caspase-activated p21-activated protein kinase 2 (PAK2) potentiates late apoptotic events. *Proc Natl Acad Sci USA* 2006;**103**:6542–7.

12. Zhang FL, Casey PJ. Protein prenylation: molecular mechanisms and functional consequences. *Ann Rev Biochem* 1996;**65**:241–70.

13. Lane KT, Beese LS. Thematic review series: lipid posttranslational modifications, Structural biology of protein farnesyltransferase and geranylgeranyltransferase type I. *J Lipid Res* 2006;**47**:681–99.

14. Maurer-Stroh S, Eisenhaber F. Refinement and prediction of protein prenylation motifs. *Genome Biol* 2005;**6**:R55.

15. Maurer-Stroh S, Koranda M, Benetka W, Schneider G, Sirota FL, Eisenhaber F. Towards complete sets of farnesylated and geranylgeranylated proteins. *PLoS Comput Biol* 2007;**3**:e66.

16. Wright LP, Philips MR. Thematic review series: lipid posttranslational modifications. CAAX modification and membrane targeting of Ras. *J Lipid Res* 2006;**47**:883–91.

17. Smotrys JE, Linder ME. Palmitoylation of intracellular signaling proteins: regulation and function. *Ann Rev Biochem* 2004;**73**:559–87.

18. Ren J, Wen L, Gao X, Jin C, Xue Y, Yao X. CSS-Palm 2.0: an updated software for palmitoylation sites prediction. *Protein Eng Des Sel* 2008.

19. Mitchell DA, Vasudevan A, Linder ME, Deschenes RJ. Protein palmitoylation by a family of DHHC protein S-acyltransferases. *J Lipid Res* 2006;**47**:11,188–227.

20. Roth AF, Wan J, Bailey AO, Sun B, Kuchar JA, Green WN, et al. Global analysis of protein palmitoylation in yeast. *Cell* E2006;**125**:1003–13.

21. Linder ME, Deschenes RJ. Palmitoylation: policing protein stability and traffic. *Nat Rev Mol Cell Biol* 2007;**8**:74–84.

22. Kleuss C, Krause E. G$_{as}$ is palmitoylated at the N-terminal glycine. *EMBO J* 2003;**22**:826–32.

23. Paulick MG, Bertozzi CR. The glycosylphosphatidylinositol anchor: a complex membrane-anchoring structure for proteins. *Biochemistry* 2008;**47**:6991–7000.

24. Eisenhaber B, Bork P, Eisenhaber F. Prediction of potential GPI-modification sites in proprotein sequences. *J Mol Biol* 1999;**292**:741–58.

25. Mann RK, Beachy PA. Novel lipid modifications of secreted protein signals. *Ann Rev Biochem* 2004;**73**:891–923.

26. Willert K, Brown JD, Danenberg E, Duncan AW, Weissman IL, Reya T, et al. Wnt proteins are lipid-modified and can act as stem cell growth factors. *Nature* 2003;**423**:448–52.

27. Takada R, Satomi Y, Kurata T, Ueno N, Norioka S, Kondoh H, et al. Monounsaturated fatty acid modification of Wnt protein: its role in Wnt secretion. *Dev Cell* 2006;**11**:791–801.

28. Miura GI, Treisman JE. Lipid modification of secreted signaling proteins. *Cell Cycle* 2006;**5**:1184–8.

29. Hofmann K. A superfamily of membrane-bound O-acyltransferases with implications for Wnt signaling. *Trends in Biochem Sci* 2000;**25**:111–12.

30. Buglino JA, Resh MD. Hhat is a palmitoylacyltransferase with specificity for N-palmitoylation of Sonic Hedgehog. *J Biol Chem* 2008;**283**:22,076–88.

31. Kojima M, Kangawa K. Ghrelin: structure and function. *Physiol Rev* 2005;**85**:495–522.

32. Yang J, Brown MS, Liang G, Grishin NV, Goldstein JL. Identification of the acyltransferase that octanoylates ghrelin, an appetite-stimulating peptide hormone. *Cell* 2008;**132**:387–96.

33. Gutierrez JA, Solenberg PJ, Perkins DR, Willency JA, Knierman MD, Jin Z, et al. Ghrelin octanoylation mediated by an orphan lipid transferase. *Proc Natl Acad Sci USA* 2008;**105**:6320–5.

34. Bhatnagar RS, Gordon JI. Understanding covalent modifications of proteins by lipids: where cell biology and biophysics mingle. *Trends Cell Biol* 1997;**7**:14–20.

35. Shahinian S, Silvius JR. Doubly-lipid-modified protein sequence motifs exhibit long-lived anchorage to lipid bilayer membranes. *Biochemistry* 1995;**34**:3813–22.

36. Murray D, McLaughlin S, Honig B. The role of electrostatic interactions in the regulation of the membrane association of G-protein βγ heterodimers. *J Biol Chem* 2001;**276**:45,153–9.

37. Magee A, Marshall C. New insights into the interaction of Ras with the plasma membrane. *Cell* 1999;**98**:9–12.

38. Schroeder H, Leventis R, Shahinian S, Walton PA, Silvius JR. Lipid-modified, cysteinyl-containing peptides of diverse structures are efficiently S-acylated at the plasma membrane of mammalian cells. *J Cell Biol* 1996;**134**:647–60.

39. Rocks O, Peyker A, Kahms M, Verveer PJ, Koerner C, Lumbierres M, et al. An acylation cycle regulates localization and activity of palmitoylated Ras isoforms. *Science* 2005;**307**:1746–52.

40. Goodwin JS, Drake KR, Rogers C, Wright L, Lippincott-Schwartz J, Philips MR, et al. Depalmitoylated Ras traffics to and from the Golgi complex via a nonvesicular pathway. *J Cell Biol* 2005;**170**:261–72.

41. Mor A, Philips MR. Compartmentalized Ras/MAPK signaling. *Ann Rev Immunol* 2006;**24**:771–800.

42. Pike LJ. Rafts defined: a report on the Keystone Symposium on Lipid Rafts and Cell Function. *J Lipid Res* 2006;**47**:1597–8.

43. Brown D, London E. Structure and function of sphingolipid- and cholesterol-rich membrane rafts. *J Biol Chem* 2000;**275**(17):221.

44. Escriba PV, Wedegaertner PB, Goni FM, Vogler O. Lipid-protein interactions in GPCR-associated signaling. *Biochim Biophys Acta* 2007;**1768**:836–52.

45. Hayashi T, Rumbaugh G, Huganir RL. Differential regulation of AMPA receptor subunit trafficking by palmitoylation of two distinct sites. *Neuron* 2005;**47**:709–23.

46. Petaja-Repo UE, Hogue M, Leskela TT, Markkanen PM, Tuusa JT, Bouvier M. Distinct subcellular localization for constitutive and agonist-modulated palmitoylation of the human delta opioid receptor. *J Biol Chem* 2006;**281**:15,780–9.

47. Abrami L, Kunz B, Iacovache I, van der Goot FG. Palmitoylation and ubiquitination regulate exit of the Wnt signaling protein LRP6 from the endoplasmic reticulum. *Proc Natl Acad Sci USA* 2008;**105**:5384–9.

48. Abrami L, Leppla SH, van der Goot FG. Receptor palmitoylation and ubiquitination regulate anthrax toxin endocytosis. *J Cell Biol* 2006;**172**:309–20.

49. Kho Y, Kim SC, Jiang C, Barma D, Kwon SW, Cheng J, et al. A tagging-via-substrate technology for detection and proteomics of farnesylated proteins. *Proc Natl Acad Sci USA* 2004;**101**:12,479–84.

50. Kostiuk MA, Corvi MM, Keller BO, Plummer G, Prescher JA, Hangauer MJ, et al. Identification of palmitoylated mitochondrial proteins using a bio-orthogonal azido-palmitate analogue. *FASEB J* 2008;**22**:721–32.

51. Zhang Jr J, Planey SL, Ceballos C, Stevens SM, Keay SK, Zacharias DA. Identification of CKAP4/p63 as a major substrate of the palmitoyl acyltransferase DHHC2, a putative tumor suppressor, using a novel proteomics method. *Mol Cell Proteomics* 2008;**7**:1378–88.

Chapter 20

Structural Principles of Lipid Second Messenger Recognition

Roger L. Williams

Medical Research Council, Laboratory of Molecular Biology, Cambridge, England, UK

INTRODUCTION

Structural analyses have shown that domains with a variety of different folds can recognize a single type of lipid second messenger, and that a single type of fold can evolve different binding sites and alternative modes of interaction for the same lipid second messenger. Specificity in lipid recognition is achieved by both electrostatic and shape complementarity. A common theme suggested by the structures of the lipid-modifying enzymes and the specific recognition modules is that secondary, non-specific membrane interactions cooperate with specific lipid recognition to increase membrane avidity. Most binding domains have evolved mechanisms such as partial membrane penetration to bind lipids without removing them from the bilayer.

A wide range of lipid second messengers that are generated in response to external signals has been characterized in terms of their molecular biology. This chapter will focus on underlying structural principles involved in recognizing these messengers both by the enzymes that produce or consume them and by the downstream effector domains. In most cases, the lipid-modifying enzymes are structurally homologous to enzymes that catalyze an analogous reaction using soluble substrates, suggesting that the constraints imposed by the catalytic chemistry are a stronger determinant of fold than specific lipid binding. For example, the lipid kinases are homologous to protein kinases [1, 2]. The phosphoinositide phosphatases are homologous to protein phosphatases and endonucleases [3, 4]. The phosphoinositide-specific phospholipases C have a catalytic domain with a TIM-barrel fold similar to many other enzymes and an arrangement of catalytic residues similar to nucleases [5]. A variety of domains present in downstream effector proteins also specifically recognize the lipid second messengers. In contrast to the metabolizing enzymes, these effector domains typically bind lipids with higher affinity and have unique folds.

PHOSPHOLIPID SECOND MESSENGER RECOGNITION BY ACTIVE SITES OF ENZYMES

The phosphoinositides are the most diverse family of lipid messengers. All of them share a phosphatidyl D-myo-inositol (PtdIns) scaffold that can be phosphorylated at all possible combinations of the 3-, 4-, and 5-hydroxyls to generate lipid messengers with specific roles in intracellular signaling. Several generalizations can be made regarding the recognition of phosphoinositides by proteins. The enzymes that recognize phosphoinositides as substrates tend to envelope the headgroup and make Van der Waals contacts with both faces of the inositol ring (Figure 20.1). In contrast, the domains that have evolved simply to bind the phosphoinositides, such as PH domains, tend to form interactions with some or all of the phosphates but to leave one or both faces of the ring exposed (Figure 20.2). Presumably, the tendency for the enzymes to more fully bury the headgroup arises from a necessity to exclude water from the active site or to more precisely position the reactive moieties in the active site. Within a family of domains, the affinity of phosphoinositide headgroup binding generally correlates with the number of hydrogen bonds between the phosphoinositide and the protein. The enzymes generally have a lower affinity for the phosphoinositide headgroup than the highest affinity binding modules – as would be expected from the role of an enzyme to preferentially recognize the transition state rather than the substrate or the product.

Phosphoinositide 3-Kinase (PI3K)

PI3Ks catalyze the phosphorylation of phosphoinositides at the 3-OH, giving rise to the second messengers PtdIns(3)P, PtdIns(3,4)P_2, and PtdIns(3,4,5)P_3. The structure of

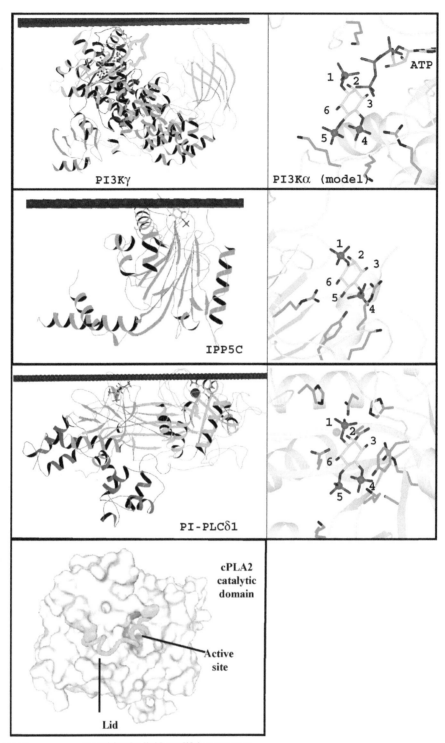

FIGURE 20.1 Lipid second messenger recognition by lipid-modifying enzymes.
The left panels show the overall folds of the phosphoinositide-modifying enzymes with putative membrane-interacting regions placed in contact with a schematic membrane represented by a layer of spheres. The bound phosphoinositides are shown in stick representation. In the right panels, close-up views of the phosphoinositide/protein interactions are shown. The structures were optimally superimposed on the inositide moieties to present a common view. The molecular surface of cPLA$_2$'s catalytic domain is shown in the lower panel.

PI3Kγ, representative of both PI3- and PI4-kinases, has a catalytic domain with an N-terminal lobe consisting of a five-stranded β-sheet closely related to protein kinases and a C-terminal lobe that is predominantly helical and more distantly related to protein kinases. The primary determinant of substrate preference for both PI3Ks is a region in the C-terminal lobe analogous to the activation loop of protein kinases [6, 7]. Models of substrate binding proposed

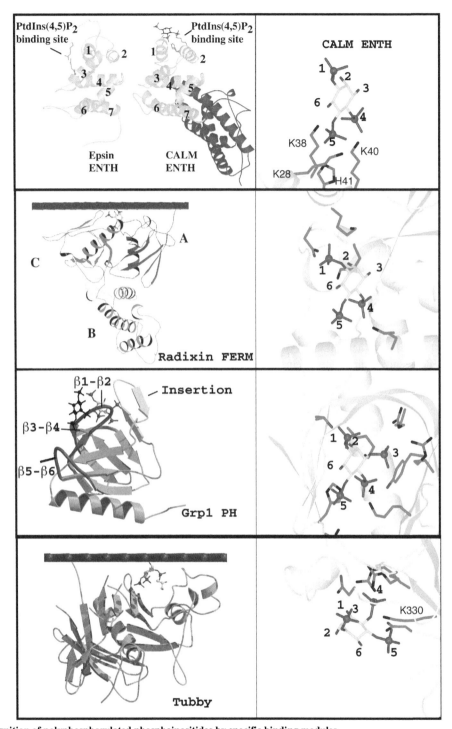

FIGURE 20.2 Recognition of polyphosphorylated phosphoinositides by specific binding modules.
The representations are as in Figure 20.1. In the first pair of panels, the PtdIns(4,5)P$_2$-binding sites of the epsin and CALM ENTH domains are illustrated. The structurally similar regions of the two domains are colored yellow (color shown in online version). Part of the epsin PtdIns(4,5)P$_2$ site involves N-terminal residues that have been modeled (dotted lines). In the third pair of panels, Ins(1,3,4,5)P$_4$ bound to the Grp1 PH domain is shown (magenta phosphates). To illustrate the differences in the locations of the phosphoinositide binding pockets in β-spectrin and other PH domains such as Grp1, an Ins(1,4,5)P$_3$ (black) has been placed on the Grp1 domain in a location analogous to the β-spectrin binding pocket.

for PI3Ks place the phosphoinositide in a shallow pocket (Figure 20.1) so that the 4- and 5-phosphates interact with basic residues in the activation loop, and the 1-phosphate contacts a Lys in a loop analogous to the glycine-rich loop of protein kinases (but without glycines in PI3Ks) [2,7]. As with PLCδ1, the location of the active site and accessory domains for membrane binding suggest that the enzyme interacts with the membrane in such a manner that substrate lipids do not have to be removed from the lipid bilayer (Figure 20.1).

Phosphatidylinositol Phosphate 4- and 5-Kinases (PIPkins)

PtdIns(4,5)P$_2$-mediated signal transduction is essential for cytoskeletal organization and dynamics, membrane trafficking, and apoptosis. Synthesis of PtdIns(4,5)P$_2$ is catalyzed by PIPkins [8]. The type IIβ?PIPkin has an N-terminal lobe with a seven-stranded anti-parallel β-sheet structurally related to protein kinases and a C-terminal lobe consisting of a smaller five-stranded β-sheet [1]. The PIPkins have a requirement for phosphorylated phosphoinositides due to a cluster of four conserved, basic residues in a putative phosphoinositide-binding pocket. The binding pocket is surprisingly shallow and open and suggests that there are few or no contacts with the 2- and 3-OH of the headgroup. The specificity of the enzyme for PtdIns(4)P versus PtdIns(5)P is completely dictated by a loop in the C-terminal lobe analogous to the activation loop of PI3K and protein kinases, and a single point mutation in this loop can swap the specificity [9].

PTEN, a 3-Phosphoinositde Phosphatase

Essential to any signal transduction system is a mechanism to produce second messengers and a mechanism to eliminate them. PTEN has a critical role in cells to antagonize the action of PI 3-kinases by catalyzing the dephosphorylation of the 3-phosphate from PtdIns(3,4,5)P$_3$. The structure of PTEN has a fold and active site configuration similar to the dual-specificity protein phosphatases [3]. A model for substrate binding places His 93 and Lys 128 as ligands of the 5-phosphate [3]. Although the 4-phosphate is deeply buried in the PTEN active site, there is no basic residue present with which it would associate. This is consistent with the ability of the enzyme to dephosphorylate PtdIns(3)P, PtdIns(3,4)P$_2$, and PtdIns(3,4,5)P$_3$. The 3-phosphate is also deeply buried, but, consistent with the presence of the scissile bond on this group, there is a basic residue, Arg 130, interacting with it.

Inositol Polyphosphate 5-Phosphatase (IPP5P)

IPP5P plays an essential role in signaling by utilizing both inositol phosphates and phosphatidylinositol polyphosphates as substrates. The 5-phosphatases regulate the levels of both the soluble Ins(1,4,5)P$_3$ and the membrane-resident PtdIns(4,5)P$_2$. The structure of the catalytic domain of the synaptojanin IPP5P from *S. pombe* bound to the product of the reaction, Ins(4)P, shows an active site located at the bottom of a funnel-shaped depression containing the histidine

essential for catalysis [4]. The catalytic mechanism is closely related to those of nucleases such as DNase I and DNase III. The 4-phosphate of the Ins(4)P interacts with three basic groups in the active site and makes water-mediated interactions with the divalent metal co-factor (Figure 20.1). The product of the reaction binds in a catalytically non-productive manner with the 4-phosphate remote from the catalytic histidine, thus showing why this family of enzymes is not able to use Ins(1,4)P$_2$ as a substrate.

Phosphoinositide-Specific Phospholipase C (PI-PLC)

PtdIns(4,5)P$_2$ is hydrolyzed by PI-PLC. The catalytic domain of the mammalian PLCδ1 consists of a (β/α)$_8$ barrel [5], a common architecture for enzymes in general. Principles of PtdIns(4,5)P$_2$ headgroup recognition by PI-PLC have been inferred from a complex of PLCδ1 with the product of the reaction, Ins(1,4,5)P$_3$. With the exception of the 6-OH of the headgroup, all of the hydroxyls of the bound inositide are stereospecifically recognized by the enzyme. The PtdIns(4,5)P$_2$ headgroup lodges edge-on in the binding pocket with the 3-OH at the bottom and the 1-OH at the top. This places the 1-OH at the level of the putative membrane-binding surface, suggesting that the enzyme does not remove substrate from the membrane during the catalytic cycle, similarly to most of the phosphoinositide-recognizing enzymes and binding domains (Figure 20.1).

Cytosolic Phospholipase A$_2$

The phospholipase A$_2$ (PLA$_2$) family of enzymes hydrolyzes the sn-2 bond of phospholipids to generate free fatty acids and lysophospholipids. The cytosolic PLA$_2$ (cPLA$_2$) selectively hydrolyzes phospholipids with an sn-2 arachidonic acid and therefore has a key role in supplying the precursor for eicosanoid biosynthesis. cPLA$_2$ has an N-terminal C2 domain that is important for Ca^{2+}-dependent membrane translocation and a catalytic domain. The enzyme has a central β-sheet with an active-site nucleophile located in a portion of the structure analogous to the nucleophilic elbow of other phospholipases having an α/β hydrolase fold [10]. Apart from this feature, however, cPLA$_2$ has a quite divergent fold. Residues in the active site that are buried by a flexible lid accomplish recognition of the substrate. Upon binding to the membrane interface, this lid undergoes a conformational change to expose a wide hydrophobic platform surrounding a funnel-shaped pocket that cradles the substrate (Figure 20.1). Even though the structure suggests that the catalytic domain partially penetrates into the hydrophobic portion of the lipid membrane, the cleft leading to the active-site nucleophile is deep enough to require that the substrate be removed from the lipid bilayer [10].

PHOSPHOINOSITIDE-BINDING DOMAINS

Polyphosphorylated Phosphoinositide-Binding Domains

ENTH Domain

Several proteins involved in endocytosis have an N-terminal domain of about 140 residues known as the ENTH domain, which is necessary for binding to PtdIns $(4,5)P_2$. The ENTH domains of CALM [11], AP180 [12], and epsin [13-15] consist of helices wound into a solenoid reminiscent of other helical domains such as armadillo and TPR. The PtdIns(4,5)P_2 binding sites in the CALM and epsin ENTH domains differ significantly (Figure 20.2). The unique binding site of the CALM ENTH domain is on an exposed surface with the PtdIns(4,5)P_2 headgroup poised at the tips of three lysines (K28, K38, K40) and a histidine (H41) in helices α1 and α2 and the loop between them (Figure 20.2). The residues involved in the interaction define a KX$_9$KX(K/R)(H/Y) motif that is present in other AP180 homologs but not in epsin [11]. The binding site in epsin involves basic residues in helices α3, α4 and a disordered N-terminal region that changes conformation upon lipid binding (Figure 20.2) [14,15]. As was observed for PH domains, similarity in domain fold does not imply that the same region of the fold is used to interact with phosphoinositides.

The FERM Domain

The FERM domain is found in the ezrin/radixin/moesin (ERM) family of proteins as well as in talin, the erythrocyte band 4.1 protein, several tyrosine kinases and phosphatases, and the tumor suppressor merlin. Members of the ERM family of proteins have three structural domains, and the N-terminal FERM domain binds to PtdIns(4,5)P_2-containing membranes. Phospholipid binding is masked by an intra or intermolecular interaction between the C-terminal domain and the FERM domain. The FERM domain consists of three compact modules, A, B, and C [16–18]. Although the C module has an overall fold similar to PH domains that are known to bind PtdIns(4,5)P_2, the crystal structure of the radixin complex with the Ins(1,4,5)P_3 shows that the phosphoinositide binds between the A and C modules [19]. Two basic residues from the A module interact with the 4- and 5-phosphates and one from the C module interacts with the 1-phosphate (Figure 20.2). The binding site is more open than most PtdIns(4,5)P_2 binding sites but less open than the ENTH-type PtdIns(4,5)P_2 binding site. PtdIns(4,5)P_2 binding causes conformational changes in the C module that prevent a self-association with the C-terminal tail of the protein and enable the N-terminal domain to interact with the cytosolic regions of integral membrane proteins. Mutagenesis suggests that the β5–β6 and β6–β7 loops in the C module may constitute a second PtdIns(4,5)P_2-binding site [20,21]. The phosphoinositide

binding pocket defines part of a basic surface that is likely to be juxtaposed to the lipid bilayer, leaving an acidic groove between subdomains B and C free to interact with integral membrane adhesion proteins (Figure 20.2) [19].

Tubby C-Terminal DNA-Binding Domain

A common feature among the tubby family proteins is the presence of a C-terminal DNA-binding domain with a unique fold consisting of a 12-stranded anti-parallel β-barrel and a hydrophobic helix running through the barrel [22]. PtdIns(4,5)P_2 binding to the C-terminal domain causes Tubby to be localized to the plasma membrane until the levels of PtdIns(4,5)P_2 fall in response to receptor-mediated activation of PLC-β [22]. Loss of plasma-membrane localization is accompanied by nuclear translocation of the protein. The complex of the C-terminal domain of Tubby with glycerophosphoinositol 4,5-bisphosphate shows the PtdIns(4,5)P_2 headgroup in a shallow pocket that involves residues from three adjacent β-strands [22] and is located at one edge of the putative DNA-binding surface. The side chain of a single Lys (330) intercalates between the 4- and 5-phosphates in a manner that is unique to Tubby and the CALM-N ENTH domains (Figure 20.2). In these domains, Lys side chain approaches the 4- and 5-phosphates approximately parallel to the plane of the inositol ring. In Tubby, the Lys makes an unusually close (2.1-Å) contact with the 5-phosphate. An additional Arg that coordinates the 4-pho s phate is also positioned so that 3-phosphorylated lipids could interact with it, which may account for the PtdIns(3,4,5)P_3 and PtdIns(3,4)P_2 binding observed *in vitro* [23].

PH Domains

PH domains are among the most common phosphoinositide-binding modules present in mammalian genomes and show a wide range of phosphoinositide affinities and specificities. They consist of two orthogonal β-sheets curving to form a barrel-like structure closed off by a C-terminal α-helix [8,24]. High-affinity binding to phosphoinositides is achieved using residues in the β1–β2 (VL1), β3–β4 (VL2), and β6–β7 (VL3) loops. The PtdIns(4,5)P_2-specific PH domain of PLC-δ1 [25] differs from other PH domains in that the orientation of the bound inositide is flipped by 1808 so that the position occupied by the 5-phosphate in PLC-δ1 is occupied by the 3-phosphate in the 3-phosphoinositide-specific PH domains that have been characterized (Figure 20.2).

Among PH domains recognizing 3-phosphoinositides, three types of specificities are apparent: PtdIns(3,4,5)P_3-specificity such as the PH domain of Grp1 and Btk, dual PtdIns (3,4,5)P_3/PtdIns(3,4)P_2-specificity such as the PH domain of DAPP1 and PKB and PtdIns(3,4)P_2-specificity such as the C-terminal PH domain of TAPP1. The PtdIns(3,4,5)

P_3 specificity is achieved by enveloping the 5-phosphate by using insertions in either the $\beta6$–$\beta7$ loop (as in GRP1 [26,27]) or in the $\beta1$–$\beta2$ loop (as in Btk, [28]). DAPP1 makes more interactions with the 4-phosphate while the 5-phosphate is largely exposed. The PtdIns(3,4)P_2 specificity of the TAPP1 PH domain arises from steric clashes of the 5-phosphate with residues in the $\beta1$–$\beta2$ loop [29]. The analogous region of the closely related DAPP1 PH domain has a Gly that makes space to accommodate the 5-phosphate of PtdIns(3,4,5)P_3. Basic and hydrophobic residues in the $\beta1$–$\beta2$ loop of Grp1 and Btk suggest that these PH domains may have additional, non-specific interactions with lipid bilayers that enhance membrane avidity (Figure 20.2) [27].

Other modes of phosphoinositide binding have been shown for PH domains. The PH domain of β-spectrin uses the $\beta5$–$\beta6$ loop and the side of the $\beta1$–$\beta2$ loop opposite that used by PLC-$\delta1$ to interact with Ins(1,4,5)P_3 [30], showing that the same fold can be adapted to several different binding modes (Figure 20.2). The PH domain from β-spectrin is an example of a PH domain with low affinity and little specificity for lipid binding. More recent analyses of the genome suggest that this may be characteristic of the vast majority of PH domains [24].

PtdIns(3)P-Binding Domains

PtdIns(3)P is present in mammalian cells at fairly high concentrations relative to such transient lipid second messengers as PtdIns(3,4,5)P_3. Its distribution in cells is restricted mainly to endosomal membranes. PtdIns(3)P levels can increase rapidly during certain processes such as receptor-mediated phagocytosis [31]. Two structurally unrelated domain types, FYVE and PX, are capable of specifically binding PtdIns(3)P [32].

FYVE Domains

The FYVE domains are found in many proteins involved in membrane transport [33]. The FYVE domains from Vps27 [34], Hrs [35], and EEA1 [36,37] consist of two small β-sheets stabilized by two Zn^{2+} ions and a C-terminal α-helix. The PtdIns(3)P forms hydrogen bonds with the protein by using the 1- and 3-phosphates and the 4-, 5-, and 6-OH groups [37]. The close approach of these hydrogen-bonding partners precludes polyphos phorylated phosphoinositides from binding (Figure 20.3). The 3-phosphate forms a hydrogen bond with the last arginine in the (R/K) (R/K)HHCR

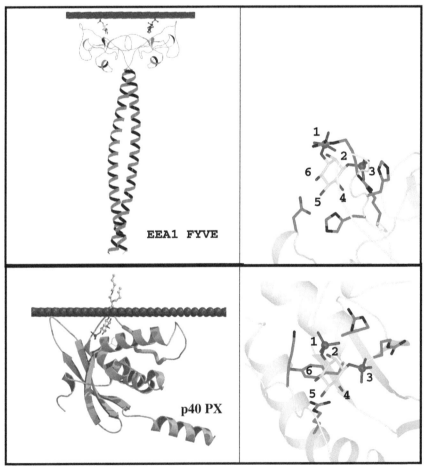

FIGURE 20.3 PtdIns(3)P recognition by specific binding modules. The representations are as in Fig. 20.1.

signature motif characteristic of the FYVE domains. The 1-phosphate interacts with the protein via the first Arg of this motif. Like the PH domains, the FYVE domain buries only one face of the bound phosphoinositide. For EEA1, the face with the axial 2-OH is exposed to solution. The presence of the coiled-coil region preceding the EEA1 FYVE domain helps to unambiguously define the mode of membrane interaction and suggests that a loop flanking the PtdIns(3)P pocket, the "turret" loop, penetrates into the lipid bilayer (Figure 20.3). Biophysical measurements indicate that this partial membrane penetration follows rather than precedes specific PtdIns(3)P binding [38].

PX Domains

PX domains are found in a wide range of proteins including many involved in lipid modification, intracellular signaling, and vesicle trafficking [39]. They consist of a three-stranded β-sheet subdomain and an α-helical subdomain that are joined by a conserved RR(Y/F) motif [40,41]. The structure of the PX domain from the p40 cytosolic subunit of the NADPH oxidase in a complex with PtdIns(3)P shows that the first Arg from the RR(Y/F) motif has a structural role in the core of the protein, while the second Arg and the Tyr residue interact with the 3-phosphate and the face of the inositide ring, respectively [42] (Figure 20.3). The PX domain buries the face of the inositide adjacent to the axial 2-OH, leaving the opposite face largely exposed. The mode of membrane binding of the PX domain is suggested by the diacylglycerol moiety of the bound PtdIns(3)P and hydrophobic residues adjacent to the phosphoinositide binding pocket (Figure 20.3).

NON-PHOSPHOINOSITIDE LIPID MESSENGER RECOGNITION

C1 Domains

The C1 domain is essential for membrane localization and activation of many proteins involved in signal transduction, including the protein kinase C isozymes [42]. The C1 domains are 50-residue modules containing two small β-sheets and a short C-terminal helix. The domains have been classified into two groups, the "typical" domains that fit a profile derived for phorbol ester or diacylglycerol (DAG) binding and the "atypical" domains that do not [43]. The phorbol ester sits in a groove that is formed by a splaying of adjacent β-strands in a sheet [44,45]. Hydrophilic groups on the phorbol ester intercalate between the strands and make backbone interactions with their exposed main-chain atoms. Once the phorbol ester is bound, the entire end of the domain presents a hydrophobic surface that penetrates into the lipid bilayer. Available binding data are consistent with a model in which the DAG fits into the same groove as the phorbol ester, forming hydrogen bonds with the main-chain atoms of the strands using its 3-OH.

FUTURE DIRECTIONS

Although much progress has been made in defining the nature of the interactions of lipid second messengers with proteins, many questions remain unanswered. Several lipid second messengers have been characterized for which there is no structural information about specific binding modules; e.g., PtdIns(3,5)P$_2$ and phosphatidic acid. A dimension of response to lipid-messenger recognition that remains largely unexplored is the effect of membrane binding on membrane structure during processes such as formation of multivesicular bodies. Many proteins use multiple weak interactions to bind to membranes in response to lipid second messengers, but an analysis of the energetics of the individual interactions is often lacking. Although membrane translocation in response to lipid second messengers is common, the nature and extent of allosteric responses mediated by membrane interactions are not clear. With methodologies that have emerged in the wake of genomic studies, we can look forward to answers to many of these questions in the near future.

ACKNOWLEDGEMENTS

I apologize to colleagues whose work I was unable to cite given the wide scope of the review and the severe limitations on space. Marketa Zvelebil is thanked for coordinates of the PI3Kα model.

REFERENCES

1. Rao VD, Misra S, Boronenkov IV, Anderson RA, Hurley JH. Structure of type IIβ phosphatidylinositol phosphate kinase: a protein kinase fold flattened for interfacial phosphorylation. *Cell* 1998;**94**:829–39.
2. Walker EH, Perisic O, Ried C, Stephens L, Williams RL. Structural insights into phosphoinositide 3-kinase catalysis and signalling. *Nature* 1999;**402**:313–20.
3. Lee JO, Yang H, Georgescu MM, Di Cristofano A, Maehama T, et al. Crystal structure of the PTEN tumor suppressor: implications for its phosphoinositide phosphatase activity and membrane association. *Cell* 1999;**99**:323–34.
4. Tsujishita Y, Guo S, Stolz LE, York JD, Hurley JH. Specificity determinants in phosphoinositide dephosphorylation: crystal structure of an archetypal inositol polyphosphate 5-phosphatase. *Cell* 2001;**105**:379–89.
5. Essen L-O, Perisic O, Cheung R, Katan M, Williams RL. Crystal structure of a mammalian phosphoinositide-specific phospholipase Cδ. *Nature* 1996;**380**:595–602.
6. Bondeva T, Pirola L, Bulgarelli-Leva G, Rubio I, Wetzker R, et al. Bifurcation of lipid and protein kinase signals of PI3Kγ to the protein kinases PKB and MAPK. *Science* 1998;**282**:293–6.

7. Pirola L, Zvelebil MJ, Bulgarelli-Leva G, Van Obberghen E, Waterfield MD, et al. Activation loop sequences confer substrate specificity to phosphoinositide 3-kinase α (PI3Kα). *J Biol Chem* 2001;**276**:21,544–54.

8. Hurley JH, Misra S. Signaling and subcellular targeting by membrane-binding domains. *Annu. Rev. Biophys. Biomol. Struct.* 2000;**29**:49–79.

9. Kunz J, Fuelling A, Kolbe L, Anderson RA. Stereo-specific substrate recognition by phosphatidylinositol phosphate kinases is swapped by changing a single amino acid residue. *J. Biol. Chem.* 2002;**277**:5611–19.

10. Dessen A, Tang J, Schmidt H, Stahl M, Clark JD, et al. Crystal structure of human cytosolic phospholipase A₂ reveals a novel topology and catalytic mechanism. *Cell* 1999;**97**:349–60.

11. Ford MGJ, Pearse BMF, Higgins MK, Vallis Y, Owen DJ, et al. Simultaneous binding of PtdIns(4,5)P₂ and clathrin by AP180 in the nucleation of clathrin lattices on membranes. *Science* 2001;**291**:1051–5.

12. Mao Y, Chen J, Maynard JA, Zhang B, Quiocho FA. A novel all helix fold of the AP180 amino-terminal domain for phosphoinositide binding and clathrin assembly in synaptic vesicle endocytosis. *Cell* 2001;**104**:433–40.

13. Hyman J, Chen H, Di Fiore PP, De Camilli P, Brunger AT. Epsin 1 undergoes nucleocytosolic shuttling and its eps15 interactor NH(2)-terminal homology (ENTH) domain, structurally similar to Armadillo and HEAT repeats, interacts with the transcription factor promyelocytic leukemia Zn^{(2)+} finger protein (PLZF). *J. Cell Biol.* 2000;**149**:537–46.

14. Itoh T, Koshiba S, Kigawa T, Kikuchi A, Yokoyama S, et al. Role of the ENTH domain in phosphatidylinositol-4,5-bisphosphate binding and endocytosis. *Science* 2001;**291**:1047–51.

15. Ford MG, Mills IG, Peter BJ, Vallis Y, Praefcke GJ, Evans PR, McMahan HT. Curvature of clathrin-coated pits driven by epsin. *Nature* 2002;**419**:361–6.

16. Shimizu T, Seto A, Maita N, Hamada K, Tsukita S, et al. Structural basis for neurofibromatosis type 2: crystal structure of the merlin FERM domain. *J Biol Chem* 2001;**277**:10,332–36.

17. Edwards SD, Keep NH. The 2.7 A crystal structure of the activated FERM domain of moesin: an analysis of structural changes on activation. *Biochemistry* 2001;**40**:7061–8.

18. Pearson MA, Reczek D, Bretscher A, Karplus PA. Structure of the ERM protein moesin reveals the FERM domain fold masked by an extended actin binding tail domain. *Cell* 2000;**101**:259–70.

19. Hamada K, Shimizu T, Matsui T, Tsukita S, Tsukita S, et al. Structural basis of the membrane-targeting and unmasking mechanisms of the radixin FERM domain. *EMBO J* 2000;**19**:4449–62.

20. Barret C, Roy C, Montcourrier P, Mangeat P, Niggli V. Mutagenesis of the phosphatidylinositol 4,5-bisphosphate (PIP(2)) binding site in the NH(2)-terminal domain of ezrin correlates with its altered cellular distribution. *J Cell Biol* 2000;**151**:1067–80.

21. Niggli V. Structural properties of lipid-binding sites in cytoskeletal proteins. *Trends Biochem Sci* 2001;**26**:604–11.

22. Boggon TJ, Shan WS, Santagata S, Myers SC, Shapiro L. Implication of tubby proteins as transcription factors by structure-based functional analysis. *Science* 1999;**286**:2119–25.

23. Santagata S, Boggon TJ, Baird CL, Gomez CA, Zhao J, et al. G-protein signaling through tubby proteins. *Science* 2001;**292**:2041–50.

24. Lemmon MA, Ferguson KM, Abrams CS. Pleckstrin homology domains and the cytoskeleton. *FEBS Letts* 2002;**513**:71–6.

25. Ferguson KM, Lemmon MA, Schlessinger J, Sigler PB. Structure of the high affinity complex of inositol trisphosphate with a phospholipase C pleckstrin homology domain. *Cell* 1995;**83**:1037–46.

26. Ferguson KM, Kavran JM, Sankaran VG, Fournier E, Isakoff SJ, et al. Structural basis for discrimination of 3-phosphoinositides by pleckstrin homology domains. *Mol Cell* 2000;**6**:373–84.

27. Lietzke SE, Bose S, Cronin T, Klarlund J, Chawla A, et al. Structural basis of 3-phosphoinositide recognition by pleckstrin homology domains. *Mol Cell* 2000;**6**:385–94.

28. Baraldi E, Carugo KD, Hyvonen M, Surdo PL, Riley AM, et al. Structure of the PH domain from Bruton's tyrosine kinase in complex with inositol 1,3,4,5-tetrakisphosphate. *Structure Fold Des* 1999;**7**:449–60.

29. Thomas CC, Dowler S, Deak M, Alessi DR, van Aalten DM. Crystal structure of the phosphatidylinositol 3,4-bisphosphate-binding pleckstrin homology (PH) domain of tandem PH-domain-containing protein 1 (TAPP1): molecular basis of lipid specificity. *Biochem. J.* 2001;**358**:287–94.

30. Hyvonen M, Macias MJ, Nilges M, Oschkinat H, Saraste M, et al. Structure of the binding site for inositol phosphates in a PH domain. *EMBO J* 1995;**14**:4676–85.

31. Vieira OV, Botelho RJ, Rameh L, Brachmann SM, Matsuo T, et al. Distinct roles of class I and class III phosphatidylinositol 3-kinases in phagosome formation and maturation. *J. Cell Biol.* 2001;**155**:19–25.

32. Misra S, Miller GJ, Hurley JH. Recognizing phosphatidylinositol 3-phosphate. *Cell* 2001;**107**:559–62.

33. Gillooly DJ, Simonsen A, Stenmark H. Cellular functions of phosphatidylinositol 3-phosphate and FYVE domain proteins. *Biochem J* 2001;**355**:249–58.

34. Misra S, Hurley JH. Crystal structure of a phosphatidylinositol 3-phosphate-specific membrane-targeting motif, the FYVE domain of Vps27p. *Cell* 1999;**97**:657–66.

35. Mao Y, Nickitenko A, Duan X, Lloyd TE, Wu MN, et al. Crystal structure of the VHS and FYVE tandem domains of Hrs, a protein involved in membrane trafficking and signal transduction. *Cell* 2000;**100**:447–56.

36. Kutateladze T, Overduin M. Structural mechanism of endosome docking by the FYVE domain. *Science* 2001;**291**:1793–6.

37. Dumas JJ, Merithew E, Sudharshan E, Rajamani D, Hayes S, et al. Multivalent endosome targeting by homodimeric EEA1. *Mol. Cell* 2001;**8**:947–58.

38. Stahelin RV, Long F, Diraviyam K, Bruzik KS, Murray D, et al. Phosphatidylinositol 3-phosphate induces the membrane penetration of the FYVE domains of Vps27p and Hrs. *J Biol Chem* 2002;**277**:26,379–88.

39. Wishart MJ, Taylor GS, Dixon JE. Phoxy lipids: revealing PX domains as phosphoinositide binding modules. *Cell* 2001;**105**:817–20.

40. Hiroaki H, Ago T, Ito T, Sumimoto H, Kohda D. Solution structure of the PX domain, a target of the SH3 domain. *Nat Struct Biol* 2001;**8**:526–30.

41. Bravo J, Karathanassis D, Pacold CM, Pacold ME, Ellson CD, et al. The crystal structure of the PX domain from p40^{phox} bound to phosphatidylinositol 3-phosphate. *Mol Cell* 2001;**8**:829–39.

42. Ono Y, Fujii T, Igarashi K, Kuno T, Tanaka C, et al. Phorbol ester binding to protein kinase C requires a cysteine-rich zinc-finger-like sequence. *Proc Natl Acad Sci USA* 1989;**86**:4868–71.

43. Hurley JH, Newton AC, Parker PJ, Blumberg PM, Nishizuka Y. Taxonomy and function of C1 protein kinase C homology domains. *Protein Sci* 1997;**6**:477–80.

44. Zhang G, Kazanietz MG, Blumberg PM, Hurley JH. Crystal structure of the cys2 activator-binding domain of protein kinase C delta in complex with phorbol ester. *Cell* 1995;**81**:917–24.

45. Xu RX, Pawelczyk T, Xia TH, Brown SC. NMR structure of a protein kinase C-gamma phorbol-binding domain and study of protein-lipid micelle interactions. *Biochemistry* 1997;**36**:10,709–17.

Chapter 21

Pleckstrin Homology (PH) Domains

Mark A. Lemmon

Department of Biochemistry and Biophysics, University of Pennsylvania Medical Center, Philadelphia, Pennsylvania

IDENTIFICATION AND DEFINITION OF PH DOMAINS

The 100–120-amino acid pleckstrin homology (PH) domain was first named in 1993 [1–3] as a region of sequence similarity that occurs twice in pleckstrin [4] and is shared by a large number of other proteins. Levels of sequence identity between PH domains are generally low, lying between around 10 (or less) to 30 percent, and there is no conserved motif that identifies PH domains. Rather, PH domains are defined by a pattern of sequence similarity that suggests a common fold, and may therefore share structural similarity in the absence of functional relatedness. The majority of PH domain-containing proteins require membrane association for some aspect of their function. These proteins participate in cellular signaling, cytoskeletal organization, membrane trafficking, and/or phospholipid modification. Sequences encoding PH domains occur in some 252 genes in the first draft of the human genome sequence [5], making this the eleventh most populous domain family in humans. PH domains occur in 77 genes in *D. melanogaster*, 71 genes in *C. elegans*, and 27 in *S. cerevisiae* [5]. Understanding the functions of these common domains has therefore been a subject of considerable interest.

THE STRUCTURE OF PH DOMAINS

Structures of 15 different PH domains have been determined by NMR and/or X-ray crystallography [6–19]. At the core of each PH domain is the same seven-stranded β-sandwich of two near-orthogonal β-sheets containing four- and three-strands respectively (Figure 21.1). A characteristic C-terminal α-helix (αC) closes off one "splayed" or open corner [20] of the β-sandwich (top in Figure 21.1), while three interstrand loops (the most variable in PH domains) close off the opposite splayed corner (abutting the membrane surface in Figure 21.1). This core fold has also been seen in several other classes of domain that share no significant sequence similarity with PH domains [21]. These include the phosphotyrosine binding (PTB) domain [22, 23], the Enabled/VASP homology 1 (EVH1) domain [24, 25], a Ran binding domain [26], and the FERM domain (for band 4.1, ezrin, radixin, moesin homology domain) [27]. The basic β-sandwich structure has been termed the PH domain "superfold" by Saraste and colleagues [28]. The frequent occurrence of this fold probably reflects its adaptability to multiple functions by creating a stable structural scaffold that can bear loops with quite different recognition properties.

Beyond the conserved β-sandwich fold, one characteristic shared by all PH domains of known structure (except the *C. elegans* Unc89 PH domain [7]) is a marked electrostatic sidedness. Each PH domain is electrostatically polarized, with a positively charged face that coincides with the three most variable loops in the PH domain [9, 29]. This positively-charged face abuts the membrane in Figure 21.1, and its existence provided part of the motivation for initial tests of PH domain binding to (negatively charged) membrane surfaces.

PH DOMAINS AS PHOSPHOINOSITIDE-BINDING MODULES

The Fesik laboratory was the first to point out that PH domains can bind membranes containing phosphoinositides [30]. Specifically, they showed that the N-terminal PH domain from pleckstrin binds phosphatidylinositol-(4,5)-bisphosphate (PtdIns(4,5)P_2) with a K_D of approximately

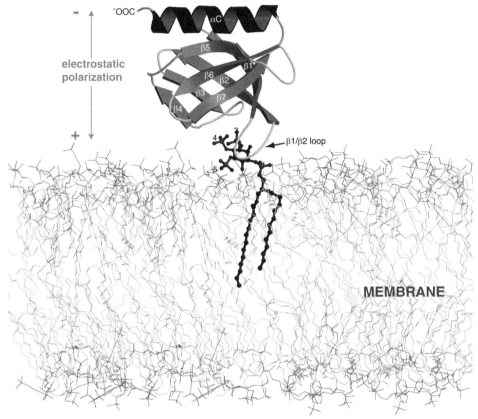

FIGURE 21.1 Hypothetical view of how the DAPP1 PH domain binds to PtdIns(3,4,5)P₃ in a membrane.
The X-ray crystal structure of the DAPP1 PH domain [11] is shown in a ribbon representation with bound Ins(1,3,4,5)P₄. The β-sandwich structure of the PH domain can be seen, with strands β1 though β4 forming a sheet behind the plane of the paper, and strands 5 through 7 forming a β-sheet in front of the plane of the page. The characteristic C-terminal α-helix (αC) is also labeled (and caps the upper splayed corner of the β-sandwich). The direction of electrostatic polarization of PH domains is depicted schematically on the left. The positive face abuts the membrane in this orientation. A diacylglycerol molecule has been attached to the Ins(1,3,4,5)P₄ molecule to generate a hypothetical view of PtdIns(3,4,5)P₃ bound to the DAPP1 PH domain. The PtdIns (3,4,5)P₃ is embedded in a stick model of a phosphatidylcholine bilayer to guide thinking as to how the PH domain might bind the lipid headgroup in this context. MOLSCRIPT [98] was used to generate this figure.

30 μM. NMR analyses demonstrated that the positively charged face of the domain (shown to abut the membrane in Figure 21.1) is the site at which the lipid binds [30]. A large number of subsequent studies have shown that phosphoinositide binding is a characteristic shared *in vitro* by nearly all PH domains, and a view has emerged that phosphoinositide binding is a conserved and likely physiologically relevant function for most PH domains [21, 31, 32]. For several PH domains, phosphoinositide binding has been convincingly demonstrated to be an important (and perhaps the only) function. In these cases the PH domain specifically recognizes the headgroup of a particular phosphoinositide, and this interaction plays an important role in targeting the PH domain-containing protein to cellular membranes [21, 33]. However, PH domains in this category are rare. The majority – perhaps over 90 percent of PH domains – bind phosphoinositides with only low affinity and specificity [21, 34–36]. How these PH domains participate in membrane targeting (if indeed they do) is not yet clear.

HIGHLY SPECIFIC RECOGNITION OF PHOSPHOINOSITIDES (AND INOSITOL PHOSPHATES) BY PH DOMAINS

PH Domain Binding to Phosphatidylinositol-4,5-Bisphosphate

The phospholipase C-δ₁ (PLC-δ₁) PH domain was the first shown to recognize a specific phosphoinositide with high affinity [37–39]. The PLC-δ₁ PH domain recognizes both PtdIns(4,5)P₂ (which it binds with a K_D of approximately 2 μM) and its isolated soluble headgroup, inositol-(1,4,5)-trisphosphate (Ins(1,4,5)P₃), with which it forms a 1 : 1 complex (K_D = 210 nM) [39]. An X-ray crystal structure of the Ins(1,4,5)P₃/PLC-δ₁ PH domain complex [10] showed that the three variable loops on the positively-charged face of the PH domain form the PtdIns(4,5)P₂/Ins(1,4,5)P₃ binding site. The detailed structure of this binding site also provided clear explanations for the strong Ins(1,4,5)

P_3-specificity of the PLC-δ_1 PH domain (it binds Ins $(1,4,5)P_3$ at least 15-fold more strongly than any other inositol polyphosphate). When expressed as a green fluorescent protein (GFP) fusion, or analyzed by indirect immunofluorescence, the PLC-δ_1 PH domain shows clear plasma membrane localization [40–43]. GFP fusion proteins of this PH domain have been used to identify the location of PtdIns $(4,5)P_2$ in living cells, and to monitor PtdIns$(4,5)P_2$ dynamics and/or Ins$(1,4,5)P_3$ accumulation in response to different agonists [41–45].

Recognition of Phosphatidylinositol 3-Kinase Products

Following the realization that some PH domains recognize specific phosphoinositides, it was found that protein kinase B (PKB, also known as Akt), a serine/threonine kinase with an N-terminal PH domain, is a downstream effector of phosphatidylinositol 3-kinase (PI3-kinase) [46, 47]. Mutations in the PKB PH domain prevent its PI3-kinase-dependent activation, indicating that the PH domain itself plays a critical role in this step [47]. The PKB PH domain specifically recognizes both PtdIns$(3,4,5)P_3$ and PtdIns$(3,4)P_2$, the major products of agonist-stimulated PI3-kinase, but does not bind strongly to PtdIns$(4,5)P_2$ or other phosphoinositides [48–50]. As discussed elsewhere in this volume, PtdIns$(3,4,5)P_3$ and PtdIns$(3,4)P_2$ are all but undetectable in quiescent cells but accumulate transiently in the plasma membrane following stimulation of cells with a variety of agonists (to an estimated local concentration of $150\,\mu M$ [51]). A PH domain that fails to bind PtdIns$(4,5)P_2$ (present constitutively in the plasma membrane), but which binds strongly to PtdIns$(3,4,5)P_3$ and/or PtdIns$(3,4)P_2$, will be recruited to the plasma membrane specifically when these PI3-kinase-generated lipid second messengers are present. The PH domain from PKB has these binding characteristics, and can be shown (as a GFP fusion protein) to be recruited efficiently to the plasma membrane of mammalian cells following growth factor stimulation [52, 53]. As discussed elsewhere in this volume, once recruited by its PH domain to PI3-kinase products at the plasma membrane, PKB is activated at this location by phosphorylation at two sites [54]. One phosphorylation event is performed by a serine/threonine kinase named PDK1 (for phosphoinositide-dependent kinase-1), which also has a PH domain that can recruit it to the plasma membrane in a PI3-kinase-dependent manner [55, 56].

Other PH domains that specifically recognize PI3-kinase products include that from Bruton's tyrosine kinase (Btk) [34, 57–59] and the PH domain from the Arf–guanine nucleotide exchanger Grp1 (general receptor for phosphoinositides-1) [60]. Both of these PH domains bind exclusively (and strongly) to PtdIns$(3,4,5)P_3$ or its headgroup Ins$(1,3,4,5)P_4$ [35, 59, 60]. A point mutation (at arginine-28) in the Btk PH domain, which leads to agammaglobulinemia in humans and mice [61, 62], abolishes PtdIns$(3,4,5)$ P_3/Ins$(1,3,4,5)P_4$ binding [34, 57, 58]. The effects of this Btk mutation on B cell signaling provided the first clue that PH domains may play a role in signal transduction. Like the PKB PH domain, the Btk and Grp1 PH domains are recruited directly to the plasma membrane upon PI3-kinase activation [53, 63–65].

Skolnik and colleagues identified more than 12 different PH domains capable of driving PI3-kinase-dependent plasma membrane recruitment using a novel yeast-based assay [66]. Where studied, each of these PH domains binds *in vitro* to PtdIns$(3,4,5)P_3$ (or Ins$(1,3,4,5)P_4$) with a K_D in the 10–100 nM range, and selects for PtdIns$(3,4,5)$ P_3 over PtdIns$(4,5)P_2$ by a factor of 20 or more [21]. PH domains in this group share a sequence motif centered on the $\beta 1/\beta 2$ loop that links the first two β-strands of the PH domain sandwich. Several crystal structures of PH domains bound to Ins$(1,3,4,5)P_4$ have shown how this motif defines a specific binding site for the PtdIns$(3,4,5)P_3$ [6, 11, 67] (Figure 21.2). The structural details of the binding site are remarkably well conserved across different structures and bear a strong resemblance (in structure and sequence) to the Ins$(1,4,5)P_3$ binding site of the PLC-δ_1 PH domain. The sequence motif identified by Skolnik and colleagues [66] serves as a strong and reliable predictor of which PH domains specifically recognize PI3-kinase products.

Ptdins$(3,4,5)P_3$ Versus PtdIns$(3,4)P_2$

Among the PH domains with the sequence motif shown in Figure 21.2, some bind equally well to both PtdIns$(3,4,5)P_3$ and PtdIns$(3,4)P_2$ (e.g., the PKB and DAPP1 PH domains), while others bind only PtdIns$(3,4,5)P_3$ (e.g., the Grp1 and Btk PH domains). PH domains that recognize only PtdIns $(3,4,5)P_3$ tend either to have extended $\beta 1/\beta 2$ loops or (as in the Grp1 PH domain) insertions elsewhere in the structure that can make specific contacts with the 5-phosphate group. In the complex between the DAPP1 (dual-specific) PH domain and Ins$(1,3,4,5)P_4$ there are no hydrogen bonds between PH domain side chains and the 5-phosphate group [11], providing one explanation for why this PH domain binds equally well to Ins$(1,3,4,5)P_4$/PtdIns$(3,4,5)P_3$ and Ins$(1,3,4)P_3$/PtdIns$(3,4,5)P_2$.

There is no currently known PH domain that binds exclusively to PtdIns$(3,4)P_2$. Alessi and colleagues identified the C-terminal PH domain from TAPP1 (for tandem PH domain-containing protein-1) as a PH domain that prefers PtdIns$(3,4)P_2$ over other phosphoinositides according to protein–lipid overlay studies [68]. However, Ferguson and colleagues [11] showed clearly that this PH domain (called AA054961 in that study) binds with high affinity to the headgroups of both PtdIns$(3,4)P_2$ and PtdIns$(3,4,5)P_3$. In several other assays, using isolated headgroups and intact lipids, it has been shown that the

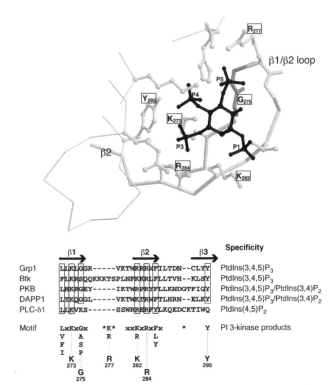

FIGURE 21.2 Close-up view of Ins(1,3,4,5)P$_4$ in the binding site of the Grp1 PH domain, depicting the side chains of residues in the sequence motif that predicts PI3-kinase product specificity. The β1/β2 loop of the Grp1 PH domain is shown to "cradle" the Ins(1,3,4,5)P$_4$ molecule, with several residues marked forming side-chain hydrogen bonds with the bound lipid headgroup. The motif that predicts specificity for PI3-kinase product binding is shown in the lower part of the figure, imposed upon the sequences of the N-terminal portions of the Grp1, Btk, PKB, and DAPP1 PH domains. The motif positions corresponding to the residues highlighted in the structural figure are also shown. K273, in strand β1 of Grp1, forms a hydrogen bond with both the 3- and 4-phosphates of Ins(1,3,4,5)P$_4$. G275 must be small in order to allow space for the inositol ring in this binding configuration. K282 forms a hydrogen bond with the Ins(1,3,4,5)P$_4$ 1-phosphate. R284 forms a critical hydrogen bond with the 3-phosphate. This is equivalent to the arginine at which mutations in Btk cause agammaglobulinemias. Y295, in strand β3, is a conserved feature of PH domains that bind PI3-kinase products, and its side-chain forms a hydrogen bond with the 4-phosphate group. These 5 motif characteristics are conserved in all PH domains that recognize PtdIns(3,4,5)P$_3$ and/or PtdIns(3,4)P$_2$, and several (but not all) are conserved in PtdIns(4,5)P$_2$ binding by the PLC-δ$_1$ PH domain [10]. Equivalents to the additional interaction of R277 with the 5-phosphate of Ins(1,3,4,5)P$_4$ are seen only in PtdIns(3,4,5)P$_3$-specific PH domains (Grp1 and Btk), and not those that also bind PtdIns(3,4)P$_2$ [11]. The extra long β1/β2 loop of the Btk PH domain contributes to 5-phosphate interactions, as does the β6/β7 loop insertion (not shown) of the Grp1 PH domain.

C-terminal TAPP1 PH domain does prefer PtdIns(3,4)P$_2$, but binds to this phosphoinositide only four-fold more strongly than to PtdIns(3,4,5)P$_3$ [V. J. Sankaran and M. A. Lemmon, unpublished data]. In spite of this weak selectivity for PtdIns(3,4)P$_2$ over PtdIns(3,4,5)P$_3$, Alessi and colleagues have provided some evidence to suggest that the TAPP1 PH domain is recruited to the plasma membrane *in vivo* when PtdIns(3,4)P$_2$ production is stimulated but

not when PtdIns(3,4,5)P$_3$ is thought to accumulate without PtdIns(3,4)P$_2$ production [69]. Whether other proteins exist that are regulated exclusively by PtdIns(3,4)P$_2$ remains to be seen.

PH Domains with other Phosphoinositide-Binding Specificities

Dowler and colleagues [68] recently identified several PH domains that appear from protein–lipid overlay assays to have novel phosphoinositide specificities. Their strategy was to identify PH domains with sequences that match (or closely resemble) the PI3-kinase product-binding motif presented in Figure 21.2, and to assess phosphoinositide-binding specificity. The C-terminal TAPP1 PH domain was identified as a target for PI3-kinase products with this approach, as were examples of PH domains that appear in overlay assays to recognize PtdIns-4-P, PtdIns-3-P, or PtdIns(3,5)P$_2$ specifically. It should be stressed that these phosphoinositide-binding specificities have not yet been confirmed via quantitative approaches *in vitro* or with localization studies *in vivo*, and that several of them appear from surface plasmon resonance (SPR) studies to have rather low affinities [68]. As well as PH domains with apparently novel specificities, Dowler *et al.* found several PH domains (with sequences related to the motif in Figure 21.2) that interact with all phosphoinositides tested [68]. These results argue that the PtdIns(3,4,5)P$_3$-specific binding site depicted in Figure 21.2 can be "remodeled" with only a handful of mutations to generate binding sites that instead recognize only the PtdIns(4,5)P$_2$ headgroup (the PLC-δ$_1$ PH domain), or perhaps only the PtdIns-4-P, PtdIns-3-P or PtdIns(3,5)P$_2$ headgroup. Remodeling of a different nature can alternatively generate a binding site that accommodates any phosphoinositide headgroup, so that binding is promiscuous.

Non-Specific Phosphoinositide Binding by PH Domains: the Majority Occupation?

Although this fact may not be immediately clear from a reading of the PH domain literature, by far the majority (> 90 percent) of PH domains do not have a sequence that significantly resembles the motif shown in Figure 21.2. Nonetheless, most PH domains lacking the motif do appear capable of phosphoinositide binding, although binding is weak and non-specific in almost every case [21, 35, 36]. Where K$_D$ values have been reported for phosphoinositide binding by this class of PH domains, they have ranged from around 30 μM to 4 mM or weaker [12, 13, 30, 34, 36, 57, 70–73]. In one case, that of the β-spectrin PH domain, a crystal structure of the PH domain with a weakly bound Ins(1,4,5)P$_3$ was reported [13]. Ins(1,4,5)P$_3$ binds to the surface of this electrostatically polarized PH domain, in the center of its positively charged face. NMR studies have similarly located the site of weak phosphoinositide binding in other PH domains to the variable loops on the positively-charged face [30, 57, 73, 74], most likely driven by delocalized electrostatic attraction to the negatively charged ligand.

Although the physiological relevance of specific, high-affinity, phosphoinositide binding by PH domains has been well established in several cases, it remains unclear in most cases whether weak and promiscuous binding of phosphoinositides to the majority of PH domains plays any physiological role. It has been shown that the low-affinity, non-specific binding of phosphoinositides to the PH domain of dynamin is essential for this protein's function in receptor-mediated endocytosis [75–77]. Similarly, the low affinity (and usually promiscuous) binding of PH domains from Dbl-family members to phosphoinositides [72] appears to be critical for their Rac/Rho exchange activity *in vivo* and their ability to transform cells [78–81]. Despite intensive study, and three crystal structures of DH/PH fragments from Dbl-family proteins [14, 15, 17], it remains unclear how low-affinity binding of phosphoinositides to the PH domains of these proteins influences the exchange activity of the adjacent DH (Dbl homology) domain. For many other proteins with PH domains in this class it has been demonstrated that the PH domain is critical for *in vivo* function, but it has not been established whether or not phosphoinositide binding is a physiologically relevant feature of the PH domain. Many more studies are required to address this question for other "promiscuous" PH domains.

BINDING OF PH DOMAINS TO NON-PHOSPHOINOSITIDE LIGANDS

Since the first description of PH domains, many potential protein binding-partners have been reported. The first were $\beta\gamma$-subunits of heterotrimeric G-proteins, which were suggested to bind all PH domains [82] but now appear only to participate in membrane targeting of a small subset, which includes the PH domains from β-adrenergic receptor kinases (βARKs) [83, 84]. Other reported protein targets for PH domains include protein kinase C (PKC) isoforms [85, 86], the product of the TCL1 (for T cell leukemia) oncogene (which binds the PKB PH domain) [87–89], the receptor for activated PKC (RACK1) [90], $G_{12}\alpha$ [91], a protein called BAP-135 (reported to bind the Btk PH domain) [92], filamentous actin [93], acidic motifs found in proteins such as nucleolin (shown to bind the PH domains of IRS1 and IRS2) [94], and several others (reviewed in [21, 95]). Although not all of these PH domain/protein interactions have been demonstrated to have physiological relevance, there is no doubt that some do, and that protein binding by PH domains cannot be ignored. Despite the relative wealth of reported protein targets, however, no

common themes emerge from the described PH domain/protein interactions. This should not be surprising given the observed diversity in the modes of protein-target recognition by the structurally related EVH1, PTB, and Ran-binding domains [21].

POSSIBLE ROLES OF NON-PHOSPHOINOSITIDE PH LIGANDS

PH domains for which protein targets have been reported include both examples that bind phosphoinositides weakly and promiscuously (e.g., the βARK, IRS-1, and dynamin PH domains), as well as PH domains that bind strongly and specifically to particular phosphoinositides (e.g., the Btk and PKB PH domains). It can therefore not be argued that the protein targets described for PH domains are simply alternatives to, or surrogates for, the well-studied (but rare) specific phosphoinositide ligands. Rather, it appears likely that some PH domains bind multiple ligands.

Cooperation of Multiple Ligands in Membrane Recruitment of PH Domains

A requirement for simultaneous PH domain binding to two different ligands was first demonstrated for membrane targeting by the βARK PH domain [83]. The βARK PH domain binds very weakly to PtdIns(4,5)P_2 ($K_D > 200\,\mu$M) [12]. It also binds rather weakly to the $\beta\gamma$-subunits of heterotrimeric G-proteins [82]. Neither of these weak interactions alone is sufficient for high-affinity targeting of βARK to membranes, but the two interactions can cooperate to recruit βARK efficiently to relevant membrane surfaces [83].

Golgi Targeting of PH Domains by Multiple Ligands

The PH domain from oxysterol binding protein (OSBP), as well as several other related PH domains, is targeted specifically to the Golgi through interactions that appear to require both phosphoinositides and another unidentified (Golgi-specific) component [96]. These Golgi-targeted PH domains, which include those from FAPP1 and the Goodpasture antigen binding protein (GPBP), are highly promiscuous in their phosphoinositide binding (and are not PtdIns-4-P-specific) [34, 96], arguing that phosphoinositide recognition alone cannot possibly determine their Golgi targeting. Phosphoinositide binding by these PH domains is several-fold weaker than PtdIns(4,5)P_2 binding by the PLC-δ_1 PH domain ([96], and D. Keleti, V. J. Sankaran, and M. A. Lemmon, unpublished), further suggesting that it may not be strong enough to drive membrane targeting of the OSBP PH domain independently. Studies in a series of yeast mutants have demonstrated that

Golgi targeting of the OSBP and FAPP1 PH domains is dependent on PtdIns-4-P and not on PtdIns(4,5)P$_2$ production [96, 97], but that the activity of Arf1p is also important [96]. It is therefore hypothesized that the presence of two binding partners in the Golgi is responsible for specific targeting of the OSBP, FAPP1, and GPBP PH domains to that organelle. On its own, phosphoinositide binding by these PH domains is not strong enough to drive membrane targeting *in vivo*, and would certainly not provide targeting specificity. The second (so far unidentified) target of these PH domains is thought to be Golgi-specific, but does not bind to the PH domains tightly enough to achieve Golgi targeting on its alone. Rather both phosphoinositide and this unknown component must be present in the same membrane (the Golgi) in order to recruit the OSBP and other PH domains to that compartment with high affinity and specificity. According to this model [96], PtdIns-4-P is implicated in Golgi targeting of the OSBP, FAPP1, and other PH domains not because of headgroup recognition, but because this happens to be the most abundant phosphoinositide in the membranes that contain the second PH domain ligand.

CONCLUSIONS

The eleventh most populous domain family in humans is now rather well understood structurally and lends its name to the PH domain superfold that includes proteins involved in binding to variety of phosphoinositide and protein ligands. Ligand binding by a small subgroup of PH domains – those that bind phosphoinositide headgroups with high affinity and specificity – is now understood quite well, although it remains possible that some PH domains from this class have additional, as yet unidentified, binding partners. PH domains that do not bind phosphoinositides with high affinity or specificity constitute the majority – perhaps 90 percent. The interactions driven by these PH domains are far less well understood. In many cases it even remains unclear whether phosphoinositide binding observed *in vitro* has any relevance *in vivo*. How weak and non-specific phosphoinositide binding could contribute to membrane binding is a question that has yet to be fully addressed. It may do so though cooperation of multiple ligands that bind to a single PH domain (as discussed for the βARK and OSBP PH domains). Alternatively, the PH domain may be one of several domains within a multidomain protein or oligomer that cooperate with one another in driving membrane targeting. In these cases, specificity of membrane targeting may be defined not by the precise nature of the individual interactions (as with PH domains that bind PI3-kinase products), but rather by the available combinations of interactions. Recruitment to a specific membrane may require that two or more PH domain targets coexist in that membrane.

REFERENCES

1. Haslam RJ, Koide HB, Hemmings BA. Pleckstrin domain homology. *Nature* 1993;**363**:309–10.

2. Mayer BJ, Ren R, Clark KL, Baltimore D. A putative modular domain present in diverse signaling molecules. *Cell* 1993;**73**:629–30.

3. Musacchio A, Gibson T, Rice P, Thompson J, Saraste M. The PH domain: a common piece in a patchwork of signalling proteins. *Trends Biochem Sci* 1993;**18**:343–8.

4. Tyers M, Rachubinski RA, Stewart MI, Varrichio AM, Shorr RGL, Haslam RJ, Harley CB. Molecular cloning and expression of the major protein kinase C substrate of platelets. *Nature* 1988;**333**:470–3.

5. Consortium IHGS. Initial sequencing and analysis of the human genome. *Nature* 2001;**409**:860–921.

6. Baraldi E, Djinovic Carugo K, Hyvönen M, Lo Surdo P, Riley AM, Potter BVL, O'Brien R, Ladbury JE, Saraste M. Structure of the PH domain from Bruton's tyrosine kinase in complex with inositol 1,3,4,5-tetrakisphosphate. *Structure* 1999;**7**:449–60.

7. Blomberg N, Baraldi E, Sattler M, Saraste M, Nilges M. Structure of a PH domain from the C. elegans muscle protein UNC–89 suggests a novel function. *Structure Fold Des* 2000;**8**:1079–87.

8. Dhe-Paganon S, Ottinger EA, Nolte RT, Eck MJ, Shoelson SE. Crystal structure of the pleckstrin homology-phosphotyrosine binding (PH-PTB) targeting region of insulin receptor substrate 1. *Proc Natl Acad Sci USA* 1999;**96**:8378–83.

9. Ferguson KM, Lemmon MA, Schlessinger J, Sigler PB. Crystal structure at 2.2-Å resolution of the pleckstrin homology domain from human dynamin. *Cell* 1994;**79**:199–209.

10. Ferguson KM, Lemmon MA, Schlessinger J, Sigler PB. Structure of a high affinity complex between inositol-1,4,5-trisphosphate and a phospholipase C pleckstrin homology domain. *Cell* 1995;**83**:1037–46.

11. Ferguson KM, Kavran JM, Sankaran VG, Fournier E, Isakoff SJ, Skolnik EY, Lemmon MA. Structural basis for discrimination of 3-phosphoinositides by pleckstrin homology domains. *Mol Cell* 2000;**6**:373–84.

12. Fushman D, Najmabadi-Kaske T, Cahill S, Zheng J, LeVine H, Cowburn D. The solution structure and dynamics of the pleckstrin homology domain of G protein-coupled receptor kinase 2 (β-adrenergic receptor kinase 1): A binding partner of G{ΣB}βγ{/ΣB} subunits. *J Biol Chem* 1998;**273**:2835–43.

13. Hyvönen M, Macias MJ, Nilges M, Oschkinat H, Saraste M, Wilmanns M. Structure of the binding site for inositol phosphates in a PH domain. *EMBO J* 1995;**14**:4676–85.

14. Rossman KL, Worthylake DK, Snyder JT, Siderovski DP, Campbell SL, Sondek J. A crystallographic view of interactions between Dbs and Cdc42: PH domain-assisted guanine nucleotide exchange. *EMBO J* 2002;**21**:1315–26.

15. Soisson SM, Nimnual AS, Uy M, Bar-Sagi D, Kuriyan J. Crystal structure of the Dbl and pleckstrin homology domains from the human Son of Sevenless protein. *Cell* 1998;**95**:259–68.

16. Thomas CC, Dowler S, Deak M, Alessi DR, van Aalten DM. Crystal structure of the phosphatidylinositol 3,4-bisphosphate-binding pleckstrin homology (PH) domain of tandem PH-domain-containing protein 1 (TAPP1): molecular basis of lipid specificity. *Biochem J* 2001;**358**:287–94.

17. Worthylake DK, Rossman KL, Sondek J. Crystal structure of Rac1 in complex with the guanine nucleotide exchange region of Tiam1. *Nature* 2000;**408**:682–8.

18. Yoon HS, Hajduk PJ, Petros AM, Olejniczak ET, Meadows RP, Fesik SW. Solution structure of a pleckstrin-homology domain. *Nature* 1994;**369**:672–5.

19. Zhang P, Talluri S, Deng H, Branton D, Wagner G. Solution structure of the pleckstrin homology domain of Drosophila beta-spectrin. *Structure* 1995;**3**:1185–95.

20. Chothia C. Principles that determine the structure of proteins. *Annu Rev Biochem* 1984;**53**:537–72.

21. Lemmon MA, Ferguson KM. Signal-dependent membrane targeting by pleckstrin homology (PH) domains. *Biochem J* 2000;**350**:1–18.

22. Eck MJ, Dhe-Paganon S, Trüb T, Nolte RT, Shoelson SE. Structure of the IRS-1 PTB domain bound to the juxtamembrane region of the insulin receptor. *Cell* 1996;**85**:695–705.

23. Zhou M-M, Ravichandran KS, Olejniczak ET, A.M. P, Meadows RP, Sattler M, Harlan JE, Wade WS, Burakoff SJ, Fesik SW. Structure and ligand recognition of the phosphotyrosine binding domain of Shc. *Nature* 1995;**378**:584–92.

24. Beneken J, Tu JC, Xiao B, Nuriya M, Yuan JP, Worley PF, Leahy DJ. Structure of the Homer EVH1 domain-peptide complex reveals a new twist in polyproline recognition. *Neuron* 2000;**26**:143–54.

25. Prehoda KE, Lee DJ, Lim WA. Structure of the enabled/VASP homology 1 domain–peptide complex: a key component in the spatial control of actin assemble. *Cell* 1999;**97**:471–80.

26. Vetter IR, Nowak C, Nishimotot T, Kuhlmann J, Wittinghofer A. Structure of a Ran-binding domain complexed with Ran bound to a GTP analogue: implications for nuclear transport. *Nature* 1999;**398**:39–46.

27. Pearson MA, Reczek D, Bretscher A, Karplus PA. Structure of the ERM protein moesin reveals the FERM domain fold masked by an extended actin binding tail domain. *Cell* 2000;**101**:259–70.

28. Blomberg N, Baraldi E, Nilges M, Saraste M. The PH superfold: a structural scaffold for multiple functions. *Trends Biochem Sci* 1999;**24**:441–5.

29. Macias MJ, Musacchio A, Ponstingl H, Nilges M, Saraste M, Oschkinat H. Structure of the pleckstrin homology domain from β-spectrin. *Nature* 1994;**369**:675–7.

30. Harlan JE, Hajduk PJ, Yoon HS, Fesik SW. Pleckstrin homology domains bind to phosphatidylinositol 4,5-bisphosphate. *Nature* 1994;**371**:168–70.

31. Hurley JH, Misra S. Signaling and subcellular targeting by membrane-binding domains. *Annu Rev Biophys Biomol Struct* 2000;**29**:49–79.

32. Bottomley MJ, Salim K, Panayotou G. Phospholipid-binding domains. *Biochim Biophys Acta* 1998;**1436**:165–83.

33. Rameh LE, Cantley LC. The role of phosphoinositide 3-kinase lipid products in cell function. *J Biol Chem* 1999;**274**:8347–50.

34. Rameh IIILE, Arvidsson A-K, Carraway KL, Couvillon AD, Rathbun G, Cromptoni A, VanRenterghem B, Czech MP, Ravichandran KS, Burakoff SJ, Wang D-S, Chen C-S, Cantley LC. A comparative analysis of the phosphoinositide binding specificity of pleckstrin homology domains. *J Biol Chem* 1997;**272**:22,059–22,066.

35. Kavran JM, Klein DE, Lee A, Falasca M, Isakoff SJ, Skolnik EY, Lemmon MA. Specificity and promiscuity in phosphoinositide binding by pleckstrin homology domains. *J Biol Chem* 1998;**273**: 30,497–30,508.

36. Takeuchi H, Kanematsu T, Misumi Y, Sakane F, Konishi H, Kikkawa U, Watanabe Y, Katan M, Hirata M. Distinct specificity in the binding of inositol phosphates by pleckstrin homology domains of pleckstrin, RAC-protein kinase, diacylglycerol kinase and a new 130kDa protein. *Biochim Biophys Acta* 1997;**1359**:275–85.

37. Yagisawa H, Hirata M, Kanematsu T, Watanabe Y, Ozaki S, Sakuma K, Tanaka H, Yabuta N, Kamata H, Hirata H, Nojima H. Expression and characterization of an inositol 1,4,5-trisphosphate binding domain of phosphatidylinositol-specific phospholipase C-delta 1. *J Biol Chem* 1994;**26**(9):20,179–20,188.

38. Garcia P, Gupta R, Shah S, Morris AJ, Rudge SA, Scarlata S, Petrova V, McLaughlin S, Rebecchi MJ. The pleckstrin homology domain of phospholipase C-delta 1 binds with high affinity to phosphatidylinositol 4,5-bisphosphate in bilayer membranes. *Biochemistry* 1995;**34**:16,228–16,234.

39. Lemmon MA, Ferguson KM, O'Brien R, Sigler PB, Schlessinger J. Specific and high-affinity binding of inositol phosphates to an isolated pleckstrin homology domain. *Proc Natl Acad Sci USA* 1995;**92**: 10,472–10,476.

40. Paterson HF, Savopoulos JW, Perisic O, Cheung R, Ellis MV, Williams RL, Katan M. Phospholipase C delta 1 requires a pleckstrin homology domain for interaction with the plasma membrane. *Biochem J* 1995;**312**:661–6.

41. Hirose K, Kadowaki S, Tanabe M, Takeshima H, Iino M. Spatiotemporal dynamics of inositol 1,4,5-trisphosphate that underlies complex Ca^{2+} mobilization patterns. *Science* 1999;**284**:1527–30.

42. Stauffer TP, Ahn S, Meyer T. Receptor-induced transient reduction in plasma membrane PtdIns(4,5)P_2 concentration monitored in living cells. *Curr Biol* 1998;**8**:343–6.

43. Varnai P, Balla T. Visualization of phosphoinositides that bind pleckstrin homology domains: calcium- and agonist-induced dynamic changes and relationship to myo-[^{3}H]inositol-labeled phosphoinositide pools. *J Cell Biol* 1998;**143**:501–10.

44. Tall EG, Spector I, Pentyala SN, Bitter I, Rebecchi MJ. Dynamics of phosphatidylinositol 4,5-bisphosphate in actin-rich structures. *Curr Biol* 2000;**10**:743–6.

45. Botelho RJ, Teruel M, Dierckman R, Anderson R, Wells A, York JD, Meyer T, Grinstein S. Localized biphasic changes in phosphatidylinositol-4,5-bisphosphate at sites of phagocytosis. *J Cell Biol* 2000;**151**:1353–68.

46. Burgering BM, Coffer PJ. Protein kinase B (c-Akt) in phosphatidylinositol-3-OH kinase signal transduction. *Nature* 1995;**376**:599–602.

47. Franke TF, Yang SI, Chan TO, Datta K, Kazlauskas A, Morrison DK, Kaplan DR, Tsichlis PN. The protein kinase encoded by the Akt proto-oncogene is a target of the PDGF-activated phosphatidylinositol 3-kinase. *Cell* 1995;**81**:727–36.

48. Franke TF, Kaplan DR, Cantley LC, Toker A. Direct regulation of the Akt proto-oncogene product by phosphatidyl inositol-3,4-bisphosphate. *Science* 1997;**275**:665–8.

49. Frech M, Andjelkovic M, Ingley E, Reddy KK, Falck JR, Hemmings BA. High affinity binding of inositol phosphates and phosphoinositides to the pleckstrin homology domain of RAC/ protein kinase B and their influence on kinase activity. *J Biol Chem* 1997;**272**:8474–81.

50. Klippel A, Kavanaugh WM, Pot D, Williams LT. A specific product of phosphatidylinositol 3-kinase directly activates the protein kinase Akt through its pleckstrin homology domain. *Mol Cell Biol* 1997;**17**:338–44.

51. Stephens LR, Jackson TR, Hawkins PT. Agonist-stimulated synthesis of phosphatidylinositol 3,4,5-trisphosphate: a new intracellular signaling system? *Biochim Biophys Acta* 1993;**1179**:27–75.

52. Watton SJ, Downward J. Akt/PKB localisation and 3′ phosphoinositide generation at sites of epithelial cell-matrix and cell–cell interaction. *Curr Biol* 1999;**9**:433–6.

53. Gray A, Van der Kaay J, Downes CP. The pleckstrin homology domains of protein kinase B and GRP1 (general receptor for phosphoinositides-1) are sensitive and selective probes for the cellular detection of phosphatidylinositol 3,4-bisphosphate and/or phosphatidylinositol 3,4,5-trisphosphate *in vivo*. *Biochem J* 1999;**344**:929–36.

54. Vanhaesebroeck B, Alessi DR. The PI3K–PDK1 connection: more than just a road to PKB. *Biochem J* 2000;**346**:561–76.

55. Anderson KE, Coadwell J, Stephens LR, Hawkins PT. Translocation of PDK–1 to the plasma membrane is important in allowing PDK-1 to activate protein kinase B. *Curr Biol* 1998;**8**:684–91.

56. Currie RA, Walker KS, Gray A, Deak M, Casamayor A, Downes CP, Cohen P, Alessi DR, Lucocq J. Role of phosphatidylinositol 3,4,5-trisphosphate in regulating the activity and localization of 3-phosphoinositide-dependent protein kinase-1. *Biochem J* 1999;**337**: 575–83.

57. Salim K, Bottomley MJ, Querfurth E, Zvelebil MJ, Gout I, Scaife R, Margolis RL, Gigg R, Smith CIE, Driscoll PC, Waterfield MD, Panayotou G. Distinct specificity in the recognition of phosphoinositides by the pleckstrin homology domains of dynamin and Bruton's tyrosine kinase. *EMBO J* 1996;**15**:6241–50.

58. Fukuda M, Kojima T, Kabayama H, Mikoshiba K. Mutation of the pleckstrin homology domain of Bruton's tyrosine kinase in immunodeficiency impaired inositol 1,3,4,5-tetrakisphosphate binding capacity. *J Biol Chem* 1996;**271**:30,303–30,306.

59. Kojima T, Fukuda M, Watanabe Y, Hamazato F, K. M. Characterization of the pleckstrin homology domain of Btk as an inositol polyphosphate and phosphoinositide binding domain. *Biochem Biophys Res Commun* 1997;**236**:333–9.

60. Klarlund JK, Guilherme A, Holik JJ, Virbasius A, Czech MP. Signaling by 3,4,5-phosphoinositide through proteins containing pleckstrin and Sec7 homology domains. *Science* 1997;**275**:1927–30.

61. Rawlings DJ, Saffran DC, Tsukada S, Largaespada DA, Grimaldi JC, Cohen L, Mohr RN, Bazan JF, Howard M, Copeland NG. Mutation of unique region of Bruton's tyrosine kinase in immunodeficient XID mice. *Science* 1993;**261**:358–61.

62. Thomas JD, Sideras P, Smith CI, Vorechovsky I, Chapman V, Paul WE. Colocalization of X-linked agammaglobulinemia and X-linked immunodeficiency genes. *Science* 1993;**261**:355–8.

63. Varnai P, Rother KI, Balla T. Phosphatidylinositol 3-kinase-dependent membrane association of the Bruton's tyrosine kinase pleckstrin homology domain visualized in single living cells. *J Biol Chem* 1999;**274**:10,983–10,989.

64. Nagel W, Zeitlmann L, Schilcher P, Geiger C, Kolanus J, Kolanus W. Phosphoinositide 3-OH kinase activates the beta2 integrin adhesion pathway and induces membrane recruitment of cytohesin-1. *J Biol Chem* 1998;**273**:14,853–14,861.

65. Venkateswarlu K, Gunn-Moore F, Oatey PB, Tavare JM, Cullen PJ. Nerve growth factor- and epidermal growth factor-stimulated translocation of the ADP-ribosylation factor-exchange factor GRP1 to the plasma membrane of PC12 cells requires activation of phosphatidylinositol 3-kinase and the GRP1 pleckstrin homology domain. *Biochem J* 1998;**335**:139–46.

66. Isakoff SJ, Cardozo T, Andreev J, Li Z, Ferguson KM, Abagyan R, Lemmon MA, Aronheim A, Skolnik EY. Identification and analysis of PH domain-containing targets of phosphatidylinositol 3-kinase using a novel *in vivo* assay in yeast. *EMBO J* 1998;**17**:5374–87.

67. Lietzke SE, Bose S, Cronin T, Klarlund J, Chawla A, Czech MP, Lambright DG. Structural basis of 3-phosphoinositide recognition by pleckstrin homology domains. *Mol Cell* 2000;**6**:385–94.

68. Dowler S, Currie RA, Campbell DG, Deak M, Kular G, Downes CP, Alessi DR. Identification of pleckstrin-homology-domain-containing proteins with novel phosphoinositide-binding specificities. *Biochem J* 2000;**351**:19–31.

69. Kimber WA, Trinkle-Mulcahy L, Cheung PC, Deak M, Marsden LJ, Kieloch A, Watt S, Javier RT, Gray A, Downes CP, Lucocq JM,

Alessi DR. Evidence that the tandem-pleckstrin-homology-domain-containing protein TAPP1 interacts with Ptd(3,4)P$_2$ and the multi-PDZ-domain-containing protein MUPP1 *in vivo*. *Biochem J* 2002;**361**:525–36.

70. Klein DE, Lee A, Frank DW, Marks MS, Lemmon MA. The pleckstrin homology domains of dynamin isoforms require oligomerization for high affinity phosphoinositide binding. *J Biol Chem* 1998;**273**:27,725–27,733.

71. Koshiba S, Kigawa T, Kim JH, Shirouzu M, Bowtell D, Yokoyama S. The solution structure of the pleckstrin homology domain of mouse Son-of-sevenless 1 (mSos1). *J Mol Biol* 1997;**20**:579–91.

72. Snyder JT, Rossman KL, Baumeister MA, Pruitt WM, Siderovski DP, Der CJ, Lemmon MA, Sondek J. Quantitative analysis of the effect of phosphoinositide interactions on the function of Dbl family proteins. *J Biol Chem* 2001;**276**:45,868–45,875.

73. Zheng J, Cahill SM, Lemmon MA, Fushman D, Schlessinger J, Cowburn D. Identification of the binding site for acidic phospholipids on the PH domain of dynamin: Implications for stimulation of GTPase activity. *J Mol Biol* 1996;**255**:14–21.

74. Zheng J, Chen R-H, Corbalan-Garcia S, Cahill SM, Bar-Sagi D, Cowburn D. The solution structure of the pleckstrin homology domain of human SOS1. A possible structural role for the sequential association of diffuse B cell lymphoma and pleckstrin homology domains. *J Biol Chem* 1997;**272**:30,340–30,344.

75. Achiriloaie M, Barylko B, Albanesi JP. Essential role of the dynamin pleckstrin homology domain in receptor-mediated endocytosis. *Mol Cell Biol* 1999;**19**:1410–15.

76. Lee A, Frank DW, Marks MS, Lemmon MA. Dominant-negative inhibition of receptor-mediated endocytosis by a dynamin-1 mutant with a defective pleckstrin homology domain. *Curr Biol* 1999;**9**:261–4.

77. Vallis Y, Wigge P, Marks B, Evans PR, McMahon HT. Importance of the pleckstrin homology domain of dynamin in clathrin-mediated endocytosis. *Curr Biol* 1999;**9**:257–60.

78. Booden MA, Campbell SL, Der CJ. Critical but distinct roles for the pleckstrin homology and cysteine-rich domains as positive modulators of Vav2 signaling and transformation. *Mol Cell Biol* 2002;**22**:2487–97.

79. Russo C, Gao Y, Mancini P, Vanni C, Porotto M, Falasca M, Torrisi MR, Zheng Y, A. E. Modulation of oncogenic DBL activity by phosphoinositol binding to pleckstrin homology domains. *J Biol Chem* 2001;**276**:19,524–19,531.

80. Han J, Luby-Phelps K, Das B, Shu X, Xi Y, Mosteller RD, Krishna UM, Falck JR, White MA, Broek D. Role of substrates and products of PI 3-kinase in regulating activation of Rac-related guanine triphosphatases by Vav. *Science* 1998;**279**:558–60.

81. Nimnual AS, Yatsula BA, Bar-Sagi D. Coupling of the Ras and Rac guanine triphosphatases through the Ras exchanger Sos. *Science* 1998;**279**:560–3.

82. Touhara K, Inglese J, Pitcher JA, Shaw G, Lefkowitz RJ. Binding of G protein beta gamma-subunits to pleckstrin homology domains. *J Biol Chem* 1994;**269**:10,217–10,220.

83. Pitcher JA, Touhara K, Payne ES, Lefkowitz RJ. Pleckstrin homology domain-mediated membrane association and activation of the β-adrenergic receptor kinase requires coordinate interaction with G$_{βγ}$ subunits and lipid. *J Biol Chem* 1995;**270**:11,707–11,710.

84. Jamora C, Yamanouye N, Van Lint J, Laudenslager J, Vandenheede JR, Faulkner DJ, Malhotra V. Gβγ-mediated regulation of Golgi organization is through the direct activation of protein kinase D. *Cell* 1999;**98**:59–68.

85. Yao L, Suzuki H, Ozawa K, Deng J, Lehel C, Fukamachi H, Anderson WB, Kawakami Y, Kawakami T. Interactions between protein kinase

C and pleckstrin homology domains. Inhibition by phosphatidylinositol 4,5-bisphosphate and phorbol 12-myristate 13-acetate. *J Biol Chem* 1997;**272**:13,033–13,039.

86. Konishi H, Kuroda S, Tanaka M, Matsuzaki H, Ono Y, Kameyama K, Haga T, Kikkawa U. Molecular cloning and characterization of a new member of the RAC protein kinase family: association of the pleckstrin homology domain of three types of RAC protein kinase with protein kinase C subspecies and beta gamma subunits of G proteins. *Biochem Biophys Res Comm* 1995;**216**:526–34.

87. Kunstle G, Laine J, Pierron G, Kagami SS, Nakajima H, Hoh F, Roumestand C, Stern MH, Noguchi M. Identification of Akt association and oligomerization domains of the Akt kinase coactivator TCL1. *Mol Cell Biol* 2002;**22**:1513–25.

88. Laine J, Kunstle G, Obata T, Sha M, Noguchi M. The protooncogene TCL1 is an Akt kinase coactivator. *Mol Cell* 2000;**6**:395–407.

89. Pekarsky Y, Koval A, Hallas C, Bichi R, Tresini M, Malstrom S, Russo G, Tsichlis P, Croce CM. Tcl1 enhances Akt kinase activity and mediates its nuclear translocation. *Proc Natl Acad Sci USA* 2000;**97**:3028–33.

90. Rodriguez MM, Ron D, Touhara K, Chen C-H, Mochly-Rosen D. RACK1, a protein kinase C anchoring protein, coordinates the binding of activated protein kinase C and select pleckstrin homology domains *in vitro*. *Biochemistry* 1999;**38**:13,787–13,794.

91. Jiang Y, Ma W, Wan Y, Kozasa T, Hattori S, Huang XY. The G protein $G_{\alpha 12}$ stimulates Bruton's tyrosine kinase and a rasGAP through a conserved PH/BM domain. *Nature* 1998;**395**:808–13.

92. Yang W, Desiderio S. BAP-135, a target for Bruton's tyrosine kinase in response to B cell receptor engagement. *Proc Natl Acad Sci USA* 1997;**94**:604–9.

93. Yao L, Janmey P, Frigeri LG, Han W, Fujita J, Kawakami Y, Apgar JR, Kawakami T. Pleckstrin homology domains interact with filamentous actin. *J Biol Chem* 1999;**274**:19,752–19,761.

94. Burks DJ, Wang J, Towery H, Ishibashi O, Lowe D, Riedel H, White MF. IRS pleckstrin homology domains bind to acidic motifs in proteins. *J Biol Chem* 1998;**273**:31,061–31,067.

95. Maffucci T, Falasca M. Specificity in pleckstrin homology (PH) domain membrane targeting: a role for a phosphoinositide-protein co-operative mechanism. *FEBS Letts* 2001;**506**:173–9.

96. Levine TP, Munro S. Targeting of Golgi-specific pleckstrin homology domains involves both PtdIns 4-kinase-dependent, and independent, components. *Curr Biol* 2002;**12**:695–704.

97. Stefan CJ, Audhya A, Emr SD. The yeast synaptojanin-like proteins control the cellular distribution of phosphatidylinositol (4,5)-bisphosphate. *Mol Biol Cell* 2002;**13**:542–57.

98. Kraulis PJ. MOLSCRIPT: A program to produce both detailed and schematic plots of protein structures. *J Appl Crystallog* 1991;**24**:946–50.

Chapter 22

PX Domains

Christian D. Ellson and Michael B. Yaffe

Koch Institute for Integrated Cancer Research, Massachusetts Institute of Technology, Cambridge, Massachusetts

HISTORY AND OVERVIEW OF PX DOMAINS

The phox (*ph*agocyte *ox*idase) homology (PX) domain, containing ~100 amino acids, was initially identified by sequence profiling as a conserved region present in the C2 domain-containing class of PI 3-kinases (PI3Ks) and in the N-terminal region of the p40phox and p47phox subunits of the NADPH oxidase [1]. For nearly 5 years following their discovery, the function of PX domains remained obscure. A conserved polyproline motif conforming to the consensus sequence PXXP, where X denotes any amino acid, was noted within most PX domain sequences. This observation, coupled with the presence of one or more SH3 domains in numerous PX domain-containing proteins (Figure 22.1), led to speculation that one function of PX domains might involve binding to SH3 domains [1]. Subsequent work has shown that many PX domains function as modular signaling domains that recognize one or more phosphoinositide species, especially the lipid products of PI3Ks, similar to PH domains and FYVE domains.

PX domain-containing proteins are found in all eukaryotes from yeast to human, and can be loosely divided into two groups. The first group consists of a large family of cytoplasmic and/or para-membrane proteins known as sorting nexins (SNX), including SNX1 through SNX29 in higher eukaryotes [2–7], and Vam7p, Vps5p, Mvp1p, and Grd19p in yeast [5, 8]. Eleven mammalian SNXs contain no additional recognizable domain other than the PX domain, nine also have a BAR domain, and the remainder have a variety of other domains of various functions [5, 6] (Figure 22.1). Collectively, all members of the SNX family are believed to be involved in vesicular trafficking (Table 22.1) [5–7]. A second group of PX domain-containing proteins all contain one or more domains of known function in addition to the PX domain and are not classed as SNXs on the grounds that they are not thought to be involved in

vesicular trafficking. These co-associating domains include protein–protein interaction domains such as SH3 domains, PDZ domains, and RGS domains; protein–lipid binding domains such as C2 domains and PH domains; and catalytic domains such as the lipid kinase domain of PI3K or the phospholipase domain of phospholipase D (PLD) (Figure 22.1a). These additional domains play an important role in defining the functions of their constituent proteins, whose intracellular localization is often determined, in part, by the PX domain (Table 22.1).

LIPID-BINDING SPECIFICITY AND STRUCTURE OF PX DOMAINS

The observation that many PX domain-containing proteins are vesicle- or membrane-associated suggested that their ligands might be specific phospholipids. Furthermore, it was noted that the vesicular localization of some PX domain-containing proteins was dependent on the activity of PI3K, an enzyme responsible for the generation of 3-phosphorylated phosphoinositides [9, 10]. This was experimentally explored via protein overlay assays on solid-phase immobilized phospholipids, and by solution-phase binding assays using phospholipid-containing synthetic liposomes. Both types of experiment demonstrated that the ligands for many PX domains were specific phosphoinositide products of PI3K [9–12]. Different PX domains show distinct specificity for different phosphoinositides. The PX domains of p40phox and p47phox, for example, bind to phosphatidylinositol-3-phosphate [PtdIns(3)P] and phosphatidylinositol-3,4-bisphosphate [PtdIns (3,4)P_2], respectively [11, 12]. The majority of PX domains tested to date demonstrate phosphoinositide binding specificity towards PtdIns(3)P, including the PX domains of p40phox, Vam7p [9], SNX3 [10], cytokine-independent survival kinase (CISK) [13], and all high-affinity PX domains from the budding yeast *Saccharomyces cerevisiae* [14].

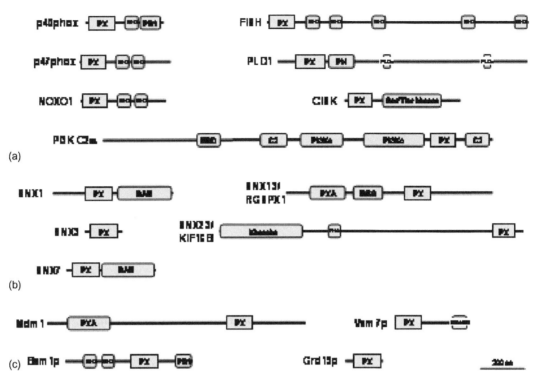

FIGURE 22.1 Domain profiles of PX-domain-containing proteins with known phosphoinositide binding specificity.
Protein sequences were collated using the NCBI Entrez Protein interface and domain structure was determined with the Simple Modular Architecture Research Tool (SMART). (a) Mammalian proteins that function as enzymes or adaptors; (b) mammalian sorting nexins (SNXs); (c) Yeast proteins (*Saccharomyces cerevisiae*). Domain abbreviations are: SH3, Src homology 3; PB1, phox and Bem1p; PH, pleckstrin homology; PLDc, phospholipase D catalytic; RBD, Ras binding domain; C2, PKC conserved 2; PI3Ka/c, phosphoinositide 3-kinase accessory/catalytic; BAR, Bin/Amphiphysin/Rvs; PXA, PX associated; RGS, regulator of G-protein signaling; FHA, Forkhead associated.

The PX domain fold is composed of three anti-parallel β-strands packed against a helical subdomain containing four α-helices and a short stretch of 3_{10} helix (Figure 22.2) [9,15,16]. Insight into the structural basis of phosphoinositide-binding specificity of PX domains first emerged from the NMR structure of the Vam7p PX domain [9], and the high-resolution X-ray crystal structure of the p40phox PX domain bound to PtdIns(3)P [15]. The understanding of the structural features of the PX domain fold and the phosphoinositide binding pocket as observed in these initial studies has been confirmed and extended by the subsequently solved crystal structures of the PX domains of p47phox [17], Bem1p [18], CISK [19], PI3K-C2α [20], and Grd19p (both free and bound to PtdIns(3)P) [21]. The PX domain of p40phox in complex with PtdIns(3)P is illustrated in Figure 22.1 as the example [15]. There are four key residues within the PX domain of p40phox that are required for, and determine, phosphoinositide binding and specificity; Arg58, Tyr59, Lys92, and Arg105 (Figure 22.2, Table 22.1). Arg58 in the p40phox PX domain forms two salt-bridges with the 3-phosphate of the lipid in the p40phoxPX:PtdIns(3)P crystal structure. A basic residue (Arg or Lys) at this position is a common feature in all PX domains that have, to date, been found to bind to 3-phosphorylated phosphoinositides (Table I). Notably, all

PX domains that have been found to bind non-3-phosphorylated phosphoinositides–i.e., the PX domains of PI3K-C2α (PtdIns(4,5)P$_2$), NOXO1 (PtdIns(4)P, PtdIns(5)P, and PtdIns(3,5)P$_2$), and Bem1p (PtdIns(4)P) all lack a basic residue at this position (Table 22.1), suggesting that a residue equivalent to Arg58 is specific to PX domains that bind to 3-phosphorylated phosphoinositides. The 4- and 5-hydroxyl groups of PtdIns(3)P form hydrogen bonds with Arg105 in the p40phox structure (Figure 22.2). Phosphorylation on either the 4- or 5-hydroxyl would sterically impinge on the R105 side-chain, rationalizing the PtdIns(3)P binding specificity of the p40phox PX domain. However, a basic residue at the position equivalent to Arg105 in p40phox is also conserved in PX domains that bind 3-phosphoinositides other than PtdIns(3)P, including those of p47phox and PLD1. Presumably, either alterations in the loops surrounding this residue relieve this steric clash and allow direct interactions of R105 with the lipid phosphates in the 4- and 5- positions, in place of the interaction with the 4- and 5- hydroxyl groups seen in the p40phox structure, or the phosphoinositide must bind in an alternative orientation [17]. Tyrosine-59 and Lys92 in the p40phoxPX:PtdIns(3)P crystal structure are also important amino acids for phosphoinositide binding. Tyr59 forms the floor of the lipid-binding pocket through interactions between its aromatic

TABLE 22.1 PX domain-containing proteins with known phosphoinositide binding specificity

Protein	Key residues[1]	PX domain lipid target[2]	Protein function	Reference
p40phox	RY, K, R	3P	Activation of NADPH oxidase	[11, 12, 42–44]
p47phox	RF, K, R	3,4P$_2$/PA	Activation of NADPH oxidase	[12, 17]
FISH	RY, K, R	3P/3,4P$_2$	Podosome formation	[47–49]
PLD1	KF, R, R	3,4,5P$_3$/phospholipids	Cell division, signal transduction, and vesicle trafficking	[23, 34, 50]
PI3K C2-α	TF, R, K	4,5P$_2$	EGF receptor signaling	[20, 51]
CISK	RY, K, R	3P	Cell survival, chemokine receptor degradation	[13, 52, 53]
NOXO1	SW, R, R	4P/5P/3,5P$_2$	Activation of non-phagocyte NADPH oxidase	[54, 55]
RGS-PX1/SNX13	RY, K, K	3P/3,4P$_2$	Gα-specific GAP; Endosomal trafficking	[56, 57]
KIF16B/SNX23	RY, K, R	3P	Early endosome motility	[58]
SNX1	RF, K, R	3P/3,5P$_2$	Retrieval of CI-M6PR from sorting endosomes	[4, 59, 60]
SNX3	RY, K, R	3P	Regulates endosomal function	[10]
SNX7	RY, K, R	3P	Vesicle trafficking	[10]
Mdm1p	RY, K, K	3P	Yeast nuclear/mitochondrial inheritance	[14, 61]
Bem1p	YY, P, R	4P	Yeast cell polarity, vacuole homeostasis/ fusion	[18, 62–64]
Vam7p	RY, R, R	3P	Yeast vesicle-vacuole fusion	[9, 14, 65]
Grd19p	RY, K, R	3P	Retrieval of proteins from yeast prevacuole to late Golgi	[21, 66]

[1]Residues important for phosphoinositide binding and specificity are shown and are equivalent to Arg58, Tyr59, Lys92, and Arg105 in the PX domain of p40phox. Equivalent residues were identified from manually-adjusted sequence alignments and structural information where available. A basic amino acid (R or K) in the Arg58 equivalent sites predict for 3-phosphorylated phosphoinositide specificity. Non-basic residues are underlined and correlate with specificity towards phosphoinositides that are not phosphorylated in the 3-position. Standard single letter amino acid code is used.
[2]Phosphoinositide species are abbreviated as standard, where 3P=PtdIns(3)P, 3,4P$_2$- PtdIns(3,4)P$_2$, etc. PA=phosphatidic acid.

side-chain and the inositol ring, and an aromatic residue at this position (Tyr, Trp or Phe) is conserved in all phosphoinositide-binding PX domains (Table 22.1). In the p40phox PX:PtdIns(3)P structure, Lys92 forms a salt-bridge with the non-bridging oxygens of the PtdIns(3)P 1-phosphoryl group (as does the side-chain Arg60) to stabilize the interaction between the domain and the membrane proximal portion of the inositol lipid. Both of these residues are conserved in some, but not all (e.g., Bem1p), PX domains, suggesting that other residues also participate.

Binding of PX domains to phosphophoinositides in lipid bilayers involves two processes; membrane binding is initiated by non-specific electrostatic interaction between the cationic membrane binding surface of the domain and the anionic membrane surface, followed by membrane penetration of hydrophobic residues on loop regions surrounding the phosphoinositide binding pocket, including on the membrane insertion loop (MIL–Figure 22.2) [18,20, 22-24]. This membrane penetration is triggered by specific phosphoinositide binding, although in the case of Bem1p, significant membrane binding can occur in the absence of its PtdIns(4)P ligand [18].

In the cases of the PX domains of p47phox and PLD1, some of the surface cationic residues have been proposed to form a secondary lipid binding site, distinct from the phosphoinositide-binding site, that interacts with anionic phospholipids, most notably the PLD enzyme product phosphatidic acid (PA) and also phosphatidylserine [17, 23].

In common with many other PX domains, the p40phox and p47phox PX domains contain a conserved PXXP motif that forms a type II polyproline (PP$_{II}$) helix (Figure 22.2). Since type II polyproline helices are well known to bind to SH3 domains, and both p40phox and p47phox also contain SH3 domains (Figure 22.1), this structural observation

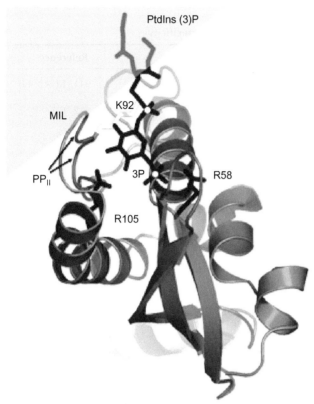

FIGURE 22.2 Structure of the p40phox PX domain in complex with PtdIns(3)P (PDB ID:1H6H) [15].
The structure is oriented such that the acyl chains of the PtdIns(3)P and the membrane insertion loop (MIL) would penetrate the membrane positioned in the top left of the figure (shaded). Residues that play key roles in phosphoinositide binding are depicted: Arg58 coordinates the 3-phosphate; Lys92 hydrogen bonds with the 1-phosphate; Arg105 interacts with the 4- and 5-hydroxyl groups.

suggests that some PX domains may form intramolecular interactions with their co-associating SH3 domains. To date, no intramolecular association of this type has been conclusively demonstrated–an initial report of the PX domain of p47phox binding to its own SH3 domain through the PP$_{II}$ motif [16] has since proven to be physiologically unlikely [25]. However, a PXXP-dependent intermolecular interaction has been demonstrated between the PX domain of PLD2 and the SH3 domain of PLCγ [26] (see below). Furthermore, a number of PX domains have been reported to interact with other proteins: the PX domains of p40phox and p47phox interact with the FERM domain of moesin [27]; that of PLD2 interacts with PKCζ [28], Syk [29] and dynamin [30], as well as PLCγ; the PX domain of Vam7p interacts with the yeast HOPS complex [31]; and a genome-wide yeast two-hybrid screen demonstrated interaction of several yeast PX domains with members of the Yip1 family of proteins [32]. Together these studies establish the PX domain not only as a phosphoinositide-binding module, but also as a protein–protein interaction domain.

PHYSIOLOGICAL FUNCTION OF PX DOMAINS

The function of several PX domain-containing proteins is reasonably well understood, although the exact role fulfilled by the PX domain is not yet well defined. In common with other phosphoinositide-binding modules, such as the PH domain, lipid binding of the PX domain can serve alternative functions. Subcellular and/or suborganellar localization of the PX domain containing protein can be dictated by the specific site and timing of the generation of the target phosphoinositide; this spatio-temporal level of regulation may also be co-defined by additional protein–membrane and/or protein–protein interactions of the protein. This allows for the local concentration of the PX protein on the appropriate membrane, and is likely to define the functional importance of the PX domain in mammalian SNXs and yeast PtdIns(3)P-binding PX domains. For example, the PX domain of Vam7p localizes it to the PtdIns(3)P-rich yeast vacuole and, in complex with other proteins, promotes vacuole-vesicle fusion [9]. However, introduction of a point mutation that prevents PtdIns(3)P binding does not completely ablate the function of Vam7p; rather, it reduces its efficiency by increasing the Km of the reaction [9, 33]. Similarly, complete deletion of the PX domain also lowers the Km of the reaction, but the defect in vesicle-vacuole fusion caused by this can be overcome with increased expression levels of the truncated mutant [9, 33]. This demonstrates that it is not PtdIns(3)P binding *per se* that the PX domain contributes, but the mechanism by which Vam7p is locally concentrated at the appropriate cellular site.

Another mechanism of action of phosphoinositide binding modules, including PX domains, is to effect a steric change in the target molecule. For example, the PtdIns(3,4,5)P$_3$-interacting PX domain of PLD1 not only induces its association with the plasma membrane but also increases the basal phospholipase activity of the enzyme, as evidenced by the fact that enzymatic activity is increased when PtdIns(3,4,5)P$_3$ is added to an *in vitro* assay [34]. Steric effects are, however, notoriously difficult to prove conclusively experimentally, as interaction with the target phosphoinositide also increases the local concentration of the protein (as discussed above).

As previously noted, a number of PX domains can participate in protein–protein interactions, and this represents an additional mode of PX domain action. In contrast to PLD1, the PX domain of PLD2 does not bind to phosphoinositides [34]. However, protein–protein interactions have been reported between the PX domain of PLD2 and the signaling enzymes PLCγ, PKCζ, and Syk [26,28,29], and in each case appear to increase the activity and downstream signaling of the partner protein, independently of the lipase activity of PLD2. Furthermore, the PX domains of both PLD1 and PLD2 interact with the GTP-bound form of dynamin [30]. Upon direct binding to dynamin's

GTPase domain, the PLD-PXs act as GTPase-activating proteins (GAPs) and stimulate the GTPase activity of the dynamin molecule. This effect is dependent on a putative arginine finger in the PLD-PXs that is spatially distinct from the phosphoinositide binding pocket of the domains [30]. As dynamin GTPase activity positively regulates EGFR endocytosis, the PLD-PXs enhance the rate of this process under physiological conditions. As the PX domain of PLD1 is capable of binding PtdIns(3,4,5)P$_3$, as well as binding dynamin, this demonstrates that phosphoinositide- and protein-binding are not necessarily mutually exclusive properties of PX domains, although the interplay between these two binding events has not been explored.

The most extensively studied example of PX domain function comes from the p40phox and p47phox proteins, after which the PX domain was named. The phox proteins are subunits of the NADPH oxidase, the heme-containing enzyme responsible for superoxide production and killing of microorganisms by phagocytic cells. In resting phagocytes, the inactive NADPH oxidase is separated into a set of cytoplasmic subunits including p47phox, p67phox, and p40phox and a membrane-bound heme-containing flavocytochrome b$_{558}$, which consists of gp91phox and p22phox. Upon phagocyte activation, an assembled trimer of the cytoplasmic components docks with the membrane-bound subunits to form a catalytically active enzyme that can transfer electrons from NADPH to oxygen to form reactive oxygen species (ROS), such as superoxide [35]. Of the cytosolic components, p67phox is most essential for NADPH oxidase activity, but has no intrinsic membrane-localization mechanism [36]. Production of superoxide in response to most stimuli requires the activity of PI3K, suggesting that specific PI3K lipid products may be directly involved in regulating oxidase assembly. The different lipid-binding specificities observed for the PX domains of p47phox (PtdIns(3,4)P$_2$) and p40phox (PtdIns(3)P) may target these proteins, along with co-associated p67phox, to specific membrane compartments that contain the appropriate PI3K-derived lipids [37]; PtdIns(3)P is rapidly synthesized on pathogen-containing phagosomes [38,39], and PtdIns(3,4)P$_2$ is generated both at the phagocytic cup during phagocytosis [38] and at the plasma membrane following stimulation with soluble stimuli [40]. In contrast to their isolated PX domains, neither full-length p40phox nor p47phox display significant interaction with their target phosphoinositides. The crystal structure [41] and truncational analysis [37] of full-length p40phox revealed an intramolecular interaction that masked the PX domain's ability to bind phosphoinositides. The mechanism by which this intramolecular interaction is relieved is not yet understood. However, the requirement for the PX domain of p40phox to interact with PtdIns(3)P, resulting in activation of the NADPH oxidase and ROS production, is well-established. Mutation of Arg58 results in loss of PtdIns(3)P-binding and a concomitant defect in NADPH oxidase activity in numerous assay systems [42–44]. Furthermore, mice carrying an Arg58Ala mutation demonstrate a reduced ability to clear bacterial infection, underscoring the *in vivo* physiological significance of the interaction of the PX domain with PtdIns(3)P [44]. In addition to its PtdIns(3)P-binding activity, the PX domain in p40phox also performs important lipid-independent functions that are required for oxidase assembly. Deletion of the PX domain in full-length p40phox prevents translocation of the co-associated p67phox subunit to the cytoskeleton/membrane, whereas p40phox containing an intact PX domain that is unable to bind to lipids, is capable of driving oxidase assembly but deficient in activating ROS production [42]. This, and other work, has lead to a model whereby p40phox recruits p67phox to the phagosome in a PX domain-dependent, but PtdIns(3)P-independent manner, possibly through interaction with moesin [42,45,46]. Subsequent binding of the p40phox PX to PtdIns(3)P stabilizes and/or enhances the phagosomal localization of p67phox, and may act as an allosteric switch to trigger increased ROS production [42,45,46]. As with p40phox, the PX domain of p47phox is also masked in the context of the full-length protein [17]. It is not known exactly how this masking occurs, but it can be relieved by extensive phosphorylation of serine residues in the C-terminus of the protein [17,25], although the contribution to NADPH oxidase activity that phosphoinositide binding by the p47phox PX domain has not been demonstrated.

In summary, PX domains are members of an expanding family of established phosphoinositide-binding domains that includes PH domains, FYVE domains, C2 domains, ENTH domains, and tubby domains. The large differences in structure between these domains suggest that their lipid-binding function arose through the convergent evolution of different structures for the same biological function of lipid binding. Additionally, the PX domain has been shown to participate in a variety of lipid binding-independent functions by acting as a protein–protein interaction module. The examples presented in this chapter show how our understanding of the functional roles that PX domains play in the regulation of their parent proteins has progressed, and it will be important for future work to further elucidate their various mechanisms of action.

REFERENCES

1. Ponting CP. Novel domains in NADPH oxidase subunits, sorting nexins, and PtdIns 3-kinases: binding partners of SH3 domains? *Protein Sci* 1996;**5**(11):2353–7.

2. Kurten RC, Cadena DL, Gill GN. Enhanced degradation of EGF receptors by a sorting nexin, SNX1. *Science* 1996;**272**(5264):1008–10.

3. Teasdale RD, Loci D, Houghton F, Karlsson L, Gleeson PA. A large family of endosome-localized proteins related to sorting nexin 1. *Biochem J* 2001;**358**(Pt 1):7–16.

4. Haft CR, de la Luz Sierra M, Barr VA, Haft DH, Taylor SI. Identification of a family of sorting nexin molecules and characterization of their association with receptors. *Mol Cell Biol* 1998;**18**(12):7278–87.

5. Carlton J, Bujny M, Rutherford A, Cullen P. Sorting nexins–unifying trends and new perspectives. *Traffic* 2005;**6**(2):75–82.

6. Seet LF, Hong W. The Phox (PX) domain proteins and membrane traffic. *Biochim Biophys Acta* 2006;**1761**(8):878–96.

7. Worby CA, Dixon JE. Sorting out the cellular functions of sorting nexins. *Nature Rev Mol Cell Biol* 2002;**3**(12):919–31.

8. Pelham HR. Insights from yeast endosomes. *Curr Opin Cell Biol* 2002;**14**(4):454–62.

9. Cheever ML, Sato TK, de Beer T, Kutateladze TG, Emr SD, Overduin M. Phox domain interaction with PtdIns(3)P targets the Vam7 t-SNARE to vacuole membranes. *Nature Cell Biol* 2001;**3**(7):613–18.

10. Xu Y, Hortsman H, Seet L, Wong SH, Hong W. SNX3 regulates endosomal function through its PX-domain-mediated interaction with PtdIns(3)P. *Nat Cell Biol* 2001;**3**(7):658–66.

11. Ellson CD, Gobert-Gosse S, Anderson KE, Davidson K, Erdjument-Bromage H, Tempst P, Thuring JW, Cooper MA, Lim ZY, Holmes AB, Gaffney PR, Coadwell J, Chilvers ER, Hawkins PT, Stephens LR. PtdIns(3)P regulates the neutrophil oxidase complex by binding to the PX domain of p40(phox). *Nat Cell Biol* 2001;**3**(7):679–82.

12. Kanai F, Liu H, Field SJ, Akbary H, Matsuo T, Brown GE, Cantley LC, Yaffe MB. The PX domains of p47phox and p40phox bind to lipid products of PI(3)K. *Nat Cell Biol* 2001;**3**(7):675–8.

13. Virbasius JV, Virbasius JV, Song X, Pomerleau DP, Zhan Y, Zhou GW, Czech MP. Activation of the Akt-related cytokine-independent survival kinase requires interaction of its phox domain with endosomal phosphatidylinositol 3-phosphate. *Proc Natl Acad Sci USA* 2001;**98**(23):12,908–12,913.

14. Yu JW, Lemmon MA. All phox homology (PX) domains from Saccharomyces cerevisiae specifically recognize phosphatidylinositol 3-phosphate. *J Biol Chem* 2001;**276**(47):44,179–44184.

15. Bravo J, Karathanassis D, Pacold CM, Pacold ME, Ellson CD, Anderson KE, Butler PJ, Lavenir I, Perisic O, Hawkins PT, Stephens L, Williams RL. The crystal structure of the PX domain from p40(phox) bound to phosphatidylinositol 3-phosphate. *Mol Cell* 2001;**8**(4):829–839.

16. Hiroaki H, Ago T, Ito T, Sumimoto H, Kohda D. Solution structure of the PX domain, a target of the SH3 domain. *Nat Struct Biol* 2001;**8**(6):526–530.

17. Karathanassis D, Stahelin V, Bravo J, Perisic O, Pacold CM, Cho W, Williams RL. Binding of the PX domain of p47(phox) to phosphatidylinositol 3,4-bisphosphate and phosphatidic acid is masked by an intramolecular interaction. *EMBO J* 2002;**21**(19):5057–5068.

18. Stahelin RV, Karathanassis D, Murray D, Williams RL, Cho W. Structural and membrane binding analysis of the Phox homology domain of Bem1p: basis of phosphatidylinositol 4-phosphate specificity. *J Biol Chem* 2007;**282**(35):25,737–25747.

19. Xing Y, Liu D, Zhang R, Joachimiak A, Songyang Z, Xu W. Structural basis of membrane targeting by the Phox homology domain of cytokine-independent survival kinase (CISK-PX). *J Biol Chem* 2004;**279**(29):30,662–30669.

20. Stahelin RV, Karathanassis D, Bruzik KS, Waterfield MD, Bravo J, Williams RL, Cho W. Structural and membrane binding analysis of the Phox homology domain of phosphoinositide 3-kinase-C2alpha. *J Biol Chem* 2006;**281**(51):39,396–39406.

21. Zhou CZ, de La Sierra-Gallay IL, Quevillon-Cheruel S, Collinet B, Minard P, Blondeau K, Henckes G, Aufrere R, Leulliot N, Graille M, Sorel I, Savarin P, de la Torre F, Poupon A, Janin J, van Tilbeurgh H. Crystal structure of the yeast Phox homology (PX) domain protein Grd19p complexed to phosphatidylinositol-3-phosphate. *J Biol Chem* 2003;**278**(50):50,371–50376.

22. Stahelin RV, Burian A, Bruzik KS, Murray D, Cho W. Membrane binding mechanisms of the PX domains of NADPH oxidase p40phox and p47phox. *J Biol Chem* 2003;**278**(16):14,469–14479.

23. Stahelin RV, Ananthanarayanan B, Blatner NR, Singh S, Bruzik KS, Murray D, Cho W. Mechanism of membrane binding of the phospholipase D1 PX domain. *J Biol Chem* 2004;**279**(52):54,918–54926.

24. Lee SA, Kovacs J, Stahelin RV, Cheever ML, Overduin M, Setty TG, Burd CG, Cho W, Kutateladze TG. Molecular mechanism of membrane docking by the Vam7p PX domain. *J Biol Chem* 2006;**281**(48):37,091–37101.

25. Groemping Y, Lapouge K, Smerdon SJ, Rittinger K. Molecular basis of phosphorylation-induced activation of the NADPH oxidase. *Cell* 2003;**113**(3):343–355.

26. Jang IH, Lee S, Park JB, Kim JH, Lee CS, Hur EM, Kim IS, Kim KT, Yagisawa H, Suh PG, Ryu SH. The direct interaction of phospholipase C-gamma 1 with phospholipase D2 is important for epidermal growth factor signaling. *J Biol Chem* 2003;**278**(20):18,184–18190.

27. Wientjes FB, Reeves EP, Soskic V, Furthmayr H, Segal AW. The NADPH oxidase components p47(phox) and p40(phox) bind to moesin through their PX domain. *Biochem Biophys Res Commun* 2001;**289**(2):382–388.

28. Kim JH, Ohba M, Suh PG, Ryu SH. Novel functions of the phospholipase D2-Phox homology domain in protein kinase Czeta activation. *Mol Cell Biol* 2005;**25**(8):3194–3208.

29. Lee JH, Kim YM, Kim NW, Kim JW, Her E, Kim BK, Kim JH, Ryu SH, Park JW, Seo DW, Han JW, Beaven MA, Choi WS. Phospholipase D2 acts as an essential adaptor protein in the activation of Syk in antigen-stimulated mast cells. *Blood* 2006;**108**(3):956–964.

30. Lee CS, Kim IS, Park JB, Lee MN, Lee HY, Suh PG, Ryu SH. The phox homology domain of phospholipase D activates dynamin GTPase activity and accelerates EGFR endocytosis. *Nat Cell Biol* 2006;**8**(5):477–484.

31. Stroupe C, Collins KM, Fratti RA, Wickner W. Purification of active HOPS complex reveals its affinities for phosphoinositides and the SNARE Vam7p. *EMBO J* 2006;**25**(8):1579–1589.

32. Vollert CS, Uetz P. The phox homology (PX) domain protein interaction network in yeast. *Mol Cell Proteomics* 2004;**3**(11):1053–1064.

33. Fratti RA, Wickner W. Distinct targeting and fusion functions of the PX and SNARE domains of yeast vacuolar Vam7p. *J Biol Chem* 2007;**282**(17):13,133–13138.

34. Lee JS, Kim JH, Jang IH, Kim HS, Han JM, Kazlauskas A, Yagisawa H, Suh PG, Ryu SH. Phosphatidylinositol (3,4,5)-trisphosphate specifically interacts with the phox homology domain of phospholipase D1 and stimulates its activity. *J Cell Sci* 2005;**118**(19):4405–4413.

35. Lambeth JD. NOX enzymes and the biology of reactive oxygen. *Nat Rev Immunol* 2004;**4**(3):181–189.

36. Koshkin V, Lotan O, Pick E. The cytosolic component p47(phox) is not a sine qua non participant in the activation of NADPH oxidase but is required for optimal superoxide production. *J Biol Chem* 1996;**271**(48):30,326–30,329.

37. Ueyama T, Tatsuno T, Kawasaki T, Tsujibe S, Shirai Y, Sumimoto H, Leto TL, Saito N. A regulated adaptor function of p40phox: distinct p67phox membrane targeting by p40phox and by p47phox. *Mol Biol Cell* 2007;**18**(2):441–454.

38. Vieira OV, Botelho RJ, Rameh L, Brachmann SM, Matsuo T, Davidson HW, Schreiber A, Backer JM, Cantley LC, Grinstein S. Distinct roles of class I and class III phosphatidylinositol 3-kinases in phagosome formation and maturation. *J Cell Biol* 2001;**155**(1):19–25.

39. Ellson CD, Anderson KE, Morgan G, Chilvers ER, Lipp P, Stephens LR, Hawkins PT. Phosphatidylinositol 3-phosphate is generated in phagosomal membranes. *Curr Biol* 2001;**11**(20):1631–1635.

40. Andrews S, Stephens LR, Hawkins PT. PI3K class IB pathway in neutrophils. *Sci STKE* 2007;**2007**(407):cm3.

41. Honbou K, Minakami R, Yuzawa S, Takeya R, Suzuki NN, Kamakura S, Sumimoto H, Inagaki F. Full-length p40phox structure suggests a basis for regulation mechanism of its membrane binding. *EMBO J* 2007;**26**(4):1176–1186.

42. Bissonnette SA, Glazier CM, Stewart MQ, Brown GE, Ellson CD, Yaffe MB. Phosphatidylinositol 3-phosphate-dependent and -independent functions of p40phox in activation of the neutrophil NADPH oxidase. *J Biol Chem* 2008;**283**(4):2108–2119.

43. Suh CI, Stull ND, Li XJ, Tian W, Price MO, Grinstein S, Yaffe MB, Atkinson S, Dinauer MC. The phosphoinositide-binding protein p40phox activates the NADPH oxidase during FcgammaIIA receptor-induced phagocytosis. *J Exp Med* 2006;**203**(8):1915–1925.

44. Ellson C, Davidson K, Anderson K, Stephens LR, Hawkins PT. PtdIns3P binding to the PX domain of p40phox is a physiological signal in NADPH oxidase activation. *EMBO J* 2006;**25**(19):4468–4478.

45. Chen J, He R, Minshall RD, Dinauer MC, Ye RD. Characterization of a mutation in the Phox homology domain of the NADPH oxidase component p40phox identifies a mechanism for negative regulation of superoxide production. *J Biol Chem* 2007;**282**(41):30,273–30,284.

46. Tian W, Li XJ, Stull ND, Ming W, Suh CI, Bissonnette SA, Yaffe MB, Grinstein S, Atkinson SJ, Dinauer MC. Fc{gamma}R-stimulated activation of the NADPH oxidase: Phosphoinositide binding protein p40phox regulates NADPH oxidase activity after enzyme assembly on the phagosome. *Blood* 2008. in press.

47. Abram CL, Seals DF, Pass I, Salinsky D, Maurer L, Roth TM, Courtneidge SA. The adaptor protein fish associates with members of the ADAMs family and localizes to podosomes of Src-transformed cells. *J Biol Chem* 2003;**278**(19):16,844–16,851.

48. Lock P, Abram CL, Gibson T, Courtneidge SA. A new method for isolating tyrosine kinase substrates used to identify fish, an SH3 and PX domain-containing protein, and Src substrate. *Embo J* 1998;**17**(15):4346–4357.

49. Seals Jr DF, Azucena EF, Pass I, Tesfay L, Gordon R, Woodrow M, Resau JH, Courtneidge SA. The adaptor protein Tks5/Fish is required for podosome formation and function, and for the protease-driven invasion of cancer cells. *Cancer Cell* 2005;**7**(2):155–165.

50. Liscovitch M, Czarny M, Fiucci G, Tang X. Phospholipase D: molecular and cell biology of a novel gene family. *Biochem J* 2000;**345**(3):401–415.

51. Song X, Xu W, Zhang A, Huang G, Liang X, Virbasius JV, Czech MP, Zhou GW. Phox homology domains specifically bind phosphatidylinositol phosphates. *Biochemistry* 2001;**40**(30):8940–8944.

52. Slagsvold T, Marchese A, Brech A, Stenmark H. CISK attenuates degradation of the chemokine receptor CXCR4 via the ubiquitin ligase AIP4. *Embo J* 2006;**25**(16):3738–3749.

53. Xu J, Liu D, Gill G, Songyang Z. Regulation of cytokine-independent survival kinase (CISK) by the Phox homology domain and phosphoinositides. *J Cell Biol* 2001;**154**(4):699–705.

54. Cheng G, Lambeth JD. NOXO1, regulation of lipid binding, localization, and activation of Nox1 by the Phox homology (PX) domain. *J Biol Chem* 2004;**279**(6):4737–4742.

55. Banfi B, Clark RA, Steger K, Krause KH. Two novel proteins activate superoxide generation by the NADPH oxidase NOX1. *J Biol Chem* 2003;**278**(6):3510–3513.

56. Zheng B, Ma YC, Ostrom RS, Lavoie C, Gill GN, Insel PA, Huang XY, Farquhar MG. RGS-PX1, a GAP for GalphaS and sorting nexin in vesicular trafficking. *Science* 2001;**294**(5548):1939–1942.

57. Zheng B, Tang T, Tang N, Kudlicka K, Ohtsubo K, Ma P, Marth JD, Farquhar MG, Lehtonen E. Essential role of RGS-PX1/sorting nexin 13 in mouse development and regulation of endocytosis dynamics. *Proc Natl Acad Sci USA* 2006;**103**(45):16,776–16,781.

58. Blatner NR, Wilson MI, Lei C, Hong W, Murray D, Williams RL, Cho W. The structural basis of novel endosome anchoring activity of KIF16B kinesin. *EMBO J* 2007;**26**(15):3709–3719.

59. Carlton J, Bujny M, Peter BJ, Oorschot VM, Rutherford A, Mellor H, Klumperman J, McMahon HT, Cullen PJ. Sorting nexin-1 mediates tubular endosome-to-TGN transport through coincidence sensing of high- curvature membranes and 3-phosphoinositides. *Curr Biol* 2004;**14**(20):1791–1800.

60. Cozier GE, Carlton J, McGregor AH, Gleeson PA, Teasdale RD, Mellor H, Cullen PJ. The phox homology (PX) domain-dependent, 3-phosphoinositide-mediated association of sorting nexin-1 with an early sorting endosomal compartment is required for its ability to regulate epidermal growth factor receptor degradation. *J Biol Chem* 2002;**277**(50):48,730–48,736.

61. Fisk HA, Yaffe MP. Mutational analysis of Mdm1p function in nuclear and mitochondrial inheritance. *J Cell Biol* 1997;**138**(3):485–494.

62. Ago T, Takeya R, Hiroaki H, Kuribayashi F, Ito T, Kohda D, Sumimoto H. The PX domain as a novel phosphoinositide- binding module. *Biochem Biophys Res Commun* 2001;**287**(3):733–738.

63. Han BK, Bogomolnaya LM, Totten JM, Blank HM, Dangott LJ, Polymenis M. Bem1p, a scaffold signaling protein, mediates cyclin-dependent control of vacuolar homeostasis in *Saccharomyces cerevisiae*. *Genes Dev* 2005;**19**(21):2606–2618.

64. Xu H, Wickner W. Bem1p is a positive regulator of the homotypic fusion of yeast vacuoles. *J Biol Chem* 2006;**281**(37):27,158–27,166.

65. Sato TK, Darsow T, Emr SD. Vam7p, a SNAP-25-like molecule, and Vam3p, a syntaxin homolog, function together in yeast vacuolar protein trafficking. *Mol Cell Biol* 1998;**18**(9):5308–5319.

66. Voos W, Stevens TH. Retrieval of resident late-Golgi membrane proteins from the prevacuolar compartment of *Saccharomyces cerevisiae* is dependent on the function of Grd19p. *J Cell Biol* 1998;**140**(3):577–590.

Chapter 23

FYVE Domains in Membrane Trafficking and Cell Signaling

Christopher Stefan, Anjon Audhya and Scott D. Emr
Weill Institute for Cell and Molecular Biology, Department of Molecular Biology and Genetics, Cornell University, Ithaca, New York

INTRODUCTION

The recruitment of cytoplasmic proteins to specific membrane compartments is important for a diverse spectrum of cellular processes including intracellular protein trafficking, cytokine and growth factor receptor signaling, actin cytoskeleton organization and apoptosis [1–4]. Many proteins are localized to membranes through tightly regulated interactions with membrane-associated factors. Derivatives of phosphatidylinositol (PtdIns) that can be reversibly phosphorylated at different positions of the inositol ring are ideally suited for this function. A set of well-conserved specific lipid kinases [5] generate different phosphoinositide (PI) species phosphorylated at the 3′, 4′, or 5′ positions of the inositol headgroup that each can recruit/activate a specific subset of cytoplasmic effector proteins. The activity of these target proteins can then be attenuated through the action of PI-specific phosphatases and lipases [6–9]. Several studies have now identified multiple, well-conserved PI-binding motifs, each of which can recognize particular PI isoforms with a high degree of specificity [10]. In this chapter, we discuss the structural basis of a specialized zinc finger that binds PtdIns 3-phosphate [PI(3)P], termed the FYVE domain, and the roles played by several proteins harboring this motif in membrane trafficking and cell signaling.

ROLE FOR PI(3)P IN MEMBRANE TRAFFICKING AND IDENTIFICATION OF THE FYVE DOMAIN

A role for PI(3)P in vesicular transport was first discovered in the study of Golgi to vacuole transport in yeast [11]. *Saccharomyces cerevisiae* expresses one PtdIns 3-kinase isoform, Vps34 [5]. Deletion of the *VPS34* gene resulted in a lack of PI(3)P synthesis and defects in endosomal membrane trafficking from the Golgi and plasma membrane to the lysosome-like vacuole [12]. Likewise, PI(3)P has been shown to play important roles in several membrane trafficking pathways to mammalian lysosomes [13]. The fungal metabolite wortmannin, an inhibitor of PI3-kinase activity, has been shown to impair homotypic endosome fusion *in vitro* and the transport of enzymes such as cathepsin D to lysosomes *in vivo* [4, 14, 15]. Accordingly, the human homolog of the yeast Vps34 PtdIns 3-kinase has been identified and found to be sensitive to wortmannin [3, 5].

Several proteins have been implicated as downstream effectors of PI(3)P in vesicle transport. One of these, mammalian EEA1 (early endosome antigen 1) has been shown to localize to endosomal membranes in a wortmannin-sensitive manner [16, 17]. Consequently, deletion of the FYVE domain of EEA1 was shown to diminish its endosomal association, suggesting that this domain may directly bind PI(3)P [18]. The FYVE domain, named after the first four proteins found to contain this motif (*F*ab1, *Y*OTB, *V*ac1 and *E*EA1), was originally identified as a RING-finger family member that coordinates two Zn^{++} ions through eight cysteine/histidine residues spaced in a conserved manner (CX$_2$CX$_{9-39}$CX$_{1-3}$(C/H)X$_{2-3}$CX$_2$CX$_{4-48}$CX$_2$C) (Figure 23.1) [18, 19]. Importantly, subsequent studies demonstrated the ability of FYVE domains to specifically bind PI(3)P *in vitro*, as recombinant EEA1 FYVE sedimented with liposomes containing PI(3)P but not other PI species [20–22]. The identification of modular protein domains that bind PI(3)P with high affinity and specificity, such as the FYVE domain, has been a crucial step in further understanding the roles of this lipid in membrane trafficking events, as described in greater detail below.

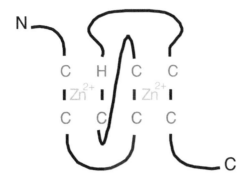

Fab1	WMK DESSKE CFS C GKT FNTFR RKHH C RI CX₄CX₂CX₁₅ RV CYN CYE
YDR313	WQA DEEAHS CFQ C KTN FSFLV RRHH C RI CX₄CX₂CX₂₇ RT CNE CYD
Vac1	WRD DRSVLF C NI C SEP FGLLL RKHH C RL CX₄CX₆ CX₃₂ RL C SH C ID
EEA1	WAE DNEVQN CMA C GKG FSVTV RRHHC RQCX₄CX₂CX₁₄ RV C DA C FN

FIGURE 23.1 The FYVE domain is a conserved RING finger domain.

(Top) Schematic cartoon of the RING finger FYVE domain. The conserved cysteine/histidine residues that coordinate two Zn⁺⁺ atoms are shown. (Bottom) Sequence alignment of the conserved cysteine/histidine motifs in the FYVE domains of Fab1, Pib1 (YDR313c), Vac1, and EEA1. Identical residues are shown in boldface.

STRUCTURAL BASIS FOR PI(3)P RECOGNITION BY THE FYVE DOMAIN

Insight into the molecular mechanisms that mediate the interaction between FYVE domains and PI(3)P is provided by several structural studies on this motif [23, 24]. The FYVE domain, as mentioned, is an approximately 80 amino-acid sequence containing eight conserved cysteine/histidine residues that coordinate two Zn^{++} atoms. In addition, several other residues are conserved, most notably a highly basic R(R/K)HHCR patch adjacent to the third cysteine residue, an amino-terminal WxxD motif, and a conserved hydrophobic region upstream of the basic patch (Figure 23.1). As first determined from the crystal structure of the yeast Vps27 FYVE domain, the basic patch is localized within β1 of two double-stranded anti-parallel β-sheets (comprised of β1/β2 and β3/β4), which are stabilized by the two zinc atoms and a C-terminal α-helix [25]. Molecular modeling suggested that the inositol head group of PI(3)P fits into a pocket created by the backbone of the first β-sheet, and the 3′-phosphate group contacts side groups of the final histidine and arginine residues found in the basic patch (Figure 23.2) [25]. Additionally, from this model, the 1′-phosphate of PI(3)P is poised to form a salt-bridge with the first arginine in the conserved basic patch [25]. In combination, these interactions are specific for PI(3)P, as additional or other phosphate groups on the inositol ring would prohibit interaction with the FYVE domain due to spatial constraints, consistent with previous *in vitro* binding studies indicating this motif does not bind other phosphoinositides.

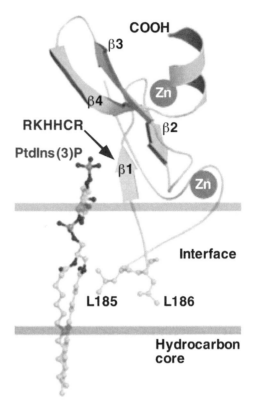

FIGURE 23.2 The FYVE domain is a modular PI(3)P-binding motif.

The crystal structure of the FYVE domain of yeast Vps27 [27, 28]. The ribbon depicts four β strands followed by a carboxy-terminal α-helix. The two Zn^{++} atoms are shown. The highly conserved positively charged residues (RKHHCR) in the β1 strand predicted to make contacts with the 3′-phosphate group of PI(3)P are indicated. In addition, residues in a hydrophobic loop upstream of the first β-sheet predicted to penetrate into the membrane bilayer are indicated [25]. The membrane layer is divided into an interfacial region (lipid headgroups and the hydrophobic interface) and a hydrocarbon core (lipid acyl chains).Reprinted from [27] (Kutateladze *et al.*, 1999), with permission; © 1999 Cell Press.

Although a similar structure was proposed for the FYVE domain of *Drosophila* Hrs, a homolog of Vps27, a different model for PI(3)P binding was suggested [26]. The major difference involved an anti-parallel association of two Hrs FYVE monomers to generate a homodimer with two ligand-binding pockets. Residues from β1 including the conserved basic patch, together with a hydrophobic strand from β4 of the opposite FYVE monomer line each pocket. In addition, this model differed from that of the Vps27 FYVE structure in regard to the orientation of the FYVE domain with respect to the membrane. Thus, even though both models were consistent in regard to the overall structure of the FYVE domain, further studies were required to help resolve the true nature of the interaction between the FYVE domain and PI(3)P containing membranes.

To gain further insight into the interaction of the FYVE domain with PI(3)P, NMR studies of the EEA1 FYVE domain were performed [27, 28]. As expected, these studies highlighted the importance of the basic residues found in the first β-sheet as they displayed large chemical shift changes in the presence of PI(3)P. In addition, residues in a hydrophobic loop upstream of the first β-sheet displayed chemical shifts, but only in the presence of micelle-embedded PI(3)P, suggesting these residues contact the membrane non-specifically. Similar to the membrane orientation predicted by the Vps27 FYVE structure (see Figure 23.2), this hydrophobic loop may extend into the cytoplasmic side of the membrane bilayer, perhaps directly interacting with hydrophobic acyl chains.

However, when fused to GFP or GST, the EEA1 FYVE domain alone failed to efficiently localize to cellular membranes [20–22]. Residues from an additional coiled-coil region adjacent to the FYVE domain were required. Examination of the crystal structure of the EEA1 FYVE domain including these additional residues has revealed the formation of stable homodimers that could bind two molecules of inositol 1,3-bisphosphate, a soluble mimic of PI(3)P [29]. However, unlike the model proposed from studies of the Hrs FYVE domain, each EEA1 FYVE domain independently bound inositol 1,3-bisphosphate. Dimerization of the EEA1 FYVE domains was mediated primarily through interactions between the coiled-coil domains. Taken together, these data suggest that dimerization enhances binding of individual FYVE domains to membrane-restricted PI(3)P. While the EEA1 FYVE structures have provided accurate models for PI(3)P binding, dimerization, and hydrophobic insertion into the lipid bilayer, additional structural and biochemical studies have identified factors involved in targeting and stabilization of FYVE domains at cellular membranes [30–33]. These include interactions with acidic phospholipids and the cytosolic pH, in part to mediate protonation of the conserved histidine residues. Interestingly, the residues adjacent to the EEA1 FYVE domain required for membrane localization are required for EEA1 to bind Rab5, a small GTPase that functions on membranes in the endocytic pathway [17,34]. Together, these data suggest that a combination of protein–lipid, both PI(3)P and acidic phospholipids, and protein–protein (e.g., FYVE domain dimerization and Rab5) interactions are essential for specific and stable localization of FYVE domain containing proteins to particular cellular membranes.

CONSERVATION OF THE FYVE DOMAIN AND LOCALIZATION OF PI(3)P

To date, analysis of the human genome has uncovered a total of more than 30 FYVE domain-containing proteins, while the *Caenorhabditis elegans* genome contains 12 and the *S. cerevisiae* genome harbors 5. Thus, while the FYVE domain itself has been well conserved through the course of evolution, there appears to have been a significant expansion in the roles played by this lipid-binding motif. As described earlier, a major role for PI(3)P, and by extension its FYVE domain-containing effectors, involves endocytic membrane transport. However, several findings also show that PI(3)P has roles in growth factor signaling and actin cytoskeleton organization through the recruitment/activation of other FYVE domain-containing proteins.

Initial studies of PI(3)P localization were carried out using a GFP fusion to the FYVE domain of EEA1 in yeast [20]. Results indicated that the fusion co-localized with prevacuolar endosomes and weakly labeled the vacuolar membrane. Importantly, this localization was dependent on Vps34 PtdIns 3-kinase activity, demonstrating a requirement for PI(3)P in mediating membrane association *in vivo* [20]. Two FYVE domains in tandem fused to GFP similarly localized to endosomal structures in fibroblasts [35]. However, due to the limitations of light microscopy, a more detailed analysis of PI(3)P localization required more extensive studies of the recombinant FYVE domain dimer. Using an electron microscopic labeling approach, PI(3)P was found to be highly enriched on endosomes as expected from previous work, but the lipid was also observed in the nucleolus, and in the internal vesicles of multivesicular bodies (MVBs) [35]. Consistent with the presence of PI(3)P on vesicles inside the lumen of MVBs, the efficient turnover of PI(3)P in yeast was shown to be dependent in part on hydrolase-mediated degradation in the vacuole [36]. It is likely that most if not PI(3)P effectors are recruited to and/or activated at endosomal, lysosomal, or autophagosomal membranes, the major sites of PI(3)P accumulation in cells.

FYVE DOMAIN-CONTAINING PROTEINS IN MEMBRANE TRAFFICKING

Studies of EEA1 have been instrumental in defining the localization of PI(3)P. Closer examination of the protein reveals it is a large coiled-coil protein (Figure 23.3a) that can bind to the GTP-bound form of Rab5, a GTPase required for endosomal membrane fusion [17]. Together with PI(3)P, Rab5-GTP recruits EEA1 to endosomal membranes where it functions in membrane fusion. Consistent with this hypothesis, depletion of EEA1 inhibits homotypic endosome fusion *in vitro*, while excess EEA1 stimulates fusion [17, 34, 37]. Furthermore, studies suggest that EEA1 may engage in oligomeric complexes during membrane fusion, which may tether Rab5-positive endosomes, thus facilitating pairing of SNARE proteins to drive membrane fusion [38, 39].

Similar to EEA1, another FYVE domain-containing protein, Rabenosyn-5 (Table 23.1), is an effector of Rab5-GTP. Its localization to the endosome is dependent on PI(3)

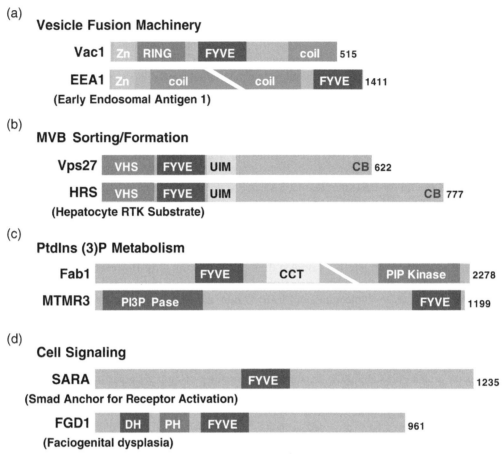

FIGURE 23.3 Schematic representation of protein motifs found within FYVE domain proteins. Several FYVE domain proteins have additional domains that bind protein targets. Together, these PI(3)P–protein and protein–protein interactions define the specific function of each FYVE domain protein. Examples of yeast and mammalian FYVE domain proteins that act in various cellular processes are shown. (a) FYVE domain proteins implicated in vesicle targeting and fusion events; (b) FYVE domain proteins involved in endosomal/MVB sorting; (c) FYVE domain proteins containing enzymatic activities implicated in PI synthesis and turnover; (d) FYVE domain proteins involved in cell signaling. Other abbreviations: Zn, Zn++ finger domain; RING, Zn++ finger domain; coil, coiled-coil domain; VHS, conserved domain found in Vps27, Hrs, and STAM; SH3, Src homology 3 domain; UIM, ubiquitin-interacting motif; CB, clathrin box binding motif; CCT, chaperonin-like region; PIP kinase, PI(3)P 5-kinase catalytic domain; PI3P Pase, myotubularin-related PI(3)P phosphatase catalytic domain; DH, Dbl homology domain; PH, pleckstrin homology domain.

P binding and is required for endosome fusion [40]. While EEA1 appears to interact with specific SNARES including syntaxin-13 [39], Rabenosyn-5 directly interacts with the Sec1-like protein hVps45, suggesting distinct roles for these Rab5/PI(3)P effectors in endosome fusion [40]. Similarly, the yeast homolog of Rabenosyn-5, Vac1, also contains a FYVE domain (Figure 23.3a) and interacts with Vps21 and Vps45, homologs of Rab5 and Sec1 [41, 42]. Deletion of *VAC1* results in an accumulation of vesicles destined for the endosome and a defect in protein sorting to the lysosome like vacuole, suggesting that Vac1 is evolutionarily conserved and required for endosome docking and/or fusion [43]. Substitutions in the FYVE domain of Vac1 also result in defects in vacuolar protein sorting, indicating a requirement for this domain in Vac1 function. However, these Vac1 mutants can still associate with membranes, suggesting additional factors are involved in Vac1 localization [42]. Nevertheless, the FYVE domains found

in EEA1 and Vac1/Rabenosyn-5 may still be required for concentration of these proteins on PI(3)P rich endosomes, while Rab binding may further drive specificity of these interactions. Once on endosomal membranes, EEA1 and Vac1/Rabenosyn-5 appear to intimately participate in the machinery that drives endosome fusion (Figure 23.4).

In addition to Rab5, another small GTPase (Rab4) that has been implicated in the recycling of internalized receptors back to the plasma membrane regulates a FYVE domain-containing effector, Rabip4 (Table 23.1). Like EEA1 and Vac1/Rabenosyn-5, Rabip4 localizes to endosomes and can affect endosomal morphology [44]. Moreover, overproduction of Rabip4 leads to the intracellular retention of normally recycled transporters such as Glut1 [44] . These data suggest that FYVE domain-containing proteins and thus PI(3)P are not only involved in anterograde trafficking to lysosomes but also endosomal membrane recycling to the plasma membrane.

TABLE 23.1 FYVE domain-containing proteins discussed in this review As well as binding to PI(3)P, many FYVE domain-containing proteins have other domains that interact with protein targets; together, these interactions likely play important roles in defining protein function

Protein	Cellular function	Domains	Targets	Reference(s)
Yeast				
Vac1	Golgi to endosome transport	FYVE, RING, COIL	PI(3)P, Vps21, Vps45, Pep12	41–43
Vps27	MVB sorting/formation (ESCRT pathway)	VHS, FYVE, UIM	PI(3)P, ubiquitin	46–49
Fab1	PI(3,5)P$_2$ synthesis	FYVE, PIP kinase	PtdIns(3)P	67–69
Pib1	Ubiquitin ligase	FYVE, E3 ubiquitin ligase	PI(3)P, unknown	20, 60
Pib2	Unknown	FYVE	PI(3)P?	20
Mammalian				
EEA1	Endosome fusion	FYVE, COIL	PI(3)P, Rab5, syntaxins	17–23, 34–39
Rabenosyn-5	Endosome fusion	FYVE	PI(3)P, Rab5, and hVps45	40
Rabip4	Endosomal membrane recycling	FYVE	PI(3)P, Rab4	44
Hrs	MVB sorting/formation (ESCRT pathway)	VHS, FYVE, UIM, CB	PI(3)P, ubiquitin, clathrin	50–52
PIKfyve	PI(3,5)P$_2$ synthesis	FYVE, PIP kinase	PI(3)P	70
MTMR4	PI(3)P turnover	FYVE, PI3P phosphatase	PI(3)P	71–74
Endofin	TGFβ signaling	FYVE	PI(3)P, Smad4	80–83
Frabin	Actin cytoskeleton	FYVE, PH, GEF	PI isoforms, Cdc42, Rac	85, 86
DFCP1	Autophagy	FYVE	PI(3)P	56, 57

Another well-conserved FYVE domain-containing protein that has been implicated in membrane trafficking is Hrs (Figure 23.3b), a *h*epatocyte growth factor *r*eceptor tyrosine kinase *s*ubstrate [45]. Studies of its yeast homolog Vps27 also demonstrate a requirement for this PI(3)P effector in endosomal sorting, functioning after Vac1 in the ESCRT-mediated multivesicular body (MVB) sorting pathway [46]. Specifically, inactivation of Vps27 results in a defect in the generation of intralumenal vesicles within MVBs and the vacuole [47,48], as Vps27/Hrs recruits the downstream ESCRT complexes that mediate MVB vesicle formation [49]. Similarly, mouse embryos that lack Hrs exhibit defects in endosomal morphogenesis [50]. Strikingly, Hrs and EEA1 are localized to different regions on endosomes. Whereas EEA1 co-localizes with Rab5 on endosomes, Hrs localizes to clathrin-enriched endosomal domains and can bind to clathrin via a carboxy-terminal clathrin interacting motif [51]. Disruption of the PI(3) P-FYVE interaction in Hrs by treatment with the PtdIns 3-kinase inhibitor wortmannin results in loss of both Hrs and clathrin localization to endosomes, again demonstrating the importance of the FYVE domain in endosomal function [52]. Likewise, inactivation of the PtdIns 3-kinase

Vps34 in yeast disrupts Vps27 endosomal membrane localization [49]. Further studies are required to precisely determine what other requirements may be necessary for Hrs/Vps27 to associate with endosomes in addition to PI(3)P.

In addition to regulation of endosome fusion, recycling, and MVB vesicle formation, FYVE domain-containing proteins have been implicated in other membrane trafficking pathways. In *C. elegans*, depletion of the WD-40 and FYVE domain containing protein, WDFY2, by siRNA resulted in endocytic internalization defects [53]. WDFY2 localized to early endosomes near the plasma membrane distinct from Rab5 and EEA1 positive endosomes that may comprise an early step in sorting of internalized cargo [53]. Similarly, mammalian orthologs of WDFY2 have been implicated in transferrin uptake [53] and recently been shown to associate with the protein kinases Akt and PKCζ as well as the transcription factor Foxo1 [54, 55], suggesting a potential regulatory link between membrane trafficking and cell signaling pathways.

PI(3)P and FYVE domain-containing proteins have also been implicated in autophagy. For example, the double FYVE domain-containing protein, DFCP1 (Table 23.1), initially localized to the endoplasmic reticulum and Golgi

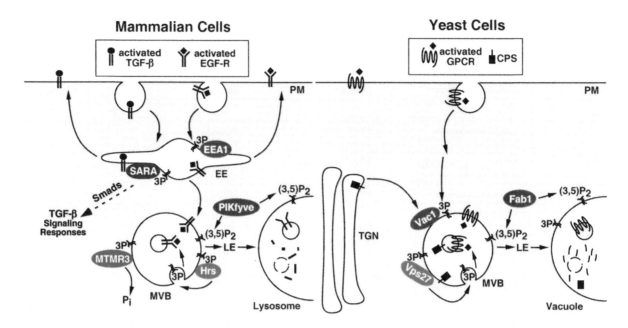

FIGURE 23.4 Cellular localization and functions of FYVE domain-containing proteins in mammalian and yeast cells.
In mammalian and yeast cells, PI(3)P (designated as 3P) recruits the Rab effectors EEA1 or Vac1 to endosomes where they participate in vesicle fusion in Golgi (TGN) to vacuole transport and endocytic trafficking. The FYVE domain-containing orthologs Hrs and Vps27 are required for MVB sorting pathways that transport PI(3)P and cargo proteins, such as carboxypeptidase S (CPS) and internalized cell surface receptors, to the vacuole lumen where they are degraded. The PI(3)P 5-kinases Fab1/PIKfyve and the PI(3)P-specific phosphatase MTMR3 terminate PI(3)P signaling by converting PI(3)P to PI(3,5)P$_2$ or PtdIns, respectively. PI(3)P 5-kinase signaling via PI(3,5)P$_2$ (designated as 3,5P$_2$) is also required for MVB sorting. In mammalian cells, the FYVE domain containing-protein SARA recruits Smad proteins, effectors of transforming growth factor beta (TGFβ) signaling, to the endosome.

[56] has recently been shown to translocate to PI(3)
P-enriched autophagosomal compartments in a Vps34 dependent fashion under starvation conditions [57]. In similar fashion, the autophagy-linked FYVE domain protein, Alfy, and its ortholog Bchs (blue cheese) in *Drosophila* have been shown to bind PI(3)P and relocalize from the ER and Golgi to autophagosomes upon stress conditions [58, 59].

In addition to Vac1 and Vps27, yeast harbor another FYVE domain-containing protein that localizes to the endo-some and vacuole, Pib1 (Table 23.1) [20]. Localization of Pib1 to these structures is dependent on its FYVE domain through an interaction with PI(3)P [60]. In addition to its FYVE domain, Pib1 contains a RING domain that pos-sesses E2-dependent ubiquitin ligase activity *in vitro* [60]. In light of previous studies indicating a role for ubiqui-tin modification in the sorting of transmembrane proteins into multivesicular bodies [47, 61–63], the finding that Pib1 is a RING-type E3 ubiquitin ligase that localizes to the endosome via a PI(3)P-FYVE interaction is interest-ing. However, deletion of *PIB1* fails to result in a defect in sorting of known ubiquitinated cargo through the MVB pathway [60], indicating that another E3 ubiquitin ligase acts in this process. Consistent with this, the HECT domain E3 ubiquitin ligase Rsp5 has been shown to mediate ubiq-uitination of MVB cargo proteins [64]. Alternatively, Pib1 may be responsible for ubiquitination of a specific subset

of substrates that have not yet been examined. Interestingly, mammalian cells also encode for a RING E3 ubiquitin ligase and FYVE domain-containing protein, CARP-2/ Rififylin, that localizes to endosomes and has been impli-cated in recycling [65], as well as down-regulation of sign-aling by the TNF (tumor necrosis factor) receptor complex at endosomes [66].

FYVE DOMAIN-CONTAINING PROTEINS INVOLVED IN PI(3)P METABOLISM

Additional FYVE domain-containing proteins found in both yeast and mammalian cells involved in membrane trafficking are the PI(3)P 5-kinases, Fab1 and PIKfyve (Figure 23.3c). Deletion of *FAB1* in yeast results in loss of PI(3,5)P$_2$ synthesis and consequent MVB sorting defects and drastically enlarged vacuoles [48, 67]. This abnormal vacuole morphology may also be in part due to defects in the recycling and/or turnover of membranes deposited at the vacuole. Recently, a downstream target effector, Atg18, for PI(3, 5)P$_2$ generated by Fab1 in this process has been suggested [68]. Interestingly, Fab1 has been shown to local-ize to both prevacuolar and vacuolar membranes, similar to the distribution of PI(3)P, suggesting that its amino-terminal FYVE domain may have a role in localization and/or activity of Fab1 [67]. However, deletion of an amino-terminal fragment of Fab1 including its FYVE domain fails to significantly perturb its function, since this form of Fab1 can rescue a temperature sensitive *fab1* mutant, suggesting that other determinants for Fab1 localization exist [67]. Indeed, a recent study has shown that Fab1, through its chaperonin-like domain, forms a vacuole-associated signaling complex with the Vac14 and Fig4 proteins [69]. This Fab1 complex is tethered to the vacuole via an interaction between the FYVE domain in Fab1 and PI(3)P on the vacuole. Similarly, the FYVE domain of mammalian PIKfyve is absolutely critical for its localization to membranes of the late endocytic pathway [70].

In addition to PI kinases, previous studies have uncovered a set of PI phosphatases that contain a FYVE domain, such as MTMR3 (Figure 23.3c) and MTMR4 (Table 23.1). Both are members of the myotubularin family of phosphatases that were originally shown to dephosphorylate serine/threonine and tyrosine residues *in vitro*, but were subsequently shown to act upon PI(3)P as their primary substrate [9,71]. Interestingly, various myotubularin members have been implicated in multiple disorders including myotubular myopathy [71], which involves defects in muscle differentiation and Charcot-Marie-Tooth disease [72], a condition caused by defects in myelin development. MTMR3, MTMR4, and other myotubularin isoforms have been localized to endosomes, and their overexpression results in reduced levels of endosomal PI(3)P dependent on their FYVE domains [73, 74]. More importantly however, the identification of both PI(3)P 5-kinases and PI(3)P phosphatases that contain FYVE domains raises the possibility that the FYVE domain may serve a regulatory role in the control of these enzyme activities when lipid is bound (Figure 23.4). Recruitment of either a PI(3)P 5-kinase or PI(3)P-specific phosphatase could play a role in terminating PI(3)P signaling by converting PI(3)P to PI(3,5)P_2 or PtdIns, respectively. Future work on these members of the FYVE domain family will be informative in shedding light on this question.

FYVE DOMAIN-CONTAINING PROTEINS IN SIGNALING

In addition to membrane trafficking and phosphoinositide metabolism, FYVE domains are also found in proteins required for other cellular processes, such as growth factor signaling. The FYVE domain containing-protein SARA (Figure 23.3d) recruits Smad proteins, effectors of transforming growth factor beta (TGFβ) signaling, to the endosome (Figure 23.4) [75]. There, bound TGFβ receptors can phosphorylate Smad2 and Smad3 via their cytoplasmic

serine/threonine kinase domain [75]. Phosphorylated Smads can then bind to Smad4, and this resulting complex is able to translocate to the nucleus and activate transcription of target genes [75]. Interestingly, SARA provides an excellent example in which the trafficking of cell-surface receptors is intimately coupled to intracellular signaling [76, 77]. In this case, the FYVE domain of SARA spatially regulates TGFβ signaling, restricting it to endosomes that contain both PI(3)P and activated TGFβ receptors. This spatial control permits the cell to prevent inappropriate activation of Smad signaling and allows for a large range of separation between the on and off states of this pathway. Consistent with this, treatment with the PtdIns 3-kinase inhibitor wortmannin results in mislocalization of SARA and leads to defects in Smad phosphorylation and downstream transcriptional activation [78, 79]. Similar to SARA, a protein named endofin (Table 23.1) also localizes to endosomes in a PI(3)P-dependent manner [80]. Although 50 percent identical to SARA, endofin fails to interact with Smad2 [80]; however a role in TGFβ signaling has been suggested in a more recent study through Smad4 interactions [81]. Unlike SARA, endofin has also been demonstrated to recruit clathrin to early endosomes through the VHS domain containing protein TOM1 [82, 83], suggesting yet another link between signaling pathways and membrane trafficking at the endosome in a PI(3)P-dependent manner.

Although less clear, FYVE domains are also found in proteins that regulate the actin cytoskeleton. These include Fgd1, *faciogenital dysplasia gene product implicated in the developmental disease Aarskog-Scott syndrome (Figure 23.3d), and Frabin (Table 23.1), that act as guanine nucleotide exchange factors (GEFs) for the small GTPases Cdc42 and Rac. Fgd1 specifically activates Cdc42, which in turn regulates actin cytoskeleton organization [84]. Frabin, through the action of Cdc42-dependent and independent pathways, has been implicated in filopodia and lamellipodia formation respectively [85]. However, these events likely do not involve endocytic trafficking, since early studies have indicated that this family of FYVE domain containing proteins does not localize to the endosome [86]. Closer examination of these GEFs shows they also contain PH domains that can bind other PI species [85], and their FYVE domains lack a well conserved tryptophan residue that is conserved in most other FYVE domains. Further studies are required to determine if these regulators of actin cytoskeleton organization actually bind PI(3)P through their FYVE domains or bind another ligand that may be structurally related to PI(3)P.

In addition to membrane trafficking, growth factor signaling, and actin cytoskeletal organization, FYVE domains have also been found in proteins required for other cellular processes, such as apoptosis [87]. The PH and FYVE domain-containing family of proteins, the Phafins, have recently been shown to localize to lysosomes in a manner dependent on their FYVE domains and PI(3)P. Following translocation to lysosomes, the Phafins have implicated in caspase-independent induction of apoptosis through the lysosomal-mitochondrial membrane permeabilization and TNF signaling pathways [87].

FYVE-LIKE DOMAINS

Comparative structural and biochemical studies have been informative in determining the molecular basis for PI(3)P binding and subcellular targeting of various FYVE domains [23,28,30,31,88]. In addition to the highly conserved FYVE domain of *bona fide* PI(3)P effectors, such as EEA1 and Hrs, several other FYVE-like domains have previously been identified. For example, the role of the FYVE-like domain in CARP-2/Rififylin has been unclear, as the conserved basic patch bears substitutions and the WxxD motif is lacking. It remains to be determined how this might effect PI(3)P binding or if CARP-2 localization is dependent on PI(3)P *in vivo*. Consistent with this, the crystal structure of the FYVE-like domain from CARP-2 indicated that it may not bind PI(3)P [88]. However, a more recent study demonstrated that CARP-2 localized to endosomal membranes, mediated by its FYVE-like domain [66]. Future studies aimed at determining the ligand binding specificities of additional FYVE-like domains should help to resolve these apparent discrepancies.

CONCLUSIONS

The identification of the FYVE domain as a specific PI(3)P binding motif has significantly impacted the field of membrane trafficking and shed light on additional cellular functions of PI(3)P. Through localization studies of the FYVE domain using both conventional light microscopy and high resolution electron microscopy, PI(3)P has been found to exist in endosomal membranes, on intralumenal vesicles contained within MVBs, autophagosomes, and in vacuolar/lysosomal membranes. In addition, GFP-FYVE fusions have been used to observe PI(3)P on phagosomes [89]. However, additional PI(3)P interacting proteins are involved in this process, as recent studies indicate that another lipid binding motif, the PX domain, specifically recognizes PI(3)P [90]. Accordingly, it is likely that new effectors of this lipid will continue to emerge.

The question still remains whether the FYVE domain alone provides sufficient specificity for protein localization. Previous studies would favor a significant but not singular role for the FYVE domain in this regard. Instead, PI(3)P-FYVE interactions coupled with protein–protein interactions likely ensure specific membrane recruitment of these proteins. This level of specificity would help to ensure appropriate membrane restricted responses and functions. Further studies are required to determine the validity of this concept and whether this is a general principle or may only apply to a certain subset of FYVE-domain containing proteins.

ACKNOWLEDGEMENTS

We thank members of the Emr laboratory, past and present, for helpful discussions and comments on this manuscript. Scott D. Emr is the Frank H.T. Rhodes Class of '56 Professor and Director of the Joan and Sanford I. Weill Institute for Cell and Molecular Biology at Cornell University.

REFERENCES

1. Janmey PA. Phosphoinositides and calcium as regulators of cellular actin assembly and disassembly. *Annu Rev Physiol* 1994;**56**:169–91.
2. Odorizzi G, Babst M, Emr SD. Phosphoinositide signaling and the regulation of membrane trafficking in yeast. *Trends Biochem Sci* 2000;**25**:229–35.
3. Rameh LE, Cantley LC. The role of phosphoinositide 3-kinase lipid products in cell function. *J Biol Chem* 1999;**274**:8347–50.
4. Simonsen A, Wurmser AE, Emr SD, Stenmark H. The role of phosphoinositides in membrane transport. *Curr Opin Cell Biol* 2001;**13**:485–92.
5. Fruman DA, Meyers RE, Cantley LC. Phosphoinositide kinases. *Annu Rev Biochem* 1998;**67**:481–507.
6. Berridge MJ. Phosphatidylinositol hydrolysis: a multifunctional transducing mechanism. *Mol Cell Endocrinol* 1981;**24**:115–40.
7. Hughes WE, Cooke FT, Parker PJ. Sac phosphatase domain proteins. *Biochem J* 2000;**350**:337–52.
8. Majerus PW, Kisseleva MV, Norris FA. The role of phosphatases in inositol signaling reactions. *J Biol Chem* 1999;**274**:10,669–10,672.
9. Wishart MJ, Taylor GS, Slama JT, Dixon JE. PTEN and myotubularin phosphoinositide phosphatases: bringing bioinformatics to the lab bench. *Curr Opin Cell Biol* 2001;**13**:172–81.
10. Hurley JH, Meyer T. Subcellular targeting by membrane lipids. *Curr Opin Cell Biol* 2001;**13**:146–52.
11. Wurmser AE, Gary JD, Emr SD. Phosphoinositide 3-kinases and their FYVE domain-containing effectors as regulators of vacuolar/lysosomal membrane trafficking pathways. *J Biol Chem* 1999;**274**:9129–32.
12. Stack JH, Herman PK, Schu PV, Emr SD. A membrane-associated complex containing the Vps15 protein kinase and the Vps34 PI3-kinase is essential for protein sorting to the yeast lysosome-like vacuole. *EMBO J* 1993;**12**:2195–204.
13. Corvera S. Phosphatidylinositol 3-kinase and the control of endosome dynamics: new players defined by structural motifs. *Traffic* 2001;**2**:859–66.
14. Brown WJ, DeWald DB, Emr SD, Plutner H, Balch WE. Role for phosphatidylinositol 3-kinase in the sorting and transport of newly synthesized lysosomal enzymes in mammalian cells. *J Cell Biol* 1995;**130**:781–96.
15. Li G, D'Souza-Schorey C, Barbieri MA, Roberts RL, Klippel A, Williams LT, Stahl PD. Evidence for phosphatidylinositol 3-kinase as a regulator of endocytosis via activation of Rab5. *Proc Natl Acad Sci USA* 1995;**92**:10,207–10,211.

16. Patki V, Virbasius J, Lane WS, Toh BH, Shpetner HS, Corvera S. Identification of an early endosomal protein regulated by phosphatidylinositol 3-kinase. *Proc Natl Acad Sci USA* 1997;**94**:7326–30.

17. Simonsen A, Lippe R, Christoforidis S, Gaullier JM, Brech A, Callaghan J, Toh BH, Murphy C, Zerial M, Stenmark H. EEA1 links PI(3)K function to Rab5 regulation of endosome fusion. *Nature* 1998;**394**:494–8.

18. Stenmark H, Aasland R, Toh BH, D'Arrigo A. Endosomal localization of the autoantigen EEA1 is mediated by a zinc- binding FYVE finger. *J Biol Chem* 1996;**271**:24,048–24,054.

19. Mu FT, Callaghan JM, Steele-Mortimer O, Stenmark H, Parton RG, Campbell PL, McCluskey J, Yeo JP, Tock EP, Toh BH. EEA1, an early endosome-associated protein. EEA1 is a conserved alpha-helical peripheral membrane protein flanked by cysteine "fingers" and contains a calmodulin-binding IQ motif. *J Biol Chem* 1995;**270**:13,503–13,511.

20. Burd CG, Emr SD. Phosphatidylinositol(3)-phosphate signaling mediated by specific binding to RING FYVE domains. *Mol Cell* 1998;**2**:157–62.

21. Patki V, Lawe DC, Corvera S, Virbasius JV, Chawla A. A functional PI(3)P-binding motif. *Nature* 1998;**394**:433–4.

22. Gaullier JM, Simonsen A, D'Arrigo A, Bremnes B, Stenmark H, Aasland R. FYVE fingers bind PI(3)P. *Nature* 1998;**394**:432–3.

23. Misra S, Miller GJ, Hurley JH. Recognizing phosphatidylinositol 3-phosphate. *Cell* 2001;**107**:559–62.

24. Fruman DA, Rameh LE, Cantley LC. Phosphoinositide binding domains: embracing 3-phosphate. *Cell* 1999;**97**:817–20.

25. Misra S, Hurley JH. Crystal structure of a phosphatidylinositol 3-phosphate-specific membrane-targeting motif, the FYVE domain of Vps27p. *Cell* 1999;**97**:657–66.

26. Mao Y, Nickitenko A, Duan X, Lloyd TE, Wu MN, Bellen H, Quiocho FA. Crystal structure of the VHS and FYVE tandem domains of Hrs, a protein involved in membrane trafficking and signal transduction. *Cell* 2000;**100**:447–56.

27. Kutateladze TG, Ogburn KD, Watson WT, de Beer T, Emr SD, Burd CG, Overduin M. Phosphatidylinositol 3-phosphate recognition by the FYVE domain. *Mol Cell* 1999;**3**:805–11.

28. Kutateladze T, Overduin M. Structural mechanism of endosome docking by the FYVE domain. *Science* 2001;**291**:1793–6.

29. Dumas JJ, Merithew E, Sudharshan E, Rajamani D, Hayes S, Lawe D, Corvera S, Lambright DG. Multivalent endosome targeting by homodimeric EEA1. *Mol Cell* 2001;**8**:947–58.

30. Blatner NR, Stahelin RV, Diraviyam K, Hawkins PT, Hong W, Murray D, Cho W. The molecular basis of the differential subcellular localization of FYVE domains. *J Biol Chem* 2004;**279**:53,818–53,827.

31. Hayakawa A, Hayes SJ, Lawe DC, Sudharshan E, Tuft R, Fogarty K, Lambright D, Corvera S. Structural basis for endosomal targeting by FYVE domains. *J Biol Chem* 2004;**279**:5958–66.

32. Kutateladze TG, Capelluto DG, Ferguson CG, Cheever ML, Kutateladze AG, Prestwich GD, Overduin M. Multivalent mechanism of membrane insertion by the FYVE domain. *J Biol Chem* 2004;**279**:3050–7.

33. Lee SA, Eyeson R, Cheever ML, Geng J, Verkhusha VV, Burd C, Overduin M, Kutateladze TG. Targeting of the FYVE domain to endosomal membranes is regulated by a histidine switch. *Proc Natl Acad Sci USA* 2005;**102**:13,052–13,057.

34. Lawe DC, Patki V, Heller-Harrison R, Lambright D, Corvera S. The FYVE domain of early endosome antigen 1 is required for both phosphatidylinositol 3-phosphate and Rab5 binding. Critical role of this dual interaction for endosomal localization. *J Biol Chem* 2000;**275**:3699–705.

35. Gillooly DJ, Morrow IC, Lindsay M, Gould R, Bryant NJ, Gaullier JM, Parton RG, Stenmark H. Localization of phosphatidylinositol 3-phosphate in yeast and mammalian cells. *EMBO J* 2000;**19**:4577–88.

36. Wurmser AE, Emr SD. Phosphoinositide signaling and turnover: PI(3)P, a regulator of membrane traffic, is transported to the vacuole and degraded by a process that requires lumenal vacuolar hydrolase activities. *EMBO J* 1998;**17**:4930–42.

37. Lawe DC, Chawla A, Merithew E, Dumas J, Carrington W, Fogarty K, Lifshitz L, Tuft R, Lambright D, Corvera S. Sequential roles for phosphatidylinositol 3-phosphate and Rab5 in tethering and fusion of early endosomes via their interaction with EEA1. *J Biol Chem* 2002;**277**:8611–17.

38. Christoforidis S, McBride HM, Burgoyne RD, Zerial M. The Rab5 effector EEA1 is a core component of endosome docking. *Nature* 1999;**397**:621–5.

39. McBride HM, Rybin V, Murphy C, Giner A, Teasdale R, Zerial M. Oligomeric complexes link Rab5 effectors with NSF and drive membrane fusion via interactions between EEA1 and syntaxin 13. *Cell* 1999;**98**:377–86.

40. Nielsen E, Christoforidis S, Uttenweiler-Joseph S, Miaczynska M, Dewitte F, Wilm M, Hoflack B, Zerial M. Rabenosyn-5, a novel Rab5 effector, is complexed with hVPS45 and recruited to endosomes through a FYVE finger domain. *J Cell Biol* 2000;**151**:601–12.

41. Peterson MR, Burd CG, Emr SD. Vac1p coordinates Rab and phosphatidylinositol 3-kinase signaling in Vps45p-dependent vesicle docking/fusion at the endosome. *Curr Biol* 1999;**9**:159–62.

42. Burd CG, Peterson M, Cowles CR, Emr SD. A novel Sec18p/NSF-dependent complex required for Golgi-to-endosome transport in yeast. *Mol Biol Cell* 1997;**8**:1089–104.

43. Weisman LS, Emr SD, Wickner WT. Mutants of *Saccharomyces cerevisiae* that block intervacuole vesicular traffic and vacuole division and segregation. *Proc Natl Acad Sci USA* 1990;**87**:1076–80.

44. Cormont M, Mari M, Galmiche A, Hofman P, Le Marchand-Brustel Y. A FYVE-finger-containing protein, Rabip4, is a Rab4 effector involved in early endosomal traffic. *Proc Natl Acad Sci USA* 2001;**98**:1637–42.

45. Komada M, Kitamura N. Growth factor-induced tyrosine phosphorylation of Hrs, a novel 115- kilodalton protein with a structurally conserved putative zinc finger domain. *Mol Cell Biol* 1995;**15**:6213–21.

46. Piper RC, Cooper AA, Yang H, Stevens TH. VPS27 controls vacuolar and endocytic traffic through a prevacuolar compartment in *Saccharomyces cerevisiae*. *J Cell Biol* 1995;**131**:603–17.

47. Shih SC, Katzmann DJ, Schnell JD, Sutanto M, Emr SD, Hicke L. Epsins and Vps27p/Hrs contain ubiquitin-binding domains that function in receptor endocytosis. *Nat Cell Biol* 2002;**4**:389–93.

48. Odorizzi G, Babst M, Emr SD. Fab1p PI(3)P 5-kinase function essential for protein sorting in the multivesicular body. *Cell* 1998;**95**:847–58.

49. Katzmann DJ, Stefan CJ, Babst M, Emr SD. Vps27 recruits ESCRT machinery to endosomes during MVB sorting. *J Cell Biol* 2003;**162**:413–23.

50. Komada M, Soriano P. Hrs, a FYVE finger protein localized to early endosomes, is implicated in vesicular traffic and required for ventral folding morphogenesis. *Genes Dev* 1999;**13**:1475–85.

51. Raiborg C, Bache KG, Mehlum A, Stang E, Stenmark H. Hrs recruits clathrin to early endosomes. *EMBO J* 2001;**20**:5008–21.

52. Raiborg C, Bremnes B, Mehlum A, Gillooly DJ, D'Arrigo A, Stang E, Stenmark H. FYVE and coiled-coil domains determine the specific localisation of Hrs to early endosomes. *J Cell Sci* 2001;**114**:2255–63.

53. Hayakawa A, Leonard D, Murphy S, Hayes S, Soto M, Fogarty K, Standley C, Bellve K, Lambright D, Mello C, Corvera S. The WD40 and FYVE domain containing protein 2 defines a class of early endosomes necessary for endocytosis. *Proc Natl Acad Sci USA* 2006;**103**:11,928–11,933.

54. Fritzius T, Burkard G, Haas E, Heinrich J, Schweneker M, Bosse M, Zimmermann S, Frey AD, Caelers A, Bachmann AS, Moelling K. A WD-FYVE protein binds to the kinases Akt and PKCzeta/lambda. *Biochem J* 2006;**399**:9–20.

55. Fritzius T, Moelling K. Akt- and Foxo1-interacting WD-repeat-FYVE protein promotes adipogenesis. *EMBO J* 2008;**27**:1399–410.

56. Ridley SH, Ktistakis N, Davidson K, Anderson KE, Manifava M, Ellson CD, Lipp P, Bootman M, Coadwell J, Nazarian A, Erdjument-Bromage H, Tempst P, Cooper MA, Thuring JW, Lim ZY, Holmes AB, Stephens LR, Hawkins PT. FENS-1 and DFCP1 are FYVE domain-containing proteins with distinct functions in the endosomal and Golgi compartments. *J Cell Sci* 2001;**114**:3991–4000.

57. Axe EL, Walker SA, Manifava M, Chandra P, Roderick L, Habermann A, Griffiths G, Ktistakis N. Autophagosome formation from membrane compartments enriched in phosphatidylinositol 3-phosphate and dynamically connected to the endoplasmic reticulum. *J Cell Biol* 2008;**182**:685–701.

58. Simonsen A, Birkeland HC, Gillooly DJ, Mizushima N, Kuma A, Yoshimori T, Slagsvold T, Brech A, Stenmark H. Alfy, a novel FYVE-domain-containing protein associated with protein granules and autophagic membranes. *J Cell Sci* 2004;**117**:4239–51.

59. Simonsen A, Cumming RC, Finley KD. Linking lysosomal trafficking defects with changes in aging and stress response in Drosophila. *Autophagy* 2007;**3**:499–501.

60. Shin ME, Ogburn KD, Varban OA, Gilbert PM, Burd CG. FYVE domain targets Pib1p ubiquitin ligase to endosome and vacuolar membranes. *J Biol Chem* 2001;**276**:41,388–41,393.

61. Bishop N, Horman A, Woodman P. Mammalian class E vps proteins recognize ubiquitin and act in the removal of endosomal protein–ubiquitin conjugates. *J Cell Biol* 2002;**157**:91–101.

62. Katzmann DJ, Babst M, Emr SD. Ubiquitin-dependent sorting into the multivesicular body pathway requires the function of a conserved endosomal protein sorting complex, ESCRT-I. *Cell* 2001;**106**:145–55.

63. Amerik AY, Nowak J, Swaminathan S, Hochstrasser M. The Doa4 deubiquitinating enzyme is functionally linked to the vacuolar protein-sorting and endocytic pathways. *Mol Biol Cell* 2000;**11**:3365–80.

64. Katzmann DJ, Sarkar S, Chu T, Audhya A, Emr SD. Multivesicular body sorting: ubiquitin ligase Rsp5 is required for the modification and sorting of carboxypeptidase S. *Mol Biol Cell* 2004;**15**:468–80.

65. Coumailleau F, Das V, Alcover A, Raposo G, Vandormael-Pournin S, Le Bras S, Baldacci P, Dautry-Varsat A, Babinet C, Cohen-Tannoudji M. Over-expression of Rififylin, a new RING finger and FYVE-like domain-containing protein, inhibits recycling from the endocytic recycling compartment. *Mol Biol Cell* 2004;**15**:4444–56.

66. Liao W, Xiao Q, Tchikov V, Fujita K, Yang W, Wincovitch S, Garfield S, Conze D, El-Deiry WS, Schutze S, Srinivasula SM. CARP-2 is an endosome-associated ubiquitin ligase for RIP and regulates TNF-induced NF-kappaB activation. *Curr Biol* 2008;**18**:641–9.

67. Gary JD, Wurmser AE, Bonangelino CJ, Weisman LS, Emr SD. Fab1p is essential for PI(3)P 5-kinase activity and the maintenance of vacuolar size and membrane homeostasis. *J Cell Biol* 1998;**143**:65–79.

68. Efe JA, Botelho RJ, Emr SD. Atg18 regulates organelle morphology and Fab1 kinase activity independent of its membrane recruitment by phosphatidylinositol 3,5-bisphosphate. *Mol Biol Cell* 2007;**18**:4232–44.

69. Botelho RJ, Efe JA, Teis D, Emr SD. Assembly of a Fab1 phosphoinositide kinase signaling complex requires the Fig4 phosphoinositide phosphatase. *Mol Biol Cell*. 2008.

70. Sbrissa D, Ikonomov OC, Shisheva A. Phosphatidylinositol 3-phosphate-interacting domains in PIKfyve. Binding specificity and role in PIKfyve. Endomenbrane localization. *J Biol Chem* 2002;**277**:6073–9.

71. Taylor GS, Maehama T, Dixon JE. Inaugural article: myotubularin, a protein tyrosine phosphatase mutated in myotubular myopathy, dephosphorylates the lipid second messenger, phosphatidylinositol 3-phosphate. *Proc Natl Acad Sci USA* 2000;**97**:8910–15.

72. Kim SA, Taylor GS, Torgersen KM, Dixon JE. Myotubularin and MTMR2, phosphatidylinositol 3-phosphatases mutated in myotubular myopathy and type 4B Charcot-Marie-Tooth disease. *J Biol Chem* 2002;**277**:4526–31.

73. Lorenzo O, Urbe S, Clague MJ. Analysis of phosphoinositide binding domain properties within the myotubularin-related protein MTMR3. *J Cell Sci* 2005;**118**:2005–12.

74. Lorenzo O, Urbe S, Clague MJ. Systematic analysis of myotubularins: heteromeric interactions, subcellular localisation and endosome related functions. *J Cell Sci* 2006;**119**:2953–9.

75. Tsukazaki T, Chiang TA, Davison AF, Attisano L, Wrana JL. SARA, a FYVE domain protein that recruits Smad2 to the TGFbeta receptor. *Cell* 1998;**95**:779–91.

76. Miura S, Takeshita T, Asao H, Kimura Y, Murata K, Sasaki Y, Hanai JI, Beppu H, Tsukazaki T, Wrana JL, Miyazono K, Sugamura K. Hgs (Hrs), a FYVE domain protein, is involved in Smad signaling through cooperation with SARA. *Mol Cell Biol* 2000;**20**:9346–55.

77. Panopoulou E, Gillooly DJ, Wrana JL, Zerial M, Stenmark H, Murphy C, Fotsis T. Early endosomal regulation of Smad-dependent signaling in endothelial cells. *J Biol Chem* 2002;**277**:18,046–18,052.

78. Itoh F, Divecha N, Brocks L, Oomen L, Janssen H, Calafat J, Itoh S, Dijke Pt P. The FYVE domain in Smad anchor for receptor activation (SARA) is sufficient for localization of SARA in early endosomes and regulates TGF-beta/Smad signalling. *Genes Cells* 2002;**7**:321–31.

79. Itoh S, ten Dijke P. Negative regulation of TGF-beta receptor/Smad signal transduction. *Curr Opin Cell Biol* 2007;**19**:176–84.

80. Seet LF, Hong W. Endofin, an endosomal FYVE domain protein. *J Biol Chem* 2001;**276**:42,445–42,454.

81. Chen YG, Wang Z, Ma J, Zhang L, Lu Z. Endofin, a FYVE domain protein, interacts with Smad4 and facilitates transforming growth factor-beta signaling. *J Biol Chem* 2007;**282**:9688–95.

82. Seet LF, Hong W. Endofin recruits clathrin to early endosomes via TOM1. *J Cell Sci* 2005;**118**:575–87.

83. Seet LF, Liu N, Hanson BJ, Hong W. Endofin recruits TOM1 to endosomes. *J Biol Chem* 2004;**279**:4670–9.

84. Zheng Y, Fischer DJ, Santos MF, Tigyi G, Pasteris NG, Gorski JL, Xu Y. The faciogenital dysplasia gene product FGD1 functions as a Cdc42Hs-specific guanine-nucleotide exchange factor. *J Biol Chem* 1996;**271**:33,169–33,172.

85. Obaishi H, Nakanishi H, Mandai K, Satoh K, Satoh A, Takahashi K, Miyahara M, Nishioka H, Takaishi K, Takai Y. Frabin, a novel FGD1-related actin filament-binding protein capable of changing cell shape and activating c-Jun N-terminal kinase. *J Biol Chem* 1998;**273**:18,697–18,700.

86. Kim Y, Ikeda W, Nakanishi H, Tanaka Y, Takekuni K, Itoh S, Monden M, Takai Y. Association of frabin with specific actin and membrane structures. *Genes Cells* 2002;**7**:413–20.

87. Chen W, Li N, Chen T, Han Y, Li C, Wang Y, He W, Zhang L, Wan T, Cao X. The lysosome-associated apoptosis-inducing protein

containing the pleckstrin homology (PH) and FYVE domains (LAPF), representative of a novel family of PH and FYVE domain-containing proteins, induces caspase-independent apoptosis via the lysosomal-mitochondrial pathway. *J Biol Chem* 2005;**280**:40,985–40,995.

88. Tibbetts III MD, Shiozaki EN, Gu L, McDonald ER, El-Deiry WS, Shi Y. Crystal structure of a FYVE-type zinc finger domain from the caspase regulator CARP2. *Structure* 2004;**12**:2257–63.

89. Ellson CD, Anderson KE, Morgan G, Chilvers ER, Lipp P, Stephens LR, Hawkins PT. Phosphatidylinositol 3-phosphate is generated in phagosomal membranes. *Curr Biol* 2001;**11**:1631–5.

90. Sato TK, Overduin M, Emr SD. Location, location, location: membrane targeting directed by PX domains. *Science* 2001;**294**:1881–5.

Type I Phosphatidylinositol 4-Phosphate 5-Kinases (PI4P 5-kinases)

K.A. Hinchliffe and R.F. Irvine

Department of Pharmacology University of Cambridge, Cambridge, England, UK

INTRODUCTION

Type I phosphatidylinositol 4-phosphate 5-kinases (PI4P5Ks) phosphorylate phosphatidylinositol 4-phosphate (PI4P) in the 5-position to form phosphatidylinositol 4,5-bisphosphate (PI45P$_2$). Because metabolic evidence suggests that *in vivo* the major route of synthesis of PI45P2 in animal cells is by the 5-phosphorylation of PI4P, both in the plasma membrane [1, 2] and the nucleus [3] , type I PI4P5Ks are obviously the enzymes primarily responsible for regulating levels of this multifunctional lipid. In the test tube, type I PI4P5Ks have been reported to catalyze other reactions. For example, both Iα and Iβ isoforms can convert PI into PI5P [4], and PI3P into PI34P$_2$ [4, 5] or PI35P$_2$ [4], or even eventually PI345P$_3$ [4, 5]. A PI4P5K from Arabadopsis shows a similar flexibility when expressed in insect cells [6]. However, the 5-phosphorylation of PI4P is the major activity of the type I enzymes, and the physiological significance (or even natural occurrence) of these other reactions remains unclear. The exception is the demonstration that an endogenous type I PI4P5K (isoform unknown) makes a physiologically significant contribution to the synthesis of PI345P$_3$ from PI34P$_2$ in response to cell stress [7].

Several fuller reviews have discussed these enzymes directly or indirectly (see, for example, [8–10]).

BASIC PROPERTIES

Cloning

Our current understanding of type I PI4P5Ks is that there are three distinct mammalian isoforms, and no other obvious candidate emerges from a scan of the current human genome database. Nomenclature is rather confusing, as the type Iβ cloned from mouse [11] and the human isoform called type Iα cloned shortly afterwards by Loijens and Anderson [12] are exact orthologs (and similarly, mouse type Iα and human type Iβ). As the type Iγ [13] has come from the same species (mouse) and lab as the original cloning of the Iα and Iβ, we have in Figure 24.1 used the mouse nomenclature. The isoform that Carvajal *et al.* [14] identified as the STM7 gene, mapping close to the Friedrich's ataxia gene (X25), is the human type Iβ isoform.

Loijens and Anderson [12] reported two splice variants of the human type Iα and one of the type Iβ, and the mouse type Iγ also has at least two splice variants [13].

Structure

The lineup in Figure 24.1 tells a superficially simple story of a highly conserved central core, which consists of the catalytic site interspersed with some loops of significant variation between isoforms, and then virtually no sequence similarity whatsoever between the isoforms at the C-and N-termini. The latter, in turn, implies diverse isoform-specific regulation, which has implications in the discussions about regulation, below.

There is also a close similarity in the catalytic "core" with the yeast gene Mss4 [11, 15], but again no similarity in other parts of the sequence between the yeast and the mammalian enzymes. There is a limited amount of similarity with the members of the Fab1 family, both yeast and mammalian; these are PI3P 5-kinases, now given the name type III PIP kinases. Also, there is similarity in the catalytic core with the type II PI5P4Ks, and from this, and from the X-ray structure of the type IIβ PI5P4K [16], some deductions can been made about probable crucial residues for catalytic activity in the type I enzymes. Ishihara *et al.* showed that lysine 138 of the type Iα PI4P 5 K, which they identified as being in the putative ATP-binding site, is essential for catalytic activity [13].

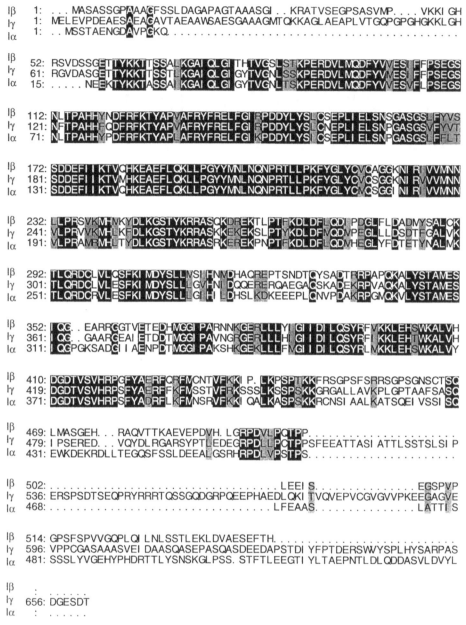

Iβ   1: ...MASASSGPAAACFSSLDAGAPAGTAAASGI..KRATVSEGPSASVMP....VKKIGH
Iγ   1: MELEVPDEAESAEACAVTAEEAAWSAESGAAAGMTQKKAGLAEAPLVTGQPGPGHGKKLGH
Iα   1: ..MSSTAENGDAVPCKQ..

Iβ 52: RSVDSSGETTYKKTTSSALKGAIQLGITHTVGSLSTKPERDVLMQDFYVVESIFFPSEGS
Iγ 61: RGVDASGETTYKKTTSSTLKGAIQLGIGYTVGNLSSKPERDVLMQDFYVVESIFFPSEGS
Iα 15: .....NEEKTYKKTASSAIKGAIQLGIGYTVGNLTSKPERDVLMQDFYVVESVFLPSEGS

Iβ 112: NLTPAHHYNDFRFKTYAPVAFRYFRELFGIRPDDYLYSLCSEPLIELSNSGASGSLFYVS
Iγ 121: NFTPAHHFCDFRFKTYAPVAFRYFRELFGIRPDDYLYSLCNEPLIELSNPGASGSVFYYT
Iα 71: NLTPAHHYPDFRFKTYAPLAFRYFRELFGIKPDDYLYSICSEPLIELSNPGASGSLFFLT

Iβ 172: SDDEFIIKTVCHKEAEFLQKLLPGYYMNLNQNPRTLLPKFYGLYCVQAGGKNIRIVVMNN
Iγ 181: SDDEFIIKTVMHKEAEFLQKLLPGYYMNLNQNPRTLLPKFYGLYCVQSGGKNIRVVVMNN
Iα 131: SDDEFIIKTVCHKEAEFLQKLLPGYYMNLNQNPRTLLPKFYGLYCMQSGGININIRIVVMNN

Iβ 232: LLPRSVKMHVKYDLKGSTYKRRASCKDREKTLPTFKDLDFLQDIPDGLFLDADMYSALCK
Iγ 241: VLPRVVKMHLKFDLKGSTYKRRASKKEKEKSLPTYKDLDFMQDNPEGLLLDSDTFGALVK
Iα 191: VLPRAMRMHLTYDLKGSTYKRRASRKEREKPNPTFKDLDFLQDVHEGLYFDTETYNALMK

Iβ 292: TLQRDCLVLCSFKIMDYSLLVSIFNMDHAQREPTSNDTQYSADTRRPAPCKALYSTAMES
Iγ 301: TLQRDCLVLESFKIMDYSLLLGVHNIDQQERERQAEGACSKADEKRPVAQKALYSTAMES
Iα 251: TLQRDCRVLESFKIMDYSLLLGIHILDHSLKDKEEEPLQNVPDAKRPGMQKVLYSTAMES

Iβ 352: ICG..EARRCGTVETEDHVGGIPARNNKGERLLLYIGIIDILQSYRFVKKLEHSWKALVH
Iγ 361: ICG..GAARCEAIETDDTVGGIPAVNGRGERLLLHIGIIDILQSYRFIKKLEHIWKALVH
Iα 311: ICGPGKSADCIIAENPDTVGGIPAKSHKGEKLLLFMGIIDILQSYRLMKKLEHSWKALVY

Iβ 410: DGDTVSVHRPGFYAERFQRFMCNTVFKKIP.LKPSPTKKFRSGPSFSRRSGPSGNSCTSC
Iγ 419: DGDTVSVHRPSFYAERFFKFMSSTVFRKSSSLKSSPSKKGRGALLAVKPLGPTAAFSASC
Iα 371: DGDTVSVHRPSFYADRFLKFMNSRVFKKIQALKASPSKKRCNSIAALKATSQEIVSSISC

Iβ 469: LMASGEH...RAQVTTKAEVEPDVH.LGRPDVLPCTPP...................
Iγ 479: IPSERED...VQYDLRGARSYPTLEDEGRPDLLPCTPPSFEEATTASIATTLSSTSLSIP
Iα 431: EWKDEKRDLLTEGQSFSSLDEEALGSRHRPDLVPSTPS...................

Iβ 502: ....LEEIS...........EGSPVP
Iγ 536: ERSPSDTSEQPRYRRRTQSSGQDGRPQEEPHAEDLQKITVQVEPVCGVGVVPKEEGAGVE
Iα 468: ....LFEAAS...........LATTIS

Iβ 514: GPSFSPVVGQPLQILNLSSTLEKLDVAESEFTH..........................
Iγ 596: VPPCGASAAASVEIDAASQASEPASQASDEEDAPSTDIYFPTDERSWYYSPLHYSARPAS
Iα 481: SSSLYVGEHYPHDRTTLYSNSKGLPSS.STFTLEEGTIYLTAEPNTLDLQDDASVLDVYL

Iβ   : ....
Iγ 656: DGESDT
Iα   : ....

FIGURE 24.1 The sequences of the mouse type I PI4P5Ks are shown (Genebank accession numbers: Iβ, NM_008847; Iγ, NM_008844; Iα, NM_008846).

Substrate Specificity

Kunz et al. [17] have shown that the substrate specificity of the type I and II PIP kinases is dictated largely by their "activation loop," that is, transferring the activation loop from type IIβ PIPK into type Iβ (human) converted the type I enzyme into a PI5P4K activity (the activation loop of the orthologous mouse type Iα is residues 347–387 in Figure 24.1). The converse (converting a type II PIPK into a PI4P5K activity by inserting a type I loop) was also observed. These observations have recently been taken a stage further by some elegant site-directed changes in this

loop [18]. A remarkable finding is that changing a single residue, glutamate 362 in human type Iβ (equivalent to E362 in murine Iα, see Figure 24.1), to an alanine transformed its substrate specificity to that resembling a type II activity in that it would use PI5P as a substrate (though its activity against PI4P was diminished rather than lost). Kunz et al. have suggested that the activation loop might fold into an α-helix in vivo [18]. The structure of type IIβ PI5P4K [16] suggests how the activation loop might lie adjacent to the presumed active site where the PI5P substrate head-group binds. Although the activation loop did not crystallize [16], it seems likely that it will influence the orientation of two loops that link contiguous β strands; both of these loops contribute residues that interact with the inositol 1,5 bisphosphate moiety of the substrate [16].

Localization

These latter studies on substrate specificity [18] have implications also for the localization of the type I enzymes. So far, localization studies have suggested that they are all primarily in the plasma membrane (though see below for some possible regulation of this). The combined data of the two papers on the influence of the activation loop [17, 18] demonstrated that changing the substrate specificity from favoring PI4P to PI5P also changes the localization of the type Iβ enzyme from plasma membrane to cytosol. This in turn implies that the plasma membrane localization is governed primarily by interaction with the PI4P substrate. This conclusion is subject to the caveat that these studies use transfection, which might saturate endogenous (protein) binding sites in locations other than the plasma membrane, and it is then the "excess" that is being visualized, bound to its substrate.

Chatah and Abrams have reported a different localization of human type Iβ PI4P5K, that is, perinuclear, with a translocation to the plasma membrane after prolonged activation of the cells [19]. Our own experience is that there is some variation in subcellular localization of transfected type I PI4P5Ks between cell types, and also some dependence on culture conditions and length of transfection time. There is still a lot to learn about the localization *in vivo* of endogenous type I PI4P5Ks and how it is controlled.

REGULATION

Given their self-evident role in cell regulation, it is not surprising that type I PI4P5Ks have been found to be subject to a variety of regulatory influences. Only a brief summary of the literature to the end of 2001 is possible here.

Phosphatidic Acid

This lipid has long been known to be a potent stimulator of type I (but not type II) PIPKs [20]. Under some circumstances it can be essential – for example, Honda *et al.* could only see the effects of Arf-6 (below) if PA was supplied [21]. Jones *et al.* [22] have produced evidence that endogenous PA may be a significant regulator of type I PI4P5Ks *in vivo*. PA is of course the product of PLD, itself an enzyme frequently tied in with PI45P₂ and with type I PI4P5Ks (e.g. [23]), and it may be that the two enzymes have a complex inter-regulatory relationship.

Monomeric G Proteins

There is abundant evidence that members of the Rho and Arf family can regulate type I PI4P5Ks, though to a significant degree we do not know the physiological veracity of these events, nor the isoform involved. The clear difference between the three isoforms (Figure 24.1) raises the possibility that *in vivo* there may be significant specificity in the G-protein–PI4P5K interaction.

Arguably the strongest evidence supports regulation by members of the Arf family. For example, Honda *et al.* [21] purified from brain cytosol the major GTPγS-dependent activator of murine type Iα PI4P5K, and found it to be Arf-1. They went on to show that its localization in HeLa cells was not consistent with its being a natural regulator of type Iα PI4P5K (Arf-1 being predominantly in the Golgi in these cells), but rather that Arf-6 fitted the bill under all the criteria they addressed. Martin *et al.* [24] also thought that Arf and not Rho (see below) was the endogenous regulator of a type I PI4P5K (isoform unknown). Brown *et al.* [25] have implicated Arf-6 in endosome formation, and showed that human type Iα PI4P5K can mimic the effects of a constitutively active Arf-6 (though again the endogenous type I PI4P5K is unknown). Arf-1 may regulate PI45P₂ synthesis in the Golgi, though in these experiments it most likely recruited the type I PI4P5K from the cytosol [26, 27].

There is also a reasonable case for type I PI4P5K activation by Rho family members, though it is sometimes confusing. Thus using the Rho-specific C3 Botulinum toxin, Chong *et al.* [28] implied that Rho regulates a type I PI4P5K activity in fibroblasts, whereas others have failed to see a Rho-type I PI4P5K interaction in experiments where it did interact directly with Rac [29]. Rac interaction with type I PI4P5Ks has been suggested in other experiments [30, 31], and there are convincing data placing type Iα or Iβ PI4P5Ks in the signaling pathway from the thrombin receptor, via Rac, to actin polymerization [32]. For the most part the evidence for Rho involvement still remains indirect [33]. Some of these simplistic contradictions may be due to differences in isoforms, though Honda *et al.* [21] stated that this was unlikely to be the reason they could not see an effect of Rho in their experiments. An interaction with RhoGDI has also been reported [31], and we think that a fair summary of the state of play is that the involvement of monomeric G-proteins in regulation of type I PI4P5Ks is real, and important, but incompletely understood.

Phosphorylation

Several protein kinases have been reported to associate with or regulate type I PI4P5Ks; for example, casein kinase I in *S. pombe* [34], Rho-kinase [35] (which might explain some of the contradictions about Rho, though see ref [33]), and PKCμ (also known as PKD) [36]. Also, Park *et al.* [37] showed that all three type I PI4P5Ks can be negatively regulated by PKA, and also suggested that receptor activation led to a dephosphorylation (and thus activation) by an uncharacterized mechanism that may involve PKC. Wenk *et al.* [38] have shown a stimulation-dependent dephosphorylation of type Iγ PI4P5K in synapses (where it is the major type I isoform). Another intriguing possible regulatory mechanism has been suggested by Itoh *et al.* [39]. All three isoforms of type I PI4P5K are capable of autophosphorylation, an activity that is stimulated by PI and that leads to an inhibition of the enzyme's activity against PI4P. The physiological relevance of this awaits further study, as does the even more intriguing (and as yet untested) possibility that, like some of the type I PI3Ks [40], type I PI4P5Ks might phosphorylate other proteins.

Other Regulation Mechanisms

Mejillano *et al.* [41] have suggested that human type Iα PI4P5K is cleaved by caspase during apoptosis – an event that, because they also suggest PI45P$_2$ to be anti-apoptotic, serves as part of the amplification of the apoptotic process once it has started. Recently, Barbieri *et al.* have shown an isomeric specificity for the involvement of type I PI4P5Ks in EGF receptor-mediated endocytosis [42], in that the mouse type Iβ PI4P5K was required but the type Iα was not. How the type Iβ is regulated in this process is an intriguing question for further exploration.

FUNCTION

The physiological role of type I PI4P 5Ks is self-evidently well established (in contrast with the more enigmatic type II PI5P 4-kinases; see Chapter 131 of Handbook of Cell Signaling, Second Edition), because their primary function is to synthesize PI45P$_2$. Thus the question, what is the function of type I PI4P 5Ks, is essentially the same as the question, what is the function of PI45P$_2$. This is now a huge topic, with upwards of 20 suggested physiological functions (see, for example, [8, 10] for reviews), and therefore is outside the scope of this short review.

ACKNOWLEDGEMENTS

K. A. Hinchliffe is supported by the MRC, and R. F. Irvine is supported by the Royal Society.

REFERENCES

1. King CE, Stephens LR, Hawkins PT, Guy GR, Michell RH. Multiple metabolic pools of phosphoinositides and phosphatidate in human erythrocytes incubated in a medium that permits rapid transmembrane exchange of phosphate. *Biochem J* 1987;**244**:209–17.

2. Stephens LR, Hughes KT, Irvine RF. Pathway of phosphatidylinositol (3,4,5)-trisphosphate synthesis in activated neutrophils. *Nature* 1991; **351**:33–9.

3. Vann LR, Wooding FB, Irvine RF, Divecha N. Metabolism and possible compartmentalization of inositol lipids in isolated rat-liver nuclei. *Biochem J* 1997;**327**:569–76.

4. Tolias KF, Rameh LE, Ishihara H, Shibasaki Y, Chen J, Prestwich GD, Cantley LC, Carpenter CL. Type I phosphatidylinositol-4-phosphate 5-kinases synthesize the novel lipids phosphatidylinositol 3,5-bisphosphate and phosphatidylinositol 5-phosphate. *J Biol Chem* 1998;**273**:18,040–6.

5. Zhang X, Loijens JC, Boronenkov IV, Parker GJ, Norris FA, Chen J, Thum O, Prestwich GD, Majerus PW, Anderson RA. Phosphatidylinositol-4-phosphate 5-kinase isozymes catalyze the synthesis of 3-phosphate-containing phosphatidylinositol signaling molecules. *J Biol Chem* 1997;**272**:17,756–61.

6. Elge S, Brearley C, Xia HJ, Kehr J, Xue HW, Mueller-Roeber B. An Arabidopsis inositol phospholipid kinase strongly expressed in procambial cells: synthesis of PtdIns(4,5)P2 and PtdIns(3,4,5)P3 in insect cells by 5-phosphorylation of precursors. *Plant J* 2001;**26**:561–71.

7. Halstead JR, Roefs M, Ellson CD, D'Andrea S, Chen C, D'Santos CS, Divecha N. A novel pathway of cellular phosphatidylinositol(3,4,5)-trisphosphate synthesis is regulated by oxidative stress. *Curr Biol* 2001;**11**:386–95.

8. Hinchliffe KA, Ciruela A, Irvine RF. PIPkins, their substrates and their products: new functions for old enzymes. *Biochim Biophys Acta* 1998;**1436**:87–104.

9. Anderson RA, Boronenkov IV, Doughman SD, Kunz J, Loijens JC. Phosphatidylinositol phosphate kinases, a multifaceted family of signaling enzymes. *J Biol Chem* 1999;**274**:9907–10.

10. Martin TF. Phosphoinositide lipids as signaling molecules: common themes for signal transduction, cytoskeletal regulation, and membrane trafficking. *Annu Rev Cell Dev Biol* 1998:14,231–64.

11. Ishihara H, Shibasaki Y, Kizuki N, Katagiri H, Yazaki Y, Asano T, Oka Y. Cloning of cDNAs encoding two isoforms of 68-kDa type I phosphatidylinositol-4-phosphate 5-kinase. *J Biol Chem* 1996;**271**:23,611–14.

12. Loijens JC, Anderson RA. Type I phosphatidylinositol-4-phosphate 5-kinases are distinct members of this novel lipid kinase family. *J Biol Chem* 1996;**271**:32,937–43.

13. Ishihara H, Shibasaki Y, Kizuki N, Wada T, Yazaki Y, Asano T, Oka Y. Type I phosphatidylinositol-4-phosphate 5-kinases. Cloning of the third isoform and deletion/substitution analysis of members of this novel lipid kinase family. *J Biol Chem* 1998;**273**:8741–8.

14. Carvajal JJ, Pook MA, dos Santos M, et al. The Friedreich's ataxia gene encodes a novel phosphatidylinositol-4-phosphate 5-kinase. *Nat Genet* 1996;**14**:157–62.

15. Yoshida S, Ohya Y, Nakano A, Anraku Y. Genetic interactions among genes involved in the STT4-PKC1 pathway of *Saccharomyces cerevisiae*. *Mol Gen Genet* 1994;**242**:631–40.

16. Rao VD, Misra S, Boronenkov IV, Anderson RA, Hurley JH. Structure of type II beta phosphatidylinositol phosphate kinase: a protein kinase fold flattened for interfacial phosphorylation. *Cell* 1998;**94**:829–39.

17. Kunz J, Wilson MP, Kisseleva M, Hurley JH, Majerus PW, Anderson RA. The activation loop of phosphatidylinositol phosphate kinases determines signaling specificity. *Mol Cell* 2000;**5**:1–11.

18. Kunz J, Fuelling A, Kolbe L, Andeson RA. Stereospecific substrate recognition by phosphatidylinositol phosphate kinases is swapped by changing a single amino acid residue. *J Biol Chem* 2002;**277**: 5611–19.

19. Chatah NE, Abrams CS. G-protein-coupled receptor activation induces the membrane translocation and activation of phosphatidylinositol-4-phosphate 5-kinase I alpha by a Rac- and Rho-dependent pathway. *J Biol Chem* 2001;**276**:34,059–65.

Chapter | 24 Type I Phosphatidylinositol 4-Phosphate 5-Kinases (PI4P 5-kinases) 273

20. Jenkins GH, Fisette PL, Anderson RA. Type I phosphatidylinositol 4-phosphate 5-kinase isoforms are specifically stimulated by phosphatidic acid. *J Biol Chem* 1994;**269**:11,547–54.

21. Honda A, Nogami M, Yokozeki T, Yamazaki M, et al. Phosphatidylinositol 4-phosphate 5-kinase alpha is a downstream effector of the small G protein ARF6 in membrane ruffle formation. *Cell* 1999;**99**:521–32.

22. Jones DR, Sanjuan MA, Merida I. Type I alpha phosphatidylinositol 4-phosphate 5-kinase is a putative target for increased intracellular phosphatidic acid. *FEBS Letts4* 2000;**76**:160–5.

23. Divecha N, Roefs M, Halstead JR, et al. Interaction of the type I alpha PIP kinase with phospholipase D: a role for the local generation of phosphatidylinositol 4,5-bisphosphate in the regulation of PLD2 activity. *EMBO J* 2000;**19**:5440–9.

24. Martin A, Brown FD, Hodgkin MN, Bradwell AJ, Cook SJ, Hart M, Wakelam MJ. Activation of phospholipase D and phosphatidylinositol 4-phosphate 5-kinase in HL60 membranes is mediated by endogenous Arf but not Rho. *J Biol Chem* 1996;**271**:17,397–403.

25. Brown FD, Rozelle AL, Yin HL, Balla T, Donaldson JG. Phosphatidylinositol 4,5-bisphosphate and Arf6-regulated membrane traffic. *J Cell Biol* 2001;**154**:1007–17.

26. Godi A, Pertile P, Meyers R, et al. ARF mediates recruitment of PtdIns-4-OH kinase-beta and stimulates synthesis of PtdIns(4,5)P2 on the Golgi complex. *Nat Cell Biol* 1999;**1**:280–7.

27. Jones DH, Morris JB, Morgan CP, Kondo H, Irvine RF, Cockcroft S. Type I phosphatidylinositol 4-phosphate 5-kinase directly interacts with ADP-ribosylation factor 1 and is responsible for phosphatidylinositol 4,5-bisphosphate synthesis in the golgi compartment. *J Biol Chem* 2000;**275**:13,962–6.

28. Chong LD, Traynor-Kaplan A, Bokoch GM, Schwartz MA. The small GTP-binding protein Rho regulates a phosphatidyl inositol 4-phosphate 5-kinase in mammalian cells. *Cell* 1994;**79**:507–13.

29. Tolias KF, Cantley LC, Carpenter CL. Rho family GTPases bind to phosphoinositide kinases. *J Biol Chem* 1995;**270**:17,656–9.

30. Hartwig JH, Bokoch GM, Carpenter CL, Janmey PA, Taylor LA, Toker A, Stossel TP. Thrombin receptor ligation and activated Rac uncap actin filament barbed ends through phosphoinositide synthesis in permeabilized human platelets. *Cell* 1995;**82**:643–53.

31. Tolias KF, Couvillon AD, Cantley LC, Carpenter CL. Characterization of a Rac1- and RhoGDI-associated lipid kinase signaling complex. *Mol Cell Biol* 1998;**18**:762–70.

32. Tolias KF, Hartwig JH, Ishihara H, Shibasaki Y, Cantley LC, Carpenter CL. Type Ialpha phosphatidylinositol-4- phosphate 5-kinase mediates Rac-dependent actin assembly. *Curr Biol* 2000;**10**:153–6.

33. Matsui T, Yonemura S, Tsukita S. Activation of ERM proteins *in vivo* by Rho involves phosphatidyl-inositol 4-phosphate 5-kinase and not ROCK kinases. *Curr Biol* 1999;**9**:1259–62.

34. Vancurova I, Choi JH, Lin H, Kuret J, Vancura A. Regulation of phosphatidylinositol 4-phosphate 5-kinase from Schizosaccharomyces pombe by casein kinase I. *J Biol Chem* 1999;**274**:1147–55.

35. Oude Weernink PA, Schulte P, Guo Y, et al. Stimulation of phosphatidyl inositol-4-phosphate 5-kinase by Rho-kinase. *J Biol Chem* 2000;**275**:10,168–74.

36. Nishikawa K, Toker A, Wong K, Marignani PA, Johannes FJ, Cantley LC. Association of protein kinase Cmu with type II phosphatidylinositol 4-kinase and type I phosphatidylinositol-4-phosphate 5-kinase. *J Biol Chem* 1998;**273**:23,126–33.

37. Park SJ, Itoh T, Takenawa T. Phosphatidylinositol 4-phosphate 5-kinase type I is regulated through phosphorylation response by extracellular stimuli. *J Biol Chem* 2001;**276**:4781–7.

38. Wenk MR, Pellegrini L, Klenchin VA, et al. Pip kinase igamma is the major pi(4,5)p(2) synthesizing enzyme at the synapse. *Neuron* 2001;**32**:79–88.

39. Itoh T, Ishihara H, Shibasaki Y, Oka Y, Takenawa T. Autophosphorylation of type I phosphatidylinositol phosphate kinase regulates its lipid kinase activity. *J Biol Chem* 2000;**275**:19,389–94.

40. Bondeva T, Pirola L, Bulgarelli Leva G, Rubio I, Wetzker R, Wymann MP. Bifurcation of lipid and protein kinase signals of PI3Kgamma to the protein kinases PKB and MAPK. *Science* 1998;**282**:293–6.

41. Mejillano M, Yamamoto M, Rozelle AL, Sun HQ, Wang X, Yin HL. Regulation of apoptosis by phosphatidylinositol 4,5-bisphosphate inhibition of caspases, and caspase inactivation of phosphatidylinositol phosphate 5-kinases. *J Biol Chem* 2001;**276**:1865–72.

42. Barbieri MA, Heath CM, Peters EM, Wells A, Davis JN, Stahl PD. Phosphatidylinositol-4-phosphate 5-kinase-1beta is essential for epidermal growth factor receptor-mediated endocytosis. *J Biol Chem* 2001;**276**:47,212–16.

Chapter 25

Phosphoinositide 3-Kinases

David A. Fruman

Department of Molecular Biology & Biochemistry, and Center for Immunology, University of California, Irvine, California

INTRODUCTION

This chapter is intended to provide a brief summary of PI3K signaling, with an emphasis on recent advances. Length restrictions prevent a comprehensive review of this broad and rapidly advancing research field. The reader is referred in the text to several excellent reviews that provide more detail on specific topics.

THE ENZYMES

Phosphoinositide 3-kinase (PI3K) isoforms have been divided into three classes that differ in subunit structure, substrate selectivity, and regulation [1, 2]. Class I PI3Ks exist as heterodimers with a tightly bound regulatory (also termed adaptor) subunit (Figure 25.1). Each class I catalytic subunit possesses an adaptor binding domain, a Ras binding domain, a C2 domain, a helical domain, and a kinase domain. The class IA subgroup (p110α, p110β and p110δ) associates with regulatory subunits (p85α, p55α, p50α, p85β or p55γ) that have multiple modular protein–protein interaction domains (Figure 25.1). Class IA PI3Ks function downstream of receptors with intrinsic or associated tyrosine kinase (TyrK) activity. TyrKs activate class IA PI3Ks through at least two mechanisms: tyrosine phosphorylation (pTyr) and increased GTP loading of Ras [3, 4]. Class IA PI3Ks bind to pTyr-containing proteins via Src homology-2 (SH2) domains in the regulatory subunits, and Ras-GTP binds to the catalytic subunits. For many transmembrane receptors, mutation of specific tyrosine residues that bind PI3K impairs receptor function. Ras-induced tumorigenesis is diminished in mice with a knock-in mutation in the Ras-binding domain of p110α [5], illustrating the importance of PI3K as a Ras effector.

The class IA regulatory subunits have complex roles in PI3K function. These subunits stabilize the catalytic subunits but maintain them in a low activity conformation in the basal state (Figure 25.2). Binding of the Src-homology 2 (SH2) domains of the regulatory subunit to pTyr-containing peptides releases the intramolecular inhibition; since the pTyr-containing proteins are often membrane-associated, this interaction also increases access of the catalytic subunit to substrates and to Ras-GTP. Other domains of the regulatory subunits may also contribute to activation or localization. On the other hand, various modular domains in the regulatory subunits participate in negative regulatory circuits that dampen activation of PI3K and other crucial signaling components [6].

The class IB enzyme (p110γ) interacts with a distinct regulatory subunit, either p101 or p84 (Figure 25.1) [7]. The class IB enzyme is activated by βγ subunits of heterotrimeric G proteins following engagement of G-protein-coupled receptors (GPCRs). p110γ is enriched in leukocytes and its catalytic activity is required for many functions in the immune system [7]. Both p101 and the Ras-binding function of p110γ are required for neutrophil migration in response to chemokines and bacterial products [8]. p110γ also has non-catalytic scaffolding functions that are important for cardiac function [9]. By a mechanism that is unclear, the p110β isoform of the class IA subgroup also functions downstream of GPCRs, and has an unexpectedly limited role in TyrK signaling [10–12].

Crystal structures have been solved for p110α in complex with portions of p85α, and for p110γ [13–15]. The p110γ structure showed that the helical domain acts as a scaffold or spine on which the other three functional domains are organized [15]. The kinase domain is similar in structure to protein kinases, as predicted from primary sequence comparison and limited protein kinase activity of PI3K enzymes. The shape of the substrate-binding pocket helps to explain the selectivity for phosphoinositide recognition and the likely basis for differential recognition of singly and multiply phosphorylated substrates by different PI3K classes. Subsequent structural work has clarified the mode of binding of various PI3K inhibitors to the active

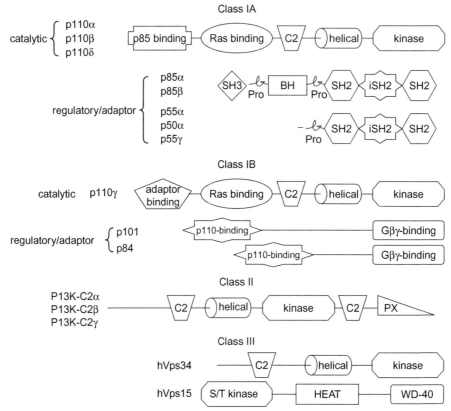

FIGURE 25.1 Schematic diagram of the domain structure of mammalian PI3Ks.
The common names of the proteins are listed at the left of each structure (hVps34 and hVps15 are the human homologs of class III PI3K (Vps34) and its associated serine/threonine kinase (Vps15), originally cloned from *S. cerevisiae*). Different modular domains are denoted by distinct shapes and/or by their common abbreviations. The three class IA catalytic subunits associated with each of the five regulatory isoforms without any apparent preference. p85α, p55α and p50α are alternative transcripts of a single gene. The class IB regulatory isoform p84 is sometimes referred to as p87, and its domain structure is inferred by homology to structural information obtained from a study of p101 [129]. Abbreviations: SH3, Src-homology 3; BH, BCR homology (also known as RhoGAP homology); pro, proline-rich motif; SH2, Src-homology 2; iSH2, inter-SH2 domain; PX, Phox homology.

site of p110γ, and has provided a model for allosteric activation by Ras [16–18]. Structural studies of the p110α isoform bound to fragments of p85α provided long-sought answers to how the iSH2 domain of p85α interacts with the N-terminal region of p110α, and how cancer-associated mutations might increase kinase activity [13, 14]. One finding was that mutations affecting the adaptor-binding domain (ABD) are in residues of a surface that contacts the kinase domain of the same (catalytic) subunit, rather than at the interface with iSH2 of the regulatory subunit. These structures also revealed that the catalytic and regulatory subunits make at least two important contacts other than the ABD–iSH2 interaction: the helical domain of p110α interacts with the N-SH2 domain of p85α, and the C2 domain of p110α contacts the iSH2 domain of p85α (Figure 25.2). Mutations that weaken these intersubunit contacts or the ABD–kinase domain interaction are likely to alter the conformation of the kinase domain to increase enzyme activity.

Class II PI3Ks are distinguished by the presence of an additional C-terminal C2 domain and a Phox homology (PX) domain (Figure 25.1). Although comparatively little is known about class II PI3K regulation, there is growing evidence that these enzymes can be activated by extracellular signals [19, 20]. Genes for class I and class II enzymes have been found in all multicellular animals. Class III PI3Ks are found in all eukaryotes from yeast to humans, and associate with a serine kinase in both organisms (yeast vps15p and human p150). These enzymes appear to have a housekeeping function related to vesicular transport and protein sorting [2, 21]. In addition, class III PI3Ks participate in nutrient sensing pathways and regulation of autophagy [22].

THE PRODUCTS

PI3Ks phosphorylate the 3′-hydroxyl of the D-*myo*-inositol ring of phosphatidylinositol (PtdIns) (Figure 25.3a). Four distinct 3′-phosphoinositides (3′PIs) exist in mammalian cells: PtdIns(3)P, PtdIns(3, 4)P_2, PtdIns(3, 5)P_2, and PtdIns(3, 4, 5)P_3. The pathways of synthesis and degradation

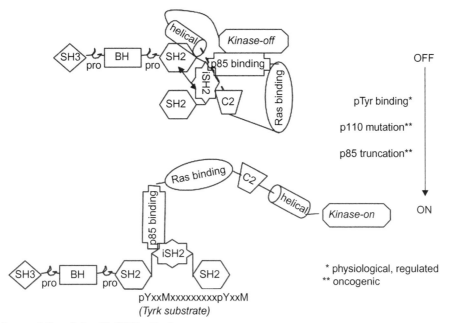

FIGURE 25.2 Model for regulation of class IA PI3K activation.
There is a constitutive, high-affinity interaction between the p85-binding domain in p110 (also known as the adaptor-binding domain, ABD) and the iSH2 domain of the regulatory subunit (shown as p85 in this figure). There are also several intermolecular and intramolecular contacts that regulate kinase activity. In the low activity state, the p110 helical domain contacts the p85 N-SH2 domain, the p110 C2 domain contacts the p85 iSH2 domain, and within p110 the p85-binding domain contacts the kinase domain. An intramolecular interaction in p85 between the N-SH2 and iSH2 has also been reported (two-way arrow) [130]. Release of these interactions switches the enzyme to the high activity state. There are at least three mechanisms to accomplish this "OFF–ON" switch, as indicated. Binding of SH2 domains to pTyr peptides is a physiological mechanism during tyrosine kinase signaling. Oncogenic mutations in p85 (truncation, iSH2 deletions) or p110 (point mutation in helical domain, p85-binding domain or C2 domain) can also promote the active conformation. The N-terminal domains of p85 might also regulate kinase activity directly, or indirectly via localization. Further details of the three-dimensional structures can be found in references [10, 11].

of these lipids have been reviewed in detail [2, 23] and are summarized in Figure 3B. PtdIns(3, 4, 5)P$_3$ is produced only by class I PI3Ks using PtdIns(4, 5)P$_2$ as a substrate. PtdIns(3, 4)P$_2$ can be generated from PtdIns(4)P by class I or class II PI3Ks, or by 5′-phosphatase action on PtdIns(3, 4, 5)P$_3$, or by a PtdIns(3)P-4-kinase. The signaling functions of PtdIns(3, 4, 5)P$_3$ and PtdIns(3, 4)P$_2$ are well studied, and will be discussed further below. Although PtdIns(3)P can be produced by all PI3Ks *in vitro*, the majority of PtdIns(3)P in cells appears to be made by class III PI3Ks and is detected primarily in endosomal vesicles [24]. PtdIns(3, 5)P$_2$ is generated *in vivo* by a PtdIns(3) P-5-kinase called PIKfyve, and like PtdIns(3)P is involved in vesicle trafficking [25, 26].

PHOSPHATASES

The membrane-targeting signal provided by 3′PIs can be modulated by the action of phosphoinositide phosphatases (PPases). PTEN (*P*hosphatase and *TEN*sin homology deleted on chromosome 10) hydrolyzes the 3′-phosphate of PtdIns(3, 4, 5)P$_3$ and PtdIns(3, 4)P$_2$, effectively reversing the action of PI3K (Figure 25.3b) [2]. SHIP1 and SHIP2 are related 5′-PPases that contain N-terminal SH2 domains

(*SH*2-containing *I*nositol polyphosphate 5-*P*hosphatase). SHIPs can remove the 5′-phosphate from PtdIns(3, 4, 5)P$_3$ to produce PtdIns(3, 4)P$_2$ (Figure 25.3b) [2, 27]. Hence, these enzymes may alter the spectrum of PI3K effectors recruited to the membrane rather than simply turning the signal off. The SH2 domains of SHIP1 and SHIP2 are selective for phosphotyrosines within a particular sequence context known as the immunoreceptor tyrosine-based inhibitory motif (ITIM). ITIMs are found in a number of receptors (e.g., FcγRIIB) whose ligation attenuates signaling through antigen receptors [27]. The myotubularin family of enzymes, called MTMRs (*Myo*Tubular *My*opathy *R*elated), selectively remove the 3′-phosphate from PtdIns(3)P and PtdIns(3, 5)P$_2$ [28].

LIPID BINDING DOMAINS

Pleckstrin homology (PH) domains are small (60–100aa) protein modules that mediate protein–lipid and protein–protein interactions. A subset of PH domains binds selectively to 3′PIs [29–31]. It is important to note that PtdIns(4)P and PtdIns(4, 5)P$_2$ are much more abundant in cellular membranes compared to 3′PIs; thus, for a PH domain to be considered 3′PI-specific, the binding affinity for 3′PIs lipids must be considerably higher than the

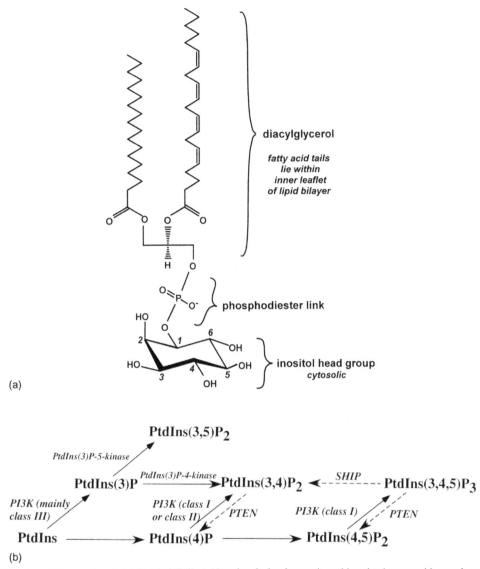

(a)

(b)

FIGURE 25.3 (a) Structure of D-*myo*-phosphatidylinositol (PtdIns). Note that the head group is positioned to interact with cytoplasmic molecules. Although free hydroxyls exist at positions 2–6, phosphorylation *in vivo* has only been detected at positions 3, 4, and 5. Reprinted, with permission, from the *Annual Review of Biochemistry,* Volume 70 ©2001, by Annual Reviews www.annualreviews.org (b) Pathways of synthesis and degradation of 3′PIs. The major enzymes responsible for particular reactions are indicated. For simplicity, many of the enzymes involved in metabolism of other phosphoinositides are omitted.

affinity for PtdIns(4)P or PtdIns(4, 5)P$_2$. Subgroups of PH domains show greater affinity for either PtdIns(3, 4, 5)P$_3$ or PtdIns(3, 4)P$_2$, while others bind these lipids comparably. 3′PI-selective PH domains are found in a variety of proteins involved in signal transduction (Figure 25.4), some of which are discussed in the next section.

PX domains are found in diverse proteins involved in vesicle trafficking, protein sorting, and signal transduction [30] (Figure 25.4). Like PH domains, different PX domains exhibit selectivity for different phosphoinositides. The crystal structure of the PX domain of p40phox bound to PtdIns(3)P shows that the lipid binds in a positively charged pocket and suggests how phosphoinositide binding specificity is determined [32].

The FYVE domain, originally identified in several yeast proteins, is a protein module that binds selectively to PtdIns(3)P [30]. Although some mammalian FYVE domain-containing proteins are involved in signal transduction, the primary role of most mammalian and all yeast proteins with this module is to mediate membrane trafficking (Figure 25.4).

Additional modular domains that can bind selectively to 3′PIs include the Sec14 homology domain and the PROPPIN domain [30, 33, 34].

The identification of 3′PI-selective modular domains has been exploited to create a panel of fluorescent bioprobes for spatiotemporal study of PI3K activation by fluorescence microscopy [35]. PH domains fused to green

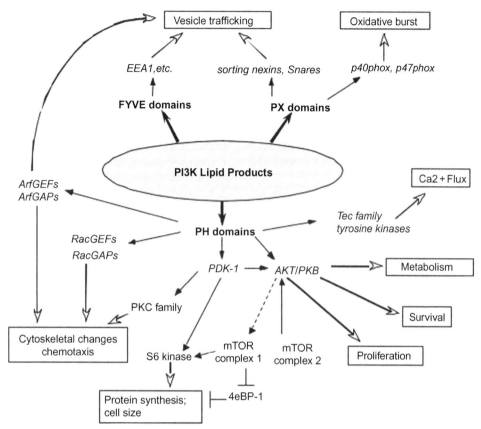

FIGURE 25.4 Overview of the effector proteins and signaling pathways regulated by PI3K lipid products.
Common 3'PI lipid-binding modules are in bold print; functional responses are in boxes. The diagram shows selected effector proteins (in italics) whose lipid-binding domains have been well studied and whose activation has been linked to particular responses. The dashed arrow indicates indirect activation of mTOR complex-1 by AKT through at least two mechanisms [50]. GEF: guanine nucleotide exchange factor; GAP: GTPase-activating protein.

fluorescent protein (GFP) are commonly used to visualize the location and kinetics of class I PI3K activity in transfected cells. Similarly, FYVE domains fused to fluorescent proteins are helpful for detecting PtdIns(3)P in live cells. Expression of the AKT PH domain fused to GFP in all cells of a transgenic mouse strain has enabled studies of PI3K signaling dynamics in primary immune cells [36–38].

EFFECTORS AND RESPONSES

PI3K activation has been linked to distinct cellular responses downstream of different receptors [2, 39]. For example, PI3K is required for proliferation induced by numerous growth factors and cytokines, for glucose uptake triggered by insulin, and for cell migration in response to chemoattractants. A major challenge in PI3K research has been to determine how specificity in signaling is achieved. With all the factors than can trigger increases in 3'PIs, how is it that different stimuli evoke distinct responses through PI3K?

There are several answers to this puzzle. One level of specificity is conferred by differential expression of PI3K

isoforms in distinct tissues/cell types. For example, the p110δ isoform is mainly expressed in leukocytes, and the phenotypes of p110δ knockout mice are mainly confined to the immune system [40]. Similarly, some PI3K effectors are differentially expressed. An example is Btk, a Tec family tyrosine kinase that contains a PH domain and is expressed primarily in B lymphocytes and mast cells. Mice lacking either Btk or the predominant class IA regulatory isoform, p85α, exhibit similar defects in B cell activation [41–43]. Another factor in signaling specificity could be the compartmentalization of PI3K activation. In other words, distinct localization of receptors in membrane subdomains affects the pool of PI3K substrates and effectors utilized. 3'PI lipids accumulate at the leading edge of cells migrating in response to chemoattractants, leading to localized activation of PI3K effectors [7, 44]. Finally, full activation of a given PI3K effector often requires synergy with other signals, which may be differentially provided by distinct receptors. For example, the guanine nucleotide exchange factor p-Rex1 is fully activated by the combined action of PtdIns(3, 4, 5)P$_3$ and G protein βγ subunits [45].

Figure 25.4 summarizes current knowledge of the linkage of certain PI3K effectors to distinct responses. It is important to note that this diagram is simplified for clarity, and some effectors have been linked to additional functions. A central player in many responses to PI3K activation is phosphoinositide-dependent kinase-1 (PDK-1) [46]. This serine/threonine kinase has a PH domain that binds both PtdIns(3, 4, 5)P$_3$ and PtdIns(3, 4)P$_2$. Current evidence suggests that PDK-1 is constitutively active but only gains access to certain substrates upon binding 3'PI lipids. Phosphorylation by PDK-1 contributes to the activation of many downstream kinases, including AKT, S6 kinase and some protein kinase C isoforms.

There are three AKT (also known as PKB) serine/threonine kinases in mammals [47]. Each has a 3'PI-selective PH domain and phosphorylates many substrates in a PI3K-dependent manner. Through this phosphorylation program, AKT controls many cell fate decisions, including cell growth (size increase), proliferation, survival, nutrient uptake, and metabolism [48]. One group of AKT substrates discussed in this chapter is the FOXO family of transcription factors, which activate cell-specific target genes in a manner opposed by AKT phosphorylation [49]. The three AKTs have partly overlapping functions based in part on different tissue distribution [47]. Full AKT activation requires phosphorylation of a threonine in the activation loop by PDK-1, together with phosphorylation of a serine residue in a hydrophobic motif near the C-terminus. The kinase responsible for hydrophobic motif phosphorylation, often referred to as PDK2, is thought to be the mammalian target of rapamycin (mTOR) complex-2 in many cell systems [50] (Figure 25.4). The interplay of AKT, mTOR complexes 1 and 2, and their substrates is complex, and the topic of several recent reviews [50–52]. In response to DNA damage signals, DNA-dependent protein kinase (DNA-PK) can serve as PDK-2 [53].

GENETICS

Natural and engineered mutations in PI3Ks and lipid phosphatases have established the fundamental role of PI3K signaling in promoting the growth, proliferation, and survival of normal and transformed cells. This section will begin with a survey of some of the seminal results from animal models, followed by a discussion of the rapidly expanding evidence for PI3K pathway alterations in human cancer. The section concludes with a brief overview of the role of PI3K in innate immunity and inflammation.

It was shown more than 10 years ago that the catalytic and regulatory subunits of class IA PI3K are proto-oncogenes. The transforming oncogene of an avian sarcoma virus, ASV16, encodes a membrane-targeted variant of p110 whose expression in cells causes accumulation of 3'PIs [54]. A truncated variant of p85α (termed p65), first isolated from a T cell lymphoma, cooperates with v-raf in fibroblast transformation assays and promotes lymphoproliferation

when expressed as a transgene in T cells [55, 56]. Mice heterozygous for a disrupted PTEN gene exhibit a similar lymphoproliferative disorder [57, 58]. Mice with homozygous loss of PTEN specifically in T cells develop autoimmune symptoms associated with spontaneously activated T cells that are resistant to apoptosis [59]. Mice lacking SHIP1 develop a myeloproliferative disorder and have lower activation thresholds for a variety of immune cell stimuli [27]. Mutations downstream of PI3K can also cause cell transformation, such as activating mutations in AKT or loss of FOXO function [60, 61].

In addition to these examples of increased PI3K signaling promoting proliferation and tumorigenesis, there are also examples of decreased PI3K signaling causing impaired proliferation and transformation. Forced expression of PTEN in PTEN-deficient cells of various tissue origin impairs growth by inducing cell cycle arrest and/or apoptosis. Loss of p110α in fibroblasts attenuates transformation by several oncogenes [62], and deletion of class IA regulatory subunits impairs pre-B cell transformation by BCR-ABL [63, 64]. Deletion of the class IA regulatory isoform p85α in mice abrogates B lymphocyte proliferation in response to antigen receptor engagement, and reduces stem cell factor-driven mast cell growth [39]. These mice lack gastrointestinal mast cells and show impaired immune responses to T cell-independent antigens, bacteria, and parasitic worms [43, 65]. Deletion or inactivation of the p110δ catalytic isoform in mice also impairs various aspects of B cell and mast cell development and function [39, 40]. In T cells, maximal proliferation requires p110δ as well as redundant functions of the regulatory subunits p85α/p55α/p50α/p85β [40, 66–68]. However, reduced PI3K signaling in T cells also leads to autoimmunity, likely due to impaired function of regulatory T cells [67, 69].

Recent data indicate that PI3K signaling regulates the survival and self-renewal of stem cells [70, 71]. Loss of PTEN in hematopoietic stem cells (HSC) causes short-term expansion but long-term depletion of the HSC pool, with increased percentages of cells with the phenotype of leukemic stem cells [72]. Combined deletion of three FOXO family members in HSC also increases cell cycling and reduces long-term repopulating ability [73], suggesting that FOXO inactivation is an important outcome of PI3K signaling that is opposed by PTEN in HSC. The balance of PI3K and PTEN activity also regulates multipotency and self-renewal in embryonic stem cells and neural stem cells [71, 74].

How does PI3K activation drive cell proliferation and survival? It is well established that AKT phosphorylates numerous substrates to promote cell cycle progression and oppose apoptotic signals [48]. However, an emerging theme is that PI3K/AKT signaling also drives cellular metabolic changes downstream of growth factor receptors [48, 52]. In C. elegans, a class I PI3K functions downstream of the insulin receptor homolog in a pathway that

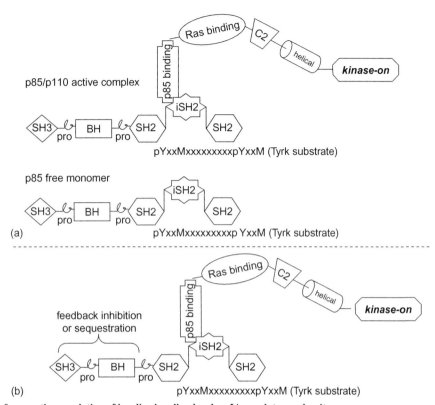

FIGURE 25.5 Models for negative regulation of insulin signaling by class IA regulatory subunits.
(a) Free monomers of p85 (or p55/p50) compete with active heterodimeric PI3K. (b) N-terminal domains of regulatory subunits engage in feedback inhibitory pathways, or sequester signaling complexes.

regulates both metabolism and lifespan [75]. This pathway involves the worm orthologs of PDK-1 and AKT, and is antagonized by PTEN. In mice, SHIP2 phosphatase is a critical modulator of insulin signaling, as SHIP2-deficient mice show increased insulin sensitivity [76]. The p110α isoform of class IA PI3K plays a dominant role in signaling downstream of insulin receptors and other receptors that influence both cellular and organismal metabolism [16, 77]. Inactivation of FOXO family of transcription factors by AKT-dependent phosphorylation appears to be an important point of regulation in metabolic changes following growth factor signaling [48, 78].

Based on these findings, it was expected that mice deficient in class IA regulatory subunits would show insulin resistance. However, in most cases examined, the opposite result has been observed. Mice lacking p85α, p85β, p85α/p55α/p50α or p55α/p50α exhibit hypoglycemia with increased insulin sensitivity and decreased glucose tolerance [79–83]. One model (Figure 25.5) is that regulatory subunits are expressed in excess of catalytic subunits in insulin-responsive cells, such that free monomers compete with and limit the recruitment of p85/p110 dimers; in knockout cells, this competition is reduced [84]. This model has been challenged recently by a mass spectrometry study that failed to detect free monomers [85]. Another model is that regulatory subunits participate in feedback

inhibition of PI3K signaling, in part through activation of the stress kinase JNK and the lipid phosphatase PTEN [6, 86, 87]. A third model is that monomeric p85 subunits sequester insulin-regulated signaling complexes in an intracellular location distinct from the plasma membrane [88]. Regardless of the mechanism of negative regulation, it is now clear that class IA regulatory subunits do also have a positive role in insulin signaling. Combined deletion of *Pik3r1* and *Pik3r2* in muscle or liver cells causes insulin resistance and whole-body glucose intolerance [89, 90]. A reasonable model is that negative regulation of insulin signaling is mediated by adaptor functions and sequestration effects, whereas positive regulation is controlled by the PI3K catalytic activation capacity of regulatory subunits. The latter effect is dominant when both *Pik3r1* and *Pik3r2* are absent in insulin-responsive tissues, therefore causing resistance to insulin.

PI3K and AKT signaling also promote cell growth (increased cell size). The pathway was first delineated in *Drosophila*, in which class I PI3K acts downstream of the insulin receptor ortholog to control cell size [91]. Overexpression of PI3K in wing imaginal discs increases cell size and yields enlarged wings in the adult fly; mutation of *Drosophila* PTEN has a similar effect. Conversely, mutation of class I PI3K genes or expression of dominant-negative PI3K reduces cell and wing size. The critical downstream effectors of PI3K in the *Drosophila* system are AKT and dTOR, the ortholog of mTOR [91, 92]. PI3K/AKT signaling also regulates the size of mouse cardiac myocytes [93–95]. As in the fruit fly, TOR activation in mammalian cells is a critical control point in cell size regulation downstream of PI3K/AKT (Figure 25.4). Importantly, mTOR integrates signals both from PI3K/AKT and from nutrient availability to determine the cellular capacity for growth [51, 52, 96].

The first evidence for increased PI3K signaling in human cancer was the discovery that the tumor suppressor PTEN is a lipid phosphatase specific for 3′-phosphorylated phosphoinositides [97]. Inherited mutations in PTEN cause the autosomal dominant cancer syndrome Cowden syndrome in humans, and are associated with other benign hamartomatous diseases [98]. Moreover, loss of PTEN function is seen in a large fraction of sporadic human cancers [99, 100]. The p110α isoform likely has a unique role in control of cell proliferation and survival, since many human tumors possess activating mutations in *PIK3CA*, the gene encoding p110α [101, 102]. Most of these mutations are gain-of-function and increase PI3K activity *in vitro* and *in vivo* [103–106]. Structural studies have begun to identify the molecular mechanisms by which these mutations augment p110α enzyme activity [13, 14]. Mutations affecting class IA regulatory subunits also occur in human cancer [107–109]. Furthermore, upstream oncogenes can lead to dysregulated PI3K signaling in tumors where PTEN is intact and PI3K genes are not mutated [99, 110]. Downstream of PI3K, gain-of-function mutations in AKT are also tumorigenic [60, 111].

Both class IA and IB PI3K are important for responses of innate immune cells and elevated PI3K signals contribute to inflammatory diseases [112]. p110γ, the class IB isoform, and the p110δ class IA isoform are expressed primarily in leukocytes. Deletion of mouse p110γ causes defects in inflammatory responses that correlate with defective chemotaxis to GPCR ligands such as f-Met–Leu–Phe and C5a [7]. Loss of p110γ or treatment with a selective inhibitor ameliorates disease in mouse models of arthritis and lupus [113–115]. The essential function of p110δ in T cell expansion and differentiation, and in mast cell responses, has established a central role for this isoform in allergic disease [40, 116]. p110δ also functions in human neutrophils in a second wave of PI3K activation following initial GPCR-mediated p110γ activation [117]. A reciprocal loop occurs in mast cells, where FcεRI-mediated class IA PI3K activation is followed by release of autocrine factors that activate class IB signaling through GPCRs [118]. In T cell development, p110δ and p110γ have mostly redundant functions, likely due to overlapping survival signals initiated by receptors linked either to TyrKs or G proteins [119, 120].

Genetic studies have shown that lymphocyte chemotaxis, homing, and basal motility are partially dependent on PI3K activation [121]. In T cells, p110γ is required for chemokine-dependent responses as predicted from the linkage of class IB PI3K to GPCR signaling. In B cells, p110γ is dispensable whereas p110δ and possibly other class IA isoforms contribute to chemokine responses. The mechanism by which GPCRs activate class IA PI3K in B cells is not known. The class IA regulatory subunits p85α and p85β have non-redundant functions in basal lymphocyte motility within living lymphoid tissue [122].

Loss-of-function mutations that eliminate the phosphatase activity of MTMRs are the cause of human myopathy and neuropathy syndromes [28].

PHARMACOLOGY

The first generation of PI3K inhibitors included the natural product wortmannin and the synthetic compound LY294002 [1, 2]. These agents inhibit proliferation and chemotaxis in a variety of cell systems. Although wortmannin and LY294002 have been used in thousands of papers, they are not definitive tools, since they inhibit all PI3K classes, as well as related kinases and other cellular enzymes. Hence, these compounds inhibit some cellular responses even when PI3K signaling is abrogated genetically [66]. The combined efforts of pharmaceutical companies and academic labs have yielded a growing set of pharmacological agents, whose selectivity is sufficient to test the physiological roles of individual PI3K enzymes and subgroups. For example, a selective p110δ inhibitor causes similar effects in B cells as genetic inactivation of p110δ kinase activity, and has no demonstrable effects on p110δ-deficient cells [123]. A landmark paper in 2006 mapped the structure–activity relationships of several chemotypes of PI3K inhibitors against a large panel of PI3Ks and related enzymes [16]. This work, together with growing knowledge of PI3K structure, points the way towards further refinement of targeted agents [124, 125]. Several studies have indicated that "second-generation" PI3K inhibitors are effective therapeutic agents in animal models of cancer and inflammation, without severe toxicity [114, 115, 126, 127]. Indeed, clinical trials of some pharmaceutical PI3K inhibitors have been initiated [128], with more likely on the horizon.

SYNOPSIS

Signaling through PI3K is an evolutionarily conserved process that enables reversible membrane localization of cytoplasmic proteins. Modular domains (including PH, PX, and FYVE) that interact with 3′PIs are broadly distributed among proteins of different function. The recruitment and activation of distinct subsets of PI3K isoforms and effectors in a receptor-specific and cell type-specific manner allows PI3K activation to be linked to different functional responses. Altered activity of the PI3K signaling pathway is characteristic of many human disease states. The ability to generate isoform-selective inhibitors of PI3Ks or 3-phosphatases might lead to development of novel therapies for cancer, diabetes, and immune dysfunction.

REFERENCES

1. Fruman DA, Meyers RE, Cantley LC. Phosphoinositide kinases. *Annu Rev Biochem* 1998;**67**:481–507.

2. Vanhaesebroeck B, Leevers SJ, Ahmadi K, Timms J, Katso R, Driscoll PC, Woscholski R, Parker PJ, Waterfield MD. Synthesis and function of 3-phosphorylated inositol lipids. *Annu Rev Biochem* 2001;**70**:535–602.

3. Jimenez C, Hernandez C, Pimentel B, Carrera AC. The p85 regulatory subunit controls sequential activation of phosphoinositide 3-kinase by Tyr kinases and Ras. *J Biol Chem* 2002;**277**:41,556–62.

4. Rodriguez-Viciana P, Warne PH, Dhand R, Vanhaesebroeck B, Gout I, Fry MJ, Waterfield MD, Downward J. Phosphatidylinositol-3-OH kinase as a direct target of Ras. *Nature* 1994;**370**:527–32.

5. Gupta S, Ramjaun AR, Haiko P, Wang Y, Warne PH, Nicke B, Nye E, Stamp G, Alitalo K, Downward J. Binding of ras to phosphoinositide 3-kinase p110alpha is required for ras-driven tumorigenesis in mice. *Cell* 2007;**129**:957–68.

6. Taniguchi CM, Aleman JO, Ueki K, Luo J, Asano T, Kaneto H, Stephanopoulos G, Cantley LC, Kahn CR. The p85alpha regulatory subunit of phosphoinositide 3-kinase potentiates c-Jun N-terminal kinase-mediated insulin resistance. *Mol Cell Biol* 2007;**27**:2830–40.

7. Andrews S, Stephens LR, Hawkins PT. PI3K class IB pathway. *Sci STKE 2007* 2007;**cm2**.

8. Suire S, Condliffe AM, Ferguson GJ, Ellson CD, Guillou H, Davidson K, Welch H, Coadwell J, Turner M, Chilvers ER, Hawkins PT, Stephens L. Gbetagammas and the Ras binding domain of p110gamma are both important regulators of PI(3)Kgamma signalling in neutrophils. *Nat Cell Biol* 2006;**8**:1303–9.

9. Patrucco E, Notte A, Barberis L, Selvetella G, Maffei A, Brancaccio M, Marengo S, Russo G, Azzolino O, Rybalkin SD, Silengo L, Altruda F, Wetzker R, Wymann MP, Lembo G, Hirsch E. PI3Kgamma modulates the cardiac response to chronic pressure overload by distinct kinase-dependent and -independent effects. *Cell* 2004;**118**:375–87.

10. Ciraolo E, Iezzi M, Marone R, Marengo S, Curcio C, Costa C, Azzolino O, Gonella C, Rubinetto C, Wu H, Dastru W, Martin EL, Silengo L, Altruda F, Turco E, Lanzetti L, Musiani P, Ruckle T, Rommel C, Backer JM, Forni G, Wymann MP, Hirsch E. Phosphoinositide 3-kinase p110β activity: key role in metabolism and mammary gland cancer but not development. *Sci Signaling* 2008;**1**:ra3.

11. Guillermet-Guibert J, Bjorklof K, Salpekar A, Gonella C, Ramadani F, Bilancio A, Meek S, Smith AJ, Okkenhaug K, Vanhaesebroeck B. The p110beta isoform of phosphoinositide 3-kinase signals downstream of G protein-coupled receptors and is functionally redundant with p110gamma. *Proc Natl Acad Sci USA* 2008;**105**:8292–7.

12. Jia S, Liu Z, Zhang S, Liu P, Zhang L, Lee SH, Zhang J, Signoretti S, Loda M, Roberts TM, Zhao JJ. Essential roles of PI(3)K-p110beta in cell growth, metabolism and tumorigenesis. *Nature* 2008;**454**:776–9.

13. Huang CH, Mandelker D, Schmidt-Kittler O, Samuels Y, Velculescu VE, Kinzler KW, Vogelstein B, Gabelli SB, Amzel LM. The structure of a human p110alpha/p85alpha complex elucidates the effects of oncogenic PI3Kalpha mutations. *Science* 2007;**318**:1744–8.

14. Miled N, Yan Y, Hon WC, Perisic O, Zvelebil M, Inbar Y, Schneidman-Duhovny D, Wolfson HJ, Backer JM, Williams RL. Mechanism of two classes of cancer mutations in the phosphoinositide 3-kinase catalytic subunit. *Science* 2007;**317**:239–42.

15. Walker EH, Perisic O, Ried C, Stephens L, Williams RL. Structural insights into phosphoinositide 3-kinase catalysis and signalling. *Nature* 1999;**402**:313–20.

16. Knight ZA, Gonzalez B, Feldman ME, Zunder ER, Goldenberg DD, Williams O, Loewith R, Stokoe D, Balla A, Toth B, Balla T, Weiss WA, Williams RL, Shokat KM. A pharmacological map of the PI3-K family defines a role for p110alpha in insulin signaling. *Cell* 2006;**125**:733–47.

17. Pacold ME, Suire S, Perisic O, Lara-Gonzalez S, Davis CT, Walker EH, Hawkins PT, Stephens L, Eccleston JF, Williams RL. Crystal structure and functional analysis of Ras binding to its effector phosphoinositide 3-kinase gamma. *Cell* 2000;**103**:931–43.

18. Walker EH, Pacold ME, Perisic O, Stephens L, Hawkins PT, Wymann MP, Williams RL. Structural determinants of phosphoinositide 3-kinase inhibition by wortmannin, LY294002, quercetin, myricetin, and staurosporine. *Mol Cell* 2000;**6**:909–19.

19. Falasca M, Hughes WE, Dominguez V, Sala G, Fostira F, Fang MQ, Cazzolli R, Shepherd PR, James DE, Maffucci T. The role of phosphoinositide 3-kinase C2alpha in insulin signaling. *J Biol Chem* 2007;**282**:28,226–36.

20. Falasca M, Maffucci T. Role of class II phosphoinositide 3-kinase in cell signalling. *Biochem Soc Trans* 2007;**35**:211–14.

21. Birkeland HC, Stenmark H. Protein targeting to endosomes and phagosomes via FYVE and PX domains. *Curr Top Microbiol Immunol* 2004;**282**:89–115.

22. Backer JM. The regulation and function of Class III PI3Ks: novel roles for Vps34. *Biochem J* 2008;**410**:1–17.

23. Tolias KF, Cantley LC. Pathways for phosphoinositide synthesis. *Chem Phys Lipids* 1999;**98**:69–77.

24. Gillooly DJ, Morrow IC, Lindsay M, Gould R, Bryant NJ, Gaullier JM, Parton RG, Stenmark H. Localization of phosphatidylinositol 3-phosphate in yeast and mammalian cells. *EMBO J* 2000;**19**:4577–88.

25. Michell RH, Heath VL, Lemmon MA, Dove SK. Phosphatidylinositol 3, 5-bisphosphate: metabolism and cellular functions. *Trends Biochem Sci* 2006;**31**:52–63.

26. Shisheva A. PIKfyve: partners, significance, debates and paradoxes. *Cell Biol Intl* 2008;**32**:591–604.

27. Rohrschneider LR, Fuller JF, Wolf I, Liu Y, Lucas DM. Structure, function, and biology of SHIP proteins. *Genes Dev* 2000;**14**:505–20.

28. Robinson FL, Dixon JE. Myotubularin phosphatases: policing 3-phosphoinositides. *Trends Cell Biol* 2006;**16**:403–12.

29. Ferguson KM, Kavran JM, Sankaran VG, Fournier E, Isakoff SJ, Skolnik EY, Lemmon MA. Structural basis for discrimination of 3-phosphoinositides by pleckstrin homology domains. *Mol Cell* 2000;**6**:373–84.

30. Lemmon MA. Membrane recognition by phospholipid-binding domains. *Nat Rev Mol Cell Biol* 2008;**9**:99–111.

31. Lietzke SE, Bose S, Cronin T, Klarlund J, Chawla A, Czech MP, Lambright DG. Structural basis of 3-phosphoinositide recognition by pleckstrin homology domains. *Mol Cell* 2000;**6**:385–94.

32. Bravo J, Karathanassis D, Pacold CM, Pacold ME, Ellson CD, Anderson KE, Butler PJ, Lavenir I, Perisic O, Hawkins PT, Stephens L, Williams RL. The crystal structure of the PX domain from p40(phox) bound to phosphatidylinositol 3-phosphate. *Mol Cell* 2001;**8**:829–39.

33. Saito K, Tautz L, Mustelin T. The lipid-binding SEC14 domain. *Biochim Biophys Acta* 2007;**1771**:719–26.

34. Stromhaug PE, Reggiori F, Guan J, Wang CW, Klionsky DJ. Atg21 is a phosphoinositide binding protein required for efficient lipidation

and localization of Atg8 during uptake of aminopeptidase I by selective autophagy. *Mol Biol Cell* 2004;**15**:3553–66.

35. Varnai P, Balla T. Live cell imaging of phosphoinositide dynamics with fluorescent protein domains. *Biochim Biophys Acta* 2006;**1761**:957–67.

36. Ferguson GJ, Milne L, Kulkarni S, Sasaki T, Walker S, Andrews S, Crabbe T, Finan P, Jones G, Jackson S, Camps M, Rommel C, Wymann M, Hirsch E, Hawkins P, Stephens L. PI(3)Kgamma has an important context-dependent role in neutrophil chemokinesis. *Nat Cell Biol* 2007;**9**:86–91.

37. Garcon F, Patton DT, Emery JL, Hirsch E, Rottapel R, Sasaki T, Okkenhaug K. CD28 provides T-cell costimulation and enhances PI3K activity at the immune synapse independently of its capacity to interact with the p85/p110 heterodimer. *Blood* 2008;**111**:1464–71.

38. Nishio M, Watanabe K, Sasaki J, Taya C, Takasuga S, Iizuka R, Balla T, Yamazaki M, Watanabe H, Itoh R, Kuroda S, Horie Y, Forster I, Mak TW, Yonekawa H, Penninger JM, Kanaho Y, Suzuki A, Sasaki T. Control of cell polarity and motility by the PtdIns(3, 4, 5)P3 phosphatase SHIP1. *Nat Cell Biol* 2007;**9**:36–44.

39. Deane JA, Fruman DA. Phosphoinositide 3-kinase: diverse roles in immune cell activation. *Annu Rev Immunol* 2004;**22**:563–98.

40. Okkenhaug K, Ali K, Vanhaesebroeck B. Antigen receptor signalling: a distinctive role for the p110delta isoform of PI3K. *Trends Immunol* 2007;**28**:80–7.

41. Donahue AC, Fruman DA. PI3K signaling controls cell fate at many points in B lymphocyte development and activation. *Semin Cell Dev Biol* 2004;**15**:183–97.

42. Fruman DA, Snapper SB, Yballe CM, Davidson L, Yu JY, Alt FW, Cantley LC. Impaired B cell development and proliferation in absence of phosphoinositide 3-kinase p85alpha. *Science* 1999;**283**:393–7.

43. Suzuki H, Terauchi Y, Fujiwara M, Aizawa S, Yazaki Y, Kadowaki T, Koyasu S. Xid-like immunodeficiency in mice with disruption of the p85alpha subunit of phosphoinositide 3-kinase. *Science* 1999;**283**:390–2.

44. Stephens L, Ellson C, Hawkins P. Roles of PI3Ks in leukocyte chemotaxis and phagocytosis. *Curr Opin Cell Biol* 2002;**14**:203–13.

45. Hill K, Krugmann S, Andrews SR, Coadwell WJ, Finan P, Welch HC, Hawkins PT, Stephens LR. Regulation of P-Rex1 by phosphatidylinositol (3, 4, 5)-trisphosphate and Gbetagamma subunits. *J Biol Chem* 2005;**280**:4166–73.

46. Storz P, Toker A. 3'-phosphoinositide-dependent kinase-1 (PDK-1) in PI 3-kinase signaling. *Front Biosci* 2002;**7**:d886–902.

47. Dummler B, Hemmings BA. Physiological roles of PKB/Akt isoforms in development and disease. *Biochem Soc Trans* 2007;**35**:231–5.

48. Manning BD, Cantley LC. AKT/PKB signaling: navigating downstream. *Cell* 2007;**129**:1261–74.

49. Calnan DR, Brunet A. The FoxO code. *Oncogene* 2008;**27**:2276–88.

50. Guertin DA, Sabatini DM. Defining the role of mTOR in cancer. *Cancer Cell* 2007;**12**:9–22.

51. Bhaskar PT, Hay N. The two TORCs and Akt. *Dev Cell* 2007;**12**:487–502.

52. Wullschleger S, Loewith R, Hall MN. TOR signaling in growth and metabolism. *Cell* 2006;**124**:471–84.

53. Bozulic L, Surucu B, Hynx D, Hemmings BA. PKBalpha/Akt1 acts downstream of DNA-PK in the DNA double-strand break response and promotes survival. *Mol Cell* 2008;**30**:203–13.

54. Chang HW, Aoki M, Fruman D, Auger KR, Bellacosa A, Tsichlis PN, Cantley LC, Roberts TM, Vogt PK. Transformation of chicken cells by the gene encoding the catalytic subunit of PI 3-kinase. *Science* 1997;**276**:1848–50.

55. Borlado LR, Redondo C, Alvarez B, Jimenez C, Criado LM, Flores J, Marcos MA, Martinez AC, Balomenos D, Carrera AC. Increased phosphoinositide 3-kinase activity induces a lymphoproliferative disorder and contributes to tumor generation inI vivo. *Faseb J* 2000;**14**:895–903.

56. Jimenez C, Jones DR, Rodriguez-Viciana P, Gonzalez-Garcia A, Leonardo E, Wennstrom S, von Kobbe C, Toran JL, L RB, Calvo V, Copin SG, Albar JP, Gaspar ML, Diez E, Marcos MA, Downward J, Martinez AC, Merida I, Carrera AC. Identification and characterization of a new oncogene derived from the regulatory subunit of phosphoinositide 3-kinase. *EMBO J* 1998;**17**:743–53.

57. Di Cristofano A, Kotsi P, Peng YF, Cordon-Cardo C, Elkon KB, Pandolfi PP. Impaired Fas response and autoimmunity in Pten+/− mice. *Science* 1999;**285**:2122–5.

58. Podsypanina K, Ellenson LH, Nemes A, Gu J, Tamura M, Yamada KM, Cordon-Cardo C, Catoretti G, Fisher PE, Parsons R. Mutation of Pten/Mmac1 in mice causes neoplasia in multiple organ systems. *Proc Natl Acad Sci USA* 1999;**96**:1563–8.

59. Suzuki A, Yamaguchi MT, Ohteki T, Sasaki T, Kaisho T, Kimura Y, Yoshida R, Wakeham A, Higuchi T, Fukumoto M, Tsubata T, Ohashi PS, Koyasu S, Penninger JM, Nakano T, Mak TW. T cell-specific loss of Pten leads to defects in central and peripheral tolerance. *Immunity* 2001;**14**:523–34.

60. Bellacosa A, Testa JR, Staal SP, Tsichlis PN. A retroviral oncogene, akt, encoding a serine-threonine kinase containing an SH2-like region. *Science* 1991;**254**:274–7.

61. Paik JH, Kollipara R, Chu G, Ji H, Xiao Y, Ding Z, Miao L, Tothova Z, Horner JW, Carrasco DR, Jiang S, Gilliland DG, Chin L, Wong WH, Castrillon DH, DePinho RA. FoxOs are lineage-restricted redundant tumor suppressors and regulate endothelial cell homeostasis. *Cell* 2007;**128**:309–23.

62. Zhao JJ, Cheng H, Jia S, Wang L, Gjoerup OV, Mikami A, Roberts TM. The p110alpha isoform of PI3K is essential for proper growth factor signaling and oncogenic transformation. *Proc Natl Acad Sci USA* 2006;**103**:16,296–300.

63. Kharas MG, Deane JA, Wong S, O'Bosky KR, Rosenberg N, Witte ON, Fruman DA. Phosphoinositide 3-kinase signaling is essential for ABL oncogene-mediated transformation of B-lineage cells. *Blood* 2004;**103**:4268–75.

64. Kharas MG, Janes MR, Scarfone VM, Lilly MB, Knight ZA, Shokat KM, Fruman DA. Ablation of PI3K blocks BCR-ABL leukemogenesis in mice, and a dual PI3K/mTOR inhibitor prevents expansion of human BCR-ABL+ leukemia cells. *J Clin Invest* 2008;**118**:3038–50.

65. Fukao T, Yamada T, Tanabe M, Terauchi Y, Ota T, Takayama T, Asano T, Takeuchi T, Kadowaki T, Hata Ji J, Koyasu S. Selective loss of gastrointestinal mast cells and impaired immunity in PI3K-deficient mice. *Nat Immunol* 2002;**3**:295–304.

66. Deane JA, Kharas MG, Oak JS, Stiles LN, Luo J, Moore TI, Ji H, Rommel C, Cantley LC, Lane TE, Fruman DA. T-cell function is partially maintained in the absence of class IA phosphoinositide 3-kinase signaling. *Blood* 2007;**109**:2894–902.

67. Fruman DA. The role of class I phosphoinositide 3-kinase in T-cell function and autoimmunity. *Biochem Soc Trans* 2007;**35**:177–80.

68. Okkenhaug K, Bilancio A, Farjot G, Priddle H, Sancho S, Peskett E, Pearce W, Meek SE, Salpekar A, Waterfield MD, Smith AJ, Vanhaesebroeck B. Impaired B and T cell antigen receptor signaling in p110delta PI 3-kinase mutant mice. *Science* 2002;**297**:1031–4.

69. Oak JS, Fruman DA. Role of phosphoinositide 3-kinase signaling in autoimmunity. *Autoimmunity* 2007;**40**:433–41.

70. Rossi DJ, Weissman IL. Pten, tumorigenesis, and stem cell self-renewal. *Cell* 2006;**125**:229–31.

71. Welham MJ, Storm MP, Kingham E, Bone HK. Phosphoinositide 3-kinases and regulation of embryonic stem cell fate. *Biochem Soc Trans* 2007;**35**:225–8.

72. Yilmaz OH, Valdez R, Theisen BK, Guo W, Ferguson DO, Wu H, Morrison SJ. Pten dependence distinguishes haematopoietic stem cells from leukaemia-initiating cells. *Nature* 2006;**441**:475–82.

73. Tothova Z, Kollipara R, Huntly BJ, Lee BH, Castrillon DH, Cullen DE, McDowell EP, Lazo-Kallanian S, Williams IR, Sears C, Armstrong SA, Passegue E, DePinho RA, Gilliland DG. FoxOs are critical mediators of hematopoietic stem cell resistance to physiologic oxidative stress. *Cell* 2007;**128**:325–39.

74. Groszer M, Erickson R, Scripture-Adams DD, Lesche R, Trumpp A, Zack JA, Kornblum HI, Liu X, Wu H. Negative regulation of neural stem/progenitor cell proliferation by the Pten tumor suppressor gene in vivo. *Science* 2001;**294**:2186–9.

75. Kenyon C. A conserved regulatory system for aging. *Cell* 2001;**105**:165–8.

76. Clement S, Krause U, Desmedt F, Tanti JF, Behrends J, Pesesse X, Sasaki T, Penninger J, Doherty M, Malaisse W, Dumont JE, Le Marchand-Brustel Y, Erneux C, Hue L, Schurmans S. The lipid phosphatase SHIP2 controls insulin sensitivity. *Nature* 2001;**409**:92–7.

77. Foukas LC, Claret M, Pearce W, Okkenhaug K, Meek S, Peskett E, Sancho S, Smith AJ, Withers DJ, Vanhaesebroeck B. Critical role for the p110alpha phosphoinositide-3-OH kinase in growth and metabolic regulation. *Nature* 2006;**441**:366–70.

78. Gross DN, van den Heuvel AP, Birnbaum MJ. The role of FoxO in the regulation of metabolism. *Oncogene* 2008;**27**:2320–36.

79. Chen D, Mauvais-Jarvis F, Bluher M, Fisher SJ, Jozsi A, Goodyear LJ, Ueki K, Kahn CR. p50alpha/p55alpha phosphoinositide 3-kinase knockout mice exhibit enhanced insulin sensitivity. *Mol Cell Biol* 2004;**24**:320–9.

80. Fruman DA, Mauvais-Jarvis F, Pollard DA, Yballe CM, Brazil D, Bronson RT, Kahn CR, Cantley LC. Hypoglycaemia, liver necrosis and perinatal death in mice lacking all isoforms of phosphoinositide 3-kinase p85 alpha. *Nat Genet* 2000;**26**:379–82.

81. Mauvais-Jarvis F, Ueki K, Fruman DA, Hirshman MF, Sakamoto K, Goodyear LJ, Iannacone M, Accili D, Cantley LC, Kahn CR. Reduced expression of the murine p85alpha subunit of phosphoinositide 3-kinase improves insulin signaling and ameliorates diabetes. *J Clin Invest* 2002;**109**:141–9.

82. Terauchi Y, Tsuji Y, Satoh S, Minoura H, Murakami K, Okuno A, Inukai K, Asano T, Kaburagi Y, Ueki K, Nakajima H, Hanafusa T, Matsuzawa Y, Sekihara H, Yin Y, Barrett JC, Oda H, Ishikawa T, Akanuma Y, Komuro I, Suzuki M, Yamamura K, Kodama T, Suzuki H, Yamamura K, Kodama T, Suzuki H, Koyasu S, Aizawa S, Tobe K, Fukui Y, Yazaki Y, Kadowaki T. Increased insulin sensitivity and hypoglycaemia in mice lacking the p85 alpha subunit of phosphoinositide 3-kinase. *Nat Genet* 1999;**21**:230–5.

83. Ueki K, Yballe CM, Brachmann SM, Vicent D, Watt JM, Kahn CR, Cantley LC. Increased insulin sensitivity in mice lacking p85beta subunit of phosphoinositide 3-kinase. *Proc Natl Acad Sci USA* 2002;**99**:419–24.

84. Ueki K, Fruman DA, Brachmann SM, Tseng YH, Cantley LC, Kahn CR. Molecular balance between the regulatory and catalytic subunits of phosphoinositide 3-kinase regulates cell signaling and survival. *Mol Cell Biol* 2002;**22**:965–77.

85. Geering B, Cutillas PR, Nock G, Gharbi SI, Vanhaesebroeck B. Class IA phosphoinositide 3-kinases are obligate p85-p110 heterodimers. *Proc Natl Acad Sci USA* 2007;**104**:7809–14.

86. Taniguchi CM, Tran TT, Kondo T, Luo J, Ueki K, Cantley LC, Kahn CR. Phosphoinositide 3-kinase regulatory subunit p85alpha suppresses insulin action via positive regulation of PTEN. *Proc Natl Acad Sci USA* 2006;**103**:12, 093–7.

87. Ueki K, Fruman DA, Yballe CM, Fasshauer M, Klein J, Asano T, Cantley LC, Kahn CR. Positive and negative roles of p85 alpha and p85 beta regulatory subunits of phosphoinositide 3-kinase in insulin signaling. *J Biol Chem* 2003;**278**:48,453–66.

88. Luo J, Field SJ, Lee JY, Engelman JA, Cantley LC. The p85 regulatory subunit of phosphoinositide 3-kinase down-regulates IRS-1 signaling via the formation of a sequestration complex. *J Cell Biol* 2005;**170**:455–64.

89. Luo J, Sobkiw CL, Hirshman MF, Logsdon MN, Li TQ, Goodyear LJ, Cantley LC. Loss of class IA PI3K signaling in muscle leads to impaired muscle growth, insulin response, and hyperlipidemia. *Cell Metab* 2006;**3**:355–66.

90. Taniguchi CM, Kondo T, Sajan M, Luo J, Bronson R, Asano T, Farese R, Cantley LC, Kahn CR. Divergent regulation of hepatic glucose and lipid metabolism by phosphoinositide 3-kinase via Akt and PKClambda/zeta. *Cell Metab* 2006;**3**:343–53.

91. Weinkove D, Leevers SJ. The genetic control of organ growth: insights from Drosophila. *Curr Op Genet Dev* 2000;**10**:75–80.

92. Neufeld TP. Genetic analysis of TOR signaling in *Drosophila*. *Curr Top Microbiol Immunol* 2004;**279**:139–52.

93. Luo J, McMullen JR, Sobkiw CL, Zhang L, Dorfman AL, Sherwood MC, Logsdon MN, Horner JW, Depinho RA, Izumo S, Cantley LC. Class IA Phosphoinositide 3-kinase regulates heart size and physiological cardiac hypertrophy. *Mol Cell Biol* 2005;**25**:9491–502.

94. Shioi T, Kang PM, Douglas PS, Hampe J, Yballe CM, Lawitts J, Cantley LC, Izumo S. The conserved phosphoinositide 3-kinase pathway determines heart size in mice. *EMBO J* 2000;**19**:2537–48.

95. Shioi T, McMullen JR, Kang PM, Douglas PS, Obata T, Franke TF, Cantley LC, Izumo S. Akt/protein kinase B promotes organ growth in transgenic mice. *Mol Cell Biol* 2002;**22**:2799–809.

96. Edinger AL. Controlling cell growth and survival through regulated nutrient transporter expression. *Biochem J* 2007;**406**:1–12.

97. Maehama T, Dixon JE. The tumor suppressor, PTEN/MMAC1, dephosphorylates the lipid second messenger, phosphatidylinositol 3, 4, 5-trisphosphate. *J Biol Chem* 1998;**273**:13,375–8.

98. Gustafson S, Zbuk KM, Scacheri C, Eng C. Cowden syndrome. *Semin Oncol* 2007;**34**:428–34.

99. Cully M, You H, Levine AJ, Mak TW. Beyond PTEN mutations: the PI3K pathway as an integrator of multiple inputs during tumorigenesis. *Nat Rev Cancer* 2006;**6**:184–92.

100. Vivanco I, Sawyers CL. The phosphatidylinositol 3-kinase AKT pathway in human cancer. *Nat Rev Cancer* 2002;**2**:489–501.

101. Bader AG, Kang S, Zhao L, Vogt PK. Oncogenic PI3K deregulates transcription and translation. *Nat Rev Cancer* 2005;**5**:921–9.

102. Samuels Y, Ericson K. Oncogenic PI3K and its role in cancer. *Curr Opin Oncol* 2006;**18**:77–82.

103. Bader AG, Kang S, Vogt PK. Cancer-specific mutations in PIK3CA are oncogenic in vivo. *Proc Natl Acad Sci USA* 2006;**103**:1475–9.

104. Gymnopoulos M, Elsliger MA, Vogt PK. Rare cancer-specific mutations in PIK3CA show gain of function. *Proc Natl Acad Sci USA* 2007;**104**:5569–74.

105. Isakoff SJ, Engelman JA, Irie HY, Luo J, Brachmann SM, Pearline RV, Cantley LC, Brugge JS. Breast cancer-associated PIK3CA mutations are oncogenic in mammary epithelial cells. *Cancer Res* 2005;**65**:10,992–11,000.

106. Samuels Jr Y, Diaz LA, Schmidt-Kittler O, Cummins JM, Delong L, Cheong I, Rago C, Huso DL, Lengauer C, Kinzler KW, Vogelstein B, Velculescu VE. Mutant PIK3CA promotes cell growth and invasion of human cancer cells. *Cancer Cell* 2005;**7**:561–73.

107. Network, T. C. G. A. R. Comprehensive genomic characterization defines human glioblastoma genes and core pathways. *Nature* 2008;**455**:1061–8.

108. Parsons Jr DW, Jones S, Zhang X, Lin JC, Leary RJ, Angenendt P, Mankoo P, Carter H, Siu IM, Gallia GL, Olivi A, McLendon R, Rasheed BA, Keir S, Nikolskaya T, Nikolsky Y, Busam DA, Tekleab H, Diaz LA, Hartigan J, Smith DR, Strausberg RL, Marie SK, Shinjo SM, Yan H, Riggins GJ, Bigner DD, Karchin R, Papadopoulos N, Parmigiani G, Vogelstein B, Velculescu VE, Kinzler KW. An integrated genomic analysis of human glioblastoma multiforme. *Science* 2008;**321**:1807–12.

109. Philp AJ, Campbell IG, Leet C, Vincan E, Rockman SP, Whitehead RH, Thomas RJ, Phillips WA. The phosphatidylinositol 3′-kinase p85alpha gene is an oncogene in human ovarian and colon tumors. *Cancer Res* 2001;**61**:7426–9.

110. Kharas MG, Fruman DA. ABL oncogenes and phosphoinositide 3-kinase: mechanism of activation and downstream effectors. *Cancer Res* 2005;**65**:2047–53.

111. Carpten JD, Faber AL, Horn C, Donoho GP, Briggs SL, Robbins CM, Hostetter G, Boguslawski S, Moses TY, Savage S, Uhlik M, Lin A, Du J, Qian YW, Zeckner DJ, Tucker-Kellogg G, Touchman J, Patel K, Mousses S, Bittner M, Schevitz R, Lai MH, Blanchard KL, Thomas JE. A transforming mutation in the pleckstrin homology domain of AKT1 in cancer. *Nature* 2007;**448**:439–44.

112. Rommel C, Camps M, Ji H. PI3K delta and PI3K gamma: partners in crime in inflammation in rheumatoid arthritis and beyond? *Nat Rev Immunol* 2007;**7**:191–201.

113. Barber DF, Bartolome A, Hernandez C, Flores JM, Fernandez-Arias C, Rodriguez-Borlado L, Hirsch E, Wymann M, Balomenos D, Carrera AC. Class IB-phosphatidylinositol 3-kinase (PI3K) deficiency ameliorates IA-PI3K-induced systemic lupus but not T cell invasion. *J Immunol* 2006;**176**:589–93.

114. Barber DF, Bartolome A, Hernandez C, Flores JM, Redondo C, Fernandez-Arias C, Camps M, Ruckle T, Schwarz MK, Rodriguez S, Martinez AC, Balomenos D, Rommel C, Carrera AC. PI3Kgamma inhibition blocks glomerulonephritis and extends lifespan in a mouse model of systemic lupus. *Nat Med* 2005;**11**:933–5.

115. Camps M, Ruckle T, Ji H, Ardissone V, Rintelen F, Shaw J, Ferrandi C, Chabert C, Gillieron C, Francon B, Martin T, Gretener D, Perrin D, Leroy D, Vitte PA, Hirsch E, Wymann MP, Cirillo R, Schwarz MK, Rommel C. Blockade of PI3Kgamma suppresses joint inflammation and damage in mouse models of rheumatoid arthritis. *Nat Med* 2005;**11**:936–43.

116. Ali K, Bilancio A, Thomas M, Pearce W, Gilfillan AM, Tkaczyk C, Kuehn N, Gray A, Giddings J, Peskett E, Fox R, Bruce I, Walker C, Sawyer C, Okkenhaug K, Finan P, Vanhaesebroeck B. Essential role for the p110delta phosphoinositide 3-kinase in the allergic response. *Nature* 2004;**431**:1007–11.

117. Condliffe AM, Davidson K, Anderson KE, Ellson CD, Crabbe T, Okkenhaug K, Vanhaesebroeck B, Turner M, Webb L, Wymann MP, Hirsch E, Ruckle T, Camps M, Rommel C, Jackson SP, Chilvers ER, Stephens LR, Hawkins PT. Sequential activation of class IB and class IA PI3K is important for the primed respiratory burst of human but not murine neutrophils. *Blood* 2005;**106**:1432–40.

118. Laffargue M, Calvez R, Finan P, Trifilieff A, Barbier M, Altruda F, Hirsch E, Wymann MP. Phosphoinositide 3-kinase gamma is an essential amplifier of mast cell function. *Immunity* 2002;**16**:441–51.

119. Swat W, Montgrain V, Doggett TA, Douangpanya J, Puri K, Vermi W, Diacovo TG. Essential role of PI3Kdelta and PI3Kgamma in thymocyte survival. *Blood* 2006;**107**:2415–22.

120. Webb LM, Vigorito E, Wymann MP, Hirsch E, Turner M. Cutting edge: T cell development requires the combined activities of the p110gamma and p110delta catalytic isoforms of phosphatidylinositol 3-kinase. *J Immunol* 2005;**175**:2783–7.

121. Oak JS, Matheu MP, Parker I, Cahalan MD, Fruman DA. Lymphocyte cell motility: the twisting, turning tale of phosphoinositide 3-kinase. *Biochem Soc Trans* 2007;**35**:1109–13.

122. Matheu MP, Deane JA, Parker I, Fruman DA, Cahalan MD. Class IA phosphoinositide 3-kinase modulates basal lymphocyte motility in the lymph node. *J Immunol* 2007;**179**:2261–9.

123. Bilancio A, Okkenhaug K, Camps M, Emery JL, Ruckle T, Rommel C, Vanhaesebroeck B. Key role of the p110delta isoform of PI3K in B-cell antigen and IL-4 receptor signaling: comparative analysis of genetic and pharmacologic interference with p110delta function in B cells. *Blood* 2006;**107**:642–50.

124. Ruckle T, Schwarz MK, Rommel C. PI3Kgamma inhibition: towards an "aspirin of the 21st century"? *Nat Rev Drug Discov* 2006;**5**:903–18.

125. Workman P, Clarke PA, Guillard S, Raynaud FI. Drugging the PI3 kinome. *Nat Biotech* 2006;**24**:794–6.

126. Engelman JA, Chen L, Tan X, Crosby K, Guimaraes AR, Upadhyay R, Maira M, McNamara K, Perera SA, Song Y, Chirieac LR, Kaur R, Lightbown A, Simendinger J, Li T, Padera RF, Garcia-Echeverria C, Weissleder R, Mahmood U, Cantley LC, Wong KK. Effective use of PI3K and MEK inhibitors to treat mutant Kras G12D and PIK3CA H1047R murine lung cancers. *Nat Med* 2008.

127. Maira SM, Stauffer F, Brueggen J, Furet P, Schnell C, Fritsch C, Brachmann S, Chene P, De Pover A, Schoemaker K, Fabbro D, Gabriel D, Simonen M, Murphy L, Finan P, Sellers W, Garcia-Echeverria C. Identification and characterization of NVP-BEZ235, a new orally available dual phosphatidylinositol 3-kinase/mammalian target of rapamycin inhibitor with potent *in vivo* antitumor activity. *Mol Cancer Ther* 2008;**7**:1851–63.

128. Garcia-Echeverria C, Sellers WR. Drug discovery approaches targeting the PI3K/Akt pathway in cancer. *Oncogene* 2008;**27**:5511–26.

129. Voigt P, Brock C, Nurnberg B, Schaefer M. Assigning functional domains within the p101 regulatory subunit of phosphoinositide 3-kinase gamma. *J Biol Chem* 2005;**280**:5121–7.

130. Shekar SC, Wu H, Fu Z, Yip SC, Nagajyothi., Cahill SM, Girvin ME, Backer JM. Mechanism of constitutive phosphoinositide 3-kinase activation by oncogenic mutants of the p85 regulatory subunit. *J Biol Chem* 2005;**280**:27,850–5.

Chapter 26

Inositol Pentakisphosphate: A Signal Transduction Hub

Stephen B. Shears

Laboratory of Signal Transduction, National Institute of Environmental Health Sciences, Research Triangle Park, North Carolina

INTRODUCTION

The phosphorylated inositol moiety is viewed as a fundamental signaling entity that the cell utilizes to generate combinatorially complex arrays of communication pathways [1]. Thus, there are a large number of different inositol phosphates inside cells, several of which are thought to have distinct signaling activities. This chapter focuses on Ins(1,3,4,5,6)P_5, and discusses the evidence that it serves a number of signaling roles, both by itself, and also as a precursor pool for other biologically active inositol polyphosphates (Figure 26.1).

SYNTHESIS OF INS(1,3,4,5,6)P_5

There are two metabolic pathways by which Ins(1,3,4,5,6)P_5 can be synthesized (Figure 26.1a). One involves a two-step phosphorylation of Ins(1,4,5)P_3 by an inositol polyphosphate multikinase (IPMK) that is also known as IPK2 [2]. This is the only means by which yeast cells can synthesize Ins(1,3,4,5,6)P_5 [3]. Higher animals have an additional, more protracted pathway. This also utilizes the promiscuous IPMK, in this case for its Ins(1,3,4,6)P_4 5-kinase activity [4, 5]. This particular metabolic route is distinctive in recruiting the Ins(1,3,4)P_3 6-kinase activity of ITPK1 [4, 6] (Figure 26.1a). There is some debate in the literature concerning which of these two alternative pathways makes the major contribution to synthesis of Ins(1,3,4,5,6)P_5 *de novo*. We can be sure that the meandering route that requires ITPK1 can make a significant contribution, because RNAi against ITPK1 in HeLa cells reduced their Ins(1,3,4,5,6)P_5 levels by nearly 90 percent [7]. On the other hand, the *Drosophila* genome does not encode an ITPK1 homolog, yet flies still synthesize Ins(1,3,4,5,6)P_5 [8]. In general, both metabolic pathways probably make a contribution, likely to varying degrees depending upon the cell type. Another contentious issue concerns the relative contributions of the Ins(1,4,5)P_3 3-kinases vs IPMK for the phosphorylation of Ins(1,4,5)P_3 (Figure 26.1). Again, this may be a situation that probably varies between cell types, but mouse embryo fibroblasts, at least, can synthesize Ins(1,3,4,5,6)P_5 independently of the Ins(1,4,5)P_3 kinases [9]. Studies into the intracellular distribution of the enzymes that synthesize Ins(1,3,4,5,6)

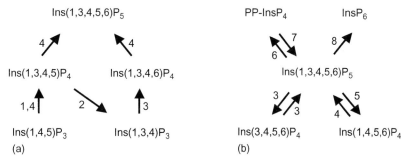

FIGURE 26.1 Synthesis and metabolism of Ins(1,3,4,5,6)P_5.
(a) The steps involved in *de novo* Ins(1,3,4,5,6)P_5; (b) how Ins(1,3,4,5,6)P_5 is metabolized. **1**, Ins(1,4,5)P_3 3-kinase; **2**, Ins(1,3,4,5)P_4 5-phosphatase; **3**, ITPK1; **4**, IPMK/IPK2; **5**, PTEN; **6**, "InsP$_6$ kinase"; **7**, DIPP; **8**, Ins(1,3,4,5,6)P_5 kinase/IPK1.

P_5 has revealed that some of this polyphosphate must be generated in the nucleus [10]. This has been suggested to facilitate the proposed role of $Ins(1,3,4,5,6)P_5$ in chromatin remodeling (see below).

All nucleated cells contain approximately 30–50 μM $Ins(1,3,4,5,6)P_5$ [11, 12]. There are some synchronized fluctuations in cellular levels of $Ins(1,3,4,5,6)P_5$ as cells transit through the cell cycle [13–15] and during cellular differentiation [16]. However, the steady-state levels of $Ins(1,3,4,5,6)P_5$ are not generally considered to respond acutely to receptor activation. A recently described notable exception to the latter rule is the three-fold increase in $Ins(1,3,4,5,6)P_5$ levels seen following Wnt3a addition to mouse F9 teratocarcinoma cells that overexpress the rat Fz1 receptor [17]. However, as the receptor was overexpressed, there should be some reservations concerning the extent to which endogenous Fz1 receptors might regulate $Ins(1,3,4,5,6)P_5$ synthesis.

FUNCTIONS OF INS(1,3,4,5,6)P₅ AS A PRECURSOR POOL

There are several routes by which $Ins(1,3,4,5,6)P_5$ can be metabolized (Figure 26.1b). Two of its metabolites, $Ins(3,4,5,6)P_4$ and $PP-InsP_4$ (see below), are present in cells at much lower concentrations than is the $Ins(1,3,4,5,6)$ P_5 itself. Thus, small changes in the metabolism of the $Ins(1,3,4,5,6)P_5$ precursor can institute relatively large alterations in the concentration of the product. As we shall see, this can be an important signaling mechanism. $Ins(1,3,4,5,6)P_5$ is also a precursor for $InsP_6$, which fulfils several important roles in cells (see below).

DEPHOSPHORYLATION OF INS(1,3,4,5,6)P₅ TO INS(3,4,5,6)P₄

$Ins(3,4,5,6)P_4$ is an especially well-defined and important biologically active metabolite of $Ins(1,3,4,5,6)P_5$ (Figure 26.1b). $Ins(3,4,5,6)P_4$ has been shown to inhibit CaMKII-activated Cl^- channels in the plasma membrane [18–20]. The molecularly distinct process by which Ca^{2+} directly activates some Cl^- channels is also blocked by $Ins(3,4,5,6)P_4$ in certain situations [21] although not in others [20]. The efficacy of $Ins(3,4,5,6)P_4$ ($IC_{50} = 3$–7 μM) [18, 19] ensures that its actions occur within a physiologically-relevant concentration range (1 μM in resting cells, rising to 4–10 μM after receptor activation [22, 23]).

The conductance of ion channels or transporters is frequently regulated by a dynamic balance between competing stimulatory and inhibitory signals. This is because a small shift in the cellular concentrations of either an activator or an inhibitor can yield an "all or nothing" response for ion movements: a switch in the conductance between its "on"

and "off" states. This can influence the transmembrane movement of millions of ions per second [24]. A increase in cellular levels of $Ins(3,4,5,6)P_4$ and the resulting impact upon Cl^- channel conductance can therefore have a substantial biological effect. CaMKII-activated Cl^- channels are present in many different cell types, where, for example, they regulate salt and fluid secretion [22, 23, 25], cell volume homeostasis [26], and electrical excitability in neurones and smooth muscle [27, 28]. To date, the only one of these biological activities so far demonstrated to be inhibited by $Ins(3,4,5,6)P_4$ is salt and fluid secretion [22, 23, 25]. The molecular identity of the responsible channel(s) still remains to be determined.

Intracellular acidic vesicles, such as endosomes, lysosomes, and secretory vesicles, also contain proteins (members of the ClC family) that transport Cl^- ions across their delimiting membranes [29]. These were originally considered to be ion channels, and it was thought that they allowed Cl^- to move into the vesicle [29]. Lately, they have emerged to be Cl^-/H^+ antiporters [30, 31]. Although ClCs are believed to assist in the acidification of the vesicle lumen, their new status as antiporters has rather muddied our understanding of the exact molecular mechanisms involved [30]. Nevertheless, the acidification role of at least one of these proteins, probably ClC3, is inhibited by $Ins(3,4,5,6)P_4$ [32]. This action of $Ins(3,4,5,6)P_4$ can prevent the proper acidification of secretory granules in pancreatic cells that is necessary to facilitate insulin release [32]. Pathological amplification of the inhibition of insulin secretion by $Ins(3,4,5,6)P_4$ might contribute to Type 2 diabetes [32].

Specificity is one of the hallmarks of an efficient cellular signal; this is certainly the case here. Cl^- channel conductance is unaffected by $Ins(1,3,4)P_3$, $Ins(3,4,5)P_3$, $Ins(3,4,6)P_3$, $Ins(4,5,6)P_3$, $Ins(3,5,6)P_3$, $Ins(1,3,4,6)P_4$, $Ins(1,3,4,5)P_4$, $Ins(1,4,5,6)P_4$, and $Ins(1,3,4,5,6)P_5$ [18, 19, 21, 33]. Another important feature is signal amplification, whereby quite small changes in $Ins(3,4,5,6)P_4$ levels dramatically alter ion channel activity. This is seen in the precipitous dose–response curve that describes the highly cooperative manner with which $Ins(3,4,5,6)P_4$ inhibits Cl^- channels [18, 19, 34]. Unfortunately, the mechanism of action of $Ins(3,4,5,6)P_4$ is still to be determined; it does not appear to be a channel blocker [20].

What controls the steady-state levels of $Ins(3,4,5,6)P_4$? The enzyme that synthesizes $Ins(3,4,5,6)P_4$ from $Ins(1,3,4,5,6)P_5$ is actually ITPK1 [33, 35]. This is possible because ITPK1 is not just an inositol phosphate kinase. This enzyme also acts as a phosphotransferase, removing the 1-phosphate group from $Ins(1,3,4,5,6)P_5$ and transferring it directly to the 6-OH of $Ins(1,3,4)P_3$ [35]. By accepting this phosphate group, $Ins(1,3,4)P_3$ phosphorylation actually enhances the rate of dephosphorylation of $Ins(1,3,4,5,6)P_5$ to $Ins(3,4,5,6)P_4$ [35]. This is how receptor-dependent, PLC-initiated increases in levels of $Ins(1,3,4)P_3$

can elevate $Ins(3,4,5,6)P_4$ levels [36]. That is, cellular levels of $Ins(3,4,5,6)P_4$ are under the supervision of cell-surface receptors. Once PLC activity returns to its pre-stimulated state, $Ins(1,3,4)P_3$ levels decline and the phosphotransferase activity of ITPK1 gives way to its $Ins(3,4,5,6)P_4$ 1-kinase activity, thereby restoring $Ins(3,4,5,6)P_4$ levels to their lower, basal level. All of these catalytic activities of ITPK1 take place at a single active site [35, 37]. Interestingly, in polarized epithelial cells, ITPK1 is concentrated at the apical pole of the cell, presumably so that it can supervise $Ins(3,4,5,6)P_4$ synthesis and metabolism proximal to the Cl^- channels that are inhibited by this polyphosphate [25].

PHOSPHORYLATION OF INS(1,3,4,5,6)P$_5$ TO PP-INSP$_4$

There is a family of kinases that converts $Ins(1,3,4,5,6)P_5$ to a diphosphorylated derivative (PP-$InsP_4$) [38]. These particular kinases generally receive more attention for phosphorylating $InsP_6$ [39] (in fact they are widely described in the literature as "$InsP_6$ kinases"), but $Ins(1,3,4,5,6)P_5$ is also a physiologically relevant substrate [40]. The cellular levels of PP-$InsP_4$ are relatively low, usually around 1 percent of $Ins(1,3,4,5,6)P_5$ levels [41].

PP-$InsP_4$ and the other diphosphorylated inositol phosphates are considered "high energy" phosphate donors, and recent work shows they can profitably use this energy to phosphorylate proteins in a kinase-independent manner, *in vitro* at least [42, 43]. However, it is clear that inositol pyrophosphates must also have other mechanisms of action *in vivo* [44].

Two independent groups [45, 46] who work with yeasts recently described an interesting phenotype following genetic manipulation of the degree of "$InsP_6$ kinase" expression. Both laboratories demonstrated that telomere length is inversely proportional to the cellular levels of PP-$InsP_4$. Telomeres are the nucleoprotein complexes that occur at the ends of eukaryotic linear chromosomes. These chromosomal caps prevent nucleolytic degradation and provide a mechanism for cells to distinguish natural termini from broken ends; the latter are signs of DNA damage (for reviews, see [47, 48]). However, it has still to be determined what might be the biological significance of the cell apparently deploying PP-$InsP_4$ as a regulator of telomere length. The mechanism of action of PP-$InsP_4$ has also yet to be directly demonstrated. Finally, these genetic experiments manipulated PP-$InsP_4$ levels outside their physiologically-relevant range of values. Thus, it is still necessary to show that this correlation between PP-$InsP_4$ and telomere length has some biological meaning.

There is rapid, ongoing metabolic cycling between $Ins(1,3,4,5,6)P_5$ and PP-$InsP_4$ *in vivo* [38]. Such cycles are often the focus of regulatory mechanisms. However, little

is known concerning how PP-$InsP_4$ synthesis might be regulated, which would typically be expected if PP-$InsP_4$ were a second messenger. There is some fairly old evidence that this turnover is inhibited by Ca^{2+} [41], but this work has not been pursued further. More work is needed in order to pursue the possibility that PP-$InsP_4$ turnover has a cell-signaling function.

PHOSPHORYLATION OF INS(1,3,4,5,6)P$_5$ TO INSP$_6$

Animal cells cannot survive in the absence of the cellular synthesis of $InsP_6$ (and/or the inositol pyrophosphates made from it) [49]. $InsP_6$ is usually the most abundant inositol polyphosphate in cells, except when, very occasionally, it is exceeded in concentration by $Ins(1,3,4,5,6)P_5$ itself [50]. Thus, changes in $Ins(1,3,4,5,6)P_5$ metabolism have a limited ability to impact upon total cellular $InsP_6$ pool size, unlike the situation with regards to $Ins(3,4,5,6)P_4$ and PP-$InsP_4$ (see above). However, there may be some compartmentalized synthesis of $InsP_6$ at certain subcellular locations. This is evidenced by the recent discovery that the $Ins(1,3,4,5,6)P_5$ kinase is concentrated in discrete cellular foci within the nucleolus (the site of ribosome biosynthesis) [51], and in transcriptionally active euchromatin [51]. Foci of $Ins(1,3,4,5,6)P_5$ kinase have also been detected in the centrosome (which nucleates and organizes the various microtubule arrays that form the mitotic spindle, cilia, and interphase cytoplasmic microtubule network), and in the ciliary body itself [52]. In fact, there is defective ciliary beating in the Kupffer's vesicle in zebrafish embryos from which the $Ins(1,3,4,5,6)P_5$ kinase is genetically eliminated [52].

The importance in these various subcellular zones of $InsP_6$ (or possibly some other activity of the kinase itself) remains to be established. This is despite the fact that more biological actions have been attributed to $InsP_6$ than any other inositol phosphate. This intellectual vacuum may reflect the ease with which experimental artifacts can lead to mis-identification of the mechanisms of action of $InsP_6$ [53]. Nevertheless, several groups are vigorously promoting the idea that $InsP_6$ is a multifunctional intracellular signal, regulating protein kinases, protein phosphatases, and ion channels [54–56]. While this topic continues to be debated, few would dispute that more recent genetically based, *in vivo* experiments have led to convincing demonstrations that $InsP_6$ regulates mRNA export [57, 58], and acts as a structural co-factor for adenosine deaminases acting on RNA [59]. Nevertheless, the latter functions can be fulfilled by sub-physiological levels of $InsP_6$ that are well outside the concentration range within which $InsP_6$ normally fluctuates. In these contexts, $InsP_6$ is viewed as a non-regulated co-factor, albeit an essential one, but not as a signaling molecule.

OTHER FUNCTIONS FOR INS(1,3,4,5,6)P$_5$: REGULATION OF PTEN

We [60, 61] have proposed that Ins(1,3,4,5,6)P$_5$ regulates the activities of PTEN, a tumor suppressor. PTEN has classically been recognized as a PtdIns(3,4,5)P$_3$ 3-phosphatase [62] that downregulates the lipid's signaling activities, such as cell proliferation and Akt-dependent cell survival [62]. To achieve this effect, PTEN must associate with cellular membranes. Nevertheless, it is well known that most of the cell's endogenous PTEN is not localized to membranes, but is instead freely distributed around the cytoplasm and in the nucleus [63–66]. The functional significance of the large reservoir of soluble PTEN has puzzled workers in this field; indeed, some have even argued that cytosolic PTEN is inactive [66]. My laboratory discovered that PTEN is also a high-affinity Ins(1,3,4,5,6)P$_5$ 3-phosphatase [61]. We believe that competition from soluble Ins(1,3,4,5,6)P$_5$ substrate will temper the ability of PTEN to bind and dephosphorylate PtdIns(3,4,5)P$_3$ and further restrict PTEN's already weak protein phosphatase activity [67]. Ins(1,3,4,5,6)P$_5$ may therefore be viewed as "clamping" PTEN activity, the significance being that overall regulation of a signaling system is much tighter, and permits greater amplification, if it has to be de-inhibited (i.e., PTEN escapes Ins(1,3,4,5,6)P$_5$) as well as activated (i.e., PTEN locates PtdIns(3,4,5)P$_3$). This hypothesis does not require PTEN to be very active as an Ins(1,3,4,5,6)P$_5$ phosphatase. Indeed, it may not be; increases in the degree of PTEN expression have been shown to *increase* rather than *decrease* Ins(1,3,4,5,6)P$_5$ levels (the mechanism is unknown) [60]. In any case, for our hypothesis to be correct, it is just necessary that the Ins(1,3,4,5,6)P$_5$ binds to the active site of PTEN–something it is clearly capable of [61].

Any Ins(1,3,4,5,6)P$_5$ that is dephosphorylated by PTEN to Ins(1,4,5,6)P$_4$ will be replenished by Ins(1,4,5,6)P$_4$ 3-kinase activity of IPMK [68, 69]. Indeed, many years ago we demonstrated the dynamic nature of Ins(1,3,4,5,6)P$_5$ 3-phosphatase/Ins(1,4,5,6)P$_4$ 3-kinase metabolic cycling *in vivo* [11, 50]. The latter results indicate the extent to which there may be ongoing dephosphorylation of Ins(1,3,4,5,6)P$_5$ by PTEN.

SopB, a virulence factor in *Salmonella*, dephosphorylates host cell Ins(1,3,4,5,6)P$_5$ to Ins(1,4,5,6)P$_4$ [70, 71]. Other work [72] has indicated that this might have pathological consequences, with the newly-synthesized Ins(1,4,5,6)P$_4$ acting as an antagonist of PtdIns(3,4,5)P$_3$, which normally inhibits Cl$^-$ secretion across the intestinal epithelium. The net result: Ins(1,4,5,6)P$_4$ may enhance diarrhetic salt and fluid secretion [72]. It has not since been considered whether Ins(1,4,5,6)P$_4$ might have broader functions as an PtdIns(3,4,5)P$_3$ antagonist. Mammalian cells do possess another, rather active, Ins(1,3,4,5,6)P$_5$ 3-phosphatase

in the form of a multiple inositol polyphosphate phosphatase (MINPP) [73]. However, this enzyme resides in the lumen of the endoplasmic reticulum, with, apparently, highly restricted access to its substrates [74]. Overall, the idea that the dephosphorylation of Ins(1,3,4,5,6)P$_5$ to Ins(1,4,5,6)P$_4$ is a physiological cell-signaling event would benefit from additional work.

AN EXPANDING LIST OF FURTHER PROPOSED FUNCTIONS FOR INS(1,3,4,5,6)P$_5$

There are reports that Ins(1,3,4,5,6)P$_5$ serves a number of signaling roles in mammalian cells. For example, Ins(1,3,4,5,6)P$_5$ has been shown to activate casein kinase 2 [17, 56] and to inhibit glycogen synthase kinase 3β [17]. These effects are proposed to underlie some of the actions of Wnt ligands that regulate essential aspects of early development, since Wnt3a-dependent activation of the Fz1 receptor acutely elevated Ins(1,3,4,5,6)P$_5$ levels [17]. However, as mentioned above, the latter data were obtained in an overexpression model, and so their biological significance may be considered somewhat in question.

There are genetic experiments that suggest Ins(1,3,4,5,6)P$_5$ can regulate chromatin remodeling [75]. In an effort to study the mechanisms involved, Wu and colleagues [76] studied nucleosome movement along a *Drosophila hsp70* DNA fragment driven by the yeast SWI/SNF chromatin remodeling complex. It was reported that 500 μM Ins(1,3,4,5,6)P$_5$ stimulated this *in vitro* assay for nucleosome sliding [76]. However, as explained previously ([77], and see above), this concentration is well beyond that which prevails in the cell, and probably unattainable even if there were to be compartmentalization of Ins(1,3,4,5,6)P$_5$ synthesis. Thus, the physiological significance of the observation must be seriously in doubt.

Other *in vitro* studies indicate that Ins(1,3,4,5,6)P$_5$ regulates viral assembly [78], L-type Ca^{2+} channels [79], cellular proliferation [80, 81], and apoptotic responses [81]. In yeast, a genomic screen has implicated Ins(1,3,4,5,6)P$_5$ as being an inhibitor of the cellular pathway that identifies and degrades aberrant mRNA transcripts [82]. These various forays into different aspects of cellular biology would benefit from some follow-up studies.

CONCLUDING STATEMENT

This chapter has shown that there are a large number of proposed functions for Ins(1,3,4,5,6)P$_5$ and its metabolites, some well-defined, others somewhat preliminary. Overall, however, there is a strong case that Ins(1,3,4,5,6)P$_5$ is a multifunctional signaling hub.

REFERENCES

1. York JD, Hunter T. Unexpected mediators of protein phosphorylation. *Science* 2004;**306**:2053–5.

2. Fujii M, York JD. A role for rat inositol polyphosphate kinases, rIpk2 and rIpk1, in inositol pentakisphosphate and inositol hexakisphosphate production in Rat-1 cells. *J Biol Chem* 2005;**280**:1156–64.

3. Odom AR, Stahlberg A, Wente SR, York JD. A role for nuclear inositol 1,4,5-trisphosphate kinase in transcriptional control. *Science* 2000;**287**:2026–9.

4. Shears SB. The pathway of myo-inositol 1,3,4-trisphosphate phosphorylation in liver, Identification of myo-inositol 1,3,4-trisphosphate 6-kinase, myo-inositol 1,3,4-trisphosphate 5-kinase, and myo-inositol 1,3,4,6-tetrakisphosphate 5-kinase. *J Biol Chem* 1989;**264**:19,879–86.

5. Chang S-C, Miller AL, Feng Y, Wente SR, Majerus PW. The human homologue of the rat inositol phosphate multikinase is an inositol 1,3,4,6-tetrakisphosphate 5-kinase. *J Biol Chem* 2002;**277**:43,836–43.

6. Wilson MP, Majerus PW. Isolation of inositol 1,3,4-trisphosphate 5/6-kinase, cDNA cloning, and expression of recombinant enzyme. *J Biol Chem* 1996;**271**:11,904–910.

7. Verbsky JW, Chang SC, Wilson MP, Mochizuki Y, Majerus PW. The pathway for the production of inositol hexakisphosphate in human cells. *J Biol Chem* 2005;**280**:1911–20.

8. Seeds AM, Sandquist JC, Spana EP, York JD. A molecular basis for inositol polyphosphate synthesis in Drosophila melanogaster. *J Biol Chem* 2004;**279**:47222.

9. Leyman A, Pouillon V, Bostan A, Schurmans S, Erneux C, Pesesse X. The absence of expression of the three isoenzymes of the inositol 1,4,5-trisphosphate 3-kinase does not prevent the formation of inositol pentakisphosphate and hexakisphosphate in mouse embryonic fibroblasts. *Cell Signal* 2007;**19**:1497–504.

10. Seeds AM, York JD. Inositol polyphosphate kinases: regulators of nuclear function. *Biochem Soc Symp* 2007:183–97.

11. Oliver Jr. KG, Putney JW, Obie JF, Shears SB The interconversion of inositol 1,3,4,5,6-pentakisphosphate and inositol tetrakisphosphates in AR4-2J cells. *J Biol Chem* 1992;**267**:21,528–34.

12. Pittet D, Schlegel W, Lew DP, Monod A, Mayr GW. Mass changes in inositol tetrakis- and pentakisphosphate isomers induced by chemotactic peptide stimulation in HL-60 cells. *J Biol Chem* 1989;**264**:18,489–93.

13. Barker CJ, Wright J, Hughes PJ, Kirk CJ, Michell RH. Complex changes in cellular inositol phosphate complement accompany transit through the cell cycle. *Biochem J* 2004;**380**:465–73.

14. Guse AH, Greiner E, Emmrich F, Brand K. Mass changes of inositol 1,3,4,5,6-pentakisphosphate and inositol hexakisphosphate during cell cycle progression in rat thymocytes. *J Biol Chem* 1993;**268**:7129–33.

15. Balla T, Sim SS, Baukal AJ, Rhee SG, Catt KJ. Inositol polyphosphates are not increased by overexpression of ins(1,4,5)P3 3-kinase but show cell-cycle dependent changes in growth factor-stimulated fibroblasts. *Mol Biol Cell* 1994;**5**:17–28.

16. Mountford JC, Bunce CM, French PJ, Michell RH, Brown G. Intracellular concentrations of inositol, glycerophosphoinositol and inositol pentakisphosphate increase during haematopoietic cell differentiation. *Biochim Biophys Acta* 1994;**1222**:101–8.

17. Gao Y, Wang HY. Inositol pentakisphosphate mediates Wnt/beta-catenin signaling. *J Biol Chem* 2007;**282**:26,490–502.

18. Xie W, Kaetzel MA, Bruzik KS, Dedman JR, Shears SB, Nelson DJ. Inositol 3,4,5,6-tetrakisphosphate inhibits the calmodulin-dependent protein kinase II-activated chloride conductance inT84 colonic epithelial cells. *J Biol Chem* 1996;**271**:14,092–14,097.

19. Ho MWY, Shears SB, Bruzik KS, Duszyk M, French AS. Inositol 3,4,5,6-tetrakisphosphate specifically inhibits a receptor-mediated Ca^{2+}-dependent Cl^- current in CFPAC-1 cells. *Am J Physiol* 1997;**272**:C1160–8.

20. Ho MWY, Kaetzel MA, Armstrong DL, Shears SB. Regulation of a human chloride channel: a paradigm for integrating input from calcium, CaMKII and Ins(3,4,5,6)P4. *J Biol Chem* 2001;**276**:18,673–18,680.

21. Ismailov II, Fuller CM, Berdiev BK, Shlyonsky VG, Benos DJ, Barrett KE. A biologic function for an "orphan" messenger: D-myo-Inositol 3,4,5,6-tetrakisphosphate selectively blocks epithelial calcium-activated chloride current. *Proc Natl Acad Sci USA* 1996;**93**:10,505–10,509.

22. Vajanaphanich M, Schultz C, Rudolf MT, Wasserman M, Enyedi P, Craxton A, Shears SB, Tsien RY, Barrett KE, Traynor-Kaplan AE. Long-term uncoupling of chloride secretion from intracellular calcium levels by Ins(3,4,5,6)P4. *Nature* 1994;**371**:711–14.

23. Carew MA, Yang X, Schultz C, Shears SB. Ins(3,4,5,6)P4 inhibits an apical calcium-activated chloride conductance in polarized monolayers of a cystic fibrosis cell-line. *J Biol Chem* 2000;**275**:26,906–13.

24. Clapham D. How to lose your hippocampus by working on chloride channels. *Neuron* 2001;**29**:1–6.

25. Yang L, Reece J, Gabriel SE, Shears SB. Apical localization of ITPK1 enhances its ability to be a modifier gene product in a murine tracheal cell model of cystic fibrosis. *J Cell Sci* 2006;**119**:1320–8.

26. Nilius B, Prenen J, Voets T, Eggermont J, Bruzik KS, Shears SB, Droogmans G. Inhibition by inositoltetrakisphosphates of calcium- and volume-activated Cl^- currents in macrovascular endothelial cells. *Pflügers Arch Eur J Physiol* 1998;**435**:637–44.

27. Frings S, Reuter D, Kleene SJ. Neuronal Ca^{2+}-activated Cl^- channels–homing in on an elusive channel species. *Progr Neurobiol* 2000;**60**:247–89.

28. Wang XQ, Deriy LV, Foss S, Huang P, Lamb FS, Kaetzel MA, Bindokas V, Marks JD, Nelson DJ. CLC-3 channels modulate excitatory synaptic transmission in hippocampal neurons. *Neuron* 2006;**52**:321–33.

29. Estévez R, Jentsch TJ. CLC chloride channels: correlating structure with function. *Curr Opin Struct Biol* 2002;**12**:531–9.

30. Jentsch TJ. Chloride and the endosomal-lysosomal pathway: emerging roles of CLC chloride transporters. *J Physiol* 2007;**578**:633–40.

31. Miller Jr. FJ, Filali M, Huss GJ, Stanic B, Chamseddine A, Barna TJ, Lamb FS Cytokine activation of nuclear factor kappa B in vascular smooth muscle cells requires signaling endosomes containing Nox1 and ClC-3. *Circ Res* 2007;**101**:663–71.

32. Renström E, Ivarsson R, Shears SB. Ins(3,4,5,6)P4 inhibits insulin granule acidification and fusogenic potential. *J Biol Chem* 2002;**277**:26,717–20.

33. Ho MWY, Yang X, Carew MA, Zhang T, Hua L, Kwon Y-U, Chung S-K, Adelt S, Vogel G, Riley AM, Potter BVL, Shears SB. Regulation of Ins(3456)P4 signaling by a reversible kinase/phosphatase. *Curr Biol* 2002;**12**:477–82.

34. Xie W, Solomons KRH, Freeman S, Kaetzel MA, Bruzik KS, Nelson DJ, Shears SB. Regulation of Ca^{2+}-dependent Cl^- conductance in T84 cells: cross-talk between Ins(3,4,5,6)P4 and protein phosphatases. *J Physiol (Lond)* 1998;**510**:661–73.

35. Chamberlain PP, Qian X, Stiles AR, Cho J, Jones DH, Lesley SA, Grabau EA, Shears SB, Spraggon G. Integration of inositol phosphate signaling pathways via human ITPK1. *J Biol Chem* 2007;**282**:28,117–25.

36. Yang X, Rudolf M, Yoshida M, Carew MA, Riley AM, Chung S-K, Bruzik KS, Potter BVL, Schultz C, Shears SB. Ins(1,3,4)P_3 acts in vivo as a specific regulator of cellular signaling by Ins(3,4,5,6)P_4. *J Biol Chem* 1999;**274**:18,973–18,980.

37. Miller GJ, Wilson MP, Majerus PW, Hurley JH. Specificity determinants in inositol polyphosphate synthesis: crystal structure of inositol 1,3,4-trisphosphate 5/6-kinase. *Mol Cell* 2005;**18**:201–12.

38. Menniti Jr. FS, Miller RN, Putney JW, Shears SB Turnover of inositol polyphosphate pyrophosphates in pancreatoma cells. *J Biol Chem* 1993;**268**:3850–6.

39. Bennett M, Onnebo SM, Azevedo C, Saiardi A. Inositol pyrophosphates: metabolism and signaling. *Cell Mol Life Sci* 2006;**63**:552–64.

40. Saiardi A, Caffrey JJ, Snyder SH, Shears SB. The inositol hexakisphosphate kinase family: catalytic flexibility, and function in yeast vacuole biogenesis. *J Biol Chem* 2000;**275**:24,686–24,692.

41. Glennon MC, Shears SB. Turnover of inositol pentakisphosphates, inositol hexakisphosphate and diphosphoinositol polyphosphates in primary cultured hepatocytes. *Biochem J* 1993;**293**:583–90.

42. Bhandari Jr. R, Saiardi A, Ahmadibeni Y, Snowman AM, Resnick AC, Kristiansen TZ, Molina H, Pandey A, Werner JK, Juluri KR, Xu Y, Prestwich GD, Parang K, Snyder SH Protein pyrophosphorylation by inositol pyrophosphates is a posttranslational event. *Proc Natl Acad Sci USA* 2007;**104**:15,305–10.

43. Saiardi A, Bhandari A, Resnick R, Cain A, Snowman AM, Snyder SH. Inositol pyrophosphate: physiologic phosphorylation of proteins. *Science* 2004;**306**:2101–5.

44. Majerus PW. A discrete signaling function for an inositol pyrophosphate. *Sci STKE* 2007;**2007**:e72.

45. York SJ, Armbruster BN, Greenwell P, Petes TD, York JD. Inositol diphosphate signaling regulates telomere length. *J Biol Chem* 2005;**280**:4264–9.

46. Saiardi A, Resnick AC, Snowman AM, Wendland B, Snyder SH. Inositol pyrophosphates regulate cell death and telomere length via PI3K-related protein kinases. *Proc Natl Acad Sci USA* 2005;**102**:1911–14.

47. d'Adda DF, Teo SH, Jackson SP. Functional links between telomeres and proteins of the DNA-damage response. *Genes Dev* 2004;**18**:1781–99.

48. Smogorzewska A, de Lange T. Regulation of telomerase by telomeric proteins. *Annu Rev Biochem* 2004;**73**:177–208.

49. Verbsky J, Lavine K, Majerus PW. Disruption of the mouse inositol 1,3,4,5,6-pentakisphosphate 2-kinase gene, associated lethality, and tissue distribution of 2-kinase expression. *Proc Natl Acad Sci USA* 2005;**102**:8448–53.

50. Menniti Jr. FS, Oliver KG, Nogimori K, Obie JF, Shears SB, Putney JW Origins of *myo*-inositol tetrakisphosphates in agonist-stimulated rat pancreatoma cells, Stimulation by bombesin of *myo*-inositol 1,3,4,5,6-pentakisphosphate breakdown to *myo*-inositol 3,4,5,6-tetrakisphosphate. *J Biol Chem* 1990;**265**:11,167–11,176.

51. Brehm MA, Schenk TM, Zhou X, Fanick W, Lin H, Windhorst S, Nalaskowski MM, Kobras M, Shears SB, Mayr GW. Intracellular localization of human inositol 1,3,4,5,6-pentakisphosphate 2-kinase. *Biochem J* 2007;**408**:335–45.

52. Sarmah B, Winfrey VP, Olson GE, Appel B, Wente SR. A role for the inositol kinase Ipk1 in ciliary beating and length maintenance. *Proc Natl Acad Sci USA* 2007;**104**:19,843–19,848.

53. Shears SB. Assessing the omnipotence of inositol hexakisphosphate. *Cell Signal* 2001;**13**:151–8.

54. Barker CJ, Leibiger IB, Leibiger B, Berggren PO. Phosphorylated inositol compounds in beta -cell stimulus–response coupling. *Am J Physiol Endocrinol Metab* 2002;**283**:E1113–22.

55. Irvine RF, Schell M. Back in the water: the return of the inositol phosphates. *Nat Rev Mol Cell Biol* 2001;**2**:327–38.

56. Solyakov L, Cain K, Tracey BM, Jukes R, Riley AM, Potter BV, Tobin AB. Regulation of casein kinase-2 (CK2) activity by inositol phosphates. *J Biol Chem* 2004;**279**:43,403–14.

57. Alcázar-Román AR, Tran EJ, Guo S, Wente SR. Inositol hexakisphosphate and Gle1 activate the DEAD-box protein Dbp5 for nuclear mRNA export. *Nat Cell Biol* 2006;**8**:645–7.

58. York JD, Odom AR, Murphy R, Ives EB, Wente SR. A phospholipase C-dependent inositol polyphosphate kinase pathway required for efficient messenger RNA export. *Science* 1999;**285**:96–100.

59. Macbeth MR, Schubert HL, Vandemark AP, Lingam AT, Hill CP, Bass BL. Inositol hexakisphosphate is bound in the ADAR2 core and required for RNA editing. *Science* 2005;**309**:1534–9.

60. Deleu S, Choi K, Pesesse X, Cho J, Sulis ML, Parsons R, Shears SB. Physiological levels of PTEN control the size of the cellular Ins(1,3,4,5,6)P(5) pool. *Cell Signal* 2006;**18**:488–98.

61. Caffrey JJ, Darden T, Wenk MR, Shears SB. Expanding coincident signaling by PTEN through its inositol 1,3,4,5,6-pentakisphosphate 3-phosphatase activity. *FEBS Letts* 2001;**499**:6–10.

62. Di Cristofano A, Pandolfi PP. The multiple roles of PTEN in tumor suppression. *Cell* 2000;**100**:387–90.

63. McMenamin ME, Soung P, Perera S, Kaplan I, Loda M, Sellers WR. Loss of PTEN expression in paraffin-embedded primary prostate cancer correlates with high Gleason score and advanced stage. *Cancer Res* 1999;**59**:4291–6.

64. Perren A, Komminoth P, Saremaslani P, Matter C, Feurer S, Lees JA, Heitz PU, Eng C. Mutation and expression analysis reveal differential subcellular compartmentalization of PTEN in endocrine pancreatic tumors compared to normal islet cells. *Am J Pathol* 2000;**157**:1097–103.

65. Leslie NR, Bennett D, Gray A, Pass I, Hoang-Xuan K, Downes CP. Targeting mutants of PTEN reveal distinct subsets of tumour suppressor functions. *Biochem J* 2001;**357**:427–35.

66. Li Z, Dong X, Wang Z, Liu W, Deng N, Ding Y, Tang L, Hla T, Zeng R, Li L, Wu D. Regulation of PTEN by Rho small GTPases. *Nat Cell Biol* 2005;**7**:399–404.

67. Myers MP, Stolarov JP, Eng C, Li J, Wang SI, Wigler MH, Parsons R, Tonks NK. PTEN, the tumor suppressor from human chromosome 10q23, is a dual specificity phosphatase. *Proc Natl Acad Sci USA* 1997;**94**:9052–7.

68. Riley AM, Deleu S, Qian X, Mitchell J, Chung SK, Adelt S, Vogel G, Potter BV, Shears SB. On the contribution of stereochemistry to human ITPK1 specificity: Ins(1,4,5,6)P(4) is not a physiologic substrate. *FEBS Letts* 2006;**580**:324–30.

69. Chang SC, Majerus PW. Inositol polyphosphate multikinase regulates inositol 1,4,5,6-tetrakisphosphate. *Biochem Biophys Res Commun* 2006;**339**:209–16.

70. Norris FA, Wilson MP, Wallis TS, Galyov EE, Majerus PW. SopB, a protein required for virulence of *Salmonella dublin*, is an inositol phosphate phosphatase. *Proc Natl Acad Sci USA* 1998;**95**:14,057–9.

71. Zhou D, Chen L-M, Hernandez L, Shears SB, Galán JE. A *Salmonella* inositol polyphosphatase acts in conjunction with other bacterial effectors to promote host-cell actin cytoskeleton rearrangements and bacterial internalization. *Mol Microbiol* 2001;**39**:248–59.

72. Eckmann L, Rudolf MT, Ptasznik A, Schultz C, Jiang T, Wolfson N, Tsien R, Fierer J, Shears SB, Kagnoff MF, Traynor-Kaplan A. D-*myo*-inositol 1,4,5,6-tetrakisphosphate produced in human intestinal epithelial cells in response to *Salmonella* invasion inhibits phosphoinositide 3-kinase signaling pathways. *Proc Natl Acad Sci USA* 1997;**94**:14,456–60.

73. Nogimori Jr. K, Hughes PJ, Glennon MC, Hodgson ME, Putney JW, Shears SB Purification of an inositol (1,3,4,5)-tetrakisphosphate 3-phosphatase activity from rat liver and its substrate specificity. *J Biol Chem* 1991;**266**:16,499–506.

74. Ali N, Craxton A, Shears SB. Hepatic Ins(1,3,4,5)P$_4$ 3-phosphatase is compartmentalized inside endoplasmic reticulum. *J Biol Chem* 1993;**268**:6161–7.

75. Steger DJ, Haswell ES, Miller AL, Wente SR, O'Shea EK. Regulation of chromatin remodeling by inositol polyphosphates. *Science* 2003;**299**:114–16.

76. Shen X, Xiao H, Ranallo R, Wu W-H, Wu C. Modulation of ATP-dependent chromatin-remodeling complexes by inositol polyphosphates. *Science* 2003;**299**:112–14.

77. Shears SB. How versatile are inositol phosphate kinases? *Biochem J* 2004;**377**:265–80.

78. Campbell S, Fisher RJ, Towler EM, Fox S, Issaq HJ, Wolfe T, Phillips LR, Rein A. Modulation of HIV-like particle assembly *in vitro* by inositol phosphates. *Proc Natl Acad Sci USA* 2001;**98**:10,875–10,879.

79. Quignard JF, Rakotoarisoa L, Mironneau J, Mironneau C. Stimulation of L-type Ca^{2+} channels by inositol pentakis- and hexakisphosphates in rat vascular smooth muscle cells. *J Physiol* 2003;**549**:729–37.

80. Orchiston EA, Bennett D, Leslie NR, Clarke RG, Winward L, Downes CP, Safrany ST. PTEN M-CBR3, a versatile and selective regulator of inositol 1,3,4,5,6-pentakisphosphate (Ins(1,3,4,5,6)P$_5$). Evidence for Ins(1,3,4,5,6)P$_5$ as a proliferative signal. *J Biol Chem* 2004;**279**:1116–22.

81. Piccolo E, Vignati S, Maffucci T, Innominato PF, Riley AM, Potter BV, Pandolfi PP, Broggini M, Iacobelli S, Innocenti P, Falasca M. Inositol pentakisphosphate promotes apoptosis through the PI 3-K/Akt pathway. *Oncogene* 2004;**23**:1754–65.

82. Wilson MA, Meaux S, van Hoof A. A genomic screen in yeast reveals novel aspects of nonstop mRNA metabolism. *Genetics* 2007;**177**:773–84.

Chapter 27

IP₃ Receptors

Colin W. Taylor and Zhao Ding

Department of Pharmacology, University of Cambridge, Cambridge, England, UK

INTRODUCTION

Inositol 1,4,5-trisphosphate receptors (IP₃Rs) are large proteins: the native receptor is some 20 nm across, extends more than 10 nm from the membrane of the endoplasmic reticulum (ER), and each of its four subunits comprises about 2700 amino acid residues [1]. Their close relatives, ryanodine receptors (RyRs), are even bigger [2]. Size is important for these intracellular Ca^{2+} channels because it allows opening of the channels to be controlled by many different intracellular stimuli, and it allows them selectively to direct the Ca^{2+} they release to specific target proteins or organelles. These features endow both families of channels with the ability to integrate diverse intracellular signals and so to transduce these signals into cytosolic Ca^{2+} signals that can specifically regulate almost every aspect of cellular behavior.

IP₃Rs play an essential role in linking the many receptors that stimulate IP₃ formation to release of Ca^{2+} from intracellular stores [3]. However, they are also important in mediating the regenerative propagation of intracellular Ca^{2+} signals by means of Ca^{2+}-induced Ca^{2+} release [4]: Ca^{2+} released by one channel can stimulate the opening of a neighbor. Controlling this process determines whether IP₃-evoked Ca^{2+} signals remain local, allowing selective regulation of only certain proteins, or propagate regeneratively, and provide a Ca^{2+} signal to the entire cell. Finally, by emptying intracellular Ca^{2+} stores, IP₃Rs cause Ca^{2+} to dissociate from the ER Ca^{2+}-sensing protein, STIM, which then activates store-operated Ca^{2+} channels and so Ca^{2+} entry across the plasma membrane [5]. IP₃R are thus responsible for initiating and sustaining Ca^{2+} signals, and interactions between IP₃Rs determine whether Ca^{2+} signals are local or global.

Both IP₃Rs and RyRs are expressed predominantly in the ER (or sarcoplasmic reticulum), but each may also be expressed in the membranes of other intracellular organelles (e.g., Golgi apparatus and secretory vesicles) and, at least in some cells, within the plasma membrane [6].

DIVERSITY OF IP₃RS

IP₃Rs are expressed in most animal cells. Three genes encode closely related subtypes in vertebrates, but there is only a single gene in invertebrates [7–9]. IP₃Rs are also expressed in unicellular organisms, including *Dictyostelium* [10] and *Paramecium* [11]. It remains unclear whether IP₃Rs (or RyRs) similar to those found in animals are expressed in plants [12,13]. At least two of the vertebrate subtypes (IP₃R1 and IP₃R2) are alternatively spliced [14]. Because the functional channel is a tetramer, which can assemble from the same or different subunits, and most mammalian cells express more than one receptor subtype, there is considerable scope for IP₃R diversity [15]. The different subtypes and their splice variants are differentially expressed in mature tissues [7], during development [9] they respond differently to chronic stimulation, and their assembly into heterotetramers is regulated. But the physiological significance of IP₃R heterogeneity is not yet entirely clear. There are differences in the affinities of the subtypes for IP₃, in their modulation by other intracellular stimuli [16, 17], and, most notably, in the proteins with which they interact [18–20]. Furthermore, knocking out different IP₃R subtypes produces very different phenotypes [21–23]. The properties shared by all IP₃Rs are, however, more striking than these differences. All IP₃Rs are tetrameric cation channels with large conductances and only modest selectivity for Ca^{2+} over K^+ [6]. The latter is not a problem for IP₃Rs within the ER, because the activity of Ca^{2+} pumps ensures that, of the cations that permeate IP₃R, only Ca^{2+}

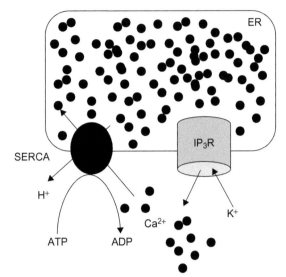

FIGURE 27.1 Ca^{2+} uptake by and release from the ER.
The SR/ER Ca^{2+}-ATPase (SERCA) uses ATP to drive the active uptake of Ca^{2+} into the ER. Hydrolysis of a single ATP drives the uptake of 2 Ca^{2+} in exchange for 2 or 3 H^+ [62]. The steep Ca^{2+} gradient across the ER membrane then provides the driving force for a rapid release of Ca^{2+} via IP$_3$R when its channel opens. The electrogenic release of Ca^{2+} must be compensated by movements of other ions if Ca^{2+} release is to continue. TRIC channels, which are selectively permeable to K^+ [24] (not shown), and the IP$_3$R itself provide routes for a compensating inward movement of K^+.

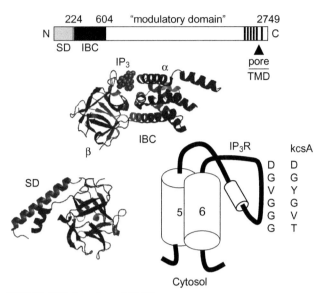

FIGURE 27.2 Structure of IP$_3$R1.
The linear sequence of a single subunit of IP$_3$R1 is shown highlighting the suppressor domain (SD, PDB accession number 1XZZ) [30], the IP$_3$-binding core (IBC, PDB accession number 1N4K) [27], the modulatory domain, and the C-terminal region with its six transmembrane domains (TMD). The proposed structure of the pore is shown with the putative selectivity filter compared with that of the bacterial K^+ channel KcsA [63].

has a significant electrochemical gradient across the ER membrane. Indeed, the ability of the IP$_3$Rs to allow K^+ to permeate may facilitate rapid Ca^{2+} release by providing a route for a K^+ flux (perhaps in concert with the recently discovered TRIC proteins [24]) into the ER to electrically compensate the release of Ca^{2+} (Figure 27.1). All IP$_3$Rs have similar primary structures, Ca^{2+} and IP$_3$ control the opening of all IP$_3$Rs, and they are all modulated by additional signals.

STRUCTURE OF IP$_3$R

Key structural features of a single subunit of the type 1 IP$_3$R are shown in Figure 27.2. The IP$_3$-binding site lies close to the N-terminal, and towards the C-terminal there are six transmembrane domains (TMDs). The last pair of these (TMD5–6), together with the intervening luminal loop (P-loop), from each of the four subunits form the pore of the channel [25]. The overall organization of the pore is probably similar to that of K^+ channels, with the two TMDs and the pore helix holding a relatively conserved selectivity filter (Figure 27.2). However, the detailed structure of the IP$_3$R pore must differ from that of K^+ channels to account for its larger conductance and its selectively (albeit modest) for divalent over monovalent cations. By analogy with RyR [26], the large conductance probably results from an increase in the diameter of the open pore,

and negatively charged residues flanking the luminal loop may contribute to the modest selectivity for divalent cations. The large stretch of residues separating the IP$_3$-binding site from the TMD region has been termed the "modulatory region" (Figure 27.2), although it is now clear that many modulators exert their effects through sites that lie elsewhere in the IP$_3$R.

Each of the four subunits of the IP$_3$R has an IP$_3$-binding site lying towards the N-terminal and formed by two distinct domains (α and β) linked by a short stretch of residues that includes the S1 splice site (Figure 27.2). A high-resolution structure of this IP$_3$-binding core (IBC; residues 224–604) shows IP$_3$ nestled in a positively-charged cleft between the two domains [27], leading to the suggestion that IP$_3$ binding might cause the clam-like IBC to close [1] in a manner analogous to activation of glutamate receptors [28]. Immediately N-terminal to the IBC is the suppressor domain (SD; residues 1–223) [29], so called because it reduces the affinity of the IBC for IP$_3$. The structure of the SD has also been resolved [30] (Figure 27.2), though not yet its relationship with the IBC. The IBC alone binds IP$_3$ with such high affinity that it can be used experimentally as an "IP$_3$ sponge" to define the role of IP$_3$ in intact cells [31]. This is a useful tool, because the only antagonists of IP$_3$R (heparin, xestospongin, and 2-aminoethyldiphenylborane) are notoriously lacking in specificity. More importantly, the SD probably plays an essential role in transmitting the conformational changes that follow IP$_3$ binding from the IBC to the pore. The details are not yet resolved, but a plausible scheme suggests that after IP$_3$ binding to the IBC, the SD interacts with the short cytosolic loop that links TMD4 and TMD5 [32,33], relieving its constrictive force on the channel gate and leading to opening of the channel.

REGULATION OF IP$_3$RS BY IP$_3$ AND CA^{2+}

Despite earlier controversy, it is now clear that all IP$_3$Rs are biphasically regulated by cytosolic Ca^{2+}. This is an important feature that allows IP$_3$Rs to respond to Ca^{2+} released by neighboring IP$_3$Rs or by other Ca^{2+} channels, and thereby regeneratively to propagate Ca^{2+} signals across cells [4]. Luminal Ca^{2+} has also been reported to regulate IP$_3$Rs, but, as with RyRs, it has proven difficult to resolve whether this really results from Ca^{2+} regulating from the luminal surface of the receptor or at its cytosolic surface after Ca^{2+} has passed through the open channel. Regulation of IP$_3$Rs by luminal Ca^{2+} may be unresolved, but there is no such uncertainty about their regulation by cytosolic Ca^{2+}: all IP$_3$Rs are stimulated by Ca^{2+}, and all are inhibited by higher concentrations of Ca^{2+} [18, 34, 35]. The details of how this biphasic regulation by cytosolic Ca^{2+} occurs have not been resolved. It is accepted that IP$_3$ and Ca^{2+} must both bind to the IP$_3$R before the channel can open, and that binding of IP$_3$ regulates Ca^{2+} binding. In other words, IP$_3$ works by tuning the Ca^{2+} sensitivity of the IP$_3$R. One scheme suggests that the major effect of IP$_3$ is to relieve Ca^{2+} inhibition by decreasing the affinity of an inhibitory Ca^{2+}-binding site [36], while another suggests that IP$_3$ binding causes a stimulatory Ca^{2+}-binding site to be exposed and an inhibitory one to be concealed [37, 38]. Whether this Ca^{2+} regulation involves Ca^{2+} binding to the IP$_3$R itself or to associated proteins remains unresolved. The role of calmodulin, in particular, as a possible mediator of Ca^{2+} inhibition remains contentious [34, 39]. There are numerous Ca^{2+}-binding sites within the IP$_3$R [40], but the most persuasive evidence for a regulatory role of a Ca^{2+}-binding site implicates a sequence centered on Glu-2100 of IP$_3$R1 [16, 35, 41]. Homologous sequences are well conserved in all IP$_3$Rs and RyRs [17]. Mutation of the conserved Glu within this region decreases the Ca^{2+}-sensitivity of IP$_3$R activation [16], and perhaps also the inhibition. Furthermore, switching the putative Ca^{2+}-sensor region between IP$_3$R subtypes appears sufficient to switch the sensitivity of the Ca^{2+} regulation [35]. However, the link between IP$_3$ binding and control of Ca^{2+} binding remains unclear, and nor is it clear whether additional Ca^{2+}-binding sites contribute to the biphasic regulation of IP$_3$Rs. In particular, the possible role of two putative Ca^{2+}-binding sites within the IBC (Ca-I and Ca-II) that may be regulated by IP$_3$ [42] deserves further attention.

In summary, co-regulation of IP$_3$Rs by Ca^{2+} and IP$_3$ is ubiquitous and important in controlling the spatial propagation of Ca^{2+} signals, but the underlying mechanisms remain incompletely defined.

MODULATION OF IP$_3$R

IP$_3$Rs are modulated by many additional pathways. ATP binds to sites within the modulatory domain and increases IP$_3$ sensitivity [43], but its effects differ profoundly between IP$_3$R subtypes [44]. Many protein kinases (e.g., PKA, PKG, PKC, CaMKII, ERK, tyrosine kinases, etc.), often with the involvement of anchoring proteins, phosphorylate and modulate IP$_3$Rs. Cyclic GMP-dependent protein kinase (PKG), for example, can both phosphorylate IP$_3$R and a protein (IRAG) that associates tightly with the IP$_3$R to inhibit gating [45]. Chemicals that modify sulfydryl groups, including reactive oxygen species, increase IP$_3$ sensitivity by modifying conserved cysteine residues towards the C-terminal. A protein within the ER lumen, ERp44, associates specifically with the TMD region of IP$_3$R1 and mediates regulation of the IP$_3$R by luminal Ca^{2+}, pH, and redox state [46]. These mechanisms allow the redox state of the cell to regulate IP$_3$R sensitivity. IP$_3$Rs are also subject to proteolytic cleavage, which contributes to both regulation of IP$_3$R levels [47] and to their regulation, because cleavage of IP$_3$R1 by caspase-3 causes its activation and thereby accelerates apoptosis [48, 49].

PROTEIN INTERACTIONS WITH IP$_3$R

More than 300 proteins have been suggested to interact with IP$_3$Rs, and about 20 of them have been studied in detail [18–20]. They include proteins that regulate IP$_3$Rs, Ca^{2+}-sensing proteins that respond to the active IP$_3$Rs, and proteins that control the subcellular distribution of IP$_3$Rs. Examples include Bcl-X$_L$ [50] and the mutant huntingtin Htt^{exp} [51], which bind near the C-terminal and increase sensitivity to IP$_3$. IRBIT binds to the IBC and inhibits IP$_3$Rs by competing with IP$_3$ [52], while receptor for activated C kinase-1 (RACK1), calmodulin, and Ca^{2+}-binding protein (CaBP) bind within the SD and modulate binding of IP$_3$ [53–55]. Another example is chromogranin, which binds to the third luminal of the IP$_3$R loop to potentiate responses to IP$_3$ [56]. Ca^{2+}-calmodulin-dependent protein kinase II (CaMKII) binds directly to IP$_3$R2 and thereby responds specifically to Ca^{2+} released by it [57].

Most IP$_3$Rs are found in the membranes of the ER, but they also occur in the Golgi apparatus, secretory vesicles, nuclear envelope, and plasma membrane [6]. Even within the ER, the distribution of IP$_3$Rs is far from uniform. On a molecular scale, IP$_3$Rs occur in clusters [58], allowing the Ca^{2+} released by one receptor rapidly to influence its neighbors [59]. At a cellular level, they can be concentrated in discrete areas of the ER – at the apical pole of pancreatic acinar cells, for example. There are also important functional associations between IP$_3$Rs in the ER and other membranes: there are, for example, close associations between IP$_3$Rs and mitochondria [60]. We are only just beginning to unravel the mechanisms responsible for putting IP$_3$Rs into the right places, but scaffolding proteins are likely to be important. A single example illustrates the likely complexity of the interactions between scaffold proteins and IP$_3$Rs. The Homer proteins form a family of dimeric scaffold proteins that assemble signaling proteins at excitatory synapses. The N-terminal of Homer binds to IP$_3$R, type-1 metabotropic glutamate receptors, and to Shank, another scaffold protein which is targeted by its PDZ domain to the postsynaptic density and which itself binds further signaling proteins [61]. This chain of protein–protein interactions both targets IP$_3$Rs to the dendritic spines of hippocampal neurones and brings them into intimate association with other signaling proteins, including receptors that stimulate IP$_3$ formation and channels that mediate Ca^{2+} entry.

In summary, IP$_3$Rs are ubiquitous intracellular Ca^{2+} channels that are regulated by both IP$_3$ and cytosolic Ca^{2+}, allowing them to control both local and global Ca^{2+} signals. Their modulation by many addition signals and association with many different proteins allows them effectively to integrate diverse signals before returning them to the cell as a Ca^{2+} signal.

REFERENCES

1. Taylor CW, da Fonseca PCA, Morris EP. IP$_3$ receptors: the search for structure. *Trends Biochem Sci* 2004;**29**:210–19.

2. Sitsapesan R, Williams AJ. *The structure and function of ryanodine receptors*. London: Imperial College Press; 1998. pp. 1–325.

3. Berridge MJ, Bootman MD, Roderick HL. Calcium signalling: dynamics, homeostasis and remodelling. *Nat Rev Mol Cell Biol* 2003;**4**:517–29.

4. Bootman MD, Berridge MJ, Lipp P. Cooking with calcium: the recipes for composing global signals from elementary events. *Cell* 1997;**91**:367–73.

5. Hogan PG, Rao A. Dissecting I$_{CRAC}$, a store-operated calcium current. *Trends Biochem Sci* 2007;**32**:235–45.

6. Dellis O, Dedos S, Tovey SC, Rahman T-U, Dubel SJ, Taylor CW. Ca^{2+} entry through plasma membrane IP$_3$ receptors. *Science* 2006;**313**:229–33.

7. Taylor CW, Genazzani AA, Morris SA. Expression of inositol trisphosphate receptors. *Cell Calcium* 1999;**26**:237–51.

8. Zhang D, Boulware MJ, Pendleton MR, Nogi T, Marchant JS. The inositol 1,4,5-trisphosphate receptor (Itpr) gene family in *Xenopus*: identification of type 2 and type 3 inositol 1,4,5-trisphosphate receptor subtypes. *Biochem J* 2007;**404**:383–91.

9. Ashworth R, Devogelaere B, Fabes J, et al. Molecular and functional characterization of inositol trisphosphate receptors during early zebrafish development. *J Biol Chem* 2007;**282**:13,984–93.

10. Traynor D, Milne JLS, Insall RH, Kay RR. Ca^{2+} signalling is not required for chemotaxis in Dictyostelium. *EMBO J* 2000;**19**:4846–54.

11. Ladenburger EM, Korn I, Kasielke N, Wassmer T, Plattner H. An Ins(1,4,5)P$_3$ receptor in *Paramecium* is associated with the osmoregulatory system. *J Cell Sci* 2006;**119**:3705–17.

12. Sanders D, Pelloux J, Brownlee C, Harper JF. Calcium at the crossroads of signaling. *Plant Cell* 2002;**14**(Suppl):S401–17.

13. Nagata T, Iizumi S, Satoh K, et al. Comparative analysis of plant and animal calcium signal transduction element using plant full-length cDNA data. *Mol Biol Evol* 2004;**21**:1855–70.

14. Patel S, Joseph SK, Thomas AP. Molecular properties of inositol 1,4,5-trisphosphate receptors. *Cell Calcium* 1999;**25**:247–64.

15. Taylor CW, Genazzani AA, Morris SA. Expression of inositol trisphosphate receptors. *Cell Calcium* 1999;**26**:237–51.

16. Miyakawa T, Mizushima A, Hirose K, et al. Ca^{2+}-sensor region of IP3 receptor controls intracellular Ca^{2+} signaling. *EMBO J* 2001;**20**:1674–80.

17. Bezprozvanny I. The inositol 1,4,5-trisphosphate receptors. *Cell Calcium* 2005;**38**:261–72.

18. Foskett JK, White C, Cheung KH, Mak DO. Inositol trisphosphate receptor Ca^{2+} release channels. *Physiol Rev* 2007;**87**:593–658.

19. Patterson RL, Boehning D, Snyder SH. Inositol 1,4,5-trisphosphate receptors as signal integrators. *Annu Rev Biochem* 2004;**73**:437–65.

20. Choe C, Ehrlich BE. The inositol 1,4,5-trisphosphate receptor (IP$_3$R) and its regulators: sometimes good and sometimes bad teamwork. *Sci STKE* 2006;**Re15**.

21. Matsumoto M, Nagata E. Type 1 inositol 1,4,5-trisphosphate receptor knock-out mice: their phenotypes and their meaning in neuroscience and clinical practice. *J Mol Med* 1999;**77**:406–11.

22. Futatsugi A, Nakamura T, Yamada MK, et al. IP$_3$ receptor types 2 and 3 mediate exocrine secretion underlying energy metabolism. *Science* 2005;**309**:2232–4.

23. Li X, Zima AV, Sheikh F, Blatter LA, Chen J. Endothelin-1-induced arrhythmogenic Ca^{2+} signaling is abolished in atrial myocytes of inositol-1,4,5-trisphosphate (IP$_3$)-receptor type 2-deficient mice. *Circ Res* 2005;**96**:1274–81.

24. Yazawa M, Ferrante C, Feng J, et al. TRIC channels are essential for Ca^{2+} handling in intracellular stores. *Nature* 2007;**448**:78–82.

25. Ramos-Franco J, Galvan D, Mignery GA, Fill M. Location of the permeation pathway in the recombinant type-1 inositol 1,4,5-trisphosphate receptor. *J Gen Physiol* 1999;**114**:243–50.

26. Tinker A, Williams AJ. Probing the structure of the conduction pathway of the sheep cardiac sarcoplasmic reticulum calcium-release channel with permeant and impermeant organic cations. *J Gen Physiol* 1993;**102**:1107–29.

27. Bosanac I, Alattia J-R, Mal TK, et al. Structure of the inositol 1,4,5-trisphosphate receptor binding core in complex with its ligand. *Nature* 2002;**420**:696–701.

28. Madden DR. The structure and function of glutamate receptor ion channels. *Nat Rev Neurosci* 2002;**3**:91–101.

29. Yoshikawa F, Uchiyama T, Iwasaki H, et al. High efficient expression of the functional ligand binding site of the inositol 1,4,5-trisphosphate receptor in Escherichia coli. *Biochem Biophys Res Commun* 1999;**257**:792–7.

30. Bosanac I, Yamazaki H, Matsu-ura T, Michikawa M, Mikoshiba K, Ikura M. Crystal structure of the ligand binding suppressor domain of type 1 inositol 1,4,5-trisphosphate receptor. *Mol Cell* 2005;**17**:193–203.

31. Uchiyama T, Yoshikawa F, Hishida A, Furuichi T, Mikoshiba K. A novel recombinant hyper-affinity inositol 1,4,5-trisphosphate (IP$_3$) absorbent traps IP$_3$, resulting in specific inhibition of IP$_3$-mediated calcium signaling. *J Biol Chem* 2002;**277**:8106–13.

32. Boehning D, Joseph SK. Direct association of ligand-binding and pore domains in homo- and heterotetrameric inositol 1,4,5-trisphosphate receptors. *EMBO J* 2000;**19**:5450–9.

33. Schug ZT, Joseph SK. The role of the S4–S5 linker and C-terminal tail in inositol 1,4,5-trisphosphae receptor function. *J Biol Chem* 2006;**281**:24,431–40.

34. Taylor CW, Laude AJ. IP$_3$ receptors and their regulation by calmodulin and cytosolic Ca^{2+}. *Cell Calcium* 2002;**32**:321–34.

35. Tu H, Wang Z, Bezprozvanny I. Modulation of mammalian inositol 1,4,5-trisphosphate receptor isoforms by calcium: a role of calciun sensor region. *Biophys J* 2005;**88**:1056–69.

36. Mak D-OD, McBride S, Foskett JK. Inositol 1,4,5-tris-phosphate activation of inositol tris-phosphate receptor Ca2+ channel by ligand tuning of Ca2+ inhibition. *Proc Natl Acad Sci USA* 1998;**95**:15,821–5.

37. Marchant JS, Taylor CW. Cooperative activation of IP$_3$ receptors by sequential binding of IP$_3$ and Ca^{2+} safeguards against spontaneous activity. *Curr Biol* 1997;**7**:510–18.

38. Adkins CE, Taylor CW. Lateral inhibition of inositol 1,4,5-trisphosphate receptors by cytosolic Ca^{2+}. *Curr Biol* 1999;**9**:1115–18.

39. Michikawa T, Hirota J, Kawano S, et al. Calmodulin mediates calcium-dependent inactivation of the cerebellar type 1 inositol 1,4,5-trisphosphate receptor. *Neuron* 1999;**23**:799–808.

40. Sienaert I, Missiaen L, De Smedt H, Parys JB, Sipma H, Casteels R. Molecular and functional evidence for multiple Ca^{2+}-binding domains on the type 1 inositol 1,4,5-trisphosphate receptor. *J Biol Chem* 1997;**272**:25,899–906.

41. Tu H, Nosyreva E, Miyakawa T, et al. Functional and biochemical analysis of the type 1 inositol (1,4,5)-trisphosphate receptor calcium sensor. *Biophys J* 2003;**85**:290–9.

42. Bosanac I, Michikawa T, Mikoshiba K, Ikura M. Structural insights into the regulatory mechanism of IP$_3$ receptor. *Biochim Biophys Acta* 2004;**1742**:89–102.

43. Miyakawa T, Maeda A, Yamazawa T, Hirose K, Kurosaki T, Iino M. Encoding of Ca^{2+} signals by differential expression of IP$_3$ receptor subtypes. *EMBO J* 1999;**18**:1303–8.

44. Tu H, Wang Z, Nosyreva E, De Smedt H, Bezprozvanny I. Functional characterization of mammalian inositol 1,4,5-trisphosphate receptor isoforms. *Biophys J* 2005;**88**:1046–55.

45. Ammendola A, Geiselhöringer A, Hofmann F, Schlossmann J. Molecular determinants of the interaction between the inositol 1,4,5-trisphosphate receptor-associated cGMP kinase substrate (IRAG) and cGMP kinase Ib. *J Biol Chem* 2001;**276**:24,153–9.

46. Higo T, Hattori M, Nakamura T, Natsume T, Michikawa T, Mikoshiba K. Subtype-specific and ER lumenal environment-dependent regulation of inositol 1,4,5-trisphosphate receptor type 1 by ERp44. *Cell* 2005;**120**:85–98.

47. Wojcikiewicz RJ, Ernst SA, Yule DI. Secretagogues cause ubiquitination and down-regulation of inositol 1,4,5-trisphosphate receptors in rat pancreatic acinar cells. *Gastroenterology* 1999;**116**:1194–201.

48. Hirota J, Furuichi T, Mikoshiba K. Inositol 1,4,5-trisphosphate receptor type 1 is a substrate for caspase-3 and is cleaved during apoptosis in a caspase-3-dependent manner. *J Biol Chem* 1999;**274**:34,433–7.

49. Assefa Z, Bultynck G, Szlufcik K, et al. Caspase-3-induced truncation of type 1 inositol trisphosphate receptor accelerates apoptotic cell death and induces inositol trisphosphate-independent calcium release during apoptosis. *J Biol Chem* 2004;**279**:43,227–36.

50. Li C, Wang X, Vais H, Thompson CB, Foskett JK, White C. Apoptosis regulation by Bcl-xL modulation of mammalian inositol 1,4,5-trisphosphate receptor channel isoform gating. *Proc Natl Acad Sci USA* 2007;**104**:12,565–70.

51. Tang TS, Tu H, Chan EY, et al. Huntingtin and huntingtin-associated protein 1 influence neuronal calcium signaling mediated by inositol-(1,4,5) triphosphate receptor type 1. *Neuron* 2003;**39**:227–39.

52. Devogelaere B, Nadif Kasri N, Derua R, et al. Binding of IRBIT to the IP$_3$ receptor: determinants and functional effects. *Biochem Biophys Res Commun* 2006;**343**:49–56.

53. Patterson RL, van Rossum DB, Barrow RK, Snyder SH. RACK1 binds to inositol 1,4,5-trisphosphate receptors and mediates Ca^{2+} release. *Proc Natl Acad Sci USA* 2004;**101**:2328–32.

54. Nadif Kasri N, Holmes AM, Bultynck G, et al. Regulation of InsP$_3$ receptor activity by neuronal Ca^2-binding proteins. *EMBO J* 2004;**23**:312–21.

55. Nadif Kasri N, Bultynck G, Sienaert I, et al. The role of calmodulin for inositol 1,4,5-trisphosphate receptor function. *Biochim Biophys Acta* 2002;**1600**:19–31.

56. Thrower EC, Choe CU, So SH, Jeon SH, Ehrlich BE, Yoo SH. A functional interaction between chromogranin B and inositol 1,4,5-trisphosphate receptor/Ca^{2+} channel. *J Biol Chem* 2003;**278**:49,699–708.

57. Bare DJ, Kettlun CS, Liang M, Bers DM, Mignery GA. Cardiac type-2 inositol 1,4,5-trisphosphate receptor: interaction and modulation by CaMKII. *J Biol Chem* 2005;**280**:15,912–20.

58. Mak D-OD, Foskett JK. Single-channel kinetics, inactivation, and spatial distribution of inositol trisphosphate (IP3) receptors in Xenopus oocyte nucleus. *J Gen Physiol* 1997;**109**:571–87.

59. Berridge MJ, Lipp P, Bootman MD. The versatility and universality of calcium signalling. *Nat Rev Mol Cell Biol* 2000;**1**:11–21.

60. Rutter GA, Rizzuto R. Regulation of mitochondrial metabolism by ER Ca^{2+} release: an intimate connection. *Trends Biochem Sci* 2000;**25**:215–21.

61. Sala C, Piëch V, Wilson NR, Passafaro M, Liu G, Sheng M. Regulation of dendritic spine morphology and synaptic function by shank and homer. *Neuron* 2001;**31**:115–30.

62. Toyoshima C, Inesi G. Structural basis of ion pumping by Ca^{2+}-ATPase of the sarcoplasmic reticulum. *Annu Rev Biochem* 2004;**73**:2269–92.

63. Doyle DA, Cabral JM, Pfuetzner RA, et al. The structure of the potassium channel: molecular basis of K$^+$ conduction and selectivity. *Science* 1998;**280**:69–76.

Chapter 28

PTEN

Lloyd C. Trotman

Cancer Center, Cold Spring Harbor Laboratory, Cold Spring Harbor, New York

INTRODUCTION

The PTEN phosphatase has an active site motif characteristic for the protein tyrosine phosphatase superfamily, yet its major substrate is the membrane lipid phosphatidylinositol (3,4,5) trisphosphate (PIP_3). PTEN is thus the major antagonist of PI3 kinase signaling, which explains its supreme role in tumor suppression, development, and growth control. One decade of research on PTEN has provided us with fundamental insights into many aspects of biology, such as phospholipid metabolism and the downstream signaling networks, and regulation of basic cellular processes such as migration or senescence; finally, it has even allowed us to redefine genetic principles of tumor suppression. As the first mechanisms of PTEN regulation and the role of PTEN in disease recurrence after targeted therapy are now becoming clear, a conceptual framework for a successful cancer therapy is emerging. Since such efforts will require a comprehensive understanding of its biology, this chapter will integrate the various facets of PTEN research.

PTEN DISCOVERY AND FUNCTION

Analysis of the genomic 10q23 region, which frequently displays loss of heterozygosity in human tumors, simultaneously led two research teams to the cloning of the tumor suppressor gene which they named *PTEN* (for *P*hosphatase and *TEN*sin homolog deleted on chromosome *TEN*) [1], or *MMAC1* (for *M*utated in *M*ultiple *A*dvanced *C*ancers *1*) [2] (see also timeline in Figure 28.1). This discovery yielded the first direct evidence of the long suspected role of phosphatases in tumor suppression through their ability to reverse oncogenic kinase function. While initial experiments soon established PTEN target preferences on synthetic peptides [3], the highly acidic nature of the optimal substrate (polyGlu$_4$Tyr$_1$) and the unusual dual specificity

for both phosphorylated Tyr and Ser/Thr residues prompted research on the ability of PTEN to target phospholipids, especially the highly charged phosphatidyl-inositol second messengers, which resulted in the identification of PIP_3 as the major cellular target of PTEN [4]. PTEN specifically dephosphorylates the 3-position on the inositol ring to PI-$(4,5)P_2$ and thus effectively reverses PI-3 kinase (PI3K) function (see Figure 28.2, below).

Since PIP_3 is an essential second messenger for recruitment and activation of PH-domain containing kinases such as AKT (also known as PKB) to the plasma membrane, the discovery of this target revealed that a phosphatase is at the top control level of a signaling pathway for intercellular communication, which is highly conserved among metazoans [5]. *PTEN* mutation in *Drosophila*, for example, provided the first link between PTEN and control of cell/organ size [6], which can be rescued by mutations in *dAKT*. But the first genetic proof of this signaling hierarchy between PI3K, PTEN, and AKT came with the availability of *Pten*-deficient mouse embryonic fibroblasts and activation specific phospho-AKT antibodies [7]. To date, phospho-Akt specific western blotting and immunohistochemistry remain the most common readouts of PTEN activity, as quantification of PIP_3 levels requires HPLC or radiolabeling and thin layer chromatography.

PTEN AND CANCER

Together with the discovery of the *PTEN* gene in a loss of heterozygosity (LOH) -prone region of chromosome 10, mutational analysis immediately revealed frequent PTEN point mutations in commonly used glioblastoma, prostate and breast cancer cell lines, and primary tumor samples [1, 2], confirming the notion that PTEN is the critical target gene of the 10q23 region. Up to 40 percent of glioblastoma samples show *PTEN* loss, and in endometrial cancers

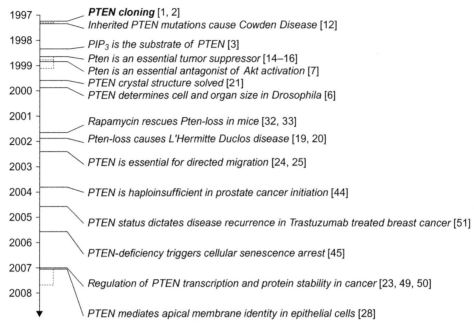

FIGURE 28.1 A timeline of landmark discoveries in PTEN biology.
References are given in brackets.

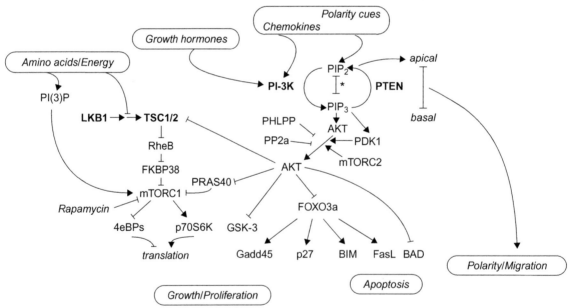

FIGURE 28.2 The PTEN/PI3-kinase signaling pathway connects extracellular inputs (boxed, top) with basic cellular functions (boxed, bottom).
Only a selection of molecular mediators are shown (see also text). The most frequently targeted cancer genes are indicated in bold print, arrows denote stimulation, blunted arrows inhibition.

mutation frequency reaches up to 80 percent (depending on the subtype), while other epithelial tumors also commonly present with *PTEN* deletions: LOH is frequent in prostate cancer (up to 60 percent), melanoma (30–40 percent), and in gastric (>40 percent) and breast cancers (>20%) [5, 8–11]. However, these numbers are subject to refinement, as both the resolution of CGH array based copy number analysis as well as the feasibility of routinely sequencing cancer genes in patient samples are ever increasing. Most notably, though, immunohistochemical (IHC) analysis of PTEN levels and activity in tumor tissue will provide the needed comprehensive analysis to complement the genetic and expression data. To illustrate, while *PTEN* LOH frequency is relatively low in breast cancer, and mutation virtually never seen, loss of PTEN protein by IHC is found in close to 50 percent of cases [11]. Taken together, the spectrum and frequency of PTEN somatic alterations in cancer is steadily increasing, and rival those of the p53 tumor suppressor.

Another major line of evidence for the important tumor-suppressive function of PTEN was published almost simultaneously with its discovery; namely, the realization that *PTEN* germline mutations are linked to cancer susceptibility syndromes [12]. Cowden disease (CD, MIM 158350) is an autosomal dominant cancer susceptibility syndrome with strong predisposition for breast, thyroid, and skin tumors, featuring large numbers of initially benign hamartomas in these tissues. Another inherited disorder, L'Hermitte Duclos disease, which is characterized by macrocephaly, seizures, and ataxia, that co-segregates with CD, was also found to harbor *PTEN* mutations, thus confirming the role of PTEN in these two syndromes. Indeed, the linkage with *PTEN* mutation has proven so strong that several phenotypically related syndromes are today collectively referred to as PTEN Hamartoma Tumor Syndromes (PTHS), with CD, Proteus-like- and Bannayan-Riley-Ruvalcaba syndromes showing the highest PTEN mutation frequency (80 percent, 60 percent, and 60 percent, respectively) [13].

Even though a firm connection between *PTEN* and cancer could soon be established, there appeared to be seemingly contradictory data regarding the ability of *PTEN* loss to initiate tumorigenesis as opposed to a proposed role only in advanced cancers. In certain tumor types (e.g., gliomas) loss mainly occurs at late stages; in contrast, germline mutations evidently are able to initiate neoplasia in various tissues. Clearly, to answer such questions, strong model systems for cancer were needed, and they were delivered mostly by recapitulating and dissecting *Pten*-mutant tumorigenesis in the mouse.

MOUSE MODELS FOR PTEN FUNCTION

Murine knockout models for *Pten* unambiguously demonstrated its potential to trigger cancer. While *Pten*-null animals displayed embryonic lethality, *Pten*-heterozygous animals were readily obtained and developed lesions in a wide variety of tissues, including breast, prostate, colon, and endometrium, as well as the thyroid and adrenal glands [14–16]. In addition, complete conditional inactivation of *Pten* using the *Cre/Lox* technology has confirmed roles for *Pten* in regulating the proliferation of skin, heart, and brain tissues, as well as cell and organ sizes in the latter two [17]. In summary, the *in vivo* analysis of *Pten* activity revealed four core biological functions: the ability of Pten to curb proliferation as seen in epithelial tumors; the essential function of Pten in mediating apoptosis, as seen in B and T lymphocytes [18]; the control of cell size, polarity and migration to maintain tissue architecture, as seen in

epithelial tumors; and a brain-specific phenotype [19, 20]. The last example is particularly instructive in demonstrating the power of mouse models for our understanding of human disease, as they revealed that abnormal size and migration of *Pten*-deficient neurons in the developing cerebellum, and not hyperproliferation, underlie human L'Hermitte Duclos disease.

PTEN STRUCTURE

The crystal structure of PTEN was also solved soon after its discovery [21], and it revealed an N-terminal catalytic domain with a wide active site pocket that accomplishes accommodation of the phosphoinositide head group and a C-terminal C2-domain, which regulates plasma membrane association and stabilizes the phosphatase domain. This structure also revealed how cancer-associated mutations target the catalytic domain directly or more indirectly – for example, via mutation of critical residues in the C2 domain that lie at the interface with the catalytic domain. Notably, PTEN contains three unstructured regions, located at the N- and C-termini and in an internal loop of the C2 domain, all of which are subject to posttranslational modifications that regulate PTEN stability, cellular localization, and activity in human cancer [22, 23].

THE PTEN SIGNALING PATHWAY

Our understanding of the PTEN pathway is expanding fast, and has greatly profited from its high conservation in model organisms such as *Dictyostelium*, *Caenorhabditis*, and *Drosophila*. As shown in Figure 28.2, there are three basic cell functions for which PTEN is essential.

To enable directional migration, small G proteins such as Ras and Rho regulate localized production and hydrolysis of PIP_3 through their effects on PI3K and PTEN, respectively, in a coordinated process which was discovered by studying chemoattractant responses in the slime mold *Dictyostelium discoideum* [24, 25]. This results in PIP_3-dependent accumulation of actin-remodeling factors to the chemoattractant-facing "front" side of the cell, and PTEN/PIP_2-dependent establishment of the cells back-side identity in migration [26, 27]. Similarly, to establish and maintain cell polarity in epithelial glands during organogenesis, it has been shown that PTEN and PIP_2 are defining the apical plasma membrane via recruitment of Cdc42 and aPKC, but, in contrast to migration, the upstream mediators of this event are still unknown [28].

A second function of PTEN signaling converges on its ability to induce apoptosis – a characteristic that is most strikingly visible in *Pten* heterozygous mice, which mostly succumb to an autoimmune disorder due to failure of lymphocyte apoptosis. Collectively, PTEN seems to mediate pro-apoptotic functions via inactivation of AKT, which can directly inactivate the pro-apoptotic BAD protein as well as inhibit FOXO3a-dependent transcription of apoptosis genes such as Fas-ligand and Bim, to name but a few [29].

The third functional network negatively regulated by PTEN controls cell growth, proliferation, and metabolism, and is also chiefly mediated by AKT. As above, AKT inhibits key targets through their phosphorylation: GSK3-phosphorylation thus links PTEN to metabolism and cell cycle control (via glycogen synthase and MYC, respectively); FOXO3a-phosphorylation leads to decreased transcription of cell cycle inhibitors such as $p27^{Kip1}$ and GADD45 [30]; and activation of the mTOR complex 1 (mTORC1) either indirectly via phosphorylation of the Tuberous Sclerosis Complex tumor-suppressor TSC2 or more directly via phosphorylation of the mTORC1 inhibitor PRAS40 leads to increased protein synthesis [31]. As illustrated in Figure 28.2, mTOR kinase therefore couples protein translation with sensing of the cells' energy levels via TSC (PTEN-AKT-dependent sensor) and to amino acid availability through a PI-(3)P signaling pathway (PTEN-AKT-independent sensor) that is conserved in yeast [32], while PIP_3 signaling is not. It thus appears as if the requirement for multicellular organisms to conditionally integrate nutrient availability with growth hormone controlled proliferation is achieved by introduction of the PIP_3 signaling pathway of which PTEN is the major antagonist. This realization, and the discovery of the naturally occurring mTORC1 inhibitor rapamycin, resulted in tremendous interest in therapeutic effects of that compound. Indeed, successful preclinical trials with rapamycin in *Pten* mutant mice fostered high hopes for the potential of this drug in clinical cancer therapy [33, 34]. Yet early results from ongoing clinical trials seem to indicate high variability of results, with specific effects in only few cancer types, in contrast to trials using rapamycin in Tuberous Sclerosis Complex patients, which are more promising [35].

It will therefore be of particular interest to see whether, in *PTEN*-deficient epithelial tumorigenesis, signaling through the AKT-FOXO3a axis is more important than that through the TSC-mTOR axis (see Figure 28.2). Mouse modeling so far highlights the importance of signaling through the former axis in *Pten* heterozygous prostate and colon cancer, upon introduction of mutations that affect FOXO3a and p27 [36, 37].

The p110α subunit of PI3K can also be activated by Ras, and this interaction was found to be an essential component for mutant Ras-mediated tumorigenesis [38], subjecting yet another layer of extracellular growth signals to PTEN activity. Finally, neo-angiogenesis can be promoted both through mTOR activity [39] as well as through inactivation of Forkhead transcription factors [40] which link PTEN to yet another exclusive characteristic of cancer.

Overall, the current picture of cellular PTEN function attributes a central role to its regulation of AKT kinase.

However, the frequency of *AKT* alterations in cancer is far below that of *PTEN*. Therefore, notwithstanding additional AKT-independent PTEN functions, our current knowledge strongly suggests that the PIP_3/PIP_2-dependent maintenance of cellular architecture, which is achieved independently of AKT by PI3K and PTEN, is one essential factor in PTEN-mediated suppression of cancerous transformation. This notion is consistent with findings in an *in vitro* breast cancer model [41], and also confirmed by human cancer genetics, since, in contrast to AKT, mutations of PI3K are increasingly appreciated as frequent events in human cancer [5].

Taken together, PTEN has emerged as a principle negative regulator of the cell's signaling pathways for proliferation, apoptosis, and architecture, and its ability to antagonize mTOR enables PTEN to overrule sensors which signal nutrient availability and recruitment of vasculature. PTEN alteration therefore enables a cell to acquire most, if not all, of the traits previously described as "hallmarks of the cancer cell" [42].

HAPLOINSUFFICIENCY AND SENESCENCE IN CANCER

Early reports on PTEN loss in epithelial cancers of breast and prostate presented researchers with a paradox: while LOH analysis of the *PTEN* locus revealed frequent deletion of one copy, mutation of the remaining allele was rarely found, as Cowden-disease patients frequently showed retention of their remaining normal *PTEN* copy in tumor tissues [8]. These observations raised the question of whether PTEN heterozygosity might be insufficient for prevention of certain tumor types – a principle termed *haploinsufficiency*, which does not conform to the classic "two-hit hypothesis." Originally defined for the retinoblastoma tumor suppressor (and essential for the realization of the role of tumor-opposing genes in cancer), this hypothesis states that it is tumor-suppressor nullizygosity that invariably leads to cancer [43]. As mouse models allow precise genetic dissection of disease progression, the question of *Pten* haploinsufficiency in cancer initiation was solved in that system. Initial results clearly demonstrated *Pten* haploinsufficiency for the lethal autoimmunity phenotype in hetereozygous lymphocytes, and for haploinsufficiency of *Pten* in prostate cancer if *Pten* heterozygosity was either combined with expression of an oncogenic transgene in prostate (resulting in accelerated tumor formation) or with loss of *p27^{kip1}* [36, 44]. Yet the first evidence that *Pten* would be haploinsufficient not in combination, and its hemizygous loss able to initiate cancer formation, came with the establishment of a hypomorphic series of *Pten*-expressing mice, which revealed that Pten protein levels and the corresponding Akt activity determine tumor onset and rate of progression in the prostate [45]. In addition, the realization that complete loss of *Pten* leads to a p53-dependent cellular senescence response that blocks tumorigenesis as long as the p53 pathway is intact, demonstrated that there is even a cell intrinsic antitumoral response that protects from tumorigenesis by the "two-hit" paradigm [46]. These findings collectively explain (1) the high frequency of incomplete *PTEN* loss in early tumors, and (2) the high frequency of complete combined *PTEN* and *p53* disruption in late-stage tumors. Notably, the senescence response, which has been found to counteract tumor formation in various tissues [47], can be bypassed in cancers of mice where one or both copies of *p27*^kip1 or the promyelocytic leukemia (*Pml*) gene are lost in combination with *Pten*, as these genetic combinations cause prostate and also lethal colon carcinomas (which form neither in *Pml*- nor in *p27*-loss alone) [36, 37].

Since PML and p27 degradation at the protein level have been associated with cancer progression in tissues that show frequent *PTEN* LOH [48, 49], it is possible that the senescence and haploinsufficiency paradigms direct emergence of tumors to tissues that are most likely to combine *PTEN* heterozygosity with the loss of such secondary tumor suppressors.

PTEN IN CANCER THERAPY

The above insights into tumor suppression strongly suggest that understanding PTEN regulation harbors great therapeutic potential, since many tumors are diagnosed while they still retain one copy of the gene. Thus, the identification of the first ubiquitin E3-ligase of PTEN, NEDD4-1, represents a great step towards elucidation of how tumor cells degrade PTEN, and how this posttranslational modification could be inhibited therapeutically [50]. Intriguingly, monoubiquitination of PTEN is essential for its nuclear import and stability, as revealed in a patient family where mutation of one ubiquitination site leads to nuclear PTEN exclusion and familial Cowden disease. Thus, PTEN degradation and nuclear transport are closely linked processes, and interventions favoring the latter should be of great therapeutic potential, as nuclear PTEN localization has been generally associated with better disease outcomes (see [23] and references therein). Inhibition of *PTEN* at the transcriptional level has also been associated with T cell leukemia, as a NOTCH1 target was found to repress *PTEN* in a manner sensitive to the clinical NOTCH1 inhibitor γ-secretase (GSI) [51].

Indeed, GSI resistance in *NOTCH1*-driven leukemia was also found to be associated with *PTEN* mutation, demonstrating the importance of the PTEN pathway in such malignancies, and highlighting the need for means to rescue PTEN function in order to prevent resistance to targeted therapy.

Additional examples for a similar role of PTEN in targeted therapy have been reported, collectively linking PTEN/PI3K alterations with resistance to the HER2 inhibitors Herceptin/Trastuzumab in breast cancer [52], and resistance to the EGFR inhibitors Erlotinib and Gefitinib in glioblastoma patients [53].

In conclusion, a picture is emerging in which the neoplastic process requires fine-tuned regulation of PTEN that dictates tumor formation, tumor progression to malignancy, and tumor sensitivity to targeted therapy. Our continued efforts to understand PTEN regulation are therefore of critical impact to the future of cancer treatment.

REFERENCES

1. Li J, Yen C, Liaw D, et al. PTEN, a putative protein tyrosine phosphatase gene mutated in human brain, breast, and prostate cancer. *Science* 1997;**275**:1943–7.
2. Steck PA, Pershouse MA, Jasser SA, et al. Identification of a candidate tumour suppressor gene, MMAC1, at chromosome 10q23.3 that is mutated in multiple advanced cancers. *Nat Genet* 1997;**15**:356–62.
3. Myers MP, Stolarov JP, Eng C, et al. P-TEN, the tumor suppressor from human chromosome 10q23, is a dual-specificity phosphatase. *Proc Natl Acad Sci USA* 1997;**94**:9052–7.
4. Maehama T, Dixon JE. The tumor suppressor, PTEN/MMAC1, dephosphorylates the lipid second messenger, phosphatidylinositol 3,4,5-trisphosphate. *J Biol Chem* 1998;**273**:13,375–8.
5. Engelman JA, Luo J, Cantley LC. The evolution of phosphatidylinositol 3-kinases as regulators of growth and metabolism. *Nat Rev Genet* 2006;**7**:606–19.
6. Huang H, Potter CJ, Tao W, et al. PTEN affects cell size, cell proliferation and apoptosis during Drosophila eye development. *Development* 1999;**126**:5365–72.
7. Stambolic V, Suzuki A, de la Pompa JL, et al. Negative regulation of PKB/Akt-dependent cell survival by the tumor suppressor PTEN. *Cell* 1998;**95**:29–39.
8. Sansal I, Sellers WR. The biology and clinical relevance of the PTEN tumor suppressor pathway. *J Clin Oncol* 2004;**22**:2954–63.
9. Byun DS, Cho K, Ryu BK, et al. Frequent monoallelic deletion of PTEN and its reciprocal associatioin with PIK3CA amplification in gastric carcinoma. *Intl J Cancer* 2003;**104**:318–27.
10. Wu H, Goel V, Haluska FG. PTEN signaling pathways in melanoma. *Oncogene* 2003;**22**:3113–22.
11. Saal LH, Johansson P, Holm K, et al. Poor prognosis in carcinoma is associated with a gene expression signature of aberrant PTEN tumor suppressor pathway activity. *Proc Natl Acad Sci USA* 2007;**104**:7564–9.
12. Liaw D, Marsh DJ, Li J, et al. Germline mutations of the PTEN gene in Cowden disease, an inherited breast and thyroid cancer syndrome. *Nat Genet* 1997;**16**:64–7.
13. Zbuk KM, Eng C. Cancer phenomics: RET and PTEN as illustrative models. *Nature. Rev Cancer* 2007;**7**:35–45.
14. Di Cristofano A, Pesce B, Cordon-Cardo C, Pandolfi PP. Pten is essential for embryonic development and tumour suppression. *Nat Genet* 1998;**19**:348–55.
15. Podsypanina K, Ellenson LH, Nemes A, et al. Mutation of Pten/Mmac1 in mice causes neoplasia in multiple organ systems. *Proc Natl Acad Sci USA* 1999;**96**:1563–8.
16. Suzuki A, de la Pompa JL, Stambolic V, et al. High cancer susceptibility and embryonic lethality associated with mutation of the PTEN tumor suppressor gene in mice. *Curr Biol* 1998;**8**:1169–78.

17. Kishimoto H, Hamada K, Saunders M, et al. Physiological functions of Pten in mouse tissues. *Cell Struct Funct* 2003;**28**:11–21.

18. Di Cristofano A, Kotsi P, Peng YF, Cordon-Cardo C, Elkon KB, Pandolfi PP. Impaired Fas response and autoimmunity in Pten mice. *Science* 1999;**285**:2122–5.

19. Kwon CH, Zhu X, Zhang J, et al. Pten regulates neuronal soma size: a mouse model of Lhermitte-Duclos disease. *Nat Genet* 2001;**29**:404–11.

20. Backman SA, Stambolic V, Suzuki A, et al. Deletion of Pten in mouse brain causes seizures, ataxia and defects in soma size resembling Lhermitte-Duclos disease. *Nat Genet* 2001;**29**:396–403.

21. Lee JO, Yang H, Georgescu MM, et al. Crystal structure of the PTEN tumor suppressor: implications for its phosphoinositide phosphatase activity and membrane association. *Cell* 1999;**99**:323–34.

22. Tamguney T, Stokoe D. New insights into PTEN. *J Cell Sci* 2007;**120**:4071–9.

23. Trotman LC, Wang X, Alimonti A, et al. Ubiquitination regulates PTEN nuclear import and tumor suppression. *Cell* 2007;**128**:141–56.

24. Funamoto S, Meili R, Lee S, Parry L, Firtel RA. Spatial and temporal regulation of 3-phosphoinositides by PI 3-kinase and PTEN mediates chemotaxis. *Cell* 2002;**109**:611–23.

25. Iijima M, Devreotes P. Tumor suppressor PTEN mediates sensing of chemoattractant gradients. *Cell* 2002;**109**:599–610.

26. Charest PG, Firtel RA. Big roles for small GTPases in the control of directed cell movement. *Biochem J* 2007;**401**:377–90.

27. Li Z, Dong X, Wang Z, et al. Regulation of PTEN by Rho small GTPases. *Nat Cell Biol* 2005;**7**:399–404.

28. Martin-Belmonte F, Gassama A, Datta A, et al. PTEN-mediated apical segregation of phosphoinositides controls epithelial morphogenesis through Cdc42. *Cell* 2007;**128**:383–97.

29. Franke TF, Hornik CP, Segev L, Shostak GA, Sugimoto C. PI3K/Akt and apoptosis: size matters. *Oncogene* 2003;**22**:8983–98.

30. The many forks in FOXO's road (trans. Brunet A, Griffith EC, and Greenberg ME). *Sci STKE* 2003.

31. Guertin DA, Sabatini DM. Defining the role of mTOR in cancer. *Cancer Cell* 2007;**12**:9–22.

32. Nobukuni T, Joaquin M, Roccio M, et al. Amino acids mediate mTOR/raptor signaling through activation of class 3 phosphatidylinositol 3OH-kinase. *Proc Natl Acad Sci USA* 2005;**102**:14,238–43.

33. Podsypanina K, Lee RT, Politis C, et al. An inhibitor of mTOR reduces neoplasia and normalizes p70/S6 kinase activity in Pten mice. *Proc Natl Acad Sci USA* 2001;**98**:10,320–5.

34. Neshat MS, Mellinghoff IK, Tran C, et al. Enhanced sensitivity of PTEN-deficient tumors to inhibition of FRAP/mTOR. *Proc Natl Acad Sci USA* 2001;**98**:10,314–19.

35. Franz DN, Leonard J, Tudor C, et al. Rapamycin causes regression of astrocytomas in tuberous sclerosis complex. *Ann Neurol* 2006;**59**:490–8.

36. Di Cristofano A, De Acetis M, Koff A, Cordon-Cardo C, Pandolfi PP. Pten and p27KIP1 cooperate in prostate cancer tumor suppression in the mouse. *Nat Genet* 2001;**27**:222–4.

37. Trotman LC, Alimonti A, Scaglioni PP, Koutcher JA, Cordon-Cardo C, Pandolfi PP. Identification of a tumour suppressor network opposing nuclear Akt function. *Nature* 2006;**441**:523–7.

38. Gupta S, Ramjaun AR, Haiko P, et al. Binding of ras to phosphoinositide 3-kinase p110alpha is required for ras-driven tumorigenesis in mice. *Cell* 2007;**129**:957–68.

39. Majumder PK, Febbo PG, Bikoff R, et al. mTOR inhibition reverses Akt-dependent prostate intraepithelial neoplasia through regulation of apoptotic and HIF-1-dependent pathways. *Nat Med* 2004;**10**:594–601.

40. Paik JH, Kollipara R, Chu G, et al. FoxOs are lineage-restricted redundant tumor suppressors and regulate endothelial cell homeostasis. *Cell* 2007;**128**:309–23.

41. Aranda V, Haire T, Nolan ME, et al. Par6-aPKC uncouples ErbB2 induced disruption of polarized epithelial organization from proliferation control. *Nat Cell Biol* 2006;**8**:1235–45.

42. Hanahan D, Weinberg RA. The hallmarks of cancer. *Cell* 2000;**100**:57–70.

43. Knudson Jr. AG. Mutation and cancer: statistical study of retinoblastoma. *Proc Natl Acad Sci USA* 1971;**68**:820–3.

44. Kwabi-Addo B, Giri D, Schmidt K, et al. Haploinsufficiency of the Pten tumor suppressor gene promotes prostate cancer progression. *Proc Natl Acad Sc i USA* 2001;**98**:11,563–8.

45. Trotman LC, Niki M, Dotan ZA, et al. Pten dose dictates cancer progression in the prostate. *PLoS Biol* 2003;**1**:E59.

46. Chen Z, Trotman LC, Shaffer D, et al. Crucial role of p53-dependent cellular senescence in suppression of Pten-deficient tumorigenesis. *Nature* 2005;**436**:725–30.

47. Narita M, Lowe SW. Senescence comes of age. *Nat Med* 2005;**11**:920–2.

48. Cordon-Cardo C, Koff A, Drobnjak M, et al. Distinct altered patterns of p27KIP1 gene expression in benign prostatic hyperplasia and prostatic carcinoma. *J Natl Cancer Inst* 1998;**90**:1284–91.

49. Gurrieri C, Capodieci P, Bernardi R, et al. Loss of the tumor suppressor PML in human cancers of multiple histologic origins. *J Natl Cancer Inst* 2004;**96**:269–79.

50. Wang X, Trotman LC, Koppie T, et al. NEDD4-1 is a proto-oncogenic ubiquitin ligase for PTEN. *Cell* 2007;**128**:129–39.

51. Palomero T, Sulis ML, Cortina M, et al. Mutational loss of PTEN induces resistance to NOTCH1 inhibition in T-cell leukemia. *Nat Med* 2007;**13**:1203–10.

52. Nagata Y, Lan KH, Zhou X, et al. PTEN activation contributes to tumor inhibition by trastuzumab, and loss of PTEN predicts trastuzumab resistance in patients. *Cancer Cell* 2004;**6**:117–27.

53. Mellinghoff IK, Wang MY, Vivanco I, et al. Molecular determinants of the response of glioblastomas to EGFR kinase inhibitors. *N Engl J Med* 2005;**353**:2012–24.

Chapter 29

PTEN/MTM Phosphatidylinositol Phosphatases

Knut Martin Torgersen[1], Soo-A Kim[1] and Jack E. Dixon[2]

[1]*The Life Science Institute and Department of Biological Chemistry, University of Michigan Medical School, Ann Arbor, Michigan*

[2]*Pharmacology/Cellular and Molecular Medicine/Chemistry and Biochemistry, University of California, San Diego, La Jolla, California*

PTEN

Introduction

PTEN (phosphatase and tensin homolog deleted on chromosome 10) was first identified as a tumor suppressor gene localized on chromosome 10q23. PTEN mutations are found at high frequencies in certain tumors, including endometrial carcinomas, gliomas, and breast and prostate cancers. Furthermore, germline mutations in the PTEN gene are found in the related autosomal disorders Cowden disease, and Lhermitte-Duclos and Bannayan-Zonana syndromes. Biochemical and genetic analyses of PTEN and its role in these diseases have placed it in a group of gatekeeper genes essential for controlling cell growth and development [1–3].

Activity and Function

PTEN is a member of the protein tyrosine phosphatase (PTP) superfamily of enzymes characterized by the invariant Cys–x_5–Arg (Cx$_5$R) active site motif (Figure 29.1). However, unlike other PTP superfamily enzymes, PTEN utilizes the lipid second messenger phosphatidylinositol 3,4,5-trisphosphate (PIP$_3$) as its substrate [4]. This places PTEN as a negative regulator of phosphatidyl inositol 3-kinase (PI3K) signaling [5]. PTEN has been reported to regulate signaling through Akt/PKB, PDK1, SGK1, and Rho GTPases and therefore as a modulator of a broad range of cellular processes [6, 7]. Loss of PTEN function can lead to tumor development through defects in cell cycle regulation, apoptosis, or angiogenesis.

Homozygous PTEN$^{-/-}$ mice die before birth, and embryos display regions of increased proliferation and disturbed developmental patterning. Heterozygous PTEN$^{+/-}$ mice are viable, but spontaneously develop various types of tumors [3]. Cells from both PTEN$^{-/-}$ mice and PTEN$^{+/-}$ have constitutively activated Akt and are resistant to apoptotic stimuli. A direct role of PTEN and its lipid phosphatase activity in the regulation of Akt has been demonstrated in several tumor cell lines [2, 3]. Furthermore, PTEN$^{+/-}$ mice have a tendency of developing both T cell lymphomas and autoimmune disorders. A role for PTEN in this postulated link between autoimmune disorders and cancers is further supported by studies of mice where PTEN is conditionally targeted in T cells [8].

Mice in which PTEN is conditionally deleted in neuronal brain cells develop macrocephaly as a result of increased cell numbers, decreased cell death, and enlarged soma size [9, 10]. Targeted deletion early in brain development suggests a role for PTEN in controlling the proliferation and potency of stem cells, whereas restricted deletion of PTEN in postmitotic neurons does not result in increased cell proliferation, but rather causes a progressive enlargement of soma size resulting in enlarged cerebellum and seizures. It is of note that the abnormal phenotype of these mice resembles that of Lhermitte-Duclos disease, suggesting that loss of PTEN function is sufficient to cause this disease in humans.

A conserved role for PTEN as a PIP$_3$ phosphatase and negative regulator of PI3K signaling has also been demonstrated by genetic studies of *D. melanogaster* (dPTEN) and *C. elegans* (Daf-18). By balancing signals from the insulin receptor, dPTEN controls cell size and number in flies, whereas Daf-18 regulates metabolism and longevity in worms [11, 12].

The crystal structure of PTEN has revealed several features that contribute to its unique substrate specificity [13]. A four-residue insertion in one loop of the PTP domain results in the widening and extension of the catalytic

	Name	Active site	C-terminal domains	Length (aa)	Chromosome	Disease
PTEN	PTEN	IHCKAGKGRT	PDZ-binding	403	10q23.3	Multiple cancers, Cowden and Lhermitte-Duclos
	PTENP1	IHCKAGKGRT	PDZ-binding	>405	9p21	—
	PTENR1	IHCKGGKGRT	—	>408	—	—
	TPIP	IHCKGGKGRT	—	445	13	—
	TPTE	IHCKGGTDRT	—	551	21 and others	—
MTM	MTM1	VHCSDGWDRT	PDZ-binding	603	Xq28	Myotybular myopathy
	MTMR1	VHCSDGWDRT	PDZ-binding	669	Xq28	—
	MTMR2	VHCSDGWDRT	PDZ-binding	643	11q22	Charcot-Marie-Tooth 4B
	MTMR3	VHCSDGWDRT	FYVE	1199	22q12.2	—
	MTMR4	VHCSDGWDRT	FYVE	1195	17q22-23	—
	MTMR6	VHCSDGWDRT	—	>567	13q12	—
	MTMR7	VHCSDGWDRT	—	>574	8p22	—
	MTMR8	VHCSDGWDRT	—	704	Xq11.2-12	—
	MTMR5 (SBF1)	VGLEDGWDIT	PH	1930	22pter	—
	MTMR9 (LIP-STYX)	IHGTEGTDST	—	549	8	—
	MTMR10	LQEEEGRDLS	—	>451	15	—
	MTMR11	LQERGDRDLN	—	710	1	—
	MTMR12 (3-PAP)	LLEENASDLC	—	>637	5	—

FIGURE 29.1 PTEN and myotubularin-related genes in human.
Length of predicted protein products (amino acids) and chromosomal localization are listed for each gene. Conserved amino acids within predicted active site sequences are presented, including the catalytic cysteine (yellow) and arginine (light blue), and non-catalytic basic (blue) and acidic (red) residues. Non-catalytic domains predicted for carboxy-terminal regions as well as related diseases are also listed.

pocket and enough space for the bulky PIP$_3$ headgroup. In addition, the two lysines (Lys125 and Lys128) within the Cx$_5$R active site sequence, as well as an upstream histidine (His93), coordinate the D1 and D5 phosphate groups of the inositol ring. Hence, the specificity of PTEN toward PIP$_3$ is generated by a larger active site pocket combined with the conserved residues within the Cx$_5$R active site. C-terminal to the PTP catalytic domain PTEN contains a Ca^{2+} independent C2 domain, two PEST sequences, and a PDZ-binding motif (Figure 29.2). These domains are likely to play important roles in PTEN regulation (see below).

The human genome contains several PTEN-related genes, but so far little is known about their function. Most of these genes exhibit restricted expression pattern and/or subcellular localization different from PTEN, and do not appear to regulate Akt phosporylation. In that respect it is interesting to note that these genes have a different active site sequence, which might suggest a different substrate specificity [14, 15].

Regulation

The crystal structure of PTEN revealed an extensive interface between its PTP domain and C2 domain, suggesting

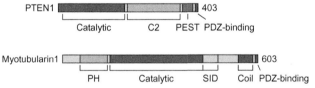

FIGURE 29.2 Structural features of PTEN and myotubularin phosphoinositide phosphatases.
PTEN and myotubularin contain a catalytic domain that encompasses the CX$_5$R active site motif of PTP. In addition, both proteins possess several other domains/motifs that are likely to facilitate membrane association and protein–protein interactions. The C2 domain of PTEN is required for binding to lipid vesicles, whereas the carboxy-terminal PDZ-binding motif mediates interaction with PDZ-containing proteins. Phosphorylation in this region inhibits PDZ binding. Myotubularin contains a PH domain that may function to regulate membrane association. Furthermore, myotubularin contains a coiled coil motif as well as a putative PDZ-binding motif.

that membrane targeting and lipid phosphatase activity are interdependent [13]. This is further supported by the observation that mutations affecting this interface are frequently found in cancers [3]. *In vitro*, the C2 domain of PTEN binds phospholipids independent of Ca^{2+} and its structural characteristics predict a direct membrane association. Mutations in critical lipid binding residues inhibit the ability of PTEN to function as a tumor suppressor and cannot be rescued by artificial membrane targeting. Hence, both structural and functional analysis suggest that the C2 domain plays a dual role of both membrane recruitment and positioning of the PTP domain.

The extreme C-terminus of PTEN contains tandem PEST sequences and a consensus PDZ-binding domain. Whereas the regulatory role of the PEST sequences remain elusive, the PDZ-binding motif has been demonstrated to associate with several PDZ-domain containing proteins [3, 16]. The identification of phosphorylation sites in the C-terminal tail of PTEN regulating PDZ-binding and complex formation (Figure 29.2) suggests additional levels of PTEN regulation [3, 16].

Finally, several regulatory elements have recently been identified in the PTEN promoter, including binding sites for the tumor suppressors p53, early growth response-1 (Egr-1), and the perioxisome proliferator-activator receptor γ (PPARγ) [17–19]. The inducible transactivation of PTEN by these genes leads to reduced Akt activity and increased cell survival.

MYOTUBULARIN: A NOVEL FAMILY OF PHOSPHATIDYLINOSITOL PHOSPHATASES

Myotubularin-related proteins constitute one of the largest and most highly conserved protein tyrosine phosphatase (PTP) subfamilies in eukaryotes [14, 19]. The MTM family includes at least eight catalytically active proteins as well as five catalytically inactive proteins in humans [14, 20]. Phylogenetic analysis of MTM family proteins allow a division of myotubularin family into six subgroups, which include the catalytically active MTM1/MTMR1/MTMR2, MTMR3/MTMR4, and MTMR6/MTMR7/MTMR8 enzymes, as well as the MTMR5 (Sbf1), MTMR9 (LIP-STYX), and MTMR10/MTMR11/MTMR12 (3-PAP) inactive forms [14, 20]. One gene from *D. melanogaster* and *C. elegans* corresponding to each of these subfamilies has been identified [14].

Phosphatase Activity

Myotubularin (MTM1), the first characterized member of this novel family, utilizes the lipid second messenger phosphatidylinositol 3-phosphate (PI(3)P) as a physiological substrate [21, 22]. In addition, recent findings demonstrate that other MTM-related phosphatases, MTMR1, MTMR2, MTMR3, MTMR4, and MTMR6, also dephosphorylate PI(3)P, a finding that suggests that activity toward this substrate is common to all active myotubularin family enzymes [23, 24].

The consensus CX_5R active site motif of PTP/DSP (dual specificity protein phosphatase) is found in the myotubularin family proteins, and the sequence CSDGWDR is invariant within all members of the active phosphatase subgroups. Unlike PTEN, in which two lysine residues within

its active site (CKAGKGR) contribute to substrate specificity by interacting with the D1 and D5 phosphates of PIP$_3$, two aspartic acid residues are found in myotubularin family phosphatases. It is possible that interactions between the active site aspartic acid residues and phoshoryl groups at either the D4 or D5 position of the inositol ring may contribute to the high degree of specificity for PI(3)P found in MTM family enzymes.

One of the most notable characteristics of the human MTM family is the existence of at least five catalytically inactive forms, which contain germline substitution in catalytically essential residues within the PTP active site motif (Figure 29.1). Myotubularin-related inactive forms may function to regulate PI(3)P levels by opposing the actions of myotubularin phosphatases, or directly affect the activity and/or subcellular localization of their active MTM counterparts (see Chapter 100 of Handbook of Cell Signaling, Second Edition).

Myotubularin Family and Human Diseases

To date, two myotubularin-related proteins have been associated with human disease. The myotubularin gene on chromosome Xq28, MTM1, is mutated in X-linked myotubular myopathy (XLMTM), a severe congenital muscular disorder characterized by hypotonia and generalized muscle weakness in newborn males [25]. Myogenesis in affected individuals is arrested at a late stage of differentiation/maturation following myotube formation, and the muscle cells have characteristically large, centrally located nuclei [25].

Mutations in a second MTM family member, MTMR2 on chromosome 11q22, have recently been shown to cause the neurodegenerative disorder, type 4B Charcot-Marie-Tooth disease (CMT4B) [26]. CMT4B is an autosomal recessive demyelinating neuropathy characterized by abnormally folded myelin sheaths and Schwann cell proliferation in peripheral nerves.

Because these two highly similar genes, MTM1 and MTMR2 (64 percent identity, 76 percent similarity), utilize the same physiologic substrate, have a ubiquitous expression pattern, and are mutated in diseases with different target tissues and pathological characteristics, myotubularin and MTMR2 may be subjected to differential regulatory mechanisms that preclude functional redundancy. Although their specific physiological roles are not known, a recent study has shown that developmental expression and subcellular localization of myotubularin and MTMR2 are differentially regulated, resulting in their utilization of specific cellular pools of PI(3)P [23].

Structural Features

In addition to the phosphatase domain, myotubularin-related proteins possess several motifs known to mediate protein–protein interactions and lipid binding. A PH domain, which was previously defined as a GRAM domain in myotubularin, is present in the N-terminal region of all myotubularin family members, including the catalytically inactive MTMs (Figure 29.2). Although the physiologic relevance of this domain is not known, its presence in the myotubularin family lipid phosphatases suggests a role in membrane targeting of these proteins. A coiled-coil motif is also present in all family members (Figure 29.2), and may play a role in the regulation of MTM proteins through interactions with protein effectors and/or subcellular location. Some myotubularin family members have additional lipid-binding domains. For example, MTMR3 and MTMR4 contain a FYVE domain, and MTMR5 has an additional PH domain in its C-terminal region (Figure 29.1).

Although the roles of the PH and FYVE domains in MTM function have yet to be determined, it is possible that they serve as targeting motifs to direct the lipid phosphatase domains to specific subcellular environments where PI(3)P is abundant. The physiologic function of myotubularin and related proteins in cell development and signaling processes remains unknown. Studies directed toward clarifying the regulation of myotubularin-related enzymes, as well as identifying downstream effectors, will be of significant value in understanding their roles in cell signaling and development.

REFERENCES

1. Cantley LC, Neel BG. New insights into tumor suppression: PTEN suppresses tumor formation by restraining the phosphoinositide 3-kinase/Akt pathway. *Proc Natl Acad Sci USA* 1999;**96**:4240–5.
2. Di Cristofano A, Pandolfi PP. The multiple roles of PTEN in tumor suppression. *Cell* 2000;**100**:387–90.
3. Maehama T, Taylor GS, Dixon JE. PTEN and myotubularin: novel phosphoinositide phosphatases. *Annu Rev Biochem* 2001;**70**:247–79.
4. Maehama T, Dixon JE. The tumor suppressor PTEN/MMAC1, dephosphorylates the lipid second messenger, phosphatidylinositol 3,4,5-trisphosphate. *J Biol Chem* 1998;**273**:13,375–8.
5. Vanhaesebroeck B, Leevers SJ, Ahmadi K, Timms J, Katso R, Driscoll PC, Woscholski R, Parker PJ, Waterfield MD. Synthesis and function of 3-phosphorylated inositol lipids. *Annu Rev Biochem* 2001;**70**:535–602.
6. Datta SR, Brunet A, Greenberg ME. Cellular survival: a play in three Akts. *Genes Dev* 1999;**13**:2905–27.
7. Toker A, Newton AC. Cellular signaling: pivoting around PDK-1. *Cell* 2000;**103**:185–8.
8. Suzuki A, Yamaguchi MT, Ohteki T, Sasaki T, Kaisho T, Kimura Y, Yoshida R, Wakeham A, Higuchi T, Fukumoto M, Tsubata T, Ohashi PS, Koyasu S, Penninger JM, Nakano T, Mak TW. T cell-specific loss of Pten leads to defects in central and peripheral tolerance. *Immunity* 2001;**14**:523–34.
9. Penninger JM, Woodgett J. Stem cells. PTEN-coupling tumor suppression to stem cells? *Science* 2001;**294**:2116–18.
10. Morrison SJ. Pten-uating neural growth. *Nat Med* 2002;**8**:1618.
11. Edgar BA. From small flies come big discoveries about size control. *Nat Cell Biol* 1999;**1**:E191–3.
12. Guarente L, Kenyon C. Genetic pathways that regulate ageing in model organisms. *Nature* 2000;**408**:255–62.
13. Lee J-O, Yang H, Georgescu M-M, Di Cristofano A, Maehama T, Shi Y, Dixon JE, Pandolfi P, Pavletich NP. Crystal structure of the PTEN tumor suppressor: implications for its phosphoinositide phosphatase activity and membrane association. *Cell* 1999;**99**:323–34.
14. Wishart MJ, Taylor GS, Slama JT, Dixon JE. PTEN and myotubularin phosphoinositide phosphatases: bringing bioinformatics to the lab bench. *Curr Opin Cell Biol* 2001;**13**:172–81.
15. Leslie NR, Downes CP. PTEN: the down side of PI 3-kinase signalling. *Cell Signal* 2002;**14**:285–95.
16. Stambolic V, MacPherson D, Sas D, Lin Y, Snow B, Jang Y, Benchimol S, Mak TW. Regulation of PTEN transcription by p53. *Mol Cell* 2001;**8**:317–25.
17. Virolle T, Adamson ED, Baron V, Birle D, Mercola D, Mustelin T, de Belle I. The Egr-1 transcription factor directly activates PTEN during irradiation-induced signalling. *Nat Cell Biol* 2001;**3**:1124–8.
18. Patel L, Pass I, Coxon P, Downes CP, Smith SA, Macphee CH. Tumor suppressor and anti-inflammatory actions of PPARγ agonists are mediated via upregulation of PTEN. *Curr Biol* 2001;**11**:764–8.
19. Laporte J, Blondeau F, Buj-Bello A, Tentler D, Kretz C, Dahl N, Mandel J-L. Characterization of the myotubularin dual specificity phosphatase gene family from yeast to human. *Hum Mol Genet* 1998;**7**:1703–12.
20. Laporte J, Blondeau F, Buj-Bello A, Mandel J-L. The myotubularin family: from genetic disease to phosphoinositide metabolism. *Trends Genet* 2001;**17**:221–8.
21. Taylor GS, Maehama T, Dixon JE. Myotubularin, a protein tyrosine phosphatase mutated in myotubular myopathy, dephosphorylates the lipid second messenger, phosphatidylinositol 3-phosphate. *Proc Natl Acad Sci USA* 2000;**97**:8910–15.
22. Blondeau F, Laporte J, Bodin S, Superti-Furga G, Payrastre B, Mandel J-L. Myotubularin, a phosphatase deficient in myotubular myopathy, acts on phosphatidylinositol 3-kinase and phosphatidylinositol 3-phosphate pathway. *Hum Mol Genet* 2000;**9**:2223–9.
23. Kim S-A, Taylor GS, Torgersen KM, Dixon JE. Myotubularin and MTMR2, phosphatidylinositol 3-phosphatases mutated in myotubular myopathy and type 4B Charcot-Marie-Tooth disease. *J Biol Chem* 2002;**277**:4526–31.
24. Laporte J, Liaubet L, Blondeau F, Tronchere H, Mandel JL, Payrastre B. Functional redundancy in the myotubularin family. *Biochem Biophys Res Commun* 2002;**291**:305–12.
25. Laporte J, Biancalana V, Tanner SM, Kress W, Schneider V, Wallgren-Pettersson C, Herger F, Buj-Bello A, Blondeau F, Liechti-Gallati S, Mandel J-L. MTM1 mutations in X-linked myotubular myopathy. *Hum Mutation* 2000;**15**:393–409.
26. Bolino A, Muglia M, Conforti FL, LeGuern E, Salih MAM, Georgiou D-M, Christodoulou K, Hausmanowa-Petrusewicz I, Mandich P, Schenone A, Gambardella A, Bono F, Quattrone A, Devoto M, Monaco AP. Charcot-Marie-Tooth type 4B is caused by mutations in the gene encoding myotubularin-related protein-2. *Nat Genet* 2000;**25**:17–19.

Chapter 30

Diacylglycerol Kinases

Matthew K. Topham[1] and Steve M. Prescott[2]

[1]*The Huntsman Cancer Institute and Department of Internal Medicine, University of Utah, Salt Lake City, Utah*

[2]*Oklahoma Medical Research Foundation, Oklahoma City, Oklahoma*

ABBREVIATIONS

DAG, diacylglycerol; DGK, diacylglycerol kinase; PKC, protein kinase C; PA, phosphatidic acid; RasGRP, Ras guanyl nucleotide releasing protein; MuLK, multisubstrate lipid kinase; AGK, acylglycerol kinase; PH, pleckstrin homology; SAM, sterile alpha motif; MARCKS, myristoylated alanine rich C kinase substrate; TCR, T cell receptor; EGFR, epidermal growth factor receptor.

INTRODUCTION

Many signaling cascades are initiated by phospholipase C isozymes. One product of this reaction is diacylglycerol (DAG), a prolific second messenger that activates proteins involved in a variety of signaling cascades. The protein kinase Cs (PKCs) are the best characterized DAG-activated proteins [1–3], but diacylglycerol also activates other proteins, including the four RasGRP guanine nucleotide-exchange factors [4–10] and some transient receptor potential channels [11]. DAG also recruits several proteins (chimaerins, protein kinase D, and the Munc13 proteins [12]) to membrane compartments.

Because it can associate with a diverse set of proteins, DAG potentially activates numerous signaling cascades. Thus, its accumulation needs to be strictly regulated. Diacylglycerol kinases (DGKs), which phosphorylate DAG, are widely considered to be responsible for terminating diacylglycerol signaling. However, the product of the DGK reaction, phosphatidic acid (PA), also can be a signal: it can activate phosphatidylinositol 4-phosphate 5-kinases and PKCζ, it helps recruit Raf1 and SOS to the plasma membrane, and it is involved in vesicle trafficking. By manipulating both DAG and PA levels, the DGKs potentially regulate numerous signaling events.

THE DGK FAMILY

DGKs have been identified in most organisms that have been studied, and it appears that they have gained specialization in more complex species. For example, bacteria express only one DGK, which is an integral membrane protein capable of phosphorylating DAG and other lipids such as ceramide. This DGK does not appear to have structural elements that allow regulation of its activity, indicating that the limiting factor is access to its substrates. With the exception of yeast, in which no DGKs have been identified, higher organisms appear to have several DGKs that can be grouped by common structural elements into five subfamilies (Figure 30.1). The mammalian DGKs are the best characterized, and 10 of them have been identified [13–15]. A recently identified enzyme called multisubstrate lipid kinase (MuLK) [16] or acylglycerol kinase (AGK) [17], which phosphorylates diacylglycerol and other lipids, is not structurally similar to the DGKs and is not included in this review. All of the mammalian DGKs have two common structural features: a catalytic domain, and at least two C1 domains which are thought to bind diacylglycerol. The C1 domain furthest from the amino terminus in each DGK has an extended motif of unknown significance. The catalytic domains are composed of accessory and catalytic subunits– which are separated in the type II DGKs δ, η, and κ–and each catalytic subunit has an ATP binding site with the sequence Gly–X–Gly–X–X–Gly, which is similar to protein kinase catalytic domains. Mutation of the second glycine in this motif to an aspartate or alanine renders the DGK catalytically inactive [18–20]. Other structural domains, which form the basis of the five subtypes, appear to have regulatory roles. For example, type I DGKs, α, β, and γ, have calcium-binding EF hand motifs that make these enzymes more active in the presence of calcium. Type II DGKs, δ, η, and κ, have pleckstrin homology (PH) domains

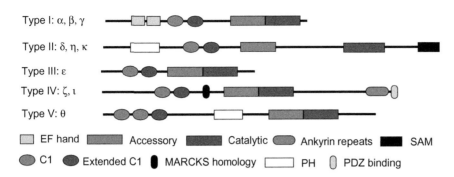

FIGURE 30.1 The 10 members of the mammalian diacylglycerol kinase family are grouped by sequence homology into five subtypes. Shown are protein motifs common to several DGKs.

near their amino termini and DGKs δ and η have a sterile alpha motif (SAM) at their carboxy terminus. These SAM domains appear to be sites of protein–protein interaction and might allow these DGKs to form oligomers. The only type III DGK, ε, does not have identifiable structural motifs outside of its C1 and catalytic domains. Interestingly, this is the only DGK that displays specificity toward acyl chains of DAG–it dramatically prefers DAGs with an arachidonoyl group at the *sn*-2 position. Type IV DGKs, ζ and ι, have a motif enriched in basic amino acids that acts as a nuclear localization signal and is a substrate for conventional PKCs. This motif is homologous to the phosphorylation site domain of the myristoylated alanine-rich C kinase substrate (MARCKS) protein. DGKs ζ and ι also have four ankyrin repeats and a PDZ binding motif at their carboxy termini that may be sites of protein–protein interactions. The only type V DGK, θ, is distinguished by three C1 domains, a PH domain, and a Ras-association domain within the PH domain. To date, no binding partners for the PH and Ras-association domains have been identified. Alternative splicing of DGKs β [21], γ [22], δ [23], ζ [24], η [25], and ι [26] greatly expand the structural diversity of the DGK family, and indicate that these enzymes likely have specific roles dictated by their unique structural motifs.

REGULATION OF DGKS

Activation of DGKs is complex, requiring translocation to a membrane compartment as well as binding to appropriate co-factors. Additional regulation of their activity occurs by posttranslational modifications. This complexity allows tissue- or cell-specific regulation, depending on the availability of co-factors and the type of stimulus that the cell receives.

DGKα demonstrates the complex regulation of DGKs. In T lymphocytes it translocates to at least two membrane compartments, depending upon the agonist used to activate the cells. For example, stimulation of T cells with IL-2 causes DGKα to translocate from the cytosol to a perinuclear region [27], but activation of the T cell receptor (TCR) results in the localization of DGKα to the plasma membrane [18]. At the membrane, the activity of DGKα

can be modified by the availability of several co-factors. Calcium is known to bind to the EF-hand structures and stimulates DAG kinase activity *in vitro*, and lipids modify its activity: phosphatidylinositol 3-kinase lipid products, phosphatidylserine, and sphingosine activate DGKα *in vitro* and likely *in vivo* [28, 29]. Finally, DGKα can be phosphorylated by several protein kinases, including PKC isoforms and Src. Although the consequences of these phosphorylations are not clear, evidence suggests that phosphorylation by Src enhances DAG kinase activity [30]. Thus, several events can modify the activity of DGKα, and combinations of them likely allow titration of its activity depending upon the cellular context.

Like DGKα, other DGK isotypes appear to be regulated by access to DAG through membrane translocation and by the availability of lipid or protein co-factors. Members of each DGK subfamily appear to be regulated similarly, although there probably are subtle differences between subfamily members owing to tissue-specific expression patterns, unique binding partners, alternative splicing, and subcellular localization. Type II DGKs, for example, have a PH domain, and this motif in DGKδ binds to phosphatidylinositols, which do not activate DGKδ but might provide a localization cue [31, 32]. The activity of types III and IV DGKs can be modified by lipids: DGKε is inhibited by phosphatidylinositols and by phosphatidylserine, while type IV DGKs are activated by these anionic phospholipids [33]. Type IV DGKs are also strongly regulated by subcellular translocation. They are imported into the nucleus, which requires their MARCKS homology domain, a nuclear localization signal that is regulated by PKC phosphorylation [20, 34]. There is also evidence that the syntrophin family of scaffolding proteins regulates nuclear import of DGKζ by associating with its carboxy-terminal PDZ binding domain to sequester DGKζ in the cytoplasm [35]. And we have observed that DGKζ has a strong nuclear export signal (M. K. Topham, unpublished). Thus, nuclear accumulation of type IV DGKs is exquisitely regulated, suggesting an important nuclear function for these isozymes. Finally, DGKθ, a type V DGK, is inhibited when it associates with active RhoA [36], and it translocates to the plasma membrane when PKCs ε and η phosphorylate it [37].

PARADIGMS OF DGK FUNCTION

Although there is substantial information regarding their regulation, little is known of the biologic functions of the individual DGKs. However, recently a few paradigms have emerged.

Regulating Ras Signaling by Spatially Modulating DAG Levels

Evidence suggests that DGKs selectively bind and regulate DAG-activated proteins. Van der Bend and colleagues [38] initially tested this concept either by initiating spatially restricted DAG synthesis through receptor activation or by causing non-specific, global DAG generation with exogenous PLC. They observed DAG kinase activity–measured by generation of PA–following receptor activation, but not after treating the cells with exogenous PLC. Their data demonstrate that DGKs are active only in spatially restricted compartments following physiologic generation of DAG. Recently, more specific examples of spatially restricted DAG kinase function have emerged. We found that DGKζ associated with RasGRP1, a guanine nucleotide-exchange factor for Ras [39]. Their association was enhanced in the presence of phorbol esters, which are slowly metabolized DAG analogs. Since RasGRP1 requires DAG to function, we hypothesized that DGKζ associated with it metabolize DAG spatially, and consequently to regulate the function of RasGRP1. Indeed, we found that catalytically inactive DGKζ did not inhibit RasGRP1. Demonstrating the specificity of this regulation, we found that five other DAG kinases also did not significantly inhibit RasGRP1.

When we tested DGKι, which is structurally similar to DGKζ, we found that it did not affect RasGRP1, but instead could bind and inhibit RasGRP3 [40]. This RasGRP activates both Ras and Rap1 and we discovered that DGKι predominantly inhibited its activation of Rap1. Surprisingly, we found that deleting DGKι in mice reduced Ras signaling and inhibited tumor formation, which was the opposite of the effects of deleting DGKζ [40]. The attenuated Ras signaling caused by DGKι deficiency likely resulted from unopposed activation of RasGRP3. The activated RasGRP3 caused excessive activation of Rap1, which then formed non-productive complexes with Raf-1 to inhibit Ras signaling. Collectively, these results demonstrate that DGK function is spatially restricted, and that the biological outcome resulting from their activity depends upon the signaling proteins that they bind and regulate.

Regulation of Neuronal Signaling

Inositol phospholipids, including PIP_2, a precursor of DAG, are enriched in arachidonate at the *sn*-2 position. Some DAG targets, including PKCs, are specifically activated by diacylglycerol containing unsaturated fatty acids, such as arachidonate. Consequently, to maintain the integrity of some DAG-activated signaling cascades, it is important that phosphatidylinositols maintain a proper fatty acid composition. Because DGKε selectively phosphorylates arachidonoyl-DAG, the first step in resynthesis of phosphatidylinositols, it may be responsible for their enrichment with arachidonate. Inositol lipid signaling is an important component of neuronal transmission. In a collaborative effort we examined seizure susceptibility in mice with targeted deletion of DGKε [41], and found that the null mice were resistant to seizures induced by electroconvulsive shock. Examination of brain lipids revealed that compared to wild-type mice, DGKε-deficient mice had reduced levels of arachidonate in both PIP_2 and DAG. This lipid profile demonstrated a critical role for DGKε in maintaining a proper balance of arachidonate-enriched inositol phospholipids. Thus, through its selectivity for arachidonoyl-DAG, DGKε regulates lipid signaling events and, consequently, seizure susceptibility.

DGKθ might also regulate neuronal signaling. Nurrish and colleagues [42] isolated a *Caenorhabditis elegans* strain resistant to serotonin-induced inhibition of locomotion. The mutated gene responsible for the effect, *dgk-1*, is homologous to DGKθ. Their data suggested a model in which DGK-1 metabolizes DAG, which is produced upon activation of the serotonin receptor. By reducing DAG levels, DGK-1 limits membrane translocation of UNC-13, a protein recruited by DAG that may mediate acetylcholine release. These data have not been tested in a vertebrate model, but suggest that, in humans, DGKθ might regulate signaling events downstream of the serotonin receptor.

Nuclear DGKs

There is a nuclear phosphatidylinositol cycle regulated separately from its plasma membrane/cytosolic counterpart [43]. DAG is present in nuclear preparations and appears to fluctuate with the cell cycle, but the specific pattern of its accumulation is not clear because of the many different methods used to isolate nuclei. DAG kinases have also been observed in nuclei, and appear to have a prominent role there. Some DGKs, like DGKs α, γ, ζ and ι, translocate to the nucleus, while others, like DGKθ, are constitutively located there [44]. These DGKs are confined to specific compartments within the nucleus. For example, both DGKθ and DGKζ appear in a speckled pattern within the nucleus, while DGKα associates with the nuclear envelope [45]. This compartmentalization suggests that DGK isotypes have specific roles in the nucleus. Movement of proteins in the nucleus largely occurs by random diffusion, so overexpression of one DGK isotype may interfere with the function of another DGK. This fact, combined with the lack of specific DGK inhibitors, has made it difficult to study the nuclear function of the different DGK isotypes. However, they likely affect nuclear signaling, either by terminating DAG signals or by generating PA. For example, in T lymphocytes, the PA produced by nuclear DGKα appears to be necessary for IL-2-mediated progression to the S phase of the cell cycle [27]. Conversely, nuclear DGKζ inhibits exit from the G1 phase of the cell cycle by metabolizing DAG [20]. These data indicate both the complexity and importance of lipid signaling and DGK function in the nucleus.

Visual Signal Transduction

A clear role for DGK function has been demonstrated in *Drosophila*. A mutant strain, *rdgA*, undergoes rapid retinal degeneration after birth. The defect results from a deficiency in retinal DGK activity because of a point mutation that inactivates dDGK2, a DAG kinase very similar to mammalian type IV DGKs. Although there are differences in photoreceptor signaling between *Drosophila* and vertebrates, these observations indicate that DGKs may be functionally important in the vertebrate retina. Three mammalian DGK isoforms, γ, ε, and ι, have been definitively localized to the retina, but their functions there have not yet been identified [29]. There is, though, evidence of light-dependent activation of PIP_2 hydrolysis and generation of PA in vertebrate retina, indicating a role there for DGK activity.

Regulation of Cell Motility

DGKs γ and ζ appear to be important for cell motility, and both DGKs seem to mediate their effects, in part, through Rac1. However, they have opposing effects on its activity: DGKζ activates Rac1 signaling [46], while DGKγ is an inhibitor [47]. Although both DGKs bind Rac1, neither of them appears to directly regulate Rac1. Instead, through spatial metabolism of DAG or production of PA, they exert their effects on proteins upstream of Rac1 that then regulate its activity. Both DGKs modulate alterations in cytoskeleton dynamics that affect membrane ruffling, lamellipodium formation, and/or neurite outgrowth. The physiological relevance of their functions is not yet clear, and no abnormalities attributable to altered cell motility are yet apparent in DGKζ knockout mice.

Regulation of Immune Function

Based on inhibitor studies, DGKs have long been know to modulate leukocyte functions, but the DGKs responsible for these effects were not known. Recently, however, studies in knockout mice have demonstrated important roles for DGKs α and ζ in leukocytes. Zhong and colleagues deleted the gene encoding DGKζ and found that lymphocytes

from the mice had excessive Ras signaling following TCR stimulation, and consequently were hyperproliferative [48]. DGKζ also appears to function in mast cells downstream of the immunoglobulin E receptor and in macrophages downstream of Toll-like receptors [49, 50]. Like DGKζ, DGKα also functions downstream of the TCR, and deletion of its gene in mice or inhibitors of DGKα enhanced TCR signaling and promoted proliferation [51, 52]. Ultimately, DGKα deficiency caused T lymphocytes to become resistant to anergy. Other DGK isoforms are expressed in leukocytes, but their specific function *in vivo* remain to be established.

Regulation of Receptor Tyrosine Kinase Signaling

DGKs δ and α regulate receptor tyrosine kinase signaling events. We found that DGKδ knockout mice have a phenotype very similar to epidermal growth factor receptor (EGFR) null mice [53]. Among the known DGK knockout mice, this phenotype was unique to DGKδ. We found that DGKδ deficiency led to excessive DAG accumulation, which activated PKCs that then reduced EGFR expression and specific activity. The changes in EGFR activity were, in part, caused by phosphorylation of a threonine residue (T654) in EGFR, while the effects on the expression levels of EGFR are still under investigation.

DGKα appears to function downstream of at least two receptor tyrosine kinases. Upon activation of the hepatocyte growth factor receptor, DGKα is phosphorylated by Src and promotes cell migration through Rac [54]. Downstream of vascular endothelial growth factor receptor, DGKα is phosphorylated by Src and affects angiogenic signaling [55]. Although it affects two important tyrosine kinase signaling pathways, DGKα mice appear normal, indicating that it might be required for pathologic responses rather than homeostatic signaling.

CONCLUSIONS

Diacylglycerol kinases are expressed in all multicellular organisms that have been studied. Their structural diversity and complexity indicate that they are functionally important in a variety of cellular signaling events. Since they can affect both DAG and PA signals, DGK activity plays a central role in many lipid signaling pathways.

REFERENCES

1. Newton AC. Regulation of protein kinase C. *Cell Biol* 1997;**9**:161–7.

2. Parekh DB, Ziegler W, Parker PJ. Multiple pathways control protein kinase C phosphorylation. *EMBO J* 2000;**19**:496–503.

3. Toker A. Signaling through protein kinase C. *Frontiers Biosci* 1998;**3**: d1134–47.

4. Ebinu JO, Bottorff DA, Chan EYW, Stang SL, Dunn RJ, Stone JC. RasGRP, a ras guanyl nucleotide-releasing protein with calcium-and diacylglycerol-binding motifs. *Science* 1998;**280**:1082–6.

5. Tognon CE, Kirk HE, Passmore LA, Whitehead IP, Der CJ, Kay RJ. Regulation of RasGRP via a phorbol ester-responsive C1 domain. *Mol Cell Biol* 1998;**18**:6995–7008.

6. Kawasaki H, Springett GM, Shinichiro T, et al. A Rap guanine nucleotide exchange factor enriched highly in the basal ganglia. *Proc Natl Acad Sci USA* 1998;**95**:13,278–13,283.

7. Yamashita S, Mochizuki N, Ohba Y, et al. CalDAG-GEFIII Activation of Ras, R-Ras, and Rap1. *J Biol Chem* 2000;**275**:25,488–25,493.

8. Clyde-Smith J, Silins G, Gartside M, et al. Characterization of RasGRP2, a plasma membrane-targeted, dual specificity Ras/Rap exchange factor. *J Biol Chem* 2000;**275**:32,260–32,267.

9. Yang Y, Wong GW, Krilis SA, Madhusudhan MS, Sali A, Stevens RL. RasGRP4, a new mast cell-restricted Ras guanine nucleotide-releasing protein with calcium- and diacylglycerol-binding motifs. *J Biol Chem* 2002;**277**:25,756–25,774.

10. Reuther GW, Lambert QT, Rebhun JF, Caligiuri MA, Quilliam LA, Der CJ. RasGRP4 Is a Novel Ras Activator Isolated from Acute Myeloid Leukemia. *J Biol Chem* 2002;**277**:30,508–30,514.

11. Lucas P, Ukhanov K, Leinders-Zufall T, Zufall F. A diacylglycerol-gated cation channel in vomeronasal neuron dendrites is impaired in TRPC2 mutant mice. *Neuron* 2003;**40**:551–61.

12. Ron D, Kazanietz MG. New insights into the regulation of protein kinase C and novel phorbol ester receptors. *FASEB J* 1999;**13**:1658–76.

13. Topham MK, Prescott SM. Diacylglycerol kinases: regulation and signaling roles. *Thromb Haemost* 2003;**88**:912–18.

14. Luo B, Regier DS, Prescott SM, Topham MK. Diacylglycerol kinases. *Cell Signal* 2004;**16**:983–9.

15. Imai H, Kai M, Yasuda S, Kanoh H, Sakane F. Identification and characterization of a novel type II diacylglycerol kinase, DGK kappa. *J Biol Chem* 2005;**280**:39,870–39,881.

16. Waggoner DW, Johnson LB, Mann PC, Morris V, Guastella J, Bajjalieh SM. MuLK, a eukaryotic multi-substrate lipid kinase. *J Biol Chem* 2004;**279**:38,228–38,235.

17. Bektas M, Payne SG, Liu H, Goparaju S, Milstien S, Spiegel S. A novel acylglycerol kinase that produces lysophosphatidic acid modulates cross talk with EGFR in prostate cancer cells. *J Cell Biol* 2005;**169**:801–11.

18. Sanjuan MA, Jones DA, Izquierdo M, Merida I. Role of diacylglycerol kinase a in the attenuation of receptor signaling. *J Cell Biol* 2001;**153**:207–19.

19. Nagaya H, Wada I, Jia Y-J, Kanoh H. Diacylglycerol kinase δ suppresses ER-to-Golgi traffic via its SAM and PH domains. *Mol Biol Cell* 2002;**13**:302–16.

20. Topham MK, Bunting M, Zimmerman GA, McIntyre TM, Blackshear PJ, Prescott SM. Protein kinase C regulates the nuclear localization of diacylglycerol kinase-zeta [see comments]. *Nature* 1998;**394**:697–700.

21. Caricasole A, Bettini E, Sala C, et al. Molecular cloning and characterization of the human diacylglycerol kinase beta (DGKbeta) gene: alternative splicing generates DGKbeta isotypes with different properties. *J Biol Chem* 2001;**277**:4790–6.

22. Kai M, Sakane F, Imai S-I, Wada I, Kanoh H. Molecular cloning of a diacylglycerol kinase isozyme predominantly expressed in human retina with a truncated and inactive enzyme expression in most other human cells. *J Biol Chem* 1994;**269**:18492–8.

23. Sakane F, Imai S, Yamada K, Murakami T, Tsushima S, Kanoh H. Alternative splicing of the human diacylglycerol kinase δ gene generates

two isoforms differing in their expression patterns and in regulatory functions. *J Biol Chem* 2002;**277**:43,519–43,526.

24. Ding L, Bunting M, Topham MK, McIntyre TM, Zimmerman GA, Prescott SM. Alternative splicing of the human diacylglycerol kinase ζ gene in muscle. *Proc Natl Acad Sci USA* 1997;**94**:5519–24.

25. Murakami T, Sakani F, Imai S, Houkin K, Kanoh H. Identification and characterization of two splice variants of diacylglycerol kinase η. *J Biol Chem* 2003;**278**:34,364–34,372.

26. Ito T, Hozumi Y, Sakane F, et al. Cloning and characterization of diacylglycerol kinase iota splice variants in rat brain. *J Biol Chem* 2004;**279**:23,317–23,326.

27. Flores I, Casaseca T, Martinez-A C, Kanoh H, Merida I. Phosphatidic acid generation through interleukin 2 (IL-2)-induced a-diacylglycerol kinase activation is an essential step in IL-2-mediated lymphocyte proliferation. *J Biol Chem* 1996;**271**:10,334–10,340.

28. Cipres A, Carrasco S, Merino E, et al. Regulation of diacylglycerol kinase α by phosphoinositide 3-kinase lipid products. *J Biol Chem* 2003;**278**:35,629–35,635.

29. Topham MK, Prescott SM. Mammalian diacylglycerol kinases, a family of lipid kinases with signaling functions. *J Biol Chem* 1999;**274**:11,447–11,450.

30. Baldanzi G, Cutrupi S, Chianale F, et al. Diacylglycerol kinase-alpha phosphorylation by Src on Y335 is required for activation, membrane recruitment and Hgf-induced cell motility. *Oncogene* 2007;**27**:942–56.

31. Sakane F, Imai SI, Kai M, Wada I, Kanoh H. Molecular cloning of a novel diacylglycerol kinase isozyme with a pleckstrin homology domain and a C-terminal tail similar to those of the EPH family of protein tyrosine kinases. *J Biol Chem* 1996;**271**:8394–401.

32. Takeuchi H, Kanematsu T, Misumi Y, et al. Distinct specificity in the binding of inositol phosphates by pleckstrin homology domains of pleckstrin, RAC-protein kinase, diacylglycerol kinase and a new 130-kDa protein. *Biochim Biophys Acta* 1997;**1359**:275–85.

33. Thirugnannam S, Topham MK, Epand RM. Physiological implications of the contrasting modulation of the activities of the ε- and ζ-isoforms of diacylglycerol kinases. *Biochemistry* 2001;**40**:10,607–10,613.

34. Ding L, Traer E, McIntyre TM, Zimmerman GA, Prescott SM. The cloning and characterization of a novel human diacylglycerol kinase, DGKι. *J Biol Chem* 1998;**273**:32,746–32,752.

35. Hogan A, Shepherd L, Chabot J, et al. Interaction of γ1-syntrophin with diacylglycerol kinase-ζ. *J Biol Chem* 2001;**276**:26,526–26,533.

36. Los AP, van Baal J, de Widt J, Divecha N, van Blitterswijk WJ. Structure–activity relationship of diacylglycerol kinase theta. *Biochim Biophys Acta* 2004;**1636**:169–74.

37. van Baal J, de Widt J, Divecha N, van Blitterswijk WJ. Translocation of diacylglycerol kinase theta from cytosol to plasma membrane in response to activation of G-protein-coupled receptors and protein kinase C. *J Biol Chem* 2005;**280**:9870–8.

38. van der Bend RL, de Widt J, Hilkmann H, van Blitterswijk WJ. Diacylglycerol kinase in receptor-stimulated cells converts its substrate in a topologically restricted manner. *J Biol Chem* 1994;**269**:4098–102.

39. Topham MK, Prescott SM. Diacylglycerol kinase z regulates Ras activation by a novel mechanism. *J Cell Biol* 2001;**152**:1135–43.

40. Regier DS, Higbee J, Lund KM, Sakane F, Prescott SM, Topham MK. Diacylglycerol kinase iota regulates ras guanyl-releasing protein 3 and inhibits Rap1 signaling. *Proc Natl Acad Sci USA* 2005;**102**:7595–600.

41. Rodriquez de Turco EB, Tang W, Topham MK, et al. Diacylglycerol kinase ε regulates siezure susceptibility and long-term potentiation through arachidonoyl-inositol lipid signaling. *Proc Natl Acad Sci USA* 2001;**98**:4740–5.

42. Nurrish S, Segalat L, Kaplan JM. Serotonin inhibition of synaptic transmission: Ga_o decreases the abundance of UNC-13 at release sites. *Neuron* 1999;**24**:231–42.

43. Divecha N, Banfic H, Irvine RF. Inositides and the nucleus and inositides in the nucleus. *Cell* 1993;**74**:405–7.

44. Matsubara T, Shirai Y, Miyasaka K, et al. Nuclear transportation of diacylglycerol kinase gamma and its possible function in the nucleus. *J Biol Chem* 2006;**281**:6152–64.

45. van Blitterswijk WJ, Houssa B. Properties and functions of diacylglycerol kinases. *Cell Signal* 2000;**12**:595–605.

46. Yakubchyk Y, Abramovici H, Maillet J, et al. Regulation of neurite outgrowth in N1E-115 cells through PDZ-mediated recruitment of diacylglycerol kinase zeta. *Mol Cell Biol* 2005;**25**:7289–302.

47. Tsushima S, Kai M, Yamada K, et al. Diacylglycerol kinase gamma serves as an upstream suppressor of Rac1 and lamellipodium formation. *J Biol Chem* 2004;**279**:28,603–28,613.

48. Zhong X, Hainey EA, Olenchock BA, et al. Enhanced T cell responses due to diacylglycerol kinase z deficiency. *Nat Immunol* 2003;**4**:882–90.

49. Olenchock BA, Guo R, Silverman MA, et al. Impaired degranulation but enhanced cytokine production after Fc epsilonRI stimulation of diacylglycerol kinase zeta-deficient mast cells. *J Exp Med* 2006;**203**:1471–80.

50. Liu CH, Machado FS, Guo R, et al. Diacylglycerol kinase zeta regulates microbial recognition and host resistance to *Toxoplasma gondii*. *J Exp Med* 2007;**204**:781–92.

51. Olenchock BA, Guo R, Carpenter JH, et al. Disruption of diacylglycerol metabolism impairs the induction of T cell anergy. *Nat Immunol* 2006;**7**:1174–81.

52. Zha Y, Marks R, Ho AW, et al. T cell anergy is reversed by active Ras and is regulated by diacylglycerol kinase-alpha. *Nat Immunol* 2006;**7**:1166–73.

53. Crotty T, Cai J, Sakane F, Taketomi A, Prescott SM, Topham MK. Diacylglycerol kinase d regulates protein kinase C and epidermal growth factor receptor signaling. *Proc Natl Acad Sci USA* 2006;**103**:15,485–15,490.

54. Chianale F, Cutrupi S, Rainero E, et al. Diacylglycerol kinase-α mediates hepatocyte growth factor-induced epithelial cell scatter by regulating Rac activation and membrane ruffling. *Mol Biol Cell* 2007;**18**:4859–71.

55. Baldanzi G, Mitola S, Cutrupi S, et al. Activation of diacylglycerol kinase alpha is required for VEGF-induced angiogenic signaling *in vivo*. *Oncogene* 2004;**23**:4828–38.

Chapter 31

Phospholipase C

Hong-Jun Liao and Graham Carpenter

Department of Biochemistry, Vanderbilt University School of Medicine, Nashville, Tennessee

INTRODUCTION

The phospholipase C enzymes that hydrolyze phosphatidylinositol 4,5–bisphosphate in mammalian cells are subdivided into four families, denoted β, γ, δ, and ε, based on sequence similarities. Each family has a unique organization of regulatory sequence motifs or domains that facilitate protein–protein and/or protein–phospholipid interactions. Utilizing these motifs, each family responds to distinct hormonal signals or intracellular cues to produce the second messenger molecules inositol 1,4,5–trisphosphate and diacyl glycerol. These metabolites in turn control intracellular levels of free Ca^{2+} and protein kinase C activity, respectively. This review, in addition to discussing molecular structure/ function and activation mechanisms for phospholipase C enzymes, presents the physiologic consequences of PLC genetic knockouts.

This chapter is focused on the phosphoinositide-specific phospholipase C (PLC) isozymes expressed in mammalian cells. This family of isozymes is defined on the basis of sequence similarities and the capacity to mediate the hydrolysis of phosphatidylinositol 4,5–bisphosphate (PI 4,5–P_2) to the second messenger molecules inositol 1,4,5–trisphosphate and diacylglycerol. The former provokes mobilization of intracellular Ca^{2+} by regulating the release of stored Ca^{2+} from within intracellular organelles into the cytosol and nucleus. The latter functions as an endogenous and required activator of protein kinase C isozymes. Hence, this enzyme uniquely activates two second messengers, which in turn may control a variety of signaling pathways and thereby influence a panoply of cellular events. This review is constrained by space, and readers are referred to other recent reviews [1–3] for additional information and pertinent references.

PLC ANATOMY

The eleven mammalian PI 4,5–P_2 specific PLCs are divided into four subgroups (designated β, γ, δ, ε) based on

sequence similarities that produce an organization of structural motifs unique to each subgroup [1]. The organization of these motifs or domains is illustrated in Figure 31.1. All PLC isozymes have motifs designated X and Y, which in the native protein are folded together to constitute the catalytic domain. An X-ray structure of PLC-δ1 provides the clearest picture of exactly how this enzymatic center is organized and suggests potential catalytic mechanisms [2]. In addition to conserved catalytic function, each PLC subgroup is characterized by additional motifs that are involved in regulating aspects of enzyme function, such as topological localization within the cell and sensitivity to protein:protein and protein:lipid interactions.

PLC ACTIVATION MECHANISMS

PLC-β

The activity of four PLC-β isozymes is regulated by hormones that bind to G-protein-coupled receptors (GPCRs) [1]. These receptors typically have multiple membrane spanning domains, have no intrinsic catalytic activity, and utilize heterotrimeric G proteins to communicate with downstream second messenger producing enzymes, such as PLC isozymes. When GPCRs are stimulated by hormone binding, G protein complexes containing α, β, γ subunits are activated with the following characteristics: GDP bound to the α subunit is replaced by GTP, dissociating the trimeric complex into two active species – a GTP-bound free α subunit and a βγ dimeric complex. Both of these act as signal transducers to activate PLC-β isoforms in a manner that may depend on both for maximal activation.

The α·GTP subunit interacts with a region in PLC-β that includes a portion of the C2 domain, while the βγ complex interacts with part of the PH domain. The region within PLC-β that binds the α·GTP subunit also appears to function as a dimerization interface, suggesting that PLC-β

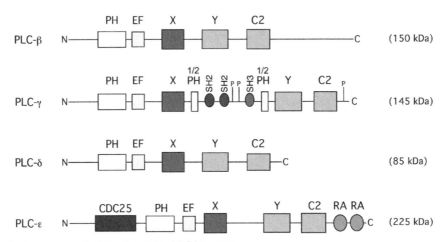

FIGURE 31.1 Schematized arrangement of domains within PLC isozymes.
Functional domains or motifs (PH, EF, SH2, SH3, C2, RA, CDC25, X, Y) are explained in the text. Tyrosine phosphorylation sites are depicted by Y in the γ isozymes.

may function as a dimer [4, 5]. Since α subunits and βγ complexes are constitutively anchored by lipid modifications to the cytoplasmic face of the plasma membrane, one consequence of these interactions is to promote a catalytically competent association of PLC-β with the plasma membrane. In unstimulated cells, PLC-βs are usually found loosely associated with the plasma membrane in what can be termed a catalytically incompetent association.

There is evidence for most PLCs that formation of a highly specific plasma membrane association is necessary for hydrolysis of the plasma membrane-localized substrate PI 4,5–P_2. Productive membrane association by PLC-β is further facilitated by interaction of its PH domain with phosphatidyl inositol 3–phosphate. Separate regions of the PH domain accommodate this phosphoinositide and βγ complexes. Although the C2 domain in PLC-δ does mediate a Ca^{2+}-dependent phospholipid interaction, there is no evidence for this in PLC-β. While interactions of the C2 and PH domains of PLC-β with membrane-bound molecules might seem sufficient to explain formation of a productive membrane:enzyme complex, it is unclear whether this association *per se* is sufficient for increased catalytic activity or whether these interactions also provoke changes within the X/Y catalytic domain.

Signal transduction mechanisms are, by definition, reactions that are readily reversible. In the case of PLC-β, the most readily reversible component resides in the α·GTP subunit, which can be rapidly converted by intrinsic GTPase activity to α·GTP.

PLC-γ

This subgroup contains two members, γ1 and γ2 [1, 3]. Initially, the cloning and sequencing of the isozymes indicated significant differences in the COOH terminal sequences; however, more recent data indicate that these

apparent differences resulted from sequencing errors for γ2 [6]. As shown in Figure 31.1, PLC-γ uniquely contains motifs known as SH2 and SH3 domains in addition to the PH and C2 domains present in other PLC subgroups. The SH2 motifs, in particular, are important to facilitate activation of γ isozymes by growth factor receptor tyrosine kinases (RTKs). RTKs possess a single transmembrane domain separating the ligand-binding ectodomain from a cytoplasmic domain that contains sequences encoding tyrosine kinase activity [7]. Ligand binding facilitates dimerization of RTKs and this in turn facilitates activation of the tyrosine kinase domain. Substrates for the tyrosine kinase include the receptor itself and other proteins, such as PLC-γ.

The initial step in growth factor-dependent activation of PLC-γ involves the recognition of autophosphorylation sites in a RTK by the SH2 domains of PLC-γ1 [1, 3]. This recognition event is a prerequisite for tyrosine phosphorylation of PLC-γ by the RTK, which constitutes a major step in the activation mechanism. Receptor association may also relocalize PLC-γ1 from the cytosol to the cytoplasmic face of the plasma membrane, much like the association of PLC-β with membrane-anchored G-protein subunits.

Compared to other PLCs, PLC-γ has an elongated linker segment between the X and Y domains. This linker contains not only the SH2 and SH3 domains, but also a split PH domain and at least two important tyrosine phosphorylation sites, which are close together between the C-SH2 and SH3 domains and are conserved in the γ1 and γ2 isoforms. It is possible that modulation of the structure of this linker region, by protein: protein interaction and/or by tyrosine phosphorylation, contributes to activation of the catalytic site [3]. One additional site of tyrosine phosphorylation is located C-terminal to C2 domain in both the γ1 and γ2 isoforms [1, 3], while a fourth site in γ2 is located between the Y and C2 domains [8].

Evidence has been presented that in some cell systems PI-3 kinase activity and the formation of phosphatidylinositol 3,4,5–trisphosphate (PI 3,4,5–P_3) is necessary for maximal PLC-γ1 activation [1, 3]. The site of action of PI 3,4,5–P_3 is most likely the N-terminal PH domain of PLC-γ1. However, not all reports are in agreement on this and data have been presented to support an interaction of PI 3,4,5–P_3 with the C-SH2 domain of PLC-γ1. The possible contributions of the SH3 and C2 domains of PLC-γ1 to the activation mechanism are unclear. Evidence has been presented to indicate that the SH3 domain of PLC-γ1 can associate with PIKE, a nuclear GTPase, and stimulate its activation [9].

In hematopoietic cells, adaptor proteins, such as LAT in T cells, are localized in specialized membrane microdomains termed lipid rafts and are also tyrosine phosphorylated following antigen activation [1, 3]. In T cells, phosphorylated LAT becomes associated with PLC-γ1, and this interaction is necessary for PLC-γ1 activation. Whether similar membrane components are involved in PLC-γ1 activation in non-hematopoietic cells is unknown. However, there is evidence that RTK activation provokes the preferential association of tyrosine phosphorylated PLC-γ with caveolae [10, 11], which resemble raft membrane microdomains but contain the protein caveolin.

PLC-δ

The activation mechanisms for this subgroup of PLC isozymes, which are not activated by GPCRs or RTKs, are perhaps least understood. Studies *in vitro* indicate that while all PLCs require free Ca^{2+}, the δ isozymes are the most sensitive to free Ca^{2+} levels [1]. This has led to the notion that this enzymes activity may be enhanced by increased levels of intracellular Ca^{2+}, and there is evidence consistent with this in a few cell-based systems. Available data also indicate that the PH and C2 domains facilitate membrane association of the δ isozymes. The PH domain of PLC-δ recognizes PI 4,5–P2 and may not only tether the enzyme to the plasma membrane but may also facilitate a processive mechanism of hydrolysis. The C2 domain mediates membrane association by recognition of a Ca^{2+}-phospholipid complex in the membrane. It is interesting that the PH domain of PLC-δ also binds inositol 1,4,5–P_3, the product of PI 4,5–P_2 hydrolysis, and this may represent a mechanism to decrease PLC-δ activity when product levels become high.

PLC-ε

This is the most recent addition to the mammalian PLC family and was foreshadowed by the isolation of a *C. elegans* cDNA that has the same organization [1]. Within the PLC family, PLC-ε has novel protein:protein interaction motifs that indicate that its activation is directly controlled by the G protein Ras, which is a particularly important molecule in signal transduction pathways initiated by growth factor RTKs.

Near its N-terminus, PLC-ε contains a sequence identified as a CDC25 or a RasGEF (guanine nucleotide exchange factor) motif, which in other proteins facilitates the activation of Ras by mediating the exchange of GDP for GTP. Evidence indicates that expression of exogenous PLC-ε in cells does promote increased levels of Ras$\cong$GTP. This would place PLC-ε upstream of Ras in a signal transduction pathway. PLC-ε also has RA or Ras association motifs near its C-terminus. This sequence motif allows recognition of PLC-ε by the GTP-bound, or activated, form of Ras. Evidence shows that PLC-ε is indeed recognized with high affinity by Ras·GTP and is not recognized by Ras GTP. This predicts that Ras is an activator of PLC-ε and would place this PLC isoform downstream of activated Ras. Also, recognition of PLC-ε by activated Ras could also facilitate membrane translocation of PLC-ε, as Ras is constitutively membrane-localized by the presence of covalent lipid constituents. The complex relationship of PLC-ε to Ras is analogous to the reported observation that PLC-β1 acts as a GAP (GTPase activating protein) toward the α·GTP subunit that is its direct activator [12]. Ras is a prototype for a large family of single subunit GTPases, and there are data suggesting that PLC-ε also participates in signaling dependent on Rap 1 [13] and Rap2B [14], members of the Ras superfamily.

It also appears that PLC-ε can be activated by heterotrimeric G-protein subunits, including $\beta\gamma$ complexes [15] and at least one α·GTP subunit [16]. The former proceeds through recognition of a PH domain in PLC-ε. Hence, PLC-ε may be activated by growth factor RTKs through Ras or by GPCRs through heterotrimeric G proteins. This finding raises the prospect that the PLC activity downstream of these receptors may represent contributions by more than one PLC subgroup.

PLC PHYSIOLOGY

While structural and biochemical questions regarding PLC isozymes have yielded significant information regarding the molecular mechanisms by which these enzymes are activated in cells, there is much less information available regarding the role of these PLC isozymes in physiologic or pathophysiologic processes. Given the ubiquitous occurrence of PLC isozymes and the pleiotropic potential of the second messengers that they generate, this may seem either too obvious or too complex a question to resolve. In view of the multiplicity of isoforms in each PLC subgroup, one might expect substantial functional redundancy to exist, although this expectation is partially offset by the

TABLE 31.1 Phenotypes resulting from targeted disruption of PLC genes in mice

Gene (protein)	Phenotype	Reference
Plcb1 (PLC-β1)	Death 2–6 weeks after birth, increased level of recurrent seizures due to decrease in muscarinic acetylcholine signaling	17
Plcb2 (PLC-β2)	Normal lifespan, increased neutrophil chemotactic response	18
Plcb3 (PLC-β3)	Embryonic lethal E2.5	19
	Normal lifespan, decreased opoid-dependent behavioral responses	20
	Increased skin ulcers	21
Plcb4 (PLC-β4)	Normal life span, locomotor ataxia due to metabotropic glutamate receptor signaling	17
	Impaired visual response	22
	Decreased climbing fiber elimination	23
	Decreased long-term depression, decreased conditioned motor learning	24
Plcg1 (PLC-γ1)	Embryonic lethal E9.0, impaired erythrogenesis, vasculogenesis	25
	Chimeric mice (plcg1$^{-/-}$ and $^{+/+}$) mice, impaired hematopoiesis, polycystic kidney	26
Plcg2 (PLC-γ2)	Normal lifespan, decreased mast cell function, decreased B cell numbers	27
Plcd4 (PLC-δ4)	Normal lifespan, male infertility due to deficiency in acrosome reaction	28

The results, in some cases, represent phenotypes obtained at the first crucial point in development when a particular PLC isoform becomes required for further development of the organism.

In the case of Plcb3 knockouts, discordant results have been reported that may reflect the manner in which the gene was actually disrupted. When Plcb3 genomic information corresponding to exons encoding the last one-third of the X domain plus the first two-thirds of the Y domain was deleted, embryonic lethality was produced at approximately 2.5 days in gestation [19]. In the second knockout [20], a genomic deletion representing one exon encoding some residues in the X domain was produced and the mice were normal as far as development and growth are concerned. At this time the discrepancy between these two studies has not been resolved. It is possible that in one knockout a mutant protein was produced that acted as a dominant-negative molecule affecting other pathways. Alternatively, one of the knockouts may represent only a partial loss of PLC-β1 function.

REFERENCES

1. Rhee SG. Regulation of phosphoinositide-specific phospholipase C. *Annu Rev Biochem* 2001;**70**:281–312.
2. Williams RL. Mammalian phosphoinositide-specific phospholipase C. *Biochim Biophys Acta* 1999;**1441**:255–67.
3. Carpenter G, Ji Q-S. Phospholipase C-γ as a signal transducing element. *Exp Cell Res* 1999;**253**:15–24.
4. Singer AU, Waldo GL, Harden TK, Sondek J. A unique fold of phospholipase C-β mediates dimerization and interaction with αq. *Nat Struct Biol* 2002;**9**:32–6.
5. Ilkaeva O, Kinch LN, Paulssen RH, Ross EM. Mutations in the carboxyl-terminal domain of phospholipase C-β1 delineate the dimer interface and a potential Gα$_q$ interaction site. *J Biol Chem* 2002;**277**:4294–300.
6. Ozdener F, Kunapuli SP, Daniel JL. Carboxyl terminal sequence of human phospholipase C-γ2. *Platelets* 2001;**12**:121–3.
7. Schlessinger J, Ullrich A. Growth factor signaling by receptor tyrosine kinases. *Neuron* 1992;**9**:383–91.
8. Watanabe D, Hashimoto S, Ishiai M, Matsushita M, Baba Y, Kishimoto T, Kurosaki T, Tsukada S. Four tyrosine residues in phospholipase C-γ2, identified as Btk-dependent phosphorylation sites, are required for B cell antigen receptor-coupled calcium signaling. *J Biol Chem* 2001;**276**:38,595–601.
9. Ye K, Aghdasi B, Luo HR, Moriarity JL, Wu FY, Hong JJ, Hurt KJ, Bae SS, Suh P-G, Snyder SH. Phospholipase Cγ1 is a physiological guanine nucleotide exchange factor for the nuclear GTPase PIKE. *Nature* 2002;**415**:541–4.
10. Wang X-J, Liao H-J, Chattopadhyay A, Carpenter G. EGF-dependent translocation of green fluorescent protein-tagged PLC-γ1 to the plasma membrane and endosomes. *Exp Cell Res* 2001;**267**:28–36.
11. Jang I-H, Kim JH, Lee BD, Bae SS, Park MH, Suh P-G, Ryu SH. Localization of phospholipase C-γ1 signaling in caveolae: importance of BGF-induced phosphoinositide hydrolysis but not in tyrosine phosphorylation. *FEBS Letts* 2001;**491**:4–9.

differing patterns of expression for each isoform. Also at play is the extent to which PLC-dependent signal transduction is necessary, sufficient, or dispensable for any given cell response. These issues can to some extent be addressed by selective abrogation of each PLC isozyme through targeted gene disruption technology. The contents of Table 31.1 describe results that have been obtained to date by the application of this technology to PLC genes.

12. Berstein G, Blank JL, Jhon D-Y, Exton JH, Rhee SG, Ross EM. Phospholipase C-β is a GTPase-activating protein form Gq/11, its physiological regulator. *Cell* 1992;**70**:411–18.

13. Jin T-G, Satoh T, Liao Y, Song C, Gao X, Kariya K-i, Hu C-D, Kataoka T. Role of the CDC25 homology domain of phospholipase Cε in amplification of Rap1-dependent signaling. *J Biol Chem* 2001;**276**:30,301–7.

14. Schmidt M, Evellin S, Weernink PAO, vom Dorp F, Rehmann H, Lomasney JW, Jakobs KH. A new phospholipase C-calcium signalling pathway mediated by cyclic AMP and a Rap GTPase. *Nat Cell Biol* 2001;**3**:1020–4.

15. Wing MR, Houston D, Kelley GG, Der CJ, Siderovski DP, Harden TH. Activation of phospholipase C-ε by heterotrimeric G protein βγ-subunits. *J Biol Chem* 2001;**276**:48,257–61.

16. Lopez I, Mak EC, Ding J, Hamm HE, Lomasney JW. A novel bifunctional phospholipase C that is regulated by Gα12 and stimulates the Ras/mitogen-activated protein kinase pathway. *J Biol Chem* 2001;**276**:2758–65.

17. Kim, D., Jun, K. I. S., Lee, S. B., Kang, N.-G., Min, D. S., Kim, Y.-H., Ryu, S. H., Suh, P.-G., and Shin, H.-S. Nature 389, 290–3.

18. Jiang H, Kuang Y, Wu Y, Xie W, Simon MI, Wu D. Roles of phospholipase Cβ2 in chemoattractant-elicited responses. *Proc Natl Acad Sci USA* 1997;**94**:7971–5.

19. Wang S, Gebre-Medhin S, Betsholtz C, Stålberg P, Zhou Y, Larsson C, Weber G, Feinstein R, Öberg K, Gobl A, Skogseid B. Targeted disruption of the mouse phospholipase Cβ3 gene results in early embryonic lethality. *FEBS Letts* 1998;**441**:261–5.

20. Xie W, Samoriski GM, McLaughlin JP, Romoser VA, Smrcka A, Hinkle PM, Bidlack JM, Gross RA, Jiang H, Wu D. Genetic alteration of phospholipase Cβ3 expression modulates behavioral and cellular responses to μ opioids. *Proc Natl Acad Sci USA* 1999;**96**:10,385–90.

21. Li Z, Jiang H, Xie W, Zhang Z, Smrcka AV, Wu D. Roles of PLC-β2 and -β3 and PI3Kγ in chemoattractant-mediated signal transduction. *Science* 2000;**287**:1046–9.

22. Jiang H, Lyubarsky A, Dodd R, Vardi N, Pugh E, Baylor D, Simon MI, Wu D. Phospholipase Cβ4 is involved in modulating the visual response in mice. *Proc Natl Acad Sci USA* 1996;**93**:14,598–601.

23. Kano M, Hashimoto K, Watanabe M, Kurihara H, Offermanns S, Jiang H, Wu Y, Jun K, Shin H-S, Ionue Y, Simon MI, Wu D. Phospholipase Cβ4 is specifically involved in climbing fiber synapse elimination in the developing cerebellum. *Proc Natl Acad Sci USA* 1998;**95**:15,724–9.

24. Miyata M, Kim H-T, Hashimoto K, Lee T-W, Cho S-Y, Jiang H, Wu Y, Jun K, Wu D, Kano M, Shin H-S. Deficient long-term synaptic depression in the rostral cerebellum correlated with impaired motor learning in phospholipase Cβ4 mutant mice. *Eur J Neurosci* 2001;**13**:1945–54.

25. Ji Q-S, Winnier GE, Niswender KD, Horstman D, Wisdom R, Magnuson MA, Carpenter G. Essential role of the tyrosine kinase substrate phospholipase C-γ1 in mammalian growth and development. *Proc Natl Acad Sci USA* 1997;**94**:2999–3003.

26. Shirane M, Sawa H, Kobayashi Y, Nakano T, Ktajima K, Shinkai Y, Nagashima K, Negishi I. Deficiency of phospholipase C-γ1 impairs renal development and hematopoiesis. *Development* 2001;**128**:5173–80.

27. Wang D, Feng J, Wen R, Marine J-C, Sangster MY, Parganas E, Hoffmeyer A, Jackson CW, Cleveland JL, Murray PJ, Ihle JN. Phospholipase Cγ2 is essential in the functions of B cell and several Fc receptors. *Immunity* 2000;**13**:25–35.

28. Fukami K, Nakao K, Inoue T, Kataoka Y, Kurokawa M, Fissore RA, Nakamura K, Katsuki M, Mikoshiba K, Yoshida N, Takenawa T. Requirement of phospholipase Cδ4 for the zna pellucida-induced acrosome reaction. *Science* 2001;**292**:920–3.

Chapter 32

Phospholipase D

Wenjuan Su and Michael A. Frohman

Center for Developmental Genetics, Graduate Program in Molecular and Cellular Pharmacology, and the Department of Pharmacology, Stony Brook University, Stony Brook, New York, New York

INTRODUCTION

Phospholipase D (PLD) is most broadly associated with hydrolysis of the phosphodiester bond of the glycerolipid phosphatidylcholine (PC) to form phosphatidic acid (PA) and choline (Figure 32.1, reviewed in [1]).

More generally, members of the PLD superfamily catalyze a transphosphatidylation reaction that, depending on the specific gene product, can employ a wide variety of primary alcohol-like nucleophiles such as water, ethanol, 1-butanol, glycerol, and diacylglycerol to synthesize PA,

phosphatidyl alcohols, and other lipids [2]. Also, some PLD superfamily members hydrolyze other substrates, including cardiolipin [3], DNA [4], and covalently-linked stalled DNA-topoisomerase complexes [5]. PLD genes are found in all species, from viruses and bacteria to plants and animals. Canonical animal PLD, which exhibits basal activity but is best known for being stimulated during signal transduction events involving G-protein-coupled receptors (GPCRs) and receptor tyrosine kinases (RTKs), has been implicated in a wide range of cell biological processes, including regulation of receptor activation, regulation

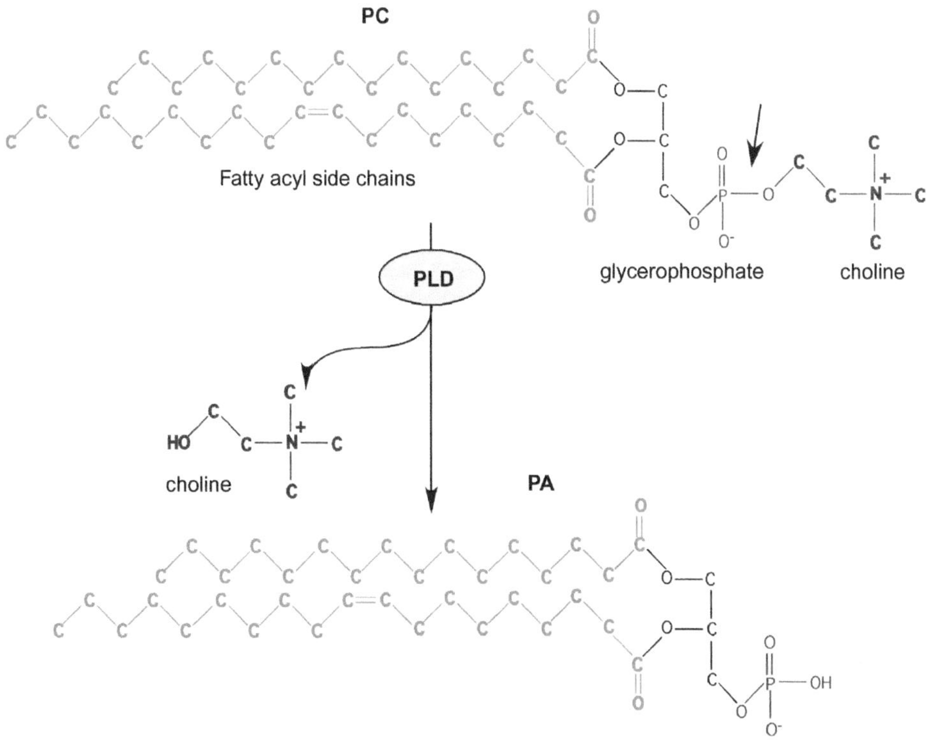

FIGURE 32.1 PLD substrate and product. PLD hydrolyzes PC to produce PA and free choline.

of Ras and Raf1 signaling pathways [6, 7], organization of the actin cytoskeleton [8, 9], endoplasmic reticulum and Golgi vesicle trafficking [10–13], receptor endocytosis [14] and secretory granule exocytosis [15–18], phagocytosis [19, 20], and chemotaxis and the oxidative burst response [21, 22]. This wide spectrum of roles has correspondingly implicated PLD in a number of disease-associated processes, including diabetes [23, 24], atherogenesis [25], obesity [26], oncogenesis [27–29], immunological responses [7, 20], and neuroendocrine function [11, 17].

THE PLD GENE FAMILY

PLD activity was defined in plants in 1947, but PLD genes were not identified until 1993, when N-terminal protein sequence was generated from a PLD enzyme purified from castor bean, making possible the cloning of a PLD cDNA [30]. The plant sequence then revealed that Spo14, a yeast gene already known to be required for sporulation, was in fact a PLD [31], and led to the cloning of mammalian PLDs [9, 32] as well as to the identification of PLD homologs in multiple species including *C. elegans* [32], *Drosophila* [32–34], bacteria [32, 35], and viruses [2], which gave rise to the definition of a PLD superfamily (Figure 32.2).

Mammals have two classical PLD genes, PLD1 [32] and PLD2 [9], and six more related genes that use non-classical substrates [3] or for which no activity has yet been defined. PLD1 and PLD2 are approximately 50 percent identical on the amino acid level, and contain the same protein domain organization. The non-classical family members share in common only the well-characterized "HKD" catalytic domain. PLD1 and PLD2 encode proteins approximately

100 kDa in size, and are found in association with the cytoplasmic leaflet of membranes in a variety of subcellular locations. In contrast, the non-classical gene PLD3 localizes to the lumen of the ER and is anchored in the membrane by a C-terminal membrane-spanning domain [36], and the non-classical gene MitoPLD localizes to the outer surface of the mitochondria and is anchored in the membrane by an N-terminal membrane-spanning domain [3].

PLD activity is relatively low in cells maintained in culture in the basal setting. Most cells express both PLD1 and PLD2, albeit to varying levels, and the basal activity that is observed is most commonly ascribed to PLD2. Upon stimulation by a wide variety of agonists, including serum, growth factors, and cytokines, both PLDs increase in activity, with PLD1 being the more dramatically responsive isoform [9]. Study of the factors controlling activation of PLD1, using recombinant purified enzyme, revealed that it is required that the membranes containing the target substrate PC also contain the lipid PtdIns(4,5)P$_2$ as a co-factor, and that soluble stimulatory proteins also be present [32, 37]. The most potent stimulators are members of the ARF and Rho families of GTPases, and classical isoforms of protein kinase C [38]. Activation of ARF and Rho by loading of GTP is required for them to stimulate PLD1. PKC, in contrast, stimulates PLD1 through a phosphorylation-independent, regulatory domain-mediated interaction that is facilitated by but does not require activation of PKC by diacylglycerol [39]. Binding sites for PtdIns(4,5)P$_2$, ARF, Rho, and PKC on PLD1 have all been identified with varying degrees of precision, and localize to different sites on the enzyme (Figure 32.2, reviewed in [1]). Taken together, these four groups of factors together elevate the level of PLD1 activity 8000-fold over its basal activity

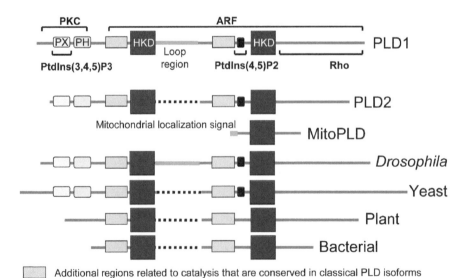

FIGURE 32.2 Schematic of motifs and domains for representative PLD superfamily members. Mammalian PLD is shown in alignment with PLD homologs from lower organisms and related HKD-containing proteins. The Phox homology (PX) domain, pleckstrin homology (PH) domain, and PtdIns(4,5)P$_2$-binding motif mediate membrane association. The PtdIns(4,5)P$_2$-binding motif and the HKD catalytic domain are essential for catalysis. Regions where protein and lipid regulators are known to interact with PLD1 are indicated.

in vitro [37]. Levels of activation *in vivo* are unlikely ever to be as dramatic, since the lipid and protein stimulators are never completely absent in the cell, and nor, during signal transduction events, do they likely rise to the level at which PLD1 would become maximally stimulated. PLD2 also exhibits a strong requirement for $PtdIns(4,5)P_2$ for enzymatic activity, but, in contrast to PLD1, otherwise exhibits a potent basal activity *in vitro* that is not increased by addition of ARF, Rho, or PKC [9]. PLD2 similarly exhibits potent activity in resting cells when overexpressed, which has led to the proposal that it is controlled at normal levels of expression via inhibitory mechanisms that are suppressed during signal transductions events – but the precise mechanisms controlling PLD2 activity *in vivo* remain unsettled. PLD1 and PLD2 undergo serine and tyrosine phosphorylation under some circumstances, and this may have modest positive or negative effects on PLD activity, but is probably not a major regulatory mechanism for this gene family.

One of the non-classical PLD isoforms has also been characterized. Mitochondrial PLD (MitoPLD) contains a HKD domain, but uses as substrate cardiolipin instead of PC. MitoPLD is required for the process of mitochondrial fusion through promoting trans-mitochondrial membrane adherence via its product PA [3], although the precise function of PA in this setting remains to be defined.

PLD Protein Functional Domains

The defining feature of PLD is the catalytic domain (PLD-c), which contains the "HKD" motif $HxK(x)_4D(x)_6GSxN$. The current model of PC catalysis involving a two-step mechanism is based on the crystallization of a bacterial PLD [40]. Water, the *in vivo* nucleophile, attacks the diester phosphate group of PC, protonating the HKD histidine. PC then enters the active pocket of the enzyme. Liberation of the proton releases free choline through the formation of a PLD-PA intermediate; re-acquisition of a proton from water then attacks the PLD-PA intermediate, causing the release of PA [2, 41]. PLD1 and PLD2 can similarly perform this reaction using ethanol or 1-butanol to generate the relatively stable compounds phosphatidylethanol or phosphatidylbutanol, which are thus useful as readouts for cellular PLD activity.

PLD1 and PLD2 also encode several lipid binding domains, including a pleckstrin homology (PH) domain, a $PtdIns(4,5)P_2$ binding motif, and a phox consensus sequence (PX). The PH domain mediates entry of PLD1 into lipid rafts, which facilitates the recycling of the enzyme to endosomes after its translocation to the plasma membrane upon PMA stimulation [42]. The $PtdIns(4,5)P_2$ binding motif consists of a sequence of conserved basic and aromatic amino acids that promotes interaction with negatively charged lipids [43, 44]. Mutations to this site both eliminate enzymatic activity and interaction with the plasma membrane [42], the site of the greatest concentration of $PtdIns(4,5)P_2$. PX domains mediate protein–protein interactions or bind to phosphatidylinositol phosphates (PIPs) [45]. The PX domain of PLD has been reported to bind to $PtdIns(3,4,5)P_3$ and $PtdIns(5)P$ and to facilitate internalization of PLD1 from the plasma membrane [42]. The PX domain has also been reported to mediate interaction of PLD1 with a variety of receptors and other proteins [46].

Subcellular Distribution

PLD1 has been reported to have a perinuclear localization in many cell lines, consistent with a Golgi, endoplasmic reticulum, late endosome, or storage vesicle distribution [9, 47–49]. In some cell lines, however, in particular neuroendocrine ones, PLD1 localizes to the plasma membrane [17]. These differences probably reflect variations of normal PLD1 subcellular tracking in the different cell lines. Upon stimulation, PLD1 located in peri-nuclear vesicles translocates to the plasma membrane and then recycles to sorting/ recycling endosomes and early endosomes [15, 23, 42]. PLD2 is most frequently reported to localize to the plasma membrane [9, 14, 50], but has also been described to exhibit cytoplasmic distribution and co-localize with β-actin [51, 52] or the Golgi [13, 53], presumably reflecting patterns of cycling analogous to that reported for PLD1 [54]. Translocation of PLD2 to membrane ruffles occurs in Hela cells in response to serum and epidermal growth factor (EGF) stimulation [50].

Functions for PLD and the Second Messenger it Generates, PA

PA has been linked to a diverse range of cellular processes using methodology that relies on the use of primary alcohols to divert PLD to generate phosphatidylalcohols instead of PA, or on a variety of relatively non-specific inhibitors such as ceramide or neomycin, expression of dominant-negative isoforms, and RNAi. Knockout mice and more specific inhibitors are in development, and should in the future offer more powerful tools that better define and validate the current hypotheses concerning PLD function in cells and physiologically. The processes currently proposed to be regulated by PLD activity include intracellular membrane trafficking (vesicle transport, endocytosis, exocytosis) and signaling pathways that regulate cell growth, survival differentiation, and shape (Ras activation, mammalian target of rapamycin (mTOR) signaling, and stimulation of $PtdIns(4,5)P_2$ synthesis). PA can also be converted to other second messengers, such as DAG and the mitogen LPA, and affect additional processes.

VESICLE TRAFFICKING

PLD and PA have been linked to multiple types of vesicle trafficking [55], including Golgi transport [10–13], endocytosis [14, 56–58], Fcγ receptor-mediated phagocytosis [20, 59], and exocytosis [17, 18].

Membrane Vesicle Production

PLD1 has been proposed to play a role in vesicular transport between the endoplasmic reticulum and Golgi apparatus by recruiting coatomer to budding membranes [12]; supporting this hypothesis, primary alcohol blocks protein transport from the endoplasmic reticulum to Golgi in a manner that can be alleviated by exogenous provision of PA [10, 12, 55]. PA production by PLD has also been reported to be required for structural integrity of Golgi [60] and secretory vesicle formation from the trans-Golgi network [11], and a very recent report has described a role for PLD2 in facilitating the action of the protein BARS to mediate scissioning of budded vesicles from the *trans*-Golgi [13]. Roles for both PLD1 and PLD2 have been proposed in ER and Golgi trafficking; however, neither enzyme predominantly localizes to the ER or Golgi, raising the issue of whether these functions are performed by one or both enzymes as they transit through the ER and Golgi in cycling pathways, or whether a small subset of cellular PLD resides there.

ENDOCYTOSIS

Multiple studies have described roles for PLD in receptor endocytosis. PA levels increase when EGF binds to its receptor [61], and EGF receptor (EGFR) internalization is blocked by 1-butanol [46]. EGFR internalization is also decreased by expression of catalytically-inactive PLD1 or PLD2, and elevated by overexpression of wild-type PLD [62]. PLD2 regulates internalization of the μ-opioid receptor MOR1, the metabolic glutamate receptor, and the angiotensin II receptor, as has been demonstrated using primary alcohol, expression of catalytically-inactive PLD2, and siRNA approaches [14, 56–58].

EXOCYTOSIS

The role of PLD derived PA is best established in exocytosis (Figure 32.3). PLD enzymatic activity promotes exocytosis in adipocytes [23], neuroendocrine [11, 17, 18], neuronal [63], mast [64], neutrophil [65] and pancreatic β cells [24]. In neuroendocrine cells, PLD1 acts downstream of ARF6 to regulate fusion of dense core granules into the plasma membrane in a PtdIns(4,5)P$_2$-dependent manner [17, 18]. In adipocytes [23], PLD1 present on storage vesicles containing the GLUT4 glucose transporter passively translocates to the plasma membrane in response to insulin stimulation. PLD activity is not required for the storage vesicles to dock at the plasma membrane or engage SNARE complex proteins; however, in the absence of PLD activity, complete fusion of the storage vesicles with the plasma membrane is greatly decreased, and evidence has been presented that the block occurs at the stage of hemifusion – i.e., after the proximal leaflets of the vesicle and plasma membranes have fused, but before the distal ones have done so.

Precisely how PLD activity manages to regulate all of these processes remains an area of active investigation. PA can act as a lipid anchor and recruit proteins that regulate these processes; it can also stimulate enzymes such as type I PtdIns(4)P 5-kinase (PIPKI) to form PtdIns(4,5)P$_2$, another lipid that regulates many facets of membrane

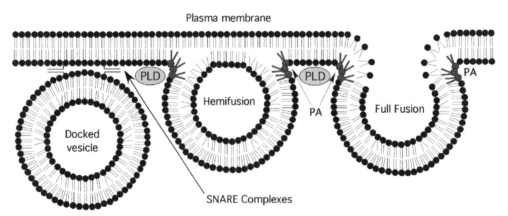

FIGURE 32.3 A proposed model for PA-mediated vesicle fusion.
Vesicles attach to the membrane and undergo SNARE-complex mediated fusion of the contacting (proximal) membrane leaflets (hemi-fusion). PLD1 generates PA (gray), which then acts as a fusogenic lipid to lower the energy required to progress towards pore formation and expansion, e.g. fusion of the distal membrane leaflets. PA may act through promoting negative membrane curvature at vesicle–plasma membrane junctions as shown, through facilitating SNARE complex action, through promoting PtdIns(4,5)P$_2$ production and CAPS recruitment, through conversion to DAG, another fusogenic lipid, or additional mechanisms.

trafficking (66). For example, PtdIns(4,5)P$_2$ produced by PA activation of PIPK recruits the protein CAPS, which triggers fusion of dense core vesicles into the plasma membrane [66, 67]. PA has also been proposed to act as a fusogenic lipid that induces negative membrane curvature and bilayer instability [17, 68], or to facilitate the function of SNARE complex proteins, in particular subsequent to hemifusion of the vesicles and the plasma membrane [69]. Similarly, PA generated by MitoPLD facilitates the mitochondrial fusion mediated by Mfn, the mitochondrial analog to SNARE complex proteins, although the mechanism undertaken by PA in this setting is unknown (3). Finally, PA can be converted to other fusogenic lipids, including DAG [70], which is quite fusogenic as well [70, 71].

Taken together, PLD and PA have been proposed to affect numerous cell biological events related to membrane vesicle trafficking; however, some of the studies are more rigorous than others, and it will be interesting to see in the next few years which of these roles remain of interest as more powerful tools for analysis of PLD function become widespread in use.

SIGNAL TRANSDUCTION

PLD has been linked to many signaling pathways, including ones involving the role in TNFα signaling, stimulation of PIPKI to generate PtdIns(4,5)P$_2$, activation of mTOR, and regulation of Ras activation (Figure 32.4).

TNFα Signaling

Tumor necrosis factor-α (TNFα) is a pleiotropic inflammatory cytokine that has been associated with a wide range of diseases, including infection, autoimmune disease, allergy, and tumorigenesis [72–74]. Binding of TNFα to its transmembrane receptors TNFR I and II leads to the activation of multiple signaling cascades, including ones involving MAPK and NFκB [75]. Several pieces of evidence implicate the involvement of PLD in TNFα signaling: PLD activity increases at early time points during TNFα-induced cell death, and TNFα-induced cell death is significantly decreased when PLD1 is overexpressed [76, 77]. Conversely, cells treated with siRNA to knockdown PLD exhibit higher susceptibility to TNFα-induced cell death. These suggest a protective role for PLD against TNFα-induced apoptosis. However, the molecular mechanisms underlying this relationship are not clear. PLD1 has recently been reported to be required for the activation of sphingosine kinase, cytosolic calcium signals, and NFκB and ERK1/2 pathways in human monocytic cells upon TNFα stimulation [78]. Interestingly, PLD1 translocates from the cytoplasm to the plasma membrane upon TNFα stimulation, indicating that PLD1 may affect TNFα-receptor links to downstream signaling pathways at the plasma membrane. PLD1 expression is also critical for TNFα-induced production of several cytokines, such as IL-1β, IL-6, and IL-13. Taken together, PLD activity may facilitate some downstream TNFα pathways (e.g., cytokine release) while blocking others (e.g., apoptosis), suggesting potential clinical relevance with respect to the possibility of manipulating PLD activity to promote the beneficial and inhibit the pathological outcomes mediated by TNFα.

PIPKI and PtdIns(4,5)P$_2$ Production

The activation of PIPKI to generate PtdIns(4,5)P$_2$ from PtdIns(4)P is a key signaling effect mediated by PLD formation of PA [79, 80–82]. PtdIns(4,5)P$_2$ is an essential factor in many cellular processes, in particular for membrane trafficking and cytoskeleton reorganization [83]. PLD2 has been reported to translocate to ruffling membranes to activate PIPKIα in response to EGF [50]. However, overexpression of PLD1 and PLD2 inhibits PMA-induced membrane ruffling and cortactin translocation, suggesting a complex role for PLD in this process [84]. Given the different stimuli (EGF versus PMA), the role of PLD in membrane ruffling may be context-dependent. As described earlier, PtdIns(4,5)P$_2$ is also a key regulator of PLD activity; thus, reciprocal stimulation of PLD and PIPKI may form a positive-feedback loop for localized production of PA and

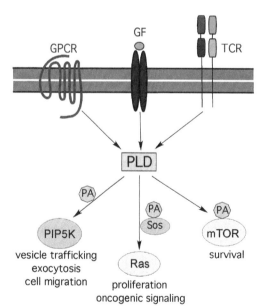

FIGURE 32.4 PA produced by PLD acts as a second messenger. Extracellular signaling via G-protein-coupled receptors (GPCR), growth factor (GF)-activation of tyrosine receptor kinases, or the T cell receptor (TCR), activates PLD, resulting in PA production. PA binds to and activates mTOR, which functions to promote cell survival. PA also activates PIP5-kinase (PIPKI), which generates PtdIns(4,5)P$_2$, a key player in vesicle trafficking and other cellular processes involving cytoskeleton reorganization. Finally, PA functions as an important upstream regulator of Ras activation by recruiting Sos or RasGRP1 to the plasma membrane.

PtdIns(4,5)P$_2$, helping to spatiotemporally regulate many types of cellular signaling events (reviewed in [83]).

mTOR Activation and Cell Survival

Several lines of evidence point to a role of PLD in cell survival [85]. mTOR, a serine/threonine kinase, appears to be a likely downstream target through which PLD-generated survival signals are generated [28]. mTOR regulates cell proliferation and growth, and acts downstream of PtdIns 3-kinase/Akt survival signaling [86]. Activation of mTOR and its downstream effector S6 kinase in HEK293 cells is blocked by 1-butanol in a manner that can be rescued by provision of exogenous PA, suggesting PLD involvement [87]. PA directly interacts with the FK506-binding protein 12-rapamycin-binding (FRB) domain of mTOR and this interaction is blocked by pretreatment with rapamycin, an inhibitor of mTOR, suggesting overlap of the PA and rapamycin binding sites [87]. In support of this idea, elevated expression of PLD2 and concomitant elevated production of PA confers rapamycin resistance to MCF-7 cells, while introduction of a catalytically-inactive PLD2 mutant increases rapamycin sensitivity in MDA-MB-231 cells [88]. However, other studies have shown that mTOR activation is dependent on PLD1 [89–91], and siRNA targeted against PLD1, but not PLD2, almost completely blocks S6 kinase activity upon serum stimulation [89]. Although it remains unsettled as to which PLD isoform is most important in mTOR activation, mTOR appears to be a well-defined target of PLD-generated PA, and this activation mechanism provides a pathway that functions in parallel to the PtdIns 3-kinase/Akt survival signaling that was previously thought to be the sole regulator of mTOR activity.

Ras Activation

An unexpected role has been described for PLD2 in Ras activation in EGF- and TCR-regulated signaling pathways [6, 7]. Receptor tyrosine-kinase (RTK)-Ras signaling plays an essential role in regulating a broad range of cellular processes, including cell proliferation, differentiation, and survival [92]. Son of sevenless (Sos), a guanine nucleotide-exchange factor (GEF), links RTKs to Ras activation, and converts Ras-GDP to Ras-GTP once Sos translocates to the plasma membrane upon ligand stimulation. Co-localization of Sos and Ras on the plasma membrane is critical for Ras activation to occur. For many years, it was thought that the driving force for Sos translocation was interaction of its SH2 domain with tyrosine-phosphorylated receptors [93, 94]. However, recent studies have shown that the SH2 domain is biochemically and genetically dispensable, and that it is production of PA by PLD2 at the plasma that triggers Sos translocation there [6]. PA production by

PLD2 increases when cells are stimulated by growth factors, and the PA interacts with a defined site on the Sos PH domain. Overexpression of PLD2 suffices to recruit Sos to plasma membrane to activate Ras, whereas knockdown of PLD2 by shRNA blocks EGF-induced Sos translocation and Ras activation. In T cells, Ras activation following TCR engagement is required for T cell development, selection, and function [95]. Mor and colleagues have shown that PLD2 similarly regulates another Ras-GEF, RasGRP1, in response to activation of the integrin LFA-1 [7]. In this case, the PA generated by PLD2 is converted to DAG, which is the signaling lipid that then directly recruits RasGRP1 to the plasma membrane to activate Ras.

Taken together, PLD's roles in TNFα signaling, PtdIns(4,5)P$_2$ production, mTOR activation, and Ras activation all point to a general pro-survival role for this gene family, which is increasingly becoming intriguing in light of genetic studies related to cancer, as discussed below.

PLD, A POTENTIAL DRUG TARGET

PLD has been demonstrated to be important for many cellular processes, as described thus far. Immune responses and tumorigenesis, as discussed below, represent two key physiological areas for which PLD may be a good drug target through development of specific inhibitors.

PLD in Immune Function

PLD enzymes are expressed in almost all leukocytes, including monocytes, macrophages, basophils, eosinophils, dendritic cells, lymphocytes, and NK cells (reviewed in [22]). In particular, PLD activity has been implicated in two critical leukocyte functions: phagocytosis with oxidative burst, and chemotaxis [19, 96–98].

In contrast to the low levels of PLD activity observed in resting cells, macrophages and neutrophils exhibit 10 times more PLD activity when challenged with virulent or killed microorganisms [19, 96, 99]. PLD concentrates at forming phagosomes, and PA is produced concomitantly [20, 59]. How PLD activity facilitates phagocytosis remains incompletely understood. However, the factors involved in phagocytosis, such as PtdIns(4,5)P$_2$, PKC, and small GTPases, all connect to PLD activity in a manner that has been demonstrated to be important for cellular changes in phagocytosis, such as actin reorganization and membrane trafficking. PLD has also been linked to oxidative burst and neutrophil degranulation [100–102]. Deficiencies in PLD-driven production of PA impair respiratory burst and microbial killing in macrophages and neutrophils [20, 59].

PLD has also been implicated in regulating chemotaxis, another important function of leukocytes and, in particular, neutrophils. PLD activity increases dramatically in neutrophils exposed to chemoattractants such as ENA-78, IL-8, and fMLP [98, 103, 104]. fMLP is produced via enteric flora, and is known to be a contributor to inflammatory bowel disease. Very recent work has shown that knockdown of PLD by siRNA abolishes basal and chemokine-induced chemotaxis, while overexpression of PLD enhances chemotaxis [98]. PLD activity appears to affect cell polarity and directionality in response to IL-8, suggesting leads for further studies on this topic.

Phagocytosis, oxidative burst, and chemotaxis are key steps in host immunity. Validation of PLD as an important player in the regulation of these processes would provide enthusiasm for exploring whether inhibition of PLD could be beneficial in settings in which excess immune activity is deleterious, such as arthritis and other autoimmune diseases.

PLD and Cancer

Increased PLD activity has been implicated in many types of human cancer, including breast, colon, gastric, and kidney [28]. PLD2 expression levels correlate inversely with survival of patients with colorectal carcinoma [105], and PLD2 point mutations and deletions have been found in tumors from patients with breast cancer [29]. Studies using breast cancer cell lines have shown that elevated PLD activity generates a survival signal, as discussed previously. Inhibiting PLD activity in MDA-MB-231 cells that have elevated PLD activity results in apoptosis [106]. Conversely, elevating PLD activity in MDA-MB-231 cells leads to rapamycin resistance [88], indicating that mTOR signaling is affected by the elevated PLD activity. PLD-generated increases in PA that activate mTOR may generate a PtdIns 3-kinase/Akt-independent survival signal [107]. PLD activity has also been implicated in cell transformation by the oncogenes v-Src, v-Ras, and v-Raf [108]. Moreover, elevating PLD activity in transformed cells drives cell proliferation, as shown in the context of PLD2-generated PA stimulating Ras activation [6]. Taken together, these data strongly argue for a role of PLD, especially PLD2, in oncogenic signaling in cancer cells.

PLD activity has been shown to be important in cell motility. PLD2 stimulates cell protrusion in v-Src-transformed cells [109], and is required for EGF-induced membrane ruffling [50]. PLD activity has also been implicated in tumor invasion [110, 111]. Recent data have shown that the ability of MDA-MB-231 cells to migrate on and invade into Matrigel is dependent on PLD and mTOR [112]. Taken together, elevated PLD activity, which is found in many human tumors, may contribute to tumor metastasis via regulation of cell migration and invasion. Lowering PLD activity could be an effective way to prevent metastasis, which is the major challenge of cancer therapy.

In summary, elevated PLD activity plays a critical part in many aspects of tumorigenesis including cell survival,

proliferation, and transformation, as well as migration. These observations reinforce the hypothesis that PLD is a promising drug target, and that inhibiting PLD activity is clinically attractive. However, the most commonly used PLD inhibitor is primary alcohol, which is not specific and cannot be used in patients. Effort has been taken to find specific PLD inhibitors, and a small-molecule PLD2 inhibitor has been reported recently [113]. However, its specificity and potency need to be further characterized *in vivo*, and its clinical implications evaluated.

ACKNOWLEDGEMENTS

We apologize to the many excellent publications of our colleagues that we were unable to cite for reasons of space. This review was supported by NIH awards GM071520 and DK64166.

REFERENCES

1. Jenkins GM, Frohman MA. Phospholipase D: a lipid centric review. *Cell Mol Life Sci* 2005;**62**:2305–16.

2. Sung TC, Roper RL, Zhang Y, et al. Mutagenesis of phospholipase D defines a superfamily including a trans-Golgi viral protein required for poxvirus pathogenicity. *EMBO J* 1997;**16**:4519–30.

3. Choi SY, Huang P, Jenkins GM, Chan DC, Schiller J, Frohman MA. A common lipid links Mfn-mediated mitochondrial fusion and SNARE-regulated exocytosis. *Nat Cell Biol* 2006;**8**:1255–62.

4. Ponting CP, Kerr ID. A novel family of phospholipase D homologues that includes phospholipid synthases and putative endonucleases. *Protein Sci* 1996;**5**:914–22.

5. Interthal H, Pouliot JJ, Champoux JJ. The tyrosyl-DNA phosphodiesterase Tdp1 is a member of the phospholipase D superfamily. *Proc Natl Acad Sci USA* 2001;**98**:12,009–12,014.

6. Zhao C, Du G, Skowronek K, Frohman MA, Bar-Sagi D. Phospholipase D2-generated phosphatidic acid couples EGFR stimulation to Ras activation by Sos. *Nat Cell Biol* 2007;**9**:706–12.

7. Mor A, Campi G, Du G, et al. The lymphocyte function-associated antigen-1 receptor costimulates plasma membrane Ras via phospholipase D2. *Nat Cell Biol* 2007;**9**:713–19.

8. Cross MJ, Roberts S, Ridley AJ, et al. Stimulation of actin stress fibre formation mediated by activation of phospholipase D. *Curr Biol* 1996;**6**:588–97.

9. Colley WC, Sung TC, Roll R, et al. Phospholipase D2, a distinct phospholipase D isoform with novel regulatory properties that provokes cytoskeletal reorganization. *Curr Biol* 1997;**7**:191–201.

10. Bi K, Roth MG, Ktistakis NT. Phosphatidic acid formation by phospholipase D is required for transport from the endoplasmic reticulum to the Golgi complex. *Curr Biol* 1997;**7**:301–7.

11. Chen YG, Siddhanta A, Austin CD, et al. Phospholipase D stimulates release of nascent secretory vesicles from the trans-Golgi network. *J Cell Biol* 1997;**138**:495–504.

12. Ktistakis NT, Brown HA, Waters MG, Sternweis PC, Roth MG. Evidence that phospholipase D mediates ADP ribosylation factor-dependent formation of Golgi coated vesicles. *J Cell Biol* 1996;**134**:295–306.

13. Yang J-S, Gad H, Lee S, et al. COPI vesicle fission: a role for phosphatidic acid and insight into Golgi maintenance. *Nat Cell Biol* 2008. in press.

14. Du G, Huang P, Liang BT, Frohman MA. Phospholipase D2 localizes to the plasma membrane and regulates angiotensin II receptor endocytosis. *Mol Biol Cell* 2004;**15**:1024–30.

15. Brown FD, Thompson N, Saqib KM, et al. Phospholipase D1 localises to secretory granules and lysosomes and is plasma-membrane translocated on cellular stimulation. *Curr Biol* 1998;**8**:835–8.

16. Siddhanta A, Backer JM, Shields D. Inhibition of phosphatidic acid synthesis alters the structure of the Golgi apparatus and inhibits secretion in endocrine cells. *J Biol Chem* 2000;**275**:12,023–12,031.

17. Vitale N, Caumont AS, Chasserot-Golaz S, et al. Phospholipase D1: a key factor for the exocytotic machinery in neuroendocrine cells. *EMBO J* 2001;**20**:2424–34.

18. Vitale N, Chasserot-Golaz S, Bader MF. Regulated secretion in chromaffin cells: an essential role for ARF6-regulated phospholipase D in the late stages of exocytosis. *Ann NY Acad Sci* 2002;**971**:193–200.

19. Kusner DJ, Hall CF, Schlesinger LS. Activation of phospholipase D is tightly coupled to the phagocytosis of Mycobacterium tuberculosis or opsonized zymosan by human macrophages. *J Exp Med* 1996;**184**:585–95.

20. Corrotte M, Chasserot-Golaz S, Huang P, et al. Dynamics and function of phospholipase D and phosphatidic acid during phagocytosis. *Traffic* 2006;**7**:365–77.

21. Waite KA, Wallin R, Qualliotine-Mann D, McPhail LC. Phosphatidic acid-mediated phosphorylation of the NADPH oxidase component p47-phox. Evidence that phosphatidic acid may activate a novel protein kinase. *J Biol Chem* 1997;**272**:15,569–15,578.

22. Gomez-Cambronero J, Di Fulvio M, Knapek K. Understanding phospholipase D (PLD) using leukocytes: PLD involvement in cell adhesion and chemotaxis. *J Leukoc Biol* 2007;**82**:272–81.

23. Huang P, Altshuller YM, Hou JC, Pessin JE, Frohman MA. Insulin-stimulated plasma membrane fusion of Glut4 glucose transporter-containing vesicles is regulated by phospholipase D1. *Mol Biol Cell* 2005;**16**:2614–23.

24. Hughes WE, Elgundi Z, Huang P, Frohman MA, Biden TJ. Phospholipase D1 regulates secretagogue-stimulated insulin release in pancreatic beta-cells. *J Biol Chem* 2004;**279**:27,534–27,541.

25. Zheng XL, Gui Y, Du G, Frohman MA, Peng DQ. Calphostin-C induction of vascular smooth muscle cell apoptosis proceeds through phospholipase D and microtubule inhibition. *J Biol Chem* 2004;**279**:7112–18.

26. Andersson L, Bostrom P, Rutberg M, et al. Extracellular signal-regulated kinase 2 (ERK2) and Phospholipase D1 regulate the amount of cytosolic lipid droplets. *J Cell Sci* 2006;**119**:2245–6.

27. Welsh CJ, Yeh GC, Phang JM. Increased phospholipase D activity in multidrug resistant breast cancer cells. *Biochem Biophys Res Commun* 1994;**202**:211–17.

28. Foster DA. Regulation of mTOR by phosphatidic acid? *Cancer Res* 2007;**67**:1–4.

29. Wood LD, Parsons DW, Jones S, et al. The genomic landscapes of human breast and colorectal cancers. *Science* 2007;**318**:1108–13.

30. Wang X, Xu L, Zheng L. Cloning and expression of phosphatidylcholine-hydrolyzing phospholipase D from Ricinus communis. *J Biol Chem* 1994;**269**:20,312–20,317.

31. Rose K, Rudge SA, Frohman MA, Morris AJ, Engebrecht J. Phospholipase D signaling is essential for meiosis. *Proc Natl Acad Sci USA* 1995;**92**:12,151–12,155.

32. Hammond SM, Altshuller YM, Sung TC, et al. Human ADP-ribosylation factor-activated phosphatidylcholine-specific phospholipase D defines a new and highly conserved gene family. *J Biol Chem* 1995;**270**:29,640–29,643.

33. LaLonde M, Janssens H, Yun S, et al. A role for Phospholipase D in *Drosophila* embryonic cellularization. *BMC developmental biology* 2006;**6**:60–72.

34. LaLonde MM, Janssen H, Rosenbaum E, et al. Regulation of phototransduction responsiveness and retinal degeneration by a phospholipase D-generated signaling lipid. *J Cell Biol* 2005;**169**:471–9.

35. Zhao Y, Stuckey JA, Lohse DL, Dixon JE. Expression, characterization, and crystallization of a member of the novel phospholipase D family of phosphodiesterases. *Protein Sci* 1997;**6**:2655–8.

36. Munck A, Bohm C, Seibel NM, Hashemol Hosseini Z, Hampe W. Hu-K4 is a ubiquitously expressed type 2 transmembrane protein associated with the endoplasmic reticulum. *FEBS J* 2005;**272**:1718–26.

37. Hammond SM, Jenco JM, Nakashima S, et al. Characterization of two alternately spliced forms of phospholipase D1. Activation of the purified enzymes by phosphatidylinositol 4,5-bisphosphate, ADP-ribosylation factor, and Rho family monomeric GTP-binding proteins and protein kinase C-alpha. *J Biol Chem* 1997;**272**:3860–8.

38. Morris AJ, Hammond SM, Colley C, et al. Regulation and functions of phospholipase D. *Biochem Soc Trans* 1997;**25**:1151–7.

39. Singer WD, Brown HA, Jiang X, Sternweis PC. Regulation of phospholipase D by protein kinase C is synergistic with ADP-ribosylation factor and independent of protein kinase activity. *J Biol Chem* 1996;**271**:4504–10.

40. Leiros I, Secundo F, Zambonelli C, Servi S, Hough E. The first crystal structure of a phospholipase D. *Structure Fold Des* 2000;**8**:655–67.

41. Stuckey JA, Dixon JE. Crystal structure of a phospholipase D family member. *Nat Struct Biol* 1999;**6**:278–84.

42. Du G, Altshuller YM, Vitale N, et al. Regulation of phospholipase D1 subcellular cycling through coordination of multiple membrane association motifs. *J Cell Biol* 2003;**162**:305–15.

43. Morris AJ. Regulation of phospholipase D activity, membrane targeting and intracellular trafficking by phosphoinositides. *Biochem Soc Symp* 2007:247–57.

44. Papayannopoulos V, Co C, Prehoda KE, Snapper S, Taunton J, Lim WA. A polybasic motif allows N-WASP to act as a sensor of PIP(2) density. *Mol Cell* 2005;**17**:181–91.

45. Xu Y, Seet LF, Hanson B, Hong W. The Phox homology (PX) domain, a new player in phosphoinositide signalling. *Biochem J* 2001;**360**:513–30.

46. Lee CS, Kim IS, Park JB, et al. The phox homology domain of phospholipase D activates dynamin GTPase activity and accelerates EGFR endocytosis. *Nat Cell Biol* 2006;**8**:477–84.

47. Sung TC, Zhang Y, Morris AJ, Frohman MA. Structural analysis of human phospholipase D1. *J Biol Chem* 1999;**274**:3659–66.

48. Freyberg Z, Sweeney D, Siddhanta A, Bourgoin S, Frohman M, Shields D. Intracellular localization of phospholipase D1 in mammalian cells. *Mol Biol Cell* 2001;**12**:943–55.

49. Lucocq J, Manifava M, Bi K, Roth MG, Ktistakis NT. Immunolocalisation of phospholipase D1 on tubular vesicular membranes of endocytic and secretory origin. *Eur J Cell Biol* 2001;**80**:508–20.

50. Honda A, Nogami M, Yokozeki T, et al. Phosphatidylinositol 4-phosphate 5-kinase alpha is a downstream effector of the small G protein ARF6 in membrane ruffle formation. *Cell* 1999;**99**:521–32.

51. Divecha N, Roefs M, Halstead JR, et al. Interaction of the type Ialpha PIPkinase with phospholipase D: a role for the local generation of phosphatidylinositol 4, 5-bisphosphate in the regulation of PLD2 activity. *EMBO J* 2000;**19**:5440–9.

52. Lee S, Park JB, Kim JH, et al. Actin directly interacts with phospholipase D, inhibiting its activity. *J Biol Chem* 2001;**276**:28,252–28,260.

53. Freyberg Z, Bourgoin S, Shields D. Phospholipase D2 is localized to the rims of the Golgi apparatus in mammalian cells. *Mol Biol Cell* 2002;**13**:3930–42.

54. Sciorra VA, Rudge SA, Wang J, McLaughlin S, Engebrecht J, Morris AJ. Dual role for phosphoinositides in regulation of yeast and mammalian phospholipase D enzymes. *J Cell Biol* 2002;**159**:1039–49.

55. Jones D, Morgan C, Cockcroft S. Phospholipase D and membrane traffic. Potential roles in regulated exocytosis, membrane delivery and vesicle budding. *Biochim Biophys Acta* 1999;**1439**:229–44.

56. Koch T, Brandenburg LO, Schulz S, Liang Y, Klein J, Hollt V. ADP-ribosylation factor-dependent phospholipase D2 activation is required for agonist-induced mu-opioid receptor endocytosis. *J Biol Chem* 2003;**278**:9979–85.

57. Koch T, Brandenburg LO, Liang Y, et al. Phospholipase D2 modulates agonist-induced mu-opioid receptor desensitization and resensitization. *J Neurochem* 2004;**88**:680–8.

58. Bhattacharya M, Babwah AV, Godin C, et al. Ral and phospholipase D2-dependent pathway for constitutive metabotropic glutamate receptor endocytosis. *J Neurosci* 2004;**24**:8752–61.

59. Iyer SS, Barton JA, Bourgoin S, Kusner DJ. Phospholipases D1 and D2 coordinately regulate macrophage phagocytosis. *J Immunol* 2004;**173**:2615–23.

60. Sweeney DA, Siddhanta A, Shields D. Fragmentation and re-assembly of the Golgi apparatus in vitro. A requirement for phosphatidic acid and phosphatidylinositol 4,5-bisphosphate synthesis. *J Biol Chem* 2002;**277**:3030–9.

61. Slaaby R, Jensen T, Hansen HS, Frohman MA, Seedorf K. PLD2 complexes with the EGF receptor and undergoes tyrosine phosphorylation at a single site upon agonist stimulation. *J Biol Chem* 1998;**273**:33,722–33,727.

62. Shen Y, Xu L, Foster DA. Role for phospholipase D in receptor-mediated endocytosis. *Mol Cell Biol* 2001;**21**:595–602.

63. Humeau Y, Vitale N, Chasserot-Golaz S, et al. A role for phospholipase D1 in neurotransmitter release. *Proc Natl Acad Sci USA* 2001;**98**:15,300–15,305.

64. Peng Z, Beaven MA. An essential role for phospholipase D in the activation of protein kinase C and degranulation in mast cells. *J Immunol* 2005;**174**:5201–8.

65. Stutchfield J, Cockcroft S. Correlation between secretion and phospholipase D activation in differentiated HL60 cells. *Biochem J* 1993;**293**(Pt 3):649–55.

66. Grishanin RN, Kowalchyk JA, Klenchin VA, et al. CAPS acts at a pre-fusion step in dense-core vesicle exocytosis as a PIP2 binding protein. *Neuron* 2004;**43**:551–62.

67. Olsen HL, Hoy M, Zhang W, et al. Phosphatidylinositol 4–kinase serves as a metabolic sensor and regulates priming of secretory granules in pancreatic beta cells. *Proc Natl Acad Sci USA* 2003;**100**:5187–92.

68. Kooijman EE, Chupin V, de Kruijff B, Burger KN. Modulation of membrane curvature by phosphatidic acid and lysophosphatidic acid. *Traffic* 2003;**4**:162–74.

69. Vicogne J, Vollenweider D, Smith JR, Huang P, Frohman MA, Pessin JE. Asymmetric phospholipid distribution drives in vitro reconstituted SNARE-dependent membrane fusion. *Proc Natl Acad Sci USA* 2006;**103**:14761–6.

70. Jun Y, Fratti RA, Wickner W. Diacylglycerol and its formation by phospholipase C regulate Rab- and SNARE-dependent yeast vacuole fusion. *J Biol Chem* 2004;**279**:53,186–53,195.

71. Barona T, Byrne RD, Pettitt TR, Wakelam MJ, Larijani B, Poccia DL. Diacylglycerol induces fusion of nuclear envelope membrane precursor vesicles. *J Biol Chem* 2005;**280**:41,171–41,177.

72. Tracey KJ, Cerami A. Tumor necrosis factor, other cytokines and disease. *Annu Rev Cell Biol* 1993;**9**:317–43.

73. Tracey KJ, Cerami A. Tumor necrosis factor: a pleiotropic cytokine and therapeutic target. *Annu Rev Med* 1994;**45**:491–503.

74. Palladino MA, Bahjat FR, Theodorakis EA, Moldawer LL. Anti-TNF-alpha therapies: the next generation. *Nat Rev* 2003;**2**:736–46.

75. Vilcek J, Lee TH. Tumor necrosis factor. New insights into the molecular mechanisms of its multiple actions. *J Biol Chem* 1991;**266**:7313–16.

76. Birbes H, Zeiller C, Komati H, Nemoz G, Lagarde M, Prigent AF. Phospholipase D protects ECV304 cells against TNFalpha-induced apoptosis. *FEBS Letts* 2006;**580**:6224–32.

77. De Valck D, Vercammen D, Fiers W, Beyaert R. Differential activation of phospholipases during necrosis or apoptosis: a comparative study using tumor necrosis factor and anti-Fas antibodies. *J Cell Biochem* 1998;**71**:392–9.

78. Sethu S, Mendez-Corao G, Melendez AJ. Phospholipase D1 plays a key role in TNF-alpha signaling. *J Immunol* 2008;**180**:6027–34.

79. Skippen A, Jones DH, Morgan CP, Li M, Cockcroft S. Mechanism of ADP-ribosylation factor-stimulated phosphatidylinositol 4,5-bisphosphate synthesis in HL60 cells. *J Biol Chem* 2002;**277**:5823–31.

80. Moritz A, De Graan PN, Gispen WH, Wirtz KW. Phosphatidic acid is a specific activator of phosphatidylinositol-4-phosphate kinase. *J Biol Chem* 1992;**267**:7207–10.

81. Jenkins GH, Fisette PL, Anderson RA. Type I phosphatidylinositol 4-phosphate 5-kinase isoforms are specifically stimulated by phosphatidic acid. *J Biol Chem* 1994;**269**:11547–54.

82. Jones DR, Sanjuan MA, Merida I. Type Ialpha phosphatidylinositol 4-phosphate 5-kinase is a putative target for increased intracellular phosphatidic acid. *FEBS Letts* 2000;**476**:160–5.

83. Oude Weernink PA, Lopez de Jesus M, Schmidt M. Phospholipase D signaling: orchestration by PIP2 and small GTPases. *Naunyn Schmiedebergs Arch Pharmacol* 2007;**374**:399–411.

84. Hiroyama M, Exton JH. Studies of the roles of ADP–ribosylation factors and phospholipase D in phorbol ester–induced membrane ruffling. *J Cell Physiol* 2005;**202**:608–22.

85. McDermott M, Wakelam MJ, Morris AJ. Phospholipase D. *Biochem Cell Biol* 2004;**82**:225–53.

86. Yang Q, Guan KL. Expanding mTOR signaling. *Cell Res* 2007;**17**:666–81.

87. Fang Y, Vilella-Bach M, Bachmann R, Flanigan A, Chen J. Phosphatidic acid-mediated mitogenic activation of mTOR signaling. *Science* 2001;**294**:1942–5.

88. Chen Y, Zheng Y, Foster DA. Phospholipase D confers rapamycin resistance in human breast cancer cells. *Oncogene* 2003;**22**:3937–42.

89. Fang Y, Park IH, Wu AL, et al. PLD1 regulates mTOR signaling and mediates Cdc42 activation of S6K1. *Curr Biol* 2003;**13**:2037–44.

90. Kam Y, Exton JH. Role of phospholipase D1 in the regulation of mTOR activity by lysophosphatidic acid. *FASEB J* 2004;**18**:311–19.

91. Shi M, Zheng Y, Garcia A, Xu L, Foster DA. Phospholipase D provides a survival signal in human cancer cells with activated H-Ras or K-Ras. *Cancer Letts* 2007;**258**:268–75.

92. Schubbert S, Shannon K, Bollag G. Hyperactive Ras in developmental disorders and cancer. *Nature. Rev Cancer* 2007;**7**:295–308.

93. McCollam L, Bonfini L, Karlovich CA, et al. Functional roles for the pleckstrin and Dbl homology regions in the Ras exchange factor Son-of-sevenless. *J Biol Chem* 1995;**270**:15,954–15,957.

94. Byrne JL, Paterson HF, Marshall CJ. p21Ras activation by the guanine nucleotide exchange factor Sos, requires the Sos/Grb2 interaction and a second ligand-dependent signal involving the Sos N-terminus. *Oncogene* 1996;**13**:2055–65.

95. Scheele JS, Marks RE, Boss GR. Signaling by small GTPases in the immune system. *Immunol Rev* 2007;**218**:92–101.

96. Serrander L, Fallman M, Stendahl O. Activation of phospholipase D is an early event in integrin-mediated signalling leading to phagocytosis in human neutrophils. *Inflammation* 1996;**20**:439–50.

97. Gelas P, Von Tscharner V, Record M, Baggiolini M, Chap H. Human neutrophil phospholipase D activation by N-formylmethionyl-leucylphenylalanine reveals a two-step process for the control of phosphatidylcholine breakdown and oxidative burst. *Biochem J* 1992;**287**(1):67–72.

98. Lehman N, Di Fulvio M, McCray N, Campos I, Tabatabaian F, Gomez-Cambronero J. Phagocyte cell migration is mediated by phospholipases PLD1 and PLD2. *Blood* 2006;**108**:3564–72.

99. Fallman M, Gullberg M, Hellberg C, Andersson T. Complement receptor-mediated phagocytosis is associated with accumulation of phosphatidylcholine-derived diglyceride in human neutrophils. Involvement of phospholipase D and direct evidence for a positive feedback signal of protein kinase. *J Biol Chem* 1992;**267**:2656–63.

100. Melendez AJ, Bruetschy L, Floto RA, Harnett MM, Allen JM. Functional coupling of FcgammaRI to nicotinamide adenine dinucleotide phosphate (reduced form) oxidative burst and immune complex trafficking requires the activation of phospholipase D1. *Blood* 2001;**98**:3421–8.

101. Palicz A, Foubert TR, Jesaitis AJ, Marodi L, McPhail LC. Phosphatidic acid and diacylglycerol directly activate NADPH oxidase by interacting with enzyme components. *J Biol Chem* 2001;**276**:3090–7.

102. Sergeant S, Waite KA, Heravi J, McPhail LC. Phosphatidic acid regulates tyrosine phosphorylating activity in human neutrophils: enhancement of Fgr activity. *J Biol Chem* 2001;**276**:4737–46.

103. Cui Y, English DK, Siddiqui RA, Heranyiova M, Garcia JG. Activation of endothelial cell phospholipase D by migrating neutrophils. *J Investig Med* 1997;**45**:388–93.

104. Bacon KB, Flores-Romo L, Life PF, et al. IL-8-induced signal transduction in T lymphocytes involves receptor-mediated activation of phospholipases C and D. *J Immunol* 1995;**154**:3654–66.

105. Saito M, Iwadate M, Higashimoto M, Ono K, Takebayashi Y, Takenoshita S. Expression of phospholipase D2 in human colorectal carcinoma. *Oncol Rep* 2007;**18**:1329–34.

106. Zhong M, Shen Y, Zheng Y, Joseph T, Jackson D, Foster DA. Phospholipase D prevents apoptosis in v-Src-transformed rat fibroblasts and MDA-MB-231 breast cancer cells. *Biochem Biophys Res Commun* 2003;**302**:615–19.

107. Chen Y, Rodrik V, Foster DA. Alternative phospholipase D/mTOR survival signal in human breast cancer cells. *Oncogene* 2005;**24**:672–9.

108. Foster DA, Xu L. Phospholipase D in cell proliferation and cancer. *Mol Cancer Res* 2003;**1**:789–800.

109. Shen Y, Zheng Y, Foster DA. Phospholipase D2 stimulates cell protrusion in v-Src-transformed cells. *Biochem Biophys Res Commun* 2002;**293**:201–6.

110. Imamura F, Horai T, Mukai M, Shinkai K, Sawada M, Akedo H. Induction of in vitro tumor cell invasion of cellular monolayers by lysophosphatidic acid or phospholipase D. *Biochem Biophys Res Commun* 1993;**193**:497–503.

111. Pai JK, Frank EA, Blood C, Chu M. Novel ketoepoxides block phospholipase D activation and tumor cell invasion. *Anticancer Drug Des* 1994;**9**:363–72.

112. Zheng Y, Rodrik V, Toschi A, et al. Phospholipase D couples survival and migration signals in stress response of human cancer cells. *J Biol Chem* 2006;**281**:15,862–15,868.

113. Monovich L, Mugrage B, Quadros E, et al. Optimization of halopemide for phospholipase D2 inhibition. *Bioorg Med Chem Letts* 2007;**17**:2310–11.

Chapter 33

Role of Phospholipase A2 Forms in Arachidonic Acid Mobilization and Eicosanoid Generation

Jesús Balsinde[1] and Edward A. Dennis[2]

[1]*Institute of Molecular Biology and Genetics, Spanish National Research Council (CSIC) and University of Valladolid School of Medicine, Valladolid, Spain*

[2]*Department of Chemistry and Biochemistry, School of Medicine and Revelle College, University of California at San Diego, La Jolla, CA*

INTRODUCTION

Phospholipase A_2 (PLA_2) plays a central role in lipid signaling, and consequently in a variety of inflammatory conditions. PLA_2 cleaves the sn-2 ester bond of cellular phospholipids, producing a free fatty acid and a lysophospholipid, both of which are implicated in lipid signaling. The free fatty acid produced is frequently arachidonic acid (AA, 5,8,11,14-eicosatetraenoic acid), the biosynthetic precursor of the eicosanoid family of potent inflammatory mediators that includes the prostaglandins, thromboxane, leukotrienes, and lipoxins. The other product of PLA_2 action on phospholipids is a lysophospholipid, which, depending on its molecular composition, may be converted into platelet-activating factor, another potent inflammatory mediator.

Phospholipase A_2 (PLA_2) constitutes a superfamily of enzymes that catalyzes the hydrolysis of the sn-2 ester bond in phospholipids, generating a free fatty acid and a lysophospholipid. This reaction is of the utmost importance in the context of cellular signaling, since it constitutes the main pathway by which arachidonic acid (AA) is liberated from phospholipids. Free AA may play itself a role in signaling, but it is also the precursor of a large family of compounds known as the eicosanoids, which includes the cyclooxygenase-derived prostaglandins and the lipoxygenase-derived leukotrienes [1]. If the other product of PLA_2 action on phospholipids is a choline-containing lysophospholipid possessing an alkyl linkage in the sn-1 position, then an acetyltransferase can act upon it to produce platelet-activating factor (PAF, 1-O-alkyl-2-acetyl-sn-3-phosphocholine).

The importance of the eicosanoids and platelet-activating factor as key mediators of inflammation as well as other pathophysiological conditions is now universally accepted [1–3]. Aspirin and other non-steroidal anti-inflammatory drugs (NSAIDs) are well established as cyclooxygenase inhibitors, and are widely used in clinical practice. Similarly, the pharmaceutical industry has been actively pursuing lipoxygenase inhibitors and receptor antagonists for both leukotrienes and PAF. Note that since prostaglandins, leukotrienes and PAF all derive from the action of a PLA_2, direct inhibition of such an enzyme would have the potential of blocking all three of the pathways at once, which could be of therapeutic advantage in certain settings.

The above approach is hampered, however, by the fact that cells in general contain multiple PLA_2 enzymes. For example, in humans, no less than 22 different proteins have been identified to possess PLA_2 activity [4–6]. Thus, a first step in a rational PLA_2 drug design strategy is to define the different PLA_2 classes present in cells as well as their putative roles in eicosanoid and PAF synthesis.

PLA_2 GROUPS

PLA_2s have been systematically classified according to their nucleotide sequence [4–6]. The latest update to this classification, published in 2006 [6], included 15 groups, most of them with several subgroups. Only PLA_2s whose nucleotide sequence has been determined were included in the classification. However, this obvious criterion is the cause of some confusion, since many reports have appeared that erroneously link certain enzyme activities and functions to particular PLA$_2$ groups without it having been verified that such an association actually exists. This chapter summarizes details found in these previous more extensive reviews [4–6].

A parallel classification of the PLA$_2$s on the basis of biochemical properties is also frequently used, and it has value in describing PLA$_2$ activities for which sequence data are not available. This classification contemplates five main PLA$_2$ classes, based on whether the enzyme is secreted (sPLA$_2$), cytosolic Ca^{2+}-dependent (cPLA$_2$), cytosolic Ca^{2+}-independent (iPLA$_2$), specific for PAF (PAF acetyl hydrolases), or located in lysosomes (acidic PLA$_2$). One must be aware of the fact that this classification is not devoid of problems either – for example, the group IVC PLA$_2$ is sometimes referred to as cPLA$_2\gamma$, despite it being a Ca^{2+}-independent enzyme. In addition, the PAF acetylhydrolase PLA$_2$s (groups VII and VIII) also distribute among these categories.

Generally, the sPLA$_2$s (groups I, II, III V, IX, X, XI, XII, XIII, XIV) require millimolar levels of Ca^{2+} for activity, have low molecular masses, and lack specificity for arachidonate-containing phospholipids. The cPLA$_2$s (group IV, comprising six subgroups) have higher molecular masses, require Ca^{2+} for translocation to membranes but not for activity, and are selective for arachidonate-containing phospholipids. Finally, the iPLA$_2$s (group VI, comprising six subgroups, and also group IVC – see above), which are Ca^{2+} independent, have high molecular masses but are not selective for arachidonate-containing phospholipids [4–6]. The PAF acetylhydrolases (groups VII and VIII, comprising two subgroups each) exhibit specificity for PAF or oxidized phospholipids as substrates, and can be found inside or outside the cells. Finally, the lysosomal PLA$_2$ (group XV) is a Ca^{2+}-independent, low-molecular mass acidic enzyme.

CELLULAR FUNCTION IN AA RELEASE

A myriad of agents that exert effects on cells via receptor-dependent or independent pathways elicits a series of signals that ultimately lead to increased PLA$_2$ activity. Elucidation of these signals has been the subject of much effort for the past 10 years [7–9]. The situation is further complicated by the evidence that most cells contain several PLA$_2$ forms, and that all of them may eventually participate in the signaling process. With regard to the regulation of AA mobilization and eicosanoid production, two are the PLA$_2$ forms that have been generally implicated, namely cPLA$_2\alpha$ (group IVA PLA$_2$) and a sPLA$_2$, which, depending on cell type and stimulus may belong to groups IIA, V or X. cPLA$_2\alpha$ is universally recognized as the initiator and major regulator of the AA release, whereas the sPLA$_2$ serves an accessory role, and amplifies the cPLA$_2$-regulated response by a variety of mechanisms.

In general, cells respond to receptor agonists by liberating AA by two temporally distinct phases, namely an immediate, acute response, which usually occurs at the expense of pre-existing intracellular effectors, and a delayed response that takes several hours to develop and may require protein synthesis. In either case, the foremost event is the translocation and activation cPLA$_2\alpha$ in an intracellular compartment, generally in the vicinity of the nuclear membrane. The mechanism of translocation and activation of this enzyme by intracellular Ca^{2+} elevations and phosphorylation at Ser^{505} by mitogen-activated protein kinase cascades has been well defined [10, 11]. However, an important aspect of the biochemistry of cPLA$_2\alpha$ that remains to be elucidated is the mechanism of activation of this enzyme under circumstances that do not involve a rise in the intracellular Ca^{2+} level. A prototypic agonist in this regard is the bacterial lipopolysaccharide, which activates the cPLA$_2\alpha$-mediated AA release and eicosanoid production in cells of the innate immune system via engagement of toll-like receptor 4 [12, 13]. In in vitro assay systems, the Ca^{2+}-dependence of cPLA$_2\alpha$ activity can be drastically reduced if phosphatidylinositol bisphosphate is incorporated into the substrate [14–16]. There is increasing evidence that cPLA$_2\alpha$ translocation and activity can also be regulated in intact cells by phosphatidylinositol bisphosphate at Ca^{2+} levels similar to those found in resting unstimulated cells [17–19].

cPLA$_2\alpha$ has been found to interact with a number of intracellular proteins, including the nuclear protein PLIP [20], vimentin [21], p11 (calpactin I light chain) [22], and p47phox and p67phox [23]. The physiological significance of these interactions and their role in regulating AA mobilization remain to be elucidated. A recent paper suggests that phosphorylation of the enzyme at Ser^{727} – a phosphorylation reaction that does not increase cPLA$_2\alpha$'s specific activity – is necessary for proper interaction with p11 [24].

During the AA mobilization response of immunoinflammatory cells such as macrophages or mast cells to various stimuli, a sPLA$_2$ frequently participates in the process, thereby creating an amplification loop that results in a greatly enhanced release of AA for eicosanoid synthesis. However, the mode how sPLA$_2$ participates in this process is still a very controversial issue.

Exogenous sPLA$_2$, particularly that belonging to groups IIA, V, and X, or sPLA$_2$ overexpressed in various cells is capable of augmenting the release of AA and other fatty acids under different experimental settings [25–28], thereby amplifying the essential role of cPLA$_2\alpha$ in the biosynthesis of eicosanoids. Regarding the endogenous enzyme, studies using mice in which the gene encoding group V sPLA$_2$ was deleted have provided unequivocal evidence for the role of this enzyme in eicosanoid production by macrophages and mast cells in vivo [29,30]. Interestingly, the augmenting effect of group V sPLA$_2$ is observed in cells on a C57BL/6 genetic background, but not in cells on a BALB/c background [30]. These data provide evidence that, depending on the genetic background, two different phenotypes may exist in cells regarding the involvement of sPLA$_2$ in eicosanoid generation. Whether these two phenotypes may also manifest in cells depending on culture conditions is unknown at present.

Group V sPLA$_2$ might potentially release AA after secretion through two pathways, one involving re-internalization via the caveolae system [31, 32] and the other involving direct interaction with the phosphatidylcholine-rich membranes to directly release AA from various cellular membranes [31, 33, 34]. Either of these mechanisms would require the sPLA$_2$ to exit the cell as a consequence of the activation process, and to subsequently re-associate and/or undergo re-internalization for action in an autocrine or paracrine manner. However, recently it has been suggested that the human group IIA and group X enzymes release AA primarily during the secretion process on the inside rather than on the outside of the cell, with the involvement of cPLA$_2\alpha$ [35].

CROSS-TALK BETWEEN CPLA$_2\alpha$ AND SPLA$_2$

During AA release cPLA$_2\alpha$ and sPLA$_2$ often act coordinately, the activity of one modifying the activity of the other, or *vice versa*. Depending on stimulation conditions, cPLA$_2\alpha$ may modulate sPLA$_2$ at a gene level or at the level of enzyme activity itself [36–38], although the mechanisms involved remain obscure. sPLA$_2$ regulation of cPLA$_2\alpha$ has also been found and investigated in detail in some instances. In murine mesangial cells, an adenoviral infection technique was used to stably express group IIA and/or group V sPLA$_2$ into the cells [39]. cPLA$_2\alpha$ was found to be responsible for AA release and, when present, both group IIA and V sPLA$_2$s amplified the cPLA$_2\alpha$-mediated response, this resulting in an increased AA release response [39]. In these studies, a scenario in which sPLA$_2$ enhances the activity of cPLA$_2\alpha$, which then acts to release AA, was proposed on the basis of experiments demonstrating preferential release of AA over oleic acid in response to H$_2$O$_2$ in cells expressing both kinds of PLA$_2$s. Moreover, a correlation was found to exist between the expression level of cPLA$_2\alpha$ and the magnitude of AA release. Such a correlation did not appear to occur between the expression level of sPLA$_2$ and the extent of AA release.

A very recent study in mouse mast cells from mice lacking sPLA$_2$ (group V) by genetic deletion has strengthened further the concept of sPLA$_2$ acting to enhance the activity of cPLA$_2\alpha$ [40]. In these cells, group V sPLA$_2$ was found to regulate AA mobilization and eicosanoid generation upon activation with TLR2 agonists through amplification of cPLA$_2\alpha$ phosphorylation by the extracellular-regulated kinases. Interestingly, such a regulation was not appreciated when the cells were stimulated to release AA through

c-kit or FcεRI [40]. Thus the positive effects of sPLA$_2$ on AA mobilization appear to be not only cell-specific [30] but also receptor-specific [40].

In general agreement with these data, it has also been shown that sPLA$_2$s, acting on the outer membrane of human neutrophils, release fatty acids and lysoPC, which in turn can act on the cell to increase cytosolic free calcium levels and phosphorylation of cPLA$_2\alpha$, resulting in cPLA$_2\alpha$-dependent leukotriene biosynthesis [41]. Nonetheless, sPLA$_2$s may also act to release AA in a cPLA$_2\alpha$-independent manner, as demonstrated by studies in cells from mice lacking cPLA$_2\alpha$ by genetic disruption [42].

iPLA$_2$ ROLE IN AA RELEASE

The group VIA PLA$_2$ enzyme (iPLA$_2$-VIA) is ubiquitously expressed in all types of cells, and has the potential to participate in AA release under some conditions. However, the role of iPLA$_2$-VIA in AA release is controversial, which is mostly due to the fact that most studies implicating iPLA$_2$-VIA have relied on the use of bromoenol lactone (BEL) – a compound that manifests high selectivity for iPLA$_2$-VIA *in vitro*, but whose selectivity *in vivo* is low [43]. In some studies BEL was found to inhibit the AA release, but in others, notably in phagocytes, no significant effect was detected [5]. It seems likely that the involvement of iPLA$_2$-VIA in AA release is markedly cell- and stimulus-dependent, as are most of the roles attributed to this enzyme in cell physiology [44]. Since various iPLA$_2$-VIA splice variants co-exist in cells [44, 45], it is plausible that the enzyme be regulated by multiple regulatory mechanisms that differ among cell types and stimulation conditions. This could explain the multiplicity of functions that this enzyme appears to serve depending on cell type.

Recently, mice with targeted disruption of the gene encoding for iPLA$_2$-VIA have been generated [46]. Use of cells from these animals has made it clear again that the involvement of iPLA$_2$-VIA in AA mobilization may notably differ depending on cell type and stimulation conditions. Thus, peritoneal macrophages from iPLA$_2$-null mice appear to release AA in response to zymosan in a manner that is indistinguishable from that of cells from wild-type animals [47]. In contrast, however, iPLA$_2$-VIA appears to be crucial for the eicosanoid response of macrophages stimulated via class A scavenger receptors [48].

In some macrophage-like cell lines such as P388D$_1$ and U937 [49, 50], and also in human neutrophils [51], iPLA$_2$-VIA has been proposed to mediate phospholipid reacylation reactions by regulating the steady-state level of lysophosphatidylcholine. The lysophospholipids produced by iPLA$_2$-VIA can be used to re-incorporate some of the fatty acids (including AA) that have previously been released by its Ca^{2+}-dependent counterparts. Thus, in some cell types all three kinds of PLA$_2$ (sPLA$_2$, cPLA$_2$, iPLA$_2$) may serve important but distinct functions in AA metabolism in cells.

REFERENCES

1. Smith WL, De Witt DL, Garavito RM. Cyclooxygenases: structural, cellular, molecular biology. *Annu Rev Biochem* 2000;**69**:145–82.

2. Bonventre JV, Huang Z, Taheri MR, O'Leary E, Li E, Moskowitz MA, Sapirstein A. Reduced fertility and postischaemic brain injury in mice deficient in cytosolic phospholipase A$_2$. *Nature* 1997;**390**:622–5.

3. Uozumi N, Kume K, Nagase T, Nakatani N, Ishii S, Tashiro F, Komagata Y, Maki K, Ikuta K, Ouichi Y, Miyazaki J, Shimizu T. Role of cytosolic phospholipase A$_2$ in allergic response and parturition. *Nature* 1997;**390**:618–22.

4. Six DA, Dennis EA. The expanding superfamily of phospholipase A$_2$ enzymes: classification and characterization. *Biochim Biophys Acta* 2000;**1488**:1–19.

5. Balsinde J, Winstead MV, Dennis EA. Phospholipase A$_2$ regulation of arachidonic acid mobilization. *FEBS Lett* 2002;**531**:2–6.

6. Schaloske R, Dennis EA. The phospholipase A$_2$ superfamily and its group numbering system. *Biochim Biophys Acta* 2006;**1761**:1246–59.

7. Fitzpatrick FA, Soberman R. Regulated formation of eicosanoids. *J Clin Invest* 2001;**107**:1347–51.

8. Murakami M, Kudo I. Phospholipase A$_2$. *J Biochem (Tokyo)* 2002;**131**: 285–92.

9. Balsinde J, Balboa MA, Insel PA, Dennis EA. Regulation and inhibition of phospholipase A2. *Annu Rev Pharmacol Toxicol* 1999;**39**:175–89.

10. Dessen A. Structure and mechanism of human cytosolic phospholipase A$_2$. Bio. *Biochim phys Acta* 2000;**1488**:40–7.

11. Ghosh M, Tucker DE, Burchett SA, Leslie CC. Properties of the Group IV phospholipase A$_2$ family. *Pro Lipid Res* 2006;**45**:487–510.

12. Qi HY, Shelhamer JH. Toll-like receptor 4 signaling regulates cytosolic phospholipase A$_2$ activation and lipid generation in lipopolysaccharide-stimulated macrophages. *J Biol Chem* 2005;**280**:38,969–38,975.

13. Buczynski MW, Stephens DL, Bowers-Gentry RC, Grkovich A, Deems RA, Dennis EA. TLR-4 and sustained calcium agonists synergistically produce eicosanoids independent of protein synthesis in RAW264.7 cells. *J Biol Chem* 2007;**282**:22,834–22,847.

14. Mosior M, Six DA, Dennis EA. Group IV cytosolic phospholipase A$_2$ binds with high affinity and specificity to phosphatidylinositol 4,5-bisphosphate resulting in dramatic increases in activity. *J Biol Chem* 1998;**273**:2184–91.

15. Six DA, Dennis EA. Essential Ca^{2+}-independent role of the group IVA cytosolic phospholipase A$_2$. C2 domain for interfacial activity. *J Biol Chem* 2003;**278**:23,450–842.

16. Das S, Cho W. Roles of catalytic domain residues in interfacial binding and activation of group IV cytosolic phospholipase A$_2$. *J Biol Chem* 2002;**277**:23,838–23,846.

17. Balsinde J, Balboa MA, Li W, Llopis J, Dennis EA. Cellular regulation of cytosolic group IV phospholipase A2 by phosphatidylinositol bisphosphate levels. *J Immunol* 2000;**164**:5398–402.

18. Casas J, Gijón MA, Vigo AG, Crespo MS, Balsinde J, Balboa MA. Phosphatidylinositol 4,5-bisphosphate anchors cytosolic group IVA phospholipase A$_2$ to perinuclear membranes and decreases its calcium requirement for translocation in live cells. *Mol Biol Cell* 2006;**17**:155–62.

19. Subramanian, P., Vora, M., Gentile, L. B., Stahelin, R. V., Chalfant, C. E. Anionic lipids activate group IVA cytosolic phospholipase A$_2$ via distinct and separate mechanisms. J. Lipid Res. 48: 2701–2708.

20. Sheridan AM, Force T, Yoon HJ, O'Leary E, Choukroun G, Taheri MR, Bonventre JV. PLIP, a novel splice variant of Tip60, interacts with group IV cytosolic phospholipase A2, induces apoptosis, and potentiates prostaglandin production. *Mol Cell Biol* 2001;**21**:4470–81.

21. Nakatani Y, Tanioka T, Sunaga S, Murakami M, Kudo I. Identification of a cellular protein that functionally interacts with the C2 domain of cytosolic phospholipase A$_2$α. *J Biol Chem* 2000;**275**:1161–8.

22. Wu T, Angus CW, Yao XL, Logun C, Shelhamer JH. P11, a unique member of the S100 family of calcium-binding proteins, interacts with and inhibits the activity of the 85-kDa cytosolic phospholipase A$_2$. *J Biol Chem* 1997;**272**:17,145–17,153.

23. Shmelzer Z, Haddad N, Admon E, Pessach I, Leto TL, Eitan-Hazan Z, Hershfinkel M, Levy R. Unique targeting of cytosolic phospholipase A$_2$ to plasma membranes mediated by the NADPH oxidase in phagocytes. *J Cell Biol* 2003;**162**:683–92.

24. Tian W, Wijewickrama GT, Kim JH, Das S, Tun MP, Gokhale N, Jung JW, Kim KP, Cho W. Mechanism of regulation of group IVA phospholipase A$_2$ activity by Ser727 phosphorylation. *J Biol Chem* 2008;**283**:3960–71.

25. Murakami M, Kambe T, Shimbara S, Higashino K, Hanasaki K, Arita H, Horiguchi M, Arita M, Arai H, Inoue K, Kudo I. Different functional aspects of the group II subfamily (types IIA and V) and type X secretory phospholipase A$_2$s in regulating arachidonic acid release and prostaglandin generation. *J Biol Chem* 1999;**274**:31,435–31,444.

26. Balsinde J, Balboa MA, Dennis EA. Functional coupling between secretory phospholipase A2 and cyclooxygenase-2 and its regulation by cytosolic group IV phospholipase A2. *Proc Natl Acad Sci USA* 1998;**95**:7951–6.

27. Shirai Y, Balsinde J, Dennis EA. Localization and functional interrelationships among cytosolic Group IV, secreted Group V, and Ca^{2+}-independent Group VI phospholipase A$_2$s in P388D$_1$ macrophages using GFP/RFP constructs. *Biochim Biophys Acta* 2005;**1735**:119–29.

28. Wijewickrama GT, Kim JH, Kim YJ, Abraham A, Oh Y, Ananthanarayanan B, Kwatia M, Ackerman SJ, Cho W. Systematic evaluation of transcellular activities of secretory phospholipases A$_2$. High activity of group V phospholipases A$_2$ to induce eicosanoid biosynthesis in neighboring inflammatory cells. *J Biol Chem* 2006;**281**:10,935–10,944.

29. Satake Y, Díaz BL, Balestrieri B, Lam BK, Kanaoka Y, Grusby MJ, Arm JP. Role of group V phospholipase A$_2$ in zymosan-induced eicosanoid generation and vascular permeability revealed by targeted gene disruption. *J Biol Chem* 2004;**279**:16,488–16,494.

30. Diaz BL, Satake Y, Kikawada E, Balestrieri B, Arm JP. Group V secretory phospholipase A$_2$ amplifies the induction of cyclooxygenase 2 and delayed prostaglandin D$_2$ generation in mouse bone marrow culture-derived mast cells in a strain-dependent manner. *Biochim Biophys Acta* 2006;**1761**:1489–97.

31. Murakami M, Koduri RS, Enomoto A, Shimbara S, Seki M, Yoshihara K, Singer A, Valentin E, Ghomashchi F, Lambeau G, Gelb MH, Kudo I. Distinct arachidonate-releasing functions of mammalian secreted phospholipase A$_2$s in human embryonic kidney 293 and rat mastocytoma RBL-2H3 cells through heparan sulfate shuttling and external plasma membrane mechanisms. *J Biol Chem* 2001;**276**:10,083–10,096.

32. Balboa MA, Shirai Y, Gaietta G, Ellisman MH, Balsinde J, Dennis EA. Localization of group V phospholipase A2 in caveolin-enriched granules in activated P388D1 macrophage-like cells. *J Biol Chem* 2003;**278**:48,059–48,065.

33. Han SK, Kim KP, Koduri R, Bittova L, Muñoz M, Leff AR, Wilton MH, Gelb MH, Cho W. Roles of Trp31 in high membrane binding and proinflammatory activity of human group V phospholipase A$_2$. *J Biol Chem* 1999;**274**:11,881–11,188.

34. Kim YJ, Kim P, Rhee HJ, Das S, Rafter JD, Oh YS, Cho W. Internalized group V secretory phospholipase A$_2$ acts on the perinuclear membranes. *J Biol Chem* 2002;**277**:9358–65.

35. Mounier C, Ghomashchi F, Lindsay MR, James S, Singer AG, Parton RG, Gelb MH. Arachidonic acid release from mammalian cells transfected with human groups IIA and X secreted phospholipase A$_2$??occurs predominantly during the secretory process and with the involvement of cytosolic phospholipase A$_2\alpha$. *J Biol Chem* 2004;**279**:25,024–25,038.

36. Balsinde J, Shinohara H, Lefkowitz LJ, Johnson CA, Balboa MA, Dennis EA. Group V phospholipase A2-dependent induction of cyclooxygenase-2 in macrophages. *J Biol Chem* 1999;**274**:25,967–25,970.

37. Balboa MA, Pérez R, Balsinde J. Amplification mechanisms of inflammation: paracrine stimulation of arachidonic acid mobilization by secreted phospholipase A$_2$ is regulated by cytosolic phospholipase A$_2$-derived hydroperoxyeicosatetraenoic acid. *J Immunol* 2003;**171**:989–94.

38. Kuwata H, Nonaka T, Muralami M, Kudo I. Search of factors that intermediate cytokine-induced group IIA phospholipase A2 expression through the cytosolic phospholipase A2- and 12/15-lipoxygenase-dependent pathway. *J Biol Chem* 2005;**280**:25,830–25,839.

39. Han WK, Sapirstein A, Hung CC, Alessandrini A, Bonventre JV. Cross-talk between cytosolic phospholipase A$_2\alpha$ (cPLA2α) and secretory phospholipase A$_2$ (sPLA$_2$) in hydrogen peroxide-induced arachidonic acid release in murine mesangial cells: sPLA$_2$ regulates cPLA$_2\alpha$ activity that is responsible for arachidonic acid release. *J Biol Chem* 2003;**278**:24,153–24,163.

40. Kikawada E, Bonventre JV, Arm JP. Group V secretory PLA$_2$ regulates TLR2-dependent eicosanoid generation in mouse mast cells through amplification of ERK and cPLA$_2\alpha$ activation. *Blood* 2007;**110**:561–7.

41. Kim YJ, Kim SP, Han SK, Muñoz NM, Zhu X, Sano A, Leff AR, Cho W. Group V phospholipase A$_2$ induces leukotriene biosynthesis in human neutrophils through the activation of group IVA phospholipase A$_2$. *J Biol Chem* 2002;**277**:36,479–36,488.

42. Muñoz NM, Kim YJ, Meliton AY, Kim KP, Han SK, Boetticher E, O'Leary E, Myou S, Zhu X, Bonventre JV, Leff AR, Cho W. Human group V phospholipase A$_2$ induces group IVA phospholipase A$_2$-independent cysteinyl leukotriene synthesis in human eosinophils. *J Biol Chem* 2003;**278**:38,813–38,820.

43. Balsinde J, Pérez R, Balboa MA. Calcium-independent phospholipase A$_2$ and apoptosis. *Biochim Biophys Acta* 2006;**1761**:1344–50.

44. Balsinde J, Balboa MA. Cellular regulation and proposed biological functions of group VIA calcium-independent phospholipase A$_2$ in activated cells. *Cell Signal* 2005;**17**:1052–62.

45. Winstead M, Balsinde J, Dennis EA. Ca^{2+}-independent phospholipase A$_2$: structure and function. *Biochimm Biophys Acta* 2000;**1488**:28–39.

46. Bao S, Miller DJ, Ma Z, Woltmann M, Eng G, Ramanadham S, Molley K, Turk J. Male mice that do not express group VIA phospholipase A$_2$ produce spermatozoa with impaired motility and have greatly reduced fertility. *J Biol Chem* 2004;**279**:38,194–38,200.

47. Bao S, Li Y, Lei X, Wohltmann M, Jin W, Bohrer A, Semenkovich CF, Ramanadham S, Tabas. I. Turk,J. Attenuated free cholesterol loading-induced apoptosis but preserved phospholipid composition of peritoneal macrophages from mice that do not express group VIA phospholipase A$_2$. *J Biol Chem* 2007;**282**:27,100–27,114.

48. Nikolic DM, Gong MC, Turk J, Post SR. Class A scavenger receptor-mediated macrophage adhesion requires coupling of calcium-independent phospholipase A$_2$ and 12/15-lipoxygenase to Rac and Cdc42 activation. *J Biol Chem* 2007;**282**:33,405–33,411.

49. Balsinde J, Balboa MA, Dennis EA. Antisense inhibition of group VI Ca2+-independent phospholipase A$_2$ blocks phospholipid fatty acid remodeling in murine P388D$_1$ macrophages. *J Biol Chem* 1997;**272**:29,317–29,321.

50. Pérez R, Melero R, Balboa MA, Balsinde J. Role of group VIA calcium-independent phospholipase A$_2$ in arachidonic acid release, phospholipid fatty acid incorporation, and apoptosis in U937 cells responding to hydrogen peroxide. *J Biol Chem* 2004;**279**:40,385–40,391.

51. Daniele JJ, Fidelio GD, Bianco ID. Calcium dependency of arachidonic acid incorporation into cellular phospholipids of different cell types. *Prostag Oth Lipid M* 1999;**5–6**:341–50.

Chapter 34

Prostaglandin Mediators

Emer M. Smyth and Garret A. FitzGerald

Institute for Translational Medicine and Therapeutics, University of Pennsylvania, Philadelphia, Pennsylvania

INTRODUCTION

Arachidonic acid (AA), a 20-carbon unsaturated fatty acid containing four double bonds ($\Delta 5,8,11,14:C20:4$), circulates in plasma in both free and esterified forms and is a natural constituent of the phospholipid domain of cell membranes. AA is mobilized for release by phospholipases (PLs) A_2, particularly type IV cytosolic (c) PLA_2 [1], following its calcium-dependent translocation to the nuclear membrane and the endoplasmic reticulum (Figure 34.1). Three major groups of enzymes, prostaglandin G/H synthase (PGHS), lipoxygenase, or cytochrome p450, then catalyze the formation of the prostaglandins (PGs) and thromboxane A_2 (TxA_2), the leukotrienes, or the epoxyeicosatrienoic acids, respectively. Collectively, these products are known as eicosanoids. A parallel family of free radical catalyzed isomers, the isoeicosanoids, are formed by direct peroxidation of AA *in situ* in cell membranes [2]. This chapter will focus on the PGs and TxA_2, collectively termed the prostanoids.

THE CYCLOOXYGENASE PATHWAY

Prostanoids are formed by the action of PGHS, or cyclooxygenase (COX), on AA to form bisenoic products containing two double bonds, denoted by a subscript 2 (e.g., PGE_2 [3]). COX-1 or COX-2 dimers [4], homotypically inserted into the ER membrane, contain both cyclooxygenase and hydroperoxidase activities [3]. AA is sequentially transformed into the unstable cyclic endoperoxides, PGG_2 and PGH_2, for delivery to downstream isomerases and synthases to generate TxA_2 and D, E, F, and I series PGs (Figure 34.1). It is presently not understood either how AA is delivered specifically to COX or how PGH_2 is presented to downstream enzymes. Two COX genes have been identified: COX-1 is expressed constitutively in most cells,

while COX-2 is upregulated by cytokines, shear stress, and tumor promoters [3]. These observations suggest housekeeping functions, such as gastric epithelial cytoprotection and hemostasis, for COX-1-derived prostanoids, although it appears that both enzymes contribute to the generation of autoregulatory prostanoids. Conversely, the inducible COX-2 is considered the dominant source of prostanoid formation in inflammation and cancer, although both isozymes can contribute to prostanoid formation in syndromes of human inflammation, including atherosclerosis [5] and rheumatoid arthritis [6]. COX-1 and -2 can also utilize eicosapentaenoic acid (EPA), an ω-3 fatty acid abundant in fish oils, to generate the 3-series of prostaglandins (e.g., PGE_3) although AA remains the preferred substrate for both isozymes [7].

COX-1 and COX-2 are closely related in their amino acid sequence [8] and crystal structure [9]. Although both isozymes demonstrate similar subcellular distribution [10], preference for downstream enzymes is sometimes evident in heterologous expression systems and apparently *in vivo*. COX-1 preferentially couples with TxS, PGFS [11], and the cytosolic (c) PGES isozymes [12]. COX-2 prefers PGIS [11] and the microsomal (m) PGES isozymes, which are induced by cytokines and tumor promoters [13]. Two forms of PGDS [14, 15] and PGFS [16, 17] have been identified, underscoring the diversity of the isomerases and synthases.

COX Deletion

Deletion of COX-2 results in multiple defects of implantation and reproduction, leading to breeding difficulties [18]. Offspring have variably revealed cardiac fibrosis, renal defects, and impaired inflammatory responses; however, the extent to which these phenotypes are modulated by genetic background is presently unclear. Impaired inflammatory responses [19] and delayed parturition [20] secondary to

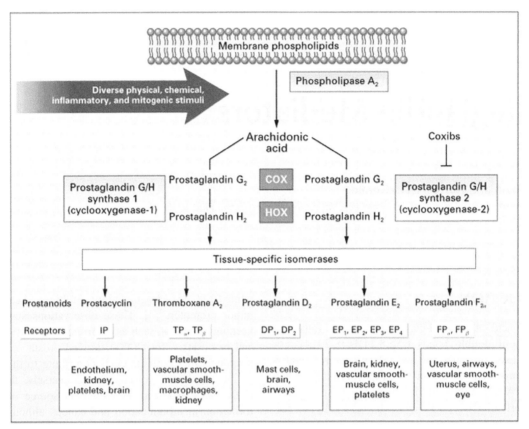

FIGURE 34.1 Production and actions of prostanoids: Arachidonic acid, a 20-carbon fatty acid containing four double bonds, is lib-erated from the sn2 position in membrane phospholipids by PLA2. COX converts arachidonic acid to the unstable intermediateprostaglandin H2, which is converted by tissue-specific isomerases to multiple prostanoids. These bioactive lipids activate specificcell-membrane receptors of the superfamily of GPCRs.

COX-1 deletion have been reported. Deletion of the COX-2, but not the COX-1, gene increases the frequency of patent ductus arteriosus (PDA) in newborn pups [21]. Coincidental deletion of COX-1 increases the frequency of the COX-2 knockout PDA phenotype [21]. It is the absence of PGE_2 that underlies this phenotype; deletion of 15-hydroxyprostaglandin deydrogenase, the major inactivating enzyme of PGE_2, produces sustained high levels of PGE_2 throughout the perinatal period and results in ductal closure [22]. Decreased PGE_2 generation following deletion of COX-2 also underlies reduced disease in mouse models of multiple cancers, including intestinal, colonic, and mammary cancers [23]. The role of COX-1 in cancer is less clear, but it is likely that both isozymes contribute to oncogenic processes [23]. Both COX-1 and COX-2 are subject to developmental regulation, and interference with their developmental expression may condition adult phenotypes [24]. COX-1 and COX-2 are expressed in a spatially and temporally segregated manner during thymic development, where they may influence T cell maturation [25].

activating membrane receptors at (or close to) the site of their formation. Specific G protein-coupled receptors (GPCRs) have been cloned for all the prostanoids [26]. A single gene product has been identified for prostacyclin (the IP), $PGF_{2\alpha}$ (the FP), and TxA_2 (the TP), while four distinct PGE_2 receptors (the EP_{1-4}) and two PGD_2 (DP_1 and DP_2) have been cloned. The prostanoid receptors appear to derive from an ancestral EP receptor, and share high homology. Phylogenetic comparison of this family reveals three subclusters [26]: first, the EP_2, EP_4, IP, and DP_1, the relaxant receptors, which increase cAMP generation; second, EP_1, FP, and TP, the contractile receptors, which increase intracellular calcium levels; and third, the EP_3, which elevates intracellular calcium and decreases cAMP. The DP_2 (also known as CTRH2 and GPR44), a member fMLP receptor superfamily [27, 28], is the exception to this characterization. Differential mRNA splicing gives rise to additional isoforms of the TP (α and β) [29], FP (A and B) [30], and EP_3 (I, II, III, IV, V, VI, e and f) [31]. The prostanoid receptors have been reviewed thoroughly elsewhere [26, 32].

PROSTANOID RECEPTORS

Due to their short half-lives (seconds to minutes), prostanoids act as autacoids, rather than circulating hormones, by

THROMBOXANE A₂ (TxA₂)

TxA_2, the major product of platelet COX-1, is a potent vasoconstrictor [33], mitogen [34], and platelet activator

[35]. Despite the diversity of platelet agonists, inhibition of platelet TxA_2 formation apparently accounts for cardioprotection from aspirin [36], reflecting the importance of TxA_2 as an amplification signal for more potent agonists, such as thrombin and ADP [35]. TxA_2, also a major product of macrophage COX-2, contributes to atherogenesis in mouse models [37]. Analogous to its role in vascular proliferation (see below), TxA_2 may also mediate cellular hypertrophy [38]. Tumor growth and angiogenesis of colon cancer cells *in vivo* was increased when TxA_2 synthase was overexpressed [39].

Two forms of the platelet TP have been segregated pharmacologically, one mediating shape change, the other aggregation [40]. However, the cloned human TP splice variants (splice variants of the mouse TP are not apparent) do not account for this distinction, and $TP\alpha$ is apparently the sole isoform expressed in platelets [41, 42]. Recognized differences between the splice variants are limited to G protein activation in heterologous expression systems [43, 44] and agonist-induced desensitization and sequestration [45, 46]. Given the identification of distinct low homology GPCRs mediating ADP-induced platelet shape change and aggregation [47], it seems likely that at least one more distinct TP remains to be identified. In this regard, evidence suggests that $iPF_{2\alpha}$-III and iPE_2III, both isoprostanes, act *in vivo* at the TP [48] but do not bind to either isoform *in vitro* [49], further suggesting the existence of another TP. Distinct receptor sites can be generated through GPCR heterodimerization [50]. Interestingly, dimerization of $TP\alpha$ and $TP\beta$ augments $iPF_{2\alpha}$-III and iPE_2III signaling [49], suggesting that the $TP\alpha TP\beta$ heterodimer constitutes an alternative TP. The extent to which associations between $TP\alpha$ with $TP\beta$, and/or other prostanoid receptors, might contribute to the family of prostanoid receptors remains to be examined. The cloned TPs couple via G_q, G_{11}, $G_{12/13}$, and G_h (which is also tissue transglutaminase II) to activate PLC-dependent inositol phosphate generation and elevate intracellular calcium [26, 32]. Activation of the TP isoforms may also activate or inhibit adenylyl cyclase, via G_s ($TP\alpha$) or G_i($TP\beta$), respectively, and signal via G_q and related proteins to MAP kinase signaling pathways. Heterodimerization of $TP\alpha$, the receptor for prostacyclin (see below), facilitates $TP\alpha$-adenylyl cyclase activation [51].

TP mRNAs are expressed widely in lung, liver, kidney, heart, uterus, and vascular cells, with $TP\alpha$ usually the predominant isoform [52]. Despite reports of abundant TP expression in thymus, the role of TxA_2 in lymphocyte development and function is presently unclear. A naturally occurring mutation in the first intracellular loop of the TP is associated with a mild bleeding disorder and platelet resistance to TP agonists [42], while a polymorphism in the TP has been linked to bronchodilator resistance in asthma [53].

TP Deletion/Overexpression

Deletion of the TP reveals a mild hemostatic defect, resistance to AA-induced platelet activation [54] and decreased atherogenesis [55], reflecting the role of TxA_2 in vascular biology. TP-null mice have reduced proliferative responses to vascular injury, while the opposite is true of mice engineered to overexpress $TP\beta$ in the vasculature [56]. $TP\beta$ overexpressors also develop a syndrome reminiscent of intrauterine growth retardation, probably secondary to placental ischemia [57].

PROSTACYCLIN (PGI_2)

A major product of COX-2 in healthy individuals [58], PGI_2 is a potent vasodilator, inhibitor of platelet aggregation by all recognized agonists [59], and inhibitor of cell proliferation *in vitro* [60]. PGI_2 biosynthesis is increased in syndromes of platelet activation [61, 62], perhaps as a homeostatic response to accelerated platelet–vascular interactions. Chronic PGI_2 treatment can reduce pulmonary vascular resistance in patients with primary pulmonary hypertension [63]. Delivery of the gene for PGIS *in vivo* diminishes vascular smooth muscle cell proliferation and migration in response to injury [64], while transplanted colon adenocarcinoma cells overexpressing PGIS formed smaller and less vascular tumors in mice [39]. PGIS polymorphs have been associated with essential hypertension [65] and myocardial infarction [66], while PGI_2 attenuates angiotensin II-induced renal vasoconstriction and systemic hypertension [67]. PGI_2 also attenuates the response to thrombotic stimuli in dogs [68], and specifically limits the effects on TxA_2 on platelets and the vessel wall in mice [56].

The sole identified IP couples to activation of adenylyl cyclase via G_s, although it can also activate phospholipase C via G_q. IP mRNA is abundantly expressed in kidney, where PGI_2 may regulate renal blood flow, renin release, and glomerular filtration rate, and in lung, where PGI_2 can modulate vascular tone [26, 32]. The IP is also expressed in the spinal column, where PGI_2 plays a role in pain perception, and in the liver, where its role is unknown. Expression within the cardiovascular system is most abundant in the aorta, consistent with the major biological role of PGI_2 in platelet and macrovascular homeostasis. IP expression in the heart, together with reports that COX-2-dependent PGI_2 formation limits oxidant-induced injury in cardiomyocytes [69], suggests a possible protective role for PGI_2 in cardiac tissue. A major difference between humans and rodents is the marked expression of IP in the thymus [70], although the functional relevance of this observation is not clear.

It is likely that at least one other IP remains to be identified. Pharmacologically distinct IP sites in the brain and kidney are not attributable to the cloned IP [71, 72]. In addition, PGI_2 may activate the peroxisome proliferator activated receptors (PPARs). However, while both PGI_2 and iloprost activated $PPAR\alpha$ and $PPAR\delta$ *in vitro*, another PGI_2 analog, cicaprost, did not [73], and it is as yet unclear whether PGI_2 activation of PPARs occurs *in vivo*. The loss of

PGI$_2$-mediated PPARδ activation was thought to underlie the implantation defect in COX-2-deficient mice, although no implantation defect was evident in PPARδ-deficient mice [74]. A signalling deficient IP polymorph is associated with accelerated human cardiovascular disease with some alterations in ligand binding and signaling in overexpression systems, have been identified [75].

IP Deletion

Although results in IP-deficient mice have implicated PGI$_2$ in the mediation of pain and inflammation [76], these consequences seem conditioned by genetic background. Platelets of IP-null mice are resistant to disaggregation by IP agonists [76] and the thrombotic and proliferative response to vascular injury is enhanced [56, 77], as is hypoxia-induced pulmonary hypertension and remodeling [78]. IP deficient mice develop salt-sensitive hypertension, cardiac hypertrophy and cardiac fibrosis [79]. Lung fibrosis is also elevated in IP$^{-/-}$ mice [80], as is pulmonary inflammation, reflecting the inhibitory effects of PGI$_2$ on T and dendritic cell function [81]. Despite its expression in murine thymus, disordered T cell function secondary to IP deletion has not been reported, and the IP appears to play a minor role, if any, in murine T cell maturation [25]. Deletion of the IP undermines the atheroprotective effect of female gender in LDL receptor deficient mice [82], a possible consequence of estrogen-mediated upregulation of PGIS [83] and/or its protection from free radical attack. In contrast to the IP knockouts, PGIS deficient mice develop renal vascular disorders [84]. The divergence in IP$^{-/-}$ and PGIS$^{-/-}$ phenotypes suggests the existence of a second IP; however, elevations of both PGE$_2$ and TxA$_2$ in PGIS knockouts may underlie their pathology [84].

PROSTAGLANDIN D$_2$ (PGD$_2$)

PGD$_2$, the major COX product formed by mast cells, is released during allergic responses, including asthma and systemic mastocytosis [85, 86]. Under physiological conditions, COX-1 is the predominant source of PGD$_2$ [87]. LPS-evoked changes in biosynthesis of PGD$_2$ in humans are coordinated with the acute inflammatory response, and fall rapidly during the resolution phase [87]. Infusion of PGD$_2$ in humans results in flushing, nasal stuffiness, and hypotension [88]. In mice, overexpression of lipocain-like PGDS increases response to bronchial challenge with ovalbumin [89]. The hematopoietic PGDS is expressed abnormally in patients with coronary disease [90], and a polymorphic variant has been linked to human asthma [91]. PGD$_2$, an abundant COX product in brain, is considered an important regulator of sleep–wake cycles [92]. Deletion of PGDS abolishes allodynia (touch-evoked pain) in mice [93], demonstrating a role for PGD$_2$ in pain perception.

The DP$_1$ is coupled positively to adenylyl cyclase through G$_s$ [26, 32], which directs PDG$_2$-induced inhibition of platelet aggregation, bronchodilation, and vasodilation. Among the prostanoid receptors the DP$_1$ is the least abundant, with minor expression reported in mouse ileum, lung, stomach, and uterus, and expression in the CNS limited specifically to the leptomeninges. A chemoattractant receptor-homologous molecule (CRTH2), expressed on T-helper (H) type-2 cells [28], was classified as the DP$_2$. This receptor is distinct from other prostanoid receptors, couples to increased intracellular Ca^{2+}, and directs PGD$_2$-induced chemotaxis and migration of TH2 cells. Both DPs integrate coordinately the effects of PGD$_2$ on eosinophils and basophils, modulating chemokinesis, degranulation, and apoptosis [94]. Despite their distinctions, the two DPs have apparent complementary roles in initiation and maintenance of allergic reactions [95]. DP$_2$ and PGDS are coordinately expressed at the fetal/maternal interface in human deciduas, where they may participate in lymphocyte recruitment [96].

PGD$_2$ may be metabolized to F-ring and J-ring metabolites [87]. These include 15-deoxy Δ(12,14) PGJ$_2$, a selective DP$_2$ ligand [97], and possible natural ligand for PPARγ regulating adipogenesis, inflammation, tumorigenesis, and immunity [98]. However, while PGJ$_2$, and its metabolite, can activate the nuclear receptor *in vitro* [99], it is presently unclear whether sufficient concentrations are formed *in vivo*.

DP Deletion

Deletion of the DP$_1$ sharply reduces ovalbumin-induced lymphocyte and eosinophil infiltration and airway hyperreactivity, reflecting PGD$_2$'s apparent role in asthma [100]. Work with these mice demonstrates the action of PGD$_2$ on arachnoid trabecular cells in the basal forebrain to increase extracellular adenosine, which in turn facilitates induction of sleep [101]. DP$_2$-null mice have reduced allergic skin reactions [102] but, surprisingly, increased allergic airway inflammation [103]. Antagonism or deletion of DP$_1$, but not DP$_2$, reduced the flushing response to the hypolipedimic drug niacin [104], raising the possibility that co-administration of a DP$_1$ antagonist might reduce this complication and enhance compliance.

PROSTAGLANDIN E$_2$ (PGE$_2$)

PGE$_2$ regulates diverse biological processes, including cell growth, inflammation, reproduction, sodium homeostasis, and blood pressure [105]. Its biological effects are complex and often opposing; vasodilation in the arterial and venous systems [106], but constriction of smooth muscle in the trachea, gastric fundus, and ileum [107]. The COX-1-cPGES axis is considered the predominant source of homeostatic PGE$_2$ [12, 108]. The inducible mPGES-1 isoform, which is frequently co-regulated and co-localized with COX-2 [108], is the major source of PGE$_2$ associated with inflammatory and pyretic responses. Deletion of mPGES-1 in mice diminished inflammatory responses and offset inflammatory disease, including experimentally-induced arthritis [109], atherosclerosis [110], and abdominal aortic aneurysm formation [111]. Constitutive expression of mPGES-1 has been reported in some tissues, with abundant levels in the kidney [108]. COX-2-derived prostanoids may differentially regulate salt excretion and glomerular circulation in volume overload or depletion [112]. PGE_2, along with PGI_2, apparently derived from COX-2, maintains renal blood flow and salt excretion [113]–effects that may be counterbalanced by COX-1-derived TxA_2 [67]. mPGES-1-deficient mice display increased salt sensitivity [114], although in most studies blood pressure is unaltered [77, 111].

Both the EP_2 and the EP_4 activate adenylyl cyclase via G_s [101]. Differences in agonist-induced desensitization may be one reason for the presence of such similar receptors for PGE_2. The EP_1, via an unclassified G protein, and the EP_3, via G_q, are coupled to PLC activation [26, 32]. A splice variant of the EP_1 in the rat may antagonize coupling of other EPs [100]. EP_3 variants couple to G_s-mediated activation, and G_i-mediated inhibition, of adenylyl cyclase.

The mRNAs for all four EPs are widely expressed; however, the limited distribution of EP_1 and EP_2, compared with EP_3 and EP_4, together with induction of EP_2 in response to inflammatory stimuli, suggests specialized functions for the different EPs [115]. The biological actions of PGE_2 may be conditioned by this differential receptor expression and/or PGE_2 levels. EP_4 directs platelet inhibition at low PGE_2 concentrations, while increased PGE_2 levels in, for example, inflammation lead to EP_3-mediated platelet aggregation [116]. Indeed, high concentrations of PGE_2 condition platelet responses through EP_3- and IP-mediated regulation of intracellular cAMP [116]. Despite higher renal EP_4 expression [32], evidence supports EP_2-mediated renal vasodilation and salt handling [117], while an EP_1-directed increase in salt excretion may contribute to PGE_2-dependent natriuresis [32]. PGE_2 may also directly stimulate renin and angiotensin II generation in the kidney [118], or directly constrict the renal vasculature [119], leading to hypertension. In the gastrointestinal tract, cytoprotective effects are mediated by EP_1 in stomach [120], but by EP_3 and EP_4 in the intestine [121]. EP_1 and EP_3 appear responsible for myometrial contractility caused by PGE analogs, such as misoprost, used to induce labor [122], while selective EP_2-mediated inhibition of myometrial contractility [123] may be useful against preterm labor. EP_2- [124] and EP_3-mediated [125] interactions with growth factors may underlie the proliferative and angiogenic actions of PGE_2 in cancer. In the immune system, activation of the EP_2 inhibits T cell proliferation, while both EP_2 and EP_4 receptors regulate antigen-presenting function *in vivo* [126]. All four EPs are implicated in promoting tumorogenesis through modulation of proliferation, migration, angiogenesis, apoptosis, and immune function [23], although the precise role of the individual receptors

in cancer is likely conditioned on context. Circulating levels of interleukin-1β induce coordinate COX-2-mPGES expression at the blood–brain barrier, permitting activation of the central EPs [127]. Localized infusions of PGE_2 into the third ventricle induce wakefulness via the EP_1 and EP_2, while EP_4 activation in the subarachnoid space induces sleep [128]. Pyrexial responses may mediated through the EP_3 [129], while the EP_1 and the EP_3 increase neuronal excitability and pain perception [129, 130].

EP Deletion

Knockout mouse models have been generated for all the EPs. EP_1-deficient mice have reduced nociceptive perception, while male, but not female, knockouts have reduced systolic blood pressure accompanied by elevated renin-angiotensin activity [130]. EP_2-deficient mice are normotensive at baseline but demonstrate increased salt- and pressor hormone-induced hypertension, although this is modified by genetic background [117]. The EP_2 knockouts also demonstrate a preimplantation defect, which may underlie some of the breeding difficulties seen in the COX-2 knockouts (see above). EP_3-deficient mice are resistant to pyrogen-induced fever [129]. However, despite its abundant expression in the kidney, there is no renal phenotype in EP_3 knockouts [131]. Deletion of the EP_4 results in PDA and neonatal death [132]. Most studies with EP knockouts indicate that EP_1, EP_2, or EP_4 contribute to early carcinogenesis, with EP_3 playing no role or a protective role [133].

PROSTAGLANDIN $F_{2\alpha}$ ($PGF_{2\alpha}$)

$PGF_{2\alpha}$ actions include luteolysis [134] and smooth muscle contraction across a variety of tissues [135, 136]. PGFS catalyzes the reduction of PGH_2 to $PGF_{2\alpha}$, PGD_2 to 9α 11β $PGF_{2\alpha}$ [137], and retinal to retinol [138]. It exists in at least two isoforms, identified initially in liver [17] and lung [137], and is also expressed in lymphocytes [137] and spinal cord [139]. $PGF_{2\alpha}$ induces cardiac myocyte hypertrophy and induction of myofibrillar genes, independent of muscle contraction [140], suggesting a role for this eicosanoid during development, in compensatory hypertrophy, and/or in recovery of the heart from injury.

Thus far, one GPCR for $PGF_{2\alpha}$, the FP, which couples via G_q activation of PLC, has been cloned [26, 32]. Stimulation of FP also activates Rho kinase, leading to the formation of actin stress fibers, phosphorylation of p125 focal adhesion kinase, and cell rounding [141]. Similar to the EP_3 and TP, carboxy terminal splice variants, FP_A and FP_B [30], have been identified. These are indistinguishable in their ligand-binding properties and signaling, but may differ in their constitutive activity [30] and rates of desensitization [142]. FP_A and FP_B also differ in coupling to the Tcf/β catenin-signaling pathway, which may underlie the prolonged cytoskeletal effects mediated thought FP_B [143]. The FP is expressed in kidney, heart, lung, and stomach; however, it is most abundant in the corpus luteum, where its expression varies during the estrus cycle, consistent with the role for $PGF_{2\alpha}$ in luteolysis. The FP is also expressed in the ciliary body of the eye, where FP agonists have clinical utility in the treatment of raised intraocular pressure in patients with glaucoma [144]. Although activation of the FP results in vaso- and broncho-constriction [136], cell proliferation [145], and cardiomyocyte hypertrophy [146], the role of this prostanoid in cardiopulmonary disease is poorly characterized. Similarly, activation of the FP blocks pre-adipocyte differentiation *in vitro* [147], but the role of the FP, if any, in obesity is poorly understood.

FP Deletion

Mice deficient in the FP do not deliver at term, resulting from a failure to induce the oxytocin receptor and lack of the normal decline in elevated progesterone levels [148]. Ovariectomy restores responsiveness to oxytocin and permits successful parturition. COX-1-derived $PGF_{2\alpha}$ in these mice appears important for luteolysis, consistent with delayed parturition in COX-1-deficient mice [20]. Subsequent upregulation of COX-2 generates prostanoids, including $PGF_{2\alpha}$ and TxA_2, important in the final stages of parturition [149]. Mice lacking both COX-1 and oxytocin underwent normal parturition, demonstrating the critical interplay between $PGF_{2\alpha}$ and oxytocin in onset of labor [20].

CONCLUDING REMARKS

The cyclooxygenase pathway of arachidonic acid metabolism generates a family of evanescent mediators with wide and varied physiological and pathophysiological actions. Understanding the biological role of the prostanoids requires examination of the biosynthetic pathways that lead to their temporal and tissue-specific generation, together with the array of signaling pathways activated by their multiple receptors.

REFERENCES

1. Leslie CC. Properties and regulation of cytosolic phospholipase A2. *J Biol Chem* 1997;**272**:16,709–12.
2. Patrono C, FitzGerald GA. Isoprostanes: potential markers of oxidant stress in atherothrombotic disease. *Arterioscler Thromb Vasc Biol* 1997;**17**:2309–15.
3. Herschman HR. Prostaglandin synthase 2. *Biochim Biophys Acta* 1996;**1299**:125–40.
4. Garavito RM, Picot D, Loll PJ. The 3.1 Å X-ray crystal structure of the integral membrane enzyme prostaglandin H2 synthase-1. *Adv Prostaglandin Thromboxane Leukot Res* 1995;**23**:99–103.
5. Wijeyaratne SM, Abbott CR, Homer-Vanniasinkam S, Mavor AI, Gough MJ. Differences in the detection of cyclo-oxygenase 1 and 2 proteins in symptomatic and asymptomatic carotid plaques. *Br J Surg* 2001;**88**:951–7.
6. Iniguez MA, Pablos JL, Carreira PE, Cabre F, Gomez-Reino JJ. Detection of COX-1 and COX-2 isoforms in synovial fluid cells from inflammatory joint diseases. *Br J Rheumatol* 1998;**37**:773–8.
7. Wada M, DeLong CJ, Hong YH, Rieke CJ, Song I, Sidhu RS, Yuan C, Warnock M, Schmaier AH, Yokoyama C, Smyth EM, Wilson SJ, FitzGerald GA, Garavito RM, Sui de X, Regan JW, Smith WL. Enzymes and receptors of prostaglandin pathways with arachidonic acid-derived versus eicosapentaenoic acid-derived substrates and products. *J Biol Chem* 2007;**282**:22,254–66.
8. Smith WL, Garavito RM, DeWitt DL. Prostaglandin endoperoxide H synthases (cyclooxygenases)-1 and -2. *J Biol Chem* 1996;**271**:33,157–60.
9. FitzGerald GA, Loll P. COX in a crystal ball: current status and future promise of prostaglandin research. *J Clin Invest* 2001;**107**:1335–7.
10. Spencer AG, Woods JW, Arakawa T, Singer I, Smith WL. Subcellular localization of prostaglandin endoperoxide H synthases-1 and -2 by immunoelectron microscopy. *J Biol Chem* 1998;**273**:9886–93.
11. Ueno N, Murakami M, Tanioka T, Fujimori K, Tanabe T, Urade Y, Kudo I. Coupling between cyclooxygenase, terminal prostanoid synthase, and phospholipase A2. *J Biol Chem* 2001;**276**:34,918–27.
12. Tanioka T, Nakatani Y, Semmyo N, Murakami M, Kudo I. Molecular identification of cytosolic prostaglandin E2 synthase that is functionally coupled with cyclooxygenase-1 in immediate prostaglandin E2 biosynthesis. *J Biol Chem* 2000;**275**:32,775–82.
13. Murakami M, Naraba H, Tanioka T, Semmyo N, Nakatani Y, Kojima F, Ikeda T, Fueki M, Ueno A, Oh S, Kudo I. Regulation of prostaglandin E2 biosynthesis by inducible membrane-associated prostaglandin E2 synthase that acts in concert with cyclooxygenase-2. *J Biol Chem* 2000;**275**:32,783–92.
14. Nagata A, Suzuki Y, Igarashi M, Eguchi N, Toh H, Urade Y, Hayaishi O. Human brain prostaglandin D synthase has been evolutionarily differentiated from lipophilic-ligand carrier proteins. *Proc Natl Acad Sci USA* 1991;**88**:4020–4.
15. Kanaoka Y, Ago H, Inagaki E, Nanayama T, Miyano M, Kikuno R, Fujii Y, Eguchi N, Toh H, Urade Y, Hayaishi O. Cloning and crystal structure of hematopoietic prostaglandin D synthase. *Cell* 1997;**90**:1085–95.
16. Watanabe K, Fujii Y, Nakayama K, Ohkubo H, Kuramitsu S, Kagamiyama H, Nakanishi S, Hayaishi O. Structural similarity of bovine lung prostaglandin F synthase to lens epsilon-crystallin of the European common frog. *Proc Natl Acad Sci USA* 1988;**85**:11–15.
17. Suzuki T, Fujii Y, Miyano M, Chen LY, Takahashi T, Watanabe K. cDNA cloning, expression, and mutagenesis study of liver-type prostaglandin F synthase. *J Biol Chem* 1999;**274**:241–8.
18. Lim H, Paria BC, Das SK, Dinchuk JE, Langenbach R, Trzaskos JM, Dey SK. Multiple female reproductive failures in cyclooxygenase 2-deficient mice. *Cell* 1997;**91**:197–208.
19. Langenbach R, Morham SG, Tiano HF, Loftin CD, Ghanayem BI, Chulada PC, Mahler JF, Lee CA, Goulding EH, Kluckman KD, et al. Prostaglandin synthase 1 gene disruption in mice reduces arachidonic acid-induced inflammation and indomethacin-induced gastric ulceration. *Cell* 1995;**83**:483–92.
20. Gross GA, Imamura T, Luedke C, Vogt SK, Olson LM, Nelson DM, Sadovsky Y, Muglia LJ. Opposing actions of prostaglandins and oxytocin determine the onset of murine labor. *Proc Natl Acad Sci USA* 1998;**95**:11,875–9.

21. Loftin CD, Trivedi DB, Tiano HF, Clark JA, Lee CA, Epstein JA, Morham SG, Breyer MD, Nguyen M, Hawkins BM, Goulet JL, Smithies O, Koller BH, Langenbach R. Failure of ductus arteriosus closure and remodeling in neonatal mice deficient in cyclooxygenase-1 and cyclooxygenase-2. *Proc Natl Acad Sci USA* 2001;**98**:1059–64.

22. Coggins KG, Latour A, Nguyen MS, Audoly L, Coffman TM, Koller BH. Metabolism of PGE2 by prostaglandin dehydrogenase is essential for remodeling the ductus arteriosus. *Nat Med* 2002;**8**:91–2.

23. Cha YI, DuBois RN. NSAIDs and cancer prevention: targets downstream of COX-2. *Annu Rev Med* 2007;**58**:239–52.

24. Grosser T, Yusuff S, Cheskis E, Pack MA, FitzGerald GA. Developmental Expression of Functional Cyclooxygenases in Zebrafish. *Proc Natl Acad Sci USA* 2002;**14**:14.

25. Rocca B, Spain LM, Pure E, Langenbach R, Patrono C, FitzGerald GA. Distinct roles of prostaglandin H synthases 1 and 2 in T-cell development. *J Clin Invest* 1999;**103**:1469–77.

26. Narumiya S, Sugimoto Y, Ushikubi F. Prostanoid receptors: structures, properties, and functions. *Physiol Rev* 1999;**79**:1193–226.

27. Nagata K, Hirai H, Tanaka K, Ogawa K, Aso T, Sugamura K, Nakamura M, Takano S. CRTH2, an orphan receptor of T-helper-2-cells, is expressed on basophils and eosinophils and responds to mast cell-derived factor(s). *FEBS Letts* 1999;**459**:195–9.

28. Hirai H, Tanaka K, Yoshie O, Ogawa K, Kenmotsu K, Takamori Y, Ichimasa M, Sugamura K, Nakamura M, Takano S, Nagata K. Prostaglandin D2 selectively induces chemotaxis in T helper type 2 cells, eosinophils, and basophils via seven-transmembrane receptor CRTH2. *J Exp Med* 2001;**193**:255–61.

29. Raychowdhury MK, Yukawa M, Collins LJ, McGrail SH, Kent KC, Ware JA. Alternative splicing produces a divergent cytoplasmic tail in the human endothelial thromboxane A2 receptor. *J Biol Chem* 1994;**269**:19,256–61.

30. Pierce KL, Bailey TJ, Hoyer PB, Gil DW, Woodward DF, Regan JW. Cloning of a carboxyl-terminal isoform of the prostanoid FP receptor. *J Biol Chem* 1997;**272**:883–7.

31. Schmid A, Thierauch KH, Schleuning WD, Dinter H. Splice variants of the human EP3 receptor for prostaglandin E2. *Eur J Biochem* 1995;**228**:23–30.

32. Breyer RM, Bagdassarian CK, Myers SA, Breyer MD. Prostanoid receptors: subtypes and signaling. *Annu Rev Pharmacol Toxicol* 2001;**41**:661–90.

33. Dorn II GW, Sens D, Chaikhouni A, Mais D, Halushka PV. Cultured human vascular smooth muscle cells with functional thromboxane A2 receptors: measurement of U46619-induced 45calcium efflux. *Circ Res* 1987;**60**:952–6.

34. Pakala R, Willerson JT, Benedict CR. Effect of serotonin, thromboxane A2, and specific receptor antagonists on vascular smooth muscle cell proliferation. *Circulation* 1997;**96**:2280–6.

35. FitzGerald GA. Mechanisms of platelet activation: thromboxane A2 as an amplifying signal for other agonists. *Am J Cardiol* 1991;**68**:11B–15B.

36. Patrono C. Aspirin as an antiplatelet drug. *N Engl J Med* 1994;**330**:1287–94.

37. Cayatte AJ, Du Y, Oliver-Krasinski J, Lavielle G, Verbeuren TJ, Cohen RA. The thromboxane receptor antagonist S18886 but not aspirin inhibits atherogenesis in apo E-deficient mice: evidence that eicosanoids other than thromboxane contribute to atherosclerosis. *Arterioscler Thromb Vasc Biol* 2000;**20**:1724–8.

38. Ali II S, Davis MG, Becker MW, Dorn GW. Thromboxane A2 stimulates vascular smooth muscle hypertrophy by up-regulating the synthesis and release of endogenous basic fibroblast growth factor. *J Biol Chem* 1993;**268**:17,397–403.

39. Pradono P, Tazawa R, Maemondo M, Tanaka M, Usui K, Saijo Y, Hagiwara K, Nukiwa T. Gene transfer of thromboxane A(2) synthase and prostaglandin I(2) synthase antithetically altered tumor angiogenesis and tumor growth. *Cancer Res* 2002;**62**:63–6.

40. Dorn II GW, DeJesus A. Human platelet aggregation and shape change are coupled to separate thromboxane A2-prostaglandin H2 receptors. *Am J Physiol* 1991;**260**:H327–H34.

41. Habib A, FitzGerald GA, Maclouf J. Phosphorylation of the thromboxane receptor alpha, the predominant isoform expressed in human platelets. *J Biol Chem* 1999;**274**:2645–51.

42. Hirata T, Kakizuka A, Ushikubi F, Fuse I, Okuma M, Narumiya S. Arg60 to Leu mutation of the human thromboxane A2 receptor in a dominantly inherited bleeding disorder. *J Clin Invest* 1994;**94**:1662–7.

43. Vezza R, Habib A, FitzGerald GA. Differential signaling by the thromboxane receptor isoforms via the novel GTP-binding protein, Gh. *J Biol Chem* 1999;**274**:12,774–9.

44. Hirata T, Ushikubi F, Kakizuka A, Okuma M, Narumiya S. Two thromboxane A2 receptor isoforms in human platelets. Opposite coupling to adenylyl cyclase with different sensitivity to Arg60 to Leu mutation. *J Clin Invest* 1996;**97**:949–56.

45. Yukawa M, Yokota R, Eberhardt RT, von Andrian L, Ware JA. Differential desensitization of thromboxane A2 receptor subtypes. *Circ Res* 1997;**80**:551–6.

46. Parent JL, Labrecque P, Orsini MJ, Benovic JL. Internalization of the TXA2 receptor alpha and beta isoforms. Role of the differentially spliced cooh terminus in agonist-promoted receptor internalization. *J Biol Chem* 1999;**274**:8941–8.

47. Takasaki J, Kamohara M, Saito T, Matsumoto M, Matsumoto S, Ohishi T, Soga T, Matsushime H, Furuichi K. Molecular cloning of the platelet P2T(AC) ADP receptor: pharmacological comparison with another ADP receptor, the P2Y(1) receptor. *Mol Pharmacol* 2001;**60**:432–9.

48. Audoly LP, Rocca B, Fabre JE, Koller BH, Thomas D, Loeb AL, Coffman TM, FitzGerald GA. Cardiovascular responses to the isoprostanes iPF(2alpha)-III and iPE(2)-III are mediated via the thromboxane A(2) receptor in vivo. *Circulation* 2000;**101**:2833–40.

49. Wilson SJ, McGinley K, Huang AJ, Smyth EM. Heterodimerization of the alpha and beta isoforms of the human thromboxane receptor enhances isoprostane signaling. *Biochem Biophys Res Commun* 2007;**352**:397–403.

50. Devi LA. Heterodimerization of G-protein-coupled receptors: pharmacology, signaling and trafficking. *Trends Pharmacol Sci* 2001;**22**:532–7.

51. Wilson SJ, Roche AM, Kostetskaia E, Smyth EM. Dimerization of the human receptors for prostacyclin and thromboxane facilitates thromboxane receptor-mediated cAMP generation. *J Biol Chem* 2004;**279**:53,036–47.

52. Miggin SM, Kinsella BT. Expression and tissue distribution of the mRNAs encoding the human thromboxane A2 receptor (TP) alpha and beta isoforms. *Biochim Biophys Acta* 1998;**1425**:543–59.

53. Leung TF, Tang NL, Lam CW, Li AM, Chan IH, Ha G. Thromboxane A2 receptor gene polymorphism is associated with the serum concentration of cat-specific immunoglobulin E as well as the development and severity of asthma in Chinese children. *Pediatr Allergy Immunol* 2002;**13**:10–17.

54. Thomas DW, Mannon RB, Mannon PJ, Latour A, Oliver JA, Hoffman M, Smithies O, Koller BH, Coffman TM. Coagulation defects and altered hemodynamic responses in mice lacking receptors for thromboxane A2. *J Clin Invest* 1998;**102**:1994–2001.

55. Kobayashi T, Tahara Y, Matsumoto M, Iguchi M, Sano H, Murayama T, Arai H, Oida H, Yurugi-Kobayashi T, Yamashita JK, Katagiri H, Majima M, Yokode M, Kita T, Narumiya S. Roles of thromboxane A(2) and prostacyclin in the development of atherosclerosis in apoE-deficient mice. *J Clin Invest* 2004;**114**:784–94.

56. Cheng Y, Austin SC, Rocca B, Koller BH, Coffman TM, Grosser T, Lawson JA, FitzGerald GA. Role of prostacyclin in the cardiovascular response to thromboxane A2. *Science* 2002;**296**:539–41.

57. Rocca III B, Loeb AL, Strauss JF, Vezza R, Habib A, Li H, FitzGerald GA. Directed vascular expression of the thromboxane A2 receptor results in intrauterine growth retardation. *Nat Med* 2000;**6**:219–21.

58. Catella-Lawson F, McAdam B, Morrison BW, Kapoor S, Kujubu D, Antes L, Lasseter KC, Quan H, Gertz BJ, FitzGerald GA. Effects of specific inhibition of cyclooxygenase-2 on sodium balance, hemodynamics, and vasoactive eicosanoids. *J Pharmacol Exp Ther* 1999;**289**:735–41.

59. Moncada S, Vane JR. Prostacyclin: homeostatic regulator or biological curiosity? *Clin Sci (Colch)* 1981;**61**:369–72.

60. Zucker TP, Bonisch D, Hasse A, Grosser T, Weber AA, Schror K. Tolerance development to antimitogenic actions of prostacyclin but not of prostaglandin E1 in coronary artery smooth muscle cells. *Eur J Pharmacol* 1998;**345**:213–20.

61. Fitzgerald DJ, Roy L, Catella F, FitzGerald GA. Platelet activation in unstable coronary disease. *N Engl J Med* 1986;**315**:983–9.

62. Fitzgerald DJ, Doran J, Jackson E, FitzGerald GA. Coronary vascular occlusion mediated via thromboxane A2-prostaglandin endoperoxide receptor activation *in vivo*. *J Clin Invest* 1986;**77**:496–502.

63. McLaughlin VV, Genthner DE, Panella MM, Rich S. Reduction in pulmonary vascular resistance with long-term epoprostenol (prostacyclin) therapy in primary pulmonary hypertension. *N Engl J Med* 1998;**338**:273–7.

64. Numaguchi Y, Naruse K, Harada M, Osanai H, Mokuno S, Murase K, Matsui H, Toki Y, Ito T, Okumura K, Hayakawa T. Prostacyclin synthase gene transfer accelerates reendothelialization and inhibits neointimal formation in rat carotid arteries after balloon injury. *Arterioscler Thromb Vasc Biol* 1999;**19**:727–33.

65. Nakayama T, Soma M, Rahmutula D, Tobe H, Sato M, Uwabo J, Aoi N, Kosuge K, Kunimoto M, Kanmatsuse K, Kokubun S. Association study between a novel single nucleotide polymorphism of the promoter region of the prostacyclin synthase gene and essential hypertension. *Hypertens Res* 2002;**25**:65–8.

66. Nakayama T, Soma M, Rehemudula D, Takahashi Y, Tobe H, Satoh M, Uwabo J, Kunimoto M, Kanmatsuse K. Association of 5′ upstream promoter region of prostacyclin synthase gene variant with cerebral infarction. *Am J Hypertens* 2000;**13**:1263–7.

67. Qi Z, Chuan-Ming H, Langenbach RI, Breyer RM, Redha R, Morrow JD, Breyer MD. Opposite effects of cyclooxygenases 1 and 2 activity on the pressor response to angiotensin II. *J Clin Invest.* 2002;**110**:61–9.

68. Hennan JK, Huang J, Barrett TD, Driscoll EM, Willens DE, Park AM, Crofford LJ, Lucchesi BR. Effects of selective cyclooxygenase-2 inhibition on vascular responses and thrombosis in canine coronary arteries. *Circulation* 2001;**104**:820–5.

69. Adderley SR, Fitzgerald DJ. Oxidative damage of cardiomyocytes is limited by extracellular regulated kinases 1/2-mediated induction of cyclooxygenase-2. *J Biol Chem* 1999;**274**:5038–46.

70. Namba T, Oida H, Sugimoto Y, Kakizuka A, Negishi M, Ichikawa A, Narumiya S. cDNA cloning of a mouse prostacyclin receptor. Multiple signaling pathways and expression in thymic medulla. *J Biol Chem* 1994;**269**:9986–9992.

71. Hebert RL, Regnier L, Peterson LN. Rabbit cortical collecting ducts express a novel prostacyclin receptor. *Am J Physiol* 1995;**268**:F145–F154.

72. Takechi H, Matsumura K, Watanabe Y, Kato K, Noyori R, Suzuki M. A novel subtype of the prostacyclin receptor expressed in the central nervous system. *J Biol Chem* 1996;**271**:5901–6.

73. Forman BM, Chen J, Evans RM. Hypolipidemic drugs, polyunsaturated fatty acids, and eicosanoids are ligands for peroxisome proliferator-activated receptors alpha and delta. *Proc Natl Acad Sci USA* 1997;**94**:4312–7.

74. Peters JM, Lee SS, Li W, Ward JM, Gavrilova O, Everett C, Reitman ML, Hudson LD, Gonzalez FJ. Growth, adipose, brain, and skin alterations resulting from targeted disruption of the mouse peroxisome proliferator-activated receptor beta(delta). *Mol Cell Biol* 2000;**20**:5119–28.

75. Arehart E, Stitham J, Asselbergs FW, Douville K, MacKenzie T, Fetalvero KM, Gleim S, Kasza Z, Rao Y, Martel L, Segel S, Robb J, Kaplan A, Simons M, Powell RJ, Moore JH, Rimm EB, Martin KA, Hwa J. Acceleration of cardiovascular disease by a dysfunctional prostacyclin receptor mutation: potential implications for cyclooxygenase-2 inhibition. *Circ Res* 2008;**102**:986–93.

76. Murata T, Ushikubi F, Matsuoka T, Hirata M, Yamasaki A, Sugimoto Y, Ichikawa A, Aze Y, Tanaka T, Yoshida N, Ueno A, Oh-ishi S, Narumiya S. Altered pain perception and inflammatory response in mice lacking prostacyclin receptor. *Nature* 1997;**388**:678–82.

77. Cheng Y, Wang M, Yu Y, Lawson J, Funk CD, Fitzgerald GA. Cyclooxygenases, microsomal prostaglandin E synthase-1, and cardiovascular function. *J Clin Invest* 2006;**116**:1391–9.

78. Hoshikawa Y, Voelkel NF, Gesell TL, Moore MD, Morris KG, Alger LA, Narumiya S, Geraci MW. Prostacyclin receptor-dependent modulation of pulmonary vascular remodeling. *Am J Respir Crit Care Med* 2001;**164**:314–8.

79. Francois H, Athirakul K, Howell D, Dash R, Mao L, Kim HS, Rockman HA, Fitzgerald GA, Koller BH, Coffman TM. Prostacyclin protects against elevated blood pressure and cardiac fibrosis. *Cell Metab* 2005;**2**:201–7.

80. Lovgren AK, Jania LA, Hartney JM, Parsons KK, Audoly LP, Fitzgerald GA, Tilley SL, Koller BH. COX-2-derived prostacyclin protects against bleomycin-induced pulmonary fibrosis. *Am J Physiol Lung Cell Mol Physiol* 2006;**291**:L144–L156.

81. Zhou W, Hashimoto K, Goleniewska K, O'Neal JF, Ji S, Blackwell TS, Fitzgerald GA, Egan KM, Geraci MW, Peebles Jr RS. Prostaglandin I2 analogs inhibit proinflammatory cytokine production and T cell stimulatory function of dendritic cells. *J Immunol* 2007;**178**:702–10.

82. Egan K, Austin S, Smyth EM, FitzGerald GA. Accelerated atherogenesis in protacyclin receptor deficient mice. *Circulation* 2000;**102**:234.

83. Seeger H, Mueck AO, Lippert TH. Effect of estradiol metabolites on prostacyclin synthesis in human endothelial cell cultures. *Life Sci* 1999;**65**:L167–L170.

84. Yokoyama C, Yabuki T, Shimonishi M, Wada M, Hatae T, Ohkawara S, Takeda J, Kinoshita T, Okabe M, Tanabe T. Prostacyclin-deficient mice develop ischemic renal disorders, including nephrosclerosis and renal infarction. *Circulation* 2002;**106**:2397–403.

85. Sladek II K, Sheller JR, FitzGerald GA, Morrow JD, Roberts LJ. Formation of PGD2 after allergen inhalation in atopic asthmatics. *Adv Prostaglandin Thromboxane Leukot Res* 1991:433–6.

86. Roberts II LJ, Sweetman BJ, Lewis RA, Austen KF, Oates JA. Increased production of prostaglandin D2 in patients with systemic mastocytosis. *N Engl J Med* 1980;**303**:1400–4.

87. Song WL, Wang M, Ricciotti E, Fries S, Yu Y, Grosser T, Reilly M, Lawson JA, FitzGerald GA. Tetranor PGDM, an abundant urinary metabolite reflects biosynthesis of prostaglandin D2 in mice and humans. *J Biol Chem* 2008;**283**:1179–88.

88. Heavey DJ, Lumley P, Barrow SE, Murphy MB, Humphrey PP, Dollery CT. Effects of intravenous infusions of prostaglandin D2 in man. *Prostaglandins* 1984;**28**:755–67.

89. Fujitani Y, Kanaoka Y, Aritake K, Uodome N, Okazaki-Hatake K, Urade Y. Pronounced eosinophilic lung inflammation and Th2 cytokine release in human lipocalin-type prostaglandin D synthase transgenic mice. *J Immunol* 2002;**168**:443–9.

90. Inoue T, Takayanagi K, Morooka S, Uehara Y, Oda H, Seiki K, Nakajima H, Urade Y. Serum prostaglandin D synthase level after coronary angioplasty may predict occurrence of restenosis. *Thromb Haemost* 2001;**85**:165–70.

91. Noguchi E, Shibasaki M, Kamioka M, Yokouchi Y, Yamakawa-Kobayashi K, Hamaguchi H, Matsui A, Arinami T. New polymorphisms of haematopoietic prostaglandin D synthase and human prostanoid DP receptor genes. *Clin Exp Allergy* 2002;**32**:93–6.

92. Hayaishi O. Molecular mechanisms of sleep–wake regulation: a role of prostaglandin D2. *Philos Trans R Soc Lond B Biol Sci* 2000;**355**:275–80.

93. Eguchi N, Minami T, Shirafuji N, Kanaoka Y, Tanaka T, Nagata A, Yoshida N, Urade Y, Ito S, Hayaishi O. Lack of tactile pain (allodynia) in lipocalin-type prostaglandin D synthase-deficient mice. *Proc Natl Acad Sci USA* 1999;**96**:726–30.

94. Monneret G, Gravel S, Diamond M, Rokach J, Powell WS. Prostaglandin D2 is a potent chemoattractant for human eosinophils that acts via a novel DP receptor. *Blood* 2001;**98**:1942–8.

95. Pettipher R, Hansel TT, Armer R. Antagonism of the prostaglandin D2 receptors DP1 and CRTH2 as an approach to treat allergic diseases. *Nat Rev Drug Discov* 2007;**6**:313–25.

96. Michimata T, Tsuda H, Sakai M, Fujimura M, Nagata K, Nakamura M, Saito S. Accumulation of CRTH2-positive T-helper 2 and T-cytotoxic 2 cells at implantation sites of human decidua in a prostaglandin D(2)-mediated manner. *Mol Hum Reprod* 2002;**8**:181–7.

97. Monneret G, Li H, Vasilescu J, Rokach J, Powell WS. 15-Deoxy-delta 12,14-prostaglandins D2 and J2 are potent activators of human eosinophils. *J Immunol* 2002;**168**:3563–9.

98. Harris SG, Padilla J, Koumas L, Ray D, Phipps RP. Prostaglandins as modulators of immunity. *Trends Immunol* 2002;**23**:144–50.

99. Forman BM, Tontonoz P, Chen J, Brun RP, Spiegelman BM, Evans RM. 15–Deoxy-delta 12, 14-prostaglandin J2 is a ligand for the adipocyte determination factor PPAR gamma. *Cell* 1995;**83**:803–12.

100. Matsuoka T, Hirata M, Tanaka H, Takahashi Y, Murata T, Kabashima K, Sugimoto Y, Kobayashi T, Ushikubi F, Aze Y, Eguchi N, Urade Y, Yoshida N, Kimura K, Mizoguchi A, Honda Y, Nagai H, Narumiya S. Prostaglandin D2 as a mediator of allergic asthma. *Science* 2000;**287**:2013–7.

101. Mizoguchi A, Eguchi N, Kimura K, Kiyohara Y, Qu WM, Huang ZL, Mochizuki T, Lazarus M, Kobayashi T, Kaneko T, Narumiya S, Urade Y, Hayaishi O. Dominant localization of prostaglandin D receptors on arachnoid trabecular cells in mouse basal forebrain and their involvement in the regulation of non-rapid eye movement sleep. *Proc Natl Acad Sci USA* 2001;**98**:11,674–9.

102. Satoh T, Moroi R, Aritake K, Urade Y, Kanai Y, Sumi K, Yokozeki H, Hirai H, Nagata K, Hara T, Utsuyama M, Hirokawa K, Sugamura K, Nishioka K, Nakamura M. Prostaglandin D2 plays an essential role in chronic allergic inflammation of the skin via CRTH2 receptor. *J Immunol* 2006;**177**:2621–9.

103. Chevalier E, Stock J, Fisher T, Dupont M, Fric M, Fargeau H, Leport M, Soler S, Fabien S, Pruniaux MP, Fink M, Bertrand CP, McNeish J, Li B. Cutting edge: chemoattractant receptor-homologous molecule expressed on Th2 cells plays a restricting role on IL-5 production and eosinophil recruitment. *J Immunol* 2005;**175**:2056–60.

104. Cheng K, Wu TJ, Wu KK, Sturino C, Metters K, Gottesdiener K, Wright SD, Wang Z, O'Neill G, Lai E, Waters MG. Antagonism of the prostaglandin D2 receptor 1 suppresses nicotinic acid-induced vasodilation in mice and humans. *Proc Natl Acad Sci USA* 2006;**103**:6682–7.

105. Dubois RN, Abramson SB, Crofford L, Gupta RA, Simon LS, Van De Putte LB, Lipsky PE. Cyclooxygenase in biology and disease. *Faseb J* 1998;**12**:1063–73.

106. Lydford SJ, McKechnie KC, Dougall IG. Pharmacological studies on prostanoid receptors in the rabbit isolated saphenous vein: a comparison with the rabbit isolated ear artery. *Br J Pharmacol* 1996;**117**:13–20.

107. Coleman RA, Kennedy I, Humphrey PPA, Bunce K, Lumley P. Prostanoids and their receptors. In: Hansch C, editor. *Comprehensive medicinal chemistry*. Oxford: Pergamon Press; 1990. p. 643–714.

108. Samuelsson B, Morgenstern R, Jakobsson PJ. Membrane prostaglandin E synthase-1: a novel therapeutic target. *Pharmacol Rev* 2007;**59**:207–24.

109. Trebino CE, Stock JL, Gibbons CP, Naiman BM, Wachtmann TS, Umland JP, Pandher K, Lapointe JM, Saha S, Roach ML, Carter D, Thomas NA, Durtschi BA, McNeish JD, Hambor JE, Jakobsson PJ, Carty TJ, Perez JR, Audoly LP. Impaired inflammatory and pain responses in mice lacking an inducible prostaglandin E synthase. *Proc Natl Acad Sci USA* 2003;**100**:9044–9.

110. Wang M, Zukas AM, Hui Y, Ricciotti E, Pure E, FitzGerald GA. Deletion of microsomal prostaglandin E synthase-1 augments prostacyclin and retards atherogenesis. *Proc Natl Acad Sci USA* 2006;**103**:14,507–12.

111. Wang M, Lee E, Song W, Ricciotti E, Rader DJ, Lawson JA, Pure E, Fitzgerald GA. Microsomal prostaglandin E synthase-1 deletion suppresses oxidative stress and angiotensin II-induced abdominal aortic aneurysm formation. *Circulation* 2008. in press.

112. Yang T, Singh I, Pham H, Sun D, Smart A, Schnermann JB, Briggs JP. Regulation of cyclooxygenase expression in the kidney by dietary salt intake. *Am J Physiol* 1998;**274**:F481–F489.

113. Breyer MD, Breyer RM. Prostaglandin E receptors and the kidney. *Am J Physiol Renal Physiol* 2000;**279**:F12–F23.

114. Jia Z, Zhang A, Zhang H, Dong Z, Yang T. Deletion of microsomal prostaglandin E synthase-1 increases sensitivity to salt loading and angiotensin II infusion. *Circ Res* 2006;**99**:1243–51.

115. Sugimoto Y, Narumiya S. Prostaglandin E receptors. *J Biol Chem* 2007;**282**:11,613–7.

116. Fabre JE, Nguyen M, Athirakul K, Coggins K, McNeish JD, Austin S, Parise LK, FitzGerald GA, Coffman TM, Koller BH. Activation of the murine EP3 receptor for PGE2 inhibits cAMP production and promotes platelet aggregation. *J Clin Invest* 2001;**107**:603–10.

117. Kennedy CR, Zhang Y, Brandon S, Guan Y, Coffee K, Funk CD, Magnuson MA, Oates JA, Breyer MD, Breyer RM. Salt-sensitive hypertension and reduced fertility in mice lacking the prostaglandin EP2 receptor. *Nat Med* 1999;**5**:217–20.

118. Jensen BL, Schmid C, Kurtz A. Prostaglandins stimulate renin secretion and renin mRNA in mouse renal juxtaglomerular cells. *Am J Physiol* 1996;**271**:F659–F669.

119. Inscho EW, Carmines PK, Navar LG. Prostaglandin influences on afferent arteriolar responses to vasoconstrictor agonists. *Am J Physiol* 1990;**259**:F157–F163.

120. Araki H, Ukawa H, Sugawa Y, Yagi K, Suzuki K, Takeuchi K. The roles of prostaglandin E receptor subtypes in the cytoprotective action of prostaglandin E2 in rat stomach. *Aliment Pharmacol Ther* 2000;**14**(Suppl. 1):116–24.

121. Kunikata T, Tanaka A, Miyazawa T, Kato S, Takeuchi K. 16,16-Dimethyl prostaglandin E2 inhibits indomethacin-induced small intestinal lesions through EP3 and EP4 receptors. *Dig Dis Sci* 2002;**47**:894–904.

122. Asboth G, Phaneuf S, Europe-Finner GN, Toth M, Bernal AL. Prostaglandin E2 activates phospholipase C and elevates intracellular calcium in cultured myometrial cells: involvement of EP1 and EP3 receptor subtypes. *Endocrinology* 1996;**137**:2572–9.

123. Tani K, Naganawa A, Ishida A, Egashira H, Sagawa K, Harada H, Ogawa M, Maruyama T, Ohuchida S, Nakai H, Kondo K, Toda M. Design and synthesis of a highly selective EP2-receptor agonist. *Bioorg Med Chem Letts* 2001;**11**:2025–8.

124. Sonoshita M, Takaku K, Sasaki N, Sugimoto Y, Ushikubi F, Narumiya S, Oshima M, Taketo MM. Acceleration of intestinal polyposis through prostaglandin receptor EP2 in Apc(Delta 716) knockout mice. *Nat Med* 2001;**7**:1048–51.

125. Pai R, Soreghan B, Szabo IL, Pavelka M, Baatar D, Tarnawski AS. Prostaglandin E2 transactivates EGF receptor: a novel mechanism for promoting colon cancer growth and gastrointestinal hypertrophy. *Nat Med* 2002;**8**:289–293.

126. Nataraj C, Thomas DW, Tilley SL, Nguyen MT, Mannon R, Koller BH, Coffman TM. Receptors for prostaglandin E(2) that regulate cellular immune responses in the mouse. *J Clin Invest* 2001;**108**:1229–1235.

127. Ek M, Engblom D, Saha S, Blomqvist A, Jakobsson PJ, Ericsson-Dahlstrand A. Inflammatory response: pathway across the blood-brain barrier. *Nature* 2001;**410**:430–1.

128. Yoshida Y, Matsumura H, Nakajima T, Mandai M, Urakami T, Kuroda K, Yoneda H. Prostaglandin E (EP) receptor subtypes and sleep: promotion by EP4 and inhibition by EP1/EP2. *Neuroreport* 2000;**11**:2127–31.

129. Ushikubi F, Segi E, Sugimoto Y, Murata T, Matsuoka T, Kobayashi T, Hizaki H, Tuboi K, Katsuyama M, Ichikawa A, Tanaka T, Yoshida N, Narumiya S. Impaired febrile response in mice lacking the prostaglandin E receptor subtype EP3. *Nature* 1998;**395**:281–4.

130. Stock JL, Shinjo K, Burkhardt J, Roach M, Taniguchi K, Ishikawa T, Kim HS, Flannery PJ, Coffman TM, McNeish JD, Audoly LP. The prostaglandin E2 EP1 receptor mediates pain perception and regulates blood pressure. *J Clin Invest* 2001;**107**:325–31.

131. Fleming EF, Athirakul K, Oliverio MI, Key M, Goulet J, Koller BH, Coffman TM. Urinary concentrating function in mice lacking EP3 receptors for prostaglandin E2. *Am J Physiol* 1998;**275**:F955–F961.

132. Nguyen M, Camenisch T, Snouwaert JN, Hicks E, Coffman TM, Anderson PA, Malouf NN, Koller BH. The prostaglandin receptor EP4 triggers remodelling of the cardiovascular system at birth. *Nature* 1997;**390**:78–81.

133. Fulton AM, Ma X, Kundu N. Targeting prostaglandin E EP receptors to inhibit metastasis. *Cancer Res* 2006;**66**:9794–7.

134. Horton EW, Poyser NL. Uterine luteolytic hormone: a physiological role for prostaglandin F2alpha. *Physiol Rev* 1976;**56**:595–651.

135. Dong YJ, Jones RL, Wilson NH. Prostaglandin E receptor subtypes in smooth muscle: agonist activities of stable prostacyclin analogues. *Br J Pharmacol* 1986;**87**:97–107.

136. Barnard JW, Ward RA, Taylor AE. Evaluation of prostaglandin F2 alpha and prostacyclin interactions in the isolated perfused rat lung. *J Appl Physiol* 1992;**72**:2469–74.

137. Suzuki-Yamamoto T, Nishizawa M, Fukui M, Okuda-Ashitaka E, Nakajima T, Ito S, Watanabe K. cDNA cloning, expression and characterization of human prostaglandin F synthase. *FEBS Letts* 1999;**462**:335–40.

138. Endo K, Fukui M, Mishima M, Watanabe K. Metabolism of vitamin A affected by prostaglandin F synthase in contractile interstitial cells of bovine lung. *Biochem Biophys Res Commun* 2001;**287**:956–61.

139. Vanegas H, Schaible HG. Prostaglandins and cyclooxygenases [correction of cycloxygenases] in the spinal cord. *Prog Neurobiol* 2001;**64**:327–63.

140. Adams JW, Migita DS, Yu MK, Young R, Hellickson MS, Castro-Vargas FE, Domingo JD, Lee PH, Bui JS, Henderson SA. Prostaglandin F2 alpha stimulates hypertrophic growth of cultured neonatal rat ventricular myocytes. *J Biol Chem* 1996;**271**:1179–86.

141. Pierce KL, Fujino H, Srinivasan D, Regan JW. Activation of FP prostanoid receptor isoforms leads to Rho-mediated changes in cell morphology and in the cell cytoskeleton. *J Biol Chem* 1999;**274**:35,944–9.

142. Fujino H, Srinivasan D, Pierce KL, Regan JW. Differential regulation of prostaglandin F(2alpha) receptor isoforms by protein kinase C. *Mol Pharmacol* 2000;**57**:353–8.

143. Fujino H, Regan JW. FP prostanoid receptor activation of a T-cell factor/beta -catenin signaling pathway. *J Biol Chem* 2001;**276**:12,489–92.

144. Kunapuli P, Lawson JA, Rokach J, FitzGerald GA. Functional characterization of the ocular prostaglandin f2alpha (PGF2alpha) receptor. Activation by the isoprostane, 12-iso-PGF2alpha. *J Biol Chem* 1997;**272**:27,147–54.

145. Hesketh TR, Moore JP, Morris JD, Taylor MV, Rogers J, Smith GA, Metcalfe JC. A common sequence of calcium and pH signals in the mitogenic stimulation of eukaryotic cells. *Nature* 1985;**313**:481–4.

146. Kunapuli P, Lawson JA, Rokach JA, Meinkoth JL, FitzGerald GA. Prostaglandin F2alpha (PGF2alpha) and the isoprostane, 8, 12-iso-isoprostane F2alpha-III, induce cardiomyocyte hypertrophy. Differential activation of downstream signaling pathways. *J Biol Chem* 1998;**273**:22,442–52.

147. Casimir DA, Miller CW, Ntambi JM. Preadipocyte differentiation blocked by prostaglandin stimulation of prostanoid FP2 receptor in murine 3T3-L1 cells. *Differentiation* 1996;**60**:203–10.

148. Sugimoto Y, Yamasaki A, Segi E, Tsuboi K, Aze Y, Nishimura T, Oida H, Yoshida N, Tanaka T, Katsuyama M, Hasumoto K, Murata T, Hirata M, Ushikubi F, Negishi M, Ichikawa A, Narumiya S. Failure of parturition in mice lacking the prostaglandin F receptor. *Science* 1997;**277**:681–3.

149. Tsuboi K, Sugimoto Y, Iwane A, Yamamoto K, Yamamoto S, Ichikawa A. Uterine expression of prostaglandin H2 synthase in late pregnancy and during parturition in prostaglandin F receptor-deficient mice. *Endocrinology* 2000;**141**:315–24.

Chapter 35

Leukotriene Mediators

Jesper Z. Haeggström and Anders Wetterholm

Department of Medical Biochemistry and Biophysics, Division of Chemistry II, Karolinska Institutet, Stockholm, Sweden

INTRODUCTION

The leukotrienes (LT) constitute a group of bioactive lipids derived from the metabolism of polyunsaturated fatty acids, e.g., arachidonic acid [1]. In two consecutive reactions, arachidonic acid is transformed into an unstable epoxide compound, LTA4. This intermediate is either hydrolyzed into the dihydroxy acid LTB4 or conjugated with glutathione to form LTC4. The latter compound together with its metabolites LTD4 and LTE4 are referred to as the cysteinyl-containing leukotrienes (cys-LTs).

Leukotrienes possess a wide range of biological activities elicited via specific, G-protein-coupled, cell surface receptors [2]. LTB4 is a very potent chemoattractant for neutrophils, and recruits inflammatory cells to the site of injury. This compound also induces chemokinesis and increases leukocyte adhesion to the endothelial cells of the vessel wall. The cys-LTs are potent constrictors of smooth muscle, particularly in the airways, leading to bronchoconstriction. In the microcirculation, the cys-LTs constrict arterioles and increase the permeability of the post-capillary venules, which results in extravasation of plasma. Due to their potent biological activities, leukotrienes are considered to be chemical mediators in a number of inflammatory and allergic disorders, such as rheumatoid arthritis, inflammatory bowel disease, and bronchial asthma [3].

5-LIPOXYGENASE

Five-lipoxygenase (5-LO) catalyzes the first two steps in leukotriene biosynthesis [4] (Figure 35.1). Free arachidonic acid is oxygenated into the hydroperoxide 5-HPETE, which is subsequently dehydrated to yield the unstable epoxide intermediate LTA4. The enzyme, which predominantly is found in bone marrow derived cells, is stimulated by Ca^{2+} and ATP. Furthermore, it contains one atom of non-heme iron that is involved in catalysis [5]. Mutagenetic analysis has demonstrated that His-372, His-550, and the C-terminal isoleucin Ile-673 are iron ligands [6]. The gene encoding human 5-LO, as well as the promoter, has been characterized, and some important features are listed in Table 35.1, together with data for the other key enzymes in the leukotriene cascade.

The only crystal structure of a mammalian lipoxygenase that has been determined is rabbit 15-LO [7]. This enzyme contains an N-terminal β-barrel domain, a structure also found in the C-terminal domain of lipases. The role of this domain for lipoxygenases is presently unclear, but for 5-LO it has been shown to bind Ca^{2+}, which stimulates enzyme activity and presumably facilitates its association of 5-LO with membranes during catalysis (see following section) [8]. 5-LO is also a substrate for p38 kinase-dependent MAPKAP kinases *in vitro*, suggesting that phosphorylation may be one additional factor, which determines 5-LO translocation and enzyme activity [9].

5-Lipoxygenase Activating Protein (FLAP) and Cellular Leukotriene Biosynthesis

Cellular 5-LO activity is dependent on a small membrane protein, 5-lipoxygenase activating protein (FLAP), which presumably presents or transfers arachidonic acid to 5-LO [10].

Early studies showed that upon cell stimulation leading to an increase in Ca^{2+}, 5-LO is activated and translocates to a membrane compartment [11]. Of particular interest was the discovery that FLAP is localized to the nuclear envelope of neutrophils and that 5-LO, upon cell activation, translocates to the same compartment [12] (Figure 35.2). Further analysis revealed that 5-LO can also be present in the nucleus of resting cells associated with the

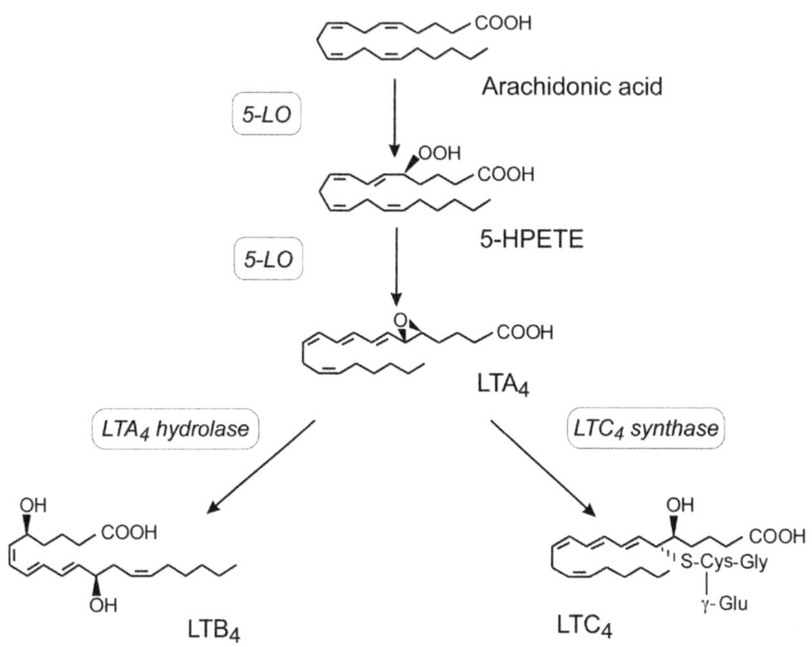

FIGURE 35.1 Enzymes and intermediates in the leukotriene cascade.

TABLE 35.1 Properties of enzymes and receptors in leukotriene biosynthesis and action[a]

	Protein size (no. of amino acids) [b]	Prosthetic group[c]	Gene size (kb)	Exon no.	Putative cis-elements of promoter regions	Chromosomal location	Gene-deficient mice
5-Lipoxygenase	673	Fe	> 82	14	Sp1, AP-2, NF-κB	10	+
FLAP	160	—	> 31	5	TATA, AP-2, GRE	13	+
LTA$_4$ hydrolase	610	Zn	> 35	19	XRE, AP-2	12	+
LTC$_4$ synthase	149	—	2.5	5	Sp1, AP-1, AP-2	5	+
BLT$_1$	351	—	5.5	3	Sp1, CpG site, NFκB AP-1	14	+
BLT$_2$[d]	357	—	ND[c]	1[c]		14	−
CysLT$_1$	336	—	ND	ND		X	+
CysLT$_2$	345	—	ND	ND		13	−

[a]Data refer to human proteins. ND, not determined.
[b]Initial methionine excluded.
[c]1 mol metal per mol protein.
[d]The ORF of BLT$_2$ is included in the promoter of the BLT$_1$ gene.

nuclear euchromatin, a site from which it translocates to the nuclear envelope. In addition, 5-LO has been shown to associate with growth factor receptor-binding protein 2 (Grb2), an "adaptor" protein for tyrosine kinase-mediated cell signaling, through Src homology 3 (SH3) domain interactions [13]. It is interesting that inhibitors of tyrosine kinase activity, a determinant of SH3 interactions, also inhibited the catalytic activity of 5-LO and its translocation during cellular activation. In addition, an internal bipartite nuclear localization sequence, spanning Arg-638–Lys-655, has been shown to be necessary for the redistribution of 5-LO to the nuclear compartment [14, 15]. Moreover, recent data indicate that also the N-terminal β-barrel domain in 5-LO plays a role in this process [16].

Not only 5-LO and FLAP are associated with the cell nucleus and nuclear membrane. Thus, LTC4 synthase (see section entitled Leukotriene C4 synthase) resides in this compartment, and recent data suggest that the enzyme is located on the outer nuclear membrane and peripheral endoplasmic reticulum [17]. It is interesting that the soluble LTA$_4$ hydrolase (see section entitled Leukotriene A$_4$ hydrolase) was also reported to reside in the nucleus of rat basophilic leukemia cells and rat alveolar macrophages [18].

Together, these findings imply that leukotriene biosynthesis is carried out by a complex of enzymes assembled at the nuclear membrane (*cf.* Figure 35.2). This conclusion in turn suggests that these enzymes and/or their products may have additional intracellular and intranuclear functions, perhaps related to signal transduction or gene regulation. In line with this notion, it has been reported that LTB4 is a natural ligand to the nuclear orphan receptor PPARα, suggesting that LTB4 may have intranuclear functions [19]. It was also reported that 5-LO can interact with several

cellular proteins, including coactosine-like protein (CLP) and transforming growth factor type β-receptor-I-associated protein (TRAP-1) [20]. In addition, 5-LO interacts with a human homologue of the protein "Dicer," a member of the RNase III family of nucleases, which is implicated in the RNA interference mechanism of gene regulation [20, 21].

Leukotriene A$_4$ Hydrolase

Leukotriene A$_4$ hydrolase catalyzes the final step in the biosynthesis of the proinflammatory compound LTB$_4$ (Figure 35.1). In contrast to 5-LO, LTA$_4$ hydrolase is widely distributed and has been detected in almost all mammalian cells, organs, and tissues examined. The enzyme has been purified from several mammalian sources, and cDNAs encoding the human, mouse, rat, and guinea-pig enzymes have been cloned and sequenced [22].

Sequence comparison with certain zinc metalloenzymes revealed the presence of a zinc-binding motif (HEXXH–X$_{18}$–E) in LTA4 hydrolase [23]. Accordingly, LTA4 hydrolase was found to contain a catalytic zinc. The three proposed zinc-binding ligands, His-295, His-299, and Glu-318, were verified by mutagenetic analysis. Furthermore, the enzyme was found to exhibit a chloride-activated peptidase activity. Based on its zinc signature, sequence homology, and aminopeptidase activity, LTA4 hydrolase has been classified as a member of the M1 family of the MA clan of metallopeptidases [24].

Identification of Catalytically Important Amino Acid Residues and Crystal Structure of LTA$_4$ Hydrolase

In addition to the zinc-binding ligands, several amino acid residues of catalytic importance have been identified by site-directed mutagenesis. Thus, mutagenetic replacements of Glu-296 in LTA$_4$ hydrolase abrogated only the peptidase activity, a finding that suggests a direct catalytic role for Glu-296 in the peptidase reaction, possibly as a general base [25]. Furthermore, sequence comparisons and mutational analysis have demonstrated that Tyr-383 plays an important role in the peptidase reaction of LTA$_4$ hydrolase, presumably as a proton donor [26].

Typically, LTA$_4$ hydrolase undergoes "suicide" inactivation with a concomitant covalent modification of the enzyme by its substrate LTA4 [27]. Mutational analysis has demonstrated that Tyr-378 is a major structural determinant for suicide inactivation [28]. Mutated proteins, carrying a Gln or Phe residue in position 378, were neither inactivated nor covalently modified by LTA$_4$.

Recently, the X-ray crystal structure of LTA$_4$ hydrolase in complex with the competitive inhibitor bestatin was determined [29]. The protein molecule is folded into an N-terminal, a catalytic, and a C-terminal domain, packed in a

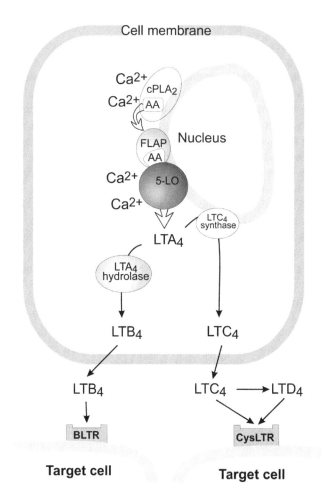

FIGURE 35.2 Leukotriene biosynthesis at the nuclear membrane of an activated leukocyte.

flat triangular arrangement. Although the three domains pack closely and make contact with each other, a deep cleft is created between them. At the bottom of the interdomain cleft, the zinc site is located. As predicted from previous work, the metal is bound to the three amino acid ligands, His-295, His-299, and Glu-318. In the vicinity of the prosthetic zinc, the catalytic residues Glu-296 and Tyr-383 are located at positions that are commensurate with their proposed roles as general base and proton donor in the aminopeptidase reaction.

Close to the catalytic zinc, a glutamic acid residue (Glu-271), belonging to a conserved GXMEN motif in the M1 family of zinc peptidases, was identified [29]. By mutational analysis and crystallography it was shown that Glu-271 is necessary for both catalytic activities of LTA_4 hydrolase [30]. Presumably, the carboxylate of the glutamic acid residue participates in the opening of the epoxide moiety of LTA_4 and formation of a carbocation intermediate. In the peptidase reaction, the role of Glu-271 may be to serve as an N-terminal recognition site and to stabilize the transition state during turnover of peptide substrates.

The crystal structure, in combination with site-directed mutagenesis studies, also suggested that Asp-375 is a critical determinant for the introduction of the 12R-hydroxyl group of LTB4 [31].

Leukotriene C_4 Synthase

Leukotriene C_4 synthase catalyzes the committed step in the biosynthesis of cys-LTs through conjugation of LTA_4 with glutathione. The enzyme is a membrane-bound homodimer with a subunit molecular mass of 18 kDa [32]. LTC_4 synthase has been cloned and sequenced [33, 34]. Two consensus sequences for protein kinase C phosphorylation were found, and subsequent studies have shown that phosphorylation reduces the LTC_4 synthase activity [35]. Sequence comparisons of LTC_4 synthase and FLAP demonstrated a surprising 31 percent identity between the two proteins. In addition, recent work has identified two microsomal GSH transferases (MGST2 and MGST3) that possess LTC_4 synthase activity and exhibit a high degree of similarity to both LTC_4 synthase and FLAP [36, 37].

Leukotriene Receptors

For LTB4, two types of surface receptors are known (BLT_1 and BLT_2). The BLT_1-receptor has been cloned and characterized as a 43-kDa, G-protein-coupled receptor with seven transmembrane-spanning domains (7TM) [38]. The BLT_1 receptor is only expressed in inflammatory cells [39] and shows a high degree of specificity for LTB_4 with a K_d of 0.15–1 nM [38, 40].

A second G-protein-coupled 7TM receptor for LTB_4, BLT_2, has recently been identified [40–42]. This receptor is homologous to the BLT_1 receptor but has a higher K_d value for LTB_4 (23 nM) [43]. In contrast to the BLT_1 receptor, BLT_2 is ubiquitously expressed in various tissues.

The cys-LTs are recognized by at least two receptor types ($CysLT_1$ and $CysLT_2$), both of which have been cloned and characterized as G-protein-coupled 7TM receptors [44–48]. The $CysLT_1$ receptor mRNA is found in, for example, spleen, peripheral blood leukocytes, lung tissue, smooth muscle cells, and tissue macrophages [45, 47]. The preferred ligands for the $CysLT_1$ receptor are LTD_4 followed by LTC_4 and LTE4, in decreasing order of potency.

The $CysLT_2$ receptor contains 345 amino acids with approximately 40 percent sequence identity to the CysLT1 receptor [44, 46, 48]. This receptor binds LTC_4 and LTD_4 equally well, whereas LTE_4 shows low affinity to the receptor. Studies on the tissue distribution of the $CysLT_2$ receptor show high levels of mRNA in heart, brain, peripheral blood leukocytes, spleen, placenta, and lymph nodes, whereas only small amounts are found in the lung. The functional role(s) of the $CysLT_2$ receptor is presently unclear, but its wide tissue distribution suggests many possibilities, including regulation of brain and/or cardiac functions.

Gene Targeting of Enzymes and Receptors in the Leukotriene Cascade

The roles of the key enzymes and two of the receptors (BLT_1 and $CysLT_1$) in the leukotriene cascade have been studied by gene targeting. 5-LO-deficient mice are more resistant to lethal effects of PAF-induced shock, and also show a marked reduction in the ear inflammatory response to exogenous arachidonic acid [49]. Furthermore, 5-LO-null mice are more susceptible to infections with Klebsiella pnemoniae [50], exhibit a reduced airway reactivity in response to methacholine, and have lower levels of serum immunoglobulins [51].

FLAP deficient mice, like the 5-LO$^{-/-}$ mice, showed a blunted response to topical arachidonic acid, had increased resistance to PAF induced shock, and responded with less edema in zymosan-induced peritonitis [52]. Furthermore, the severity of collagen-induced arthritis was substantially reduced in FLAP$^{-/-}$ mice, thereby indicating a role for leukotrienes in this model of inflammation [53].

LTA_4 hydrolase$^{-/-}$ mice are resistant to the lethal effects of systemic shock induced by PAF, thus identifying LTB_4 as a key mediator of this reaction [54]. In zymosan A-induced peritonitis, LTB_4 modulates only the cellular component of the response, whereas the LTC_4 synthase$^{-/-}$ mice displayed a reduced plasma protein extravasation in this type of inflammation [55]. Furthermore, the LTC_4 synthase$^{-/-}$ mice were less prone to develop passive cutaneous anaphylaxis. Recently, the role of LTC_4 in plasma protein extravasation following zymosan A-induced peritonitis and IgE-mediated passive cutaneous anaphylaxis was confirmed in mice lacking the $CysLT_1$ receptor gene [56].

Finally, the role of the BLT1 receptor has also been studied by targeted gene disruption [57, 58]. The receptor was necessary to elicit the physiological effects of LTB$_4$ (e.g., chemotaxis, calcium mobilization, and adhesion to endothelium) and important for the recruitment of leukocytes in an *in vivo* model of peritonitis. As also observed in mice lacking 5-LO, FLAP, or LTA$_4$ hydrolase, BLT$_1^{-/-}$ mice were protected from the lethal effects of PAF-induced anaphylaxis.

ACKNOWLEDGEMENTS

This work was supported by the Swedish Medical Research Council (O3X-10350), the European Union (QLG1-CT-2001-01521), the Vårdal Foundation, the Swedish Foundation for Strategic Research, and Konung Gustav V:s 80-Årsfond.

REFERENCES

1. Funk CD. Prostaglandins and leukotrienes: advances in eicosanoid biology. *Science* 2001;**294**:1871–5.
2. Izumi T, Yokomizu T, Obinata H, Ogasawara H, Shimizu T. Leukotriene receptors: classification, gene expression, and signal transduction. *J Biochem* 2002;**132**:1–6.
3. Lewis RA, Austen KF, Soberman RJ. Leukotrienes and other products of the 5-lipoxygenase pathway. *New Engl J Med* 1990; **323**:645–55.
4. Rouzer CA, Matsumoto T, Samuelsson B. Single protein from human leukocytes possesses 5-lipoxygenase and leukotriene A$_4$ synthase activities. *Proc Natl Acad Sci USA* 1986;**83**:857–61.
5. Percival MD. Human 5-lipoxygenase contains an essential iron. *J Biol Chem* 1991;**266**:10,058–10,061.
6. Rådmark O. Mutagenesis studies of mammalian lipoxygenases. In: Serhan CN, Ward PA, editors. *Molecular and Cellular Basis of Inflammation*. Totowa, NJ: Humana Press Inc; 2000. p. 93–108.
7. Gillmor SA, Villaseñor A, Fletterick R, Sigal E, Browner M. The structure of mammalian 15-lipoxygenase reveals similarity to the lipases and the determinants of substrate specificity. *Nat Struct Biol* 1997;**4**:1003–9.
8. Hammarberg T, Provost P, Persson B, Radmark O. The N-terminal domain of 5-lipoxygenase binds calcium and mediates calcium stimulation of enzyme activity. *J Biol Chem* 2000;**275**:38,787–38,793.
9. Werz O, Klemm J, Samuelsson B, Radmark O. 5-lipoxygenase is phosphorylated by p38 kinase-dependent MAPKAP kinases. *Proc Natl Acad Sci USA* 2000;**97**:5261–6.
10. Ford-Hutchinson AW, Gresser M, Young RN. 5-Lipoxygenase. *Annu Rev Biochem* 1994;**63**:383–417.
11. Rouzer CA, Kargman S. Translocation of 5-lipoxygenase to the membrane in human leukocytes challenged with ionophore A23187. *J Biol Chem* 1988;**263**:10,980–10,988.
12. Peters-Golden M, Brock TG. Intracellular compartmentalization of leukotriene synthesis: unexpected nuclear secrets. *FEBS Letts* 2001;**487**:323–6.
13. Lepley RA, Fitzpatrick FA. 5-Lipoxygenase contains a functional Src homology 3-binding motif that interacts with the Src homology 3 domain of Grb2 and cytoskeletal proteins. *J Biol Chem* 1994;**269**:24,163–24,168.
14. Lepley RA, Fitzpatrick FA. 5-Lipoxygenase compartmentalization in granulocytic cells is modulated by an internal bipartite nuclear localizing sequence and nuclear factor kappa B complex formation. *Arch Biochem Biophys* 1998;**356**:71–6.
15. Healy AM, Peters-Golden M, Yao JP, Brock TG. Identification of a bipartite nuclear localization sequence necessary for nuclear import of 5-lipoxygenase. *J Biol Chem* 1999;**274**:29,812–29,818.
16. Chen XS, Funk CD. The N-terminal "beta-barrel" domain of 5-lipoxygenase is essential for nuclear membrane translocation. *J Biol Chem* 2001;**276**:811–18.
17. Christmas P, Weber BM, McKee M, Brown D, Soberman RJ. Membrane localization and topology of leukotriene C$_4$ synthase. *J Biol Chem* 2002;**277**:28,902–28,908.
18. Brock TG, Maydanski E, McNish RW, Peters-Golden M. Co-localization of leukotriene A$_4$ hydrolase with 5-lipoxygenase in nuclei of alveolar macrophages and rat basophilic leukemia cells but not neutrophils. *J Biol Chem* 2001;**276**:35,071–35,077.
19. Devchand PR, Keller H, Peters JM, Vazquez M, Gonzalez FJ, Wahli W. The PPARα-leukotriene B$_4$ pathway to inflammation control. *Nature* 1996;**384**:39–43.
20. Provost P, Samuelsson B, Rådmark O. Interaction of 5-lipoxygenase with cellular proteins. *Proc Natl Acad Sci USA* 1999;**96**:1881–5.
21. Bernstein E, Caudy AA, Hammond SM, Hannon GJ. Role for a bidentate ribonuclease in the initiation step of RNA interference. *Nature* 2001;**409**:363–6.
22. Wetterholm A, Blomster M, Haeggström JZ. Leukotriene A$_4$ hydrolase: a key enzyme in the biosynthesis of leukotriene B$_4$. In: Folco G, Samuelsson B, Maclouf J, Velo GP, editors. *Eicosanoids: From Biotechnology to Therapeutic Applications*. New York: Plenum Press; 1996. p. 1–12.
23. Haeggstrom JZ. Structure, function, and regulation of leukotriene A$_4$ hydrolase. *Am J Resp Crit Care Med* 2000;**161**:S25–31.
24. Barret AJ, Rawlings ND, Woessner JF. Family M1 of membrane alanyl aminopeptidase. In: Barret AJ, Rawlings ND, Woessner JF, editors. *Handbook of Proteolytic Enzymes*. London, San Diego: Academic Press; 1998. p. 994–6.
25. Wetterholm A, Medina JF, Rådmark O, Shapiro R, Haeggström JZ, Vallee BL, Samuelsson B. Leukotriene A$_4$ hydrolase: abrogation of the peptidase activity by mutation of glutamic acid–296. *Proc Natl Acad Sci USA* 1992;**89**:9141–5.
26. Blomster M, Wetterholm A, Mueller MJ, Haeggström JZ. Evidence for a catalytic role of tyrosine 383 in the peptidase reaction of leukotriene A$_4$ hydrolase. *Eur J Biochem* 1995;**231**:528–34.
27. Orning L, Gierse J, Duffin K, Bild G, Krivi G, Fitzpatrick FA. Mechanism-based inactivation of leukotriene A$_4$ hydrolase/ aminopeptidase by leukotriene A$_4$, Mass spectrometric and kinetic characterization. *J Biol Chem* 1992;**267**:22,733–22,739.
28. Mueller MJ, Blomster M, Oppermann UCT, Jörnvall H, Samuelsson B, Haeggstrom JZ. Leukotriene A$_4$ hydrolase-protection from mechanism-based inactivation by mutation of tyrosine-378. *Proc Natl Acad Sci USA* 1996;**93**:5931–5.
29. Thunnissen MGM, Nordlund P, Haeggström JZ. Crystal structure of human leukotriene A$_4$ hydrolase, a bifunctional enzyme in inflammation. *Nat Struct Biol* 2001;**8**:131–5.
30. Rudberg PC, Tholander F, Thunnissen MMGM, Haeggström JZ. Leukotriene A$_4$ hydrolase/aminopeptidase: glutamate 271 is a catalytic residue with specific roles in two distinct enzyme mechanisms. *J Biol Chem* 2002;**277**:1398–404.
31. Rudberg PC, Tholander F, Thunnissen MMGM, Samuelsson B, Haeggström JZ. Leukotriene A$_4$ hydrolase: selective abrogation of

leukotriene formation by mutation of aspartic acid 375. *Proc Natl Acad Sci USA* 2002;**99**:4215–20.

32. Nicholson DW, Ali A, Vaillancourt JP, Calaycay JR, Mumford RA, Zamboni RJ, Ford-Hutchinson AW. Purification to homogeneity and the N-terminal sequence of human leukotriene C_4 synthase: a homodimeric glutathione S-transferase composed of 18-kDa subunits. *Proc Natl Acad Sci USA* 1993;**90**:2015–19.

33. Lam BK, Penrose JF, Freeman GJ, Austen KF. Expression cloning of a cDNA for human leukotriene C_4 synthase, an integral membrane protein conjugating reduced glutathione to leukotriene A4. *Proc Natl Acad Sci USA* 1994;**91**:7663–7.

34. Welsch DJ, Creely DP, Hauser SD, Mathis KJ, Krivi GG, Isakson PC. Molecular cloning end expression of human leukotriene C4 synthase. *Proc Natl Acad Sci USA* 1994;**91**:9745–9.

35. Ali A, Ford-Hutchinson AW, Nicholson DW. Activation of protein kinase C down-regulates leukotriene C_4 synthase activity and attenuates cysteinyl leukotriene production in an eosinophilic substrain of HL-60 cells. *J Immunol* 1994;**153**:776–88.

36. Jakobsson PJ, Mancini JA, Ford-Hutchinson AW. Identification and characterization of a novel human microsomal glutathione S-transferase with leukotriene C_4 synthase activity and significant sequence identity to 5-lipoxygenase-activating protein and leukotriene C_4 synthase. *J Biol Chem* 1996;**271**:22,203–22,210.

37. Jakobsson PJ, Mancini JA, Riendeau D, Ford-Hutchinson AW. Identification and characterization of a novel microsomal enzyme with glutathione-dependent transferase and peroxidase activities. *J Biol Chem* 1997;**272**:22,934–22,939.

38. Yokomizo T, Izumi T, Chang K, Takuwa Y, Shimizu T. A G-protein-coupled receptor for leukotriene B_4 that mediates chemotaxis. *Nature* 1997;**387**:620–4.

39. Kato K, Yokomizo T, Izumi T, Shimizu T. Cell-specific transcriptional regulation of human leukotriene B(4) receptor gene. *J Exp Med* 2000;**192**:413–20.

40. Yokomizo T, Kato K, Terawaki K, Izumi T, Shimizu T. A second leukotriene B_4 receptor, BLT2: a new therapeutic target in inflammation and immunological disorders. *J Exp Med* 2000;**192**:421–31.

41. Kamohara M, Takasaki J, Matsumoto M, Saito T, Ohishi T, Ishii H, Furuichi K. Molecular cloning and characterization of another leukotriene B_4 receptor. *J Biol Chem* 2000;**275**:27,000–27,004.

42. Tryselius Y, Nilsson NE, Kotarsky K, Olde B, Owman C. Cloning and characterization of cDNA encoding a novel human leukotriene B_4 receptor. *Biochem Biophys Res Commun* 2000;**274**:377–82.

43. Yokomizo T, Kato K, Hagiya H, Izumi T, Shimizu T. Hydroxyeicosanoids bind to and activate the low affinity leukotriene B_4 receptor BLT_2. *J Biol Chem* 2001;**276**:12,454–12,459.

44. Heise CE, O'Dowd BF, Figueroa DJ, Sawyer N, Nguyen T, Im D-S, Stocco R, Bellefeuille JN, Abramovitz M, Cheng JR, Williams R, Zeng Z, Liu Q, Ma L, Clements MK, Coulombe N, Liu Y, Austin CP, George SR, O'Neill GP, Metters KM, Lynch KP, Evans JF. Characterization of the human cysteinyl leukotriene 2 (CysLT2) receptor. *J Biol Chem* 2000;**275**:30,531–30,536.

45. Lynch Jr KR, O'Neill GP, Liu Q, Im DS, Sawyer N, Metters KM, Coulombe N, Abramovitz M, Figueroa DJ, Zeng Z, Connolly BM, Bai C, Austin CP, Chateauneuf A, Stocco R, Greig GM, Kargman S, Hooks SB, Hosfield E, Williams DL, Ford-Hutchinson AW, Caskey CT,

Evans JF. Characterization of the human cysteinyl leukotriene $CysLT_1$ receptor. *Nature* 1999;**399**:789–93.

46. Nothacker HP, Wang ZW, Zhu YH, Reinscheid RK, Lin SHS, Civelli O. Molecular cloning and characterization of a second human cysteinyl leukotriene receptor: discovery of a subtype selective agonist. *Mol Pharmacol* 2000;**58**:1601–8.

47. Sarau HM, Ames RS, Chambers J, Ellis C, Elshourbagy N, Foley JJ, Schmidt DB, Muccitelli RM, Jenkins O, Murdock PR, Herrity NC, Halsey W, Sathe G, Muir AI, Nuthulaganti P, Dytko GM, Buckley PT, Wilson S, Bergsma DJ, Hay DW. Identification, molecular cloning, expression, and characterization of a cysteinyl leukotriene receptor. *Mol Pharmacol* 1999;**56**:657–63.

48. Takasaki J, Kamohara M, Matsumoto M, Saito T, Sugimoto T, Ohishi T, Ishii H, Ota T, Nishikawa T, Kawai Y, Masuho Y, Isogai T, Suzuki Y, Sugano S, Furuichi K. The molecular characterization and tissue distribution of the human cysteinyl leukotriene $CysLT_2$ receptor. *Biochem Biophys Res Commun* 2000;**274**:316–22.

49. Chen XS, Sheller JR, Johnson EN, Funk CD. Role of leukotrienes revealed by targeted disruption of the 5-lipoxygenase gene. *Nature* 1994;**372**:179–82.

50. Bailie MB, Standiford TJ, Laichalk LL, Coffey MJ, Strieter R, Peters-Golden M. Leukotriene-deficient mice manifest enhanced lethality from Klebsiella pneumonia in association with decreased alveolar macrophage phagocytic and bactericidal activities. *J Immunol* 1996;**157**:5221–4.

51. Irvin CG, Tu YP, Sheller JR, Funk CD. 5-lipoxygenase products are necessary for ovalbumin-induced airway responsiveness in mice. *Am J Physiol* 1997;**16**:L1053–8.

52. Byrum RS, Goulet JL, Griffiths RJ, Koller BH. Role of the 5-lipoxygenase-activating protein (FLAP) in murine acute inflammatory responses. *J Exp Med* 1997;**185**:1065–75.

53. Griffiths RJ, Smith MA, Roach ML, Stock JL, Stam EJ, Milici AJ, Scampoli DN, Eskra JD, Byrum RS, Koller BH, McNeish JD. Collagen-induced arthritis is reduced in 5-lipoxygenase-activating protein-deficient mice. *J Exp Med* 1997;**185**:1123–9.

54. Byrum RS, Goulet JL, Snouwaert JN, Griffiths RJ, Koller BH. Determination of the contribution of cysteinyl leukotrienes and leukotriene B4 in acute inflammatory responses using 5-lipoxygenase- and leukotriene A_4 hydrolase-deficient mice. *J Immunol* 1999;**163**:6810–19.

55. Kanaoka Y, Maekawa A, Penrose JF, Austen KF, Lam BK. Attenuated zymosan-induced peritoneal vascular permeability and IgE dependent passive cutaneous anaphylaxis in mice lacking leukotriene C_4 synthase. *J Biol Chem* 2001;**276**:22,608–22,613.

56. Maekawa A, Austen KF, Kanaoka Y. Targeted gene disruption reveals the role of cysteinyl leukotriene 1 receptor in the enhanced vascular permeability of mice undergoing acute inflammatory responses. *J Biol Chem* 2002;**277**:20,820–20,824.

57. Haribabu B, Verghese MW, Steeber DA, Sellars DD, Bock CB, Snyderman R. Targeted disruption of the leukotriene B_4 receptor in mice reveals its role in inflammation and platelet-activating factor-induced anaphylaxis. *J Exp Med* 2000;**192**:433–8.

58. Tager AM, Dufour JH, Goodarzi K, Bercury SD, von Andrian UH, Luster AD. BLTR mediates leukotriene B_4-induced chemotaxis and adhesion and plays a dominant role in eosinophil accumulation in a murine model of peritonitis. *J Exp Med* 2000;**192**:439–46.

Chapter 36

Sphingosine-1-Phosphate Receptors: An Update

Michael Maceyka and Sarah Spiegel

Department of Biochemistry and Molecular Biology, Virginia Commonwealth University School of Medicine, Richmond, Virginia

INTRODUCTION

Sphingosine-1-phosphate (S1P) has emerged over the past decade as a potent lipid mediator involved in a variety of signaling pathways that play major roles in diverse physiological processes, including angiogenesis, immune cell trafficking, and development. S1P is formed by phosphorylation of sphingosine catalyzed by one of two sphingosine kinase isoenzymes (SphK1 and SphK2) [1, 2]. Some evidence suggests that S1P has intracellular actions that promote cell growth and inhibit apoptosis, though to date no intracellular proteins have been found to be direct targets of S1P. However, almost a decade ago it was shown that S1P binds to a cell surface G-protein-coupled receptor, now known as $S1P_1$ [3]. Currently there are five generally accepted S1P receptors, $S1P_{1-5}$, which are evolutionarily related and bind S1P (and dihydro-S1P) with high affinity [4]. Other receptors have been proposed to be S1P receptors; for example, gpr6 [5, 6]. However, given the absence of direct binding assays and the ability of S1P to transactivate other receptors (see below), convincing evidence for existence of other members of the S1PR family is still lacking.

The S1PRs have been shown to regulate a diverse set of signaling pathways due to their differential coupling to G proteins [4]. Initial work characterized the roles of these receptors in cell physiology, including motility, cytoskeletal rearrangements, NO production, and calcium mobilization [7]. Recently, great strides in determining the physiological and pathophysiological functions of the S1PRs have been made following studies of knockout mice, tissue-specific $S1P_1$ knockouts (the general $S1P_1$ knockout is embryonic lethal [8]), and development of increasingly selective agonists and antagonists. A summary of commercially available $S1P_{1-3}$-specific reagents is provided in Table 36.1.

Agonist/antagonist development continues, and some success has recently been reported in the development of $S1P_4$- and $S1P_5$-specific agonists [9].

TRANSACTIVATION OF S1PRS

A recurring theme in S1PR signaling is "inside-out" signaling loops that function in an autocrine and/or paracrine fashion (reviewed in [10]). A diverse number of stimuli, including hormones, immunoglobulin receptor ligation, growth factors, and cytokines, stimulate SphK activity, typically that of SphK1, leading to the formation of S1P, which then activates cell surface S1PRs present on the same or neighboring cells. Indeed, many of the downstream effects of these stimuli require the formation of S1P and the activation of one or more S1PRs. For example, estradiol increases ERK1/2 phosphorylation in MCF-7 breast cancer cells, though the mechanism requires at least two autocrine signaling loops [11]. Estradiol activates SphK1, leading to the production of S1P and subsequent activation of $S1P_3$. $S1P_3$ then activates a matrix metalloproteinase that releases EGF, which in turn stimulates its cell surface receptor, finally leading to ERK1/2 activation. This signaling loop is also intriguing, because SphK1 is cytosolic, and the S1P that is produced signals through the exoplasmic side of cell surface receptors. While it has been suggested that SphK1 proteins themselves may be secreted and produce S1P extracellulary [12], it has been convincingly shown that ABC transporters mediate S1P secretion from some cells [13–15]. It should be noted that such inside-out signaling loops are not limited to S1PRs, as it has been shown that chemotactic signals for neutrophils induce secretion of ATP that locally activates cell surface nucleotide receptors to coordinate directed cell migration [16].

TABLE 36.1 S1PR agonists and antagonists

Compound	Receptor targeted	Type	Commercial source*	Reference
SEW2871	$S1P_1$	Agonist	Cayman (USA), Biomol (USA), Sigma (USA), Calbiochem (USA)	[25]
VPC 23019	$S1P_1/S1P_3$	Antagonist	Avanti (USA)	[73]
VPC 24191	$S1P_1/S1P_3$	Agonist	Avanti (USA)	[74]
VPC 23153	$S1P_1/S1P_3$	Agonist	Avanti (USA)	[75]
FTY720 (fingolimod) pro-drug)	$S1P_1$, $S1P_{3-5}$	Agonist	Cayman (USA), Biozol Diagnostica (Germany)	[18]
KRP-203 (pro-drug)	$S1P_1$, $S1P_{4-5}$	Agonist	Kyorin Pharmaceutical Co. (Japan)	[48]
JTE013	$S1P_2$	Antagonist	Tocris (USA), Cayman (USA)	[76]
W146	$S1P_1$	Antagonist	Avanti (USA)	[59]

Not a comprehensive list.

Many recent studies of functions of S1PRs have been carried out in mice using genetic knockouts and agonists/antagonists. Thus, we will focus this review on two of the well-studied systems in which S1PRs have been shown to play a key role: the immune system and the cardiovascular system.

S1PRS AND THE IMMUNE SYSTEM

B and T Cell Trafficking

Several rounds of screening for immunosuppressive drugs led to the discovery that FTY720, also called fingolimod, induced lymphopenia and suppressed allograft rejection (reviewed in [17]). Further work revealed that FTY720 prevented B and T cells from leaving lymphoid organs [18]. Clues to the molecular mechanisms of FTY720 came from the observation that it resembles sphingosine structurally. Indeed, like sphingosine, FTY720 was found to be phosphorylated more efficiently by SphK2 than SphK1 on its 1-OH [19]. In agreement, expression of SphK2 but not SphK1 was required for FTY720-induced lymphopenia *in vivo* [20–22]. These results indicated that FTY720 was a pro-drug and FTY720-phosphate was the inducer of lymphopenia, which was confirmed by the demonstration that FTY720-phosphate is a potent agonist of all S1PRs except $S1P_2$ [18,23]. The S1P receptor involved in mediating lymphopenia was definitively determined to be $S1P_1$ by several experimental approaches. First, when mature $S1P_1$-deficient T and B cells were used to reconstitute the immune systems of wild-type mice, these cells were unable to exit secondary lymphoid organs [24]. Second, the $S1P_1$ agonist SEW2871 was also demonstrated to prevent egress of lymphocytes from second-

ary lymphoid organs [25, 26]. That $S1P_1$ agonist treatment would have the same phenotype as $S1P_1$ knockouts suggested that $S1P_1$ may rapidly be internalized in lymphocytes. Indeed, FTY720 treatment decreased $S1P_1$ expression in lymphoid tissue [24, 27], and induced rapid internalization and degradation of the receptor in many different cell types [24, 28–30]. Moreover, thymocytes internalize all of their surface $S1P_1$ receptors within 20 minutes of treatment with 1-nM S1P [31]. This raises an intriguing issue, as the plasma concentrations of S1P are typically in the 100-nM range [32], well above the K_d values of the S1PRs. Indeed, circulating T cells do not express $S1P_1$ on their surface [31]. Conversely, the concentration of S1P is lower in secondary lymphoid organs, and T cells do express $S1P_1$ at the cell surface in these organs. These results led to the hypothesis that surface $S1P_1$ receptors are required on lymphocytes to move up the S1P concentration gradient from lymph nodes into the circulation [31, 33]. It is thought that FTY720-phosphate, which is more stable than S1P *in vivo* [18], suppresses $S1P_1$ surface expression by chronic activation and thus prevents lymphocyte egress. Another hypothesis, which is not mutually exclusive, is that, as $S1P_1$ also plays a role in vascular permeability (see below); its expression on endothelial cells may serve as a gatekeeper for lymphocytes to be able to move into the circulation [34, 35].

Natural Killer (NK) Cell Trafficking

Intriguingly, a similar requirement for an S1PR in immune cell trafficking has been observed for NK cells, though in this case the receptor involved is $S1P_5$ [36]. NK cells are formed in the bone marrow, and mature as they cycle through lymphoid and non-lymphoid organs. In $S1P_5$ knockout mice, the maturation of NK cells was unchanged; however, the steady-state distribution of NK cells was shifted, with fewer cells in the blood, spleen, and lung, and more NK cells in the bone marrow and lymph nodes [36]. Moreover, $S1P_5$ was required for NK cell recruitment to inflamed tissue. When comparing wild-type to $S1P_5$ knockouts, NK cells depended entirely on $S1P_5$ for migration towards S1P. These data suggest that NK cells require $S1P_5$ to respond to S1P gradients to move from sites of maturation or to sites of inflammation. Of note, FTY720 did not affect the numbers of circulating NK cells, and neither did it dramatically reduce the ability of NK cells to migrate towards S1P *ex vivo*. This suggests that, unlike its effect on $S1P_1$, FTY720 may not chronically downregulate $S1P_5$.

S1PRs and Other Immune Cells

The discovery of FTY720 has biased the literature towards the importance of $S1P_1$ in immune regulation. However, other S1PRs also have key roles in the immune system, such as the requirement for $S1P_2$ in mast cell degranulation and motility [37–39]. Although the specific receptors involved have not been identified, S1P release accompanies many inflammatory responses, such as eosinophil chemotaxis, neutrophil invasion, dendritic cell homing, and Th2 stimulation. A discussion of these is beyond the scope of this review, and readers are directed to recent excellent reviews elsewhere [40–42].

S1PRS AND THE CARDIOVASCULAR SYSTEM

Heart Rate

FTY720 holds enormous promise as a pro-drug to treat autoimmune disorders, and indeed has entered Phase III clinical trials for treatment of multiple sclerosis [43, 44]. However, one of the drawbacks to FTY720 treatment is that it can induce bradycardia [45]. Mechanistically, S1P treatment activates the inwardly rectifying atrial potassium channel, IKACh, in cardiac myocytes [46]. Initial investigations linked the bradycardia to activation of $S1P_3$. This was due to the observations that intravenous injection of S1P also induced bradycardia; the $S1P_1$-specific agonist SEW2871 induced lymphopenia but not bradycardia; and FTY720 did not induce bradycardia in $S1P_3$ knockout mice [25, 26]. However, KRP-203, a phosphorylatable FTY720 analog that activates $S1P_1$, but not $S1P_3$, does induce bradychardia, albeit requiring higher doses than FTY720 [47, 48]. Moreover, recent results suggest a role for $S1P_1$ in regulating negative ionotropy through both IKACh and inhibition of L-type calcium channels [49]. Differing methodological

details and choice of animal models may account for the differences in results of these experiments.

Ischemia

S1P plays a protective role in ischemia/reperfusion (I/R) injury [50]. It has been shown in isolated hearts that production of reactive oxygen species leads to the degradation of SphK1 and subsequent decrease in S1P [51]. One possible consequence of the degradation of SphK1 is decreased conversion of sphingosine to S1P resulting in increased ceramide levels that can then directly induce cardiomyocyte apoptosis. A second not mutually exclusive possibility is that secreted S1P, which may normally protect cardiomyocytes from apoptosis via stimulation of S1PRs, is decreased due to SphK1 degradation. Indeed, there is recent evidence that indicates that S1PRs can act at several levels *in vivo* to protect the heart from damage following I/R. First, exogenously added S1P reduced I/R-induced cardiomyocyte apoptosis, and this protection was linked to inhibition of $S1P_3$-dependent recruitment of polymorphonuclear leukocytes to the infarct area [52]. Activation of $S1P_3$ receptors also led to increased NO production, as the protective effects of S1P were inhibited by the NO scavenger L-NAME. Although $S1P_3$ may mediate cardioprotection when S1P is given intravenously, another group has reported that neither $S1P_2$ nor $S1P_3$ single knockout mice had larger infarct sizes than wild-type controls [53]. However, the infarct size in $S1P_2$/$S1P_3$ double-knockout animals was increased more than 50 percent [53]. The increased I/R injury was attributed to an inability to activate Akt, as hearts from double-knockouts have lower phospho-Akt levels in response to I/R. This is intriguing, as the double-knockout hearts as well as cardiomyocytes cultured from them still expressed $S1P_1$, which has been shown in other systems to couple to activation of Akt [54, 55]. These data are at odds with another report, in which it was demonstrated that the $S1P_1$ agonist SEW2871 protected cardiomyocytes from apoptosis, an effect that was blocked by the $S1P_1$ antagonist VPC 23019 or by phosphatidylinositol 3-kinase inhibitors [56]. The apparent discrepancy between these studies further highlights the need for tissue specific $S1P_1$ knockouts. Still, these reports highlight the importance of S1PR agonism in protection from I/R injury.

Endothelial Barrier Function

Endothelial cells provide a selectively permeable barrier for ions, proteins, and cells between the blood and tissues. Current evidence suggests that $S1P_1$, $S1P_2$, and $S1P_3$ act in concert to maintain the endothelial barrier. First, $S1P_1$ promotes the formation of the endothelial barrier by targeting vascular-endothelial cadherin to adherens junctions between endothelial cells and stimulation of cortical actin formation [57, 58]. Moreover, $S1P_1$ inhibition promotes capillary leakage *in vivo* [59], while in various animal models, S1P$_1$ agonism reduces endothelial leakage [60,61]. S1P$_1$-induced enhancement of barrier function has been linked to activation of Rac and α-actinin-induced cortical actin rearrangement [58]. Conversely, activation of Rho typically enhances stress fiber formation and disrupts cortical actin, thus decreasing endothelial barrier function. Consistent with this notion, S1P$_2$ and S1P$_3$, both of which can couple to the G12/13-Rho pathway, decrease endothelial barrier function [58,62]. Indeed, inhibition of S1P$_2$ with JTE013 inhibited vascular permeability in a perfused rat lung model [63]. Together, these results suggest that S1P$_1$ barrier enhancement is balanced by S1P$_2$ and S1P$_3$ to maintain normal vascular permeability. The role of S1PRs in endothelial barrier function likely extends to epithelial barriers as well, as it has been shown that S1P$_3$ is required for S1P-induced loss of tight junctions and of lung barrier integrity *in vivo* [64].

Previously, it was suggested that FTY720-phosphate downregulates S1P$_1$ on lymphocytes, thus preventing them from moving up the S1P gradient from the lymph node into the blood. However, if S1P$_1$ promotes endothelial barrier function, might FTY720-phosphate downregulation of endothelial S1P$_1$ receptors "close the gate," preventing lymphocyte egress? In an elegant series of experiments, T cell migration was assessed in lymph-node explants. S1P$_1$ agonism with SEW2871 reduced T cell egress within 5 minutes, suggesting that S1P$_1$ activation closed a normally open stromal gate [59]. However, systemic antagonism of S1P$_1$ did not increase the number of circulating lymphocytes [59], indicating that this gate is usually open. Thus, S1P$_1$ may function in both the lymphocyte and the endothelium to control lymphocyte trafficking.

Angiogenesis

S1P plays a significant role in regulation of vasculogenesis and angiogenesis. S1P has long been known to induce endothelial tube formation *in vitro* and angiogenesis *in vivo* [57, 65–67]. Moreover, an S1P-specific monoclonal antibody effectively inhibits tumor angiogenesis in xenograft models [68]. Consistent with these results, S1P$_1$ knockout mice die *in utero* due to massive hemorrhaging as pericytes, the vascular smooth muscle cells, fail to migrate and ensheath embryonic arteries [8]. Moreover, FTY720 also inhibits S1P-induced angiogenesis, likely through downregulating S1P$_1$ [67]. Intriguingly, the requirement for S1P$_1$ in angiogenesis seems mainly to lie with the endothelial cells themselves, as endothelial cell-targeted S1P$_1$ knockouts have a similar type of embryonic hemorrhagic lethality as the S1P$_1$ global knockouts [69].

Several lines of evidence suggest that other S1PRs are also involved in angiogenesis. First, both S1P$_1$ and S1P$_3$ were found to be required for neovascular growth into Matrigel plugs implanted in mice [57]. Second, the S1P$_2$/S1P$_3$ double knockout only leads to partial embryonic lethality, and the S1P$_1$ knockout embryos die earlier if S1P$_2$ and/or S1P$_3$ is also knocked out [70]. Third, S1P$_2$ knockout mice are protected from pathological angiogenesis of the retina induced by ischemia, although it is not yet clear if endothelial S1P$_2$ is involved [71]. Together, these results suggest that S1P$_{1-3}$ play partially overlapping roles in angiogenesis at different stages of development, in different tissues, and in different pathophysiological states.

CONCLUDING REMARKS

Although we have only reviewed recent evidence for roles of the S1PRs in the immune system and the cardiovascular system, these receptors are also important in other physiological systems and in several other pathophysiological conditions. In the nervous system, for example, S1PRs promote differentiation, neurogenesis, neurite growth, and astrocyte migration, among other functions (reviewed in [72]). Moreover, very little is still known of the physiological functions of S1P$_4$ and S1P$_5$, the "newest" S1PRs. As new generations of more specific S1PR agonists and antagonists are developed, together with cell- and tissue-specific knockout and "knock-in" models, more roles for S1PRs will undoubtedly be uncovered.

ACKNOWLEDGEMENTS

Supported by NIH grant R37GM043880 (Sarah Spiegel). We thank Dr S. Milstien for helpful comments and suggestions.

REFERENCES

1. Maceyka M, Payne SG, Milstien S, Spiegel S. Sphingosine kinase, sphingosine-1-phosphate, and apoptosis. *Biochim Biophys Acta* 2002;**1585**:193–201.

2. Spiegel S, Milstien S. Functions of the multifaceted family of sphingosine kinases and some close relatives. *J Biol Chem* 2007;**282**:2125–9.

3. Lee MJ, Van Brocklyn JR, Thangada S, Liu CH, Hand AR, Menzeleev R, Spiegel S, Hla T. Sphingosine-1-phosphate as a ligand for the G-protein-coupled receptor EDG-1. *Science* 1998;**279**:1552–5.

4. Spiegel S, Milstien S. Functions of a new family of sphingosine-1-phosphate receptors. *Biochim Biophys Acta* 2000;**1484**:107–16.

5. Ignatov A, Lintzel J, Kreienkamp HJ, Schaller HC. Sphingosine-1-phosphate is a high-affinity ligand for the G protein-coupled receptor GPR6 from mouse and induces intracellular Ca^{2+} release by activating the sphingosine-kinase pathway. *Biochem Biophys Res Commun* 2003;**311**:329–36.

6. Lobo MK, Cui Y, Ostlund SB, Balleine BW, Yang XW. Genetic control of instrumental conditioning by striatopallidal neuron-specific S1P receptor Gpr6. *Nat Neurosci* 2007;**10**:1395–7.

7. Spiegel S, Milstien S. Sphingosine-1-phosphate: an enigmatic signalling lipid. *Nat Rev Mol Cell Biol* 2003;**4**:397–407.

8. Liu Y, Wada R, Yamashita T, Mi Y, Deng CX, Hobson JP, Rosenfeldt HM, Nava VE, Chae SS, Lee MJ, Liu CH, Hla T, Spiegel S, Proia RL. Edg-1, the G protein-coupled receptor for sphingosine-1-phosphate, is essential for vascular maturation. *J Clin Invest* 2000;**106**:951–61.

9. Hanessian S, Charron G, Billich A, Guerini D. Constrained azacyclic analogues of the immunomodulatory agent FTY720 as molecular probes for sphingosine 1-phosphate receptors. *Bioorg Med Chem Letts* 2007;**17**:491–4.

10. Alvarez SE, Milstien S, Spiegel S. Autocrine and paracrine roles of sphingosine-1-phosphate. *Trends Endocrinol Metab* 2007;**18**:300–7.

11. Sukocheva O, Wadham C, Holmes A, Albanese N, Verrier E, Feng F, Bernal A, Derian CK, Ullrich A, Vadas MA, Xia P. Estrogen transactivates EGFR via the sphingosine 1-phosphate receptor Edg-3: the role of sphingosine kinase-1. *J Cell Biol* 2006;**173**:301–10.

12. Venkataraman K, Thangada S, Michaud J, Oo ML, Ai Y, Lee YM, Wu M, Parikh NS, Khan F, Proia RL, Hla T. Extracellular export of sphingosine kinase-1a contributes to the vascular S1P gradient. *Biochem J* 2006;**397**:461–71.

13. Mitra P, Oskeritzian CA, Payne SG, Beaven MA, Milstien S, Spiegel S. Role of ABCC1 in export of sphingosine-1-phosphate from mast cells. *Proc Natl Acad Sci USA* 2006;**103**:16,394–16,399.

14. Kobayashi N, Nishi T, Hirata T, Kihara A, Sano T, Igarashi Y, Yamaguchi A. Sphingosine 1-phosphate is released from the cytosol of rat platelets in a carrier-mediated manner. *J Lipid Res* 2006;**47**:614–21.

15. Sato K, Malchinkhuu E, Horiuchi Y, Mogi C, Tomura H, Tosaka M, Yoshimoto Y, Kuwabara A, Okajima F. Critical role of ABCA1 transporter in sphingosine 1-phosphate release from astrocytes. *J Neurochem* 2007;**103**:2610–9.

16. Chen Y, Corriden R, Inoue Y, Yip L, Hashiguchi N, Zinkernagel A, Nizet V, Insel PA, Junger WG. ATP release guides neutrophil chemotaxis via P2Y2 and A3 receptors. *Science* 2006;**314**:1792–5.

17. Schwab SR, Cyster JG. Finding a way out: lymphocyte egress from lymphoid organs. *Nat Immunol* 2007;**8**:1295–301.

18. Mandala S, Hajdu R, Bergstrom J, Quackenbush E, Xie J, Milligan J, Thornton R, Shei GJ, Card D, Keohane C, Rosenbach M, Hale J, Lynch CL, Rupprecht K, Parsons W, Rosen H. Alteration of lymphocyte trafficking by sphingosine-1-phosphate receptor agonists. *Science* 2002;**296**:346–9.

19. Paugh SW, Payne SG, Barbour SE, Milstien S, Spiegel S. The immunosuppressant FTY720 is phosphorylated by sphingosine kinase type 2. *FEBS Letts* 2003;**554**:189–93.

20. Allende ML, Sasaki T, Kawai H, Olivera A, Mi Y, van Echten-Deckert G, Hajdu R, Rosenbach M, Keohane CA, Mandala S, Spiegel S, Proia RL. Mice deficient in sphingosine kinase 1 are rendered lymphopenic by FTY720. *J Biol Chem* 2004;**279**:52,487–52,492.

21. Kharel Y, Lee S, Snyder AH, Sheasley-O'neill S L, Morris MA, Setiady Y, Zhu R, Zigler MA, Burcin TL, Ley K, Tung KS, Engelhard VH, Macdonald TL, Lynch KR. Sphingosine kinase 2 is required for modulation of lymphocyte traffic by FTY720. *J Biol Chem* 2005;**280**:36,865–36,872.

22. Zemann B, Kinzel B, Muller M, Reuschel R, Mechtcheriakova D, Urtz N, Bornancin F, Baumruker T, Billich A. Sphingosine kinase type 2 is essential for lymphopenia induced by the immunomodulatory drug FTY720. *Blood* 2006;**107**:1454–8.

23. Brinkmann V, Davis MD, Heise CE, Albert R, Cottens S, Hof R, Bruns C, Prieschl E, Baumruker T, Hiestand P, Foster CA, Zollinger M, Lynch KR. The immune modulator, FTY720, targets sphingosine 1-phosphate receptors. *J Biol Chem* 2002;**277**:21,453–21,457.

24. Matloubian M, Lo CG, Cinamon G, Lesneski MJ, Xu Y, Brinkmann V, Allende ML, Proia RL, Cyster JG. Lymphocyte egress from thymus and peripheral lymphoid organs is dependent on S1P receptor 1. *Nature* 2004;**427**:355–60.

25. Sanna MG, Liao J, Jo E, Alfonso C, Ahn MY, Peterson MS, Webb B, Lefebvre S, Chun J, Gray N, Rosen H. Sphingosine 1-phosphate (S1P) receptor subtypes S1P1 and S1P3, respectively, regulate lymphocyte recirculation and heart rate. *J Biol Chem* 2004;**279**:13,839–13,848.

26. Forrest M, Sun SY, Hajdu R, Bergstrom J, Card D, Hale J, Keohane CA, Meyers C, Milligan J, Mills S, Nomura N, Rosenbach M, Shei GJ, Singer II, Tian M, West S, White V, Xie J, Rosen H, Proia R, Doherty G, Mandala S. Immune cell regulation and cardiovascular efects of sphingosine 1-phosphate receptoragonists in rodents are mediated via distinct receptor sub-types. *J Pharmacol Exp Ther* 2004;**309**:758–68.

27. Cyster JG. Chemokines, sphingosine-1-phosphate, and cell migration in secondary lymphoid organs. *Annu Rev Immunol* 2005;**23**:127–59.

28. Graler MH, Goetzl EJ. The immunosuppressant FTY720 downregulates sphingosine 1-phosphate G protein-coupled receptors. *FASEB J* 2004;**18**:551–3.

29. Oo ML, Thangada S, Wu MT, Liu CH, Macdonald TL, Lynch KR, Lin CY, Hla T. Immunosuppressive and anti-angiogenic sphingosine 1-phosphate receptor-1 (S1P1) agonists induce ubiquitinylation and proteosomal degradation of the receptor. *J Biol Chem* 2007;**282**(12):9082–9.

30. Gonzalez-Cabrera PJ, Hla T, Rosen H. Mapping pathways downstream of sphingosine 1-phosphate subtype 1 by differential chemical perturbation and proteomics. *J Biol Chem* 2007;**282**:7254–64.

31. Schwab SR, Pereira JP, Matloubian M, Xu Y, Huang Y, Cyster JG. Lymphocyte sequestration through S1P lyase inhibition and disruption of S1P gradients. *Science* 2005;**309**:1735–9.

32. Pyne S, Pyne NJ. Sphingosine 1-phosphate signalling in mammalian cells. *Biochem J* 2000;**349**:385–402.

33. Pappu R, Schwab SR, Cornelissen I, Pereira JP, Regard JB, Xu Y, Camerer E, Zheng YW, Huang Y, Cyster JG, Coughlin SR. Promotion of lymphocyte egress into blood and lymph by distinct sources of sphingosine-1-phosphate. *Science* 2007;**316**:295–8.

34. Rosen H, Goetzl EJ. Sphingosine 1-phosphate and its receptors: an autocrine and paracrine network. *Nat Rev Immunol* 2005;**5**:560–70.

35. Rosen H, Sanna MG, Cahalan SM, Gonzalez-Cabrera PJ. Tipping the gatekeeper: S1P regulation of endothelial barrier function. *Trends Immunol* 2007;**28**:102–7.

36. Walzer T, Chiossone L, Chaix J, Calver A, Carozzo C, Garrigue-Antar L, Jacques Y, Baratin M, Tomasello E, Vivier E. Natural killer cell trafficking in vivo requires a dedicated sphingosine 1-phosphate receptor. *Nat Immunol* 2007;**8**:1337–44.

37. Jolly PS, Bektas M, Olivera A, Gonzalez-Espinosa C, Proia RL, Rivera J, Milstien S, Spiegel S. Transactivation of sphingosine-1-phosphate receptors by FcεRI triggering is required for normal mast cell degranulation and chemotaxis. *J Exp Med* 2004;**199**:959–70.

38. Yokoo E, Yatomi Y, Takafuta T, Osada M, Okamoto Y, Ozaki Y. Sphingosine 1-phosphate inhibits migration of RBL-2H3 cells via S1P2: cross-talk between platelets and mast cells. *J Biochem (Tokyo)* 2004;**135**:673–81.

39. Jolly PS, Bektas M, Watterson KR, Sankala H, Payne SG, Milstien S, Spiegel S. Expression of SphK1 impairs degranulation and motility of RBL-2H3 mast cells by desensitizing S1P receptors. *Blood* 2005;**105**:4736–42.

40. Chun J, Rosen H. Lysophospholipid receptors as potential drug targets in tissue transplantation and autoimmune diseases. *Curr Pharm Des* 2006;**12**:161–71.

41. Goetzl EJ, Wang W, McGiffert C, Liao JJ, Huang MC. Sphingosine 1-phosphate as an intracellular messenger and extracellular mediator in immunity. *Acta Paediatr (Suppl)* 2007;**96**:49–52.

42. Rivera J, Olivera A. Src family kinases and lipid mediators in control of allergic inflammation. *Immunol Rev* 2007;**217**:255–68.

43. Kappos L, Antel J, Comi G, Montalban X, O'Connor P, Polman CH, Haas T, Korn AA, Karlsson G, Radue EW. Oral fingolimod (FTY720) for relapsing multiple sclerosis. *N Engl J Med* 2006;**355**:1124–40.

44. Brinkmann V. Sphingosine 1-phosphate receptors in health and disease: mechanistic insights from gene deletion studies and reverse pharmacology. *Pharmacol Ther* 2007;**115**:84–105.

45. Budde K, Schmouder RL, Brunkhorst R, Nashan B, Lucker PW, Mayer T, Choudhury S, Skerjanec A, Kraus G, Neumayer HH. First human trial of FTY720, a novel immunomodulator, in stable renal transplant patients. *J Am Soc Nephrol* 2002;**13**:1073–83.

46. Koyrakh L, Roman MI, Brinkmann V, Wickman K. The heart rate decrease caused by acute FTY720 administration is mediated by the G protein-gated potassium channel I. *Am J Transplant* 2005;**5**:529–36.

47. Shimizu H, Takahashi M, Kaneko T, Murakami T, Hakamata Y, Kudou S, Kishi T, Fukuchi K, Iwanami S, Kuriyama K, Yasue T, Enosawa S, Matsumoto K, Takeyoshi I, Morishita Y, Kobayashi E. KRP-203, a novel synthetic immunosuppressant, prolongs graft survival and attenuates chronic rejection in rat skin and heart allografts. *Circulation* 2005;**111**:222–9.

48. Fujishiro J, Kudou S, Iwai S, Takahashi M, Hakamata Y, Kinoshita M, Iwanami S, Izawa S, Yasue T, Hashizume K, Murakami T, Kobayashi E. Use of sphingosine-1-phosphate 1 receptor agonist, KRP-203, in combination with a subtherapeutic dose of cyclosporine A for rat renal transplantation. *Transplantation* 2006;**82**:804–12.

49. Landeen LK, Aroonsakool N, Haga JH, Hu BS, Giles WR. Sphingosine-1-phosphate receptor expression in cardiac fibroblasts is modulated by *in vitro* culture conditions. *Am J Physiol Heart Circ Physiol* 2007;**292**:H2698–711.

50. Maceyka M, Milstien S, Spiegel S. Shooting the messenger: oxidative stress regulates sphingosine-1-phosphate. *Circ Res* 2007;**100**:7–9.

51. Pchejetski D, Kunduzova O, Dayon A, Calise D, Seguelas MH, Leducq N, Seif I, Parini A, Cuvillier O. Oxidative stress-dependent sphingosine kinase-1 inhibition mediates monoamine oxidase A-associated cardiac cell apoptosis. *Circ Res* 2007;**100**:41–9.

52. Theilmeier G, Schmidt C, Herrmann J, Keul P, Schafers M, Herrgott I, Mersmann J, Larmann J, Hermann S, Stypmann J, Schober O, Hildebrand R, Schulz R, Heusch G, Haude M, von Wnuck Lipinski K, Herzog C, Schmitz M, Erbel R, Chun J, Levkau B. High-density lipoproteins and their constituent, sphingosine-1-phosphate, directly protect the heart against ischemia/reperfusion injury *in vivo* via the S1P3 lysophospholipid receptor. *Circulation* 2006;**114**:1403–9.

53. Means CK, Xiao CY, Li Z, Zhang T, Omens JH, Ishii I, Chun J, Brown JH. Sphingosine 1-phosphate S1P2 and S1P3 receptor-mediated Akt activation protects against *in vivo* myocardial ischemia-reperfusion injury. *Am J Physiol Heart Circ Physiol* 2007;**292**:H2944–51.

54. Balthasar S, Samulin J, Ahlgren H, Bergelin N, Lundqvist M, Toescu EC, Eggo MC, Tornquist K. Sphingosine 1-phosphate receptor expression profile and regulation of migration in human thyroid cancer cells. *Biochem J* 2006;**398**:547–56.

55. Morales-Ruiz M, Lee MJ, Zollner S, Gratton JP, Scotland R, Shiojima I, Walsh K, Hla T, Sessa WC. Sphingosine 1-phosphate activates Akt, nitric oxide production, and chemotaxis through a Gi protein/phosphoinositide 3-kinase pathway in endothelial cells. *J Biol Chem* 2001;**276**:19,672–19,677.

56. Zhang J, Honbo N, Goetzl EJ, Chatterjee K, Karliner JS, Gray MO. Signals from type 1 sphingosine 1-phosphate receptors enhance adult mouse cardiac myocyte survival during hypoxia. *Am J Physiol Heart Circ Physiol* 2007;**293**:H3150–8.

57. Lee MJ, Thangada S, Claffey KP, Ancellin N, Liu CH, Kluk M, Volpi M, Sha'afi RI, Hla T. Vascular endothelial cell adherens junction assembly and morphogenesis induced by sphingosine-1-phosphate. *Cell* 1999;**99**:301–12.

58. Singleton PA, Dudek SM, Chiang ET, Garcia JG. Regulation of sphingosine 1-phosphate-induced endothelial cytoskeletal rearrangement and barrier enhancement by S1P1 receptor, PI3 kinase, Tiam1/Rac1, and alpha-actinin. *FASEB J* 2005;**19**:1646–56.

59. Sanna MG, Wang SK, Gonzalez-Cabrera PJ, Don A, Marsolais D, Matheu MP, Wei SH, Parker I, Jo E, Cheng WC, Cahalan MD, Wong CH, Rosen H. Enhancement of capillary leakage and restoration of lymphocyte egress by a chiral S1P(1) antagonist *in vivo*. *Nat Chem Biol* 2006;**2**:434–41.

60. Sanchez T, Estrada-Hernandez T, Paik JH, Wu MT, Venkataraman K, Brinkmann V, Claffey K, Hla T. Phosphorylation and action of the immunomodulator FTY720 inhibits VEGF-induced vascular permeability. *J Biol Chem* 2003;**278**:47,281–47,290.

61. Peng X, Hassoun PM, Sammani S, McVerry BJ, Burne MJ, Rabb H, Pearse D, Tuder RM, Garcia JG. Protective effects of sphingosine 1-phosphate in murine endotoxin-induced inflammatory lung injury. *Am J Resp Crit Care Med* 2004;**169**:1245–51.

62. Singleton PA, Dudek SM, Ma SF, Garcia JG. Transactivation of sphingosine 1-phosphate receptors is essential for vascular barrier regulation. Novel role for hyaluronan and CD44 receptor family. *J Biol Chem* 2006;**281**:34,381–34,393.

63. Sanchez T, Skoura A, Wu MT, Casserly B, Harrington EO, Hla T. Induction of vascular permeability by the sphingosine-1-phosphate receptor-2 (S1P2R) and its downstream effectors ROCK and PTEN. *Arterioscler Thromb Vasc Biol* 2007;**27**:1312–18.

64. Gon Y, Wood MR, Kiosses WB, Jo E, Sanna MG, Chun J, Rosen H. S1P3 receptor-induced reorganization of epithelial tight junctions compromises lung barrier integrity and is potentiated by TNF. *Proc Natl Acad Sci USA* 2005;**102**:9270–5.

65. Wang F, Van Brocklyn JR, Hobson JP, Movafagh S, Zukowska-Grojec Z, Milstien S, Spiegel S. Sphingosine 1-phosphate stimulates cell migration through a G(i)- coupled cell surface receptor, Potential involvement in angiogenesis. *J Biol Chem* 1999;**274**:35,343–35,350.

66. Lee OH, Kim YM, Lee YM, Moon EJ, Lee DJ, Kim JH, Kim KW, Kwon YG. Sphingosine 1-phosphate induces angiogenesis: its angiogenic action and signaling mechanism in human umbilical vein endothelial cells. *Biochem Biophys Res Commun* 1999;**264**:743–50.

67. LaMontagne K, Littlewood-Evans A, Schnell C, O'Reilly T, Wyder L, Sanchez T, Probst B, Butler J, Wood A, Liau G, Billy E, Theuer A, Hla T, Wood J. Antagonism of sphingosine-1-phosphate receptors by FTY720 inhibits angiogenesis and tumor vascularization. *Cancer Res* 2006;**66**:221–31.

68. Visentin B, Vekich JA, Sibbald BJ, Cavalli AL, Moreno KM, Matteo RG, Garland WA, Lu Y, Yu S, Hall HS, Kundra V, Mills GB, Sabbadini RA. Validation of an anti-sphingosine-1-phosphate antibody as a potential therapeutic in reducing growth, invasion, and angiogenesis in multiple tumor lineages. *Cancer Cell* 2006;**9**:225–38.

69. Allende ML, Yamashita T, Proia RL. G-protein coupled receptor S1P1 acts within endothelial cells to regulate vascular maturation. *Blood* 2003;**102**:3665–7.

70. Kono M, Mi Y, Liu Y, Sasaki T, Allende ML, Wu YP, Yamashita T, Proia RL. The sphingosine-1-phosphate receptors S1P1, S1P2, and S1P3 function coordinately during embryonic angiogenesis. *J Biol Chem* 2004;**279**:29,367–29,373.

71. Skoura A, Sanchez T, Claffey K, Mandala SM, Proia RL, Hla T. Essential role of sphingosine 1-phosphate receptor 2 in pathological angiogenesis of the mouse retina. *J Clin Invest* 2007;**117**:2506–16.

72. Dev KK, Mullershausen F, Mattes H, Kuhn RR, Bilbe G, Hoyer D, Mir A. Brain sphingosine-1-phosphate receptors: Implication for FTY720 in the treatment of multiple sclerosis. *Pharmacol Ther* 2008;**117**:77–93.

73. Davis MD, Clemens JJ, Macdonald TL, Lynch KR. Sphingosine 1-phosphate analogs as receptor antagonists. *J Biol Chem* 2005;**280**:9633–41.

74. Clemens JJ, Davis MD, Lynch KR, Macdonald TL. Synthesis of para-alkyl aryl amide analogues of sphingosine-1-phosphate: discovery of potent S1P receptor agonists. *Bioorg Med Chem Letts* 2003;**13**:3401–4.

75. Clemens JJ, Davis MD, Lynch KR, Macdonald TL. Synthesis of benzimidazole based analogues of sphingosine-1-phosphate: discovery of potent, subtype-selective S1P4 receptor agonists. *Bioorg Med Chem Letts* 2004;**14**:4903–6.

76. Osada M, Yatomi Y, Ohmori T, Ikeda H, Ozaki Y. Enhancement of sphingosine 1-phosphate-induced migration of vascular endothelial cells and smooth muscle cells by an EDG-5 antagonist. *Biochem Biophys Res Commun* 2002;**299**:483–7.

Cyclic Nucleotides

Chapter 37

Adenylyl Cyclases

Adam J. Kuszak and Roger K. Sunahara

Department of Pharmacology, University of Michigan Medical School, Ann Arbor, Michigan

INTRODUCTION

Cyclic adenosine monophosphate (3′,5′-cyclic AMP or cAMP) is a key component of prokaryotic and eukaryotic intracellular signaling pathways, and the enzyme responsible for its synthesis is adenylyl cyclase (AC). In eukaryotes cAMP primarily activates protein kinase A (PKA), which regulates many enzymes, secondary kinases, transcription factors, receptors, and channels through protein phosphorylation [1]. Cyclic AMP also directly activates exchange factors of small molecular weight G proteins [2, 3], activates cyclic nucleotide-gated channels, and regulates the activity of some cyclic GMP-specific phosphodiesterases [4]. Adenylyl cyclases and cAMP signaling pathways are conserved through both eukaryotes and prokaryotes. In bacteria cAMP binds directly to transcription factors and is responsible for repression of genes involved in metabolism, also serving as a feedback mechanism [5].

In mammals, 10 genes have been cloned that encode either membrane-bound (AC1 to AC9) or a soluble form (sAC) of AC. In higher eukaryotes, receptor-activated G proteins, Ca^{2+}-activated calmodulin (CaM), protein kinases, and bicarbonate ions are thought to be the native modulators of AC activity [6–9]. AC function may also be affected by cellular stress [10, 11], as well as by exogenous small molecules such as the diterpene forskolin [12]. Although multiple AC isoforms may be expressed in a single cell, individual isoforms are often selectively regulated by specific factors and organized by compartmentalization [13].

With multiple isoforms, regulators, and extensive roles in cellular signaling pathways, ACs have been the subject of a vast amount of research. Biochemical characterization of ACs has revealed their catalytic mechanism at the atomic level [14–16]. While extensive knowledge of how ACs are regulated has been gained, new modes of regulation are still being discovered. Finally, recent work has found novel links between ACs and physiological functions such as learning and memory, drug withdrawal, cardiac stress, and longevity [17–22].

This chapter will focus on the mammalian ACs, but will make note of some differences between mammalian ACs and those of bacteria and prokaryotic toxins. We will summarize the structure and catalytic mechanism of ACs, the many modes of AC regulation, and discuss normal and aberrant AC function in physiology and disease states.

CLASSIFICATION/STRUCTURE/FUNCTION

Catalytic Core

Nucleotidyl cyclases (NCs) have been broadly organized into three classes. Class I NCs are found only in gram-negative, enteric, γ-proteobacteria such as *Escherichia*, *Yersinia*, and *Pseudomonas*. Studies on Class I NCs are limited, but they are believed to contain two domains: an N-terminal catalytic domain and a C-terminal regulatory domain [23, 24]. Class II NCs are a highly divergent group of enzymes comprised of exotoxins secreted by pathogenic prokaryotes like *B. pertussis*, *B. anthracis*, and *P. aeroginosa* [25–27]. Recent advances have elucidated much about the structure and mechanism of Class II cyclases, specifically those of the *B. pertussis* AC, CyaA, and the *B. anthracis* AC, edema factor [28, 29]. Finally, Class III NCs include all eukaryotic ACs currently identified, some prokaryotic ACs, as well as all guanylyl cyclases (GCs) and diguanylyl cyclases. Class III NCs are therefore very diverse, yet the mammalian ACs isoforms display 50–90 percent sequence identity. The 9 membrane-bound forms (tmACs) are composed of 12 transmembrane (TM) segments and 2 large cytoplasmic domains (C_1 and C_2), sharing ~60 percent homology between isoforms (Figure 37.1b). These integral membrane proteins exist as tandem repeats of 6-TM segments followed by a large cytoplasmic loop (Figure 37.1a). The cytoplasmic loops share considerable sequence similarity with GCs, to the degree that a few point mutations may be introduced into AC to convert it to a functional GC, and *vice versa* [30, 31]. Unlike mammalian ACs, prokaryotic Class III ACs differ

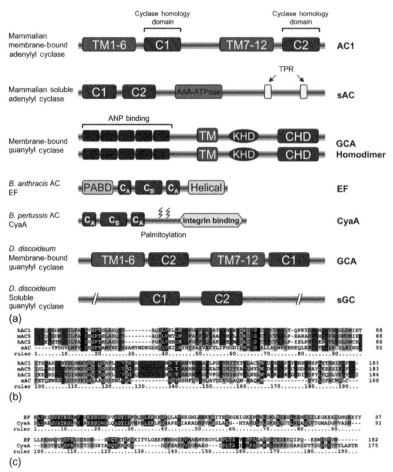

FIGURE 37.1 (a) Illustration of the domain structure of both eukaryotic and prokaryotic adenylyl and guanylyl cyclases. Cyclase homology domains (CHD, C_1, C_2, C_A, and C_B), transmembrane domains (TM), AAA ATPase, tetratricopeptide repeats (TPR), atrial natriuretic peptide (ANP) binding domains, kinase homology domains (KHD), anthrax protective antigen binding domain (PABD) and helical domain, and putative palmitoylation sites are shown. The functional enzymes are organized as homo- or heterodimers of the catalytic domains. (b) An alignment of the amino acid sequences of mammalian adenylyl cyclases. Shown are the first CHD of human AC1 (aa 294–476, GI: 62512172), human AC2 (aa 281–464, GI: 118572617), mouse AC5 (aa 461–643, GI: 122065123), and human sAC (aa 42–197, GI: 8272648). The selected cyclases were chosen as representative of ACs which are differentially regulated. "*" indicates a fully conserved residue; ":" indicates a strongly conserved residue; "." denotes a weakly conserved residue. Catalytically active aspartates are denoted by ‡. Residues conserved among membrane cyclases are denoted by "^". (c) An alignment of the CHDs of the bacterial ACs EF (aa 342–518; GI: 729244) and CyaA (aa 49–223; GI: 34978355). Symbols as in (b); catalytically important Asp and His residues are marked ‡.

greatly in length and domain structure, and often only have one catalytic domain [32].

The X-ray crystal structure of the cytoplasmic domains has provided atomic detail of the active site and the mechanism of catalysis (see Figure 37.3, below) [16]. The catalytic core of mammalian AC is composed of a heterodimer of the C_1 and C_2 domains, which are related to each other by a two-fold pseudo-symmetry (Figure 37.2). The active site of catalysis, where nucleotides bind, is also located at the domain interface, but is pseudo-symmetrically related to the forskolin binding site (Figure 37.3). Two aspartic acid residues in the active site support ATP binding through coordinating two magnesium ions. These residues are invariant amongst most Cyclase Homology Domains (CHDs) of both ACs and GCs [33, 34]. In a manner similar to DNA and RNA polymerases, RNA spliceosomes, and reverse transcriptases, ACs utilize divalent cations (e.g., Mg^{2+} or Ca^{2+}) to deprotonate the 3′-hydroxyl of the ribose ring of ATP (Figure 37.3d) [16]. This key step is necessary for the subsequent nucleophilic attack on the 5′ α-phosphate by the newly formed oxyanion. The end products are cAMP and the leaving group pyrophosphate (PPi). Similarity between the AC reaction to DNA polymerases is so strong, in fact, that the anti-viral DNA polymerase inhibitor foscarnet was found to suppress forskolin-stimulated catalysis of purified AC catalytic domains [35]. The stimulatory G protein, Gs, binds on the CHD surface, contacting both domains, while forskolin binds to a hydrophobic pocket at the interface of the two domains (Figures 37.2a, 37.3c).

Interestingly, the catalytic domain sequence of soluble adenylyl cyclase (sAC) is more closely related to some bacterial ACs than mammalian ACs [36]. The sAC

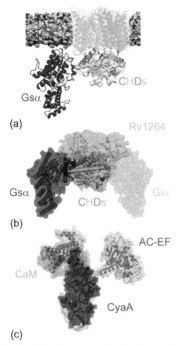

(a)

(b)

(c)

FIGURE 37.2 Illustration of the overall structural features of adenylyl cyclases.

(a) Ribbon structure of mammalian membrane-bound AC interacting with Gsα. Transmembrane (tm) AC contains 12 TM domains and two 40-kDa, cytoplasmic, cyclase homology domains (CHDs), C_1 and C_2 (PDB 1CJK). The complex of the homologous C_1 and C_2 domains is pseudo-symmetrically arranged. Gsα stimulates cyclase activity via an interaction with C_2. The transmembrane domains (helices within space-filling density) are modeled using the ABC transporter structure (PDB 2NQ2). (b) Regulators of CHDs interact at multiple surfaces. A representative CHD structure is shown (PDB 1CJK). Gsα (bound to ATPαS) switch II regulates AC by binding a groove between C_2 α1 and α3 helices. Giα (olive, modeled after Gsα) does not compete with Gsα to inhibit AC, but rather binds the homologous helical groove on C_1. *Mycobacterium tuberculosis* Rv1264 (PDB 1Y11) is a two-domain enzyme with an N-terminal regulatory domain and a C-terminal CHD which functions as a homodimer. Each regulatory domain makes extensive contacts with the CHD helices of the other dimer partner. (c) Comparison of bacterial AC structures. *Bacillus anthracis* EF (PDB 1XFV) and *Bordetella pertussis* CyaA (PDB 1YRU) are superimposed, showing the structural similarity. Calmodulin (CaM, PDB 1XFV), the activator of EF, is also shown to indicate its regulatory surface contacts. Structural models were generated and rendered using Pymol (Delano Scientific, Palo Alto, CA).

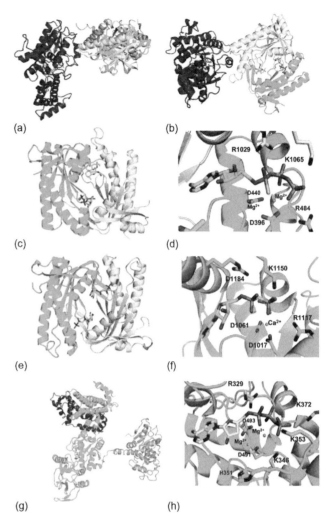

(a)   (b)

(c)   (d)

(e)   (f)

(g)   (h)

FIGURE 37.3 Comparison of eukaryotic and prokaryotic adenylyl cyclase active sites.

(a) Mammalian tmAC C_1 and C_2 with bound ATPαS and forskolin, interacting with Gsα (PDB 1CJK). (b) 90° rotated view of the C_1, C_2, and Gsα complex, where the dimeric nature of the CHD is more apparent. (c, d) Domain organization and detail of the mammalian tmAC CHD active site. The molecular detail of ATPαS illustrates the coordination of two Mg^{2+} ions by conserved aspartates D396 and D440. (e, f) Domain organization and detail of the mammalian soluble AC CHD active site (PDB 1WC0). The similarity of the sAC CHD to tmAC is apparent; however, sAC utilizes the two active site aspartates to coordinate a single Ca^{2+} for deprotonation of the ATP 3′ hydroxyl ion. (g, h) Domain organization and detail of the anthrax EF and calmodulin, and the CHD active site. In EF the C_A and C_B domains create a CHD which is structurally different from tmACs, yet contains aspartates that coordinate two catalytically important metals in a similar fashion, as well as a histidine (H351) utilized for stabilizing the reactive hydroxyl ion.

ortholog CyaC from cyanobacterium *S. platensis* shares an overall active site architecture that is similar to the tmACs, but contains two cooperatively-related active sites [28]. In addition, sACs utilize Ca^{2+} rather than Mg^{2+} to coordinate ATP binding and catalysis (Figure 37.3f).

New insight into non-mammalian AC structure and catalytic mechanism has been achieved in recent years with the elucidation of the X-ray crystal structure of multiple additional ACs. The crystal structure of AC from both *Bacillus anthracis* (the Edema Factor, EF, of anthrax) and of *Bordetella pertussis* (CyaA adenylyl cyclase domain) and catalytic mechanisms have been delineated [29, 37, 38]. Class II ACs have no clear structural homology with Class

III ACs, have only one cyclase homology domain, and are presumably monomeric. However the catalytic mechanisms seem to converge between the Class II and Class III cyclases. EF has two domains which comprise the catalytic core, C_A and C_B, and a third helical domain (Figures 37.1a, 37.2c, 37.3 g. Within the active site, two aspartates (Asp 491 and Asp 493) coordinate metal ions, similar to mammalian ACs (Figure 37.3 h [37]. The active site also contains similar positively-charged residues for stabilizing the transition site. Unlike mammalian ACs, however, a histidine (His 351) on the opposite side of the nucleotide ribose is positioned to stabilize an OH^- ion near the ribose 3′OH group (Figure 37.3 h) [29]. EF was originally thought to utilize a single catalytic metal ion to promote loss of the PPi [37], but it is now believed to use a two-metal-ion catalysis analogous to Class III cyclases [29]. While CyaA of *B. pertussis* shares ~25 percent sequence homology with EF, CyaA is structurally very similar, containing domains that correlate to the C_A and C_B domains [38]. Also like EF, the CyaA active site contains Mg^{2+} ions coordinated by two aspartate residues (D188 and D190) and a catalytically important histidine (H298) (Figure 37.1c).

Oligomerization

Although the functional role of the membrane domains of tmACs remains controversial, it has been suggested that tmACs can dimerize through these domains [39]. Intramolecular dimerization of the two membrane domains of AC8 was observed using fluorescence resonance energy transfer (FRET) and co-immunoprecipitation studies [39]. In addition, isoform-specific dimerization was found to be important for membrane trafficking of the enzyme [39]. FRET studies between cyan fluorescent protein (CFP)- and yellow fluorescent protein (YFP)-tagged TM domains suggest that dimerization of AC8 occurs via the second TM domains [40]. Although biochemical evidence for tmAC dimerization is increasing, more research is required before an understanding of the functional relevance is appreciated.

ENDOGENOUS AND EXOGENOUS REGULATION

G Proteins

Not only has our understanding of AC structure and mechanism expanded, but so has the characterization of how ACs are regulated by cellular signaling molecules. Hormonal and neurotransmitter regulation of ACs occurs primarily through heterotrimeric G proteins in both vertebrates and invertebrates (Table 37.1, Figures 37.2b, 37.4). Cell surface

TABLE 37.1 Summary of the regulatory properties of mammalian adenylyl cyclases[#]

AC isoform	Tissue distribution	Gαs	Gβγ	Gαi	Protein kinases	Calcium	Forskolin	Notes
AC1	Brain, adrenal (medulla)	↑	↓	↓(Gαo)	↑PKC ↓CaMKIV	↑CAM	↑	
AC2	Brain, skeletal muscle, lung (heart)	↑	↑*		↑PKC		↑	
AC3	Brain, olfactory epithelium (heart)	↑	↓		↑PKC ↓CaMKII	↑CaM	↑	
AC4	Brain (heart, kidney, liver, lung, BAT, uterus	↑	↑*		↑PKC		↑	
AC5	Heart, brain, kidney, liver, lung, uterus, adrenal, BAT	↑	↑*	↓ ↓PKA	↑PKCα, ζ	↓	↑	
AC6	Ubiquitous	↑	↑*	↓	↑PKC	↓	↑	
AC7	Ubiquitous, high in brain (heart)	↑	↑*		↑PKC		↑	
AC8	Brain, lung (testis, adrenal, uterus, heart)	↑	↓			↑CaM	↑	
AC9	Brain, skeletal muscle (heart)	↑		↓(weak)			↑(weak)	
sAC	Testis							Activated by bicarbonate

Gβγ stimulation of AC isoforms is conditional upon Gsα co-activation.
[#]*Upward arrows indicate a stimulation of activity, downward arrows denote an inhibition of activity.*

G protein-coupled receptor (GPCR) activation by extra-cellular stimuli in turn leads to activation of G proteins by initiating the exchange of GDP for GTP. The α subunit of the stimulatory G protein (Gsα) activates all nine tmACs in a nucleotide-dependent fashion [7, 8, 41, 42]. Gsα activates tmACs by binding to a site that is shared by the C_1 and C_2 domains, priming the active site for ATP binding and catalysis [15]. Gsα activation of ACs is terminated by GTP hydrolysis to GDP. The α subunits of the inhibitory family of G proteins, Gi$_{1,2,3}$, Go, and Gz [43–45], attenuate AC activity in an isoform-dependent manner (Table 37.1) [6–8,41]. Giα inhibition of AC5 and AC6 is not a result of competition with Gsα, but rather through binding to a site that is pseudo-symmetrically related to the Gsα site (Figure 37.2b) [46]. The G$\beta\gamma$ subunits of heterotrimeric G proteins can also stimulate AC2, AC4, and AC7 activity, though it is dependent upon co-activation by Gsα [43, 47]. G$\beta\gamma$ interacts with AC2's C_1 domain to exert its stimulation [48]. In contrast, G$\beta\gamma$ has the opposite effect on AC1, AC3, and AC8, causing an inhibition of cAMP accumulation [6, 48, 49]. G$\beta\gamma$ has been found to inhibit AC5 and AC6 in cell culture experiments [50], but not in *in vitro* experiments [51]. The *in vitro* stimulation by G$\beta\gamma$ was Gsα-dependent, similar to the AC2 family of cyclases. These opposing findings may result from G$\beta\gamma$ effects on other signaling molecules, and will have to be resolved in future research.

In the past decade, Regulators of G protein Signaling (RGS) molecules, which catalyze the GTP hydrolysis of heterotrimeric G proteins, have been shown to play a critical role in controlling G protein activity. Recently, RGS proteins have been implicated in directly regulating tmACs as well. RGS2 decreases Gs-stimulated increases in cAMP [52, 53], and binds purified AC5 C_1 [54]. In addition, bioluminescence resonance energy transfer (BRET) experiments suggest that RGS2 may interact directly with AC2 and AC6 [55]. It remains to be seen if RGS interaction at the level of ACs represents a new form of RGS influence on G-protein signaling.

Ca^{2+} and Calmodulin (CaM)

CaM is a ubiquitous Ca^{2+} sensor protein and is a potent activator of several mammalian tmAC isoforms: AC1 [56], AC8 [57], and perhaps AC3 [58]. The primary source of calcium ions is thought to be capacitative entry through Ca^{2+} channels [59,60]. CaM is also implicated in the pathology of the bacterial exotoxins mentioned above, as it is the principle exotoxin AC activator (Figure 37.2c) [29,38]. Ca^{2+}/CaM inhibits AC1 and AC3 indirectly through the activity of CaM-dependent protein kinase II and IV. Phosphorylation of AC1 and AC3 by CaM kinases inhibits the cyclase by blocking the binding of activators. In this sense, posttranslational modification of ACs by phosphorylation, which can be caused by either PKA or protein kinase C (PKC), is generally inhibitory. PKA supports a negative-feedback mechanism whereby the more cAMP that is produced by ACs, the more PKA is activated and thus the more ACs are phosphorylated and inhibited. Cellular Ca^{2+} levels can also regulate AC activity directly, as all AC isoforms are inhibited by high (upper micromolar to millimolar) concentrations [61]. However, submicromolar concentrations within the physiologic range and compatible with capacitative entry effectively and specifically inhibit AC5, AC6, and anthrax EF [61–63].

Bicarbonate

The sole soluble mammalian AC isoform (sAC) is unique in its regulation; it is not affected by classic AC modulators like G proteins, CaM, or forskolin [64]. Instead, sAC is activated by bicarbonate [9], and the bicarbonate ion induces a closure of the active site similar to the structural changes induced by Gsα on tmACs [28]. Prokaryotic soluble ACs are also stimulated by bicarbonate, suggesting that sAC may be an evolutionarily conserved bicarbonate sensor [10].

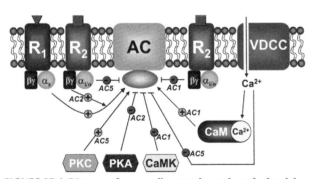

FIGURE 37.4 Diagram of mammalian membrane-bound adenylyl cyclase regulators.

The stimulatory (circles with pluses) or inhibitory (circles with minuses) influences of hormone-receptor-mediated G-protein regulation (R1 and R2), protein kinase regulation, and Ca^{2+} and/or calmodulin (CaM) effects on AC activity are summarized. AC isoform-specific effects are indicated in italics next to the stimulatory and inhibitory signs. For example, the effect of G$\beta\gamma$ is inhibitory on the AC1 family of cyclases (AC1, AC3, and AC8) and stimulatory on the AC2 family (AC2, AC4, and AC7), though the stimulation is dependent on Gsα co-activation. Ca^{2+}-CaM directly activates the AC1 family of cyclases, whereas Ca^{2+} alone inhibits the AC5 family (AC5 and AC6). Ca^{2+} is believed to enter the cell by capacitative entry via volgate-dependent Ca^{2+} channels (VDCC). Protein kinase A (PKA) activity can inhibit AC2, while phosphorylation by PKC can stimulate AC5. Phosphorylation of AC1 and AC3 by CaM kinases (CaMK) prevents binding of AC activators. Not shown is forskolin, which potently stimulates all membrane-bound AC isoforms except AC9.

Other Modulators

The small molecule forskolin (isolated from the plant *Choleus forskohlii*) is a potent activator of all mammalian tmACs except AC9, which is weakly activated [65]. While the binding site of forskolin is within the conserved catalytic domain, the stimulatory actions appear to be selective for membrane-bound vertebrate and invertebrate forms [65]. On the other side of exogenous modulation, ACs are inhibited by a class of adenosine analogs known as P-site inhibitors [66]. P-site inhibitors and deoxy-adenosine analogs are thought to be derived from RNA metabolism [67]. These small molecules act by binding to a conformation of the enzyme that closely resembles the posttransitional, product-bound state [34, 68]. Inhibition is enhanced in the presence of PPi. The ability to exogenously modulate AC function is intriguing, as specific ACs are increasingly being associated with certain physiological roles and disease states.

ROLES IN PHYSIOLOGY AND DISEASES

Our understanding of the physiological role of ACs has recently been dramatically advanced through a series of elegant genetic studies. These studies have revealed unexpected roles in physiological development and disease states. One example is the elucidation of sAC's role in sperm function. Spermatozoa are not immediately able to fertilize an egg; they must undergo a reorganization of plasma membrane lipids and proteins and enter a hyperactive motility state, collectively termed capacitation. Capacitation is stimulated by bicarbonate, and is dependent on high concentrations of cAMP. Originally, changes in pH levels were thought to link bicarbonate stimulation to elevated cAMP levles. However, it was recently determined that sAC, which is most highly expressed in the testis, provides a direct link between bicarbonate and cAMP in sperm capacitation [9]. Furthermore, mice lacking sAC are infertile due to a lack of sperm motility [69,70]. The discovery that sAC is required for male fertility has lead to the intriguing idea of targeting sAC therapeutically as a male contraceptive.

Adenylyl cyclase function has also proven to be important in learning. The *D. melanogaster* mutant *rutabaga* is deficient in a calcium-activated AC thought to be the ortholog of the mammalian AC1 subfamily of ACs [71]. Deficiency of this AC causes these flies to avoid a trained odor, indicating that AC is important in memory and learning [72]. Similar observations were later reported with AC1 (and also the related AC8) gene disruption in mice [73]. Recently, Abel and coworkers created transgenic mice expressing a constitutively active mutant of Gsα. Although the constantly active Gsα promoted high AC activity, cAMP levels were low as a result of compensatory phosphodiesterase activity. These mice exhibited impairments in spatial and fear conditioning learning tasks [17], supporting AC's important role in mammalian learning. Similarly, Martel and colleagues found that hippocampal levels of cAMP are important for enhancing memory processes [18].

ACs have been also been implicated in olfactory responses in mice. While AC2, AC3, and AC4 are all expressed in the olfactory cilia, knockout studies with AC3-null mice illustrated a complete loss of odorant responses [74]. Further investigating the role of AC3 in mouse odorant perception, researchers found that AC3-null mice do not respond to mouse milk, urine, or pheromones, and male sexual behaviors were absent [75]. Cyclic AMP production by AC3 has also been shown to be necessary for establishing proper axonal identity and for axonal projections in glomeruli of olfactory sensory neurons [76, 77]. Thus, AC3 has proven to be crucial for olfactory function.

In addition to being critical for development of learning and olfactory pathways, ACs are important players in disease states. For example, ACs are theorized to have a role in developing drug dependencies. Following chronic treatment with opiate drugs like morphine, several ACs are upregulated and become sensitized to additional stimulation by either Gsα or forskolin [78, 79]. Although originally observed in response to opiate drugs, sensitization of AC (also called heterologous sensitization and supersensitization) may be a characteristic response to chronic stimulation of all G protein-coupled receptors that promote Gi/o activity. Not all ACs can undergo sensitization, however. Specifically, AC1, AC5, AC6, and AC8 are sensitized, while AC2, AC3, AC4, and AC7 are not [80–82]. Multiple lines of evidence suggest that sensitization involves phosphorylation of the ACs by protein kinases including PKC, PKA, and Raf-1 kinase [83–87]. However, protein kinase activity has also been shown to decrease AC sensitization [85, 88], and a thorough understanding of the molecular mechanism and consequences of AC sensitization on the cellular and whole system level remains to be determined.

In recent years, the use of β-adrenergic receptor (βAR) antagonists (β-blockers) for treating heart failure has gained widespread acceptance [89]. β-blockers prevent βAR stimulation of AC, which causes less sympathetic stimulation in an overworked heart, resulting in decreased remodeling of the heart and less myocardial damage. AC5-null mice exhibited ventricular hypertrophy to a similar extent as wild-type mice following induction of pressure overload [90]. However, AC5-null mice had preserved left ventricle function as well as decreased levels of myocardial apoptosis [20]. Thus, suppressing AC5 activity by disrupting the AC5 gene, or through the use of β-blockers, results in a higher tolerance to pressure overload. Moreover, the AC5-selective P-site inhibitor, PMC-6 (a derivative of the classic AC inhibitor 9-(cyclopentyl)adenine) exhibited a dose-dependent decrease in cAMP accumulation in cardiac myocytes, prevented isoproterenol-induced cardiac myocyte apoptosis, but did not have an effect on myocyte contractility [21].

Taken together these findings suggest that inhibition of AC5 function is beneficial in the failing heart. It should be noted that these data are in slight disagreement with studies overexpressing AC5 using the transgenic activated Gqα mouse model of cardiac hypertrophy [91]. Overexpression of AC5 in the Gqα overexpression background resulted in a recovery of cAMP levels to normal as well as a recovery of contractility, although the hypertrophy remained [92, 93]. Therefore, it is currently unclear whether AC5 activity is beneficial or detrimental in the failing heart. The contributions of AC5 to hypertrophy may be difficult to assess in this complex phenotypic model of cardiac hypertrophy (i.e., Gqα overexpression background), and deserves further characterization.

Another surprising finding in AC5-null mice was an ∼30 percent increase in median lifespan [20, 22]. The AC5 gene disruption resulted in preserved bone density, lower body weights, and less incidence of age-induced cardiomyopathy. In addition, Raf-1/MEK1/ERK signaling activity increased in AC5-null mice, resulting in increased in anti-apoptosis, cell survival, and anti-oxidative stress signaling. These findings lend themselves to a model where the lack of AC5 signaling results in less PKA activity, which in turn allows for higher levels of Raf-1/MEK1/ERK activity [22].

In light of an increased understanding of AC roles in physiological and disease states, there is currently a strong effort to discover compounds which can be used therapeutically to alter AC function. Model-based searches for AC isoform-specific small molecule forskolin derivatives, as well as small molecule non-nucleosides that mimic P-site inhibitors, have been reported [94]. This screen found several compounds with increased AC isoform specificity. Importantly, a 6-(3-dimethylaminopropionyl) modification of forskolin resulted in a compound with enhanced activity towards AC5. In addition, ribose-substituted P-site inhibitors were found to have higher AC5 selectivity, leading to the discovery of 2-amino-7-(2-furanyl)-7,8-dihydro-5(6H)-quinazolinone, or NKY80, as a novel AC5 inhibitor [94]. Such advances in pharmacological manipulation of specific AC function are promising at a time when ACs roles in various disease states are being elucidated at a fast pace.

SUMMARY

Adenylyl cyclases (ACs) are vital in both eukaryotes and prokaryotes, where they generate the cyclic AMP required for a multitude of cellular functions. Although mammalian and bacterial ACs differ structurally, they appear to have converged on a common catalytic mechanism shared by RNA and DNA polymerases. The extensive cellular regulation of AC function through G proteins and calmodulin has been well characterized over the years, yet new methods of altering AC function are still being discovered. In addition, transgenic models have revealed roles for specific AC isoforms in diverse areas such as learning and memory, olfaction, cardiac function, and aging. These isoform-specific effects, together with their differential expression patterns, reveal their potential as therapeutic targets. Recently, derivatives of forskolin, nucleosides, and P-site inhibitors have been developed which

show promising isoform specificity and display strong therapeutic potential. Indeed, there remains much to be learned about ACs as new regulators and physiological roles emerge, and the potential of targeting specific AC isoforms therapeutically will guide future research.

REFERENCES

1. Taylor SS, Buechler JA, Yonemoto W. cAMP-dependent protein kinase: framework for a diverse family of regulatory enzymes. *Annu Rev Biochem* 1990;**59**:971–1005.
2. de Rooij J, Zwartkruis FJ, Verheijen MH, Cool RH, Nijman SM, Wittinghofer A, Bos JL. Epac is a Rap1 guanine-nucleotide-exchange factor directly activated by cyclic AMP. *Nature* 1998;**396**:474–7.
3. Kawasaki H, Springett GM, Mochizuki N, Toki S, Nakaya M, Matsuda M, Housman DE, Graybiel AM. A family of cAMP-binding proteins that directly activate Rap1. *Science* 1998;**282**:2275–9.
4. Broillet MC, Firestein S. Cyclic nucleotide-gated channels. Molecular mechanisms of activation. *Ann NY Acad Sci* 1999;**868**:730–40.
5. Daniel PB, Walker WH, Habener JF. Cyclic AMP signaling and gene regulation. *Annu Rev Nutr* 1998;**18**:353–83.
6. Sunahara RK, Dessauer CW, Gilman AG. Complexity and diversity of mammalian adenylyl cyclases. *Annu Rev Pharmacol Toxicol* 1996;**36**:461–80.
7. Smit MJ, Iyengar R. Mammalian adenylyl cyclases. *Adv Sec Mess Phosph Res* 1998;**32**:1–21.
8. Patel TB, Du Z, Pierre S, Cartin L, Scholich K. Molecular biological approaches to unravel adenylyl cyclase signaling and function. *Gene* 2001;**269**:13–25.
9. Chen Y, Cann MJ, Litvin TN, Iourgenko V, Sinclair ML, Levin LR, Buck J. Soluble adenylyl cyclase as an evolutionarily conserved bicarbonate sensor. *Science* 2000;**289**:625–8.
10. Zippin JH, Levin LR, Buck J. CO(2)/HCO(3)(-)-responsive soluble adenylyl cyclase as a putative metabolic sensor. *Trends Endocrinol Metab* 2001;**12**:366–70.
11. van Es S, Virdy KJ, Pitt GS, Meima M, Sands TW, Devreotes PN, Cotter DA, Schaap P. Adenylyl cyclase G, an osmosensor controlling germination of Dictyostelium spores. *J Biol Chem* 1996;**271**(23):623–5.
12. Seamon KB, Padgett W, Daly JW. Forskolin: unique diterpene activator of adenylate cyclase in membranes and in intact cells. *Proc Natl Acad Sci USA* 1981;**78**:3363–7.
13. Cooper DM. Compartmentalization of adenylate cyclase and cAMP signalling. *Biochem Soc Trans* 2005;**33**:1319–22.
14. Zhang G, Liu Y, Ruoho AE, Hurley JH. Structure of the adenylyl cyclase catalytic core. *Nature* 1997;**386**:247–53.
15. Tesmer JJ, Sunahara RK, Gilman AG, Sprang SR. Crystal structure of the catalytic domains of adenylyl cyclase in a complex with Gsalpha. GTPgammaS. *Science* 1997;**278**:1907–16.
16. Tesmer JJ, Sprang SR. The structure, catalytic mechanism and regulation of adenylyl cyclase. *Curr Opin Struct Biol* 1998;**8**:713–19.
17. Bourtchouladze R, Patterson SL, Kelly MP, Kreibich A, Kandel ER, Abel T. Chronically increased Gsalpha signaling disrupts associative and spatial learning. *Learn Mem* 2006;**13**:745–52.
18. Martel G, Millard A, Jaffard R, Guillou JL. Stimulation of hippocampal adenylyl cyclase activity dissociates memory consolidation processes for response and place learning. *Learn Mem* 2006;**13**:342–8.
19. Watts VJ, Neve KA. Sensitization of adenylate cyclase by Galpha i/o-coupled receptors. *Pharmacol Ther* 2005;**106**:405–21.

20. Okumura S, Takagi G, Kawabe J, Yang G, Lee MC, Hong C, Liu J, Vatner DE, Sadoshima J, Vatner SF, Ishikawa Y. Disruption of type 5 adenylyl cyclase gene preserves cardiac function against pressure overload. *Proc Natl Acad Sci USA* 2003;**100**:9986–90.

21. Iwatsubo K, Minamisawa S, Tsunematsu T, Nakagome M, Toya Y, Tomlinson JE, Umemura S, Scarborough RM, Levy DE, Ishikawa Y. Direct inhibition of type 5 adenylyl cyclase prevents myocardial apoptosis without functional deterioration. *J Biol Chem* 2004;**279**(40):938–45.

22. Yan L, Vatner DE, O'Connor JP, Ivessa A, Ge H, Chen W, Hirotani S, Ishikawa Y, Sadoshima J, Vatner SF. Type 5 adenylyl cyclase disruption increases longevity and protects against stress. *Cell* 2007;**130**:247–58.

23. Holland MM, Leib TK, Gerlt JA. Isolation and characterization of a small catalytic domain released from the adenylate cyclase from *Escherichia coli* by digestion with trypsin. *J Biol Chem* 1988;**263**(14):661–8.

24. Reddy P, Hoskins J, McKenney K. Mapping domains in proteins: dissection and expression of Escherichia coli adenylyl cyclase. *Anal Biochem* 1995;**231**:282–6.

25. Ladant D, Ullmann A. *Bordetella pertussis* adenylate cyclase: a toxin with multiple talents. *Trends Microbiol* 1999;**7**:172–6.

26. Leppla SH. Anthrax toxin edema factor: a bacterial adenylate cyclase that increases cyclic AMP concentrations of eukaryotic cells. *Proc Natl Acad Sci USA* 1982;**79**:3162–6.

27. Yahr TL, Vallis AJ, Hancock MK, Barbieri JT, Frank DW. ExoY, an adenylate cyclase secreted by the *Pseudomonas aeruginosa* type III system. *Proc Natl Acad Sci USA* 1998;**95**(13):899–904.

28. Steegborn C, Litvin TN, Levin LR, Buck J, Wu H. Bicarbonate activation of adenylyl cyclase via promotion of catalytic active site closure and metal recruitment. *Nat Struct Mol Biol* 2005;**12**:32–7.

29. Shen Y, Zhukovskaya NL, Guo Q, Florian J, Tang WJ. Calcium-independent calmodulin binding and two-metal-ion catalytic mechanism of anthrax edema factor. *EMBO J* 2005;**24**:929–41.

30. Sunahara RK, Beuve A, Tesmer JJ, Sprang SR, Garbers DL, Gilman AG. Exchange of substrate and inhibitor specificities between adenylyl and guanylyl cyclases. *J Biol Chem* 1998;**273**(16):332–8.

31. Beuve A. Conversion of a guanylyl cyclase to an adenylyl cyclase. *Methods* 1999;**19**:545–50.

32. Sinha SC, Sprang SR. Structures, mechanism, regulation and evolution of class III nucleotidyl cyclases. *Rev Physiol Biochem Pharmacol* 2006;**157**:105–40.

33. Liu Y, Ruoho AE, Rao VD, Hurley JH. Catalytic mechanism of the adenylyl and guanylyl cyclases: modeling and mutational analysis. *Proc Natl Acad Sci USA* 1997;**94**(13):414–19.

34. Dessauer CW, Tesmer JJ, Sprang SR, Gilman AG. The interactions of adenylate cyclases with P-site inhibitors. *Trends Pharmacol Sci* 1999;**20**:205–10.

35. Kudlacek O, Mitterauer T, Nanoff C, Hohenegger M, Tang WJ, Freissmuth M, Kleuss C. Inhibition of adenylyl and guanylyl cyclase isoforms by the antiviral drug foscarnet. *J Biol Chem* 2001;**276**:3010–16.

36. Buck J, Sinclair ML, Schapal L, Cann MJ, Levin LR. Cytosolic adenylyl cyclase defines a unique signaling molecule in mammals. *Proc Natl Acad Sci USA* 1999;**96**:79–84.

37. Drum CL, Yan SZ, Bard J, Shen YQ, Lu D, Soelaiman S, Grabarek Z, Bohm A, Tang WJ. Structural basis for the activation of anthrax adenylyl cyclase exotoxin by calmodulin. *Nature* 2002;**415**:396–402.

38. Guo Q, Shen Y, Lee YS, Gibbs CS, Mrksich M, Tang WJ. Structural basis for the interaction of Bordetella pertussis adenylyl cyclase toxin with calmodulin. *EMBO J* 2005;**24**:3190–201.

39. Gu C, Sorkin A, Cooper DM. Persistent interactions between the two transmembrane clusters dictate the targeting and functional assembly of adenylyl cyclase. *Curr Biol* 2001;**11**:185–90.

40. Gu C, Cali JJ, Cooper DM. Dimerization of mammalian adenylate cyclases. *Eur J Biochem* 2002;**269**:413–21.

41. Hanoune J, Defer N. Regulation and role of adenylyl cyclase isoforms. *Annu Rev Pharmacol Toxicol* 2001;**41**:145–74.

42. Sunahara RK, Tesmer JJ, Gilman AG, Sprang SR. Crystal structure of the adenylyl cyclase activator Gsalpha. *Science* 1997;**278**:1943–7.

43. Tang WJ, Gilman AG. Type-specific regulation of adenylyl cyclase by G protein beta gamma subunits. *Science* 1991;**254**:1500–3.

44. Taussig R, Tang WJ, Hepler JR, Gilman AG. Distinct patterns of bidirectional regulation of mammalian adenylyl cyclases. *J Biol Chem* 1994;**269**:6093–100.

45. Kozasa T, Gilman AG. Purification of recombinant G proteins from Sf9 cells by hexahistidine tagging of associated subunits. Characterization of alpha 12 and inhibition of adenylyl cyclase by alpha z. *J Biol Chem* 1995;**270**:1734–41.

46. Dessauer CW, Tesmer JJ, Sprang SR, Gilman AG. Identification of a Gialpha binding site on type V adenylyl cyclase. *J Biol Chem* 1998;**273**(25):831–9.

47. Gao BN, Gilman AG. Cloning and expression of a widely distributed (type IV) adenylyl cyclase. *Proc Natl Acad Sci USA* 1991;**88**:10,178–10,182.

48. Diel S, Klass K, Wittig B, Kleuss C. Gbetagamma activation site in adenylyl cyclase type II. Adenylyl cyclase type III is inhibited by Gbetagamma. *J Biol Chem* 2006;**281**:288–94.

49. Steiner D, Saya D, Schallmach E, Simonds WF, Vogel Z. Adenylyl cyclase type-VIII activity is regulated by G(betagamma) subunits. *Cell Signal* 2006;**18**:62–8.

50. Bayewitch ML, Avidor-Reiss T, Levy R, Pfeuffer T, Nevo I, Simonds WF, Vogel Z. Inhibition of adenylyl cyclase isoforms V and VI by various Gbetagamma subunits. *FASEB J* 1998;**12**:1019–25.

51. Gao X, Sadana R, Dessauer CW, Patel TB. Conditional stimulation of type V and VI adenylyl cyclases by G protein betagamma subunits. *J Biol Chem* 2007;**282**:294–302.

52. Roy AA, Lemberg KE, Chidiac P. Recruitment of RGS2 and RGS4 to the plasma membrane by G proteins and receptors reflects functional interactions. *Mol Pharmacol* 2003;**64**:587–93.

53. Sinnarajah S, Dessauer CW, Srikumar D, Chen J, Yuen J, Yilma S, Dennis JC, Morrison EE, Vodyanoy V, Kehrl JH. RGS2 regulates signal transduction in olfactory neurons by attenuating activation of adenylyl cyclase III. *Nature* 2001;**409**:1051–5.

54. Salim S, Sinnarajah S, Kehrl JH, Dessauer CW. Identification of RGS2 and type V adenylyl cyclase interaction sites. *J Biol Chem* 2003;**278**(15):842–9.

55. Roy AA, Baragli A, Bernstein LS, Hepler JR, Hebert TE, Chidiac P. RGS2 interacts with Gs and adenylyl cyclase in living cells. *Cell Signal* 2006;**18**:336–448.

56. Krupinski J, Coussen F, Bakalyar HA, Tang WJ, Feinstein PG, Orth K, Slaughter C, Reed RR, Gilman AG. Adenylyl cyclase amino acid sequence: possible channel- or transporter-like structure. *Science* 1989;**244**:1558–64.

57. Cali JJ, Zwaagstra JC, Mons N, Cooper DM, Krupinski J. Type VIII adenylyl cyclase. A Ca^{2+}/calmodulin-stimulated enzyme expressed in discrete regions of rat brain. *J Biol Chem* 1994;**269**(12):190–5.

58. Choi EJ, Xia Z, Storm DR. Stimulation of the type III olfactory adenylyl cyclase by calcium and calmodulin. *Biochemistry* 1992;**31**:6492–8.

59. Fagan KA, Mahey R, Cooper DM. Functional co-localization of transfected Ca(2+)-stimulable adenylyl cyclases with capacitative Ca2+ entry sites. *J Biol Chem* 1996;**271**(12):438–44.

60. Fagan KA, Graf RA, Tolman S, Schaack J, Cooper DM. Regulation of a Ca^{2+}-sensitive adenylyl cyclase in an excitable cell. Role of voltage-gated versus capacitative Ca^{2+} entry. *J Biol Chem* 2000;**275**: 40187–94.

61. Cooper DM. Molecular and cellular requirements for the regulation of adenylate cyclases by calcium. *Biochem Soc Trans* 2003;**31**:912–15.

62. Gu C, Cooper DM. Ca(2+), Sr(2+), and Ba(2+) identify distinct regulatory sites on adenylyl cyclase (AC) types VI and VIII and consolidate the apposition of capacitative cation entry channels and Ca(2+)-sensitive ACs. *J Biol Chem* 2000;**275**:6980–6.

63. Guillou JL, Nakata H, Cooper DM. Inhibition by calcium of mammalian adenylyl cyclases. *J Biol Chem* 1999;**274**(35):539–45.

64. Neer EJ. Physical and functional properties of adenylate cyclase from mature rat testis. *J Biol Chem* 1978;**253**:5808–12.

65. Hacker BM, Tomlinson JE, Wayman GA, Sultana R, Chan G, Villacres E, Disteche C, Storm DR. Cloning, chromosomal mapping, and regulatory properties of the human type 9 adenylyl cyclase (ADCY9). *Genomics* 1998;**50**:97–104.

66. Londos C, Wolff J. Two distinct adenosine-sensitive sites on adenylate cyclase. *Proc Natl Acad Sci USA* 1977;**74**:5482–6.

67. Bushfield M, Shoshani I, Johnson RA. Tissue levels, source, and regulation of 3′-AMP: an intracellular inhibitor of adenylyl cyclases. *Mol Pharmacol* 1990;**38**:848–53.

68. Dessauer CW, Gilman AG. The catalytic mechanism of mammalian adenylyl cyclase. Equilibrium binding and kinetic analysis of P-site inhibition. *J Biol Chem* 1997;**272**(27):787–95.

69. Esposito G, Jaiswal BS, Xie F, Krajnc-Franken MA, Robben TJ, Strik AM, Kuil C, Philipsen RL, van Duin M, Conti M, Gossen JA. Mice deficient for soluble adenylyl cyclase are infertile because of a severe sperm-motility defect. *Proc Natl Acad Sci USA* 2004;**101**: 2993–8.

70. Hess KC, Jones BH, Marquez B, Chen Y, Ord TS, Kamenetsky M, Miyamoto C, Zippin JH, Kopf GS, Suarez SS, Levin LR, Williams CJ, Buck J, Moss SB. The "soluble" adenylyl cyclase in sperm mediates multiple signaling events required for fertilization. *Dev Cell* 2005;**9**:249–59.

71. Levin LR, Han PL, Hwang PM, Feinstein PG, Davis RL, Reed RR. The *Drosophila* learning and memory gene rutabaga encodes a Ca^{2+}/calmodulin-responsive adenylyl cyclase. *Cell* 1992;**68**:479–89.

72. Zars T, Wolf R, Davis R, Heisenberg M. Tissue-specific expression of a type I adenylyl cyclase rescues the rutabaga mutant memory defect: in search of the engram. *Learn Mem* 2000;**7**:18–31.

73. Liauw J, Wu LJ, Zhuo M. Calcium-stimulated adenylyl cyclases required for long-term potentiation in the anterior cingulate cortex. *J Neurophysiol* 2005;**94**:878–82.

74. Wong ST, Trinh K, Hacker B, Chan GC, Lowe G, Gaggar A, Xia Z, Gold GH, Storm DR. Disruption of the type III adenylyl cyclase gene leads to peripheral and behavioral anosmia in transgenic mice. *Neuron* 2000;**27**:487–97.

75. Wang Z, Balet Sindreu C, Li V, Nudelman A, Chan GC, Storm DR. Pheromone detection in male mice depends on signaling through the type 3 adenylyl cyclase in the main olfactory epithelium. *J Neurosci* 2006;**26**:7375–9.

76. Chesler AT, Zou DJ, Le Pichon CE, Peterlin ZA, Matthews GA, Pei X, Miller MC, Firestein S. A G protein/cAMP signal cascade is required for axonal convergence into olfactory glomeruli. *Proc Natl Acad Sci USA* 2007;**104**:1039–44.

77. Zou DJ, Chesler AT, Le Pichon CE, Kuznetsov A, Pei X, Hwang EL, Firestein S. Absence of adenylyl cyclase 3 perturbs peripheral olfactory projections in mice. *J Neurosci* 2007;**27**:6675–83.

78. Sharma SK, Klee WA, Nirenberg M. Dual regulation of adenylate cyclase accounts for narcotic dependence and tolerance. *Proc Natl Acad Sci USA* 1975;**72**:3092–6.

79. Chakrabarti S, Wang L, Tang WJ, Gintzler AR. Chronic morphine augments adenylyl cyclase phosphorylation: relevance to altered signaling during tolerance/dependence. *Mol Pharmacol* 1998;**54**:949–53.

80. Avidor-Reiss T, Nevo I, Saya D, Bayewitch M, Vogel Z. Opiate-induced adenylyl cyclase superactivation is isozyme-specific. *J Biol Chem* 1997;**272**:5040–7.

81. Watts VJ, Neve KA. Sensitization of endogenous and recombinant adenylate cyclase by activation of D2 dopamine receptors. *Mol Pharmacol* 1996;**50**:966–76.

82. Thomas JM, Hoffman BB. Isoform-specific sensitization of adenylyl cyclase activity by prior activation of inhibitory receptors: role of beta gamma subunits in transducing enhanced activity of the type VI isoform. *Mol Pharmacol* 1996;**49**:907–14.

83. Gordon AS, Yao L, Jiang Z, Fishburn CS, Fuchs S, Diamond I. Ethanol acts synergistically with a D2 dopamine agonist to cause translocation of protein kinase C. *Mol Pharmacol* 2001;**59**:153–60.

84. Oak JN, Lavine N, Van Tol HH. Dopamine D(4) and D(2L) Receptor stimulation of the mitogen-activated protein kinase pathway is dependent on trans-activation of the platelet-derived growth factor receptor. *Mol Pharmacol* 2001;**60**:92–103.

85. Johnston CA, Beazely MA, Vancura AF, Wang JK, Watts VJ. Heterologous sensitization of adenylate cyclase is protein kinase A-dependent in Cath.a differentiated (CAD)-D2L cells. *J Neurochem* 2002;**82**:1087–96.

86. Varga EV, Rubenzik MK, Stropova D, Sugiyama M, Grife V, Hruby VJ, Rice KC, Roeske WR, Yamamura HI. Converging protein kinase pathways mediate adenylyl cyclase superactivation upon chronic delta-opioid agonist treatment. *J Pharmacol Exp Ther* 2003;**306**: 109–15.

87. Varga EV, Yamamura HI, Rubenzik MK, Stropova D, Navratilova E, Roeske WR. Molecular mechanisms of excitatory signaling upon chronic opioid agonist treatment. *Life Sci* 2003;**74**:299–311.

88. Iwami G, Kawabe J, Ebina T, Cannon PJ, Homcy CJ, Ishikawa Y. Regulation of adenylyl cyclase by protein kinase A. *J Biol Chem* 1995;**270**(12):481–4.

89. Wild DM, Kukin M. Beta-blockers to prevent symptomatic heart failure in patients with stage A and B heart failure. *Curr Heart Fail Rep* 2007;**4**:99–102.

90. Okumura S, Kawabe J, Yatani A, Takagi G, Lee MC, Hong C, Liu J, Takagi I, Sadoshima J, Vatner DE, Vatner SF, Ishikawa Y. Type 5 adenylyl cyclase disruption alters not only sympathetic but also parasympathetic and calcium-mediated cardiac regulation. *Circ Res* 2003;**93**:364–71.

91. D'Angelo 2nd DD, Sakata Y, Lorenz JN, Boivin GP, Walsh RA, Liggett SB, Dorn GW. Transgenic Galphaq overexpression induces cardiac contractile failure in mice. *Proc Natl Acad Sci USA* 1997;**94**: 8121–6.

92. Roth DM, Gao MH, Lai NC, Drumm J, Dalton N, Zhou JY, Zhu J, Entrikin D, Hammond HK. Cardiac-directed adenylyl cyclase expression improves heart function in murine cardiomyopathy. *Circulation* 1999;**99**:3099–102.

93. Tepe NM, Liggett SB. Transgenic replacement of type V adenylyl cyclase identifies a critical mechanism of beta-adrenergic receptor dysfunction in the G alpha q overexpressing mouse. *FEBS Letts* 1999;**458**:236–40.

94. Onda T, Hashimoto Y, Nagai M, Kuramochi H, Saito S, Yamazaki H, Toya Y, Sakai I, Homcy CJ, Nishikawa K, Ishikawa Y. Type-specific regulation of adenylyl cyclase. Selective pharmacological stimulation and inhibition of adenylyl cyclase isoforms. *J Biol Chem* 2001;**276**(47):785–93.

Chapter 38

Phosphodiesterase Families

James Surapisitchat and Joseph A. Beavo

Department of Pharmacology, University of Washington School of Medicine, Seattle, Washington

INTRODUCTION

Since its discovery over 50 years ago by Rall and Sutherland, interest in cAMP and its companion, cGMP, has spawned a field of research that has impacted almost every area of biomedical research and has been the basis of five different Nobel Prizes [1]. Many hormones and neurotransmitters act as "first messengers" by binding to receptors and activating a cascade of signaling events via the production of the "second messengers" cAMP and cGMP. These cyclic nucleotides in turn regulate a large number of processes, including proliferation, chemotaxis, differentiation, contraction, gene transcription, and inflammation. These second messengers are produced by adenylyl cyclases (cAMP) and guanylyl cyclases (cGMP), and are utilized by nearly all eukaryotes from amoebae to man. cAMP and cGMP act on a variety of enzymes including protein kinase A and G (PKA and PKG), cyclic nucleotide gated ion channels and guanine-nucleotide exchange factors (GEFs) to regulate numerous downstream signaling cascades. The effects and regulation of cyclic nucleotide signaling are not only controlled by the stimuli that activate cyclic nucleotide synthesis and the downstream effector molecules, but also by the activity of specific cyclic nucleotide phosphodiesterases (PDEs) that hydrolyze cyclic nucleotides, turning off its signaling. Thus, the amplitude and duration of cyclic nucleotide signaling is controlled not only by their production by the cyclases, but also by their hydrolysis by PDEs.

To date, there are 11 known PDE gene families, each with their own distinct characteristics (Figure 38.1). "Cyclic nucleotide phosphodiesterase" was first described as a widely distributed enzyme that could catalyze the hydrolysis of cAMP and cGMP to their respective 5′ monophosphates [2]. The initial studies of PDE activity used either tissue homogenates or partially purified preparations of these enzymes from various tissues. The characteristics of PDE activity from these studies varied greatly depending on the PDE source. It was unclear whether these differences were a consequence of the purification scheme of these enzymes (that is, the presence of different

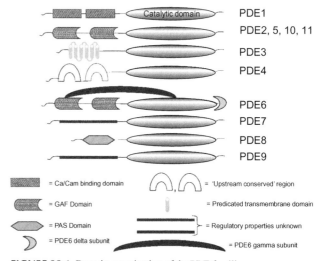

FIGURE 38.1 Domain organization of the PDE families.

contaminating proteins), or of the existence of multiple forms of PDE. PDEs were therefore referred to in terms of the tissue from which they were purified (for example, rat liver PDE or bovine brain PDE). Later, anion exchange chromatography experiments demonstrated the presence of several PDE activities in an individual tissue or cell type. These observations were later confirmed in experiments by use of immunocytochemistry, immunoblotting, and *in situ* hybridization. With the purification of multiple enzymes to apparent homogeneity and more stringent characterization of their properties, PDEs were subsequently named according to their regulatory properties and substrate specificities (for example, calcium/calmodulin-stimulated PDE or cGMP-stimulated PDE). With the advent of molecular biology there was a virtual explosion of new information, including the cloning of the previously known and many new PDE genes, as well the identification of a number of new splice variants. Nucleotide sequence data for the PDEs have also allowed for their organization into gene families according to homology. Fortunately, this organization rather closely agrees with the earlier organizations by regulatory and kinetic properties. As more data emerge regarding the distribution, characteristics, and roles of the many PDE isozymes, it is clear that the regulation of cyclic nucleotide signaling by PDEs is far more complex than could have been imagined when they were first studied in the 1960s.

THE GENE FAMILIES

There currently exist 11 PDE gene families. The nomenclature for a PDE contains, in order, two letters to indicate species, then PDE, followed with a number indicating gene family, a letter to represent an individual gene, and finally a number to identify the splice variant. For example, HSPDE7A1 represents *Homo sapiens* PDE gene family 7, gene A, splice variant 1. For a correlation between older and current PDE nomenclature, see Beavo *et al.* [3]. The kinetic properties, substrate specificities, and drug sensitivities of these families (Table 38.1) have been described extensively elsewhere [3–6], and will not be discussed at length here. In general, the phosphodiesterases share the same organizational structure. Each protein has an N-terminal domain that confers regulatory properties to the protein, followed by a more C-terminal ~270 amino acid catalytic domain and a short C-terminal tail. The sequence identity in the catalytic domain between genes is only about 35 percent, yet all PDEs possess the signature sequence $H–D–X_2–H–X_4–N$ [7]. The substrate specificities of the different PDE families run the gamut from dual-specificity PDEs to those that are highly specific for either cAMP or cGMP. Further, the relative substrate specificity can vary even between members of a gene family. For example, within the PDE1 family, PDE1A2 has a K_m for cGMP that is approximately 20-fold lower than that for cAMP, yet PDE1C2 has a K_m that is equal for both. In addition to variation in specificity, the activity of a PDE toward one nucleotide may depend upon the concentration of the other. For instance, PDE2s hydrolyze cAMP and cGMP with relatively similar K_m values. However, the presence of a small amount of cGMP (which allosterically binds to PDE2) stimulates the enzymes' catalytic activity toward cAMP several-fold [8]. To make things more complex, there are also PDEs for which the cyclic nucleotides are competitive inhibitors for one another. Cyclic AMP is a competitive inhibitor of cGMP hydrolysis by PDE10 [9], and cGMP is a potent competitive inhibitor of cAMP hydrolysis for PDE3s [10]. This variety and flexibility in substrate specificities of the PDEs makes them a family of enzymes with tremendous diversity, suitable for the fine tuning of many cyclic nucleotide-mediated signaling systems.

As mentioned above, most of the PDEs also possess domains within their N-termini that regulate the activity of the catalytic site (Figure 38.1). The PDE1 proteins have two Ca^{2+}/calmodulin binding domains, and binding of calmodulin

TABLE 38.1 Characteristics of the PDE families

PDE family	Genes	# Splice variants	Regulatory domains, role	Phosphorylation	Substrates	Commonly used inhibitors
PDE1	1A, 1B, 1C	12	CaM, activation	PKA	cGMP, cAMP	Vinpocetine, IC86340
PDE2	2A	3	GAF, activation	Unknown	cAMP, cGMP	EHNA, Bay 60-7550
PDE3	3A, 3B	4	Transmembrane domains, membrane targeting	PKB, PKA	cAMP>cGMP (low Vmax)	Milrinone, Cilostamide, Trequinsin
PDE4	4A, 4B, 4C, 4D	>20	UCR1, UCR2, unclear	ERK, PKA	cAMP	Rolipram
PDE5	5A	3	GAF, activation	PKG, PKA	cGMP	Sildenafil, Vardenafil, Tadalafil
PDE6	6A, 6B, 6C	1 each	GAF, activation	PKC, PKA	cGMP	Dipyrimadole, Zaprinast
PDE7	7A, 7B	7	Unknown	Unknown	cAMP	None Identified
PDE8	8A, 8B	9	PAS, unknown	Unknown	cAMP	None Identified
PDE9	9A	20	Unknown	Unknown	cGMP	None Identified
PDE10	10A	10	GAF, unknown	PKA	cAMP, cGMP	None Identified
PD11	11A	4	GAF, unknown	PKA, PKG	cAMP, cGMP	None Identified

to these PDEs stimulates their activity [11]. The PDE2, PDE5, PDE6, PDE10, and PDE11 proteins all have allosteric, cyclic nucleotide-binding domains known to be part of the larger GAF domain family [12]. The consequence of binding of cGMP or cAMP to these domains varies with the PDE. As discussed above, cGMP binding to the GAF domain of PDE2 stimulates activity. For PDE5, binding of cGMP to this domain also increases its activity, thus making PDE5 a central player in a negative feedback loop in regulating cGMP signaling. For further discussion of the GAF domains in PDEs, see Chapter 187 of Handbook of Cell Signaling, Second Edition. The longer variants of the PDE3 family proteins have six predicted transmembrane segments in their amino terminal domains, consistent with the observation that PDE3 activity is at least partially membrane-associated. The PDE4 family, a large family of enzymes with four genes and many splice variants, is responsible for the majority of basal cAMP-hydrolyzing activity in many cell types. For further discussion of the PDE4 family, see Chapter 174 of Handbook of Cell Signaling, Second Edition. The N-terminus of PDE8 contains a PAS domain, a domain that generally is found in proteins that are involved in sensing and responding to the cellular environment (for example, redox state, light levels, or energy levels) [13]. The PAS domains in many proteins bind small molecules such as heme, NAD, or chromaphores, and can also serve as a site for homodimerization. It will be interesting to see whether some small molecules also bind the PDE8 PAS domain, what effect this may have on activity, and also whether the PAS domain of PDE8 serves to dimerize the protein. PDE7 and PDE9 have substantial N-terminal segments that bear no resemblance to known proteins. What roles these domains of PDE7 and PDE9 have in regulation of the protein remain to be seen.

In addition to different regulatory domains, different PDEs are subject to protein phosphorylation by a variety of kinases that can alter PDE activity. In pancreatic β cells, PDE3B can be phosphorylated and activated by PKB in response to leptin stimulation [14]. Various members of the PDE4 family can be phosphorylated and regulated by PKA and ERK. PDE5 can be phosphorylated by PKG, stabilizing it ability to bind and be activated by cGMP. Thus, in addition to the great variation in nucleotide specificity and kinetic properties, the PDE superfamily comprises a complex set of enzymes that can provide cross-talk between the cGMP and cAMP pathways, and with Ca^{2+}/CaM-dependent pathways and various kinase pathways, and allow the cell exquisite control of cyclic nucleotide dynamics.

IMPLICATIONS OF MULTIPLE GENE FAMILIES/SPLICE VARIANTS

In addition to the multiple means of regulation of the various PDEs by their distinctive regulatory domains, another remarkable feature of the PDE superfamily is the highly individual expression and localization patterns of its members. Individual PDEs, within gene families, and even between splice variants, have unique tissue, cellular, and even subcellular expression patterns. For example, the PDE1 genes PDE1A, PDE1B, and PDE1C are all expressed in brain. However, *in situ* localization studies have shown that PDE1A is expressed primarily in the cortex, PDE1B in the striatum, and PDE1C in the cerebellum [15]. The expression pattern of PDE1C splice variants has been further broken down. PDE1C5 is highly expressed in testis; PDE1C1 is more generally distributed, being found in heart, testis, cerebellum, and olfactory epithelium; and PDE1C2 is primarily expressed in the olfactory neuroepithelium.

The PDE3 genes provide another example of the complex localization of family members. PDE3A and PDE3B can be found in distinct cell types, with PDE3A expressed in platelets and PDE3B in adipose cells and hepatocytes. However, both are apparently expressed in vascular smooth muscle cells, cardiomyocytes, and endothelial cells, although probably in different compartments. The expression and localization of PDE3A and 3B among and within cell types adds additional layers of complexity to cyclic nucleotide regulation.

The ability to produce multiple N-terminal variants allows for specific differential targeting of the PDE2 family members. The three PDE2A splice variants differ only at their extreme N-terminus. However, the PDE2A2 N-terminus contains a putative transmembrane domain, and the PDE2A3 variant contains an N-terminal myristoylation site, both of which probably allow for targeting to membrane compartments of the cell and are responsible for the membrane-associated forms of PDE2 activity observed in tissue homogenates [16].

All of the above examples of differential localization of PDE genes/splice variants imply that each of these PDEs probably plays specialized roles in the regulation of cyclic nucleotide signaling in cells. Perhaps the definitive example of precise localization of a PDE to achieve specialized function is in the case of the PDE6 gene family. In the photoreceptor cells of the retina, a visual signal is generated through the activation of a cascade of proteins that ultimately result in the activation of PDE6. PDE6 rapidly hydrolyzes the cGMP in the cell, the resident cGMP-gated cation channels close, and the cells become hyperpolarized. All of the proteins involved in this cascade, including PDE6, are expressed primarily in the retina and specifically targeted to the membrane disks of the photoreceptors. Further discussion and specific details of the role of PDE6 in phototransduction can be found in Chapter 178 of Handbook of Cell Signaling, Second Edition. Although the PDE6 family is the first known and best characterized example of a highly specialized PDE, there are certainly others. A prime candidate example is PDE1C2. As mentioned previously, PDE1C2 is highly expressed in the olfactory neuroepithelium. The other major PDE expressed there is PDE4A. However, expression of PDE1C2 is restricted to the cilia of the epithelium, where it co-localizes with adenylate cyclase III. PDE4A is not expressed in the cilia of the neurons, but rather throughout the remainder of the neuronal layer [17]. Clearly, in the olfactory neuroepithelium, PDE1C2 and PDE4A are playing different roles in regulating cAMP during olfaction.

A demonstration of different PDEs playing different functional roles was also shown in insulin secreting pancreatic β cells. Inhibition of either PDE3 or PDE4 can elevate cAMP levels in β cells. However, only inhibition of PDE3 blocked leptin inhibition of insulin secretion [14]. These results suggest that PDE3 and PDE4 regulate different functional pools of cAMP in β cells. Analogous examples have been shown in cardiac myocyctes, and will be discussed in Chapter 195 of Handbook of Cell Signaling, Second Edition.

While the different roles of PDE3 and PDE4 in β cells is one of several emerging examples of specific functional pools of cAMP regulated by one specific PDE, there are examples in which a single pool of cAMP may be regulated by multiple PDEs [18]. In cardiomyocytes and endothelial cells, cGMP, via its differential regulation of PDE2A (activation) and PDE3A (inhibition), has a biphasic effect on cAMP-mediated processes, such as L-type Ca^{2+} channel phosphorylation by PKA in cardiomyocytes and inhibition of thrombin-induced permeability in endothelial cells [10, 19]. Not only are these examples of cGMP/cAMP crosstalk, but they also provide examples in which single functional pools of cAMP may be regulated by multiple PDEs.

The concept of compartmentalization of cyclic nucleotide signaling to discrete "pools" is a topic of increasing interest despite being first raised nearly 25 years ago by Buxton and Brunton [20]. Compartmentalization explains how a second messenger as ubiquitous as cAMP and cGMP can mediate different effects within a cell [21, 22]. The idea that there are different cyclic nucleotide "pools" available only to certain effectors such as kinases, channels, and exchange factors localized by scaffolding proteins such as AKAPs is critical to the theory of compartmentalization. Increasing evidence points to PDEs being vitally important and critical for maintaining these pools. The fact that there are 11 different PDE families, each with their unique characteristics, allows for differential regulation of these cyclic nucleotide pools. In the simplest scenario, one PDE may be localized to a certain region of the cell to regulate one pool of cyclic nucleotide(s), while another, due to differential splicing or expression of another PDE gene altogether, may be targeted to regulate a different cyclic nucleotide pool. However, due to the 11 different PDE families that can be differentially regulated by other signaling pathways such as phosphorylation, intracellular calcium, and other cyclic nucleotides, the regulation of one pool of cyclic nucleotide often may be under the influence of more than one PDE. This would allow for one function to be differentially regulated by different signaling pathways. Other scenarios can also be imagined where a certain cyclic nucleotide pool may be regulated in layers or

where different PDEs act as sequential barriers like dams in a river for control of cyclic nucleotide flux to specific effector molecules and cellular functions. Thus, our increasing understanding of the roles of the 11 different PDE families and the over 50 different splice variants is becoming progressively more important for our understanding of cyclic nucleotide and cellular signaling.

ALTERED PDE EXPRESSION IN PATHOLOGICAL STATES

Alteration in PDE expression in response to various stimuli is a common mechanism that many cells and tissues use to alter cyclic nucleotide signaling. For example, nitroglycerine (NTG) is used in the treatment of hypertension for its vasorelaxing effects, but its therapeutic use is limited due to the development of nitrate tolerance. In rats treated with NTG, PDE1A expression and activity is upregulated [23]. Inhibition of PDE1A leads to partial restoration of smooth muscle responsiveness to nitrates. PDE1C is found to be upregulated in proliferating smooth muscle, suggesting a potential target in the treatment of atherosclerosis or restenosis after angioplasty [24]. PDE7 and PDE8 are upregulated during T cell activation by CD3 and CD28 [25]. Other examples of PDE upregulation include the upregulation of PDE1B in monocyte to macrophage differentiation, and PDE5 in vascular smooth muscle in response to angiotensin II stimulation and in hypertrophied hearts [26–28]. Few examples of downregulation of PDEs have been re-ported to our knowledge. PDE3A expression and activity has been found to be decreased in patients with heart failure [29]. In 3T3-L1 adipocytes, TNFα decreased PDE3B expression, implicating a mechanism by which TNFα regulates lipolysis [30]. Recently, it has been shown that TNFα can increase PDE2 (cGMP-stimulated) and decrease PDE3A (cGMP-inhibited) in endothelial cells altering their response to cGMP signaling [10, 31]. This growing list of examples in which alteration of PDE expression and activity in response to physiological and pathological stimuli leading to altered cellular responses to cyclic nucleotide signaling must be taken into account in developing treatments that manipulate cyclic nucleotide signaling.

PDE INHIBITORS AS THERAPEUTIC AGENTS

The wide variety of PDE isozymes implies that inhibition or activation of PDEs has tremendous therapeutic potential. Indeed, some of the oldest drugs used by man (for example, caffeine, ginseng) are PDE inhibitors! Some PDE inhibitors such as theophylline, papaverine, and dipyridamole were in fact used before their mechanism of action was known. However, with a new appreciation of the complexity of the

PDE signaling systems and the availability of more sophisticated endpoint assays, new generations of PDE-inhibiting drugs are being developed. It seems likely that this will be extended to PDE activators, although to our knowledge this has not yet happened.

Non-selective PDE inhibitors have long been known to have anti-inflammatory properties, and have been used for the treatment of asthma, stroke, and chronic obstructive pulmonary disease (COPD). Early treatments for these diseases with selective PDE4 inhibitors have unfortunately been hampered by the side effect of emesis. Novel PDE4 inhibitors such as Ariflo from GlaxoSmithKline, now in phase III clinical trials for the treatment of asthma and COPD, may have milder emetic side effects [32]. Milrinone, a PDE3 inhibitor that has been used to treat patients in congestive heart failure, has recently been shown *in vitro* to increase conductance of the CFTR transporter, indicating some promise for the treatment of cystic fibrosis [22, 33]. Cilostazol, also a PDE3 inhibitor, is currently approved for the treatment of patients with intermittent claudication, and clinical trials suggest that cilostazol also may be useful in the prevention of restenosis after angioplasty [34, 35]. Dipyridamole can inhibit PDE activity in platelets and is commonly used in combination with aspirin (ASA) to reduce clotting, despite the initial lack of clinical data demonstrating an added benefit of dipyridamole over ASA alone [36]. However, the European Stroke Prevention Study (ESPS2) clearly demonstrated an additive effect of dipyridamole and ASA in the prevention of subsequent stroke [4]. Thus, dipyridamole (Aggrenox®) continues to be recommended to be prescribed in combination with ASA [37]. Finally, Viagra® (sildenafil), Levitra® (vardenafil), and Cialis® (tadalafil), all PDE5-specific inhibitors, have been used successfully for the treatment of male erectile dysfunction with minimal side effects. More recently, it has been shown that PDE5-specific inhibitors are useful in the treatment of pulmonary hypertension [38], and may be of use for treatment of cardiac hypertrophy and even jet lag [39, 40].

Recent studies using PDE inhibitors and cyclic nucleotide analogs have suggested a re-evaluation of previously reported findings. The recent identification and characterization of PDE8 has shown that this PDE, unlike most others, is insensitive to the widely used non-selective inhibitor, IBMX [41]. Moreover, *in vivo* studies have shown that PDE8 is important in regulating Leydig cell function [42] and insulin secretion [43]. Past and future work using IBMX thus needs to be carefully analyzed. Other useful tools in studying cyclic nucleotide signaling are cyclic nucleotide analogs, in part because of their reported (or assumed) resistance to hydrolysis by PDEs. However, recent findings have shown that this might not be the case (Poppe and colleagues, unpublished information), since several analogs have been shown to be substrates for PDEs and many are inhibitors of PDEs. One such example

is 8-pCPT-2OMe-cAMP, an analog used to study Epac function due to its ability to activate Epac but not PKA. Findings that a function is dependent on Epac based on the use of this analog need to be re-evaluated, since inhibition of a PDE by 8-pCPT-2OMe-cAMP may lead to increased endogenous cAMP and activation of PKA. The use of cyclic nucleotide analogs and their effects on PDEs will be discussed in detail later in this volume (see Chapter 191 of Handbook of Cell Signaling, Second Edition).

WHERE DO WE GO FROM HERE?

Although it is certainly possible that more PDEs exist that bear little resemblance in linear sequence to the currently known proteins, it is also certain that most if not all of the proteins responsible for the major PDE activities in tissues have been identified and their genes cloned. Therefore the major focus of future research on PDEs likely will be on the *in vivo* function of PDEs, especially the newly cloned PDEs such as PDE8, 9, 10, and 11. With the field of cyclic nucleotide signaling and signal transduction in general focusing on compartmentalized signaling domains, PDEs will undoubtedly play a critical role in these compartments. Several PDEs have already been shown to be critical in establishing cAMP signaling pools [18]. With such a large family of genes, each differentially expressed, localized, and regulated, cyclic nucleotide compartmentalization and signaling can potentially be very complex. With the added factor of altered PDE expression under normal versus pathological conditions resulting in altered cyclic nucleotide signaling, the need for specific inhibitors as well as techniques to measure cyclic nucleotides and PDE activity in real-time and space is even more crucial in enabling understanding of PDE function and treating cyclic nucleotide-based diseases. Nearly 50 years after the discovery of cAMP, cGMP, and PDEs, the field of PDE research is still ripe for potential new discoveries, and particularly as an area for drug development.

ACKNOWLEDGEMENTS

This work was supported by NIH grant DK21723 and the Foundation Leducq.

REFERENCES

1. Beavo JA, Brunton LL. Cyclic nucleotide research – still expanding after half a century. *Nat Rev* 2002;**3**:710–18.
2. Cheung WY. Cyclic nucleotide phosphodiesterase. *Adv Biochem Psychopharmacol* 1970;**3**:51–65.
3. Beavo JA. Cyclic nucleotide phosphodiesterases: functional implications of multiple isoforms. *Physiol Revi* 1995;**75**:725–48.
4. Redman AR, Ryan GJ. Analysis of trials evaluating combinations of acetylsalicylic acid and dipyridamole in the secondary prevention of stroke. *Clin Ther* 2001;**23**:1391–408.

5. Houslay MD. PDE4 cAMP-specific phosphodiesterases. *Prog Nucleic Acid Res Mol Biology* 2001;**69**:249–315.

6. Soderling SH, Beavo JA. Regulation of cAMP and cGMP signaling: new phosphodiesterases and new functions. *Curr Opin Cell Biol* 2000;**12**:174–9.

7. Charbonneau H, Beier N, Walsh KA, Beavo JA. Identification of a conserved domain among cyclic nucleotide phosphodiesterases from diverse species. *Proc Natl Acad Sci USA* 1986;**83**:9308–12.

8. Martins TJ, Mumby MC, Beavo JA. Purification and characterization of a cyclic GMP-stimulated cyclic nucleotide phosphodiesterase from bovine tissues. *J Biol Chem* 1982;**257**:1973–9.

9. Soderling SH, Bayuga SJ, Beavo JA. Isolation and characterization of a dual-substrate phosphodiesterase gene family: PDE10A. *Proc Natl Acad Sci USA* 1999;**96**:7071–6.

10. Surapisitchat J, Jeon KI, Yan C, Beavo JA. Differential regulation of endothelial cell permeability by cGMP via phosphodiesterases 2 and 3. *Circ Res* 2007;**101**:811–18.

11. Kakkar R, Raju RV, Sharma RK. Calmodulin-dependent cyclic nucleotide phosphodiesterase (PDE1). *Cell Mol Life Sci* 1999;**55**:1164–86.

12. Aravind L, Ponting CP. The GAF domain: an evolutionary link between diverse phototransducing proteins. *Trends Biochem Sci* 1997;**22**:458–9.

13. Taylor BL, Zhulin IB. PAS domains: internal sensors of oxygen, redox potential, and light. *Microbiol Mol Biol Rev* 1999;**63**:479–506.

14. Zhao AZ, Bornfeldt KE, Beavo JA. Leptin inhibits insulin secretion by activation of phosphodiesterase 3B. *J Clin Invest* 1998;**102**:869–73.

15. Yan C, Bentley JK, Sonnenburg WK, Beavo JA. Differential expression of the 61-kDa and 63-kDa calmodulin-dependent phosphodiesterases in the mouse brain. *J Neurosci* 1994;**14**:973–84.

16. Beavo JA, Hardman JG, Sutherland EW. Stimulation of adenosine 3′,5′-monophosphate hydrolysis by guanosine 3′,5′-monophosphate. *J Biol Chem* 1971;**246**:3841–6.

17. Juilfs DM, Fulle HJ, Zhao AZ, Houslay MD, Garbers DL, Beavo JA. A subset of olfactory neurons that selectively express cGMP-stimulated phosphodiesterase (PDE2) and guanylyl cyclase-D define a unique olfactory signal transduction pathway. *Proc Natl Acad Sci USA* 1997;**94**:3388–95.

18. Fischmeister R, Castro LR, Abi-Gerges A, et al. Compartmentation of cyclic nucleotide signaling in the heart: the role of cyclic nucleotide phosphodiesterases. *Circ Res* 2006;**99**:816–28.

19. Vandecasteele G, Verde I, Rucker-Martin C, Donzeau-Gouge P, Fischmeister R. Cyclic GMP regulation of the L-type Ca(2+) channel current in human atrial myocytes. *J Physiol* 2001;**533**:329–40.

20. Buxton IL, Brunton LL. Compartments of cyclic AMP and protein kinase in mammalian cardiomyocytes. *J Biol Chem* 1983;**258**:10,233–10,239.

21. Hayes JS, Brunton LL, Mayer SE. Selective activation of particulate cAMP-dependent protein kinase by isoproterenol and prostaglandin E1. *J Biol Chem* 1980;**255**:5113–19.

22. Kelley TJ, Thomas K, Milgram LJ, Drumm ML. In vivo activation of the cystic fibrosis transmembrane conductance regulator mutant deltaF508 in murine nasal epithelium. *Proc Natl Acad Sci USA* 1997;**94**:2604–8.

23. Kim D, Rybalkin SD, Pi X, et al. Upregulation of phosphodiesterase 1A1 expression is associated with the development of nitrate tolerance. *Circulation* 2001;**104**:2338–43.

24. Rybalkin SD, Bornfeldt KE, Sonnenburg WK, et al. Calmodulin-stimulated cyclic nucleotide phosphodiesterase (PDE1C) is induced in human arterial smooth muscle cells of the synthetic, proliferative phenotype. *J Clin Invest* 1997;**100**:2611–21.

25. Li L, Yee C, Beavo JA. CD3- and CD28-dependent induction of PDE7 required for T cell activation. *Science* 1999;**283**:848–51.

26. Bender AT, Ostenson CL, Wang EH, Beavo JA. Selective up-regulation of PDE1B2 upon monocyte-to-macrophage differentiation. *Proc Natl Acad Sci USA* 2005;**102**:497–502.

27. Kim D, Aizawa T, Wei H, et al. Angiotensin II increases phosphodiesterase 5A expression in vascular smooth muscle cells: a mechanism by which angiotensin II antagonizes cGMP signaling. *J Mol Cell Cardiol* 2005;**38**:175–84.

28. Nagendran J, Archer SL, Soliman D, et al. Phosphodiesterase type 5 is highly expressed in the hypertrophied human right ventricle, and acute inhibition of phosphodiesterase type 5 improves contractility. *Circulation* 2007;**116**:238–48.

29. Ding B, Abe J, Wei H, et al. Functional role of phosphodiesterase 3 in cardiomyocyte apoptosis: implication in heart failure. *Circulation* 2005;**111**:2469–76.

30. Rahn Landstrom T, Mei J, Karlsson M, Manganiello V, Degerman E. Down-regulation of cyclic-nucleotide phosphodiesterase 3B in 3T3-L1 adipocytes induced by tumour necrosis factor alpha and cAMP. *Biochem J* 2000;**346**:337–43.

31. Seybold J, Thomas D, Witzenrath M, et al. Tumor necrosis factor-alpha-dependent expression of phosphodiesterase 2: role in endothelial hyperpermeability. *Blood* 2005;**105**:3569–76.

32. Kroegel C, Foerster M. Phosphodiesterase-4 inhibitors as a novel approach for the treatment of respiratory disease: cilomilast. *Expert Opin Invest Drugs* 2007;**16**:109–24.

33. Smith SN, Middleton PG, Chadwick S, et al. The *in vivo* effects of milrinone on the airways of cystic fibrosis mice and human subjects. *Am J Respir Cell Mol Biol* 1999;**20**:129–34.

34. Eberhardt RT, Coffman JD. Drug treatment of peripheral vascular disease. *Heart Dis (Hagerstown)* 2000;**2**:62–74.

35. El-Beyrouty C, Spinler SA. Cilostazol for prevention of thrombosis and restenosis after intracoronary stenting. *Ann Pharmacother* 2001;**35**:1108–13.

36. Gibbs CR, Lip GY. Do we still need dipyridamole?. *Br J Clin Pharmacol* 1998;**45**:323–8.

37. Sudlow C. Dipyridamole with aspirin is better than aspirin alone in preventing vascular events after ischaemic stroke or TIA. *Br Med J (Clin Res ed)* 2007;**334**:901.

38. Galie N, Ghofrani HA, Torbicki A, et al. Sildenafil citrate therapy for pulmonary arterial hypertension. *N Engl J Med* 2005;**353**:2148–57.

39. Takimoto E, Champion HC, Li M, et al. Chronic inhibition of cyclic GMP phosphodiesterase 5A prevents and reverses cardiac hypertrophy. *Nature Med* 2005;**11**:214–22.

40. Agostino PV, Plano SA, Golombek DA. Sildenafil accelerates reentrainment of circadian rhythms after advancing light schedules. *Proc Natl Acad Sci USA* 2007;**104**:9834–9.

41. Soderling SH, Bayuga SJ, Beavo JA. Cloning and characterization of a cAMP-specific cyclic nucleotide phosphodiesterase. *Proc Natl Acad Sci USA* 1998;**95**:8991–6.

42. Vasta V, Shimizu-Albergine M, Beavo JA. Modulation of Leydig cell function by cyclic nucleotide phosphodiesterase 8A. *Proc Natl Acad Sci USA* 2006;**103**:19,925–19,930.

43. Dov A, Abramovitch E, Warwar N, Nesher R. Diminished PDE8B potentiates biphasic insulin response to glucose. *Endocrinology* 2007;**149**:741–8.

cAMP-Dependent Protein Kinase

Susan S. Taylor [1, 2] and Elzbieta Radzio-Andzelm[2]

[1]Department of Chemistry and Biochemistry

[2]Department of Pharmacology, Howard Hughes Medical Institute, University of California at San Diego, La Jolla, California

INTRODUCTION

cAMP-dependent protein kinase (PKA) is one of the best characterized members of the large protein kinase super-family. The catalytic subunit (C) serves as a structural prototype for the entire family. The inactive holoenzyme comprises a regulatory (R) subunit dimer and two catalytic subunits. Binding of cAMP to the R subunits unleashes the active C subunits. In this chapter, the structure of the C subunit is described and correlated with its function. The structure of the dimerization/docking domain of RIIα and the cAMP binding domains of RIα and RIIβ are also described and correlated with the dynamic properties of the R subunits.

cAMP-dependent protein kinase (PKA) was one of the first protein kinases to be discovered [1], the first to be sequenced [2], the first to be cloned [3], and the first protein kinase for which a crystal structure was solved [4]. It thus serves in many ways as a prototype for the entire protein kinase superfamily, which represents approximately 2 percent of the human genome. cAMP is an ancient stress response signal; for example, it is a universal indicator of glucose deprivation. Whereas in bacteria the cAMP second messenger is linked to the catabolite gene activator protein, in mammals it is linked primarily to the activation of PKA. PKA is ubiquitous in mammalian cells, and regulates many diverse pathways.

As stated above, the inactive holoenzyme complex consists of a regulatory (R) subunit dimer and two catalytic (C) subunits. Binding of cAMP to the R subunits unleashes the active C subunits, thus allowing them to phosphorylate a variety of protein substrates, both cytosolic and nuclear [5, 6]. In addition to serving as inhibitors of PKA activity and receptors for cAMP, the R subunits also serve as adapters that tether the C subunits to specific cellular locations by binding to A kinase anchoring proteins (AKAPs) [7]. PKI, another inhibitor of the C subunit that is independent of cAMP [8], also contributes to trafficking of the free C subunits between the cytoplasm and the nucleus [9]. The inhibitors of PKA activity are both modular and multifunctional proteins. A review of PKA structure thus must include the diverse set of proteins that contribute overall to PKA regulation. The structures of the C subunit and its inhibitors, both the R subunits and PKI, are described here, as well as more recent structures of the RIα and RIIα holoenzyme complexes [10-12].

CATALYTIC SUBUNIT

In mammals, three isoforms of the C subunit have been identified: α, β, and γ [3, 13, 14]. The Cα subunit is expressed constitutively in all cells, whereas expression of Cβ is tissue-specific, and especially prominent in brain. Cγ is found primarily in testes. Several splice variants of both Cα [15] and Cβ [16] also exist; all differ in the first exon. In the primary form of Cα, β, and γ, exon I codes for 14 amino acids that include an N-terminal myristylation site [17]. The other C subunit splice variants are typically not myristylated. In addition to co-translational myristylation, the C subunit is phosphorylated posttranslationally at two essential sites [18]. Phosphorylation at Thr197 in the activation loop is essential for efficient catalysis [19]. Phosphorylation at Ser338 is essential for stability, and is very likely to be an important part of the maturation of the initial transcript into an active enzyme [20]. Phosphorylation at Ser10 [21] and deamidation of Asn2 [22] are other posttranslational modifications that have been identified.

Catalytic Properties

The C subunit is a highly concerted enzyme; all its energy is focused on transferring the γ-phosphate of ATP to an appropriate substrate protein [23]. There are two general recognition motifs for PKA substrates [24]: Arg–Arg–X–Ser/Thr–Hyd, and Arg–X–X–Arg–X–X–Ser/Thr–Hyd, where X is any residue and Hyd is a hydrophobic residue. The mechanism for catalysis has been carefully defined by Adams [25]. Pre-steady-state kinetics established that the actual rate of phosphoryl transfer is very fast (>500 per second) whereas the kcat is only 20 per second. For PKA, the k_{cat} correlates, in general, with the release of ADP and the conformational changes that allow for its release [26]. The K_m for the heptapeptide, kemptide (Leu–Arg–Arg–Ala–Ser–Leu–Gly), is 10–20 μM; however, this K_m does not reflect a true binding affinity. The K_d (200–300 μM) more accurately reflects affinity [27].

STRUCTURE OF THE CATALYTIC SUBUNIT

Multiple forms of the C subunit have been crystallized, and these structures provide a molecular understanding of nucleotide binding, peptide binding, and conformational flexibility [4, 28–30]. The C subunit comprises a highly conserved core containing a smaller ATP binding domain (residues 40–126) that is dominated by β structure, and a larger, mostly helical lobe (residues 127–300) that provides a docking site for peptides/proteins as well as several essential residues that contribute to catalysis (Figure 179.1). The adenine ring of ATP is buried at the base of the cleft between the two lobes, and the peptide docks to the surface of the large lobe at the edge of the cleft. This core is conserved in all protein kinases that phosphorylate Ser, Thr, or Tyr [31].

Conserved Core

As recognized initially by Hanks and Hunter [32], the conserved kinase core consists of a set of sequence motifs that span the entire core (Figure 179.1). Although affinity labeling provided clues about the roles of some of these motifs [6, 33, 34], the first crystal structure revealed the unique architecture that brings most of these conserved motifs to the active site cleft, where they contribute primarily to the binding of ATP and phosphoryl transfer [35]. The detailed characterization of the C subunit is reviewed in Johnson *et al.* [36].

Small Lobe

In general, the small lobe is more "loosely" structured than the large lobe (Figure 179.2, left). One of the most essential features of this enzyme is the glycine-rich loop that links β strands 1 and 2. In most of the crystal structures, this loop is disordered or ordered poorly [37]. Only in the ternary complex where ATP, or an ATP analog, and an inhibitor peptide, PKI (5–24), are bound [29, 38], and in a recently solved aluminum fluoride complex that mimics a transition state intermediate, is the tip of the loop firmly anchored (Figure 179.2, center) [39].

The hydrogen bond between the backbone amide of Ser53 and the γ-phosphate of ATP is probably the driving force for catalysis [40]. It positions the γ-phosphate for transfer to a peptide or protein substrate. The two other essential residues in the small lobe are Lys72 in β−strand 3, which anchors the α- and β-phosphates of ATP, and Glu91 in the C helix, which interacts with Lys72. All crystal structures of the C subunit so far have been of the active, fully phosphorylated protein. Phosphorylation decreases the K_m (ATP) 50-fold, and increases the rate of phosphoryl transfer from $500\,s^{-1}$ to 20^{-1} [19]. For many other protein kinases, the proper orientation of the C helix is dependent

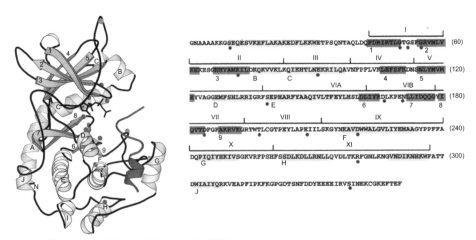

FIGURE 179.1 Structure and sequence of the catalytic subunit of PKA.
A ribbon diagram of the mouse C subunit bound to ATP and an inhibitor peptide PKI(5-24) is on the left [29]. b strands are in medium gray; a-helices are light gray. PKI(5-24) is dark gray. Conserved residues are indicated as dark-gray balls, phosphorylation sites as medium-gray balls. On the right is the sequence with the same color coding.

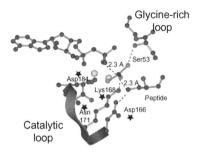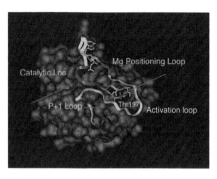

FIGURE 179.2 The two lobes converge at the active site cleft to mediate catalysis.
The dynamic small lobe (N-lobe), shown on the left, binds to ATP and positions the g-phosphate for phosphoryl transfer. The large lobe (C-lobe) is very stable and provides a docking surface for the catalytic machinery (right). Also shown is the Thr197, which is phosphorylated in the active form of PKA. The aluminum fluoride complex, shown in the center, mimics a transition state intermediate where we see the convergence of active site residues in a complex of C subunit with ADP, AlF3, and a substrate peptide. This structure reveals how many of the conserved residues cluster around the active site cleft and contribute to the transfer of the g-phosphate of ATP to the peptide substrate. The two Mg21 ions are shown as spheres.

upon phosphorylation of the activation loop in the large lobe [41]. Most likely, this is also true for the C subunit when the protein is unphosphorylated.

Large Lobe

Although the large lobe is mostly helical, there is also an extended β-sheet that lies at the active site cleft. Most of the conserved residues in the large lobe are localized on this β-sheet, and the β-sheet is anchored firmly through hydrophobic interactions to the rest of the helical large lobe (Figure 179.2, right). The catalytic loop linking β-strands 6 and 7 contains three conserved residues; Asp166 and Asn171 are universally conserved, whereas Lys168 is conserved in all Ser/Thr specific kinases. Although Asp166 is positioned to be a catalytic base, it contributes only minimally to phosphoryl transfer and is thought to be used primarily for orienting the peptide hydroxyl moiety rather than contributing significantly to the nucleophilic properties of the attacking group [42]. Asn171 binds to the second metal ion that interacts with the α- and γ-phosphates of ATP. It also hydrogen bonds to the backbone carbonyl of Asp166, thereby stabilizing the backbone of the catalytic loop. The magnesium-positioning loop, residues 184–187, links β strands 8 and 9. Asp184 binds the activating Mg ion that bridges the β- and γ-phosphates of ATP. β-strand 9 is followed by the activation loop, which is positioned for optimal phosphorylation by the phosphorylation of Thr197. When expressed in *E. coli*, Thr 197 can be autophosphorylated. However, Thr197 is also an excellent substrate for PDK1 [43], and in mammalian cells it is more likely that the C subunit is phosphorylated by a heterologous protein kinase, not by autophosphorylation [44]. Thr197 is followed by the P + 1 loop, named because three residues (Leu198, Pro202, and Leu205) fold inward and form a docking site for hydrophobic P + 1 residue. In fact, however, this loop can be more appropriately referred to as the peptide-organizing loop, since almost every residue contributes to some aspect

of peptide recognition. Gly200 and Thr201 are essential and conserved in all Ser/Thr protein kinases. Gly200 abuts the backbone of the P-site residue and forms a hydrogen bond to the P + 1 backbone amide. In contrast, the side-chain of Thr201 interacts directly with catalytic loop residues, where it is wedged between and hydrogen bonds to the side-chains of Lys168, which positions the γ-phosphate of ATP and Asp166, which positions the hydroxyl acceptor in the peptide substrate. The bridging role of Thr201 is seen most clearly in the structure of ADP, AlF3, and a substrate peptide (Figure 179.2, center) [39]. Glu203 in PKA provides a docking site for the P-6 Arg and Tyr204 hydrogen bonds to Glu230, a primary recognition site for the P-2 Arg. The aromatic ring of Tyr204 also contributes to peptide binding.

Flanking Tails

In PKA, as seen in Figure 179.3, the core is flanked by 40 additional residues, referred to as the N-tail, at the N-terminus, which begins with a myristyl moiety attached to the N-terminal Gly. This is followed by an amphipathic helix that is anchored by hydrophobic interactions to both the small and large lobe of the core [45]. The core is followed by a 50-residue C-terminal tail (C-tail) that is anchored to the large lobe (residues 301–318), has a flexible anionic "gate" that draws basic peptides to the active site cleft, and terminates with a hydrophobic motif at the C terminus (Phe–Ser–Glu–Phe) [37, 46]. This hydrophobic motif is anchored to a hydrophobic pocket on the small lobe and probably helps orient the C helix into its active conformation.

The AGC subfamily of the kinome tree includes PKA, PKC, and PKA, as well as other kinases such as Akt, PDK1, and RSK. A comprehensive analysis of the AGC subfamily has established that the C-tail is a conserved feature of all AGC kinases, and that it contains many conserved motifs that allow it to interact not only with the kinase core but also with other proteins [47]. There are three well-defined regions that are defined on the basis of their interactions

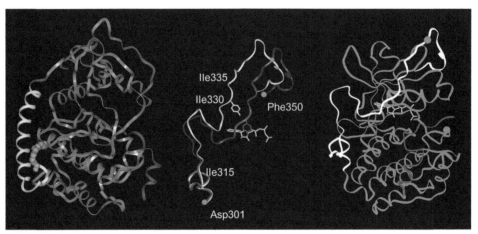

FIGURE 179.3 The N and C terminal tails of the catalytic subunit.
The structure of the myristylated N terminus (residues 1–40) was observed in the mammalian C subunit (left); this structure represents an open conformation. On the right is a structure of a ternary complex of the recombinant C subunit with the C terminal tail highlighted in white. In the center is shown the conformation of the C terminal tail in an "open" and "closed" conformation. Tyr330 in the closed conformation forms a nucleation site by interacting with the ribose of ATP, the linker through Glu127, and the P-3 Arg through a water molecule. Replacement of Tyr330 with Ala leads to significant loss of activity. In the absence of ATP, the tail tends to be disordered.

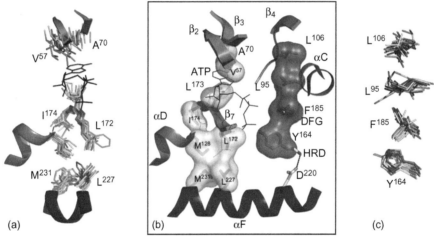

FIGURE 179.4 The kinase core contains two hydrophobic spines that are comprised of spatially conserved but non-linear elements.
The regulatory spine (right) is assembled only in active kinases [48] while the catalytic spine (left) is completed by the adenine ring of ATP [49]. The F helix serves as a hydrophobic scaffold onto which the rest of the molecule assembles.

with the core. The C-lobe tether (CLT: residues 301–318 in PKA) is anchored to the large lobe, while the N-lobe tether (NLT: residues 328–350 in PKA) is firmly anchored to the small lobe through the hydrophobic motif. The variable and highly dynamic region is anchored to the active site in the closed conformation, and is an integral part of the ATP binding pocket. It is thus referred to as the Active Site Tether (AST). Based on this analysis of the AGC family of protein kinases, the C-tail is now considered to be an essential *cis*-regulatory element for all of the AGC kinases.

Conserved Non-Linear Motifs

Because there are now many protein kinase structures available, we can begin to understand some of the global

features that describe the architecture and organization of this enzyme family. The catalytic subunit of PKA continues to serve as a prototype for the family. Using a graph theory-based method, Local Spatial Pattern (LSP) alignment, we have discovered a defining principle for the kinase core. A comparison of active and inactive kinases established first that there is a regulatory spine (R-spine) comprised of hydrophobic residues from the large and small lobes [48]. In the absence of activation, typically mediated by the phosphorylation of the activation loop, the spine is broken (Figure 179.4). Further analysis of active kinases established that there is a second hydrophobic spine referred to as a catalytic spine (C-spine) [49]. This second spine is again built from residues in the large and small lobes; however, in this case the two lobes are linked by the adenine ring of ATP (Figure 179.4). The hydrophobic F helix that spans the large lobe serves as the organizing unit for the entire kinase core. The R-spine in anchored to the N-terminus of the F helix, while the C-spine is anchored to the C-terminus of the F helix. The highly unusual hydrophobic F helix that extends across the middle of the large lobe provides a docking surface for all of the functional elements of the large lobe, including the catalytic loop, the activation segment, and the αH–αI loop.

PROTEIN KINASE INHIBITOR

PKI contains a 20-residue inhibitor domain that binds to the free C subunit ($K_d = 2$ nM). In solution, PKI, which contains 75 amino acids, is mostly disordered, with the exception of two helical regions [50]. The first helical region provides high-affinity binding for PKI to the C subunit [8]. This amphipathic helix precedes the consensus site, which for PKI contains an Ala at the P site. The high-affinity binding of PKI requires ATP. While the consensus site segment of PKI binds to the active site cleft region, the high-affinity binding of PKI requires the amphipathic helix that docks to a hydrophobic pocket composed of Tyr^{235}–Pro–Pro–Phe–Phe [28]. The second helix in the C subunit lies in the region that harbors the nuclear export signal [9].

REGULATORY SUBUNITS

As seen in Figure 179.5, the R subunits are modular proteins that are multifunctional and highly flexible [51]. There are two major isoforms, types I and II, with α and β subtypes in each class. RIα and RIIα are expressed in most mammalian tissues, whereas the expression of the β isoforms is more tissue-specific. There are also unique isoform distribution patterns: RIα is expressed predominantly in growing and transformed cells, and RIIα predominates in differentiated cells [52]. The isoforms are clearly not functionally redundant. The only R subunit that is essential is RIα. Deletion of RIα is embryonically lethal, and leads to cardiac defects [53]. Knockouts of other isoforms give unique phenotypes but are not lethal, and RIα tends to compensate when other R subunits are deleted [54]. For example, deletion of RIIβ gives a lean phenotype with a resistance to alcohol toxicity [53, 55]. Myxomas and Carney disease are caused by premature stop codons in RIα [56].

Clearly, there is still much to be learned about the physiological importance of the PKA isoforms. RIα requires ATP and $2Mg^{2+}$ ions to form a tight complex with the C subunit [57]; the high-affinity binding of ATP (60 nM) and C (0.2 nM) are synergistic. Type I holoenzyme is activated at lower levels of cAMP than type II holoenzyme [58, 59]. RII binding to the C subunit is independent of MgATP; instead, RII subunits are autophosphorylated at the consensus inhibitor site by the C subunit.

Molecular Architecture

All mammalian R subunits share the same organization. At the N-terminus is a dimerization/docking domain that locks the enzyme into a stable dimer. In the RI subunits, the two protomers are actually linked by a disulfide bond [60]. This is followed by a flexible and variable linker region that also contains a pseudo-substrate inhibitor site. In the absence of cAMP, this inhibitor site binds to the active site cleft of the C subunit, thus blocking access of other substrates. At the C terminus lie two stable, tandem cAMP-binding domains. In RIα, the first cAMP-binding domain also contributes to the docking of the C subunit [61]. Domain A thus shuttles between two conformations: a C bound form associated with the holoenzyme, and a cAMP-bound conformation. The

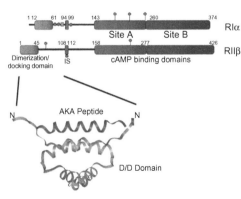

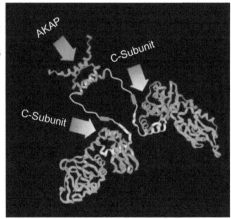

FIGURE 179.5 Domain organization of the regulatory subunits.
The modular organization of RI and RII is shown on the top left, and a model of the subunit showing a flexible linker is on the right. On the bottom left is a structure of the dimerization/docking domain of RIIa bound to an AKAP peptide [69–71]. The figure on the left is drawn by Ashton D. Taylor.

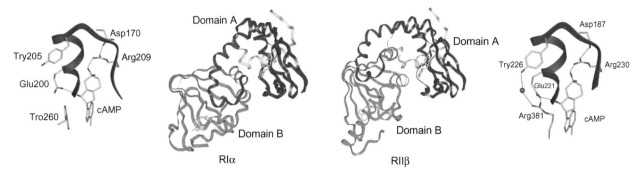

FIGURE 179.6 Structures of the regulatory subunits.
In the center are the structures of the cAMP-binding domains of RIa [68] and RIIb [67]. On the far left and right are the phosphate binding cassettes (PBCs) of RIa and RIIb, respectively.

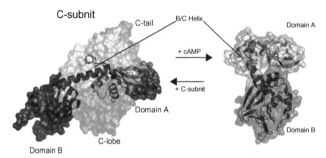

FIGURE 179.7 The RIa subunit undergoes major conformational changes when it forms a holoenzyme complex.
The compact cAMP binding domains in the cAMP-bound conformation of RIa [68] are separated in the holoenzyme complex by an extended B/C helix [10]. Domain A and the extended B/C helix wrap around the large lobe (C-lobe) of the catalytic subunit, while the inhibitor site of RIa docks to the active site cleft.

second cAMP-binding domain serves as a gatekeeper and regulates access of cAMP to site A [62].

cAMP is an ancient signaling molecule that has been conserved from bacteria to man. The cAMP-binding domain that serves to shield the cyclic phosphate from solvent and from phosphodiesterases is also ancient [62, 63]. In bacteria, the cAMP-binding domain is linked to a DNA-binding domain in the catabolite gene activator protein, whereas in mammals it is linked to protein kinase activation and is also found in cyclic nucleotide-gated channels [64] and in a cGMP exchange factor, EPAC [65]. The highly conserved phosphate-binding cassette that surrounds the cyclic phosphate is the recognition motif for this [63]. One side of this motif interacts with cAMP, while the other side interacts with the rest of the domain and is the heart of an extended network of interactions that reach to both the C-subunit docking site and the B domain [66]. Although the motif and the overall domain are highly conserved in RI and RII, the network of interactions (Figure 179.6) that lead to the cooperative binding of cAMP and the release of the catalytic subunit are remarkably different in RIα and RIIβ [67, 68].

The D/D domains are composed of a very stable four-helix bundle (Figure 179.5), but once again there are striking differences between RI and RII [69–72]. The AKAP-binding

surface is formed at the dimer interface, and dimerization is essential for AKAP binding. An amphipathic helix from the AKAP docks to this dimer interface [70].

Dynamics

In the absence of C subunit, the region that links the D/D domain to cAMP-binding domain A is quite mobile, as demonstrated by time-resolved fluorescence anisotropy [73]. Even in the holoenzyme, the linker remains quite mobile. Small angle X-ray scattering reveals a highly asymmetric structure [74], and once again there are significant differences between RI and RII isoforms [75–77]. Hydrogen/deuterium exchange in the presence and absence of C subunit and in the presence and absence of cAMP has provided a glimpse of the dynamic network that links cAMP binding to the release of the C subunit [66, 78–80].

HOLOENZYME COMPLEXES

The most recent advances in our understanding of PKA come from structures of the holoenzyme complexes [10–12]. These complexes reveal striking differences between the catalytic and the regulatory subunits. The catalytic subunit serves as stable scaffold for docking of the regulatory subunits. As expected, the inhibitor site of the regulatory subunit (the P − 3 to P + 1 site) docks to the active site cleft in a manner that is completely analogous to PKI. However, in order to achieve high affinity, the regulatory subunits dock to a large surface on the catalytic subunit that is distinct from the site where the amphipathic helix of PKI docks. The regulatory subunits also undergo a dramatic conformational change as they release cAMP and then bind to the catalytic subunit. As seen in Figure 179.7, the two cyclic nucleotide binding domains are split apart by the extension of the B/C helix in cyclic nucleotide binding domain A, which becomes extended into one long helix. These structures allow us for the first time to understand how the catalytic subunit is actually inhibited by the regulatory subunits, and how the holoenzyme complexes are activated by cAMP.

ACKNOWLEDGEMENTS

We would like to thank Sventja von Daake for her help in preparing the updated version of this manuscript. This work was supported by grants from the National Institutes of Health and by the Howard Hughes Medical Institute.

REFERENCES

1. Walsh DA, Perkins JP, Krebs EG. An adenosine 3′,5′-monophosphate-dependent protein kinase from rabbit skeletal muscle. *J Biol. Chem* 1968;**243**:3763–5.

2. Shoji S, Ericsson LH, Walsh KA, Fischer EH, Titani K. Amino acid sequence of the catalytic subunit of bovine type II adenosine cyclic 3′,5′-phosphate dependent protein kinase. *Biochemistry* 1983;**22**:3702–9.

3. Uhler MD, Carmichael DF, Lee DC, Chivia JC, Krebs EG, McKnight GS. Isolation of cDNA clones for the catalytic subunit of mouse cAMP-dependent protein kinase. *Proc Natl Acad Sci USA* 1986;**83**:1300–4.

4. Knighton DR, Zheng J, Ten Eyck LF, et al. Crystal structure of the catalytic subunit of cAMP-dependent protein kinase. *Science* 1991;**253**:407–14.

5. Francis SH, Corbin JD. Structure and function of cyclic nucleotide-dependent protein kinases. *Annu Rev Physiol* 1994;**56**:237–72.

6. Taylor SS, Buechler JA, Yonemoto W. cAMP-dependent protein kinase: framework for a diverse family of regulatory enzymes. *Annu Rev Biochemistry* 1990;**59**:971–1005.

7. Michel JJ, Scott JD. AKAP mediated signal transduction. *Annu Rev Pharmacol Toxicol* 2002;**42**:235–57.

8. Walsh DA , Angelos KL , Van Patten SM , Glass DB , Garetto LP Kemp BE, editor. *Peptides and protein phosphorylation*. Boca Raton: CRC Press; 1990. p. 43–84.

9. Wen W, Meinkoth JL, Tsien RY, Taylor SS. Identification of a signal for rapid export of proteins from the nucleus. *Cell* 1995;**82**:463–73.

10. Kim C, Cheng CY, Saldanha SA, Taylor SS. PKA-I holoenzyme structure reveals a mechanism for cAMP-dependent activation. *Cell* 2007;**130**:1032–43.

11. Wu J, Brown SH, von Daake S, Taylor SS. PKA type IIalpha holoenzyme reveals a combinatorial strategy for isoform diversity. *Science* 2007;**318**:274–9.

12. Kim C, Xuong NH, Taylor SS. Crystal structure of a complex between the catalytic and regulatory (RIalpha) subunits of PKA. *Science* 2005;**307**:690–6.

13. Beebe SJ, Oyen O, Sandberg M, Froysa A, Hansson V, Jahnsen T. Molecular cloning of a tissue-specific protein kinase (C gamma) from human testis–representing a third isoform for the catalytic subunit of cAMP-dependent protein kinase. *Mol Endocrinol* 1990;**4**:465–75.

14. Showers MO, Maurer RA. A cloned bovine cDNA encodes an alternate form of the catalytic subunit of cAMP-dependent protein kinase. *J Biol Chem* 1986;**261**:16,288–16,291.

15. Agustin JT, Wilkerson CG, Witman GB. The unique catalytic subunit of sperm cAMP-dependent protein kinase is the product of an alternative Calpha mRNA expressed specifically in spermatogenic cells. *Mol Biol Cell* 2000;**11**:3031–44.

16. Guthrie CR, Skalhegg BS, McKnight GS. Two novel brain-specific splice variants of the murine Cbeta gene of cAMP-dependent protein kinase. *J Biol Chem* 1997;**272**:29,560–29,565.

17. Carr SA, Biemann K, Shoji S, Parmalee DC, Titani K. n-Tetradecanoyl in the NH2 terminal blocking group of the catalytic subunit of the cyclic AMP-dependent protein kinase from bovine cardiac muscle. *Proc Natl Acad Sci USA* 1982;**79**:6128–31.

18. Shoji S, Titani K, Demaille JG, Fischer EH. Sequence of two phosphorylated sites in the catalytic subunit of bovine cardiac muscle adenosine 3′,5′-monophosphate-dependent protein kinase. *J Biol Chem* 1979;**254**:6211–14.

19. Adams JA, McGlone ML, Gibson RM, Taylor SS. Phosphorylation modulates catalytic function and regulation in the cAMP-dependent protein kinase. *Biochemistry* 1995;**34**:2447–54.

20. Yonemoto W, McGlone ML, Grant B, Taylor SS. Autophosphorylation of the catalytic subunit of cAMP-dependent protein kinase in *Escherichia coli*. *Protein Eng* 1997;**10**:915–25.

21. Toner-Webb J, van Patten SM, Walsh DA, Taylor SS. Autophosphorylation of the Catalytic Subunit of cAMP-Dependent Protein Kinase. *J Biol Chem* 1992;**267**:25,174–25,180.

22. Kinzel V, Konig N, Pipkorn R, Bossemeyer D, Lehmann WD. The amino terminus of PKA catalytic subunit – a site for introduction of posttranslational heterogeeities by deamidation: D-Asp2 and D-isoAsp2 containing isozymes. *Protein Sci* 2000;**11**:2269–77.

23. Li F, Gangal M, Juliano C, Gorfain E, Taylor SS, Johnson DA. Evidence for an internal entropy contribution to phosphoryl transfer: a study of domain closure, backbone flexibility, and the catalytic cycle of cAMP-dependent protein kinase. *J Mol Biol* 2002;**315**:459–69.

24. Zetterquist OZ , Ragnarson U , Engtrom L Kemp BE, editor. *Peptides and protein phosphorylation*. Boca Raton: CRC Press; 1990. p. 171–87.

25. Adams JA. Kinetic and catalytic mechanisms of protein kinases. *Chem Rev* 2001;**101**:2271–90.

26. Adams JA, Taylor SS. The energetic limits of phosphotransfer in the catalytic subunit of cAMP-dependent protein kinase as measured by viscosity experiments. *Biochemistry* 1992;**31**:8516–22.

27. Adams JA, Taylor SS. Phosphorylation of peptide substrates for the catalytic subunit of cAMP-dependent protein kinase. *J Biol Chem* 1993;**268**:7747–52.

28. Knighton DR, Zheng J, Ten Eyck LF, Xuong NH, Taylor SS, Sowadski JM. Structure of a peptide inhibitor bound to the catalytic subunit of cyclic adenosine monophosphate-dependent protein kinase. *Science* 1991;**253**:414–20.

29. Zheng J, Knighton DR, Ten Eyck LF, et al. Crystal structure of the catalytic subunit of cAMP-dependent protein kinase complexed with MgATP and peptide inhibitor. *Biochemistry* 1993;**32**:2154–61.

30. Zheng J, Knighton DR, Xuong N-h, Taylor SS, Sowadski JM, Ten Eyck LF. Crystal structures of the myristylated catalytic subunit of cAMP-dependent protein kinase reveal open and closed conformations. *Protein Sci* 1993;**2**:1559–73.

31. Taylor SS, Knighton DR, Zheng J, Ten Eyck LF, Sowadski JM. Structural framework for the protein kinase family. *Annu Rev Cell Biol* 1992;**8**:429–62.

32. Hanks SK, Hunter T. Protein kinases 6. The eukaryotic kinase superfamily: kinase (catalytic) domain structure and classification. *FASEB J* 1995;**8**:576–96.

33. Buechler JA, Taylor SS. Identification of Asp 184 as an essential residue in the catalytic subunit of cAMP-dependent protein kinase.. *Biochemistry* 1988;**27**:7356–61.

34. Zoller MJ, Taylor SS. Affinity labeling of the nucleotide binding site of the catalytic subunit of cAMP-dependent protein kinase using *p*-fluorosulfonyl-[^{14}C]benzoyl 5′-adenosine: identification of a modified lysine residue. *J Biol Chem* 1979;**254**:8363–8.

35. Taylor SS, Knighton DR, Zheng J, Sowadski JM, Gibbs CS, Zoller MJ. A template for the protein kinase family. *Trends Biochem Sci* 1993;**18**:84–9.

36. Johnson DA, Akamine P, Radzio-Andzelm E, Madhusudan M, Taylor SS. Dynamics of cAMP-dependent protein kinase. *Chem Rev* 2001;**101**:2243–70.

37. Narayana N, Cox S, Xuong NH, Ten Eyck LF, Taylor SS. A binary complex of the catalytic subunit of cAMP-dependent protein kinase and adenosine further defines conformational flexibility. *Structure* 1997;**5**:921–35.

38. Bossemeyer D, Engh RA, Kinzel V, Ponstingl H, Huber R. Phosphotransferase and substrate binding mechanism of the cAMP-dependent protein kinase catalytic subunit from porcine heart as deduced from the 2.0-Å structure of the complex with Mn^{2+} adenyl imidodiphosphate and inhibitor peptide PKI(5-24). *EMBO J* 1993;**12**:849–59.

39. Madhusudan M, Akamine P, Xuong NH, Taylor SS. Crystal structure of a transition state mimic of the catalytic subunit of cAMP-dependent protein kinase. *Nature Struct Biol* 2002;**9**:273–7.

40. Aimes RT, Hemmer W, Taylor SS. Serine-53 at the tip of the glycine-rich loop of cAMP-dependent protein kinase: role in catalysis, P-site specificity, and interaction with inhibitors. *Biochemistry* 2000;**39**:8325–32.

41. Johnson L, Noble M, Owen D. Active and inactive protein kinases: structural basis for regulation. *Cell* 1996;**85**:149–58.

42. Zhou J, Adams JA. Is there a catalytic base in the active site of cAMP-dependent protein kinase? *Biochemistry* 1997;**36**:2977–84.

43. Cheng X, Ma Y, Moore M, Hemmings BA, Taylor SS. Phosphorylation and activation of cAMP-dependent protein kinase by phosphoinositide-dependent protein kinase. *Proc Natl Acad Sci USA* 1998;**95**:9849–54.

44. Cauthron RD, Carter KB, Liauw S, Steinberg RA. Physiological phosphorylation of protein kinase A at Thr-197 is by a protein kinase A kinase. *Mol Cell Biol* 1998;**18**:1416–23.

45. Veron M, Radzio-Andzelm E, Tsigelny I, Ten Eyck LF, Taylor SS. A conserved helix motif complements the protein kinase core. *Proc Natl Acad Sci USA* 1993;**90**:10,618–10,622.

46. Batkin M, Schvartz I, Shaltiel S. Snapping of the carboxyl terminal tail of the catalytic subunit of PKA onto its core: characterization of the sites by mutagenesis. *Biochemistry* 2000;**39**:5366–73.

47. Kannan N, Haste N, Taylor SS, Neuwald AF. The hallmark of AGC kinase functional divergence is its C-terminal tail, a cis-acting regulatory module. *Proc Natl Acad Sci USA* 2007;**104**:1272–7.

48. Kornev AP, Haste NM, Taylor SS, Eyck LF. Surface comparison of active and inactive protein kinases identifies a conserved activation mechanism. *Proc Natl Acad Sci USA* 2006;**103**:17,783–17,788.

49. Kornev AP, Taylor SS, Ten Eyck LF. A helix scaffold for the assembly of active protein kinases. *Proc Natl Acad Sci USA* 2008;**105**:14,377–14,382.

50. Hauer JA, Barthe P, Taylor SS, Parello J, Padille A. Two well defined motifs in the cAMP-dependent protein kinase inhibitior (PKIa) correlate with inhibitory and nuclear export function. *Protein Sci* 1999;**8**:545–53.

51. Li F, Gangal M, Jones JM, et al. Consequences of cAMP and catalytic-subunit binding on the flexibility of the A-kinase regulatory subunit. *Biochemistry* 2000;**39**:15,626–15,632.

52. Stratakis CA, Cho-Chung YS. Protein kinase A and human disease. *Trends Endocrinol Metab* 2002;**13**:50–2.

53. Cummings DE, Brandon EP, Planas JV, Motamed K, Idzerda RL, McKnight GS. Genetically lean mice result from targeted disruption of the RII beta subunit of protein kinase A [see comments]. *Nature* 1996;**382**:622–6.

54. Amieux PS, Cummings DE, Motamed K, et al. Compensatory regulation of RIa protein levels in protein kinase A mutant mice. *J Biol Chem* 1997;**272**:3993–8.

55. Thiele TE, Willis B, Stadler J, Reynolds JG, Bernstein IL, McKnight GS. High ethanol consumption and low sensitivity to ethanol-induced sedation in protein kinase A-mutant mice. *J Neurosci* 2000;**20**:RC75.

56. Kirschner LS, Carney JA, Svetlana DP, et al. Mutations of the gene encoding the protein kinase A type I – a regulatory subunit in patients with Carney complex. *Nat Genet* 2000;**26**:89.

57. Herberg FW, Doyle ML, Cox S, Taylor SS. Dissection of the nucleotide and metal-phosphate binding sites in cAMP-dependent protein kinase. *Biochemistry* 1999;**38**:6352–60.

58. Cadd GG, Uhler MD, McKnight GS. Holoenzymes of cAMP-dependent protein kinase containing the neural form of type I regulatory subunit have an increased sensitivity to cyclic nucleotides. *J Biol Chem* 1990;**265**:19,502–19,506.

59. Steinberg SF, Brunton LL. Compartmentation of G-protein-coupled signaling pathways in cardiac myocytes. *Annu Rev Pharmacol Toxicol* 2001;**41**:751–73.

60. Bubis J, Neitzel JJ, Saraswat LD, Taylor SS. A point mutation abolishes binding of cAMP to site A in the regulatory subunit of cAMP-dependent protein kinase. *J Biol Chem* 1988;**263**:9668–73.

61. Huang LJ, Taylor SS. Dissecting cAMP binding domain A in the RIa subunit of cAMP-dependent protein kinase: distinct subsites for recognition of cAMP and the catalytic subunit. *J Biol Chem* 1998;**273**:26,739–26,746.

62. Ogreid D, Doskeland SO. Cyclic nucleotides modulate the release of [3H] adenosine cyclic 3′,5′-phosphate bound to the regulatory moiety of protein kinase I by the catalytic subunit of the kinase. *Biochemistry* 1983;**22**:1686–96.

63. Canaves JM, Taylor SS. Classification and phylogenetic analysis of cAMP-dependent protein kinase regulatory subunit family. *J Mol Evolution* 2002;**52**:17–29.

64. Bonigk W, Bradley J, Muller F, et al. The native rat olfactory cyclic nucleotide-gated channel is composed of three distinct subunits. *J Neurosci* 1999;**19**:5332–47.

65. de Rooij J, Zwartkruis FJ, Verheijen MH, et al. Epac is a Rap1 guanine-nucleotide-exchange factor directly activated by cyclic AMP. *Nature* 1998;**396**:474–7.

66. Anand GS, Hughes CA, Jones JM, Taylor SS, Komives EA. Amide H/2H exchange reveals communication between the cAMP and

catalytic subunit-binding sites in the R(I)alpha subunit of protein kinase A. *J Mol Biol* 2002;**323**:377–86.

67. Diller TC, Xuong NH, Taylor SS. Type IIB regulatory subunit of cAMP-dependent protein kinase: purification strategies to optimize crystallization. *Protein Expr Purif* 2000;**20**:357–64.

68. Su Y, Dostmann WRG, Herberg FW, et al. Regulatory (RIa) Subunit of protein kinase A: structure of deletion mutant with cAMP binding domains. *Science* 1995;**269**:807–19.

69. Banky P, Roy M, Newlon MG, et al. Related protein-protein interaction modules present drastically different surface topographies despite a conserved helical platform. *J Mol Biol* 2003;**330**:1117–29.

70. Newlon MG, Roy M, Morikis D, et al. A novel mechanism of PKA anchoring revealed by solution structures of anchoring complexes. *EMBO J* 2001;**7**:1651–62.

71. Newlon MG, Roy M, Morikis D, et al. The molecular basis for protein kinase A anchoring revealed by solution NMR. *Nature Struct Biol* 1999;**6**:222–7.

72. Banky P, Newlon MG, Roy M, Garrod S, Taylor SS, Jennings PA. Isoform-specific differences between the type Ialpha and IIalpha cyclic AMP-dependent protein kinase anchoring domains revealed by solution NMR. *J Biol Chem* 2000;**275**:35,146–35,152.

73. Gangal M, Li F, Jones JM, et al. Consequences of cAMP and catalytic subunit binding on the flexibility of the A-kinase of the regulatory subunit. *Biochemistry* 2000;**39**:15,626–15,632.

74. Tung CS, Walsh DA, Trewhella J. A structural model of the catalytic subunit-regulatory subunit dimeric complex of the cAMP-dependent protein kinase. *J Biol Chem* 2002;**277**:12,423–12,431.

75. Vigil D, Blumenthal DK, Heller WT, et al. Conformational differences among solution structures of the type Ialpha, IIalpha and IIbeta protein kinase A regulatory subunit homodimers: role of the linker regions. *J Mol Biol* 2004;**337**:1183–94.

76. Vigil D, Blumenthal DK, Taylor SS, Trewhella J. The conformationally dynamic C helix of the RIalpha subunit of protein kinase A mediates isoform-specific domain reorganization upon C subunit binding. *J Biol Chem* 2005;**280**:35,521–35,527.

77. Vigil D, Blumenthal DK, Taylor SS, Trewhella J. Solution scattering reveals large differences in the global structures of type II protein kinase A isoforms. *J Mol Biol* 2006;**357**:880–9.

78. Anand GS, Hotchko M, Brown SH, Ten Eyck LF, Komives EA, Taylor SS. R-subunit isoform specificity in protein kinase A: distinct features of protein interfaces in PKA types I and II by amide H/2H exchange mass spectrometry. *J Mol Biol* 2007;**374**:487–99.

79. Hamuro Jr Y, Zawadzki KM, Kim JS, Stranz DD, Taylor SS, Woods VL. Dynamics of cAPK type IIbeta activation revealed by enhanced amide H/2H exchange mass spectrometry (DXMS). *J Mol Biol* 2003;**327**:1065–76.

80. Zawadzki KM, Hamuro Y, Kim JS, et al. Dissecting interdomain communication within cAPK regulatory subunit type IIbeta using enhanced amide hydrogen/deuterium exchange mass spectrometry (DXMS). *Protein Sci* 2003;**12**:1980–90.

Chapter 40

Cyclic GMP-Dependent Protein Kinase: Targeting and Control of Expression

Thomas M. Lincoln, Hassan Sellak, Nupur Dey, Chung-Sik Choi and Felricia Brown

Department of Physiology, College of Medicine, University of South Alabama, Mobile, Alabama

INTRODUCTION

In this brief review, we will address three aspects of PKG function that are of recent interest: the role of PKG-I targeting to subcellular proteins especially in smooth muscle cells (SMC), the role of PKG-I in regulating vascular SMC (VSMC) gene expression, and the regulation of the expression of PKG-I.

CYCLIC GMP-DEPENDENT PROTEIN KINASE: STRUCTURE AND FUNCTION

Cyclic GMP interacts with three known "receptor" proteins in cells: cyclic nucleotide-gated ion channels, cGMP binding phosphodiesterases, and PKG (recently reviewed in [1, 2]). The only recognized and physiologically relevant receptor protein for cGMP in VSMC is PKG. There are only two mammalian PKG (known as PRKG) genes: PKG-I and PKG-II [2, 3]. PKG-I is the more widely expressed enzyme, and is found in abundance in VSMC, platelets, and cerebellar Purkinje cells, while being less abundant in cardiomyocytes, endocrine secretory cells, endothelial cells, and certain neurons. The PKG-II enzyme is localized in intestinal epithelial cells, glomerular juxtamedullary cells, chondrocytes, and various parts of the brain, but not in VSMC. PKG is not expressed, or at least not highly expressed, in various cultured cells such as Cos7, CHO, HeLa, and many other tumor cell lines [4, 5].

The subunit structure of the PKG isoforms is well known. The kinase exists as a homodimer (subunit Mr for PKG-I is approximately 79 kDa whereas the PKG-II is approximately 90 kDa). As shown in Figure 40.1, the protein structure can be divided into two general domains: an N-terminal regulatory domain that contains the cGMP binding sites, and a C-terminal catalytic domain. PKG-I is expressed as two isoforms as a result of alternate mRNA splicing of the first and second exons: a PKG-Iα and a PKG-Iβ. These two isoforms of PKG are different only in the first approximate 90 amino acids, and from Figure 40.1 it can be seen that this region of the enzyme encodes an N-terminal leucine/isoleucine zipper (LZ) region (for dimerization of the holoenzyme), an autoinhibitory domain, and an area of autophosphorylation. Hence, although PKG-Iα and -Iβ have differences in the N-terminal sequences (dimerization, autoinhibitory, and autophosphorylation regions), they are identical in sequence with regard to cGMP binding regions and catalytic domains. PKG-II is not known to exist as isoforms. On the other hand, PKG-II is myristolyated on the N-terminal amino acid, thus targeting the kinase to hydrophobic sites in the cell such as lipid raft domains [6].

Targeted deletion of either the PKG-I or PKG-II genes does not result in embryonic lethality, and the pups appear to be born phenotypically normal. However, PKG-I null animals die within 5 weeks due to vascular and gastrointestinal distress [7], whereas PKG-II null mice are dwarfed as a result of chondrocyte abnormalities [8]. PKG-I null animals predictably, demonstrate loss of endothelial-dependent relaxation mechanisms, but, interestingly, are not hypertensive until near death at approximately 5 weeks of age [7]. Conditional null PKG-I knockout animals have been described where Cre is activated by a tamoxifen-driven SM22α promoter to induce smooth muscle-specific knockouts. Curiously, these animals do not demonstrate any phenotype (e.g., hypertension, gastrointestinal abnormalities, or erectile dysfunction). When crossed with atherosclerotic apoE null mice, however, the PKG-I-deficient animals demonstrate decreased atherosclerosis after 16 weeks [9]. This finding led the authors to conclude that the cGMP–PKG pathway in mice at least is proatherogenic – a notion that is surprising, given the widely accepted anti-atherogenic role of NO itself [10].

Handbook of Cell Signaling, Three-Volume Set 2 ed.

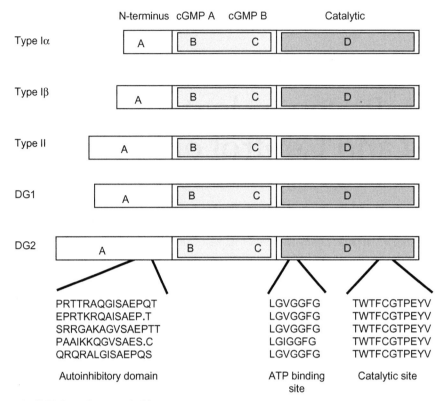

FIGURE 40.1 Structure of cGMP-dependent protein kinases.
The two mammalian genes encoding PKG (Type I and Type II) and two *Drosophila* genes (DG1 and DG2) are shown for comparison. PKG proteins are divided into four separate regions, A, B, C, and D. The catalytic domains (D) for the two isoforms of Type I PKG are identical, and highly homologous with Type II and the DG isoforms. The binding sites for PKG are identified as sites B and C. Site A is the N-terminus, which contains the leucine zipper (LZ) and the autoinhibitory domain. Shown in the figure are the different autoinhibitory domains of all five PKG proteins. Each autoinhibitory sequence is different, suggesting different regulatory mechanisms for controlling PKG activity. In general, there is a substrate "analog" sequence in each autoinhibitory domain; for example, PKG-Iα contains RAQG while PKG-Iβ is KRQA. PKG-II is the more conventional RRGA. The ATP binding site consists of the "glycine loop" which is rather strictly conserved in all protein kinases. The catalytic site for PKG proteins is identical and differs.

TARGETING OF PKG-I IN VSMC

Much has been speculated about (1) targeting of PKG-I to subcellular structures or proteins in cells, and (2) the mechanism of PKG-I targeting. Some earlier studies suggested that PKG-I was targeted to cytoskeletal structures such as vimentin [11] or troponin [12]. In the case of vimentin, it was found that the N-terminal domain of the PKG-I dimer was most likely involved in targeting, similar to that seen for the PKA regulatory subunit targeting to A-kinase Anchoring Proteins (AKAPs). These early studies suggested that the different N-terminal domains of PKG-I isoforms may function to target PKG-I to different proteins in smooth muscle cells.

Overall, little is known about the physiological significance of the different N-terminal regions, although there is much speculation regarding the different biological roles of PKG-Iα and PKG-Iβ. It is known that the apparent K_{ACT} for cGMP is approximately 10-fold higher for PKG-Iβ than PKG-Iα, despite having identical cGMP binding sites [13, 14]. It is believed that the lower sensitivity to activation of the Iβ isoform is due to the fact that the autoinhibitory

domain for PKG-Iβ exerts a greater inhibitory influence on the catalytic domain, thus requiring saturation of cGMP binding sites to produce full activation of PKG-Iβ [15]. PKG-Iα, on the other hand, is more sensitive to activation by cGMP, and is in fact partially active under conditions where cGMP is not elevated in the cell and when only one cGMP binding site is occupied [14, 15].

The role of the different N-terminal sequences (i.e., the LZ domain) of the PKG-I isoforms to direct different cellular localization of the two isoforms was first confirmed in the targeting of PKG-I to the smooth muscle myosin light-chain phosphatase. PKG-Iα, but not PKG-Iβ, has been shown to bind specifically to the Myosin Binding Subunit (MBS) of the smooth muscle myosin light-chain phosphatase [16]. Functionally, PKG-Iα activates the phosphatase but only if the kinase is bound to the MBS. The physiologic relevance of this targeting was confirmed in studies where it was shown that MBS is expressed as different alternately spliced isoforms [17–19]. When a specific exon (exon Y) is spliced into the sequence, this creates a frame shift resulting in termination of translation before the PKG-binding domain on the C-terminus is completed. Hence, the spliced-in isoform does not bind PKG-Iα. When exon Y is spliced out of the coding region, termination of reading does not occur until after the PKG-binding sequence is completed. Only the spliced out MBS binds PKG-Iα and more importantly only in SMC expressing only the spliced out MBS does cGMP promote smooth muscle relaxation [19].

With regard to PKG-Iβ, this isoform, but not PKG-Iα, is bound to a sarcoplasmic reticulum (SR) protein known as IRAG for (*I*nositol phosphate *R*eceptor *A*ssociated *G*-kinase binding protein [20, 21]). It has been known for several years that PKG-I catalyzes the phosphorylation of SR proteins such as phospholamban and the IP$_3$ receptor [22–24], and inhibits Ca release from the SR as a component of the Ca-lowering effects of NO-cGMP signaling in VSMC [22, 23]. PKG-dependent phosphorylation of both phospholamban and the IP$_3$ receptor requires the presence of IRAG that serves to "scaffold" the IP$_3$ receptor, phospholamban, and PKG-I into a complex that can be regulated by cGMP [24–26].

It is apparent, therefore, that the seemingly minor sequence differences in PKG-Iα and PKG-Iβ may relate to functional differences in the two proteins. The expression of the two isoforms in various smooth muscle tissues is also different. Generally, PKG-Iα is more highly expressed than PKG-Iβ at the protein level in *vascular* SMC, whereas PKG-Iβ seems to be the more highly expressed isoform in *non-vascular* SMC. In pulmonary tissue (which contains VSMC and airway SMC), only PKG-Iα is expressed. In platelets, on the other hand, only PKG-Iβ is expressed. How the differential distribution of PKG-I isoforms in different smooth muscle tissue accounts for functional differences in cGMP signaling is unclear at this time. PKG-I also is known to activate membrane ion channels, particularly the "big K" Ca^{2+}-activated K$^+$ channel (BK channel) in smooth muscle. At least two reports indicate that PKG-I catalyzes the phosphorylation of the BK channel in SMC [27, 28]. The questions to be asked are: is there a preference for PKG-Iα or Iβ in BK channel phosphorylation, and is this due to specific targeting of a specific isoform to the channel? Likewise, PKG-Iα catalyzes the phosphorylation of the G-protein regulatory protein, RGS-2, leading to an inhibition of receptor-Gq signaling in SMC [29]. Does targeting of PKG-Iα, but not Iβ, result in specific regulation of receptor signaling in SMC? Our overall view of the role of PKG-I targeting and signaling is shown in Figure 40.2.

ROLE OF PKG-I IN THE REGULATION OF VSMC PROLIFERATION AND PHENOTYPE

Recent studies have shown that PKG-I is involved in the regulation of VSMC phenotype. Our laboratory first demonstrated that cultured VSMC deficient in expression of PKG due to repetitive passaging assumed a more contractile

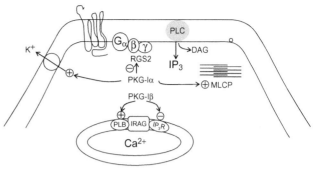

FIGURE 40.2 Mechanisms of PKG-I isoform action in vascular smooth muscle.
PKG-Iα and PKG-Iβ are believed to be separately targeted via the N-terminal domains to different regions of the cell. PKG-Iα is known to be specifically targeted to myosin light-chain phosphatase (MLCP) and regulator of G-protein signaling-2 (RGS2) to activate the phosphatase and inhibit G-protein-dependent PLC activity. PKG-Iα, and possibly PKG-Iβ, is also believed to activate large conductance Ca-dependent K channels, although the targeting mechanisms are not known. On the other hand, PKG-Ib seems to be more specifically localized to the sarcoplasmic reticulum to the inositol phosphate receptor associated G-kinase binding protein, IRAG. There, PKG-Iβ catalyzes the phosphorylation of phospholamban (PLB), which in turn activates the Ca-ATPase, SERCA. Also, PKG-Ib catalyzes the phosphorylation of the IP$_3$R to inhibit Ca gating. Both effects decrease intracellular free Ca concentrations.

phenotype upon restoration of PKG expression [30–33]. Restored PKG expression resulted in increased expression of smooth muscle specific marker proteins such as smooth muscle myosin heavy chain (SMMHC), calponin, smoothelin, heavy caldesmon, and α-actin [30–33]. Furthermore, restoration of PKG expression decreased the expression in VSMC of extracellular matrix proteins such as collagen, osteopontin, and thrombospondin [31, 34]. Other laboratories have shown that PKG expression increases the expression of cyclin kinase inhibitors such as p27^{kip} and decreases the expression of inflammatory adhesion molecules [35–37]. The mechanism of regulation of smooth muscle cell-specific gene expression is unknown at this time. Pilz and co-workers [38–40] found that PKG-I expression in VSMC increased serum response factor (SRF)-dependent gene expression from promoters of smooth muscle specific genes such as SM22α. On the other hand, PKG-I inhibited SRF-dependent expression of growth related genes such as c-fos. The latter findings were consistent with the observation that NO and cGMP are growth inhibitory to VSMC. Work from these investigators suggest that at least part of the growth inhibitory effects of PKG on growth gene expression is due to the phosphorylation and inhibition of RhoA activity.

On the other hand, the effects of PKG-I on smooth muscle specific gene expression and differentiation are also SRF-dependent. SRF interacts with numerous co-transcriptional regulators, including myocardin, a smooth muscle specific co-transcription factor and member of the larger myocardin-related factor (MRF) family of proteins [41–43]. Pilz's laboratory demonstrated that PKG-I enhanced myocardin-stimulated SRF activity in part through the phosphorylation of a myocardin regulatory protein, cysteine-rich protein-2 (CRP-2) in VSMC [39–40]. Our laboratory has also observed enhanced myocardin-dependent SRF activity in a non-smooth muscle cell line (i.e., Cos7) when transfected with myocardin and PKG-I cDNA (unpublished observations). Furthermore, using quantitative chromatin immunoprecipitation (ChIP), we observed that PKG-I increased SRF binding to smooth muscle-specific promoter regions of SMMHC and SM22α (unpublished observations). Thus, PKG-I appears to play a modulatory role in myocardin-SRF dependent gene expression although the precise mechanisms are unknown at this time.

REGULATION OF PKG-I EXPRESSION

It has been appreciated for decades that the expression patterns of PKG are highly variable in cells and tissues, especially when compared with PKA [44, 45]. Hence, not only is PKG not as widely expressed as PKA, but the levels of PKG expressed in either cell culture systems or in tissues studied under different physiological and pathophysiological conditions vary greatly. In studies conducted in our own laboratory, for instance, PKG-I levels decrease markedly upon passaging of rat aortic SMC [46, 47]. Notably, as rat aortic SMC are passaged and lose PKG, their growth properties improve and their phenotype becomes less "contractile" or "differentiated." Furthermore, in animal models where balloon catheters are used to injure vessels *in situ* to produce a proliferative response of the VSMC, PKG expression decreases in these proliferating cells [48, 49]. Others have shown that PKG levels decrease under various conditions such as hypoxia or normoxia followed by hypoxia, diabetes, and aging [50–54]. These findings are consistent with our general hypothesis that PKG may function as a "brake" on VSMC proliferation and to induce differentiation. Adaptation to culture or to wound-healing conditions *in vivo* may necessitate downregulation of the cGMP-PKG pathway for cell survival.

REGULATION OF PKG-I MRNA LEVELS

The PKG-I mRNA level is very low in the cell. Using real-time PCR, we have estimated that the levels of PKG-I in primary cultures of bovine aortic SMC are approximately 1000–2000 copies per cell [55]. In fact, northern blot analysis of PKG-I mRNA expression is exceedingly difficult to accomplish due to the low copy number. To determine whether physiological mechanisms exist in the regulation of PKG-I mRNA expression, we have defined the sequences in the proximal 5'-upstream non-transcribed promoter region from the Iα exon that regulate PKG-I mRNA expression. Briefly, both upstream stimulatory factor

(USF) transcription factors and Sp1 proteins appear to be important in regulating PKG-I mRNA expression [56, 57]. Recently, Pilz's laboratory has shown that the Kruppel-like factor (KLF-4) binds to the Sp1 sites on the PKG-I proximal promoter and regulates gene transcription [58]). In addition, these investigators have made the very interesting and important observation that the small molecular weight G proteins, RhoA and rac, regulated KLF-4 binding and PKG-I gene expression. In particular, the role of rac to increase PKG-I gene expression may be an important link between increased cell density and PKG-I expression in cultured cells [47], and the possible involvement of the cytoskeleton and signaling pathways that communicate with the cell nucleus.

Our laboratory reported that inflammatory, atherogenic cytokines such as IL-1β and TNFα decreased PKG-I mRNA and protein levels in primary cultures of bovine aortic VSMC [55]. One mechanism that appeared responsible for this event was a cytokine-dependent increase in iNOS expression, NO biosynthesis, and a decrease in Sp1 binding to the PKG-I promoter. Suppression of iNOS activity or sGC activity inhibited the downregulation of PKG-I mRNA induced by the cytokines [55]. Interestingly, PKA inhibition also suppressed the effects of cytokines on PKG-I mRNA expression, suggesting that high elevations in cGMP in response to iNOS expression cross-activate PKA and lead to decreased Sp1 protein binding to the PKG-I promoter [55]. Whether RhoA activity and KLF-4 binding to the PKG-I promoter are affected by inflammatory cytokines is not known at this time.

Going back to an earlier discussion, it is interesting that the Iα form of the kinase, in contrast to the Iβ form, exists as a partially active species not requiring elevations in cGMP for activity. As suggested by Dostmann's laboratory, PKG-Iα may serve to maintain differentiated characteristics of smooth muscle in the absence of cGMP elevation [59]. Thus, that PKG-Iα in particular would be susceptible to downregulation upon sustained elevations in cGMP (for example, in response to iNOS expression in inflammation) makes physiological sense: cytokines increase iNOS expression, cGMP levels are persistently elevated, PKG-Iα is downregulated, and the cells dedifferentiate – if only temporarily – to accomplish wound-healing functions through proliferation and fibrotic activity. In this regard, a new mode for the regulation of PKG-I expression was recently uncovered in our laboratory. In response to robust elevations in cGMP, PKG-Iα (but not PKG-Iβ) undergoes ubiquitination [60]. The mechanisms responsible are not well-understood, but may be dependent upon specific cGMP-induced autophosphorylation of PKG-Iα (whose autophosphorylation sites are different than those of PKG-Iβ). Thus, it appears that perhaps a more significant mechanism regulating PKG-I expression, and especially the vascular isoform PKG-Ia, is downregulation via the ubiquitin–proteasome pathway. Such a mechanism is reminiscent of the role of phorbol esters that activate protein kinase C and also downregulate the kinase by proteasome degradation.

FINAL THOUGHTS

Atherosclerosis and other vascular diseases are now known to be related to, if not outright caused by, inflammatory mediators that activate growth and de-differentiation programs in VSMC (see [61] for a review). TNFα, IL-1, and even lipopolysaccharide (LPS) stimulate VSMC proliferation and phenotypic modulation, and these effects are enhanced in the presence of growth factors such as PDGF that activate tyrosine-kinase receptors in VSMC [62–64]. The inflammatory response may be initiated by overt injury, dyslipidemias, hyperglycemia, infectious agents, and perhaps a wide variety of other risk factors associated with atherosclerosis, stroke, and cardiovascular disorders.

Given the importance of NO signaling in maintaining normal vascular function, it is not surprising that there is abundant literature relating dysfunction of NO signaling to the aforementioned disorders. For instance, oxidative stress – brought on by inflammatory pathway-dependent activation

of NADPH oxidase systems as well as mitochondrial dysregulation – produces reactive oxygen species (ROS) such as superoxide that react with NO to produce peroxynitrite [65–67]. Peroxynitrite in turn is a potent oxidizing agent that has been proposed to induce cellular damage through nitration of tyrosine residues in proteins and lipid peroxidation [68–70]. However, it has been the view of our laboratory that because peroxynitrite production is not an amplifying signal transduction pathway *per se* but rather a stoichiometric phenomenon (e.g., protein nitration), peroxynitrite alone probably does not account for the myriad of changes in VSMC phenotype and function, such as the changes that occur in gene expression under inflammatory conditions [10]. For the pronounced phenotypic changes to occur under these conditions, we proposed that there should be disruptions in NO signaling *downstream* from NO production. In the case for PKG-I, we found that relatively low concentrations of TNFα and IL-β inhibit PKG mRNA and protein expression in primary, contractile phenotype VSMC [55–56]. At least a major part of the mechanism for these effects is related to the induction of type II NO synthase (iNOS). Thus, downregulation of PKG coupled with the direct effects of growth factors and cytokines is hypothesized to lead to phenotypic modulation of VSMC to a fibroproliferative state that is characteristic of vascular lesions and disorders (Figure 40.3).

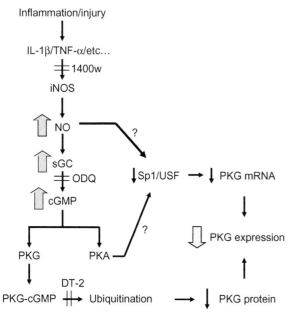

FIGURE 40.3 Control of PKG-I expression.
PKG-I is regulated at both the mRNA and protein level. Transcription factors binding to Sp1 sites and upstream stimulatory factor (USF) sites in the 3′-promoter region of PKG-I stimulate transcription. NO and cGMP inhibit binding of Sp1 proteins to the Sp1 site. The inhibition of Sp1 binding may be due in part to cGMP-dependent activation of PKA owing to the robust elevation of cGMP in response to inflammatory cytokine-induced expression of type II NO synthase (iNOS) and the resultant activation of sGC. Inhibition of iNOS by 1400W or sGC by ODQ inhibit the downregulation of PKG-I mRNA. At the protein level, cGMP binding to PKG-I stimulates ubiquitination and degradation of PKG-I. This effect is blocked by the PKG-I catalytic inhibitor, DT-2, suggesting that autophosphorylation may be responsible for ubiquitination of PKG-I.

REFERENCES

1. Hofmann F, Feil R, Kleppesch T, Schlossmann J. Function of cGMP-dependent protein kinases as revealed by gene deletion. *Physiol Rev* 2006;**86**:1–23.
2. Lincoln TM, Wu X, Sellak H, Dey N, Choi CS. Regulation of vascular smooth muscle cell phenotype by cyclic GMP and cyclic GMP-dependent protein kinase. *Front Biosci* 2006;**11**:356–67.
3. Hofmann F. The biology of cyclic GMP-dependent protein kinases. *J Biol Chem* 2005;**280**:1–4.
4. Hou Y, Gupta N, Schoenlein P, Wong E, Martindale R, Ganapathy V, Browning D. An anti-tumor role for cGMP-dependent protein kinase. *Cancer Letts* 2005;**240**:60–8.
5. Hou Y, Wong E, Martin J, Schoenlein PV, Dostmann WE, Browning D. A role for cyclic-GMP dependent protein kinase in anoikis. *Cell Signal* 2006;**18**:882–8.
6. El-Husseini AE, Williams J, Reiner RB, Pelech S, Vincent SR. Localization of the cGMP-dependent protein kinase in relation to nitric oxide signaling in the brain. *J Chem Neuroanat* 1999;**17**:44–55.
7. Pfeifer A, Klatt P, Massberg S, Ny L, Sausbier M, Hirneiss C, Wang GX, Korth M, Aszodi A, Andersson KE, Krombach F, Mayerhofer A, Ruth P, Fassler R, Hofmann F. Defective smooth muscle regulation in cGMP kinase I-deficient mice. *EMBO J* 1998;**17**:3045–51.
8. Pfeifer A, Aszodi A, Seidler U, Ruth P, Hofmann F, Fassler R. Intestinal secretory defects and dwarfism in mice lacking cGMP-dependent protein kinase II. *Science* 1996;**274**:2082–6.
9. Wolfsgruber W, Feil S, Brummer S, Kuppinger O, Hofmann F, Feil R. A proatherogenic role for cGMP-dependent protein kinase in vascular smooth muscle cells. *Proc Nat Acad Sci USA* 2003;**100**:13,519–13,524.

10. Lincoln TM, Dey N, Sellak H. Signal transduction in smooth muscle invited review: cGMP-dependent protein kinase signaling in smooth muscle: from the regulation of tone to gene expression. *J Appl Physiol* 2001;**91**:1421–30.

11. MacMillan-Crow LA, Lincoln TM. High-affinity binding and localization of the cyclic GMP-dependent protein kinase with the intermediate filament protein vimentin. *Biochemistry* 1994;**33**:8035–43.

12. Yuasa K, Michibata H, Omori K, Yanaka N. A novel interaction of cGMP-dependent protein kinase I with Troponin T. *J Biol Chem* 1999;**274**:37,429–37,434.

13. Wolfe L, Corbin JD, Francis SH. Characterization of a novel isozyme of cGMP-dependent protein kinase from bovine aorta. *J Biol Chem* 1989;**264**:7734–41.

14. Ruth P, Landgraf W, Keilbach A, May B, Egleme C, Hofmann F. The activation of expressed cGMP-dependent protein kinase isozymes Iα and Iβ is determined by the different amino-termini. *Eur J Biochem* 1991;**202**:1339–44.

15. Busch JL, Bessay EP, Francis SH, Corbin JD. A conserved serine juxtaposed to the psuedosubstrate site of Type I cGMP-dependent protein kinase contributes strongly to autoinhibition and lower cGMP affinity. *J Biol Chem* 2002;**277**:34,048–34,054.

16. Surks HK, Mochizuki N, Kasai Y, Georgescu SP, Tang KM, Ito M, Lincoln TM, Mendelsohn ME. Regulation of myosin phosphatase by a specific interaction with cGMP-dependent protein kinase I-alpha. *Science* 1999;**86**:1583–7.

17. Khatri JJ, Joyce KM, Borzovich FV, Fisher SA. Role of myosin phosphatase isoforms in cGMP-mediated smooth muscle relaxation. *J Biol Chem* 2001;**276**:37,250–37,257.

18. Huang QQ, Fisher SA, Brozovich FV. Unzipping the role of myosin light chain phosphatase in smooth muscle cell relaxation. *J Biol Chem* 2003;**279**:597–603.

19. Given AM, Ogut O, Brozovich FV. MYPT1 mutants demonstrate the importance of aa 888–928 for the interaction with PKG-I alpha. *Am J Physiol (Cell* 2007;**292**:C432–9.

20. Schlossmann J, Ammendola A, Ashman K, Zong X, Huber A, Neubauer G, Wang GX, Allescher HD, Korth M, Wilm M, Hofmann F, Ruth P. Regulation of intracellular calcium by a signalling complex of IRAG, IP₃ receptor and cGMP kinase Ibeta. *Nature* 2000;**404**:197–201.

21. Geiselhöringer A, Werner M, Sigl K, Smital P, Worner R, Acheo L, Stieber J, Weinmeister P, Feil R, Feil S, Wegener J, Hofmann F, Schlossmann J. IRAG is essential for relaxation of receptor-triggered smooth muscle contraction by cGMP kinase. *EMBO J* 2004;**23**:4222–31.

22. Komalavilas P, Lincoln TM. Phosphorylation of the inositol 1, 4, 5-trisphosphate receptor protein by cyclic GMP-dependent protein kinase. *J Biol Chem* 1994;**269**:8701–7.

23. Komalavilas P, Lincoln TM. Phosphorylation of the inositol 1, 4, 5-trisphosphate receptor: cyclic GMP-dependent protein kinase mediates cAMP and cGMP dependent phosphorylation in the intact rat aorta. *J Biol Chem* 1996;**271**:21,933–21,938.

24. Koller A, Schlossmann J, Ashman K, Uttenweiler-Joseph S, Ruth P, Hofmann F. Association of phospholamban with a cGMP kinase signaling complex. *Biochem Biophys Res Commun* 2003;**300**:155–60.

25. Geiselhoringer A, Werner M, Sigl K, Smital P, Warner R, Acheo L, Steiber J, Weinmeister P, Feil R, Feil S, Wegener J, Hofmann F, Schlossmann J. IRAG is essential for relaxation of receptor-triggered smooth muscle contraction by cGMP kinase. *EMBO J* 2004;**23**:4222–31.

26. Geiselhöringer A, Gaisa M, Hofmann F, Schlossmann J. Distribution of IRAG and cGKI-isoforms in murine tissues. *FEBS Letts* 2004;**575**:19–22.

27. Alioua A, Tanaka Y, Wallner M, Hofmann F, Ruth P, Meera P, Toro L. The large conductance, voltage-dependent and calcium-sensitive K+ channel, Hslo, is a target of cGMP-dependent protein kinase phosphorylation in vivo. *J Biol Chem* 1998;**273**:32,950–32,956.

28. Fukao M, Mason HS, Britton FC, Kenyon JL, Horowitz B, Keef KD. Cyclic GMP-dependent protein kinase activates cloned BKCa channels expressed in mammalian cells by direct phosphorylation at serine 1072. *J Biol Chem* 1999;**274**:10,927–10,935.

29. Tang KM, Wang GR, Lu P, Karas RH, Aronovitz M, Heximer SP, Kalternbronn KM, Blumer KJ, Siderovski DP, Zhu Y, Mendelsohn ME. Regulator of G-protein signaling-2 mediates vascular smooth muscle relaxation and blood pressure. *Nat Med* 2003;**9**:1506–9.

30. Boerth NJ, Dey NB, Cornwell TL, Lincoln TM. Cyclic GMP-dependent protein kinase regulates vascular smooth muscle cell phenotype. *J Vasc Res* 1997;**34**:245–59.

31. Dey NB, Boerth NJ, Murphy-Ullrich JE, Chang PL, Prince CW, Lincoln TM. Cyclic GMP-dependent protein kinase inhibits osteopontin and thrombospondin production in rat aortic smooth muscle cells. *Circ Res* 1998;**82**:139–46.

32. Brophy CM, Woodrum DA, Pollock J, Dickinson M, Komalavilas P, Cornwell TL, Lincoln TM. Cyclic GMP-dependent protein kinase expression restores contractile function in cultured vascular smooth muscle cells. *J Vasc Res* 2002;**39**:95–103.

33. Dey NB, Foley KF, Lincoln TM, and Dostmann WEG. Inhibition of cGMP-dependent protein kinase reverses phenotypic modulation of vascular smooth muscle cells. *J Cardiovasc Pharmacol* 2005;**45**:404–13.

34. Wang S, Wu X, Lincoln TM, Murphy-Ullrich JE. Expression of a constitutively active cGMP-dependent protein kinase prevents glucose stimulation of thrombospondin-1 expression and TGF-β activity. *Diabetes* 2003;**52**:2144–50.

35. Gu M, Lynch J, Brecher P. Nitric oxide increases p21 (Waf1/Cip1) expression by a cyclic GMP-dependent pathway that includes activation of extracellular signal-regulated kinase and p70 (S6K). *J Biol Chem* 2000;**275**:11,389–11,396.

36. Tanner FC, Meier P, Greutert H, Champion C, Nabel EG, Luscher TF. Nitric oxide modulates expression of cell cycle regulatory proteins: a cytostatic strategy for inhibition of human vascular smooth muscle cell proliferation. *Circulation* 2000;**101**:1982–9.

37. Wong D, Prameya R, Dorovini-Zis K, Vincent SR. Nitric oxide regulates interactions of PMN with human brain microvessel endothelial cells. *Biochem Biophys Res Commun* 2004;**323**:142–8.

38. Zhang T, Zhuang S, Casteel E, Looney DJ, Boss GR, Pilz RB. A cysteine-rich LIM-only protein mediates regulation of smooth muscle-specific gene expression by cGMP-dependent protein kinase. *J Biol Chem* 2007;**282**:33,367–33,380.

39. Pilz RB, Casteel DE. Regulation of gene expression by cyclic GMP. *Circ Res* 2003;**93**:1034–46.

40. Gudi T, Chen JC, Casteel DE, Seasholtz TM, Boss GR, Pilz RB. cGMP-dependent protein kinase inhibits serum-response element transcription by inhibiting Rho activation and functions. *J Biol Chem* 2002;**277**:37,382–37,393.

41. Miano JM. Serum response factor: toggling between disparate programs of gene expression. *J Mol Cell Cardiol* 2003;**35**:577–93.

42. Yoshida T, Sinha S, Dandre F, Wamgoff BR, Hoofnagle MH, Kremer BE, Wang DZ, Olson EN, Owens GK. Myocardin is a key regulator of CArG-dependent transcription of multiple smooth muscle marker genes. *Circ Res* 2003;**92**:856–64.

43. Wang Z, Wang DZ, Teg Pipes GC, Olson EN. Myocardin is a master regulator of smooth muscle gene expression. *Proc Natl Acad Sci USA* 2003;**100**:7129–34.

44. Lincoln TM, Corbin JD. Characterization and biological role of the cGMP-dependent protein kinase. *Adv Cyclic Nucleo Res* 1983;**15**:139–92.

45. Walter U. Distribution of cyclic GMP-dependent protein kinase in various rat tissues and cell lines determined by a sensitive and specific radioimmunoassay. *Eur J Biochem* 1981;**118**:339–46.

46. Cornwell TL, Lincoln TM. Regulation of intracellular Ca levels in cultured vascular smooth muscle cells: reduction of Ca by atriopeptin and 8-bromo-cyclic GMP is mediated by cGMP-dependent protein kinase. *J Biol Chem* 1989;**264**:1146–55.

47. Cornwell TL, Soff GA, Traynor AE, Lincoln TM. Regulation of the expression of cyclic GMP-dependent protein kinase by cell density in vascular smooth muscle cells. *J Vasc Res* 1994;**31**:330–7.

48. Anderson PG, Boerth NJ, Liu M, McNamara DB, Cornwell TL, Lincoln TM. Cyclic GMP-dependent protein kinase expression in coronary arterial smooth muscle in response to balloon catheter injury. *Arterioscler Thromb Vasc Biol* 2000;**20**:2192–7.

49. Sinnaeve P, Chiche JD, Gillijns H, van Pelt N, Wirthlin D, van de Werf F, Collen D, Bloch KD, Janssens SP. Overexpression of a constitutively active protein kinase G mutant reduces neointima formation and in-stent restenosis. *Circulation* 2002;**105**:2911–16.

50. Gao Y, Dhamakoti S, Trevino EM, Sander FC, Portugal AM, Raj JU. Effect of oxygen on cyclic GMP-dependent protein kinase-mediated relaxation in ovine fetal pulmonary arteries and veins. *Am J Physiol* 2003;**285**:L611–18.

51. Jernigan NL, Walker BR, Resta TC. Pulmonary PKG-1 is upregulated following chronic hypoxia. *Am J Physiol* 2003;**285**:L634–42.

52. Jacob A, Smolenski A, Lohmann SM, Begum N. MKP-1 expression and stabilization and cGK Ialpha prevent diabetes- associated abnormalities in VSMC migration. *Am J Physiol* 2004;**287**:C1077–86.

53. Lin CS, Liu X, Tu R, Chow S, Lue TF. Age-related decrease of protein kinase G activation in vascular smooth muscle cells. *Biochem Biophys Res Commun* 2001;**287**:244–8.

54. Scavone C, Munhoz CD, Kawamoto EM, Glezer I, de Sa Lima L, Marcourakis T, Markus RP. Age-related changes in cyclic GMP and PKG-stimulated cerebellar Na, K-ATPase activity. *Neurobiol Aging* 2005;**26**:907–16.

55. Browner N, Sellak H, Lincoln TM. Down-regulation of cGMP-dependent protein kinase expression by inflammatory cytokines in vascular smooth muscle cells. *Am J Physiol* 2004;**287**:C88–96.

56. Sellak H, Yang X, Cao X, Cornwell T, Soff GA, Lincoln TM. Sp1 transcription factor as a molecular target for nitric oxide- and cyclic nucleotide-mediated suppression of cGMP-dependent protein kinase-I— expression in vascular smooth muscle cells. *Circ Res* 2002;**90**:405–12.

57. Sellak H, Choi CS, Browner N, Lincoln TM. Upstream stimulatory factors (USF-1/USF-2) regulate human cGMP-dependent protein kinase I gene expression in vascular smooth muscle cells. *J Biol Chem* 2005;**280**:18,425–18,433.

58. Zeng Y, Zhuang S, Gloddek J, Tseng CC, Boss GR, Pilz RB. Regulation of cGMP-dependent protein kinase expression by Rho and Kruppel-like transcription factor-4. *J Biol Chem* 2006;**281**:16,951–16,961.

59. Taylor MS, Okwuchukwuasanya C, Nicki C, Tegge W, Brayden JE, Dostmann WRG. Inhibition of cGMP-dependent protein kinase by the cell permeable peptide DT-2 reveals a novel mechanism of vasoregulation. *Mol Pharmacol* 2004;**65**:1111–19.

60. Dey NB, Busch JL, Francis SH, Corbin JD, Lincoln TM. Cyclic GMP specifically suppresses Type 1α cGMP-dependent protein kinase expression by ubiquitination. *Cell Signaling* 2009.

61. Libby P. What have we learned about the biology of atherosclerosis? The role of inflammation. *Am J Cardiol* 2001;**88**:3J–6J.

62. Beasley D, Cooper A. Constitutive expression of interleukin-1a precursor promotes human vascular smooth muscle cell proliferation. *Am J Physiol* 1999;**276**:H901–12.

63. Taylor AM, McNamara CA. Regulation of vascular smooth muscle cell growth: targeting the final common pathway. *Arterioscler Thromb Vasc Biol* 2003;**23**:1717–20.

64. Anwar A, Zahid AA, Scheidegger KJ, Brink M, Delafontaine P. Tumor necrosis factor-alpha regulates insulin-like growth factor-1 and insulin-like growth factor binding protein-3 expression in vascular smooth muscle. *Circulation* 2002;**105**:1220–5.

65. Beckman JS, Beckman TW, Chen J, Marshall PA, Freeman BA. Apparent hydroxy radical production by peroxynitrite: implications for endothelial injury from nitric oxide and superoxide. *Proc Nat Acad Sci USA* 1990;**87**:1620–4.

66. Hogg N, Darley-Usmar V, Graham A, Moncada S. Peroxynitrite and atherosclerosis. *Biochem Soc Trans* 1993;**21**:358–62.

67. Eiserich JP, Hristova M, Cross CE, Jones AD, Freeman BA, Halliwell B, van der Vliet A. Formation of nitric oxide-derived inflammatory oxidants by myeloperoxidase in neutrophils. *Nature* 1998;**391**:393–7.

68. Ishiropoulos H, Zhu L, Chen J, Tsai M, Martin JC, Smith CD, Beckman JS. Peroxynitritie-mediated tyrosine nitration catalyzed by superoxide dismutase. *Arch Biochem Biophys* 1992;**298**:431–7.

69. Radi R, Beckman JS, Bush KM, Freeman BA. Peroxynitrite oxidation of sulfhydryls: the cytotoxic potential of superoxide and nitric oxide. *J Biol Chem* 1991;**266**:4244–50.

70. O'Donnell VB, Freeman BA. Interactions between nitric oxide and lipid oxidation pathways: implications for vascular disease. *Circ Res* 2001;**88**:12–21.

Chapter 41

Substrates of Cyclic Nucleotide-Dependent Protein Kinases

Neil F. W. Saunders[1], Ross I. Brinkworth[1], Bruce E. Kemp[2] and Bostjan Kobe[1]

[1]*School of Chemistry and Molecular Biosciences and Institute for Molecular Bioscience, University of Queensland, Brisbane, Queensland, Australia*

[2]*St Vincent's Institute & Department of Medicine, University of Melbourne, St Vincent's Hospital, Fitzroy, Victoria, Australia*

INTRODUCTION

The cyclic nucleotide-dependent protein kinases cAMP- and cGMP-dependent protein kinase (PKA and PKG) are closely-related enzymes belonging to subgroup I of the AGC serine-threonine protein kinase family. PKA in particular has been studied extensively, with many known substrates. Our previous review focused on the use of peptide libraries as a tool to elucidate the specificity of PKA and PKG [1]. In this chapter, we examine the available data concerning known substrates of PKA/PKG, and the features of the kinase sequences and structures that account for their peptide specificities, before highlighting some bioinformatics tools and databases that can be used to analyze PKA/PKG and their substrates.

CLASSIFICATION AND NOMENCLATURE

The catalytic domains of PKA and PKG are closely related, sharing approximately 50 percent identity. Several schemes exist to classify protein kinases according to sequence similarity in the catalytic domain. In the Kinase Sequence Database [2], PKA and PKG belong to KSD family 17 (PKC/PKA) and family 105 (cAMP-I, PKG-I, PKG-II), respectively. The PANTHER database [3] classifies PKA as cAMP-dependent protein kinase catalytic subunit (PKA C, PTHR22985:SF84) and PKG as cGMP-dependent protein kinase (PTHR22985:SF90). In the classification system used by the kinase.com resource for the human kinome [4], PKA and PKG belong to group AGC, family PKA, and group AGC, family PKG, respectively. In addition, numerous synonyms for the genes that encode PKA/PKG gene products are used in the public sequence databases.

Gene names and synonyms from the current release of the UniProt database [5] are summarized in Table 41.1.

SUBSTRATES OF PKA AND PKG

The PredikinDB database stores information about phosphorylation sites obtained from UniProt annotations (http://predikin.biosci.uq.edu.au) [5a]. The current release of PredikinDB contains 487 and 16 protein sequences annotated as substrates of PKA and PKG, respectively (Table 41.2). Between them, the substrates contain 382 (PKA) and 21 (PKG) non-redundant heptapeptide phosphorylation sites, of which in both cases around 70 percent are annotated with high confidence ("by similarity" or "certain"). The number of known substrates for PKA is in fact the largest for any protein kinase; by contrast, PKG substrates are rather poorly characterized.

Non-redundant high confidence phosphorylation sites can be used to construct frequency matrices for PKA and PKG substrates (Table 41.3). This illustrates the key feature of PKA/PKG substrates; namely, a strong preference for arginine or lysine residues at the $P-3$ and $P-2$ positions, relative to the phosphorylation site (P0; residues in the substrate sequence are termed $P-1$ to $P-n$ N-terminal or $P+1$ to $P+n$ C-terminal to P0). The PKA/PKG substrates consensus motif $[RK] [RK]-X-[ST]$ is in fact one of only three motifs for substrates of serine-threonine protein kinase families, along with casein kinase II and protein kinase C, that is conserved enough to warrant an entry (PDOC00004) in the PROSITE database [6]. Positions $P-1$ and $P+1$ to $P+3$ are rather less conserved, although PKA exhibits some preference for aliphatic residues at $P+1$ and Ala or Ser at $P+2$ and $P+3$. The structural features of PKA/PKG that determine these preferences are discussed further

TABLE 41.1 Gene names and synonyms of PKA and PKG in the UniProt database[1]

Kinase	Gene name	Synonyms
PKA	TPK1	PKA1, SRA3, YJL164C, J0541
	PRKACB	PKACB, Prkacb, Pkacb
	pka1	tpk, SPBC106.10
	TPK2	PKA2, YKR1, YPL203W
	PRKACA	Prkaca, PKACA, Pkaca
	PRKACG	
	CAPK	
	kin-1	ZK909.2
	pkaC	PK2, PK3
	Pka-C1	CdkA, DC0, CG4379
	TPK3	PKA3, YKL166C, YKL630
	Pka-C2	DC1, CG12066
	Pka-C3	DC2, CG6117
	PRKX	PKX1, Prkx, Pkare
	PRKY	
PKG	PRKG1	Prkg1, PRKGR1A, PRKG1B, Prkg1b, PRKGR1B, Prkgr1b
	Pkg21D	DG1, CG3324
	for	DG2, PGK2, Pkg24A, CG10033
	PRKG2	Prkg2, PRKGR2, Prkgr2

[1]*Names and synonyms were retrieved from the PredikinDB database of phosphorylation sites and associated protein kinases (http://predikin.biosci.uq.edu.au). Another resource of kinase sequence information is the Kinase Sequence Database http://sequoia.ucsf.edu/ksd/*

TABLE 41.2 Substrates of PKA and PKG annotated in the UniProt database[1]

	PKA	PKG
Substrates	487	16
Sites[2] (non-redundant sites)		
Certain	161 (137)	11 (8)
By similarity	337 (139)	9 (6)
Probable	7 (6)	4 (3)
Potential	195 (100)	4 (4)
Total	700 (382)	28 (21)

[1]*Other information on phosphorylation sites include PhosphoSite (http://www.phosphosite.org/; Cell Signalling Technology) and PhosphoELM (http://phospho.elm.eu.org/).*
[2]*Sites are labeled according to annotation confidence (with the degree of confidence decreasing in the order certain, by similarity, probable or potential). The numbers in parenthesis refer to "non-redundant" heptapeptide sequences. The difference is that in the "redundant" sites, the same heptapeptide phosphorylation site sequence occurs in more than one protein – i.e., several proteins may have the same local phosphorylation site sequence.*

the more limited roles for PKG substrates are a reflection of the far smaller number of known substrates in the public databases, as compared with PKA substrates. Although the precise substrates of PKG are often not well documented, it is known to modulate a number of important physiological processes [9]. Of particular interest are the regulation of NO-mediated smooth muscle relaxation [10], intestinal secretion [11], and the development of the nervous system [12]. Tissue-specific expression of PKG isoforms also appears to be an important factor in determining its physiological roles. For example, PKG-Iβ, but not -Iα, is recruited to the endoplasmic reticulum by an inositol 1,4,5-trisphosphate receptor-associated substrate [13]. Alternative splicing of PKG, induced by angiotensin, also modulates the nitrate tolerance of smooth muscle cells in hypertension [14].

STRUCTURAL AND SEQUENCE FEATURES RELEVANT TO PEPTIDE SPECIFICITY

Protein crystal structures have revealed that protein kinases have extended grooves to accommodate the protein substrate sequence around the target phosphorylation site (Figure 41.2). Residues in the substrate sequence make contact with amino acid residues in the protein kinase substrate binding groove termed substrate-determining residues (SDRs) (Figure 41.2), and so determine which substrates are preferred to bind to the kinase [15, 15a]. SDRs can be located by aligning a query protein sequence

under "Structural and sequence features relevant to peptide specificity" (see below).

Substrates of PKA and PKG are involved in many diverse biological processes. This can be illustrated using a Gene Ontology (GO) analysis [7], in which all known substrates of PKA and PKG in UniProt were submitted to the GO annotation services at AgBase [8]. GO annotations for the biological processes assigned to substrates of PKA and PKG are illustrated in Figure 41.1. Several categories – signal transduction, transcription, and transport – are common to substrates of both kinases. Substrates of PKA appear to be involved with a wider range of biological processes than those of PKG, and some processes, such as nucleic acid metabolism, are more prevalent in PKG substrates. However, at present it is not clear to what degree

TABLE 41.3 Amino acid frequencies[1] at positions P−3 to P+3 for substrates of PKA and PKG in the UniProt database

PKA (220 unique sites)																			
	A	C	D	E	F	G	H	I	K	L	M	N	P	Q	R	S	T	V	W	Y
P−3	3	0	1	1	1	1	2	3	31	8	1	1	6	2	146	4	5	3	0	1
P−2	0	1	1	1	1	1	6	1	37	1	0	2	0	6	131	12	17	2	0	0
P−1	11	2	4	1	9	20	2	7	12	31	9	18	17	9	22	22	4	16	1	3
P0	0	0	0	0	0	0	0	0	0	0	0	0	0	0	0	196	24	0	0	0
P+1	10	1	10	5	13	10	2	19	8	35	4	7	7	7	9	29	13	24	6	1
P+2	25	8	12	5	10	7	2	9	10	17	1	15	17	8	7	31	17	13	1	5
P+3	24	0	6	20	2	13	3	2	16	15	4	9	17	14	16	28	18	9	3	1
PKG (10 unique sites)																			
	A	C	D	E	F	G	H	I	K	L	M	N	P	Q	R	S	T	V	W	Y
P−3	0	0	0	0	0	0	0	0	0	0	0	0	0	0	9	0	0	1	0	0
P−2	0	0	0	0	0	0	0	0	6	0	0	0	0	0	4	0	0	0	0	0
P−1	3	0	0	0	0	0	0	3	1	0	0	0	0	0	1	0	0	2	0	0
P0	0	0	0	0	0	0	0	0	0	0	0	0	0	0	9	1	0	0	0	0
P+1	0	0	0	0	1	1	0	0	2	0	0	2	0	3	1	0	0	0	0	0
P+2	2	0	0	0	0	1	0	0	0	0	0	0	0	1	0	3	1	1	1	0
P+3	3	0	0	1	0	3	0	0	1	0	0	1	0	0	0	1	0	0	0	0

[1]*Frequencies were calculated from non-redundant heptapeptide phosphorylation site sequences annotated for PKA/PKG phosphorylation sites with confidence "certain" or "by similarity" in UniProt.*

with a profile Hidden Markov model [16] of the kinase catalytic domain obtained from a database such as Pfam [17] or SMART [18]. Figure 41.3 illustrates an alignment of human PKA and PKG using the SMART HMM S_TKc and the HMMER program hmmalign. The major SDRs are summarized in Table 41.4. The conserved motifs (in bold) simply act as anchors in the alignment reference sequence (HMM) from which the positions of SDRs in the query sequence can be calculated.

The sequences of the catalytic domains in PKA and PKG are 50 percent identical and 70 percent similar. This translates to the major SDRs from each kinase, which are identical at 9/16 positions and similar at a further 6/16. The SDRs that affect substrate positions P−3 and P−2, where Arg and Lys are strongly preferred, are identical for the P−2 position and conserved for the P−3 position (Figure 41.2). The P−3 and P−2 substrate residues make more contacts (H bonds, electrostatic bridges, and van der Waals contacts)

with the kinase catalytic domain than any other substrate side-chains, and so provide the bulk of the interactions required for peptide specificity.

The most significant differences in SDRs are in those that bind to the P+1 and P+2 positions in the substrate heptapeptide. V525 in PKG substitutes for L206 in PKA. The shorter valine side-chain makes fewer contacts with the substrate, suggesting a weaker preference for hydrophobic residues at the P+1 position in PKG substrates, compared with those of PKA. The P+2 position is influenced by the "KE loop" region of serine-threonine protein kinases, found between the AMK motif and a Glu residue C-terminal of AMK (Figure 41.2). In PKG, D399 and T400 are substituted for K82 and L83 in PKA. However, specificity at the P+2 position is rather broad in most serine-threonine protein kinases, and the limited number of known PKG substrates makes it difficult to ascertain the effect of these substitutions. It may

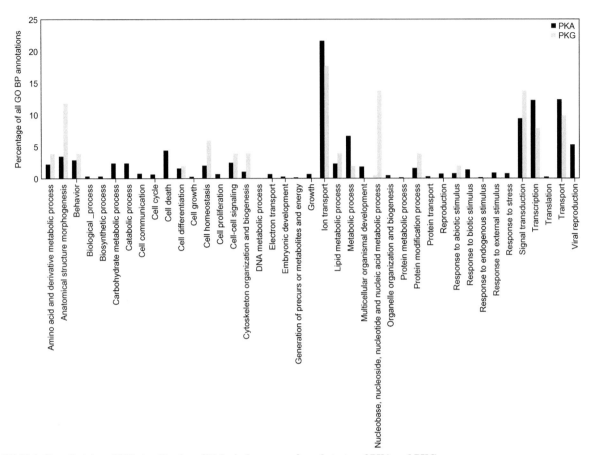

FIGURE 41.1 Gene Ontology (GO) classification of biological processes for substrates of PKA and PKG.
Substrates of PKA and PKG were classified by GO biological process using the tools GORetriever and GOSlimViewer at the AgBase server.

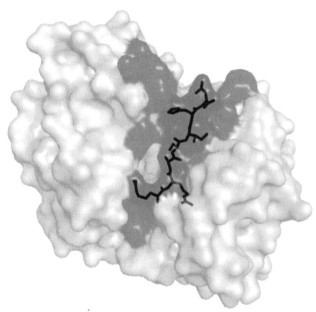

FIGURE 41.2 Surface representation of PKA showing bound substrate peptide and location of substrate-determining residues.
The major SDRs are shown in dark gray. Peptide substrate RRASIHD is shown as black sticks. Figure generated from PDB structure 1JBP.

be instructive to compare known substrates of both PKA and PKG to cases where only one or the other is known to phosphorylate a substrate. For example, PKA and PKG both phosphorylate S660, S700, S737, and S813 in the cystic fibrosis transmembrane conductance regulator at the sequences RRNSILT, RKNSILN, RRLSLVP, and RRLSQET (phosphorylation site shown in bold) [19], whereas the sequence RTLSVSS in glycogen synthase is a substrate for PKA, but not PKG [20]. This suggests that within certain sequences, positions P+2 and P+3 can play an important role in determining the differential peptide specificity in PKA and PKG. Alternatively, the differences between PKA and PKG in the pocket binding the P−3 residue (Phe and Ser in PKA vs. Trp and Thr in PKG; Table 41.4) may account for differentially recognizing the Arg–Thr pair of residues at the P−3 and P−2 positions in glycogen synthase.

SDRs are used by the Predikin method [15, 15a] to predict phosphorylation sites for a kinase given only its sequence. Predikin weight matrices for substrates of PKA and PKG at positions P − 3 to P + 3 are presented in Table 41.5. Broadly, the matrices agree well

```
KAPCA_HUMAN  FERIKTLGTGSFGRVMLVKHKE.TGNHYAMKILDkqkvvklkqiEHTLNE
KGP1A_HUMAN  FNIIDTLGVGGFGRVELVQLKSeESKTFAMKILKkrhivdtrqqEHIRSE
                    *.                     *.
                    -1                     +2
                    -1                     +2
                                           +2


KAPCA_HUMAN  KRILQAvNFPFLVKLEFSFKDnSNLYMVMEYVPG--GEMFSHLRRIGR--
KGP1A_HUMAN  KQIMQGaHSDFIVRLYRTFKDsKYLYMLMEACLG--GELWTILRDRGS--
                                            *  ::
                                            -3
                                            -3
                                            -3


KAPCA_HUMAN  ----FSEpHARFYAAQIVLTFEYLHSLDLIYRDLKPENLLIDQqgYIQVT
KGP1A_HUMAN  ----FEDsTTRFYTACVVEAFAYLHSKGIIYRDLKPENLILDHrgYAKLV


KAPCA_HUMAN  DFGFAKRVkg..rtWTLCGTPEYLAPEIIlskGYNK-AVDWWALGVLIYE
KGP1A_HUMAN  DFGFAKKIgfgkktWTFCGTPEYVAPEIIlnkGHDI-SADYWSLGILMYE
                *              *.  ****:
               +1             +3    -2
                              +3    -2
                                    -2
                                      +1
                                      +1


KAPCA_HUMAN  MAAGYPPFFadQPIQIYEKIVsgkvr..fpshfSSDLKDLLRNLLQVDLT
KGP1A_HUMAN  LLTGSPPFSgpDPMKTYNIILrgidmiefpkkiAKNAANLIKKLCRDNPS


KAPCA_HUMAN  KRFGNlKNGV--NDIKNHKWF
KGP1A_HUMAN  ERLGNlKNGV--KDIQKHKWF
```

FIGURE 41.3 Alignment of catalytic domains from PKA (KAPCA_HUMAN) and PKG (KGP1A_HUMAN).
Catalytic domains were aligned using the HMMER program hmmalign and the profile HMM S_TKc (SMART database accession SM00220). The "anchor motifs" used to locate substrate-determining residues (SDRs) in the alignment are shown in bold. SDRs are underlined. Identity (*) and similarity (: or .) between SDRs in PKA and PKG is indicated. The "KE loop" region is shown in italics. Numbers below the alignment indicate the position in the substrate relative to the phosphorylation site which interacts with the corresponding SDR.

TABLE 41.4 Major substrate determining residues in PKA and PKG

Position relative to p[ST]	SDRs[1]	Residue in PKA	Residue in PKG
P−3	GEL+1	E128	E445
	GEL +3	F130	W447
	GEL +4	S131	T448
P−2	APE−2	Y205	Y524
	APE −3	E204	E523
	APE −5	T202	T521
P−1	GXG +2	G53	G369
	GXG +3	S54	G370
P+1	APE −1	L206	V525
	APE −4	P203	P522
	DFG +3	F188	F505
P+2	AMK +10	V81	V398
	AMK +11	K82	D399
	AMK +12	L83	T400
P+3	APE −8	L199	F518
	APE −9	T198	T517

[1]The SDRs column indicates the position of the SDR relative to the anchor motifs GXG, AMK, GEL, DFG and APE, used to locate SDRs in a sequence alignment.

with observed PKA/PKG substrates, with the exception that a strong preference for hydrophobic residues (Phe, Val, and particularly Met) is predicted at the P+1 position. Ile, Leu, and Val are found at P+1 in approximately 40 percent of known PKA substrate heptapeptides, but Met and Phe are not common. One reason for this is that the regions surrounding a phosphorylation site need to be solvent-exposed and bind the kinase in an extended conformation; an excess of large hydrophobic residues would therefore be unfavorable [21].

Given the similar peptide specificity of PKA and PKG, it is clear that other factors are important in determining the substrates that they phosphorylate and the pathways that they regulate. Collectively, these factors are described by the term "recruitment," which encompasses any process that brings a protein kinase and its substrate together in the cell [22]. Recruitment processes include (1) co-expression of kinase and substrate, and (2) co-localization of kinase and substrate by mechanisms including docking or scaffolding interactions.

Subcellular targeting of PKA is a well-characterized process involving A-kinase anchoring proteins (AKAPs) [23, 24]. AKAPs bind to a high-affinity docking domain on the N-terminus of the PKA regulatory subunit, forming complexes in which PKA is localized with substrates that participate in the cAMP activation pathway. There is now

TABLE 41.5 Predikin predicted weight matrix at positions P −3 to P+3 for substrates of PKA and PKG based on substrate-determining residues

PKA(P17612)

	A	C	D	E	F	G	H	I	K	L	M	N	P	Q	R	S	T	V	W	Y
P−3	0.92	−1.89	−3.21	−3.89	−2.89	−3.47	−2.47	−0.16	1.84	−2.43	−1.89	−2.89	−0.42	0.21	2.47	−3.89	1.30	0.24	−0.89	−2.47
P−2	1.11	−0.76	−0.60	−0.69	0.56	0.09	−0.50	−1.82	−0.21	−0.76	0.24	−2.08	−0.86	−0.71	1.46	−0.18	0.88	0.33	−2.08	−2.67
P−1	−0.37	−0.75	−0.91	−0.95	1.10	0.86	−0.70	−3.78	0.43	−0.83	0.80	0.22	−0.37	−0.91	0.40	0.86	0.29	−1.84	−0.53	0.87
P0	−5.13	−3.55	−4.87	−5.55	−4.55	−5.13	−4.13	−4.87	−5.13	−5.87	−3.55	−4.55	−5.13	−4.87	−5.13	2.94	2.83	−5.13	−2.55	−3.36
P+1	0.10	−1.96	−3.28	0.32	1.62	−3.54	−2.54	1.43	0.10	−1.30	3.19	−2.96	−3.54	−3.28	−0.57	−0.98	0.36	0.73	−0.96	−2.54
P+2	1.27	2.31	−2.39	−3.07	−2.07	−2.65	−1.65	0.99	−2.65	−3.39	−1.07	0.45	−2.65	2.88	−0.14	−3.07	−2.39	−0.14	−0.07	−1.65
P+3	−1.61	−0.49	0.20	0.40	−1.49	1.97	−1.75	0.32	0.96	−0.95	−0.49	−1.49	0.29	−2.49	−4.09	0.54	−0.48	−0.98	−1.50	−1.07

PKG (Q13976)

	A	C	D	E	F	G	H	I	K	L	M	N	P	Q	R	S	T	V	W	Y
P−3	−0.90	−1.88	−3.20	−3.88	−2.88	−3.46	−2.46	−0.64	1.86	−2.42	−1.88	−2.88	−0.40	0.23	2.49	−3.88	1.32	0.26	−0.88	−2.46
P−2	1.11	−0.76	−0.60	−0.69	0.56	0.09	−0.50	−1.82	−0.21	−0.76	0.24	−2.08	−0.86	−0.71	1.46	−0.18	0.88	0.33	−2.08	−2.67
P−1	−2.01	−2.47	1.28	−0.72	0.58	0.46	0.32	−3.79	0.13	0.07	2.04	−0.62	−0.68	−0.66	−0.15	−0.06	−0.42	−0.92	1.38	0.85
P0	−5.13	−3.55	−4.87	−5.55	−4.55	−5.13	−4.13	−4.87	−5.13	−5.87	−3.55	−4.55	−5.13	−4.87	−5.13	2.94	2.83	−5.13	−2.55	−3.36
P+1	0.10	−1.96	−3.28	0.32	1.62	−3.54	−2.54	1.43	0.10	−1.30	3.19	−2.96	−3.54	−3.28	−0.57	−0.98	0.36	0.73	−0.96	−2.54
P+2	0.16	2.13	−0.90	0.43	−2.60	−3.18	−0.16	−2.92	−0.35	−0.20	−1.60	−2.60	−3.18	−2.92	1.49	1.51	−2.92	0.54	−0.60	−2.18
P+3	0.36	0.09	1.03	−0.24	0.39	1.75	−0.12	−3.11	−1.12	−1.02	−1.78	0.55	−0.20	−0.12	−1.12	0.52	−0.22	−2.32	−0.78	−3.47

growing evidence for the localization of PKG to subcellular scaffolding complexes ("GKAPs"), although they have proved more difficult to identify than the AKAPs [25].

CONCLUSIONS

PKA is an intensively studied protein kinase, and is well characterized in terms of sequence analysis, structure, and peptide specificity. PKG is less well characterized, with far fewer documented substrates. However, peptide specificity predictions based on substrate-determining residues indicate that PKG has a similar specificity to PKA. Improved database annotation of PKG phosphorylation sites is required to resolve the subtle differences in peptide specificity between PKA and PKG. Given the similarities in peptide specificity between PKA and PKG, it seems likely that substrate recruitment and tissue-specific expression profiles will remain important factors in determining the physiological roles of these kinases. The accelerating use of mass spectrometry in phosphoproteomics at the tissue level [26], coupled with the increasing knowledge of signaling pathways, means that we can look forward to an avalanche of data for all protein kinases, including the classical kinases such as PKA and PKG.

REFERENCES

1. Brinkworth R, Kobe B, Kemp B. Peptide substrates of cyclic nucleotide-dependent protein kinases. In: Bradshaw R, Dennis E, editors. *Handbook of cell signaling*. San Diego, CA: Academic Press; 2003. p. 495–9.

2. Buzko O, Shokat K. A kinase sequence database: sequence alignments and family assignment. *Bioinformatics* 2002;**18**:1274–5.

3. Mi H, Lazareva-Ulitsky B, Loo R, Kejariwal A, Vandergriff J, Rabkin S, Guo N, Muruganujan A, Doremieux O, Campbell MJ, Kitano H, Thomas PD. The PANTHER database of protein families, subfamilies, functions and pathways. *Nucleic Acids Res* 2005;**33**:D284–8.

4. Manning G, Whyte DB, Martinez R, Hunter T, Sudarsanam S. The protein kinase complement of the human genome. *Science* 2002;**298**:1912–34.

5. The UniProt Consortium. The Universal Protein Resource (UniProt). *Nucleic Acids Res* 2007;**35**:D193–7.

5a. Saunders NF, Kobe B. The Predikin webserver: improved prediction of protein kinase peptide specificity using structural information. *Nucleic Acids Res* 2008;**36**:W286–290.

6. Hulo N, Bairoch A, Bulliard V, Cerutti L, Castro ED, Langendijk-Genevaux PS, Pagni M, Sigrist CJA. The PROSITE database. *Nucleic Acids Res* 2006;**34**:D227–30.

7. Gene Ontology Consortium. The Gene Ontology (GO) project in 2006. *Nucleic Acids Res* 2006;**34**:D322–6.

8. McCarthy FM, Bridges SM, Wang N, Magee GB, Williams WP, Luthe DS, Burgess SC. AgBase: a unified resource for functional analysis in agriculture. *Nucleic Acids Res* 2007;**35**:D599–603.

9. Schlossmann J, Feil R, Hofmann F. Insights into cGMP signalling derived from cGMP kinase knockout mice. *Front Biosci* 2005;**10**:1279–89.

10. Schlossmann J, Ammendola A, Ashman K, Zong X, Huber A, Neubauer G, Wang GX, Allescher HD, Korth M, Wilm M, Hofmann F, Ruth P. Regulation of intracellular calcium by a signalling complex of IRAG, IP3 receptor and cGMP kinase I-beta. *Nature* 2000;**404**:197–201.

11. Kim BJ, Lee JH, Jun JY, Chang IY, So I, Kim KW. Vasoactive intestinal polypeptide inhibits pacemaker activity via the nitric oxide-cGMP-protein kinase G pathway in the interstitial cells of Cajal of the murine small intestine. *Mol Cells* 2006;**21**:337–42.

12. Feil R, Hofmann F, Kleppisch T. Function of cGMP-dependent protein kinases in the nervous system. *Rev Neurosci* 2005;**16**:23–41.

13. Geiselhöringer A, Gaisa M, Hofmann F, Schlossmann J. Distribution of IRAG and cGKI-isoforms in murine tissues. *FEBS Letts* 2004;**575**:19–22.

14. Gerzanich V, Ivanov A, Ivanova S, Yang JB, Zhou H, Dong Y, Simard JM. Alternative splicing of cGMP-dependent protein kinase I in angiotensin-hypertension: novel mechanism for nitrate tolerance in vascular smooth muscle. *Circ Res* 2003;**93**:805–12.

15. Brinkworth RI, Breinl RA, Kobe B. Structural basis and prediction of substrate specificity in protein serine/threonine kinases. *Proc Natl Acad Sci USA* 2003;**100**:74–9.

15a. Saunders NF, Brinkworth RI, Huber T, Kemp BE, Kobe B. Predikin and PredikinDB: a computational framework for the prediction of protein kinase peptide specificity and an associated database of phosphorylation sites. *BMC Bioinformatics* 2008;**9**:245.

16. Eddy SR. Profile hidden Markov models.. *Bioinformatics* 1998;**14**:755–63.

17. Finn RD, Mistry J, Schuster-Böckler B, Griffiths-Jones S, Hollich V, Lassmann T, Moxon S, Marshall M, Khanna A, Durbin R, Eddy SR, Sonnhammer ELL, Bateman A. Pfam: clans, web tools and services. *Nucleic Acids Res* 2006;**34**:D247–51.

18. Letunic I, Copley RR, Pils B, Pinkert S, Schultz J, Bork P. SMART 5: domains in the context of genomes and networks. *Nucleic Acids Res* 2006;**34**:D257–60.

19. Picciotto MR, Cohn JA, Bertuzzi G, Greengard P, Nairn AC. Phosphorylation of the cystic fibrosis transmembrane conductance regulator. *J Biol Chem* 1992;**267**(12):742–52.

20. Flotow H, Graves PR, Wang AQ, Fiol CJ, Roeske RW, Roach PJ. Phosphate groups as substrate determinants for casein kinase I action. *J Biol Chem* 1990;**265**:14,264–14,269.

21. Iakoucheva LM, Radivojac P, Brown CJ, O'Connor TR, Sikes JG, Obradovic Z, Dunker AK. The importance of intrinsic disorder for protein phosphorylation. *Nucleic Acids Res* 2004;**32**:1037–49.

22. Zhu G, Liu Y, Shaw S. Protein kinase specificity. a strategic collaboration between kinase peptide specificity and substrate recruitment. *Cell Cycle* 2005;**4**:52–6.

23. Beene DL, Scott JD. A-kinase anchoring proteins take shape. *Curr Opin Cell Biol* 2007;**19**:192–8.

24. Taylor SS, Kim C, Cheng CY, Brown SHJ, Wu J, Kannan N. Signaling through cAMP and cAMP-dependent protein kinase: diverse strategies for drug design. *Biochim Biophys Acta* 2007;**17**:16–26.

25. Dodge-Kafka KL, Langeberg L, Scott JD. Compartmentation of cyclic nucleotide signaling in the heart: the role of A-kinase anchoring proteins. *Circ Res* 2006;**98**:993–1001.

26. Villén J, Beausoleil SA, Gerber SA, Gygi SP. Large-scale phosphorylation analysis of mouse liver. *Proc Natl Acad Sci USA* 2007;**104**:1488–93.

Chapter 42

Physiological Substrates of PKA and PKG

Anja Ruppelt[1, 2], Nikolaus G. Oberprieler[1, 2], George Magklaras[2] and Kjetil Taskén[1, 2]

[1]Biotechnology Centre of Oslo

[2]Centre for Molecular Medicine Norway, Nordic EMBL Partnership, University of Oslo, Oslo, Norway

INTRODUCTION

The cAMP- and cGMP-dependent protein kinases (PKA and PKG, respectively) belong to the ACG subclass of Ser/Thr-specific protein kinases and generally prefer the phosphate acceptor residue preceded by a row of basic residues. S is also the favored phosphate acceptor, when taking into account the 12-fold higher frequency of S over T in eukaryotic proteins. Based on an extensive body of work with peptide substrates *in vitro* (see Chapter 182 of Handbook of Cell Signaling, Second Edition) and mapping of phosphorylation sites in physiological substrates *in vivo*, PKA is well known to phosphorylate substrates with the general motif $R(R/K)X(S/T)$ [1–6], whereas, the consensus for PKG substrates is $(R/K_{2-3})(X K)(S/T)$ and includes more basic residues than the PKA consensus [7] (Table 42.1). However, considerable overlap of sites phosphorylated by both kinases is observed *in vivo*. By analysis of physiological substrates (Table 42.2), the preference for PKA in the $P-3$ and $P-$ positions is $RR=RK>>KR=KK$, and there are weaker preferences for small residues (S, G, P) at $P-1$, for a basic residue (R) at $P-4$ to $P-7$, and for a hydrophobic residue (F, I, L, V) at $P+1$ [8]. Substrate specificity of PKG is similar to that of PKA, but in physiological substrates (Table 42.3) a stronger preference for $R>>K$ at $P-3$ (at position $P-2$ $K=R$), a slight preference for basic or neutral residues (K, R, S) in $P-4$, and an increased frequency of neutral and hydrophobic residues (S, V, A) at $P-1$ and (S, L, A) at $P+1$ is observed (this report).

Here, we present data regarding the total availability of PKA and PKG consensus sites in the human proteome, estimate frequencies of phosphorylation of different motifs, and attempt to give an overview of physiological substrates of both kinases that meet a set of eligibility criteria. However, mechanisms whereby the phosphorylation event alters function of each individual substrate and thereby regulates its physiological role are not included in this short overview.

ABUNDANCE OF PKA AND PKG PHOSPHORYLATION SITES IN THE HUMAN PROTEOME

We have focused our search on the human subset of the International Protein Index database [9]. Version 3.34 of the IPI human subset (October 2007) contained 67,556 minimally redundant yet maximally complete human proteins. A small PERL (Practical Extraction and Report Language; PERL Portal) script was written to generate all the permutations of the canonical PKA and PKG motif $(R/K)(R/K)X(S/T)$, and search each of the produced motif combinations against the IPI human database using the "patmatdb" EMBOSS protein motif search application [10]. A little more than 57,000 motifs were found in human proteins (Table 42.1a). Based on the total abundance of canonical motifs (phosphorylated and non-phosphorylated) in a limited set of approximately 100 physiological substrates for PKA where the phosphorylation sites had been mapped, Shabb estimated the probability of *in situ* phosphorylation of permutations of the canonical sequence, which was $RRXS$ $(0.8)>RRXT$, $RKX(S/T)$ $(0.5–0.3)>KKX(S/T)$, $KRX(S/T)$ (<0.2) [8]. A similar analysis of frequency of PKG phosphorylation (this report, from substrates listed in Table 42.3) shows that probability of phosphorylation *in vivo* by PKG is estimated to be $R(R/K)XS$ $(0.7)>RKXT$ $(0.5)>KRXS$, $RRXT$ $(0.3)>KKXS$ (<0.2). Using these probabilities, we estimate that PKA, PKG, or both can phosphorylate approximately 21,000 sites of the 57,000 putative sites in the human proteome *in vivo*. When both frequency of different sites

TABLE 42.1 Abundance of PKA and PKG motifs[1]

(a) Frequency and probability of phosphorylation of canonical substrate sequences

Consensus	*Homo sapiens*	Estimated phosphorylated motifs of PKA[2]	Estimated phosphorylated motifs of PKG[3]
RRXS	11592	9273	8346
RKXS	5626	2645	3769
KRXS	8678	1563	2864
KKXS	4730	757	804
RRXT	8740	3321	2186
RKXT	6089	1827	3045
KRXT	7703	1079	0
KKXT	4301	301	0

(b) Frequency of high-affinity motifs for PKA[4,5] and PKG[5] identified in physiological substrates[4] or by peptide library screens[5]

PKA consensus	*Homo sapiens*	PKG consensus	*Homo sapiens*
RRSS(L, V, I, F)[1]	696	RKKS	1348
RRGS(L, V, I, F)[1]	413	KRKKS	274
RRPS(L, V, I, F)[1]	315	KARKXS	59
RR(S, G, P)SF[1]	158	KARKKS	11
RRAS[3]	1755	AKRKKS	24
AERRAS[33]	3	KRKKSL	15
RAERRASI[3]	0	KXRKKSL	21
LRRASLG[3] (kemptide)	0	KARKKSL	5
		TQAKRKKSLA	0

[1]All available human protein sequences including predicted proteins translated from the full draft of the human genome were analyzed (for Table 42.1a: October 2007, 67,756 sequences in the International Protein Index (IPI) from EMBL-European Bioinformatics Institute, Hinxton, Cambridge, UK, which provides a minimally redundant yet maximally complete set of human proteins assembled from SWISS-PROT, TrEMBL, RefSeq and Ensembl, see http://www.ebi.ac.uk/IPI/IPIhelp.html; for Table 42.1b: September 2007, 5,507,867 sequences in the Basic Local Alignment Search Tool version BLASTP 2.2.17 (Aug-26-2007) National Center for Biotechnology Information, US National Library of Medicine, 8600 Rockville Pike, Bethesda, MD 20894, which provides non-redundant GenBank CDS translations+PDB+SwissProt+PIR+PRF excluding environmental samples from WGS projects, see http://ncbi.nlm.nih.gov/blast/beta/
[2]Probability for the motif being phosphorylated by PKA[2] [8] or PKG[3] (this study) in vivo.
[3]Probability for the motif being phosphorylated by PKA[2] [8] or PKG[3] (this study) in vivo.
[4]Preferred in vivo motifs according to [8].
[5]High-affinity substrates from peptide library screens [12, 13, 7].

and probability of phosphorylation are taken into account, RRXS, followed by RRXT and RKXS, stand out as the most abundant *in vivo* sites for both kinases (Table 42.1a). Notably, however, this analysis also predicts phosphorylation of significant numbers of substrates with less prevalent motifs such as K(R/K)X(S/T) with lower affinity, but which may be physiologically relevant, especially in a context where kinase and substrate are co-localized [8]. In contrast, preferred substrates defined by more detailed analysis of physiological substrates (Tables 42.2, 42.3) or by *in vitro* phosphorylation of generations of peptide libraries with PKA and PKG are clearly less abundant due

to their more restricted motifs as identified by homology searching (BLAST) (Table 42.1b). High-affinity substrates such as LRRASLG (Kemptide, [11]) and RAERRASI [7] for PKA and TQAKRKKSLA for PKG [12] used *in vitro* in enzyme assays were not found in the human proteome.

PHYSIOLOGICAL SUBSTRATES

General criteria for identification of physiological substrates of protein kinases were originally outlined by Krebs and

TABLE 42.2 Physiological substrates of PKA

Substrate	*In vivo/in situ* site	Sequence	Accession no./species	Ref.
Autophosphorylation				
cAMP-dependent protein kinase regulatory subunit type IIα	Ser-95[1]	PGRFD<u>RRVS</u>VCAET	P00515 (bovine)	8
Receptor mediated signaling				
β₂-adrenergic receptor	Ser-262[3]	TGHGL<u>RRSS</u>KFCLK	P07550 (human)	8
	Ser-345[3]	QELLC<u>LRRSS</u>LKAY		
	Ser-346[3]	ELLCL<u>RRSS</u>LKAYG		
Prostacyclin receptor (IP)	Ser-357[3]	QAPLS<u>RPAS</u>GRRDP	A54416 (mouse)	15
Regulator of protein signaling RGS9-1	Ser-427[3]	EPQGT<u>TRKS</u>SSLPF	O46469 (bovine)	16
	Ser-428[3]	PQGTT<u>RKSS</u>SLPFM		
Regulator of G protein signaling (RGS10)	Ser-168[3]	AQTAA<u>KRAS</u>RIYNT	Q9CQE5 (mouse)	17
G-protein-coupled receptor kinase 1 (GRK1)	Ser-21[3]	AFIA<u>ARGS</u>FDGSSSQ	Q15835 (human)	18
G-protein-coupled receptor kinase 2 (GRK2)	Ser-685[3]	VPLV<u>QRGS</u>ANGL	P25098 (human)	19
G-protein-coupled receptor kinase 7 (GRK7)	Ser-23[3]	YLQA<u>RKPS</u>DCDSKEL	Q8WTQ7 (human)	18
	Ser-36[3]	ELQR<u>RRRS</u>LALPGLQ		
G α13	Thr-203[3]	ILLA<u>RRPT</u>KGIHEYD	Q14344 (human)	20
Histamine H1 receptor, HRH1	Ser-398[3]	WKRL<u>RSHS</u>RQYVSGL	P35367 (human)	21
Integrin α4	Ser-1027[3]	QEEN<u>RRDS</u>WSYIN	P13612 (human)	22
Integrin β4	Ser-1364[3]	PSGSQ<u>RPS</u>VSDDTGC	P16144 (human)	23
Cardiac ryanodine receptor	Ser-2808[3]	YNRT<u>RRIS</u>QTSQV	Q92736 (human)	24
Regulator of G-protein signaling 4 (RGS4)	Ser-52[3]	VVIC<u>QRVS</u>QEEVKKW	Q0R4E4 (rabbit)	19
cAMP signaling				
cAMP-specific phosphodiesterase PDE4D3	Ser-54[1,3]	FVHSQ<u>RRES</u>FLYRS	P14270 (rat)	8
cGMP inhibited phosphodiesterase PDE3B	Ser-302[1,3]	SGKMF<u>RRPS</u>LPCIS	Q63085 (rat)	8
D1 dopamine receptor	Thr-268[3]	FKMSF<u>KRET</u>KVLKT	P18901 (rat)	8
Metabotropic glutamate receptor subunit mGluR2	Ser-843[1,3]	FGSAA<u>PRAS</u>ANLGQ	P31421 (rat)	8
Phosphodiesterase 4D5 (PDE4D5)	Ser-126[1,3]	FVHSQ<u>RRES</u>FLYRS	AAC00069 (human)	25
Human phosphodiesterases 3B (HSPDE3B)	Ser-73[3]	PQQP<u>RRCS</u>PFCRARL	Q13370 (human)	26

(Continued)

TABLE 42.2 (Continued)

Substrate	*In vivo/in situ* site	Sequence	Accession no./species	Ref.
	Ser-269[3]	IRPRRRSSCVSLGET	Q13370 (human)	
	Ser-318[3]	CKIFRRPSLPCISRE	Q13370 (human)	
A-kinase anchoring protein 13 (AKAP-Lbc/AKAP-79)	Ser-1565[3]	LSPFRRHSWGPGK	Q12802 (human)	27,28
	Ser-2733[1,2]	SVSPKRNSISRT		
Yotiao (AKAP9)	Ser-55[3]	QKKKRKTSSSKHDVS	Q99996 (human)	29
cGMP signaling				
Phosducin	Ser-73[1]	KERMSRKMSIQEYE	P20942 (rat)	8
Phosphoinositide and calcium signaling				
Elongation factor-2 kinase	Ser-499[3]	SRLHLPRPSAVALE	P70531 (rat)	8
Inositol 1,4,5-trisphosphate Type I receptor	Ser-1589[3]	ARNAARRDSVLAAS	P29994 (rat)	8
	Ser-1755[1]	IRPSGRRESLTSFG		
Phospholipase C-γ1	Ser-1248[1]	FHVRAREGSFEARY	P10686 (rat)	8
Phospholipase C-β3	Ser1105[1,3]	ILDRKRHNSISEAK	Q01970 (human)	8
Thromboxane A$_2$ receptor TPα	Ser-329[3]	PRLSTRPRSLSLQP	NP 001051 (human)	8
CaM-kinase kinase α	Thr-108[3]	SPRAWRRPTIESHH	BAA75246 (rat)	30
	Ser-458[3]	VKSMLRKRSFGNPF		
Arachidonate 5 lipooxygenase	Ser-524[3]	GMRGRKSSGFPKS	P09917 (human)	31
Rho signaling				
RhoA small GTP-binding protein	Ser-188[1,3]	QARRGKKKSGCLVL	P06749 (human)	8
Rho guanine nucleotide exchange factor 7	Ser-516[3]	RKPERKPSDEEF	Q14155 (human)	32
	Thr-526[3]	EFASRKSTAALE		
T cell receptor signaling				
COOH-terminal Src kinase (Csk)	Ser-364[1,3]	EALREKKFSTKSDV	P41240 (human)	8
Mitogen-activated protein kinase signaling				
Hematopoeietic protein tyrosine phosphatase (hePTP)	Ser-23[1,3]	VRLQERRGSNVALM	P35236 (human)	8
Mammalian STE20-like kinase 3 b isoform (MST3b)	Thr-18[3]	LALNKRRATLPHPG	AAD42039 (human)	8
v-Mos	Ser-56[1,3]	PSVDSRSCSIPLVA	P00538 (maloney murine sarcoma virus)	8
	Ser-102[1,3]	VCLMHRLGSGGFGS		
	Ser-263[1,3]	QDLRGRQASPPHIG		

Substrate	*In vivo/in situ* site	Sequence	Accession no./species	Ref.
GTPase activating protein specific for Rap1 (rap1GAP)	Ser-490[1,3]	GKSPTRKKSGPFGS	P47736 (human)	8
	Ser-499[1,3]	GPFGSRRSSAIGIE		
Guanine nucleotide exchange factor Ras-GRF1	Ser-916[1,3]	NKEVFRRMSLANTG	P27671 (mouse)	8
Protein tyrosine phosphatase-SL (PTP-SL)	Ser-231[3]	IGLQERRGSNVSLT	NP 035347 (human)	8
Raf-1 serine/threonine protein kinase	Ser-43[1,3]	QFGYQRRASDDGKL	P11345 (rat)	8
Rap1b low molecular weight GTP/GDP-bindingprotein	Ser-179[3]	VPGKARKKSSCQLL	P09526 (human)	8
Striatal-enriched protein tyrosine phosphatase, 61 kDa (STEP) STEP$_{61}$ and 46-kDa STEP$_{46}$ splice variant	Ser-160[1,3]	LPPEDRRQSVSRQP	P54830 (mouse)	8
	Ser221[1,3,5]	MGLQERRGSNVSLT		
Ras related protein 1A (RAP1A GTPase)	Ser-180[3]	PVEKKKPKKKSCLLL	P62834 (human)	33
GTPase activation protein (RGS14)	Ser-260[3]	ANAALRRESQGSL	O43566 (human)	34
	Thr-495[3]	SATGKRQTCDIE		

Modulators of protein phosphatase 1

Glycogen binding (G) subunit of protein phosphatase 1	Ser-46[1]	SPQPSRRGSDSSED	NP 002702 (human)	8
	Ser-65[1]	PSSGTRRVSFADSF		
Inhibitor-1of (I-1) protein phosphatase I	Thr-35[1]	EQIRRRRPTPATLV	P01099 (rabbit)	8
Dopamine and cAMP-regulated phosphoprotein, 32 kDa (DARPP-32)	Thr-34[1,2]	EMIRRRRPTPAMLF	P07516 (bovine)	8
Neurabin II protein, protein phosphatase 1 regulatory subunit 9B (spinophilin)	Ser-94[1,3]	SERGVRLSLPRASS	Q96SB3 (human)	35
	Ser-100[1,3]	LSLPRASSLNEN		
Protein phosphatase 1 regulatory subunit 16A (Myosin phosphatase-targeting subunit 3) (MYPT3)	Ser-340[3]	SLLRRRTSSAGSR	Q96I34 (human)	36
	Ser-341[3]	LLRRRTSSAGSRG		
	Ser-353[3]	KVVRRVSLTQRT		

(Continued)

TABLE 42.2 (Continued)

Substrate	*In vivo/in situ* site	Sequence	Accession no./species	Ref.
Transcriptional regulation				
Cyclic AMP response element binding protein (CREB)	Ser133[1,2,3]	REILS<u>RRPS</u>YRKIL	P15337 (rat)	8
Cyclic AMP responsive element modulator (CREMτ)	Ser-117[3]	REILS<u>RRPS</u>YRKIL	P27699 (mouse)	8
Nuclear factor of activated T cells 3 (NFAT3)	Ser-272[1,3]	SPCGK<u>RRYS</u>SSGTP	Q14934 (human)	8
	Ser-289[1,3]	SPALS<u>RRGS</u>LGEEG		
Nuclear factor κB (NFκB)	Ser-276[1,3]	VSMQL<u>RRPS</u>DRELS	Q04207 (mouse)	8
Retinoic acid receptor-α (RARα)	Ser-369[1,3]	VYVRK<u>RRPS</u>RPHMF	P11416 (mouse)	8
Retinoid X receptor-α (RXRα)	Ser-27[3]	LTSPT<u>GRGS</u>MAAPS	P19793 (human)	8
Sex determining region of Y gene product (SRY protein)	Ser-32[3]	NIPAL<u>RRSS</u>SFLCT	Q05066 (human)	8
SRY-box related transcription factor SOX9	Ser-64[3]	GEPDL<u>KKES</u>EEDKF	P48436 (human)	8
	Ser-181[2,3]	KYQPR<u>RRKS</u>VKNGQ		
Steroidogenic factor-1 (SF-1)	Ser-430[3]	CLVEV<u>RALS</u>MQAKE	P33242 (mouse)	8
Thyroid hormone receptor α1	Ser-28[1,3]	LDGKRK<u>RKSS</u>QCLV	P04625 (chick)	8
	Ser-29[1,3]	DGKRK<u>RKSS</u>QCLVK		
Vasoactive intestinal polypeptide receptor transcriptional repressor protein (VIPR-RP)	Ser-245[3]	KTKKA<u>RKDS</u>EEGES	AAC40192 (rat)	8
	Ser-361[3]	KGSPT<u>KRES</u>VSPED		
Class II transactivator (CIITA)	Ser-834[3]	VQELP<u>GRLS</u>FLGTR	AAA88861 (human)	37
	Ser1050[3]	LAASL<u>LRLS</u>LYNNC		
Thyroid transcription factor (TTF1)	Ser-337[3]	PDLAH<u>HAAS</u>PAALQ	P23441 (rat)	38
HAND1 transcription factor	Ser-98[1]	GRLGR<u>RKGS</u>GPKKE	O96004 (human)	39
	Thr-107[1]	PKKE<u>RRRT</u>ESINS		
	Ser-109[1]	KERR<u>RTES</u>INSAF		
Histone deacetylase 8, HDAC8	Ser-39[3]	AKIP<u>KRAS</u>MVHSL	Q9BY41 (human)	40
NFE2L1, transcription co-factor	Ser-599[3]	PSAL<u>KKGS</u>KEKQA	Q14494 (human)	41

Substrate	*In vivo/in situ* site	Sequence	Accession no./species	Ref.
Nuclear factor NFκB p105 subunit	Ser-337[3]	FVQL<u>RRKS</u>DLETSE	P19838 (human)	42
Polypyrimidine tract-binding protein 1, PTBP1	Ser-16[3]	IAVGT<u>KRGS</u>DELFS	P26599 (human)	43
Histones				
Histone H1c	Ser-37[1]	PAGVR<u>RKAS</u>GPPVS	P15864 (mouse)	8
Histone H3	Ser-10[1,2,3]	RTKQT<u>ARKS</u>TGGKA	P16106 (human)	8
Apoptosis and cell survival				
Bcl-2/Bcl-X$_L$-antagonist, causing cell death (BAD)	Ser155[1,2,3]	YGREL<u>RRMS</u>DEFEG	Q61337 (mouse)	8
Glycogen synthase kinase-3α(GSK-3α)	Ser-21[2]	GSGRA<u>RTSS</u>FAEPG	P18265 (rat)	8
Glycogen synthase kinase-3β(GSK-3β)	Ser-9[2,3]	MSGRP<u>RTTS</u>FAESC	P18266 (rat)	8
Interleukin receptor-3 β$_c$ chain	Ser-585[2,3]	YLGPP<u>HSRS</u>LPDIL	NP_000386 (human)	8
Inhibitor of apoptosis protein (IAP) (Survivin)	Ser-20[3]	PFLKD<u>HRIS</u>TFKN	O15392 (human)	44
Proteasome subunit Rpt6	Ser-120[3]	L<u>RNDS</u>YTLH	O13563 (human/yeast)	45
Ligand-gated ion channels				
GABA$_A$ receptor β1 subunit	Ser-409[3]	KGRIR<u>RRAS</u>QLKVK	P50571 (mouse)	8
GABA$_A$ receptor β3 subunit	Ser-408[3]	KTHLR<u>RRSS</u>QLKIK	P15433 (mouse)	8
	Ser-409[3]	KKTHL<u>RRRS</u>SQLKI		
Glutamate receptor GluR1 subunit (AMPA receptor)	Ser845[1,2,3]	RTSTL<u>PRNS</u>GAGAS	P19490 (rat)	8
Glutamate receptor GluR4 subunit (AMPA receptor)	Ser-842[1,3]	AIRNK<u>ARLS</u>ITGSV	P48058 (human)	8
Glutamate receptor GluR6 subunit (kainate receptor)	Ser-684[3]	AFMSS<u>RRQS</u>VLVKS	P42260 (rat)	8
Glutamate receptor NR1A subunit (NMDA receptor)	Ser897[1,2,3]	SSFKR<u>RRSS</u>KDTST	P35439 (rat)	8
Nicotinic acetylcholine receptor δ subunit	Ser-361[1,3]	NDLKL<u>RRSS</u>SVGYI	P02718 (T. californica)	8
	Ser-362[1,3]	DLKLR<u>RSSS</u>VGYIS		
P$_{2X2}$ purinoreceptor	Ser-431[3]	AVQSP<u>RPCS</u>ISALT	P49653 (rat)	8
Sodium ion movement				
Na$^+$ H$^+$ exchanger 3 (NHE3)	Ser-552[1,3]	VAEGE<u>RRGS</u>LAFIR	P26433 (rat)	8
	Ser-605[1,3]	QSLEQ<u>RRRS</u>IRDTE		
Na$^+$,K$^+$ ATPase α1 subunit	Ser943[1,2,3]	VICKT<u>RRNS</u>VFQQG	P06685 (rat)	8
Serum and glucocorticoid regulated kinase (Sgk)	Thr-369[3]	DDLIN<u>KKIT</u>PPFNP	O00141 (human)	8
Voltage sensitive Na$^+$ channel (Rat brain type IIA) α subunit	Ser-573[1,3]	SLFSP<u>RRNS</u>RASLF	P04775 (rat)	8

(*Continued*)

TABLE 42.2 (Continued)

Substrate	*In vivo/in situ* site	Sequence	Accession no./species	Ref.
	Ser-610[1,3]	EDNDSRRDSLFVPH		
	Ser-623[1,3]	HRHGERRPSNVSQA		
	Ser-687[1,3]	TEIRKRRSSSYHVS		
Voltage-sensitive Na+ channel (cardiac type H1) α subunit	Ser526[1,2,3]	RTSMRPRSSRGSIF	P15389 (rat)	8
	Ser529[1,2,3]	MRPRSSRGSIFTFR		
Brain sodium channel 2	Ser-479[3]	QKEAKRSSADKGVA	P78348 (human)	46
Sodium-dependent vitamin C transporter (SVCT) 2	Ser-402[3]	YYACARLSCAPP	Q9EPR4 (mouse)	47
	Ser-639[3]	ENSRSSDKDSQ		

Chloride conductance

Substrate	*In vivo/in situ* site	Sequence	Accession no./species	Ref.
Cystic fibrosis transmembrane conductance regulator (CFTR)	Ser-660[1,3]	QFSAERRNSILTET	P13569 (human)	8
	Ser-700[1]	EFGEKRKNSILNPI		
	Ser-737[3]	DEPLERRLSLVPDS		
	Ser-795[3]	TTASTRKVSLAPQA		
	Ser-813[3]	IDIYSRRLSQETGL		
Phospholemman	Ser-68[1,3]	FRSSIRRLSTRRR	O08589 (rat)	8

Potassium channels

Substrate	*In vivo/in situ* site	Sequence	Accession no./species	Ref.
Shaker K+ channel	Ser-507[3]	TLGQHMKKSSLSES	P08511 (*Drosophila*)	8
	Ser-508[3]	LGQHMKKSSLSESS		
Slo KCa channel splice variant A1C2E1G3I0	Ser-942[2,3]	PIVLQRRGSVYGAN	JH0697[4] (*Drosophila*)	8
hSlo BKCa α subunit of large conductance Ca2+-dep. K+ channel (maxi-K)	Ser-869[3]	VHGMLRQPSITTGV	NP 002238 (human)	8
Kv1.1 α subunit of the Shaker RCK1 voltage-gated K+ channel	Ser-446[3]	DSDLSRRSSSTISK	P10499 (rat)	8
Kvβ1.3 subunit of the Kv1.5 K+ channel Ikur	Ser-24[3]	ENTKLRRQSGFSVA	AAC41926.1 (human)	8
Kv4.2 α subunit of the Shal-type K+ channel	Thr-38[1,2]	PPRQERKRTQDALI	NP 062671 (mouse)	8
	Ser-552[1,2]	NVSGSHRGSVQELS		
Kir 1.1 Renal outer medullary K+ channel 1, 2 (ROMK1, ROMK2)	Ser-25[1,3]	SRQRARLVSKEGRC	P35560 (rat)	8
	Ser-200[1,3]	IRVANLRKSLLIGS		

Substrate	*In vivo/in situ* site	Sequence	Accession no./species	Ref.
	Ser-294[1,3]	SATCQVRTSYVPEE		
Kir2.1	Ser-425[3]	EPRPLRRESEI	Q64273 (rat)	8
Kir2.3 Inward rectifier K+ channel (IRK)	Ser-440[1,3]	DNISYRRESAI	P48050 (human)	8
Kir6.2 subunit of the ATP-sensitive K+ channel (K$_{ATP}$)	Ser-372[3]	ARGPLRKRSVPMAK	Q14654 (human)	8
	Thr-224[3]	HMQVVRKTTSPEGE		
SUR1 subunit of the ATP-sensitive K+ channel (K$_{ATP}$)	Ser-1571[3]	EKLLSRKDSVFASF	Q09428 (human)	8
Kir2.2 (ATP-sensitive inward rectifier potassium channel 12)	Ser-431[3]	LEQRPYRRESEI	Q14500 (human)	48
Kir6.1 subunit of the ATP-sensitive K+ channel	Ser-385[3]	SRKRNSMRRNN	Q15842 (human)	49
	Thr-234[1,3]	QVVKKTTTPEGEV	Q15842 (human)	
SUR2B subunit of the ATP-sensitive K+ channel	Thr-635[1,3]	FESCKKHTGVQPK	O60706 (human)	49
	Ser-1468[1,3]	RAFVRKSSILIM	O60706 (human)	
Potassium voltage-gated channel subfamily H member 2	Ser-283[1,3]	SCASVRRASSADDI	Q12809 (human)	50
	Ser-1137[1,3]	EGPTRRLSLPGQLG		
Calcium channels				
L-type calcium channel Cav1.2 (CACNA1C)	Ser-1928[3]	ASLGRRASFHLECLK	P15381 (rabbit)	51
Ryanodine receptor calcium release channel (RyR2)	Ser-2808[2,3]	YNRTRRISQTSQVSV	Q92736 (human)	52
Water homeostasis				
Aquaporin-2	Ser-256[2,3]	EREVRRRQSVELHS	P34080 (rat)	8
Other transporters				
P-glycoprotein mdr1b	Ser-665[1,3]	SKSPLIRRSIYRSV	P06795 (mouse)	8
	Ser-681[1,3]	KQDQERRLSMKEAV		
Steroidogenic acute regulatory protein (StAR)	Ser-57[3]	INQVRRRSSLLGSR	P49675 (human)	8
	Ser-195[3]	VRCAKRRGSTCVLA		
Extracellular proteins				
Atrial natriuretic peptide	Ser-104[1]	GPRSLRRSSCFGGR	P01161 (rat)	8
Vitronectin	Ser-378[1]	RNQNSRRPSRATWL	P04004 (human)	8

(Continued)

TABLE 42.2 (Continued)

Substrate	*In vivo/in situ* site	Sequence	Accession no./species	Ref.
Trafficking and motility				
Actin bundling protein L-plastin	Ser-5[1,3]	MARGSVSDEEMMEL	P13796 (human)	8
Low-density lipoprotein receptor-related protein (LRP)	Ser-4520[3]	MGGHGSRHSLASTD	Q07954 (human)	8
Myosin light-chain kinase (MLCK) and telokin splice variant	Ser-1005[1,3,66]	SGLSGRKSSTGSPT	P29294 (rabbit)	8
Protein tyrosine phosphatase-PEST (PTP-PEST)	Ser-39[1]	DFMRLRRLSTKYRT	NP 002826 (human)	8
	Ser-435[1]	DKKLERNLSFEIKK		
Small heat shock-related protein HSP20	Ser-16[1,2]	QPSWLRRASAPLPG	O14558 (human)	8
Synapsin I	Ser-9[1]	MNYLRRRLSDSNFM	P09951 (rat)	8
Vasodilator-stimulated phosphoprotein (VASP)	Ser-157[1,3]	SEHIERRVSNAGGP	P50552 (human)	8
	Ser-239[1,2,3]	AGAKLRKVSKQEEA		
	Thr-278[1,3]	MLARRRKATQVGEK		
Snapin	Ser-50[3]	SHVHAVRESQVELR	XP_057189 (human)	53
Sso1 t-SNARE	Ser-79[3]	EQASHLRHSLDNFV	NP_015092 (yeast)	54
Cysteine string protein (csp)	Ser-10[3]	DCQRQRSLSTSGES	Q29455 (bovine)	55
Claudin 3	Thr-192[3]	PPREKKYTATKVVYS	O15551 (human)	56
Synaptotagmin-12	Ser-97[3]	GPPSRKGSLSIEDTF	P97610 (rat)	57
Striated muscle contraction				
Myosin-binding protein-C cardiac isoform	Ser-275[1,3]	LLSAFRRTSLAGGG	Q14896 (human)	8
	Ser-284[1,3]	LAGGGRRISDSHED		
	Ser-304[1,3]	SSLLKKRDSFRTPR		
Phospholamban	Ser-16[1,2]	TRSAIRRASTIEMP	P26678 (human)	8
Ryanodine receptor type 2 (sarcoplasmic reticulum Ca^{2+} release channel)	Ser-2809[3]	LYNRTRRISQTSQV	P30957 (rabbit)	8
Troponin I	Ser-23[1]	APAPIRRRSSNYRA	P19429 (human)	8
	Ser-24[1]	PAPIRRRSSNYRAY		
Voltage-sensitive L-type Ca^{2+} channel (skeletal muscle) α1 subunit	Ser-1757[1,3]	PERGQRRTSLTGSL	P07293 (rabbit)	8
	Ser-1854[1,3]	PGSLSRRSSLGSLD		

Substrate	*In vivo/in situ* site	Sequence	Accession no./species	Ref.
Voltage-sensitive L-type Ca^{2+} channel (cardiac) α1 subunit	Ser-1928[1,2,3]	SASLGRRASFHLEC	P15381 (rabbit)	8
Voltage-sensitive L-type Ca^{2+} channel (cardiac) β$_{2a}$ subunit	Ser-459[1,3]	DRSAPRSASQAEEE	A42044 (rat)	8
	Ser-478[1,3]	VKKSQHRSSSATHQ		
	Ser-479[1,3]	KKSQHRSSSATHQN		
Metabolism and respiration				
ATP citrate lyase	Ser-454[1]	TPAPSRTASFSESR	P16638 (rat)	8
Cytochrome P450 CYP2E1	Ser-129[1,3]	TWKDVRRFSLSILR	P05182 (rat)	8
Glycogen synthase (muscle type)	Ser-7[1,3]	MPLSRTLSVSSLPG	AAB69872 (rabbit)	8
	Ser-697[1,3]	APEWPRRASCTSSS		
	Ser-710[1,3]	SSSGGSKRSNSVDT		
Hormone-sensitive lipase	Ser-563[1,3]	RLTESMRRSVSEAA	P15304 (rat)	8
	Ser-659[1,3]	PDGFHPRRSSQGVL		
	Ser-660[1,3]	DGFHPRRSSQGVLH		
Phenylalanine hydroxylase	Ser-16[1,2]	NPGLGRKLSDFGQE	P00439 (human)	8
Phosphorylase kinase α subunit (muscle type)	Ser-1018[1]	KQVEFRRLSISTES	P18688 (rabbit)	8
Phosphorylase kinase β subunit (muscle type)	Ser-26[1]	RARTKRSGSVYEPL	P12798 (rabbit)	8
6-phosphofructo-2-kinase-fructose-2, 6-bisphosphatase liver isozyme 1	Ser-32[1]	SVLQRRRGSSIPQF	P07953 (rat)	8
6-phosphofructo-1-kinase, isozyme A (muscle type)	Ser-376[1]	EAMKLRGRSFMNNW	P00511 (rabbit)	8
Pyruvate kinase (liver type)	Ser-43[1]	PAGYLRRASVAQLT	P12928 (rat)	8
Tyrosine hydroxylase	Ser-40[1,2,3]	PRFIGRRQSLIEDA	P04177 (rat)	8
Nuclear-encoded subunit of complex I (NDUFS4)	Ser-131[1]	ANFSWNKRTRVSTK	Q02375 (bovine)	58
GTPase activation protein (RGS14)	Ser-260[3]	ANAALRRESQGSLN	O43566 (human)	34
	Thr-495[3]	SATGKRQTCDIEGL		
GTPase activation protein (RGS14)	Ser-260[3]	ANAALRRESQGSLN	O43566 (human)	34
	Thr-495[3]	SSATGKRQTCDIEG		
GTPase activation protein (RGS14)	Ser-260[3]	NAALRRESQGSLNS	O43566 (human)	34
	Thr-495[3]	SATGKRQTCDIEGL		
Perilipin	Ser-492[3]	EKPARRVSDSFFRPS	Q8CGN5 (mouse)	59

(*Continued*)

TABLE 42.2 (Continued)

Substrate	*In vivo/in situ* site	Sequence	Accession no./species	Ref.
	Ser-517[3]	YSQLRKKS		
Pyruvate kinase 1 (Pyk1)	Ser-22[3]	AGSDLRRTSIIGTI	P00549 (yeast)	60
Micellaneous				
cAMP-regulated phosphoprotein 16/19 kDa (ARPP16/19)	Ser-104[1,2,7]	QDLPQRKPSLVASK	P56211 (human)	8
cAMP-regulated phosphoprotein, 21 kDa (ARPP-21)	Ser-55[1,2]	AQNQERRKSKSGAG	A34957 (bovine)	8
Serine/threonine protein kinase LKB1	Ser-431[2,3]	SSNKIRRLSACKQQ	NP_035622 (mouse)	8
Phogrin	Ser-680[3]	GPHTSRINSVSSQL PSPSARSSTSSWSE	CAA90600 (rat)	61
	Thr-699[3]			
14-3-3 zeta/delta, protein kinase C inhibitor protein-1, Factor activating exoenzyme S, FAS))	Ser-58[3]	VVGARRSSWRVVSS	P63104 (human)	62
Brefeldin A-inhibited guanine nucleotide-exchange proteins (BIG1)	Ser-883[3]	EIAGKKISMKETKEL	Q9Y6D6 (human)	63
Casein kinase 1 delta (CK1)	Ser-370[3]	RERKVSMRLH	P48730 (human)	64
CTP synthetase (CTPS)	Thr-455[3]	MRLGKRRTLFQTKNS	P17812 (human)	65
Cullin 5	Ser-730[3]	IMKMRKKISNAQLQ	Q93034 (human)	66
LASP1, LIM and SH3 domain protein 1	Ser-146[3]	MEPERRDSQDGSSY	Q14847 (human)	67
Merlin	Ser-518[3]	DTDMKRLSMEIEKE	P35240 (human)	68
Src kinase, c-src	Ser-17[2,3]	DASQRRRSLEPAEN	P12931 (human)	69
Rabphilin	Ser-234[1]	HGPPTRRASEARM	P47709 (rat)	70

[1]*Direct sequencing and/or phosphopeptide mapping.*
[2]*Phospho/dephospho-specific antibodies.*
[3]*Site-directed mutagenesis.*
[4]*Ser-952 in this splice variant.*
[5]*STEP$_{61}$ Ser-221 is equivalent to STEP$_{46}$ Ser-49.*
[6]*MLCK Ser-1005 is equivalent to telokin Ser-13.*
[7]*ARPP-19 Ser-104 is equivalent to ARPP-16 Ser-88.*

Beavo [13], reviewed by Shabb [8], and can be summarized as follows:

1. The target protein should be phosphorylated stoichometrically and dephosphorylated by phosphatase *in vitro* at significant kinetic rates
2. Functional properties of the substrate should change in correlation with the degree of phosphorylation
3. Phosphorylation of the substrate should be demonstrated *in vivo* or in intact cells with accompanying functional changes

4. The cellular levels of protein kinase should correspond to the extent of phosphorylation of the substrate
5. The *in situ* phosphorylation sequence should be identified (new, adds stringency).

Newly available technologies such as deletion/mutation mapping and mass spectrometry have led to a current consensus. The primary evidence that *in vivo* phosphorylation occurs in response to elevated cAMP or cGMP levels – criteria (3) to (5) – is considered a more stringent approach when compared to *in vitro* mapping of PKA and

TABLE 42.3 Physiological substrates of PKG

Substrate	*In vivo/in vitro* site	Sequence	Accession no./species	Ref.
Autophosphorylation				
Autophosphorylation	Ser-63	ATQQAQKQSASTLQ	P14619 (human)	71
	Ser-79	PRTKRQAISAEPTA		
Regulation of smooth muscle tone				
Cardiac troponin I	Ser-23	APAPIRRRSSNYRAY	P19429 (human)	72, 73
(cTN1)	Ser-24	APAPIRRRSSNYRAY		
(CRP2) Cystein-rich Protein 2	Ser-104	VRTEERKTSGPPKGP	P36201 (rat)	74, 75
Hsp20 (Heat-shock 20-kD-like protein P20)		QPSWLRRASAPLPG	P97541 (rat)	76
Regulation of smooth muscle tone (by regulation of intracell. Ca level)				
Cardiac phospholamban	Ser-16	TRSAIRRASTIEMP	P26678 (human)	77
L-type Ca $^{2+}$ channel α1c subunit	Ser-533	HRISKSKFSRYWRR	P15381 (rabbit)	78
Calcium-acivated maxi K$^+$ channel (BK$_{Ca}$) (α-subunit)	Ser-1072[1]	SQSSSKKSSSHVS	AAA84000 (canine)	79, 80
Ins (1,4,5) P3 receptor type I	Ser-1756	IRPSGRRESLTSFG	P29994 (rat)	81
Platelet aggregation				
Heat shock protein 27 (Hsp27)	Ser-15	PFSLLRGPSWDPFR	XP_004991	82
	Ser-78	APAYSRALSRQLSSG	(*Homo sapiens*)	
	Ser-82	SRALSRQLSSGVSEI		
	Thr-143	SRCFTRKYTLPPGV		
VASP (vasodilator stimulated phosphoprotein)	Ser-157	SEHIERRVSNAGGP	CAA86523 (human)	83, 84
	Ser-239	AGAKLRKVSKQEEA		
	Thr-278	MLARRRKATQVGEK		
Neuronal function				
G-septin	Ser-91	KSQVSRKASSWNRE	AAD21035	85, 86
GABA$_A$ receptor β2 subunit	Ser-410[1]	KSRLRRRASQLKIT	(rat)	
GABA$_A$ receptor β3-subunit	Ser-409	KTHLRRRSSQLKIK	P15432 (mouse)	87
G substrate	Thr-68	Qkkprrkdtpalhi	1095220 (mouse)	87
	Thr-119	QKKPRRKDTPALHM	AAD13030 (human)	88
Dopamin/DARPP-32	Thr-34	EMIRRRRPTPAMLF	P07516 (bovine)	89
Thromboxane receptor alpha (TPα)	Ser-331[1]	LSTRPRSLSLQPQL	NP_001051 (human)	90, 91
Nucleus				
Splicing factor SF1	Ser-20	FPSKKRKRSRWNQD	CAA03883 (human)	92

(Continued)

TABLE 42.3 (Continued)

Substrate	*In vivo/in vitro* site	Sequence	Accession no./species	Ref.
Histone H2B	Ser-32	KDGKKRKRSRKESY	XP_059791(human)	93
Germ cell development				
GKAP42	Ser-106[1]	SPNPAQKESREENW	BAA92254(mouse)	94
Metabolic enzymes				
Tyrosine hydroxylase	Ser-40	PRFIGRRQSLIEDA	P04177(rat)	95
Nitric-oxide synthase NOS-III	Ser-633	SWRRKRKESSNTDS	P29474 (human)	96
	Ser-1177	VTSRIRTQSSTPQF		
6-Phospho-fructo-2-kinase	Ser-32	SVLQRRRGSSIPQF	P07953(rat)	97
Regulation of other signaling pathways				
cGMP-binding cGMP-specific phosphodiesterase	Ser-92	PGTPTRKISASEFD	Q28156 (bovine)	98
(CGB-PDE)	Ser-43[1]	QFGYQRRASDDGKL	P04049 (human)	99
c-Raf1	Ser-26[1]	VVKTLRRGSKFIKW	P51432 (mouse)	100
PLC-beta3	Ser-1105[1]	ILDRKRNNSISEAK		
HIV-1 replication and infectivity				
Vif (one of the HIV-1 Proteins	Ser-144	QAGHNKVGSLQYLA	AAA44202	101
	Thr-188	TKGHRGSHTMNGH	(HIV type I)	
Filamentprotein				
Vimetin	Ser-26[2]	PGTASRPSSTRSYV	P48616 (bovine)	102, 103
Cytoskeletal proteins				
LASP1, LIM and SH3 domain protein 1	Ser-146[1]	GMEPERRDSQDGSS	Q14847 (human)	67
Formin homology domain-containing protein, FHOD1	Ser-1131[1]	AARERKRSRGNRKSL	Q9Y613 (human)	104

[1]*Site-directed mutagenesis.*
[2]*Potential phosphorylation site because the serine residue meets the consensus sequence.*

PKG sites (typically the method of use in older literature), where a pitfall is that less stringent phosphorylations may occur. However, application of these criteria excludes phosphorylations occurring *in vivo* that are either silent (i.e., no functional change) or where the function has not yet been mapped. Conversely, substrates implicated in physiological pathways where the precise mechanism or site has not been identified to date are also excluded.

Table 42.2 presents a comprehensive list of PKA substrates that meet the above eligibility criteria identified through review of the literature. This listing is based on a review by Shabb [8], the previous version of this chapter, and has been revised by adding new published data. This brings the number of identified and listed physiological substrates of PKA that meet the eligibility criteria to 163, and the number of analyzed motifs to 229. Furthermore, we present a corresponding list of 31 PKG substrates, representing 40 analyzed motifs that have been identified by systematic literature search and application of the same set of criteria (Table 42.3). Interestingly, 13 of those motifs are also phosphorylated by PKA (Tables 42.2, 42.3), indicating at least 30 percent overlap in substrates. During these searches we additionally revealed a number of potential substrates that do not yet fulfill a convincing combination of the general criteria. Furthermore, the discrepancy between identified substrates that meet the criteria (Tables 42.2, 42.3) and the estimated number of substrates (Table 42.1) indicates that approximately 99 percent of human PKA and

PKG substrates are as yet unidentified. The identification and functional analysis of consequences of phosphorylation constitute a major task to be addressed in the future.

CONCLUDING REMARKS

The specificity of a substrate is determined not only by the primary sequence, but also by several other factors that affect the degree of phosphorylation of a given target. The tertiary structure of the substrate affects the function and kinetics of the kinase, such as, for example, the catalytic subunit of cAMP-kinase that in part acquires its substrate specificity from the conserved F at position $P-11$ [14]. Organization of the microenvironment around a phosphorylation event has clear impact. In that respect, anchoring proteins (AKAPs, GKAPs) play an important role by locating PKA and PKG in close vicinity to their substrates, and demonstrate how low-affinity substrates may become physiologically relevant.

REFERENCES

1. Glass DB, Krebs EG. Comparison of the substrate specificity of adenosine 3′,5′-monophosphate- and guanosine 3′,5′′-monophosphate-dependent protein kinases. Kinetic studies using synthetic peptides corresponding to phosphorylation sites in histone H2B. *J Biol Chem* 1979;**254**(19):9728–38.

2. Glass DB. Differential responses of cyclic GMP-dependent and cyclic AMP-dependent protein kinases to synthetic peptide inhibitors. *Biochem J* 1983;**213**(1):159–64.

3. Glass DB, Cheng HC, Mende-Mueller L, Reed J, Walsh DA. Primary structural determinants essential for potent inhibition of cAMP-dependent protein kinase by inhibitory peptides corresponding to the active portion of the heat-stable inhibitor protein. *J Biol Chem* 1989;**264**(15):8802–10.

4. Kennelly PJ, Krebs EG. Consensus sequences as substrate specificity determinants for protein kinases and protein phosphatases. *J Biol Chem* 1991;**266**(24):15,555–15,558.

5. Kemp BE, Pearson RB, House CM. Pseudosubstrate-based peptide inhibitors. *Methods Enzymol* 1991;**201**:287–304.

6. Kemp BE, Faux MC, Means AR, House CM, Tiganis T, Hu SH, Mitchelhill KI. Structural aspects: pseudosubstrate and substrate interactions. In: Woodgett JR, editor. *Protein kinases*. Oxford: IRL Press and Oxford University Press; 1994. p. 30–67.

7. Tegge W, Frank R, Hofmann F, Dostmann WR. Determination of cyclic nucleotide-dependent protein kinase substrate specificity by the use of peptide libraries on cellulose paper. *Biochemistry* 1995;**34**(33):10,569–10,577.

8. Shabb JB. Physiological substrates of cAMP-dependent protein kinase. *Chem Rev* 2001;**101**(8):2381–411.

9. Kersey PJ, Duarte J, Williams A, Karavidopoulou Y, Birney E, Apweiler R. The International Protein Index: an integrated database for proteomics experiments. *Proteomics* 2004;**4**(7):1985–8.

10. Rice P, Longden I, Bleasby A. EMBOSS: the European Molecular Biology Open Software Suite. *Trends Genet* 2000;**16**(6):276–7.

11. Kemp BE, Graves DJ, Benjamini E, Krebs EG. Role of multiple basic residues in determining the substrate specificity of cyclic AMP-dependent protein kinase. *J Biol Chem* 1977;**252**(14):4888–94.

12. Dostmann WR, Nickl C, Thiel S, Tsigelny I, Frank R, Tegge WJ. Delineation of selective cyclic GMP-dependent protein kinase Ialpha substrate and inhibitor peptides based on combinatorial peptide libraries on paper. *Pharmacol Ther* 1999;**82**(2–3):373–87.

13. Krebs EG, Beavo JA. Phosphorylation–dephosphorylation of enzymes. *Annu Rev Biochem* 1979;**48**:923–59.

14. Knighton DR, Zheng JH, Ten Eyck LF, Xuong NH, Taylor SS, Sowadski JM. Structure of a peptide inhibitor bound to the catalytic subunit of cyclic adenosine monophosphate-dependent protein kinase. *Science* 1991;**253**(5018):414–20.

15. Lawler OA, Miggin SM, Kinsella BT. Protein kinase A-mediated phosphorylation of serine 357 of the mouse prostacyclin receptor regulates its coupling to G(s)-, to G(i)-, and to G(q)-coupled effector signaling. *J Biol Chem* 2001;**276**(36):33,596–33,607.

16. Balasubramanian N, Levay K, Keren-Raifman T, Faurobert E, Slepak VZ. Phosphorylation of the regulator of G protein signaling RGS9-1 by protein kinase A is a potential mechanism of light- and Ca^{2+}-mediated regulation of G protein function in photoreceptors. *Biochemistry* 2001;**40**(42):12,619–12,627.

17. Burgon PG, Lee WL, Nixon AB, Peralta EG, Casey PJ. Phosphorylation and nuclear translocation of a regulator of G protein signaling (RGS10). *J Biol Chem* 2001;**276**(35):32,828–32,834.

18. Horner TJ, Osawa S, Schaller MD, Weiss ER. Phosphorylation of GRK1 and GRK7 by cAMP-dependent protein kinase attenuates their enzymatic activities. *J Biol Chem* 2005;**280**(31):28,241–28,250.

19. Huang J, Zhou H, Mahavadi S, Sriwai W, Murthy KS. Inhibition of Galphaq-dependent PLC-beta1 activity by PKG and PKA is mediated by phosphorylation of RGS4 and GR K2. *Am J Physiol Cell Physiol* 2007;**292**(1):C200–8.

20. Manganello JM, Huang JS, Kozasa T, Voyno-Yasenetskaya TA, Le Breton GC. Protein kinase A-mediated phosphorylation of the Galpha13 switch I region alters the Galphabetagamma13-G protein-coupled receptor complex and inhibits Rho activation. *J Biol Chem* 2003;**278**(1):124–30.

21. Horio S, Ogawa M, Kawakami N, Fujimoto K, Fukui H. Identification of amino acid residues responsible for agonist-induced down-regulation of histamine H(1) receptors. *J Pharmacol Sci* 2004;**94**(4):410–19.

22. Han J, Liu S, Rose DM, Schlaepfer DD, McDonald H, Ginsberg MH. Phosphorylation of the integrin alpha 4 cytoplasmic domain regulates paxillin binding. *J Biol Chem* 2001;**276**(44):40,903–40,909.

23. Wilhelmsen K, Litjens SH, Kuikman I, Margadant C, van Rheenen J, Sonnenberg A. Serine phosphorylation of the integrin {beta}4 subunit is necessary for EGF receptor-induced hemidesmosome disruption. *Mol Biol Cell* 2007;**18**(9):3512–22.

24. Xiao B, Sutherland C, Walsh MP, Chen SR. Protein kinase A phosphorylation at serine-2808 of the cardiac Ca2+-release channel (ryanodine receptor) does not dissociate 12.6-kDa FK506-binding protein (FKBP12.6). *Circ Res* 2004;**94**(4):487–95.

25. Baillie G, MacKenzie SJ, Houslay MD. Phorbol 12-myristate 13-acetate triggers the protein kinase A-mediated phosphorylation and activation of the PDE4D5 cAMP phosphodiesterase in human aortic smooth muscle cells through a route involving extracellular signal regulated kinase (ERK). *Mol Pharmacol* 2001;**60**(5):1100–11.

26. Palmer D, Jimmo SL, Raymond DR, Wilson LS, Carter RL, Maurice DH. Protein kinase A phosphorylation of human phosphodiesterase 3B promotes 14-3-3 protein binding and inhibits phosphatase-catalyzed inactivation. *J Biol Chem* 2007;**282**(13):9411–9.

27. Diviani D, Abuin L, Cotecchia S, Pansier L. Anchoring of both PKA and 14–3-3 inhibits the Rho-GEF activity of the AKAP-Lbc signaling complex. *Embo J* 2004;**23**(14):2811–20.

28. Carnegie GK, Smith FD, McConnachie G, Langeberg LK, Scott JD. AKAP-Lbc nucleates a protein kinase D activation scaffold. *Mol Cell* 2004;**15**(6):889–99.

29. Chen L, Kurokawa J, Kass RS. Phosphorylation of the A-kinase-anchoring protein Yotiao contributes to protein kinase A regulation of a heart potassium channel. *J Biol Chem* 2005;**280**(36):31,347–31,352.

30. Kitani T, Okuno S, Fujisawa H. Regulation of Ca(2+)/calmodulin-dependent protein kinase kinase alpha by cAMP-dependent protein kinase: II, Mutational analysis. *J Biochem (Tokyo)* 2001;**130**(4):515–25.

31. Luo M, Jones SM, Flamand N, Aronoff DM, Peters-Golden M, Brock TG. Phosphorylation by protein kinase a inhibits nuclear import of 5-lipoxygenase. *J Biol Chem* 2005;**280**(49):40,609–40,616.

32. Chahdi A, Miller B, Sorokin A. Endothelin 1 induces beta 1Pix translocation and Cdc42 activation via protein kinase A-dependent pathway. *J Biol Chem* 2005;**280**(1):578–84.

33. Hu CD, Kariya K, Okada T, Qi X, Song C, Kataoka T. Effect of phosphorylation on activities of Rap1A to interact with Raf-1 and to suppress Ras-dependent Raf-1 activation. *J Biol Chem* 1999;**274**(1):48–51.

34. Hollinger S, Ramineni S, Hepler JR. Phosphorylation of RGS14 by protein kinase A potentiates its activity toward G alpha i. *Biochemistry* 2003;**42**(3):811–19.

35. Hsieh-Wilson LC, Benfenati F, Snyder GL, Allen PB, Nairn AC, Greengard P. Phosphorylation of spinophilin modulates its interaction with actin filaments. *J Biol Chem* 2003;**278**(2):1186–94.

36. Yong J, Tan I, Lim L, Leung T. Phosphorylation of myosin phosphatase targeting subunit 3 (MYPT3) and regulation of protein phosphatase 1 by protein kinase A. *J Biol Chem* 2006;**281**(42):31,202–31,211.

37. Li G, Harton JA, Zhu X, Ting JP. Downregulation of CIITA function by protein kinase a (PKA)-mediated phosphorylation: mechanism of prostaglandin E, cyclic AMP, and PKA inhibition of class II major histocompatibility complex expression in monocytic lines. *Mol Cell Biol* 2001;**21**(14):4626–35.

38. Feliciello A, Allevato G, Musti AM, De Brasi D, Gallo A, Avvedimento VE, Gottesman ME. Thyroid transcription factor 1 phosphorylation is not required for protein kinase A-dependent transcription of the thyroglobulin promoter. *Cell Growth Differ* 2000;**11**(12):649–54.

39. Firulli BA, Howard MJ, McDaid JR, McIlreavey L, Dionne KM, Centonze VE, Cserjesi P, Virshup DM, Firulli AB. PKA, PKC, and the protein phosphatase 2 A influence HAND factor function: a mechanism for tissue-specific transcriptional regulation. *Mol Cell* 2003;**12**(5):1225–37.

40. Lee H, Rezai-Zadeh N, Seto E. Negative regulation of histone deacetylase 8 activity by cyclic AMP-dependent protein kinase A. *Mol Cell Biol* 2004;**24**(2):765–73.

41. Narayanan K, Ramachandran A, Peterson MC, Hao J, Kolsto AB, Friedman AD, George A. The CCAAT enhancer-binding protein (C/EBP)beta and Nrf1 interact to regulate dentin sialophosphoprotein (DSPP) gene expression during odontoblast differentiation. *J Biol Chem* 2004;**279**(44):45,423–45,432.

42. Guan H, Hou S, Ricciardi RP. DNA binding of repressor nuclear factor-kappaB p50/p50 depends on phosphorylation of Ser337 by the protein kinase A catalytic subunit. *J Biol Chem* 2005;**280**(11):9957–62.

43. Xie J, Lee JA, Kress TL, Mowry KL, Black DL. Protein kinase A phosphorylation modulates transport of the polypyrimidine tract-binding protein. *Proc Natl Acad Sci USA* 2003;**100**(15):8776–81.

44. Dohi T, Xia F, Altieri DC. Compartmentalized phosphorylation of IAP by protein kinase A regulates cytoprotection. *Mol Cell* 2007;**27**(1):17–28.

45. Zhang F, Hu Y, Huang P, Toleman CA, Paterson AJ, Kudlow JE. Proteasome function is regulated by cyclic AMP-dependent protein kinase through phosphorylation of RPT6. *J Biol Chem* 2007;**282**:22,460–22,471.

46. Leonard AS, Yermolaieva O, Hruska-Hageman A, Askwith CC, Price MP, Wemmie JA, Welsh MJ. cAMP-dependent protein kinase phosphorylation of the acid-sensing ion channel-1 regulates its binding to the protein interacting with C-kinase-1. *Proc Natl Acad Sci USA* 2003;**100**(4):2029–34.

47. Wu X, Zeng LH, Taniguchi T, Xie QM. Activation of PKA and phosphorylation of sodium-dependent vitamin C transporter 2 by prostaglandin E2 promote osteoblast-like differentiation in MC3T3-E1 cells. *Cell Death Differ* 2007.

48. Zitron E, Kiesecker C, Luck S, Kathofer S, Thomas D, Kreye VA, Kiehn J, Katus HA, Schoels W, Karle CA. Human cardiac inwardly rectifying current IKir2.2 is upregulated by activation of protein kinase A. *Cardiovasc Res* 2004;**63**(3):520–7.

49. Quinn KV, Giblin JP, Tinker A. Multisite phosphorylation mechanism for protein kinase A activation of the smooth muscle ATP-sensitive K$^+$ channel. *Circ Res* 2004;**94**(10):1359–66.

50. Kagan A, Melman YF, Krumerman A, McDonald TV. 14-3-3 amplifies and prolongs adrenergic stimulation of HERG K+ channel activity. *EMBO J* 2002;**21**(8):1889–98.

51. Wang M, Berlin JR. Channel phosphorylation and modulation of L-type Ca^{2+} currents by cytosolic Mg^{2+} concentration. *Am J Physiol Cell Physiol* 2006;**291**(1):C83–92.

52. Wehrens XH, Lehnart SE, Reiken S, Vest JA, Wronska A, Marks AR. Ryanodine receptor/calcium release channel PKA phosphorylation: a critical mediator of heart failure progression. *Proc Natl Acad Sci USA* 2006;**103**(3):511–18.

53. Chheda MG, Ashery U, Thakur P, Rettig J, Sheng ZH. Phosphorylation of Snapin by PKA modulates its interaction with the SNARE complex. *Nat Cell Biol* 2001;**3**(4):331–8.

54. Marash M, Gerst JE. t-SNARE dephosphorylation promotes SNARE assembly and exocytosis in yeast. *EMBO J* 2001;**20**(3):411–21.

55. Evans GJ, Wilkinson MC, Graham ME, Turner KM, Chamberlain LH, Burgoyne RD, Morgan A. Phosphorylation of cysteine string protein by protein kinase A. Implications for the modulation of exocytosis. *J Biol Chem* 2001;**276**(51):47,877–47,885.

56. D'Souza T, Agarwal R, Morin PJ. Phosphorylation of claudin-3 at threonine 192 by cAMP-dependent protein kinase regulates tight junction barrier function in ovarian cancer cells. *J Biol Chem* 2005;**280**(28):26,233–26,240.

57. Maximov A, Shin OH, Liu X, Sudhof TC. Synaptotagmin-12, a synaptic vesicle phosphoprotein that modulates spontaneous neurotransmitter release. *J Cell Biol* 2007;**176**(1):113–24.

58. Technikova-Dobrova Z, Sardanelli AM, Speranza F, Scacco S, Signorile A, Lorusso V, Papa S. Cyclic adenosine monophosphate-dependent phosphorylation of mammalian mitochondrial proteins: enzyme and substrate characterization and functional role. *Biochemistry* 2001;**40**(46):13,941–13,947.

59. Miyoshi II H, Perfield JW, Souza SC, Shen WJ, Zhang HH, Stancheva ZS, Kraemer FB, Obin MS, Greenberg AS. Control of adipose triglyceride lipase action by serine 517 of perilipin A globally regulates protein kinase A-stimulated lipolysis in adipocytes. *J Biol Chem* 2007;**282**(2):996–1002.

60. Portela P, Moreno S, Rossi S. Characterization of yeast pyruvate kinase 1 as a protein kinase A substrate, and specificity of the phosphorylation site sequence in the whole protein. *Biochem J* 2006;**396**(1):117–26.

61. Wasmeier C, Hutton JC. Secretagogue-dependent phosphorylation of the insulin granule membrane protein phogrin is mediated by cAMP-dependent protein kinase. *J Biol Chem* 2001;**276**(34):31,919–31,928.

62. Gu YM, Jin YH, Choi JK, Baek KH, Yeo CY, Lee KY. Protein kinase A phosphorylates and regulates dimerization of 14-3-3 epsilon. *FEBS Letts* 2006;**580**(1):305–10.

63. Citterio C, Jones HD, Pacheco-Rodriguez G, Islam A, Moss J, Vaughan M. Effect of protein kinase A on accumulation of brefeldin A-inhibited guanine nucleotide-exchange protein 1 (BIG1) in HepG2 cell nuclei. *Proc Natl Acad Sci USA* 2006;**103**(8):2683–8.

64. Giamas G, Hirner H, Shoshiashvili L, Grothey A, Gessert S, Kuhl M, Henne-Bruns D, Vorgias CE, Knippschild U. Phosphorylation of casein kinase 1 delta: Identification of Ser370 as the major phosphorylation site targeted by PKA *in vitro* and *in vivo. Biochem J* 2007;**406**(3):389–98.

65. Choi MG, Carman GM. Phosphorylation of human CTP synthetase 1 by protein kinase A: identification of Thr455 as a major site of phosphorylation. *J Biol Chem* 2007;**282**(8):5367–77.

66. Burnatowska-Hledin M, Zhao P, Capps B, Poel A, Parmelee K, Mungall C, Sharangpani A, Listenberger L. VACM-1, a cullin gene family member, regulates cellular signaling. *Am J Physiol Cell Physiol* 2000;**279**(1):C266–73.

67. Butt E, Gambaryan S, Gottfert N, Galler A, Marcus K, Meyer HE. Actin binding of human LIM and SH3 protein is regulated by cGMP- and cAMP-dependent protein kinase phosphorylation on serine 146. *J Biol Chem* 2003;**278**(18):15,601–15,607.

68. Alfthan K, Heiska L, Gronholm M, Renkema GH, Carpen O. Cyclic AMP-dependent protein kinase phosphorylates merlin at serine 518 independently of p21-activated kinase and promotes merlin-ezrin heterodimerization. *J Biol Chem* 2004;**279**(18):18,559–18,566.

69. Schmitt JM, Stork PJ. PKA phosphorylation of Src mediates cAMP's inhibition of cell growth via Rap1. *Mol Cell* 2002;**9**(1):85–94.

70. Lonart G, Sudhof TC. Characterization of rabphilin phosphorylation using phospho-specific antibodies. *Neuropharmacology* 2001;**41**(6):643–9.

71. de Jonge HR, Rosen OM. Self-phosphorylation of cyclic guanosine 3′,5′-monophosphate-dependent protein kinase from bovine lung. Effect of cyclic adenosine 3′,5′-monophosphate, cyclic guanosine 3′,5′-monophosphate and histone. *J Biol Chem* 1977;**252**(8):2780–3.

72. Vallins WJ, Brand NJ, Dabhade N, Butler-Browne G, Yacoub MH, Barton PJ. Molecular cloning of human cardiac troponin I using polymerase chain reaction. *FEBS Letts* 1990;**270**(1–2):57–61.

73. Yuasa K, Michibata H, Omori K, Yanaka N. A novel interaction of cGMP-dependent protein kinase I with troponin T. *J Biol Chem* 1999;**274**(52):37,429–37,434.

74. Okano I, Yamamoto T, Kaji A, Kimura T, Mizuno K, Nakamura T. Cloning of CRP2, a novel member of the cysteine-rich protein family with two repeats of an unusual LIM/double zinc-finger motif. *FEBS Letts* 1993;**333**(1–2):51–5.

75. Huber A, Neuhuber WL, Klugbauer N, Ruth P, Allescher HD. Cysteine-rich protein 2, a novel substrate for cGMP kinase I in enteric neurons and intestinal smooth muscle. *J Biol Chem* 2000;**275**(8):5504–11.

76. Beall AC, Kato K, Goldenring JR, Rasmussen H, Brophy CM. Cyclic nucleotide-dependent vasorelaxation is associated with the phosphorylation of a small heat shock-related protein. *J Biol Chem* 1997;**272**(17):11,283–11,287.

77. Raeymaekers L, Hofmann F, Casteels R. Cyclic GMP-dependent protein kinase phosphorylates phospholamban in isolated sarcoplasmic reticulum from cardiac and smooth muscle. *Biochem J* 1988;**252**(1):269–73.

78. Jiang LH, Gawler DJ, Hodson N, Milligan CJ, Pearson HA, Porter V, Wray D. Regulation of cloned cardiac L-type calcium channels by cGMP-dependent protein kinase. *J Biol Chem* 2000;**275**(9):6135–43.

79. Alioua A, Huggins JP, Rousseau E. PKG-I alpha phosphorylates the alpha-subunit and upregulates reconstituted GKCa channels from tracheal smooth muscle. *Am J Physiol* 1995;**268**(6 Pt 1):L1057–63.

80. Fukao M, Mason HS, Britton FC, Kenyon JL, Horowitz B, Keef KD. Cyclic GMP-dependent protein kinase activates cloned BKCa channels expressed in mammalian cells by direct phosphorylation at serine 1072. *J Biol Chem* 1999;**274**(16):10,927–10,935.

81. Komalavilas P, Lincoln TM. Phosphorylation of the inositol 1,4,5-trisphosphate receptor. Cyclic GMP-dependent protein kinase mediates cAMP and cGMP dependent phosphorylation in the intact rat aorta. *J Biol Chem* 1996;**271**(36):21,933–21,938.

82. Butt E, Immler D, Meyer HE, Kotlyarov A, Laass K, Gaestel M. Heat shock protein 27 is a substrate of cGMP-dependent protein kinase in intact human platelets: phosphorylation-induced actin polymerization caused by HSP27 mutants. *J Biol Chem* 2001;**276**(10):7108–13.

83. Butt E, Abel K, Krieger M, Palm D, Hoppe V, Hoppe J, Walter U. cAMP- and cGMP-dependent protein kinase phosphorylation sites of the focal adhesion vasodilator-stimulated phosphoprotein (VASP) in vitro and in intact human platelets. *J Biol Chem* 1994;**269**(20):14,509–14,517.

84. Haffner C, Jarchau T, Reinhard M, Hoppe J, Lohmann SM, Walter U. Molecular cloning, structural analysis and functional expression of the proline-rich focal adhesion and microfilament-associated protein VASP. *EMBO J* 1995;**14**(1):19–27.

85. Xue J, Wang X, Malladi CS, Kinoshita M, Milburn PJ, Lengyel I, Rostas JA, Robinson PJ. Phosphorylation of a new brain-specific septin, G-septin, by cGMP-dependent protein kinase. *J Biol Chem* 2000;**275**(14):10,047–10,056.

86. Ammendola A, Geiselhoringer A, Hofmann F, Schlossmann J. Molecular determinants of the interaction between the inositol 1,4,5-trisphosphate receptor-associated cGMP kinase substrate (IRAG) and cGMP kinase Ibeta. *J Biol Chem* 2001;**276**(26):24,153–24,159.

87. McDonald BJ, Moss SJ. Conserved phosphorylation of the intracellular domains of GABA(A) receptor beta2 and beta3 subunits by cAMP-dependent protein kinase, cGMP-dependent protein kinase protein kinase C and Ca^{2+}/calmodulin type II-dependent protein kinase. *Neuropharmacology* 1997;**36**(10):1377–85.

88. Endo S, Suzuki M, Sumi M, Nairn AC, Morita R, Yamakawa K, Greengard P, Ito M. Molecular identification of human G-substrate, a possible downstream component of the cGMP-dependent protein kinase cascade in cerebellar Purkinje cells. *Proc Natl Acad Sci USA* 1999;**96**(5):2467–72.

89. Hemmings Jr HC, Williams KR, Konigsberg WH, Greengard P. DARPP-32, a dopamine- and adenosine 3':5'-monophosphate-regulated neuronal phosphoprotein. I. Amino acid sequence around the phosphorylated threonine. *J Biol Chem* 1984;**259**(23):14,486–14,490.

90. Wang GR, Zhu Y, Halushka PV, Lincoln TM, Mendelsohn ME. Mechanism of platelet inhibition by nitric oxide: in vivo phosphorylation of thromboxane receptor by cyclic GMP-dependent protein kinase. *Proc Natl Acad Sci USA* 1998;**95**(9):4888–93.

91. Yamamoto S, Yan F, Zhou H, Tai HH. Serine 331 is the major site of receptor phosphorylation induced by agents that activate protein kinase G in HEK 293 cells overexpressing thromboxane receptor alpha. *Arch Biochem Biophys* 2001;**393**(1):97–105.

92. Wang X, Robinson PJ. Cyclic GMP-dependent protein kinase and cellular signaling in the nervous system. *J Neurochem* 1997;**68**(2):443–56.

93. Glass DB, Krebs EG. Phosphorylation by guanosine 3′,5′-monophosphate-dependent protein kinase of synthetic peptide analogs of a site phosphorylated in histone H2B. *J Biol Chem* 1982;**257**(3):1196–200.

94. Yuasa K, Omori K, Yanaka N. Binding and phosphorylation of a novel male germ cell-specific cGMP-dependent protein kinase-anchoring protein by cGMP-dependent protein kinase Ialpha. *J Biol Chem* 2000;**275**(7):4897–905.

95. Rodriguez-Pascual F, Ferrero R, Miras-Portugal MT, Torres M. Phosphorylation of tyrosine hydroxylase by cGMP-dependent protein kinase in intact bovine chromaffin cells. *Arch Biochem Biophys* 1999;**366**(2):207–14.

96. Butt E, Bernhardt M, Smolenski A, Kotsonis P, Frohlich LG, Sickmann A, Meyer HE, Lohmann SM, Schmidt HH. Endothelial nitric-oxide synthase (type III) is activated and becomes calcium independent upon phosphorylation by cyclic nucleotide-dependent protein kinases. *J Biol Chem* 2000;**275**(7):5179–87.

97. Murray KJ, El-Maghrabi MR, Kountz PD, Lukas TJ, Soderling TR, Pilkis SJ. Amino acid sequence of the phosphorylation site of rat liver 6-phosphofructo-2-kinase/fructose-2,6-bisphosphatase. *J Biol Chem* 1984;**259**(12):7673–81.

98. Thomas MK, Francis SH, Corbin JD. Substrate- and kinase-directed regulation of phosphorylation of a cGMP-binding phosphodiesterase by cGMP. *J Biol Chem* 1990;**265**(25):14,971–14,978.

99. Suhasini M, Li H, Lohmann SM, Boss GR, Pilz RB. Cyclic-GMP-dependent protein kinase inhibits the Ras/Mitogen-activated protein kinase pathway. *Mol Cell Biol* 1998;**18**(12):6983–94.

100. Xia C, Bao Z, Yue C, Sanborn BM, Liu M. Phosphorylation and regulation of G-protein-activated phospholipase C-beta 3 by cGMP-dependent protein kinases. *J Biol Chem* 2001;**276**(23):19,770–19,777.

101. Yang X, Goncalves J, Gabuzda D. Phosphorylation of Vif and its role in HIV-1 replication. *J Biol Chem* 1996;**271**(17):10,121–10,129.

102. Wyatt TA, Lincoln TM, Pryzwansky KB. Vimentin is transiently co-localized with and phosphorylated by cyclic GMP-dependent protein kinase in formyl-peptide-stimulated neutrophils. *J Biol Chem* 1991;**266**(31):21,274–21,280.

103. MacMillan-Crow LA, Lincoln TM. High-affinity binding and localization of the cyclic GMP-dependent protein kinase with the intermediate filament protein vimentin. *Biochemistry* 1994;**33**(26):8035–43.

104. Wang Y, El-Zaru MR, Surks HK, Mendelsohn ME. Formin homology domain protein (FHOD1) is a cyclic GMP-dependent protein kinase I-binding protein and substrate in vascular smooth muscle cells. *J Biol Chem* 2004;**279**(23):24,420–24,426.

Chapter 43

Inhibitors of Cyclic AMP- and Cyclic GMP-Dependent Protein Kinases

Wolfgang R. Dostmann and Christian K. Nickl

Department of Pharmacology, University of Vermont, College of Medicine, Burlington, Vermont

INTRODUCTION

The cyclic nucleotide-dependent protein kinases PKA and PKG serve as primary targets for the second messengers cAMP and cGMP, respectively. Both kinases have served as Rosetta Stones in our understanding of a vast number of intracellular signaling mechanisms ranging from smooth muscle cell relaxation to neuronal synaptic plasticity (for reviews, see [1–6]). Therefore, the search for potent inhibitors

of these kinases has been extensively investigated. However, the structural similarities of PKA and PKG have posed a formidable obstacle in the design of selective inhibitors that specifically target cyclic nucleotide-dependent protein kinases and show little inhibitory potency to other basophilic Ser/Thr-kinases.

The domain structures of PKA and PKG dictate key target sites for putative inhibitors. Figure 43.1 compares the domain structures of PKA and PKG and illustrates

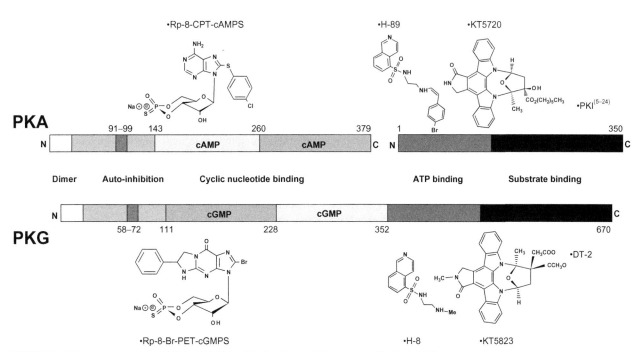

FIGURE 43.1 Domain structures of PKA and PKG indicating three distinct target sites for inhibitors.
Rp-thiophosphoryl analogs of cAMP and cGMP target the cyclic nucleotide binding sites, napthalenesulfonyl derivatives (H-series) and staurosporine analogs (K-series) are ATP-site directed, and peptide inhibitors such as PKI and DT-2 are specific for the substrate binding sites.

three distinct classes of inhibitors and their various sites of actions. The regulatory components of cyclic nucleotide-dependent protein kinases each harbor two tandem cyclic nucleotide binding sites that allow allosteric and cooperative control of kinase activity [1, 7–11]. Of a particular class of cAMP/cGMP derivatives, (Rp)-phosphorothioates are the only known inhibitors that bind to the cyclic nucleotide binding sites [12–15]. Although their mode of action is still not completely understood, studies have indicated that the binding of these derivatives fails to induce the conformational changes essential for releasing catalytic activity [16–18]. A large pool of derivatives, moderate selectivity, and cell membrane permeability are regarded as the major advantages of Rp-cyclic nucleotide derivatives as tools in intact cell studies [15, 19, 20]. However, partial antagonism and limited selectivity restrict their versatility [14, 15, 21]. The catalytic components of cyclic nucleotide-dependent protein kinases contain two target sites for inhibitors: the ATP-binding site, and the substrate binding site. Compounds mimicking ATP represent a diverse class of inhibitors, as has been known for all other major families of protein kinases [22–24]. In contrast, peptide inhibitors designed to block the substrate binding site of PKG have long remained elusive, partly because the sequence requirements for PKG inhibition do not follow a classic consensus sequence and the kinase appears to non-specifically favor positively charged amino acids [25–27]. Recently, we have developed a new class of potent and cell-membrane permeable PKG peptide inhibitors [25]. We utilized SPOT-based combinatorial peptide libraries [28–32] to identify PKG selective substrate and inhibitor peptides, of which DT-2 is the most potent and selective PKG inhibitor known today [33]. DT-2 shows unsurpassed PKG specificity and, due to its membrane-permeable segment from HIV-1 tat[47–59], remarkable cellular translocation characteristics [34].

act as kinase agonists [39–43], prompted the synthesis and subsequent analysis of an entirely new class of cyclic nucleotide analogs [12, 20, 44] in an effort to identify selective and isozyme-specific inhibitors (for a comprehensive overview and a complete list of references, visit www.biolog. de). Tables 43.1 and 43.2 provide examples of PKA- and PKG-specific Rp-cyclic nucleotide phosphorothioates (Rp-cNMPS). These studies revealed that the competitive binding of Rp-cAMPS to both cAMP binding sites in the PKA holoenzyme is thought to prevent dissociation of the regulatory subunits [16–18]. Likewise, it is believed that binding of Rp-cGMPS and analogs to PKG are unable to induce the conformational change needed to expose the enzyme's catalytic cleft.

Recent studies have demonstrated that Rp-cAMPS analogs can function as partial agonists dependent on the presence of MgATP and enzyme concentration [14, 18]. These findings raise questions concerning the usefulness of inhibition constants when dealing with Rp-cAMPS analogs. However, this phenomenon is not observed for PKG, partly because the holoenzyme complex does not dissociate into regulatory and catalytic subunits. Another important finding is that the sulfur substitution in Rp-cNMPS increases the relative lipophilicity and hence cell-membrane permeability compared to their cNMP counterparts [21, 45] (see also http://www.biolog.de/technical-info/lipophilicity-data/). In addition, Rp-cNMPS exhibits complete resistance to phosphodiesterases [46]. These properties have markedly contributed to the diverse applications that cyclic nucleotide binding site-targeted inhibitors have enjoyed in intracellular signaling research [15]. It should be noted that the Rp-cGMPS derivatives Rp-8-CPT-cGMPS and Rp-8-Br-PET-cGMPS, as PKG inhibitors (Table 43.2) with high selectivity, low toxicity, and superior membrane permeability, have gained supremacy in their class [12, 13].

CYCLIC NUCLEOTIDE BINDING SITE-TARGETED INHIBITORS

The cyclic-nucleotide binding sites of PKA and PKG show significant sequence similarity with respect to the recognition motif of the nucleotide phosphodiester (FGE…RAA and FGE…RTA for all PKA and PKG isoforms, respectively). The crystal structures of PKA RIIβ [35] and RIα [36] revealed the architecture of the cyclic nucleotide binding pocket with an invariant arginine situated at its base. It is thought that chelation of the phosphate moiety by the invariant arginine is the first step in cAMP binding, although the conformational constraints of cyclic nucleotide binding appear much more complex [7, 10, 37, 38]. The discovery that sulfur substitution of the axial exocyclic phosphate oxygen (Rp-) of cyclic nucleotides resulted in inhibitors of PKA and PKG, while cyclic nucleotide analogs carrying the equivalent apical sulfur substitution (Sp-)

ATP BINDING SITE-TARGETED INHIBITORS

Synthetic protein kinase inhibitors that are competitive with ATP and specific for PKA and/or PKG represent a structurally diverse group of small ligand compounds [24, 47]. Polycyclic aromatics, such as isoquinolinesulfonyl and napthalenesulfonyl compounds ("H-series") and naturally occurring molecules, such as staurosporine analogs ("K-series") have served primarily as valuable inhibitors of AGC-type protein kinases, notably PKC [22, 24, 47–51]. In fact, the inhibitory potency against PKC is a defining property of most ATP-site inhibitors of the above series. However, a subset of compounds including the H-series H89 and KT5720, and the K-series H8 and KT5823, comprises moderately specific inhibitors for PKA and PKG, respectively (see Tables 43.1, 43.2). The relatively straightforward chemical modifications of isoquinoline-derivatives in particular produced a wealth of selective protein kinase inhibitors with

TABLE 43.1 Selected cyclic nucleotide-dependent protein kinase inhibitors

PKA inhibitors	K_i/IC$_{50}$ (µM)	Isoform selectivity	Comments[1]	Refs	PKG inhibitors	K_i/IC$_{50}$ (µM)	Isoform selectivity	Comments[1]	Refs
Kemptide	376	PKA Iα	pep, PKA	73, 76	H2B[(29–35)]	86, PKG Iα	PKG Iα, Iβ	pep, non	94
Rp-cAMPS	12.5/7.9	PKA I	cyc, lip, PKA, PDE	40, 43, 113	Rp-cGMPS	20, PKG Iα	PKG Iα, Iβ	cyc, lip, PKG, PDE	44
	3.7	PKA II							
H-8	1.2	—	ATP, lip	22	WW21	7.5, PKG Iα	PKG Iα	pep, PKG	28
Rp-8-Br-cAMPS	—	PKA I	cyc, lip, PDE	14, 15, 19, 114	HA1004	1.4, PKG Iα	—	ATP, lip, AGC, CaMK CMGC	49
Rp-8-CPT-cAMPS	—	PKA II	cyc,, lip, PDE	14, 15, 19, 115	Rp-8-Br-cGMPS	4.0, PKG Iα	PKG Iα, Iβ	cyc lip, PKG, PDE	116
Rp-8-PIP-cAMPS	—	PKA II, site B	cyc,, lip, PDE	20	Rp-8-CPT-cGMPS	0.5, PKG Iα	PKG II	cyc lip, PKG, PDE	12
KT5720	0.06, PKA	—	ATP, lip	47, 50	**H-8**	0.5, PKG Iα	—	ATP, lip, PKG	22
H-89	0.048, PKA	—	ATP, lip	47, 48, 50	KT5823	0.234, PKG Iα	—	ATP, lip, PKG	47, 50
4-cyano-3-methyl-isoquinoline	0.03, PKA	—	ATP, AGC	51	K-252b	0.1, PKG Iα	—	ATP, lip, AGC, CaMK	50
K-252a	0.018,	—	ATP, lip	50	Rp-8-Br-PET-cGMPS	0.035, PKG Iα	PKG Iα	cyc, lip, PKG, PDE	13
						0.03, PKG Iβ			
staurosporine	0.008, PKA	—	ATP, non	117–119	K-252a	0.02, PKG Iα	—	ATP, lip, AGC, CaMK	50
balanol	0.004 PKA	PKA I	ATP, PKA	56, 57	GRTGR(PTR)NAI	3.8, PKG Iα	—	pep	112
PKI[(5–24)]	0.002, PKA	PKA Iα	pep, PKA	28, 77, 81, 92	DT-3	0.025, PKG Iα	PKG Iα	pep, MTS, PKG	25
AdcAhx(D-Arg)$_6$	0.008, PKA	—	ATP, MTS	63	DT-2	0.012, PKG Iα	PKG Iα	pep, MTS, PKG	25
H9-(CH$_2$)$_5$C(O) (D-Arg)$_6$-NH$_2$	0.0053, PKA	—	ATP, MTS	63					

[1]The following abbreviations are used: AGC, specificity for AGC subfamily protein kinases; ATP, ATP-binding site inhibitor; CaMK, specificity for CaMK subfamily protein kinases; CMGC, specificity for CMGC subfamily protein kinases; cyc, cyclic nucleotide binding site inhibitor; lip, lipophilic, cell permeable; MTS, membrane translocation signal; non, non-specific; PDE, resistant against PDE hydrolysis; pep, peptide binding site inhibitor; PKA, high selectivity for PKA; PKC, high selectivity for PKC; PKG, high selectivity for PKG.

TABLE 43.2 Representative commercially available inhibitors with increasing PKA/PKG selectivity

	K_i (µM)	PKA/PKG selectivity	References
Cyclic nucleotide inhibitors			
Rp-cAMPS	7.9 PKA I/II	1.5×10^{-1}	40, 42–44, 113
	52, PKG Iα		
Rp-cGMPS	20, PKA II	1×10^{0}	12, 44
	20, PKG Iα		
Rp-8-CPT-cGMPS	8.3, PKA II	1.7×10^{1}	12
	0.5, PKG Iα		
Rp-8-Cl-cGMPS	100, PKA II	6.7×10^{1}	44
	1.5, PKG		
Rp-8-Br-PET-cGMPS	11, PKA II	3.1×10^{2}	13
	0.035, PKG Iα	3.7×10^{2}	
	0.030, PKG Iβ		
ATP analogs			
KT5720	0.06, PKA I	3.0×10^{-2}	50
	>2, PKG Iα		
H89	0.048, PKA I	1.0×10^{-1}	24, 48
	0.48, PKG Iα		
H8	1.2, PKA I	2.5×10^{1}	22, 24
	0.48, PKG Iα		
KT5823	>10, PKA I	$>4.2\times10^{1}$	47, 50
	0.234, PKG Iα		
Peptide inhibitors			
PKI[5−24]	0.002, PKA	1.8×10^{-5}	92, 120
TTYADFIASGRTGRRNAIHD	111, PKG		
PKI[14−22]	0.073, PKA	1.5×10^{-3}	28, 77
GRTGRRNAI	47, PKG		
Ala-Kemptide	376, PKA	4.7×10^{-1}	73, 76, 90
LRRAALG	800, PKG		
[A[32]]-H2B[29−35]	550, PKA	6.4×10^{0}	90
RKRARKE	86, PKG		
PKGI	1020, PKA	6.9×10^{1}	99
GRTGRRN(D-Ala)I	14.8, PKG		
WW21	750, PKA	1.0×10^{2}	28
TQAKRKKALAMA	7.5, PKG		
DT-3	493, PKA	1.97×10^{4}	28, 105
RQIKIWFQNRRMKWKKLRKKKKKH	0.025, PKG		
DT-2	16.5, PKA	1.32×10^{3}	28, 105
YGRKKRRQRRRPPLRKKKKKH	0.012, PKG		

potential for clinical applications [22, 24, 48–51]. In addition, the cell-membrane permeability of these compounds further amplified their versatility in dissecting signaling pathways involving protein kinase signaling. However, concerns regarding toxicity and reports of problems in *in vivo* conditions, using KT5823 and KT5720 as specific PKG and PKA inhibitors, have questioned their usefulness in intact cell preparations [52–55] (for a detailed discussion, see: http://www.biolog.de/technical-info/h-89-kt-5720-pitfalls/).

Recently, it was observed that the natural product balanol and its derivatives inhibit PKA and PKC of the AGC-subfamily with high potency [56–58]. The crystal structure of balanol in complex with the catalytic subunit of PKA [59] confirmed a structural peculiarity of most ATP-site targeted inhibitors: the molecules satisfy essential interaction within the ATP-binding site, but utilize unique interactions with the enzyme, thus gaining selectivity and specificity. It has been shown that analogs of balanol display variability in protein kinase inhibition, and the structural determinants of their protein kinase selectivity can now be elucidated with computational methods [60–62].

Another novel class of ATP-mimetic inhibitors utilizes a bisubstrate approach by conjugation of oligoarginine peptides with adenine derivatives [63–67]. By combining adenine, adenosine, and isoquinolinesulfonamide aromatic structures with oligoarginine of various lengths through a series of conformationally flexible linkers, the resulting kinase inhibitors were shown to exhibit considerable potencies towards PKA and PKC. Interestingly, hexa-(L-arginine) derivatives were more potent than tetra-(L-arginine) analogs, and Hexa-(D-arginine) conjugates showed further enhanced inhibitory potencies [63]. It is believed that these novel inhibitors target the ATP, as well as the substrate binding sites [65, 68]. Future studies directed against the linker region will show whether higher specificity towards a selected kinase of the AGC family can be achieved. An additional advantage using poly-arginine conjugates is their effective translocation through cellular membranes, which should further enhance the utility of these compounds [69–71].

Although inhibitors directed against the ATP binding site show a high degree of diversity, recent studies have suggested that they all utilize similar residues at the specific positions important for binding [72]. This finding may contribute significantly in the design of novel kinase inhibitors and kinase targeted drugs with pre-defined selectivity profiles.

PEPTIDE BINDING SITE-TARGETED INHIBITORS

The observation that relatively short peptides corresponding to the regulatory subunit's auto-phosphorylation site were effective substrates for PKA [73] and the discovery of the protein kinase inhibitor PKI [74, 75] prompted a comprehensive search of PKA inhibitory peptides [74, 76–81] and presented

a prime example for the concept of consensus sequence-targeted inhibitor design [27, 74, 82, 83]. However, full appreciation of the intricate structural web existing between kinase and inhibitor occurred only after the crystal structure of the catalytic subunit of the PKA : PKI adduct was solved [84, 85]. Peptides derived from PKI isoforms α and β [86–88], namely PKI$^{5–24}$ and PKI$^{5–22}$, are still the most potent and, more importantly, the most selective PKA inhibitors known today (Tables 43.1, 43.2). However, their use in intact cell studies is limited (i.e., to patch-clamp techniques) due to their inability to cross the plasma membrane. Unfortunately, fusion peptides of PKI$^{15–22}$ with membrane translocation signal (MTS) peptides derived from the antennapedia-homeo domain or *HIV-1* tat [28, 89] showed a profound loss in PKA selectivity (W. Dostmann, unpublished results).

Attempts to identify PKG selective inhibitor peptides based on the auto-inhibitory domain of the enzyme or *in vivo* substrates have been tedious at best, due to the lack of a well-defined consensus sequence [90–96]. Only a relative preference for basic residues surrounding the phosphate acceptor site has been established [26, 80, 90, 93, 94, 97]. Various synthetic peptides have been used to analyze and optimize the sequence requirements for PKG substrates and inhibitors [98–102]. Recently, the identification of selective inhibitors of PKG by a novel peptide library screen specifically designed to select for tight binding peptides was reported [25, 28, 31]. Cellular internalization of the peptides was accomplished by N-terminal fusion to the membrane translocation sequences, from either the HIV-1 Tat protein [22, 48, 49, 51–54, 56, 57, 59, 61, 103, 104], DT-2, or from the *Drosophila* antennapedia homeo-domain [22, 24, 46–54, 56, 57, 59, 103, 104], DT-3 [25, 105]. Surprisingly, these fusion peptides result in an extraordinary synergism with respect to PKG inhibition (Tables 43.1, 43.2). It was shown that DT-2 in particular effectively inhibited NO-induced vasodilation, further emphasizing the central role for PKG in the modulation of vascular contractility [105]. Isolated pressurized cerebral arteries were utilized to further evaluate the potential functional impact of this unique inhibition by DT-2 [33]. These findings indicated that DT-2 not only effectively inhibits cGMP-stimulated PKG activity, but also reduces basal PKG activity both *in vitro* and *in vivo*. The validity of DT-2 as a superior inhibitor of PKG in terms of potency, selectivity, and membrane permeability has been further demonstrated [106–111]. Additional studies suggest that the uptake of DT-2 is multifaceted. Cell morphology appears to be a strong determinant of the uptake pattern of the inhibitor [34]. These results suggest that the cell membrane permeability of DT-2 and DT-3, combined with enormous PKG selectivity, will significantly advance our experimental ability to dissect PKG-mediated intracellular pathways from PKA and other kinases. A potentially new class of peptide mimetic inhibitors makes use of the poly-L-proline type II (PPII) conformation found in substrates of PKA and PKG (Table 43.1) [112]. These pseudosubstrate PPII peptides may serve as the basis for the design of selective, high-affinity protein kinase inhibitors.

CONCLUSIONS

Rp-phosphorotioate derivatives of cAMP and cGMP competitively inhibit cyclic nucleotide-dependent protein kinases by stabilizing the enzymes in their inactive holoenzyme states. A large pool of derivatives, moderate selectivity, and cell membrane permeability are regarded as their advantages as tools in intact cell studies. However, partial antagonism and limited potencies restrict their versatility. ATP-analogs are a highly resourceful group of protein kinase inhibitors. Cell membrane permeability and limited selectivity highlight their advantages and disadvantages, respectively. Peptide-derived inhibitors present the most potent and selective group of PKA and PKG blockers. Low cell membrane permeability remains their main obstacle in cellular research. Recently, a subset of PKG-selective peptide inhibitors employing membrane translocation sequences as a means of cellular delivery has overcome this problem.

ACKNOWLEDGEMENTS

This work was supported by grants from the NIH (HL68891) and the Totman Trust for Medical Research.

REFERENCES

1. Kim C, Vigil D, Anand G, Taylor SS. Structure and dynamics of PKA signaling proteins. *Eur J Cell Biol* 2006;**85**:651–4.
2. Taylor SS, Kim C, Cheng CY, Brown SH, Wu J, Kannan N. Signaling through cAMP and cAMP-dependent protein kinase: diverse strategies for drug design. *Biochim Biophys Acta* 2008;**1784**:16–26.
3. Feil R, Feil S, Hofmann F. A heretical view on the role of NO and cGMP in vascular proliferative diseases. *Trends Mol Med* 2005;**11**:71–5.
4. Hofmann F, Feil R, Kleppisch T, Schlossmann J. Function of cGMP-dependent protein kinases as revealed by gene deletion. *Physiol Rev* 2006;**86**:1–23.
5. Beene DL, Scott JD. A-kinase anchoring proteins take shape. *Curr Opin Cell Biol* 2007;**19**:192–8.
6. Lincoln TM, Wu X, Sellak H, Dey N, Choi CS. Regulation of vascular smooth muscle cell phenotype by cyclic GMP and cyclic GMP-dependent protein kinase. *Front Biosci* 2006;**11**:356–67.
7. Berman HM, Ten Eyck LF, Goodsell DS, Haste NM, Kornev A, Taylor SS. The cAMP binding domain: an ancient signaling module. *Proc Natl Acad Sci USA* 2005;**102**:45–50.
8. Das R, Esposito V, Abu-Abed M, Anand GS, Taylor SS, Melacini G. cAMP activation of PKA defines an ancient signaling mechanism. *Proc Natl Acad Sci USA* 2007;**104**:93–8.
9. Kannan N, Haste N, Taylor SS, Neuwald AF. The hallmark of AGC kinase functional divergence is its C-terminal tail, a cis-acting regulatory module. *Proc Natl Acad Sci USA* 2007;**104**:1272–7.
10. Kim C, Cheng CY, Saldanha SA, Taylor SS. PKA-I holoenzyme structure reveals a mechanism for cAMP-dependent activation. *Cell* 2007;**130**:1032–43.
11. Pfeifer A, Ruth P, Dostmann W, Sausbier M, Klatt P, Hofmann F. Structure and function of cGMP-dependent protein kinases. *Rev Physiol Biochem Pharmacol* 1999;**135**:105–49.
12. Butt E, Eigenthaler M, Genieser HG. (Rp)-8-pCPT-cGMPS, a novel cGMP-dependent protein kinase inhibitor. *Eur J Pharmacol* 1994;**269**:265–8.
13. Butt E, Pohler D, Genieser HG, Huggins JP, Bucher B. Inhibition of cyclic GMP-dependent protein kinase-mediated effects by (Rp)-8-bromo-PET-cyclic GMPS. *Br J Pharmacol* 1995;**116**:3110–16.
14. Gjertsen BT, Mellgren G, Otten A, Maronde E, Genieser HG, Jastorff B, Vintermyr OK, McKnight GS, Doskeland SO. Novel (Rp)-cAMPS analogs as tools for inhibition of cAMP-kinase in cell culture. Basal cAMP-kinase activity modulates interleukin-1 beta action. *J Biol Chem* 1995;**270**(20):599–607.
15. Schwede F, Maronde E, Genieser H, Jastorff B. Cyclic nucleotide analogs as biochemical tools and prospective drugs. *Pharmacol Ther* 2000;**87**:199–226.
16. Dostmann WR. (RP)-cAMPS inhibits the cAMP-dependent protein kinase by blocking the cAMP-induced conformational transition. *FEBS Letts* 1995;**375**:231–4.
17. Wu J, Jones JM, Nguyen-Huu X, Ten Eyck LF, Taylor SS. Crystal structures of RIalpha subunit of cyclic adenosine 5′-monophosphate (cAMP)-dependent protein kinase complexed with (Rp)-adenosine 3′,5′-cyclic monophosphothioate and (Sp)-adenosine 3′,5′-cyclic monophosphothioate, the phosphothioate analogues of cAMP. *Biochemistry* 2004;**43**:6620–9.
18. Dostmann WR, Taylor SS. Identifying the molecular switches that determine whether (Rp)-cAMPS functions as an antagonist or an agonist in the activation of cAMP-dependent protein kinase I. *Biochemistry* 1991;**30**:8710–16.
19. Dostmann WR, Taylor SS, Genieser HG, Jastorff B, Doskeland SO, Ogreid D. Probing the cyclic nucleotide binding sites of cAMP-dependent protein kinases I and II with analogs of adenosine 3′,5′-cyclic phosphorothioates. *J Biol Chem* 1990;**265**(10):484–91.
20. Ogreid D, Dostmann W, Genieser HG, Niemann P, Doskeland SO, Jastorff B. Rp)- and (Sp)-8-piperidino-adenosine 3′,5′-(cyclic)thiophosphates discriminate completely between site A and B of the regulatory subunits of cAMP-dependent protein kinase type I and II. *Eur J Biochem* 1994;**221**:1089–94.
21. Poppe H, Rybalkin SD, Rehmann H, Hinds TR, Tang XB, Christensen AE, Schwede F, Genieser HG, Bos JL, Doskeland SO, Beavo JA, Butt E. Cyclic nucleotide analogs as probes of signaling pathways. *Nat Methods* 2008;**5**:277–8.
22. Hidaka H, Inagaki M, Kawamoto S, Sasaki Y. Isoquinolinesulfonamides, novel and potent inhibitors of cyclic nucleotide dependent protein kinase and protein kinase C.. *Biochemistry* 1984;**23**:5036–41.
23. Hidaka H, Kobayashi R. Protein kinase inhibitors. *Essays Biochem* 1994;**28**:73–97.
24. Ono-Saito N, Niki I, Hidaka H. H-series protein kinase inhibitors and potential clinical applications. *Pharmacol Ther* 1999;**82**:123–31.
25. Dostmann WR, Taylor MS, Nickl CK, Brayden JE, Frank R, Tegge WJ. Highly specific, membrane-permeant peptide blockers of cGMP-dependent protein kinase Ialpha inhibit NO-induced cerebral dilation. *Proc Natl Acad Sci USA* 2000;**97**(14):772–7.

26. Glass DB, Krebs EG. Phosphorylation by guanosine 3′,5′-monophos-phate-dependent protein kinase of synthetic peptide analogs of a site phosphorylated in histone H2B. *J Biol Chem* 1982;**257**:1196–200.

27. Kennelly PJ, Krebs EG. Consensus sequences as substrate specificity determinants for protein kinases and protein phosphatases. *J Biol Chem* 1991;**266**(15):555–8.

28. Dostmann WR, Nickl C, Thiel S, Tsigelny I, Frank R, Tegge WJ. Delineation of selective cyclic GMP-dependent protein kinase Ialpha substrate and inhibitor peptides based on combinatorial peptide libraries on paper. *Pharmacol Ther* 1999;**82**:373–87.

29. Frank R. Spot-synthesis: an easy thechnique for the positionally addressable, parallel chemical synthesis on a membrane support.. *Tetrahedron* 1992;**48**:9217–32.

30. Martens W, Greiser-Wilke I, Harder TC, Dittmar K, Frank R, Orvell C, Moennig V, Liess B. Spot synthesis of overlapping peptides on paper membrane supports enables the identification of linear monoclonal antibody binding determinants on morbillivirus phosphoproteins. *Vet Microbiol* 1995;**44**:289–98.

31. Tegge W, Frank R, Hofmann F, Dostmann WR. Determination of cyclic nucleotide-dependent protein kinase substrate specificity by the use of peptide libraries on cellulose paper.. *Biochemistry* 1995;**34**:10,569–10,577.

32. Tegge WJ, Frank R. Analysis of protein kinase substrate specificity by the use of peptide libraries on cellulose paper (SPOT-method). *Methods Mol Biol* 1998;**87**:99–106.

33. Taylor MS, Okwuchukwuasanya C, Nickl CK, Tegge W, Brayden JE, Dostmann WR. Inhibition of cGMP-dependent protein kinase by the cell-permeable peptide DT-2 reveals a novel mechanism of vasoregulation. *Mol Pharmacol* 2004;**65**:1111–19.

34. Foley KF, De Frutos S, Laskovski KE, Tegge W, Dostmann WR. Culture conditions influence uptake and intracellular localization of the membrane permeable cGMP-dependent protein kinase peptide inhibitor DT-2. *Front Biosci* 2005;**10**:1302–12.

35. Diller TC, Madhusudan., Xuong NH, Taylor SS. Molecular basis for regulatory subunit diversity in cAMP-dependent protein kinase: crystal structure of the type II beta regulatory subunit. *Structure* 2001;**9**:73–82.

36. Su Y, Dostmann WR, Herberg FW, Durick K, Xuong NH, Ten Eyck L, Taylor SS, Varughese KI. Regulatory subunit of protein kinase A: structure of deletion mutant with cAMP binding domains. *Science* 1995;**269**:807–13.

37. Kornev AP, Taylor SS, Ten Eyck LF. A generalized allosteric mechanism for cis-regulated cyclic nucleotide binding domains. *PLoS Comput Biol* 2008;**4**. e1000056.

38. Masterson LR, Mascioni A, Traaseth NJ, Taylor SS, Veglia G. Allosteric cooperativity in protein kinase A. *Proc Natl Acad Sci USA* 2008;**105**:506–11.

39. Botelho LH, Rothermel JD, Coombs RV, Jastorff B. cAMP analog antagonists of cAMP action. *Methods Enzymol* 1988;**159**:159–72.

40. de Wit RJ, Hekstra D, Jastorff B, Stec WJ, Baraniak J, Van Driel R, Van Haastert PJ. Inhibitory action of certain cyclophosphate derivatives of cAMP on cAMP-dependent protein kinases. *Eur J Biochem* 1984;**142**:255–60.

41. de Wit RJ, Hoppe J, Stec WJ, Baraniak J, Jastorff B. Interaction of cAMP derivatives with the 'stable' cAMP-binding site in the cAMP-dependent protein kinase type I. *Eur J Biochem* 1982;**122**:95–9.

42. Hofmann F, Gensheimer HP, Landgraf W, Hullin R, Jastorff B. Diastereomers of adenosine 3′,5′-monothionophosphate (cAMP[S]) antagonize the activation of cGMP-dependent protein kinase. *Eur J Biochem* 1985;**150**:85–8.

43. Rothermel JD, Parker Botelho LH. A mechanistic and kinetic analysis of the interactions of the diastereoisomers of adenosine 3′,5′-(cyclic)phosphorothioate with purified cyclic AMP-dependent protein kinase. *Biochem J* 1988;**251**:757–62.

44. Butt E, van Bemmelen M, Fischer L, Walter U, Jastorff B. Inhibition of cGMP-dependent protein kinase by (Rp)-guanosine 3′,5′-monophosphorothioates. *FEBS Letts* 1990;**263**:47–50.

45. Krass J, Jastorff B, Genieser HG. Determination of Lipophilicity by Gradient Elution High-Performance Liquid Chromatography. *Anal Chem* 1997;**69**:2575–81.

46. Erneux C, Miot F. Cyclic nucleotide analogs used to study phosphodiesterase catalytic and allosteric sites. *Methods Enzymol* 1988;**159**:520–30.

47. Hidaka H, Kobayashi R. Pharmacology of protein kinase inhibitors. *Annu Rev Pharmacol Toxicol* 1992;**32**:377–97.

48. Chijiwa T, Mishima A, Hagiwara M, Sano M, Hayashi K, Inoue T, Naito K, Toshioka T, Hidaka H. Inhibition of forskolin-induced neurite outgrowth and protein phosphorylation by a newly synthesized selective inhibitor of cyclic AMP-dependent protein kinase, N-[2-(p-bromocinnamylamino)ethyl]-5-isoquinolinesulfonamide (H-89), of PC12D pheochromocytoma cells. *J Biol Chem* 1990;**265**:5267–72.

49. Ishikawa T, Inagaki M, Watanabe M, Hidaka H. Relaxation of vascular smooth muscle by HA-1004, an inhibitor of cyclic nucleotide-dependent protein kinase. *J Pharmacol Exp Ther* 1985;**235**:495–9.

50. Kase H, Iwahashi K, Nakanishi S, Matsuda Y, Yamada K, Takahashi M, Murakata C, Sato A, Kaneko M. K-252 compounds, novel and potent inhibitors of protein kinase C and cyclic nucleotide-dependent protein kinases. *Biochem Biophys Res Commun* 1987;**142**:436–40.

51. Lu ZX, Quazi NH, Deady LW, Polya GM. Selective inhibition of cyclic AMP-dependent protein kinase by isoquinoline derivatives. *Biol Chem Hoppe Seyler* 1996;**377**:373–84.

52. Burkhardt M, Glazova M, Gambaryan S, Vollkommer T, Butt E, Bader B, Heermeier K, Lincoln TM, Walter U, Palmetshofer A. KT5823 inhibits cGMP-dependent protein kinase activity in vitro but not in intact human platelets and rat mesangial cells. *J Biol Chem* 2000;**275**:33,536–33,541.

53. Komalavilas P, Shah PK, Jo H, Lincoln TM. Activation of mitogen-activated protein kinase pathways by cyclic GMP and cyclic GMP-dependent protein kinase in contractile vascular smooth muscle cells. *J Biol Chem* 1999;**274**:34,301–34,309.

54. Wyatt TA, Pryzwansky KB, Lincoln TM. KT5823 activates human neutrophils and fails to inhibit cGMP-dependent protein kinase phosphorylation of vimentin. *Res Commun Chem Pathol Pharmacol* 1991;**74**:3–14.

55. Davies SP, Reddy H, Caivano M, Cohen P. Specificity and mechanism of action of some commonly used protein kinase inhibitors. *Biochem J* 2000;**351**:95–105.

56. Koide K, Bunnage ME, Gomez Paloma L, Kanter JR, Taylor SS, Brunton LL, Nicolaou KC. Molecular design and biological activity of potent and selective protein kinase inhibitors related to balanol. *Chem Biol* 1995;**2**:601–8.

57. Setyawan J, Koide K, Diller TC, Bunnage ME, Taylor SS, Nicolaou KC, Brunton LL. Inhibition of protein kinases by balanol: specificity within the serine/threonine protein kinase subfamily. *Mol Pharmacol* 1999;**56**:370–6.

58. Akamine P, Madhusudan., Brunton LL, Ou HD, Canaves JM, Xuong NH, Taylor SS. Balanol analogues probe specificity determinants and the conformational malleability of the cyclic 3′,5′-adenosine monophosphate-dependent protein kinase catalytic subunit. *Biochemistry* 2004;**43**:85–96.

59. Narayana N, Diller TC, Koide K, Bunnage ME, Nicolaou KC, Brunton LL, Xuong NH, Ten Eyck LF, Taylor SS. Crystal structure of the potent natural product inhibitor balanol in complex with the catalytic subunit of cAMP-dependent protein kinase.. *Biochemistry* 1999;**38**:2367–76.

60. Pande V, Ramos MJ, Gago F. The protein kinase inhibitor balanol: structure–activity relationships and structure-based computational studies. *Anticancer Agents Med Chem* 2008;**8**:638–45.

61. Wong CF, Hunenberger PH, Akamine P, Narayana N, Diller T, McCammon JA, Taylor S, Xuong NH. Computational analysis of PKA-balanol interactions. *J Med Chem* 2001;**44**:1530–9.

62. Wong CF, Kua J, Zhang Y, Straatsma TP, McCammon JA. Molecular docking of balanol to dynamics snapshots of protein kinase A. *Proteins* 2005;**61**:850–8.

63. Enkvist E, Lavogina D, Raidaru G, Vaasa A, Viil I, Lust M, Viht K, Uri A. Conjugation of adenosine and hexa-(D-arginine) leads to a nanomolar bisubstrate-analog inhibitor of basophilic protein kinases. *J Med Chem* 2006;**49**:7150–9.

64. Enkvist E, Raidaru G, Vaasa A, Pehk T, Lavogina D, Uri A. Carbocyclic 3′-deoxyadenosine-based highly potent bisubstrate-analog inhibitor of basophilic protein kinases.. *Bioorg Med Chem Lett* 2007;**17**:5336–9.

65. Kuznetsov A, Uri A, Raidaru G, Jarv J. Kinetic analysis of inhibition of cAMP-dependent protein kinase catalytic subunit by the peptide–nucleoside conjugate AdcAhxArg6. *Bioorg Chem* 2004;**32**:527–35.

66. Loog M, Uri A, Raidaru G, Jarv J, Ek P. Adenosine-5′-carboxylic acid peptidyl derivatives as inhibitors of protein kinases.. *Bioorg Med Chem Lett* 1999;**9**:1447–52.

67. Ricouart A, Gesquiere JC, Tartar A, Sergheraert C. Design of potent protein kinase inhibitors using the bisubstrate approach. *J Med Chem* 1991;**34**:73–8.

68. Viht K, Schweinsberg S, Lust M, Vaasa A, Raidaru G, Lavogina D, Uri A, Herberg FW. Surface-plasmon-resonance-based biosensor with immobilized bisubstrate analog inhibitor for the determination of affinities of ATP- and protein-competitive ligands of cAMP-dependent protein kinase. *Anal Biochem* 2007;**362**:268–77.

69. Raagel H, Lust M, Uri A, Pooga M. Adenosine-oligoarginine conjugate, a novel bisubstrate inhibitor, effectively dissociates the actin cytoskeleton. *FEBS J* 2008;**275**:3608–24.

70. Uri A, Raidaru G, Subbi J, Padari K, Pooga M. Identification of the ability of highly charged nanomolar inhibitors of protein kinases to cross plasma membranes and carry a protein into cells.. *Bioorg Med Chem Lett* 2002;**12**:2117–20.

71. Viht K, Padari K, Raidaru G, Subbi J, Tammiste I, Pooga M, Uri A. Liquid-phase synthesis of a pegylated adenosine-oligoarginine conjugate, cell-permeable inhibitor of cAMP-dependent protein kinase.. *Bioorg Med Chem Lett* 2003;**13**:3035–9.

72. Sheinerman FB, Giraud E, Laoui A. High affinity targets of protein kinase inhibitors have similar residues at the positions energetically important for binding. *J Mol Biol* 2005;**352**:1134–56.

73. Kemp BE, Graves DJ, Benjamini E, Krebs EG. Role of multiple basic residues in determining the substrate specificity of cyclic AMP-dependent protein kinase. *J Biol Chem* 1977;**252**:4888–94.

74. Scott JD, Fischer EH, Demaille JG, Krebs EG. Identification of an inhibitory region of the heat-stable protein inhibitor of the cAMP-dependent protein kinase. *Proc Natl Acad Sci USA* 1985;**82**:4379–83.

75. Scott JD, Fischer EH, Takio K, Demaille JG, Krebs EG. Amino acid sequence of the heat-stable inhibitor of the cAMP-dependent protein kinase from rabbit skeletal muscle. *Proc Natl Acad Sci USA* 1985;**82**:5732–6.

76. Bhatnagar Jr. D, Glass DB, Roskoski R, Lessor RA, Leonard NJ Synthetic peptide analogues differentially alter the binding affinities of cyclic nucleotide dependent protein kinases for nucleotide substrates. *Biochemistry* 1988;**27**:1988–94.

77. Glass DB, Cheng HC, Mende-Mueller L, Reed J, Walsh DA. Primary structural determinants essential for potent inhibition of cAMP-dependent protein kinase by inhibitory peptides corresponding to the active portion of the heat-stable inhibitor protein. *J Biol Chem* 1989;**264**:8802–10.

78. Glass DB, Lundquist LJ, Katz BM, Walsh DA. Protein kinase inhibitor-(6-22)-amide peptide analogs with standard and nonstandard amino acid substitutions for phenylalanine Inhibition of cAMP-dependent protein kinase. *J Biol Chem* 1989;**264**:14,579–14,584.

79. Kemp BE, Pearson RB. Intrasteric regulation of protein kinases and phosphatases. *Biochim Biophys Acta* 1991;**1094**:67–76.

80. Kemp BE, Pearson RB, House CM. Pseudosubstrate-based peptide inhibitors. *Methods Enzymol* 1991;**201**:287–304.

81. Scott JD, Glaccum MB, Fischer EH, Krebs EG. Primary-structure requirements for inhibition by the heat-stable inhibitor of the cAMP-dependent protein kinase. *Proc Natl Acad Sci USA* 1986;**83**:1613–16.

82. Kemp BE, Parker MW, Hu S, Tiganis T, House C. Substrate and pseudosubstrate interactions with protein kinases: determinants of specificity. *Trends Biochem Sci* 1994;**19**:440–4.

83. Pearson RB, Kemp BE. Protein kinase phosphorylation site sequences and consensus specificity motifs: tabulations. *Methods Enzymol* 1991;**200**:62–81.

84. Knighton DR, Zheng JH, Ten Eyck LF, Xuong NH, Taylor SS, Sowadski JM. Structure of a peptide inhibitor bound to the catalytic subunit of cyclic adenosine monophosphate-dependent protein kinase. *Science* 1991;**253**:414–20.

85. Gibson RM, Ji-Buechler Y, Taylor SS. Identification of electrostatic interaction sites between the regulatory and catalytic subunits of cyclic AMP-dependent protein kinase. *Protein Sci* 1997;**6**:1825–34.

86. Kumar P, Van Patten SM, Walsh DA. Multiplicity of the beta form of the cAMP-dependent protein kinase inhibitor protein generated by post-translational modification and alternate translational initiation. *J Biol Chem* 1997;**272**:20,011–20,020.

87. Kumar P, Walsh DA. A dual-specificity isoform of the protein kinase inhibitor PKI produced by alternate gene splicing. *Biochem J* 2002;**362**:533–7.

88. Van Patten SM, Ng DC, Th'ng JP, Angelos KL, Smith AJ, Walsh DA. Molecular cloning of a rat testis form of the inhibitor protein of cAMP-dependent protein kinase. *Proc Natl Acad Sci USA* 1991;**88**:5383–7.

89. Wadia JS, Dowdy SF. Protein transduction technology. *Curr Opin Biotechnol* 2002;**13**:52–6.

90. Glass DB. Differential responses of cyclic GMP-dependent and cyclic AMP-dependent protein kinases to synthetic peptide inhibitors. *Biochem J* 1983;**213**:159–64.

91. Glass DB. Substrate specificity of the cyclic GMP-dependent protein kinase. In: Kemp BE, editor. *Peptides and protein phosphorylation.* Boca Raton, FL: CRC Press, Inc; 1990. p. 209–38.

92. Glass DB, Cheng HC, Kemp BE, Walsh DA. Differential and common recognition of the catalytic sites of the cGMP-dependent and cAMP-dependent protein kinases by inhibitory peptides derived from the heat-stable inhibitor protein. *J Biol Chem* 1986;**261**(12):166–71.

93. Glass DB, Krebs EG. Comparison of the substrate specificity of adenosine 3′,5′-monophosphate- and guanosine 3′,5′-monophosphate-dependent protein kinases. Kinetic studies using synthetic peptides corresponding to phosphorylation sites in histone H2B. *J Biol Chem* 1979;**254**:9728–38.

94. Glass DB, Smith SB. Phosphorylation by cyclic GMP-dependent protein kinase of a synthetic peptide corresponding to the autophosphorylation site in the enzyme. *J Biol Chem* 1983;**258**(14):797–803.

95. Mitchell RD, Glass DB, Wong CW, Angelos KL, Walsh DA. Heat-stable inhibitor protein derived peptide substrate analogs: phosphorylation by cAMP-dependent and cGMP-dependent protein kinases.. *Biochemistry* 1995;**34**:528–34.

96. Poteet-Smith CE, Corbin JD, Francis SH. The pseudosubstrate sequences alone are not sufficient for potent autoinhibition of cAMP- and cGMP-dependent protein kinases as determined by synthetic peptide analysis. *Adv Second Messenger Phosphoprotein Res* 1997;**31**:219–35.

97. Zeilig CE, Langan TA, Glass DB. Sites in histone H1 selectively phosphorylated by guanosine 3′,5′-monophosphate-dependent protein kinase. *J Biol Chem* 1981;**256**:994–1001.

98. Butt E, Abel K, Krieger M, Palm D, Hoppe V, Hoppe J, Walter U. cAMP- and cGMP-dependent protein kinase phosphorylation sites of the focal adhesion vasodilator-stimulated phosphoprotein (VASP) in vitro and in intact human platelets. *J Biol Chem* 1994;**269**:14,509–14,517.

99. Lev-Ram V, Jiang T, Wood J, Lawrence DS, Tsien RY. Synergies and coincidence requirements between NO, cGMP, and Ca^{2+} in the induction of cerebellar long-term depression.. *Neuron* 1997;**18**:1025–38.

100. Werner DS, Lee TR, Lawrence DS. Is protein kinase substrate efficacy a reliable barometer for successful inhibitor design? *J Biol Chem* 1996;**271**:180–5.

101. Wood JS, Yan X, Mendelow M, Corbin JD, Francis SH, Lawrence DS. Precision substrate targeting of protein kinases. The cGMP- and cAMP-dependent protein kinases. *J Biol Chem* 1996;**271**:174–9.

102. Yan X, Corbin JD, Francis SH, Lawrence DS. Precision targeting of protein kinases. An affinity label that inactivates the cGMP– but not the cAMP-dependent protein kinase. *J Biol Chem* 1996;**271**:1845–8.

103. Gustafsson AB, Brunton LL. Differential and selective inhibition of protein kinase A and protein kinase C in intact cells by balanol congeners. *Mol Pharmacol* 1999;**56**:377–82.

104. Hunenberger PH, Helms V, Narayana N, Taylor SS, McCammon JA. Determinants of ligand binding to cAMP-dependent protein kinase. *Biochemistry* 1999;**38**:2358–66.

105. Dostmann WR, Tegge W, Frank R, Nickl CK, Taylor MS, Brayden JE. Exploring the mechanisms of vascular smooth muscle tone with highly specific, membrane-permeable inhibitors of cyclic GMP-dependent protein kinase Ialpha. *Pharmacol Ther* 2002;**93**:203–15.

106. Bove PF, Wesley UV, Greul AK, Hristova M, Dostmann WR, van der Vliet A. Nitric oxide promotes airway epithelial wound repair through enhanced activation of MMP-9. *Am J Respir Cell Mol Biol* 2007;**36**:138–46.

107. Dey NB, Foley KF, Lincoln TM, Dostmann WR. Inhibition of cGMP-dependent protein kinase reverses phenotypic modulation of vascular smooth muscle cells. *J Cardiovasc Pharmacol* 2005;**45**:404–13.

108. Honda A, Moosmeier MA, Dostmann WR. Membrane-permeable cygnets: rapid cellular internalization of fluorescent cGMP-indicators. *Front Biosci* 2005;**10**:1290–301.

109. Hou Y, Wong E, Martin J, Schoenlein PV, Dostmann WR, Browning DD. A role for cyclic-GMP dependent protein kinase in anoikis. *Cell Signal* 2006;**18**:882–8.

110. Krieg T, Philipp S, Cui L, Dostmann WR, Downey JM, Cohen MV. Peptide blockers of PKG inhibit ROS generation by acetylcholine and bradykinin in cardiomyocytes but fail to block protection in the whole heart.. *Am J Physiol Heart Circ Physiol* 2005;**288**:H1976–81.

111. Zhu CB, Carneiro AM, Dostmann WR, Hewlett WA, Blakely RD. p38 MAPK activation elevates serotonin transport activity via a trafficking-independent, protein phosphatase 2A-dependent process. *J Biol Chem* 2005;**280**:15,649–15,658.

112. Zhang R, Mamai A, Flemer S, Natarajan A, Madalengoitia JS, Nickl CK, Dostmann WR. Poly-L-proline Type II peptide ,imics as probes of the active site occupancy requirements of cGMP dependent protein kinase. *J Peptide Res* 2006;**66**:151–9.

113. Van Haastert PJ, Van Driel R, Jastorff B, Baraniak J, Stec WJ, De Wit RJ. Competitive cAMP antagonists for cAMP-receptor proteins. *J Biol Chem* 1984;**259**:10,020–10,024.

114. Schaap P, van Ments-Cohen M, Soede RD, Brandt R, Firtel RA, Dostmann W, Genieser HG, Jastorff B, van Haastert PJ. Cell-permeable non-hydrolyzable cAMP derivatives as tools for analysis of signaling pathways controlling gene regulation in *Dictyostelium*. *J Biol Chem* 1993;**268**:6323–31.

115. Constantinescu A, Gordon AS, Diamond I. cAMP-dependent protein kinase types I and II differentially regulate cAMP response element-mediated gene expression: implications for neuronal responses to ethanol. *J Biol Chem* 2002;**277**:18,810–18,816.

116. Wei JY, Cohen ED, Genieser HG, Barnstable CJ. Substituted cGMP analogs can act as selective agonists of the rod photoreceptor cGMP-gated cation channel. *J Mol Neurosci* 1998;**10**:53–64.

117. Nakano H, Kobayashi E, Takahashi I, Tamaoki T, Kuzuu Y, Iba H. Staurosporine inhibits tyrosine-specific protein kinase activity of Rous sarcoma virus transforming protein p60. *J Antibiot (Tokyo)* 1987;**40**:706–8.

118. Ruegg UT, Burgess GM. Staurosporine, K-252 and UCN-01: potent but nonspecific inhibitors of protein kinases. *Trends Pharmacol Sci* 1989;**10**:218–20.

119. Tamaoki T, Nomoto H, Takahashi I, Kato Y, Morimoto M, Tomita F. Staurosporine, a potent inhibitor of phospholipid/Ca++ dependent protein kinase. *Biochem Biophys Res Commun* 1986;**135**:397–402.

120. Cheng HC, Kemp BE, Pearson RB, Smith AJ, Misconi L, Van Patten SM, Walsh DA. A potent synthetic peptide inhibitor of the cAMP-dependent protein kinase. *J Biol Chem* 1986;**261**:989–92.

Chapter 44

AKAP Transduction Units: Context dependent Assembly of Signaling Complexes

John D. Scott and Lorene K. Langeberg

Howard Hughes Medical Institute, University of Washington, Seattle, Washington

INTRODUCTION

The dissemination of cellular signals often involves the positioning of signaling proteins in proximity to their upstream activators and downstream targets. In fact, the clustering of receptors, G proteins and enzymes with their substrates is believed to contribute significantly to the specificity of signaling. This sophisticated degree of organization may also help to prevent the indiscriminate activation of related signaling complexes that are close by. This is of particular importance for second messenger-dependent signaling pathways that lead to the activation of broad specificity enzymes such as the cAMP dependent protein kinase (PKA), protein kinase C (PKC), and a variety of calmodulin-dependent kinases (Cam kinases). Compartmentalization of these enzymes is often achieved through their association with scaffolding proteins that simultaneously coordinate the location of several enzymes [1–5].

A-Kinase Anchoring Proteins (AKAPs) are a family of scaffolding proteins which package PKA and other signaling enzymes into multiprotein complexes [6, 7]. As discussed in Chapter 165 of Handbook of Cell Signaling, Second Edition, each AKAP contains both a conserved amphipathic helix that binds to the R subunit dimer with high affinity and a targeting domain that directs the PKA–AKAP complex to specific subcellular compartments [8–11]. A now well-defined biological role for AKAPs is to place PKA in the proximity of enzymes such as phosphatases and phosphodiesterases that terminate cAMP signaling events [12–17]. The focus of this chapter is to highlight advances in our understanding of AKAP signaling complexes and their role in facilitating this bidirectional control of various signaling events.

G-PROTEIN SIGNALING THROUGH AKAP SIGNALING COMPLEXES

A shared property of several AKAPs is to position enzymes in microenvironments where they can respond to upstream signals. Clearly, there are potential advantages of anchoring PKA in proximity to primary transduction elements such as G-protein-coupled receptors and the cAMP synthesis machinery. Two anchoring proteins, gravin/AKAP250 and AKAP79/150, maintain kinase complexes that bind to the β2-adrenergic (β2-AR) receptor [18, 19]. The AKAP79 complex binds to regions within the third cytoplasmic loop and C-terminal tail of the β2-AR in an agonist-independent manner, whereas gravin/AKAP250 recruits PKA and PKC to the receptor in an agonist-dependent manner [19–21] (Figure 44.1). These receptor-based AKAP complexes also contribute to β2-AR phosphorylation, desensitization, and indirect activation of MAP kinase pathways that emanate from the receptor [22]. Furthermore, dephosphorylation of β2-AR and the receptor kinase GRK2 are likely to be important signal termination events in this process and could be mediated by an anchored pool of PP-2B that associates with AKAP79. Another anchoring protein, MAP2, seems to nucleate a membrane-associated signaling complex that includes β2-AR, adenylyl cyclase, PKA, PP-2B, and a substrate for the kinase the class C L-type Ca^{2+} channel [23, 24]. The identification of such a signaling complex emphasizes the notion that receptors, effectors, kinases and their substrates are spatially coordinated. However, it also confirms the view put forward by a number of investigators that in some cases cAMP may not have to diffuse very far from its site of synthesis to activate the PKA holoenzyme.

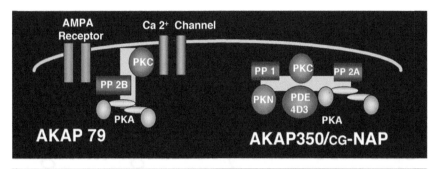

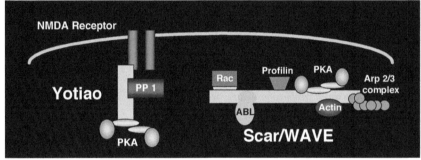

FIGURE 44.1 AKAP signaling complexes: a schematic representation of certain AKAP signaling complexes discussed in this article. Each interacting protein is labeled.

This latter notion is further supported by two recent publications. Willoughby and colleagues used a combination of co-immunoprecipitation and live cell imaging techniques to demonstrate that gravin maintains a signaling complex that includes protein kinase A (PKA) and the type 4 phosphodiesterase PDE4D3 or PDE4D5 [25]. These authors propose that gravin-associated PDE4s provide a means to rapidly terminate cAMP signals just below the plasma membrane with concomitant effects on localized ion channel, transporter, or enzyme activities. Bauman and colleagues have recently described an interaction between AKAP79/150 and adenylyl cyclases V and VI [26]. PKA anchoring facilitates the preferential phosphorylation of adenylyl cyclase to inhibit cAMP synthesis. Using real-time cellular imaging experiments, they show that anchoring of PKA with the cAMP synthesis machinery ensures rapid termination of cAMP signaling upon activation of the kinase [26]. These findings suggest that such a protein configuration permits the formation of a negative feedback loop to temporally regulate cAMP production at the plasma membranes.

Other classes of G proteins have been implicated in the channeling of signals through AKAP complexes, although not necessarily via cAMP-dependent mechanisms [4]. Scar/WAVE-1 is a member of the Wiskott-Aldrich syndrome family of scaffolding proteins that binds PKA, the Abl tyrosine kinase, and the Arp2/3 complex, a group of seven proteins that control actin remodeling [27–29] (Figure 44.1). The dynamic assembly of this complex at sites of lamellapodial extension occurs in response to growth factor signals that activate the low molecular weight GTPase Rac

[27]. Consequently, Scar/WAVE may direct PKA and Abl toward cytoskeletal substrates, and synchronize cell movement by ensuring efficient transmission of Rac-mediated signals to the actin remodeling machinery. Analogous AKAP signaling networks participate in the formation of actin stress fibers. AKAP-Lbc, a splice variant of the Lbc oncogene, encodes a chimeric molecule that anchors PKA and functions as a Rho-selective guanine nucleotide exchange factor [30]. Application of lysophosphatydic acid or selective expression of Gα12 enhances cellular AKAP-Lbc activation and leads to the formation of actin stress fibers in fibroblasts [30]. This provides an example where the spatial organization of heterotrimeric and small molecular weight G proteins may involve interactions with the same AKAP. However, AKAP–Lbc complexes may participate in the regulation of other signaling processes. Work published concurrently by the Diviani and the Pawson groups have shown that PKA phosphorylation of serine 1565 on AKAP–Lbc generates a binding site for 14–3–3 proteins that suppress AKAP–Lbc Rho GEF activity [31, 32]. Other work has shown that the synergistic actions anchored PKA and PKC also contribute to PKD activation. PKC phosphorylation primes this kinase for activation, whereas PKA phosphorylation of serine 2737 on AKAP–Lbc releases the PKD from the activation complex [33]. The above examples of highly localized phosphorylation events that occur within AKAP signaling complexes highlight the utility of kinase anchoring as a means to restrict the substrate accessibility of broad spectrum enzymes such as PKA, PKC, and PKD.

KINASE/PHOSPHATASE SIGNALING COMPLEXES

Several AKAP signaling complexes include both signal-transduction and signal-termination enzymes. This generates a locus to regulate the forward and backward steps of a given signaling process. One example of this is the clustering of second-messenger regulated kinases and phosphatases at the excitatory synapses of neurons by the AKAP79/150 family of anchoring proteins [34].

The human form, AKAP79, and its bovine and murine counterparts, AKAP75 and AKAP150 respectively, are enriched in the synaptosomal and postsynaptic density fractions of neuronal lysates, and are present in dendritic spines [35, 36]. In 1995, the A subunit of the calcium/calmodulin-dependent phosphatase PP2B was identified in a two-hybrid screen using AKAP79 as bait [37]. Subsequent biochemical and cellular analyses defined the phosphatase binding site and demonstrated that both enzymes simultaneously associate with AKAP79/150 in neurons [37]. A year later, it was demonstrated that PKC is also a component of the AKAP79 signaling complex [38]. At that time it was postulated that the simultaneous anchoring of these three signaling enzymes generated a locus for the integration of distinct second-messenger signals at the postsynaptic membranes [39]. Functional studies have largely confirmed this notion showing that the AKAP79/150 signaling complex controls the phosphorylation status and facilitates the regulation of a variety of ion channels, including L-type calcium channels, KCNQ potassium channels, aquiporin water channels, and AMPA-type glutamate receptor ion channels [40–43] (Figure 44.1).

The most detailed studies have dissected the phosphorylation events that occur on the cytoplasmic tail of AMPA-type glutamate receptors. This channel is present at the terminals of excitatory synapses, and is gated by the release of glutamate across the synaptic cleft [44, 45]. A series of reports has shown that the AKAP79 signaling complex is recruited into a larger transduction unit with the GluR1 subunit of the AMPA type glutamate receptor (reviewed by Dodge and Scott [34]). Simultaneous association with the membrane-associated guanylate kinase bridging protein SAP97 brings the channel and the signaling complex together [46]. Functional studies indicate that AKAP79-bound PKA enhances GluR1 phosphorylation on serine 845 in the cytoplasmic tail of the channel subunit, an important site for the regulation of channel function during the induction of long-term synaptic depression (LTD) [47–49]. These findings extend an earlier report showing that perfusion of anchoring inhibitor peptides into cultured hippocampal neurons antagonizes PKA anchoring and causes rundown of synaptic AMPA-type glutamate receptor activity [50]. Since this phenomenon occurs with a time-course that is similar to the inhibition of the kinase, it was initially assumed that disruption of PKA anchoring displaced the kinase from the proximity the AMPA receptor. However, more recent

studies indicate that the phosphatase PP2B may play a prominent role in the downregulation of channel activity. Electrophysiological recordings suggest that the proximity of the AKAP79-bound phosphatase to sites of calcium entry ensures that the enzyme is rapidly activated upon synaptic elevation of intracellular calcium, and responsible for the dephosphorylation of serine 845. Interestingly, serine 845 is phosphorylated by PKA upon elevation of synaptic cAMP levels [51, 52]. Thus AKAP79/150 maintains a kinases and a phosphatase in close proximity to the channel in a manner that allows second-messenger dependent changes in the phosphorylation status and activity of GluR1 (Figure 44.1).

Other synaptic AKAPs also maintain kinase/phosphatase complexes. For example, yotiao interacts with the NR1a subunit of synaptic NMDA glutamate receptor ion channels and anchors PKA and protein phosphatase 1 (PP-1) [13, 53–56]. The modulation of NMDA receptors containing NR1a requires interactions with the scaffolding protein, as peptide-mediated displacement of either PKA or PP-1 causes changes in the modulation of channel activity [13]. Thus, yotiao maintains a signaling complex that is directly attached to the substrate, the NMDA receptor (Figure 44.1). While this example certainly highlights the role of AKAPs to ensure the rapid preferential phosphorylation of substrates, the compartmentalization of enzymes in the yotiao complex may also contribute to the segregation of signals at excitatory synapses where the GluR1/AKAP79 complex is in the immediate vicinity.

Two AKAP additional signaling complexes are found at the centrosome. AKAP350/CG-NAP, a large centrosomal AKAP of unknown function, has been reported to bind three kinases (PKA, PKC, and PKN) and two phosphatases (PP-1 and PP2A; Figure 44.1) [28, 57–59]. Likewise pericentrin, an integral component of the centreolar machinery, anchors PKA and other enzymes, presumably for a role in the coordination of centrosomal phosphorylation events. Interestingly, both AKAP350.CG-NAP and pericentrin contain a C-terminal PACT domain that is responsible for targeting each anchoring protein to the centrosome. Expression of this 100 amino acid region alone is sufficient to promote centrosomal targeting of GFP [60]. This raises the intriguing possibility that the PACT domains of these anchoring proteins might interact with the same structure in the centrosome in a mutually exclusive manner. This could represent one mechanism to generate greater diversity, as distinct signaling complexes could be tethered to the same cellular locus. Thus, the possibilities for coordinated phosphorylation and dephosphorylation events mediated by association with AKAPs are increased.

CAMP SIGNALING UNITS

Another way to exert tight control of PKA phosphorylation events is to compartmentalize the kinase with enzymes that control the intracellular concentrations of its activator, cAMP. In fact, two recent reports have demonstrated that phosphodiesterases, the enzymes that catalyze cAMP degradation, are components of AKAP–PKA signaling complexes [15, 16]. These findings add to the complexity of cAMP signaling, as they point toward a role for anchored pools of phosphodiesterase in the tight control of local second-messenger concentrations. This in turn controls where and when PKA becomes active. For example, a muscle-selective anchoring protein mAKAP directly binds PKA and a splice variant of the cAMP-specific type 4 phosphodiesterase PDE4D3, and targets them to the perinuclear membranes of cardiomyocytes [15]. Yet in Sertoli cells the PDE4D3 interacts with AKAP350/CG-NAP, one of the large centrosomal AKAPs discussed above (Figure 44.1) [16].

Two important regulatory factors that are built into these cAMP-signaling modules favor the signal termination process. First, the tethered PDE is constitutively active and will rapidly restore basal cAMP levels when the flow of second messenger is turned off from its site of synthesis at the plasma membrane. Secondly, elegant experiments have demonstrated that PKA phosphorylation of PDE4D3 on serine 54 increases the Vmax of the enzyme two- to three-fold over basal conditions [61–65]. Phosphorylation of PDE4D3 increases cAMP degradation to favor reformation of the PKA holoenzyme. PKA anchoring is a unique and critical element in this PKA–PDE4D3 feedback loop, as displacement of the kinase with the anchoring inhibitor peptide Ht31 prevents cAMP-dependent stimulation of the mAKAP associated PDE4D activity [15]. This emphasizes the importance of PDE localization to maintain the balance of intracellular cAMP levels. This notion is also supported by recent imaging studies using intermolecular FRET which have shown that micro-gradients of cAMP emanate from sites of synthesis at the plasma membrane. Hormonal stimulation of cardiomyocytes induced changes in the rate and magnitude of local cAMP gradients with the concomitant effect on the activation of anchored PKA pools [66]. Thus, multiple regulatory processes are involved in controlling where and when cellular PKA activation occurs.

Although PDE4D3 is a substrate for the kinase, it is clear that there are other PKA substrates associated with the mAKAP scaffold. For example, the regulation of ryanodine receptor (RyR) phosphorylation is important for maintaining contractility in response to β-adrenergic signaling and increases in intracellular Ca^{2+} concentration in the heart. Hyperphosphorylation of sarcoplasmic reticulum RyR leads to increased Ca^{2+} sensitivity of the channel and decreased sensitivity to β-adrenergic stimulation [67–69]. These changes are manifest in human heart tissue undergoing heart failure where changes in RyR phosphorylation are also detected (67). Atypical regulation of RyR function may be due to several factors that regulate cAMP/PKA signaling in heart, including loss of phosphatase activity from the RyR complex [67] and defects in regulation of cAMP

levels by PDE activity associated with the complex [15]. Interestingly two groups have detected PP1 and PP2A phosphatase subunits in the mAKAP signaling complex. Given the myriad of binding partners for mAKAP, it is plausible to suggest that the composition of this signaling network may be altered in response to different intracellular stimuli and in disease states.

CONCLUSIONS AND PERSPECTIVES

AKAPs provide the platforms for the assembly of multi-protein signaling complexes in a variety of cellular compartments. Two factors contribute to the diversity of these signaling units. First, individual AKAP complexes may control distinct signaling events within the same subcellular compartment. This may be best exemplified by WAVE-1 and AKAP–Lbc, which nucleate the formation intracellular signaling cascades to catalyze distinct forms of cytoskeletal reorganization. In both cases, receptor occupancy at the plasma membrane triggers the assembly of signaling complexes that transmit distinct signals to the actin cytoskeleton [27,30]. Likewise, AKAP350 and pericentrin may synchronize different phosphorylation events at the centrosome, and three anchoring proteins, D-AKAP-1/sAKAP82, D-AKAP-2, and Rab32, are involved in mitochondrial signaling processes [70–72]. Secondly, the recruitment and release of individual enzymes from the signaling complex provides a dynamic component to the composition of a given protein network. For example, Ca^{2+}/calmodulin antagonizes PKC anchoring by competing for binding to AKAP79. The calcium influx from ion channels within the synaptic membrane releases PKC from its anchor, changing the activity status of this kinase. Presumably the soluble enzyme is more available to propagate calcium/phospholipid signaling events as it has a less restricted access to its substrates. An additional level of complexity may be present, as biochemical and proteomic experiments have detected each PKC isoform in AKAP79/150 complexes isolated from rat brain [73,74]. This implies that at these synapses a variety of AKAP79/150 signaling complexes exist which contain conventional, novel or atypical PKC isozymes. Again this adds to the diversity of AKAP signaling, as each PKC class responds to different combinations of phospholipid activator. These examples not only highlight the sophisticated degree of spatial organization achieved by anchoring proteins, but also emphasize the degree of specificity that can be generated through the combinatorial assembly of unique AKAP signaling complexes.

ACKNOWLEDGEMENTS

John D. Scott and Lorene K. Langeberg are supported in part by NIH grant GM48231.

REFERENCES

1. Pawson T, Scott JD. Signaling through scaffold, anchoring, and adaptor proteins. *Science* 1997;**278**:2075.

2. Jordan JD, Landau EM, Iyengar R. Signaling networks: the origins of cellular multitasking. *Cell* 2000;**103**:193.

3. Hunter T. Signaling–2000 and beyond. *Cell* 2000;**100**:113.

4. Bauman AL, Scott JD. Kinase- and phosphatase-anchoring proteins: harnessing the dynamic duo. *Nat Cell Biol* 2002;**4**:E203.

5. Smith FD, Langeberg LK, Scott JD. The wheres and whens of kinase anchoring. *Trends Biochem Sci* 2006;**31**:316.

6. Wong W, Scott JD. AKAP Signalling complexes: focal points in space and time. *Nat Rev Mol Cell Biol* 2004;**5**:959.

7. Tasken K, Aandahl EM. Localized effects of cAMP mediated by distinct routes of protein kinase A. *Physiol Rev* 2004;**84**:137.

8. Carr DW, Stofko-Hahn RE, Fraser C, Bishop SM, Acott TS, Brennan RG, Scott JD. Interaction of the regulatory subunit (RII) of cAMP-dependent protein kinase with RII-anchoring proteins occurs through an amphipathic helix binding motif. *J Biol Chem* 1991;**266**:14,188.

9. Newlon MG, Roy M, Morikis D, Carr DW, Westphal R, Scott JD, Jennings PA. A novel mechanism of PKA anchoring revealed by solution structures of anchoring complexes. *EMBO J* 2001;**20**:1651.

10. Alto NM, Soderling SH, Hoshi N, Langeberg LK, Fayos R, Jennings PA, Scott JD. Bioinformatic design of A-kinase anchoring protein in silico: a potent and selective peptide antagonist of type II protein kinase A anchoring. *Proc Natl Acad Sci USA* 2003;**100**:4445.

11. Gold MG, Lygren B, Dokurno P, Hoshi N, McConnachie G, Tasken K, Carlson CR, Scott JD, Barford D. Molecular basis of AKAP specificity for PKA regulatory subunits. *Mol Cell* 2006;**24**:383.

12. Smith FD, Scott JD. Signaling complexes: junctions on the intracellular information super highway. *Curr Biol* 2002;**12**:R32.

13. Westphal RS, Tavalin SJ, Lin JW, Alto NM, Fraser ID, Langeberg LK, Sheng M, Scott JD. Regulation of NMDA receptors by an associated phosphatase-kinase signaling complex. *Science* 1999;**285**:93.

14. Tavalin SJ, Colledge M, Hell JW, Langeberg LK, Huganir RL, Scott JD. Regulation of GluR1 by the A-kinase anchoring protein 79 (AKAP79) signaling complex shares properties with long-term depression. *J Neurosci* 2002;**22**:3044.

15. Dodge KL, Khouangsathiene S, Kapiloff MS, Mouton R, Hill EV, Houslay MD, Langeberg LK, Scott JD. mAKAP assembles a protein kinase A/PDE4 phosphodiesterase cAMP signaling module. *EMBO J* 2001;**20**:1921.

16. Tasken KA, Collas P, Kemmner WA, Witczak O, Conti M, Tasken K. Phosphodiesterase 4D and protein kinase a type II constitute a signaling unit in the centrosomal area. *J Biol Chem* 2001;**276**:21,999.

17. Dodge-Kafka KL, Soughayer J, Pare GC, Carlisle Michel JJ, Langeberg LK, Kapiloff MS, Scott JD. The protein kinase A anchoring protein mAKAP coordinates two integrated cAMP effector pathways. *Nature* 2005;**437**:574.

18. Shih M, Lin F, Scott JD, Wang HY, Malbon CC. Dynamic complexes of beta2-adrenergic receptors with protein kinases and phosphatases and the role of gravin. *J Biol Chem* 1999;**274**:1588.

19. Fraser ID, Cong M, Kim J, Rollins EN, Daaka Y, Lefkowitz RJ, Scott JD. Assembly of an A kinase-anchoring protein-beta(2)-adrenergic receptor complex facilitates receptor phosphorylation and signaling. *Curr Biol* 2000;**10**:409.

20. Lin F, Wang H, Malbon CC. Gravin-mediated formation of signaling complexes in beta 2-adrenergic receptor desensitization and resensitization. *J Biol Chem* 2000;**275**:19,025.

21. Ferguson SS. Evolving concepts in G protein-coupled receptor endocytosis: the role in receptor desensitization and signaling. *Pharmacol Rev* 2001;**53**:1.

22. Daaka Y, Luttrell LM, Lefkowitz RJ. Switching of the coupling of the beta2-adrenergic receptor to different G proteins by protein kinase A. *Nature* 1997;**390**:88.

23. Davare MA, Avdonin V, Hall DD, Peden EM, Burette A, Weinberg RJ, Horne MC, Hoshi T, Hell JW. A beta2 adrenergic receptor signaling complex assembled with the Ca^{2+} channel Cav1.2. *Science* 2001;**293**:98.

24. Davare MA, Dong F, Rubin CS, Hell JW. The A-kinase anchor protein MAP2B and cAMP-dependent protein kinase are associated with class C L-type calcium channels in neurons. *J Biol Chem* 1999;**274**:30,280.

25. Willoughby D, Wong W, Schaack J, Scott JD, Cooper DM. An anchored PKA and PDE4 complex regulates subplasmalemmal cAMP dynamics. *EMBO J* 2006;**25**:2051.

26. Bauman AL, Soughayer J, Nguyen BT, Willoughby D, Carnegie GK, Wong W, Hoshi N, Langeberg LK, Cooper DM, Dessauer CW, Scott JD. Dynamic regulation of cAMP synthesis through anchored PKA-adenylyl cyclase V/VI complexes. *Mol Cell* 2006;**23**:925.

27. Westphal RS, Soderling SH, Alto NM, Langeberg LK, Scott JD. Scar/WAVE-1, a Wiskott-Aldrich syndrome protein, assembles an actin-associated multi-kinase scaffold. *EMBO J* 2000;**19**:4589.

28. Diviani D, Scott JD. AKAP signaling complexes at the cytoskeleton. *J Cell Sci* 2001;**114**:1431.

29. Machesky LM, Insall RH. Scar1 and the related Wiskott-Aldrich syndrome protein, WASP, regulate the actin cytoskeleton through the Arp2/3 complex. *Curr Biol* 1998;**8**:1347.

30. Diviani D, Soderling J, Scott JD. AKAP-Lbc anchors protein kinase A and nucleates Galpha 12-selective Rho-mediated stress fiber formation. *J Biol Chem* 2001;**276**:44,247.

31. Diviani D, Abuin L, Cotecchia S, Pansier L. Anchoring of both PKA and 14–3–3 inhibits the Rho-GEF activity of the AKAP-Lbc signaling complex. *EMBO J* 2004;**23**:2811.

32. Jin J, Smith FD, Stark C, Wells CD, Fawcett JP, Kulkarni S, Metalnikov P, O'Donnell P, Taylor P, Taylor L, Zougman A, Woodgett JR, Langeberg LK, Scott JD, Pawson T. Proteomic, functional, and domain-based analysis of in vivo 14–3–3 binding proteins involved in cytoskeletal regulation and cellular organization. *Curr Biol* 2004;**14**:1436.

33. Carnegie GK, Smith FD, McConnachie G, Langeberg LK, Scott JD. AKAP-Lbc nucleates a protein kinase D activation scaffold. *Mol Cell* 2004;**15**:889.

34. Dodge K, Scott JD. AKAP79 and the evolution of the AKAP model. *FEBS Letts* 2000;**476**:58.

35. Carr DW, Stofko-Hahn RE, Fraser C, Cone RD, Scott JD. Localization of the cAMP-dependent protein kinase to the postsynaptic densities by A-kinase anchoring proteins: characterization of AKAP79. *J Biol Chem* 1992;**24**:16,816.

36. Sik A, Gulacsi A, Lai Y, Doyle WK, Pacia S, Mody I, Freund TF. Localization of the A kinase anchoring protein AKAP79 in the human hippocampus. *Eur J Neurosci* 2000;**12**:1155.

37. Coghlan VM, Perrino BA, Howard M, Langeberg LK, Hicks JB, Gallatin WM, Scott JD. Association of protein kinase A and protein phosphatase 2B with a common anchoring protein. *Science* 1995;**267**:108.

38. Klauck TM, Faux MC, Labudda K, Langeberg LK, Jaken S, Scott JD. Coordination of three signaling enzymes by AKAP79, a mammalian scaffold protein. *Science* 1996;**271**:1589.

39. Faux MC, Scott JD. Molecular glue: kinase anchoring and scaffold proteins. *Cell* 1996;**70**:8.

40. Fraser ID, Scott JD. Modulation of ion channels: a "current" view of AKAPs. *Neuron* 1999;**23**:423.

41. Gao T, Yatani A, Dell'Acqua ML, Sako H, Green SA, Dascal N, Scott JD, Hosey MM. cAMP-dependent regulation of cardiac L-type Ca^{2+} channels requires membrane targeting of PKA and phosphorylation of channel subunits. *Neuron* 1997;**19**:185.

42. Jo I, Ward DT, Baum MA, Scott JD, Coghlan VM, Hammond TG, Harris HW. AQP2 is a substrate for endogenous PP2B activity within an inner medullary AKAP-signaling complex. *Am J Physiol Renal Physiol* 2001;**281**:F958.

43. Potet F, Scott JD, Mohammad-Panah R, Escande D, Baro II. AKAP proteins anchor cAMP-dependent protein kinase to KvLQT1/IsK channel complex. *Am J Physiol Heart Circ Physiol* 2001;**280**:H2038.

44. Mayer ML, Westbrook GL. The physiology of excitatory amino acids in the vertebrate central nervous system. *Prog Neurobiol* 1987;**28**:197.

45. Jahr CE, Lester RAJ. Synaptic excitation mediated by glutamate-gated ion channels. *Curr Opin Neurobiol* 1992;**2**:395.

46. Colledge M, Dean RA, Scott GK, Langeberg LK, Huganir RL, Scott JD. Targeting of PKA to glutamate receptors through a MAGUK-AKAP complex. *Neuron* 2000;**27**:107.

47. Kameyama K, Lee HK, Bear MF, Huganir RL. Involvement of a post-synaptic protein kinase A substrate in the expression of homosynaptic long-term depression. *Neuron* 1998;**21**:1163.

48. Lee HK, Barbarosie M, Kameyama K, Bear MF, Huganir RL. Regulation of distinct AMPA receptor phosphorylation sites during bidirectional synaptic plasticity. *Nature* 2000;**405**:955.

49. Banke TG, Bowie D, Lee H, Huganir RL, Schousboe A, Traynelis SF. Control of GluR1 AMPA receptor function by cAMP-dependent protein kinase. *J Neurosci* 2000;**20**:89.

50. Rosenmund C, Carr DW, Bergeson SE, Nilaver G, Scott JD, Westbrook GL. Anchoring of protein kinase A is required for modulation of AMPA/kainate receptors on hippocampal neurons. *Nature* 1994;**368**:853.

51. Raymond LA, Blackstone CD, Huganir RL. Phosphorylation and modulation of recombinant GluR6 glutamate receptors by cAMP-dependent protein kinase. *Nature* 1993;**361**:637.

52. Swope SL, Moss SI, Raymond LA, Huganir RL. Regulation of ligand-gated ion channels by protein phosphorylation. *Adv Second Messenger Phosphoprotein Res* 1999;**33**:49.

53. Lin JW, Wyszynski M, Madhavan R, Sealock R, Kim JU, Sheng M. Yotiao, a novel protein of neuromuscular junction and brain that interacts with specific splice variants of NMDA receptor subunit NR1. *J Neurosci* 1998;**18**:2017.

54. Raman IM, Tong G, Jahr CE. β-adrenergic regulation of synaptic NMDA receptors by cAMP-dependent protein kinase. *Neuron* 1996;**16**:415.

55. Wang L-Y, Orser BA, Brautigan DL, Macdonald JF. Regulation of NMDA receptors in cultured hippocampal neurons by protein phosphatases 1 and 2A. *Nature* 1994;**369**:230.

56. Snyder GL, Fienberg AA, Huganir RL, Greengard P. A dopamine/D1 receptor/protein kinase A/dopamine- and cAMP-regulated phosphoprotein (Mr 32 kDa)/protein phosphatase-1 pathway regulates dephosphorylation of the NMDA receptor. *J Neurosci* 1998;**18**:10297.

57. Takahashi M, Mukai H, Oishi K, Isagawa T, Ono Y. Association of immature hypo-phosphorylated protein kinase C epsilon with an anchoring protein CG-NAP. *J Biol Chem* 2000.

58. Schmidt PH, Dransfield DT, Claudio JO, Hawley RG, Trotter KW, Milgram SL, Goldenring JR. AKAP350: a multiply spliced A-kinase anchoring protein associated with centrosomes. *J Biol Chem* 1999;**274**:3055.

59. Witczak O, Skalhegg BS, Keryer G, Bornens M, Tasken K, Jahnsen T, Orstavik S. Cloning and characterization of a cDNA encoding an A-kinase anchoring protein located in the centrosome, AKAP450. *EMBO J* 1999;**18**:1858.

60. Gillingham AK, Munro S. The PACT domain, a conserved centrosomal targeting motif in the coiled-coil proteins AKAP450 and pericentrin. *EMBO Rep* 2000;**1**:524.

61. Sette C, Conti M. Phosphorylation and activation of a cAMP-specific phosphodiesterase by the cAMP-dependent protein kinase. *J Biol Chem* 1996;**271**:16,526.

62. Lim J, Pahlke G, Conti M. Activation of the cAMP-specific phosphodiesterase PDE4D3 by phosphorylation. Identification and function of an inhibitory domain. *J Biol Chem* 1999;**274**:19,677.

63. Oki N, Takahashi SI, Hidaka H, Conti M. Short term feedback regulation of cAMP in FRTL-5 thyroid cells. Role of PDE4D3 phosphodiesterase activation. *J Biol Chem* 2000;**275**:10831.

64. Conti M. Phosphodiesterases and cyclic nucleotide signaling in endocrine cells. *Mol Endocrinol* 2000;**14**:1317.

65. Hoffmann R, Wilkinson IR, McCallum JF, Engels P, Houslay MD. cAMP-specific phosphodiesterase HSPDE4D3 mutants which mimic activation and changes in rolipram inhibition triggered by protein kinase A phosphorylation of Ser-54: generation of a molecular model. *Biochem J* 1998;**333**:139.

66. Zaccolo M, Pozzan T. Discrete microdomains with high concentration of cAMP in stimulated rat neonatal cardiac myocytes. *Science* 2002;**295**:1711.

67. Marx SO, Reiken S, Hisamatsu Y, Jayaraman T, Burkhoff D, Rosemblit N, Marks AR. PKA phosphorylation dissociates FKBP12.6 from the calcium release channel (ryanodine receptor): defective regulation in failing hearts. *Cell* 2000;**101**:365.

68. Fink MA, Zakhary DR, Mackey JA, Desnoyer RW, Apperson-Hansen C, Damron DS, Bond M. AKAP-mediated targeting of protein kinase a regulates contractility in cardiac myocytes. *Circ Res* 2001;**88**:291.

69. Zakhary DR, Moravec CS, Bond M. Regulation of PKA binding to AKAPs in the heart: alterations in human heart failure. *Circulation* 2000;**101**:1459.

70. Huang LJ, Durick K, Weiner JA, Chun J, Taylor SS. Identification of a novel dual specificity protein kinase A anchoring protein, D-AKAP1. *J Biol Chem* 1997;**272**:8057.

71. Huang LJ, Durick K, Weiner JA, Chun J, Taylor SS. D-AKAP2, a novel protein kinase A anchoring protein with a putative RGS domain. *Proc Natl Acad Sci USA* 1997;**94**:11,184.

72. Alto NM, Soderling J, Scott JD. Rab32 is an A-kinase anchoring protein and participates in mitochondrial dynamics. *J Cell Biol* 2002;**158**:659.

73. Faux MC, Rollins EN, Edwards AS, Langeberg LK, Newton AC, Scott JD. Mechanism of A-kinase-anchoring protein 79 (AKAP79) and protein kinase C interaction. *Biochem J* 1999;**343**:443.

74. Husi H, Ward MA, Choudhary JS, Blackstock WP, Grant SG. Proteomic analysis of NMDA receptor-adhesion protein signaling complexes [see comments]. *Nat Neurosci* 2000;**3**:661.

Chapter 45

Cyclic Nucleotide Specificity and Cross-Activation of Cyclic Nucleotide Receptors

John B. Shabb

Department of Biochemistry and Molecular Biology, University of North Dakota School of Medicine and Health Sciences, Grand Forks, North Dakota

MOLECULAR BASIS FOR CAMP/CGMP SELECTIVITY OF PKA AND PKG

The cyclic nucleotide-binding domains (CNBDs) of cAMP-dependent protein kinase (PKA) and cGMP-dependent protein kinase (PKG) are members of an evolutionarily conserved family of regulatory modules that are also found in the bacterial catabolite gene activator protein, cyclic nucleotide-regulated guanine nucleotide exchange factors (Epac1 and Epac2), and cyclic nucleotide-gated (CNG) and hyperpolarization-activated cyclic nucleotide-modulated (HCN) channels [1, 2]. In-depth descriptions of these individual cyclic nucleotide-dependent proteins are found in other chapters within this section of this handbook. The typical CNBD contains about 124 residues. At its core is an eight-stranded anti-parallel β-barrel which harbors the conserved contacts for the cyclic phosphate ribose moiety of cyclic nucleotides. A less-conserved C-terminal α-helix forms a hinged hydrophobic lid over the cyclic nucleotide purine moiety. The cAMP/cGMP selectivity of each CNBD-containing protein is determined in large part by specific structural features within their cyclic nucleotide-binding pockets.

The affinities of PKA and PKG for their physiological cyclic nucleotides are at least 100-fold greater than for the opposing cyclic nucleotides [3–5]. The middle residues of the RAA and RTA motifs found in PKA and PKG CNBDs, respectively, are critical for determining if the protein kinase will bind cGMP with high or low affinity. This Ala/Thr difference, which is conserved among all PKAs and PKGs [6], is bracketed by two of the invariant CNBD residues. The Arg interacts with the equatorial oxygen of the cyclic phosphate, and the invariant Ala forms one side of the cyclic nucleotide binding pocket. Mutation of the middle Ala residue in the RAA motif in each of the two CNBDs of the type

Iα regulatory subunit of PKA increases its affinity 200-fold for cGMP with only minor effects on cAMP affinity [7, 8]. Likewise, mutation of the Thr residue in the RTA motifs of both cGMP-binding domains of PKG Iα [9], PKG Iβ [10], and PKG II [11] reduces their affinities for cGMP relative to cAMP. These mutagenesis studies are consistent with cyclic nucleotide analog mapping [3–5], molecular modeling [6, 12], and crystallographic [13, 14] studies of PKA and PKG CNBDs. All support the interpretation that hydrogen bonding between the 2-amino position of the cGMP purine ring and the Thr hydroxyl of the RTA motif confers cGMP selectivity to PKG CNBDs when the cyclic nucleotide is bound in its *syn* conformation. The absence of this hydrogen-bonding potential in PKA CNBDs is a major factor in a much reduced affinity of PKA for cGMP.

Identification of CNBD residues responsible for selective high-affinity cAMP-binding in PKA is much more problematic. There is little structural or sequence conservation within the part of the CNBD that surrounds the purine moiety in PKA [13, 14], the bacterial catabolite gene activator protein [15], Epac2 [16], HCN channel [17], or the bacterial LotiK potassium channel [18]. All mammalian CNBDs show aromatic or hydrophobic stacking between the purine moiety and a residue within the C-terminal α-helix. This could be a source of cAMP-selectivity, but mutagenesis studies in this region of the PKA RIα cyclic nucleotide-binding B domain reveal only minor contributions to cAMP/cGMP selectivity [19].

One way in which PKG may overcome its cAMP selectivity barrier is through autophosphorylation. The *in vitro* cGMP-dependent slow-phase autophosphorylation within the autoinhibitory regions of the PKG Iα and PKG Iβ isoforms increases their sensitivities to cGMP and cAMP at least two-fold [20, 21]. Phosphomimetic mutation of one such site in PKG Iβ yields a constitutively activated kinase in a cultured cell line [22]. Similarly, phosphomimetic mutation of a PKC-dependent site in PKG Iα makes this isoform more sensitive to cGMP [23]. If these types of phosphorylation events occur *in vivo*, they may sensitize PKG to cAMP cross-activation. This may be of special relevance in tissues such as vascular smooth muscle, in which cAMP levels may be as much as five times higher than cGMP [24].

CONFOUNDING FACTORS IN CROSS-ACTIVATION STUDIES

Experimentally proving cross-activation of PKA and PKG in cells and tissues is challenging. This is in part because of the close structural and functional relationships between PKA and PKG. The two kinases often target the same phosphorylation sites on the same physiological effector (see Chapter 183 of Handbook of Cell Signaling, Second Edition for the physiological substrates of PKA and PKG). The more ubiquitous PKA is almost always co-expressed with the more specialized PKG isoforms. These subtleties make it critical that pharmacological inhibitors of each kinase show both potency and specificity (see Chapter 181 of Handbook of Cell Signaling, Second Edition for more on PKA and PKG inhibitors). The heat-stable protein kinase inhibitor and peptide derivatives are the most specific competitive inhibitors of PKA. Even so, the specificity of the membrane-permeable myristoylated version may elicit unspecific effects [25]. The PKG inhibitor KT5823 has the added problem of not always being active [26, 27]. The PKA inhibitors H89 and KT5720 are also potent inhibitors of at least a half-dozen other protein kinases [28, 29]. Furthermore, the *in vitro* behavior of these inhibitors cannot be generalized to cell- and tissue-based assays. Cohen and colleagues describe prudent steps investigators should take to minimize erroneous interpretation of protein kinase inhibitors [28].

The wide array of cyclic nucleotide analogs and their Rp-cAMPS and Rp-cGMPS phosphorothioate derivatives represent another useful class of activators and inhibitors for testing cross-activation hypotheses (for more on cyclic nucleotide analogs, see Chapter 191 of Handbook of Cell Signaling, Second Edition). Each has its distinct *in vitro* affinity for each CNBD, lipophilicity, and resistance to phosphodiesterases. The phosphorothioate analogs inhibit kinase activation by competing with cyclic nucleotides for binding to the CNBDs of PKA and PKG. High concentrations of the Rp-cGMPS analogs can cause unspecific effects in cell- and tissue-based assays [25, 30].

The enzymes that hydrolyze cAMP and cGMP are themselves subject to regulation by cyclic nucleotides. For example, the cGMP-inhibited cyclic nucleotide phosphodiesterase PDE3 has similar substrate binding affinities for cAMP and cGMP, but the k_{cat} for cGMP is one-tenth that of cAMP (see Chapter 175 of Handbook of Cell Signaling, Second Edition). Intracellular elevation of cGMP reduces the activity of PDE3 by competing with cAMP for its active site. The result is cGMP-stimulated elevation of cAMP and potential activation of PKA.

The best studied CNG channels are those from photoreceptors and olfactory neurons, or the HCN channel from the cardiac sinoatrial node (see Chapter 185 of Handbook of Cell Signaling, Second Edition), but CNG channel subunits are also detectable in many other tissue types. Rod and cone CNG channels are cGMP selective, whereas olfactory CNG channels respond to cAMP and cGMP equally. Many of the same cyclic nucleotide analogs that are used to dissect PKA and PKG functions in cells are also used to characterize CNG channels. For example, the PKG agonists 8-Br-cGMP and 8-pCPT-cGMP also activate rod CNG channels [31, 32], and the PKA-selective 8-pCPT-cAMP is a potent olfactory CNG channel agonist [33]. The PKA and PKG Rp-cGMPS and Rp-cAMPS antagonists either activate or inhibit CNG channels, depending on the isotype being studied [32, 34]. The potential interference of CNG channels in PKA or PKG cross-activation should therefore be considered.

The Epac1 and Epac2 isoforms are implicated in the regulation of cell adhesion and insulin secretion through the modulation of Rap1 activity (see Chapter 186 of Handbook of Cell Signaling, Second Edition). As with the CNG channels, 8-substituted cAMP analogs commonly used to activate PKA are effective Epac agonists [35, 36]. Likewise, Rp-cAMPS analogs that prevent PKA activation also competitively inhibit cAMP activation of Epac [35, 36]. Interestingly, the cAMP-selective Epac1 binds to but is not activated by cGMP [35, 36]. The Epac-selective activator 8-pCPT-2′-O-Me-cAMP [37] has been invaluable in discerning between Epac and PKA cellular actions.

CAMP CROSS-ACTIVATION OF PKG

The paradigm for cyclic nucleotide cross-activation is cAMP/PKG-mediated smooth muscle relaxation. Early observations suggested that intracellular elevation of cAMP or cGMP could mediate smooth muscle relaxation. Even though vascular smooth muscle contains high levels of both PKA and PKG [24], a series of studies in two independent laboratories came to the same conclusion: that both cyclic nucleotides elicited their responses through the activation of PKG. Francis, Corbin, and co-workers observed that the potencies of an array of cAMP and cGMP analogs in eliciting relaxation of pig coronary arteries and guinea-pig tracheal smooth muscle correlated more strongly with their *in vitro* affinities for PKG than for PKA [24]. They subsequently demonstrated that moderate cAMP elevation in coronary arteries was sufficient to activate PKG [38]. Lincoln and co-workers capitalized on the observation that cultured rat aortic smooth muscle cells lost the expression of PKG over time [39]. The blunted forskolin inhibition of Ca^{2+} uptake in these PKG-depleted cells was reversed by introduction of purified PKG into these cells, but not PKA [40].

Since these early studies, a number of other groups have observed varying degrees of cAMP/PKG-mediated impairment of smooth muscle contractile response initiated by an array of physiological and pharmacological agents [41–45]. Of particular note is the study by Pelligrino, which suggests an indirect route of cAMP/PKG cross-activation through inhibition of cGMP efflux [44]. One of the main routes for cGMP efflux is through the multidrug resistance protein MRP5. This study showed that antisense oligonucleotide knockdown of MRP5 in the rat cerebral vasculature blocked cAMP potentiated nitric oxide (NO)-induced vasodilation. Further studies are needed to corroborate this provocative finding.

Other cAMP/PKG cross-activation studies focus on specific targets of PKA or PKG involved in smooth muscle tone. Activation of certain large-conductance calcium-activated potassium channel (BK_{Ca}) isoforms leads to inhibition of the L-type Ca^{2+} channel and relaxation. The cAMP-dependent activation of BK_{Ca} channels appears to require PKG in a variety of vascular smooth muscle cells [46–49]. The L-type Ca^{2+} channel itself may be inhibited by cAMP/PKG cross-activation in smooth muscle cells [50, 51], but only at supraphysiological cAMP levels. Likewise, partial cAMP/PKG cross-activation may inhibit store-operated Ca^{2+} channel activity in porcine tracheal smooth muscle cells [52]. Forskolin-induced phosphorylation of the inositol 1,4,5-trisphosphate (IP_3) receptor is prevented at least partially by inhibition of PKG in rat aortas [53] or gastric smooth muscle cells [54], suggesting yet another potential site of cAMP/PKG cross-activation. Interpretation of all of the above studies relies on the pharmacological use of PKA- and PKG-selective inhibitors and activators or the *in vitro* measurement of protein kinase activation ratios.

Mouse models deficient in specific components of the cAMP or cGMP signaling pathway provide powerful independent tools for assessing the physiological relevance of cross-activation. Aortic smooth muscle from mice deficient in PKG I exhibits unperturbed cAMP-induced relaxation relative to that from wild-type mice [55]. This argues strongly against the physiological relevance of cAMP/PKG cross-activation in this tissue. Genetically engineered BK_{Ca}-deficient mice, on the other hand, exhibit increased sensitivity to β-adrenergic-induced tracheal smooth muscle relaxation when compared to wild-type mice [56]. This increased sensitivity is correlated with a compensatory increase in smooth muscle PKG but not PKA. Results from the BK_{Ca} model do not confirm or refute a role for cAMP/PKG cross-activation of BK_{Ca} channels, but rather demonstrate a potential physiological link between cross-activation and some aspect of BK_{Ca} regulation of smooth muscle tone.

CGMP CROSS-ACTIVATION OF PKA

The preponderance of evidence supports an antiproliferative role for PKG I in vascular smooth muscle [57]. Nevertheless, evidence persists that there may also be a PKG-independent role for cGMP through cross-activation

of PKA. Initial studies showed that the cGMP-dependent inhibition of cultured rat aortic smooth muscle cells could be blocked by Rp-cAMPS but not Rp-cGMPS [58]. Subsequent *in vitro* studies demonstrate a role for cGMP inhibition of PDE3 leading to activation of PKA [59, 60]. Mice deficient in smooth muscle PKG I provide further support for a PKG-independent pathway of NO-induced inhibition of atherogenesis [61]. The mouse studies provocatively suggest that low NO levels, working through PKG, actually stimulate smooth muscle proliferation, whereas high NO levels lead to the more classic antiproliferative effects which are mediated through PKA and not PKG [62].

It is possible that cGMP activation of PKA may also contribute to smooth muscle relaxation. The loss of NO-responsive vasodilation in PKG I-deficient mice supports the central role of PKG I in NO-induced smooth muscle relaxation, and argues against a physiological role for cross-activation in this process [55]. Under exceptional conditions when cGMP levels are very high, however, cross-activation of PKA can contribute to NO-induced relaxation in isolated smooth muscle preparations from these same PKG I-deficient mice [63–65].

Mice deficient in PKG I have erectile dysfunction, suggesting that PKG I is important for normal functioning of the corpus cavernosum and the cAMP/PKA pathway cannot compensate for this loss [66]. Other studies suggest that penile erection by PKA activation occurs through a circuitous cross-activation path where PDE3 is inhibited by cGMP which is elevated upon sildenafil inhibition of the cGMP-binding, cGMP-specific PDE5 [67, 68]. Sildenafil reversal of erectile dysfunction in PKG I-deficient mice has not been reported yet.

The primary action of activated platelet inhibition by NO is through PKG I (see Chapter 192 of Handbook of Cell Signaling, Second Edition). Consistent with this, activated platelets derived from PKG I-deficient mice are unresponsive to NO treatment even though the platelets have an intact cAMP/PKA pathway [69]. Nevertheless, platelet PKA may be activated in response to NO under certain conditions [70–73] mediated through cGMP inhibition of PDE3 [71–73]. Some of the effects attributed to PKA may be due to unspecific actions of PKG activators and inhibitors in platelets [25]. Mice deficient in PDE3A show delayed collagen-induced thrombosis and death, firmly implicating a role for this PDE3 isoform in inhibition of platelet activation [74]. Nitric oxide-induced platelet inhibition has not been explored with this mouse model yet.

The *Escherichia coli* heat-stable enterotoxin (STa) induces supraphysiological accumulation of cGMP, resulting in increased chloride secretion in human intestinal epithelium. This action is likely mediated through PKA in the T84 human intestinal epithelial cell line, since little or no PKG is detectable in these cells [75]. Other similar studies support this conclusion [76, 77]. The cGMP/PKA cross-activation observed in permanent cell lines has yet to be substantiated in a more physiological model. Mice deficient in PKG II do not exhibit STa-induced fluid accumulation in the gut, as is seen in wild-type mice. The functional cAMP/PKA pathway in intestinal epithelia is therefore not able to compensate for the loss of PKG II [78].

Elevated cGMP suppresses TNFα-induced apoptosis of rat hepatocytes. This effect is blocked by inhibitors of PKA but not inhibitors of PKG. The cAMP-mediated suppression of apoptosis may itself be only partially PKA-dependent [79]. Moreover, the PKA-independent effect appears to be mediated through Epac, since 8-pCPT-2′-O-Me-cAMP suppresses apoptosis and activates Rap1 without activating PKA [80]. These observations suggest a potential for cGMP/Epac cross-activation in apoptosis suppression.

The site of associative learning in the honeybee is the antennal lobe. PKA activity is enhanced in this organ upon photorelease of NO, and this effect is blocked upon inhibition of soluble guanylyl cyclase or PKA but not PKG or PDE3 [81]. Modest synergistic activation of purified honeybee PKA by low levels of cAMP and cGMP provides further support for direct cGMP/PKA cross-activation in this system [82]. In a cricket model for learning and memory, induction of long-term memory enhancement with 8-Br-cGMP is completely blocked by inhibition of adenylyl cyclase, calmodulin, or CNG channels, but not by inhibiton of PKG. This and other evidence suggest that cGMP/PKA cross-activation begins with cGMP activation of CNG channels followed by an influx of Ca^{2+} sufficient to activate Ca^{2+}-calmodulin-dependent adenylyl cyclase, which in turn increases cAMP levels [83]. This may be the first example of cGMP/PKA cross-activation via CNG channel activation.

REFERENCES

1. Shabb JB, Corbin JD. Cyclic nucleotide-binding domains in proteins having diverse functions. *J Biol Chem* 1992;**267**:5723–6.

2. Rehmann H, Wittinghofer A, Bos JL. Capturing cyclic nucleotides in action, snapshots from crystallographic studies. *Nat Rev Mol Cell Biol* 2007;**8**:63–73.

3. Døskeland SO, Øgreid D, Ekanger R, Sturm PA, Miller JP, Suva RH. Mapping of the two intrachain cyclic nucleotide binding sites of adenosine cyclic 3′, 5′-phosphate dependent protein kinase I. *Biochemistry* 1983;**22**:1094–101.

4. Øgreid D, Ekanger R, Suva RH, Miller JP, Døskeland SO. Comparison of the two classes of binding sites (A and B) of type I and type II cyclic-AMP-dependent protein kinases by using cyclic nucleotide analogs. *Eur J Biochem* 1989;**181**:19–31.

5. Corbin JD, Øgreid D, Miller JP, Suva RH, Jastorff B, Døskeland SO. Studies of cGMP analog specificity and function of the two intrasubunit binding sites of cGMP-dependent protein kinase. *J Biol Chem* 1986;**261**:1208–14.

6. Weber IT, Shabb JB, Corbin JD. Predicted structures of the cGMP binding domains of the cGMP-dependent protein kinase: a key alanine/threonine difference in evolutionary divergence of cAMP and cGMP binding sites. *Biochemistry* 1989;**28**:6122–7.

7. Shabb JB, Ng L, Corbin JD. One amino acid change produces a high affinity cGMP-binding site in cAMP-dependent protein kinase. *J Biol Chem* 1990;**265**:16,031–16,034.

8. Shabb JB, Buzzeo BD, Ng L, Corbin JD. Mutating protein kinase cAMP-binding sites into cGMP-binding sites. Mechanism of cGMP selectivity. *J Biol Chem* 1991;**266**:24,320–24,326.

9. Reed RB, Sandberg M, Jahnsen T, Lohmann SM, Francis SH, Corbin JD. Structural order of the slow and fast intrasubunit cGMP-binding sites of type Iα cGMP-dependent protein kinase. *Adv Sec Mess Phosph Res* 1997;**31**:205–17.

10. Reed RB, Sandberg M, Jahnsen T, Lohmann SM, Francis SH, Corbin JD. Fast and slow cyclic nucleotide-dissociation sites in cAMP-dependent protein kinase are transposed in type Iβ cGMP-dependent protein kinase. *J Biol Chem* 1996;**271**:17,570–17,575.

11. Taylor MK, Uhler MD. The amino-terminal cyclic nucleotide binding site of the type II cGMP-dependent protein kinase is essential for full cyclic nucleotide-dependent activation. *J Biol Chem* 2000;**275**:28,053–28,062.

12. Weber IT, Steitz TA, Bubis J, Taylor SS. Predicted structures of cAMP binding domains of type I and II regulatory subunits of cAMP-dependent protein kinase. *Biochemistry* 1987;**26**:343–51.

13. Su Y, Dostmann WR, Herberg FW, Durick K, Xuong NH, Ten Eyck L, Taylor SS, Varughese KI. Regulatory subunit of protein kinase A: structure of deletion mutant with cAMP binding domains. *Science* 1995;**269**:807–13.

14. Diller TC, Madhusudan., Xuong NH, Taylor SS. Molecular basis for regulatory subunit diversity in cAMP-dependent protein kinase: crystal structure of the type I Iβ?regulatory subunit. *Structure* 2001;**9**:73–82.

15. Weber IT, Steitz TA. Structure of a complex of catabolite gene activator protein and cyclic AMP refined at 2.5 A resolution. *J Mol Biol* 1987;**198**:311–26.

16. Rehmann H, Prakash B, Wolf E, Rueppel A, de Rooij J, Bos JL, Wittinghofer A. Structure and regulation of the cAMP-binding domains of Epac2. *Nat Struct Biol* 2003;**10**:26–32.

17. Zagotta WN, Olivier NB, Black KD, Young EC, Olson R, Gouaux E. Structural basis for modulation and agonist specificity of HCN pacemaker channels. *Nature* 2003;**425**:200–5.

18. Clayton GM, Silverman WR, Heginbotham L, Morais-Cabral JH. Structural basis of ligand activation in a cyclic nucleotide regulated potassium channel. *Cell* 2004;**119**:615–27.

19. Kapphahn MA, Shabb JB. Contribution of the carboxyl-terminal regional of the cAMP-dependent protein kinase type Iα regulatory subunit to cyclic nucleotide interactions. *Arch Biochem Biophys* 1997;**348**:347–56.

20. Landgraf W, Hullin R, Göbel C, Hofmann F. Phosphorylation of cGMP-dependent protein kinase increases the affinity for cAMP. *Eur J Biochem* 1986;**154**:113–17.

21. Smith JA, Francis SH, Walsh KA, Kumar S, Corbin JD. Autophosphorylation of the type Iβ?cGMP-dependent protein kinase increases basal catalytic activity and enhances allosteric activation by cGMP or cAMP. *J Biol Chem* 1996;**271**:20,756–20,762.

22. Collins SP, Uhler MD. Cyclic AMP- and cyclic GMP-dependent protein kinases differ in their regulation of cyclic AMP response element-dependent gene transcription. *J Biol Chem* 1999;**274**:8391–404.

23. Hou Y, Lascola J, Dulin NO, Ye RD, Browning DD. Activation of cGMP-dependent protein kinase by protein kinase C. *J Biol Chem* 2003;**278**:16,706–16,712.

24. Francis SH, Noblett BD, Todd BW, Wells JN, Corbin JD. Relaxation of vascular and tracheal smooth muscle by cyclic nucleotide analogs that preferentially activate purified cGMP-dependent protein kinase. *Mol Pharmacol* 1988;**34**:506–17.

25. Gambaryan S, Geiger J, Schwarz UR, Butt E, Begonja A, Obergfell A, Walter U. Potent inhibition of human platelets by cGMP analogs independent of cGMP-dependent protein kinase. *Blood* 2004;**103**:2593–600.

26. Burkardt M, Glazova M, Gambaryan S, Vollkommer T, Butt E, Bader B, Heermeier K, Lincoln TM, Walter U, Palmetshofer A. KT5823 inhibits cGMP-dependent protein kinase activity *in vitro* but not in intact human platelets and rat mesangial cells. *J Biol Chem* 2000;**275**:33,536–33,541.

27. Bain J, McLauchlan H, Elliott M, Cohen P. The specificities of protein kinase inhibitors: an update. *Biochem J* 2003;**371**:199–204.

28. Davies SP, Reddy H, Matilde C, Cohen P. Specificity and mechanism of action of some commonly used protein kinase inhibitors. *Biochem J* 2000;**351**:95–105.

29. Bain J, Plater L, Elliott M, Shpiro N, Hasties J, McLauchlan H, Klevernic I, Arthur S, Alessi D, Cohen P. The selectivity of protein kinase inhibitors; a further update. *Biochem J* 2007;**408**:297–315.

30. Smolenski A, Burkhardt AM, Eigenthaler M, Butt E, Gambaryan S, Lohmann SM, Walter U. Functional analysis of cGMP-dependent protein kinases I and II as mediators of NO/cGMP effects. *N S Arch Pharmacol* 1998;**358**:134–9.

31. Zimmerman AL, Yamanaka G, Eckstein F, Baylor DA. Interaction of hydrolysis-resistant analogs of cyclic GMP with the phosphodiesterase and light-sensitive channel of retinal rod outer segments. *Proc Natl Acad Sci USA* 1985;**82**:8813–17.

32. Wei JY, Cohen ED, Genieser HG, Barnstable CJ. Substituted cGMP analogs can act as selective agonists of the rod photoreceptor cGMP-gated cation channel. *J Mol Neurosci* 1998;**10**:53–64.

33. Frings S, Lynch JW, Lindemann B. Properties of cyclic nucleotide-gated channels mediating olfactory transduction. Activation, selectivity, and blockage. *J Gen Physiol* 1992;**100**:45–67.

34. Kramer RH, Tibbs GR. Antagonists of cyclic nucleotide-gated channels and molecular mapping of their site of action. *J Neurosci* 1996;**16**:1285–93.

35. Christensen AE, Selheim F, de Rooij J, Dremier S, Schwede F, Dao KK, Martinez A, Maenhaut C, Bos JL, Genieser HG, Døskeland SO. cAMP analog mapping of Epac1 and cAMP kinase. Discriminating analogs demonstrate that Epac and cAMP kinase act synergistically to promote PC-12 cell neurite extension. *J Biol Chem* 2003;**278**:35,394–35,402.

36. Rehmann H, Schwede F, Døskeland SO, Wittinghofer A, Bos JL. Ligand-mediated activation of the cAMP-responsive guanine nucleotide exchange factor Epac. *J Biol Chem* 2003;**278**:38,548–38,556.

37. Enserink JM, Christensen AE, de Rooij J, van Triest M, Schwede F, Genieser HG, Døskeland SO, Blank JL, Bos JL. A novel Epac-specific cAMP analogue demonstrates independent regulation of Rap1 and ERK. *Nat Cell Biol* 2002;**4**:901–6.

38. Jiang H, Colbran JL, Francis SH, Corbin JD. Direct evidence for cross-activation of cGMP-dependent protein kinase by cAMP in pig coronary arteries. *J Biol Chem* 1992;**267**:1015–19.

39. Cornwell TL, Lincoln TM. Regulation of intracellular Ca^{2+} levels in cultured vascular smooth muscle cells. Reduction of Ca^{2+} by atriopeptin and 8-Br-cyclic GMP is mediated by cyclic GMP-dependent protein kinase. *J Biol Chem* 1989;**264**:1146–55.

40. Lincoln TM, Cornwell TL, Taylor AE. cGMP-dependent protein kinase mediates the reduction of Ca^{2+} by cAMP in vascular smooth muscle cells. *Am J Physiol* 1990;**258**:C399–407.

41. Murthy KS, Makhlouf GM. Interaction of cA-kinase and cG-kinase in mediating relaxation of dispersed smooth muscle cells. *Am J Physiol Cell Physiol* 1994;**268**:C171–80.

42. Dhanakoti SN, Gao Y, Nguyen MQ, Raj U. Involvement of cGMP-dependent protein kinase in the relaxation of ovine pulmonary arteries to cGMP and cAMP. *J Appl Physiol* 2000;**88**:1637–42.

43. Keung W, Vanhoutte PM, Man RYK. Acute impairment of contractile responses by 17β-estradiol is cAMP and protein kinase G dependent in vascular smooth muscle cells of the porcine coronary arteries. *Br J Pharmacol* 2005;**144**:71–9.

44. Xu H-L, Wolde HM, Gavrilyuk V, Baughman VL, Pelligrino DA. cAMP modulates cGMP-mediated cerebral arteriolar relaxation *in vivo*. *Am J Physiol Heart Circ Physiol* 2004;**287**:H2501–9.

45. Leffler CW, Fedinec AL, Parfenova H, Jaggar JH. Permissive contributions of NO and prostacyclin in CO-induced cerebrovascular dilation in piglets. *Am J Physiol Heart Circ Physiol* 2005;**289**:H432–8.

46. White RE, Kryman JP, El-Mowafy AM, Han G, Carrier GO. cAMP-dependent vasodilators cross-activate the cGMP-dependent protein kinase to stimulate BK$_{Ca}$ channel activity in coronary artery smooth muscle cells. *Circ Res* 2000;**86**:897–905.

47. Zhu S, Han G, White RE. PGE$_2$ action in human coronary artery smooth muscle: role of potassium channels and signaling cross-talk. *J Vasc Res* 2002;**39**:477–88.

48. Barman SA, Zhu S, Han G, White RE. cAMP activates BK$_{Ca}$ channels in pulmonary arterial smooth muscle via cGMP-dependent protein kinase. *Am J Physiol Lung Cell Mol Physiol* 2003;**284**:L1004–11.

49. Burnette JO, White RE. PGI$_2$ opens potassium channels in retinal pericytes by cyclic AMP-stimulated cross-activation of PKG. *Exp Eye Res* 2006;**83**:1359–65.

50. Koh SD, Sanders KM. Modulation of Ca^{2+} current in canine colonic myocytes by cyclic nucleotide-dependent mechanisms. *Am J Physiol Cell Physiol* 1996;**271**:C794–803.

51. Ishikawa T, Hume JR, Keef KD. Regulation of Ca^{2+} channels by cAMP and cGMP in vascular smooth muscle cells. *Circ Res* 1993;**73**:1128–37.

52. Ay B, Iyanoye A, Sieck GC, Prakash YS, Pabelick CM. Cyclic nucleotide regulation of store-operated Ca^{2+} influx in airway smooth muscle. *Am J Physiol Lung Cell Mol Physiol* 2006;**290**:L278–83.

53. Komalavilas P, Lincoln TM. Phosphorylation of the inositol 1, 4, 5-trisphosphate receptor by cyclic GMP-dependent protein kinase. *J Biol Chem* 1994;**269**:8701–7.

54. Murthy KS, Zhou H. Selective phosphorylation of the IP$_3$R-I *in vivo* by cGMP-dependent protein kinase in smooth muscle. *Am J Physiol Gastr L Physiol* 2003;**284**:G221–30.

55. Pfeifer A, Klatt P, Massberg S, Ny L, Sausbier M, Hirneiss C, Wang G-X, Korth M, Aszódi A, Andersson K-E, Krombach F, Mayerhofer A, Ruth P, Fässler R, Hofmann F. Defective smooth muscle regulation in cGMP kinase I-deficient mice. *EMBO J* 1998;**17**:3045–51.

56. Sausbier M, Zhou XB, Beier C, Sausbier U, Wolpers D, Maget S, Martin C, Dietrich A, Ressmeyer AR, Renz H, Schlossman J, Hofmann F, Neuhuber W, Gudermann T, Uhlig S, Korth M, Ruth P. Reduced rather than enhanced cholinergic airway constriction in mice with ablation of the large conductance Ca^{2+}-activated K$^+$ channel. *FASEB J* 2007;**21**:812–22.

57. Lincoln TM, Wu X, Sellak H, Dey N, Choi C-S. Regulation of vascular smooth muscle cell phenotype by cyclic GMP and cyclic GMP-dependent protein kinase. *Front Biosci* 2006;**11**:356–67.

58. Cornwell TL, Arnold E, Boerth NJ, Lincoln TM. Inhibition of smooth muscle cell growth by nitric oxide and activation of cAMP-dependent protein kinase by cGMP. *Am J Physiol* 1992;**267**:C1405–13.

59. Osinski MT, Rauch BH, Schrör K. Antimitogenic actions of organic nitrates are potentiated by sildenafil and mediated *via* activation of protein kinase A. *Mol Pharmacol* 2001;**59**:1044–50.

60. Aizawa T, Wei H, Miano JM, Abe J, Berk BC, Yan C. Role of phosphodiesterase 3 in NO/cGMP-mediated anti-inflammatory effects in vascular smooth muscle cells. *Circ Res* 2003;**93**:406–13.

61. Wolfsgruber W, Feil S, Brummer S, Kuppinger O, Hofmann F, Feil R. A proatherogenic role for cGMP-dependent protein kinase in vascular smooth muscle cells. *Proc Natl Acad Sci USA* 2003;**100**:13,519–13,524.

62. Feil R, Feil S, Hofmann F. A heretical view on the role of NO and cGMP in vascular proliferative diseases. *Trends Mol Med* 2005;**11**:71–5.

63. Sausbier M, Schubert R, Voigt V, Hirneiss C, Pfeifer A, Korth M, Kleppisch T, Ruth P, Hofmann F. Mechanisms of NO/cGMP-dependent vasorelaxation. *Circ Res* 2000;**87**:825–30.

64. Wörner R, Lukowski R, Hofmann F, Wegener JW. cGMP signals mainly through cAMP kinase in permeabilized murine aorta. *Am J Physiol Heart Circ Physiol* 2006;**292**:H237–44.

65. Bonnevier J, Fässler R, Somlyo AP, Somlyo AV, Arner A. Modulation of Ca^{2+} sensitivity by cyclic nucleotides in smooth muscle from protein kinase G-deficient mice. *J Biol Chem* 2004;**279**:5146–51.

66. Hedlund P, Aszódi A, Pfeifer A, Alm P, Hofmann F, Ahmad M, Fässler R, Andersson KE. Erectile disfunction in cyclic GMP-dependent kinase I-deficient mice. *Proc Natl Acad Sci USA* 2000;**97**:2349–54.

67. Stief CG, Ückert S, Becker AJ, Harringer W, Truss MC, Forssmann WG, Jonas U. Effects of sildenafil on cAMP and cGMP levels in isolated human cavernous and cardiac tissue. *Urology* 2000;**55**:146–50.

68. Prieto D, Rivera L, Recio P, Ruiz Rubio JL, Hernández M, Garcia-Sacristán A. Role of nitric oxide in the relaxation elicited by sildenafil in penile resistance arteries. *J Urol* 2006;**175**:1164–70.

69. Massberg S, Sausbier M, Klatt P, Bauer M, Pfeifer A, Siess W, Fässler R, Ruth P, Krombach F, Hofmann F. Increased adhesion and aggregation of platelets lacking cyclic guanosine 3′,5′-monophosphate kinase I. *J Exp Med* 1999;**189**:1255–63.

70. Li Z, Ajdic J, Eigenthaler M, Du X. A predominant role for cAMP-dependent protein kinase in the cGMP-induced phosphorylation of vasodilator-stimulated phosphoprotein and platelet inhibition in humans. *Blood* 2003;**101**:4423–9.

71. Maurice DH, Haslam RJ. Molecular basis of the synergistic inhibition of platelet function by nitrovasodilators and activators of adenylate cyclase: inhibition of cyclic AMP breakdown by cyclic GMP. *Mol Pharmacol* 1990;**37**:671–81.

72. Nolte C, Eigenthaler M, Horstrup K, Hönig-Liedl P, Walter U. Synergistic phosphorylation of the focal adhesion assoicated vasodilator-stimulated phosphoprotein in intact human platelets in response to cGMP- and cAMP-elevating platelet inhibitors. *Biochem Pharmacol* 1994;**48**:1569–75.

73. Jensen BO, Selheim F, Døskeland SO, Gear ARL, Holmsen H. Protein kinase A mediates inhibition of the thrombin-induced platelet shape change by nitric oxide. *Blood* 2004;**104**:2775–82.

74. Sun B, Li H, Shakur Y, Hensley J, Hockman S, Kambayashi J, Manganiello VC, Liu Y. Role of phosphodiesterase type 3A and 3B in regulating platelet and cardiac function using subtype-selective knockout mice. *Cell Signal* 2007;**19**:1765–71.

75. Forte LR, Thorne PK, Eber SL, Krause WJ, Freeman RH, Francis SH, Corbin JD. Stimulation of intestinal Cl^- transport by heat-stable enterotoxin: activation of cAMP-dependent protein kinase by cGMP. *Am J Physiol Cell Physiol* 1992;**263**:C607–15.

76. Chao AC, de Sauvage FJ, Dong YJ, Wagner JA, Goeddel DV, Gardner P. Activation of intestinal CFTR Cl^- channel by heat-stable enterotoxin and guanylin *via* cAMP-dependent protein kinase. *EMBO J* 1994;**13**:1065–72.

77. Tien XY, Brasitus TA, Kaetzel MA, Dedman JR, Nelson DJ. Activation of the cystic fibrosis transmembrane conductance regulator by cGMP in the human colonic cancer cell line Caco-2. *J Biol Chem* 1994;**269**:51–4.

78. Pfeifer A, Aszódi A, Seidler U, Ruth P, Hofmann F, Fässler R. Intestinal secretory defects and dwarfism in mice lacking cGMP-dependent protein kinase II. *Science* 1996;**274**:2082–6.

79. Li J, Yang S, Billiar TR. Cyclic nucleotides suppress tumor necrosis factor α-mediated apoptosis by inhibiting caspase activation and cytochrome c release in primary hepatocytes *via* a mechanism independent of Akt activation. *J Biol Chem* 2000;**275**:13,026–13,034.

80. Cullen KA, McCool J, Anwer MS, Webster CRL. Activation of cAMP-guanine exchange factor confers PKA-independent protection from hepatocyte apoptosis. *Am J Physiol Gastrointest Liver Physiol* 2004;**287**:G334–43.

81. Müller U. Prolonged activation of cAMP-dependent protein kinase during conditioning induces long-term memory in honeybees. *Neuron* 2000;**27**:159–68.

82. Leboulle G, Muller U. Synergistic activation of insect cAMP-dependent protein kinase A (type II) by cyclic AMP and cyclic GMP. *FEBS Letts* 2004;**576**:216–20.

83. Matsumoto Y, Unoki S, Aonuma H, Mizunami M. Critical role of nitric oxide-cGMP cascade in the formation of cAMP-dependent long-term memory. *Learn Mem* 2006;**13**:35–44.

G-Proteins

Signal Transduction by G Proteins: Basic Principles, Molecular Diversity, and Structural Basis of Their Actions

Lutz Birnbaumer

Laboratory of Neurobiology, National Institute of Environmental Health Sciences, Research Triangle Park, North Carolina

INTRODUCTION

Cells do not live in isolation, and respond to extracellular stimuli either with specific tasks such as neurotransmission, activation of substrate uptake, or fatty acid release, or with adaptive changes that insure homeostatic cohabitation with other cell types as is needed in multicellular organisms to insure better survival (stress responses). Extracellular cues include nutrients, intoxicants, and a multitude of signaling molecules such as autacoids, growth factors, and hormones. Signaling molecules are either membrane-permeant (e.g., steroid hormones) or -impermeant (peptide hormones, biogenic amines), for which cells have evolved separate response mechanisms. For membrane-permeant signals, receptors are for the most part intracellular, cytosolic, or nuclear, and the changes they elicit are frequently modulation of gene expression with direct participation of the receptor in the regulatory complex that transcribes response genes. For membrane-impermeant signals, nature has evolved a repertoire of mechanisms by which binding of the ligand to its receptor on the cell surface leads to intracellular changes of one or more enzymatic activities, and to activation or inhibition of regulatory signaling pathways. In many instances, the action or actions of a receptor involve promotion or disruption of multimeric protein complexes. Some of the signaling pathways activated by receptors in response to membrane-impermeant ligands are wholly cytosolic or submembranous; others affect nuclear gene expression. The process by which the extracellular ligand–receptor interaction leads to changes inside the cell is commonly referred to as "signal transduction" – the ligand is the extracellular signal whose message is transduced into an intracellular signal of a different chemical nature.

The signals generated in this way are second messengers, such as (1) ions entering through ligand-gated ion channels; (2) signaling molecules derived from the receptor itself, such as the cytosolic domain of Notch which is proteolytically cleaved off and translocates into the nucleus to regulate gene expression; (3) products of the activation of the receptor's intrinsic enzymatic activities, such as cGMP formed by the guanylyl cyclase activity of the atrial natriuretic peptide guanylin receptors; (4) tyrosine phosphorylation, first of self and then of proteins recruited to the phosphorylated receptor, as happens when EGF, PDGF, and insulin bind to their respective homodimeric receptors; and (5) sequential serine/threonine phosphorylation, also first of self (receptor II phosphorylating receptor I) and then of cytosolic transducing proteins such as R-SMADs by phosphorylated receptor I, as it happens in response to interaction of the TGFβ and bone morphogenic protein (BMP) family of signaling molecules with their heteromeric receptors. In the case of R-SMADs, their phosphorylation exposes a nuclear localization signal as well as a heterodimerization domain, leading ultimately to translocation into the nucleus and changes in gene expression. For general references, see *Molecular Cell Biology* (6th edition) by Lodish and colleagues [1].

The use of heterotrimeric G proteins as signal transducers constitutes yet a different type of signal transduction process. In this case evolution has led a structurally related superfamily of receptors, the seven-transmembrane receptors (also referred to as G-Protein-Coupled Receptors or GPCRs) to acquire the ability to recruit and regulate the intrinsic signaling capacity of a family of structurally related heterotrimeric regulatory GTPases, the G proteins (Figure 46.1). With ~800 members (Table 46.1), GPCRs

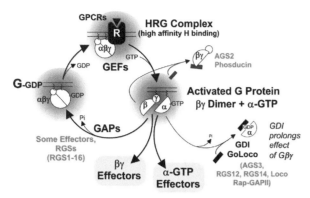

FIGURE 46.1 Hormone receptor complexes act as guanine nucleotide exchange factors (GEFs) to catalyze activation of the trimeric G protein with formation of α-GTP plus $\beta\gamma$, both of which modulate effector functions. Spontaneous decay of α-GTP to α-GDP can be modulated by GTPase activating proteins (GAPs) such as RGSs and some effectors, and by activators of G-protein signaling (AGSs) that tilt the balance between effects through α-GTP vs those through $\beta\gamma$.

TABLE 46.1 Human and mouse GPCR families*

Receptor	Homo sapiens	Mus musculus
1. Rhodopsin (Class A)	**659**	**1318**
(non-olfactory	(271)	(281)
olfactory	(388)	1037
olfactory pseudogenes)	(~450)	(?)
alpha (opsin family) opsin prostaglandin amine melatonin melacortin (MC)	101	105
beta	43	46
gamma (peptides) SOG MCH chemokine	64	67
delta MAS-related glycoprotein purine	63	63
olfactory	388	1037
2. Secretin (Class B)	**15**	**15**
3. Glutamate (Class C)	**22**	**79**
4. Adhesion	**33**	**31**
5. Frizzled	**11**	**11**
6. Taste type 2 (TAS2)	**25**	**34**
7. Vomeronasal type 1 (VR1)	**3**	**165**
8. Other	**23**	**25**
Total	**780**	**1687**

MAS, MAS oncogene; MCH, melanin concdentrating hormone; SOG, somatostatin, opioid, galanin.
From [42] (Bjarnadottir et al., 2006).

constitute the largest superfamily of evolutionary related proteins encoded in the mammalian genome. Its members recognize and are activated by a multitude of chemically diverse signaling molecules that include peptide and glycoprotein hormones (e.g., ACTH, somatostatin, GnRH, glucagon, cholecystokinin, luteinizing hormone, neurokinins), neurotransmitters (e.g., norepinephrine, serotonin, GABA, acetylcholine), amino acids (e.g., glutamate), lipids (e.g., prostacyclins, leukotrienes), phospholipids (phosphatidylcholine, sphingosine-1-phosphate), and sensory inputs such as light, taste-eliciting compounds, odors and pheromones.

This chapter deals with the mechanism of activation/deactivation, and the actions of the heterotrimeric G proteins responsible for transducing the binding of a signaling molecule to its seven-transmembrane receptor into cellular responses. Activation of a G protein by a receptor, first proposed by Rodbell and collaborators to be a two-step process in which a GTP-dependent transducer regulates an enzyme or amplifier, without clear knowledge of the number of molecular components involved [2], is now known to be a multi-step process that includes GDP/GTP exchange and subunit dissociation of the $\alpha\beta\gamma$ G protein, followed by spontaneous deactivation as a consequence of the hydrolysis of the activating GTP to GDP, and reassociation of the separated subunits in preparation for reactivation by a new round of GDP/GTP exchange promoted by the activating hormone receptor (HR) complex (Figure 46.1, bottom panel). As shown in this figure, the basic cycle is affected by modulators of G-protein signaling, the RGSs (Regulators of G protein Signaling), which increase the rate at which α-GTP self-deactivates, and the AGS2- and AGS3-type modulators (Activators of G protein Signaling) which sequester either α-GDP or G$\beta\gamma$ and thereby bias signaling through α-GTP vs signaling through G$\beta\gamma$ (see below).

RAS, THE PROTOTYPIC REGULATORY GTPASE

As the name indicates, regulatory GTPases are proteins that bind and hydrolyze GTP. Their regulatory power lies in the fact that their conformation differs when occupied by GTP or GDP. They are also referred to as molecular switches. Due to their intrinsic GTPase activity, they carry a built-in deactivation timer that prevents the GTP state from being long-lived. The crystal structure of the 180 amino-acid regulatory GTPase *ras*, in its GTP- and GDP-liganded forms, shows two principal regions that differ in the GTP state

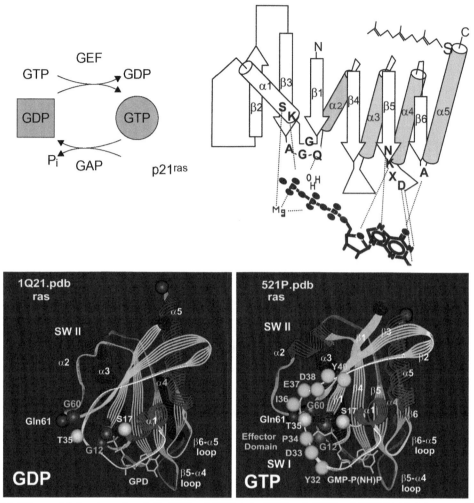

FIGURE 46.2 **Model of the crystal structure of** *ras* **in its GDP and GTP states (PDB code** *1Q21* **and** *521P***) highlighting secondary structures (α-helices and β-strands) as well as switch-1 (effector domain) and switch-2 regions.**
The top left panel depicts the basic GTPase regulatory cycle and its two principal modulators: GEFs, responsible for activation, and GAPs, responsible for deactivation of the regulatory GTPases. The top right panel shows a two-dimensional representation of the three-dimensional features of the basic GTPase fold found in all regulatory GTPases.

as compared to the GDP state, referred to as switch I and switch II [3]. Switch I, aa32–40, is a large loop connecting α-helix 1 (α1) to β-strand 2 (β2). Mutations in this region interfere with activation of several of the downstream effectors of *ras*. Switch II, aa 60–75, changes conformation even more drastically than switch I, to the extent that upon binding of GTP, aa66–74 rearrange into a well ordered α-helix (α2). switches I and II are required for activation of downstream effectors by *ras*-GTP, the activated form of *ras*. *Ras*-GDP appears to be neutral. Mutations of Gln61 (at the start of switch II after β3), and of Gly12 (in the loop between β1 and α1–referred to as the P-loop), reduce the intrinsic GTPase activity of *ras*, prolonging the lifespan of the activated GTP state. As is the case for most (if not all) regulatory GTPases, the *ras* GTPase is under regulation of two types of proteins, a GEF (Guanine Nucleotide Exchange Factor) responsible for promoting the transition

of *ras*-GDP to *ras*-GTP, and several GAPs (GTPase Activating Proteins) which, as their name indicates, shorten the lifespan of *ras*-GTP by increasing the catalytic efficacy of its intrinsic GTPase (see [4] and references therein). SOS (Son Of Sevenless) and *ras*-GAP1, are the prototypes of *ras* GEF and *ras* GAP, respectively. Mutation of Ser17 to Asn17, close to the amino end of α2, interferes with the nucleotide exchange reaction and locks *ras*-GDP into an inactive conformation in which it may still bind to proteins regulated by *ras*-GTP but without affecting their activity. By reason of occupying the site to which *ras*-GTP should bind to activate effectors, Asn-17 *ras* interferes with signaling by wild-type *ras*-GTP and is referred to as a dominant negative form of *ras* [5]. Figure 46.2 depicts *ras* in its GTP and GDP conformations and highlights the locations of Gly12 (G12), Ser17 (S17), Gln61 (Q61), as well as switch I and switch II. The upper left panel of Figure 46.2

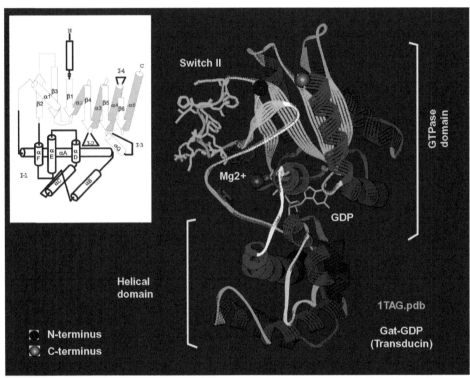

FIGURE 46.3 Crystal structure of an α subunit of a heterotrimeric G protein in its GDP state (PDB code *1TAG*).
Its GTPase domain is oriented in the same way as that of *ras* in Figure 46.2. Atoms of amino acid side-chains of Switch II and GDP are displayed as sticks. Mg is a light-blue ball. GTPase and helical domains are indicated. *Upper left inset*: two-dimensional diagram of the three-dimensional features of a heterotrimeric G protein a subunit. Additions to the *ras* structure are in heavy lines.

presents this basic GTPase cycle in schematic form, with GEF and GAP regulating the lifetimes of the GTP and GDP states. The upper right panel is a schematic two-dimensional representation of the main three-dimensional features of this regulatory GTPase (adapted from [3]).

HETEROTRIMERIC G PROTEINS

Trimeric G proteins are responsible for transducing the effects of seven-transmembrane receptors, and are regulatory GTPases which, while more complex than *ras*, nevertheless preserve the basic features of the *ras*-type regulatory GTPases. G proteins activated by seven-transmembrane receptors are αβγ trimers, of which the α subunit is the GTPase-bearing subunit. The β and γ subunits form a dimer that exists free, or bound to effector proteins (see below), or in association with α-GDP. Gβ and Gγ have never been found isolated as individual proteins.

Subunit Structure

α subunits were crystallized in the laboratories of Paul Siegler at Yale University, and of Steven Sprang at the University of Texas at Dallas, in collaboration with Heidi Hamm and Alfred Gilman, respectively [6, 7]. From a

structural viewpoint, α subunits of heterotrimeric G proteins are made up of a *ras*-like GTPase domain and a helical domain with 6 α-helices (αA through αF) inserted in the center of what would be *ras*'s switch I (Figures 46.3, 46.4). In their GDP-liganded form, Gαs tend to have a disorganized switch-II region and exhibit high affinity for Gβγ. Gα-GDPs are therefore found associated with Gβγ dimers as heterotrimers. Gβγ locks GDP into its binding site on Gα, causing its dissociation rate to drop by a factor of 300 [8]. At the same time, Gβγ also shields switch II from interacting with possible effectors (Figure 46.4). Consequently, activation of a heterotrimeric G protein requires two events: (1) exchange of GTP for GDP, and (2) dissociation of Gβγ from Gα. There is some discussion as to whether dissociation is an absolute requirement for deployment of the signaling capacity of Gα-GTP. At a minimum, the activation process has to involve the opening of a βγ complex similar to the opening of the valves of a clam. While in most cases the opening process includes separation of the subunits, there are situations in which the valves may remain connected at the region forming the hinge. The amino acids forming this hypothetical hinge have not been identified. Subunit dissociation has not been measured in intact cells in an unequivocal way, but is readily seen *in vitro*. Dissociation exposes the regulating surface of Gα-GTP and allows for effector regulation with direct

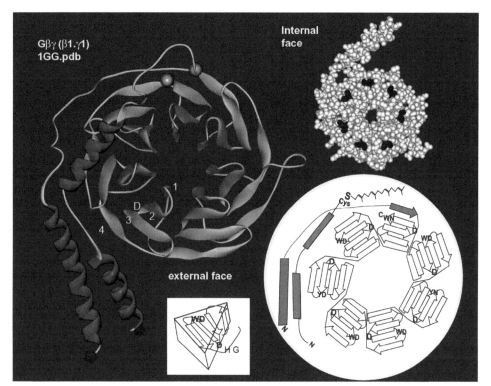

FIGURE 46.4 View of a seven-bladed Gβ propeller from the "external" side determined by its orientation when complexed with α-GDP. *Upper right inset*: Space-filling CPK representation of the same propeller viewed from the side facing a-GDP in its trimeric form. Blue: conserved aspartic acids (D) present in each propeller blade. *Bottom right*: two-dimensional diagram of the three-dimensional features of a Gβ propeller highlighting the "zipping," outermost β strand of blade seven, and location of conserved D residues shown in blue in the space-filling representation of the model. Note the coiled-coil interaction of the γ subunit with the N-terminus of the Gβ subunit.

involvement of switch II ([9]; PDB code *1CS4*). The converse also applies. That is, Gβγ is a signaling molecule able to regulate effector functions, and its association with Gα-GDP results in occlusion of its Gβγ's signaling surface.

The structure of Gβγ deserves separate comment ([10, 11]; Figures 46.4, 46.5). Gβ is a seven-bladed propeller, of which each blade is made up of four antiparallel β-strands running from the center to the periphery. The innermost β-strand runs parallel to the axis of rotation of the propeller. The next two β-strands change pitch to approach the orientation of the outermost strand, which runs along the periphery of the propeller and is co-planar with the circle described by the rotating propeller. Blade seven is made up of three β-strands contributed by the very C-terminus of Gβ, and a fourth (outermost) "zipping" β-strand recruited from the sequence immediately preceding those that create the first blade. Preceding the "zipping" β-strand is an extended N-terminus with a long α-helix that interacts with the Gγ by forming a coiled coil. Gβγ dimers have so far been crystallized only in association with Gα subunits or with a regulatory protein, phosducin, found in photoreceptor cells of the retina, but not in isolation. Conformational changes in Gβ upon dissociating from Gα, if they occur, have not yet been observed.

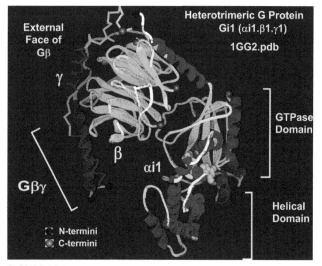

FIGURE 46.5 Model of G$_i$1 heterotrimer with bound GDP deduced from crystal X-ray diffraction studies (PDB code *1GG2*). Note that the internal face of Gβ shields the Gα subunit's switch II from being accessible to effectors. Blue balls, N-terminal α carbons; brown balls, C-terminal α carbons; gray balls on Gβ, α carbons of conserved aspartic acids.

Lipid Modifications

Gα and Gβγ subunits engaged in signal transduction are membrane bound by virtue of lipid modifications. Gβγ dimers are anchored to membranes through a C15 or C20 polyisoprene attached in thioether inkage to the Cys of a C-terminal CAAX motif of the Gγ subunit (where C is cysteine, A is an aliphatic amino acid, and X is any amino acid). Posttranslational processing not only attaches the polyisoprene, but also removes the last three amino acids and methylates the new C-terminus. Prenylation is not necessary for association of Gγ with Gβ, but is required for association of Gβγ to Gα-GDP and for regulation of effector – e.g., adenylyl cyclase. Further, prenylation contributes to the association of the Gβγ dimer to membranes.

Most Gα subunits engaged in signal transduction are palmitoylated in thioester linkage to a cysteine near their N-terminus. Palmitoylation facilitates their anchoring to the plasma membrane. Gα subunits of the G_i/G_o family are also myristoylated in amide linkage at their N-terminal glycines (Gly2 of the primary transcript). Gα myristoylation increases affinity for Gβγ dimers. Removal of the myristic acid by Gly2 to Ala2 mutation renders G_iα subunits inactive as inhibitors of adenylyl cyclase. As a C16 aliphatic acid, palmitic acid is a stronger membrane anchor than the C14 myristic acid. Some but not all non-palmitoylated Gα subunits fail to localize to membranes, and are found in the cytosol. Lipid modification of Gα and Gγ subunits are therefore essential for their normal biological activity (reviewed in [12]).

Molecular Diversity of G Proteins

Each of the subunits that make up a heterotrimeric G protein is encoded by a family of structurally homologous genes. There are 16 Gα (one with two splice variants), 5 Gβ and 11 Gγ genes, raising the theoretical possibility of close to 1000 distinct heterotrimers. Gα subunits are the longest of the three subunits, ranging from 350 to 390. Their sequence similarities vary from almost identical when G_i1α is compared to G_i3α (86 percent identical) to up to 60 percent different when G_sα is compared to G16α. Amino acid sequence alignments show Gβs to be a structurally very closely related family, with β1–β4 being each 350 aa long and differing by no more than 17 percent in their amino acid sequence. Gβ5, with 395 aa, exhibits the same degree of similarity, differing primarily by a 45-aa N-terminal extension. Gγs are the shortest (68–75 aa) and the most diverse of the three subunits, differing in amino acid sequence between 40 and 65 percent. GGL (Gγ-like) domains of RGS 6, 7, 9, and 11 (for RGS see below) constitute an additional group of Gγ subunits that interact with Gβ5's atypical N-terminal extension.

The actual number of G protein isoforms in any given cell is much lower than 1000, for two reasons: (1) there is

no cell known that expresses all G protein subunit genes, and (2) there are structural limitations that do not allow all βγ dimers to form – for example, while β1 interacts with γ1, β2 and β3 do not; β3 does not interact with γ1 or γ2, but partners with γ4. Conversely, γ1 partners with β1 but not with either β2 or β3, and γ2 partners with β1 and β2, failing to do so with β3. The complete spectrum of permissible interactions among the 5 β and 11 γ subunits still needs to be worked out. One Gβ, Gβ5, interacts preferentially with the GGL domain of RGS 6, 7, 9, and 11. It is quite possible that even for biochemically permissible interactions, there may be βγ dimers that never form because they are not co-expressed in the same cell.

G proteins are named after their α subunits. This has as its origin in the fact that for the first two G proteins discovered, G_s and transducin (G_t), it was established that the major (then sole) signaling function of the G protein resided in their α subunits. Gαs activates adenylyl cyclases (ACs) and Gα$_t$ subunits activate visual phosphodiesterase (PDE, a tetramer of one α, one β, and two inhibitory γs). Activation of visual PDE results from the association of αt-GTP subunits with PDEγ subunits, thereby suppressing their inhibitory effects on PDEαβ. A phylogenetic tree of G-protein α subunits (Figure 46.6) clusters their sequences into four subfamilies: G_s, G_q, G12/13, and the PTX-sensitive Gα-subunit subfamily. The latter includes three α subunits that play roles in light and taste perception, plus the α subunits of the G_i/G_o family. PTX-sensitive α subunits not only show higher sequence similarity to each other than to the remaining α subunits, but also, as their name indicates, are substrates for the ADP-ribosyl-transferase activity of the S1 subunit of pertussis toxin. The ADP-ribosylated amino acid

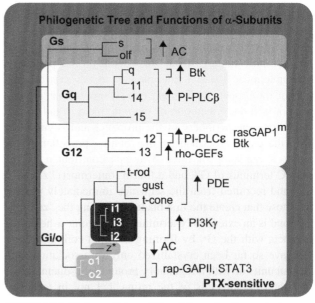

FIGURE 46.6 Phylogenetic tree of G-protein α subunits reveals structural subdivisions that have functional correlates.

is a Cys at position -4 from their C-termini. PTX uncouples this group of G proteins from activation by receptors by virtue of creating of a steric hindrance to the G protein–receptor interaction. From a biochemical point of view, even though G proteins ADP-ribosylated by PTX become non-functional, the modification does not inactivate these α subunits. Under *in vitro* conditions, they can still be activated by non-hydrolyzable GTP analogs. Included in this structural group is the α subunit of G_z (α_z), which lacks a Cys at -4 from the C-terminus and is PTX-insensitive. It is functionally a G_i, its closest structural homolog. As shown in Figure 46.6, and listed in more detail in Table 46.2, α subunits modulate the activity of a large and diverse group of enzymes, including ACs, phospholipase Cβs (PLCβs), phosphatidylinositol 3-OH kinases (PI3Ks), as well as type-6 visual and type-3 gustatory PDEs and Rho-GEFs for regulation of cytoskeletal remodeling. An analysis of

the evolution of the intron–exon structure of G-protein α subunits (Figure 46.7) clusters the genes into the same groups: G_s, $G_q/11$, G12/13, and $G_{i/o}$ and $G_{taste/vision}$. This type of analysis places G_z in the G_i/G_o group solely on the basis of open reading frame-sequence similarity, but shows no similarity in its intron–exon organization to those of any G-protein α subunit as it has only three exons (two coding) instead of the eight or nine exons of the $G_{i/o}$ family of α subunits. The loss of introns is highly suggestive of reactivation of a pseudo-gene with attendant chance mutation of the Cys at -4 from the C-terminus.

Both Gα subunits and G$\beta\gamma$ dimers regulate, positively or negatively, a diverse set of cellular functions (reviewed in [13, 14]). In some instances the effects of G$\beta\gamma$ dimers are in concert with those of α subunits, in others regulation of effector by G$\beta\gamma$ is unrelated to regulation by a Gα. It is not clear at this time whether effectors distinguish G$\beta\gamma$ isoforms.

TABLE 46.2 G-protein α-subunit effector systems identified by functional or direct interaction

α Subunits		Effector/binging partner	Effect–comment
$G_s\alpha$		AC I through VII, also AC IX	Stimulation
		Caveolin	Subcellular localization
$G_{olf}\alpha$		Most likely all ACs	
$G_i1\alpha$, $G_i2\alpha$ $G_i3\alpha$		All ACs except AC II & IV	Inhibition
	αi2	AC IX not tested	Stimulation by suppression of inhibition by ATP
	αi3	Kir 6::SUR	Anti-autophagy
	αi1	?	Stimulation
		PI3 kinase γ	
$G_o\alpha$	αo	AC I (not AC V or VI)	Inhibition
	αi/o	rap1-GAPII (GoLoco domain)	Promotes degradation of Rap1-GAPII
	αo	PKAcat α & β	Sequestration in extranuclear compartment
		?	Vesicular monoamine and glutamate transporters (VMAT and VGLUT)
$G_t\alpha$ rod, cone		PDE6γ subunit	Releases inhibition of GMP-specific PDE$\alpha\beta$ by DEγ
$G_t3\alpha$ (gustducin)		cAMP-specific Ca/CaM-stimulated PDE 1A	Stimulation (mechanism:?)
$G_q\alpha$	αq, α11, α14 & α16	Phospholipase Cβ1 through β4	Stimulation
	αq	Btk	Stimulation
G13α		p115-RhoGEF	Stimulation
		PDZ-RhoGEF	Stimulation
		LARG	Stimulation
Gα12		LARG tyrosine-phosphorylated by TEC kinase	Stimulation
		Btk	Stimulation
		ras-GAP1^m	Stimulation

AC, adenylyl cyclase; Btk, Brunton's tyrosine kinase; CaM, calmodulin; GAP, GTPase activating protein; GEF, Guanine nucleotide exchange factor; Kir, inwardly rectifying K$^+$ channel; LARG, Leukemia-associated Rho-GEF; PDE, phosphodieatserase; PI3K, phopshatidylinositol 3 kinase; PKA, cAMP-stimulated protein kinase.
TEC kinase, a member of a family of non-receptor cytosolic tyrosine kinases, founding member: TEC, other members: Btk, Itk.

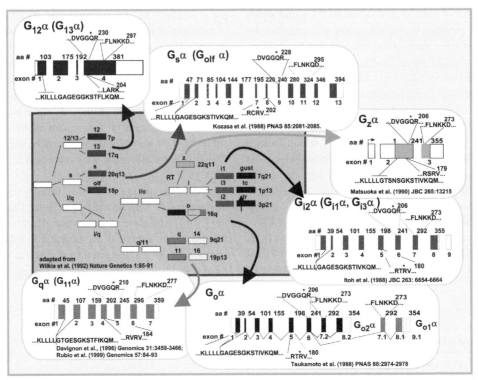

FIGURE 46.7 Analysis of the intron – exon structure, and sequence similarities among proteins encoded in the exons, allows for development of a model for the evolution of α-subunit genes in which a series of gene duplications, and loss of one intergenic region including most of the exons of the downstream gene followed by rescue of a pseudogene, lead to the present-day G protein α subunits.

The general layout of the figure was adapted from a similar layout [43] (Wilkie *et al.*, 1992). Other references on the figure refer to the publications in which the intron–exon structures shown were first described.

As shown in Table 46.3, the gamut of Gβγ effectors is as complex and as diverse as that of the α subunits.

Among the regulated functions worth mentioning are inhibition of type-1 AC, but co-stimulation with Gsα of type-2 and type-4 ACs; stimulation of type-3 PLCβ independently of co-existing stimulation by the Gq/11 group of α subunits; co-stimulation with Gαi2 (possibly also αi1 and αi3) of PI3Kγ; and co-stimulation of PI3Kβ with tyrosine phosphorylated p85, the regulatory subunit of PI3K. *In vitro* reconstitution experiments in which PI3Kβ was incubated with Gβγ and a tyrosine phosphorylated peptide corresponding to the tyrosine phosphorylated sequence of p85, showed that stimulation by each was 3- to 5-fold, but became 100-fold when the tyrosine phosphorylated peptide and the Gβγ were added together. This type of cross-dependence on dual inputs gives rise to complex G-protein signaling networks, such as that illustrated in Figure 46.8 for regulation of adenylyl cylcases by G proteins and in Figure 46.9 for regulation of phospholipid and Ca²⁺ signaling.

Gβγ dimers also regulate a variety of ion channels (Table 46.3), some positively and others negatively, and thereby provide a nexus between regulation of second-messenger formation by enzymes and regulation of cell excitability by voltage, which Gβγ dimers tend to dampen

(activation of potassium channels and inhibition of presynaptic Ca²⁺ channels).

In conclusion, signal transduction by G proteins is the result of structurally similar receptors activating structurally very similar G proteins, which then regulate, positively or negatively, the activity of a diverse gamut of structurally unrelated cellular functions that affect intracellular levels of cAMP, inositol tris-phosphate (IP3), Ca²⁺, diacylglycerol (DAG), phosphatidyl inositol 3 phosphates (PIP3s), as well as the cytoskelton and ion channel activity. Through their interactions with Rho-GEFs (Figure 46.9), the G12/13 subfamily of G proteins regulates lammelipodia and filopodia and the attendant cytoskeletal changes. Moreover, since second messengers such as cAMP, DAG, and Ca²⁺ affect protein kinases, it is unlikely that there is a cellular function that in one way or another is not under the controlling or modulatory influence that emanates from activation of heterotrimeric G proteins. Indeed, as shown in Tables 46.2 and 46.3, and Figures 46.8–10, cellular responses include the activation of PI3K and its sequels, including PDK-Akt-NFκB anti-apoptotic responses, the effects of Gβγ on dynamin, Golgi vesiculation and vesicle budding, and the effects of Gβγ resulting from interaction with SNAP-25 affecting neurotransmission [15, 16].

TABLE 46.3 G-protein βγ-dimer effectors identified by functional or direct interaction

Effector/binding partner		Effect–comment
AC I		Inhibition
AC II & IV		Stimulation; dependent on GTP-αs
PLCβ1–4, not PLCβ4		Stimulation
Kir3 (GIRKs)	Kir3.1::Kir3.2 Kir3.2a::Kir3.2c Kir3.1::Kir3.4 (CIR) [Kir3.4]₂	Stimulation, dependent on PIP2
Kir6 (Kir,ATP; BIR)::SUR		Stimulation [suppression of inhibition by ATP]; the interaction appears to be with SUR
	Kir6.2::SUR2A	
GRK2 & -3 (βARK1 & -2)		Gβγ acts as scaffold and as an activator
Ca channels		
	CaV2.1-2.-3 (α1C-B-E)	Inhibition - relieved by depolarization (e.g. prepulse)
PKD (PKCμ)		Stimulation – affects vesicle budding processes
Dynamin		Inhibits GTPase – affects dynamin dependent vesicle budding
Shc		βγ acts as scaffold to nucleate Shc-Grb2-SOS-src
PI3Kγ (p101::p110)		Stimulation
PI3Kβ (p110::p85)		Stimulation – depends on pY interacting with p85
Glucocorticoid receptor		Inhibits activation of transcription by GR in the nucleus
Brunton's kinase		Interaction with PH domain - facilitates activation by Gqα and G12α
SNAP25		Inhibition of neurosecretion

name::name, *denotes functional unit formed of the indicated subunits.*
AC, adenylylcyclase; βARK, β-adeneric receptor; BIR, β cell Kir; CIR, cardiac Kir; GIRK, G-protein-regulated inwardly rectifying K⁺ channel; GR, glucocorticoid receptor; GRK, GPCR kinase; Kir, inwardly rectifying K⁺ channel; PI3K, phosphatidylinositol 3 kinase; PIP2, phosphatidylinositol 4,5-bisphosphate; PK, protein kinase; PLC, phospholipase C; pY, phosphotyrosine; SUR, sulfonylurea receptor.

MECHANISM OF G-PROTEIN ACTIVATION BY RECEPTORS

Receptors acting through heterotrimeric G proteins play the role of GEFs in the regulatory GTPase cycle. Ligand binding to GPCRs has, as its final effect, the GDP–GTP exchange with attendant subunit dissociation into Gα-GTP plus Gβγ (Figure 46.1). The final effect of a hormone acting through a GPCR on any given cell depends on the type of G protein activated by the receptor, and the repertoire of effectors (i.e., regulated enzymes, ion channels, and other affected molecules) in the target cell.

Mechanism of Activation of a G Protein by a Receptor

At the molecular level, activation of a G protein by a GPCR is still only partially understood. This is because of lack of knowledge regarding which amino acids of the receptor make contact with which amino acids of the G protein. In contrast, the regions of each molecule important for productive interaction are well known, as are some of the kinetic and molecular state changes that occur when a receptor under the influence of an activating ligand (i.e. an agonist) interacts with and activates a G protein. Thus, binding of an agonist

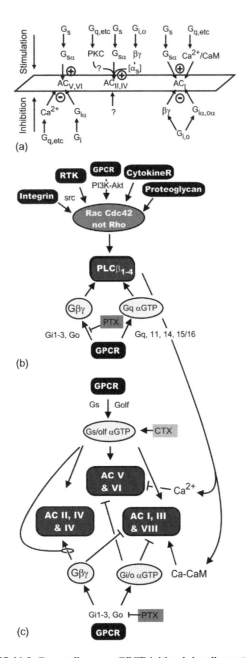

(a)

(b)

(c)

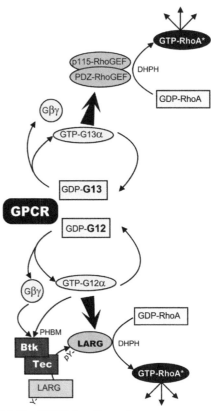

FIGURE 46.9 The G12/13 signaling pathway.
Rho-GEF activities involve their DHPH domains. DH, Dbl-domain; PH, pleckstrin homology domain; BM, Btk domain; Y, tyrosine; pY, phospho-tyrosine. G12α and Gβγ are shown as activators of the Tec kinases Tec and Btk. The interaction has been shown to involve the PHBM domain of the Tec kinase [46–48]. Adapted from [14] (Birnbaumer, 2007).

FIGURE 46.8 Cross-talk among GPCR initiated signaling pathways involving the G$_s$ and the G$_q$/11 G proteins.
(a) Adenylyl cyclases are targets of multiple regulatory signaling pathways and respond differently, depending on which group they belong to. The figure shows these differences as summarized in 1994 (adapted from [44] (Taussig *et al.*, 1994)). (b) The β-type phosphoinositide-specific phospholipases C are targets of three different signaling pathways (for signals impinging on the Rac-Cdc42 GTPases, see [45] (Cerione, 2004) and references therein), the G$_q$α family of α subunits (Table 46.2), and Gβγ dimers (Table 46.3). (c) Example of cross-talk from the G$_q$ (G$_q$α), G$_i$ (Gβγ), and Cdc42-Rac signaling pathways to the G$_s$ signaling pathway may generate increases or decreases in cAMP levels depending on the subset of adenylyl cyclases expressed in the target cell. Adapted from [14] (Birnbaumer, 2007).

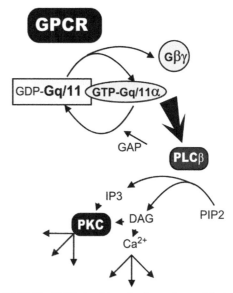

FIGURE 46.10 The G$_q$/11 signal transduction pathway. Adapted from [14] (Birnbaumer, 2007).

to a GPCR in the absence of guanine nucleotide (GTP or GDP), as can be done *in vitro* with purified membranes, has two consequences: (1) a shift of the equilibrium between two states of the receptor, from being mostly in state I (inactive) characterized by having low affinity for agonist as well as for the G protein, to mostly in state II (active) having higher affinity for the activating ligand and also for the unoccupied, nucleotide-free trimeric G protein. This leads to the stable association of the agonist–receptor complex to the G protein. The latter causes the G protein to reduce its affinity for GDP. Bound GDP, or pre-bound [3H]GDP, will thus dissociate under these conditions. Mg ion has to be present if receptor is to cause GDP dissociation, as Mg, by an as yet unknown mechanism, reverses the affinity-increasing effect of Gβγ. Addition at this point of GTP or a GTP analog such as GTPγS or GMP-P(NH)P leads to its binding in place of GDP (binding that is of much higher affinity than that of GDP owing to the presence of Mg, the co-activator of G proteins) and the attendant activation of the G protein. For most of the cases where this has been studied, a high concentration of Mg ion, ~50 mM, mimics the action of the agonist-activated receptor in facilitating dissociation of GDP. With a purified G protein, incubation with Mg ion and GTPγS or GMP-P(NH)P leads not only to accumulation of Gα-bound guanine nucleotide but also to subunit dissociation – i.e., formation of Gα-GTPγS plus free Gβγ, Dissociation is evident in several ways, the easiest being by a shift in sedimentation velocity from that corresponding to an approximately Mr 100,000 protein (Gαβγ) to that of two co-sedimenting proteins of Mr ~50,000 (Ga-GTP + Gβγ). The Mr values of α subunits are in the 40,000–50,000 range, and those of Gβγ complexes are also approximately 50,000.

In intact membranes, where activation of a G protein of the αβγ type by agonist occupancy of a receptor can be measured in terms of stimulation of the activity of an effector (e.g., adenylyl cyclase, phospholipase C, visual phosphodiesterase, or an inwardly rectifying potassium channel), the net effect of receptor activity is thus facilitation of the activation of the G proteins by Mg-GTP. This comes about as a consequence of a receptor-induced shift in the apparent K_m for Mg ion from high millimolar to low micromolar. In other words, a receptor appears to act by reducing the concentration of Mg required for activation of the G protein by GTP from being above physiologic to being below physiologic. Free cytosolic Mg is in the order of 0.5 mM. The effect of glucagon (receptor) shifting the concentration of Mg ion required for activation of liver G_s (the stimulatory regulatory component of adenylyl cyclase) by GTPγS is illustrated in Figure 46.11.

Molecular Activation Sequence of a GTPase by GTP and Mg^{2+}

The structures of crystallized trimers, and α subunits in the presence of GDP, GDP-Mg, or stable GTP analogs in

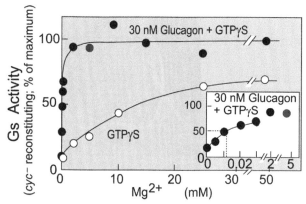

FIGURE 46.11 Hormone-stimulated activation of a G protein by GTP-Mg.
At the biochemical level, activated HR complexes have the overriding effect of forming Mg^{2+} binding pockets, as shown, for example, by the reduction in the concentration of Mg ions required for G-protein activation by GTP. The results of a three-step reaction are shown. In the first step (not shown), the adenylyl cyclase enzyme of liver membranes was inactivated by treatment with N-ethyl maleimide, leaving an intact receptor–G protein system in its natural membrane environment. The second step tested for the effect of varying Mg ions on GDP/GTPγS exchange in the absence and presence of the hormone glucagon. In the third step, the G_sα-GTPγS complexes formed were extracted and quantified in a standard reconstitution assay. The figure shows the effect of the glucagon-activated receptor on the Mg required for G-protein activation. Note that (1) hormone was not necessary for G_s activation, as long as a high enough level of (supra-physiologic) Mg was present during incubation with GTPγS, and (2) that in the presence of the hormone, the Mg required to activate G_s was ~1000-fold lower than in its absence. *Inset*: Same as main panel, but with an expanded Mg concentration scale. Adapted from [49] (Iyengar and Birnbaumer, 1982).

the presence of Mg^{2+}, have provided a clear picture of the activation sequence for a G-protein α subunit. Not surprisingly, both the nucleotide and Mg^{2+} ion play key roles. Thus, both *ras*-like regulatory GTPases and heterotrimeric G-protein α subunits acquire their active or "effector-plus" conformation by the combined binding of GTP and Mg^{2+}. Biochemical analyses have shown that while the equilibrium dissociation constant (K_d) for the dissociation of Mg from Mg-GTP is 60 μM, that for the dissociation of Mg from the α subunit–Mg-GTP complex is 10–20 nM [8, 17, 18]. This indicates that the protein contributes to the binding of Mg^{2+} in major ways and, conversely, that Mg^{2+} acts as a keystone to lock the GTPase fold into the effector-plus conformation. The effector-plus conformation cannot be stabilized in absence of Mg, as it cannot be stabilized in the absence of GTP. The molecular mechanism by which this comes about is essentially the same for the *ras* GTPase as for the heterotrimeric α subunits. Crystals of *ras* and α subunits of trimeric G proteins with bound GDP-Mg and GTP-Mg show that the six coordination bonds that hold Mg^{2+} in place in the protein – Mg-GDP complex are given by one oxygen of the β-phosphate of GDP, one oxygen from what in *ras* is Ser17, and the oxygens of four water molecules. One of these water molecules is stabilized by a

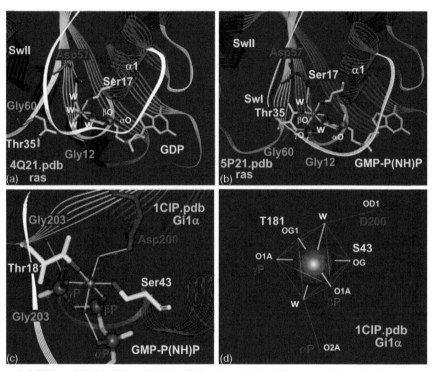

FIGURE 46.12 GTP-liganded GTPase folds bind Mg with high affinity due to their ability to coordinate Mg through two non-water linkages: that of an oxygen of the γ-phosphate and that of the γ1 oxygen of a Thr*(Thr35 in ras).
(a), (b) Coordination of Mg as seen in GDP-*ras* and GTP-*ras* crystals. (c) Coordination of Mg as seen in a $G_i1\alpha$ crystal. Note the similarity between the *ras*-GTPase fold and the Gα subunit GTPase fold. Numerical equivalences between *ras* and α-subunit amino acids are shown in Table 46.4. (d) Octahedral coordination of Mg in GTP–protein complexes as seen in the crystal of $G_i1\alpha$-GTP-Mg.

hydrogen bond to a carboxyl oxygen of an Asp (OD1, Asp57 in *ras*) and the other is stabilized by a hydrogen bond to one of the oxygens of the α-phosphate of the GDP (O1A). In the protein – Mg-GTP complex, Mg is essentially in the same place in space as in the protein – Mg-GDP complex [18], but its position is stabilized further – and hence its affinity is increased – by replacement of two of the coordinating water molecules by what in *ras* is the oxygen of Thr35 (OD1) and by one of the oxygens of the γ-phosphate (OG2). The two remaining water molecules are those stabilized by hydrogen bonds to OD1 of Asp and O1A of the α-phosphate. To allow for coordination of Mg by Thr35, the protein has to change the conformation not only of switch II, but also of switch I (Figure 46.12, Table 46.4).

Structural Determinants of the Receptor – G Protein Interaction

On the GPCR side, mutational analysis has shown that amino acids in the third intracellular loop of GPCRs are involved in the ability of a GPCR to activate a G protein. Swapping intracellular loops between receptors of different G-protein preference, such as between M1 and M2 muscarinic receptors, or β- and αl-adrenergic receptors, also points to the third intracellular loop as responsible for defining G protein specificity. Further, most of the receptor mutations that are of the gain-of-function type are in the distal (C-terminal) end of the third intracellular loop. It is not known, however, why some mutations are activating and others inactivating. Moreover, it is also not known whether these amino acids actually contact the G protein and, if so, which G-protein subunit they contact (for discussion, see [19–22]).

Mutational analysis and sequence-swapping experiments with G-protein α subunits indicate that receptors interact with the very C-terminus of the G-protein α subunit. Indeed, swapping as few as 3 of the last 10 amino acids between two α subunits can lead to a switch in the type of receptor that activates the G protein [23]. The C-terminus is not the only region of interaction of an α subunit with a receptor. Multiple sites have been identified by mutational analysis, including the α3β5 and α4β6 loops of the GTPase domain (see insets of Figures 46.2 and 46.3; [21, 24]). The α-subunit C-terminus, and the α3β5 and α4β6 loops, are part of the same face of the molecule presumed to be immediately juxtamembranous and to be recognized by the receptor.

The crystal structure of rhodopsin was solved by Palczewki and collaborators (reviewed in [25]). It is a ligand-inhibited GPCR in which the inactivating ligand is the 11-cis retinal covalently bound in Schiff-base linkage to Lys296 of the apo-protein opsin. Ligand-free opsin is a constitutively active GPCR (Ops*). Ops* has been crystallized in isolation

TABLE 46.4 Coordination bonds of Mg^{2+} and conformation-stabilizing hydrogen bonds involved in activation of a regulatory GTPase by GTP amd Mg^{2+}

Bond type	Ras	Gsα	Gi1α	Tα
A. Coordination bonds				
Ser OD1 to Mg	Ser17	Ser54	Ser43	Ser43
Thr OG1 to Mg	Thr35	Thr204	Thr181	Thr177
GTP OB1 to Mg				
GTP OG1 to Mg				
HOH1 O to Mg	Asp57	Asp223	Asp200	Asp196
HOH2 O to Mg				
B. Hydrogen bonds				
Asp OD1 to HOH1				
αP 0A1 to HOH2				
γP OG2 to NH	Gly60	Gly226	Gly203	Gly199

[26] and as a complex with the 11 C-terminal amino acids of transducin α (GαCT) [27]. Together with spectroscopic studies which described the changes occurring at the receptor during the transition from inactive to active states [28, 29], the crystal structures describe a model in which the receptor attracts the C-terminal α5-helix of the Gα subunit to enter into a pocket formed by transmembrane segments (TM) 7, 6, 5, and 3, and in so doing causes the α5-helix to move along its longitudinal axis. This movement is thought to cause a change in the positioning of the β6α5 loop, causing GDP to dissociate [27, 29, 30]. The pocket with the GαCT is depicted in Figure 46.13aii.

The β6α5 loop amino acids of G-protein α subunits (Thr–Cys–Ala–Thr–Thr–Asp–Thr or TCATDT) are highly conserved. Of these, A (Ala326 in G$_i$1α and Ala366 in G$_s$$\alpha$) is invariant. Ala and its neighboring Cys are part of a lid that prevents GDP from dissociating (Figure 46.13b & c). [Ala366Ser]G$_s$$\alpha$ is a naturally occurring mutation that results in spontaneous activation of G$_s$$\alpha$ with attendant very early-onset puberty in men. The reason for spontaneous activation of the Ala366Ser mutant is an augmented spontaneous dissociation rate of GDP allowing entry of the activating GTP and Mg [31]. Taken together, the crystal structures, spectroscopic measurements, and docking of the C-terminus of the α5-helix in a pocket created upon receptor activation points to a change in the conformation of the β6α5 loop at the other end of the α5-helix as the mechanism by which receptors trigger activation of G proteins.

Receptors do not only interact with the α subunit of the trimeric G proteins. Free Gα subunits are not recognized by receptors; they are only recognized in the context of the heterotrimer. In agreement with this idea, injection of subunit-specific antisense oligonucleotides, or subunit-specific antibodies, suppresses receptor-mediated effector regulation, not only upon suppressing Gα [32, 33] but also either Gβ or Gγ subunits [34, 35], all in an isoform-specific manner. It has been shown that in pituitary cells the M4 muscarinic receptor activates a G$_o$ G protein of subunit composition αo1β3γ1, while the somatostatin receptor activates a G$_o$ of subunit composition αo2β1γ3. It follows that receptors "proofread" the subunit subtypes that make up the particular trimer they enter in contact with.

Figure 46.4 shows that only one of the two faces of the Gβ propeller is exposed to the milieu while the other faces ("looks at") the a subunit's switch-II region. The exposed face and the sides of the propeller are therefore available for interaction with receptor. In turn, since receptor interacts with Gβ and Gγ, it can reasonably be expected that receptors may affect the G$\beta$$\gamma$ interaction with Gα. Regardless of the final outcome of the events responsible for G-protein activation by a receptor at the submolecular level, the overall reaction for a receptor activating a G protein is to facilitate the exit of GDP with automatic replacement with GTP. This replacement is driven energetically by the fact that GTP concentration is roughly 10 times that of GDP, and that binding of GTP and Mg^{2+} are cooperative

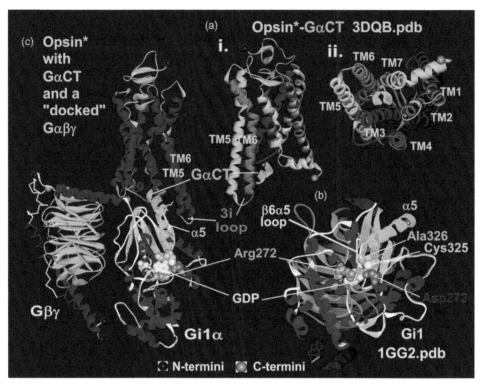

FIGURE 46.13 Model of the mechanism by which a GPCR activates a heterotrimeric G protein.
(ai), (aii) Two views of the active conformation of the ligand-free active conformation of opsin (Opsin*) occupied by the 11-amino acid C-terminus of transducin (GaCT, green). Note that GaCT sits in a pocket formed by transmembrane segments (TMs) 7, 6, 5, and 3 (aii). (b) View of the Gα subunit from the "bottom," illustrating how amino acids of the β6α5-loop and elsewhere prevent exit of the GDP from its binding site. (c) The trimeric G protein G$_i$1 has been docked manually to the Opsin*–GaCT complex. The crystal structure of G$_i$1 has an eight amino-acid C-terminal truncation so that only three amino acids overlap between G$_i$α1 and GaCT. Details and angles between Opsin* and G$_i$1 are tentative, but illustrate the most likely mechanism by which a GPCR activates a heterotrimeric G protein.

and essentially irreversible, as described above. Acquisition of the effector-plus conformation and subunit dissociation are therefore obligatory consequences of the association of GTP and Mg to the α subunit of the trimer. Reversal requires GTP hydrolysis.

MODULATION OF ACTIVITY BY SHORTENING OR EXTENSION OF THE ACTIVE STATE

RGSs or Regulators of G-Protein Signaling

As there are receptors that, by virtue of their GEF activity, promote activation of the heterotrimeric G proteins, there are also GAPs, GTPase activating proteins, that accelerate the GTPase activity of activated, MgGTP-liganded Gα subunits. Two types of Gα GAPs have been identified. One type is the RGSs, or Regulators of G-protein signaling. RGSs accelerate GTPase activity by 100-fold or more, and exhibit Gα-subunit selectivity. Sixteen RGSs are known, and many of them are multidomain, multifunction proteins, and thus

not only affect the Gα subunits but also aid in the organization of multicomponent "signaling complexes" [36]. The second type of GAPs are some of the effectors regulated by the α subunits. GAP activities of effectors increase k$_{cat}$ of the Gα-GTP complexes by only 10- to 20-fold. In both instances, increased GTP hydrolysis insures not only prompt turn-off of the signaling protein, but also a faster approach to equilibrium, therefore increasing the rate at which responses to an extracellular stimulus can be obtained. Indeed, for RGS proteins, faster on/off rates of the regulated function may be a primary *raison d'être*. In contrast, when effectors act as GAPs, the primary purpose may be to insure that they are indeed affected by the activated α-GTP complex. The intrinsic GTPase activity of Gα subunits is very low, in the order of 4–8 s^{-1}, giving them a rather long half-life, insuring that they "find" their effector(s) while still in their GTP state. Once the effector has been found and the receptor message has been delivered, the recipient of the message "kills" the messenger through activation of its deactivating mechanism. Thus, continued stimulation of effector, if this is desirable, requires continued presence of receptor agonist and constant reactivation of the G protein.

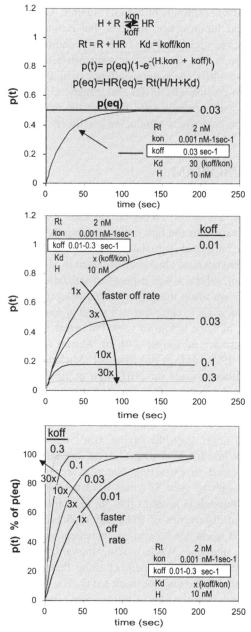

FIGURE 46.14 Approach to equilibrium of a reversible bimolecular reaction,

$$\text{H} + \text{R} \underset{\text{koff}}{\overset{\text{kon}}{\Longleftrightarrow}} \text{HR}$$

HR = 0 at t_0. The panels illustrate the effect of increasing the koff rate, equivalent to the effect of introducing a GTPase-activating RGS into a regulatory GTPase cycle, on the rate at which equilibrium is reached and on the proportion of the total binding molecules activated.

Figure 46.14 presents simulations that illustrate the effect of GAPs in accelerating the rate at which equilibrium is established in a simple on/off reaction, such as binding of a hormone H to its receptor R. The three panels show (1) the basic rate at which a bi-molecular reaction reaches equilibrium, (2) the effect of increasing the koff rate on the rate at which equilibrium is reached, expressed

as HR formed in absolute concentrations, and (3) the fact that while the number of complexes at equilibrium decreases with increasing values of koff, the rate at which equilibrium is reached increases with increases in koff. It follows that, given the low intrinsic GTPase activity of Gα subunits, the need for a rapid response can only be satisfied by existence of both a GAP and a very high concentration of reactants, so that the amplitude of the read-out signal (regulated effector) is large enough upon activation of only a small fraction of the GTPase. This is in fact the case with the activation of transducin (heterotrimeric G protein) by rhodopsin. The GTPase of transducin is stimulated by RGS9, a GGL RGS. This insures that the physiological rapid turn-off occurs within a tenth of a second, as opposed to having a half-life of 10 seconds ($t_{1/2} = \ln2/koff = 10\,s$, where koff corresponds to published intrinsic GTPase activity of Gα subunits of 4 per minute) [37]. However, to insure that sufficient active transducin will be formed, Mother Nature endowed the visual system with the highest known concentrations of receptor (rhodopsin; 40 percent of disc membrane protein) and regulated G protein (transducin; 10 percent of disc protein) (reviewed in [14]).

AGSs, Activators of G-Protein Signaling – the GoLoco Domain

As described above, the free Gα-GTP and the Gβγ released upon G-protein activation are both active regulators of effector systems. RGSs and effectors acting as GAPs shorten the lifetime of the active Gα with formation of Gα-GDP. Due to the high affinity of Gα-GDP for Gβγ, a GAP activity accelerates not only de-activation of Gα-regulated effectors, but also that of the Gβγ-regulated effectors from which Gβγ is sequestered by Gα-GDP. AGSs, or activators of G-protein signaling, were uncovered in a search for molecules that potentiate the effect of α-factor in baker's yeast [38]. The yeast α-mating factor acts by activating an αβγ heterotrimeric G protein and initiating a Gβγ-mediated cascade of reactions that lead to growth arrest and preparation for mating to the opposite mating type. One of the AGSs, AGS3, was found to potentiate Gβγ signaling by binding preferentially to Gα-GDP, and to do so via a domain found in unrelated proteins and referred to as GoLoco. In mammalian systems, GoLoco domains act as they do in yeast: they bind preferentially to GDP-liganded Gα subunit of the G_i/G_o type, thus prolonging Gβγ signaling. In addition to AGS3, proteins with a GoLoco domain include several RGSs and Rap-GAPII. This hints at a role that transcends its direct function of binding the GDP forms of G_i/G_o and involves participation in integration of multicomponent signaling pathways (reviewed in [39]).

Even though they were identified in the same type of bioassay, AGS1, AGS2, and AGS3 differ in their mode of action. AGS1 (also Rasdex1) is a *ras*-related protein that appears to act as a GEF, whereas AGS2 interacts with Gβγ and AGS3 binds to α-GDP. Why a Gβγ-interacting protein (AGS2) would enhance the effectiveness of either a Gβγ-or a Gα-GTP remains to be determined. Another Gβγ interacting protein, phosducin, serves to attenuate the action of transducin in the retina. The existence of Gβγ- and Gα-interacting proteins (RGSs, GoLoco proteins, phosducin, and non-GoLoco AGSs) points to the fact that proper cell homeostasis requires fine tuning of the basic regulatory G-protein cycle. These fine-tuning mechanisms are therefore responsible for the ultimate ability of a cell to live a productive life that is in concert with the needs of the whole organism.

FUTURE DIRECTION

While research into the activation of rhodopsin and coupling to transducin α has made fantastic advances, the story is still incomplete. Figure 46.15 shows the same "docked" complex depicted in Figure 46.13c, but instead of being viewed from its side it is shown as it would be seen from the top. Clearly, the footprint of the GPCR is too small to account for its proofreading ability whereby it checks for the isoform identity of all three G-protein subunits. Is it that receptors have two activities – one to trigger GDP dissociation by the mechanism described above, and the second to proofread the subunits? Evidence has accumulated, for some GPCRs, that they form functional dimers. Could such dimers cooperate, with one molecule doing the proofreading and in so doing positioning the other molecule to

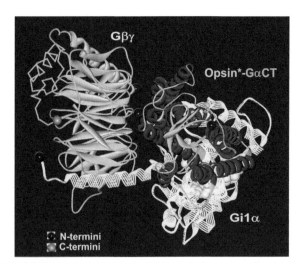

FIGURE 46.15 The footprint of a GPCR–opsin–is much smaller than that of a heterotrimeric G protein–G$_i$1α.Gβγ. The R–G complex is viewed from the outside of the cell (top down).through the plasma membrane. The GPCR (red) creates a pocket for insertion from within the cell of the C-terminal end (green) of the Gα (beige). It is difficult to visualize how a single GPCR molecule can interact with all three subunits of the heterotrimeric G protein. The participation of two GPCRs (dimers?) in the activation of one trimeric G protein has been suggested.

engage the C-terminus of the Gα subunit? Optical fluorescence and luminscence methods that evaluate interacting molecules in real time have been proposed to study GPCR dimerization, but are not with controversy (see comments by Lohse [40]). A major challenge will therefore be to obtain co-crystals of GPCRs and trimeric G proteins. Such crystals will hopefully shed light on the next unknown of this fascinating signal-tranducing machinery.

IN MEMORIAM

This is the revision of a chapter written for the 2002 edition of the *Handbook of Cell Signaling*. That version was written 32 years after the first data regarding the GTP requirement in hormonal stimulation of liver adenylyl cyclase (then adenyl cyclase) had been obtained in Martin Rodbell's laboratory, where I was a postdoctoral fellow [2]. It was then that we made the proposal that receptor may be acting on adenylyl cyclase through a signal transducer driven by GTP, in some unknown way. Two great thinking minds of that era are no longer with us. Martin Rodbell passed away in 1998 [41], as did Michael (Mickey) Schramm in 2002. Mickey visited the NIH often in the late 1960s, and fed us ideas that were incorporated into our thinking without us even realizing it. Seldom, if ever, have we properly given credit to Mickey's influence on our thinking. Better late than never, goes the saying. But I wish I had done it earlier. While life is destined to end, our task as researchers dedicated to extracting Nature's secrets never ends. Even though by now 38 years have passed, signal transduction through G proteins has not been fully resolved, as the recent discoveries of RGS and AGS proteins, and the continued quest for co-crystals between receptors and G proteins, illustrate. While looking forward to new discoveries, we should also remember those that contributed in major ways to the way we think today. The present chapter was written with the idea of introducing "signaling through G proteins" to the next generation of investigators. I hope not to have failed too badly.

ACKNOWLEDGEMENTS

This work was supported by the Intramural Research Program of the NIH, National Institute of Environmental Health Sciences (Z01-ES101643).

REFERENCES

1. Lodish H, Berk A, Kaiser CA, Krieger M, Scott MP, Bretshcer A, Ploeg H, Matsudaira P. *Molecular cell biology*. 6th ed. New York, NY: H. Freeman and Co; 2008 p. 623–712.
2. Rodbell M, Birnbaumer L, Pohl SL, Krans HMJ. Properties of the adenyl cyclase systems in liver and adipose cells: the mode of action of hormones. *Acta Diabetol Lat* 460;**7**(Suppl. 1):9–57.

3. Kim SH, Priveé GG, Milburn MV. Conformational switch and structural basis offor oncogenic mutations of Ras proteins. In: Dickey B, Birnbaumer L, editors. *Handbook of experimental pharmacology, vol. 108/I, GTPases in biology.* Heidelberg: Springer Verlag; 1993. p. 177–93.

4. Dickey B, Birnbaumer L, editors. *Handbook of experimental pharmacology, vol. 108/I, GTPases in biology.* Heidelberg: Springer Verlag; 1993.

5. Feig LA. Dominant inhibitory Ras mutants: tools for eucidading Ras function. In: Dickey B, Birnbaumer L, editors. *Handbook of experimental pharmacology, vol. 108/I, GTPases in biology.* Heidelberg, Germany: Springer Verlag; 1993. p. 289–98.

6. Noel JP, Hamm HE, Sigler PB. The 2.2-Å crystal structure of transducin a GTP^S. *Nature* 1993;**366**:654–63.

7. Coleman DE, Berghuis AM, Lee E, Lindner ME, Gilman AG, Sprang SR. Structures of active conformations of $G_i\alpha 1$ and the mechanism of GTP hydrolysis. *Science* 1994;**265**:1405–12.

8. Higashijima T, Ferguson KM, Sternweis PC, Smigel MD, Gilman AG. Effects of Mg^{2+} and the $\beta\gamma$ subunit complex on the interactions of guanine nucleotides with G proteins. *J Biol Chem* 1987;**262**:762–6.

9. Tesmer JJ, Sunahara RK, Gilman AG, Sprang SR. Crystal structure of the catalytic domains of adenylyl cyclase in a complex with $G_s\alpha$. GTP^S. *Science* 1997;**278**:1907–16.

10. Lambright DG, Noel JP, Hamm HE, Sigler PB. Structral determinants for the activation of the a subunit of a heterotrimeric G protein. *Nature* 1994;**239**:621–8.

11. Wall MA, Coleman DE, Lee E, IniguezLluhi JA, Posner BA, Gilman AG, Sprang SR. The structure of the G protein heterotrimer $G_i\alpha 1$ $\beta 1$ $\gamma 2$. *Cell* 1995;**83**:1047–58.

12. Casey PJ. Protein lipidation in cell signaling. *Science* 1995;**268**:221–5.

13. Birnbaumer L. The discovery of signal transduction by G proteins. A personal account and an overview of the initial findings and contributions that led to our present understanding. *Biochim Biophys Acta Biomembranes* 2007;**1768**:756–71.

14. Birnbaumer L. Expansion of signal transduction by G proteins from 3 to 16 a subunits plus $\beta\gamma$ dimers. The second 15 years or so. *Biochim Biophys Acta Biomembranes* 2007;**1768**:772–93.

15. Gerachshenko T, Blackmer T, Yoon EJ, Bartleson C, Hamm HE, Alford S. $G\beta\gamma$ acts at the C terminus of SNAP-25 to mediate presynaptic inhibition. *Nat Neurosci* 2005;**8**:597–605.

16. Yoon EJ, Hamm HE, Currie KP. G protein $\beta\gamma$ subunits modulate the number and nature of exocytotic fusion events in adrenal chromaffin cells independent of calcium entry. *J Neurophysiol* 2008. 14 September, epub ahead of print.

17. Raw AS, Coleman DE, Gilman AG, Sprang SR. Structural and biochemical characterization of the GTP^S-, GDP.Pi-, and GDP-bound forms of a GTPase-deficient Gly42 → Val mutant of $G_i\alpha 1$. *Biochemistry* 1997;**36**:15,660–15,669.

18. John J, Rensland H, Schlichting I, Vetter I, Borasio GD, Goody RS, Wittinghofer A. Kinetic and structural analysis of the Mg^{2+}-binding site of the guanine nucleotide-binding protein p21H-ras. *J Biol Chem* 1993;**268**:923–9.

19. Bourne HR. How receptors talk to trimeric G proteins. *Curr Opin Cell Biol* 1995;**9**:134–42.

20. Wess J. Molecular basis of receptor/G protein coupling selectivity. *Pharmacol Ther* 1998;**80**:231–64.

21. Grishina G, Berlot CH. A surface exposed region of $G_s\alpha$ in which substitutions decrease receptor mediated activation and increase receptor affinity. *Mol Pharmacol* 2000;**57**:1081–92.

22. Berlot CH. A highly effective dominant negative as construct containing mutations that affect distinct functions inhibits multiple G_s-coupled receptor signalling pathways. *J Biol Chem* 2002;**277**:21,080–21,085.

23. Conklin BR, Farfel Z, Lustig KD, Julius D, Bourne HR. Substitution of three amino acids switches receptor specificity of $G_q\alpha$ to that of $G_i\alpha$. *Nature* 1993;**363**:274–6.

24. Hamm HE. How activated receptors couple to G propteins. *Proc Natl Acad Sci USA* 2002;**98**:4819–21.

25. Palczewski K. G protein-coupled receptor rhodopsin. *Annu Rev Biochem* 2006;**75**:743–67.

26. Park JH, Scheerer P, Hofmann KP, Choe HW, Ernst OP. Crystal structure of the ligand-free G-protein-coupled receptor opsin. *Nature* 2008;**454**:183–7.

27. Scheerer P, Park JH, Hildebrand PW, Kim YJ, Krauss N, Choe HW, Hofmann KP, Ernst OP. Crystal structure of opsin in its G-protein-interacting conformation. *Nature* 2008;**455**:497–502.

28. Van Eps N, Oldham WM, Hamm HE, Hubbell WL. Structural and dynamical changes in an alpha-subunit of a heterotrimeric G protein along the activation pathway. *Proc Natl Acad Sci USA* 2006;**103**:16,194–16,199.

29. Oldham WM, Van Eps N, Preininger AM, Hubbell WL, Hamm HE. Mapping allosteric connections from the receptor to the nucleotide-binding pocket of heterotrimeric G proteins. *Proc Natl Acad Sci USA* 2007;**104**:7927–32.

30. Schwartz TW, Hubbell WL. Structural biology: a moving story of receptors. *Nature* 2008;**455**:473–4.

31. Iiri T, Herzmark P, Nakamoto JM, van Dop C, Bourne HR. Rapid GDP release from G_s alpha in patients with gain and loss of endocrine function. *Nature* 1994;**371**:164–8.

32. Kleuss C, Hescheler J, Ewel C, Rosenthal W, Schultz G, Wittig B. Assignment of G-protein subtypes to specific receptors inducing inhibition of calcium currents. *Nature* 1991;**353**:43–8.

33. Chen C, Clarke IJ. Go2 Protein mediates the reduction in Ca^{2+} current by somatostatin in cultured ovine somatotrophs. *J Physiol* 1996;**491**:21–9.

34. Kleuss C, Scherübl H, Hescheler J, Schultz G, Wittig B. Different β-subunits determine G protein interaction with transmembrane receptors. *Nature* 1992;**358**:424–6.

35. Kleuss C, Scherübl H, Hescheler J, Schultz G, Wittig B. Selectivity in signal transduction determined by γ subunits of heterotrimeric G proteins. *Science* 1993;**259**:832–4.

36. DeVries L, Zheng B, Fischer T, Elenko E, Farquhar MG. The regulator of G protein signaling family. *Annu Rev Pharmacol Toxicol* 2000;**40**:235–71.

37. Chen CK, Burns ME, He W, Wensel TG, Baylor DA, Simon MI. Slowed recovery of rod photoresponse in mice lacking the GTPase accelerating protein RGS9-1. *Nature* 2000;**403**:557–60.

38. Cismowski MJ, Takesono A, Bernard ML, Duzic E, Lanier SM. Receptor-independent activators of heterotrimeric G proteins. *Life Sci* 2001;**68**:2301–8.

39. Kimple RJ, Willard FS, Siderovski DR. The GoLoco motif: heralding a new tango between G protein signaling and cell division. *Mol Interventions* 2002;**2**:88–100.

40. Lohse MJ. G protein-coupled receptors: too many dimers? *Nat Methods* 2006;**3**:972–3.

41. Birnbaumer L. Martin Rodbell (1925 1998) In memoriam. *Science 283* 1999;**1656**.

42. Bjarnadóttir TK, Gloriam DE, Hellstrand SH, Kristiansson H, Fredriksson R, Schiöth HB. Comprehensive repertoire and phylogenetic analysis of the G protein-coupled receptors in human and mouse. *Genomics* 2006;**88**:263–73.

43. Wilkie TM, Gilbert DJ, Olsen AS, Chen XN, Amatruda TT, Korenberg JR, Trask BJ, de Jong P, Reed RR, Simon MI, Jenkins NA, Copeland NG. Chromosomal evolution of the G protein a subunit multigene family. *Nat Genet* 1992;**1**:85–91.

44. Taussig R, Tang WJ, Hepler JR, Gilman AG. Distinct patterns of bidirectional regulation of mammalian adenylyl cyclases. *J Biol Chem* 1994;**269**:6093–100.

45. Cerione RA. Cdc42: new roads to travel. *Trends Cell Biol* 2004;**14**:127–32.

46. Tsukada S, Simon MI, Witte ON, Katz A. Binding of βγ subunits of heterotrimeric G proteins to the PH domain of Bruton tyrosine kinase. *Proc Natl Acad Sci USA* 1994;**91**:11,256–11,260.

47. Langhans-Rajasekaran SA, Wan Y, Huang XY. Activation of Tsk and Btk tyrosine kinases by G protein βγ subunits. *Proc Natl Acad Sci USA* 1995;**92**:8601–5.

48. Jiang Y, Ma W, Wan Y, Kozasa T, Hattori S, Huang XY. The G protein G121α stimulates Bruton's tyrosine kinase and a rasGAP through a conserved PH/BM domain. *Nature* 1998;**395**:808–13.

49. Iyengar R, Birnbaumer L. Hormone receptor modulates the regulatory component of adenylyl cyclases by reducing its requirement for Mg ion and enhancing its extent of activation by guanine nucleotides. *Proc Natl Acad Sci USA* 1982;**79**:5179–83.

Chapter 47

G-Protein-Coupled Receptors, Signal Fidelity, and Cell Transformation

Todd R. Palmby, Hans Rosenfeldt* and J. Silvio Gutkind

Oral and Pharyngeal Cancer Branch, National Institute of Dental and Craniofacial Research, National Institutes of Health, Bethesda, Maryland

Current address: Food and Drug Administration, Silver Spring, Maryland

INTRODUCTION

G protein-coupled receptors (GPCRs) make up the largest group of transmembrane proteins implicated in signal transduction. These receptors are activated by a large variety of ligands, including cytokines and chemokines, peptide and non-peptide neurotransmitters, hormones, growth factors, odorant molecules and light. Such a wide spectrum of sensitivity is reflected in the number of genes that encode GPCRs in animal genomes, including *Drosophila* (1 percent of total genes), *Caenorhabditus elegans* (>5 percent of all genes), and even humans, where more than 2 percent of human genes are responsible for over 800 proteins with a structure similar to GPCRs [1, 2]. Numerous GPCRs have been implicated in physiology and in the progression of hereditary diseases [3], and this link has transformed the investigation of new therapeutic drugs: GPCRs and the signaling pathways that they control have become a major focus of pharmaceutical development [1, 2, 4, 5]. Indeed, several approved drugs, designed to treat a wide range of indications, directly interact with GPCRs as agonists or antagonists. Seven-transmembrane domain-containing receptors are called G-protein-coupled receptors due to their characterized relationship to heterotrimeric G proteins (α, β, and γ subunits). GPCRs become stabilized in an active conformation upon ligand-binding, allowing them to catalyze the exchange of GDP for GTP bound to the G-protein α subunit. Gα and G$\beta\gamma$ subunits subsequently activate effector proteins.

The downstream signaling events initiated by GPCR activation depend on the G-protein-coupling specificity of each receptor. The α subunits of G proteins are divided into four subfamilies – Gα_s, Gα_i, Gα_q, and G$\alpha_{12/13}$ – and a single GPCR can couple to either one or more families of Gα proteins. Each G protein activates multiple downstream effectors. Typically, Gα_s stimulates adenylyl cyclase resulting in increased levels of cAMP, while Gα_i inhibits adenylyl cyclase resulting in decreased levels of cAMP. Members of the Gα_q family bind to and activate phospholipase C which cleaves phosphatidylinositol bisphosphate (PIP$_2$) into diacylglycerol and phophatidylinositol trisphosphate (IP$_3$). Receptor-mediated activation of G$\alpha_{12/13}$ family subunits leads to activation of the Rho small GTPase pathway via direct interaction between G$\alpha_{12/13}$ family subunits and the Guanidine Exchange Factors (GEFs) that regulate Rho. The Gβ subunits and Gγ subunits signal as a dimer and activate many signaling molecules, including phospholipases, ion channels, and lipid kinases.

Besides the regulation of these classical second-messenger generating systems, G proteins α and $\beta\gamma$ subunits can also control the activity of key intracellular molecules, including small GTP-binding proteins of the Ras and Rho families and members of the mitogen activated protein kinase (MAPK) family of serine-threonine kinases, including extracellular regulated kinase (ERK), c-jun N-terminal kinase (JNK), p38, and ERK5, through an intricate network of signaling events that have yet to be fully elucidated [5, 6].

In this review we will attempt to give a sense of the multiplicity of classical and novel signaling systems that transmit GPCR-mediated changes in cell behavior. We emphasize that the complexity of known GPCR-initiated signals is likely not to exist in one single cell type; that the signaling pathways available to GPCRs in a particular tissue are likely to be much more limited. However, because of the vast signaling potential of GPCRs, these critical cell-specific transduction pathways are in danger of being overridden by deregulated mechanisms that are sensitive to GPCRs. Thus, we will also focus on mechanisms that enforce GPCR signaling fidelity to physiological pathways as a barrier against GPCR-induced pathological outcomes such as cell transformation.

Handbook of Cell Signaling, Three-Volume Set 2 ed.

GPCRS AND ONCOGENESIS

Based on a collective body of work on GPCRs and their role in cancer, it has become evident that this class of cell surface receptors has a major role in the development, progression, and metastasis of cancer [7]. GPCRs have been shown to have roles in multiple stages of cancer progression, from being commandeered by cancer cells to induce proliferation, altering the detection of cancer cells by the immune system, and increasing the blood supply of a tumor, to driving cancer cells to invade surrounding tissues and travel to distant sites throughout the body [7]. There are many modes of activating GPCRs in the context of cancer, in which GPCR hyperactivation can provide a selective advantage to the cancer cell.

Occasionally, activating and inactivating GPCRs and α-subunit mutations have been identified in human diseases, including cancer [8]. For example, activating mutations of the TSH receptor are found in some thyroid carcinomas and approximately 80 percent of thyroid adenomas [9, 10]. Hypermorphic mutations have also been detected in other GPCRs, such as LH receptors, which can cause hyperplastic growth of Leydig cells in a form of familial male precocious puberty [11], and Ca^{2+}-sensing G-protein-linked receptors, which can cause autosomal dominant hypercalcemia [12] and may be involved in certain cancers [13].

Mitogens such as thrombin, lysophosphatidic acid (LPA), bombesin, endothelin, and prostaglandins can stimulate cell proliferation by binding to their cognate GPCRs in many different cell types [reviewed in 7, 14, 15–17]. In addition, the description of the GPCR encoded by the *mas* oncogene was the first evidence of a connection between cancer and GPCRs. Unlike other oncogenes, the *mas* gene-product does not contain any activating mutations when compared to other GPCRs, and requires ligand-binding for its transforming ability. This observation, taken together with other findings showing that ectopic expression of serotonin 1C and muscarinic m1, m3, and m5 receptors transforms mouse cells in an agonist-dependent fashion [18, 19], suggested that endogenous GPCRs can be tumorigenic in the presence of excess locally produced or circulating ligand, and does not require mutation to be transforming.

Neuropeptides are often involved in both autocrine and paracrine activation of GPCRs in many human carcinomas. For example, GPCR ligands such as bombesin, gastrin-releasing peptide (GRP), neuromedin B (NMB), bradykinin, cholecystokinin (CCK), galanin, neurotensin (NT), and vasopressin are secreted by small cell lung cancer cells (SCLC). These tumors also express the GPCRs sensitive to these agonists, and thus use GPCRs to stimulate their own proliferation in an autocrine or paracrine fashion (for an extensive review, see [7, 20]). A variety of neuropeptide receptors and their ligands play a role in the progression of colon adenomas and carcinomas, gastric hyperplasia and cancer, prostate cancer, and pancreatic hyperplasia and carcinoma (reviewed

in [7, 17, 20]). Thus, ectopic expression of GPCRs allows tumor cells to override the physiological signaling pathways that are inherent to the cell type of the cancer's origin, and use the unchecked potential of GPCR-controlled signaling mechanisms to drive cell proliferation.

Another way in which cellular and viral GPCRs can promote tumorigenesis is by promoting the development of blood vessels that support tumor growth. A variety of G-protein-coupled receptors have been implicated in angiogenesis, including those binding sphingosine-1 phosphate, LPA, PAF, thrombin, IL-8, GROα–γ, MCP-1, and SDF-1, have been all implicated in tumor-induced angiogenesis and vasculogenesis [21]. The sphingosine-1-phosphate (S1P) receptor S1P$_1$/EDG-1 is a particularly interesting example because of its interactions with receptor tyrosine kinases [22, 23]. This GPCR was originally cloned from endothelial cells, and supports G$_i$-dependent cell migration and Rac activation in human embryonic kidney (HEK) cells [22, 24]. Interestingly, Mouse Embryonic Factor (MEF) derived from S1P$_1$$^{-/-}$ animals not only exhibit deficits in S1P-directed Rac activation and cell migration, but also those elicited by other mitogens such as PDGF [25–27]. These results, led to a transactivation model in which S1P generated by PDGF receptor stimulation activated the EDG-1 receptor in a paracrine or autocrine way. Since cell migration is essential for blood vessel formation, this signaling relationship between the S1P$_1$/EDG-1 and the PDGF receptors might be a critical step in angiogenesis, including that promoted by tumors.

Viral GPCRs have also been implicated in pathological blood vessel formation. The identification of the Kaposi's sarcoma (KS)-associated herpes virus (KSHV) as the viral etiologic agent of KS [28] has shed a new light into the pathogenesis of this enigmatic disease. KS is the most frequent tumor arising in HIV-infected patients, and remains a significant cause of morbidity among the world's AIDS population [29]. KS is a highly angiogenic tumor, with spindle cells, likely of endothelial origin, representing the most prominent cell type. KSHV also causes two rare B-cell lymphomas whose incidences are also increased in AIDS patients, primary effusion lymphoma (PEL), and multicentric Castleman's disease (MCD) [30]. Although the KSHV genome encodes several candidate oncogenes, its virally-encoded GPCR (KSHV GPCR), which is highly related to the CXCR1 and CXCR2 chemokine receptors [31], is unique in that it is transforming, pro-angiogenic, and can promote the tumor growth of cells in a paracrine fashion [32, 33].

The KSHV GPCR is constitutively active due to the presence of a several structural deviations from closely related GPCRs, including a mutation (Asp142Val) within its DRY motif in the second intracytoplasmic loop [34]. KSHV GPCR enhances the activity of a complex signaling network whose contribution to KS is beginning to be unraveled. For example, the transforming, pro-survival and angiogenic effects of KSHV GPCR involve the activation of multiple MAPKs whose activity converges in the nucleus to control transcription factors such as hypoxia-inducible factor 1α (HIF-1α), AP-1, and NFκB. The activity of these transcription factors, in turn, promotes the expression and secretion of pro-inflammatory cytokines such as IL-1β, IL-2, IL-4, IL-6, VEGF, MIP-1/CCL3, and IL-8/CXCL8, which may participate in this unusual paracrine neoplasia [35]. Among the multiple pathways stimulated by KSHV GPCR, the Akt–mTOR signaling route has recently taken a center stage, as its pharmacological inhibition prevents KS progression in animal model systems [36] and in renal transplantation patients that develop KS upon immunosuppression [37]. KSHV GPCR activates the Akt–mTOR by stimulating PI3K through βγ-subunits released from both pertussis-toxin-sensitive and -insensitive G proteins, as well as in an autocrine fashion by upregulating the expression of the vascular endothelial growth factor (VEGF) receptor KDR2 [38] and the release of VEGF [35].

In addition to the direct involvement of GPCRs in oncogenesis, their cognate heterotrimeric G proteins have also been found to have initiating roles in cancer. A minimum 10 of the 17 Gα subunits have been described to have transforming potential (reviewed in [39]), including members of the four trimeric G-protein classes: Gα$_{12/13}$, Gα$_q$, Gα$_i$, and Gα$_s$. In many cases these proteins are similar to GPCRs in that they stimulate carcinogenesis in their intact form when they are overexpressed outside their normal cellular context. However, mutations that inhibit the basal GTPase activity of two of these Gα subunits, Gα$_s$ and Gα$_{i2}$, have been described in several tumors types (see below).

Oncogenic mutations of Gα$_q$ family members have not been found in human cancers, and research carried out with laboratory-generated active forms of these proteins has yielded contradictory data. It seems that, depending on cell type, activated Gα$_q$ is transforming when expressed at low levels [40] but can lead to apoptosis when present at high levels [39]. It is interesting to reiterate that highly transforming receptors, such as serotonin-1C, muscarinic m1, and α$_1$-adrenergic receptors, are coupled to Gα$_q$, and that the KSHV-GPCR, a constitutively active Gα$_i$/Gα$_q$-coupled receptor, has been implicated in Kaposi sarcoma progression [31], suggesting that parallel pathways emanating from Gα$_q$-coupled receptors may be necessary in addition to the activation of the Gα$_q$ subunit itself to induce cell transformation and growth.

Examples of transforming G proteins include the *gep*, *gip2*, and *gsp* oncogenes. The *gep* gene was simultaneously identified as an oncogenic sequence present in Ewing's sarcoma and as a transcript that induces strong transformation of NIH 3T3 cells [41, 42]. *gep* turned out to be a wild-type Gα$_{12}$ subunit, belonging to the Gα$_{12/13}$ family. This result has been consistent with subsequent findings: increased expression of Gα$_{12/13}$ subunits has been detected in many human cancers, but, interestingly, no mutations have been found. For example, breast, colon, and prostate

adenocarcinoma-derived cell lines express elevated levels of wild-type Gα$_{12/13}$ (reviewed in [43]). The *gip2* oncogene is a constitutively active mutant of Gα$_{i2}$. This mutation has been found in human ovarian sex cord stromal tumors and adrenal cortical tumors [44], although how often Gα$_{i2}$ mutations occur in these cancer types remains a point of controversy. Fibroblast transformation resulting from transfection of *gip2* has been suggested to result from the derepression of the Ras-ERK1/2 pathway after cAMP/PKA inhibition [45]. However, Gα$_{i2}$ can also stimulate the Ras-ERK1/2 cascade via Rap1 inhibition [46] or stimulation [47], depending on the cellular context, and by stimulating the release of βγ G-protein subunits [48]. The *gsp* oncogene codes for a GTPase-deficient mutant of Gα$_s$, and is found in thyroid toxic adenomas (30 percent), thyroid carcinomas (10 percent), growth hormone-secreting pituitary adenomas, and McCune-Albright syndrome. Interestingly, responses to *gsp* expression are cell-type specific. Increases in cAMP and activated PKA resulting from the presence of this oncogene can inhibit Raf1 and prevent transformation in some cells [49]. By contrast, the presence of *gsp* and the same downstream second messengers can lead to inhibited cell growth in other cell types such as PC12 and thyroid cells [49–51].

gsp illustrates the importance of cellular context in regard to the transformation potential of GPCRs and G proteins. Cell types that express the small GTPase Rap1 and the downstream kinase B-Raf, such as PC12 and thyroid cells, are predisposed to ERK activation in response to cAMP increases [50, 51]. Thus, the availability of the transforming ERK pathway to *gsp* depends on the organization of the signaling pathways present in a particular cell type. The cell-type dependence of *gsp* transformation is only one representative case of a more general phenomenon. For example, the Rho GTPases and MAP kinases are activated downstream of GPCRs, and may contribute to cell-type specific responses to GPCR stimulation through temporal–spatial organization via scaffolding molecules.

A MAPK SIGNALING NETWORK LINKS GPCRS TO BIOLOGICAL OUTCOMES

Various signaling pathways mediate and control the proliferative capacity of GPCRs. MAP kinases (MAPKs) are a group of highly related proline-targeted serine-threonine kinases, which are central members of the signaling pathways downstream of GPCRs. Although many members of this group have been recently discovered, we will focus on the three best-known classes of MAPKs: the extracellular signal-regulated kinases (ERKs), the c-Jun N-terminal kinases (JNKs), and the p38 kinases. These molecules are phosphorylated by a family of proteins known as the MAP kinase kinases (MAPKKs), which are phosphorylated themselves by the MAP kinase kinase kinase (MAPKKK) class of proteins. MAP kinases are typically extremely specific within a particular cascade, with only a very limited number of proteins at each step. For example, in the ERK cascade the MAPKKK Raf-1 will not phosphorylate a MAPKK other than MEK1 or MEK2, two isotypes that perform the same function in that specific MAP kinase pathway. Thus, MAP kinase molecules such as ERK are downstream of phosphorylation cascades forming separate and parallel signaling pathways.

GPCRs can signal to a variety of MAPKKKs that are linked to MAP kinases, including those of the ERK, JNK, and p38 pathways (Figure 47.1). MAPKs translocate to the nucleus upon activation, where they regulate the expression of genes that play a key role in physiological and pathological cell growth. These signaling molecules affect gene transcription by phosphorylating transcription factors that control the synthesis of these critical mRNAs [52]. Each MAP kinase cascade has a different range of intracellular targets and is therefore able to induce different cellular responses, such as cell proliferation, apoptosis, and migration. The substantial body of work surrounding the mechanism of GPCR regulation of MAPK function provides an extensive understanding about how GPCRs regulate such a diverse array of biological responses.

ERK CASCADE

GPCR stimulation of p42 and p44 MAPK (MAPK/ERK1/2) is one of the most intensely studied and comprehended signaling events, and is known to occur in many different

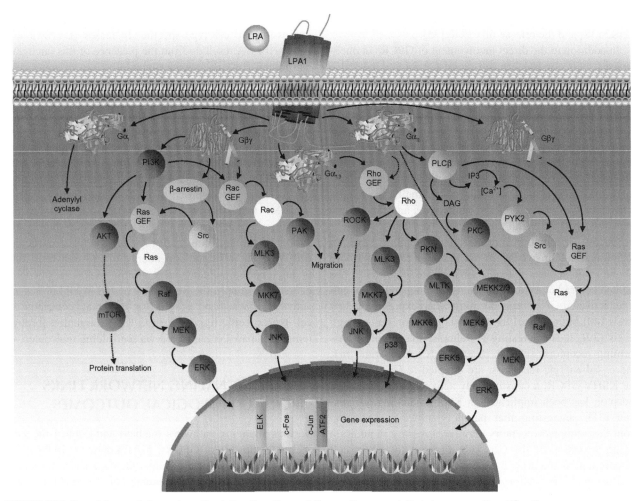

FIGURE 47.1 G-protein-coupled receptors stimulating $G\alpha_q$, $G\alpha_l$, and $G\alpha_{12}$ pathways contribute to cancer cell proliferation.
Lysophosphatidic acid (LPA) receptors are coupled to $G\alpha_q$, $G\alpha_l$, and $G\alpha_{12}$ (15). LPA signaling through $G\alpha_i$ and $G\beta\gamma$ subunits activates phosphatidylinositol 3-kinase (PI3K), which results in the stimulation of the Akt survival pathway and increased protein translation by the activation of the mammalian target of rapamycin (mTOR) signaling pathway. Activation of PI3K and Src by $G\beta\gamma$ subunits also stimulates the activity of both Ras and Rac, which lead to subsequent activation of the Raf-MEK-extracellular signal-regulated kinase (ERK) pathway controlling cell proliferation and invasion and Pak to promote cell migration and JNK to regulate gene expression. The activation of $G\alpha_{12}$ stimulates Rho guanine nucleotide exchange factors (RhoGEFs), including p115-RhoGEF, LARG and PDZ-RhoGEF, which activate Rho and downstream pathways such as Rho-associated coiled-coil containing protein kinase 1 (ROCK), which causes actin polymerization and migration, and c-jun N-terminal kinase (JNK) and p38, controlling gene expression. $G\alpha_q$ stimulation also activates Rho, promoting migration, and MAPK cascades including ERK, JNK, p38 and ERK5 through an unclear mechanism leading to a boost in gene expression [6]. DAG, diacylglycerol; IP3, inositol triphosphate.

cell types in response to a diversity of agonists (reviewed in [53]). Much of this research has focused on the molecular mechanisms regulating this signaling cascade, and has revealed an assortment of cell-specific pathways. Many of these pathways include tyrosine kinases, PI-3 kinases, and PKC as downstream targets for GPCRs that, in turn, lead to activation of the ERK cascade.

The first evidence for tyrosine kinases acting as a link between GPCRs and the Ras-ERK cascade was provided by tyrosine kinase inhibition, which reduces both activation of ERK1/2 [54] and the rapid tyrosine phosphorylation of the adapter molecule Shc, a posttranslational modification that induces Shc–GRB2 complex formation [55]. This response has been attributed to the activity of a subset of non-receptor tyrosine kinases (NRTKs) and receptor tyrosine kinases (RTKs). Src, or Src-like kinases can phosphorylate Shc upon stimulation by LPA or expression of $\beta\gamma$ subunits [56]. NRTKs such as Csk, Lyn, Btk, Pyk2, and Fak, have been implicated in signaling from $G\alpha_i$- and $G\alpha_q$-coupled receptors to the ERK cascade in multiple cell types and conditions (reviewed in [57, 58]). The variety of tyrosine kinases mediating the activation of the ERK cascade reflects the number of different cell types and conditions in which the ERK pathway is involved. Thus, multiple tyrosine kinases in separate pathways may converge on the same ERK cascade to produce different biological results.

The protein and lipid kinases of the PI3K family are also essential for GPCR signaling to ERKs. A member of this family, PI3Kγ, can be activated by binding to G$\beta\gamma$ subunits. After activation, PI3Kγ signals to tyrosine kinases that phosphorylate Shc, leading to increases in ERK activity [59]. The PI3Kβ isoform can also mediate GPCR-directed signaling to the ERK cascade [60]. Further, PI3K may potentially activate Rac and PAK in combination with Ras to stimulate Raf kinase activity [61]. The variety of pathways stimulated by PI3Kγ and -β provides an example of the complex mechanisms by which GPCRs can activate the ERK cascade. Interestingly, these pathways all converge on the ERK pathway, impinging on the MAPKKK, Raf-1, which suggests that although the GPCR-driven signals mediated by PI3K and tyrosine kinases are numerous and complex, they are specific and tightly controlled as they both proceed through the activation of the ERK cascade.

Gα_q-coupled receptors can use alternative mechanisms to activate the ERK pathway. These variations include signaling cascades that are mediated by protein kinase C, Ras, or a combination of the two. The cell type and the particular extracellular stimulus often determine which signaling mechanism will be involved. This specified response includes the second messengers that are synthesized in response to Gα_q, such as diacylglycerol and elevated levels of intracellular Ca^{2+}, that can stimulate Ras through the guanine nucleotide exchange factors (GEFs) RasGRF and RasGRP (also called CalDGEF). These two GEFs are expressed only in certain tissues, and stimulate only Ras and

Ras-related GTPases (reviewed in [57]). By contrast, the mechanism by which PKC stimulates ERK activity is not yet completely elucidated, since direct phosphorylation of Raf does not seem to be sufficient to fully activate MEK and MAPK [62]. PKC may additionally modulate other molecules that regulate the interaction between Ras and Raf.

The consistent theme in GPCR signaling to ERK1/2 is that there is a complex network of transduction pathways that are ultimately integrated by a cascade including three kinases (Raf, MEK, and ERK). The range of different signals that can impinge on this one cascade is unlikely to co-exist in one single cell type, but represents the summation of the different mechanisms by which GPCRs can connect to ERK in many cell types. In order to maintain normal physiology, it is critical for GPCRs to activate ERK in a controlled fashion, restricting the duration and intensity of ERK activation by cell-specific pathways that can ensure the fidelity of each signaling step. There is always potential, however, for GPCRs to inappropriately stimulate the ERK cascade via signal transduction pathways that are extraneous to cellular context. For example, overexpression of gastrin receptors in cancer cells leads to activation of c-Src and, in turn, the ERK pathway. Thus, inappropriate c-Src activation may contribute to the transforming effects of gastrin receptors. When mechanisms of ERK activation are placed out of their normal cellular context, they are likely to be deregulated and, given the transforming potential of the ERK cascade, to contribute to tumor progression. Thus, ensuring cell-specific signal transduction pathways with respect to the ERK cascade is a critical aspect of proper GPCR function in biology.

JNK Cascade

The detailed mechanisms by which GPCRs stimulate the MAPK cascade that terminates in the c-Jun NH2-terminal kinase (JNK) remain to be fully elucidated. This molecule, also known as stress-activated protein kinase (SAPK), has sequence similarity to ERK1/2, but is activated by GPCRs through distinct pathways. The major difference in how these two MAPK cascades are regulated is that ERK1/2 stimulation often depends on Ras, while the JNK cascade is downstream of the small G proteins Rac, Rho, and Cdc42. Constitutively active mutants of Rac and Cdc42, for example, can stimulate JNK activity [63]. Moreover, Rac and Cdc42 can mediate signaling of G$\beta\gamma$ dimers and Gα_{12}, Gα_{13}, Gα_q and Gα_i to JNK [57, 58, 64]. Little is known about how GPCRs activate JNK beyond these general constraints, but recent work greatly advanced the field by identifying the first GEFs known to be responsive to G$\beta\gamma$ and PI-3 kinase, P-REX1/2 [65–67]. In addition, recent evidence suggests that another group of related GEFs, including p63-RhoGEF and Trio, can be activated by Gα_q through a direct binding [68–70]. Trio has two catalytic

Dbl Homology (DH) domains, one of which is specifically activates Rac and the other Rho [71]. Furthermore, a substantial link between $G\alpha_{12}$ and $G\alpha_{13}$ and three GEFs, p115-RhoGEF, PDZ-RhoGEF, and LARG, has demonstrated a direct activation of Rho downstream of GPCRs coupled to these G proteins. These GEFs have been shown to interact directly with $G\alpha_{12}$ and $G\alpha_{13}$, and can lead to subsequent Rho and JNK activation [72–75]. This eloquently demonstrates the signal fidelity involved in the convergence of multiple extracellular signals through one MAPK cascade. Agonists of $G\alpha_{12/13}$-, $G\alpha_q$- and $G\alpha_i$-coupled GPCRs can all lead to similar activation of JNK through three different mechanisms to activate the same network of small GTPases. Yet cell type and conditional specificity may involve additional GEF mechanisms of GTPase activation, and subtle refinements through the involvement of differentially expressed or localized single family members of each GEF group responsible for transmitting the specific input.

Another area that remains unclear is how $G\alpha_{12/13}$ stimulates Rac/Cdc42, and the downstream JNK cascade. Current candidates that might mediate this effect include two Rac/Cdc42 GEFs, Tiam1 and Dbl, and two Ras GEFs which may also catalyze Rac GTP exchange, Ras-GRF1, and Ras-GRF2. Further, the non-receptor tyrosine kinases PYK2 and FAK, which are stimulated by stress-fiber and focal complex formation, can also stimulate the JNK cascade by interacting with the adaptor proteins Crk [76] or paxillin [77]. Crk and paxillin can, in turn, stimulate GEFs for Rac and Cdc42 (reviewed in [57]).

Still, many of the open questions concerning the activation of the JNK kinase pathway are in the end questions about how Rac and Cdc42 respond to GPCRs. Which Rac effectors stimulate the JNK pathway is also an important outstanding issue. Here again, the presence of a variety of mechanisms impinging on the JNK cascade reflects the general theme of GPCR signaling complexity, and the exquisite cell specificity that is possible from such a wide range of alternative biochemical routes.

p38 Cascade

The p38 MAPKs, like the JNK family, are stimulated by cellular stress and membrane-bound receptors [78]. There are presently four p38 MAPKs known: p38α (CSBP-1), p38β, p38γ (ERK6/SAPK3), and p38δ (SAPK4) [79]. Although the GPCRs and agonists that elicit increased activity from the p38 family of MAP kinases have been the focus of much investigation, there is no clear picture of the downstream mechanisms directly controlling p38 kinases. There have been some reports showing that $G\alpha_q$ and G$\beta\gamma$ dimers stimulate p38α [80], and that two NRTKs, BTK [81] and Src [82], are involved in this mechanism. Receptors that couple to $G\alpha_q$ can also stimulate the p38α, p38γ, and p38δ isoforms [83]. Recent work indicates that $G\alpha_q$ mediates p38 activation through the MKK3 and MKK6 MAPKKs [84]. Further, research using electrophysiological techniques suggests that p38 MAP kinases are downstream from $G\alpha_{13}$ [85], and it has been proposed that this heterotrimeric G protein can initiate the activity of the p38 pathway by stimulating Ask, a MAPKKK for this cascade [86]. It is expected that the use of novel techniques, such as RNA interference or the generation of knockout animals for molecules acting upstream of p38, will enable the molecular dissection of the mechanisms by which GPCRs and other cell surface receptors activate each member of the p38 family of MAPKs.

G PROTEIN-INDEPENDENT SIGNALING

It has now become evident that the extent of GPCR signaling reaches beyond signal transduction pathways that are downstream of Gα and G$\beta\gamma$ subunits. Recent work suggests that GPCRs interact with a wide range of signaling molecules besides heterotrimeric G proteins. Molecules containing protein–protein interaction domains such as the PDZ, SH2, and SH3 motifs, as well as polyproline-containing regions, have been reported to directly interact with GPCRs (reviewed in [87]).

These specialized domains could serve GPCRs as cell-specific bypasses from trimeric G proteins to activate intracellular signaling in some cellular contexts. Several signaling proteins containing these protein–protein interaction motifs have been shown to bind GPCRs. For example, the PDZ domain interacts with proteins containing a C-terminal S/TxV(L/I) sequence common in GPCRs. SH2-containing molecules such as the adaptor Grb2 and the SH2-containing tyrosine phosphatase SHP (reviewed in [87]) have been reported to bind the β_2-adrenergic and to AT$_{1A}$ receptors, respectively. Other domains that have been reported to bind GPCRs are the polyproline-binding domains, such as SH3, WW, and EVH domains [88, 89]. For example, metabotropic glutamate receptors (mGluRs) interact with a class of molecules that harbor Enabled/VASP Homology (EVH)-like domains, such as Homer (1a–c, 2, and 3), which binds mGluRs through a C-terminal polyproline sequence (PPXXFP) (reviewed in [87]).

This expanded view of GPCR signaling, combined with emerging results showing that the *frizzled*, *smoothened*, and *Dictyostelium* cAMP receptors elicit biological responses that are independent of heterotrimeric G proteins, suggests a re-evaluation of the "G-protein-coupled receptor/heterotrimeric G-protein-associated effector" concept of GPCR function. Thus, some workers prefer terms other than "GPCR" in order to avoid using a designation that suggests a more limited range than the vast array of signaling cascades that these receptors actually control. Alternative terms include "serpentine receptor," "seven-transmembrane receptor," or "heptahelical receptor" [90]. Such a wide range of signaling possibilities requires strict organization among the signaling molecules that are downstream of GPCRs. One emerging example of how these downstream cascades can be physically organized intracellularly is a variety of scaffolding molecules that tether multiple components of signal transduction pathways in specific configurations.

GPCR EFFECTORS ARE ORGANIZED BY SCAFFOLDING MOLECULES

GPCRs stimulate physiological cell growth *via* a variety of tightly regulated signaling modules. In the past decade, scaffolding proteins have emerged as general regulatory mechanisms ensuring the fidelity of intracellular pathways. These proteins bind components of signaling pathways, physically organizing them to enable physiological responses. The prototypical signaling scaffold is the yeast protein Ste5p, which binds the components of the yeast MAP kinase cascade, leading to the mating response after activation of the pheromone GPCR [91–94], and has been suggested to maintain the signaling fidelity of this cascade [95]. Ste5p is particularly important to the mating pathway

of yeast because, unlike multicellular organisms, MAP kinase cascades share most of their components, such as the PAK-like kinase Ste20p and the MEKK-like protein Ste11p [96–99]. Like scaffolds in multicellular organisms, however, Ste5p is expressed in a specific cell type: the haploid yeast cell. Thus, Ste5p allows a generalized MAP kinase module to elicit a specialized physiological response in a particular cellular context.

Several scaffolds binding a variety of signaling pathways have now been described in multicellular organisms. These include the various MAP kinase pathways organized by scaffolding proteins such as Kinase Suppressor of RAS (KSR) and c-Jun Terminal Kinase Interacting Protein family (JIPs), the cAMP-dependent cascades that rely on the A-kinase anchoring proteins (AKAPs) for tethering, $G\beta\gamma$ signaling involving protein complexes with WD40 repeat proteins, regulation of the actin cytoskeleton through ERM proteins, and the multiple cell signals downstream of proteins that bind molecules of the β-arrestin class (reviewed in [100]) (Figure 47.2). In many cases, overexpression of these scaffolding proteins will inhibit cell transformation, suggesting that these molecules play an important role in restricting the propagation of signaling events, and preventing signal transduction cascades from causing

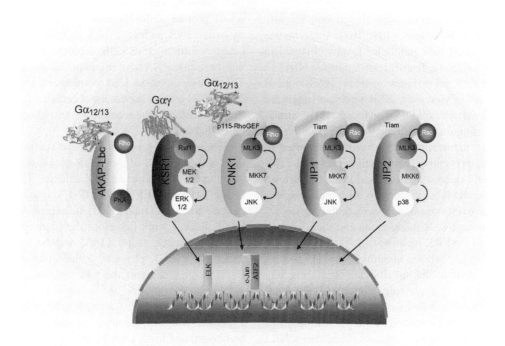

FIGURE 47.2 Signaling pathways downstream of GPCRs organized by scaffolding molecules provide fidelity to a complex network.
AKAP-Lbc is an A Kinase Anchoring Protein; A kinase anchoring proteins were first described as large protein scaffolds organizing and localizing Protein Kinase A (PKA) signaling events. This specific molecule links PKA activation by GPCRs to the regulation of Rho GTPases. Similarly, the Kinase Suppressor of Ras (KSR1) provides a direct and quick activation of the Raf/MEK/ERK pathway by binding to Gβγ. Recently, Rho Guanine Nucleotide Exchange Factors have been identified as having scaffolding properties in placing the activation of small GTPases in close proximity to the MAP kinase pathways they stimulate through binding to other scaffolding molecules that organize the MAP kinases.

pathological cell proliferation. Although there are very few published reports directly connecting G-protein-coupled receptors to these organizing molecules, we can expect that the relevance of scaffolding proteins to GPCR-induced transformation will become increasingly apparent in the foreseeable future because of their fundamental role in maintaining the integrity of biochemical routes connecting cell surface receptors to the nucleus.

KSR and CNK1

KSR was first cloned in RAS suppressor screens using *Caenorhabditis elegans* and *Drosophila* [101–103]. However, recent work has shown that KSR function is required for proper signaling between RAF-1 and ERK in both *Drosophila* [104, 105] and *C. elegans* [106]. In these invertebrate systems, ablation of KSR by either RNAi [105] or mutation [106] prevented the efficient activation of MEK and ERK by constitutively active Ras. Intriguingly, overexpression of KSR has similar negative effects on preventing Ras-V12-induced tumorigenesis in *Drosophila* imaginal discs [104]. Roy and colleagues have observed that wild-type KSR overexpression can only stimulate the ERK pathway if all other components binding KSR are concomitantly transfected, and suggest that the mechanism by which the overexpressed scaffolding protein prevents efficient signaling to ERK is through a stoichiometric excess of KSR isolating signaling components from each other.

The murine form of KSR was cloned concurrently with its invertebrate homologs [103, 107]. Experiments in mammalian systems have paralleled results obtained in *Drosophila* and *C. elegans*. Transfection studies in NIH3T3 cells showed that KSR overexpression blocks RAS-induced transformation [108]. Moreover, KSR knockout mice are resistant to polyomavirus middle T-induced tumor formation [109], paralleling results obtained with *Drosophila*. Of particular interest, mammalian KSR has been reported to bind βγ subunits, preventing ERK activation [110]. This result is particularly interesting because it is the first known link between the KSR scaffold and G-protein-coupled receptors. Further work remains to elucidate the physiological role of βγ subunit/KSR interactions and its effects on GPCR signaling to the ERK pathway.

More recent studies have identified another protein cooperating with KSR in linking Rho and RhoGEFs with JNK activation and c-jun phosphorylation. CNK1, connector enhancer of *ksr*, was first isolated in a genetic screen in *Drosophila* to identify cofactors for *kinase suppressor of ras* (*ksr*) [111]. The investigators found that CNK1 binds the DH/PH domains of two RhoGEFs, Net1 and p115RhoGEF [112]. CNK1 also binds two JNK kinase kinases, MLK2 and MLK3 (Mixed-Lineage Kinases 2 and 3), and the JNK kinase MKK7, but not MKK4. In addition, p115-RhoGEF co-immunoprecipitates with

MKK7 in the presence of CNK1, confirming that hCNK1 can act as a molecular scaffold connecting RhoGEFs to the JNK pathway. Interestingly, the formation of a complex between p115RhoGEF, CNK1, and MLKs is required to induce c-Jun phosphorylation by sphingosine-1-phosphate (S1P) activation of p115RhoGEF *via* Gα$_{12/13}$ [112]. Furthermore, data suggest that CNK1 may interact with unidentified Rac GEFs [112] and members of the Ras or Ral pathways [111, 113, 114].

JIP/IB Family of Scaffolding Molecules

The JIP/IB family of scaffolding proteins consists of three mammalian homologs JIP/IB-1, -1b, and -2 [115–117]. These molecules bind members of the mixed-lineage group of MAPKKKs (MLK) that have been reported to activate both JNK and p38 [118, 119]. In addition, both IB1/JIP1 and IB2/JIP-2 have been found to bind the JNK MAPKK, MKK7 [117, 120], while IB2/JIP-2 has also been reported to bind the p38 MAPKK MLK3 [121] and to p38γ [122].

JIP-1 was originally described as an inhibitor of the JNK pathway that could block JNK-dependent cell growth and transformation [115]. Soon thereafter IB1, an alternatively transcribed isoform of JIP-1, was found in insulin-secreting cells [116]. Like KSR, the overexpression of JIP proteins inhibits the transmission of signals through the JNK pathway. However, the physiological function of this set of protein scaffolds seems to be to temper the activity of the JNK pathway, since cells derived from JIP-1$^{+/-}$ heterozygous mice or cells depleted of endogenous JIP-1 with antisense RNA have augmented JNK activity [123–125]. These large increases lead to enhanced apoptosis in IL-1β/TNFα/IFN-treated pancreatic β cells. In fact, a mutation in the human *IB1/JIP-1* gene has recently been linked to adult onset diabetes [124].

An emerging property of these scaffolding proteins is their ability to bind certain GDP exchange factors that activate the small G proteins Rho and Rac. The Rho exchange factor p190RhoGEF binds JIP-1 through its phosphotyrosine binding domain (PTB) in neuronal and PC12 cells [126]. Recent work describes the binding of IB2/JIP-2 to the Rac exchange factor TIAM and the Rac/Ras exchange factor Ras-GRF1, physically linking known activators of Rac to components of the p38 pathway, which have previously been reported to be downstream of this small GTPase [121]. TIAM is phosphorylated in response to the G-protein-coupled receptor ligands LPA, endothelin-1, bombesin, and bradykinin [127], and this protein modification is required for the proper localization of TIAM after LPA stimulation [128]. Thus, the JIP scaffolding proteins might link G-protein-coupled receptors to the JNK and p38 pathways by physically interacting with the small GTPase exchange factors activated by GPCRs.

AKAPs

cAMP signaling was the first second-messenger described [129, 130]. Production of this nucleotide is controlled by adenylyl cyclases, a class of enzymes that are mostly controlled by GPCRs coupling to G proteins of the G_s class [131]. The best described target of cAMP is protein kinase A (PKA), a tetrameric enzyme that becomes activated upon binding to this nucleotide [132]. Activation of PKA does not occur randomly; instead, a class of scaffolding proteins, the A Kinase Anchoring Proteins (AKAPs), localizes PKA to specific sites of action [133]. Interestingly, these scaffolding proteins cannot be grouped together by homology, and are defined as AKAPs solely in terms of their capacity to bind PKA.

A common theme among AKAPs, besides their ability to bind PKA, is their multifunctional nature. A growing number of large, multivalent proteins that bind the PKA holoenzyme have been described. One example is AKAP-Lbc, which has been demonstrated to coordinate PKA signaling and RhoA activation via $G\alpha_{12/13}$. A recent study identified AKAP-Lbc as an integral component in alpha adrenergic receptor-induced cardiac hypertrophy [134]. In addition, it was shown that anchoring of PKA by AKAP-Lbc recruits 14-3-3 binding, thereby inhibiting the RhoA activation mediated by the Lbc portion of the protein [135–137]. Also, there are multiple reports of AKAP-Lbc mediating protein kinase D activation by nucleating protein kinase C, PKA and RhoA activation [138, 139].

AKAPs often serve as scaffolds for other signal transduction pathways or exhibit other functions in addition to binding PKA. For example, AKAPs now include proteins such as gravin, a protein that was first identified in patients with the autoimmune disease myasthenia gravis and that also binds protein kinase C (PKC) [140], and WAVE1-3, three orthologous members of the WASP family of proteins that interact with the ARP2/3 actin polymerization unit [141] and respond to Rac signals [135, 142]. Moreover, recent evidence suggests that WAVE3 and gravin function as tumor suppressors [143–145], and it is probable that more of these PKA-binding proteins will be shown to function as tumor suppressors.

One reason why AKAPs might serve as tumor suppressors is the fact that the $G\alpha_s$/cAMP pathway often serves as an inhibitor of cell proliferation. For example, PKA phosphorylates RAF-1 at serines 43, 259, and 621 [146]. Phosphorylation at serine 259 is critical for PKA-dependent inhibition of RAF-1 kinase activity *in vitro*, and blocks RAF-1 dependent *in vivo* activation of ERK in COS cells. Additionally, mutation of serine 43 to alanine blocks the ability of RAF-1 to bind RAS in response to cAMP increases. Thus proper organization of cAMP signaling to PKA in the correct subcellular compartment may be critical in preventing other mitogenic pathways such as the PKC or ERK cascades from becoming over-active and promoting inappropriate cell proliferation.

Proteins Containing WD40 Repeats

Gβ is part of the class of WD40 repeat proteins that contain a similar circular bladed β-propeller structure. In addition to binding to GPCRs and Gα proteins, G$\beta\gamma$ has been found to complex with other WD40 repeat proteins, such as RACK1 (Receptor for Activated C Kinase 1), dynein intermediate chain, and a protein with unknown function, AAH20044 [147–149]. RACK1, the best characterized of these interactions, is 57 percent similar at the amino acid level with Gβ1, but lacks a binding region for Gγ in the N-terminus. The RACK1 interaction is specific to the G$\beta\gamma$ and does not occur when the G$\beta\gamma$ are complexed with Gα. The interactions of G$\beta\gamma$ and RACK1 may create a mode of selective G$\beta\gamma$ inhibition. For example, G$\beta\gamma$-mediated PLC β2 and adenylyl cyclase II activation were inhibited by ectopic RACK1 both *in vitro* and *in vivo*; however, $\beta\gamma$−mediated chemotaxis and mitogen-activated protein kinase (MAPK) activation were not affected by RACK1 overexpression and RACK1 had no effect on $G\alpha_i$, $G\alpha_s$ or $G\alpha_q$-mediated signaling [148]. In addition, this interaction may facilitate the plasma membrane localization of RACK1 and scaffolding of a multiprotein complex involving additional molecules associated with RACK1 [149].

ERM Family

The ezrin-radixin-moesin (ERM) family of proteins is involved in linking proteins in the plasma membrane to the actin cytoskeleton, and thereby has a function in Rho GTPase signaling. ERM protein have been associated with cell morphology, migration, and growth (reviewed in [150, 151]), and have therefore been implicated in tumor metastasis ([152, 153]). ERM proteins are able to bind to actin and additional proteins when these binding sites are unmasked, relieving an intramolecular interaction between the N- and C-terminus [150]. The N-terminus of Radixin, a member of the ERM family of proteins, was found to bind to GTP-bound $G\alpha_{13}$, which lead to an increased binding to actin [154]. Dominant-negative mutants of radixin blocked transformation by $G\alpha_{13}$, suggesting that it functions downstream, which was further supported by a study that reported Rac1 and CAMKII activation and SRE-dependent gene transcription mediated by the C-terminus of radixin [155].

Another ERM related protein, ezrin-radixin-moesin-binding phosphoprotein 50 (EBP50), functions as a scaffolding protein for ERM proteins, ion channels, receptors, and cytoskeletal and cytoplasmic proteins (reviewed in [156]). EBP50 contains PDZ domains, which have been shown to interact with $G\alpha_q$, preferentially in its activated state, and $G\alpha_s$, but not other Gα [157]. It is still unknown whether EBP50 functionally bridges $G\alpha_q$ and $G\alpha_s$ to the other proteins with which it interacts. Interestingly, EBP50 was found to directly interact with RACK1, and this interaction was shown to be involved in the PKC regulation of the cystic fibrosis transmembrane regulator (CFTR) chloride channel [158].

Arrestin

G-protein-coupled receptors are phosphorylated by the G-protein-coupled receptor kinase (GRK) family of proteins after ligand binding [159]. The scaffolding protein arrestin subsequently interacts physically with GPCRs. While this molecule has been primarily implicated in targeting GPCRs for endocytosis, arrestin has been also shown to couple GPCRs to the activation of Src-like kinases. This process apparently involves the formation of large multi-protein complexes that can include components of the MAPK and JNK cascades [160, 161]. For example, an arrestin-tethered multi-molecular complex has been described in the Rac/Cdc42-independent activation of the JNK cascade MAPKKK ASK1 by $G\alpha_{12}$ [86]. Interestingly, arrestin only binds the neural specific JNK3 isoform, suggesting that this scaffold enforces cell-type specificity in the activation of the JNK pathway by GPCRs. In addition, there are other examples of arrestin-mediated, localized, and sustained activation of specific MAP kinase cascades. For instance, parathyroid hormone receptor activation of Erk1/2 involves both arrestin- and G-protein independent pathways [162]. Likewise, beta-arrestin2 has been shown to enhance p38 activation downstream of CXCR4, which mediates the robust chemotaxis resulting from CXCR4 activation [163]. Arrestin has also been shown to localize MAP kinase activation in a spatio-temporal fashion to facilitate protease-activated receptor 2-induced chemotaxis through a direct interaction with the receptor and MAP kinase [164]. Furthermore, more recent observations suggest that arrestin is involved in the organization of biological processes such as chemotaxis, metastasis, and apoptosis by scaffolding multiple protein complexes downstream of a variety of cell surface e receptors, in addition to classically defined GPCRs (reviewed in [165]).

CONCLUSION: GPCR BIOLOGY REQUIRES BOTH SIGNAL INTEGRATION AND FIDELITY

The essential molecular events that GPCRs use to govern such a wide range of biological events seem elusive because of their apparent complexity. Classical second-messenger generating systems are now understood to be only a subset of the mechanisms that GPCRs use in physiological and pathological contexts. At first glance, this "brave new world" of complexity looks like a highly interconnected meshwork where signals derived from a GPCR can travel to any of a wide variety of endpoints. From our current standpoint, we cannot predict with any confidence which signaling pathway, of the many possible routes, will be activated by a heptahelical receptor. Yet this situation is not the case in biology: usually, GPCR stimulation in a given context will produce a repeatable result. Moreover, an emerging concept derived from recent studies with MAP kinases and other targets of GPCRs is that scaffold proteins are organizers and keepers of specificity. It may be as important to keep signals separate between closely related cascades, such as the MAP kinases, as it is to integrate them in a coordinated response.

Although physiological functions of GPCRs, including phenotypic differentiation and cell survival or death, most likely result from the integration of a complex network of signaling cascades, it is probable that pathology induced by GPCRs, such as cancer or tissue hypertrophy, results from the breakdown of signal separation between targets downstream of these receptors. Recently gained understanding of GPCR-driven intracellular signaling networks and how they are organized by scaffolding proteins will provide a more global view of the general systems by which these receptors exert their numerous physiological roles, and elucidate their role in many pathological conditions (reviewed in [166, 167]). This new systems-level understanding may also point to novel approaches for pharmacological treatment of a variety of disease processes.

ACKNOWLEDGEMENTS

Grant support comes from the Intramural Research Program, National Institute of Dental and Craniofacial Research, NIH.

REFERENCES

1. Flower DR. Modelling G-protein-coupled receptors for drug design. *Biochim Biophys Acta* 1999;**1422**:207–34.
2. Attwood TK, Findlay JB. Fingerprinting G-protein-coupled receptors. *Protein Eng* 1994;**7**:195–203.
3. Rohrer DK, Kobilka BK. G protein-coupled receptors: functional and mechanistic insights through altered gene expression. *Physiol Rev* 1998;**78**:35–52.
4. Schwarz MK, Wells TN. New therapeutics that modulate chemokine networks. *Nature Rev Drug Discov* 2002;**1**:347–58.
5. Pierce KL, Premont RT, Lefkowitz RJ. Seven-transmembrane receptors. *Nature Rev Mol Cell Biol* 2002;**3**:639–50.
6. Marinissen MJ, Gutkind JS. G-protein-coupled receptors and signaling networks: emerging paradigms. *Trends Pharmacol Sci* 2001;**22**:368–76.
7. Dorsam RT, Gutkind JS. G-protein-coupled receptors and cancer. *Nature Rev Cancer* 2007;**7**:79–94.
8. Schoneberg T, Schulz A, Biebermann H, Hermsdorf T, Rompler H, Sangkuhl K. Mutant G-protein-coupled receptors as a cause of human diseases. *Pharmacol Ther* 2004;**104**:173–206.
9. Russo D, Arturi F, Wicker R, et al. Genetic alterations in thyroid hyperfunctioning adenomas. *J Clin Endocrinol Metab* 1995;**80**:1347–51.
10. Parma J, Duprez L, Van Sande J, et al. Diversity and prevalence of somatic mutations in the thyrotropin receptor and Gs alpha genes

as a cause of toxic thyroid adenomas. *J Clin Endocrinol Metab* 1997;**82**:2695–701.

11. Shenker A, Laue L, Kosugi S, et al. A constitutively activating mutation of the luteinizing hormone receptor in familial male precocious puberty [see comments]. *Nature* 1993;**365**:652–4.

12. Spiegel AM. Mutations in G proteins and G protein-coupled receptors in endocrine disease. *J Clin Endocrinol Metab* 1996;**81**:2434–42.

13. Hoff Jr AO, Cote GJ, Fritsche HA, et al. Calcium-induced activation of a mutant G-protein-coupled receptor causes in vitro transformation of NIH/3T3 cells. *Neoplasia* 1999;**1**:485–91.

14. Rozengurt E. Early signals in the mitogenic response. *Science* 1986;**234**:161–6.

15. Mills GB, Moolenaar WH. The emerging role of lysophosphatidic acid in cancer. *Nature Rev Cancer* 2003;**3**:582–91.

16. Young D, Waitches G, Birchmeier C, et al. Isolation and characterization of a new cellular oncogene encoding a protein with multiple potential transmembrane domains. *Cell* 1986;**45**:711–19.

17. Gutkind JS. Cell growth control by G protein-coupled receptors: from signal transduction to signal integration. *Oncogene* 1998;**17**:1331–42.

18. Julius D, Livelli TJ, Jessell TM, Axel R. Ectopic expression of the serotonin 1c receptor and the triggering of malignant transformation. *Science* 1989;**244**:1057–62.

19. Gutkind JS, Novotny EA, Brann MR, Robbins KC. Muscarinic acetylcholine receptor subtypes as agonist-dependent oncogenes. *Proc Natl Acad Sci USA* 1991;**88**:4703–7.

20. Heasley LE. Autocrine and paracrine signaling through neuropeptide receptors in human cancer. *Oncogene* 2001;**20**:1563–9.

21. Moore BB, Keane MP, Addison CL, et al. CXC chemokine modulation of angiogenesis: the importance of balance between angiogenic and angiostatic members of the family. *J Investig Med* 1998;**46**:113–20.

22. Lee MJ, Van Brocklyn JR, Thangada S, et al. Sphingosine-1-phosphate as a ligand for the G protein-coupled receptor EDG-1. *Science* 1998;**279**:1552–5.

23. Liu Y, Wada R, Yamashita T, et al. Edg-1,the G-protein-coupled receptor for sphingosine-1-phosphae, is essential for vascular maturation. *J Clin Invest* 2000;**106**:951–61.

24. Wang F, Van Brocklyn JR, Hobson JP, et al. Sphingosine 1-phosphate stimulates cell migration through a G(i)- coupled cell surface receptor. Potential involvement in angiogenesis. *J Biol Chem* 1999;**274**:35,343–35,350.

25. Hobson JP, Rosenfeldt HM, Barak LS, et al. Role of the sphingosine-1-phosphate receptor EDG-1 in PDGF-induced cell motility. *Science* 2001;**291**:1800–3.

26. Rosenfeldt HM, Hobson JP, Maceyka M, et al. EDG-1 links the PDGF receptor to Src and focal adhesion kinase activation leading to lamellipodia formation and cell migration. *FASEB J* 2001;**15**:2649–59.

27. Olivera A, Spiegel S. Sphingosine-1-phosphate as a second messenger in cell proliferation induced by PDGF and FCS mitogens. *Nature* 1993;**365**:557–60.

28. Chang Y, Cesarman E, Pessin MS, et al. Identification of herpesvirus–like DNA sequences in AIDS-associated Kaposi's sarcoma. *Science* 1994;**266**:1865–9.

29. Moore PS, Chang Y. Molecular virology of Kaposi's sarcoma-associated herpesvirus. *Philos Trans R Soc Lond* 2001;**356**:499–516.

30. Carbone A, Gloghini A. AIDS-related lymphomas: from pathogenesis to pathology. *Br J Haematol* 2005;**130**:662–70.

31. Arvanitakis L, Geras-Raaka E, Varma A, et al. Human herpesvirus KSHV encodes a constitutively active G-protein-coupled receptor linked to cell proliferation [see comments]. *Nature* 1997;**385**:347–50.

32. Bais C, Santomasso B, Coso O, et al. G-protein-coupled receptor of Kaposi's sarcoma-associated herpesvirus is a viral oncogene and angiogenesis activator [see comments] [published erratum appears in Nature 1998; 392(6672):210]. *Nature* 1998;**391**:86–9.

33. Montaner S, Sodhi A, Molinolo A, et al. Endothelial infection with KSHV genes in vivo reveals that vGPCR initiates Kaposi's sarcomagenesis and can promote the tumorigenic potential of viral latent genes. *Cancer Cell* 2003;**3**:23–36.

34. Rosenkilde MM, Kledal TN, Holst PJ, Schwartz TW. Selective elimination of high constitutive activity or chemokine binding in the human herpesvirus 8 encoded seven-transmembrane oncogene ORF74. *J Biol Chem* 2000;**275**:26,309–26,315.

35. Sodhi A, Montaner S, Gutkind JS. Viral hijacking of G-protein-coupled-receptor signalling networks. *Nature Rev Mol Cell Biol* 2004;**5**:998–1012.

36. Sodhi A, Chaisuparat R, Hu J, et al. The TSC2/mTOR pathway drives endothelial cell transformation induced by the Kaposi's sarcoma-associated herpesvirus G protein-coupled receptor. *Cancer Cell* 2006;**10**:133–43.

37. Stallone G, Schena A, Infante B, et al. Sirolimus for Kaposi's sarcoma in renal-transplant recipients. *N Engl J Med* 2005;**352**:1317–23.

38. Bais C, Van Geelen A, Eroles P, et al. Kaposi's sarcoma associated herpesvirus G protein-coupled receptor immortalizes human endothelial cells by activation of the VEGF receptor-2/ KDR. *Cancer Cell* 2003;**3**:131–43.

39. Dhanasekaran N, Tsim ST, Dermott JM, Onesime D. Regulation of cell proliferation by G proteins. *Oncogene* 1998;**17**:1383–94.

40. Kalinec G, Nazarali AJ, Hermouet S, Xu N, Gutkind JS. Mutated alpha subunit of the Gq protein induces malignant transformation in NIH 3T3 cells. *Mol Cell Biol* 1992;**12**:4687–93.

41. Xu N, Bradley L, Ambdukar I, Gutkind JS. A mutant alpha subunit of G12 potentiates the eicosanoid pathway and is highly oncogenic in NIH 3T3 cells. *Proc Natl Acad Sci USA* 1993;**90**:6741–5.

42. Chan AM, Fleming TP, McGovern ES, et al. Expression cDNA cloning of a transforming gene encoding the wild-type G alpha 12 gene product. *Mol Cell Biol* 1993;**13**:762–8.

43. Gutkind JS, Coso OA, Xu N. G12 and G13 a subunits of heterotrimeric G proteins: A novel family of oncogenes. In: Spiegel AM, editor. *G proteins, receptors, and disease.* Totowa, NJ: Humana Press; 1998. p. 101–17.

44. Lyons J, Landis CA, Harsh G, et al. Two G protein oncogenes in human endocrine tumors. *Science* 1990;**249**:655–9.

45. Miller MJ, Rioux L, Prendergast GV, Cannon S, White MA, Meinkoth JL. Differential effects of protein kinase A on Ras effector pathways. *Mol Cell Biol* 1998;**18**:3718–26.

46. Mochizuki N, Ohba Y, Kiyokawa E, et al. Activation of the ERK/MAPK pathway by an isoform of rap1GAP associated with G alpha(i) [see comments]. *Nature* 1999;**400**:891–4.

47. Schmitt JM, Stork PJ. beta 2-adrenergic receptor activates extracellular signal-regulated kinases (ERKs) via the small G protein rap1 and the serine/threonine kinase B-Raf. *J Biol Chem* 2000;**275**:25,342–25,350.

48. Crespo P, Xu N, Simonds WF, Gutkind JS. Ras-dependent activation of MAP kinase pathway mediated by G-protein beta gamma subunits [see comments]. *Nature* 1994;**369**:418–20.

49. Chen J, Iyengar R. Suppression of Ras-induced transformation of NIH 3T3 cells by activated G alpha s. *Science* 1994;**263**:1278–81.

50. Ehses JA, Pelech SL, Pederson RA, McIntosh CH. Glucose-dependent insulinotropic polypeptide (GIP) activates the Raf-Mek 1/2-ERK 1/2 module via a cyclic AMP/PKA/Rap1-mediated pathway. *J Biol Chem* 2002;**277**:37,088–37,097.

51. Erhardt P, Troppmair J, Rapp UR, Cooper GM. Differential regulation of Raf-1 and B-Raf and Ras-dependent activation of mitogen-activated protein kinase by cyclic AMP in PC12 cells. *Mol Cell Biol* 1995;**15**:5524–30.

52. Davis RJ. Transcriptional regulation by MAP kinases. *Mol Reprod Dev* 1995;**42**:459–67.

53. Gutkind JS. The pathways connecting G protein-coupled receptors to the nucleus through divergent mitogen-activated protein kinase cascades. *J Biol Chem* 1998;**273**:1839–42.

54. Hordijk PL, Verlaan I, van Corven EJ, Moolenaar WH. Protein tyrosine phosphorylation induced by lysophosphatidic acid in Rat-1 fibroblasts. Evidence that phosphorylation of map kinase is mediated by the Gi-p21ras pathway. *J Biol Chem* 1994;**269**:645–51.

55. van Biesen T, Hawes BE, Luttrell DK, et al. Receptor-tyrosine-kinase- and G beta gamma-mediated MAP kinase activation by a common signalling pathway [see comments]. *Nature* 1995;**376**:781–4.

56. Luttrell LM, Hawes BE, van Biesen T, et al. Role of c-Src tyrosine kinase in G protein-coupled receptor- and Gbetagamma subunit-mediated activation of mitogen-activated protein kinases. *J Biol Chem* 1996;**271**:19,443–19,450.

57. Gutkind JS. Regulation of mitogen-activated protein kinase signaling networks by G protein-coupled receptors. *Science STKE* 2000. http://www.stke.org/cgi/content/full/OC_sigtrans;2000/40/re1.

58. Gudermann T, Grosse R, Schultz G. Contribution of receptor/G protein signaling to cell growth and transformation. *Naunyn Schmiedebergs Arch Pharmacol* 2000;**361**:345–62.

59. Lopez-Ilasaca M, Crespo P, Pellici PG, Gutkind JS, Wetzker R. Linkage of G protein-coupled receptors to the MAPK signaling pathway through PI 3-kinase gamma. *Science* 1997;**275**:394–7.

60. Murga C, Fukuhara S, Gutkind JS. A novel role for phosphatidylinositol 3-kinase beta in signaling from G protein-coupled receptors to Akt. *J Biol Chem* 2000;**275**:12,069–12,073.

61. Sun H, King AJ, Diaz HB, Marshall MS. Regulation of the protein kinase Raf-1 by oncogenic Ras through phosphatidylinositol 3-kinase, Cdc42/Rac and Pak. *Curr Biol* 2000;**10**:281–4.

62. Macdonald SG, Crews CM, Wu L, et al. Reconstitution of the Raf-1-MEK-ERK signal transduction pathway in vitro [published erratum appears in Mol Cell Biol 1994; 14(3):2223–4]. *Mol Cell Biol* 1993;**13**:6615–20.

63. Coso OA, Chiariello M, Yu JC, et al. The small GTP-binding proteins Rac1 and Cdc42 regulate the activity of the JNK/SAPK signaling pathway. *Cell* 1995;**81**:1137–46.

64. Yamauchi J, Kawano T, Nagao M, Kaziro Y, Itoh H. G(i)-dependent activation of c-Jun N-terminal kinase in human embryonal kidney 293 cells. *J Biol Chem* 2000;**275**:7633–40.

65. Welch HC, Coadwell WJ, Ellson CD, et al. P-Rex1, a PtdIns(3,4,5)P3- and Gbetagamma-regulated guanine-nucleotide exchange factor for Rac. *Cell* 2002;**108**:809–21.

66. Rosenfeldt H, Vazquez-Prado J, Gutkind JS. P-REX2, a novel PI-3-kinase sensitive Rac exchange factor. *FEBS Letts* 2004;**572**:167–71.

67. Donald S, Hill K, Lecureuil C, et al. P-Rex2, a new guanine-nucleotide exchange factor for Rac. *FEBS Letts* 2004;**572**:172–6.

68. Lutz S, Freichel-Blomquist A, Yang Y, et al. The guanine nucleotide exchange factor p63RhoGEF, a specific link between Gq/11-coupled receptor signaling and RhoA. *J Biol Chem* 2005;**280**:11,134–11,139.

69. Rojas RJ, Yohe ME, Gershburg S, Kawano T, Kozasa T, Sondek J. Galphaq directly activates p63RhoGEF and Trio via a conserved extension of the Dbl homology-associated pleckstrin homology domain. *J Biol Chem* 2007;**282**:29,201–29,210.

70. Lutz S, Shankaranarayanan A, Coco C, et al. Structure of Galphaq-p63RhoGEF-RhoA complex reveals a pathway for the activation of RhoA by GPCRs. *Science* 2007;**318**:1923–7.

71. Debant A, Serra-Pages C, Seipel K, et al. The multidomain protein Trio binds the LAR transmembrane tyrosine phosphatase, contains a protein kinase domain, and has separate rac-specific and rho-specific

guanine nucleotide exchange factor domains. *Proc Natl Acad Sci USA* 1996;**93**:5466–71.

72. Hart MJ, Jiang X, Kozasa T, et al. Direct stimulation of the guanine nucleotide exchange activity of p115 rhoGEF by Galpha13. *Science* 1998;**280**:2112–14.

73. Fukuhara S, Murga C, Zohar M, Igishi T, Gutkind JS. A novel PDZ domain containing guanine nucleotide exchange factor links heterotrimeric G proteins to Rho. *J Biol Chem* 1999;**274**:5868–79.

74. Fukuhara S, Chikumi H, Gutkind JS. Leukemia-Associated Rho Guanine nucleotide exchange factor (LARG) links heterotrimeric G proteins of the G12 family to Rho. *FEBS Letts* 2000;**485**:183–8.

75. Lee YN, Malbon CC, Wang HY. G alpha 13 signals via p115RhoGEF cascades regulating JNK1 and primitive endoderm formation. *J Biol Chem* 2004;**279**:54,896–54,904.

76. Blaukat A, Ivankovic-Dikic I, Gronroos E, et al. Adaptor proteins Grb2 and Crk couple Pyk2 with activation of specific mitogen-activated protein kinase cascades. *J Biol Chem* 1999;**274**:14,893–14,901.

77. Igishi T, Fukuhara S, Patel V, et al. Divergent signaling pathways link focal adhesion kinase to mitogen- activated protein kinase cascades. Evidence for a role of paxillin in c- jun nh(2)-terminal kinase activation. *J Biol Chem* 1999;**274**:30,738–30,746.

78. Davis RJ. Signal transduction by the JNK group of MAP kinases. *Cell* 2000;**103**:239–52.

79. Ono K, Han J. The p38 signal transduction pathway: activation and function. *Cell Signal* 2000;**12**:1–13.

80. Yamauchi J, Nagao M, Kaziro Y, Itoh H. Activation of p38 mitogen-activated protein kinase by signaling through G protein-coupled receptors. Involvement of Gbetagamma and Galphaq/11 subunits. *J Biol Chem* 1997;**272**:27,771–27,777.

81. Bence K, Ma W, Kozasa T, Huang XY. Direct stimulation of Bruton's tyrosine kinase by G(q)-protein alpha- subunit. *Nature* 1997;**389**:296–9.

82. Nagao M, Yamauchi J, Kaziro Y, Itoh H. Involvement of protein kinase C and Src family tyrosine kinase in Galphaq/11-induced activation of c-Jun N-terminal kinase and p38 mitogen-activated protein kinase. *J Biol Chem* 1998;**273**:22,892–22,898.

83. Marinissen MJ, Chiariello M, Pallante M, Gutkind JS. A network of mitogen-activated protein kinases links G protein-coupled receptors to the c-jun promoter: a role for c-Jun NH2-terminal kinase, p38s, and extracellular signal-regulated kinase 5. *Mol Cell Biol* 1999;**19**:4289–301.

84. Yamauchi J, Tsujimoto G, Kaziro Y, Itoh H. Parallel regulation of mitogen-activated protein kinase kinase 3 (MKK3) and MKK6 in Gq-signaling cascade. *J Biol Chem* 2001;**276**:23,362–23,372.

85. Wilk-Blaszczak MA, Stein B, Xu S, Barbosa MS, Cobb MH, Belardetti F. The mitogen-activated protein kinase p38-2 is necessary for the inhibition of N-type calcium current by bradykinin. *J Neurosci* 1998;**18**:112–18.

86. Berestetskaya YV, Faure MP, Ichijo H, Voyno-Yasenetskaya TA. Regulation of apoptosis by alpha-subunits of G12 and G13 proteins via apoptosis signal-regulating kinase-1. *J Biol Chem* 1998;**273**:27,816–27,823.

87. Bockaert J, Pin JP. Molecular tinkering of G protein-coupled receptors: an evolutionary success. *EMBO J* 1999;**18**:1723–9.

88. Pawson T, Scott JD. Signaling through scaffold, anchoring, and adaptor proteins. *Science* 1997;**278**:2075–80.

89. Brzostowski JA, Kimmel AR. Signaling at zero G: G-protein-independent functions for 7-TM receptors. *Trends Biochem Sci* 2001;**26**:291–7.

90. Hall RA, Premont RT, Lefkowitz RJ. Heptahelical receptor signaling: beyond the G protein paradigm. *J Cell Biol* 1999;**145**:927–32.

91. Kranz JE, Satterberg B, Elion EA. The MAP kinase Fus3 associates with and phosphorylates the upstream signaling component Ste5. *Genes Dev* 1994;**8**:313–27.

92. Choi KY, Satterberg B, Lyons DM, Elion EA. Ste5 tethers multiple protein kinases in the MAP kinase cascade required for mating in S. cerevisiae. *Cell* 1994;**78**:499–512.

93. Marcus S, Polverino A, Barr M, Wigler M. Complexes between STE5 and components of the pheromone-responsive mitogen-activated protein kinase module. *Proc Natl Acad Sci USA* 1994;**91**:7762–6.

94. Printen Jr JA, Sprague GF. Protein–protein interactions in the yeast pheromone response pathway: Ste5p interacts with all members of the MAP kinase cascade. *Genetics* 1994;**138**:609–19.

95. Elion EA. The Ste5p scaffold. *J Cell Sci* 2001;**114**:3967–78.

96. Gustin MC, Albertyn J, Alexander M, Davenport K. MAP kinase pathways in the yeast Saccharomyces cerevisiae. *Microbiol Mol Biol Rev* 1998;**62**:1264–300.

97. O'Rourke SM, Herskowitz I. The Hog1 MAPK prevents cross talk between the HOG and pheromone response MAPK pathways in *Saccharomyces cerevisiae*. *Genes Dev* 1998;**12**:2874–86.

98. Liu H, Styles CA, Fink GR. Elements of the yeast pheromone response pathway required for filamentous growth of diploids. *Science* 1993;**262**:1741–4.

99. Roberts RL, Fink GR. Elements of a single MAP kinase cascade in *Saccharomyces cerevisiae* mediate two developmental programs in the same cell type: mating and invasive growth. *Genes Dev* 1994;**8**:2974–85.

100. Andreeva AV, Kutuzov MA, Voyno-Yasenetskaya TA. Scaffolding proteins in G-protein signaling. *J Mol Signal* 2007;**2**:13.

101. Kornfeld K, Hom DB, Horvitz HR. The ksr-1 gene encodes a novel protein kinase involved in Ras-mediated signaling in *C. elegans*. *Cell* 1995;**83**:903–13.

102. Sundaram M, Han M. The C. elegans ksr-1 gene encodes a novel Raf-related kinase involved in Ras-mediated signal transduction. *Cell* 1995;**83**:889–901.

103. Therrien M, Chang HC, Solomon NM, et al. KSR, a novel protein kinase required for RAS signal transduction. *Cell* 1995;**83**:879–88.

104. Karim FD, Rubin GM. Ectopic expression of activated Ras1 induces hyperplastic growth and increased cell death in *Drosophila* imaginal tissues. *Development* 1998;**125**:1–9.

105. Roy F, Laberge G, Douziech M, Ferland-McCollough D, Therrien M. KSR is a scaffold required for activation of the ERK/MAPK module. *Genes Dev* 2002;**16**:427–38.

106. Ohmachi M, Rocheleau CE, Church D, Lambie E, Schedl T, Sundaram MV. C. elegans ksr-1 and ksr-2 have both unique and redundant functions and are required for MPK-1 ERK phosphorylation. *Curr Biol* 2002;**12**:427–33.

107. Nehls M, Luno K, Schorpp M, et al. YAC/P1 contigs defining the location of 56 microsatellite markers and several genes across a 3.4-cM interval on mouse chromosome 11. *Mamm Genome* 1995;**6**:321–31.

108. Denouel-Galy A, Douville EM, Warne PH, et al. Murine Ksr interacts with MEK and inhibits Ras-induced transformation. *Curr Biol* 1998;**8**:46–55.

109. Nguyen A, Burack WR, Stock JL, et al. Kinase suppressor of Ras (KSR) is a scaffold which facilitates mitogen-activated protein kinase activation *in vivo*. *Mol Cell Biol* 2002;**22**:3035–45.

110. Bell B, Xing H, Yan K, Gautam N, Muslin AJ. KSR-1 binds to G-protein betagamma subunits and inhibits beta gamma-induced mitogen-activated protein kinase activation. *J Biol Chem* 1999;**274**:7982–6.

111. Therrien M, Wong AM, Rubin GM. CNK, a RAF-binding multidomain protein required for RAS signaling. *Cell* 1998;**95**:343–53.

112. Jaffe AB, Hall A, Schmidt A. Association of CNK1 with Rho guanine nucleotide exchange factors controls signaling specificity downstream of Rho. *Curr Biol* 2005;**15**:405–12.

113. Jaffe AB, Aspenstrom P, Hall A. Human CNK1 acts as a scaffold protein, linking Rho and Ras signal transduction pathways. *Mol Cell Biol* 2004;**24**:1736–46.

114. Lanigan TM, Liu A, Huang YZ, et al. Human homologue of Drosophila CNK interacts with Ras effector proteins Raf and Rlf. *FASEB J* 2003;**17**:2048–60.

115. Dickens M, Rogers JS, Cavanagh J, et al. A cytoplasmic inhibitor of the JNK signal transduction pathway. *Science* 1997;**277**:693–6.

116. Bonny C, Nicod P, Waeber G. IB1, a JIP-1-related nuclear protein present in insulin-secreting cells. *J Biol Chem* 1998;**273**:1843–6.

117. Yasuda J, Whitmarsh AJ, Cavanagh J, et al. The JIP group of mitogen-activated protein kinase scaffold proteins. *Mol Cell Biol* 1999;**19**:7245–54.

118. Rana A, Gallo K, Godowski P, et al. The mixed lineage kinase SPRK phosphorylates and activates the stress-activated protein kinase activator, SEK-1. *J Biol Chem* 1996;**271**:19,025–19,028.

119. Tibbles LA, Ing YL, Kiefer F, et al. MLK-3 activates the SAPK/JNK and p38/RK pathways via SEK1 and MKK3/6. *EMBO J* 1996;**15**:7026–35.

120. Negri S, Oberson A, Steinmann M, et al. cDNA cloning and mapping of a novel islet-brain/JNK-interacting protein. *Genomics* 2000;**64**:324–30.

121. Buchsbaum RJ, Connolly BA, Feig LA. Interaction of Rac exchange factors Tiam1 and Ras-GRF1 with a scaffold for the p38 mitogen-activated protein kinase cascade. *Mol Cell Biol* 2002;**22**:4073–85.

122. Schoorlemmer J, Goldfarb M. Fibroblast growth factor homologous factors are intracellular signaling proteins. *Curr Biol* 2001;**11**:793–7.

123. Bonny C, Oberson A, Steinmann M, Schorderet DF, Nicod P, Waeber G. IB1 reduces cytokine-induced apoptosis of insulin-secreting cells. *J Biol Chem* 2000;**275**:16,466–16,472.

124. Waeber G, Delplanque J, Bonny C, et al. The gene MAPK8IP1, encoding islet-brain-1, is a candidate for type 2 diabetes. *Nature Genet* 2000;**24**:291–5.

125. Tawadros T, Formenton A, Dudler J, et al. The scaffold protein IB1/JIP-1 controls the activation of JNK in rat stressed urothelium. *J Cell Sci* 2002;**115**:385–93.

126. Meyer D, Liu A, Margolis B. Interaction of c-Jun amino-terminal kinase interacting protein-1 with p190 rhoGEF and its localization in differentiated neurons. *J Biol Chem* 1999;**274**:35,113–35,118.

127. Fleming IN, Elliott CM, Collard JG, Exton JH. Lysophosphatidic acid induces threonine phosphorylation of Tiam1 in Swiss 3T3 fibroblasts via activation of protein kinase C. *J Biol Chem* 1997;**272**:33,105–33,110.

128. Buchanan FG, Elliot CM, Gibbs M, Exton JH. Translocation of the Rac1 guanine nucleotide exchange factor Tiam1 induced by platelet-derived growth factor and lysophosphatidic acid. *J Biol Chem* 2000;**275**:9742–8.

129. Sutherland E, Rall T. the properties of an adenine ribonucleotide produced with cellular particles, ATP, Mg++, and epinephrine or glucagon. *J Am Chem Soc* 1957;**79**:3608.

130. Sutherland EW. Studies on the mechanism of hormone action. *Science* 1972;**177**:401–8.

131. Simonds WF. G protein regulation of adenylate cyclase. *Trends Pharmacol Sci* 1999;**20**:66–73.

132. Scott JD. Cyclic nucleotide-dependent protein kinases. *Pharmacol Ther* 1991;**50**:123–45.

133. Colledge M, Scott JD. AKAPs: from structure to function. *Trends Cell Biol* 1999;**9**:216–21.

134. Appert-Collin A, Cotecchia S, Nenniger-Tosato M, et al. The A-kinase anchoring protein (AKAP)-Lbc-signaling complex mediates alpha1 adrenergic receptor-induced cardiomyocyte hypertrophy. *Proc Natl Acad Sci USA* 2007;**104**:10,140–10,145.

135. Diviani D, Scott JD. AKAP signaling complexes at the cytoskeleton. *J Cell Sci* 2001;**114**:1431–7.

136. Diviani D, Abuin L, Cotecchia S, Pansier L. Anchoring of both PKA and 14-3-3 inhibits the Rho-GEF activity of the AKAP-Lbc signaling complex. *EMBO J* 2004;**23**:2811–20.

137. Klussmann E, Edemir B, Pepperle B, et al. Ht31: the first protein kinase A anchoring protein to integrate protein kinase A and Rho signaling. *FEBS Letts* 2001;**507**:264–8.

138. Yuan J, Slice LW, Rozengurt E. Activation of protein kinase D by signaling through Rho and the alpha subunit of the heterotrimeric G protein G13. *J Biol Chem* 2001;**276**:38,619–38,627.

139. Carnegie GK, Smith FD, McConnachie G, et al. AKAP-Lbc nucleates a protein kinase D activation scaffold. *Mol Cell* 2004;**15**:889–99.

140. Gordon T, Grove B, Loftus JC, et al. Molecular cloning and preliminary characterization of a novel cytoplasmic antigen recognized by myasthenia gravis sera. *J Clin Invest* 1992;**90**:992–9.

141. Westphal RS, Soderling SH, Alto NM, et al. Scar/WAVE-1, a Wiskott-Aldrich syndrome protein, assembles an actin-associated multi-kinase scaffold. *EMBO J* 2000;**19**:4589–600.

142. Mullins RD, Machesky LM. Actin assembly mediated by Arp2/3 complex and WASP family proteins. *Methods Enzymol* 2000;**325**:214–37.

143. Sossey-Alaoui K, Su G, Malaj E, et al. WAVE3, an actin-polymerization gene, is truncated and inactivated as a result of a constitutional t(1;13)(q21;q12) chromosome translocation in a patient with ganglioneuroblastoma. *Oncogene* 2002;**21**:5967–74.

144. Gelman IH. the role of SSeCKS/Gravin/AKAP12 scaffolding proteins in the spaciotemporal control of signaling pathways in oncogenesis and development. *Front Biosci* 2002;**7**:D1782–97.

145. Wikman H, Kettunen E, Seppanen JK, et al. Identification of differentially expressed genes in pulmonary adenocarcinoma by using cDNA array. *Oncogene* 2002;**21**:5804–13.

146. Dhillon AS, Pollock C, Steen H, Shaw PE, Mischak H, Kolch W. Cyclic AMP-dependent kinase regulates Raf-1 kinase mainly by phosphorylation of serine 259. *Mol Cell Biol* 2002;**22**:3237–46.

147. Dell EJ, Connor J, Chen S, et al. The betagamma subunit of heterotrimeric G proteins interacts with RACK1 and two other WD repeat proteins. *J Biol Chem* 2002;**277**:49888–95.

148. Chen S, Dell EJ, Lin F, Sai J, Hamm HE. RACK1 regulates specific functions of Gbetagamma. *J Biol Chem* 2004;**279**:17861–8.

149. Chen S, Spiegelberg BD, Lin F, et al. Interaction of Gbetagamma with RACK1 and other WD40 repeat proteins. *J Mol Cell Cardiol* 2004;**37**:399–406.

150. Bretscher A, Edwards K, Fehon RG. ERM proteins and merlin: integrators at the cell cortex. *Nature Rev Mol Cell Biol* 2002;**3**:586–99.

151. Polesello C, Payre F. Small is beautiful: what flies tell us about ERM protein function in development. *Trends Cell Biol* 2004;**14**:294–302.

152. Khanna C, Wan X, Bose S, et al. The membrane-cytoskeleton linker ezrin is necessary for osteosarcoma metastasis. *Nature Med* 2004;**10**:182–6.

153. Yu Y, Khan J, Khanna C, et al. Expression profiling identifies the cytoskeletal organizer ezrin and the developmental homeoprotein Six-1 as key metastatic regulators. *Nature Med* 2004;**10**:175–81.

154. Vaiskunaite R, Adarichev V, Furthmayr H, et al. Conformational activation of radixin by G13 protein alpha subunit. *J Biol Chem* 2000;**275**:26,206–26,212.

155. Liu G, Voyno-Yasenetskaya TA. Radixin stimulates Rac1 and Ca^{2+}/calmodulin-dependent kinase, CaMKII: cross-talk with Galpha13 signaling. *J Biol Chem* 2005;**280**:39,042–39,049.

156. Weinman EJ, Hall RA, Friedman PA, et al. The association of NHERF adaptor proteins with g protein-coupled receptors and receptor tyrosine kinases. *Annu Rev Physiol* 2006;**68**:491–505.

157. Rochdi MD, Watier V, La Madeleine C, et al. Regulation of GTP-binding protein alpha q (Galpha q) signaling by the ezrin-radixin-moesin-binding phosphoprotein-50 (EBP50). *J Biol Chem* 2002;**277**:40,751–40,759.

158. Liedtke CM, Yun CH, Kyle N, Wang D. Protein kinase C epsilon-dependent regulation of cystic fibrosis transmembrane regulator involves binding to a receptor for activated C kinase (RACK1) and RACK1 binding to Na+/H+ exchange regulatory factor. *J Biol Chem* 2002;**277**:22,925–92,933.

159. Ferguson SS. Evolving concepts in G protein-coupled receptor endocytosis: the role in receptor desensitization and signaling. *Pharmacol Rev* 2001;**53**:1–24.

160. van Biesen T, Luttrell LM, Hawes BE, Lefkowitz RJ. Mitogenic signaling via G protein-coupled receptors. *Endocr Rev* 1996;**17**:698–714.

161. McDonald PH, Chow CW, Miller WE, et al. beta-arrestin 2: a receptor-regulated MAPK scaffold for the activation of JNK3. *Science* 2000;**290**:1574–7.

162. Gesty-Palmer D, Chen M, Reiter E, et al. Distinct beta-arrestin- and G protein-dependent pathways for parathyroid hormone receptor-stimulated ERK1/2 activation. *J Biol Chem* 2006;**281**:10,856–10,864.

163. Sun Y, Cheng Z, Ma L, Pei G. Beta-arrestin2 is critically involved in CXCR4-mediated chemotaxis, and this is mediated by its enhancement of p38 MAPK activation. *J Biol Chem* 2002;**277**:49,212–49,219.

164. Ge L, Ly Y, Hollenberg M, DeFea K. A beta-arrestin-dependent scaffold is associated with prolonged MAPK activation in pseudopodia during protease-activated receptor-2-induced chemotaxis. *J Biol Chem* 2003;**278**:34,418–34,426.

165. Lefkowitz RJ, Rajagopal K, Whalen EJ. New roles for beta-arrestins in cell signaling: not just for seven-transmembrane receptors. *Mol Cell* 2006;**24**:643–52.

166. Neves SR, Ram PT, Iyengar R. G protein pathways. *Science* 2002;**296**:1636–9.

167. Eungdamrong NJ, Iyengar R. Modeling cell signaling networks. *Biol Cell* 2004;**96**:355–62.

Chapter 48

Regulation of G Proteins by Covalent Modification

Benjamin C. Jennings and Maurine E. Linder

Department of Cell Biology and Physiology, Washington University School of Medicine, St Louis, Missouri

INTRODUCTION

Covalent modifications of G-protein subunits play key roles in their function, localization, and regulation. The first modifications known were those catalyzed by bacterial toxins. The discovery of ADP-ribosylation of G-protein α subunits by cholera and pertussis toxins provided significant insights into G-protein function, and is discussed in detail in Chapter 50 of Handbook of Cell Signaling, Second Edition. Regulation of G-protein activity by phosphorylation is also covered by Luttrell and Luttrell in Chapter 221 of Handbook of Cell Signaling, Second Edition [1]. Other posttranslational modifications reported for G-protein subunits include deamidation of $G\alpha_o$ [2, 3], ADP-ribosylation of $G\beta$ by a cellular ADP-ribosyltransferase [4], acetylation of $G\gamma$ [5], and arginylation of $G\gamma$ [6]. This chapter will focus on the covalent attachment of lipids to G proteins, modifications that anchor G-protein subunits to the membrane. Accordingly, lipidation of G proteins is essential for their ability to propagate signals from cell surface receptors to effectors. All $G\alpha$ subunits are N-terminally acylated with amide-linked myristate and/or thioester-linked palmitate. All $G\gamma$ subunits are C-terminally modified by a thioether-linked farnesyl (C15) or geranylgeranyl (C20) isoprenoid. Here, we discuss how the addition of lipids impacts protein function through roles in membrane targeting, protein trafficking, and protein–protein interactions of heterotrimeric G proteins.

N-TERMINAL ACYLATION OF Gα SUBUNITS

Gα subunits are modified with myristic and/or palmitic acid as presented in Table 48.1. N-myristoylation is the addition of myristic acid (C14:0) to a protein at an N-terminal

TABLE 48.1 Lipid modifications of heterotrimeric G proteins

Subunit	Lipid modifications	Modified sequence[a]
α_i, α_o, α_z	N-myr, S-palm	H$_2$N-M**GC**--
α_t, α_g	N-myr	H$_2$N-M**G**--
α_s	N-palm, S-Palm	H$_2$N-M**GC**--
α_{olf}	S-palm	H$_2$N-MG**C**--
α_q	S-palm	H$_2$N-MTLESIMA**CC** --
α_{11}	S-palm	H$_2$N-MTLESMMA**CC**--
α_{12}	S-palm	H$_2$N-MSGVVRTLSR**C**--
α_{13}	S-palm	H$_2$N-MAD--[14]**CFPGC**[18]--
α_{14}	S-palm	H$_2$N-MAG**CCC**LS--
α_{15} (α_{16})	S-palm	H$_2$N-MAR--[8]**RCCPWC**[13]--
γ_1, γ_9, γ_{11}	Farnesyl	--**C**aaS-COOH[b]
γ_2, γ_3, γ_4, γ_5, γ_7, γ_8, γ_{10}, γ_{12}, γ_{13}	Geranylgeranyl	--**C**aaL-COOH[b]

[a]*Modified residues of human sequences are in boldface.*
[b]*"a" is an aliphatic residue; additional processing includes proteolysis of the three C-terminal residues and carboxymethylation of the prenylated cysteine.*

glycine residue exposed after removal of the initiator methionine. Attachment of myristate is through an amide linkage, and thus not readily reversible. Myristoyl-CoA: protein N-myristoyltransferase (NMT) is associated with ribosomes, and catalyzes the co-translational addition of myristate to Gα subunits [7]. All members of the Gα_i subfamily (α_t, α_z, α_g, α_o, and α_{i1-3}) are N-myristoylated [8]. Gα_s is unique among G-protein α subunits in that it is modified by amide-linked palmitate at the amino-terminal glycine [9]. The mechanism by which this amide-linked palmitate is added to Gly2 in $G\alpha_s$ is unknown.

Palmitoylation (S-palmitoylation or S-acylation) refers to the addition of palmitic acid or other long-chain fatty acids to cysteine residues in proteins [10]. Unlike myristoylation, palmitoylation occurs posttranslationally. Palmitate is attached through a reversible thioester linkage, allowing for dynamic and regulated palmitoylation/depalmitoylation cycling in cells. All G-protein α subunits except $G\alpha_t$ and $G\alpha_g$ have cysteine residues within the first 20 amino acids that are modified with palmitate.

S-palmitoylation of G-protein subunits is thought to be carried out by one or more members of the DHHC (Asp–His–His–Cys) family of protein acyltransferases (PATs) [11]. DHHC proteins are integral membrane proteins with a highly conserved DHHC motif embedded in a cysteine-rich domain. First characterized in yeast and subsequently in mammalian cells, DHHC proteins catalyze the S-palmitoylation of proteins on the cytoplasmic face of membranes [12]. It is likely that one or several of these DHHC proteins acylate $G\alpha$ subunits using palmitoyl-coenzyme A as a source of activated palmitate. DHHC3 is a candidate PAT for the addition of thioester-linked palmitate to $G\alpha_s$, based on its ability to increase palmitoylation of $G\alpha_s$ when overexpressed in tissue culture cells [13].

Palmitoylation of G-protein α subunits is reversible, and evidence suggests that palmitate turnover is regulated by G-protein-coupled receptors. An enzyme capable of depalmitoylating $G\alpha$ subunits has been identified as acyl-protein thioesterase (APT1). Purified APT1 depalmitoylates both $G\alpha_{i1}$ and $G\alpha_s$ heterotrimer in vitro, and co-expression of APT1 with $G\alpha_s$ in tissue culture cells results in faster turnover of palmitate [14]. Saccharomyces cerevisiae lacking the APT1 gene display impaired palmitate turnover on $Gp\alpha1$ in vivo and extracts from apt1Δ yeast failed to depalmitoylate $G\alpha_{i1}$ in vitro. However, there was no effect of APT1 deletion on the activity of the pheromone response pathway [15].

C-TERMINAL MODIFICATION OF Gγ

All 12 mammalian Gγ subunits are prenylated at C-terminal CaaX motifs, where "C" represents the cysteine that is modified by the isoprenoid and "a" represents an aliphatic residue. The amino acid at position "X" specifies if a farnesyl or a geranylgeranyl group is added to the CaaX box cysteine. $G\gamma_1$, γ_9, and γ_{11} terminate with serine and are modified by farnesyltransferase (FTase). The remaining Gγ subunits terminate with leucine and are modified by geranylgeranyltransferase Type I (GGTase-I). Attachment occurs through a thioether linkage and is not reversible. FTase and GGTase-I are soluble enzymes, and prenylation occurs in the cytoplasm. Prenylated Gγ then moves to the

ER by an unknown mechanism, where the "–aaX" motif is proteolyzed and reversibly carboxylmethylated at the freshly exposed cysteine. Proteolysis and methylation are mediated by RCE1 (CaaX prenyl protease 2, FACE2) and isoprenylcysteine carboxyl methyltransferase (ICMT) [16], respectively. Both enzymes are integral membrane proteins and have been localized to the ER [17, 18]. Although most Gγ subunits undergo the full processing reaction at the C-terminus, mass spectrometry of Gγ subunits isolated from brain reveals heterogeneity in the posttranslational processing. Species with incomplete processing have been identified, as well as a novel isoform resulting from cysteinylation of the carboxyl terminus [5, 19]. The functional significance of this heterogeneity has not been explored.

STRUCTURAL AND FUNCTIONAL CONSEQUENCES OF LIPID MODIFICATIONS

Myristoylation and Palmitoylation

Myristoylation is necessary for the function of $G\alpha_i$ subunits. Gpa1, a $G\alpha_i$ ortholog in Sacchaomyces cerevisiae, is encoded by an essential gene in haploid cells. Mutation of the myristoylation site in GPA1 results in lethality [20]. In mammalian cells, loss of myristoylation by mutation of Gly2 results in a cytoplasmic distribution of the protein and failure to couple to effectors and receptors [8]. The membrane-binding function of the myristoyl group is clearly key to its function, but there is debate as to whether the myristoyl group may serve other functions.

In the inactive state, native $G\alpha$ subunits form a high-affinity complex with Gβγ. Some members of the $G\alpha_i$ subfamily still form this heterotrimer when lacking myristate (e.g., $G\alpha_{i1}$ [21], $G\alpha_{i3}$ [22], and yeast Gpa1 [23]). However, $G\alpha_t$ [24] and $G\alpha_o$ [25] show reduced Gβγ interaction without myristate. Facilitation of subunit interactions by lipid modification is thought to be indirect, occurring through enhanced membrane interaction, rather than mediated by lipid-binding sites on the protein surfaces. In studies of transducin association with detergent micelles or phospholipid vesicles, the $G\alpha_t$ myristoyl group appears to bind cooperatively with the $G\gamma_t$ farnesyl chain in the membrane lipids [24]. It is proposed that lipid–lipid interactions contribute to subunit association by favoring the orientation of protein interaction surfaces and promoting their binding [24].

When $G\alpha_i$ is dissociated from Gβγ, the myristoyl group may mediate interaction of the N-terminal helix with an intramolecular binding site on the surface of $G\alpha_i$. Electron paramagnetic resonance and fluorescent probes were used to monitor mobility and conformational changes in the N-terminus of $G\alpha_{i1}$ upon activation with aluminum fluoride or with its association with Gβγ. These experiments revealed that the myristoylated N-terminus of $G\alpha_{i1}$ was largely immobile compared to the high mobility of the same region in unmodified $G\alpha_{i1}$ [26].

Effector interactions are also influenced by fatty acylation of G proteins. Activated $G\alpha_s$ and $G\alpha_i$ work to stimulate and inhibit membrane-bound adenylyl cyclase (AC), respectively. $G\alpha_s$ modified with amide-linked palmitate stimulated membrane-bound AC activity with ~60-fold higher affinity than unmodified $G\alpha_s$ [9, 27]. Membrane localization of the effector was necessary to detect this difference. An artificial soluble AC was stimulated equivalently by N-palmitoylated $G\alpha_s$ and unmodified $G\alpha_s$ [9]. Hence, it is likely that N-palmitoylation mediates high-affinity interaction by concentrating the protein at membranes. Interestingly, however, the myristoyl group on $G\alpha_i$ directly affects the affinity for AC independent of membrane localization. Myristoylation was required for $G\alpha_i$ to interact with a soluble domain of AC in an assay free of micelles or membranes [28]. It is possible that there is a binding site on the surface of AC that accommodates the myristoyl group. Alternatively, the conformation described above in which the N-terminal helix is immobilized on a putative intramolecular binding site on $G\alpha_i$ [26] may be necessary for its interaction with effector. In contrast to N-myristoylation, S-palmitoylation does not appear to have an important role in promoting $G\alpha$–effector interactions [27, 29].

Following activation and effector interaction, $G\alpha$ subunits attenuate signaling through intrinsic GTPase activity that is accelerated following interaction with *r*egulators of *G*-protein *s*ignaling (RGS) proteins. Myristoylation of $G\alpha_z$, $G\alpha_i$, and $G\alpha_o$ increases their affinity for the RGS protein $G\alpha_z$ GAP [30, 31]. Conversely, palmitoylation of $G\alpha_z$ and $G\alpha_i$ or phosphorylation of $G\alpha_z$ reduced their responses to the GAP activity of RGS proteins [30, 32]. Thus, depalmitoylation following activation may be important for returning G proteins to their inactive state.

Prenylation

Prenylation is not required for $G\beta\gamma$ dimer formation, however, it enhances the interaction of $G\alpha$ and $G\beta\gamma$. This may be an indirect effect due to concentration on membrane surfaces. However, there is evidence to support a putative S-prenyl-binding site in $G\alpha$. S-prenyl analogs containing farnesyl or geranylgeranyl moieties inhibit heterotrimer formation competitively and reversibly [33]. $G\gamma$ subunits modified with geranylgeranyl versus farnesyl groups display higher affinity for $G\alpha$ subunits and more potent stimulation of adenylyl cyclase and phospholipase C (PLC) activity [34]. Likewise, carboxylmethylation of $G\beta_1\gamma_1$ ($T\beta\gamma$) strongly enhances activation of $PLC\beta_3$ and PI3K relative to non-methylated $G\gamma_1$ [35]. Methylation may facilitate coupling to effectors by neutralizing the negative carboxylate group at the C-terminus of $G\gamma$ subunits, thus minimizing

repulsion by the negatively charged phospholipids' head groups and facilitating insertion into the leaflet of the PM.

Prenylation and primary sequence of $G\gamma$ are both important determinants of $G\beta\gamma$ effector activation. In the case of $PLC\beta$ activation, there is evidence that the prenyl group is not simply promoting effector activation indirectly through membrane interactions, but directly mediating protein–protein interactions. Peptides corresponding to the C-terminus of $G\gamma_2$ inhibit activation of $PLC\beta$ by $G\beta\gamma$ in a prenylation-dependent manner. A fluorescence-based binding assay demonstrated a direct interaction of the prenylated peptide with $PLC\beta_2$ [36]. Precedent for a prenyl-binding site in a protein is provided by the structure of prenylated Cdc42, a Rho family GTPase, bound to Rho-GDI, a guanine nucleotide dissociation inhibitor, which regulates interaction of Rho family members with membranes. In this structure, the C-terminal geranylgeranyl group on Cdc42 inserts into a hydrophobic pocket formed by Rho-GDI [37].

A second mechanism for prenylation-dependent effector interactions is suggested by the structure of the $G\beta\gamma$-phosducin complex [38]. Phosducin binds tightly to free $G\beta\gamma$, extracting it from membranes and preventing its reassociation with $G\alpha$. Phosducin induces several local conformational changes in $G\beta$ that are not seen in the structures of free $G\beta\gamma$ or the heterotrimer, including the opening of a pocket between blades 6 and 7 [38, 39]. Based on their structure of farnesylated $G\beta\gamma$ with phosducin, Loew and colleagues proposed that the farnesyl group is sequestered in the crevice [38]. $G\beta\gamma$ may undergo a similar conformational change when bound to effectors. Consistent with this model, mutations in $G\beta$ that perturb the putative prenyl binding pocket exhibit reduced potency in effector activation assays [34].

LIPIDATION INFLUENCES G-PROTEIN TRAFFICKING AND LOCALIZATION

Lipid modifications influence the affinity of proteins for membranes, and thus their subcellular localization and trafficking. Lipid-mediated localization is discussed in general for signaling proteins in Chapter 19 of Handbook of Cell Signaling, Second Edition. This section will highlight recent findings on how G-protein localization is affected by lipidation. The role of lipid modifications during nascent protein synthesis and trafficking and their role following G-protein-coupled receptor (GPCR) activation will be discussed. The reader is referred to an excellent review for more detailed coverage of this topic [11].

Lipidation During Nascent Protein Synthesis

$G\beta$ and $G\gamma$ subunits are synthesized on soluble ribosomes. G-protein β subunits associate with chaperones to prevent $G\beta$ aggregates and to facilitate assembly with $G\gamma$ subunits

[40]. Although isoprenylated or unprocessed Gγ is able to bind Gβ in the cytoplasm, unprocessed Gγ is the preferred substrate [41, 42]. Gβγ heterodimerization is irreversible, thus within the complex, Gγ is subsequently prenylated in the cytoplasm by the relevant prenyltransferase. Through an unknown mechanism, the isoprenylated Gβγ dimer is targeted to the cytoplasmic surface of the ER, where it is proteolyzed and methylated.

Gα subunits are also synthesized on cytoplasmic ribosomes. Members of the Gα$_i$-family are co-translationally modified with myrisate at the ribosome. Exactly where Gα and Gβγ assemble into heterotrimer is unknown. However, the Golgi and ER are likely sites [11]. Palmitoylation of Gα follows heterotrimer formation, most likely by a DHHC PAT localized to the Golgi or ER. Dually acylated heterotrimer is able to localize to the PM. Golgi disruption with the fungal metabolite brefeldin A does not perturb the PM targeting of α$_s$, α$_q$, or α$_z$, or the palmitoylation of α$_q$, α$_z$, or α$_i$, suggesting that trafficking occurs via a Golgi-independent pathway. However, when a GPCR is overexpressed along with Gα and Gβγ, Gαβγ trafficking to the PM is Golgi-dependent [43]. Thus, G proteins may use different trafficking pathways depending on which proteins are available to form pre-signaling complexes.

Lipidation During Steady-State and GPCR Activation

During steady-state conditions, heterotrimers were thought to statically remain at the plasma membrane in the inactive GDP state poised for nucleotide exchange induced by an agonist-activated GPCR. However, a recent study revealed dynamic shuttling of G-protein heterotrimers under basal conditions [44]. Fluorescently tagged heterotrimers were demonstrated to shuttle between the plasma membrane and endomembranes through a diffusion-mediated pathway. Interestingly, treatment of cells with the palmitoylation inhibitor 2-bromopalmitate (2BP) blocked shuttling of heterotrimers, suggesting the shuttling may depend on the palmitoylation of Gα subunits. However, the kinetics of palmitate turnover on Gα subunits ($t_{1/2} > 60$ min for α$_i$, 90 min for α$_s$) [45, 46] is slower than the kinetics of shuttling between the plasma membrane and endomembranes (1–3 min) [44]. An alternative proposal for the effect of 2BP is that an unidentified protein may undergo an acylation/deacylation cycle that regulates trafficking of the heterotrimer. If as evidence suggests that shuttling is controlled by diffusion through the cytoplasm, a putative binding partner might be necessary to bind to the heterotrimer and mask the lipid groups. It is also possible that the effect of 2BP is unrelated to protein palmitoylation.

Following treatment with agonist, GPCRs act as guanine nucleotide exchange factors replacing GDP with GTP in Gα subunits and activating those subunits for stimulation of downstream effectors. Agonist treatment also stimulates palmitate turnover of Gα subunits [8]. The current model is that following activated-receptor-induced Gα and Gβγ dissociation, the Gα subunit becomes more susceptible to a thioesterase that removes palmitate and turnover is accelerated relative to the basal state (reviewed in [8]).

Finally, recent work has shown agonist stimulation also causes diffusion-mediated translocation of some Gβγ complexes from the PM to either the Golgi or ER and this can be reversed by antagonist treatment. The identity of the γ isoform, not whether it is modified with a farnesyl- or geranylgeranyl-group, dictates whether free Gβγ traffics to the Golgi or the ER. Interestingly, 2BP blocked this receptor-induced translocation of Gβγ, suggesting it is acylation-dependent, similar to the basal state heterotrimer trafficking described above [47].

CONCLUSIONS

Lipid modifications present on Gα and Gβγ control interactions with both membranes and other proteins. Increasing evidence is revealing an important role for lipid modifications in regulating G-protein localization and trafficking. Future studies in this area will help in identifying the enzymes that mediating Gα palmitoylation and shed light on the molecular mechanisms driving G-protein trafficking between membranes. Novel roles for G-protein lipidation will likely be revealed through these studies.

REFERENCES

1. Luttrell LM, Luttrell DK. Phophorylation of G proteins. In: Bradshaw RA, Dennis EA, editors. *Handbook of cell signaling*. New York, NY: Academic Press; 2007. p. 612–99.
2. Exner T, Jensen ON, Mann M, Kleuss C, Nurnberg B. Posttranslational modification of Galphao1 generates Galphao3, an abundant G protein in brain. *Proc Natl Acad Sci USA* 1999;**96**:1327–32.
3. McIntire WE, Schey KL, Knapp DR, Hildebrandt JD. A major G protein alpha O isoform in bovine brain is deamidated at Asn346 and Asn347, residues involved in receptor coupling. *Biochemistry* 1998;**37**:14,651–14,658.
4. Lupi R, Corda D, Di Girolamo M. Endogenous ADP-ribosylation of the G protein beta subunit prevents the inhibition of type 1 adenylyl cyclase. *J Biol Chem* 2000;**275**:9418–24.
5. Cook LA, Schey KL, Wilcox MD, Dingus J, Ettling R, Nelson T, Knapp DR, Hildebrandt JD. Proteomic analysis of bovine brain G protein gamma subunit processing heterogeneity. *Mol Cell Proteomics* 2006;**5**:671–85.
6. Hamilton MH, Cook LA, McRackan TR, Schey KL, Hildebrandt JD. Gamma 2 subunit of G protein heterotrimer is an N-end rule ubiquitylation substrate. *Proc Natl Acad Sci USA* 2003;**100**:5081–6.
7. Farazi TA, Waksman G, Gordon JI. The biology and enzymology of protein N-myristoylation. *J Biol Chem* 2001;**276**:39,501–39,504.
8. Chen CA, Manning DR. Regulation of G proteins by covalent modification. *Oncogene* 2001;**20**:1643–52.
9. Kleuss C, Krause E. Galpha(s) is palmitoylated at the N-terminal glycine. *EMBO J* 2003;**22**:826–32.

10. Smotrys JE, Linder ME. Palmitoylation of intracellular signaling proteins: regulation and function. *Annu Rev Biochem* 2004;**73**:559–87.

11. Marrari Y, Crouthamel M, Irannejad R, Wedegaertner PB. Assembly and trafficking of heterotrimeric G proteins. *Biochemistry* 2007;**46**:7665–77.

12. Mitchell DA, Vasudevan A, Linder ME, Deschenes RJ. Protein palmitoylation by a family of DHHC protein S-acyltransferases. *J Lipid Res* 2006;**47**:1118–27.

13. Fukata M, Fukata Y, Adesnik H, Nicoll RA, Bredt DS. Identification of PSD-95 palmitoylating enzymes.. *Neuron* 2004;**44**:987–96.

14. Duncan JA, Gilman AG. A cytoplasmic acyl-protein thioesterase that removes palmitate from G protein alpha subunits and p21(RAS). *J Biol Chem* 1998;**273**:15,830–15,837.

15. Duncan JA, Gilman AG. Characterization of Saccharomyces cerevisiae acyl-protein thioesterase 1, the enzyme responsible for G protein alpha subunit deacylation in vivo. *J Biol Chem* 2002;**277**:31,740–31,752.

16. Winter-Vann AM, Casey PJ. Post-prenylation-processing enzymes as new targets in oncogenesis. *Nature Rev Cancer* 2005;**5**:405–12.

17. Schmidt WK, Tam A, Fujimura-Kamada K, Michaelis S. Endoplasmic reticulum membrane localization of Rce1p and Ste24p, yeast proteases involved in carboxyl-terminal CAAX protein processing and amino-terminal a-factor cleavage. *Proc Natl Acad Sci USA* 1998;**95**:11,175–11,180.

18. Dai Q, Choy E, Chiu V, Romano J, Slivka SR, Steitz SA, Michaelis S, Philips MR. Mammalian prenylcysteine carboxyl methyltransferase is in the endoplasmic reticulum. *J Biol Chem* 1998;**273**:15,030–15,034.

19. Kilpatrick EL, Hildebrandt JD. Sequence dependence and differential expression of Ggamma5 subunit isoforms of the heterotrimeric G proteins variably processed after prenylation in mammalian cells. *J Biol Chem* 2007;**282**:14,038–14,047.

20. Stone DE, Cole GM, de Barros Lopes M, Goebl M, Reed SI. N-myristoylation is required for function of the pheromone-responsive G alpha protein of yeast: conditional activation of the pheromone response by a temperature-sensitive N-myristoyl transferase. *Genes Dev* 1991;**5**:1969–81.

21. Jones Jr TL, Simonds WF, Merendino JJ, Brann MR, Spiegel AM. Myristoylation of an inhibitory GTP-binding protein alpha subunit is essential for its membrane attachment. *Proc Natl Acad Sci USA* 1990;**87**:568–72.

22. Brand SH, Holtzman EJ, Scher DA, Ausiello DA, Stow JL. Role of myristoylation in membrane attachment and function of G alpha i-3 on Golgi membranes. *Am J Physiol* 1996;**270**:C1362–9.

23. Song J, Hirschman J, Gunn K, Dohlman HG. Regulation of membrane and subunit interactions by N-myristoylation of a G protein alpha subunit in yeast. *J Biol Chem* 1996;**271**:20,273–20,283.

24. Bigay J, Faurobert E, Franco M, Chabre M. Roles of lipid modifications of transducin subunits in their GDP-dependent association and membrane binding. *Biochemistry* 1994;**33**:14,081–14,090.

25. Linder ME, Pang IH, Duronio RJ, Gordon JI, Sternweis PC, Gilman AG. Lipid modifications of G protein subunits. Myristoylation of Go alpha increases its affinity for beta gamma. *J Biol Chem* 1991;**266**:4654–9.

26. Preininger AM, Van Eps N, Yu NJ, Medkova M, Hubbell WL, Hamm HE. The myristoylated amino terminus of Galpha(i)(1) plays a critical role in the structure and function of Galpha(i)(1) subunits in solution. *Biochemistry* 2003;**42**:7931–41.

27. Kleuss C, Gilman AG. Gsalpha contains an unidentified covalent modification that increases its affinity for adenylyl cyclase. *Proc Natl Acad Sci USA* 1997;**94**:6116–20.

28. Dessauer CW, Tesmer JJ, Sprang SR, Gilman AG. Identification of a Gialpha binding site on type V adenylyl cyclase. *J Biol Chem* 1998;**273**:25,831–25,839.

29. Hepler JR, Biddlecome GH, Kleuss C, Camp LA, Hofmann SL, Ross EM, Gilman AG. Functional importance of the amino terminus of Gq alpha. *J Biol Chem* 1996;**271**:496–504.

30. Tu Y, Wang J, Ross EM. Inhibition of brain Gz GAP and other RGS proteins by palmitoylation of G protein alpha subunits. *Science* 1997;**278**:1132–5.

31. Wang J, Tu Y, Woodson J, Song X, Ross EM. A GTPase-activating protein for the G protein Galphaz. Identification, purification, and mechanism of action. *J Biol Chem* 1997;**272**:5732–40.

32. Wang J, Ducret A, Tu Y, Kozasa T, Aebersold R, Ross EM. RGSZ1, a Gz-selective RGS protein in brain. Structure, membrane association, regulation by Galphaz phosphorylation, and relationship to a Gz gtpase-activating protein subfamily. *J Biol Chem* 1998;**273**: 26,014–26,025.

33. Dietrich A, Scheer A, Illenberger D, Kloog Y, Henis YI, Gierschik P. Studies on G-protein alpha.betagamma heterotrimer formation reveal a putative S-prenyl-binding site in the alpha subunit. *Biochem J* 2003;**376**:449–56.

34. Myung CS, Yasuda H, Liu WW, Harden TK, Garrison JC. Role of isoprenoid lipids on the heterotrimeric G protein gamma subunit in determining effector activation. *J Biol Chem* 1999;**274**:16,595–16,603.

35. Parish CA, Smrcka AV, Rando RR. Functional significance of beta gamma-subunit carboxymethylation for the activation of phospholipase C and phosphoinositide 3-kinase.. *Biochemistry* 1995;**34**:7722–7.

36. Fogg VC, Azpiazu I, Linder ME, Smrcka A, Scarlata S, Gautam N. Role of the gamma subunit prenyl moiety in G protein beta gamma complex interaction with phospholipase Cbeta. *J Biol Chem* 2001;**276**:41,797–41,802.

37. Hoffman GR, Nassar N, Cerione RA. Structure of the Rho family GTP-binding protein Cdc42 in complex with the multifunctional regulator RhoGDI. *Cell* 2000;**100**:345–56.

38. Loew A, Ho YK, Blundell T, Bax B. Phosducin induces a structural change in transducin beta gamma. *Structure* 1998;**6**:1007–19.

39. Gaudet R, Bohm A, Sigler PB. Crystal structure at 2.4 angstroms resolution of the complex of transducin betagamma and its regulator, phosducin. *Cell* 1996;**87**:577–88.

40. Kubota S, Kubota H, Nagata K. Cytosolic chaperonin protects folding intermediates of Gbeta from aggregation by recognizing hydrophobic beta-strands. *Proc Natl Acad Sci USA* 2006;**103**:8360–5.

41. Higgins JB, Casey PJ. The role of prenylation in G-protein assembly and function. *Cell Signal* 1996;**8**:433–7.

42. Higgins JB, Casey PJ. In vitro processing of recombinant G protein gamma subunits. Requirements for assembly of an active beta gamma complex. *J Biol Chem* 1994;**269**:9067–73.

43. Dupre DJ, Robitaille M, Ethier N, Villeneuve LR, Mamarbachi AM, Hebert TE. Seven transmembrane receptor core signaling complexes are assembled prior to plasma membrane trafficking. *J Biol Chem* 2006;**281**:34561–73.

44. Chisari M, Saini DK, Kalyanaraman V, Gautam N. Shuttling of G protein subunits between the plasma membrane and intracellular membranes. *J Biol Chem* 2007;**282**:24,092–24,098.

45. Chen CA, Manning DR. Regulation of galpha i palmitoylation by activation of the 5-hydroxytryptamine-1A receptor. *J Biol Chem* 2000;**275**:23,516–23,522.

46. Wedegaertner PB, Wilson PT, Bourne HR. Lipid modifications of trimeric G proteins. *J Biol Chem* 1995;**270**:503–6.

47. Saini DK, Kalyanaraman V, Chisari M, Gautam N. A family of G protein betagamma subunits translocate reversibly from the plasma membrane to endomembranes on receptor activation. *J Biol Chem* 2007;**282**:24,099–24,108.

Chapter 49

Signaling Through G$_z$

Michelle E. Kimple[1], Rainbo C. Hultman[2] and Patrick J. Casey[1,2]

[1]*Departments of Pharmacology and Cancer Biology, Duke University Medical Center, Durham, North Carolina*

[2]*Biochemistry, Duke University Medical Center, Durham, North Carolina*

INTRODUCTION

The α subunit of G$_z$ was first identified in 1988 [1, 2], yet until recently relatively little had been known about its function *in vivo*. While the similarities between G$_z$ and other G$_i$ proteins cannot be ignored, the limited tissue distribution of Gα$_z$, its unusual biochemical properties, the identification of Gα$_z$-specific effectors, and the inability of other Gα$_i$ subfamily members to substitute for Gα$_z$ *in vivo* all support unique physiologic roles for G$_z$. These subjects, along with the role of G$_z$ as a prototypical G$_i$ subfamily member, will be the focus of this chapter.

GENERAL PROPERTIES

Tissue Distribution

The tissue distribution of Gα$_z$ is quite restricted as compared to that of other Gα$_i$ subfamily members. Expression of Gα$_z$ protein has been definitively demonstrated only in the central and peripheral nervous systems, platelets, placenta, adrenal medulla, retina, and pancreatic islets [3–7], although its expression has been suggested in other select tissues [4, 8–11].

Several studies have been carried out to pinpoint specific brain regions and developmental stages where Gα$_z$ is transcribed or expressed. Gα$_z$ mRNA has been identified in hippocampal (CA1, CA3, and dentate gyrus), cortical, cerebellar, and, to a lesser extent, striatal regions of the brain [12]. Gα$_z$ appears to be expressed in most hippocampal and cortical neurons, as well as in ganglion cells of the retina and Purkinje cells of the cerebellum [3, 13]. In mice, Gα$_z$ is expressed in the Purkinje cells of the cerebellum during development and through adulthood. Interestingly, unlike other Gα subunits, Gα$_z$ is localized not only to the Purkinje cell somas, but also to their axons and dendrites [13]. Quantitative RT-PCR studies suggest that Gα$_z$ mRNA is highest in sensory neurons near birth (E16 to PND 14), corresponding to the time of tissue target innervation [14]. These findings may suggest a possible role for G$_z$ signaling in the development of the nervous system.

Covalent Modifications

Unlike other members of the Gα$_i$ subfamily, Gα$_z$ lacks a consensus Cys residue near the C-terminus that is the site of modification by pertussis toxin-catalyzed ADP-ribosylation [4, 15, 16]. This makes Gα$_z$ a candidate for pertussis toxin-insensitive, G protein-dependent, signaling processes.

Another distinct property of Gα$_z$ is that it can be phosphorylated both *in vitro* and *in situ*. In permeabilized or intact platelets, PKC activation promotes rapid and stoichiometric phosphorylation of Gα$_z$, but has no effect on other members of the Gα$_i$ subfamily or Gα$_s$ [17]. Site-directed mutagenesis of potential PKC consensus sites revealed that Ser^{16} and Ser^{27} of Gα$_z$ are responsible for nearly 80 percent of the total phosphorylation [18]. This PKC-mediated phosphorylation is effectively inhibited by the presence of βγ complex, and phosphorylated Gα$_z$ has a markedly reduced ability to interact with the βγ complex [19], suggesting that the βγ contact site is near the N-terminus of Gα$_z$. Phosphorylation of Gα$_z$ also renders the α subunit much less susceptible to RGS20 (RGSZ1) action [20]. Taken together, PKC-mediated phosphorylation of Gα$_z$ may augment G$_z$ signaling by preventing both RGS20-enhanced GTP hydrolysis and re-formation of an αβγ complex, although this hypothesis has not yet been directly tested.

Localization of $G\alpha_z$ to the plasma membrane requires two lipid modifications: co-translational myristoylation on Gly^2, and subsequent palmitoylation on Cys^3. Prevention of myristoylation by substitution of Ala for Gly^2 decreases palmitoylation on Cys^3, but palmitoylation can be rescued by overexpression of $\beta\gamma$ complex, suggesting membrane association triggers a palmitoylation event [21, 22]. Substitution of Ala for Cys^3, thus preventing palmitoylation, does not affect co-translational myristoylation. This suggests that myristoylation, an irreversible modification, directly impacts the cellular localization of $G\alpha_z$, while palmitoylation, a reversible modification, is more important for the interaction between $G\alpha_z$ and its possible effectors and regulators [22, 23].

Biochemical Properties

The rate of GDP dissociation from $G\alpha_z$ is extremely slow as compared to that of most other G-protein α subunits, and almost completely suppressed at Mg^{2+} concentrations greater than $100\,\mu M$ [15]. Furthermore, the intrinsic rate of GTP hydrolysis by $G\alpha_z$ is nearly 200-fold lower than that of other $G\alpha_i$ subfamily proteins [15]. This relatively weak GTP hydrolysis activity, possibly a result of a Ser substitution of the second Gly in a conserved "GAGES" sequence [24], indicates that $G\alpha_z$ may participate in longer-duration signaling events than other $G\alpha_i$ proteins. This low GTP hydrolysis activity also suggests that Regulator of G-protein Signaling (RGS) molecules may play crucial roles in regulating G_z signaling, due to their enhancement of G-protein GTP hydrolysis (discussed below).

REGULATORS OF G_Z SIGNALING: RGS PROTEINS

RGS proteins act as negative regulators of G-protein signaling by binding to and enhancing GTP hydrolysis by G-protein α subunits. Several RGS proteins have been identified (out of the over 20 that are known in mammals) that selectively act on all members of the $G\alpha_i$ subfamily or $G\alpha_z$ specifically. These include RGS4 [25], RGS10 [26], RGS17 (RGSZ2), RGS19 (GAIP) [27], and RGS20 (RGSZ1) [20, 28]. RGS4 and RGS10 are 100- to 700-fold less active on $G\alpha_z$ than RGS17, RGS19, and RGS20 (which all contain a unique N-terminal cysteine string motif and are together termed the RGS-Rz family) [23]; furthermore, RGS17 and RGS20 are specific for $G\alpha_z$.

It is becoming increasingly appreciated that RGS-Rz proteins function not only as negative regulators of G protein signaling, but also as effectors or adaptors linking $G\alpha_z$ to other signaling pathways. For example, the interaction of RGS20 with the stathmin family member SCG10 inhibits the ability of the stathmin to promote microtubule disassembly [29].

Another example is the direct interaction of RGS17 with both the brain μ-opioid receptor and $G\alpha_z$ [30, 31]. Sumoylated RGS17 serves a scaffolding role for the μ-opioid receptor and G_z, while free RGS17 and RGS20 act as prototypical RGS proteins, accelerating $G\alpha_z$ GTP hydrolysis [30, 31].

RECEPTORS THAT COUPLE TO G_Z

Most receptors that couple to G_i proteins can also activate G_z if the G protein and/or receptors are overexpressed in cells (for comprehensive reviews see [16, 32]). Several receptors, though, which will be described in more detail below, have been demonstrated to preferentially couple to Gz *in situ* or *in vivo*.

The μ, δ, and κ opioid receptors can all couple to G_z in overexpression systems; however, only the brain μ-opioid receptor regulates G_z [33–35]. The brain μ-opioid receptor co-precipitates with $G\alpha_z$, and this association decreases with stimulation by agonists such as morphine [31]. A functional coupling between $G\alpha_z$ and the μ-opioid receptor is also supported by the finding that RGS-Rz proteins interact with the C-terminus of the μ-opioid receptor and not those of the δ receptors, and that the RGS-Rz proteins modulate $G\alpha_z$-mediated μ-opiod receptor signaling [31, 36].

Another receptor that exhibits specificity toward G_z is the α_{2A}-adrenergic receptor. Mice lacking $G\alpha_z$ exhibit decreased platelet aggregation and impaired inhibition of cAMP formation in response to epinephrine [37, 38], which acts through α_{2A}-adrenergic receptors [39]. The preferential coupling of the α_{2A}-adrenergic receptor to G_z was also observed in experiments using PC12 cells expressing wild-type $G\alpha_z$. When these cells were challenged with a cAMP analog or nerve growth factor, treatment with a specific agonist to α_{2A}-adrenergic receptors, UK14304, attenuated PC12 cell differentiation; such an effect was not observed in these cells in the absence of $G\alpha_z$ expression (Figure 49.1, left panel) [40]. Receptor coupling to G_z, however, appears to be cell type- and context-specific. For example, in the Ins-1 (832/13) pancreatic β-cell line, UK14304 strongly inhibited glucose-stimulated insulin secretion, but this inhibition was almost completely pertussis toxin-sensitive, suggesting that the α_{2A}-adrenoreceptor preferentially signals through other G_i subfamily members in β cells [7]. This finding is supported by prior studies that also showed pertussis toxin-sensitive inhibition of glucose-stimulated insulin secretion by α_{2A}-adrenoreceptor agonists [41, 42]. However, in the same Ins-1 cell line, the inhibition of glucose-stimulated insulin secretion by prostaglandin E1 was completely insensitive to pertussis toxin pre-treatment [7]. The selective coupling of the β-cell E prostanoid receptor to G_z was confirmed by selective blockade of $G\alpha_z$ activity or expression by the RGS domain of RGSZ1 or $G\alpha_z$-specific siRNA, respectively [7].

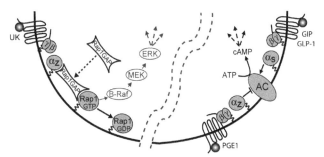

FIGURE 49.1 G$_z$ signaling pathways.

Left panel: A model for Rap1GAP involvement in G$_z$ signaling in neuronal cells. In the model shown, GTP-bound Rap1 activates a MAP kinase cascade that triggers downstream effects (dashed arrows). Concurrent activation of G$_z$ by ligation of an appropriate G protein-coupled receptor (shown is UK14304 activation of the α_{2A}-adrenoreceptor) leads to recruitment of Rap1GAP to the membrane with subsequent down-regulation of Rap1 signaling. *Right panel*: The role of G$_z$ in regulation of the insulin secretion pathway. Activation of receptors that couple to G$_s$ subfamily proteins by agonists such as gastric inhibitory polypeptide (GIP) or glucagon-like peptide 1 (GLP-1) leads to the activation of adenylyl cyclase (AC), resulting in an increase in the production of the second messenger cAMP. cAMP has been shown to augment the insulin secretory pathway at numerous points (dashed arrows). Conversely, activated G$_z$, perhaps produced by stimulation of an E prostanoid receptor, binds to and inhibits adenylyl cyclase, leading to a decrease in intracellular cAMP, thus attenuating the downstream effects of this second messenger.

A role for G$_z$ in dopaminergic signaling was first discovered when G$_z$-null mice exhibited a highly exaggerated response to cocaine [37]. Since then, studies have been performed using agonists for specific dopamine receptors to further dissect the dopaminergic receptor coupling of G$_z$. In particular, Gα_z-null mice are less sensitive to the impact of quinpirole, a D2-specific receptor agonist, on locomotor activity and dopamine release in the nuclear accumbens. These observations suggest that Gα_z endogenously couples to D2-like autoreceptors [43]. Furthermore, the Gα_z-null mice exhibited altered neuroendocrine responses to quinpirole, such as a reduction in the release of the pituitary adrenocorticotropic hormone (ACTH) and attenuated quinpirole-stimulated hypothermia [43].

EFFECTORS OF G$_z$ SIGNALING

Activated Gα_z possesses the ability to inhibit adenylyl cyclase (AC), a property shared with other members of G$_i$ subfamily [44, 45]. Gα_z and the other members of the Gα_i subfamily do, however, have different preferences for AC isoforms. Gα_z exhibits selectivity for AC I and V, while other Gα_i subfamily members inhibit AC V and VI [46, 47] [Gα_z is more potent at AC V, though, than Gα_{i2} [47]. All members of the G$_i$ subfamily, including G$_z$, have the ability to block AC II through the release of the $\beta\gamma$ complex, which binds and inhibits AC II [48].

Since expression levels of other members of the Gα_i subfamily are much higher than that of Gα_z, it was widely believed that the ability of Gα_z to inhibit AC activity was largely masked and Gα_z had other unique, and presumably important, cellular functions. In this regard, stable expression of mutationally activated Gα_z transformed Swiss3T3 and NIH3T3 cells by stimulating mitogenic pathways. Interestingly, this stimulation was apparently unrelated to the ability of Gα_z to inhibit AC [49]. Gα_z regulation of cAMP levels is functionally relevant in some cell types, though, as this appears to be the mechanism for increased insulin secretion seen in pancreatic islets lacking Gα_z [50] (Figure 201.1, right panel; see also below). Islets from Gα_z-null mice had approximately 1.9-fold increased cAMP levels, correlating well with a 1.5-fold increase in insulin secretion at a supra-threshold glucose concentration, thus explaining the phenotype observed in Gα_z-null mice of improved glucose clearance [50].

Yeast two-hybrid screens were used to identify other molecules that selectively interact with activated Gα_z: RGS20 [20], Eya2 [51], and Rap1GAP [52]. The ability of RGS-Rz subfamily proteins to act not only as negative effectors but also as effectors or scaffolding proteins for Gα_z-signaling pathways has been described above (see section, "Regulators of G$_z$ signaling: RGS proteins"). Of the other two molecules, Eya2 is a transcriptional co-activator and Rap1GAP is a GTPase activating protein for the Ras-like monomeric G protein Rap1. Eya2 interacts with transcription factors of Six family; this interaction results in a translocation of Eya2 from the cytosol to the nucleus. Expression of constitutively-active Gα_z blocks this process by competing with Six4 in an activation-dependent manner [51]. Rap1 activation induces ERK phosphorylation and cell differentiation in the PC12 line; these pathways were blocked by transfection of constitutively-activated Gα_z or receptor-mediated G$_z$ activation [40]. Co-precipitation assays revealed that Gα_z, Rap1GAP, and Rap1 form a stable complex at the membrane [52], leading to a model in which receptor-mediated activation of G$_z$ results in recruitment of Rap1GAP to the plasma membrane where it can effectively downregulate Rap1 signaling (Figure 201.1, left panel) [40].

PHENOTYPE OF Gα_z KNOCKOUT MICE

Gα_z knockout mice have been generated by two independent groups [37, 53]. Initial characterization indicated that Gα_z-null mice were viable, although they were born smaller than their littermates [37, 53]. Gα_z-null mice had significantly increased bleeding times from the tail tip as compared to wild-type mice [54]. Platelets isolated from Gα_z-null mice failed to aggregate in response to epinephrine [37, 38], although whether this contributes to protection from thrombombembolism in response to collagen injection remains unclear [37, 54]. Both groups reported that Gα_z knockout mice exhibit abnormal responses to certain psychoactive drugs, including a pronounced increase in locomotor activity in response to cocaine administration, a loss of the antidepressant effects of catecholamine reuptake inhibitors, and a reduction in the analgesic effects of morphine [37, 53].

Gα_z-null mice have increased anxiety- and depression-like behaviors, as evidenced by the results of forced swim testing [55]. This is supported by significantly increased amplitudes of 5HT-mediated potassium current and conductance in CA1 pyramidal neurons [55]. In addition to the findings described above linking G$_z$ to D2-dopamine receptor signaling [43], Gα_z-null mice have been shown to be less sensitive to pre-pulse inhibition of acoustic startle response, a process that is highly dependent upon the D2 dopamine receptor [56]. Finally, as mentioned in previous sections, Gα_z-null mice exhibit increased glucose clearance during glucose tolerance tests that is correlated with increased plasma insulin levels after glucose administration [50]. This phenotype appears to be a direct result of increase insulin secretion from the pancreatic islets [50]. Taken together, these results imply a role for G$_z$ in regulating secretory process involved in various aspects of mood, behavior, and glucose homeostasis.

CONCLUDING REMARKS

Although the precise roles of G$_z$ in cellular signaling are still being established, accumulating evidence points to the involvement of this unique G$_i$ subfamily member in several facets of cell biology. First, the temporal and spatial expression patterns of Gα_z suggest a possible role in neuron growth and/or differentiation. G$_z$ also appears to regulate platelet function through adrenergic receptor signaling. In addition, G$_z$ participates in the physiologic regulation of insulin secretion from the pancreatic β cells (Figure 201.1, right panel). Finally, G$_z$ may be involved in processes impacting on mood and/or behavior. These distinct roles further define G$_z$ as unique among the G$_i$ subfamily of heterotrimeric G proteins.

REFERENCES

1. Matsuoka M, Itoh H, Kozasa T, Kaziro Y. Sequence analysis of cDNA and genomic DNA for a putative pertussis toxin-insensitive guanine nucleotide-binding regulatory protein alpha subunit. *Proc Natl Acad Sci USA* 1988;**85**:5384–8.

2. Fong HK, Yoshimoto KK, Eversole-Cire P, Simon MI. Identification of a GTP-binding protein alpha subunit that lacks an apparent ADP-ribosylation site for pertussis toxin. *Proc Natl Acad Sci USA* 1988;**85**:3066–70.

3. Hinton DR, Blanks JC, Fong HK, Casey PJ, Hildebrandt E, Simons MI. Novel localization of a G protein, Gz-alpha, in neurons of brain and retina. *J Neurosci* 1990;**10**:2763–70.

4. Gagnon AW, Manning DR, Catani L, Gewirtz A, Poncz M, Brass LF. Identification of Gz alpha as a pertussis toxin-insensitive G protein in human platelets and megakaryocytes. *Blood* 1991;**78**:1247–53.

5. el Mabrouk M, Simoneau L, Bouvier C, Lafond J. Asymmetrical distribution of G proteins in syncytiotrophoblastic brush-border and basal-plasma membranes of human term placenta. *Placenta* 1996;**17**:471–7.

6. Zigman JM, Westermark GT, LaMendola J, Steiner DF. Expression of cone transducin, Gz alpha, and other G-protein alpha-subunit messenger ribonucleic acids in pancreatic islets. *Endocrinology* 1994;**135**:31–7.

7. Kimple ME, Nixon AB, Kelly P, et al. A role for G(z) in pancreatic islet beta-cell biology. *J Biol Chem* 2005;**280**:31,708–13.

8. Maghazachi AA, Al-Aoukaty A, Naper C, Torgersen KM, Rolstad B. Preferential involvement of Go and Gz proteins in mediating rat natural killer cell lysis of allogeneic and tumor target cells. *J Immunol* 1996;**157**:5308–14.

9. Crouch MF, Heydon K, Garnaut SM, Milburn PJ, Hendry IA. Retrograde axonal transport of the alpha-subunit of the GTP-binding protein GZ in mouse sciatic nerve: a potential pathway for signal transduction in neurons. *Eur J Neurosci* 1994;**6**:626–31.

10. Glassner M, Jones J, Kligman I, Woolkalis MJ, Gerton GL, Kopf GS. Immunocytochemical and biochemical characterization of guanine nucleotide-binding regulatory proteins in mammalian spermatozoa. *Dev Biol* 1991;**146**:438–50.

11. Paulssen EJ, Paulssen RH, Haugen TB, Gautvik KM, Gordeladze JO. Cell specific distribution of guanine nucleotide-binding regulatory proteins in rat pituitary tumour cell lines. *Mol Cell Endocrinol* 1991;**76**:45–53.

12. Friberg IK, Young AB, Standaert DG. Differential localization of the mRNAs for the pertussis toxin insensitive G-protein alpha subunits Gq, G11, and Gz in the rat brain, and regulation of their expression after striatal deafferentation. *Brain Res Mol Brain Res* 1998;**54**:298–310.

13. Schuller U, Lamp EC, Schilling K. Developmental expression of heterotrimeric G-proteins in the murine cerebellar cortex. *Histochem Cell Biol* 2001;**116**:149–59.

14. Kelleher KL, Matthaei KI, Leck KJ, Hendry IA. Developmental expression of messenger RNA levels of the alpha subunit of the GTP-binding protein, Gz, in the mouse nervous system. *Brain Res Dev Brain Res* 1998;**107**:247–53.

15. Casey PJ, Fong HK, Simon MI, Gilman AG. Gz, a guanine nucleotide-binding protein with unique biochemical properties. *J Biol Chem* 1990;**265**:2383–90.

16. Fields TA, Casey PJ. Signalling functions and biochemical properties of pertussis toxin-resistant G-proteins. *Biochem J* 1997;**321**(3):561–71.

17. Lounsbury KM, Casey PJ, Brass LF, Manning DR. Phosphorylation of Gz in human platelets. Selectivity and site of modification. *J Biol Chem* 1991;**266**:22,051–6.

18. Lounsbury KM, Schlegel B, Poncz M, Brass LF, Manning DR. Analysis of Gz alpha by site-directed mutagenesis. Sites and specificity of protein kinase C-dependent phosphorylation. *J Biol Chem* 1993;**268**:3494–8.

19. Fields TA, Casey PJ. Phosphorylation of Gz alpha by protein kinase C blocks interaction with the beta gamma complex. *J Biol Chem* 1995;**270**:23,119–25.

20. Glick JL, Meigs TE, Miron A, Casey PJ. RGSZ1, a Gz-selective regulator of G protein signaling whose action is sensitive to the phosphorylation state of Gzalpha. *J Biol Chem* 1998;**273**:26,008–13.

21. Fishburn CS, Herzmark P, Morales J, Bourne HR. Gbetagamma and palmitate target newly synthesized Galphaz to the plasma membrane. *J Biol Chem* 1999;**274**:18,793–800.

22. Wilson PT, Bourne HR. Fatty acylation of alpha z. Effects of palmitoylation and myristoylation on alpha z signaling. *J Biol Chem* 1995;**270**:9667–75.

23. Tu Y, Wang J, Ross EM. Inhibition of brain Gz GAP and other RGS proteins by palmitoylation of G protein alpha subunits. *Science* 1997;**278**:1132–5.

24. Sprang SR. G protein mechanisms: insights from structural analysis. *Annu Rev Biochem* 1997;**66**:639–78.

25. Cavalli A, Druey KM, Milligan G. The regulator of G protein signaling RGS4 selectively enhances alpha 2A-adreoreceptor stimulation of the GTPase activity of Go1alpha and Gi2alpha. *J Biol Chem* 2000;**275**:23,693–23,699.

26. Hunt TW, Fields TA, Casey PJ, Peralta EG. RGS10 is a selective activator of G alpha i GTPase activity. *Nature* 1996;**383**:175–7.

27. Woulfe DS, Stadel JM. Structural basis for the selectivity of the RGS protein, GAIP, for Galphai family members. Identification of a single amino acid determinant for selective interaction of Galphai subunits with GAIP. *J Biol Chem* 1999;**274**:17,718–17,724.

28. Wang J, Tu Y, Woodson J, Song X, Ross EM. A GTPase-activating protein for the G protein Galphaz. Identification, purification, and mechanism of action. *J Biol Chem* 1997;**272**:5732–40.

29. Nixon AB, Grenningloh G, Casey PJ. The interaction of RGSZ1 with SCG10 attenuates the ability of SCG10 to promote microtubule disassembly. *J Biol Chem* 2002;**277**:18,127–33.

30. Rodriguez-Munoz M, Bermudez D, Sanchez-Blazquez P, Garzon J. Sumoylated RGS-Rz proteins act as scaffolds for Mu-opioid receptors and G-protein complexes in mouse brain. *Neuropsychopharmacology* 2007;**32**:842–50.

31. Garzon J, Rodriguez-Munoz M, Lopez-Fando A, Sanchez-Blazquez P. The RGSZ2 protein exists in a complex with mu-opioid receptors and regulates the desensitizing capacity of Gz proteins. *Neuropsychopharmacology* 2005;**30**:1632–48.

32. Ho MK, Wong YH. G(z) signaling: emerging divergence from G(i) signaling. *Oncogene* 2001;**20**:1615–25.

33. Sanchez-Blazquez P, Gomez-Serranillos P, Garzon J. Agonists determine the pattern of G-protein activation in mu-opioid receptor-mediated supraspinal analgesia. *Brain Res Bull* 2001;**54**:229–35.

34. Garzon J, Martinez-Pena Y, Sanchez-Blazquez P. Gx/z is regulated by mu but not delta opioid receptors in the stimulation of the low Km GTPase activity in mouse periaqueductal grey matter. *Eur J Neurosci* 1997;**9**:1194–200.

35. Sanchez-Blazquez P, Juarros JL, Martinez-Pena Y, Castro MA, Garzon J. Gx/z and Gi2 transducer proteins on mu/delta opioid-mediated supraspinal antinociception. *Life Sci* 1993;**53**:PL381–6.

36. Garzon J, Rodriguez-Munoz M, Lopez-Fando A, Garcia-Espana A, Sanchez-Blazquez P. RGSZ1 and GAIP regulate mu- but not delta-opioid receptors in mouse CNS: role in tachyphylaxis and acute tolerance. *Neuropsychopharmacology* 2004;**29**:1091–104.

37. Yang J, Wu J, Kowalska MA, et al. Loss of signaling through the G protein, Gz, results in abnormal platelet activation and altered responses to psychoactive drugs. *Proc Natl Acad Sci USA* 2000;**97**:9984–9.

38. Yang J, Wu J, Jiang H, et al. Signaling through Gi family members in platelets. Redundancy and specificity in the regulation of adenylyl cyclase and other effectors. *J Biol Chem* 2002;**277**:46,035–42.

39. Hsu CY, Knapp DR, Halushka PV. The effects of alpha adrenergic agents on human platelet aggregation. *J Pharmacol Exp Ther* 1979;**208**:366–70.

40. Meng J, Casey PJ. Activation of Gz attenuates Rap1-mediated differentiation of PC12 cells. *J Biol Chem* 2002;**277**:43,417–24.

41. Lang J, Nishimoto I, Okamoto T, et al. Direct control of exocytosis by receptor-mediated activation of the heterotrimeric GTPases Gi and G(o) or by the expression of their active G alpha subunits. *Embo J* 1995;**14**:3635–44.

42. Laychock SG, Bilgin S. Alpha 2-adrenergic inhibition of pancreatic islet glucose utilization is mediated by an inhibitory guanine nucleotide regulatory protein. *FEBS Letts* 1987;**218**:7–10.

43. Leck KJ, Blaha CD, Matthaei KI, Forster GL, Holgate J, Hendry IA. Gz proteins are functionally coupled to dopamine D2-like receptors in vivo. *Neuropharmacology* 2006;**51**:597–605.

44. Taussig R, Tang WJ, Hepler JR, Gilman AG. Distinct patterns of bidirectional regulation of mammalian adenylyl cyclases. *J Biol Chem* 1994;**269**:6093–100.

45. Wong YH, Conklin BR, Bourne HR. Gz-mediated hormonal inhibition of cyclic AMP accumulation. *Science* 1992;**255**:339–42.

46. Ammer H, Christ TE. Identity of adenylyl cyclase isoform determines the G protein mediating chronic opioid-induced adenylyl cyclase supersensitivity. *J Neurochem* 2002;**83**:818–27.

47. Kozasa T, Gilman AG. Purification of recombinant G proteins from Sf9 cells by hexahistidine tagging of associated subunits. Characterization of alpha 12 and inhibition of adenylyl cyclase by alpha z. *J Biol Chem* 1995;**270**:1734–41.

48. Federman AD, Conklin BR, Schrader KA, Reed RR, Bourne HR. Hormonal stimulation of adenylyl cyclase through Gi-protein beta gamma subunits. *Nature* 1992;**356**:159–61.

49. Wong YH, Chan JS, Yung LY, Bourne HR. Mutant alpha subunit of Gz transforms Swiss 3T3 cells. *Oncogene* 1995;**10**:1927–33.

50. Kimple ME, Joseph JW, Bailey CL, et al. Gαz negatively regulates insulin secretion and glucose clearance. *J Biol Chem* 2008;**283**:4560–7.

51. Fan X, Brass LF, Poncz M, Spitz F, Maire P, Manning DR. The alpha subunits of Gz and Gi interact with the eyes absent transcription cofactor Eya2, preventing its interaction with the six class of homeodomain-containing proteins. *J Biol Chem* 2000;**275**:32,129–34.

52. Meng J, Glick JL, Polakis P, Casey PJ. Functional interaction between Galpha(z) and Rap1GAP suggests a novel form of cellular cross-talk. *J Biol Chem* 1999;**274**:36,663–9.

53. Hendry IA, Kelleher KL, Bartlett SE, et al. Hypertolerance to morphine in G(z alpha)-deficient mice. *Brain Res* 2000;**870**:10–19.

54. Kelleher KL, Matthaei KI, Hendry IA. Targeted disruption of the mouse Gz-alpha gene: a role for Gz in platelet function? *Thromb Haemost* 2001;**85**:529–32.

55. Oleskevich S, Leck KJ, Matthaei K, Hendry IA. Enhanced serotonin response in the hippocampus of Galphaz protein knock-out mice. *Neuroreport* 2005;**16**:921–5.

56. van den Buuse M, Martin S, Brosda J, Leck KJ, Matthaei KI, Hendry I. Enhanced effect of dopaminergic stimulation on prepulse inhibition in mice deficient in the alpha subunit of G(z). *Psychopharmacology (Berl)* 2005;**183**:358–67.

Chapter 50

Mono-ADP-Ribosylation of Heterotrimeric G Proteins

Maria Di Girolamo and Daniela Corda

Department of Cell Biology and Oncology, Consorzio Mario Negri Sud, Santa Maria Imbaro, Chieti, Italy

INTRODUCTION

The heterotrimeric G proteins have crucial roles in determining the specificity of cellular responses to extracellular signals. Several covalent modifications of these G proteins are relevant in the control of their functions under both normal and pathological conditions. Posttranslational modifications that occur on G protein α and $\beta\gamma$ subunits include: lipid modifications, which are required for targeting the subunits to the plasma membrane and for the interactions of α with $\beta\gamma$, its effectors, and the RGS proteins [1]; phosphorylation, which affects the interactions between α and $\beta\gamma$, and between G proteins with their receptors and effectors [2]; and finally, mono-ADP-ribosylation, the function of which has not yet been fully defined. In the following, we will focus on the enzymatic mono-ADP-ribosylation of the G-protein subunits, which is catalyzed by both bacterial toxins and eukaryotic enzymes, and discuss the mechanisms and potential roles of this reaction in mammals.

THE MONO-ADP-RIBOSYLATION REACTION

The mono-ADP-ribosylation reaction is catalyzed by mono-ADP-ribosyltransferases (EC 2.4.2.31), which transfer an ADP-ribose residue from βNAD^+ to a specific amino acid (arginine, asparagine, cysteine, diphtamide) of acceptor proteins, via *N*- or *S*-glycosidic linkages, with the release of nicotinamide (Figure 50.1). A second major reaction involving ADP-ribose transfer is poly-ADP-ribosylation, catalyzed by the poly(ADP-ribose)polymerases (PARPs) [3]. Recently, 17 human PARPs have been identified by gene database analysis [4, 5]. PARPs 1–6

are strictly polymerase enzymes, and can transfer multiple ADP-ribose residues, and even branched polymers of ADP-ribose linked by *O*-glycosidic linkages, onto target proteins. Despite this original definition based on their function, some of the recently identified PARPs also possess mono-ADP-ribosyltransferase activity [4–7].

The mono-ADP-ribosylation reaction can occur through both enzymatic and non-enzymatic mechanisms; the latter involves the covalent binding of free ADP-ribose to lysine and cysteine residues [8, 9]. Free ADP-ribose can either be released from βNAD^+ by cellular NAD^+-glycohydrolases (NADases), or it can result from ADP-ribosyl hydrolase activities that act on ADP-ribosylated substrates [7, 9]. Thus, both poly-ADP-ribose glycohydrolases, which hydrolyze poly-ADP-ribose polymers to free ADP-ribose, and mono-ADP-ribosylhydrolases, which revert mono-ADP-ribose modification hydrolyzing the protein–ADP-ribose linkage, end up in forming the cellular free ADP-ribose [10–12].

Until recently, *Marh1* was the only cloned and characterized mammalian mono-ADP-ribose arginine-hydrolase gene: it encodes a soluble, intracellular enzyme that specifically hydrolyzes the ADP-ribose-arginine bond [10]. Two additional genes, *Arh2* and *Arh3*, were then identified, but neither of their gene products, ARH2 and ARH3, can hydrolyze the ADP-ribose specifically bound to arginine, cysteine, diphtamide, or asparagine residues [13, 14]. Rather, ARH3 has been shown to possess a poly-ADP-ribose-glycohydrolase activity [13, 14], whereas the activity of ARH2 remains yet to be defined. The presence of ADP-ribosyltransferases and ADP-ribosylhydrolases in cells points at a mechanism that may regulate the activity of the substrates of this reaction, as also indicated by the examples given below.

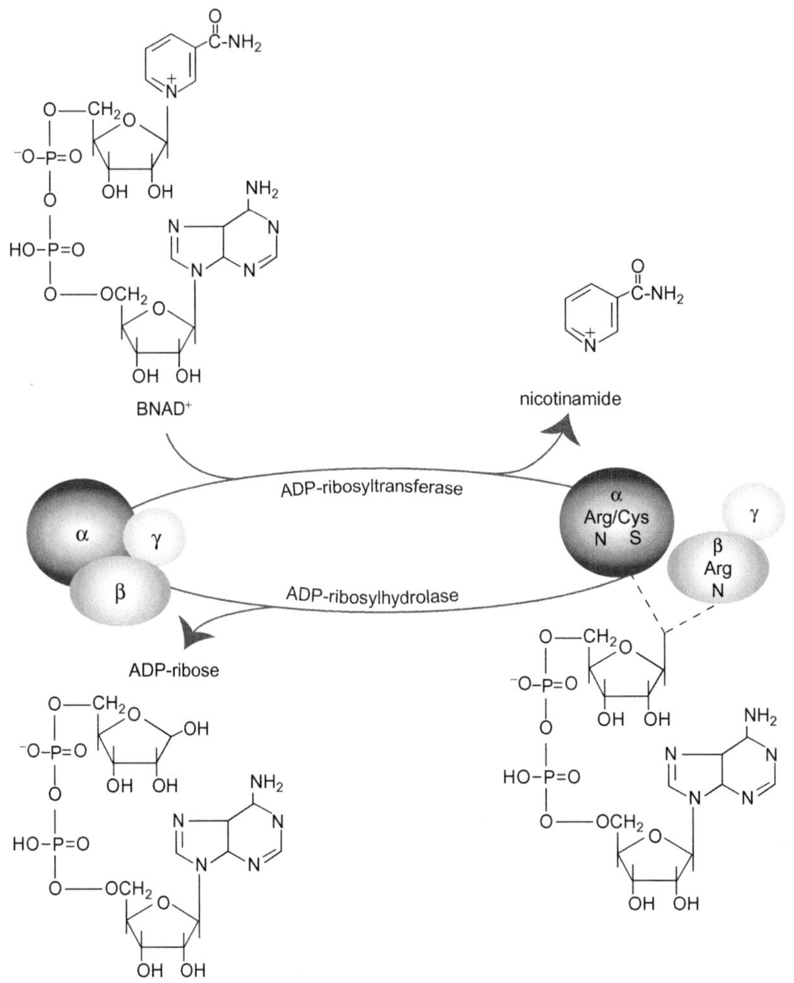

FIGURE 50.1 Scheme of the mono-ADP-ribosylation cycle.
Mono-ADP-ribosyltransferase transfers an ADP-ribose residue from βNAD^+ to an arginine (αs and β) or to a cysteine (αi, αo, and αt) of the target G protein, via *N*- or *S*-glycosidic linkages, respectively, with the release of nicotinamide. Specific ADP-ribosylhydrolases hydrolyze the aminoacid-ADP-ribose glycosidic linkage, thus regenerating free arginine or cysteine and releasing ADP-ribose.

BACTERIAL-TOXIN-INDUCED ADP-RIBOSYLATION

The mono-ADP-ribosylation of G-protein α subunits was originally described as the mechanism of cell intoxication by bacteria such as *Vibrio cholerae* (which produces cholera toxin; CT) and *Bordetella pertussis* (which produces pertussis toxin; PT) [15–17]. These toxins act specifically on G proteins (see below), and rapidly became useful tools in the identification and definition of the functional roles of these proteins (Table 50.1). Other well-characterized bacterial ADP-ribosyltransferases include the diphtheria and clostridial toxins, which act by modifying crucial proteins in the host cell, including the monomeric GTPases of the Rho family, monomeric actin and elongation factor-2, which inactivates the proteins and hence interferes with their cellular functions (reviewed in [17, 18]).

The CT and heat-labile enterotoxins LT-1 and LT-2 from *Escherichia coli* are arginine-specific ADP-ribosyltransferases that can modify Arg^{201} of the α subunit of the stimulatory G protein (G_s) and irreversibly inhibit its GTPase activity; this results in the activation of adenylyl cyclase and an increase in intracellular cyclic AMP [16, 17]. As a consequence, there is a massive efflux of salts and water from the bowel epithelial cells into the intestinal lumen, determining the diarrhea characteristic of cholera. CT is an 84-kDa oligomeric protein that consists of a monomeric A subunit, a 29-kDa globular polypeptide that has ADP-ribosyltransferase activity, and a homopentamer B subunit. The proteolytic cleavage and the reduction of the Cys187–Cys199 disulfide bond of the A subunit releases a smaller, carboxyl-terminus 7-kDa fragment (CTA2), and a larger 22-kDa catalytically active fragment (CTA1), the latter of which carries the ADP-ribosyltransferase activity of

TABLE 50.1 ADP-ribosylating/deribosylating enzymes acting on heterotrimeric G proteins

Enzyme/source	Subunit structure/ localization	Targets(s)*	Effects	Refs
Toxin ADPRT				
Cholera toxin	AB_5	G_s, G_t	Inhibition of G-protein GTPase activity	[16, 26]
Escherichia coli LT	AB_5	G_s, G_t	Inhibition of G-protein GTPase activity	[16]
Pertussis toxin	AB_5	G_i, G_o G_t	Uncoupling of G proteins from receptors	[15, 16]
Cellular ADPRT				
Human erythrocytes	Membranes	G_i	Decrease in epinephrine-mediated AC inhibition	[69, 70]
Human platelets	Membranes	G_i	Decrease in epinephrine-mediated AC inhibition	[69, 70]
Rabbit ventricles	Membranes	G_s	Increase in AC activity	[59]
Human platelets	–	G_s	Increase in AC activity	[60]
Rat brain	Homogenate	G_s, G_o	–	[61, 62]
Rabbit luteal	Membranes	G_s	Increase in AC activity	[63]
Chicken spleen	Membranes	G_s, actin	Increase in AC activity	[64]
NG108-15	Membranes	G_s	Increase in AC activity	[65]
CHO cells	Plasma membranes	$G\beta$	Inhibition of betagamma-mediated function	[73]
Cellular hydrolase				
ARH1	Cytosal	G_s	–	[10, 29]
Rat and human tissue	Cytosol	G_s	–	[10]
Human erythrocytes	Cytosol	G_i	–	[11]
CHO cells	Cytosol	$G\beta$	–	[73]

This table lists the ADP-ribosylating activities (ADPRT) identified in different systems and related to G-protein modification. The ADP-ribosylhydrolases (ARH) catalyzing the reverse reaction are also indicated. See text for other enzymes for which the action on G proteins has not been proven directly, such as the mammalian ART, Sirtuin and PARP families. AC, adenylyl cyclase; –, not determined.
* REFERRING TO $G\alpha$ SUBUNITS UNLESS SPECIFIED AS $G\beta$.

CT [19]. The delivery of the CTA1 enzyme to its target *in vivo* requires the B subunits and the CTA2 peptide. The homopentamer B subunit is a complex of five 11-kDa polypeptides that binds stoichiometrically with high affinity and specificity to five GM1 molecules on the plasma membrane [20]. To cause disease, both CT and LT co-opt molecular mechanisms of the host cell. The toxin enters by endocytosis of the toxin–receptor complexes and follows retrograde transport to the Golgi complex, where it encounters the KDEL receptor, which normally determines the retrieval of lumenal proteins of the endoplasmic reticulum (ER) that have escaped their resident compartment. The A

subunit of CT contains a KDEL sequence which binds to the specific receptor, thus promoting the retrograde transport of the toxin to the lumen of the ER [21, 22]. In the ER, the disulfide bond of the A subunit is reduced, and the CTA1 peptide is released from the rest of the toxin and translocates to the cytosol through the Sec61p channel [23]. Unfolding of the CTA1 peptide depends on the activity of the ER oxido-reductase protein disulfide isomerase (PDI) [24], which acts as a redox-dependent chaperone: in its reduced state, PDI binds and unfolds the toxin, whereas in its oxidized state, PDI releases it. The ER oxidase Ero1 is responsible for oxidation of PDI, and thus for inducing the release of the CTA1 toxin and its retrotranslocation across the ER membrane to its site of action on the inner surface of the plasma membrane [25].

The ADP-ribosyltransferase of the CTA1 peptide (and related LT-1 and LT-2) is allosterically activated in the host cells by a family of GTPases, the ADP-ribosylation factors (ARFs) [26]. The ARFs are 20-kDa guanine-nucleotide-binding proteins that act as molecular switches and are essential in membrane trafficking and actin cytoskeleton remodeling in eukaryotic cells [27]. Only ARF-GTP (the GTP-bound, active form of the ARFs) can bind effector proteins and activate CTA1. The ARFs do not interact with the entire toxin, and thus proteolysis and reduction of the disulfide bond are prerequisites for exposure of the ARF-binding site on CTA1 [26].

Despite extensive studies, the mechanism of activation of ARF with regard to A1 remained unknown until recently, when the crystal structures of the complex between CTA1 and ARF6-GTP were solved [28]. This revealed that the binding of ARF6-GTP causes major changes in the bacterial CTA1 structure, resulting in a conformation that can bind the G-protein substrate, $G\alpha_s$ [28]. The extensive toxin–ARF-GTP interface surface mimics ARF-GTP recognition of its normal cellular protein partners, which suggests that the toxin has evolved to make full use of the promiscuous binding properties of the ARFs.

Recently, it has been shown that ARH1 activity is important in the counteracting of the CT-catalyzed ADP-ribosylation of $G\alpha_s$ [29]. Thus, in cells endogenously expressing ARH1, the CT effects can be partially counteracted, whereas in ARH1$^{-/-}$ cells, the sensitivity to CT is enhanced and results in an accelerated loss of $G\alpha_s$ [29]. Thus, enzymatic cross-talk exists between bacterial toxin ADP-ribosyltransferases and the host ADP-ribosylhydrolases; clearly, in a disease state, toxin-catalyzed ADP-ribosylation overwhelms this potential host-defense system, resulting in great and prolonged ADP-ribosylation of the crucial substrates.

PT is a cysteine-specific ADP-ribosyltransferase, and it ADP-ribosylates the α subunit of the G proteins G_i, G_o, and G_t. ADP-ribosylation of the $G\alpha_i$ subunit abolishes its inhibitory activity on adenylyl cyclase, leading to uncontrolled elevated levels of cAMP production in some cell types [15, 30, 31]. A second, important cause of PT-induced disease is the uncoupling of the G_i proteins from the G-protein-coupled receptors (GPCRs): PT-catalyzed mono-ADP-ribosylation of the α subunit occurs only when this is associated with βγ, generating a modified, heterotrimeric G protein that cannot transduce its signals [32].

PT is a 119-kDa toxin that comprises a 28-kDa monomeric A component, an S1 subunit that has ADP-ribosyltransferase activity, and a B component that is a complex of five polypeptides (S2, S3, two S4s, S5) [33]. The function of the B component is to mediate the binding of PT

to sialoglycoproteins on target cells, thus initiating toxin internalization by receptor-mediated endocytosis and its trafficking through the early to late endosomes and the Golgi complex [34]. To date, no evidence has been provided that the ADP-ribosyltransferase activity of the S1 subunit requires co-factors, such as with the ARFs for CT.

ENDOGENOUS MONO-ADP-RIBOSYLATION

Vertebrate ADP-ribosyltransferase activity was first detected in turkey erythrocytes [35, 36], rat liver homogenates [37], and *Xenopus* tissues [38]. Specific enzymes have now been cloned from different sources [39–44]. The first characterized family of mammalian ADP-ribosyltransferases includes five enzymes referred to as ARTs 1–5 [45, 46]. ART1, ART2, ART3, and ART4 are glycosylphosphatidylinositol (GPI)-anchored to the cell surface. ART5 has a hydrophobic N-terminal signal sequence, and is a secretory protein [47]. ART1 and ART2 have roles in immune regulation: ART1 inhibits T lymphocyte functions, such as their cytolytic activity and proliferation, by ADP-ribosylating arginines of cell surface molecules, including the T-cell co-receptors [45, 48]. The human defensin HNP-1 is a mediator of the innate antimicrobial immunity, and is among the most recently identified substrates of ART1 [49]. ART2 is expressed in resting T cells and in natural killer cells, and has a regulatory role in rat models of autoimmune insulin-dependent diabetes mellitus, where its defective expression in T cells has been associated with increased susceptibility to this disease [45, 50]. The ART2 activity has also been characterized as a cause of T cell apoptosis, an event that is initiated by the activation of the P_2X_7 purinoceptor through its ART2-dependent ADP-ribosylation [51]. The biological functions of ART3, ART4, and ART5 remain to be defined; moreover, while human ART5 has been characterized as an arginine-specific ADP-ribosyltransferase, ART3 and ART4 appear to have lost their ADP-ribosyltransferase activities.

The catalytic domains of the mammalian ARTs are extracellular, and thus it is unlikely that the members of this family are involved in mono-ADP-ribosylation of intracellular targets, as is the case with the heterotrimeric G proteins [52]. Thus, a distinct family of intracellular ADP-ribosyltransferases, which will possibly be divergent from the known ARTs, must be active intracellularly. Indeed, recent data have indicated that at least two distinct families of such enzymes are expressed in mammalian cells. One is represented by the seven human sirtuins (sirtuins 1 to 7) that can metabolize NAD^+ and that possess ADP-ribosyltransferase activity [53, 54]. In particular, SirT4 mono-ADP-ribosylates the mitochondrial glutamate dehydrogenase GDH, thus repressing its activity [55, 56], whereas SirT6 mono-ADP-ribosylates itself [57].

The second family here includes at least some of the novel PARP proteins that are characterized by the absence of the conserved PARP1 E988 residue, which is essential for the elongation of the ADP-ribose chain [7].

Cellular ADP-ribosyltransferase activities that can catalyze the modification of G-protein arginine residues in a way similar to CT have been described in many cells and tissues (Table 50.1) [58–67]. While presenting identification of either the enzymatic activity or the G protein substrates, several of these initial reports did not fully demonstrate the mono-ADP-ribosylation reaction (no discrimination between the enzymatic and non-enzymatic ADP-ribose linkages, or the use of α-NAD$^+$ rather than β-NAD$^+$, the preferred substrate of ADP-ribosyltransferases, in the ADP-ribosylation assays); moreover, identification of the G-protein substrates remains elusive, since these reports were only based either on determination of the molecular mass or on co-migration with the bacterial toxin substrates, by polyacrylamide gel electrophoresis [58–64].

A series of more recent reports has led to a better understanding of the endogenous mono-ADP-ribosylation machinery. It has been shown that in highly purified canine cardiac sarcolemma, isoproterenol produces selective ADP-ribosylation of a single 45-kDa protein, identified as $G\alpha_s$ by immunoprecipitation. This effect correlated with the ability of NAD$^+$ to increase cyclic AMP production [66]. Similarly, a 52-kDa protein that was identified as a $G\alpha_s$ isoform is ADP-ribosylated in smooth muscle cells from bovine coronary arteries [67]. In this case, the $G\alpha_s$ modification was related to the release of eicosanoids from the endothelium, which stimulate the enzymatic reaction and thus activate K$^+$ channels [68].

Endogenous ADP-ribosylation of cysteine residues of G proteins that is comparable to the PT effect has been reported in human erythrocytes and platelets [69], where it attenuates the inhibition of adenylyl cyclase induced by epinephrine (Table 50.1) [70]. Thus, the picture emerging from these observations is that pathological modification of G proteins produced by bacterial toxins has an endogenous, and possibly physiologically relevant, counterpart which can be looked upon as an additional mechanism of regulation of G-protein-mediated functions. In addition, there are indications in *in-vitro* assays that not only the α but also the β subunit of G$_t$ is a substrate of an ADP-ribosyltransferase purified from the cytosol of turkey erythrocytes [71, 72].

Direct evidence of the functional, enzymatic modification of the G protein β subunit has been reported both in isolated plasma membranes and in intact cells [73]. The arginine-specific mono-ADP-ribosyltransferase that catalyzes this reaction is a plasma-membrane-associated, but not GPI-anchored, protein that acts intracellularly and specifically modifies residue 129 of the β subunit. The modified β subunit can then be de-ADP-ribosylated by a cytosolic ADP-ribosylhydrolase, thus revealing a cellular

ADP-ribosylation/de-ADP-ribosylation cycle that might parallel a functional activation/inactivation cycle of the $\beta\gamma$ dimer. In addition, under resting conditions, the β subunit mono-ADP-ribosylation also takes place in intact cells, thus supporting the potential physiological role of this reaction [73, 74].

The modified β subunit loses its ability to modulate effector enzymes, such as calmodulin-stimulated type 1 adenylyl cyclase and phospholipase C, indicating that the modification of arginine 129, a critical residue located in the β common-effector-binding surface [75], can indeed impair β-subunit activity [73, 74]. Mono-ADP-ribosylation can thus be considered to be a signal termination mechanism for $\beta\gamma$-mediated functions. In principle, the ADP-ribosylation of the β subunit could also affect the function of the α subunits, since it can lead to a sustained activation of α-subunit-dependent functions by sequestering the $\beta\gamma$ subunit from the signal cascade and preventing re-association of the heterotrimer.

Among the substrates of mono-ADP-ribosylating enzymes, there are other cellular proteins that are involved in cell signaling, metabolism, and organization, such as GRP78/BIP, GDH, actin, tubulin and CtBP/BARS [76–82]. Interestingly, and in line with the data summarized above for the modification of the G-protein α and β subunits, in all of these cases, ADP-ribosylation has been shown to result in inactivation of the target protein.

In conclusion, although a complete understanding of the cellular regulation of the mono-ADP-ribosylation reactions is not yet available, our increased understanding of the enzymes and substrates involved indicates that this process is relevant not only in G-protein-mediated signaling, but also in other fundamental cellular events. The recent findings on the members of the sirtuin and PARP families reported above may indeed represent the instruments to finally define the molecular players of this endogenous mono-ADP-ribosylation reaction.

ACKNOWLEDGEMENTS

The authors would like to thank Dr C. P. Berrie for editorial assistance, and acknowledge the financial support of the Italian Association for Cancer Research (AIRC, Milano, Italy), Telethon (Italy), and MIUR (Italy).

REFERENCES

1. Chen CA, Manning DR. Regulation of G proteins by covalent modification. *Oncogene* 2001;**20**:1643–52.
2. Morris AJ, Malbon CC. Physiological regulation of G protein-linked signaling. *Physiol Rev* 1999;**79**:1373–430.
3. Ueda K, Hayaishi O. ADP-ribosylation. *Annu Rev Biochem* 1985;**54**:73–100.

4. Ame JC, Spenlehauer C, de Murcia G. The PARP superfamily. *Bioessays* 2004;**26**:882–93.

5. Otto H, Reche PA, Bazan F, Dittmar K, Haag F, Koch-Nolte F. *In silico* characterization of the family of PARP-like poly(ADP-ribosyl)transferases (pARTs). *BMC Genomics* 2005;**6**:139.

6. Yu M, Schreek S, Cerni C, Schamberger C, Lesniewicz K, Poreba E, Vervoorts J, Walsemann G, Grotzinger J, Kremmer E, Mehraein Y, Mertsching J, Kraft R, Austen M, Luscher-Firzlaff J, Luscher B. PARP-10, a novel Myc-interacting protein with poly(ADP-ribose) polymerase activity, inhibits transformation. *Oncogene* 2005;**24**:1982–93.

7. Hassa PO, Haenni SS, Elser M, Hottiger MO. Nuclear ADP-ribosylation reactions in mammalian cells: where are we today and where are we going? *Microbiol Mol Biol Rev* 2006;**70**:789–829.

8. McDonald LJ, Wainschel LA, Oppenheimer NJ, Moss J. Amino acid-specific ADP-ribosylation: structural characterization and chemical differentiation of ADP-ribose-cysteine adducts formed nonenzymatically and in a pertussis toxin-catalyzed reaction. *Biochemistry* 1992;**31**:11,881–11,887.

9. Corda D, Di Girolamo M. Functional aspects of protein mono-ADP-ribosylation. *EMBO J* 2003;**22**:1953–8.

10. Takada T, Iida K, Moss J. Cloning and site-directed mutagenesis of human ADP-ribosylarginine hydrolase. *J Biol Chem* 1993;**268**:17,837–17,843.

11. Tanuma S, Endo H. Identification in human erythrocytes of mono(ADP-ribosyl) protein hydrolase that cleaves a mono(ADP-ribosyl) G_i linkage. *FEBS Letts* 1990;**261**:381–4.

12. Davidovic L, Vodenicharov M, Affar EB, Poirier GG. Importance of poly(ADP-ribose) glycohydrolase in the control of poly(ADP-ribose) metabolism. *Exp Cell Res* 2001;**268**:7–13.

13. Ono T, Kasamatsu A, Oka S, Moss J. The 39-kDa poly(ADP-ribose) glycohydrolase ARH3 hydrolyzes O-acetyl-ADP-ribose, a product of the Sir2 family of acetyl-histone deacetylases. *Proc Natl Acad Sci USA* 2006;**103**:16,687–16,691.

14. Mueller-Dieckmann C, Kernstock S, Lisurek M, von Kries JP, Haag F, Weiss MS, Koch-Nolte F. The structure of human ADP-ribosylhydrolase 3 (ARH3) provides insights into the reversibility of protein ADP-ribosylation. *Proc Natl Acad Sci USA* 2006;**103**:15,026–15,031.

15. Katada T, Ui M. Direct modification of the membrane adenylate cyclase system by islet- activating protein due to ADP-ribosylation of a membrane protein. *Proc Natl Acad Sci USA* 1982;**79**:3129–33.

16. Moss J, Vaughan M. ADP-ribosylation of guanyl nucleotide-binding regulatory proteins by bacterial toxins. *Adv Enzymol Relat Areas Mol Biol* 1988;**61**:303–79.

17. Holbourn KP, Shone CC, Acharya KR. A family of killer toxins. Exploring the mechanism of ADP-ribosylating toxins. *FEBS J* 2006;**273**:4579–93.

18. Krueger KM, Barbieri JT. The family of bacterial ADP-ribosylating exotoxins. *Clin Microbiol Rev* 1995;**8**:34–47.

19. Moss J, Stanley SJ, Lin MC. NAD glycohydrolase and ADP-ribosyltransferase activities are intrinsic to the A1 peptide of choleragen. *J Biol Chem* 1979;**254**:11,993–11,999.

20. Spangler BD. Structure and function of cholera toxin and the related Escherichia coli heat-labile enterotoxin. *Microbiol Rev* 1992;**56**:622–47.

21. Lencer WI, Constable C, Moe S, Jobling MG, Webb HM, Ruston S, Madara JL, Hirst TR, Holmes RK. Targeting of cholera toxin and *Escherichia coli* heat labile toxin in polarized epithelia: role of COOH-terminal KDEL. *J Cell Biol* 1995;**131**:951–62.

22. Chinnapen DJ, Chinnapen H, Saslowsky D, Lencer WI. Rafting with cholera toxin: endocytosis and trafficking from plasma membrane to ER. *FEMS Microbiol Letts* 2007;**266**:129–37.

23. Schmitz A, Herrgen H, Winkeler A, Herzog V. Cholera toxin is exported from microsomes by the Sec61p complex. *J Cell Biol* 2000;**148**:1203–12.

24. Tsai B, Rodighiero C, Lencer WI, Rapoport TA. Protein disulfide isomerase acts as a redox-dependent chaperone to unfold cholera toxin. *Cell* 2001;**104**:937–48.

25. Tsai B, Ye Y, Rapoport TA. Retro-translocation of proteins from the endoplasmic reticulum into the cytosol. *Nat Rev Mol Cell Biol* 2002;**3**:246–55.

26. Tsai SC, Noda M, Adamik R, Chang PP, Chen HC, Moss J, Vaughan M. Stimulation of choleragen enzymatic activities by GTP and two soluble proteins purified from bovine brain. *J Biol Chem* 1988;**263**:1768–72.

27. Inoue H, Randazzo PA. Arf GAPs and their interacting proteins. *Traffic* 2007;**8**:1465–75.

28. O'Neal CJ, Jobling MG, Holmes RK, Hol WG. Structural basis for the activation of cholera toxin by human ARF6-GTP. *Science* 2005;**309**:1093–6.

29. Kato J, Zhu J, Liu C, Moss J. Enhanced sensitivity to cholera toxin in ADP-ribosylarginine hydrolase-deficient mice. *Mol Cell Biol* 2007;**27**:5534–43.

30. Katada T, Ui M. ADP ribosylation of the specific membrane protein of C6 cells by islet-activating protein associated with modification of adenylate cyclase activity. *J Biol Chem* 1982;**257**:7210–16.

31. Bruckener KE, el Baya A, Galla HJ, Schmidt MA. Permeabilization in a cerebral endothelial barrier model by pertussis toxin involves the PKC effector pathway and is abolished by elevated levels of cAMP. *J Cell Sci* 2003;**116**:1837–46.

32. Hsia JA, Moss J, Hewlett EL, Vaughan M. ADP-ribosylation of adenylate cyclase by pertussis toxin. Effects on inhibitory agonist binding. *J Biol Chem* 1984;**259**:1086–90.

33. Tamura M, Nogimori K, Murai S, Yajima M, Ito K, Katada T, Ui M, Ishii S. Subunit structure of islet-activating protein, pertussis toxin, in conformity with the A–B model. *Biochemistry* 1982;**21**:5516–22.

34. Xu Y, Barbieri JT. Pertussis toxin-catalyzed ADP-ribosylation of G_i-2 and G_i-3 in CHO cells is modulated by inhibitors of intracellular trafficking. *Infect Immun* 1996;**64**:593–9.

35. Moss J, Stanley SJ, Watkins PA. Isolation and properties of an NAD- and guanidine-dependent ADP- ribosyltransferase from turkey erythrocytes. *J Biol Chem* 1980;**255**:5838–40.

36. West Jr. RE, Moss J Amino acid specific ADP-ribosylation: specific NAD: arginine mono-ADP- ribosyltransferases associated with turkey erythrocyte nuclei and plasma membranes. *Biochemistry* 1986;**25**:8057–62.

37. Moss J, Stanley SJ. Amino acid-specific ADP-ribosylation. Identification of an arginine- dependent ADP-ribosyltransferase in rat liver. *J Biol Chem* 1981;**256**:7830–3.

38. Godeau F, Koide SS. Xenopus mono(adenosine diphosphate ribosyl) transferase: purification, assay, and properties. *Princess Takamatsu Symp* 1983;**13**:111–18.

39. Zolkiewska A, Nightingale MS, Moss J. Molecular characterization of NAD:arginine ADP-ribosyltransferase from rabbit skeletal muscle. *Proc Natl Acad Sci USA* 1992;**89**:11,352–11,356.

40. Davis T, Shall S. Cloning of a chicken gene homologous to the rabbit mono-ADP- ribosyltransferase gene. *Biochem Soc Trans* 1995;**23**:207S.

41. Haag FA, Kuhlenbaumer G, Koch-Nolte F, Wingender E, Thiele HG. Structure of the gene encoding the rat T cell ecto-ADP- ribosyltransferase RT6. *J Immunol* 1996;**157**:2022–30.

42. Okazaki IJ, Kim HJ, Moss J. Cloning and characterization of a novel membrane-associated lymphocyte NAD:arginine ADP-ribosyltransferase. *J Biol Chem* 1996;**271**:22,052–22,057.

43. Shimoyama M, Tsuchiya M, Hara N, Yamada K, Osago H. Molecular cloning and characterization of arginine-specific ADP- ribosyltransferases from chicken bone marrow cells. *Adv Exp Med Biol* 1997;**419**:137–44.

44. Koch-Nolte F, Haag F, Braren R, Kuhl M, Hoovers J, Balasubramanian S, Bazan F, Thiele HG. Two novel human members of an emerging mammalian gene family related to mono-ADP-ribosylating bacterial toxins. *Genomics* 1997;**39**:370–6.

45. Haag F, Koch-Nolte F. Endogenous relatives of ADP-ribosylating bacterial toxins in mice and men: potential regulators of immune cell function. *J Biol Regul Homeost Agents* 1998;**12**:53–62.

46. Moss J, Balducci E, Cavanaugh E, Kim HJ, Konczalik P, Lesma EA, Okazaki IJ, Park M, Shoemaker M, Stevens LA, Zolkiewska A. Characterization of NAD:arginine ADP-ribosyltransferases. *Mol Cell Biochem* 1999;**193**:109–13.

47. Glowacki G, Braren R, Cetkovic-Cvrlje M, Leiter EH, Haag F, Koch-Nolte F. Structure, chromosomal localization, and expression of the gene for mouse ecto-mono(ADP-ribosyl)transferase ART5. *Gene* 2001;**275**:267–77.

48. Wang J, Nemoto E, Dennert G. Regulation of CTL by ecto-nictinamide adenine dinucleotide (NAD) involves ADP-ribosylation of a p56lck-associated protein. *J Immunol* 1996;**156**:2819–27.

49. Paone G, Wada A, Stevens LA, Matin A, Hirayama T, Levine RL, Moss J. ADP ribosylation of human neutrophil peptide-1 regulates its biological properties. *Proc Natl Acad Sci USA* 2002;**99**:8231–5.

50. Whalen BJ, Greiner DL, Mordes JP, Rossini AA. Adoptive transfer of autoimmune diabetes mellitus to athymic rats: synergy of CD4+ and CD8+ T cells and prevention by RT6+ T cells. *J Autoimmun* 1994;**7**:819–31.

51. Seman M, Adriouch S, Scheuplein F, Krebs C, Freese D, Glowacki G, Deterre P, Haag F, Koch-Nolte F. NAD-induced T cell death: ADP-ribosylation of cell surface proteins by ART2 activates the cytolytic P2X7 purinoceptor. *Immunity* 2003;**19**:571–82.

52. Di Girolamo M, Dani N, Stilla A, Corda D. Physiological relevance of the endogenous mono(ADP-ribosyl)ation of cellular proteins. *FEBS J* 2005;**272**:4565–75.

53. Tanny JC, Dowd GJ, Huang J, Hilz H, Moazed D. An enzymatic activity in the yeast Sir2 protein that is essential for gene silencing. *Cell* 1999;**99**:735–45.

54. Yamamoto H, Schoonjans K, Auwerx J. Sirtuin functions in health and disease. *Mol Endocrinol* 2007;**21**:1745–55.

55. Herrero-Yraola A, Bakhit SM, Franke P, Weise C, Schweiger M, Jorcke D, Ziegler M. Regulation of glutamate dehydrogenase by reversible ADP-ribosylation in mitochondria. *EMBO J* 2001;**20**:2404–12.

56. Haigis MC, Mostoslavsky R, Haigis KM, Fahie K, Christodoulou DC, Murphy AJ, Valenzuela DM, Yancopoulos GD, Karow M, Blander G, Wolberger C, Prolla TA, Weindruch R, Alt FW, Guarente L. SIRT4 inhibits glutamate dehydrogenase and opposes the effects of calorie restriction in pancreatic beta cells. *Cell* 2006;**126**:941–54.

57. Liszt G, Ford E, Kurtev M, Guarente L. Mouse Sir2 homolog SIRT6 is a nuclear ADP-ribosyltransferase. *J Biol Chem* 2005;**280**:21,313–21,320.

58. Reilly TM, Beckner S, McHugh EM, Blecher M. Isoproterenol-induced ADP-ribosylation of a single plasma membrane protein of

59. Feldman AM, Levine MA, Baughman KL, Van Dop C. NAD+-mediated stimulation of adenylate cyclase in cardiac membranes. *Biochem Biophys Res Commun* 1987;**142**:631–7.

60. Molina y Vedia L, Nolan RD, Lapetina EG. The effect of iloprost on the ADP-ribosylation of G$_s$ alpha (the alpha-subunit of G$_s$). *Biochem J* 1989;**261**:841–5.

61. Duman RS, Terwilliger RZ, Nestler EJ. Endogenous ADP-ribosylation in brain: initial characterization of substrate proteins. *J Neurochem* 1991;**57**:2124–32.

62. Matsuyama S, Tsuyama S. Mono-ADP-ribosylation in brain: purification and characterization of ADP-ribosyltransferases affecting actin from rat brain. *J Neurochem* 1991;**57**:1380–7.

63. Abramowitz J, Jena BP. Evidence for a rabbit luteal ADP-ribosyl-transferase activity which appears to be capable of activating adenylyl cyclase. *Intl J Biochem* 1991;**23**:549–59.

64. Obara S, Yamada K, Yoshimura Y, Shimoyama M. Evidence for the endogenous GTP-dependent ADP-ribosylation of the alpha-subunit of the stimulatory guanyl-nucleotide-binding protein concomitant with an increase in basal adenylyl cyclase activity in chicken spleen cell membrane. *Eur J Biochem* 1991;**200**:75–80.

65. Donnelly LE, Boyd RS, MacDermot J. G$_s$ alpha is a substrate for mono(ADP-ribosyl)transferase of NG108-15 cells. ADP-ribosylation regulates G$_s$ alpha activity and abundance. *Biochem J* 1992;**288**:331–6.

66. Quist EE, Coyle DL, Vasan R, Satumtira N, Jacobson EL, Jacobson MK. Modification of cardiac membrane adenylate cyclase activity and G$_s$ alpha by NAD and endogenous ADP-ribosyltransferase. *J Mol Cell Cardiol* 1994;**26**:251–60.

67. Li PL, Chen CL, Bortell R, Campbell WB. 11,12-Epoxyeicosatrienoic acid stimulates endogenous mono-ADP- ribosylation in bovine coronary arterial smooth muscle. *Circ Res* 1999;**85**:349–56.

68. Li PL, Campbell WB. Epoxyeicosatrienoic acids activate K$^+$ channels in coronary smooth muscle through a guanine nucleotide binding protein. *Circ Res* 1997;**80**:877–84.

69. Tanuma S, Kawashima K, Endo H. Eukaryotic mono(ADP-ribosyl)transferase that ADP-ribosylates GTP-binding regulatory G$_i$ protein. *J Biol Chem* 1988;**263**:5485–9.

70. Tanuma S, Endo H. Mono(ADP-ribosyl)ation of G$_i$ by eukaryotic cysteine-specific mono(ADP- ribosyl) transferase attenuates inhibition of adenylate cyclase by epinephrine. *Biochim Biophys Acta* 1989;**1010**:246–9.

71. Watkins PA, Kanaho Y, Moss J. Inhibition of the GTPase activity of transducin by an NAD+:arginine ADP- ribosyltransferase from turkey erythrocytes. *Biochem J* 1987;**248**:749–54.

72. Ehret-Hilberer S, Nullans G, Aunis D, Virmaux N. Mono ADP-ribosylation of transducin catalyzed by rod outer segment extract. *FEBS Letts* 1992;**309**:394–8.

73. Lupi R, Corda D, Di Girolamo M. Endogenous ADP-ribosylation of the G protein beta subunit prevents the inhibition of type 1 adenylyl cyclase. *J Biol Chem* 2000;**275**:9418–24.

74. Lupi R, Dani N, Dietrich A, Marchegiani A, Turacchio S, Berrie CP, Moss J, Gierschik P, Corda D, Di Girolamo M. Endogenous mono-ADP-ribosylation of the free Gbetagamma prevents stimulation of phosphoinositide 3 kinase-gamma and phospholipase C- beta2 and Is activated by G-protein-coupled receptors. *Biochem J* 2002;**367**(3):825–32.

75. Chen Y, Weng G, Li J, Harry A, Pieroni J, Dingus J, Hildebrandt JD, Guarnieri F, Weinstein H, Iyengar R. A surface on the G protein

beta-subunit involved in interactions with adenylyl cyclases. *Proc Natl Acad Sci USA* 1997;**94**:2711–14.

76. Leno GH, Ledford BE. ADP-ribosylation of the 78-kDa glucose-regulated protein during nutritional stress. *Eur J Biochem* 1989;**186**:205–11.

77. Mishima K, Terashima M, Obara S, Yamada K, Imai K, Shimoyama M. Arginine-specific ADP-ribosyltransferase and its acceptor protein p33 in chicken polymorphonuclear cells: co-localization in the cell granules, partial characterization, and in situ mono(ADP-ribosyl)ation. *J Biochem (Tokyo)* 1991;**110**:388–94.

78. Zolkiewska A, Moss J. Integrin alpha 7 as substrate for a glycosyl-phosphatidylinositol- anchored ADP-ribosyltransferase on the surface of skeletal muscle cells. *J Biol Chem* 1993;**268**:25,273–25,276.

79. Huang HY, Graves DJ, Robson RM, Huiatt TW. ADP-ribosyla-tion of the intermediate filament protein desmin and inhibition of desmin assembly in vitro by muscle ADP-ribosyltransferase. *Biochem Biophys Res Commun* 1993;**197**:570–7.

80. Fujita H, Okamoto H, Tsuyama S. ADP-ribosylation in adrenal glands: purification and characterization of mono-ADP-ribosyltrans-ferases and ADP-ribosylhydrolase affecting cytoskeletal actin. *Intl J Biochem Cell Biol* 1995;**27**:1065–78.

81. Di Girolamo M, Silletta MG, De Matteis MA, Braca A, Colanzi A, Pawlak D, Rasenick MM, Luini A, Corda D. Evidence that the 50-kDa substrate of brefeldin A-dependent ADP- ribosylation binds GTP and is modulated by the G-protein beta gamma subunit complex. *Proc Natl Acad Sci USA* 1995;**92**:7065–9.

82. Weigert R, Silletta MG, Spano S, Turacchio G, Cericola C, Colanzi A, Senatore S, Mancini R, Polishchuk EV, Salmona M, Facchiano F, Burger KN, Mironov A, Luini A, Corda D. CtBP/BARS induces fis-sion of Golgi membranes by acylating lysophosphatidic acid. *Nature* 1999;**402**:429–33.

Chapter 51

Specificity of G-Protein βγ Dimer Signaling

Carl A. Hansen[1], William F. Schwindinger[2] and Janet D. Robishaw[2]

[1] Department of Biological and Allied Health Sciences, Bloomsburg University, Bloomsburg, Pennsylvania

[2] Weis Center for Research, Geisinger Clinic, Danville, Pennsylvania

INTRODUCTION

The G-protein βγ dimer performs numerous roles in the signal transduction process, from stabilization and membrane targeting of the α subunit, to recognition of receptors, to direct regulation of effectors, to modulation of various proteins (e.g., GRKs, RACKs, etc.) affecting the intensity or duration of the signal [1–4]. In principal, more than 60 distinct βγ dimers can arise from combinatorial association of the 5 known β and 12 γ subtypes found in the human and mouse genomes. Adding to the structural heterogeneity, the number and type of posttranslational modifications can vary [5–6]. Although their biochemical properties have been well studied *in vitro*, key questions regarding when and where they form *in vivo* and whether they exhibit specific functions and signaling roles in the context of the organism remain to be answered.

Of particular interest, the γ subtypes are more structurally diverse than their β counterparts [2–4]. Moreover, the γ subtypes exhibit very distinct temporal and spatial patterns of expression [7–10]. Collectively, these characteristics suggest that γ subunits have heterogeneous functions in the context of the organism. However, this hypothesis has been difficult to prove since transfection and reconstitution approaches have revealed only modest differences in their biochemical properties, leading many investigators in the field to conclude they are interchangeable. Thus, more than a decade after their discovery, the vast majority of γ subtypes have not been assigned biological functions or cellular signaling roles. Here, we review recent studies aimed at elucidating the *in vivo* functions of the various γ subtypes, with a particular focus on highlighting the ways in which mouse and zebra fish can be used to identify and dissect their cellular roles.

DIVERSITY OF γ

At the genomic level, 12 γ-subunit genes have been identified in humans and named in the order of cloning [2, 11]. Despite the small sizes of the encoded proteins, the GNG genes exhibit large and complex gene structures (Figure 51.1). All GNG genes have two coding exons, and share a common splice site in the coding region, with the exception of GNG13 [12]. In addition, most GNG genes have one or more 5′ non-coding exons. For instance, alternatively spliced 5′ variants of the GNG4 exist (NCBI Accessions NM_004485.3, NM_001098721.1, and NM_001098722.1). Also, the GNG5 and GNG10 genes have 3′ non-coding exons that may affect mRNA stability or subcellular localization. Though the details remain to be worked out, the presence of additional exons may allow for multiple promoters, or alternative splicing of elements involved in regulation of translation. Several GNG genes appear to be divergently transcribed from adjacent genes, and thus may have overlapping promoter or enhancer regions that could contribute to coordinate regulation of transcription: GNGT2 is 4 kb away from ABI3; GNG3 is contained in an intron of BSCL2; GNG5 overlaps with SPATA1; GNG8 is 2 kb away from CR612700, and GNG13 is 7 kb away from LOC388199 [13]. Finally, the GNG5 and GNG10 genes seem to produce fusion proteins with upstream genes encoding a lysosomal enzyme CTBS [14] and a heat shock protein DNAJC25 (NCBI Accession NP_004116.2), respectively. In both cases, the predicted fusion proteins contain the carboxyl terminal region containing the isoprenylation site that may contribute a membrane localization signal [15–16]. Collectively, the diversity in GNG gene structure suggests variability in regulation of gene expression that may underlie specificity in function.

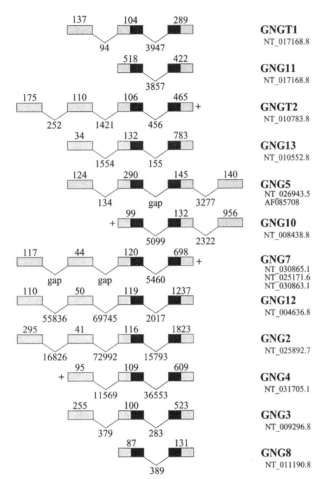

FIGURE 51.1 Structure of human G-protein γ-subunit genes.
Boxes represent exons, shaded areas represent coding region, lines represent introns. Sizes are given in base pairs. Dotted lines represent alternative splicing of 5′-untranslated region of GNG4 and distinct splice site in coding region of GNG13.

From a genomic perspective, the emergence of γ-subunit diversity appears to coincide with the evolution of chordates and vertebrates. For instance, yeast (*S. cerevisiae*) and slime mold (*D. discoidium*) have only 1 γ-subunit gene; fruit fly (*D. melanogastar*) and worm (*C. elegans*) contain 2 γ-subunit genes; sea squirt (*C. intestinalis*) has 5 γ-subunit genes; chicken (*G. gallus*) has 7 or 8 γ-subunit genes, platypus (*O. anatinus*) contains 9 γ-subunit genes; cow (*B. torus*), mouse (*M. musculus*), and human (*H. sapiens*) have 12 γ-subunit genes, and zebra fish (*D. rario*) contains 17 γ-subunit genes due to a genome duplication [17]. In higher vertebrates, GNGT1 and GNG11 apparently arose from a late gene duplication event, since zebra fish [18], frog, fish, chicken, and platypus contain only the ancestral gene. GNG7 and GNG12 likely arose from a chromosomal duplication based on the presence of DIRAS1–GNG7–GADD45B and DIRAS3–GNG12–GADD45A contigs on human chromosome 19 and chromosome 1, respectively. Presumably, increased diversity of γ- as well as α-subunit genes provides greater structural diversity at the level of

the G proteins to accommodate the more complex signaling requirements of vertebrates.

From the protein perspective, the γ subunits can be arranged into five classes based on amino acid sequence comparisons [19]: (1) γ_{t1}, γ_{11}, and γ_{t2} (62–73 percent identity); (2) γ_2, γ_3, γ_4, and γ_8 (56–75 percent identity); (3) γ_5 and γ_{10} (53% identity); (4) γ_7 and γ_{12} (76 percent identity); and (5) γ_{13}. Thus, analogous to the α subtypes, this pattern of structural diversity suggests that the γ subtypes may fall into functional divisions that have yet to be characterized. Most of the amino acid diversity is concentrated in the N-terminal third of these proteins [20–21]. This diverse region may be responsible for the specificity of interactions with α subunits and effectors. Unfortunately, corresponding structural information for this diverse region is lacking, since it is absent or disordered in the available crystal structures [22]. Additionally, the N-termini of the γ subunits may undergo important posttranslational modifications. For example, the γ_{12} subtype is phosphorylated by protein kinase C, a modification that facilitates cell spreading in response to lysophosphatidic acid [23]. Critical amino acid differences in the middle third of the γ subunits are responsible for the specificity of βγ dimer formation [24–25], while the C-terminal region directs additional posttranslational modifications [15–16]. Most γ subtypes have a –CaaL sequence that directs the addition of a geranylgeranyl group. Class 1 γ subunits have a –CaaS sequence that directs the isoprenylation with the shorter farnesyl group, and as a result, they are less tightly associated with the plasma membrane [16]. The γ_5 subtype has an unusual –CSFL motif that is incompletely processed (i.e., presence of the phenylalanine in this motif inhibits normal cleavage of the three terminal amino acids and the subsequent carboxy-methylation of the newly cleaved C-terminus), resulting in the production and detection of variably processed isoforms of the γ_5 subtype in cells [26]. Finally, the C-terminus of the γ_{13} subtype has been shown to function as a PDZ binding site [27]. The C-termini of the γ_4 and γ_5 subunits also contain putative PDZ consensus sequences, suggesting possible roles in their intracellular localization prior to processing of their –CaaX motifs.

APPLICATION OF GENETIC SYSTEMS FOR ANALYSIS OF γ SPECIFICITY

There is a saying that "*in vitro cognito sed in vivo veritas*," which, roughly translated, means "one may look for meaning *in vitro*, but the truth may only be found *in vivo*." Perhaps in no situation is this saying more applicable than identifying unique functions for the diverse γ subtypes. Since the emergence of γ diversity coincides with vertebrate evolution, elucidating the functional significance of γ diversity requires the application of genetic approaches to vertebrate model systems. Table 51.1 summarizes genetic

TABLE 51.1 Genetic studies in cells and animals published to date

Subtype	Manipulation	Phenotype/biological function	Receptor G-protein		References
Gngt1	Gngt1(S74L) knockin	Impairs light adaptation; attenuated light-dependent translocation from the outer segment to the inner region	α-effector rhodopsin	βγ-effector	
			$\alpha\beta\gamma_{T1}$		
			PDE	-	
	Gngt1(−/−) knockout	Reduction of both α_{T1} and β_1; photoreceptor degeneration; reduced sensitivity to light	rhodopsin		45
			$\alpha_1\beta_1\gamma_{T1}$		30
			PDE	-	
Gng2	Antisense knockdown	Blunted galanin-induced inhibition of voltage gated calcium channels in RINm5F and GH$_3$ cells	GAL		
			$\alpha_{1o}\beta_2\gamma_2 > \alpha_{1o}\beta_3\gamma_4$		
			-	-	
	Antisense knockdown	Blunted norepinephrine-induced increase in intracellular calcium in rat portal vein myocytes	α_1-AR		46
			$\alpha_{11}\beta_3\gamma_2$		
			Ca^{2+} influx	-	
	Antisense knockdown	Attenuated antinociception to DPDPE	DOR		47
			$\alpha\beta\gamma_2$		
			-	-	
	Antisense knockdown	Blunted activation of AC by VIP or TRH in GH$_3$ cells	VPAC, TRHR		48
			$\alpha s\beta_2\gamma_2$		
			AC	-	
	Antisense knockdown	Blunted activation of PLC by TRH in GH$_3$ cells	TRHR		49
			$\alpha_{q/11}\beta_4\gamma_2$		
			PLC	-	
	Antisense knockdown	Attenuated antinociception with deltorphin II, morphine, WIN 55212-2, clonidine, or U69,593	DOR, MOR, KOR, CB, α_2-AR		49
			$\alpha_{q/11}\beta_4\gamma_2$		
			-	-	
	Morpholino knockdown	Blocks angiogenesis in zebra fish embryos; blocks PLCγ1 and AKT phosphorylation by VEGF	-		50

(Continued)

TABLE 51.1 (Continued)

Subtype	Manipulation	Phenotype/biological function	Receptor G-protein		References
			$\alpha\beta\gamma_2$		
			-	-	
Gng3	Antisense knockdown	Blunted somatostatin-induced inhibition of voltage-gated calcium channels in RINm5F cells	SSTR2		43
			$\alpha_{o2}\beta_1\gamma_3$		
			-	-	51
	Antisense knockdown	Blunted angiotensin II-induced increase in intracellular calcium	AT_{1A}		
			$\alpha_{31}\beta_1\gamma_3$		52, 53
			-	-	
	Antisense knockdown	Blunted norepinephrine-induced increase in intracellular calcium in rat portal vein myocytes	α_1-AR		
			$\alpha_q\beta_1\gamma_3$		47
			Ca^{2+} store	-	
	Antisense knockdown	Blunted somatostatin- or dopamine induced inhibition of AC in GH$_3$ cells	SSTR, D$_2$		
			$\alpha_{i2}\beta_1\gamma_3$		49
			AC	-	
	Gng3(−/−) knockout	Increased susceptibility to seizures, reduced body weight, and decreased adiposity	-		
			$\alpha\beta\gamma_3$		
			-	-	
	Antisense knockdown	Blunted galanin-induced inhibition of voltage gated calcium channels in RINm5F and GH$_3$ cells	GAL		33
Gng4			$\alpha_{o1}\beta_2\gamma_2 > \alpha_{o1}\beta_3\gamma_4$		
			-	-	
	Antisense knockdown	Blunted dopamine-induced inhibition of PL-C in GH$_3$ cells	D$_2$		46
			$\alpha_o\beta_3\gamma_4$		
			-	PLC	
Gng5	Antisense knockdown	Blunted endothelin-induced increase in intracellular calcium	ET$_A$		49
			$\alpha_{11}\beta_3\gamma_5$		
			Ca^{2+} store, PLC	-	

)

Subtype	Manipulation	Phenotype/biological function	Receptor G-protein		References
Gng7	Ribozyme	Reduction in β_1; no reduction in α_s; attenuation of isoproterenol-, but not PGE$_1$-stimulated AC activity	β–adrenoceptor $\alpha\beta_1\gamma_7$		54
			AC	-	
	Gng7(−/−) knockout	Reductions in the α_{olf}-subunit content and dopamine-stimulated AC activity of the striatum	D$_1$		55, 56
			$\alpha_{o1f}\beta\gamma_7$		
			AC Type V	-	32
Gng11	Antisense knockdown	Increased population doublings in fibroblasts	-		
			$\alpha\beta\gamma_{11}$		57
			-	-	
Gng13	Blocking antibody	Blocked the denatonium-induced increase of inositol trisphosphate (IP$_3$) in taste tissue	TAS2R		
			$\alpha_3\beta_3\gamma_{13}$		11
			PDE	PLC-β	11

Abbreviations: AC, adenylyl cyclase; AR, adrenoreceptor; DPDPE, a δ-opioid agonist; GH$_3$, a rat pituitary cell line; PLC, phospholipase-C; RINm5F, a rat pancreatic β-cell line; TRH, thyrotrophin; U69,593, a κ-opioid agonist; VIP, vasoactive intestinal protein; WIN 55212-2, a CB1-cannabinoid agonist.

studies in cells and animals that have been published to date. In keeping with the *in vivo veritas* theme, only the animal studies are described below.

Gene targeting in mice

The heterogeneity of α, β, and γ-subunit genes creates the potential to generate hundreds of potentially diverse G-protein heterotrimers. The challenge is to identify which G-protein heterotrimers actually exist *in vivo*, and to establish what roles they play in particular receptor signaling pathways. Gene targeting in mice represents a powerful method for exploring these questions in a native system. Since the functions of the G-protein heterotrimers have been historically ascribed to their α subtypes, this approach has been extensively used to genetically ablate the various α-subunit genes and compare their resulting phenotypes [28]. However, since γ subtypes are generally considered to be interchangeable, little attention has focused on the production and analysis of γ knockout mice. Analogous to loss of the α subtype, we believe that loss of the γ subtype will provide a similar phenotype to loss of its associated α subtype by blocking the activation of a specific G-protein αβγ heterotrimer required for interaction with the upstream receptor. However, our viewpoint is not universally held;

some contend that loss of the γ subunit should result in constitutive activation of the α subunit. The only way to resolve this issue is to produce the γ knockout animals. Gratifyingly, of the three γ knockout animals (γ_{t1}, γ_3, and γ_7) made to date, loss of the γ subtype produces a phenotype that in some cases is more severe than loss of its associated α subtype. Moreover, in the process of characterizing their phenotypes, the results identify a critical requirement for the γ subunits for the assembly of particular G-protein heterotrimers.

Role of the γ_{t1} Subtype in Retina

The γ_{t1} subtype is the best studied and first γ protein to have an assigned function based on its restricted expression in the rod photoreceptor cells of the retina [29]. The rod photoreceptor cells are responsible for night vision. In these cells, the rod G protein (transducin; $\alpha_{t1}\beta_1\gamma_{t1}$) acts by coupling the light-sensitive rhodopsin receptor to stimulation of cGMP phosphodiesterase, thereby lowering cGMP levels and closing cGMP regulated cation channels that hyperpolarize these cells. In mice lacking the γ_{t1} subtype, the response to light is greatly attenuated [30], confirming that the γ_{t1} subtype is a primary player in this pathway. In this regard, the results are similar to those observed in

α_{t1} knockout mice in which the light response is largely abolished [31]. However, in other respects, the phenotype of the γ_{t1} knockout mice seems more severe than that observed in α_{t1} knockout mice. For instance, mice lacking the γ_{t1} subtype develop severe retinal degeneration at an early age, suggesting a possible role in a cell survival pathway. Moreover, in γ_{t1} knockout mice, the retinal levels of α_{t1} and β_1 proteins are dramatically reduced, suggesting an absolutely critical requirement for the γ_{t1} subunit for the expression and function of the α_{t1} subtype in particular and the rod G-protein heterotrimer in general. As discussed below, this requirement was first observed for the γ_7 and α_{olf} subtypes [32]. In both cases, the coordinate reduction in their associated α_{t1} and α_{olf} protein levels occurs post-transcriptionally, since their mRNA levels are not changed. Taken together, these results reveal a previously unrecognized requirement for the γ subtype that may be applicable to the assembly of all G-protein heterotrimers.

The existence of a phenotype in γ_{t1} knockout mice indicates that at least some of its functions are unique and cannot be replaced by related family members. The inability of other γ subunits to compensate for γ_{t1} is most likely due to the unique expression pattern of this gene in the organism. Analysis of $\gamma t1$ knockout mice suggests only modest adaptations in other γ subunits that are not sufficient to compensate for the abundant expression of the γ_{t1} protein in this tissue [30]. Therefore, rod-specific expression is likely to be a critical factor in maintaining the fidelity of G-protein mediated signaling in these cells.

Role of the γ_3 Subtype in Brain

Mice with a targeted deletion of the Gng3 gene were produced in two steps [33]. First, mice with a floxed Gng3 allele were created by using a targeting vector in which the complete coding region is flanked by loxP sites. Second, mice with a deleted Gng3 allele were produced by crossing floxed Gng3 mice with mice expressing the Cre transgene under control of the EIIa promoter to generate a whole body knockout. After initial characterization on the mixed background, mice with the deleted Gng3 allele were back-crossed for a minimum of five generations onto the C57Bl6 background prior to phenotypic analysis to reduce the possibility of background strain artifacts.

In contrast to the retinal specific expression of the γ_{t1} subtype, the γ_3 subtype is widely and abundantly expressed in most cell types in the brain. Consistent with its possible involvement in multiple neurotransmitter signaling system, mice lacking the γ_3 subtype exhibit a variety of neurological abnormalities [33]. Noted first on a mixed genetic background (i.e., FVB/N, C57Bl/6, 129 strains), γ_3 knockout mice exhibit increased frequency of spontaneous seizures and increased mortality rates. These phenotypic changes are dependent on the genetic background – both features are lost after crossing onto the seizure-resistant C57Bl/6

background for five generations, and reappear after crossing onto the seizure-sensitive FVB/N background for one generation (W. F. Schwindinger and J. D. Robishaw, unpublished). This dependence of seizure susceptibility on genetic background was also observed for the serotonin 5–HT$_{2c}$- [34] and GABA$_B$-knockout mice [35], suggesting potential signaling pathways through which disruption of γ_3 may affect neuronal excitability. Moreover, loss of γ_3 is associated with significant reductions in the levels of β_1, β_2, and α_{i3} proteins in the cortex, hippocampus, and cerebellum [33], suggesting possible G-protein heterotrimer combinations through which γ_3 may act. In future studies, the ability to limit inactivation of the Gng3 gene to specific brain regions for this animal model will allow further dissection of the functions of this gene in different neuronal regions and cell types.

Mice lacking the γ_3 subtype also exhibit a weight phenotype that is *not* dependent on the genetic background [33]. Observed first on a regular diet, the phenotype is more pronounced in γ_3 knockout mice in response to a high fat diet (WS and JDR, unpublished). Compared to their wild-type littermate controls, female and, to a lesser extent, male knockout mice gain less weight, have significantly reduced fat pad stores, and show milder hepatic steatosis. These phenotoypic changes are not associated with increased activity or energy expenditure, but rather appear to be related to reduced food consumption and decreased preference for a high-fat food. These behavioral changes are associated with an attenuated response to morphine, suggesting the μ-opioid receptor signaling system as a potential pathway through which ablation of γ_3 may adversely affect the rewarding properties of palatable foods that promote increased food intake.

The existence of two apparently unrelated phenotypes in γ_3 knockout mice suggests a requirement for γ_3 in different brain regions or distinct signaling pathways affecting neuronal excitability and food consumption that cannot be compensated for by related γ subtypes. Of the known γ subtypes, the γ_2, γ_3, and γ_7 subtypes are predominantly expressed in the brain, where they show partly overlapping patterns of mRNA expression in the cortex, striatum, hippocampus, and cerebellum [36]. Quantitative immunoblot analysis shows that loss of the γ_3 protein results in a small increase in the level of γ_2 protein (21 percent) in the cortex and the level of the γ_7 protein (31 percent) in the striatum [33]. Such adaptations may reflect a compensatory change to replace the lost γ_3 protein in the same signaling pathway. However, based on their relative levels in these brain regions, the relatively small changes in γ_2 and γ_7 proteins would not be sufficient to functionally compensate for the much larger loss of the γ_3 protein. Alternatively, such changes may reflect a secondary alteration in another signaling pathway. As discussed further below, comparative analyses of the γ_3 and γ_7 knockout mice suggest that these two γ subtypes function in different signaling pathways.

Role of the γ7 Subtype in Nervous System

Mice with a targeted deletion of the Gng7 gene were produced in the same way as described above for the Gng3 gene [32]. After initial characterization on the mixed background, mice with the deleted Gng7 allele were backcrossed for a minimum of five generations onto the C57Bl/6 background prior to phenotypic analysis to reduce the possibility of background strain artifacts.

In contrast to the broad expression of γ3 in brain, the γ7 subtype is predominantly expressed in the striatum, suggesting a possible role in control of locomotor activity. Consistent with this possibility, mice lacking γ7 exhibit specific behavioral changes including altered startle reflex and greatly attenuated locomotor responses to psychoactive drugs such as amphetamine and caffeine. The actions of these two drugs are dependent on D_1 dopamine- and A_{2a} adenosine-receptor mediated activation of the G_{olf} protein and subsequent stimulation of adenylyl cyclase activity, respectively [37]. Confirming a requirement for the γ7 subtype in these two signaling pathways, D_1 dopamine- [32] and A_{2a} adenosine- (WS and JDR, unpublished) receptor-specific agonists produce the expected increases in striatal AC activity in wild-type mice, but these responses were largely abolished in γ7 knockout mice. Moreover, loss of γ7 is associated with a specific and dramatic reduction in the level of the $α_{olf}$ protein (82 percent) in the striatum of knockout mice [32]. These results provide the first evidence that loss of a γ subunit results in loss of its α-subunit partner. With its confirmation in other γ-knockout models, this result reveals an important new signaling paradigm; namely, the level of a specific γ subtype controls the stability and/or assembly of its G-protein heterotrimer.

Because γ3 and γ7 are more than 80 percent homologous at the amino acid level and are predominantly expressed in brain, the finding that mice lacking one or the other subtype display distinctive phenotypes and different αβγ subunit associations provides conclusive proof that even closely related γ subtypes perform unique roles in separate signaling pathways or biological processes in the context of the organism. Collectively, these results strongly support the notion that the γ subunit composition contributes to the specificity of G-protein mediated signaling pathways in brain through some combination of cellular expression, subcellular localization, and/or differential protein–protein interactions.

Morpholino Antisense RNA Knockdown in Zebra Fish

Understanding the complexity of G-protein mediated signaling in vertebrate development remains a challenge. Although G proteins are used extensively to regulate development in lower eukaryotes [38–41], their importance in higher eukaryotes has been much more difficult to elucidate. In large part, this reflects the difficulty of studying

embryogenesis in vertebrate models. Traditionally, the mouse has been the vertebrate species of choice, even though it has numerous disadvantages for developmental analysis – including the time-consuming and expensive nature of knockout mice, the intrauterine development of embryos that prevents direct observation of developmental events, and the potential for embryonic lethality. More recently, the zebra fish has emerged as an alternative vertebrate species with several advantages for this type of analysis [42]. Particularly relevant to understanding γ-subunit diversity, zebra fish share a similar complexity of γ genes and show a striking degree of amino acid sequence homology compared to their mouse and human counterparts [17]. Moreover, since zebra-fish eggs are externally fertilized, various reagents (e.g., morpholino antisense oligonucleotides and RNA) can be readily introduced into eggs to manipulate gene expression. Subsequent visual and *in situ* hybridization analysis of the resulting phenotype in transparent embryos can provide a rapid survey of developmental functions for the individual γ subtypes. Finally, since cardiac function and blood circulation are not required for the first several days of development, even those embryos showing severe defects can survive long enough for morphologic and functional analysis. Attesting to both the usefulness of this model and the feasibility of this approach, targeted knockdown of the gng2 gene encoding the γ2 subtype has identified its role in regulating angiogenesis in developing zebra-fish embryos.

Role of γ2 Subtype in Angiogenesis

The zebra fish gng2 gene encodes the γ2 protein that is 94 percent identical to its mouse and human orthologs [43]. Such a high degree of sequence conservation across species suggests similar conservation of the function of the γ2 subtype. Whole mount *in situ* hybridization shows that the gng2 transcript is detected throughout embryonic development and is expressed at particularly high levels in the developing brain and vasculature [43]. Due to the transparency of zebra-fish embryos, vascular development and function can be directly observed by video microscopy in live embryos. Therefore, a morpholino antisense knockdown approach can be readily used to explore a possible role for the γ2 subtype in vascular development [43]. To achieve knockdown *in vivo*, a splice junction morpholino targeted against the first coding exon–intron boundary of gng2 is used. When injected into embryos, the splicing morpholino induces a cryptic splicing site, resulting in truncation of the coding sequence and production of a non-functional γ2 protein.

Morphologic and functional analysis of the gng2 knockdown phenotype reveals a specific defect in vascular development. Early in development, at 1.5 days post-fertilization (dpf), both control and gng2 knockdown embryos show normal vasculogenesis, as exhibited by normal development and blood circulation through the major vessels (i.e., dorsal aorta and cardinal vein). However, by 2.5 dpf, the gng2 knockdown embryos display abnormal angiogenesis, as exhibited by the absence of, or reduced formation and blood flow through, the intersomitic vessels. The gng2 knockdown phenotype is particularly interesting, since intersomitic vessels develop by angiogenic sprouting from the dorsal aorta and cardinal vein, a process that closely resembles tumor angiogenesis. Attesting to the specific nature of the angiogenic defect, a similar γ_2 knockdown phenotype is observed in embryos injected with a translation blocking morpholino targeted against gng2, the phenotype is not seen in embryos injected with five-base mismatch morpholino control, and the phenotype is partially rescued in embryos co-injected with the corresponding gng2 cDNA along with the morpholino. Collectively, these results provide conclusive evidence that the γ_2 subtype plays a critical role in angiogenesis for the first time. The identities of the α and β subtypes associated with γ_2 subtype in this process are not known. However, based on similarity to knockout mouse phenotypes, α_{13} and/or $\alpha_{q/11}$ appear to be likely candidates [28, 43].

CONCLUSION

This growing body of evidence provides conclusive proof that the G-protein γ subtypes are not functionally interchangeable in the context of the whole animal. However, studies carried out thus far represent only the tip of the iceberg, with additional phenotypes remaining to be discovered. For instance, the γ_3 subtype is induced following activation of T and B cells, and preliminary characterization of γ_3 knockout mice point to immunological defect in these animals [44]. In future studies, it will be interesting to determine whether γ-specific associations and signaling roles change from one cell type to another. Also, it will necessary to target additional γ subtypes. In this regard, the production of mice lacking the γ_5 and γ_{11} subtypes is underway.

These animal studies are the first to reveal the absolutely critical requirement of a specific γ subtype for the assembly and function of a particular G-protein heterotrimer. Just how α, β, and γ subtypes are brought together to form specific G-protein heterotrimers remains to be determined. At one level, the α, β, and γ subtypes may be expressed in a cell-specific pattern, raising the possibility that each cell contains only a subset of possible G-protein heterotrimers to participate in signaling at one time. This mechanism may be operative in sensory transduction systems, but is unlikely to be applicable to most other signaling systems that operate in tissues and cell types that must respond to many diverse stimuli. At another level, the γ subtypes may be involved in targeting the G-protein heterotrimer to specific locations within the cell. A knowledge of which G-protein heterotrimers exist and operate in

a particular signaling pathway is required to answer these types of questions, and therefore continued production and analysis of γ knockout models is needed.

ACKNOWLEDGEMENTS

This work was supported by NIH grants GM39867 and GM58191 awarded to Janet D. Robishaw.

REFERENCES

1. Smrcka AV. G protein $\beta\gamma$ subunits: central mediators of G protein coupled receptor signaling. *Cell Mol Life Sci* 2008;**65**:2191–214.

2. Yang W, Hildebrandt JD. Genomic analysis of G protein γ subunits in human and mouse – the relationship between conserved gene structure and G protein $\beta\gamma$ dimer formation. *Cell Signal* 2006;**18**:194–201.

3. Robishaw JD, Berlot CH. Translating G protein subunit diversity into functional specificity. *Curr Opin Cell Biol* 2004;**16**:206–9.

4. Robishaw JD, Schwindinger WF, Hansen CA. Specificity of G protein $\beta\gamma$ dimer signaling. In: Bradshaw RA, Dennis EA, editors. *Handbook of Cell Signaling*, Vol. 2. San Diego: Elsevier Science; 2003. p. 623–9.

5. Asano T, Morishita R, Ueda H, Asano M, Kato K. GTP-binding protein γ_{12} subunit phosphorylation by protein kinase C – identification of the phosphorylation site and factors involved in cultured cells and rat tissues *in vivo. Eur J Biochem* 1998;**251**:314–19.

6. Cook LA, Schey KL, Wilcox MD, Dingus J, Ettling R, Nelson T, Knapp DR, Hildebrandt JD. Proteomic analysis of bovine brain G protein γ subunit processing heterogeneity. *Mol Cell Proteomics* 2006;**5**:671–85.

7. Wilcox M, Dingus J, Balcueva E, McIntire W, Mehta N, Schey K, Robishaw J, Hildebrandt J. Bovine brain Go isoforms have distinct γ subunit compositions. *J Biol Chem* 1995;**270**:4189–92.

8. Hansen C, Schroering A, Carey D, Robishaw JD. Localization of a heterotrimeric G protein γ_5 subunit to focal adhesions and associated stress fibers. *J Cell Biol* 1994;**126**:811–19.

9. Ueda H, Saga S, Shinohara H, Morishita R, Kato K, Asano T. Association of the γ_{12} subunit of G proteins with actin filaments. *J Cell Sci* 1997;**110**:1503–11.

10. Saini DK, Kalyanaraman V, Chisari M, Gautam N. A family of G protein $\beta\gamma$ subunits translocate reversibly from the plasma membrane to endomembranes on receptor activation. *J Biol Chem* 2007; **282**:24,099–24,108.

11. Hurowitz E, Melnyk J, Chen Y, Kouros-Mehr H, Simon M, Shizuya H. Genomic characterization of the human heterotrimeric G protein α, β,γ subunit genes. *DNA Res* 2000;**7**:111–20.

12. Huang L, Shanker YG, Dubauskaite J, Zheng JZ, Yan W, Rosenzweig S, Spielman AI, Max M, Margolskee RF. Gγ_{13} co localizes with gustducin in taste receptor cells and mediates IP3 responses to bitter denatonium. *Nature Neurosci* 1999;**2**:1055–62.

13. Karolchik D, Kuhn RM, Baertsch R, Barber GP, Clawson H, Diekhans M, Giardine B, Harte RA, Hinrichs AS, Hsu F, Miller W, Pedersen JS, Pohl A, Raney BJ, Rhead B, Rosenbloom KR, Smith KE, Stanke M, Thakkapallayil A, Trumbower H, Wang T, Zweig AS, Haussler D, Kent WJ. The UCSC Genome Browser Database: 2008 update. *Nucleic Acids Res* 2008;**36**:D773–9.

14. Fisher K, Aronson NJ. Characterization of the cDNA and genomic sequence of a G protein γ subunit (γ5). *Mol Cell Biol* 1992;**12**:1585–91.

15. Maltese WA, Robishaw JD. Isoprenylation of C-terminal cysteine in a G protein γ subunit. *J Biol Chem* 1990;**265**:8071–4.

16. Balcueva E, Wang Q, Hughes H, Kunsch C, Yu Z, Robishaw JD. Human G protein γ₁₁ and γ₁₄ subtypes define a new functional subclass. *Exp Cell Res* 2000;**257**:310–15.

17. Cheng KC, Levenson R, Robishaw JD. Functional genomic dissection of multimeric protein families in zebra fish. *Dev Dyn* 2003;**228**(3):555–67.

18. Chen H, Leung T, Giger KE, Stauffer AM, Humbert JE, Sinha S, Horstick EJ, Hansen CA, Robishaw JD. Expression of the G protein γt1 subunit during zebra fish development. *Gene Expr Patterns* 2007;**7**:574–83.

19. Schwindinger W, Robishaw JD. Heterotrimeric G-protein βγ-dimers in growth and differentiation. *Oncogene* 2002;**20**:1653–60.

20. Rahmatullah M, Ginnan R, Robishaw JD. Specificity of G protein α–γ subunit interactions. N-terminal 15 amino acids of γ subunit specifies interaction with α subunit. *J Biol Chem* 1995;**270**:2946–51.

21. Cook LA, Schey KL, Cleator JH, Wilcox MD, Dingus J, Hildebrandt JD. Identification of a region in G protein γ subunits conserved across species but hypervariable among subunit isoforms. *Protein Sci* 2001;**10**:2548–55.

22. Sondek J, Bohm A, Lambright DG, Hamm HE, Sigler PB. Crystal structure of a G-protein βγ dimer at 2.1-Å resolution. *Nature* 1996;**379**:369–74. Erratum in Nature 379, 847.

23. Ueda H, Morishita R, Yamauchi J, Itoh H, Kato K, Asano T. Regulation of Rac and Cdc42 pathways by G(i) during lysophosphatidic acid-induced cell spreading. *J Biol Chem* 2001;**276**:6846–52.

24. Lee C, Murakami T, Simonds WF. Identification of a discrete region of the G protein γ subunit conferring selectivity in βγ complex formation. *J Biol Chem* 1995;**270**:8779–84.

25. Meister M, Dietrich A, Gierschik P. Identification of a three-amino-acid region in G protein γ₁ as a determinant of selective βγ heterodimerization. *Eur J Biochem* 1995;**234**:171–7.

26. Kilpatrick EL, Hildebrandt JD. Sequence dependence and differential expression of Gγ5 subunit isoforms of the heterotrimeric G proteins variably processed after prenylation in mammalian cells. *J Biol Chem* 2007;**282**:14,038–14,047.

27. Li Z, Benard O, Margolskee RF. Gγ13 interacts with PDZ domain-containing proteins. *J Biol Chem* 2006;**281**:11,066–11,073.

28. Wettschureck N, Offermanns S. Mammalian G proteins and their cell type specific functions. *Physiol Rev* 2005;**85**:1159–204.

29. Peng Y, Robishaw JD, Levine M, Yau K. Retinal rods and cones have distinct G protein β and subunits. *Proc Natl Acad Sci USA* 1992;**89**:10,882–10,886.

30. Lobanova ES, Finkelstein S, Herrmann R, Chen YM, Kessler C, Michaud NA, Trieu LH, Strissel KJ, Burns ME, Arshavsky VY. Transducin γ-subunit sets expression levels of α- and β-subunits and is crucial for rod viability. *J Neurosci* 2008;**28**:3510–20.

31. Calvert PD, Krasnoperova NV, Lyubarsky AL, Isayama T, Nicolo M, Kosaras B, Wong G, Gannon KS, Margolskee RF, Sidman RL, Pugh EN, Makino CL, Lem J. Phototransduction in transgenic mice after targeted deletion of the rod transducin α-subunit. *Proc Natl Acad Sci USA* 2000;**97**:13,913–13,918.

32. Schwindinger WF, Betz KS, Giger KE, Sabol A, Bronson SK, Robishaw JD. Loss of G protein γ7 alters behavior and reduces striatal αolf level and cAMP production. *J Biol Chem* 2003;**278**:6575–9.

33. Schwindinger WF, Giger KE, Betz KS, Stauffer AM, Sunderlin EM, Sim-Selley LJ, Selley DE, Bronson SK, Robishaw JD. Mice with deficiency of G protein γ3 are lean and have seizures. *Mol Cell Biol* 2004;**24**:7758–68.

34. Tecott LH, Sun LM, Akana SF, Strack AM, Lowenstein DH, Dallman MF, Julius D. Eating disorder and epilepsy in mice lacking 5-HT₂c serotonin receptors. *Nature* 1995;**374**:542–6.

35. Prosser HM, Gill CH, Hirst WD, Grau E, Robbins M, Calver A, Soffin EM, Farmer CE, Lanneau C, Gray J, Schenck E, Warmerdam BS, Clapham C, Pangalos MN. Epileptogenesis and enhanced prepulse inhibition in GABA_B1-deficient mice. *Mol Cell Neurosci* 2001;**17**:1059–70.

36. Betty M, Harnish S, Rhodes K, Cockett M. Distribution of heterotrimeric G-protein β and γ subunits in the rat brain. *Neuroscience* 1998;**85**:475–86.

37. Corvol JC, Studler JM, Schonn JS, Girault JA, Herve D. Gαolf is necessary for coupling D₁ and A₂a receptors to adenylyl cyclase in the striatum. *J Neurochem* 2001;**76**:1585–8.

38. Dohlman HG. G proteins and pheromone signaling. *Ann Rev Physiol* 2002;**64**:129–52.

39. Zhang N, Long Y, Devreotes PN. G γ in Dictyostelium: its role in localization of G βγ to the membrane is required for chemotaxis in shallow gradients. *Mol Biol* 2001;**12**:3204–13.

40. Ray K, Ganguly R. Organization and expression of the *Drosophila melanogaster* G γ₁ gene encoding the G protein γ subunit. *Gene* 1994;**148**:315–19.

41. Jansen G, Weinkove D, Plasterk RH. The G-protein γ subunit gpc-1 of the nematode *C. elegans* is involved in taste adaptation. *EMBO J* 2002;**21**:986–94.

42. Driever W, Solnica-Krezel L, Scheir AF, Neuhauss SC, Malicki J, Stemple DL, Stainier DY, Zwartkruis F, Abdelilah S, Rangine Z, Belak J, Boggs C. A genetic screen for mutations affecting embryogenesis in zebra fish. *Development* 1996;**123**:37–46.

43. Leung T, Chen H, Stauffer AM, Giger KE, Sinha S, Horstick EJ, Humbert JE, Hansen CA, Robishaw JD. Zebra fish G protein γ2 is required for VEGF signaling during angiogenesis. *Blood* 2006;**108**:160–6.

44. Dubeykovskiy A, McWhinney C, Robishaw JD. Runx-dependent regulation of G-protein γ3 expression in T-cells. *Cell Immunol* 2006;**240**:86–95.

45. Kassai H, Aiba A, Nakao K, Nakamura K, Katsuki M, Xiong WH, Yau KW, Imai H, Shichida Y, Satomi Y, Takao T, Okano T, Fukada Y. Farnesylation of retinal transducin underlies its translocation during light adaptation. *Neuron* 2005;**47**:529–39.

46. Kalkbrenner F, Degtiar VE, Schenker M, Brendel S, Zobel A, Heschler J, Wittig B, Schultz G. Subunit composition of G(o) proteins functionally coupling galanin receptors to voltage-gated calcium channels. *EMBO J* 1995;**14**:4728–37.

47. Macrez-Leprêtre N, Kalkbrenner F, Schultz G, Mironneau J. Distinct functions of G_q and G₁₁ proteins in coupling α₁-adrenoreceptors to Ca²⁺ release and Ca²⁺ entry in rat portal vein myocytes. *J Biol Chem* 1997;**272**:5261–8.

48. Hosohata K, Logan JK, Varga E, Burkey TH, Vanderah TW, Porreca F, Hruby VJ, Roeske WR, Yamamura HI. The role of the G protein γ2 subunit in opioid antinociception in mice. *Eur J Pharmacol* 2000;**392**:R9–R11.

49. Johansen PW, Lund HW, Gordeladze JO. Specific combinations of G-protein subunits discriminate hormonal signalling in rat pituitary (GH₃) cells in culture. *Cell Signal* 2001;**13**:251–6.

50. Varga EV, Hosohata K, Borys D, Navratilova E, Nylen A, Vanderah TW, Porreca F, Roeske WR, Yamamura HI. Antinociception depends on the presence of G protein γ²-subunits in brain. *Eur J Pharmacol* 2005;**508**:93–8.

51. Degtiar VE, Wittig B, Schultz G, Kalkbrenner F. A specific G(o) heterotrimer couples somatostatin receptors to voltage-gated calcium channels in RINm5F cells. *FEBS Letts* 1996;**380**:137–41.

52. Macrez-Leprêtre N, Kalkbrenner F, Morel JL, Schultz G, Mironneau J. G protein heterotrimer $\alpha_{13}\beta_1\gamma_3$ couples the angiotensin AT_{1A} receptor to increases in cytoplasmic Ca^{2+} in rat portal vein myocytes. *J Biol Chem* 1997;**272**:10,095–10,102.

53. Macrez N, Morel JL, Kalkbrenner F, Viard P, Schultz G, Mironneau J. A $\beta\gamma$ dimer derived from G_{13} transduces the angiotensin AT_1 receptor signal to stimulation of Ca^{2+} channels in rat portal vein myocytes. *J Biol Chem* 1997;**272**:23,180–23,185.

54. Macrez N, Morel JL, Mironneau J. Specific $G\alpha_{11}\beta_3\gamma_5$ protein involvement in endothelin receptor-induced phosphatidylinositol hydrolysis and Ca^{2+} release in rat portal vein myocytes. *Mol Pharmacol* 1999;**55**:684–92.

55. Wang Q, Mullah B, Hansen C, Asundi J, Robishaw JD. Ribozyme-mediated suppression of the G protein γ_7 subunit suggests a role in hormone regulation of adenylylcyclase activity. *J Biol Chem* 1997;**272**:26,040–26,048.

56. Wang Q, Mullah BK, Robishaw JD. Ribozyme approach identifies a functional association between the G protein $\beta_1\gamma_7$ subunits in the β-adrenergic receptor signaling pathway. *J Biol Chem* 1999;**274**:17,365–17,371.

57. Hossain MN, Sakemura R, Fujii M, Ayusawa D. G-protein γ subunit GNG11 strongly regulates cellular senescence. *Biochem Biophys Res Commun* 2006;**351**:645–50.

Chapter 52

Reversible Palmitoylation in G Protein Signaling

Philip B. Wedegaertner

Department of Biochemistry and Molecular Biology, Kimmel Cancer Center, Thomas Jefferson University, Philadelphia, Pennsylvania

INTRODUCTION

Numerous proteins involved in cellular signaling undergo reversible palmitoylation [1–3]. This review will focus on reversible palmitoylation of G protein α subunits (Gα), briefly discuss some more general aspects of palmitoylation and depalmitoylation, and highlight recent evidence that many Regulator of G protein Signaling (RGS) proteins are palmitoylated.

Palmitoylation is a covalent lipid modification in which the saturated 16-carbon fatty acid, palmitate, is linked through a thioester bond to a cysteine. Palmitoylation is often referred to as S-acylation or thioacylation to indicate that radio-labeled fatty acids of varying lengths can be incorporated into a protein and that recent mass spectrometry analysis has indicated that proteins can exist as multiple isoforms modified by palmitate or alternate fatty acids of longer carbon chain length, such as stearate and oleate [4, 5]. However, very few studies have addressed the identity and stoichiometry of the endogenous fatty acid covalently attached to a particular cellular protein. In fact, it remains mostly unknown if "palmitoylated" proteins exist as isoforms with different attached fatty acids, if modification by fatty acids other than palmitate imparts unique function, if specific cellular mechanisms exist to dictate a prefence for palmitate, or if palmitate is primarily used merely because of its relatively high abundance in cells.

Much of the interest in palmitoylation stems from its dynamic nature, and the resulting idea that this covalent modification is more than just a static membrane anchor. Numerous studies have demonstrated that palmitoylation is a reversible modification–i.e., attached palmitate often has a much shorter half-life than does the protein it modifies. Such reversibility implies the potential for important regulatory roles for palmitoylation.

SITES OF PALMITOYLATION IN Gα AND RGS PROTEINS

Besides the modified cysteine(s) itself, no primary amino acid consensus sequence has been identified for palmitoylation. The only consensus is that an additional hydrophobic modification or membrane targeting motif is often a prerequisite for palmitoylation [1, 3, 6, 7]. Table 52.1 lists identified sites of palmitoylation from representative Gα and RGS proteins. Based on lipid modification, Gα can be divided into two subfamilies (Table 52.1): those that are palmitoylated only, and those that undergo both myristoylation and palmitoylation. Recently, it was demonstrated that the yeast *S. cerevisiae* G protein γ subunit is palmitoylated at a cysteine adjacent to the prenylated cysteine at the extreme C-terminus [8, 9]. However, none of the 12 human Gγ subunits or the two Gγ of the model organism *C. elegans* contain a potential C-terminal palmitoylation site.

Palmitoylation has been identified in some members of the RGS family (Table 52.1) [10]. Interestingly, palmitoylation of RGS proteins can occur at one or two cysteines within short N-terminal extensions [11, 12], multiple cysteines in a cysteine string motif in the N-terminus [13], cysteines at internal sites [14, 15], or a cysteine within the RGS domain itself [16] (Table 1). Additional palmitoylated proteins involved in G-protein signaling pathways include G-protein-coupled receptors (GPCR) [17], G-protein-coupled receptor kinases (GRK) [18], and small GTPases of the ras and rho superfamily [2, 3].

TABLE 52.1 Palmitoylation of Gα and RGS proteins[a]

Gα N-termini (myristoylated and palmitoylated)[b]	
α_{i1}	M **G** C T L S A E D K A A V E R S K M I D
α_{o1}	M **G** C T L S A E E R A A L E R S K A I E
α_{z}	M **G** C R Q S S E E K E A A R R S R R I D
Gpa1	M **G** C T V S T Q T I G D E S D P F L Q N
Gα N-termini (non-myristoylated)[c]	
α_{s}	M **G** C L G N S K T E D Q R N E E K A Q R
α_{q}	M T L E S I M A C C L S E E A K E A R R
α_{14}	M A G C C C L S A E E K E S Q R I S A E
α_{16}	M A R S L R W R C C P W C L T E D E K A
α_{12}	M S G V V R T L S R C L L P A E A G A R
α_{13}	M A D F L P S R S V L S V C F P G C V L
RGS N-termini[d]	
RGS4	M C K G L A G L P A S C L R S A K D M K
RGS16	M C R T L A T F P N T C L E R A K E F K
RGS cysteine-string motif[e]	
RGS-GAIP	35 S R N P C C L C W C C C C S C S W N Q E 54
RGS internal site between DEP and GGL domains	
RGS7	123 P E N T D Y A V Y L C K R T M Q N K A R 142
RGS box palmitoylation[f]	
RGS4	86 E E N I D F W I S C E E Y K K I K S P S 105
RGS10	57 E E N V L F W L A C E D F K K M Q D K T 76
RGS16	89 E E N L E F W L A C E E F K K I R S A T 108

[a] Palmitoylated (or potentially palmitoylated) cysteines are underlined

[b] Myristoylated glycines at position 2 are in bold (myristoylation of Gα has been well-reviewed [1, 61, 62])

[c] For αs, palmitate attached through a stable amide bond to the glycine at position 2 (i.e., N-palmitoylation), after removal of the initiating methionine, has also been identified [103]

[d] Similar N-terminal sequence found in RGS5

[e] Cysteine-string motifs are found in other members of the RZ subfamily

[f] Cysteine present at similar position in RGS box of most RGS proteins.

ACTIVATION-REGULATED PALMITOYLATION OF Gα

If palmitoylation functions as a regulatory modification, then the expectation is that appropriate cellular stimuli will cause changes in the level of palmitoylation of a particular protein. Indeed, regulated changes in palmitoylation appear to be a general phenomenon for Gα [1, 7]. Regulated palmitoylation was first demonstrated for α_{s}. Palmitate attached to α_{s} turns over much more rapidly after activation by the β-adrenergic receptor (βAR) agonist isoproterenol, activation by cholera toxin, or activation by a constitutively activating mutation in α_{s} [19–21]. Time-courses of palmitate incorporation and pulse-chase analyses are consistent with a model [22, 23] in which activation leads to both more rapid depalmitoylation and more rapid subsequent

repalmitoylation of α_{s}. For Gα other than α_{s}, a thorough study showed that agonist activation of stably transfected 5-HT_{1A} receptors in CHO cells resulted in increased palmitate turnover on endogenous α_{i} [24]. In addition, stimulation of gonadotropin-releasing hormone receptors (GnRH) sin pituitary cells causes increased palmitate incorporation into α_{s}, α_{i}, and α_{q} [25, 26], stimulation of m1-muscarinic receptors in transfected cells increases the rate of palmitate incorporation into α_{q} (author's unpublished observations), stimulation of GPCRs in isolated membranes has been shown to increase palmitoylation of α_{i}, α_{o}, α_{s}, and α_{q} [25, 27, 28], and agonist-regulated palmitoylation has been observed for α_{s} and α_{q} even when they are directly fused to GPCRs [29, 30]. A similar model of regulated palmitoylation/depalmitoylation in response to agonist activation has been well-described for GPCRs [17, 31–33]. Moreover, a recent study showed regulation of palmitoylation of RGS3 by a GnRH receptor agonist [34]. We can anticipate future studies to further demonstrate physiological inputs that regulate the reversible palmitoylation of GRKs and RGS proteins.

MECHANISMS OF REVERSIBLE PALMITOYLATION

Regulated palmitoylation implies that changes in a protein's palmitoylation state are carried out by regulation of the palmitoylation and/or depalmitoylation machinery in the cell, or by changes in the accessibility of the palmitoylated protein substrate to constitutive palmitoylating and depalmitoylating activities. Although intensely studied by many investigators, the relevant enzymes and cellular pathways that regulate reversible palmitoylation remain poorly defined.

Palmitoyl Transferases

For many years, palmitoyl acyltransferases (PATs) eluded purification and identification. However, in 2002 the identification of *bona fide* PATs in the yeast *S. cerevisiae* provided a major breakthrough [35, 36]. The PATs are part of a large family of proteins that all contain a conserved Asp-His-His-Cys cysteine-rich domain (DHHC-CRD) that is required for PAT catalytic activity. Eight DHHC-CRD proteins exist in yeast, and mammals contain more than 20 DHHC-CRD proteins, or potential PATs [37]. A major challenge is to determine the specificity of the PATs for different substrates. Relevant PATs have not yet been clearly identified for Gα or other proteins involved in heterotrimeric G-protein signaling. A recent study showed that overexpression of DHHC-3 and DHHC-7 could enhance palmitoylation of co-expressed α_s [38], whereas another study showed that DHHC-9/GCP16, which is able to palmitoylate H-Ras and N-Ras, displayed no PAT activity towards α_i [39]. In addition to palmitoylation mediated by PATs, compelling evidence has been presented for non-enzymatic palmitoylation, or autoacylation, of Gα [40] and other proteins, in which palmitate is transferred directly from palmitoyl CoA to a substrate protein in the absence of an additional PAT enzyme. This somewhat controversial topic has been well-discussed recently [3], and autoacylation remains an important possibility when considering palmitoylation of G proteins.

Palmitoyl Thioesterases

Although abundant evidence exists to demonstrate that palmitoylation is a reversible modification, the depalmitoylation reaction is even less understood than the palmitoylation mechanism. However, a strong candidate for a physiologically relevant palmitoyl thioesterase, termed an acyl protein thioesterase (APT1), has been described [41]. Overexpression of APT1 in cultured cells increased the basal rate of depalmitoylation of co-expressed α_s, and a recent report showed that overexpression of APT1 in cultured intestinal epithelial cells promoted the translocation of a constitutively active α_s mutant from membranes to cytoplasm, presumably due to depalmitoylation of α_s [42]. Importantly, an APT1 ortholog was identified in yeast, and strains lacking APT1 lacked depalmitoylating activity, including an inability to depalmitoylate the yeast Gα, Gpa1 [43]. Further studies to substantiate a role for endogenous APT1 in regulated depalmitoylation of G proteins are eagerly awaited.

Proteins and Pathways that Regulate Reversible Palmitoylation

G protein $\beta\gamma$ subunits appear to regulate both palmitoylation and depalmitoylation of Gα. $\beta\gamma$ promotes palmitoylation of Gα *in vitro* [44], and allows palmitoylation of a non-myristoylated (G2A) mutant of α_i or α_z in transfected cells [45, 46]. α_s or α_q containing mutations in N-terminal $\beta\gamma$ interaction sites display greatly decreased palmitoylation [47, 48], while a mutant α_o that has an increased affinity for $\beta\gamma$ is palmitoylated to a higher level than α_o wild type [49]. One way in which $\beta\gamma$ can enhance palmitoylation of Gα is by promoting membrane targeting. Consistent with this, palmitoylation, and plasma membrane (PM) localization, of a $\beta\gamma$-binding deficient α_q is recovered when it is engineered to undergo myristoylation [47], and $\beta\gamma$-binding mutations in α_o, a subunit that is normally myristoylated *and* palmitoylated (Table 52.1), do not affect α_o palmitoylation [49]. Although myristoylation and binding to $\beta\gamma$ appear, in some cases, to function interchangeably as membrane-targeting signals for Gα, they may have additional, more specific roles in promoting palmitoylation [1, 7, 44, 49].

$\beta\gamma$ can also inhibit depalmitoylation of α_s. A mutant α_s that binds tightly to $\beta\gamma$ is refractory to activation-induced rapid depalmitoylation in cultured cells [19, 21], and purified $\beta\gamma$ inhibits depalmitoylation of α_s when assayed in cell extracts [21] or when using purified proteins [41, 50]. These results suggest that activation-induced depalmitoylation of Gα is mediated, at least in part, by its dissociation from $\beta\gamma$.

Although studies of palmitate turnover on Gα and GPCRs suggest that palmitoylation and depalmitoylation are tightly coupled, a study utilizing a βAR-α_s fusion protein was the first to separate these two activities [29]. The authors showed that palmitoylation occurred at appropriate sites in both the βAR and α_s portions of the fusion protein, and isoproterenol induced rapid depalmitoylation, as seen with the separate proteins. However, isoproterenol-induced palmitoylation, likely *re*-palmitoylation, was blocked. This result suggests that, after rapid depalmitoylation, repalmitoylation requires dissociation of the G protein from the GPCR and/or later events that do not occur normally with this fusion protein, such as desensitization or internalization.

Tools for Studying Reversible Palmitoylation

As expected from our incomplete knowledge of the proteins involved in palmitoylation and depalmitoylation, there is a dearth of tools for inhibiting these activities. The palmitate analog 2-bromopalmitate has been widely used to inhibit palmitoylation of signaling proteins, and has proved to be a valuable reagent [51, 52]. Other inhibitors of protein palmitoylation that have been used include tunicamycin and cerulenin, or cerulenin analogs [53, 54], and a recent methods paper provides an excellent discussion of these different potential inhibitors [52]. In addition, there are efforts towards developing small molecule inhibitors of PATs and APT1 [55–57].

The most widely-used method for detecting palmitoylation is incubation of cells with [^{3}H]-palmitate, followed by immunoprecipitation of the protein of interest. Palmitate-labeled proteins are then identified by gel electrophoresis and fluorography. Unfortunately, this method typically results in weak labeling that can take very long film exposures (days or weeks) to detect. Recently, an alternative method has been developed that does not require radiolabeling but instead relies on fatty-acid exchange chemistry [58, 59]. This method has made possible novel approaches and experiments for studying palmitoylation, such as a recent proteomic study that sought to identify all the palmitoylated proteins in yeast [60]. These and other technical advances will be critical for furthering our understanding of the functions and mechanisms of reversible palmitoylation.

FUNCTIONS OF REVERSIBLE PALMITOYLATION

Plasma Membrane Localization

An obvious role for palmitoylation is to tether a protein to cellular membranes, and thus regulation of this modification would allow changes in a protein's subcellular localization, either by dissociation off a cellular membrane or by transfer to different membrane domains. Palmitoylation appears to function, in a poorly understood manner, as a specific membrane targeting device that specifies localization to PM, and in some cases specialized PM microdomains [1, 61, 62]. This role of palmitoylation has been well documented and discussed in terms of a two-signal model for PM localization of signaling proteins [1, 7, 23, 61–65].

Although the consensus is that palmitoylation plays a critical role in PM localization of many proteins, including the α_i subfamily of Gα [23, 46, 66, 67] (Table 52.1), there is some disagreement regarding a role for palmitoyla-

tion in the non-myristoylated Gα (Table 52.1). Virtually all localization studies have relied on preventing palmitoylation by mutating the relevant cysteines to serines or alanines. Subcellular fractionation and immunofluorescence localization of cells transiently or stably expressing non-palmitoylated cysteine mutants of α_s, α_q, α_{11}, α_{14}, α_{16}, or α_{13} have demonstrated that such mutants display increased cytoplasmic localization and biochemical partitioning into the soluble fraction [68–73]. Moreover, overexpression of $\beta\gamma$ with palmitoylation-defective α_s or α_q cannot restore their PM localization [48], providing further support for the critical importance of palmitoylation. On the other hand, some researchers have observed that non-palmitoylated mutants of α_s, α_q, or α_{12} remain in a particulate fraction after subcellular fractionation of transiently transfected cells [20, 74–77]. However, it is important to note that in the latter cases [20, 74–77] immunofluorescence localization of the mutants has not been reported, and thus it is unknown whether these mutants arrive correctly at the PM or are mis-targeted to intracellular locations. In some cases, the major role for palmitoylation may be to localize a protein to PM microdomains, such as lipid rafts, and other membrane targeting signals in the protein may be sufficient for general PM localization.

In agreement with this, mutation of the N-terminally palmitoylated cysteines in RGS4 and RGS16 (Table 52.1) had little effect on their PM localization [11, 12, 78], but abolishing palmitoylation of RGS16 prevented localization in lipid rafts [79]. On the other hand, RGS7 can be recruited from the cytoplasm to the PM upon co-expression of active α_o, but mutation of an internal site of palmitoylation prevented the PM recruitment [15]. For GRK6, mutation of C-terminal cysteines or inhibition of palmitoylation by treatment with 2-bromopalmitate blocked PM localization [18]. Taken together, most all palmitoylation studies of Gα, RGS proteins, and GRKs support an important role for palmitoylation in PM localization and/or localization to PM microdomains.

Can Reversible Palmitoylation Regulate Changes in Subcellular Localization?

Although palmitoylation is reversible, and plenty of evidence indicates that non-palmitoylated mutants of various proteins are defective in PM localization and/or localization to PM microdomains, actually demonstrating a relationship between changes in a protein's palmitoylation and movement of that protein within a cell is a difficult problem. Nonetheless, α_s provides the best example of a correlation between reversible palmitoylation and reversible subcellular localization. As described above, activation of α_s induces rapid turnover of its attached palmitate. Similarly, activation of α_s by GPCRs, cholera toxin, or constitutively activating mutations, can promote its redistribution from PM to cytoplasm [42, 69, 80–88]. Significantly, βAR-induced redistribution of α_s appears to follow a similar timecourse as βAR-induced depalmitoylation of α_s [21, 69, 80]. In addition, replacement of the N-terminal single cysteine site of palmitoylation of α_s with other membrane targeting motifs results in mutant α_s subunits that are unable to translocate from PM to cytoplasm upon activation [86], consistent with depalmitoylation of a single palmitate playing a critical role in activation-induced subcellular redistribution of α_s. Others have speculated that regulated palmitoylation is more relevant for allowing reversible movement of Gα within PM microdomains [22, 61, 76]. Possibly, reversible palmitoylation could help to regulate the availability of G proteins at diverse subcellular locations.

The precise location of α_s during cycles of palmitoylation and depalmitoylation remains to be clearly identified. Although constitutively active mutants of α_s are found in a soluble fraction upon cell lysis, receptor-activated α_s displays only a minor decrease in partitioning into the membrane fraction [69], raising the possibility that internalized α_s is associated with membrane vesicles or endomembranes. Several studies have shown that α_s internalization does not require the clathrin coated pit-mediated endocytosis that most GPCRs utilize [7], and, consistent with α_s internalization by a unique cellular pathway, internalized α_s was not found on endosomes [87]. However, internalized α_s was found on Rab11-containing vesicles [87]. Another study showed that lipid rafts were required for activation-induced internalization of α_s [85]. More studies are needed to understand how α_s, and other G-protein subunits, traffic as a result of regulated palmitoylation. In addition, palmitoylation-dependent trafficking of G proteins needs to be considered in light of a recently described model for palmitoylated proteins, in which it was shown that H-Ras and N-Ras undergo constitutive cycles of movement back and forth from the PM to the Golgi [89, 90]; this trafficking appears to be controlled by constitutive cycles of depalmitoylation and palmitoylation at the PM and Golgi, respectively. Indeed, a recent report used several live cell imaging approaches to demonstrate constitutive and rapid shuttling of both Gα and βγ between the PM and endomembranes, and this movement was inhibited by 2-bromopalmitate [91].

Interestingly, reversible palmitoylation has been linked to PM to nuclear shuttling for RGS7 and GRK6. Not only can RGS7 itself be palmitoylated, but RGS7 also associates with another palmitoylated protein, termed RGS7 binding protein (R7BP) [92, 93]. Co-expression of R7BP with RGS7, and the other critical RGS7 interactor Gβ5, targets RGS7-β5 to the PM; as mentioned above, RGS7-β5 can also be targeted to PM by co-expression of α_o [15]. The unique aspect is that a non-palmitoylated mutant of R7BP still interacts with RGS7-β5, but the R7BP–RGS7-β5 complex is targeted to the nucleus rather than the PM [92]. In addition, treatment of cells with 2-bromopalmitate caused a displacement from the PM of the R7BP–RGS7-β5

complex and its appearance in the nucleus. Similar findings were reported for GRK6; disruption of palmitoylation by mutation or 2-bromopalmitate treatment caused GRK6 to translocate from the PM to both the cytoplasm and nucleus [18]. Other proteins have been reported to localize at the PM or nucleus depending upon whether the protein is palmitoylated or not [94–97], and thus PM–nuclear shutting may be a common pathway regulated by reversible palmitoylation. The function of nuclear RGS7 or GRK6 is currently not defined, although nuclear functions of RGS proteins have been reported [98].

Palmitoylation Affects RGS–Gα Interactions

Changes in the palmitoylation status of a protein can also affect its ability to interact with other proteins. Indeed, one of the most interesting recent advances is evidence that palmitoylation of certain Gαs influences their interaction with RGS proteins, and, *vice versa*, palmitoylation of certain RGS proteins affects their interactions with Gαs. *In vitro*, palmitoylation of α_z or α_{i1} greatly inhibited their sensitivity to the GTP hydrolysis-stimulating activity of several RGS proteins [99]. This result implies that reversible palmitoylation of Gα functions as a key switch to regulate the ability of RGS proteins to "turn off" signaling.

Palmitoylation of RGS4 or RGS10 was shown to either inhibit or accelerate their ability to stimulate GTP hydrolysis of α_z or α_{i1} [16]. The positive or negative effect of palmitoylation differed depending upon the *in vitro* GAP (GTPase activating protein) assay used, and whether palmitoylation occurred at N-terminal sites or cysteine sites in the conserved RGS box (Table 52.1). Consistent with an important role for palmitoylation of RGS proteins, when non-palmitoylated RGS16, containing N-terminal cysteine mutations (Table 52.1), was expressed in cultured cells, it failed to effectively inhibit signaling mediated by G_q or G_i [11]. Further elegant studies on RGS16 confirmed a positive role for palmitoylation in GAP activity, and provided a model in which (1) N-terminal palmitoylation is necessary for lipid raft localization; (2) raft localization facilitates internal RGS box palmitoylation; and (3) internal palmitoylation is required for GAP activity [79, 100]. On the other hand, internal palmitoylation of RGS2 appears to inhibit its GAP activity towards α_q [101]. It seems likely that the effects of palmitoylation on RGS–Gα functional interaction are mediated by promoting an optimal conformation when tethered to the membrane and/or promoting an optimal orientation for the protein–protein interaction [3, 10, 101, 102].

CONCLUSION

Palmitoylation plays an important role in membrane binding and regulating interactions of signaling proteins. Future challenges include defining cellular pathways and enzymes that regulate reversible palmitoylation, and understanding how changes in a protein's palmitoylation are translated into functional changes inside the cell.

REFERENCES

1. Chen CA, Manning DR. Regulation of G proteins by covalent modification. *Oncogene* 2001;**20**:1643–52.

2. Resh MD. Trafficking and signaling by fatty-acylated and prenylated proteins.. *Nat Chem Biol* 2006;**2**:584–90.

3. Smotrys JE, Linder ME. Palmitoylation of intracellular signaling proteins: regulation and function. *Annu Rev Biochem* 2004;**73**:559–87.

4. Liang X, Lu Y, Neubert TA, Resh MD. Mass spectrometric analysis of GAP-43/neuromodulin reveals the presence of a variety of fatty acylated species. *J Biol Chem* 2002;**277**:33,032–33,040.

5. Liang X, Lu Y, Wilkes M, Neubert TA, Resh MD. The N-terminal SH4 region of the Src family kinase Fyn is modified by methylation and heterogeneous fatty acylation: role in membrane targeting, cell adhesion, and spreading. *J Biol Chem* 2004;**279**:8133–9.

6. Escriba PV, Wedegaertner PB, Goni FM, Vogler O. Lipid–protein interactions in GPCR-associated signaling. *Biochim Biophys Acta* 2007;**1768**:836–52.

7. Marrari Y, Crouthamel M, Irannejad R, Wedegaertner PB. Assembly and trafficking of heterotrimeric G proteins. *Biochemistry* 2007;**46**:7665–77.

8. Hirschman JE, Jenness DD. Dual lipid modification of the yeast Ggamma subunit Ste18p determines membrane localization of Gbetagamma. *Mol Cell Biol* 1999;**19**:7705–11.

9. Manahan CL, Patnana M, Blumer KJ, Linder ME. Dual lipid modification motifs in G(alpha) and G(gamma) subunits are required for full activity of the pheromone response pathway in *Saccharomyces cerevisiae*. *Mol Biol Cell* 2000;**11**:957–68.

10. Jones TL. Role of palmitoylation in RGS protein function. *Methods Enzymol* 2004;**389**:33–55.

11. Druey KM, Ugur O, Caron JM, Chen CK, Backlund PS, Jones TL. Amino-terminal cysteine residues of RGS16 are required for palmitoylation and modulation of Gi- and Gq-mediated signaling. *J Biol Chem* 1999;**274**:18,836–18,842.

12. Srinivasa SP, Bernstein LS, Blumer KJ, Linder ME. Plasma membrane localization is required for RGS4 function in *Saccharomyces cerevisiae*. *Proc Natl Acad Sci USA* 1998;**95**:5584–9.

13. De Vries L, Elenko E, Hubler L, Jones TL, Farquhar MG. GAIP is membrane-anchored by palmitoylation and interacts with the activated (GTP-bound) form of G alpha i subunits. *Proc Natl Acad Sci USA* 1996;**93**:15,203–15,208.

14. Rose JJ, Taylor JB, Shi J, Cockett MI, Jones PG, Hepler JR. RGS7 is palmitoylated and exists as biochemically distinct forms. *J Neurochem* 2000;**75**:2103–12.

15. Takida S, Fischer CC, Wedegaertner PB. Palmitoylation and plasma membrane targeting of RGS7 are promoted by alpha o. *Mol Pharmacol* 2005;**67**:132–9.

16. Tu Y, Popov S, Slaughter C, Ross EM. Palmitoylation of a conserved cysteine in the regulator of G protein signaling (RGS) domain modulates the GTPase-activating activity of RGS4 and RGS10. *J Biol Chem* 1999;**274**:38,260–38,267.

17. Qanbar R, Bouvier M. Role of palmitoylation/depalmitoylation reactions in G-protein-coupled receptor function. *Pharmacol Ther* 2003;**97**:1–33.

18. Jiang X, Benovic JL, Wedegaertner PB. Plasma membrane and nuclear localization of G protein coupled receptor kinase 6A. *Mol Biol Cell* 2007;**18**:2960–9.

19. Degtyarev MY, Spiegel AM, Jones TL. Increased palmitoylation of the Gs protein alpha subunit after activation by the beta-adrenergic receptor or cholera toxin. *J Biol Chem* 1993;**268**:23,769–23,772.

20. Mumby SM, Kleuss C, Gilman AG. Receptor regulation of G-protein palmitoylation. *Proc Natl Acad Sci USA* 1994;**91**:2800–4.

21. Wedegaertner PB, Bourne HR. Activation and depalmitoylation of Gs alpha. *Cell* 1994;**77**:1063–70.

22. Mumby SM. Reversible palmitoylation of signaling proteins. *Curr Opin Cell Biol* 1997;**9**:148–54.

23. Wedegaertner PB. Lipid modifications and membrane targeting of G alpha. *Biological Signal Recept* 1998;**7**:125–35.

24. Chen CA, Manning DR. Regulation of Galpha i palmitoylation by activation of the 5-hydroxytryptamine-1A receptor. *J Biol Chem* 2000;**275**:23,516–23,522.

25. Stanislaus D, Janovick JA, Brothers S, Conn PM. Regulation of G(q/11)alpha by the gonadotropin-releasing hormone receptor. *Mol Endocrinol* 1997;**11**:738–46.

26. Stanislaus D, Ponder S, Ji TH, Conn PM. Gonadotropin-releasing hormone receptor couples to multiple G proteins in rat gonadotrophs and in GGH3 cells: evidence from palmitoylation and overexpression of G proteins. *Biol Reprod* 1998;**59**:579–86.

27. Gurdal H, Seasholtz TM, Wang HY, Brown RD, Johnson MD, Friedman E. Role of G alpha q or G alpha o proteins in alpha 1-adrenoceptor subtype-mediated responses in Fischer 344 rat aorta. *Mol Pharmacol* 1997;**52**:1064–70.

28. Bhamre S, Wang HY, Friedman E. Serotonin-mediated palmitoylation and depalmitoylation of G alpha proteins in rat brain cortical membranes. *J Pharmacol Exp Ther* 1998;**286**:1482–9.

29. Loisel TP, Ansanay H, Adam L, Marullo S, Seifert R, Lagace M, Bouvier M. Activation of the beta(2)-adrenergic receptor-Galpha(s) complex leads to rapid depalmitoylation and inhibition of repalmitoylation of both the receptor and Galpha(s). *J Biol Chem* 1999;**274**:31,014–31,019.

30. Stevens PA, Pediani J, Carrillo JJ, Milligan G. Coordinated agonist regulation of receptor and G protein palmitoylation and functional rescue of palmitoylation-deficient mutants of the G protein G11alpha following fusion to the alpha1b-adrenoreceptor: palmitoylation of G11alpha is not required for interaction with beta*gamma complex. *J Biol Chem* 2001;**276**:35,883–35,890.

31. Mouillac B, Caron M, Bonin H, Dennis M, Bouvier M. Agonist-modulated palmitoylation of beta 2-adrenergic receptor in Sf9 cells. *J Biol Chem* 1992;**267**:21,733–21,737.

32. Loisel TP, Adam L, Hebert TE, Bouvier M. Agonist stimulation increases the turnover rate of beta 2AR-bound palmitate and promotes receptor depalmitoylation. *Biochemistry* 1996;**35**:15,923–15,932.

33. Morello JP, Bouvier M. Palmitoylation: a post-translational modification that regulates signalling from G-protein coupled receptors. *Biochem Cell Biol* 1996;**74**:449–57.

34. Castro-Fernandez C, Janovick JA, Brothers SP, Fisher RA, Ji TH, Conn PM. Regulation of RGS3 and RGS10 palmitoylation by GnRH. *Endocrinology* 2002;**143**:1310–17.

35. Lobo S, Greentree WK, Linder ME, Deschenes RJ. Identification of a Ras palmitoyltransferase in *Saccharomyces cerevisiae*. *J Biol Chem* 2002;**277**:41,268–41,273.

36. Roth AF, Feng Y, Chen L, Davis NG. The yeast DHHC cysteine-rich domain protein Akr1p is a palmitoyl transferase. *J Cell Biol* 2002;**159**:23–8.

37. Mitchell DA, Vasudevan A, Linder ME, Deschenes RJ. Protein palmitoylation by a family of DHHC protein S-acyltransferases. *J Lipid Res* 2006;**47**:1118–27.

38. Fukata M, Fukata Y, Adesnik H, Nicoll RA, Bredt DS. Identification of PSD-95 palmitoylating enzymes. *Neuron* 2004;**44**:987–96.

39. Swarthout JT, Lobo S, Farh L, Croke MR, Greentree WK, Deschenes RJ, Linder ME. DHHC9 and GCP16 constitute a human protein fatty acyltransferase with specificity for H- and N-Ras. *J Biol Chem* 2005;**280**:31,141–31,148.

40. Duncan JA, Gilman AG. Autoacylation of G protein alpha subunits. *J Biol Chem* 1996;**271**:23,594–23,600.

41. Duncan JA, Gilman AG. A cytoplasmic acyl-protein thioesterase that removes palmitate from G protein alpha subunits and p21(RAS). *J Biol Chem* 1998;**273**:15,830–15,837.

42. Makita N, Sato J, Rondard P, Fukamachi H, Yuasa Y, Aldred MA, Hashimoto M, Fujita T, Iiri T. Human G(salpha) mutant causes pseudohypoparathyroidism type Ia/neonatal diarrhea, a potential cell-specific role of the palmitoylation cycle. *Proc Natl Acad Sci USA* 2007;**104**:17,424–17,429.

43. Duncan JA, Gilman AG. Characterization of *Saccharomyces cerevisiae* acyl-protein thioesterase 1, the enzyme responsible for G protein alpha subunit deacylation *in vivo*. *J Biol Chem* 2002;**277**:31,740–31,752.

44. Dunphy JT, Greentree WK, Manahan CL, Linder ME. G-protein palmitoyltransferase activity is enriched in plasma membranes. *J Biol Chem* 1996;**271**:7154–9.

45. Degtyarev MY, Spiegel AM, Jones TL. Palmitoylation of a G protein alpha i subunit requires membrane localization not myristoylation. *J Biol Chem* 1994;**269**:30,898–30,903.

46. Morales J, Fishburn CS, Wilson PT, Bourne HR. Plasma membrane localization of G alpha z requires two signals. *Mol Biol Cell* 1998;**9**:1–14.

47. Evanko DS, Thiyagarajan MM, Wedegaertner PB. Interaction with Gbetagamma is required for membrane targeting and palmitoylation of Galpha(s) and Galpha(q). *J Biol Chem* 2000;**275**:1327–36.

48. Evanko DS, Thiyagarajan MM, Siderovski DP, Wedegaertner PB. Gbeta gamma isoforms selectively rescue plasma membrane localization and palmitoylation of mutant Galphas and Galphaq. *J Biol Chem* 2001;**276**:23,945–23,953.

49. Wang Y, Windh RT, Chen CA, Manning DR. N-Myristoylation and betagamma play roles beyond anchorage in the palmitoylation of the G protein alpha(o) subunit. *J Biol Chem* 1999;**274**:37,435–37,442.

50. Iiri T, Backlund PS, Jones TL, Wedegaertner PB, Bourne HR. Reciprocal regulation of Gs alpha by palmitate and the beta gamma subunit. *Proc Natl Acad Sci USA* 1996;**93**:14,592–14,597.

51. Webb Y, Hermida-Matsumoto L, Resh MD. Inhibition of protein palmitoylation, raft localization, and T cell signaling by 2-bromopalmitate and polyunsaturated fatty acids. *J Biol Chem* 2000;**275**:261–70.

52. Resh MD. Use of analogs and inhibitors to study the functional significance of protein palmitoylation. *Methods* 2006;**40**:191–7.

53. Patterson SI, Skene JH. Inhibition of dynamic protein palmitoylation in intact cells with tunicamycin. *Methods Enzymol* 1995;**250**:284–300.

54. Lawrence DS, Zilfou JT, Smith CD. Structure-activity studies of cerulenin analogues as protein palmitoylation inhibitors. *J Medicinal Chem* 1999;**42**:4932–41.

55. Ducker CE, Griffel LK, Smith RA, Keller SN, Zhuang Y, Xia Z, Diller JD, Smith CD. Discovery and characterization of inhibitors of human palmitoyl acyltransferases. *Mol Cancer Ther* 2006;**5**:1647–59.

56. Biel M, Deck P, Giannis A, Waldmann H. Synthesis and evaluation of acyl protein thioesterase 1 (APT1) inhibitors. *Chemistry* 2006;**12**:4121–43.

57. Deck P, Pendzialek D, Biel M, Wagner M, Popkirova B, Ludolph B, Kragol G, Kuhlmann J, Giannis A, Waldmann H. Development and biological evaluation of acyl protein thioesterase 1 (APT1) inhibitors. *Angew Chem Int Ed Engl* 2005;**44**:4975–80.

58. Roth AF, Wan J, Green WN, Yates JR, Davis NG. Proteomic identification of palmitoylated proteins. *Methods* 2006;**40**:135–42.

59. Drisdel RC, Alexander JK, Sayeed A, Green WN. Assays of protein palmitoylation. *Methods* 2006;**40**:127–34.

60. Roth III AF, Wan J, Bailey AO, Sun B, Kuchar JA, Green WN, Phinney BS, Yates JR, Davis NG. Global analysis of protein palmitoylation in yeast. *Cell* 1998;**125**:1003–13.

61. Dunphy JT, Linder ME. Signalling functions of protein palmitoylation. *Biochim Biophy Acta* 1998;**1436**:245–61.

62. Resh MD. Fatty acylation of proteins: new insights into membrane targeting of myristoylated and palmitoylated proteins. *Biochim Biophys Acta* 1999;**1451**:1–16.

63. Shahinian S, Silvius JR. Doubly lipid-modified protein sequence motifs exhibit long-lived anchorage to lipid bilayer membranes. *Biochemistry* 1995;**34**:3813–22.

64. McLaughlin S, Aderem A. The myristoyl-electrostatic switch: a modulator of reversible protein-membrane interactions. *Trends Biochem Sci* 1995;**20**:272–80.

65. Cadwallader KA, Paterson H, MacDonald SG, Hancock JF. N-terminally myrisoylated ras proteins require palmitoylation or a polybasic domain for plasma membrane localization. *Mol Cell Biol* 1994;**14**:4722–30.

66. Fishburn CS, Herzmark P, Morales J, Bourne HR. Gbetagamma and palmitate target newly synthesized Galphaz to the plasma membrane. *J Biol Chem* 1999;**274**:18,793–18,800.

67. Fishburn CS, Pollitt SK, Bourne HR. Localization of a peripheral membrane protein: G beta gamma targets G alpha(z). *Proc Natl Acad Sci USA* 2000;**97**:1085–90.

68. Wedegaertner PB, Chu DH, Wilson PT, Levis MJ, Bourne HR. Palmitoylation is required for signaling functions and membrane attachment of Gq alpha and Gs alpha. *J Biol Chem* 1993;**268**:25,001–25,008.

69. Wedegaertner PB, Bourne HR, von Zastrow M. Activation-induced subcellular redistribution of $G_s\alpha$. *Mol Biol Cell* 1996;**7**:1225–33.

70. Wise A, Parenti M, Milligan G. Interaction of the G-protein G11alpha with receptors and phosphoinositidase C: the contribution of G-protein palmitoylation and membrane association. *FEBS Letts* 1997;**407**:257–60.

71. Bhattacharyya R, Wedegaertner PB. Galpha 13 requires palmitoylation for plasma membrane localization, Rho-dependent signaling, and promotion of p115-RhoGEF membrane binding. *J Biol Chem* 2000;**275**:14,992–14,999.

72. Hughes TE, Zhang H, Logothetis DE, Berlot CH. Visualization of a functional Galpha q-green fluorescent protein fusion in living cells. Association with the plasma membrane is disrupted by mutational activation and by elimination of palmitoylation sites, but not be activation mediated by receptors or AlF4. *J Biol Chem* 2001;**276**:4227–35.

73. Pedone KH, Hepler JR. The importance of N-terminal polycysteine and polybasic sequences for G14alpha and G16alpha palmitoylation, plasma membrane localization, and signaling function. *J Biol Chem* 2007;**282**:25,199–25,212.

74. Degtyarev MY, Spiegel AM, Jones TLZ. The G protein α_s subunit incorporates [^{3}H]palmitic acid and mutation of cysteine-3 prevents this modification. *Biochemistry* 1993;**32**:8057–61.

75. Hepler JR, Biddlecome GH, Kleuss C, Camp LA, Hofmann SL, Ross EM, Gilman AG. Functional importance of the amino terminus of Gq alpha. *J Biol Chem* 1996;**271**:496–504.

76. Huang C, Duncan JA, Gilman AG, Mumby SM. Persistent membrane association of activated and depalmitoylated G protein alpha subunits. *Proc Natl Acad Sci USA* 1999;**96**:412–17.

77. Jones TL, Gutkind JS. Galpha12 requires acylation for its transforming activity. *Biochemistry* 1998;**37**:3196–202.

78. Chen C, Seow KT, Guo K, Yaw LP, Lin SC. The membrane association domain of RGS16 contains unique amphipathic features that are conserved in RGS4 and RGS5. *J Biol Chem* 1999;**274**:19,799–19,806.

79. Hiol A, Davey PC, Osterhout JL, Waheed AA, Fischer ER, Chen CK, Milligan G, Druey KM, Jones TL. Palmitoylation regulates regulators of G-protein signaling (RGS) 16 function. I. Mutation of amino-terminal cysteine residues on RGS16 prevents its targeting to lipid rafts and palmitoylation of an internal cysteine residue. *J Biol Chem* 2003;**278**:19,301–19,308.

80. Levis MJ, Bourne HR. Activation of the α subunit of G$_s$ in intact cells alters its abundance, rate of degradation, and membrane avidity. *J Cell Biol* 1992;**119**:1297–307.

81. Hansen SH, Casanova JE. Gsα stimulates transcytosis and apical secretion in MDCK cells through cAMP and protein kinase A. *J Cell Biol* 1994;**126**:677–87.

82. Negishi M, Hashimoto H, Ichikawa A. Translocation of α subunits of stimulatory guanine nucleotide-binding proteins through stimulation of the prostacyclin receptor in mouse mastocytoma cells. *J Biol Chem* 1992;**267**:2367–9.

83. Ransnäs LA, Svoboda P, Jasper JR, Insel PA. Stimulation of β-adrenergic receptors of S49 lymphoma cells redistributes the α subunit of the stimulatory G protein between cytosol and membranes. *Proc Natl Acad Sci USA* 1989;**86**:7900–3.

84. Yu JZ, Rasenick MM. Real-time visualization of a fluorescent G(alpha)(s): dissociation of the activated G protein from plasma membrane. *Mol Pharmacol* 2002;**61**:352–9.

85. Allen JA, Yu JZ, Donati RJ, Rasenick MM. Beta-adrenergic receptor stimulation promotes G alpha s internalization through lipid rafts: a study in living cells. *Mol Pharmacol* 2005;**67**:1493–504.

86. Thiyagarajan MM, Bigras E, Van Tol HH, Hebert TE, Evanko DS, Wedegaertner PB. Activation-induced subcellular redistribution of G alpha(s) is dependent upon its unique N-terminus. *Biochemistry* 2002;**41**:9470–84.

87. Hynes TR, Mervine SM, Yost EA, Sabo JL, Berlot CH. Live cell imaging of Gs and the beta2-adrenergic receptor demonstrates that both alphas and beta1gamma7 internalize upon stimulation and exhibit similar trafficking patterns that differ from that of the beta2-adrenergic receptor. *J Biol Chem* 2004;**279**:44,101–44,112.

88. Sugama J, Yu JZ, Rasenick MM, Nakahata N. Mastoparan inhibits beta-adrenoceptor-G(s) signaling by changing the localization of Galpha(s) in lipid rafts. *Cell Signal* 2007;**19**:2247–54.

89. Rocks O, Peyker A, Kahms M, Verveer PJ, Koerner C, Lumbierres M, Kuhlmann J, Waldmann H, Wittinghofer A, Bastiaens PI. An acylation cycle regulates localization and activity of palmitoylated Ras isoforms. *Science* 2005;**307**:1746–52.

90. Goodwin JS, Drake KR, Rogers C, Wright L, Lippincott-Schwartz J, Philips MR, Kenworthy AK. Depalmitoylated Ras traffics to and from the Golgi complex via a nonvesicular pathway. *J Cell Biol* 2005;**170**:261–72.

91. Chisari M, Saini DK, Kalyanaraman V, Gautam N. Shuttling of G protein subunits between the plasma membrane and intracellular membranes. *J Biol Chem* 2007;**282**:24,092–24,098.

92. Drenan RM, Doupnik CA, Boyle MP, Muglia LJ, Huettner JE, Linder ME, Blumer KJ. Palmitoylation regulates plasma membrane-nuclear shuttling of R7BP, a novel membrane anchor for the RGS7 family. *J Cell Biol* 2005;**169**:623–33.

93. Martemyanov KA, Yoo PJ, Skiba NP, Arshavsky VY. R7BP, a novel neuronal protein interacting with RGS proteins of the R7 family. *J Biol Chem* 2005;**280**:5133–6.

94. Acconcia F, Ascenzi P, Bocedi A, Spisni E, Tomasi V, Trentalance A, Visca P, Marino M. Palmitoylation-dependent estrogen receptor alpha membrane localization: regulation by 17beta-estradiol. *Mol Biol Cell* 2005;**16**:231–7.

95. Li L, Haynes MP, Bender JR. Plasma membrane localization and function of the estrogen receptor alpha variant (ER46) in human endothelial cells. *Proc Natl Acad Sci USA* 2003;**100**:4807–12.

96. Rai D, Frolova A, Frasor J, Carpenter AE, Katzenellenbogen BS. Distinctive actions of membrane-targeted versus nuclear localized estrogen receptors in breast cancer cells. *Mol Endocrinol* 2005;**19**:1606–17.

97. Wiedmer T, Zhao J, Nanjundan M, Sims PJ. Palmitoylation of phospholipid scramblase 1 controls its distribution between nucleus and plasma membrane. *Biochemistry* 2003;**42**:1227–33.

98. Hepler JR. R7BP: a surprising new link between G proteins, RGS proteins, and nuclear signaling in the brain. *Sci STKE* 2005;**2005**:pe38.

99. Tu Y, Wang J, Ross EM. Inhibition of brain Gz GAP and other RGS proteins by palmitoylation of G protein alpha subunits. *Science* 1997;**278**:1132–5.

100. Osterhout JL, Waheed AA, Hiol A, Ward RJ, Davey PC, Nini L, Wang J, Milligan G, Jones TL, Druey KM. Palmitoylation regulates regulator of G-protein signaling (RGS) 16 function. II. Palmitoylation of a cysteine residue in the RGS box is critical for RGS16 GTPase accelerating activity and regulation of Gi-coupled signalling. *J Biol Chem* 2003;**278**:19,309–19,316.

101. Ni J, Qu L, Yang H, Wang M, Huang Y. Palmitoylation and its effect on the GTPase-activating activity and conformation of RGS2. *Intl J Biochem Cell Biol* 2006;**38**:2209–18.

102. Tu Y, Woodson J, Ross EM. Binding of regulator of G protein signaling (RGS) proteins to phospholipid bilayers. Contribution of location and/or orientation to Gtpase-activating protein activity. *J Biol Chem* 2001;**276**:20,160–20,166.

103. Kleuss C, Krause E. Gα$_s$ is palmitoylated at the N-terminal glycine. *EMBO J* 2003;**22**:826–32.

Chapter 53

The Influence of Intracellular Location on Function of Ras Proteins

Jodi McKay[1] and Janice E. Buss[2]

[1]Department of Biochemistry, Biophysics and Molecular Biology, Iowa State University, Ames, Iowa
[2]Department of Pharmaceutical Sciences, University of Kentucky, Lexington, Kentucky

INTRODUCTION

There are two ways that Ras proteins might come to be at multiple locations within a cell. The protein might traverse several routes, staying in one place temporarily, only to depart, perhaps casually or perhaps from the demands of a regulatory agent. Each isoform (H-, N- or K-Ras) would likely travel a separate route. This itinerant format predicts that the intracellular pool of each isoform would eventually have access to all points, and all signaling partners, on that route. Thus, understanding the vehicles, regulation of departure, and speed of Ras movement become key to understanding signaling in this scenario. A second, and not exclusive, possibility is that a significant portion of an isoform might reside more or less permanently in one location. In this case, identifying the partners that detain Ras and how they oversee its access to signaling partners becomes important. Multiple studies give us glimpses of both scenarios, and show that experimental manipulation of the location of introduced Ras proteins is a tactic that can alter Ras signaling. The challenge ahead is to find ways to apply these driving lessons to endogenous Ras proteins and learn if this helps steer Ras away from oncogenic destinations. The experiments that begin to answer these questions are described below.

DOES LOCATION ACTUALLY INFLUENCE RAS SIGNALING?

Two seminal papers have now laid a solid foundation under the concept that normal, endogenous Ras proteins, positioned at different intracellular locations, really do engage distinct signaling pathways, and in doing so produce unique biological outcomes.

Studies with the fission yeast *S. pombe* show that the single Ras1 protein (an ortholog of mammalian Ras) has only two downstream effectors – Byr2 and Scd1 – and interacts with them in two separate locations. Byr2 is part of a mitogen-activated protein kinase (MAPK) signal transduction cascade and controls yeast mating; Scd1 is part of the Cdc42 signal transduction cascade and regulates cell morphology. Ras1 is palmitoylated, and is found on both the plasma membrane and on internal membranes, similar to the distribution of palmitoylated mammalian Ras isoforms [1]. The important finding is that Ras1 which is at the plasma membrane signals to Byr2 and only to Byr2, not Scd1; Ras1 that is internal displays a reciprocal elitism and signals only to Scd1, ignoring Byr2 [1]. Thus, the distribution of Ras1 decides the allocation of signal between these two pathways.

In mammalian cells, T lymphocytes have provided an elegant story of a role of Ras location in life and death decisions. This story starts with the finding that antigenic peptides that differ only slightly in their affinity for the T cell receptor (TCR) can trigger startling differences in cellular outcome from their binding. Peptides at the less potent side of this TCR affinity boundary trigger positive selection (that is, the cells survive), while peptides that bind a bit more tightly trigger negative selection (i.e., the cells die) [2]. What is especially remarkable is that this small change in affinity builds to a large difference in the strength, kinetics, and *location* of signals that pulse through the TCR-driven cascades. The easygoing cascade of positive selection activates endogenous Ras only on Golgi membranes, while the robust signals associated with negative selection trigger Ras activation on plasma membrane as well as Golgi [2].

MODIFICATIONS BY ENDOMEMBRANE ENZYMES – NEW OPPORTUNITIES TO ABORT RAS TRAFFICKING?

Ras proteins are among the many cytosolic proteins with a sequence known as a CAAX box as their final four amino acid residues [3]. The CAAX box serves as a recognition signal for farnesyltransferase (FTase). Attachment of the farnesyl lipid initiates Ras membrane binding. In fact, the FTase may itself escort its now modified substrate to its first membrane encounter [4], which happens to be on the vast surface of the endoplasmic reticulum ER. There, two additional enzymes produce crucial C-terminal modifications before Ras proteins begin their separate journeys. The protease Ras Converting Enzyme 1 (RCE 1) recognizes the farnesylated CAAX box and removes the final three amino acids, creating a new C-terminus at the farnesylated cysteine that is the "C" of the CAAX [5]. This farnesyl cysteine then serves as the recognition site for the third enzyme, Isoprenylcysteine carboxy methyl transferase (Icmt). Icmt attaches a methyl group at the new C-terminus [6]. These three primary modifications occur on all isoforms of Ras in humans, and each eukaryotic Ras protein that has been studied. K-Ras-4B, the most abundant human K-Ras variant, does not undergo additional modifications, and, as described below, behaves quite differently (and more mysteriously) in its intracellular trafficking mechanism than other Ras isoforms. Palmitoyl groups, placed on cysteines located just N-terminal to the CAAX box, further modify the H-Ras, N-Ras, and K-Ras 4A isoforms, and enable these proteins to board the cytosolic face of both exocytic and endocytic vesicles. On H-Ras, cysteines 181 and 184 are palmitoylated, but, interestingly, the palmitates are not equivalent functionally. Cysteine 181 seems to play the dominant role in trafficking, while cysteine 184 is needed for distribution of H-Ras between microdomains of the plasma membrane [7]. The enzyme(s) responsible for palmitoylation had long eluded isolation, but now appear to have been identified as members of a family of acyl transferase enzymes with a distinctive DHHC motif [8, 9].

This quartet of modifying enzymes holds great potential for development of pharmaceutics [10, 11]. The farnesyl transferase was the first to be targeted both experimentally and clinically, and, after some initial fanfare and inevitable disappointments, is gathering momentum as a cancer chemotherapy [12] (see Chapter 222 of Handbook of Cell Signaling, Second Edition. However, it should be noted that these enzymes modify a large number of targets other than Ras proteins, and that, certainly for farnesyl transferase inhibitors, inhibition of prenylation of these other targets may be the basis of therapeutic action [12]. Nevertheless, the Rce1 endoprotease seems to be needed to enable Ras proteins to bind to membranes correctly, and has less effect on Rho family GTPases [13]. By decreasing Icmt methyl transferase expression using the *Cre-loxP* system, Bergo

and colleagues demonstrated that transformation by K-Ras was decreased [14, 15]. Methylation also appears to influence subcellular location of Ras proteins [16]. The palmitoyl transferase enzymes are much less studied, but are enticing targets for inhibition, as the acyl modification must occur repeatedly [17–19]. This offers multiple opportunities for prohibiting membrane binding, in contrast to the one-chance-is-all-you-get to prevent farnesyl addition.

RAS PROTEINS BEGIN THEIR MARCH TOWARD THE CELL SURFACE

Our appreciation of the role of intracellular residency and trafficking in the lifecycle of Ras proteins (Figure 53.1) grew from a series of papers that demonstrated that H-Ras and N-Ras proteins can be observed on endomembranes and, using GFP-tagged proteins in live cells, on vesicles which traffic outward to the cell surface [20–23]. Of special interest were the discoveries that, in Jurkat T cells, a substantial amount of N-Ras was located, at steady state, on Golgi membranes, and could be both activated on and transmit signals from that location [22, 24]. Thus, in this situation, N-Ras appears to preferentially accumulate on Golgi membranes. Whether this is because some portion of N-Ras is captured and held there by an N-Ras-specific partner, or because it revisits these membranes frequently, is not known.

Re-visiting internal membranes appears to be a fundamental part of trafficking for H-Ras and N-Ras, and is based on the cyclic removal and replacement of their palmitoyl groups (Figure 53.1) [9, 18, 25]. Outward vesicular transport requires the protein to be palmitoylated, and delivers the protein to the plasma membrane. There, H-Ras and N-Ras can encounter (unknown) deacylating enzymes that remove their palmitates and destabilize their membrane binding. If a Ras protein is rapidly re-palmitoylated by an acyltransferase at the plasma membrane [26], it will remain at that location. However, failing that rescue, H-Ras and N-Ras detach from the plasma membrane and appear to wander briefly before encountering a palmitoyltransferase on the ER or Golgi [8, 18, 19]. There they become "kinetically trapped" [27] on endomembranes, and can rejoin the vesicular pathway again. This cycle of deacylation/reacylation occurs for both active and inactive forms of Ras, but there may be differences in the rate or frequency of the cycle for GDP-bound (slower deacylation) and GTP-bound (faster deacylation) forms of H-Ras [17]. This clever repositioning provides one mechanism to replenish internal membranes with these Ras isoforms. Endocytosis, described below, may be another.

However, H-Ras and N-Ras do not always utilize vesicular vehicles. Fluorescence microscopy and photobleaching studies indicate that non-vesicular mechanisms transport H-Ras and N-Ras from plasma membrane or ER to Golgi [19]. Even the initial outward journey of newly synthesized H-Ras appears to be largely independent of

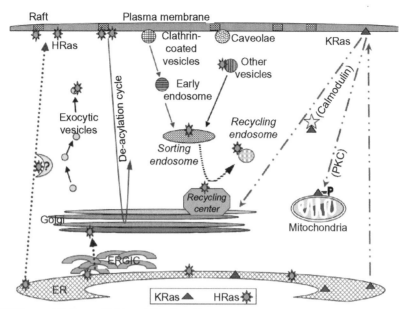

FIGURE 53.1 Pathways of H-Ras and K-Ras trafficking within the cell.
Left: Lipid modified H-Ras travels toward the plasma membrane via either a vesicular or a recently identified non-conventional, possibly non-vesicular, trafficking mechanism (dotted line). In the vesicular transport mechanism (dark arrows), H-Ras travels to the ER/Golgi Intermediate Compartment (ERGIC), then through the Golgi, and outward to the plasma membrane. Novel modes of trafficking ("rasosomes," moon shaped) have been suggested as a possible mechanism behind non-vesicular trafficking. At the plasma membrane, H-Ras preferentially partitions into lipid rafts when inactive, or cholesterol-independent regions of the bilayer when activated. H-Ras at the plasma membrane can lose its palmitoyl groups and participate in the deacylation cycle that relocates H-Ras to the Golgi (solid arrow on the left). **Center:** The mechanisms of endocytosis for H-Ras are only partially characterized. Clathrin-coated vesicles, caveolae, and other vesicles are all possibilities. Vesicles from multiple pathways merge at the sorting endosome, some depart for late endosomes and lysosomes (*not shown*) and others, carrying H-Ras, move to the endosomal recycling center and are trafficked back out to the plasma membrane for the second time via recycling endosomes. **Right:** K-Ras moves out to the plasma membrane via an unknown, but non-vesicular mechanism. K-Ras dissociates from the plasma membrane if its C-terminal polybasic region interacts with Ca^{2+}/calmodulin or is phosphorylated by PKC. Phospho-K-Ras translocates to the outer surface of mitochondria, where it interacts with Bcl-X_L. A mechanism for return of K-Ras to the plasma membrane is currently uncharacterized.

COP I- or COP II-mediated vesicular transport (Figure 53.1) [28, 29]. More specifically, when cells co-express H-Ras and a dominant negative form of Sar1, H-Ras is still able to localize to the plasma membrane. Further, when cells are treated with Brefeldin A, a pharmacological reagent that disturbs the Golgi stacks, H-Ras still localizes to the plasma membrane. This indicates that this trafficking mechanism is initiated before H-Ras reaches the Golgi, and is independent of a functional Golgi. Additionally, H-Ras trafficking continues even when microtubules are destroyed by nocodazole [29]. In cells treated with these inhibitors, a portion of H-Ras appears to enter the traditional exocytic pathway and reaches, and even accumulates, in the ERGIC (ER – Golgi intermediate compartment) [29]. However, although the alternative pathway of H-Ras transport continues to shuttle protein to the plasma membrane, the protein that entered the traditional pathway of vesicular transport before the inhibitors were added cannot de-plane and access that alternative transport system [29]. Thus, the choice of which mechanism of transport to use is a serious, possibly irreversible, decision. A similar, non-conventional secretory route of transport for yeast Ras has also been documented using genetic means to disrupt vesicular trafficking [30]. Why H-Ras simultaneously uses two methods to reach the plasma membrane is

unknown, but this duality certainly complicates hopes for interrupting transport. Caution is also warranted in studies that utilize GFP-tagged versions of H-Ras. GFP-H-Ras was found far more frequently on endomembranes than was H-Ras lacking any tags [29].

While an alternative, possibly non-vesicular, trafficking mechanism is a novel idea for H-Ras and N-Ras, non-vesicular movement had been expected for K-Ras [23]. K-Ras4B is not palmitoylated, and, once its processing by the ER-located protease and methyl transferase is complete, it very rapidly appears at the plasma membrane, with no visible vesicular intermediates (Figure 53.1) [20, 31]. The unusually lysine-rich C-terminus of K-Ras is involved in this trafficking [32, 33], but the exact mechanism behind K-Ras transport is still elusive.

RAS ACTIVATION ON ER AND GOLGI MEMBRANES

The major shift in our understanding that has come from the study of these trafficking events has been that Ras proteins are not docile travelers; in response to mitogens, they can actively signal from both cell surface and internal membranes. Study of subcellular signaling has required development of Ras proteins targeted to specific intracellular locations, and tagging of Ras and its partners with fluorescent proteins to enable real-time, live cell microscopy. The resulting images are both beautiful and compelling.

Although little endogenous Ras is found on ER membranes, artificially targeted H-Ras proteins demonstrate that this enormous intracellular membrane surface can support interactions of H-Ras and effectors, and give rise to biologically important signals. H-Ras can be targeted to the ER in two ways. When H-Ras cysteines in positions 181 and/or 184 are mutated to serines, so that palmitoylation cannot occur, the non-palmitoylated H-Ras is distributed between the ER and cytosol; a transmembrane chimera of activated H-RasV12 with N-terminal residues from the avian infectious bronchitis virus M protein is also localized on the ER [7, 21, 34]. Both styles of ER-restricted protein can bind an activation marker (the Ras binding domain of c-Raf; RBD) that can be used to indicate when the Ras protein has become GTP-bound. The non-palmitoylated version can also be activated after external treatment of the cells with EGF [21]. A portion of endogenous Ras can localize naturally to Golgi membranes, and N-Ras is particularly abundant at that location in T cells; H-Ras can be artificially tethered to the Golgi via an N-terminal KDEL receptor transmembrane sequence [24, 34]. Activation of Golgi-localized Ras proteins has been visualized by accumulation of a green fluorescent protein-tagged version of RBD (GFP-RBD) on Golgi membranes.

How can Ras that is located on endomembranes be activated by stimuli that are external? Ras activation is triggered by guanine nucleotide exchange factors (GEFs). Thus, in order for Ras that is present on an internal membrane to become GTP-bound, a GEF must either be itself activated at the plasma membrane and move to an intracellular location, or it must be pre-positioned on the membrane and become activated by a movable agent, such as Ca^{2+} [35]. Two classes of Ras GEFs meet these criteria. The RasGRF1 and RasGRF2 proteins are activated via signals from G-protein-coupled receptors or by stimuli that release calcium from internal stores in the ER, and thereby can activate H-Ras that is present on the ER [36]. For H-Ras that is positioned on Golgi membranes, the RasGRP1 protein of the separate GEF family of RasGRP proteins can translocate to that site in response to stimulation by EGF, diacylglycerol, phorbol esters, or stimulation of PLCγ [22, 37]. These treatments can activate H-Ras on Golgi membranes [22, 37].

The kinetics of Ras signaling from internal membranes are a bit slower and prolonged, compared to signal events that occur on the plasma membrane [38]. In COS-1 cells that co-express H-Ras and GFP-RBD, treatment with EGF causes the immediate and expected recruitment of the GFP-RBD to the H-Ras-GTP that is located on the plasma membrane. The H-Ras that is present on the Golgi waits several minutes before becoming GTP-bound (probably via the calcium-mediated activation of RasGRP1), and

retains this GTP for over 30 minutes. Meanwhile, the H-Ras at the plasma membrane gradually hydrolyzes its GTP, in part from the interesting reciprocity of calcium-dependent activation of the GTPase activating protein, CAPRI [21, 22]. The deacylation cycle of H-Ras may also play a role in the slow and sustained activation of Ras on the Golgi, by re-distributing more H-Ras (perhaps still in the GTP form) from the plasma membrane to that compartment. Additional mechanisms for sustaining internal H-Ras in its GTP-bound form are also possible, such as binding of Golgi-localized H-Ras-GTP to its also Golgi-localized effector, PLC-ε, which would then produce a positive cycle of diacylglycerol and IP3/calcium stimulation of RasGRP1 [39].

There are good indications that activation of H-Ras on either the ER or the Golgi produces a unique portfolio of signals from its effector pathways [40–42]. How such signals are then integrated with the other signals that emanate from the plasma membrane to produce either normal or oncogenic Ras function is not known; nor is it known in what way simply being located on a Golgi or ER membrane physically translates into favored interactions between Ras and the ERK pathway [21, 34]. Dissecting the partnerships and signals arising from a subcellular signaling pathway remains a significant challenge, as artificial Ras constructs targeted to specific compartments, especially those with N-terminal transmembrane domains, are of course altered structurally. The limited structural data available on Ras proteins that are tethered to lipid membranes (see below) suggest that membrane binding may be influenced by the GTP binding domain [43]. An indication of GTP-dependent changes in membrane interaction is also seen with palmitate turnover studies [17]. Whether differences in membrane juxtaposition contribute to differences in effector interactions will be an important future experiment.

The special localization of N-Ras on Golgi membranes in T cells offers the best current model to study endomembrane-specific signaling; and, as discussed above, the biological outcome is dramatic. Stimulation of the T cell antigen receptor with low-affinity peptides triggers Ras activation on the Golgi, while exposure to antigenic peptides with only slightly higher affinity, or with simultaneous engagement of LFA-1 (lymphocyte function-associated antigen-1), triggers Ras activation on both plasma membrane and Golgi [2, 24, 44]. This scenario suggests that specific effector pathways are triggered uniquely when only Golgi-localized Ras is activated, and that these partners underlie positive selection – a hypothesis that will surely be tested.

DESTINATION CELL SURFACE: RAS PROTEINS DISTRIBUTE AMONG SEVERAL PLASMA MEMBRANE DOMAINS

Because Ras-mediated signal transduction produces multiple, and sometimes opposing, biological and molecular responses, proper transmission of Ras mediated signal requires both specificity and precision [45]. Originally it was believed that the majority of Ras signal transduction took place at the plasma membrane. The studies described above demonstrate that this was a very incomplete description of Ras signal transduction. Ras moves between and resides on the separate membranes of intracellular organelles, and it also appears to move within minute subdomains of the plasma membrane. Especially for distribution among these microdomains, GTP loading of Ras proteins plays an important role [46].

Defining the subdomains of the plasma membrane between which Ras proteins partition has been technically daunting. In resting cells these regions appear to be too small to observe by optical microscopy, and are dynamic, probably transient, arenas of lipids and proteins. There appear to be broad areas of generic membrane in which areas of special lipid composition are embedded. These special areas have many names and defining properties (such as being enriched in glycosphingolipids and cholesterol, and resistant to detergents), and there are several kinds of subdomain [47, 48], but the most well known are popularly called "rafts" (Figure 53.1). Biochemical isolation of such tenuous entities has proven contentious. There is, however, agreement that various stimuli cause these areas to cluster and stabilize, bringing proteins into proximity to initiate the signaling event [49, 50].

The Ras isoforms differ from one another in their preference for membrane microdomains. K-Ras4B, having arrived at the plasma membrane direct from the ER, settles into areas of membrane that are not sensitive to cholesterol extraction, and abides there in times of rest as well as after events of stimulation [51]. H-Ras, whether deposited on the plasma membrane by an exocytic vesicle or the non-conventional (but unknown) mechanism, appears to reside in a cholesterol-dependent "raft" domain when inactive, but may move to another, cholesterol-independent domain, different than the type occupied by K-Ras4B, when activated [41, 46, 52]. Additional studies suggest that both H-Ras and K-Ras form nanoclusters within the membrane areas where they are located. The number of K-Ras clusters increases when cells are exposed to the mitogen EGF [53]. Active K-Ras4B can be found in regions where the outer boundaries of the nanoclusters are set by the actin cytoskeleton [54]. Recent single-particle tracking studies indicate that GDP-bound H-Ras diffuses randomly until it is activated, at which time it undergoes a short period of immobilization [55, 56]. This immobile population contains GTP-bound H-Ras, and is not affected by cholesterol extraction, indicating that it is no longer in a "raft" [55].

These movements and clustering of Ras proteins are a crucial phase of signaling [45, 47, 57]. For the Raf-1/MEK/ERK signaling pathway, both *in silico* modeling and experimental data suggest that the K-Ras clusters generate a step-like response from exposure of cells to a wide range of EGF concentrations [53]. This allows K-Ras clusters to act as "nanoswitches" that control signaling output. Future

experiments will need to address whether H-Ras clusters in cholesterol-dependent membrane domains behave similarly.

There is an additional form of clustered H-Ras (and N-Ras). Nanoparticle-bound forms ("rasosomes") of these palmitoylated Ras proteins have been observed to move rapidly and randomly through the cytosol (Figure 53.1) [58]. The connection between these particles, signaling, and perhaps the alternative pathway (described above) by which H-Ras first accesses the plasma membrane [29] remain to be explored.

NEW DESTINATIONS – MITOCHONDRIA

Although its initial outward voyage to the plasma membrane is still enigmatic, K-Ras performs a newly discovered, and unexpected, inward journey that is initiated by phosphorylation and calcium signals. K-Ras has multiple lysine residues adjacent to its farnesylated C-terminus, and these lysines are crucial for K-Ras membrane association. Serine 181 that is within this very basic region can be phosphorylated by protein kinase C (PKC), resulting in release of K-Ras from the plasma membrane [59]. This is the first clear example of regulated membrane dissociation of a Ras protein through a mechanism that is not derived from its state of lipid modification, and has been code-named the "farnesyl-electrostatic switch." It also implies that K-Ras signaling from the plasma membrane could be controlled by PKC. Of greater interest however, is the fate of the released phospho-K-Ras. It does not simply slide into the cytoplasm; it moves into the cell and associates with mitochondria (Figure 53.1), where it interacts with the apoptosis-promoting protein, Bcl-X_L [59, 60]. Yeast Ras has also been detected on mitochondria [61]. The PKC agonist bryostatin-1 inhibits growth and causes apoptosis of tumor cells in a K-Ras Ser181-dependent manner, and K-Ras with a mutation that mimics Ser181 phosphorylation causes apoptosis through a mechanism that requires Bcl-X_L [59]. These results suddenly open the door to the idea that membrane interaction of K-Ras may be controllable, and may in fact change its fundamental biological signal from growth promotion to cell death.

Neural cells seem also to make special use of K-Ras trafficking. Early studies in fibroblasts had indicated that calmodulin could bind to K-Ras, but not H-Ras, particularly when K-Ras was GTP-bound, and that this binding inhibited Raf/ERK signaling [62]. In hippocampal neurons, which contain large amounts of calmodulin, glutamate-stimulated neural activity causes K-Ras to be released from the plasma membrane and to relocate to endosomes and Golgi membranes (Figure 53.1). Calcium/calmodulin complexes formed during the neural stimulation bind to the prenylated, lysine-rich region of the K-Ras C-terminus [63]. Both that binding and the journey from plasma membrane to interior are calcium-sensitive, and reversible. However, a mechanism for return of K-Ras to the plasma membrane is currently uncharacterized. Thus, K-Ras can be extracted from the plasma membrane and directed to interior membranes by two powerful signaling mechanisms. This is also notable because H-Ras appears not to be subject to these relocation directives, and should remain at the plasma membrane, thereby spatially separating these Ras isoforms and their signals.

OLD DESTINATIONS – ENDOSOMES

K-Ras may not stopover on Golgi membranes on its way from the ER out to the cell surface, but the new studies above show that it can bind mitochondrial, endosomal, or Golgi membranes after having visited the plasma membrane. H-Ras, on the other hand, has often been seen lingering on endosomes [64]. There has been a great deal of work that now shows convincingly that endosomes are yet another membrane platform from which signaling by internalized cell surface receptors can take place [65, 66]. To what extent and through what partners Ras proteins can signal from endosomal membranes is still an open question.

The amount of H-Ras on endosomes is definitely higher than the amount of K-Ras [64, 67]. Furthermore, H-Ras is dependent on dynamin-dependent endocytosis for its ability to fully activate Raf-1 signaling [67]. Raf-1 has been found on endosomes internalized with the insulin or EGF receptors [64, 68, 69]. However, the identity of the endosomal pathway used for H-Ras internalization has not been firmly established (Figure 53.1), and it remains unclear whether H-Ras utilizes the same endosomes as its Raf effector, or whether it enters separately and joins sorting endosomes that contain its signaling partners [64, 69]. GTP-bound H-Ras can be found on these endosomes, so the possibility exists that Ras signaling may occur on endosomal membranes [64, 68]. However, GDP-bound H-Ras is also found on endosomes, indicating that internalization of H-Ras may be, in part, constitutive [64, 70]. An interesting new finding is the discovery that H-Ras (but not K-Ras), in both GTP-bound and GDP-bound forms, can be ubiquitinated [71]. This modification stabilizes H-Ras binding to endosomes, but appears to decrease its ability to recruit Raf-1 to membranes and activate the ERK pathway [71].

Despite the uncertainty of what pathways are used for H-Ras to enter the cell, H-Ras has already been seen returning to the plasma membrane from the pericentriolar endocytic recycling center [72] (Figure 53.1). The activity of Rab5 and Rab11 proteins is needed for this cycling. Even at this late stage, H-Ras may still be able to signal, as H-Ras on Rab11-positive endosomes bound GFP-RBD, indicating that it was in the GTP-bound form [72]. The key element needed now is a better understanding of the early steps of both clathrin-dependent and clathrin-independent internalization pathways to identify the classes of endosomes that merge to form sorting endosomes.

A REALITY CHECK – THE FIRST STRUCTURES OF LIPID-MODIFIED RAS ON LIPID MEMBRANES

Despite the importance of the membrane interactions of the lipidated C-terminal domain of Ras proteins, there have been no complete structures available of this region. Three-dimensional structures of several Ras proteins in isolation or in concert with a bound effector protein have been solved, but the C-terminal domain was either absent or unstructured, not lipidated, and not associated with a membrane [73]. This has left a great imaginative hole in our vision of Ras structure. The enzymatic hurdles to producing a fully lipidated and processed Ras protein are slowly being overcome, and the complexity of modeling a protein interacting with a lipid bilayer is yielding to impressive computational approaches. The structure of lipidated peptides that represent the C-terminal domain in association with artificial lipid mono- and bilayers has been studied by molecular dynamics simulations and solid state NMR [74, 75]. A combination of computational simulation and experimental validation has now provided the first, tempting model for full-length, lipidated H-Ras on a lipid bilayer [43]. The model suggests that H-Ras can assume two modes of orientation against the bilayer, and that both the GTP-binding body of the protein and the hypervariable domain, just upstream of the lipidated region, may also contact the membrane [43]. This meshes nicely with the model built from biological experiments that GTP binding affects association of H-Ras with "raft" and non-raft areas of the membrane [52].

These glimpses will soon contribute to the first clear vision of a Ras protein in action in its native membrane environment – and, if the experiments described above are any indication, that vision will need to be a high-speed movie.

REFERENCES

1. Onken B, Wiener H, Philips MR, Chang EC. Compartmentalized signaling of Ras in fission yeast. *Proc Natl Acad Sci USA* 2006;**103**:9045–50.
2. Daniels MA, Teixeiro E, Gill J, et al. Thymic selection threshold defined by compartmentalization of Ras/MAPK signalling. *Nature* 2006;**444**:724–9.
3. Wright LP, Philips MR. CAAX modification and membrane targeting of Ras. *J Lipid Res* 2006;**47**:883–91.
4. Taylor JS, Reid TS, Terry KL, Casey PJ, Beese LS. Structure of mammalian protein geranylgeranyltransferase type-1. *EMBO J* 2003; **22**:5963–74.
5. Schmidt WK, Tam A, Fujimura-Kamada K, Michaelis S. Endoplasmic reticulum membrane localization of Rce1p and Ste24p, yeast proteases involved in carboxyl-terminal CAAX protein processing and amino-terminal a-factor cleavage. *Proc Natl Acad Sci* 1998;**95**:11,175–11,180.
6. Dai Q, Choy E, Chiu V, et al. Mammalian prenylcysteine carboxyl methyltransferase is in the endoplasmic reticulum. *J Biol Chem* 1998;**273**:15,030–15,034.

7. Roy S, Plowman S, Rotblat B, et al. Individual palmitoyl residues serve distinct roles in H-Ras trafficking, microlocalization, and signaling. *Mol Cell Biol* 2005;**25**:6722–33.

8. Swarthout JT, Lobo S, Farh L, et al. DHHC9 and GCP16 constitute a human protein fatty acyltransferase with specificity for H- and N-Ras. *J Biol Chem* 2005;**280**:31,141–31,148.

9. Linder ME, Deschenes RJ. Palmitoylation: policing protein stability and traffic. *Nat Rev Mol Cell Biol* 2007;**8**:74–84.

10. Winter-Vann AM, Kamen BA, Bergo MO, et al. Targeting Ras signaling through inhibition of carboxyl methylation: an unexpected property of methotrexate. *Proc Natl Acad Sci* 2003;**100**:6529–34.

11. Winter-Vann AM, Casey PJ. Post-prenylation-processing enzymes as new targets in oncogenesis. *Nat Rev Cancer* 2005;**5**:405–12.

12. Basso AD, Kirschmeier P, Bishop WR. Thematic review series: lipid posttranslational modifications. Farnesyl transferase inhibitors. *J Lipid Res* 2006;**47**:15–31.

13. Michaelson D, Ali W, Chiu VK, et al. Postprenylation CAAX processing is required for proper localization of Ras but not Rho GTPases. *Mol Biol Cell* 2005;**16**:1606–16.

14. Bergo MO, Gavino BJ, Hong C, et al. Inactivation of Icmt inhibits transformation by oncogenic K-Ras and B-Raf. *J Clin Invest* 2004;**113**:539–50.

15. Clarke S, Tamanoi F. Fighting cancer by disrupting C-terminal methylation of signaling proteins. *J Clin Invest* 2004;**113**:513–15.

16. Chiu VK, Silletti J, Dinsell V, et al. Carboxyl methylation of Ras regulates membrane targeting and effector engagement. *J Biol Chem* 2004;**279**:7346–52.

17. Baker TL, Zheng H, Walker J, Coloff JL, Buss JE. Distinct rates of palmitate turnover on membrane-bound cellular and oncogenic H-Ras. *J Biol Chem* 2003;**278**:19,292–19,300.

18. Rocks O, Peyker A, Kahms M, et al. An acylation cycle regulates localization and activity of palmitoylated Ras isoforms. *Science* 2005;**307**:1746–52.

19. Goodwin JS, Drake KR, Rogers C, et al. Depalmitoylated Ras traffics to and from the Golgi complex via a nonvesicular pathway. *J Cell Biol* 2005;**170**:261–72.

20. Choy E, Chiu VK, Silletti J, et al. Endomembrane trafficking of Ras: the CaaX motif targets proteins to the ER and Golgi. *Cell* 1999;**98**:69–80.

21. Chiu VK, Bivona T, Hach A, et al. Ras signalling on the endoplasmic reticulum and the Golgi. *Nat Cell Biol* 2002;**4**:343–50.

22. Bivona TG, deCastro IP, Ahearn IM, et al. Phospholipase Cg activates Ras on the Golgi apparatus by means of RasGRP1. *Nature* 2003;**424**:694–8.

23. Apolloni A, Prior IA, Lindsay M, Parton RG, Hancock JF. H-ras but not K-ras traffics to the plasma membrane through the exocytic pathway. *Mol Cell Biol* 2000;**20**:2475–87.

24. deCastro IP, Bivona TG, Philips MR, Pellicer A. Ras activation in Jurkat T cells following low-grade stimulation of the T-cell receptor is specific to N-Ras and occurs only on the Golgi apparatus. *Mol Cell Biol* 2004:3485–96.

25. Greaves J, Chamberlain LH. Palmitoylation-dependent protein sorting. *J Cell Biol* 2007;**176**:249–54.

26. Dunphy JT, Greentree WK, Linder ME. Enrichment of G-protein palmitoyltransferase activity in low density membranes: *in vitro* reconstitution of Gαi to these domains requires palmitoyltransferase activity. *J Biol Chem* 2001;**76**:43,300–43,304.

27. Silvius JR. Mechanisms of Ras protein targeting in mammalian cells. *J Membr Biol* 2002;**190**:83–92.

28. Watson RT, Furukawa M, Chiang S-H, et al. The exocytotic trafficking of TC10 occurs through both classical and nonclassical secretory transport pathways in 3T3L1 adipocytes. *Mol Cell Biol* 2003;**23**:961–74.

29. Zheng H, McKay J, Buss JE. H-Ras does not need COP I- or COP II-dependent vesicular transport to reach the plasma membrane. *J Biol Chem* 2007;**282**:25,760–25,768.

30. Dong S, Mitchell DA, Lobo S, Zhao L, Bartels DJ, Deschenes RJ. Palmitoylation and plasma membrane localization of Ras2p by a non-classical trafficking pathway in *Saccharomyces cerevisiae*. *Mol Cell Biol* 2003;**23**:6574–8.

31. Hancock JF, Magee AI, Childs JE, Marshall CJ. All Ras proteins are polyisoprenylated but only some are palmitoylated. *Cell* 1989;**57**:1167–77.

32. Hancock JF, Paterson H, Marshall CJ. A polybasic domain or palmitoylation is required in addition to the CAAX motif to localize p21ras to the plasma membrane. *Cell* 1990;**63**:133–9.

33. Jackson JH, Li JW, Buss JE, Der CJ, Cochrane CG. Polylysine domain of K-ras 4B protein is crucial for malignant transformation. *Proc Natl Acad Sci* 1994;**91**:12,730–12,734.

34. Matallanas D, Sanz-Moreno V, Arozarena I, et al. Distinct utilization of effectors and biological outcomes resulting from site-specific Ras activation: Ras functions in lipid rafts and Golgi complex are dispensable for proliferation and transformation. *Mol Cell Biol* 2006;**26**:100–16.

35. Cullen PJ. Decoding complex Ca^{2+} signals through the modulation of Ras signaling. *Curr Opin Cell Biol* 2006;**18**:157–61.

36. Arozarena I, Matallanas D, Berciano MT, et al. Activation of H-Ras in the endoplasmic reticulum by the RasGRF family guanine nucleotide exchange factors. *Mol Cell Biol* 2004;**24**:1516–30.

37. Caloca MJ, Zugaza JL, Bustelo XR. Exchange factors of the RasGRP family mediate Ras activation in the Golgi. *J Biol Chem* 2003;**278**:33,465–33,473.

38. Mor A, Philips MR. Compartmentalized Ras/MAPK signaling. *Annu Rev Immunol* 2006;**24**:771–800.

39. Eungdamrong NJ, Iyengar R. Compartment-specific feedback loop and regulated trafficking can result in sustained activation of Ras at the Golgi. *Biophys J* 2007;**92**:808–15.

40. Rocks O, Peyker A, Bastiaens PIH. Spatio-temporal segregation of Ras signals: one ship, three anchors, many harbors. *Curr Opin Cell Biol* 2006;**18**:351–7.

41. Plowman SJ, Hancock JF. Ras signaling from plasma membrane and endomembrane microdomains. *Biochim Biophys Acta Mol Cell Res* 2005;**1746**:274–83.

42. Quatela SE, Philips MR. Ras signaling on the Golgi. *Curr Opin Cell Biol* 2006;**18**:162–7.

43. Gorfe AA, Hanzal-Bayer M, Abankwa D, Hancock JF, McCammon JA. Structure and dynamics of the full-length lipid-modified H-Ras protein in a 1,2-dimyristoylglycero-3-phosphocholine bilayer. *J Med Chem* 2007;**50**:674–84.

44. Mor A, Campi G, Du G, et al. The lymphocyte function-associated antigen-1 receptor costimulates plasma membrane Ras via phospholipase D2. *Nat Cell Biol* 2007;**9**:713–19.

45. Kenworthy AK. Nanoclusters digitize Ras signalling. *Nat Cell Biol* 2007;**9**:875–7.

46. Rotblat B, Prior IA, Muncke C, et al. Three separable domains regulate GTP-dependent association of H-Ras with the plasma membrane. *Mol Cell Biol* 2004;**24**:6799–810.

47. Hancock JF, Parton RG. Ras plasma membrane signalling platforms. *Biochem J* 2005;**389**:1–11.

48. Shaikh SR, Edidin MA. Membranes are not just rafts. *Chem Phys Lipids* 2006;**114**:1–3.

49. Kusumi A, Koyama-Honda I, Suzuki K. Molecular dynamics and interactions for creation of stimulation-induced stabilized rafts from small unstable steady-state rafts. *Traffic* 2004;**5**:213–30.

50. Mayor S, Rao M. Rafts: scale-dependent, active lipid organization at the cell surface. *Traffic* 2004:231–40.

51. Prior IA, Muncke C, Parton RG, Hancock JF. Direct visualization of Ras proteins in spatially distinct cell surface microdomains. *J Cell Biol* 2003;**160**:165–70.

52. Prior IA, Harding A, Yan J, Sluimer J, Parton RG, Hancock JF. GTP-dependent segregation of H-ras from lipid rafts is required for biological activity. *Nat Cell Biol* 2001;**3**:368–75.

53. Tian T, Harding A, Inder K, Plowman S, Parton RG, Hancock JF. Plasma membrane nanoswitches generate high-fidelity Ras signal transduction. *Nat Cell Biol* 2007;**9**:905–14.

54. Plowman SJ, Muncke C, Parton RG, Hancock JF. H-Ras, K-Ras, and inner plasma membrane raft proteins operate in nanoclusters with differential dependence on the actin cytoskeleton. *Proc Natl Acad Sci* 2005;**102**:15,500–15,505.

55. Murakoshi H, Iino R, Kobayashi T, et al. Single-molecule imaging analysis of Ras activation in living cells. *Proc Natl Acad Sci* 2004;**101**:7317–22.

56. Lommerse PHM, Snaar-Jagalska BE, Spaink HP, Schmidt T. Single-molecule diffusion measurements of H-Ras at the plasma membrane of live cells reveal microdomain localization upon activation. *J Cell Sci* 2005;**118**:1799–809.

57. Eisenberg S, Henis YI. Interactions of Ras proteins with the plasma membrane and their roles in signaling. *Cell Signal* 2008;**20**:31–9.

58. Rotblat B, Yizhar O, Haklai R, Ashery U, Kloog Y. Ras and its signals diffuse through the cell on randomly moving nanoparticles. *Canc Res* 2006;**66**:1974–81.

59. Bivona TG, Quatela SE, Bodemann BO, et al. PKC regulates a farnesyl-electrostatic switch on K-Ras that promotes its association with Bcl-XL on mitochondria and induces apoptosis. *Mol Cell* 2006;**21**:481–93.

60. Silvius JR, Bhagatji P, Leventis R, Terrone D. K-ras4B and prenylated proteins lacking "second signals" associate dynamically with cellular membranes. *Mol Biol Cell* 2006;**17**:192–202.

61. Wang G, Deschenes RJ. Plasma membrane localization of Ras requires class C Vps proteins and functional mitochondria in *Saccharomyces cerevisiae*. *Mol Cell Biol* 2006;**26**:3243–55.

62. Villalonga P, Lopez-Alcala C, Bosch M, et al. Calmodulin binds to K-Ras, but not to H- or N-Ras, and modulates its downstream signaling. *Mol Cell Biol* 2001;**21**:7345–54.

63. Fivaz M, Meyer T. Reversible intracellular translocation of KRas but not HRas in hippocampal neurons regulated by Ca^{2+}/calmodulin. *J Cell Biol* 2005;**170**:429–41.

64. Jiang X, Sorkin A. Coordinated traffic of Grb2 and Ras during epidermal growth factor receptor endocytosis visualized in living cells. *Mol Biol Cell* 2002;**13**:1522–35.

65. Sorkin A, vonZastrow M. Signal transduction and endocytosis: close encounters of many kinds. *Nature* 2002;**3**:600–14.

66. LeRoy C, Wrana JL. Clathrin- and nonclathrin-mediated endocytic regulation of cell signalling. *Nat Rev Mol Cell Biol* 2005;**6**:112–26.

67. Roy S, Wyse B, Hancock JF. H-Ras signaling and K-Ras signaling are differentially dependent on endocytosis. *Mol Cell Biol* 2002;**22**:5128–40.

68. Rizzo MA, Kraft CA, Watkins SC, Levitan ES, Romero G. Agonist-dependent traffic of raft-associated Ras and Raf-1 is required for activation of the MAPK cascade. *J Biol Chem* 2001;**276**:34,928–34,933.

69. Li H-S, Stolz DB, Romero G. Characterization of endocytic vesicles using magnetic microbeads coated with signalling ligands. *Traffic* 2005;**6**:324–34.

70. Rizzo MA, Shome K, Watkins SC, Romero G. The recruitment of Raf-1 to membranes is mediated by direct interaction with phosphatidic acid and is independent of association with Ras. *J Biol Chem* 2000;**275**:23,911–23,918.

71. Jura N, Scotto-Lavino E, Sobczyk A, Bar-Sagi D. Differential modification of Ras proteins by ubiquitination. *Mol Cell* 2006;**21**:679–87.

72. Gomez GA, Daniotti JL. H-Ras dynamically interacts with recycling endosomes in CHO-K1 cells: involvement of Rab5 and Rab11 in the trafficking of H-Ras to this pericentriolar endocytic compartment. *J Biol Chem* 2005;**280**:34,997–35,010.

73. Corbett KD, Alber T. The many faces of Ras: recognition of small GTP-binding proteins. *Trends Biochem Sci* 2001;**26**:710–16.

74. Gorfe AA, Pellarin R, Caflisch A. Membrane localization and flexibility of a lipidated Ras peptide studied by molecular dynamics simulations. *J Am Chem Soc* 2004;**126**:15,272–15,286.

75. Reuther G, Tan K-T, Vogel A, et al. The lipidated membrane anchor of full length N-Ras protein shows an extensive dynamics as revealed by solid-state NMR spectroscopy. *J Am Chem Soc* 2006;**128**:13,840–13,846.

Chapter 54

Role of R-Ras in Cell Growth

Gretchen A. Repasky[1], Adrienne D. Cox[2,3,4,5,6], Ariella B. Hanker[4,5,6], Natalia Mitin[2,4,6] and Channing J. Der[2,4,5,6]

[1]Birmingham-Southern College, Birmingham, Alabama

[2]Departments of Pharmacology

[3]Radiation Oncology

[4]Lineberger Comprehensive Cancer Center

[5]Curriculum in Genetics and Molecular Biology,

[6]University of North Carolina at Chapel Hill, Chapel Hill, North Carolina

INTRODUCTION

Ras proteins (H-Ras, K-Ras4A and 4B, and N-Ras) are regulators of signal transduction, mutated in 30 percent of human cancers, and targets for novel approaches for cancer treatment. Ras proteins are the founding members of a superfamily of small GTP binding and hydrolyzing proteins (GTPases). The small GTPases that share the greatest amino acid identity with Ras, such as R-Ras, Rap, and Ral, constitute members of the Ras family of proteins (Figure 54.1a). Within this family, the R-Ras subfamily proteins (R-Ras, TC21/R-Ras2, and M-Ras/R-Ras3) exhibit the strongest structural and biological similarities with Ras. Like Ras, R-Ras proteins function as regulated GDP/GTP binary switches (Figure 54.1b). While studies in experimental model systems have shown that R-Ras proteins can promote oncogenic transformation, there is only limited evidence for aberrant R-Ras function in human cancers. However, some R-Ras proteins play distinct roles in normal cell physiology. In this chapter, we first summarize the general features of R-Ras proteins that are shared with Ras proteins, and then highlight unique features of each R-Ras protein.

GENERAL PROPERTIES OF R-RAS PROTEINS: VARIATIONS ON RAS

Structure

R-Ras proteins are GTPases of approximately 200 amino acids that share significant primary (Figure 54.2a), secondary, and tertiary structural characteristics with the 21-kDa Ras GTPases. In addition, R-Ras proteins possess extended amino (10–26 residues) or carboxyl-terminal residues not present in Ras proteins that account for their larger size

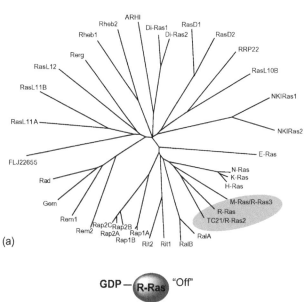

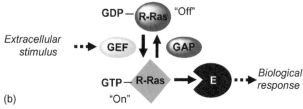

FIGURE 54.1 (a) The Ras branch of the human Ras superfamily of small GTPases. ClustalX was used to generate a dendrogram of the GTP-binding domains of human Ras subfamily members. The K-Ras2B isoform of K-Ras is shown. (b) R-Ras proteins function as GDP/GTP-regulated binary switches. Activity is most commonly associated with GEF activation. Effector (E) binding promotes downstream signaling and regulation of cell physiology.

(approximately 25kDa) and suggest functional differences with Ras proteins. Therefore, the numbering of key R-Ras protein amino acid residues relative to Ras is by the addition of 26 (R-Ras), 11 (TC21), or 10 (M-Ras) to the Ras numbering system.

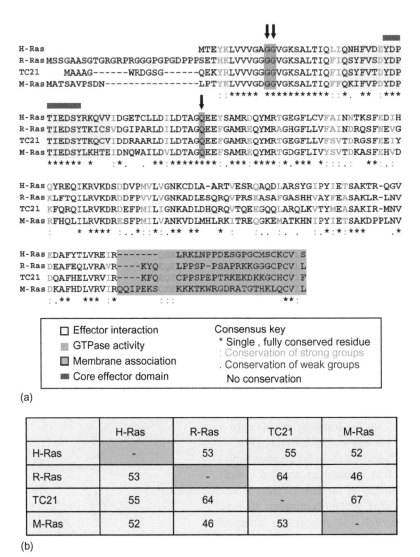

(a)

(b)

FIGURE 54.2 Amino acid sequence alignment of H-Ras and R-Ras proteins.
(a) R-Ras proteins are larger than Ras proteins due to additional amino (boxed) and carboxyl-terminal sequences. R-Ras proteins share complete sequence identity with the core Ras effector domain (H-Ras residues 32–40), but sequence differences in the extended effector domain (boxed; H-Ras residues 25–45) account for the different abilities of R-Ras proteins to interact with Ras effectors. Missense mutations at H-Ras residues G12, G13, and Q61 (indicated by arrows) have been identified in human cancers and result in constitutively activated, transforming variants of Ras proteins. Mutation of the equivalent residues in R-Ras, TC21, or M-Ras results in the generation of constitutively activated mutants of the R-Ras proteins. (b) Sequence identity between H-Ras and R-Ras proteins. ClustalX was used to align the sequences for determination of sequence identity.

Expression

The three *RAS* genes are expressed ubiquitously, although at distinct tissue-specific levels. Similarly, the three *RRAS* genes are also expressed widely. *RRAS* transcripts showed a wide range of expression levels in a variety of cell types [1]. Human TC21 protein levels were found to be highest in adult human kidney, placenta, and ovary [2]. Interestingly, *MRAS* was isolated in two separate differential gene expression cloning strategies, reflecting its regulation at the level of gene expression [3, 4]. *MRAS* gene expression is particularly high in brain and heart, with lower levels in skeletal muscle, ovary, and cells of hematopoietic origin [5–8].

Biochemistry

R-Ras proteins share 46–67 percent amino acid identity with each other, and 52–55 percent amino acid identity with H-Ras (Figure 54.2b), primarily in the consensus guanine nucleotide binding motifs and the switch I and II regions of Ras that alter conformation based on the bound guanine nucleotide (GTP or GDP), permitting function as GDP/GTP-regulated molecular switches (Figure 54.2a) [9]. These strong sequence similarities account for the fact that Ras and R-Ras proteins are regulated by an overlapping set of guanine nucleotide exchange factors (GEFs) and GTPase activating proteins (GAPs) (Table 54.1) [10].

TABLE 54.1 Regulators of Ras and R-Ras proteins

Protein	Other activities	Substrates
GEFs		
Sos1	RacGEF, feedback activation by Ras-GTP	H-Ras, N-Ras, K-Ras, R-Ras2, R-Ras3, Rac 1; not R-Ras
Sos1	RacGEF	H-Ras
Ras-GRF1	RacGEF	H-Ras, K-Ras[1], N-Ras[1], R-Ras, R-Ras2, R-Ras3, Rac1, not K-Ras[1], N-Ras[1], Rap1
Ras-GRF2	RacGEF	H-Ras, Rac1; not R-Ras
RasGRP1(CalDAG-GEFII)	DAG-,Ca^{2+}-binding	H-Ras, N-Ras, R-Ras, R-Ras2, R-Ras3; not Rap1A
RasGRP2(CalDAG-GEFI)	DAG-,Ca^{2+}-binding	N-Ras, K-Ras, R-Ras, R-Ras2, Rap1A, Rap2A; not H-Ras
RasGRP3(CalDAG-GEFIII)	DAG-,Ca^{2+}-binding	H-Ras, R-Ras, R-Ras2, R-Ras3, Rap1A, Rap2B
RasGRP4	DAG-,Ca^{2+}-binding	H-Ras
RapGEF1 (C3G)		Rap1A, Rap1B, R-Ras, R-Ras2
GAPs		
p120 RasGAP	p190 RhoGAP-binding	H-Ras, N-Ras, R-Ras, R-Ras2, R-Ras3, Rab5; not Rap proteins
Neurofibromin	GAP activity regulated by tubulin binding and arachidonic acid treatment	H-Ras, K-Ras, N-Ras, R-Ras, R-Ras2, R-Ras3; not Rap1A
SynGAP		H-Ras
GAP1[IP4BP] (R-Ras GAP)	Inositol 1,3,4,5-tetrakisphosphate-binding	H-Ras, R-Ras, R-Ras2; not R-Ras3
GAP1[m]	PIP_3-binding; $G\alpha_{12}$	H-Ras, R-Ras, R-Ras2, R-Ras3
RASAL1	PIP_2- and PIP_3-binding; Ca^{2+}-binding	H-Ras
CAPRI	Ca^{2+}-binding	H-Ras
DAB2IP (DIP1/2)	DOC-2/DAB2-binding	H-Ras, K-Ras, R-Ras, R-Ras2; not Rap1A
Plexin A1		R-Ras
Plexin B1		R-Ras

Like Ras, GTP-bound R-Ras is the active form, and mutations at positions analogous to those found in tumor-associated mutant Ras proteins (amino acids 12 and 61) also produce constitutively activated R-Ras GTPases. Finally, like Ras, effectors of R-Ras proteins typically contain Ras-binding (RBD) or Ras-association (RA) domains that are responsible for their preferential binding to the GTP-bound conformation.

Like Ras, R-Ras proteins contain a carboxyl-terminal CAAX tetrapeptide motif that directs posttranslational modifications, including the covalent addition of an isoprenoid lipid, which promotes the association of R-Ras proteins with the inner face of the plasma membrane [11] (Figure 54.3). However, whereas Ras proteins are modified by a C15 farnesyl isoprenoid, R-Ras proteins are modified by the more hydrophobic C20 geranylgeranyl isoprenoids [12].

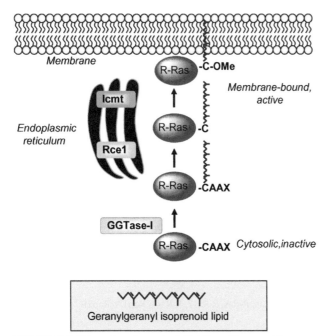

FIGURE 54.3 The carboxyl-terminal CAAX motifs present in R-Ras proteins direct posttranslational modification by protein prenyltransferases.

Whereas Ras proteins are modified by covalent addition of a C15 farnesyl isoprenoid by farnesyltransferase, R-Ras proteins are modified by covalent addition of a C20 geranylgeranyl isoprenoid by geranylgeranyltransferase-I (GGTase-I). This modification is followed by the Ras converting enzyme (Rce1)-catalyzed proteolytic removal of the AAX residues and isoprenylcysteine carboxyl methyltransferase (ICMT)-catalyzed carboxylmethylation of the now terminal isoprenylated cysteine residue. The CAAX-signaled modifications, together with specific sequences upstream of the CAAX motif, represent the two signals required to target Ras and R-Ras proteins to the plasma membrane. Like H-Ras, R-Ras and TC21 contain cysteine residues that undergo posttranslational modification by the fatty acid palmitate which facilitates membrane association. Like K-Ras4B (not shown), M-Ras contains a lysine-rich sequence in place of the palmitoylated cysteine that is important for membrane targeting.

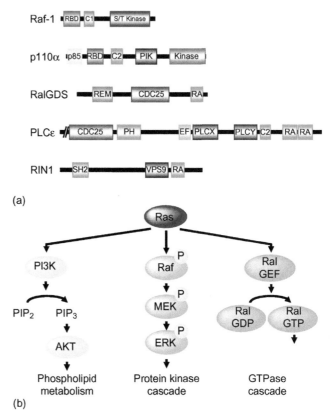

(a)

(b)

FIGURE 54.4 Downstream effectors of R-Ras proteins.

(a) Ras effectors contain RBD or RA domains. The majority of Ras effectors contain Ras-binding (RBD) or Ras association (RA) domains that promote preferential binding with the GTP-bound GTPase. Aside from this shared domain, Ras effectors exhibit significant sequence and biochemical divergence. This includes catalytic domains for protein (Raf) and lipid (PI3K) kinases, GEFs for Ras family GTPases (RalGDS and RIN1) and phospholipase (PLCe) activity. (b) Ras effector pathways involved in transformation. The three main classes of Ras effectors implicated in oncogenesis are the phosphatidylinositol 3-kinases (PI3Ks), Raf serine/threonine kinases, and GEFs specific for Ral small GTPases. The PI3K lipid kinases phosphorylate phosphoinositides and promote formation of phosphatidylinositol 3,4,5-phosphate (PIP3), which in turn facilitates the activation of other cytoplasmic signaling proteins, including the serine/threonine kinase Akt/PKB. Raf phosphorylates the MEK1/2 dual specificity protein kinases, which in turn phosphorylate the ERK1/2 serine/threonine kinases.

Isoprenoid modification of Ras and R-Ras proteins is essential for biological function. Consequently, inhibitors (FTIs) of the enzyme farnesyltransferase, which catalyzes the modification of Ras proteins, have been developed as inhibitors of oncogenic Ras function and are under clinical evaluation as anti-neoplastic drugs. Similarly, inhibitors (GGTIs) of the enzyme geranylgeranyltransferase I, which catalyzes the modification of R-Ras proteins, have also been developed, and have shown anti-tumor activity in preclinical studies [13]. TC21 can also be modified by farnesyltransferase [14]. The remaining two CAAX-signaled modifications, Rce1-catalyzed removal of the AAX peptide and Icmt-catalyzed carboxylmethylation of the prenylated cysteine residue, are also being considered for therapeutic intervention [15].

Signal Transduction and Cell Biology

Like Ras, R-Ras proteins serve as relay switches that transmit signals initiated by diverse extracellular ligands to

cytoplasmic signaling pathways. However, unlike Ras, the specific upstream signals that promote activation of endogenous R-Ras proteins are less well characterized, perhaps due in part to technical limitations and in part to limited experimental analyses. Because Ras and R-Ras proteins share common GEFs, it is not surprising many of the same signals that cause Ras activation also cause R-Ras protein family activation (Table 54.1). Once activated, Ras protein signaling is mediated by interaction with a multitude of downstream effector targets. Among these, the Raf serine/threonine protein kinases, the phosphatidylinositol 3-kinase lipid kinases (PI3Ks) and the Ral small GTPase guanine nucleotide exchange factors (RalGEFs) have been established as key mediators of Ras-mediated transformation [16] (Figure 54.4b). The contribution of each of these major Ras effectors to R-Ras family-mediated transformation is distinct, and will be discussed in the following sections. The core effector domain of Ras (amino acids 32–40) is critical for interactions with downstream signaling molecules, and R-Ras proteins share complete identity with this core effector domain (Figure 54.2a). However, flanking residues within the extended effector domain (Ras amino acids 25–45) of R-Ras proteins differ more significantly, suggesting that R-Ras subfamily members may regulate a distinct set of effectors and cytoplasmic signaling pathways.

R-Ras subfamily GTPases are implicated in regulating biological functions both similar to and distinct from those controlled by Ras proteins. For example, like Ras, constitutively activated forms of all three R-Ras members have been shown to cause growth transformation of NIH 3T3 mouse fibroblasts. Similarly, both Ras and R-Ras family proteins regulate cell survival, actin cytoskeletal organization, and differentiation. However, R-Ras proteins also cause consequences opposing those of Ras, and the biological phenotypes of R-Ras function can be significantly different from those of TC21 and M-Ras. The following sections detail the distinct properties and roles of each R-Ras subfamily member in signal transduction and cell biology.

R-RAS

GDP/GTP Regulation

R-Ras encodes a 218 amino acid protein that shares 53 percent identity overall to H-Ras (Figure 54.2) [1]. While no GEFs have been shown to be specific for R-Ras, R-Ras GDP/GTP exchange is stimulated by GEFs for some Ras and/or Rap small GTPases (e.g., RasGRF, RasGRPs, C3G) (Table 54.1) [17–23]. R-Ras interacts with some of the known Ras GAPs, and GTP hydrolysis is stimulated by p120GAP, GAP1^m, GAP1^{IP4BP} (R-Ras GAP), and DAB2IP (IDIP1/2) (Table 54.1) [24–26]. In addition, recent findings have shown that members of the Plexin family of cell surface receptors, Plexin-Al (Sema3A receptor) or Plexin-B1 (Sema4D receptor) also act as GAPs for R-Ras proteins. Plexins are receptors for Semaphorins, a large class of secreted or membrane-associated proteins that act as chemotactic cues for cell movement. Plexin-A1 and Plexin-B1 possess intrinsic R-Ras GAP activity, and ligand stimulation of these plexins causes R-Ras inactivation that is required for Plexin-B1-mediated effects on the actin cytoskeleton, cell movement, neuronal morphogenesis, and angiogenesis [27].

Signal Transduction

R-Ras fails to activate the Raf–MEK–ERK mitogen-activated protein kinase (MAPK) cascade, which is the key effector pathway required for Ras-mediated transformation

of rodent fibroblasts (Table 54.1). Ectopic expression analyses failed to find R-Ras activation of all three Raf isoforms [28]. However, R-Ras strongly activates the phophatidylinositol 3-kinase (PI3K)-Akt signaling pathway, perhaps the second most critical effector pathway for Ras-mediated transformation and survival [29–31]. Ectopic co-overexpression analyses showed that activated R-Ras can cooperate with the class I PI3K catalytic p110α, δ, and γ, but not β, subunits to promote phosphoinositide 3,4,5 phosphate formation [32]. This effector pathway has also been shown to be required for R-Ras-mediated cell survival and transformation [30, 33, 34]. R-Ras transformation of NIH 3T3 cells was found to be dependent primarily on the PI3K-AKT and not the Raf-ERK cascade, with a limited contribution by the RalGEF pathway [33], which contributes significantly to Ras-mediated transformation of human and not mouse cells [35]. Nevertheless, the GTP-bound form of each R-Ras family member can bind to isolated Ras binding domains of Raf and RalGDS, which enables the selective detection of activated R-Ras proteins [36]. However, a demonstration of binding to isolated RBDs alone is not sufficient evidence for effector activation under physiologic conditions. Ectopic co-overexpression of mutant R-Ras together with various RalGEFs did result in Ral activation in 293T cells [28]. Endogenous R-Ras located at endosomes was found to activate RalGEF and Ral to regulate calcium-triggered exocytosis in PC12 pheochromocytoma cells [32]. Finally, R-Ras did not bind to the isolated RA domains of PLC epsilon, an established Rap effector, but did interact with full-length proteins of the Rin family, and weakly to members of the RASSF1 family, of Ras effectors.

Cell Biology

Mutated *RRAS2/TC21* [37–39], but not *RRAS* or *RRAS3* genes, has been identified in human tumors. However, introduction of point mutations analogous to those that cause mutational activation of Ras (e.g., 38 V, 87 L) stimulate R-Ras transforming potential. Unlike Ras, in which mutations at positions 12 and 61 are similar in their transforming potency, in members of the R-Ras family, activating mutations analogous to Ras codon 61 (e.g., 87 L, 72 L, or 71 L) are significantly more potent than those analogous to Ras codon 12 [8, 33, 40, 41]. Although constitutively activated mutants of R-Ras cause tumorigenic transformation of NIH3T3 mouse fibroblasts, the strong morphological transformation observed in Ras-transformed rodent fibroblasts is not observed with activated mutants of R-Ras [37, 40]. Further, constitutively activated R-Ras fails to rescue the block in growth caused by expression of dominant negative Ras [42], suggesting that R-Ras induces transformation by regulation of signaling pathways distinct from those utilized by Ras. Hence, the normal biological role of R-Ras may lie in processes distinct from those that contribute to cell transformation. Like Ras, activated R-Ras also blocked anoikis in RIE-1 rat intestinal cells [43]. Ectopic expression of activated R-Ras(87L) also promoted the growth and migration of C33A cervical epithelial cells through PI3K activation [44]. Finally, a recent study in clinical samples of primary and metastatic prostate cancer identified frequent missense mutations in Plexin-B1 that hinder its R-Ras binding and GAP activity; thus, such mutations may be a common cause of inappropriate R-Ras activation in this disease [45].

R-Ras has been shown to control and promote integrin-mediated cellular adhesion [46, 47]. For example, expression of dominant negative R-Ras (S43N) in adherent cells reduced cell spreading, and ectopic expression of activated R-Ras in suspension cells promoted spreading and fibronectin assembly, suggesting that R-Ras is required for integrin-mediated cell adhesion [47]. R-Ras also promoted retinal neural outgrowth on laminin, a process dependent on integrin function [48]. R-Ras induced integrin-mediated migration and invasion of breast epithelial cells on collagen, by signaling to β2 but not β5 integrin receptors, suggesting that R-Ras induces selective activation of specific integrins [49]. R-Ras-mediated control of adhesion may be linked to both Src- and Raf-mediated pathways [50,51]. RalBP1/RLIP76, an effector of the Ral small GTPases, was also found to be an effector of R-Ras [52]. RalBP1/RLIP76, a RhoGAP and negative regulator of Rac and Cdc42, is required for adhesion-induced activation of Rac and the resulting cell spreading and migration caused by R-Ras. Phospholipase C epsilon, first identified as a Ras effector, has also been shown to be an effector of R-Ras modulation of the actin cytoskeleton and membrane protrusion [53].

R-Ras has also been shown to control apoptosis [42], although the effect on the apoptotic response varies depending on cell type. For example, R-Ras promoted myeloid cell apoptosis in response to growth factor withdrawal, an effect which was abrogated by overexpression of Bcl-2 [54]. In contrast, a greater pool of evidence from several model cell systems suggests that R-Ras more frequently blocks the apoptotic response. For example, constitutively activated R-Ras induced protection from cell death following withdrawal of anchorage or growth factors in RIE-1 rat intestinal epithelial cells, C2C12 skeletal myoblasts, and BaF3 cells, an IL-3-dependent mouse pro-B-cell line [30, 34, 43].

In summary, R-Ras is both similar to and distinct from Ras. R-Ras utilizes only some of the same effectors and signaling pathways, such as PI3K-Akt, and regulates only some of the same cellular functions, such as cell proliferation and survival, similar to Ras. Conversely, R-Ras also functions in opposing cellular regulatory roles, such as in cell adhesion. Thus, additional effectors of R-Ras likely remain to be identified.

TC21/R-RAS-2

GDP/GTP Regulation

Human TC21/R-Ras-2 is a 204 amino acid protein that shares 64 percent amino acid identity with R-Ras and 55 percent amino acid identity with H-Ras (Figure 54.2) [37, 55]. TC21 is regulated by a set of GEFs and GAPs similar to those that regulate R-Ras GDP/GTP cycling, but responds to more Ras and/or Rap regulators than does R-Ras (Table 54.1) [21]. TC21 intrinsic GTP/GDP cycling activity is activated by SOS, RasGRF1/2, C3G, and RasGRP [2, 21, 56]. Hydrolysis of TC21-bound GTP is enhanced by known Ras GAPs (p120 GAP, Gap1^m, neurofibromin, GAP1^{IP4BP}, and DAB2IP) (Table 54.1) [2,21]. Hence, extracellular signals that regulate Ras GDP/GTP cycling are also likely to cause concurrent regulation of TC21 activity.

Signal Transduction

Whether Raf is an important mediator of TC21-induced transformation remains unclear (Table 54.1). One study found that TC21 failed to interact with and activate Raf kinases *in vivo* [2,57]. Ectopic co-expression analyses showed that TC21 did not activate ERK, and only weakly activated ERK when overexpressed with c-Raf-1 but not B-Raf or A-Raf [28]. Consistent with this failure to activate Raf is the observation that, unlike Ras, TC21, as well as R-Ras, failed to induce senescence in primary fibroblasts [58]. Ras-mediated induction of senescence has been attributed to activation of the Raf/ERK pathway. However, other studies determined that TC21 interacts with and activates Raf kinases, and activates the Raf–MEK–ERK signaling pathway in transformation [56, 59]. The finding that association of 14-3-3 proteins with Raf-1 prevented its activation by R-Ras or TC21, but not by Ras, suggests that the presence of other Raf-1 interacting proteins may account for these different observations [60].

Using various screening approaches for protein–protein interactions, TC21 was found to interact with RalGDS and RalGDS-like proteins (RGL, RGL2) [61, 62]. However, whether TC21 activates RalGEFs is unclear (Table 54.1). One study showed that, although TC21 interacted with RalGDS proteins, in contrast to Ras, it did not promote activation of RalA [62]. Thus, similar to observations with R-Ras [63, 64], TC21 can bind but not activate RalGDS, suggesting that RalGDS may not be a key effector for mediating TC21-induced growth transformation. In contrast, another study [65] determined that activated TC21 stimulated activation of Ral. Finally, ectopic co-overexpression analyses showed that TC21 can cooperate with three RalGEF family members to cause transient Ral activation [32]. These differences may be attributed to experimental, cell type, or cell-strain type differences [66], or to the utilization of different expression vectors for ectopic expression of TC21 proteins [67].

In contrast to the conflicting observations regarding the involvement of Raf or RalGDS effectors in TC21-mediated transformation, there is a strong consensus that TC21 binds the p110 catalytic subunit of PI3K and stimulates PI3K and PI3K-dependent Akt activities (Table 54.1) [32, 62, 65, 68], and that this effector is critical for TC21-mediated transformation. Ectopic co-overexpression analyses showed that activated TC21 can cooperate with the class I PI3K catalytic p110α, δ, and γ, but not β, subunits to promote phosphoinositide 3,4,5 phosphate formation [32]. Inhibition of PI3K function with pharmacological inhibitors or genetic manipulation reversed TC21-mediated transformation of rodent fibroblasts [62, 65, 68]. However, activated PI3K alone does not cause transformation of fibroblasts [69], suggesting that TC21 must also utilize other effectors to mediate transformation. Like Ras, TC21 also activates phospholipase C epsilon (PLCε) (Table 54.1) [62]. Further study is clearly necessary to determine whether PLCε-mediated second-messenger activation is important for TC21-mediated transformation. Finally, TC21 does bind to Rin and RASSF1 family proteins, but not to other RA domain-containing proteins that interact with Ras [28].

Cell Biology

Of the R-Ras subfamily members, TC21 regulates biological functions most similarly to Ras. Unlike other R-Ras subfamily members, activating mutations in TC21 or over-expression of wild-type TC21 protein have been observed in several human tumors and tumor-derived cell lines [37–39, 56, 59, 70]. Similar to activated forms of Ras, activated TC21 causes strong morphologic growth and neoplastic transformation in several cell types, including rodent fibroblasts and human breast epithelial cells [41, 56, 70]. In addition, unlike R-Ras, activated TC21 can overcome the growth inhibitory actions of dominant-negative Ras, indicating that TC21 and Ras share some common functions important for normal cell proliferation [42]. However, in contrast to Ras, neither TC21 nor R-Ras caused premature senescence of primary rodent fibroblasts [58]. In addition to regulating cell proliferation and transformation, TC21 promoted cell survival of matrix-deprived intestinal epithelial cells and chemotherapeutic drug-treated NIH 3T3 cells [43, 68]. Like Ras, TC21 also promoted PC12 pheochromocytoma cell differentiation and growth cessation, and blocked serum deprivation-induced differentiation of C2 myoblasts [57]. In addition, like R-Ras, activated TC21 has been shown to induce migration and invasion of breast epithelial cells [49]. Like Ras and R-Ras, activated TC21 also blocked anoikis in RIE-1 cells [43]. TC21 caused transformation of EphH4

mouse mammary epithelial cells that was dependent on PI3K but not AKT [71]. In summary, of the three R-Ras subfamily proteins, TC21 is regulated and functions most similarly to Ras to control processes such as cell proliferation and transformation. While controversy exists over the mechanism by which TC21 causes transformation, it is clear that chronically activated versions are potent oncogenes, and contribute to various aspects of the transformation phenotype and to tumor formation in humans. Functional distinctions between the three *RAS* genes are only now beginning to be clarified. Whether *TC21* is functionally distinct from or redundant with *RAS* genes is also an issue that remains to be elucidated.

M-RAS/R-RAS-3

GDP/GTP Regulation

Human M-Ras is a 209 amino acid protein that shares 46–53 percent amino acid identity to R-Ras, TC21, and Ras. M-Ras GTP/GDP cycling is regulated more similarly to Ras than to R-Ras or TC21 (Table 54.1). For example, while M-Ras is activated by RasGRF and RasGRP, unlike R-Ras and TC21, it is also stimulated by Sos [8,21]. In addition, like Ras but unlike TC21 and R-Ras, C3G failed to upregulate GTP-bound M-Ras [21]. Similarly, M-Ras GTPase activity is stimulated by an overlapping set of GAPs (p120 Ras GAP, Neurofibromin, GAP1^m) that enhances GTP hydrolysis on Ras and other R-Ras isoforms [8, 21].

Signal Transduction

The effectors and pathways stimulated by M-Ras to cause transformation are poorly characterized, but may involve both known Ras effectors and novel mechanisms (Table 54.1). Compared to Ras proteins, M-Ras interacts only weakly with Raf kinases [8], and activated M-Ras is only a weak stimulator of the Raf–MEK–ERK MAPK cascade [3, 6, 8, 72]. Activated M-Ras did not activate ERK when expressed alone in 293T cells, but did so when co-overexpressed with ectopic Raf-1, but not B-Raf, and weakly with A-Raf [28]. Pharmacological inhibition of ERK activation caused a reduction in M-Ras-mediated fibroblast transformation, and Raf-1 cooperated with M-Ras to induce transformation, suggesting that Raf is a key mediator of M-Ras-mediated transformation [8].

In addition to Raf, M-Ras associates, to varying extents, with other known Ras effectors, such as RalGEFs, PI3K, and Rin1 [8]. Like Ras and TC21, M-Ras was found to interact strongly with RGL3, a RalGDS-like protein [73]. Similar to Ras, activated M-Ras stimulates Ral-GTP formation when co-overexpressed with ectopic RalGEFs in 293T cells [28]. In addition, M-Ras bound strongly to and activated PI3K in fibroblasts [72]. However, M-Ras only stimulated PIP3 formation significantly when co-expressed with p110α and not with p110δ, γ, or β [28]. Finally, although transient overexpression and co-precipitation analyses showed that M-Ras also interacted with other RA domain-containing proteins (Rin and RASSF proteins) as well as with the isolated RA domains of PLCε, whether these are physiologically relevant efforts of M-Ras remains to be determined.

M-Ras-GTP also interacts with the Rap GEFs, MR-GEF, and RA-GEF-2, which possess a Ras association (RA) domain [74, 75]. Co-expression of constitutively activated M-Ras with RA-GEF-2 resulted in upregulation of GTP-bound, plasma membrane-bound Rap1[75]. In contrast, activated M-Ras blocked MR-GEF-mediated GTP loading of Rap1A [74].

Cell Biology

Of the R-Ras subfamily proteins, the least is known regarding M-Ras biological function. Like R-Ras and TC21, M-Ras controls cell proliferation and survival, and may also regulate actin cytoskeletal dynamics. M-Ras is weakly transforming in NIH 3T3 fibroblasts [6], and also induced transformation of melan-a-immortalized mouse melanocytes [4]. Activated M-Ras also caused IL-3-independent proliferation of RX-6 mast cell/megakaryocyte or BaF3 mouse myeloid cells, and blocked serum withdrawal induction of C2 myoblast differentiation [3, 5, 8]. Like Ras, R-Ras, and TC21, activated M-Ras also promoted cell survival in PC12 cells, by a mechanism that required PI3K function [72]. It should be noted that these activities of M-Ras were revealed by overexpression of mutated M-Ras proteins, and hence it is not clear whether they accurately reflect functions of endogenous M-Ras activated by extracellular stimuli. Finally, although M-Ras was shown to be strongly expressed in astrocytes and neurons, and was activated by neurotrophic factors, *MRAS*-null mice were phenotypically normal [76]. This lack of phenotypic consequences is similar to that seen in *HRAS*- and/or *NRAS*-null mice, but contrasts with *KRAS*-null mice that are embryonic lethal [77], and may reflect functional overlap of M-Ras activity with Ras and other R-Ras isoforms.

CONCLUSIONS

In summary, the three R-Ras subfamily members, R-Ras, TC21/R-Ras2, and M-Ras/R-Ras3, mediate cell growth, division, differentiation, and death by utilizing both novel pathways and those regulated by Ras and other Ras-related proteins. Further, R-Ras subfamily members, especially R-Ras itself, contribute to cellular processes such as integrin-mediated cell adhesion in a manner distinct from that of Ras. Deciphering the full contribution of R-Ras, TC21, and M-Ras to cellular growth control clearly awaits further study.

ACKNOWLEDGEMENTS

We thank Lanika DeGraffenreid for expert assistance with manuscript and figure preparation. Our studies were supported by grant CA042978 from the National Cancer Institute (to Channing J. Der). Ariella B. Hanker was supported by a Department of Defense Breast Cancer Research Program predoctoral fellowship (W81XWH-06-1-0747).

REFERENCES

1. Lowe DG, Capon DJ, Delwart E, Sakaguchi AY, Naylor SL, Goeddel DV. Structure of the human and murine R-ras genes, novel genes closely related to ras proto-oncogenes. *Cell* 1987;**48**:137–46.
2. Graham SM, Vojtek AB, Huff SY, Cox AD, Clark GJ, Cooper JA, Der CJ. TC21 causes transformation by Raf-independent signaling pathways. *Mol Cell Biol* 1996;**16**:6132–40.
3. Louahed J, Grasso L, De Smet C, Van Roost E, Wildmann C, Nicolaides NC, Levitt RC, Renauld JC. Interleukin-9-induced expression of M-Ras/R-Ras3 oncogene in T-helper clones. *Blood* 1999;**94**:1701–10.
4. Wang D, Yang W, Du J, Devalaraja MN, Liang P, Matsumoto K, Tsubakimoto K, Endo T, Richmond A. MGSA/GRO-mediated melanocyte transformation involves induction of Ras expression. *Oncogene* 2000;**19**:4647–59.
5. Ehrhardt GR, Leslie KB, Lee F, Wieler JS, Schrader JW. M-Ras, a widely expressed 29-kD homologue of p21 Ras: expression of a constitutively active mutant results in factor-independent growth of an interleukin-3-dependent cell line. *Blood* 1999;**94**:2433–44.
6. Kimmelman A, Tolkacheva T, Lorenzi MV, Osada M, Chan AM. Identification and characterization of R-ras3: a novel member of the RAS gene family with a non-ubiquitous pattern of tissue distribution. *Oncogene* 1997;**15**:2675–85.
7. Matsumoto K, Asano T, Endo T. Novel small GTPase M-Ras participates in reorganization of actin cytoskeleton. *Oncogene* 1997;**15**:2409–17.
8. Quilliam LA, Castro AF, Rogers-Graham KS, Martin CB, Der CJ, Bi C. M-Ras/R-Ras3, a transforming ras protein regulated by Sos1, GRF1, and p120 Ras GTPase-activating protein, interacts with the putative Ras effector AF6. *J Biol Chem* 1999;**274**:23,850–23,857.
9. Vetter IR, Wittinghofer A. The guanine nucleotide-binding switch in three dimensions. *Science* 2001;**294**:1299–304.
10. Mitin N, Rossman KL, Der CJ. Signaling interplay in Ras superfamily function. *Curr Biol* 2005;**15**:R563–74.
11. Cox AD, Der CJ. Farnesyltransferase inhibitors and cancer treatment: targeting simply Ras? *Biochim Biophys Acta* 1997;**1333**:F51–71.
12. Reid TS, Terry KL, Casey PJ, Beese LS. Crystallographic analysis of CaaX prenyltransferases complexed with substrates defines rules of protein substrate selectivity. *J Mol Biol* 2004;**343**:417–33.
13. Philips MR, Cox AD. Geranylgeranyltransferase I as a target for anti-cancer drugs. *J Clin Invest* 2007;**117**:1223–5.
14. Carboni JM, Yan N, Cox AD, Bustelo X, Graham SM, Lynch MJ, Weinmann R, Seizinger BR, Der CJ, Barbacid M, et al. Farnesyltransferase inhibitors are inhibitors of Ras but not R-Ras2/TC21, transformation. *Oncogene* 1995;**10**:1905–13.
15. Winter-Vann AM, Casey PJ. Post-prenylation-processing enzymes as new targets in oncogenesis. *Nat Rev Cancer* 2005;**5**:405–12.
16. Repasky GA, Chenette EJ, Der CJ. Renewing the conspiracy theory debate: does Raf function alone to mediate Ras oncogenesis?. *Trends Cell Biol* 2004;**14**:639–47.

17. Gotoh T, Cai D, Tian X, Feig LA, Lerner A. p130Cas regulates the activity of AND-34, a novel Ral, Rap1, and R-Ras guanine nucleotide exchange factor. *J Biol Chem* 2000;**275**:30,118–30,123.

18. Gotoh T, Niino Y, Tokuda M, Hatase O, Nakamura S, Matsuda M, Hattori S. Activation of R-Ras by Ras-guanine nucleotide-releasing factor. *J Biol Chem* 1997;**272**:18,602–18,607.

19. Gotoh T, Tian X, Feig LA. Prenylation of target GTPases contributes to signaling specificity of Ras-guanine nucleotide exchange factors. *J Biol Chem* 2001;**276**:38,029–38,035.

20. Kawasaki H, Springett GM, Toki S, Canales JJ, Harlan P, Blumenstiel JP, Chen EJ, Bany IA, Mochizuki N, Ashbacher A, Matsuda M, Housman DE, Graybiel AM. A Rap guanine nucleotide exchange factor enriched highly in the basal ganglia. *Proc Natl Acad Sci USA* 1998;**95**:13,278–13,283.

21. Ohba Y, Mochizuki N, Yamashita S, Chan AM, Schrader JW, Hattori S, Nagashima K, Matsuda M. Regulatory proteins of R-Ras, TC21/R-Ras2, and M-Ras/R-Ras3. *J Biol Chem* 2000;**275**:20,020–20,026,.

22. Tian X, Feig LA. Basis for signaling specificity difference between Sos and Ras-GRF guanine nucleotide exchange factors. *J Biol Chem* 2001;**276**:47,248–47,256.

23. Yamashita S, Mochizuki N, Ohba Y, Tobiume M, Okada Y, Sawa H, Nagashima K, Matsuda M. CalDAG-GEFIII activation of Ras, R-ras, and Rap1. *J Biol Chem* 2000;**275**:25,488–25,493.

24. Li S, Nakamura S, Hattori S. Activation of R-Ras GTPase by GTPase-activating proteins for Ras, Gap1(m), and p120GAP. *J Biol Chem* 1997;**272**:19,328–19,332.

25. Rey I, Taylor-Harris P, van Erp H, Hall A. R-ras interacts with ras-GAP, neurofibromin and c-raf but does not regulate cell growth or differentiation. *Oncogene* 1994;**9**:685–92.

26. Yamamoto T, Matsui T, Nakafuku M, Iwamatsu A, Kaibuchi K. A novel GTPase-activating protein for R-Ras. *J Biol Chem* 1995;**270**:30,557–30,561.

27. Puschel AW. GTPases in semaphorin signaling. *Adv Exp Med Biol* 2007;**600**:12–23.

28. Rodriguez-Viciana P, Sabatier C, McCormick F. Signaling specificity by Ras family GTPases is determined by the full spectrum of effectors they regulate. *Mol Cell Biol* 2004;**24**:4943–54.

29. Marte BM, Rodriguez-Viciana P, Wennstrom S, Warne PH, Downward J. R-Ras can activate the phosphoinositide 3-kinase but not the MAP kinase arm of the Ras effector pathways. *Curr Biol* 1997;**7**:63–70.

30. Suzuki J, Kaziro Y, Koide H. An activated mutant of R-Ras inhibits cell death caused by cytokine deprivation in BaF3 cells in the presence of IGF-I. *Oncogene* 1997;**15**:1689–97.

31. Suzuki J, Kaziro Y, Koide H. Synergistic action of R-Ras and IGF-1 on Bcl-xL expression and caspase-3 inhibition in BaF3 cells: R-Ras and IGF-1 control distinct anti-apoptotic kinase pathways. *FEBS Letts* 1998;**437**:112–16.

32. Takaya A, Kamio T, Masuda M, Mochizuki N, Sawa H, Sato M, Nagashima K, Mizutani A, Matsuno A, Kiyokawa E, Matsuda M. R-Ras regulates exocytosis by Rgl2/Rlf-mediated activation of RalA on endosomes. *Mol Biol Cell* 2007;**18**:1850–60.

33. Osada M, Tolkacheva T, Li W, Chan TO, Tsichlis PN, Saez R, Kimmelman AC, Chan AM. Differential roles of Akt, Rac, and Ral in R-Ras-mediated cellular transformation, adhesion, and survival. *Mol Cell Biol* 1999;**19**:6333–44.

34. Suzuki J, Kaziro Y, Koide H. Positive regulation of skeletal myogenesis by R-Ras. *Oncogene* 2000;**19**:1138–46.

35. Hamad NM, Elconin JH, Karnoub AE, Bai W, Rich JN, Abraham RT, Der CJ, Counter CM. Distinct requirements for Ras oncogenesis in human versus mouse cells. *Genes Dev* 2002;**16**:2045–57.

36. Taylor SJ, Resnick RJ, Shalloway D. Nonradioactive determination of Ras-GTP levels using activated ras interaction assay. *Methods Enzymol* 2001;**333**:333–42.

37. Chan AM, Miki T, Meyers KA, Aaronson SA. A human oncogene of the RAS superfamily unmasked by expression cDNA cloning. *Proc Natl Acad Sci USA* 1994;**91**:7558–62.

38. Huang Y, Saez R, Chao L, Santos E, Aaronson SA, Chan AM. A novel insertional mutation in the TC21 gene activates its transforming activity in a human leiomyosarcoma cell line. *Oncogene* 1995;**11**:1255–60.

39. Barker KT, Crompton MR. Ras-related TC21 is activated by mutation in a breast cancer cell line, but infrequently in breast carcinomas in vivo. *Br J Cancer* 1998;**78**:296–300.

40. Cox AD, Brtva TR, Lowe DG, Der CJ. R-Ras induces malignant, but not morphologic, transformation of NIH3T3 cells. *Oncogene* 1994;**9**:3281–8.

41. Graham SM, Cox AD, Drivas G, Rush MG, D'Eustachio P, Der CJ. Aberrant function of the Ras-related protein TC21/R-Ras2 triggers malignant transformation. *Mol Cell Biol* 1994;**14**:4108–15.

42. Huff SY, Quilliam LA, Cox AD, Der CJ. R-Ras is regulated by activators and effectors distinct from those that control Ras function. *Oncogene* 1997;**14**:133–43.

43. McFall A, Ulku A, Lambert QT, Kusa A, Rogers-Graham K, Der CJ. Oncogenic Ras blocks anoikis by activation of a novel effector pathway independent of phosphatidylinositol 3-kinase. *Mol Cell Biol* 2001;**21**:5488–99.

44. Rincon-Arano H, Rosales R, Mora N, Rodriguez-Castaneda A, Rosales C. R-Ras promotes tumor growth of cervical epithelial cells. *Cancer* 2003;**97**:575–85.

45. Wong OG, Nitkunan T, Oinuma I, Zhou C, Blanc V, Brown RS, Bott SR, Nariculam J, Box G, Munson P, Constantinou J, Feneley MR, Klocker H, Eccles SA, Negishi M, Freeman A, Masters JR, Williamson M. Plexin-B1 mutations in prostate cancer. *Proc Natl Acad Sci USA* 2007;**104**:19,040–19,045.

46. Hughes PE, Renshaw MW, Pfaff M, Forsyth J, Keivens VM, Schwartz MA, Ginsberg MH. Suppression of integrin activation: a novel function of a Ras/Raf-initiated MAP kinase pathway. *Cell* 1997;**88**:521–30.

47. Zhang Z, Vuori K, Wang H, Reed JC, Ruoslahti E. Integrin activation by R-ras. *Cell* 1996;**85**:61–9.

48. Ivins JK, Yurchenco PD, Lander AD. Regulation of neurite outgrowth by integrin activation. *J Neurosci* 2000;**20**:6551–60.

49. Keely PJ, Rusyn EV, Cox AD, Parise LV. R-Ras signals through specific integrin alpha cytoplasmic domains to promote migration and invasion of breast epithelial cells. *J Cell Biol* 1999;**145**:1077–88.

50. Sethi T, Ginsberg MH, Downward J, Hughes PE. The small GTP-binding protein R-Ras can influence integrin activation by antagonizing a Ras/Raf-initiated integrin suppression pathway. *Mol Biol Cell* 1999;**10**:1799–809.

51. Zou JX, Liu Y, Pasquale EB, Ruoslahti E. Activated SRC oncogene phosphorylates R-ras and suppresses integrin activity. *J Biol Chem* 2002;**277**:1824–7.

52. Goldfinger LE, Ptak C, Jeffery ED, Shabanowitz J, Hunt DF, Ginsberg MH. RLIP76 (RalBP1) is an R-Ras effector that mediates adhesion-dependent Rac activation and cell migration. *J Cell Biol* 2006;**174**:877–88.

53. Ada-Nguema AS, Xenias H, Hofman JM, Wiggins CH, Sheetz MP, Keely PJ. The small GTPase R-Ras regulates organization of actin and drives membrane protrusions through the activity of PLCepsilon. *J Cell Sci* 2006;**119**:1307–19.

54. Wang HG, Millan JA, Cox AD, Der CJ, Rapp UR, Beck T, Zha H, Reed JC. R-Ras promotes apoptosis caused by growth factor deprivation via a Bcl-2 suppressible mechanism. *J Cell Biol* 1995;**129**:1103–14.

55. Drivas GT, Shih A, Coutavas E, Rush MG, D'Eustachio P. Characterization of four novel ras-like genes expressed in a human teratocarcinoma cell line. *Mol Cell Biol* 1990;**10**:1793–8.

56. Movilla N, Crespo P, Bustelo XR. Signal transduction elements of TC21, an oncogenic member of the R-Ras subfamily of GTP-binding proteins. *Oncogene* 1999;**18**:5860–9.

57. Graham SM, Oldham SM, Martin CB, Drugan JK, Zohn IE, Campbell S, Der CJ. TC21 and Ras share indistinguishable transforming and differentiating activities. *Oncogene* 1999;**18**:2107–16.

58. Lin AW, Barradas M, Stone JC, van Aelst L, Serrano M, Lowe SW. Premature senescence involving p53 and p16 is activated in response to constitutive MEK/MAPK mitogenic signaling. *Genes Dev* 1998;**12**:3008–19.

59. Rosario M, Paterson HF, Marshall CJ. Activation of the Raf/MAP kinase cascade by the Ras-related protein TC21 is required for the TC21-mediated transformation of NIH 3T3 cells. *EMBO J* 1999;**18**:1270–9.

60. Light Y, Paterson H, Marais R. 14-3-3 antagonizes Ras-mediated Raf-1 recruitment to the plasma membrane to maintain signaling fidelity. *Mol Cell Biol* 2002;**22**:4984–96.

61. Lopez-Barahona M, Bustelo XR, Barbacid M. The TC21 oncoprotein interacts with the Ral guanosine nucleotide dissociation factor. *Oncogene* 1996;**12**:463–70.

62. Murphy GA, Graham SM, Morita S, Reks SE, Rogers-Graham K, Vojtek A, Kelley GG, Der CJ. Involvement of phosphatidylinositol 3-kinase, but not RalGDS, in TC21/R-Ras2-mediated transformation. *J Biol Chem* 2002;**277**:9966–75.

63. Peterson II. SN, Trabalzini L, Brtva TR, Fischer T, Altschuler DL, Martelli P, Lapetina EG, Der CJ, White GC Identification of a novel RalGDS-related protein as a candidate effector for Ras and Rap1. *J Biol Chem* 1996;**271**:29,903–29,908.

64. Spaargaren M, Bischoff JR. Identification of the guanine nucleotide dissociation stimulator for Ral as a putative effector molecule of R-ras, H-ras, K-ras, and Rap.. *Proc Natl Acad Sci USA* 1994;**91**:12,609–12,613.

65. Rosario M, Paterson HF, Marshall CJ. Activation of the Ral and phosphatidylinositol 3′ kinase signaling pathways by the ras-related protein TC21. *Mol Cell Biol* 2001;**21**:3750–62.

66. Khosravi-Far R, White MA, Westwick JK, Solski PA, Chrzanowska-Wodnicka M, Van Aelst L, Wigler MH, Der CJ. Oncogenic Ras activation of Raf/mitogen-activated protein kinase-independent pathways is sufficient to cause tumorigenic transformation. *Mol Cell Biol* 1996;**16**:3923–33.

67. Fiordalisi II. JJ, Johnson RL, Ulku AS, Der CJ, Cox AD Mammalian expression vectors for Ras family proteins: generation and use of expression constructs to analyze Ras family function. *Methods Enzymol* 2001;**332**:3–36.

68. Rong R, He Q, Liu Y, Sheikh MS, Huang Y. TC21 mediates transformation and cell survival via activation of phosphatidylinositol 3-kinase/Akt and NF-kappaB signaling pathway. *Oncogene* 2002; **21**:1062–70.

69. Rodriguez-Viciana P, Warne PH, Khwaja A, Marte BM, Pappin D, Das P, Waterfield MD, Ridley A, Downward J. Role of phosphoinositide 3-OH kinase in cell transformation and control of the actin cytoskeleton by Ras. *Cell* 1997;**89**:457–67.

70. Clark GJ, Kinch MS, Gilmer TM, Burridge K, Der CJ. Overexpression of the Ras-related TC21/R-Ras2 protein may contribute to the development of human breast cancers. *Oncogene* 1996;**12**:169–76.

71. Erdogan M, Pozzi A, Bhowmick N, Moses HL, Zent R. Signaling pathways regulating TC21-induced tumorigenesis. *J Biol Chem* 2007; **282**:27,713–27,720.

72. Kimmelman AC, Osada M, Chan AM. R-Ras3, a brain-specific Ras-related protein, activates Akt and promotes cell survival in PC12 cells. *Oncogene* 2000;**19**:2014–22.

73. Ehrhardt GR, Korherr C, Wieler JS, Knaus M, Schrader JW. A novel potential effector of M-Ras and p21 Ras negatively regulates p21 Ras-mediated gene induction and cell growth. *Oncogene* 2001;**20**:188–97.

74. Rebhun JF, Castro AF, Quilliam LA. Identification of guanine nucleotide exchange factors (GEFs) for the Rap1 GTPase. Regulation of MR-GEF by M-Ras-GTP interaction. *J Biol Chem* 2000;**275**:34,901–34,908.

75. Gao X, Satoh T, Liao Y, Song C, Hu CD, Kariya Ki K, Kataoka T. Identification and characterization of RA-GEF-2, a Rap guanine nucleotide exchange factor that serves as a downstream target of M-Ras. *J Biol Chem* 2001;**276**:42,219–42,225.

76. Nunez Rodriguez N, Lee IN, Banno A, Qiao HF, Qiao RF, Yao Z, Hoang T, Kimmelman AC, Chan AM. Characterization of R-ras3/m-ras null mice reveals a potential role in trophic factor signaling. *Mol Cell Biol* 2006;**26**:7145–54.

77. Johnson L, Greenbaum D, Cichowski K, Mercer K, Murphy E, Schmitt E, Bronson RT, Umanoff H, Edelmann W, Kucherlapati R, Jacks T. K-ras is an essential gene in the mouse with partial functional overlap with N-ras. *Genes Dev* 1997;**11**:2468–81.

Chapter 55

The Ran GTPase: Cellular Roles and Regulation

Mary Dasso

Laboratory of Gene Regulation and Development/National Institute of Child Health and Human Development, Bethesda, Maryland

INTRODUCTION

This chapter discusses the biochemistry of the Ran pathway, how this biochemistry is utilized to promote cellular activities and recent insights into Ran pathway regulation. Similar to many other Ras-family GTPases, Ran's intrinsic rates of guanine nucleotide exchange and hydrolysis are slow. Nucleotide exchange is accelerated by a nucleotide exchange factor, called RCC1 in vertebrates, while GTP hydrolysis is activated by a single GTPase activating protein, called RanGAP1 in vertebrates (reviewed in [1]; Figure 55.1). For nomenclature in other organisms, see Table 55.1. There is an asymmetric distribution of Ran's regulators throughout interphase: RCC1 is chromatin-bound and hence nuclear, whereas RanGAP1 is predominantly cytoplasmic. The distribution of these Ran regulators leads to differential Ran-GTP distribution within cells, providing spatial cues that direct Ran-dependent processes [2, 3].

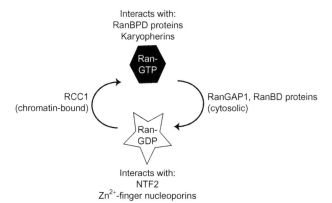

FIGURE 55.1 Core components of the Ran GTPase pathway.
RanGAP1 and RanBD-containing proteins promote cytosolic GTP hydrolysis by Ran. RCC1 catalyzes Ran guanine nucleotide exchange within the nucleus. Ran-GTP binds Karyopherins and RanBD-containing proteins. Ran-GDP binds NTF2 and zinc finger-containing nucleoporins.

These cues are interpreted by a family of Ran-GTP binding effector proteins that were originally characterized as nuclear transport receptors [4]. Here, these proteins will collectively be called Karyopherins. Karyopherins that mediate nuclear import will be called Importins, and those that mediate export will be called Exportins. Cargo loading of Karyopherins is governed by Ran-GTP levels [4, 5]: Importins bind to their cargo in the absence of Ran-GTP and dissociate upon Ran-GTP binding. Conversely, Exportins bind their cargo in ternary complexes that contain Ran-GTP, and dissociate from cargo upon Ran-GTP hydrolysis.

There are three other attributes of Ran that are generally notable. First, Ran is not isoprenylated, and is freely soluble. Second, Ran is extremely abundant, as are many of its interacting partners. Third, Ran-GTP binds to a family of proteins that possess a common structural motif (RanBD) [1]. RanBP1 is the best-studied member of this family; it does not have GAP activity, but it acts as a RanGAP1 accessory protein [6]. RanBP2 (also called Nup358) is protein with four RanBD domains that localizes to the cytoplasmic face of the nuclear pore complex (NPC) [1]. RanBP3 acts as a co-factor for the formation of complexes for nuclear export [7, 8].

STRUCTURAL ANALYSIS OF RAN PATHWAY COMPONENTS

The structure of Ran is dramatically regulated by nucleotide binding, particularly in the switch I region [9, 10]. The orientation of the switch I region precludes association between Ran-GDP and Karyopherins, providing a molecular rationale for the specificity for Karyopherins in binding to Ran-GTP only. Karyopherins are typically constructed from 18–20 HEAT repeats (*H*untingtin, *e*longation factor 3, the PR65/*A* subunit of protein phosphatase 2A

TABLE 55.1 Nomenclature for core Ran pathway components

Protein	S. cerevisiae	D. melanogaster	C. elegans	metazoan	Notes
GTPase	Gsp1/Cnr1, Gsp2/Cnr2	Ran	RAN-1	Ran	• GTPase
GAP	Rna1	Segregation Distorter	RAN-2	RanGAP1	• GTPase activating protein • Cytoplasmic during interphase in metazoans, and concentrated at NPC through RanBP2 binding • Spindle associated in mitosis for many metazoan cells • Cytoplasmic throughout cell cycle in yeast
GEF	Mtr1/Prp20/ Srm1	BJ1	RAN-3	RCC1	• Guanine nucleotide exchange factor for Ran • Localized predominantly on chromatin throughout the cell cycle
RanBP1	Yrb1/Cst20	—	—	RanBP1	• Ran-GTP binding protein • Promotes Ran-GTP release from Karyopherins, and acts as RanGAP accessory factor • Predominatly cytoplasmic during interphase • Apparently absent in some species (e.g., flies, worms)
RanBP2	Nup2?	CG11856	NPP-9	RanBP2/ Nup358	• Nuclear pore-associated Ran-binding protein • Metazoan protein contains both Ran-GTP and Ran-GDP binding motifs • Vertebrate RanGAP1 associated to RanBP2/Nup358 throughout the cell cycle
RanBP3	Yrb2p	CG10225	RAN-5	RanBP3	• Regulator of Crm1 export pathway • Increases the affinity of Crm1 for Ran-GTP and export cargo • Associates with and activates RCC1 • Promotes the association of Crm1 and RCC1 • Inhibits the association of unloaded Crm1 to the NPC
NTF2	Ntf2p	Ntf-2	RAN-4	Ntf2	• Required for efficient recycling of Ran-GDP to nucleus from cytosol in nuclear transport cycle

and the lipid kinase TOR) [4]. These repeats stack into two C-shaped arches and give Karyopherins a overall helicoidal structure, with the pitch of the superhelices varying between individual Karyopherins. This construction causes Karyopherins to be very flexible, in a manner analogous to a tightly wound spring, where each HEAT repeat represents a turn of the spring, and subtle changes in the orientation of adjacent coils substantially change the helicoidal pitch [4].

Cargos bind to relatively large interaction surfaces within the C-terminal domains of Karyopherins [4]. Cargo binding significantly changes the pitch of Karyopherins through an induced fit mechanism, in order to make optimal cargo–Karyopherin contacts. Such distortion causes Karyopherins to be functionally "spring loaded" for later release: The highly favorable interactions required for specific cargo recognition are balanced by increased internal energy as karyopherins deform during cargo binding. Karyopherins thus ensure selective cargo recognition by using large binding interfaces while simultaneously allowing cargo loading and unloading transitions to be achieved through the relatively small amount of energy released by Ran-GTP hydrolysis. Ran binds the N-terminal HEAT repeats of Karyopherins through contacts made with its the switch I and switch II regions [11–14]. At the same time, the switch I region contacts the C-terminal arch of the karyopherins. For yeast Importin-β, for instance, the latter interactions cause a change in helicoidal pitch that is incompatible with binding to its adaptor molecule (Importin-α) or other cargo, and thus promotes cargo release [12]. This might be taken as a general paradigm for Ran-GTP-mediated cargo binding or release [4].

Binding of Ran-GTP to Karyopherins does not inhibit association to RanBD-containing proteins, since they

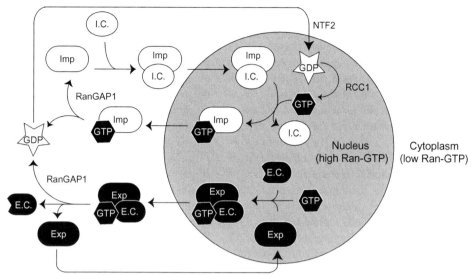

FIGURE 55.2 Ran function in nuclear trafficking.
The compartmentalization of RCC1 causes nuclear Ran-GDP (**GDP**) to be low and Ran-GTP (**GTP**) levels to be high. RanGAP1 localization results in low cytoplasmic Ran-GTP levels. After GTP hydrolysis, the NTF2 protein imports Ran-GDP. Importins (**Imp**) and their cargos (**I.C.**) form complexes in the cytoplasm, in the absence of Ran-GTP, and transit across the NPC. In the nucleus, Ran-GTP binding to importins causes cargo release, and they return to the cytosol in their Ran-GTP-bound form. In the cytoplasm, RanGAP1 and RanBP1 promote conversion of Ran-GTP to Ran-GDP, allowing receptor recycling. Exportins (**Exp**) and Ran-GTP bind to export cargo (**E.C.**) within the nucleus. These complexes transit to the cytosol, where they dissociate after Ran-GTP hydrolysis.

bind on different surfaces of Ran [10], but the interaction sites for RanGAP1 and Karyopherins significantly overlap [14, 15]. As a result, RanBD proteins are essential *in vivo* because they are required to dissociate Ran-GTP from Karyopherins to permit access of RanGAP1 [14, 16–18]. RanBD domains make extensive contacts that have been described as a "molecular embrace" of Ran [14], causing a shift in the Ran's acidic C terminus, and promoting Ran's release from bound Karyopherins. RanBP1 does not directly contact RanGAP within the RanGAP–Ran-GTP–RanBP1 ternary complex [15]. Rather, RanBP1 exerts a positive influence on GAP-mediated hydrolysis by accelerating the kinetics of Ran-GTP association to RanGAP1, probably through remodeling of the C-terminal domain of Ran [9, 19].

Ran shows an interesting divergence from other Ras-like GTPases with respect to its mechanism of GAP-mediated hydrolysis [15]. Yeast RanGAP consists of 11 leucine-rich repeats of 28–37 residues each, forming a crescent [20]. Unlike GAPs for many other ras family GTPases, RanGAP does not require an arginine residue provided by the GAP protein [15]: Structures of a Ran–RanBP1–RanGAP ternary complex in the ground state and in a transition-state mimic show that RanGAP does not provide a catalytic arginine. Rather, Ran alone provides the machinery for GTP hydrolysis, with tyrosine 39 of Ran replacing the arginine provided by other GAPs.

RCC1 is a seven-bladed beta-propeller protein [21], consistent with the earlier observation that RCC1's amino acid sequence shows of a series of seven direct repeats [22]. RCC1's interaction interface with Ran resides in

the loops between the propeller blades on one face of the protein [23].

RAN'S ROLE IN NUCLEAR TRANSPORT

Vertebrate NPCs are large (125-MDa) structures that consist of roughly 30 individual proteins, each of which is present in multiple copies [24]. While NPCs allow passive diffusion of small molecules, they exclude molecules of more than 40 kDa that are not bound to Karyopherins or other transport carriers [25]. This exclusion is maintained through a subset of nucleoporins that possess distinctive Phe–Gly (FG) sequence motifs. Several different mechanisms for regulated exclusion by FG nucleoporins have been proposed, including the idea that the FG domains form a sieve-like gel [25]. In favor of this model, saturated hydrogels formed from a single FG-repeat domain reproduce the permeability properties of NPCs, with respect to both exclusion of bulk proteins, and permeability for Importin-β and Importin-β-cargo complexes [26]. Interactions between the shallow hydrophobic cavities on the Karyopherins and the Phe side-chains of the hydrophobic FG-repeat cores may be critical for Karyopherin passage across NPCs [25]. These interactions allow Karyopherins to evade the normally exclusive properties of the NPC, but are intrinsically weak and transient, thereby allowing the rapid off-rates required for passage.

Ran functions primarily to determine the directionality of nuclear trafficking [25] (Figure 55.2): Importins bind to their cargo in the cytoplasm, where Ran-GTP is absent, and translocate readily across the NPC. In the nucleus, RCC1-generated Ran-GTP binds to Importins and releases the cargo. Conversely, Exportins and their cargo associate within the nucleus as complexes containing Ran-GTP. After these complexes translocate to the cytoplasm, RanGAP1 induces GTP hydrolysis, thereby causing the dissociation of the complexes. A small Ran-GDP-binding protein, NTF2, reimports Ran-GDP back into the nucleus, where regeneration of Ran-GTP is achieved by RCC1. Different Karyopherins show interesting variations in the way that they utilize the Ran-GTP gradient. First, while each Karyopherin can only carry a particular cargo in one direction, there is no reason why they cannot carry other cargos in the opposite direction. Indeed, bidirectional Karyopherins have been reported in yeast [27] and mammals [28]. Second, some Karyopherins bind cargos through adaptor proteins [25]. For instance, the Importin-α adaptor binds cargos with classical nuclear localization signals (cNLSs). Importin-β associates with the Importin-α–cargo complex and promotes its NPC transit. After release by Ran-GTP in the nucleus, Importin-α is re-exported in association with Ran-GTP and the Exportin CAS.

Third, import cargo unloading can be controlled by events in addition to Ran-GTP binding [29, 30], allowing Importins to direct not only nuclear import but also intranuclear targeting or assembly of cargos into macromolecular complexes. Finally, accessory proteins may control the loading of Karyopherins. For example, the Exportin Crm1 binds RanBP3, a nuclear RanBD-containing protein [31, 32]. RanBP3 increases the affinity of Crm1 for Ran-GTP and export cargo. RanBP3 also associates with RCC1 in a manner that is stimulated by Ran, and activates RCC1 as Ran GEF [7]. RanBP3 promotes the association of Crm1 and RCC1, perhaps acting as a scaffold to coordinate the loading of Ran-GTP onto Crm1. RanBP3 inhibits the association of unloaded Crm1 to the NPC in a manner that is relieved by Ran-GTP [32], suggesting that it permits Crm1 association to the NPC only after export complex assembly is complete.

RAN'S FUNCTION IN SPINDLE ASSEMBLY

In 1999, it was shown that Ran regulates spindle assembly in a manner that is independent of its nuclear transport function [33–37]. In particular, spindle assembly is severely defective when Ran-GTP levels are lowered in *Xenopus* M-phase arrested egg extracts, a mitotic system that is devoid of intact nuclei. Under these conditions, spindles are disorganized, with low densities of microtubules (MTs) [35, 37]. Conversely, increased levels of Ran-GTP in M-phase extracts cause massive polymerization of MTs in a manner that does not require chromosomes or centrosomes [34, 35, 37]. It had been previously found that mitotic chromosomes can locally stabilize MTs, probably through the action of a diffusable MT-stabilizing factor produced by a chromatin-associated enzyme (reviewed in [38]). Since

RCC1 binds to chromatin, it was natural to speculate that Ran-GTP could play a role in the localized stabilization of mitotic MTs by chromosomes. Consistent with this idea, fluorescence resonance energy transfer (FRET) experiments show that Ran-GTP concentrations are elevated in the vicinity of mitotic chromosomes [2].

A major soluble effector pathway for Ran during mitosis involves Importin-β and Importin-α. Importin-α/β bind and inhibit a growing list of spindle assembly factors (SAFs) [39]. In the vicinity of chromosomes, Ran-GTP may bind Importin-β and release these inhibitory complexes in a manner similar to its action during nuclear protein import, resulting in a local activation of the SAFs near chromosomes (Figure 55.3). At a distance from chromosomes, Ran-GTP would presumably undergo nucleotide hydrolysis and inhibition would be restored. In reality, it is likely that the site of nucleotide hydrolysis may also very important for spindle function, since vertebrate RanGAP1 is localized to the spindle in a highly regulated manner [40]. Notably, Importin-α/β may promote interphase nuclear localization of SAFs, insuring that they are not inappropriately active on MTs in cytosol outside of mitosis.

A second mitotic Ran-GTP effector pathway in vertebrates that requires the Exportin Crm1 (Figure 55.2). Crm1 localizes to kinetochores of mammalian cells throughout mitosis via a mechanism that requires neither MTs nor Ran-GTP [41]. RanGAP1 becomes linked to a small ubiquitin-like protein called SUMO-1 [42, 43], resulting in the formation of a covalently-bound conjugate (RanGAP1•SUMO-1). RanGAP1•SUMO-1 binds to RanBP2 in a complex that includes the SUMO-1 conjugating enzyme (Ubc9) [44]. This complex will be referred to as the RRSU complex, and it remains highly stable throughout the cell cycle [45]. In mitotic mammalian cells, the RRSU complex is targeted to kinetochores in a MT-dependent fashion [40]. Inhibition of Crm1 ternary complex formation using leptomycin B (LMB), a highly specific chemical inhibitor, blocks kinetochore recruitment of the RRSU complex [41], showing that its recruitment is also Crm1 dependent. It is possible that localization of the RRSU complex acts primarily to change the Ran-GTP levels in the environment of the kinetochore through the action of RanGAP1. Alternatively, RRSU recruitment may be important because it brings other components to kinetochores [46]. Notably, the extent of Crm1 and RRSU complex recruitment to kinetochores varies between different species and cell types [47], with corresponding variability in response of different mitotic cells to LMB.

Finally, Ran, Importin-β and Crm1 have all been implicated in centrosomal duplication and function. A fraction of Ran-GTP localizes to centrosomes in a manner that requires the centrosomal matrix A kinase anchoring protein (AKAP450) [48]. Depletion of AKAP450 releases Ran-GTP from centrosomes, and causes defects in MT anchorage [48]. Moreover, centrosomal cohesion can be

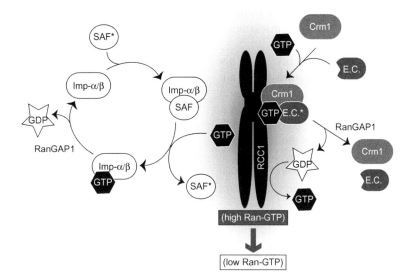

FIGURE 55.3 Dual pathways for Ran in metazoan mitosis.
Importin-α/β import receptor (***Imp-α/β***) binds active spindle assembly factors (***SAF****) in regions distant from chromosomes, causing their conversion to inhibited forms (***SAF***). Near chromosomes, chromatin-bound RCC1 generates an increased level of Ran-GTP (***GTP***). Ran-GTP disrupts Importin-α/β complexes, and activates SAFs. Crm1 is localized to kinetochores (***Crm1***), where it presumably binds to export cargo (***E.C.***) and Ran-GTP. Distal from chromosomes, these interactions would be released when Ran-GTP is converted to Ran-GDP (***GDP***) by RanGAP1 and RanBP1. Formation of Crm1 ternary complexes may render export cargo into forms with different biochemical properties (***E.C.****), or merely promotes their concentration on kinetochores.

disrupted by overexpression of RanBP1, implicating Ran-GTP pools more directly [49]. Importin-β becomes associated with spindle poles in a fashion that is dependent on MTs and TPX2, a dynein-associated SAF [50]. As with RanBP1, overexpression of Importin-β causes fragmentation of spindle poles, implicating Importin-β itself, Ran-GTP, or both, in maintenance of pole integrity. At the same time, a fraction of Crm1 has been reported to localize to centrosomes [51]; Crm1 is required for centrosomal targeting of nucleophosmin (NPM), a nucleolar phosphoprotein that has been implicated in a variety of cellular functions, including centrosomal duplication [52]. Depletion of NPM or prolonged treatment with LMB causes formation of supernumerary centrosomes [51, 52]. LMB treatment also causes nuclear accumulation of the centrosomal proteins Centrin and Pericentrin [48], suggesting Crm1 controls their shuttling between the nucleus and cytoplasm as well.

RAN'S FUNCTION IN CELL CYCLE PROGRESSION

A number of observations have implicated Ran in regulating the onset of mitosis. When arrested in S phase and then shifted to the restrictive temperature, a mutant Hamster cell line with a temperature sensitive allele of RCC1 (tsBN2 cells) progresses into mitosis prematurely, accompanied by nuclear envelope (NE) breakdown, precocious chromosome condensation, and activation of p34^{cdk1}/Cyclin B kinase [53]. These observations and others have shown that tsBN2 cells do not arrest appropriately in S phase in response to unreplicated DNA. Moreover, mutant Ran proteins can block onset of mitosis in the presence or absence of nuclei in *Xenopus* cycling extracts, which would otherwise alternate between interphase and mitosis [54, 55]. The molecular events whereby Ran regulates mitotic onset in metazoans remain unclear, and there is a lack of strong evidence from fission and budding yeast that Ran acts analogously to regulate mitosis in either of those organisms [56].

Ran plays an important role in controlling the metaphase – anaphase transition in some systems. To prevent chromosome mis-segregation, entry into anaphase is tightly controlled through the spindle assembly checkpoint (SAC), a regulatory pathway that monitors the attachment of MTs to kinetochores, and prevents activation of the anaphase-promoting complex/cyclosome (APC/C) until all chromosomes are correctly aligned onto the metaphase plate [57]. The APC/C is a ubiquitin ligase that promotes the destruction of key mitotic proteins, including B-type cyclins and Securins. In *Xenopus* egg extracts, elevated levels of Ran-GTP abrogate SAC-mediated mitotic arrest, allowing APC/C activation and dispersion of SAC components away from kinetochores [58]. These findings imply that the SAC within egg extracts is directly responsive to the concentration of Ran-GTP. *tsBN2* cells can arrest in response to the MT poison nocodazole, indicating that

SAC activation in response to fully unattached kineto-chores requires neither RCC1 nor Ran-GTP [41]. However, *tsBN2* cells undergo anaphase in the presence of inappro-priate MT-kinetochore attachments, indicating that cellular responses to these conditions are deficient.

Components downstream of Ran in the SAC have not been identified, although it is clear that neither Crm1 nor Importin-β are critical SAC effectors of Ran [41, 58]. In this context, it is interesting to note that SAC components, including the Mad1, Mad2, and Mps1 proteins, local-ize to interphase NPCs in both vertebrates and yeast [59]. Mad2 NPC association is sensitive to changes in the level of Ran-GTP [60]. It is therefore attractive to speculate that such interactions may be the target of Ran-GTP within the SAC.

RAN'S ROLE IN NUCLEAR ENVELOPE DYNAMICS

Ran-GTP plays multiple roles in nuclear envelope (NE) dynamics. First, it regulates post-mitotic NE assembly in organisms with open mitosis. Both nucleotide exchange and hydrolysis on Ran must occur to promote early events of nuclear membrane fusion during NE assembly in *Xenopus* egg extracts [61]. Remarkably, GST-Ran bound to beads can assemble structures resembling NE in egg extracts, which contain NPCs and are capable of nuclear transport [62]. Importin-β is required for NE assembly around chromatin templates or Ran on beads in *Xenopus* egg extracts [63, 64]. Moreover, concentration of Importin-β itself on beads is sufficient to induce NE assembly in this system [64]. Importin-β mutants that have a decreased affinity for nucle-oporins lack the capacity to induce NE assembly. However, an Importin-β mutant that does not interact with Importin-α is functional, suggesting that Importin-β mediates NE assembly by recruiting NPC components.

One likely target of Importin-β regulation in NE assem-bly is the Lamin B receptor (LBR), a chromatin- and lamin-binding protein of the inner NE. LBR binds to Importin-β in an Importin-α-independent fashion, and contributes to the targeting of membranes to nascent NEs around Importin-β- or Ran-coated beads [65]. It has been proposed that Importin-β initially targets membranes to chromatin through association with both LBR and chromatin-associated Ran-GTP, and that subsequent release of LBR after bind-ing of Ran-GTP and Importin-β to each other may promote NE remodeling and Importin-β recycling [65]. A second target may be the Nup107–160 complex, a group of 10 nucleoporins that localize to re-forming nuclear envelopes very early during the process of NPC assembly [63, 66]. Importin-β binding may disassemble Nup107-160 and other nucleoporins from the NPC as cells enter mitosis [67]. As cells return to interphase, Ran-GTP may reverse

this process by releasing Importin-β from the nucleoporins. Nuclear lamins have been proposed as a third NE target of Importin-α/β in egg extracts [68]. Importin-α binding prevents lamin assembly into filaments, suggesting that Ran can also control post-mitotic assembly of the nuclear lamina through Importin-α release [68]. Notably, Importin-α/β also bind proteins of the inner nuclear membrane and facilitate their import into nuclei [69], so these proteins might be additional potential targets for mitotic regulation through this receptor pathway.

Second, Ran-GTP is required for ongoing NPC assem-bly into intact NEs during interphase and for correct NE growth, even in organisms that do not undergo mitotic NE breakdown (NEBD) [70–72]. In *Xenopus* egg extracts, the insertion of new NPCs into existing NE requires Ran-GTP in a transport-independent manner, and can occur from either the inner or outer membrane [70]. The addi-tion of membranes into existing NEs was not blocked by exogenous Importin-β, and nor was nuclear transport, but assembly of NPCs was sharply inhibited. These results are consistent with the idea that Importin-β plays a role in chaperoning NPC assembly during interphase that is highly analogous to its post-mitotic function. In yeast, both Ran pathway components and Importin-β have been geneti-cally implicated in NPC assembly in a manner similar to the findings in *Xenopus* egg extracts [72, 73]. One impor-tant phenotypic difference between Ran cycle mutants and Importin-β mutants was that the former accumulated small cytoplasmic vesicles, while the latter formed elongated cytoplasmic membranes and had highly distorted NEs. These finding again suggests that Importin-β is required for NPC assembly but not NE membrane fusion, while Ran-GTP is required for both events.

Finally, high levels of Ran-GTP antagonize the final stages of NE rupture in an *in vitro* disassembly assay using permeabilized tissue culture cells and staged *Xenopus* egg extracts [74]. Inhibition of Importin-β also blocks NE rup-ture, suggesting the possibility that Importin-β is required for sequestration of NE components and disassembly of NE structures, such as the NPC and nuclear lamina. The MT network contributes to NE rupture *in vivo* [75] and within the *in vitro* disassembly assay [74], so it is also possible that Ran promotes NE rupture by remodeling MTs or alter-ing the activity of MT binding proteins and motors, in a manner related to its function in the intact mitotic spindle.

BIOLOGICAL REGULATION OF CORE RAN PATHWAY COMPONENTS

There are some indications that expression of Ran path-way components are regulated in response to developmen-tal or physiological cues [76], although a truly systematic analysis of their expression has not been made. In *Xenopus* tissue culture cells, both RanGAP1 and RanBP1 levels increase through the cell cycle, peaking in mitosis. Similar changes in RanBP1 have been observed in NIH3T3 cells: RanBP1 levels are very low in quiescent NIH3T3 cells and increase substantially as cells progress, again peaking in mitosis [77]. Overexpression of RanBP1 disrupts exit from mitosis [77], demonstrating that these changes are physiologically important.

The Ran pathway is regulated through a number of additional means, particularly through modulation of RCC1. Recently, it was shown that RanBP3 can be phosphorylated in response to activation of the Ras/ERK and PI3-kinase (PI3K) pathways [78]. This phosphorylation disrupted the interphase nuclear accumulation of Ran-GTP, probably through disturbance in RanBP3's capacity to regulate RCC1, and thereby disrupted nuclear trafficking. The association of Ran and RCC1 with chromatin is controlled through multiple mechanisms. Ran can bind to chromatin directly through association with histones H3 and H4 [79]. In *Xenopus* egg extracts, this binding occurs before acquisition of RCC1, and it facilitates RCC1 recruitment [80]. Efficient binding of RCC1 to chromatin requires its N-terminal domain [81]; RCC1 mutants lacking the N-terminal domain fail to stably associate to mitotic chromosomes and show pronounced defects in spindle assembly [82]. RCC1's association with chromatin requires interaction with core histones [83]. RCC1 binds directly to mononucleosomes and to histones H2A and H2B, and histone binding causes a modest stimulation of its catalytic activity [83]. Newly-synthesized RCC1 undergoes removal of its N-terminal methionine and mono-, di- or trimethylation of the α-amino group of its amino-terminal serine or proline. This modification promotes RCC1 binding to chromatin through electrostatic interactions between RCC1's modified N-terminal domain and DNA [84]. RCC1 thus appears to bind chromatin in a complex fashion, interacting with Ran, histones, and DNA.

RCC1's chromatin binding and catalytic activities are further regulated through interaction with Importin-α/β, phosphorylation, and mRNA splicing. The N-terminal domain of RCC1 binds to Importin-β through an importin-α isoform, importin-α3 [85]. Mitotic phosphorylation of human RCC1 at serines 2 and 11 by Cyclin B1/Cdc2 reduces importin-α/β association and accelerates the dynamics of RCC1's interaction with mitotic chromosomes [86, 87]. Human cells express at least three isoforms of RCC1, which are differentially subject to control through importin-α/β [88]. These isoforms show distinct patterns of phosphorylation, importin-α binding, and chromatin association. They show distinct profiles of expression in various tissues, and different capacities to support cell proliferation in cultured cells.

As discussed above, RanGAP1 is regulated through RRSU complex formation, promoting its association to NPCs during interphase and to spindles during mitosis.

RanGAP1 remains stably associated to the RRSU complex throughout the cell cycle [45], and there has been no demonstration that this association is responsive to physiological conditions. No comparable targeting has been described for RanGAP within yeast. Notably, however, plants maintain similar localization of RanGAP1 during interphase and mitosis through a completely distinct set of interactions. Plant RanGAP has a unique N-terminal extension (WPP domain) not found in animal or fungal RanGAPs [89], that interacts with a set of proteins that are not related to RanBP2. These observations suggest convergent evolution of to control Ran spatially.

CONCLUSIONS

Ran regulates nuclear transport, cell cycle progression, spindle assembly, and post-mitotic NE assembly, suggesting a wide role for Ran in coordinating events during the cell cycle. These facets of Ran are mechanistically linked in two ways. First, in all cases where we have a rudimentary molecular understanding of Ran function, it appears that Ran-GTP gradients provide spatial cues that serve to indicate the localization of the chromatin. This notion is particularly well supported for nuclear transport and spindle assembly. Second, gradients of Ran are interpreted through a common set of Ran-GTP-binding effectors, with the Importin-β protein playing a particularly prominent role.

REFERENCES

1. Dasso M. The Ran GTPase: theme and variations. *Curr Biol* 2002;**12**(14):R502–8.
2. Kalab P, Weis K, Heald R. Visualization of a Ran-GTP gradient in interphase and mitotic *Xenopus* egg extracts. *Science* 2002;**295**(5564):2452–6.
3. Smith AE, Slepchenko BM, Schaff JC, Loew LM, Macara IG. Systems analysis of Ran transport. *Science* 2002;**295**(5554):488–91.
4. Conti E, Muller CW, Stewart M. Karyopherin flexibility in nucleocytoplasmic transport. *Curr Opin Struct Biol* 2006;**16**(2):237–44.
5. Cook A, Bono F, Jinek M, Conti E. Structural biology of nucleocytoplasmic transport. *Annu Rev Biochem* 2007;**76**:647–71.
6. Bischoff FR, Krebber H, Smirnova E, Dong W, Ponstingl H. Co-activation of RanGTPase and inhibition of GTP dissociation by Ran-GTP binding protein RanBP1. *EMBO J* 1995;**14**(4):705–15.
7. Nemergut ME, Lindsay ME, Brownawell AM, Macara IG. Ran-binding protein 3 links Crm1 to the Ran guanine nucleotide exchange factor. *J Biol Chem* 2002;**277**(20):17,385.
8. Petosa C, Schoehn G, Askjaer P, Bauer U, Moulin M, Steuerwald U, Soler-López M, Baudin F, Mattaj I, Müller C, et al. Architecture of CRM1/Exportin1 suggests how cooperativity is achieved during formation of a nuclear export complex. *Mol Cell* 2004;**16**(5):761–75.
9. Scheffzek K, Klebe C, Fritz-Wolf K, Kabsch W, Wittinghofer A. Crystal structure of the nuclear Ras-related protein Ran in its GDP-bound form. *Nature* 1995;**374**(6520):378–81.

10. Vetter IR, Nowak C, Nishimoto T, Kuhlmann J, Wittinghofer A. Structure of a Ran–binding domain complexed with Ran bound to a GTP analogue: implications for nuclear transport. *Nature* 1999;**398**(6722):39–46.

11. Chook YM, Blobel G. Structure of the nuclear transport complex karyopherin-beta2-Ran x GppNHp. *Nature* 1999;**399**(6733):230–7.

12. Lee SJ, Matsuura SM, Liu M. Stewart, Structural basis for nuclear import complex dissociation by RanGTP. *Nature* 2005;**435**(7042):693–6.

13. Matsuura Y, Stewart M. Structural basis for the assembly of a nuclear export complex. *Nature* 2004;**432**(7019):872–7.

14. Vetter IR, Arndt A, Kutay U, Gorlich D, Wittinghofer A. Structural view of the Ran-Importin beta interaction at 2.3 A resolution. *Cell* 1999;**97**(5):635–46.

15. Seewald MJ, Korner C, Wittinghofer A, Vetter IR. RanGAP mediates GTP hydrolysis without an arginine finger. *Nature* 2002;**415**(6872):662–6.

16. Bischoff FR, Gorlich D. RanBP1 is crucial for the release of RanGTP from importin beta-related nuclear transport factors. *FEBS Letts* 1997;**419**(2–3):249–54.

17. Lounsbury KM, Macara IG. Ran-binding protein 1 (RanBP1) forms a ternary complex with Ran and karyopherin beta and reduces Ran GTPase-activating protein (RanGAP) inhibition by karyopherin beta. *J Biol Chem* 1997;**272**(1):551–5.

18. Saric M, Zhao X, Korner C, Nowak C, Kuhlmann J, Vetter IR. Structural and biochemical characterization of the Importin-beta.Ran. GTP.RanBD1 complex. *FEBS Lett* 2007;**581**(7):1369–76.

19. Seewald MJ, Kraemer A, Farkasovsky M, Körner C, Wittinghofer A, Vetter IR. Biochemical characterization of the Ran–RanBP1–RanGAP system: are RanBP proteins and the acidic tail of RanGAP required for the Ran–RanGAP GTPase reaction? *Mol Cell Biol* 2003;**23**(22):8124–36.

20. Hillig RC, Renault L, Vetter IR, et al. The crystal structure of rna1p: a new fold for a GTPase-activating protein. *Mol Cell* 1999;**3**(6):781–91.

21. Renault L, Nassar N, Vetter IR, et al. The 1.7 A crystal structure of the regulator of chromosome condensation (RCC1) reveals a seven-bladed propeller. *Nature* 1998;**392**(6671):97–101.

22. Ohtsubo M, Kai R, Furuno N, Sekiguchi T, Sekiguchi M, Hayashida H, Kuma K, Suzuki-Utsunomiya K, Mizumura H, Shefner JM, Cox GA, et al. Isolation and characterization of the active cDNA of the human cell cycle gene (RCC1) involved in the regulation of onset of chromosome condensation. *Genes Dev* 1987;**1**(6):585–93.

23. Azuma Y, Renault L, García-Ranea JA, et al. Model of the ran-RCC1 interaction using biochemical and docking experiments. *J Mol Biol* 1999;**289**(4):1119–30.

24. Cronshaw JM, Krutchinsky AN, Zhang W, Chait BT, Matunis MJ. Proteomic analysis of the mammalian nuclear pore complex. *J Cell Biol* 2002;**158**(5):915–27.

25. Stewart M. Molecular mechanism of the nuclear protein import cycle. *Nat Rev Mol Cell Biol* 2007;**8**(3):195–208.

26. Frey S, Gorlich D. A saturated FG-repeat hydrogel can reproduce the permeability properties of nuclear pore complexes. *Cell* 2007;**130**(3):512–23.

27. Yoshida K, Blobel G. The karyopherin Kap142p/Msn5p mediates nuclear import and nuclear export of different cargo proteins. *J Cell Biol* 2001;**152**(4):729–40.

28. Mingot JM, Kostka S, Kraft R, Hartmann E, Gorlich D. Importin 13: a novel mediator of nuclear import and export. *EMBO J* 2001;**20**(14):3685–94.

29. Pemberton LF, Rosenblum JS, Blobel G. Nuclear import of the TATA-binding protein: mediation by the karyopherin Kap114p

30. Senger B, Simos G, Bischoff FR, Podtelejnikov A, Mann M, Hurt E. Mtr10p functions as a nuclear import receptor for the mRNA-binding protein Npl3p. *Embo J* 1998;**17**(8):2196–207.

31. Englmeier L, Fornerod M, Bischoff FR, Petosa C, Mattaj IW, Kutay U, et al. RanBP3 influences interactions between CRM1 and its nuclear protein export substrates. *EMBO Rep* 2001;**2**(10):926–32.

32. Lindsay ME, Holaska JM, Welch K, et al. Ran-binding protein 3 is a cofactor for Crm1-mediated nuclear protein export. *J Cell Biol* 2001;**153**(7):1391–402.

33. Zhang C, Hughes M, Clarke PR. Ran-GTP stabilises microtubule asters and inhibits nuclear assembly in *Xenopus* egg extracts. *J Cell Sci* 1999;**112**(14):2453–61.

34. Wilde A, Zheng Y. Stimulation of microtubule aster formation and spindle assembly by the small GTPase Ran. *Science* 1999;**284**(5418):1359–62.

35. Kalab P, Pu RT, Dasso M. The ran GTPase regulates mitotic spindle assembly. *Curr Biol* 1999;**9**(9):481–4.

36. Carazo-Salas RE, Guarguaglini G, Gruss OJ, Segref A, Karsenti E, Mattaj IW. Generation of GTP-bound Ran by RCC1 is required for chromatin-induced mitotic spindle formation. *Nature* 1999;**400**(6740):178–81.

37. Ohba T, Nakamura M, Nishitani H, et al. Self-organization of microtubule asters induced in *Xenopus* egg extracts by GTP-bound Ran. *Science* 1999;**284**(5418):1356–8.

38. Karsenti E, Vernos I. The mitotic spindle: a self-made machine. *Science* 2001;**294**(5542):543–7.

39. Joseph J. Ran at a glance. *J Cell Sci* 2006;**119**(Pt 17):3481–4.

40. Joseph J, Tan SH, Karpova TS, McNally JG, Dasso M. SUMO-1 targets RanGAP1 to kinetochores and mitotic spindles. *J Cell Biol* 2002;**156**(4):595–602.

41. Arnaoutov A, Azuma Y, Ribbeck K, et al. Crm1 is a mitotic effector of Ran-GTP in somatic cells. *Nat Cell Biol* 2005;**7**(6):626–32.

42. Mahajan R, Delphin C, Guan T, Gerace L, Melchior F. A small ubiquitin-related polypeptide involved in targeting RanGAP1 to nuclear pore complex protein RanBP2. *Cell* 1997;**88**(1):97–107.

43. Matunis MJ, Coutavas E, Blobel G. A novel ubiquitin-like modification modulates the partitioning of the Ran-GTPase-activating protein RanGAP1 between the cytosol and the nuclear pore complex. *J Cell Biol* 1996;**135**(6 Pt 1):1457–70.

44. Saitoh H, Pu R, Cavenagh M, Dasso M. RanBP2 associates with Ubc9p and a modified form of RanGAP1. *Proc Natl Acad Sci USA* 1997;**94**(8):3736–41.

45. Joseph J, Liu ST, Jablonski SA, Yen TJ, Dasso M. The RanGAP1-RanBP2 complex is essential for microtubule-kinetochore interactions invivo. *Curr Biol* 2004;**14**(7):611–17.

46. Dasso M. Emerging roles of the SUMO pathway in mitosis. *Cell Div* 2008;**3**(1):5.

47. Arnaoutov A, Dasso M. Ran-GTP regulates kinetochore attachment in somatic cells. *Cell Cycle* 2005;**4**(9):1161–5.

48. Keryer G, Di Fiore B, Celati C, Lechtreck KF, Mogensen M, Delouvée A, Lavia P, Bornens M, Tassin A-M. Part of Ran is associated with AKAP450 at the centrosome: involvement in microtubule-organizing activity. *Mol Biol Cell* 2003;**14**(10):4260–71.

49. Di Fiore B, Ciciarello M, Mangiacasale R, Palena A, Tassin AM, Cundari E, Lavia P. Mammalian RanBP1 regulates centrosome cohesion during mitosis. *J Cell Sci* 2003;**116**(Pt 16):3399–411.

50. Ciciarello M, Mangiacasale R, Thibier C, Guarguaglini G, Marchetti E, Di Fiore B, Lavia P. Importin beta is transported to spindle poles

during mitosis and regulates Ran-dependent spindle assembly factors in mammalian cells. *J Cell Sci* 2004;**117**(26):6511–22.

51. Forgues M, Difilippantonio MJ, Linke SP, Ried T, Nagashima K, Feden J, Valerie K, Fukasawa K, Wang XW. Involvement of Crm1 in hepatitis B virus X protein-induced aberrant centriole replication and abnormal mitotic spindles. *Mol Cell Biol* 2003;**23**(15):5282–92.

52. Wang W, Budhu A, Forgues M, Wang XW. Temporal and spatial control of nucleophosmin by the Ran–Crm1 complex in centrosome duplication. *Nat Cell Biol* 2005;**7**(8):823–30.

53. Nishitani H, Ohtsubo M, Yamashita K, et al. Loss of RCC1, a nuclear DNA-binding protein, uncouples the completion of DNA replication from the activation of cdc2 protein kinase and mitosis. *Embo J* 1991;**10**(6):1555–64.

54. Clarke PR, Klebe C, Wittinghofer A, Karsenti E. Regulation of Cdc2/cyclin B activation by Ran, a Ras-related GTPase. *J Cell Sci* 1995;**108**(3):1217–25.

55. Kornbluth S, Dasso M, Newport J. Evidence for a dual role for TC4 protein in regulating nuclear structure and cell cycle progression. *J Cell Biol* 1994;**125**(4):705–19.

56. Sazer S, Dasso M. The ran decathlon: multiple roles of Ran. *J Cell Sci* 2000;**113**(7):1111–18.

57. Musacchio A, Salmon ED. The spindle-assembly checkpoint in space and time. *Nat Rev Mol Cell Biol* 2007;**8**(5):379–93.

58. Arnaoutov A, Dasso M. The Ran GTPase regulates kinetochore function. *Dev Cell* 2003;**5**(1):99–111.

59. Stukenberg PT, Macara IG. The kinetochore NUPtials. *Nat Cell Biol* 2003;**5**(11):945–7.

60. Quimby BB, Arnaoutov A, Dasso M. Ran GTPase regulates Mad2 localization to the nuclear pore complex. *Eukaryot Cell* 2005;**4**(2):274–80.

61. Hetzer M, Bilbao-Cortes D, Walther TC, Gruss OJ, Mattaj IW. GTP hydrolysis by Ran is required for nuclear envelope assembly. *Mol Cell* 2000;**5**(6):1013–24.

62. Zhang C, Clarke PR. Chromatin-independent nuclear envelope assembly induced by Ran GTPase in *Xenopus* egg extracts. *Science* 2000;**288**(5470):1429–32.

63. Harel A, Chan RC, Lachish-Zalait A, Zimmerman E, Elbaum M, Forbes DJ. Importin beta negatively regulates nuclear membrane fusion and nuclear pore complex assembly. *Mol Biol Cell* 2003;**14**(11):4387–96.

64. Zhang C, Hutchins JR, Muhlhausser P, Kutay U, Clarke PR. Role of importin-beta in the control of nuclear envelope assembly by Ran. *Curr Biol* 2002;**12**(6):498–502.

65. Ma Y, Cai S, Lv QL, et al. Lamin B receptor plays a role in stimulating nuclear envelope production and targeting membrane vesicles to chromatin during nuclear envelope assembly through direct interaction with importin beta. *J Cell Sci* 2007;**120**(3):520–30.

66. Walther TC, Alves A, Pickersgill H, et al. The conserved Nup107–160 complex is critical for nuclear pore complex assembly. *Cell* 2003;**113**(2):195–206.

67. Walther TC, Askjaer P, Gentzel M, et al. RanGTP mediates nuclear pore complex assembly. *Nature* 2003;**424**(6949):689–94.

68. Adam SA, Sengupta K, Goldman RD. Regulation of nuclear lamin polymerization by importin alpha. *J Biol Chem* 2008.

69. King MC, Lusk CP, Blobel G. Karyopherin-mediated import of integral inner nuclear membrane proteins. *Nature* 2006;**442**(7106):1003–7.

70. D'Angelo MA, Anderson DJ, Richard E, Hetzer MW. Nuclear pores form de novo from both sides of the nuclear envelope. *Science* 2006;**312**(5772):440–3.

71. Demeter J, Morphew M, Sazer S. A mutation in the RCC1-related protein pim1 results in nuclear envelope fragmentation in fission yeast. *Proc Natl Acad Sci USA* 1995;**92**(5):1436–40.

72. Ryan KJ, McCaffery JM, Wente SR. The Ran GTPase cycle is required for yeast nuclear pore complex assembly. *J Cell Biol* 2003;**160**(7):1041–53.

73. Ryan KJ, Zhou Y, Wente SR. The karyopherin Kap95 regulates nuclear pore complex assembly into intact nuclear envelopes *in vivo*. *Mol Biol Cell* 2007;**18**(3):886–98.

74. Muhlhausser P, Kutay U. An *in vitro* nuclear disassembly system reveals a role for the RanGTPase system and microtubule-dependent steps in nuclear envelope breakdown. *J Cell Biol* 2007;**178**(4):595–610.

75. Salina D, Bodoor K, Eckley DM, et al. Cytoplasmic dynein as a facilitator of nuclear envelope breakdown. *Cell* 2002;**108**(1):97–107.

76. Dasso MDS, Pu R. Multiple roles of the Ran GTPase during the cell cycle. In: Rush M, D'Eustachio P, editors. *The small GTPase Ran.* Boston, MA: Kluwer Academic Publishers; 2001. p. 105–22.

77. Battistoni A, Guarguaglini G, Degrassi F, et al. Deregulated expression of the RanBP1 gene alters cell cycle progression in murine fibroblasts. *J Cell Sci* 1997;**110**(19):2345–57.

78. Yoon SO, Shin S, Yuzhen L, et al. Ran-binding protein 3 phosphorylation links the Ras and PI3-Kinase pathways to nucleocytoplasmic transport. *Mol Cell* 2008;**29**(3):362–75.

79. Bilbao-Cortes D, Hetzer M, Langst G, Becker PB, Mattaj IW. Ran binds to chromatin by two distinct mechanisms. *Curr Biol* 2002;**12**(13):1151–6.

80. Zhang C, Goldberg MW, Moore WJ, Allen TD, Clarke PR. Concentration of Ran on chromatin induces decondensation, nuclear envelope formation and nuclear pore complex assembly. *Eur J Cell Biol* 2002;**81**(11):623–33.

81. Seino H, Hisamoto N, Uzawa S, Sekiguchi T, Nishimoto T. DNA-binding domain of RCC1 protein is not essential for coupling mitosis with DNA replication. *J Cell Sci* 1992;**102**(3):393–400.

82. Moore W, Zhang C, Clarke PR. Targeting of RCC1 to chromosomes is required for proper mitotic spindle assembly in human cells. *Curr Biol* 2002;**12**(16):1442–7.

83. Nemergut ME, Mizzen CA, Stukenberg T, Allis CD, Macara IG. Chromatin docking and exchange activity enhancement of RCC1 by histones H2A and H2B. *Science* 2001;**292**(5521):1540–3.

84. Chen T, Muratore TL, Schaner-Tooley CE, et al. N-terminal alpha-methylation of RCC1 is necessary for stable chromatin association and normal mitosis. *Nat Cell Biol* 2007;**9**(5):596–603.

85. Talcott B, Moore MS. The nuclear import of RCC1 requires a specific nuclear localization sequence receptor, karyopherin alpha3/Qip. *J Biol Chem* 2000;**275**(14):10,099–10,104.

86. Hutchins JR, Moore WJ, Hood FE, Wilson JS, Andrews PD, Swedlow JR, Clarke P. Phosphorylation regulates the dynamic interaction of RCC1 with chromosomes during mitosis. *Curr Biol* 2004;**14**(12):1099–104.

87. Li HY, Zheng Y. Phosphorylation of RCC1 in mitosis is essential for producing a high RanGTP concentration on chromosomes and for spindle assembly in mammalian cells. *Genes Dev* 2004;**18**(5):512–27.

88. Hood FE, Clarke PR. RCC1 isoforms differ in their affinity for chromatin, molecular interactions and regulation by phosphorylation. *J Cell Sci* 2007;**120**(19):3436–45.

89. Jeong SY, Rose A, Joseph J, Dasso M, Meier I. Plant-specific mitotic targeting of RanGAP requires a functional WPP domain. *Plant J* 2005;**42**(2):270–812.

Chapter 56

Cdc42 and Its Cellular Functions

Qiyu Feng and Richard A. Cerione

Department of Molecular Medicine, Department of Chemistry and Chemical Biology, Cornell University, Ithaca, New York

INTRODUCTION

Cdc42, a 22- to 25-kDa GTP-binding protein found in all eukaryotic cells, belongs to the Rho GTPase family within the superfamily of Ras-related small GTPases. The small GTPases play fundamentally important roles in cell biology, functioning as molecular switches in cellular signaling pathways and other biological processes. Cdc42 was first identified in *S. cerevisiae* as a key player in the assembly of the bud-site, and in mammals as a possible participant in EGF receptor-coupled signaling [1–3]. It was subsequently shown that Cdc42 directs actin cytoskeletal changes that are important for the formation of filopodia [4, 5] and for cell polarity-dependent processes [6], in line with its role in bud-site assembly in yeast. However, it has become increasingly appreciated that Cdc42 participates in the regulation of a wide variety of cellular functions aside from its role in the establishment of cell polarity, including gene transcription, cell-cycle progression, vesicular trafficking, and receptor endocytosis [7–14].

Like other small GTPases, Cdc42 cycles between a GDP-bound, signaling-inactive state and a GTP-bound, signaling-active state that propagates signals through its downstream targets and effector proteins. The regulation of the GTP-binding/GTP-hydrolytic cycle of Cdc42 is critical to its normal functioning in cells. Guanine nucleotide exchange factors (GEFs) act upstream from Cdc42 as positive regulators by stimulating GDP–GTP exchange and thereby promoting the formation of its signaling-active state, whereas GTPase-activating proteins (GAPs) help to "switch-off" Cdc42 by catalyzing GTP hydrolysis. Cdc42 is also maintained in either a GDP- or GTP-bound state, as well as being released from membranes, by the actions of a regulatory protein called Rho-GDI (for GDP-dissociation inhibitor) [15, 16].

The questions surrounding the cellular function of Cdc42 have now gone beyond simply identifying it as a possible participant in a specific signaling pathway, especially given its involvement in such a vast array of processes and biological activities. Rather, what is becoming of increasing interest is to determine more about the spatial and temporal activation of Cdc42 in cells, in particular the GEFs that are involved and the nature of the regulatory mechanisms that are required, as well as how the activation of Cdc42 is translated into a number of different outputs as mediated by its biological effectors, and whether many of its cellular signaling activities are coordinated. Because the answers to these questions are still far from complete, what we try to do in this chapter is provide a general appreciation for the different cellular processes that come under the regulation of Cdc42, and then allude to some of the current challenges confronting investigators in terms of the kinds of questions raised above and where they might begin to look for answers.

CDC42 REGULATES A VARIETY OF CELLULAR FUNCTIONS

In considering the different cellular functions regulated by Cdc42, the sheer number of activities that are impacted by this GTPase is striking (Figure 56.1). Thus, in the following sections, we will focus on those cellular outcomes for which a role for Cdc42 has become well established. It should be noted that in many cases it becomes rather difficult to draw a clear line that separates distinct functions. For example, cell polarity requires cytoskeletal changes and the polarized trafficking of proteins, whereas Golgi reorientation, a key step in vesicular trafficking, also depends upon cytoskeletal changes, while receptor

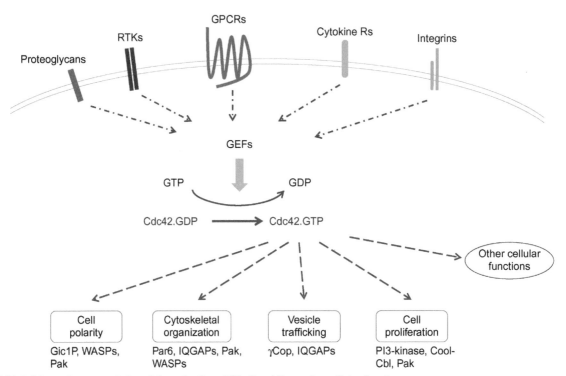

FIGURE 56.1 Schematic representation of the activation of Cdc42 and its ensuing cellular functions.
Different types of extracellular stimuli activate specific GEFs which catalyze GDP–GTP exchange on Cdc42. Activated (GTP-bound) Cdc42 stimulates a number of different downstream effector proteins and regulates a variety of cellular functions, including cell polarity, cytoskeletal organization, vesicle trafficking, and cell proliferation.

trafficking and endocytosis are important components of cell growth control.

Cell Polarity

Cell polarity probably represents the first cellular function for Cdc42 to have been firmly established. It is crucial for many aspects of cell and developmental biology, including cellular morphogenesis, differentiation, migration, and proliferation [1, 12, 17, 18]. The budding yeast, *S. cerevisiae*, is a single-celled organism in which the proper establishment of cell polarity directly underlies morphogenesis and cell division. Yeast cells become polarized as an outcome of cell-cycle control (i.e., during budding) or in response to a pheromone gradient (during mating). These stimuli result in the recruitment and activation of the yeast homolog of Cdc42, commonly designated as Cdc42p, at the site of polarized growth. Activated Cdc42p then causes actin and microtubule rearrangements, as well as influencing membrane trafficking events that help to mediate changes in cell polarity. Such changes result in the formation of a bud and, during mating, lead to the generation of a shmoo.

In multicellular organisms, cell polarity is determined primarily by external stimuli such as chemotactic signals, physical stress, and cell–cell contacts. Through different families of receptors, these signals eventually activate Cdc42. For example, receptor tyrosine kinases like the EGF receptor or the platelet-derived growth factor receptor activate Cdc42 as an early requirement for polarized cellular migration [19–21]. These growth factor-coupled signals are mediated through the Src tyrosine kinase to GEFs for Cdc42 like Vav-2 and the Cool (Cloned-out of library)/Pix (Pak-interactive exchange factor) proteins. During neutrophil responses to the chemoattractant C5a, heterotrimeric G proteins promote the activation of Cdc42 through a protein complex consisting of the $G\beta\gamma$ subunits, Pak1 (for p21-activated kinase-1), and the Cool-2/α-Pix protein [22]. In other cases, the activation of Cdc42 is induced by integrin–matrix interactions that occur near the front of the cell [23, 24]. The GEFs involved in this process are still unknown, but the Cool/Pix proteins represent attractive candidates, given that they are phosphorylated and activated by Src and Fak (focal adhesion kinase), which act downstream from integrins (Q. Feng, in preparation). During the formation of epithelial sheets upon cell–cell contact, adhesion proteins such as nectin and E-cadherin initiate signaling events that lead to the activation of Cdc42 and ultimately result in the generation of distinct apical and baso-lateral surfaces [25–28]. In each of these cases, GTP-bound Cdc42 needs to engage a variety of downstream targets and effector proteins, including members of the Pak family of serine/threonine kinases, as well as the cytoskeletal scaffold proteins WASP (Wiscott-Aldrich syndrome protein), IQGAP (for *IQ* motifs and Ras-*GAP* domain), and Par6 (partitioning-defective 6), and the trafficking protein γCOP (γ-coatomer). The possible involvements of these effectors in different activities relevant to cell polarity and other Cdc42-associated functions will be considered further below.

Cytoskeletal Organization

Another of the cellular functions of Cdc42 that was recognized relatively early involves changes in cytoskeletal architecture. In mammalian cells, these Cdc42-directed changes are manifested through the formation of filopodia [4, 5]. One mechanism by which Cdc42 directs changes in the actin cytoskeleton involves the major actin polymerization factor Arp2/3 (actin-related protein 2/3). GTP-bound Cdc42 first binds to one of its specific downstream target/effector proteins, WASP. This exposes a C-terminal Arp2/3-binding site that results in the binding of Arp2/3 to WASP and in Arp2/3 activation, which then stimulates actin polymerization. An additional influence by Cdc42 on the actin cytoskeleton may come from its ability to bind and activate members of the Pak family of serine/threonine kinases. It has been suggested that activated Pak phosphorylates and promotes the activation of the LIM kinase (where LIM is an acronym of the three gene products Lin-11, Isl-1, and Mec-3), which in turn phosphorylates ADF (actin-depolymerizing factor)/cofilin. Cofilin severs actin filaments, leading to an increase in uncapped barbed ends that serve as sites for actin polymerization and filament elongation. Cofilin also participates in filament disassembly by promoting actin monomer dissociation from the pointed ends, and both of these activities seem to be important for productive membrane protrusions [14, 29, 30].

In addition to its effects on the actin cytoskeleton, Cdc42 also regulates the microtubule cytoskeleton. In microtubule dynamics, Op18 (Oncoprotein 18)/stathmin plays an important regulatory role by promoting catastrophic disassembly and inhibiting polymerization [31]. However activated Cdc42 works through Pak to promote the phosphorylation and inactivation of Op18/stathmin, thereby resulting in the net elongation of microtubule ends [32]. An additional level of regulation comes through effects on CLIP-170 (cytoplasmic linker protein of 170 kDa), a microtubule plus-end capture protein. CLIP-170 is able to bind simultaneously to microtubules and to the scaffold protein IQGAP, a target/effector for Cdc42. Activated Cdc42 enhances the ability of CLIP-170 to bind to IQGAP, which then promotes plus-end capture [33]. Still another mode of regulation of the microtubule cytoskeleton by Cdc42 has been highlighted from work performed with migrating astrocytes. In these studies, it was suggested that the interaction of APC (adenomatosis polyposis coli) with EB1 (end-binding protein 1) at microtubule plus-ends is regulated by Cdc42 through its ability to engage the scaffold protein Par6, as well as its

associated binding partners Par3 and aPKC (atypical protein kinase C) [34].

Vesicle Trafficking

Vesicle trafficking is an essential feature of eukaryotic cells, as it involves the movement of cargo between different cellular organelles as well as between the cell and its surroundings. There have been a number of lines of evidence suggesting that Cdc42 may participate in several different aspects of vesicular trafficking, including vesicle formation at the Golgi, Golgi reorientation, targeting of exocytic vesicles, and endocytosis.

The COPI complex (also referred to as the coatomer complex) is made up of seven subunits and, together with the GTP-bound form of the small GTPase Arf1, coats Golgi-derived trafficking vesicles [35–38]. Activated Cdc42 directly associates with the γ subunit of the COPI complex (i.e., γCOP), through a pair of lysine residues located near the C-terminal end of the GTPase. It was further shown that both GTP hydrolysis-defective mutants and nucleotide-exchange-defective mutants of Cdc42 blocked the ER to Golgi transport of the viral glycoprotein VSV-G (vesicular stomatitus virus-glycoprotein). However, the constitutively activated Cdc42(F28L) mutant, which spontaneously exchanges GDP for GTP but still hydrolyzes GTP (and thus has been referred to as a "fast-cycling" mutant) [39, 40], stimulated the ER to Golgi transport of VSV-G. Thus, the ability of Cdc42 to cycle between the GDP- and GTP-bound states appeared to be linked to its ability to positively influence intracellular trafficking activity. Interestingly, it was shown that the ability of the fast-cycling Cdc42(F28L) mutant to transform NIH 3T3 cells required its interaction with the COPI complex, implying a connection between intracellular trafficking and cell growth regulation [41]. More recently, additional indications for a role for Cdc42 in intracellular trafficking events have emerged; for example, Cdc42 has been reported to regulate the retrograde transport of Shiga toxin from the Golgi to the ER, as well as post-Golgi protein sorting and transport [42].

Golgi reorientation represents a key trafficking step in polarized cells. In migrating astrocytes, the orientation of the Golgi was shown to be controlled by the Cdc42–Par6–aPKC pathway through modifications of the microtubule cytoskeleton. In NIH 3T3 cells, Cdc42, Scar2 (for suppressor of cyclic AMP receptor defects-2), and the Arp2/3 complex were also shown to influence Golgi reorientation by regulating the actin cytoskeleton [43].

Numerous studies have implicated Cdc42 in exocytosis and endocytosis. Mellman and colleagues demonstrated that in Madin-Darby canine kidney (MDCK) cells, Cdc42 controls both secretory and endocytic transport to the basolateral plasma membrane [44]. In yeast, the Cdc42-target/effector protein Iqg1p (i.e., the homolog of the mammalian IQGAPs) was shown to be localized at the plasma membrane and to associate with Sec3p, a subunit of the exocyst complex [45]. In mammalian cells, the interaction between the small GTPase RalA and another component of the exocyst complex, Sec5, was reported to be triggered by Cdc42 [46]. Moreover, Cdc42 has been suggested to facilitate the targeting of exocytic vesicles. This might require the formation of Par6–aPKC and Lgl (Lethal giant larvae)–Scribble complexes in the vicinity of the plasma membrane. During clathrin-mediated endocytosis, clathrin-coat assembly was linked to Cdc42 signaling through the Cdc42-GEF, Intersectin, which localizes to clathrin-coated pits via an Eps15 (for epidermal growth factor pathway substrate 15)-homology (EH) domain [47].

Proliferation and Transformation

The Dbl (diffuse B cell lymphoma) family of proteins consists of GEFs for Cdc42 and other Rho-related GTPases. They contain tandem Dbl-homology (DH) and Pleckstrin-homology (PH) domains that serve important functional roles (discussed further below). Many members of the Dbl family, including Dbl itself and Vav, when truncated or point-mutated, induce cellular transformation [48, 49]. These findings provided early evidence for a direct connection between the hyper-activation of Cdc42 and other Rho-related GTPases in cells, and the stimulation of excessive cell growth and/or other characteristics associated with the transformed phenotype. Activated forms of Cdc42 were subsequently shown to be capable of stimulating cell-cycle progression, whereas a dominant-negative Cdc42 mutant blocked Ras-induced transformation. A possible explanation for the latter findings came from studies showing that Cdc42, by activating Pak, could help to mediate the phosphorylation of Raf and contribute to the activation of Mek and Erk, thus helping Ras to send its full complement of transforming signals. However, as will be elaborated upon below, Cdc42 is also able to mediate cell growth regulation through mechanisms distinct from its ability to influence Ras-coupled signaling activities [50–53].

One of the more interesting and telling differences between Ras and Cdc42, with regard to their abilities to influence cell growth, is that while GTP hydrolysis-defective Ras is oncogenic and strongly stimulates cell growth, the corresponding Cdc42 mutants cause growth inhibition. On the other hand, the fast-cycling Cdc42(F28L) mutant was capable of inducing the transformation of fibroblasts in culture. Because GTP hydrolysis-defective Cdc42 mutants were also defective for stimulating the trafficking of VSV-G, this implied, as alluded to above, that the ability of Cdc42 to cycle between its GDP- and GTP-bound states is important for its role in intracellular trafficking, and this is somehow connected to its actions in cell growth regulation.

These findings then motivated efforts to identify growth factors and/or growth regulatory proteins whose cellular trafficking, and ultimately homeostasis, might be altered in cells expressing fast-cycling Cdc42. In fact, it was found that cells expressing the Cdc42(F28L) mutant contained significantly increased levels of epidermal growth factor receptors (EGFRs). The accumulation of EGFRs resulted in excessive mitogenic signaling through the Ras–Erk pathway and eventually caused cellular transformation [13, 41, 54].

What is the underlying mechanism responsible for the accumulation of EGFRs in cells expressing fast-cycling Cdc42? It turns out that in addition to Cdc42, two other proteins are involved. These are the Cool-1/β-Pix protein, which functions both as a GEF and a target for activated Cdc42, and as a binding partner for c-Cbl, which is an adaptor protein that exhibits E3 ubiquitin ligase activity [55–59]. Constitutively active, fast-cycling Cdc42 is able to persistently drive the formation of a complex between itself, Cool-1, and Cbl, resulting in the sequestration of c-Cbl away from the EGFR. This prevents the Cbl-catalyzed ubiquitination of the EGFR, leading to the accumulation of these receptors and their sustained mitogenic signaling.

SPATIAL AND TEMPORAL CONTROL OF CDC42 ACTIVITY

Given its prominent role in so many different aspects of cell biology, it is not surprising that Cdc42 is tightly regulated. To achieve the necessary spatial and temporal control of Cdc42-dependent activities in cells, mechanisms are in place to ensure the proper recruitment of Cdc42 regulators (e.g., GEFs) and target/effector proteins, together with Cdc42 itself, to the correct subcellular locations in response to specific stimuli (growth factors, hormones, stress etc.).

Protein Recruitment

Various extracellular or intracellular cues initiate the recruitment of proteins to specific sites where they are able to form signaling complexes responsible for distinct cellular and biological outputs. In the case of signaling events mediated by small GTPases, these complexes often contain both an upstream activator (e.g., GEF) and a downstream effector protein for the GTPase, thereby enabling the GTPase to respond rapidly and efficiently to different extracellular stimuli. One example is provided by the roles played by Cdc42 in budding yeast. The local activation of the yeast Cdc42 protein (Cdc42p) at the presumptive growth site is required for budding. During the G1 phase of the cell cycle, the Cdc42-GEF, Cdc24p, is sequestered within the nucleus in a complex with Far1p (factor arrest-1 protein). In late G1 phase, the cyclin-dependent kinase (CDK)-cyclin complex, Cdc28–Cln2, induces the degradation of Far1p. This results in the release of Cdc24p from the nucleus and its recruitment to the polarization site, where it binds to the bud-site selection GTPase Rsr1p (Ras-related-1 protein)/Bud1p and to the membrane adaptor protein Bem1p (bud emergence-1 protein). During mating, a pheromone gradient activates G-protein-coupled transmembrane receptors, leading to the association of Far1p with the Gβγ subunit-complex, finally causing the accumulation of Cdc24p at the site of receptor stimulation [12, 60–62]. Cdc42p and its downstream effectors are also recruited to the site of polarized growth. During cell division, Cdc42p clusters at the bud-site prior to bud emergence. The Cdc42-effector proteins Gic1p (for GTPase-interacting component-1 protein) and Gic2p help to recruit other downstream effectors such as Cla4p (Cdc42-activated signal transducing kinase of the Pak family-4 protein) and Bni1p (Bud neck-involved-1 protein) to the site of polarized growth. The scaffold protein Bem1p recruits Cdc24p, together with Cdc42p and downstream effectors, providing an actin-independent feedback mechanism to amplify stochastic fluctuations in local Cdc42p concentration.

In mammalian cells, various stimuli direct the recruitment of Cdc42, its GEFs, and its downstream effector proteins to specific cellular sites. For example, during the formation of epithelia, cadherin-mediated cell–cell contacts activate and recruit Cdc42 and Rac to cadherin–cadherin contact sites formed between neighboring cells. These recruitment events are necessary for the development of the adherens junctions. A second example is provided by the responses of T cells, as activated T cell receptors (TCRs) induce the recruitment of Cdc42, WASP, and other components of the actin polymerization machinery to specific cellular (membrane) locations [63, 64]. Still another example where the recruitment of Cdc42 to a specific cellular location is linked to a biological outcome involves cells undergoing chemotaxis in response to distinct extracellular signals. In these cases, Cdc42 is recruited to the leading edge. One pathway responsible for such recruitment is dependent upon phosphoinositide 3-kinase (PI3K) activity, and results in the activation of both Cdc42 and Rac at the leading edge of migrating cells [65–67]. Another pathway, identified in neutrophil responses to the chemoattractant C5a, involves the activation of Cdc42 by heterotrimeric G-protein-coupled receptors [22]. Chemoattractant-bound receptors activate heterotrimeric G proteins and release Gβγ subunit-complexes. These complexes bind directly to Pak1 and then activate Cdc42 through the Cool-2/α-Pix protein, which forms a tight complex with Pak1. The activated Cdc42 molecule is then able to stimulate Pak1 activity. Activated Pak1 and/or other Cdc42 effector proteins can then help to influence the localization of PTEN (phosphatase and tensin homolog) and regulate F-actin polymerization. This becomes especially important for directional sensing, persistent directional movement, and the establishment of cell polarity in a shallow chemoattractant gradient.

Finally, the recruitment of Cdc42 to the Golgi may have an important influence on its potential role in intracellular trafficking events. Cdc42 is predominantly localized to the Golgi complex in mammalian cells. This localization of Cdc42 is dependent on the activation of the Arf1 GTPase and on the ability of Cdc42 to bind to the γCOP subunit of the COPI complex. Once localized to the Golgi, activated Cdc42 triggers actin polymerization and protein transport through a WASP/Arp2/3-dependent mechanism. Activation of this pathway leads to the recruitment of a distinct set of actin-binding proteins to the Golgi [41, 42, 68–70].

Regulating the GEFs

Members of the Dbl family of GEFs contain a characteristic tandem arrangement of DH and PH domains. The DH domain is the limit functional unit necessary for GEF activity. The PH domain binds to phosphorylated phosphoinositides (PIPs) and/or proteins, and helps to target the GEF to its appropriate intracellular location, as well as in some cases modulating the catalytic activity of the DH domain. Often, GEF activation appears to be closely coupled to the re-localization of the GEF as mediated by its PH domain [71–73].

Some GEFs, such as Dbl and Vav, are subject to intramolecular (auto)inhibitory regulation. Relief from auto-inhibition can be accomplished through the phosphorylation of a specific site(s) on the GEF, or as an outcome of its binding to other proteins (see Figure 56.2). In the case of proto-Vav, upon receptor stimulation, Src- and Syk-family tyrosine kinases transiently phosphorylate this GEF at a specific site (Tyr174), leading to the activation of its GEF activity. The structure of Vav shows that Tyr174 lies at the center of an inhibitory helix that binds in a complementary cleft on the conserved face of the DH domain, sterically blocking access to the GTPase-binding site. Phosphorylation of this tyrosine residue prevents it from binding to the DH domain, as well as destabilizing the helical structure of the inhibitory arm, thus preventing the auto-inhibitory interaction and resulting in an activation of GEF activity. Lipid binding to the PH domain also influences the phosphorylation of Tyr174, thereby suggesting a cooperative role for the PH domain in relieving the auto-inhibition of Vav's GEF activity. A similar auto-regulatory interaction was demonstrated between the amino-terminal coiled-coil region of Dbl and the PH domain [16, 49, 74].

The Cool/Pix proteins provide some additional interesting examples regarding how GEFs are regulated (Figure 56.2). Cool-1/β-Pix shows little if any detectable basal GEF activity because of the presence of a negative regulatory region (called T1) that exists immediately downstream from the PH domain of the protein [59]. The T1 region does not appear to act as a classical auto-inhibitory domain, because its removal does not immediately restore

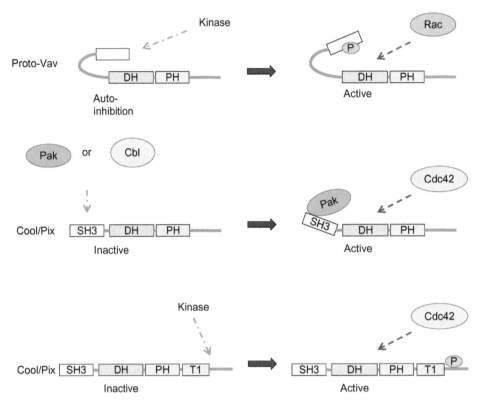

FIGURE 56.2 Schematic representation of different examples of GEF regulation.
In the case of proto-Vav, relief from auto-inhibition can be accomplished through the phosphorylation of Try174, which prevents its interaction with the DH domain and thereby allows Vav to interact with GTPases. In the case of the Cool/Pix proteins, one mode of regulation occurs when Pak or Cbl binds to the SH3 domain and enhances the interaction of Cool/Pix with Rac/Cdc42, resulting in the stimulation of GDP-GTP exchange. Another mode of regulation for Cool-1/β-Pix occurs via the EGF-dependent phosphorylation of Tyr442 which lies just downstream from its negative regulatory T1 region. This results in the activation of Cdc42-GEF activity.

GEF activity. However, the GEF activity of a Cool-1/β-Pix construct lacking the T1 region can be restored upon the binding of either Pak or Cbl to the SH3 domain of the GEF [59]. More recently, we discovered that the EGF-stimulated phosphorylation of Cool-1/β-Pix (as mediated by Src and Fak) at Tyr442 (which lies adjacent to the T1 region) also activates the GEF activity, although the resultant activity is highly specific for Cdc42 [21].

Cool-2/α-Pix, which is highly related to Cool-1, shows some additional complications regarding its regulation. Normally, Cool-2/α-Pix exists as a dimer due to interactions occurring through a carboxyl-terminal leucine zipper domain. Dimeric Cool-2 appears to be a highly specific GEF for Rac. Essential residues for the Rac-GEF activity are contributed by the DH domain from one of the Cool-2 monomers making-up the dimer, and by the PH domain from the adjacent Cool-2 monomer. However, when the Cool-2 dimer is converted to a monomer, it then acts as a GEF for both Rac and Cdc42. One way in which Cool-2 becomes monomeric is in response to signals from che-moattractant receptors. These G-protein-coupled receptors activate heterotrimeric G proteins, resulting in the generation of GTP-bound Gα subunits and Gβγ-subunit complexes. The latter associate with members of the Pak

family of serine/threonine kinases, and the resulting Gβγ-Pak complex, through its ability to bind (via Pak) to the SH3 domain, causes dimeric Cool-2 molecules to dissociate into monomers and to exhibit GEF activity [75].

COOPERATION BETWEEN CDC42 AND OTHER SMALL GTPASES IN CELLULAR FUNCTIONS

Many of the cellular functions regulated by Cdc42 are mediated through complex signaling networks. Thus, there is a good deal of potential for crosstalk between different Cdc42-coupled signaling pathways, as well as between Cdc42 and other small GTPases.

Crosstalk between Cdc42 and other small GTPases

The cooperation between various signaling pathways mediated by different small GTPases may be achieved through GEFs that can activate more than one GTPase, or through target/effectors that can serve to interface the signaling

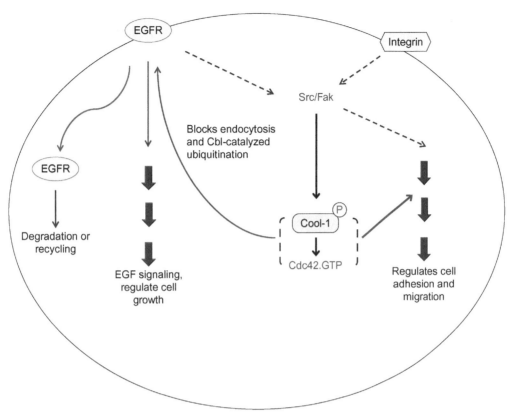

FIGURE 56.3 Schematic representation of a role for Cool-1/β-Pix in coordinating growth factor (EGF)- and integrin-coupled signaling events. Both EGF- and integrin-coupled signaling stimulate the phosphorylation of Cool-1/β-Pix, which in turn activates Cdc42. GTP-bound Cdc42, through the formation of a Cdc42–Cool-1–Cbl complex, regulates the timing of the Cbl-catalyzed ubiquitination of EGFRs. Phosphorylated Cool-1 and GTP-bound Cdc42 also help to regulate cell migration.

pathways of different GTPases. For example, Intersectin (ITSN), a Cdc42-GEF, is also able to activate Ras via the binding of Sos to the first SH3 domain of ITSN, resulting in signaling to JNK (by Ras) and to PI3K (by Ras and Cdc42). This cooperation between Cdc42 and Ras enables certain constructs of ITSN, when expressed in NIH 3T3 cells, to induce cellular transformation [76]. Another example of crosstalk between Cdc42 and Ras involves the combined regulatory effects of these GTPases on Raf activity. GTP-bound Cdc42 activates its effector Pak, which then phosphorylates both Raf1 and Mek1, thereby maximizing the activation of the Raf–Mek–Erk pathway by Ras. A third example comes from the nectin-based cell–cell contact events that lead to the c-Src-dependent activation of Cdc42. The newly generated GTP-bound Cdc42 complexes enhance the GEF activity of tyrosine-phosphorylated Vav2 toward Rac1.

Many other examples of crosstalk between Cdc42 and Rac have also been observed. STEF (Sif and Tiam1-like exchange factor)/Tiam1 (T-lymphoma invasion and metastasis-inducing protein 1), which is a Rac-specific GEF, binds to Par-3 and becomes part of a large multi-protein complex that includes Par-3, aPKC, Par-6, and GTP-

bound Cdc42. In cultured hippocampal neurons, STEF was shown to accumulate at the tip of the growing axon and to co-localize with Par-3. The spatio-temporal activation and signaling through the Cdc42–Par-6–Par-3–STEF/Tiam1–Rac complex appears to play an important role in neurite growth and axon specification [77, 78].

A functional interplay between Cdc42 and Rac was also shown to be mediated by the Cool-2/α-Pix protein [79]. In this case, activated Cdc42 binds to the DH domain of one of the Cool-2 molecules making up the Cool-2 dimer. This results in a conformational change that activates the Rac-GEF activity of Cool-2, perhaps by helping to properly align the remaining DH domain, from one of the Cool-2 monomers making up the dimer, with the adjacent (trans) PH domain.

Cooperation between Cdc42 and Signaling Pathways Regulating Cell Growth and Cell Migration

Cdc42 appears to influence different steps that are important for EGFR-coupled signaling and homeostasis (Figure 56.3).

At one level, Cdc42 appears to set the timing for EGFR downregulation and ubiquitination. This occurs through the EGF-dependent activation of Cdc42 as mediated by the phosphorylation of Cool-1/β-Pix, which stimulates its Cdc42-GEF activity (as described above; also, see Figure 56.2) and leads to the formation of a complex between GTP-bound Cdc42, phosphorylated Cool-1, and Cbl. When Cbl is part of this ternary complex, it is not able to bind to the EGFR and thus cannot help to recruit other proteins important for receptor endocytosis. Since Cbl is sequestered away from the EGFR by activated Cdc42 and phosphorylated Cool-1, it is not able to directly bind to the receptor and to catalyze its ubiquitination, which is a necessary event for the ultimate targeting of EGFRs to the lysosome. In some cases, binary complexes consisting of phosphorylated Cool-1 and Cbl can bind to the EGFR; however, the presence of Cool-1 prevents Cbl from recruiting other proteins necessary for receptor endocytosis, as well as blocking the EGF-dependent phosphorylation of Cbl which is required to activate its E3 ubiquitin ligase activity. By keeping Cbl from prematurely initiating EGFR downregulation and degradation, activated Cdc42 helps to ensure that EGF can activate Ras and generate a sufficient signal for mitogenesis. Ultimately, Cool-1 has to dissociate from Cbl in order for EGFR downregulation and degradation to occur, otherwise excessive mitogenic signaling would ensue, leading to cellular transformation. There appear to be two ways by which this can occur. One is through the de-phosphorylation of Cool-1, as only phosphorylated Cool-1 has a relatively strong affinity for Cbl. The EGF-dependent phosphorylation of Cool-1 in cells, which is a critical first step for the assembly of the Cdc42–Cool-1–Cbl complex, is transient; however, the phosphatase that catalyzes the critical de-phosphorylation of Cool-1 has still not been identified and is being actively investigated. A second way in which Cbl might "escape" from Cool-1 is through its ubiquitination, as it was recently reported that Cbl catalyzes the ubiquitination and degradation of Cool-1 [80]. This then raises an interesting possibility regarding a double feedback loop, as Cdc42 and Cool-1 would initially prevent Cbl from catalyzing EGFR ubiquitination, whereas Cbl would ultimately direct the ubiquitination and degradation of Cool-1, freeing Cbl to bind and promote the degradation of EGFRs.

The involvement of Cool-1 in the regulation of EGFR downregulation and degradation might also provide a mechanism by which mitogenic signals can be coordinated with cell migration. There are a number of lines of evidence implicating Cool-1, and one of its primary binding partners, Cat (for Cool-associated tyrosine phospho-substrate)/Git (for G protein-coupled receptor kinase interactor), a GAP for the GTPase Arf6, in cell migration [81–84]. Interestingly, we have recently found that the phosphorylation of Cool-1 at Tyr442 is not only stimulated by EGF, through Src and Fak, but also by integrins (Q. Feng, in

preparation). Moreover, we have discovered that the phosphorylation–de-phosphorylation cycle of Cool-1 is linked to the de-assembly and assembly of focal complexes, respectively, and is essential for cell migration. Thus, it is attractive to consider that the growth factor- and integrin-dependent phosphorylation of Cool-1 serves as a critical event for the convergence of signals that are important for cell growth control and for cell motility.

CONCLUSION

The initial discovery of a mammalian/human homolog for the product of the yeast *CDC42* gene suggested that this GTPase must play a fundamentally important role in cell biology. This was initially borne out by the findings that Cdc42 has a universal function in regulating cell polarity through its effects on the actin cytoskeleton. However, through the ensuing years it has become clear that Cdc42 plays essential roles in a wide range of cellular activities and biological outcomes, many of which were not predicted from its actions in budding yeast. With these findings has come the identification of a large number of GEFs and other regulatory proteins, as well as a myriad of target/effector proteins. The latter include scaffold proteins that help to interface Cdc42 with the actin and microtubule cytoskeletons, protein kinases that mediate cytoskeletal changes and transmit signals to the nucleus to influence cell cycle progression and gene expression, and a number of other target/effectors that link Cdc42 to cell growth regulation, in many cases by influencing EGFR homeostasis. Future efforts are likely to be directed at understanding how the proper GEFs and target/effectors are positioned to enable Cdc42 to efficiently trigger its many different cellular responses, and whether these various activities need to be coordinated such that the effects of Cdc42 on cell growth can be synchronized with its regulation of cell shape and movement.

REFERENCES

1. Johnson DI, Pringle JR. Molecular characterization of CDC42, a Saccharomyces cerevisiae gene involved in the development of cell polarity. *J Cell Biol* 1990;**111**:143–52.
2. Hart MJ, Polakis PG, Evans T, Cerione RA. The identification and characterization of an epidermal growth factor-stimulated phosphorylation of a specific low molecular weight GTP-binding protein in a reconstituted phospholipid vesicle system. *J Biol Chem* 1990;**265**:5990–6001.
3. Shinjo K, Koland JG, Hart MJ, Narasimhan V, Johnson DI, Evans T, Cerione RA. Molecular cloning of the gene for the human placental GTP-binding protein Gp (G25K): identification of this GTP-binding protein as the human homolog of the yeast cell-division-cycle protein CDC42. *Proc Natl Acad Sci USA* 1990;**87**:9853–7.
4. Nobes CD, Hall A. Rho, Rac, and Cdc42 GTPases regulate the assembly of multimolecular focal complexes associated with actin stress fibers, lamellipodia, and filopodia. *Cell* 1995;**81**:53–62.

5. Kozma R, Ahmed S, Best A, Lim L. The Ras-related protein Cdc42Hs and bradykinin promote formation of peripheral actin microspikes and filopodia in Swiss 3T3 fibroblasts. *Mol Cell Biol* 1995;**15**:1942–52.

6. Chant J. Cell polarity in yeast. *Annu Rev Cell Dev Biol* 1999;**15**:365–91.

7. Van Aelst L, D'Souza-Schorey C. Rho GTPases and signaling networks. *Genes Dev* 1997;**11**:2295–322.

8. Hall A. Rho GTPases and the actin cytoskeleton. *Science* 1998;**279**:509–14.

9. Bar-Sagi D, Hall A. Ras and Rho GTPases: a family reunion. *Cell* 2000;**103**:227–38.

10. Erickson JW, Cerione RA. Multiple roles for Cdc42 in cell regulation. *Curr Opin Cell Biol* 2001;**13**:153–7.

11. Etienne-Manneville S, Hall A. Rho GTPases in cell biology. *Nature* 2002;**420**:629–35.

12. Etienne-Manneville S. Cdc42--the centre of polarity. *J Cell Sci* 2004;**117**:1291–300.

13. Cerione RA. Cdc42: new roads to travel. *Trends Cell Biol* 2004;**14**:127–32.

14. Jaffe AB, Hall A. Rho GTPases: biochemistry and biology. *Annu Rev Cell Dev Biol* 2005;**21**:247–69.

15. Boguski MS, McCormick F. Proteins regulating Ras and its relatives. *Nature* 1993;**366**:643–54.

16. Schmidt A, Hall A. Guanine nucleotide exchange factors for Rho GTPases: turning on the switch. *Genes Dev* 2002;**16**:1587–609.

17. Drubin DG. Development of cell polarity in budding yeast. *Cell* 1991;**65**:1093–6.

18. Johnson DI. Cdc42: an essential Rho-type GTPase controlling eukaryotic cell polarity. *Microbiol Mol Biol Rev* 1999;**63**:54–105.

19. Chou J, Burke NA, Iwabu A, Watkins SC, Wells A. Directional motility induced by epidermal growth factor requires Cdc42. *Exp Cell Res* 2003;**287**:47–56.

20. Liu BP, Burridge K. Vav2 activates Rac1, Cdc42, and RhoA downstream from growth factor receptors but not beta1 integrins. *Mol Cell Biol* 2000;**20**:7160–9.

21. Feng Q, Baird D, Peng X, Wang J, Ly T, Guan JL, Cerione RA. Cool-1 functions as an essential regulatory node for EGF receptor- and Src-mediated cell growth. *Nat Cell Biol* 2006;**8**:945–56.

22. Li Z, Hannigan M, Mo Z, Liu B, Lu W, Wu Y, Smrcka AV, Wu G, Li L, Liu M, Huang CK, Wu D. Directional sensing requires G beta gamma-mediated PAK1 and PIX alpha-dependent activation of Cdc42. *Cell* 2003;**114**:215–27.

23. Etienne-Manneville S, Hall A. Integrin-mediated activation of Cdc42 controls cell polarity in migrating astrocytes through PKCzeta. *Cell* 2001;**106**:489–98.

24. Nobes CD, Hall A. Rho GTPases control polarity, protrusion, and adhesion during cell movement. *J Cell Biol* 1999;**144**:1235–44.

25. Rojas R, Ruiz WG, Leung SM, Jou TS, Apodaca G. Cdc42-dependent modulation of tight junctions and membrane protein traffic in polarized Madin-Darby canine kidney cells. *Mol Biol Cell* 2001;**12**:2257–74.

26. Arthur WT, Noren NK, Burridge K. Regulation of Rho family GTPases by cell–cell and cell–matrix adhesion. *Biol Res* 2002;**35**:239–46.

27. Honda T, Shimizu K, Kawakatsu T, Fukuhara A, Irie K, Nakamura T, Matsuda M, Takai Y. Cdc42 and Rac small G proteins activated by trans-interaction of nectins are involved in activation of c-Jun N-terminal kinase, but not in association of nectins and cadherin to form adherens junctions, in fibroblasts. *Genes Cells* 2003;**8**:481–91.

28. Takai Y, Nakanishi H. Nectin and afadin: novel organizers of intercellular junctions. *J Cell Sci* 2003;**116**:17–27.

29. Ghosh M, Song X, Mouneimne G, Sidani M, Lawrence DS, Condeelis JS. Cofilin promotes actin polymerization and defines the direction of cell motility. *Science* 2004;**304**:743–6.

30. Pollard TD, Borisy GG. Cellular motility driven by assembly and disassembly of actin filaments. *Cell* 2003;**112**:453–65.

31. Cassimeris L. The oncoprotein 18/stathmin family of microtubule destabilizers. *Curr Opin Cell Biol* 2002;**14**:18–24.

32. Daub H, Gevaert K, Vandekerckhove J, Sobel A, Hall A. Rac/Cdc42 and p65PAK regulate the microtubule-destabilizing protein stathmin through phosphorylation at serine 16. *J Biol Chem* 2001;**276**:1677–80.

33. Fukata M, Watanabe T, Noritake J, Nakagawa M, Yamaga M, Kuroda S, Matsuura Y, Iwamatsu A, Perez F, Kaibuchi K. Rac1 and Cdc42 capture microtubules through IQGAP1 and CLIP-170. *Cell* 2002;**109**:873–85.

34. Etienne-Manneville S, Hall A. Cdc42 regulates GSK-3beta and adenomatous polyposis coli to control cell polarity. *Nature* 2003;**421**:753–6.

35. Waters MG, Serafini T, Rothman JE. 'Coatomer': a cytosolic protein complex containing subunits of non-clathrin-coated Golgi transport vesicles. *Nature* 1991;**349**:248–51.

36. Lowe M, Kreis T. Regulation of membrane traffic in animal cells by COPI. *Biochim Biophys Acta* 1998;**1404**:53–66.

37. Nickel W, Brügger B, Wieland FT. Vesicular transport: the core machinery of COPI recruitment and budding. *J Cell Sci* 2002;**115**:3235–40.

38. Bonifacino JS, Lippincott-Schwartz J. Coat proteins: shaping membrane transport. *Nat Rev Mol Cell Biol* 2003;**4**:409–14.

39. Lin R, Bagrodia S, Cerione RA, Manor D. A novel Cdc42Hs mutant induces cellular transformation. *Curr Biol* 1997;**7**:794–7.

40. Lin R, Cerione RA, Manor D. Specific contributions of the small GTPases Rho, Rac and Cdc42 to Dbl transformation. *J Biol Chem* 1999;**274**:23,633–23,641.

41. Wu WJ, Erickson JW, Lin R, Cerione RA. The γ-subunit of the coatomer complex binds Cdc42 to mediate transformation. *Nature* 2000;**405**:800–4.

42. Luna A, Matas OB, Martínez-Menárguez JA, Mato E, Durán JM, Ballesta J, Way M, Egea G. Regulation of protein transport from the Golgi complex to the endoplasmic reticulum by CDC42 and N-WASP. *Mol Biol Cell* 2002;**13**:866–79.

43. Magdalena J, Millard TH, Etienne-Manneville S, Launay S, Warwick HK, Machesky LM. Involvement of the Arp2/3 complex and Scar2 in Golgi polarity in scratch wound models. *Mol Biol Cell* 2003;**14**:670–84.

44. Kroschewski R, Hall A, Mellman I. Cdc42 controls secretory and endocytic transport to the basolateral plasma membrane of MDCK cells. *Nat Cell Biol* 1999;**1**:8–13.

45. Osman MA, Konopka JB, Cerione RA. Iqg1p links spatial and secretion landmarks to polarity and cytokinesis. *J Cell Biol* 2002;**159**:601–11.

46. Sugihara K, Asano S, Tanaka K, Iwamatsu A, Okawa K, Ohta Y. The exocyst complex binds the small GTPase RalA to mediate filopodia formation. *Nat Cell Biol* 2002;**4**:73–8.

47. Schafer DA. Coupling actin dynamics and membrane dynamics during endocytosis. *Curr Opin Cell Biol* 2002;**14**:76–81.

48. Whitehead IP, Campbell S, Rossman KL, Der CJ. Dbl family proteins. *Biochim Biophys Acta* 1997;**1332**:F1–23.

49. Hoffman GR, Cerione RA. Signaling to the Rho GTPases: networking with the DH domain. *FEBS Letts* 2002;**513**:85–91.

50. Olson MF, Ashworth A, Hall A. An essential role for Rho, Rac, and Cdc42 GTPases in cell cycle progression through G1. *Science* 1995;**269**:1270–2.

51. Qiu RG, Abo A, McCormick F, Symons M. Cdc42 regulates anchorage-independent growth and is necessary for Ras transformation. *Mol Cell Biol* 1997;**17**:3449–58.

52. King AJ, Sun H, Diaz B, Barnard D, Miao W, Bagrodia S, Marshall MS. The protein kinase Pak3 positively regulates Raf-1 activity through phosphorylation of serine 338. *Nature* 1998;**396**:180–3.

53. Jelinek T, Catling AD, Reuter CW, Moodie SA, Wolfman A, Weber MJ. RAS and RAF-1 form a signalling complex with MEK-1 but not MEK-2. *Mol Cell Biol* 1994;**14**:8212–18.

54. Wu WJ, Tu S, Cerione RA. Activated Cdc42 sequesters c-Cbl and prevents EGF receptor degradation. *Cell* 2003;**114**:715–25.

55. Bagrodia S, Taylor SJ, Jordon KA, Van Aelst L, Cerione RA. A novel regulator of p21-activated kinases. *J Biol Chem* 1998;**273**:23,633–23,636.

56. Manser E, Loo TH, Koh CG, Zhao ZS, Chen XQ, Tan L, Tan I, Leung T, Lim L. PAK kinases are directly coupled to the PIX family of nucleotide exchange factors. *Mol Cell* 1998;**1**:183–92.

57. Levkowitz G, Waterman H, Zamir E, Kam Z, Oved S, Langdon WY, Beguinot L, Geiger B, Yarden Y. c-Cbl/Sli-1 regulates endocytic sorting and ubiquitination of the epidermal growth factor receptor. *Genes Dev* 1998;**12**:3663–74.

58. Levkowitz G, Waterman H, Ettenberg SA, Katz M, Tsygankov AY, Alroy I, Lavi S, Iwai K, Reiss Y, Ciechanover A, Lipkowitz S, Yarden Y. Ubiquitin ligase activity and tyrosine phosphorylation underlie suppression of growth factor signaling by c-Cbl/Sli-1. *Mol Cell* 1999;**4**:1029–40.

59. Feng Q, Albeck JG, Cerione RA, Yang W. Regulation of the Cool/Pix proteins: key binding partners of the Cdc42/Rac targets, the p21-activated kinases. *J Biol Chem* 2002;**277**:5644–50.

60. Shimada Y, Gulli MP, Peter M. Nuclear sequestration of the exchange factor Cdc24 by Far1 regulates cell polarity during yeast mating. *Nat Cell Biol* 2000;**2**:117–24.

61. Nern A, Arkowitz RA. A Cdc24p-Far1p Gβγ protein complex required for yeast orientation during mating. *J Cell Biol* 1999;**144**:1187–202.

62. Nern A, Arkowitz RA. Nucleocytoplasmic shuttling of the Cdc42p exchange factor Cdc24p. *J Cell Biol* 2000;**148**:1115–22.

63. Van Aelst L, Symons M. Role of Rho family GTPases in epithelial morphogenesis. *Genes Dev* 2002;**16**:1032–54.

64. Cannon JL, Burkhardt JK. The regulation of actin remodeling during T-cell–APC conjugate formation. *Immunol Rev* 2002;**186**:90–9.

65. Merlot S, Firtel RA. Leading the way: directional sensing through phosphatidylinositol 3-kinase and other signaling pathways. *J Cell Sci* 2003;**116**:3471–8.

66. Benard V, Bohl BP, Bokoch GM. Characterization of rac and cdc42 activation in chemoattractant-stimulated human neutrophils using a novel assay for active GTPases. *J Biol Chem* 1999;**274**:13,198–13,204.

67. Wang F, Herzmark P, Weiner OD, Srinivasan S, Servant G, Bourne HR. Lipid products of PI(3)Ks maintain persistent cell polarity and directed motility in neutrophils. *Nat Cell Biol* 2002;**4**:513–18.

68. Erickson JW, Zhang C, Kahn RA, Evans T, Cerione RA. Mammalian Cdc42 is a brefeldin A-sensitive component of the Golgi apparatus. *J Biol Chem* 1996;**271**:26,850–26,854.

69. Fucini RV, Chen JL, Sharma C, Kessels MM, Stamnes M. Golgi vesicle proteins are linked to the assembly of an actin complex defined by mAbp1. *Mol Biol Cell* 2002;**13**:621–31.

70. Stamnes M. Regulating the actin cytoskeleton during vesicular transport. *Curr Opin Cell Biol* 2002;**14**:428–33.

71. Rebecchi MJ, Scarlata S. Pleckstrin homology domains: a common fold with diverse functions. *Annu Rev Biophys Biomol Struct* 1998;**27**:503–28.

72. Lemmon MA, Ferguson KM. Signal-dependent membrane targeting by pleckstrin homology (PH) domains. *Biochem J* 2000;**350**:1–18.

73. Zheng Y, Zangrilli D, Cerione RA, Eva A. The pleckstrin homology domain mediates transformation by oncogenic dbl through specific intracellular targeting. *J Biol Chem* 1996;**271**:19,017–19,020.

74. Bi F, Debreceni B, Zhu K, Salani B, Eva A, Zheng Y. Autoinhibition mechanism of proto-Dbl. *Mol Cell Biol* 2001;**21**:1463–74.

75. Feng Q, Baird D, Cerione RA. Novel regulatory mechanisms for the Dbl family guanine nucleotide exchange factor Cool-2/alpha-Pix. *EMBO J* 2004;**23**:3492–504.

76. Wang JB, Wu WJ, Cerione RA. Cdc42 and Ras cooperate to mediate cellular transformation by intersectin-L. *J Biol Chem* 2005;**280**:22,883–22,891.

77. Kawakatsu T, Ogita H, Fukuhara T, Fukuyama T, Minami Y, Shimizu K, Takai Y. Vav2 as a Rac-GDP/GTP exchange factor responsible for the nectin-induced, c-Src- and Cdc42-mediated activation of Rac. *J Biol Chem* 2005;**280**:4940–7.

78. Nishimura T, Yamaguchi T, Kato K, Yoshizawa M, Nabeshima Y, Ohno S, Hoshino M, Kaibuchi K. PAR-6-PAR-3 mediates Cdc42-induced Rac activation through the Rac GEFs STEF/Tiam1. *Nat Cell Biol* 2005;**7**:270–7.

79. Baird D, Feng Q, Cerione RA. The Cool-2/alpha-Pix protein mediates a Cdc42-Rac signaling cascade. *Curr Biol* 2005;**15**:1–10.

80. Schmidt MH, Husnjak K, Szymkiewicz I, Haglund K, Dikic I. Cbl escapes Cdc42-mediated inhibition by downregulation of the adaptor molecules βPix. *Oncogene* 2006;**25**:3071–8.

81. Zhao ZS, Manser E, Loo TH, Lim L. Coupling of PAK-interacting exchange factor PIX to GIT1 promotes focal complex disassembly. *Mol Cell Biol* 2000;**20**:6354–63.

82. Lee J, Jung ID, Chang WK, Park CG, Cho DY, Shin EY, Seo DW, Kim YK, Lee HW, Han JW, Lee HY. p85 beta-PIX is required for cell motility through phosphorylations of focal adhesion kinase and p38 MAP kinase. *Exp Cell Res* 2005;**307**:315–28.

83. Brown MC, Cary LA, Jamieson JS, Cooper JA, Turner CE. Src and FAK kinases cooperate to phosphorylate paxillin kinase linker, stimulate its focal adhesion localization, and regulate cell spreading and protrusiveness. *Mol Biol Cell* 2005;**16**:4316–28.

84. Frank SR, Adelstein MR, Hansen SH. GIT2 represses Crk- and Rac1-regulated cell spreading and Cdc42-mediated focal adhesion turnover. *EMBO J* 2006;**25**:1848–59.

Chapter 57

Structure of Rho Family Targets

Helen R. Mott and Darerca Owen
Department of Biochemistry, University of Cambridge, Cambridge, England, UK

RHO SUBFAMILY PROTEIN STRUCTURES AND CONFORMATIONAL SWITCH

The Rho subfamily was originally identified as a group of proteins similar to the Ras protein (Rho – *Ras ho*mology). Extensive structural studies on the Ras superfamily, particularly Ras itself, have established the basic fold of the small G proteins in both its active and inactive forms (reviewed in [1] and [2]). The core of the G domain comprises six β-strands, five of which form a parallel β-sheet, surrounded by a single anti-parallel β-strand and five α-helices. The overall topology is similar to that of many nucleotide binding proteins. All G proteins share common features that are involved in binding and specificity for guanine nucleotides. The phosphate-binding site (P-loop) has the consensus sequence GXXXXGKS/T, where the Lys side-chain contacts the negatively charged β-phosphate oxygens and the Ser/Thr side-chain OH coordinates with Mg^{2+}. Another ligand for Mg^{2+} is provided by Thr-35 (Rac1/Cdc42 numbering). The other major consensus sequence for nucleotide binding is the guanine recognition site, which, as its name implies, binds the guanine base and has the consensus sequence N/TKXD.

The major differences between the GDP- and GTP-bound forms of small G proteins lie in two regions known as switch I (or the effector loop) and switch II. The switches change conformation dramatically when GTP or its analogs bind, due to the formation of hydrogen bonds involving Thr-35 in switch I and Gly-60 in switch II (Rac1/Cdc42 numbering) and the γ-phosphate (Figure 57.1). The switch regions are relatively mobile, and are often absent or poorly defined in X-ray and NMR-derived structures. The formation of the new interactions with GTP is sufficient to change both the dynamics and structure of the loops. This results in a reorientation of the effector loop, allowing it to adopt a conformation where it can interact with downstream effector proteins.

Structurally, the Rho family proteins are distinguished from the Ras family by the presence of an extra 10–15 amino acids between helix α3 and strand β5 (the insert region). The first Rho family protein structure solved, Rac1 bound to the GTP analog GMPPNP [3], revealed that the insert region forms two extra α-helices (α3a and α3b), which are exposed on the surface of the molecule. Subsequent structures of RhoA GDP, RhoA GTPγS [4, 5] and Cdc42 GDP [6] showed that the insert region is not sensitive to the bound nucleotide (Figure 57.1). Finally, a study of Rac1, with its insert region deleted, showed that removal of these residues had no effect on the overall structure of the protein [7]. It has been reported that removal of this insert region in Rac1 expressed *in vivo* prevented both the formation of membrane ruffles [8] and transformation [9]. Thus, it would be expected that at least one target protein of the Rho family will interact with this region, but so far no effectors have been demonstrated to bind the insert region and its function remains a mystery.

CRIB PROTEINS

ACK, PAK, and WASP

Several of the downstream effectors for the Rho family proteins Cdc42 and Rac contain a short (16 amino acid) consensus sequence known as the CRIB (for *Cdc42/Rac Interactive Binding*), which is essential for mediating interactions with the G proteins [10]. In several studies it has been shown that additional residues C-terminal to the CRIB are also necessary for tight binding, making the full GBD (G-protein binding domain) 40–45 amino acids [11, 12]. The affinities of several peptides have also been measured for two PAK homologs from *Candida albicans* [13]. Like the human homolog, the residues C-terminal to the CRIB region are required for high-affinity binding. All

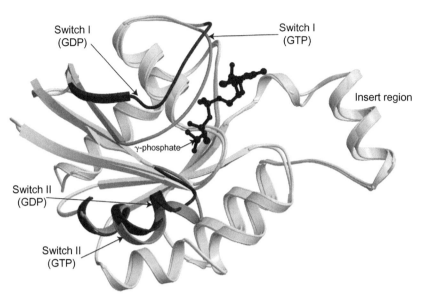

FIGURE 57.1 The GTPase switch in Rho family proteins.
The structures of RhoA · GDP and RhoA · GTPγS are shown overlaid (1a2b and 1ftn). The nucleotide is shown in a black ball-and-stick representation. When the γ-phosphate is in position, two regions of the protein, switch I and switch II, change position due to hydrogen bond formation with the phosphate oxygens. The switches in the GDP-bound form are dark gray while the GTP-bound form is light gray. All figures were made using Molscript [88] and Raster3D [89].

the CRIB proteins bind to Cdc42 · GTP, and some of them bind to Rac · GTP. Structural studies of the CRIB family proteins have addressed two fundamental questions: how do some CRIB proteins discriminate between the closely related Cdc42 and Rac, and how does binding a Rho family protein activate downstream events?

The structures of several CRIB effector fragments have been solved in complex with Rho family proteins: ACK (activated Cdc42 kinase), a tyrosine kinase, which has been implicated in integrin signaling and endocytosis [14]; WASP (Wiscott-Aldrich syndrome protein), which is thought to mediate interactions with the cytoskeleton [15]; PAKs 1, 4, and 6 (p21 activated kinase), PAK1 being a serine/threonine kinase involved in JNK signaling and cytoskeletal rearrangements ([16] and the Structural Genomics Consortium), and Par6, which is required for formation and maintenance of tight junctions and cell polarity establishment [17]. Comparison of these structures reveals interesting differences in the way that the effectors contact the G proteins (Figure 57.2). In addition, since ACK and WASP are specific for Cdc42 while PAKs bind to both Rac and Cdc42, the structures shed light on how the CRIB proteins may discriminate between two such similar molecules. In each structure, the CRIB consensus region itself binds in a similar manner, forming an intermolecular β-sheet with the β2 strand of Cdc42 and then interacting with switch I. This is reminiscent of the structure of some Ras-effector complexes, such as Raf-1 [18] and Ral-GEF [19], both of which form an intermolecular β-sheet. The regions outside the CRIB consensus all interact with the

same regions of the G protein: residues N-terminal to the CRIB interact with helix α5 and residues C-terminal to the CRIB interact with switch II, but the details of their interactions are different. ACK does not form any other secondary structure but wraps around the Cdc42, forming an irregular hairpin at the top of switch I (Figure 57.2a). WASP and PAK1 both form a β-hairpin, which interacts with both switch I and switch II of Cdc42 and is followed by a short piece of α-helix that interacts with switch II (Figure 57.2b, c). When PAK1 interacts with Rac3, in contrast, its structure more closely resembles ACK in that it does not form any secondary structure beyond the intermolecular β-sheet. In this X-ray derived structure 11 residues at the C-terminus were absent, suggesting that they are highly mobile (Figure 57.2d). These residues correspond to those in the second strand of the β-hairpin and the α-helix. In X-ray structures of Rac3-PAK4 and Cdc42-PAK6 complexes however, the residues C-terminal to the CRIB are visible and also form the β-hairpin followed by an α-helix (Figure 57.2e, f). In the four structures where they are present, the relative orientation of this hairpin and α-helix are different, as is their orientation with respect to the switch II helix. It is likely therefore that the region C-terminal to the CRIB represents a more flexible binding interface than the core CRIB region.

NMR studies on the free GBDs of the CRIB proteins have revealed no significant tertiary structure. In both PAK and WASP, there was some evidence for the formation of the short section of α-helix that is seen in the complex with Cdc42 [12, 16]. In ACK, where there is no secondary

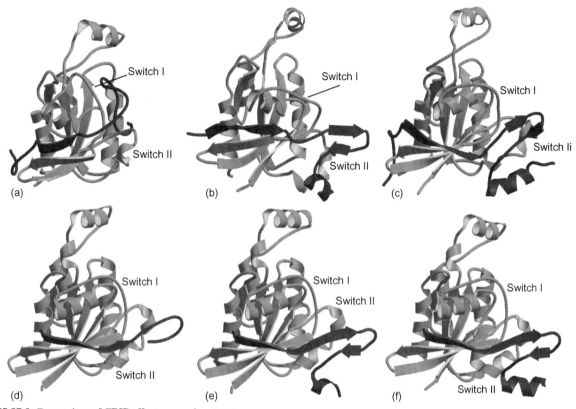

FIGURE 57.2 Comparison of CRIB effector complex structures.
In each case, the G protein is shown in pale gray and the CRIB fragment in dark gray. The insert region that defines the Rho family is at the top of each representation. The positions of the two switch regions that become fixed on effector binding are also shown. (a) Cdc42-ACK (1cf4); (b) Cdc42-PAK1 (1e0a); (c) Cdc42-WASP (1cee); (d) Rac3-PAK1 (2qme); (e) Rac3-PAK4 (2ov2); (f) Cdc42-PAK6 (2odb).

structure in the complex, none could be discerned in the free GBD [14]. Similarly, PAK proteins from *Candida albicans* have no discernible tertiary structure in the absence of Cdc42 [13].

Sequence alignments of Cdc42 and Rac reveal that switches I and II are almost completely conserved, the only difference being a single (conservative) substitution. It was therefore clear that the basis of the selectivity of CRIB effector proteins such as ACK and WASP for Cdc42 would lie outside these switches. Mutagenesis studies combined with measurements of K_d suggested that interactions with Leu-174 in helix α5 of Cdc42 contribute to binding of WASP and ACK to Cdc42, but do not contribute to PAK binding [20]. Position 174 is an Arg residue in Rac, and thus it may represent one of the points of discrimination between G proteins. Analysis of single-residue mutations may not lead us to a complete understanding of the discrimination between Cdc42 and Rac, since in all the Cdc42-effector complexes solved the buried surface area is large (2500–4000 Å^2). A recent study with two PAK homologs from *Candida albicans* dissected the roles of the core CRIB and the C-terminal residues [13]. The affinities of peptides corresponding to the core CRIB were much higher than those of the C-terminal region, and in the complexes the

C-terminal region exhibited higher mobility. It is speculated that residues in the core CRIB are important for the affinity of the complexes, but that much of the specificity of the interactions comes from the C-terminal region. This mode of binding may explain why the affinities of the CRIB-Rho family protein complexes are only 20–50 nM, even though they all bury a very large surface area. Affinities in this range are presumably necessary to ensure that the binding is readily reversible *in vivo*. A comprehensive mutagenic study of Rac1 was undertaken to determine the minimum number of changes required to allow high-affinity binding to ACK [21]. Seven residues were identified – Ser-41, Ala-42, Asn-43, Asp-47, Asn-52, Trp-56, and Arg-174 – which, when mutated in Rac1 to their Cdc42 counterparts, allowed the mutant protein to bind ACK with an affinity comparable to that of Cdc42. These residues together comprise only 5.5 percent of the total buried surface area in the Cdc42–ACK complex and only some of them directly contact Cdc42, while others appear to have a more subtle influence on the conformation of the interface.

Insight into the activation mechanisms of the CRIB proteins came when the structures of both PAK and WASP in autoinhibited forms were solved [22, 23]. WASP has no enzyme activity, but it has a region at the C-terminus that

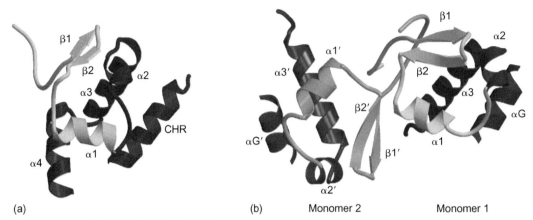

FIGURE 57.3 Structures of autoinhibited WASP and PAK fragments.
The layers of the structures are shown in different shades. In both cases, the β1/β2 and α1 elements are part of the GBD and are involved in the interaction with Cdc42; α2 and α3 are in the region immediately C-terminal to the GBD that is involved in negative regulation of the C-terminus of the molecule. (a) WASP GBD-VCA complex (1ej5), helix α4 follows α3: this helix has no counterpart in PAK. The CHR is the cofilin homology region. (b) Part of the PAK GBD–kinase complex (1f3m), αG is helix G from the kinase domain. The X-ray derived structure shows that PAK exists as a dimer. The monomer whose secondary structural elements are labeled β1, α1 etc is in the same orientation as WASP in (a). The second PAK monomer is labeled β1′, α1′, etc.

binds both the Arp2/3 complex and actin. It was found that a small region of the C-terminus that is homologous to cofilin (CHR) interacts with residues within and C-terminal to the GBD. The structure of a molecule comprising the CHR tethered by a flexible linker to the GBD extended at the C-terminus by approximately 20 residues has been solved (Figure 57.3a). This structure showed that the β-hairpin and α-helix seen in the Cdc42 complex are still present but that they now interact with three extra α-helices C-terminal to the GBD, mainly through hydrophobic contacts. These four helices form a hydrophobic surface, against which a helix from the CHR is packed. It is clear that in this form the protein cannot bind either to the Arp2/3 complex (via the CHR) or to Cdc42 (via the GBD). To bind Cdc42, the protein would have to undergo a conformational change that results in release of the CHR, allowing it to bind to other partners. The thermodynamic cost of this conformational change would be paid for by a lower binding affinity between the autoinhibited form of the protein and Cdc42 than between GBD fragments and Cdc42, as is seen, with K_ds approximately 100-fold higher for the tethered construct [23].

In Group A PAKs (PAK1–3) a region C-terminal to the GBD, known as the kinase inhibitory (or KI) domain, can bind directly to its kinase domain, inhibiting its catalytic activity. A structure of the GBD/KI bound to the kinase domain [22] reveals that the GBD/KI structure is strikingly similar to the equivalent region of WASP, although in PAK there are only three helices to WASP's four (Figure 57.3b). One of the helices from the kinase domain packs against the three helices of the KI domain in a manner closely resembling the CHR helix of WASP packing against its autoinhibitory domain. There is one striking difference in the PAK structure: it is a dimer that is held together by an interaction in part between the CRIB/KI

regions of one monomer and the kinase of the other monomer – i.e., it is inhibited in *trans* [24]. Given this structure, it would seem to be impossible to form a Cdc42–PAK complex without breaking the dimer. The interaction with the KI region locks the kinase in an inactive conformation, preventing phosphorylation of Thr-423 in the activation loop, which is necessary to achieve full kinase activation. This presents a problem for the activation mechanism, since presumably the PAK has to transphosphorylate to achieve full kinase activation, but activation by Cdc42 necessarily leads to dissolution of the dimer. The structure of the isolated kinase domain of PAK1 suggested that, even in the absence of phosphorylation of Thr-423, the activation loop was in an active conformation [25], proposing that when Cdc42 binds and breaks the dimer, PAK1 is in an intermediate semi-active state, where it can phosphorylate itself to bring about full activation. A solution study of the PAK2 kinase domain, however, suggests that it is a dimer in solution, with a K_d for dimerization of 1 μM [26]. These authors reanalyzed the PAK1 kinase structure, and suggest that there is a second dimerization interface that overlaps with the autoinhibition interface. This second dimerization interface should be capable of mediating transphosphorylation, and indeed mutants designed to block the kinase domain dimerization exhibited a much higher K_d (> 300 μM) and a concomitant decreased rate of autophosphorylation [26]. This is in agreement with an earlier study that suggested that Cdc42 binding did not break the PAK dimer [27]. The structures of the kinase domains of the group B PAKs (PAKs 4–6) have also been solved [28]. In this case, Cdc42 binding does not activate the kinases directly; rather, it may act by inducing the correct localization of the PAK. A detailed structural comparison of the Group B PAK kinase domains, in the free form and in complex with inhibitors, has allowed the conformational changes upon activation to be analyzed. The two lobes clamp together to bind ATP, and this clamping appears to be due to hinge movements in two major regions in the N-terminal lobe: the αC helix and a Gly-rich loop [28].

The activation mechanism of ACK is not known in detail, but it is also likely to involve an intramolecular inhibition. Mutation of Leu-543 of ACK, which interacts with switch II in the Cdc42–ACK complex [14], causes constitutive activation of the kinase [29]. Mutation of the equivalent Leu residue in PAK (Leu-107) or WASP (Leu-270) also disrupts their auto-inhibitory interactions [30, 31]. The activation of ACK is more complicated, since it appears that Cdc42 also disrupts an intramolecular interaction between the SH3 domain and a Pro-rich region in the C-terminal half of the protein [32]. It has been shown that neither Cdc42 nor an SH3 ligand can activate the kinase of purified ACK1 *in vitro* [33], although in HeLa cells ACK1 with a functionally defective SH3 domain that should prevent the intramolecular interaction shows enhanced autophosphorylation [34]. Furthermore, ACK1 mutants that were mutated in the CRIB region were impaired in autophosphorylation, suggesting that *in vivo* Cdc42 binding is required for ACK1 activation [34]. Cdc42 binding may activate ACK indirectly, for example by localization, since the activation loop of ACK1 is structured even in the absence of phosphorylation [35]. ACK phosphorylation does enhance the kinase activity, however [33], and there are subtle differences in the activation loop on phosphorylation [35]. The fine control of ACK activation is, therefore, not fully elucidated.

Par6

Par proteins are involved in the formation and maintenance of tight junctions and in cell polarity establishment [36, 37]. The partition-defective protein Par6 is involved in a complex with Cdc42, atypical protein kinase C (aPKC) and Par3. The region of Par6 that interacts with the Rho subfamily proteins Cdc42 and TC10 contains a "semi-CRIB" motif that includes the consensus Ile, Pro, and Phe at the N-terminus of the CRIB, but is lacking the two conserved His residues at the C-terminus of the CRIB that are necessary for interaction with Asp-38 in the conventional CRIB effector-Cdc42 complexes. The semi-CRIB alone is necessary but not sufficient for binding to Cdc42, and is adjacent to a PDZ domain that is required for high-affinity binding [36]. The crystal structure of the complex formed between Cdc42 and the semi-CRIB-PDZ fragment of Par6 [17] shows that the semi-CRIB forms an intermolecular β-sheet with the β2 strand of Cdc42, analogous to the other CRIB complexes (Figure 57.4). C-terminal to the semi-CRIB the chain reverses via an extended loop that interacts with switch II and leads into the first β-strand of the PDZ domain, which

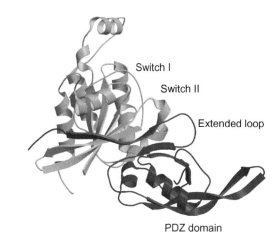

FIGURE 57.4 The Cdc42-PAR6 complex (1nf3).
Cdc42 is shown in pale gray and the switch regions are labeled. PAR6 is in dark gray. The extended loop of PAR6, between the semi-CRIB region and the PDZ domain is labeled.

forms a β-sheet with the semi-CRIB β-strand. Residues 204–208, which are part of a turn in the PDZ domain, also contact Cdc42 at both switch regions. The four-stranded β-sheet in the PDZ domain, the semi-CRIB, and the three-stranded β-sheet in Cdc42 come together to form an eight-stranded intermolecular, anti-parallel β-sheet. In the other Cdc42–CRIB complexes, the interface involves several residues from the effector and buries a surface area of at least $2500\,\text{Å}^2$. Fewer residues in Par6 contribute to the Cdc42 interaction, and this is reflected in a smaller buried surface area of only $\sim 1100\,\text{Å}^2$. NMR studies of free Par6 have shown that the semi-CRIB region is highly dynamic in the absence of Cdc42 but that once Cdc42 binds it becomes fixed in conformation [38], in agreement with fluorescence studies that indicate that there is a conformational change when Cdc42 binds to Par6 [17].

PDZ domains primarily bind to sequences at the C-terminus of target proteins, although they can also bind to short sequences within proteins (internal ligands). The Par6 PDZ domain binds to internal ligands in other proteins involved in polarity, such as Par3 and Pals1, but interestingly these interactions are not affected by Cdc42 *in vitro* [38]. In contrast, Cdc42 enhances the binding of a synthetic class I C-terminal ligand. Comparison of the free and Cdc42-bound structures of the PDZ domain from Par6 with other PDZ structures reveals that the free form of Par6 deviates from the canonical structure, and that binding of Cdc42 alters the PDZ domain so that it can become competent for binding C-terminal ligands [38].

The N-terminal region of Par6 contains a PB1 domain that binds to a PB1 domain in aPKC [39, 40]. The structure of the complex of the PB1 domains from aPKC and Par6 shows that the interaction is driven by salt-bridges that form between acidic groups in aPKC and a conserved Lys residue in Par6 [41]. Furthermore, comparison with other PB1 complexes suggests that specificity in these complexes is achieved through structural reorientations as well as *via* differences in the amino acid sequences. The interaction between aPKC and the isolated PB1 domain of Par6 activates the aPKC kinase domain, but this activation does not occur in the presence of full-length Par6 unless Cdc42 is also overexpressed [39]. This suggests that the semi-CRIB-PDZ is an autoinhibitory region for the action of the PB1 domain, and that this inhibition is relieved when Cdc42 binds. Thus, the effects of Cdc42 binding to Par6 are two-fold; it changes the conformation of the PDZ domain to allow it to bind to C-terminal ligands, and it allows the activation of aPKC.

NON-CRIB RAC EFFECTORS

p67*phox*

p67*phox* is one component of the multiprotein enzyme complex, NADPH oxidase. This complex, found in phagocytes, forms the principal defense mechanism against microbial infection in humans. Binding of Rac to p67*phox* is a critical step in the activation of the latent NADPH oxidase complex. The Rac binding region of p67*phox* had been localized to the N-terminal 200 amino acids, which contains four copies of the tetratricopeptide repeat (TPR) motif. Structures of other TPR-containing proteins had shown that each TPR motif is composed of a pair of anti-parallel α-helices (A and B), and that these repeated units pack together to form an extended structure with an amphipathic groove on the A helix face of the domain (the "TPR groove"), which mediates interactions with other proteins [42, 43]. The structure of Rac GTP complexed with the N-terminal 200 amino acids of p67*phox* revealed both an effector GBD distinct from the CRIB family of effectors and also a different use of the TPR repeats as a binding motif [44]. The N-terminus of p67*phox* contains nine α-helices: the first eight of these form four TPR motifs, while the ninth helix packs against the B helix of TPR4 (Figure 57.5a). The TPR groove is filled by a stretch of residues C-terminal to the ninth helix, which bind in an extended conformation, and thus it is not available for intermolecular interactions. Rather, contacts are made between Rac and one face of the TPR domain, which consists of a β-hairpin insertion between TPR 3 and 4 and the loops connecting TPR 1 with TRP 2 and TPR 2 with TPR 3. Contacts on the G protein side are also unusual, and include residues from helix α1 and the following loop, residues from the N-terminal end of switch I, and the loop between strand β5 and helix α5. The TPR domain does not contact all of switch I, and no contacts are seen with switch II or the insert region. This is in contrast to the complexes between Cdc42 and the CRIB effectors, where extensive contacts are made with both switches I and II. The TPR–Rac complex also differs from the CRIB effector complexes in that no intramolecular β-sheet is formed. A surface area of 1170 Å^2 is buried in

the TPR–Rac complex, less than that in the CRIB–Cdc42 complexes, possibly resulting in the lower affinity observed.

p67*phox* binds specifically to Rac rather than Cdc42. Analysis of the residues involved in the interface showed that all were conserved between Rac and Cdc42, except Gly-30. Ala-27 and Gly-30 have been defined as critical residues for the specificity of the interaction between Rac and p67*phox* mutation of these residues in Cdc42 to the corresponding residues in Rac results in a Cdc42 protein that binds p67*phox* with a relatively high affinity [44]. Ala-27 does not directly contact the TPR domain, but in Cdc42 this residue is a Lys, which would cause a steric clash, preventing binding.

Arfaptin

Arfaptin (or POR, *P*artner *o*f *R*ac) was identified independently as an effector for both the Rac and Arf small G proteins, and consequently has been proposed to be a facilitator of cross-talk between signaling pathways. Arfaptin (residues 118–341) has been crystallized alone and in complex with both Rac · GDP and Rac · GMPPNP [45]. The Arfaptin domain consists of three α helices (A–C) which form an anti-parallel α-helical bundle; two of these self-associate to give an elongated crescent-shaped dimer, with an overall length of 140 Å, which binds to one Rac molecule. Rac sits on the concave surface of the Arfaptin crescent, close to the dimer interface, and contacts are seen predominantly between switches I and II of Rac and monomer A of Arfaptin (Figure 57.5b). Switch I packs against helix αA, whilst switch II interacts with helix αB. A single contact is seen to monomer B, at His 57 in helix αA′. This interaction is sufficient to preclude the binding of another Rac molecule to Arfaptin, thus accounting for the observed stoichiometry of 1 Rac : 1 Arfaptin dimer. A solvent-accessible surface area of 1600 Å^2 is buried in the complex. Similar binding of Arfaptin affinities for both the GDP- and GTP-bound forms of Rac argues against its being a conventional effector for the G protein. In contrast, Arfaptin binds to both Arf1 and Arf6 in a GTP-dependent manner. Examination of the Rac GDP and Rac GMPPNP molecules reveals crucial similarities between the two in complex with Arfaptin. Critically, Thr-35, which coordinates to the Mg^{2+} in G-protein GTP forms, is instead in contact with Arfaptin, giving rise to a more GDP-like structure. It is predicted that a canonical Rac · GTP conformation could not be accommodated by Arfaptin [45]. Presumably, in the Arf–Arfaptin complex the G protein can take up its usual GTP-like conformation, thus allowing discrimination in that case. It is possible that Arfaptin's ability to bind both forms of Rac allows it to sequester Rac until Arf is activated, whereupon Arf displaces Rac, freeing it to signal appropriately. In this model, the function of Arf is to allow coordinated activation of Rac and Arf.

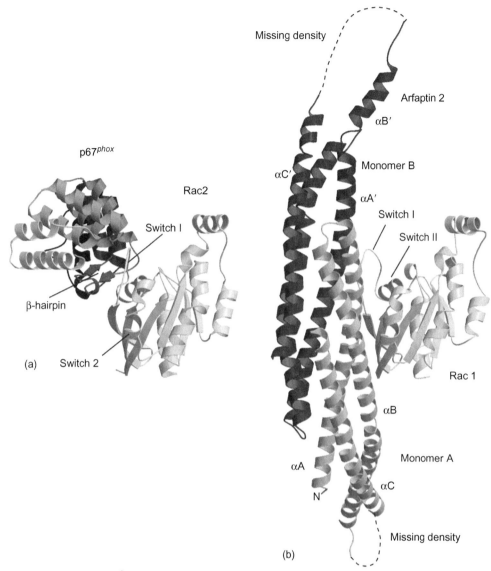

FIGURE 57.5 The structures of Rac2-p67^{phox} and Rac1-Arfaptin 2.
(a) Rac2- p67^{phox} (1e96). The layers of the TPR repeats of the p67^{phox} become darker going from the N- to the C-terminus. The switch regions of Rac and the β-hairpin of p67^{phox}, which interacts with switch I are indicated. (b) Rac1-Arfaptin2 (1i4t). The two monomers in the Arfaptin 2 dimer are shown in different shades. In the Arfaptin structure, the two regions where no electron density was observed are shown as dotted lines.

Phospholipase C-β2

Phospholipase C (PLC) isozymes hydrolyze phosphatidylinositol 4,5-bisphosphate (PI(4,5)P$_2$) into the secondary messengers diacylglycerol and inositol 1,4,5-triphosphate (IP$_3$). These products regulate both protein kinase C and the mobilization of intracellular calcium stores, and therefore control a panoply of signaling pathways. Thirteen PLC enzymes exist in humans, which are subdivided into six families: β, δ, γ, ε, ζ, and η. Generally, these isozymes share a similar architecture, containing a catalytic core that is split between two domains (X and Y), which are interrupted by an intervening sequence of 50–100 amino acids. The PLCs also contain an N-terminal PH

domain, multiple EF hands, and a C2 domain. The PLC-β family also has an extended C-terminal region that is involved in dimerization and regulation.

Avian PLC-β2 was the first member of the PLC family to be shown to be under the control of G proteins; however, it was the Gα subunit of the G$_q$ family of heterotrimeric G proteins that activated these enzymes [46]. The regulation of PLC-β by heterotrimeric G proteins was extended when it was demonstrated that the βγ subunits also activated PLC-β isoforms [47]. The Gα subunits have been shown to regulate PLC-β by interacting with their extended C-terminal region, whereas the Gβγ subunits are proposed to interact through the N-terminal PH domain [47]. More recently it has been demonstrated that Rac1 also directly activates

PLC-β, revealing an important point of confluence between heterotrimeric and small G-protein signaling in the control of the secondary messenger system [48]. Further work demonstrated that the PH domain was the site of interaction on PLC-β for Rac [49].

The structure of Rac1·GTPγS in complex with the conserved core of PLC-β2, comprising the PH, EF hands, catalytic TIM barrel and C2 domains, has been solved at 2.2-Å resolution (Figure 57.6) [50]. Rac1 binds exclusively to the PLC-β PH domain via its switch regions, burying $1200 Å^2$ of surface area. Val-36^{Rac1}, Phe-37^{Rac1} (switch I), Tyr-64^{Rac1}, and Trp-56^{Rac1} (switch II) interchelate with Arg-22$^{PLC-β}$, Gln-52$^{PLC-β}$, and Tyr-118$^{PLC-β}$ to form the central interactions, while Leu-67^{Rac1} and Leu-70^{Rac1} in the switch II α-helix are buried in a shallow hydrophobic groove lined by Ile-24$^{PLC-β}$, Pro-35$^{PLC-β}$, Val-84$^{PLC-β}$, and Tyr-118$^{PLC-β}$ at the periphery. Gln-52$^{PLC-β}$ is a crucial binding determinant, as it aids orientation of five Rac1 residues. It lies between Phe-37^{Rac1} and Trp-56^{Rac1}, forms hydrogen bonds with Asn-39^{Rac1} and Ser-71^{Rac1}, and participates in a hydrogen bonding network with Asp-57^{Rac1}. The binding determinants for PLC-β on Rac1 are conserved in both Cdc42 and RhoA; however, only Rac1 binds with high affinity to PLC-β isoforms [49]. Electrostatic surface potential differences between the small G proteins are thought to be the reason underlying the specificity. Molecular topography of the switch regions could, however, also play a part, as substitution of Trp-56 for Phe (as found in Cdc42) disrupts Rac1 binding to PLC-β. This residue has previously been found to be crucial in effector specificity determination between Cdc42 and Rac [21] as it packs behind switch I and maintains correct contouring of the G-protein surface, which some effectors seem sensitive to.

In complex with Rac1, the four domains of PLC-β form a square planar arrangement, with each domain potentially exposed to the membrane (Figure 57.6). The PH domain binds Rac1 via one surface but also makes extensive contacts with the catalytic TIM barrel domain on its opposite surface, burying 1600 Å of surface area. While the PH domains of the PLC-δ family have been well characterized for their ability to bind PI(4,5)P₂ and therefore anchor the phospholipases at the membrane in readiness to hydrolyze phosphoinositides, the PLC-β PH domains are not able to bind lipids. It is probable, therefore, that membrane associated Rac1 activates PLC-β by binding to its PH domain and orientating/stabilizing the catalytic domain relative to the membrane so it is available for efficient hydrolysis of its membrane-bound substrate [50]. It is probable that membrane-bound Gβγ subunits employ a similar activation mechanism [51], albeit via a different interface on the PH domain. The exact molecular mechanism underlying activation remains to be elucidated. The active site of uncomplexed PLC-β2 is occluded by part of the X–Y linker [47]; however, a comparison with the Rac1-bound

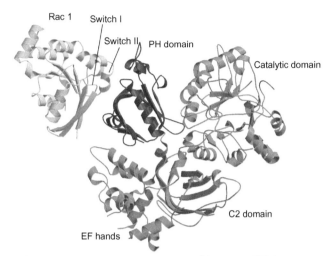

FIGURE 57.6 The Rac1-phospholipase C β2 complex (2fju). Rac1 is light gray, the PH domain of PLCβ, which forms the major contact with Rac1, is dark gray, while the other domains of PLCβ are mid-gray.

structure did not reveal any significant structural changes in PLC-β2, although this type of analysis is hampered by the lack of electron density in the unstructured X–Y linker. Further structural analysis should elucidate both the mechanism of activation of these important enzymes by small G proteins and the interplay between the heterotrimeric G proteins and small G proteins in regulating this crucial signaling node.

Plexin-B1

Plexins are a large family of transmembrane proteins that function as receptors for semaphorins [52]. They can be divided into four subfamilies, A–D, based on overall homology. While semaphorins were originally discovered as axonal guidance molecules functioning in the developing nervous system, they have more recently been shown to control adhesion and migration in many cell types [53]. Plexins have a large intracellular region that comprises two highly conserved regions which are required for semaphorin stimulated signaling, and show sequence similarity to GTPase activating proteins (GAPs) for Ras family small G proteins [54]. Plexins A1 and B1 interact directly with activated Rac1, Rnd1, and RhoD as well as with RhoA in a nucleotide-independent manner [54–60]. This gives the plexins a unique place in the Rho family effectors, as they are the first transmembrane receptor to directly interact with small G proteins.

The small G proteins all bind to the same region of the plexins, which lies in the linker region between the two GAP homology regions in the cytoplasmic domain [60]. The structure of the free GBD from plexin-B1 has been solved to 1.7Å, and consists of a 120 amino acid ubiquitin-fold

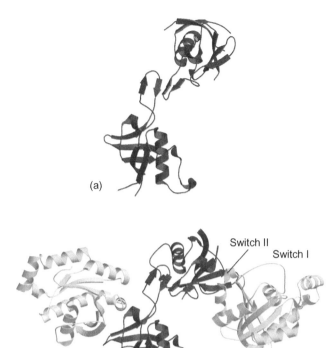

(a)

Switch II

Switch I

(b)

FIGURE 57.7 Stuctures of free and complexed plexin-B1 GBD.
(a) The free GBD of plexin-B1 (2r2o). Both monomers are dark gray. (b) The Rnd1–plexin-B1 complex (2rex). Two molecules of Rnd1 (light gray) bind to a single dimer of the plexin-B1 GDB (dark gray).

between Phe-47^{Rnd1}(switch I) and Trp-1807$^{Plexin-B1\ GBD}$. Rnd1 is essentially unchanged in the free (pdb code: 2cls) and the Plexin-B1 GBD-bound forms, and has a classic Rho small G protein structure. Contacts from both switch I and II to the plexin GBD are seen.

The GAP domain of Plexin-B1 is active towards R-Ras, a member of the Ras family of small G proteins, but only in the presence of Rnd1 [61], whose presence is indeed required for binding of Plexin- to R-Ras. Presumably, in complex with Rnd1, conformational changes in the Plexin-B1 GBD are transmitted to the adjacent GAP domain homology regions, activating both binding and catalytic activity. In a similar fashion, Plexin-B1 is also known to interact with the Rho-specific guanine nucleotide exchange factor PDZ-RhoGAP/LARG; Rnd1 has been shown to promote this interaction and to upregulate Plexin-B1-mediated RhoA activation [58]. The detailed mechanism by which Rnd1 facilitates the activities of Plexin-B1 awaits structural information on larger portions of the cytoplasmic region of the receptor.

RHO EFFECTORS

Effector proteins for Rho are divided into two classes, based on their Rho binding motifs: REM proteins (or Class 1 Rho binding motif) include the PRKs, Rhophilin, and Rhotekin, while RKH proteins (REM2 or Class 2 Rho binding motif) include the Rho kinases ROCKI and ROCKII, mDia, citron kinase, and kinectin (reviewed in [62]). Structural information is limited, at present, to one of the REM proteins, PRK1, in complexes with RhoA and Rac1, and two of the RKH family, ROCK1 and mDia.

Protein Kinase C Related Kinases

PRK1 (PKN) and -2 are highly related serine/threonine kinases with a catalytic domain homologous to that of the protein kinase C family in their C-termini and a unique regulatory domain in their N-termini [63, 64]. The N-terminus of PRK1 contains three HR1 repeats, one of which, HR1a, incorporates an inhibitory pseudo-substrate site [65]. Kinase activity is enhanced by binding of GTP-bound Rho or Rac [66–69].

The X-ray structure of RhoA in complex with the HR1a repeat of PRK1 shows that the HR1a domain is an *anti*-parallel *coiled-coil* (ACC) finger domain [70] (Figure 57.7a). The structure of the complex between HR1a and RhoA indicated two potential contacts sites on RhoA for HR1a (Figure 57.8a). The major site, contact 1, has a buried surface area of 2080Å^2, mainly involves hydrophilic interactions, and involves residues in the β2 and β3 strands of RhoA, the N-terminal part of helix α5, and residues at the ends of switch I. A second HR1a molecule contacts resi-

domain [60] (Figure 57.7a). Although the domain has the expected compact structure, several long loops are inserted into the fold. One loop, comprising residues 1818–1833, forms an anti-parallel β-sheet separate from the main fold. This in turn forms a dimerization interface with a second molecule in the X-ray structure [60], with Trp-1830 critical for dimerization.

Originally, a sequence in Plexin-B1 was identified (residues 1848–1890) that had limited homology to a CRIB motif, and this was thought to be the binding site for Rac1 [55]. However, the binding site for Rac1/Rnd1/RhoD was mapped on to the free Plexin-B1 GBD structure (using NMR chemical shift analysis) and defined as residues 1806–1818 and 1834–1848. These residues align in the structure to form an uninterrupted binding surface juxtaposed to the dimerization motif, prompting speculation that G-protein binding would disrupt dimer formation [60]. The structure of the Plexin-B1 GBD–Rnd1 complex (Structural Genomics Consortium) reveals, however, that the GBD remains as an intact dimer when bound to Rnd1, although major structural changes are seen between the bound and unbound GBD structures (Figure 57.7b). Conformational adjustments are seen in the central anti-parallel β-sheet as the GBD moves close to Rnd1. The GBD loop, 1809–1813, rotates by approximately 180° to allow an interaction

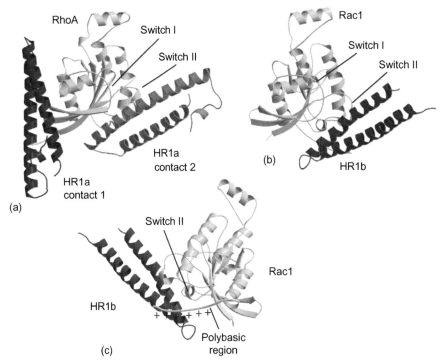

FIGURE 57.8 Structure of the RhoA-PRK1 HR1a and Rac1-PRK1 HR1b complexes.
(a) RhoA-HR1a (1cxz). The two potential HR1a binding sites are shown as Contact 1 and Contact 2. Contact 1 was defined as the primary contact site on the basis of the buried surface area, although the HR1a in Contact 2 make more interactions with the switches. RhoA is shown in pale gray and the two HR1a domains in darker gray. (b) Rac1-PRK1 (2rmk). Rac1 is pale gray and the HR1b domain is dark gray. (c) The interaction between the C-terminal tail of Rac1 with HR1b. The polybasic region of Rac1 is shown as a series of "+" signs to denote the six positive charges.

dues in switch I (Val-38 to Asn-41), strand β3 and switch II (Trp-58 and Asp-65 to Asp-76), and buries a surface area of 1640 Å^2. The size of contact 2 means that it is unlikely to be an artifact of crystal packing. It is also noteworthy that it involves more residues in the switch regions of RhoA, which, being sensitive to the bound nucleotide, would be expected to be involved in effector binding.

The HR1b domain of PRK1 also forms an anti-parallel coiled-coil, and *in vitro* studies have shown that it can bind tightly to Rac1 and, albeit at a lower affinity, to RhoA [71, 72]. The structure of the Rac1–HR1b complex reveals that HR1b contacts Rac1 at a site equivalent to the RhoA contact 2 site [73] (Figure 57.8b). As this site involves the switch regions of the G proteins, which are highly conserved between Rac1 and RhoA, it might be expected that HR1b should also tightly to RhoA. The HR1b–Rac1 structure revealed, however, that there are additional contacts between the two proteins that involve the C-terminal polybasic region (PBR) of Rac1. The Rho family proteins, like most Ras-like G proteins, are lipid-modified at the C-terminus, which allows them to be localized to membranes. Extra membrane attachment in the Rho family is provided by the PBR, which is presumed to form electrostatic interactions with membrane phospholipid head-

groups. The PBR in Rac1 is unique in that it is preceded by a triple proline motif, and also that it contains an uninterrupted stretch of six basic residues: most of the Rho family have just one or two prolines and a few basic residues interspersed with glycines and hydrophilic residues. The confluence of these motifs in Rac1 leads to a particular structure in this region: the proline motif, being conformationally rigid, locks the backbone position of the region just C-terminal to the last α-helix and this is followed by the six basic residues, which repel each other and force that section of the chain to be extended. Overall, this leads to the basic region looping back and contacting residues in switch II of Rac1 and residues in the HR1b domain (Figure 57.8c). The buried surface area in the region of the HR1b–Rac1 complex involving the Rac1 switches is 1550 Å^2, similar to the size of the same interface in the HR1a–RhoA complex, but the interactions with the C-terminus increase contribute an extra ~600 Å^2 of buried surface area. This presumably accounts for the higher affinity that HR1b has for Rac1 when compared to that for RhoA. The multiple binding sites for Rho family protein in PRK1 suggest that its regulation is more complicated than simply binding of RhoA to the HR1a domain, but the detailed mechanism of PRK1 activation and the role of binding of the HR1a and HR1b domains to Rho family proteins remains elusive.

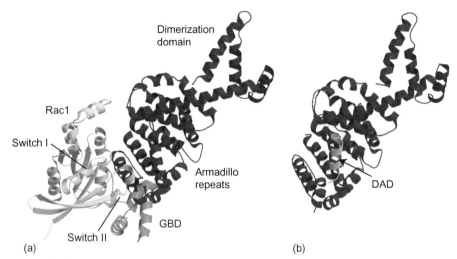

FIGURE 57. 9 The structure of active and inhibited mDia.
(a) The structure of mDia$_N$ in complex with Rac1 (1z2c). Rac1 is shown in pale gray, while mDia$_N$ is shown in dark gray. The mDia$_N$ GBD is highlighted in mid-gray and the dimerization domain and the Armadillo repeats of mDia$_N$ are labeled. b) The structure of mDia$_N$ in complex with a peptide from the mDia DAD (2bap). mDiaN is colored in dark gray and the DAD peptide in pale gray.

mDia1 (p140 Diaphanous)

The formin family of proteins comprises commanding regulators of actin cytoskeleton dynamics. They can control unbranched actin filament nucleation rate and elongation rate, and affect the actions of actin filament capping proteins. Their overall activity leads to an increase in actin filament assembly [74]. One group of formins, the Dia (diaphanous) family, is regulated by the Rho small G proteins; mDia1 is regulated by RhoA, B and C, mDia2 by RhoA, B, and Rif (RhoF), while mDia3 is known to interact with RhoD [75].

The diaphanous proteins are modular and contain a catalytic FH2 (formin homology domain 2) in their C-terminal half, preceded by an FH1 (formin homology domain1). These two domains act together to initiate actin assembly [74]. At their C-terminus, diaphanous proteins contain a DAD domain (diaphanous autoregulatory domain), which binds to sequences at the N-terminus of the protein and maintains it in an autoinhibited conformation [76]. The mDia N-terminus contains the FH3 (formin homology domain 3) region and the Rho binding domain (GBD) [77]. The functional inhibitory domain (DID) overlaps with both the GBD and the FH3. Binding of Rho and the DAD to the N-terminal regulatory sequences are mutually exclusive, and therefore Rho binding to the GBD relieves the autoinhibition on the FH2 and activates the diaphanous proteins [77]. Structures have been solved of both RhoC in complex with GBD/FH3 region of mDia1, mDia$_N$ [77], and a similar region in complex with a DAD peptide [78, 79]. This pair of structures has allowed insights into the mechanism of control of mDia1 by Rho (Figure 57.9).

The regulatory N-terminal region of mDia1 is an all-helical peptide, comprising the GBD, an armadillo repeat region (ARR), and a dimerization domain. RhoC contacts mDia1 via both the GBD and the ARR, resulting in a buried surface area in the complex of 2981 Å (Figure 57.9a). RhoC utilizes a hydrophobic patch on the surface of the switch regions, consisting of Val-38, Phe-39, and Leu-72, to form the core of the interface, which is then further stabilized by electrostatic interactions towards the periphery. Both switch regions of RhoC are therefore involved in the interaction, switch I interacting with the mDia1 GBD and switch II interacting with residues in both the GBD and the ARR (Figure 57.9a). The structure explains the GTP dependence of the Rho–mDia interaction as RhoC Val-38, Phe-39 and Glu-40 rearrange in the absence of the GTP γ-phosphate, deforming the hydrophobic core of the binding interface. Changes are also observed in Leu-69 and Tyr-66 of switch II on GDP/GTP exchange, which make additional contacts with mDia1. Indeed, the R68A and L69D mutations in Rho abolish binding to mDia1. Arg-68, of switch II, plays a critical role in the complex by wedging itself between the GBD and the first repeat of the ARR, interacting with Asn-217, Leu-163, and Asn-166 of mDia1 [78].

The structure of the N-terminal regulatory domain of mDia1 in complex with a peptide from the DAD revealed that the DAD is also helical and interacts with mDia$_N$ via mainly hydrophobic interactions. The DAD binds to the concave side of the ARR (Figure 57.9b). The DAD (residues 1180–1195) forms an amphipathic helix, which is kinked at residue 1192, that makes comprehensive contacts to the second helices of armadillo repeats 2, 3, and 4. As well as the extensive hydrophobic interactions, Asp-1183 forms a salt-bridge with Lys-213 in the second Armadillo repeat. Asp-1183 also contacts Asn-217, which is crucial for Rho binding, as it stabilizes the "arginine wedge" (Arg-68[RhoC]). Although the DAD binding site is close to the Rho interface with mDia$_N$, it is only marginally overlapping (Figure 57.9), raising the question of how Rho is able to activate mDia1. Steric clashes are postulated for Thr-1179[DAD] and Gly-1180[DAD] on RhoC binding to the GBD, along with electrostatic repulsion between Asp-1183[DAD]/Glu-1187[DAD] and Glu-64[Rho]/Asp-65[Rho]/Asp-67[Rho]. The potential importance of DAD residues outside those visible in the structure was also hypothesized, highlighting two basic regions on either side of the core DAD. More recent work has suggested that a basic region is also important (mDia1[1196]RRKR[1199]), although there is no structural information pertaining to this region to date [80]. As the GBD and the ARR are connected in a flexible manner, the autoinhibited state is thought to be able to accommodate initial, low-affinity binding of Rho to the GBD and still allow DAD binding. In a second, conformation-driven stage, Rho is able to make additional contacts to the ARR, inducing release of the DAD from the mDiaN, and formation of a high-affinity Rho–mDia complex [78]. This scenario parallels both the GEF catalyzed nucleotide exchange of Ras-like proteins, and activation of other Rho-family effectors (e.g., WASP [23] and PAK [22]). As with the majority of structural work on large multi-domain proteins, individual domains have been utilized in these studies. It is also noted that in larger constructs, Rho binding is insufficient for full activation of mDia, thus other stages may also be required to trigger mDia activity [78].

ROCK

Two ROCK isoforms have been isolated: ROCK I (also known as ROKβ and p160ROCK) and ROCKII (also known as ROKα and Rho kinase). Both isoforms contain a kinase domain in the N-terminal half of the protein, followed by a long coiled-coil region of about 600 residues that includes the Rho binding domain and finally, at the C-terminus, a PH domain that is interrupted by a C1 domain (or cysteine-rich domain). ROCKI and ROCKII are 65 percent identical overall, with 92 percent identity in the kinase domain. Myotonic dystrophy kinase, myotonic dystrophy kinase-related Cdc42-binding kinase and citron kinase all have similar domain structures to the ROCKs, and their kinase domains are homologous. The ROCK proteins are involved in several pathways downstream of Rho (reviewed in [81]), including smooth muscle contraction and stress fiber formation. ROCK is known to stabilize actin filaments by phosphorylating LIM kinase, which phosphorylates and inhibits the actin destabilizing protein cofilin.

The kinase domain of ROCK is assumed to be autoregulated by the inhibitory C-terminal half of the protein, which

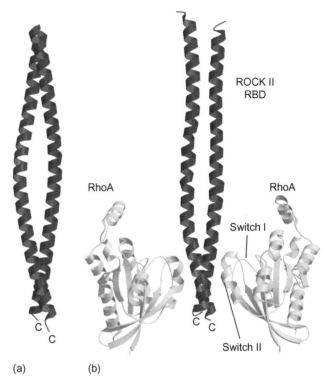

(a)　　　　　(b)

FIGURE 57.10 The structure of the ROCK GBD, both bound and free.

(a) The structure of the ROCKII GBD (1uix). The C-termini of the two α-helices of the coiled-coil are labeled. (b) The structure of the ROCKI GBD in complex with two molecules of RhoA (1s1c). The ROCKI GBD is colored dark gray and labeled as in (a). The RhoA molecules are colored pale gray.

includes the Rho binding domain and the PH domain [81]. It is also likely, however, that ROCK is at least partially regulated through dimerization. Full-length ROCKII is a dimer *in vitro* [82], and it was long assumed that dimerization was mediated through the coiled-coil domain in the center of the protein. The structure of the Rho binding domain of ROCKII was solved and indeed was shown to consist of a parallel, left-handed, coiled-coil made from two 86 amino acid long α-helices [83] (Figure 57.10a). Like most coiled-coils it is held together principally by hydrophobic interactions, although these are usually reinforced by salt-bridges at the edge of the hydrophobic interface, and in ROCKII there is only a single ionic interaction between the chains. The coiled-coil is packed tightly at the edges, but at the center it is characterized by loose packing. This results in an unusually large inter-helix distance in the center of the coiled-coil, although ordered water molecules were not observed in the resulting space. The crystallographic B factors in this region were higher than in the rest of the molecule, implying that there is some mobility in this part of the protein.

The structure of the Rho binding domain of ROCKI in complex with RhoA confirmed that the parallel coiled-coil is a feature of both isoforms of ROCK, and persists in the presence of the small G protein [84] (Figure 57.10b). The asymmetric unit includes two RhoA molecules in complex with the dimeric ROCK RBD (Rho binding domain). The minimal Rho binding motif is formed by just 13 amino acids at the C-terminus of the parallel coiled-coil, which contact both switch regions of RhoA. One of the two helices in the ROCKI dimer is opposite switch I, while the other helix faces switch II. As the coiled-coil is parallel and two RhoA molecules are interacting with the ROCK dimer, equivalent residues in the RBD contact the same residues in RhoA and there is a non-crystallographic symmetry axis running through the center of the coiled-coil. Hydrophobic residues in both RhoA switch regions contact hydrophobic residues in the RBD at the center of the interface, and these interactions are stabilized by a number of salt-bridges and hydrogen bonds at the edge of the contact surface. Four residues in RhoA, two from each switch region (Val-38, Phe-39, Asp-65, Tyr-66), are close to the RBD dimer interface, contacting both helices in the RBD and apparently stabilizing the dimer.

Like the free ROCKII coiled-coil, the center of the ROCKI coiled-coil is less tightly packed than the ends. In the ROCKI dimer there is a series of stutters in the heptad repeats that comprise the coiled-coils; these insertions of four residues into the heptad lead to a right-handed twist of coiled-coils. The consequence of these stutters in ROCKI is that the coiled-coil is left-handed (and therefore canonical) at the ends and is right-handed at the center. Although this suggests that there is a genuine difference between ROCKI and ROCKII at the center of the coiled-coil due to sequence diversity in this region, the B factors are also high for one of the helices in ROCKI in the right-handed section. The alternative conformations observed in the two X-ray structures may therefore be just two of several structures that are sampled in the full-length molecule in solution.

The mechanism of activation of ROCKs, like that of PRK1, is thought to involve relief of autoinhibition. PRK1 has multiple HR1 domains, each of which could, in principle, bind to small G proteins, suggesting that the activation mechanism may require cooperative engagement of more than one Rho-binding domain. By extension, it is possible that ROCK also has extra Rho-binding regions, and indeed sequence alignments of ROCK proteins suggested that there were other highly conserved regions within the coiled-coil. One of these regions, the RID (Rho-interacting domain), binds to RhoA with a high affinity and in a GTP-dependent manner [72] in vitro, and thus is likely to be a secondary binding site. Furthermore, a potential HR1 domain was also identified in ROCK1, between residues 420 and 550. This domain also binds to RhoA, but can interact with both the GDP- and GTP-bound forms. It is possible that the HR1 domain binds to the contact I site identified in the PRK1-RhoA structure [70], which is not expected to be nucleotide dependent, but there is no structural evidence for this as yet.

ROCK can clearly dimerize through the RBD, but a combination of light-scattering and analytical ultracentrifugation suggested the presence of a second dimerization motif N-terminal to the kinase domain [82]. This was confirmed by X-ray structures of the kinase domains of ROCKI [85] and ROCKII [86] with an N-terminal extension that includes the dimerization motif and a C-terminal extension that is essential for activity in the AGC family members and contains a hydrophobic sequence motif [87]. The structures show that the N-terminal extension from each monomer folds into three α-helices that form a head-to-head dimer and is effectively a single domain (the dimerization domain or the capped helix bundle domain). The extensive interface between the monomers is further stabilized by the C-terminal extension, which contacts the dimerization domain, interacting with both chains and contributing to the large, hydrophobic buried surface area of $\sim 4000 \text{Å}^2$. The kinase domain of ROCK appears to be active, since the activation loop, which blocks the active site in inactive kinases and prevents substrates binding, resembles an active conformation. Active kinases are often stabilized by phosphorylation of the activation loop, but ROCK does not appear to be phosphorylated and instead the loop conformation is stabilized by interactions with other residues in the protein. The hydrophobic motif in the C-terminal extension also contributes to an active kinase conformation by binding to a hydrophobic groove in the N-terminal lobe of the kinase. This correctly aligns helix αC and allows a salt bridge to form between Glu-140 in αC and Lys-121 (ROCKII numbering), which is characteristic for an active kinase. The N-terminal dimerization domain stabilizes the hydrophobic motif binding to the hydrophobic groove, so both extensions are necessary for both dimerization and a fully active conformation for the kinase.

The active sites of the two kinase domains are facing the same direction, but the orientations are such that the activation loop of one monomer is unable to be phosphorylated by the other monomer. The size and hydrophobic nature of the dimerization interface make it unlikely that it would be disrupted to allow phosphorylation to occur, suggesting that transphosphorylation is not a critical part of the activation process for ROCK. As the activation loops and the active sites of both kinase domains are on the same face of the dimer, it is likely that they can be autoinhibited in parallel by binding to the C-terminal regulatory domains. In the absence of detailed structural or biochemical information, it is not possible to determine whether they are autoinhibited in trans, in a similar manner to the PAK kinase, or by binding to the regulatory domain in the same monomer.

CONCLUDING REMARKS

We have attempted to summarize here the pertinent details of all the Rho family-effector structures determined to date. Several points emerge from this discussion. First, it is clear that the structural diversity in the Rho effectors is extensive. It is likely, for example, that the CRIB proteins, although they will all have some similarities in the way that they interact, will also differ in their details, particularly outside the short CRIB consensus sequence. The other seven effector structures are also completely different both to each other and to the CRIBs; although it is likely that families will emerge whose members interact with the G proteins in a similar, but not identical manner (e.g., coiled-coil domains). There are also, however, several other effector proteins with no sequence homology to TPR domains, Arfaptin, PH domains, Plexin-B1, PRK1, mDia, ROCK, or CRIBs. It is likely that these proteins will adopt different structures and will interact with the Rho family proteins in novel ways.

The manner in which the effectors contact the Rho family G proteins is multifarious. They utilize β-strands (the CRIBs), β-hairpins (the CRIBs and Plexin-B1), α-helices (Arfaptin, PRK-1, ROCK, mDia, and the CRIBs), inter-helical loops (p67^{phox}), and even a dimer interface (Arfaptin) to interact with the G protein. In some cases, such as the CRIB proteins, the effectors make an extensive set of contacts with the Rho family proteins, burying a large surface area (2500–4000 Å^2), while in others, such as p67^{phox}, which binds with a lower affinity, the buried surface is only ~1200 Å^2.

The region of the G proteins that interact with the downstream targets is not entirely conserved. In most cases, switch I and/or switch II are involved in the interactions, which is perhaps not surprising given that effectors bind preferentially to the GTP-bound form of the G protein. It is often the case, however, that other regions of the G protein are also involved in binding to effectors, and this may be in part to bring about specificity: the switches are relatively well-conserved within the family, while the diversity in the other regions is higher. The other regions of the Rho family that are involved in effector binding are: helix α1, the β2 and β3 strands, the β5–α5 loop, the C-terminal helix, α5, and the polybasic region at the extreme C-terminus of Rac1. In no case so far is the insert region involved in effector binding.

Structural information on the Rho family targets has moved at an exciting pace since the previous edition of this review 5 years ago. The next stage is still to determine how the binding of the Rho family protein causes the downstream effects that are seen. This point had begun to be addressed with the elegant work on PAK and WASP activation, and we now have a mechanistic insight into the activation of mDia. We can look forward to many more such breakthroughs in the future, when structures of larger portions of these effectors, encompassing the catalytic and regulatory domains, will help us to understand fully these complex systems.

REFERENCES

1. Kjeldgaard M, Nyborg J, Clark BFC. Protein motifs 10. The GTP binding motif: variations on a theme. *FASEB J* 1996;**10**:1347–68.

2. Vetter IR, Wittinghofer A. Signal transduction – the guanine nucleotide-binding switch in three dimensions. *Science* 2001;**294**:1299–304.

3. Hirshberg M, Stockley RW, Dodson G, Webb MR. The crystal structure of human rac1, a member of the rho-family complexed with a GTP analogue. *Nature Struct Biol* 1997;**4**:147–52.

4. Wei YY, Zhang Y, Derewenda U, et al. Crystal structure of RhoA-GDP and its functional implications. *Nature Struct. Biol* 1997;**4**:699–703.

5. Ihara K, Muraguchi S, Kato M, et al. Crystal structure of human RhoA in a dominantly active form complexed with a GTP analogue. *J Biol Chem* 1998;**273**:9656–66.

6. Feltham JL, Dotsch V, Raza S, et al. Definition of the switch surface in the solution structure of Cdc42Hs. *Biochemistry* 1997;**36**:8755–66.

7. Thapar R, Karnoub AE, Campbell SL. *Structural and biophysical insights into the role of the insert region in Rac1 function* 2002;**41**:3875–83.

8. Karnoub AE, Der CJ, Campbell SL. The insert region of Rac1 is essential for membrane ruffling but not cellular transformation. *Mol Cell Biol* 2001;**21**:2847–57.

9. Joneson T, Bar-Sagi D. A Rac1 effector site controlling mitogenesis through superoxide production. *J Biol, Chem* 1998;**273**:17,991–17,994.

10. Burbelo PD, Drechsel D, Hall A. A conserved binding motif defines numerous candidate target proteins for both Cdc42 and Rac GtPases. *J Biol Chem* 1995;**270**:29,071–29,074.

11. Thompson G, Owen D, Chalk PA, Lowe PN. Delineation of the Cdc42/Rac-binding domain of p21-activated kinase. *Biochemistry* 1998;**37**:7885–91.

12. Rudolph MG, Bayer P, Abo A, Kuhlmann J, Vetter IR, Wittinghofer A. The Cdc42/Rac interactive binding region motif of the Wiskott Aldrich syndrome protein (WASP) is necessary but not sufficient for tight binding to Cdc42 and structure formation. *J Biol Chem* 1998;**273**:18,067–18,076.

13. Su ZD, Osborne MJ, Xu P, Xu XL, Li Y, Ni F. A bivalent dissectional analysis of the high-affinity interactions between Cdc42 and the Cdc42/Rac interactive binding domains of signaling kinases in Candida albicans. *Biochemistry* 2005;**44**:16461–74.

14. Mott HR, Owen D, Nietlispach D, et al. Structure of the small G protein Cdc42 bound to the GTPase- binding domain of ACK. *Nature* 1999;**399**:384–8.

15. Abdul-Manan N, Aghazadeh B, Liu GA, et al. Structure of Cdc42 in complex with the GTPase-binding domain of the 'Wiskott-Aldrich syndrome' protein. *Nature* 1999;**399**:379–83.

16. Morreale A, Venkatesan M, Mott HR, et al. Structure of Cdc42 bound to the GTPase binding domain of PAK. *Nature Struct Biol* 2000;**7**:384–8.

17. Garrard SM, Capaldo CT, Gao L, Rosen MK, Macara IG, Tomchick DR. Structure of Cdc42 in a complex with the GTPase-binding domain of the cell polarity protein, Par6. *EMBO J* 2003;**22**:1125–33.

18. Nassar M, Horn G, Herrmann C, Scherer A, McCormick F, Wittinghofer A. The 2.2-Ångstrom crystal-structure of the Ras-binding domain of the serine threonine kinase C-Raf1 in complex with Rap1a and a GTP analog. *Nature* 1995;**375**:554–60.

19. Vetter IR, Linnemann T, Wohlgemuth S, et al. Structural and biochemical analysis of Ras-effector signaling via RalGDS. *FEBS Letts* 1999;**451**:175–80.

20. Owen D, Mott HR, Laue ED, Lowe PN. Residues in Cdc42 that specify binding to individual CRIB effector proteins. *Biochemistry* 2000;**39**:1243–50.

21. Elliot-Smith AE, Mott HR, Lowe PN, Laue ED, Owen D. Specificity determinants on Cdc42 for binding its effector protein ACK. *Biochemistry* 2005;**44**:12,373–12,383.

22. Lei M, Lu WG, Meng WY, et al. Structure of PAK1 in an autoinhibited conformation reveals a multistage activation switch. *Cell* 2000;**102**:387–97.

23. Kim AS, Kakalis LT, Abdul-Manan M, Liu GA, Rosen MK. Autoinhibition and activation mechanisms of the Wiskott-Aldrich syndrome protein. *Nature* 2000;**404**:151–8.

24. Parrini MC, Lei M, Harrison SC, Mayer BJ. Pak1 kinase homodimers are autoinhibited in trans and dissociated upon activation by Cdc42 and Rac1. *Mol Cell* 2002;**9**:73–83.

25. Lei M, Robinson MA, Harrison SC. The active conformation of the PAK1 kinase domain. *Structure* 2005;**13**:769–78.

26. Pirruccello M, Sondermann H, Pelton JG, et al. A dimeric kinase assembly underlying autophosphorylation in the p21 activated kinases. *J Mol Biol* 2006;**361**:312–26.

27. Buchwald G, Hostinova E, Rudolph MG, et al. Conformational switch and role of phosphorylation in PAK activation. *Mol Cell Biol* 2001;**21**:5179–89.

28. Eswaran J, Lee WH, Debreczeni JE, et al. Crystal structures of the p21-activated kinases PAK4, PAK5, and PAK6 reveal catalytic domain plasticity of active group IIPAKs. *Structure* 2007;**15**:201–13.

29. Kato J, Kaziro Y, Satoh T. Activation of the guanine nucleotide exchange factor Dbl following ACK1-dependent tyrosine phosphorylation. *Biochem Biophys Res Commun* 2000;**268**:141–7.

30. Frost JA, Khokhlatcheva A, Stippec S, White MA, Cobb MH. Differential effects of PAK1-activating mutations reveal activity-dependent and -independent effects on cytoskeletal regulation. *J Biol Chem* 1998;**273**:28,191–28,198.

31. Devriendt K, Kim AS, Mathijs G, et al. Constitutively activating mutation in WASP causes X-linked severe congenital neutropenia. *Nature Genet* 2001;**27**:313–17.

32. Yang WN, Lin Q, Guan JL, Cerione RA. Activation of the Cdc42-associated tyrosine kinase-2 (ACK-2) by cell adhesion via integrin beta(1). *J Biol Chem* 1999;**274**:8524–30.

33. Yokoyama N, Miller WT. Biochemical properties of the Cdc42-associated tyrosine kinase ACK1 – Substrate specificity, autophosphorylation, and interaction with Hck. *J Biol Chem* 2003;**278**:47,713–47,723.

34. Galisteo ML, Yang Y, Urena J, Schlessinger J. Activation of the non-receptor protein tyrosine kinase Ack by multiple extracellular stimuli. *Proc Natl Acad Sci USA* 2006;**103**:9796–801.

35. Lougheed JC, Chen RH, Mak P, Stout TJ. Crystal structures of the phosphorylated and unphosphorylated kinase domains of the Cdc42-associated tyrosine kinase ACK1. *J Biol Chem* 2004;**279**:44,039–44,045.

36. Joberty G, Petersen C, Gao L, Macara IG. The cell-polarity protein Par6 links Par3 and atypical protein kinase C to Cdc42. *Nature Cell Biol* 2000;**2**:531–9.

37. Lin D, Edwards AS, Fawcett JP, Mbamalu G, Scott JD, Pawson T. A mammalian PAR-3-PAR-6 complex implicated in Cdc42/Rac1 and aPKC signalling and cell polarity. *Nature Cell Biol* 2000;**2**:540–7.

38. Peterson FC, Penkert RR, Volkman BF, Prehoda KE. Cdc42 regulates the Par-6 PDZ domain through an allosteric CRIB-PDZ transition. *Mol Cell* 2004;**13**:665–76.

39. Yamanaka T, Horikoshi Y, Suzuki A, et al. PAR-6 regulates aPKC activity in a novel way and mediates cell– cell contact-induced formation of the epithelial junctional complex. *Genes Cells* 2001;**6**:721–31.

40. Etienne-Manneville S, Hall A. Integrin-mediated activation of Cdc42 controls cell polarity in migrating astrocytes through PKC zeta. *Cell* 2001;**106**:489–98.

41. Hirano Y, Yoshinaga S, Takeya R, et al. Structure of a cell polarity regulator, a complex between atypical PKC and Par6 PB1 domains. *J Biol Chem* 2005;**280**:9653–61.

42. Das AK, Cohen PTW, Barford D. The structure of the tetratricopeptide repeats of protein phosphatase 5: implications for TPR-mediated protein-protein interactions. *EMBO J* 1998;**17**:1192–9.

43. Scheufler C, Brinker A, Bourenkov G, et al. Structure of TPR domain-peptide complexes: critical elements in the assembly of the Hsp70-Hsp90 multichaperone machine. *Cell* 2000;**101**:199–210.

44. Lapouge K, Smith SJM, Walker PA, Gamblin SJ, Smerdon SJ, Rittinger K. Structure of the TPR domain of p67(phox) in complex with Rac center dot GTP. *Mol Cell* 2000;**6**:899–907.

45. Tarricone C, Xiao B, Justin N, et al. The structural basis of Arfaptin-mediated cross-talk between Rac and Arf signalling pathways. *Nature* 2001;**411**:215–19.

46. Rhee SG, Choi KD. Regulation of Inositol phospholipid-specific phospholipase-C isozymes. *J Biol Chem* 1992;**267**:12,393–12,396.

47. Harden TK, Sondek J. Regulation of phospholipase C isozymes by Ras superfamily GTPases. *Annu Rev Pharmacol Toxicol* 2006;**46**:355–79.

48. Illenberger D, Schwald F, Pimmer D, et al. Stimulation of phospholipase C-beta(2) by the rho GTPases Cdc42Hs and Rac1. *EMBO J* 1998;**17**:6241–9.

49. Snyder JT, Singer AU, Wing MR, Harden TK, Sondek J. The pleckstrin homology domain of phospholipase C-beta(2) as an effector site for Rac. *J Biol Chem* 2003;**278**:21,099–21,104.

50. Jezyk MR, Snyder JT, Gershberg S, Worthylake DK, Harden TK, Sondek J. Crystal structure of Rac1 bound to its effector phospholipase C-beta 2. *Nature Struct Mol Biol* 2006;**13**:1135–40.

51. Wang TL, Dowal L, El-Maghrabi MR, Rebecchi M, Scarlata S. The pleckstrin homology domain of phospholipase C-beta(2) links the binding of G beta gamma to activation of the catalytic core. *J Biol Chem* 2000;**275**:7466–9.

52. Negishi M, Oinuma I, Katoh H. Plexins: axon guidance and signal transduction. *Cell Mol Life Sci* 2005;**62**:1363–71.

53. Kruger RR, Aurandt J, Guan KL. Semaphorins command cells to move. *Nature Rev Mol Cell Biol* 2005;**6**:789–800.

54. Rohm B, Rahim B, Kleiber B, Hovatta I, Puschel AW. The semaphorin 3A receptor may directly regulate the activity of small GTPases. *FEBS Letts* 2000;**486**:68–72.

55. Vikis HG, Li WQ, He ZG, Guan KL. The semaphorin receptor plexin-B1 specifically interacts with active Rac in a ligand-dependent manner. *Proc Natl Acad Sci USA* 2000;**97**:12,457–12,462.

56. Hu HL, Marton TF, Goodman CS. Plexin B mediates axon guidance in Drosophila by simultaneously inhibiting active rac and enhancing RhoA signaling. *Neuron* 2001;**32**:39–51.

57. Driessens MHE, Hu HL, Nobes CD, et al. Plexin-B semaphorin receptors interact directly with active Rac and regulate the actin cytoskeleton by activating Rho. *Curr Biol* 2001;**11**:339–44.

58. Oinuma I, Katoh H, Harada A, Negishi M. Direct interaction of Rnd1 with plexin-B1 regulates PDZ-RhoGEF-mediated Rho activation by plexin-B1 and induces cell contraction in COS-7 cells. *J Biol Chem* 2003;**278**:25,671–25,677.

59. Zanata SM, Hovatta I, Rohm B, Puschel AW. Antagonistic effects of Rnd1 and RhoD GTPases regulate receptor activity in semaphorin 3A-induced cytoskeletal collapse. *J Neurosci* 2002;**22**:471–7.

60. Tong Y, Chugha P, Hota PK, et al. Binding of Rac1, Rnd1, and RhoD to a novel Rho GTPase interaction motif destabilizes dimerization of the plexin-B1 effector domain. *J Biol Chem* 2007;**282**:37,215–37,224.

61. Oinuma I, Ishikawa Y, Katoh H, Negishi M. The semaphorin 4D receptor plexin-B1 is a GTPase activating protein for R-Ras. *Science* 2004;**305**:862–5.

62. Bishop AL, Hall A. Rho GTPases and their effector proteins. *Biochem J* 2000;**348**:241–55.

63. Mukai H, Ono Y. A novel protein-kinase with leucine zipper-like sequences – its catalytic domain is highly homologous to that of protein-kinase-C. *Biochem Biophys Res Commun* 1994;**199**:897–904.

64. Palmer RH, Ridden J, Parker PJ. Cloning and expression patterns of 2 members of a novel protein kinase-C-related kinase family. *Eur J Biochem* 1995;**227**:344–51.

65. Kitagawa M, Shibata H, Toshimori M, Mukai H, Ono Y. The role of the unique motifs in the amino-terminal region of PKN on its enzymatic activity. *Biochem Biophys Res Commun* 1996;**220**:963–8.

66. Amano M, Mukai H, Ono Y, et al. Identification of a putative target for Rho as the serine-threonine kinase protein kinase N. *Science* 1996;**271**:648–50.

67. Lu Y, Settleman J. The Drosophila Pkn protein kinase is a Rho Rac effector target required for dorsal closure during embryogenesis. *Genes Dev* 1999;**13**:1168–80.

68. Vincent S, Settleman J. The PRK2 kinase is a potential effector target of both Rho and Rac GTPases and regulates actin cytoskeletal organization. *Mol Cell Biol* 1997;**17**:2247–56.

69. Watanabe G, Saito Y, Madaule P, et al. Protein kinase N (PKN) and PKN-related protein rhophilin as targets of small GTPase Rho. *Science* 1996;**271**:645–8.

70. Maesaki R, Ihara K, Shimizu T, Kuroda S, Kaibuchi K, Hakoshima T. The structural basis of Rho effector recognition revealed by the crystal structure of human RhoA complexed with the effector domain of PKN/PRK1. *Mol Cell* 1999;**4**:793–803.

71. Owen D, Lowe PN, Nietlispach D, et al. Molecular dissection of the interaction between the small G proteins Rac1 and RhoA and protein kinase C-related kinase 1 (PRK1). *J Biol Chem* 2003;**278**:50,578–50,587.

72. Blumenstein L, Ahmadian MR. Models of the cooperative mechanism for Rho effector recognition – Implications for RhoA-mediated effector activation. *J Biol Chem* 2004;**279**:53,419–53,426.

73. Modha R, Campbell LJ, Nietlispach D, Buhecha HR, Owen D, Mott HR. The Rac1 polybasic region is required for interaction with its effector PRK1. *J Biol Chem* 2008;**283**:1492–500.

74. Higgs HN. Formin proteins: a domain-based approach. *Trends Biochem Sci* 2005;**30**:342–53.

75. Vega FM, Ridley AJ. SnapShot: Rho family GTPases. *Cell* 2007;**129**:U1430–2.

76. Alberts AS. Identification of a carboxyl-terminal diaphanous-related formin homology protein autoregulatory domain. *J Biol Chem* 2001;**276**:2824–30.

77. Rose R, Weyand M, Lammers M, Ishizaki T, Ahmadian MR, Wittinghofer A. Structural and mechanistic insights into the interaction between Rho and mammalian Dia. *Nature* 2005;**435**:513–18.

78. Lammers M, Rose R, Scrima A, Wittinghofer A. The regulation of mDia1 by autoinhibition and its release by Rho center dot GTP. *EMBO J* 2005;**24**:4176–87.

79. Nezami AG, Poy F, Eck MJ. Structure of the autoinhibitory switch in formin mDia1. *Structure* 2006;**14**:257–63.

80. Wallar BJ, Stropich BN, Schoenherr JA, Holman HA, Kitchen SM, Alberts AS. The basic region of the diaphanous-autoregulatory domain (DAD) is required for autoregulatory interactions with the diaphanous-related formin inhibitory domain. *J Biol Chem* 2006;**281**:4300–7.

81. Riento K, Ridley AJ. Rocks: multifunctional kinases in cell behaviour. *Nature Rev Mol Cell Biol* 2003;**4**:446–56.

82. Doran JD, Liu X, Taslimi P, Saadat A, Fox T. New insights into the structure–function relationships of Rho-associated kinase: a thermodynamic and hydrodynamic study of the dimer-to-monomer transition and its kinetic implications. *Biochem J* 2004;**384**:255–62.

83. Shimizu T, Ihara K, Maesaki R, Amano M, Kaibuchi K, Hakoshima T. Parallel coiled-coil association of the RhoA-binding domain in Rho-kinase. *J Biol Chem* 2003;**278**:46,046–46,051.

84. Dvorsky R, Blumenstein L, Vetter IR, Ahmadian MR. Structural insights into the interaction of ROCKI with the switch regions of RhoA. *J Biol Chem* 2004;**279**:7098–104.

85. Jacobs M, Hayakawa K, Swenson L, et al. The structure of dimeric ROCK I reveals the mechanism for ligand selectivity. *J Biol Chem* 2006;**281**:260–8.

86. Yamaguchi H, Kasa M, Amano M, Kaibuchi K, Hakoshima T. Molecular mechanism for the regulation of Rho-kinase by dimerization and its inhibition by fasudil. *Structure* 2006;**14**:589–600.

87. Pearl LH, Barford D. Regulation of protein kinases in insulin, growth factor and Wnt signalling. *Curr Opin Struct Biol* 2002;**12**:761–7.

88. Kraulis PJ. Molscript – a program to produce both detailed and schematic plots of protein structures. *J Appl Crystallogr* 1991;**24**:946–50.

89. Merritt EA, Bacon DJ. Raster3D: photorealistic molecular graphics. *Methods in Enzymology* 1997;**277**:505–24.

Index

A

AA. *See* Arachidonic acid

Abscisic acid, signaling of, in *Arabidopsis Thaliana*, 177–178

AC 1, 457t

Acetylated residue recognition, example of, 24f

ACK, activation of, 560

Activation loop, 57–58

    cross-, of cGMP, of PKA, 443–444

    trans-, 7f, 355–356

Activators of G-protein signaling (GoLoco domain), 463–464

Acylation

    de-, 227

    N-terminal of, of Gα subunits, 481–482, 481t

    O-GlyN-, 8

    trafficking through, 227

Adaptors, types of, 25–27, 26f

Adenomatous polyposis coli (APC), 211t

Adenosine diphosphate (ADP)

    bacterial-toxin-induced, 494–496, 495t

    mono-, ribosylation reaction, 493, 494f

        endogenous, 496–497

    NADPH oxidase, PX domain and, 253

    -ribosylating/deribosylating enzymes, on GTP, 495t

    substrates for, 454–455

Adenosine triphosphate (ATP)

    analogs of, 428t

    -ase, in uptake, 296f

    binding site, 58, 425f

        targeted inhibitors of, 426, 429

    mimetic inhibitors of, 429

    mimics of, 51

Adenylyl cyclase

    bicarbonate and, 369

    Ca^{2+} and, 367f, 369

    CaM and, 367f, 369

    characteristics of, 365

    classification of, 365–368

    endogenous regulation of, 368–370, 369f

    eukaryotic active sites of, prokaryotic active sites v., 367, 367f

    exogenous regulation of, 368–370, 369f

    future of, 371

    GPCR and, 368–369, 369f

    G-proteins and, 368–369, 369f

    in human disease, 370–371

    oligomerization of, 368

    regulatory subunits of, 368t

    role of, 370

    structure of, 367, 367f

ADP. *See* Adenosine diphosphate

β-adrenergic receptor, 212t

AG 1433, 48

AGC subfamily, inhibitors of, 429

Agonists

    dependent-, 435

    EPAK, 442

    hypertrophic, 172

    kinase, 426

    MAP, 191, 222

    of S1P, 356t

    toll receptor, 133

AKAP. *See* A-kinase anchoring proteins

A-kinase anchoring proteins (AKAP)

    characteristics of, 435

    future of, 438

    with GPCR, 475

    G-proteins signaling through, 435

    schematic representation of, 436f

    signaling complexes of, 437

AKT serine/threonine kinases

    characteristics of, 280

    PKB/, inhibitors of, 50

    PTEN and, 304

Allosteric regulation, 55–56

Amino acid

    frequencies of, 401t

    $\beta 6\alpha 5$ loop, of g-proteins α subunit, 461

    residues of, LT A_4 hydrolase and, 351–352

    270, 13

Analog sensitive alleles, in kinase assays, 97, 97f

Anaphase-promoting complex/cyclosome (APC/C), 541–542

Angiogenesis

    anti-, kinase inhibitors, 48

    $G\gamma_{t1}$ subtype in, 507–508

    S1P and, 358

Animals

$G\gamma_{t1}$ subtype in, 508

    G-protein studies in, 503–505t

Antagonists

    β-blocker, 370, 413t

    dependent-, 435

    of S1P, 356t

Antennal lobe, of honeybee, 444

Anti-angiogenic kinase inhibitors, 48

Antigen receptors, signaling of, Shp2 in, 149–150

APC. *See* Adenomatous polyposis coli

APC/C. *See* Anaphase-promoting complex/cyclosome

Apoptic proteins

    cell survival and, 413t

    in PP2A, 213t

    pro-, PTK inhibitors with, 48–49

APS SH2 domain, 20f

Arabidopsis thaliana, abscisic acid signaling in, 177–178

Arachidonic acid (AA)

    actions of, 340f

    characteristics of, 339

    future of, 344

    $iPLA_2$ and, 335–336

    release of, PLA_2 and, 302f, 334–335

Arfaptin, 562, 563f

ARH1, 496

Armadillo repeat region (ARR), 567

ARR. *See* Armadillo repeat region

Arrestin

    β-,-2, 67t, 68

    in GPCR, 476

Asn-Pro-X-pTyr (NPXpY), 21

ATP. *See* Adenosine triphosphate

Autophosphorylation, 409t, 419t

Axin, 211t

B

B cells, trafficking of, S1P and, 356

Bacillus subtilis, stress signaling in, 178f

Bacterial-toxin-induced ADP, 494–496, 495t

Balanol, 427t

Bcr-Abl, in CML, 45

Bernard, Claude, 5

Bicarbonate, adenylyl cyclase and, 369

Printed and bound by CPI Group (UK) Ltd, Croydon, CR0 4YY

03/10/2024

01040311-0019